LINKING RESEARCH AND MARKETING OPPORTUNITIES FOR PULSES
IN HET 21ST CENTURY

Current Plant Science and Biotechnology in Agriculture

VOLUME 34

Aims and Scope
The book series is intended for readers ranging from advanced students to senior research scientists and corporate directors interested in acquiring in-depth, state-of-the-art knowledge about research findings and techniques related to all aspects of agricultural biotechnology. Although the previous volumes in the series dealt with plant science and biotechnology, the aim is now to also include volumes dealing with animals science, food science and microbiology. While the subject matter will relate more particularly to agricultural applications, timely topics in basic science and biotechnology will also be explored. Some volumes will report progress in rapidly advancing disciplines through proceedings of symposia and workshops while others will detail fundamental information of an enduring nature that will be referenced repeatedly.

The titles published in this series are listed at the end of this volume.

Proceedings IFLRC-III: Linking Research and Marketing Opportunities for pulses in the 21st Century.

GERMPLASM AND BIODIVERSITY

NUTRITIONAL VALUE IN FOOD AND FEED

About the Conference

The Third International Food Legume Research Conference (IFLRC III) was held at the Convention Centre in Adelaide, South Australia from the 22 to 26 September 1997. The two previous conferences had dealt with the crops, pea (*Pisum sativum*), lentil (*Lens culinaris*), faba bean (*Vicia faba*) chickpea (*Cicer arietinum*) and grasspea (*Lathyrus sativus*). Included with these crops in IFLRC III were the lupins (*Lupinus angustifolius, L albus*, and *L mutabilis*).

The title of the conference "*Linking Research and Market Opportunities for the 21st Century*" was indicative of the need, felt by many persons, that to ensure continued support from governments and funding organisations it was essential to illustrate that the research being undertaken on pulses was aimed at increasing yields and the financial return to farmers. In many countries the food legumes are perceived to warrant less support than the cereals, a perception that has arisen as a result of the considerable improvements in yield achieved in the cereals and by concerns for food sufficiency in the face of increasing populations. The advantages of including food legumes in rotations and their role in human and animal diets are often overlooked by those allocating resources for research.

Many of the food legumes are also used as feed for animals. It is fortunate these alternative uses are available as it provides a secondary and at times a more important market for the production of the crop. Lupins in Australia are an example. Lupins have been included in rotations partially because of the benefit they confer to following cereal crops but at the present time most of the production is used as animal feed. IFLRC III included the research on grain legumes as feed for ruminant and monogastric animals and also as a feed in the expanding aquaculture industries. Additionally in the conference there were papers on the use of pulses in industrial processes, a use which is likely to increase in the future.

There were 14 sessions of invited and presented papers and a final session with summaries of the conference. In addition there were poster papers submitted by participants. These papers were accepted by the conference organisers provided the legume species and the topic of the poster were relevant to the aims of the IFLRC. Mid-way through conference a day of visits to farms and research stations was arranged to illustrate local production and research on grain legumes.

Near the end of the conference a paper prepared by Professor R J Summerfield and Dr F J Muehlbauer was delivered by Professor Summerfield on the 'Continuation of the IFLRC'. In the paper they outlined the history of the IFLRCs and their past and future organisation.
Because of its comprehensive coverage the paper serves as a very good Introduction to the Adelaide conference and I have arranged for it to be printed at the beginning of these proceedings.

Every conference hopes to expand on the strengths of previous conferences. The possible changes for the next conference are included in the paper.

Also listed are the financial donors whose generous support made the conference possible and the members of the various committees who made every effort to make the Adelaide Conference a success.

International Food Legume Research Conference : concept and continuity

R.J.Summerfield[1] and F.J.Muehlbauer[2]

1 The University of Reading, Department of Agriculture, Plant Environment Laboratory, Cutbush Lane, Shinfield, Reading RG2 9AD, Berkshire and   2 US Department of Agriculture, Agricultural Research Service, 303W Johnson Hall, Washington State University, Pullman, WA 99164-6434, USA

Retrospect : the Concept

The genesis of the concept of an *International Food Legume Research Conference (IFLRC)* can be traced to an informal meeting in 1983. Those present on that evening in Aleppo, Syria (*viz.* H.L.Blain, G.C. Hawtin, F.J. Muehlbauer, A.E.Slinkard, and R.J. Summerfield) were keen to ensure that national crop improvement efforts devoted to the food legumes should not be wasteful of resources by duplicating work ongoing elsewhere. A particular concern was to build on the then unprecedented efforts in food legume research underway at ICRISAT and at ICARDA, and which had begun in 1972 and 1976, respectively. These motives were in time to be articulated as the general objective of the IFLRC, *viz.*:

> *To build communication linkages in order to promote research collaboration and the interchange of scientific and technical information on a global basis covering all aspects of research and development of cool season food legumes.*

The inaugural IFLRC became a reality in July 1986 when, under the guidance of Drs R.H. Lockerman and D.F. Bezdicek, the combined efforts of an International Advisory Board, an International Observer and an Organizing Committee came to fruition (see Summerfield, 1988).

The four hundred or so delegates from 48 countries who attended IFLRC I in Spokane, Washington, USA participated in a series of pre-planned business meetings. By their votes the delegates endorsed both the general and specific objectives of the Conference as recommended by the organizers. Activities of the IFLRC as well as the composition and functions of an International Steering Committee were also agreed with enthusiasm. These details are summarized below.

Specific Objectives of the IFLRC

- To promote research collaboration and the integration and dissemination of knowledge on cool season food legumes;
- To provide an international forum for discussion on priority problem areas requiring research attention;
- To maximise awareness and use of novel technologies and research methods;
- To encourage the evaluation and adoption of appropriate new technologies on the farm;
- To provide an international forum for promoting links between institutions with the aim of strengthening research manpower development; and
- To promote awareness of the importance of cool season food legumes among scientists, policy makers and funding agencies in order to stimulate increased research attention and financial support.

Activities of the IFLRC

The principal activity will be the International Conference to be held approximately every four to five years. Other activities may be undertaken as agreed upon by the IFLRC.

Function of the International Steering Committee

- To ensure continuity and help minimize duplication among other national and international meetings;
- To assist the host organization with logistic arrangements for the Conference;
- To formulate the Conference agenda so as to meet the objectives of the IFLRC;
- To solicit financial and logistic support from donors and other organizations for IFLRC activities; and
- To undertake any other activities approved by the IFLRC in line with Conference objectives.

Proposed Composition of the International Steering Committee

- Representative from ICARDA;
- Representative from ICRISAT;
- Chairperson of the immediate past IFLRC;
- Program Chairperson of the immediate past IFLRC;
- Host nation Chairperson (to be co-opted, once known); and
- Seven representatives from six agro-geographical regions as defined by the Food and Agriculture Organisation of the United Nations, *viz.*:

Agro-Geographical Regions		
I.	North America	USA and Canada
II.	Latin America and the Caribbean	Mexico; C. and S. America; Caribbean
III.	Europe	W. and E. Europe; USSR*; Israel
IV.	Africa	All African countries south of the Sahara
V.	Near East	N.Africa; West Asia to Afghanistan
VI.	Asia and the Pacific (including Pakistan)	India to China and Japan; Australia;
	(*Two representatives*)	New Zealand and Oceanic Isles

(*Now the Commonwealth of Independent States*)

April 1992: the Second IFLRC

During the decade which began in 1983, the use of cool season pulse crops as food and feed continued to increase, to the extent that demand exceeded supply in most countries in the West Asia and North Africa region (Oram and Belaid, 1990). It is the poorest countries, the urban poor, rural peoples and farm families that suffer most in these circumstances. Given these sorts of impetus and causes for concern, the second IFLRC was in the event to be appropriately timed and appropriately located. The International Steering Committee (ISC) elected in Spokane voted Cairo as the venue for IFLRC-II. Under the guidance of Drs A.E. Slinkard and M.C. Saxena, the ISC worked *in tandem* with the Organizing Committee to plan the Conference for April 1992. More than 200 delegates came from 38 countries to ensure "*another Major success*" [see Roberts (1994), pp. 983-988 in Muehlbauer and Kaiser (1994)].

Business of the IFLRC was conducted following agreed procedures. A ballot proved heavily in favour of the *Continuation of the IFLRC Concept* and a new ISC was voted into being.

Continuity: the IFLRC-III

The newly-elected ISC (Table 1) met for the first time on 16 April 1992. Dr F.J. Muehlbauer and Professor R.J. Summerfield were elected as Conference Chairman and Program Chairman, respectively. The Committee agreed that IFLRC-III should take place during 1997 and noted that bids to host the Conference were anticipated from:
♦ The European Association for Grain Legume Research (J. Picard);
♦ General Directorate of Agricultural Research, Turkey (I. Kusmenoglu);
♦ Indian Council of Agricultural Research (P.N. Bhal); and from an
♦ Australian Consortium of Food Legume Scientists (R.O. Rees).

As intended, the ISC had completed the voting procedures necessary to decide the location of IFLRC-III by the Spring of 1993: by unanimous vote the Conference would be held in Adelaide, Australia.

Table 1. Members of the International Steering Committee for IFLRC-III

Region	Representative	Region	Representative
1. North America	Dr F.J. Muehlbauer (Chair) USDA-ARS Washington State University, USA	2. Latin America; Caribbean	Dr J.U. Tay INIA, Chile
3. Europe	Professor R.J. Summerfield The University of Reading, UK	4. Africa (sub-Sahara)	Dr G. Bejiga Crop Science Department, Ethiopia
5. Near East	Dr H.M. Halila INRAT Tunisia	6. Asia & Pacific (inc. Pakistan)	Dr B.A. Malik NARC, Pakistan
Immediate Past Program Chair	Dr M.C. Saxena ICARDA, Syria	6. Asia & Pacific (incl. Pakistan)	Mr R.O. Rees ABARE, Australia
Immediate Past Chair	Dr A.E. Slinkard University of Saskatchewan Canada	ICARDA Representative	Dr W. Erskine ICARDA, Syria
ICRISAT Representative	Dr Y.L. Nene ICRISAT, India	Co-opted by invitation	Dr W.J. Kaiser USDA-ARS-RPIS Washington State University, USA

Adelaide, Australia : 19-26 September 1997

Mr Brian Hansen (Primary Industries, South Australia) was co-opted onto the ISC as the Chairman of the Local Organising Committee (LOC) - the composition of which is given in Table 2.

Table 2. Members of the Local Organizing Committee for IFLRC-III[+]

Name	Organization
Mr B. Hansen	Primary Industries SA
Dr R. Knight[*]	University of Adelaide
Mr T. Day	Pulse Australia
Mr R. Rees	Primary Industries SA
Mr J. Hannay	Primary Industries SA
Mr W. Hawthorne	Primary Industries SA
Dr A. Dubé	SARDI, Adelaide
Mr R. Bulfield	Sapro Marketing Pty Ltd

[+]Dr J. Hamblin and Mr H. McClelland also provided enthusiastic support.

[*]Appointed as Editor of the Proceedings, once again to be published by Kluwer Academic Publishers, The Netherlands (Knight, 1998).

A particular concern for IFLRC-III was the linking of research activities and priorities with future market opportunities and demands - hence the subtitle *"Linking Research and Market Opportunities for the 21st Century"*. The ISC, acting primarily through the efforts of the Conference and Program Chairmen (who visited Australia in March 1995 as the guests of CLIMA and the South Australia Department of Agriculture), agreed the following general objectives:

(i) Plenary sessions would address world and regional production and consumption. Other sessions would consider research concerned with breeding and varietal selection, integrated pest management, nutritional value of pulses as food for humans and feed for animals, plant architecture, biotechnology, crop protection, sustainability, farming systems, modelling and germplasm.

(ii) The marketing sessions would consider the quality and quantity of food legumes being produced, global and regional trends in supply and demand, the impact of government policies, innovations in processing and marketing, developments in the international supermarket trade and innovative pulse products.

(iii) Poster papers would be presented during the Conference in what is a magnificent Adelaide Convention Centre Exhibition Hall.

In the event more than 200 co-authors contributed to the 70 or so invited papers and 176 posters were displayed. More than 400 delegates came from 39 countries to hear what was said and to see the information on display.

Discussion sessions were wide-ranging and stimulating. Conference administration, orchestrated by Mr R. Bulfield and his SAPRO Marketing team, was flawlessly efficient. There is no doubting that IFLRC-III, which opened with generous praise from the Honourable Rob Kerin MP (Minister for Primary Industries, SA), was a resounding success.

The ISC and the LOC are very pleased to record once again our sincere thanks to the several generous sponsors on whose support the financial viability of the Conference was so crucially dependent (Table 3).

Table 3. Sponsors of the IFLRC-III

PRINCIPAL SPONSORS
Primary Industries South Australia
Grains Research & Development Corporation (GRDC)
Centre for Legumes in Mediterranean Agriculture (CLIMA)

MAJOR SPONSORS
Lief Grain PTY Ltd
SA Research & Development Institute (SARDI)
AusAid

OTHER SPONSORS*
Australian Barley Board
Food and Agriculture Organisation of the United Nations (FAO)
Bustan International
Meat Research Corporation
Pulse Australia
Rural Industries Research & Development Corporation (RIRDC)
SA Co-operative Bulk Handling Limited

SA Grains Industry Trust Fund
US Pea and Lentil Council
Ward McKenzie
Governments of Northern Territory, Queensland, Victoria and Western Australia

* The Conference Organisers also thank Ansett Australia and Malaysia Airlines for their co-operation with arrangements for the domestic and international travel for selected speakers.

THE FUTURE : IFLRC-IV

As in previous years, IFLRC business generated enthusiastic interest with keenly contested ballots for regional representations on the ISC for IFLRC-IV. The new Committee (Table 4) met for the first time on 26 September 1997 when Drs J.Hamblin (CLIMA) and W.Erskine (ICARDA) were proposed and unanimously elected to Conference Chairman and Program Chairman, respectively.

Table 4.   International Steering Committee for IFLRC-IV

Region+		Representative
• Region 1	:	B. Vandenberg (Canada)
• Region 2	:	J. Tay (Chile)
• Region 3	:	J. Wery (France)
• Region 4	:	G. Bejiga (Ethiopia)
• Region 5	:	B. Sakr (Morocco)
• Region 6	:	J. Hamblin } (Australia)
		B. Malik } (Pakistan)
• FAO Observer	:	E. Kueneman (Italy)
• ICRISAT Representative	:	W. Erskine (Syria)
• ICRISAT Representative	:	N.P.Saxena (India)
• Immediate Past IFLRC Chair	:	F.J.Muehlbauer (USA)
• Immediate Past IFLRC Program Chair	:	R.J. Summerfield (UK)
• Chair, Local Organizing Committee	:	To be coopted once known

+As in Table 1.

The Committee was pleased to record that two formal bids had been received from representatives of organisations in India and Turkey who wished to host IFLRC-IV in 2002. Both organisations have been requested to submit their final and formal bids in comprehensive detail by 30 March 1998.

Feed-back suggestions from IFLRC-III delegates were recorded for future reference, *viz*:

(i) IFLRC-IV should target the cool season food legumes and any additional legume species of major regional importance depending on the geographical location of the Conference [e.g. if in Turkey, then possibly add *Phaseolus vulgaris*; if in India, then possibly add Asiatic *Vigna* spp; *Cajanus cajan* and *Arachis hypogaea]*;

(ii)Provide more time for viewing of posters;

(iii) Include one or more Sessions targetting grain legume production for feed rather than food;

(iv)   Include a greater proportion of targetted research papers at the expense of general review papers and consider selecting topics from "offered" rather than "invited" papers;

(v)   Include broader sessions on biotechnological themes;

(vi)   Poster subjects can involve any specie of grain legume; and

(vii)   Offer awards for *"Best of"* categories, especially for younger scientists.

CONCLUDING REMARKS

The reality of the intervening decade or so since the first Conference in 1986 is that support for agricultural research in general and for that on cool season legumes in particular has declined. In the five years since the Cairo Conference this decline has accelerated not only for national programs in most countries but also at the international centres. Faced with declining support, the target of improving yields of the cool season food legumes by an average of 28 kg per hectare per year, so prominently mentioned at the Cairo Conference as required to meet the demands of an ever growing population in West Asia and North Africa, remains a daunting challenge in many regions. Nevertheless there have been exceptions to declining research support. The Centre for Legumes in Mediterranean Agriculture (CLIMA) has been established in Australia and there is strong federal and industry support provided to the national program in Canada. The scientific successes and on-farm impact of these two programmes are shining examples of the excellent return to investment in long-term research and breeding on cool season food legumes.

References

Knight, R.(ed.) 1998. *Proceedings IFLRC-III: Linking Research and Marketing Opportunities for the 21st Century.* Dordrecht: Kluwer Academic Publishers (in press).

Muehlbauer, F.J. and Kaiser, W.J. (eds) 1994. *Expanding the Production and Use of Cool Season Food Legumes,* pp. 991 Dordrecht: Kluwer Academic Publishers.

Oram, P. and Belaid, A. 1990. *Legumes in Farming Systems.* ICARDA-IFPRI Report, pp. 206.

Oram, P. 1994. *Expanding the Production and Use of Cool Season Food Legumes*, pp.3-49 (eds F.J. Muehlbauer and W.J. Kaiser). Dordrecht: Kluwer Academic Publishers.

Roberts, E.H. 1994. *Expanding the Production and Use of Cool Season Food Legumes.*pp.983-988 (eds F.J. Muehlbauer and W.J. Kaiser). Dordrecht: Kluwer Academic Publishers.

Summerfield, R.J. (ed.) 1988. *World Crops: Cool Season Food Legumes.* pp. 1179. Dordrecht: Kluwer Academic Publishers.

Editorial Notes

Not all of the 72 papers presented at the conference are included in the proceedings, as the authors of 4 presentations did not wish to submit their papers for publication. Two contributions that were invited, but for different reasons were not presented, are included.

The format of units and numbers
The contributors to the proceedings come from different countries and have expressed numbers in different ways. To avoid confusion I have adopted a common format for the papers. I apologise to those for whom this format is not the one they use customarily.
The format I have followed is one used for text, written in the English language, in which digits to the left of the decimal point are marked off in groups of three with a comma or alternatively with a space. Numbers with decimals have the decimal point marked as a period (full stop) and not as a comma. The format is most easily illustrated by examples.
56,300 and 56 300 are ways of representing fifty-six thousand three hundred.
The yield 4.35 t/ha is four decimal three five tonnes per hectare
The units used throughout are metric and the abbreviation MT is a metric tonne.

Acronyms
Some international agricultural centres and some national centres have major programs of research on the crop species of interest to the IFLRC and have been frequently referred to in these proceedings by their acronyms.
They are
ICARDA International Center for Agricultural Research in Dry Areas, Aleppo, Syria
ICRISAT International Crops Research Institute for the Semi Arid Tropics, Patancheru, India
IPGRI International Plant Genetics Resources Institute, Rome, Italy
The Australian national organisations are
CLIMA Centre for Legumes in Mediterranean Agriculture, Perth, Australia
GRDC Grains Research and Development Corporation, Canberra, Australia

In the two previous proceedings the editors provided glossaries that defined terms such as accession, cultivar, genotype and germplasm. The definitions have not changed and therefore I have not presented them again in these proceedings. Anyone wishing to refer to these definitions will find them in the Editorial Notes and Glossary of *Expanding the Production and use of Cool Season Food Legumes* (eds FJ Muehlbauer and WJ Kaiser).1994. Kluwer Academic Press.

Acknowledgments
I wish to express my gratitude to Lachlan Tailby computing officer in the Department of Plant Science, Adelaide University and my daughter Jacqueline Knight whose computing expertise greatly facilitated the production of these proceedings. The text, tables and graphs of the papers had been produced originally by the authors using different computer applications and required converting to a common format acceptable to the publishers. The help of these two persons was critical to this process.

The Pulse Economy in the Mid-1990s: A Review of Global and Regional Developments

T.G. Kelley[1], P. Parthasarathy Rao[1] and H. Grisko-Kelley[2]
1 International Crops Research Institute for the Semi-Arid Tropics, Patancheru P.O., Andhra Pradesh, India; 2 Agricultural Economist, Hyderabad, India.

Abstract

The world pulse economy seems to have stagnated during the first half of the 1990s after going through a contraction during the 1970s caused by the Green Revolution, and an expansion during the 1980s fuelled by changes in the European Community's (EC) Common Agricultural Policies (CAP) which favored production of pulses for feed. In 1996 world production stood at 57 million MT on 70.5 million ha with an average yield of 809 kg/ha. However, recent vast cutbacks by the Commonwealth of Independent States (former USSR) and reductions in EC feed production following the 1992 CAP reforms in 1993 have been offset by a sustained upward trend in Africa and a remarkable expansion in South Asia. Developing countries produce 71% of the world's pulses. India remains the largest pulse producer with 27% of world production.

Trade in pulses has increased to 14% of world production. This encompasses all pulses and regions, suggesting that trade barriers are being reduced. Imports are still concentrated in Europe, while the most important exporters are Canada, France, China, USA, and Myanmar. The latter has emerged as the largest supplier to South Asia, which it shares with inter-seasonal exports from Australia. Two-thirds of all pulses is used for food, mostly in developing countries, while about one quarter is used for feed, mostly in Europe, CIS, and Oceania (Australia). Since 1980-82, per capita food consumption of pulses declined by 6% in developing countries where relative prices of pulses have gone up while consumption of animal protein is increasing. But developed countries increased their food pulse consumption by 2.3%, if from a low base level.

Price data for pulses remain sketchy. In India, the largest consumer, prices have increased relative to other foods, pushing pulse protein out of the average diet. Declining amounts of pulse protein in developing country diets do not necessarily indicate protein deficiencies, but to a paucity of technological progress in production relative to advances in production of animal protein such as milk and poultry, which have brought down the relative prices of these foods. The same observation, coupled with a relatively strong negative price elasticity of demand for pulses, suggests that there is a significant unmet demand for pulses in developing countries which will not be satisfied at prevailing (high) prices.

Yield variability remains a challenge to a constant, abundant supply at low and stable prices. As most of the food pulses are produced and consumed in developing countries, the limitations to higher and more stable yields need to be addressed, and multi-objective frameworks of farm families need to be taken into account when developing improved cultivars and production methods.

The future of the pulse economy depends on social, dietary, economic, environmental, and infrastructural factors some of which are predictable in the process of economic growth while others -- such as government interventions and scientific breakthroughs or competing crops or protein sources -- are highly unpredictable and could rapidly change the supply or demand situation.

1. INTRODUCTION

Much of this paper is concerned with the aggregates of developing vs. developed countries or world regions, fully realizing that such aggregates conceal much[1]. The paper will not discuss in detail specific

---

[1] There is some overlap between regions and the level of development, i.e., for the purpose of the pulse economy, we can roughly equate North America, Oceania, CIS, and Europe with the developed countries group, while the remaining regions could be roughly subsumed under the developing countries umbrella. The delineations within Asia are:

South Asia: Bangladesh, Bhutan, India, Maldives, Nepal, Pakistan, Sri Lanka

East Asia: Cambodia, China, Indonesia, Japan, Korea (2), Laos, Mongolia, Myanmar, Philippines, Thailand, Viet Nam

West Asia: Afghanistan, Bahrain, Cyprus, Iran, Iraq, Israel, Jordan, Lebanon, Saudi Arabia, Syria, Turkey, Yemen.

1

R. Knight (ed.), Linking Research and Marketing Opportunities for Pulses in the 21st Century, 1–29.
© 2000 Kluwer Academic Publishers. Printed in the Netherlands.

countries or regions as they are considered in the regional reviews in these proceedings. Our analyses help explain facets of existing pulse economies from which predictions can be made for country groups.

We will refer to India, not simply because our research is done there, but because India is the world's largest pulse producing and consuming country representing 27% and 35% of world production[2], and consumption in 1996. India is also home to more than one-third of the world's poor and is predicted to replace China as the world's most populous nation by the middle of next century (World Bank, 1997).

We first review two similar papers presented at the first and second IFLRC. Next, we present current data from FAO (1997) to summarize developments in production, consumption, and trade. We use the periods 1980-82 to 1989-91 to 1994-96 (or 1993-95, depending on availability of data) for a medium- and short-term analysis. This was necessary to analyse more recent developments as they deviate from the developments in the 1980s. 1989-91 was the final triennium in the analysis of Oram and Agcaoili (1994). This paper also includes a few observations on pulse prices, but no price analyses.

The main part of the discussion revisits issues raised by Agostini and Khan (1988) and Oram and Agcaoili (1994): the protein deficits in the diets of people in developing countries, the supply constraint to pulse markets, and the problem of yield variability.

We felt it necessary to include warm season pulses, particularly pigeonpea *(Cajanus cajan)* and cowpea *(Vigna sinensis, Dolichos sinensis)*, even though the data for them are not as good as for chickpea *(Cicer arietinum)*, dry beans *(Phaseolus spp.)*, dry peas *(Pisum sativum, P. arvense)*, and lentils *(Lens esculenta, Ervum lens)*. As it turns out cowpea has been a star performer, steadily gaining ground in Africa.

At the first conference, Agostini and Khan (1988) reviewed the world grain legume scenario of the 1970s and early 80s when it was coming out of a contraction due to the rapid expansion of cereals. By 1992, when Oram delivered his address at the second conference, the world market had seen the entrance of new players: the EEC producing dry peas for feed to replace soybeans, Turkey exporting chickpea surpluses from a fallow-replacement program, and Australia producing lupins, dry pea, and even chickpeas for export.

The market has continued to change. Lupins and vetches are no longer listed separately among the pulses by FAO. Broad beans are losing significance as the area has been cut-back. The countries of the former USSR have dramatically reduced their pulse production, yet some are emerging as major exporters. The EC countries are making changes in the wake of the Common Agricultural Policy (CAP) regarding support for oilseed crops (1992), cereals and protein crops (including dry pea, faba bean, and sweet lupin) (1993), and other grain legumes (chickpea, lentil, vetches) (1995) which were causing a marked decline in production in 1994 and 1995. Canada, a new entrant, is now fifth among the world's pulse producers (15% of lentils and 10% of dry peas) and by 1995 had become the number one pulse exporter. Myanmar, after the change in government in 1988, has risen to rank fifth among the exporters and has an increasing production of dry beans and pigeonpea. Turkey and Australia, the success stories by the second conference, now rank eighth and ninth, respectively, among net exporters.

2.   SOCIO-ECONOMIC SITUATION AT THE TIME OF IFLRC I & II

When Agostini and Khan wrote their paper in 1986, they were looking back at the '70s and the decline in pulse production which had occurred due to lack of government attention which had gone to cereals in the wake of the Green Revolution. They saw a stimulation of the pulse economy in the 80s, including government interest, with Australia appearing as the world's largest lupin producer. 70% of total world production in 1984 came from developing countries, mostly as dry beans, chickpea, and faba bean, while pea and vetches and lupins were the most commonly produced pulses in the developed countries[3]. Of the 48

---

[2] India is the largest producer of dry beans (22% of world total), chickpea (74%), lentils (28%), pigeonpea (88%), and "other pulses" (15%). This latter FAOSTAT category includes data for all legumes for which FAO collects data, but that are not listed separately, i.e., vetches *(Vicia sativa)*, lupins *(Lupinus spp.)*, bambara beans *(Voandzeia subterranea)*, and others. FAO also uses "pulses NES" (not elsewhere specified) to describe these. Separate listings exist for dry beans *(Phaseolus spp.)*, broad beans (faba bean) *(Vicia faba)*, chickpeas *(Cicer arietinum)*, cowpeas *(Vigna sinensis; Dolichos sinensis)*, lentils *(Lens esculenta; Ervum lens)*, dry peas *(Pisum sativum; P. arvense)*, and pigeon peas *(Cajanus cajan)*. Data refer to the dry grain only and not to that produced or consumed as a green vegetable.

[3] All data given in this section are taken from the publications reviewed. FAO data may have been revised thereafter.

million tonnes of pulses produced world-wide, 70% went for food, with an average per capita consumption of 6.3 kg in 1983, which had fallen from 7.5 kg in 1969/71 and was seen as stabilizing in the 1980s. Feed uses, which had declined in the 1970s, were picking up, mostly in the USSR and EEC.

Domestic prices for food pulses had risen relative to cereal prices, and while world prices had fallen, they had done so less than prices for other agricultural products. Of the total world production, only 7% was traded, creating an extremely thin world market which was mostly utilized to offset shortfalls in domestic production or dispose of surplus production. This had created highly volatile prices especially for chickpea and lentil. The different pulses also fetched very different prices depending on end use and quality. Beans and peas were becoming more important as were exports of pulses in general, having witnessed an increase in volume of 79% in nearly fifteen years (from 1.9 million MT in 1969-71 to 3.4 million MT in 1984). This development went hand in hand with a reduction in the number of market players and a change in position of the group of developing countries from net-exporter to net-importer status, even though Turkey replaced the U.S. in 1982 as the world's largest pulse exporter.

For the future, Agostini and Khan predicted slow growth in global output with expanding opportunities for exports to developing countries given their inability to satisfy domestic demand as populations grew (eg India) or demand rose (eg Algeria).

The main problem, the authors discussed, was the insufficient availability of protein, especially in the poor countries who could not afford to import pulses - a problem compounded by their demand for pulses, not commonly traded, such as cowpea and pigeonpea. Agostini and Khan also understood the problem of relative, not absolute, pulse prices influencing production and they undertook to calculate some real (i.e., deflated) prices and price deviations.

By 1992 and only six years later, the scenario had changed dramatically. Oram and Agcaoili, found there had been 3.5% growth per annum for the four main cool season pulses during the 1980s, much higher in the developed than the developing countries (8% vs. 1.5%), with increases in yield and area contributing to the trend. Thus, when they reviewed FAO projections from 1986, they found FAO had overestimated output in developing countries while underestimating output in the developed countries. Actual consumption overshot the projections for feed use, while food use had been overestimated. World trade in legumes had vastly increased not only for feed use but also, unexpectedly, as food use for developing countries.

By 1989-91, world production had risen by another 17% since 1984, to 56.2 million MT, with 90% of the total concentrated in 12 countries and as much as 50% grown in India (23%), USSR (17%), and China (10%). The developing countries were still producing the lion's share of pigeonpea, cowpea, chickpea, dry beans, faba bean, and lentil, though their share had declined for six out of nine pulse crops, totalling 63.5% of world pulse production. Dry pea now contributed almost as much as dry beans to world production in pulses, 27% and 29%, respectively. (See Figure 2.)

With total production up, all uses were up, but food use had grown at only 1.3% per annum versus 9.7% for feed. These data could be disaggregated to show that the developed world was using 72% of their share for feed in 1989 (up from 51% in 1980), while the developing world's feed use was 8.5% (up only slightly from 8.2%). While total per capita consumption of food and feed was 16.4 kg in 1989 (up from 8.0 kg in 1980) for the developed countries, per capita consumption for food use only had declined in the developing countries from 7.6 kg in 1980 to 7.2 kg in 1989, mostly due to falling levels of pulse consumption in Asia and Latin America. (Note that Asia represented 54% of the world's total use of food pulses in 1988-90.)

The world market had increased by 1989/90, with 11% of production exported, but it was still a thin market with volatile prices. Peas accounted for 36% of total trade with the EC being the largest importer. The warm-season pulses especially continued to have a low share of the trade. The problem of "unsatisfactory consumption" of food pulses in developing countries pointed out earlier by Agostini and Khan persisted. Oram and Agcaoili postulated a "Year 2000" demand of 49 million MT for developing countries and 4 million MT for food use in developed countries, plus an unspecified amount for feed use in developed countries.

Given the 1992 scenario, Oram and Agcaoili put forward questions to address the postulated protein deficit in developing country diets:

1. How do we deal with yield and area instability in pulses?
2. How do we close the supply/ demand-gap?
3. Are we dealing with a supply or demand constraint?

They suggested the yield and area instabilities were functions of the production systems, use of inputs, and price signals for pulses. The supply/ demand-gap could be closed by increasing area, yield, or both, and also - for individual country deficits - by imports. And production was definitely supply-constrained. According to Oram and Agcaoili, the most promising way to increase supply was to increase yields through technological progress. So the ball was with research, and two of the issues research needed to look at were yield variability and links between socio-economic and technical research.

3.   THE CURRENT SCENARIO: 1996

3.1   Production

In 1996, world pulse production stood at 56.8 million MT, down from 58.1 million MT in 1990 (FAO, 1997). This 2.2% reduction occurred despite an increase in area, from 67.4 to 71.0 million ha, which brought the average yield down to 800 kg/ha, from 862 kg/ha in 1990, a 7.8% decrease. However, this single year comparison overstates the downswing.

The more reliable medium-term observations show a growth of 36% for total production between 1980-82 and 1994-96 and a compound growth rate of 2.1% p.a. for 1980 to 1996 (Tables 1 and 2). All the values given in tables 1 to 12, 15 & 16 are calculated from FAO statistics and are for dry grain. The totals in tables 1 & 2 are based on a phenomenal expansion of 3.6% p.a. during the 1980s, followed by near stagnation since 1990. Yields contributed 1.3% p.a. to the growth between 1980 and 1996, while expansion in area accounted for a smaller share of 0.8% p.a. But again, the higher growth rates of 1.2% p.a. for area and 2.4% p.a. for yield occurred during the 1980s. They have since come down to 0.9% p.a. for area and a reduction of 0.8% p.a. in yield during the first half of the 1990s.

Table 1. Area, Production and Yield of Total Pulses, Triennium Averages, 1980-82, 1989-91 and 1994-96, by Region

	Area ('000 ha)			Production ('000 MT)			Yield (kg/ha)		
	'80-'82	'89-'91	'94-'96	'80-'82	'89-'91	'94-'96	'80-'82	'89-'91	'94-'96
Africa	9043	11877	12572	5145	6625	7208	569	558	573
East Asia	7573	5993	6641	8282	6942	7807	1095	1152	1175
West Asia	1665	3259	3281	1640	2644	2858	986	811	871
South Asia	25490	26262	27418	11760	14877	16666	462	566	608
L. America + Caribbean	8951	8985	9364	4999	5184	6154	557	575	657
CIS	5167	5484	3165	5392	7594	4178	1060	1372	1307
Europe	2513	3048	2421	2704	7439	6204	1071	2452	2561
N. America	1138	1276	2093	1800	2249	3640	1581	1750	1738
Oceania	286	1524	2054	291	1596	2011	1034	1045	983
World	**61843**	**68421**	**70501**	**42015**	**56160**	**56995**	**679**	**821**	**809**
Developing	52545	56895	60595	31597	36998	40723	601	650	672
Developed	9297	11526	9907	10418	19162	16273	1126	1663	1642

Underlying the world picture are different pictures for country groups. In 1980-82, the "developed countries" accounted for 25% of world pulse production. By 1989-91, their share had risen to 34%. It is now down again to 29%. The fluctuation is due to production shooting up by 8.3% p.a. between 1980 and 1990, to which area expansion contributed 3.2%, and yield increases an amazing 4.9%, bringing their total production up to 19.2 million MT in 1989-91. By 1994-96, total production had shrunk to 16.3 million MT[4]. This contraction occurred at a rate of -3.8% p.a., led by a 2.6% p.a. area reduction, with yields falling at a rate of 1.2% p.a. The average yield in this group is still 1642 kg/ha.

For the "developing countries" group, growth rates have been much more even over the 1980-1996 period. In these countries, production grew at an average of 1.7% during the whole period. This sustained growth came from a 0.94% p.a. expansion in area, and 0.72% p.a. increase in yield. During the 1980s, the growth was led by yields (0.9% p.a.), while an area expansion of 1.5% p.a. has carried the growth since

[4] The triennium averages tend to underestimate the decline of 5 million MT, from 20.3 to 15.3 million MT, between 1990 and 1996.

1990. Production in 1994-96 stood at 40.7 million MT on 60.6 million hectares, with an average yield of 672 kg/ha.

Table 2. Growth Rates in Area, Production and Yield of Total Pulses, 1980-96, by Region

	Area Growth Rate (% per annum)			Production Growth Rate (% per annum)			Yield Growth Rate (% per annum)		
	'80-'96	'80-'90	'90-'96	'80-'96	'80-'90	'90-'96	'80-'96	'80-'90	'90-'96
Africa	2.45	3.22	1.12	2.58	3.03	2.22	0.12	-0.19	1.09
East Asia	-1.59	-1.64	3.80	-1.33	-0.57	3.87	0.26	1.09	0.07
West Asia	5.23	8.32	-0.06	4.15	6.76	0.15	-1.03	-1.44	0.21
South Asia	0.35	-0.04	1.29	2.04	2.33	3.13	1.68	2.37	1.83
L. America + Caribbean	0.16	0.50	0.79	1.58	0.20	2.63	1.42	-0.30	1.83
CIS	-3.95	1.64	-10.08	-2.25	5.52	-12.70	1.76	3.82	-2.91
Europe	-0.27	2.79	-3.74	6.44	13.69	-4.41	6.73	10.60	-0.69
N. America	4.96	1.89	10.08	5.77	2.28	9.31	0.78	0.38	-0.70
Oceania	13.97	22.78	5.08	13.90	22.57	4.94	-0.06	-0.17	-0.14
World	**0.80**	**1.18**	**0.85**	**2.07**	**3.63**	**0.07**	**1.26**	**2.41**	**-0.78**
Developing	0.94	0.79	1.48	1.66	1.69	1.84	0.72	0.90	0.35
Developed	0.08	3.22	-2.63	3.13	8.27	-3.79	3.05	4.88	-1.20

Regional Review South Asia was the most important region in production, with 29% of the 1994-96 total, or 17 million MT (Figure 1). There has been an average growth of 2.0% p.a. since 1980, mostly from increasing yields. East Asia, which produces 14% of the world's pulses, or 8.2 million MT, saw its production decline by 1.3% p.a. since 1980, due to area losses of 1.6% p.a. This is mostly attributable to China reducing the area under broad beans (see below). West Asia, with a world production share of 5%, or 2.9 million MT, shows a 4.2% p.a. growth in the medium run. This region had an enormous area increase of 8.3% p.a. during the 1980s, which was somewhat eroded by a yield decrease of 1.4% p.a. during the same period. Since then, slight yield increases have been balanced by very slight area losses to produce a near zero growth. In Latin America and the Caribbean, production was almost stagnant during the 1980s, but has since seen a 2.6% growth per annum, led by yield increases of 1.8% p.a. and area expansion of 0.8% p.a., which resulted in a medium-term growth rate of 1.6% p.a. and a production of 6.3 million MT in 1996, equal to 11% of the world total. This makes the region about as important as Africa, which produces 7.4 million MT, or an eighth of the world total. The medium-term growth rate for Africa has been 2.6% p.a., originating from a 3.2% p.a. area expansion during the 1980s, and a 2.2% p.a. growth rate since. Area and yield increases contribute evenly to the increase.

Figure 1. Regional Production of Pulses, 1994-96 Average

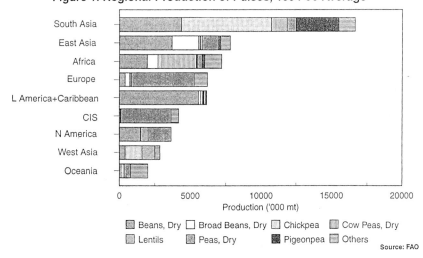

Examining the data for the developed regions, we find that Europe, which produced 11% of the world total in 1996, derived its enormous growth during the 1980s from a 10.6% p.a. yield increase fuelled by changes in the CAP which in 1978 began to give preference to protein crops (Carrouee et al., 1997). The reduction of 4.4% p.a. since 1990 has come mostly from area reductions of 3.7% p.a., which brought total production down to 5.8 million MT, but leaves the region with record yields of 2393 kg/ha. Again, the contraction is attributable to policy changes (Personal Communication, Hubertus Wolfgarten, 1997). In North America, which produces 6% of the world's total, the 5.7% p.a. growth since 1980 has been driven by a 10.1% p.a. area expansion since 1990, much higher than the 1.9% p.a. that accounted for the 1980s growth, and attributable to Canada taking up pulse production. Area under pulses in this region has thus grown from 1.3 to 2.1 million ha between 1990 and 1996. In Oceania, which produced 2.2 million MT or 4% of the world's pulses in 1996, the remarkable growth of 13.9% p.a. has been achieved entirely by area expansion: 22.8% p.a. during the 1980s, and 5.1% since. This is due mostly to Australia entering the global market for pulses.

This leaves the Commonwealth of Independent States (CIS), which produced 3.5 million MT in 1996, down to less than half of their 1990 production of 8.9 million MT. While the growth rate in the USSR had been 5.5% p.a. during the 1980s, with 3.8% p.a. growth from yield increases, and 1.6% p.a. from area expansion, it went down to -12.7% p.a. since 1990, mostly caused by an immense area reduction of 10.1% p.a. Yields fluctuate widely between a 6-year low of 1038 kg/ha in 1991 and a 6-year high of 1647 kg/ha in 1992. At the same time, the CIS lost 54.8 million MT or 30% of their cereal production, and 700,000 MT or 19% of their oilseed production. This immense decline is most likely due to restructuring, and a general economic downturn, which reduced the availability of inputs. Thompson (1997) suggests that the USSR was subsidizing both consumers and producers, leading to a steep decline especially in the livestock sector when the Union dissolved. The CIS and Europe combined, account for all of the contraction in the world pulse economy since 1990.

Major Producers Looking at single countries in 1996, India remained the most important producer (27.0%) followed by China (8.7%) and Brazil (5.0%). Together with France (4.6%), Australia (3.8%), and Canada (3.3%), the top three developed-country producers, the largest six producers in the world account for 52.5% of total production. Ukraine, formerly part of the USSR, which was the second largest pulse producer in 1989-91, occupies rank seven. Even though the broad world production picture has remained intact since the Cairo conference, a comparison with Table 4 of Oram and Agcaoili (1994) shows the changes which have occurred since. They are, the break up of the USSR and the decline in production; India increasing its share of world production as China, Brazil, and France decreased theirs; and Australia and Canada joining the top six producer group, while the United States has dropped out.

The Crops Perspective A comparison with Agostini and Khan (1988, p. 463) shows that concentration of certain pulse crops in one of the two developing/ developed country groups has increased for dry beans (88% grown in developing countries), dry peas (81% grown in developed countries), and cowpea (99% grown in developing countries), while it has decreased for chickpea (96% grown in developing countries), lentil (79% grown in developing countries), and "other pulses" (56% grown in developing countries). Broad bean (87% in developing countries) and pigeonpea (100% in developing countries) are virtually unchanged. (Table 3)

As for distribution of specific pulse crop production, dry beans are still in the lead with 32% of world total in 1996 (Figure 2), largely produced by India (23%), Brazil (16%), and China (9%). The second largest pulse crop with a 22% share is dry peas grown primarily in France (21%), Ukraine (14%), and Canada (10%). Dry beans and dry peas together still account for more than half (54%) of total world production, and they also dominate the world market for pulses, but as dry peas have lost some ground in Europe and the CIS, so has their relative importance.

Chickpea, the number three crop representing 14% of pulses in 1996, also shows an increase in production. India (73%), Turkey (9%), and Pakistan (8%) are the leading producers. "Other pulses" (mostly vetches and lupins for feed) add up to 11% of the world's production[5]. Not surprisingly, the "other pulses" are the least concentrated in any producer country: the top three producers -- India, U.K., and DPR Korea -- together account a quarter of the total, as against the most highly concentrated, pigeonpea, of which 96% is produced by India (88%), Myanmar (4%), and Malawi (3%). This warm season crop has taken over from

---

[5] In Agostini and Khan (1988), vetches and lupins were dealt with under separate entries. Oram and Agcaoili (1994) had one entry for both crops.

broad beans as the number five, representing six percent of world production[6]. The annual growth rate of 2.8% originates largely in increased area (2.4%). Certain regions, however, have deviated remarkably: East Asia shows a 10.5% growth rate for pigeonpea attributable to expansion in Myanmar, while African yields have grown by 1.4% p.a. (Table 4, 5).

Table 3. Production, as dry grain, of Major Pulses, by Region, 1994-96 Average

	World Prod.	Regional Distribution				L Amer/ Carib¹	CIS	Europe	N Amer²	Oceania	Developing	Developed
		Africa	Asia									
			East Asia	West Asia	South Asia							
	('000 mt)	(% of World Production)										
eans,	18,184	10.9	20.3	2.0	24.1	30.7	*	2.3	8.3	*	88.0	12.0
eas,	12,436	2.7	9.4	*	5.3	*	28.4	35.6	12.9	3.7	18.9	81.1
hickpeas	8,247	3.3	*	13.9	76.9	*	*	*	*	2.6	96.4	3.6
geonpeas	3,398	5.8	4.3	*	88.9	*	*	*	*	*	100.0	
road Beans,	3,331	22.5	56.6	2.9	*	4.6	*	9.6	*	2.8	87.1	12.9
ntils	2,816	2.2	4.0	31.7	39.1	*	*	*	18.2	*	79.4	20.6
wpeas,	2,546	95.8	*	*	*	*	*	*	*	*	98.7	1.3
hers	6,037	19.3	11.0	5.7	19.0	*	8.9	15.3	*	19.9	55.9	44.1
t. Pulses	**56,995**	**12.3**	**13.7**	**5.0**	**29.2**	**10.9**	**7.3**	**10.9**	**6.4**	**3.5**	**71.4**	**28.6**

Regions for which share of world total is less than 2%

1 Latin America & Caribbean   2 North America

Figure 2. Changes in Global Pulse Production by Crop, Triennium Averages

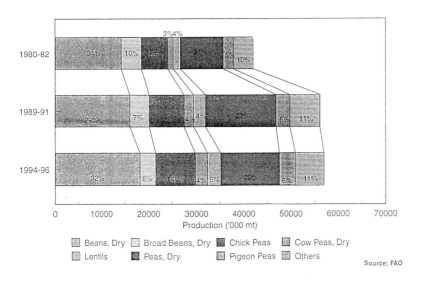

Source: FAO

Broad bean is the only pulse crop that has decreased in relative and absolute terms since 1980, down from 10% to 6% of total world production, with a growth rate of -1.6%. This was mostly attributable to area

---

[6] This data does not include Kenya, which is considered the largest producer in Africa (Muehlbauer et al, 1996).

reductions in China, which fell from 2.3 million ha in 1980 to 1.4 million in 1990 and to 1.1 million in 1996. Cyprus, Israel, and Japan have virtually stopped producing broad beans. This masks a 20.3% annual growth rate in production in Oceania (14.4% p.a. for area, 5.1% p.a. for yield). China (57%), Egypt (13%), and Ethiopia (8%) are the main producers.

Table 4. Compound growth rates (% per annum) of Production of dry grain of Major Pulses, by Region, 1980-96

	World	Africa	East Asia	West Asia	South Asia	L Amer Carib[1]	CIS	Europe	N Amer[2]	Oceania
Beans,	**1.7**	1.7	0.7	3.3	2.4	1.8	0.3	-4.2	1.6	10.7
Peas,	**2.5**	0.8	-5.9	3.6	4.1	1.5	-2.1	13.0	12.7	9.0
Chickpeas	**2.2**	0.9	-0.8	7.8	1.5	-1.4	-	-2.5	-	30.8
Pigeonpeas	**2.8**	2.5	11.3	-	2.6	1.7	-	-	-	-
Broad Beans,	**-1.6**	-1.7	-3.2	0.9	-	-1.1	-	-3.2	0.3	20.3
Lentils	**4.6**	-1.1	40.8	4.2	3.8	0.3	12.4	-6.6	11.7	30.1
Cowpeas,	**6.8**	7.2	7.1	-0.9	-3.3	-0.1	-	5.8	-10.7	-4.1
Others	**2.2**	2.0	2.8	-1.1	0.2	1.7	-3.7	4.2	-	15.1
Tot. Pulses	**2.1**	**2.6**	**-1.3**	**4.2**	**2.0**	**1.6**	**-2.3**	**6.4**	**5.8**	**13.9**

1 Latin America & Caribbean   2 North America

Table 5. Compound growth rates (% per annum) in Yield of Major Pulses, by Region, 1980-96

	World	Africa	East Asia	West Asia	South Asia	L Amer. +Carib[1]	CIS	Europe	N Amer[2]	Oceania
Beans,	**1.3**	0.6	0.5	0.4	1.8	1.5	-2.1	3.2	0.7	1.6
Peas,	**2.6**	0.4	0.02	2.1	2.7	1.4	1.8	3.3	0.1	-2.9
Chickpeas	**1.4**	-0.7	-0.9	-1.4	1.7	2.1	-	-0.3	-	-1.4
Pigeonpeas	**0.3**	1.4	0.7	-	0.2	0.4	-	-	-	-
Broad Beans,	**-0.01**	0.01	2.5	0.8	-	0.4	-	1.7	-2.3	5.1
Lentils	**1.7**	0.4	9.0	-0.7	2.2	2.1	3.9	-0.9	2.1	4.2
Cowpeas,	**2.5**	2.8	2.1	0.7	0.2	4.6	-	-0.7	2.1	-1.7
Others	**1.4**	-0.9	1.7	-0.8	2.2	-3.0	1.9	3.0	-	1.4
Tot. Pulses	**1.3**	**0.1**	**0.3**	**-1.0**	**1.7**	**1.4**	**1.8**	**6.7**	**0.8**	**-0.1**

Source: Calculations from FAO

Lentils and cowpeas account for 5% and 4% of world pulse production, up from 4% and 3%, respectively. While lentils grew at 4.6% p.a., with 2.8% p.a. in area and 1.7% p.a. in yield, cowpeas had the highest global annual growth rate at 6.8%, with 4.1% p.a. in area and 2.6% p.a. in yield. This is mostly attributable to large increases in area (4.4% p.a.) and yield (2.8% p.a.) in Africa. The top producers of this crop are Nigeria (65%), Niger (17%), and Myanmar (2%), while lentil production is dominated by India (28%), Turkey (22%), and Canada (15%).

3.2 Trade

The world market volume[7] in 1995 stood at 8.0 million MT representing 14.2% of world production; the 1993-95 average was 7.6 million MT or 13.4% of production (Table 6). The proportion of production traded is now higher than in cereals, where world trade was 244 million MT, with 12.8% of production traded. The value of all pulse exports exceeded US$ 2.5 billion in 1993-95 (Table 7).

---

[7] As measured by exports, since export data are usually more reliable than import figures.

Table 6. Imports and Exports ('000 MT) of Total Pulses, by Region, 1980-82 and 1993-95

	Imports		Exports	
	1980-82	1993-95	1980-82	1993-95
Africa	330	786	181	177
East Asia	430	634	434	1734
West Asia	177	248	454	570
South Asia	196	952	9	86
L.America+Caribbean	812	714	300	430
CIS	16	21	39	267
Europe	1055	3608	522	2052
N. America	50	124	970	1692
Oceania	19	24	69	595
World	**3102**	**7130**	**2979**	**7603**
Developing	1731	3050	1373	2988
Developed	1371	4105	1606	4645

Table 7. Trade in pulses by Volume ('000 Mt) and Value ($ Million), 1980-95

Year	Volume of Exports	Growth Rate (%)	Value of Exports	Value in real terms[*]
1980-82	2,979		1,514	1,514
1985-87	4,659	11.3	1,688	1,261
1990-92	6,522	8.0	2,477	1,383
1993-95	7,604	5.5	2,547	1,194
[*] Nominal value deflated by 6% per annum				

Virtually all pulse crops have seen an increase in export volume (Table 8, 9) as well as in trade as a share of production (Table 10), suggesting that trade barriers are being reduced. The most widely exported crop relative to its production is still lentils, which however, saw only a slight increase. The highest increase in exports has occurred in dry peas, which are now almost as widely traded as lentils, and broad beans, mostly due to the 1974-78 changes in Europe (see Gent, 1994).

Table 8. World Trade in Selected Pulse Crops, 1980-1995

	Exports (Million MT)		
	1980-82	1989-91	1993-95
Beans,	1.4	1.8	2.3
Peas,	0.6	2.5	3.2
Chickpeas	0.2	0.5	0.4
Broad Beans,	0.2	0.5	0.6
Lentils	0.4	0.5	0.7
Total	**2.9**	**6.2**	**7.6**

Table 9. Imports and Exports ('000 MT), of Major Food Grain Legumes, by Region 1980-82 and 1993-95 Averages

	Beans		Peas		Chickpeas		Broad Beans		Lentils		Total Pulses	
	80-82	93-95	80-82	93-95	80-82	93-95	80-82	93-95	80-82	93-95	80-82	93-95
					Imports							
World	**1480**	**1829**	**536**	**2870**	**134**	**467**	**144**	**582**	**354**	**650**	**3102**	**7131**
Africa	116	284	7	43	24	38	5	116	113	145	330	787
East Asia	319	472	38	89	5	6	15	25	13	13	430	634
West Asia	6	18	2	10	35	48	21	20	65	51	177	248
South Asia	119	137	6	216	8	227	-	-	13	106	196	952
L Amer+Carib	522	355	74	120	10	14	-	1	26	101	812	714
CIS	16	-	-	21	-	-	-	-	-	-	16	21
Europe	347	479	390	2356	41	119	102	418	124	227	1055	3608
N America	26	46	12	36	12	14	-	2	1	8	50	124
Oceania	7	12	6	3	-	0.2	1	0.1	0.1	2	19	24
					Exports							
World	**1410**	**2347**	**633**	**3210**	**225**	**441**	**171**	**605**	**369**	**737**	**2979**	**7604**
Africa	61	108	1	6	4	3	39	8	6	16	181	177
East Asia	373	1254	3	33	1	1	15	313	6	57	434	1734
West Asia	15	77	-	1	153	191	26	14	250	260	454	570
South Asia	0.4	1	-	1	1	2	-	-	1	37	9	87
L Amer+Carib	216	327	9	27	64	66	-	1	10	8	301	430
CIS	-	-	38	267	-	-	-	-	-	-	38	267
Europe	66	101	319	1699	2	11	83	197	22	16	522	2052
N America	678	464	201	858	-	8	2	7	75	343	970	1693
Oceania	1	17	61	319	-	162	7	68	-	2	69	595

Table 10. Imports and Exports as a percentage Share of Total Production for Major Pulses, by Region, 1980-82 and 1993-95 Averages

	Beans		Peas		Chickpeas		Broad Beans		Lentils		Tot. Pulses	
	80-82	93-95	80-82	93-95	80-82	93-95	80-82	93-95	80-82	93-95	80-82	93-95
					Imports							
World	**10.4**	**10.5**	**6.0**	**21.2**	**2.4**	**6.2**	**3.4**	**17.8**	**23.7**	**23.1**	**7.4**	**12.6**
Africa	7.5	14.6	2.3	13.1	9.9	13.7	0.5	16.3	156.8	219.0	6.4	11.3
East Asia	11.0	13.9	1.7	8.0	4.8	7.1	0.6	1.4	1715.6	13.0	5.2	8.6
West Asia	2.5	4.9	22.9	89.9	7.9	4.3	24.1	20.1	13.7	5.5	10.8	8.6
South Asia	3.9	3.1	1.7	32.8	0.2	4.0	-	-	2.0	9.7	1.7	6.0
L Amer+Carib	11.9	6.7	67.6	94.9	5.5	10.9	0.0	0.3	45.2	177.3	16.2	12.2
CIS	27.3	-	-	0.5	-	-	-	-	-	-	0.3	0.4
Europe	50.9	121.5	46.1	48.3	45.8	218.2	21.4	125.4	195.5	799.9	39.0	53.8
N America	1.9	3.3	4.1	2.4	-	-	-	11.8	0.5	1.5	2.8	3.7
Oceania	170.4	44.2	4.5	0.5	-	0.1	12.0	0.1	21.3	14.5	6.6	1.1
					Exports							
World	**9.9**	**13.5**	**7.1**	**23.7**	**4.0**	**5.8**	**4.0**	**18.5**	**24.7**	**26.2**	**7.1**	**13.4**
Africa	4.0	5.5	0.4	1.8	1.7	0.9	4.1	1.2	8.5	24.4	3.5	2.5
East Asia	12.9	37.0	0.1	3.0	1.4	1.7	0.6	17.1	765.0	55.9	5.2	23.5
West Asia	6.0	21.4	1.0	5.6	34.3	16.8	30.2	13.5	52.8	28.0	27.7	19.8
South Asia	0.0	0.0	-	0.1	0.0	0.0	-	-	0.1	3.3	0.1	0.5
L Amer+Carib	4.9	6.2	8.6	21.1	37.3	51.6	0.0	0.8	17.3	13.4	6.0	7.4
CIS	-	-	0.8	6.0	-	-	-	-	-	-	0.7	5.2
Europe	9.6	25.6	37.8	40.9	2.5	20.0	17.4	59.0	34.3	57.2	19.3	30.6
N America	49.9	33.0	71.2	58.3	-	-	38.9	48.6	52.1	68.8	53.9	49.9
Oceania	25.7	63.6	47.2	62.7	-	85.2	78.4	66.9	-	17.3	23.8	27.9

Geographically, the highest increases in export shares have occurred in East Asia and Europe, while there have been decreases in West Asia and North America (which still exports half of its production). Europe exports roughly one third of its production[8]. As for import shares, they are still concentrated in Europe and growing (54% in 1993-95, up from 39% in 1980-82), followed by Latin America and the Caribbean (12%,

---

[8] Data problems exist due to transshipments being counted, e.g. Switzerland, Pakistan, and Macao appear as major exporters.

down from 16%) and Africa (11%, up from 6%). It is interesting to note that South Asia (India), so far not much engaged in exports, has entered the lentil trade, as have Africa and Oceania (Australia), while East and West Asia have pulled out to some extent. The highest import dependency exists in Europe, particularly for lentils. South Asia is said to be a fairly self-contained trade region (Kana, 1997), but their import share has risen to 6% in 1993-95, which translates into a 486% increase in real terms, giving South Asia the position of second largest importing region. Likewise, Myanmar is not only exporting to India, Pakistan, and Bangladesh, but also to Japan, Singapore, Korea, and Malaysia (Kyi, 1996).

In 1993-95, dry peas were the most important pulse crop traded by volume (3.21 million MT or 42%) followed by dry beans (2.35 million MT or 31%) (Table 8). The dry beans trade was valued at 1.11 billion US$ as opposed to dry peas at US$ 715 million, illustrating the higher valuation of food (beans) vs. feed (peas) use. The lentil trade was 0.7 million MT (US$ 280 million). The total average value of all legumes exported in 1993-95 was 2.55 billion US$, nominally up from 1990-92 by 700 million US$, but down in real terms (Table 7).

3.3   Utilization

Looking at consumption (Table 11), the total world production is split into roughly two-thirds for food (62% in 1993-95, up from 59% in 1989-91, but still down from 69% in 1980-82), while about a quarter goes to feed (27% in 1993-95, markedly up from 19% in 1980-82 and down from 29% in 1989-91), and the remaining tenth is used for seed, waste, and stockholding (11% in 1993-95 and 1989-91, down from 13% in 1980-82).

Table 11. Food & Feed Use of Total Pulses, by Region, 1980-82 and 1993-95 Averages

	Average Quantity ('000 MT)						Share of total Pulse Utilization (%)					
	1980-82			1993-95			1980-82			1993-95		
	Food	Feed	Other	Food	Feed	Other	Food	Feed	Other	Food	Feed	Other
World	28677	7819	5358	34517	15400*	6200*	68.5	18.7	12.8	61.5	27.4	11.0
Africa	4233	119	817	5852	99	1285	81.9	2.3	15.8	80.9	1.4	17.8
East Asia	6122	1198	855	5168	1102	732	74.9	14.7	10.5	73.8	15.7	10.5
West Asia	959	159	221	1771	360	413	71.6	11.9	16.5	69.6	14.2	16.2
South Asia	9761	972	1197	13696	1294	1494	81.8	8.1	10.8	83.1	7.8	9.1
L Amer+Carib	4671	67	586	5402	29	636	87.7	1.3	11.0	89.0	0.5	10.5
CIS	771	3645	955	374*	3811*	651*	14.4	67.9	17.8	7.7	78.5	13.6
Europe	1398	1491	443	1716	5924	645	42.0	44.7	13.3	20.7	71.5	7.8
N America	735	18	88	1229	251	232	87.4	2.1	10.5	71.8	14.7	13.5
Oceania	28	150	39	88	723	194	12.8	69.3	18.0	8.7	72.0	19.3

* estimated

The aggregate figures conceal differing trends in the regions. Africa, South Asia, and Latin America and the Caribbean, for example, consume over 80% of pulses as food in the form of soft cooked beans in gravy eaten over rice or cassava; cleaned split pulse *dhals* eaten with rice; and whole cooked beans, respectively. The shares have gone down for Africa and up for South Asia and Latin America/ Caribbean. However, due to absolute increases in consumption over the past 15 years, the compound growth rates for food consumption have been 2.7%, 2.3%, and 1.1%. Feed use has declined for Africa (-2% p.a.) and Latin America and the Caribbean (-5.4% p.a.)[9], and increased in South Asia (1.7% p.a.) (Table 12).

East and West Asia and North America use between 70% and 74% of their consumption for food, with the trend pointing down, and between 14% and 16% for feed, with the trend going up. Again due to differences in the totals between these regions, the growth rates have been -1.5%, 5.1%, and 4.3%, respectively, for food consumption. Feed use declined slightly in East Asia (-0.3% p.a.), but increased sharply in West Asia (6.8% p.a.), and dramatically in North America (24% p.a.).

---

[9] Note that the large growth rate here, as in the case of feed use in Oceania, reflects a very low base level and is thus not significant in the overall scenario.

12

Table 12. Change (Compound growth rates) in Utilization of Pulse Products, by Region, 1980-95

	food	feed	other[*]
Africa	2.71	-2.03	4.15
East Asia	-1.52	-0.26	-1.50
West Asia	5.06	6.80	5.20
South Asia	2.31	1.66	1.60
L Amer + Carib	1.07	-5.43	0.67
CIS	-4.49	4.40[*]	-
Europe	1.40	11.61	2.07
N America	4.29	23.96	8.22
Oceania	7.33	11.11	12.26
World	1.50	7.97[*]	2.02[*]

[*] including seed, waste and industrial use

The regions with the highest feed use -- above 70% and trends sharply up -- are parts of the former USSR (now CIS), Europe (dominated by the EC), and Oceania (dominated by Australia). Compound growth rates for these three have been -4.5%, 1.4%, and a very high 7.3% for food use, and 4.4%, 11.6%, and 11.1% for feed use, respectively.

Examining pulse utilization rates in selected developing countries and their development over the past 15 years, only China (East Asia) shows a marked departure from the "food" group (above 70% food use), yet its new proportions (53:35 food:feed ratio) still keep it well out of the "feed" group. Colombia, Brazil, and Mexico (Latin America) and Indonesia, Malaysia, and Thailand (East Asia) all show negligible proportions of their pulse use in feed, and their food pulse shares are all above 80%. Turkey (West Asia) increased the proportions of both food and feed uses to 59:23 (from 56:20) as other uses declined.

As for per capita consumption of pulses and pulse products for food, the world average has fluctuated between a low of 5.8 kg per annum for the year 1992 and a high of 6.6 kg in 1983, around a slowly declining trend. Although aggregate consumption of food pulses at the global level rose by 1.5% per annum between 1980-95, per capita consumption declined over the entire period by 2.8%. This disaggregates into per capita consumption figures fluctuating between 2.7 kg and 3.1 kg for the developed countries group, with a 2.3% growth in per capita consumption over the entire period. Per capita consumption for the developing countries group declined by 6% over the 15-year period, with consumption fluctuating between 7.9 kg and 6.7 kg per annum. One explanation for these developments is the saturation of developed country diets with animal protein so that other foods are slowly becoming more desirable, whereas total protein consumption in developing countries is still shifting away from pulse and cereal protein towards animal protein, when affordable.

4.   REVISITING THE MAIN RESEARCH QUESTIONS

4.1   Protein Deficiencies in Developing Country Diets

While it is understood that animal protein will replace cereal and pulse protein in the human diet as incomes go up or relative prices change[10], researchers are still postulating shortages of pulse protein in developing country diets, as indicated by declining, or at best stable, per capita consumption figures (Agostini and Khan, 1988; Gowda et al., 1997). While this may be true for very poor countries and poor segments of advancing developing countries, we hold that for the most part, the reduction in pulse consumption has not contributed to a deteriorating protein nutritional status. As shown below, the data appear to substantiate this view.

Pulse Protein Overrated? Although pulses are an inexpensive source of protein (Pachico, 1993), their protein status is probably overrated. Walker (1987) cites three reasons. First, most of the nutritional

---

[10] For a detailed analysis of pigeonpea protein in Indian diets, see Mueller et al. (1990). They found that "the objective of maximizing protein consumption would be achieved not by buying more protein-rich pigeonpea, but by buying more low-protein cereals. As a consequence, the contribution of pigeonpea to total protein consumption would fall when both budget constraint and minimum pulse requirement were relaxed" (p.467).

evidence indicates that protein deficiencies are rare in poor urban and rural households in Asia. Deficiencies related to lack of calories, vitamins, and minerals are more common. Secondly, other studies have shown that a threshold amount of calories must be ingested before additional protein can be efficiently absorbed. So for energy deficient individuals, it is calories and not protein that is required. And finally, more than dietary concerns or cost of nutrients, taste preference influences food intake. Thus, even traditionally pulse-consuming societies will add vegetables, milk and dairy products (vegetarian), eggs, and seafood and meat (non-vegetarian) to their diets, and increase the share of preferred pulses, at the expense of "basic" pulses.

Competing Protein Source. Technological progress benefits products at different rates and other protein-rich foods can become relatively cheaper than pulses. For example, the use of pharmaceuticals in the poultry industry has made it possible to produce eggs and chicken on much less land, with labor-saving techniques, leading to lower prices. Fish and prawn farms are becoming alternatives to "traditional agriculture". Milk production per cow has vastly increased through breeding programmes using artificial insemination. All these technologies have been adopted by developing countries, leading to lower prices of animal protein in those products, which are acceptable to populations which observe dietary restrictions, such as vegetarians. At the same time, pulse prices have increased relative to other foods. Table 13 gives values for India.

Table 13. Expenditure and Own Price Elasticities for Selected Commodities of Consumption in India

	Expenditure elasticity		Own price elasticity	
	Model I[1]	Model II[2]	Model I[1]	Model II[2]
Rice	0.049	0.384	-0.250	-0.46
Wheat	-0.075	0.382	-0.185	-0.60
Pulses	0.283	0.647	-0.518	-1.07
Milk	0.435	1.100	-0.637	-0.66
Meat, Fish & Eggs	0.770	0.890	-0.878	-0.56

1 Kumar et al (1994)   2 Radhakrishna and Ravi (1990)

If our hypothesis is true, we can expect the following developments in protein intake per capita for developing countries:

a) an increase in the total,

b) the shares of cereal and pulse protein first increasing and later declining, or at least stabilizing, and the share of animal protein increasing, over time, at a rate determined by changes in relative prices, changes in income and dietary preferences, and

c) a reduction in intake of the less-preferred pulses, to the benefit of more highly valued ones, and

d) greater fluctuation of all the figures for poorer countries who rely less on imports and government programs to buffer their food economies, and for countries which produce mostly for their own consumption.

For protein-saturated (mainly developed) countries, we expect alternative uses of pulses such as in convenience foods (e.g., frozen foods, take-outs, semi-prepared foods, snack foods) and specialty foods (e.g., vegetarian dishes in otherwise non-vegetarian cultures, "natural" foods), and a gradual decline in protein from animal sources for reasons of health.

Visual Evidence We have chosen graphs depicting average daily per capita intake of protein from cereal, pulses, and animal sources over a 25-year period to support our hypotheses. Figure 3 shows the world per capita intake of all protein increasing over time, from 56 g/day in 1970 to 63 g/day in 1995, a 12.5% increase. At the same time, intake of protein from cereals increased moderately from 30 g to 33 g per person per day (10.0%) as against a 23.8% increase - from 21 g to 26 g - for dietary protein from animal sources, while pulse protein intake declined from 5 g to 4 g, by 20%. Thus, the protein composition was 54% cereal, 9% pulse, and 38% animal protein in 1970 as against 52% to 6% to 41% in 1995[11].

Figures 4 and 5 disaggregate the world picture into developed vs. developing countries. In the developed countries, total protein intake increased over the period, from 83 g to a peak of 90 g in 1990, and has been declining since, to 86 g in 1995. At the same time, cereal protein has been fairly stable, as has been pulse

---

[11] This assumes only negligible amounts of protein from other sources such as oilseeds and nuts, which does not hold true for countries which consume large amounts of soybean in their diet.

protein at 2 g per day. Protein from animal sources increased from 1970 to the peak in 1990, and has since declined. For the developing countries as a whole, per capita protein from all sources has gone up by 20%, from 45 g in 1970 to 56 g in 1995. Cereal protein contributed 30 g in 1970 and 34 g in 1995, while pulse protein contributed 6 g and 4 g, respectively. Animal products contributed 10 g to the total in 1970, as against 18 g in 1995, an 80% increase. To compare the two groups, the total protein intake in developing countries was 54% of that of developed in 1970, and 65% in 1995. Progress has clearly been made.

Asia (East, South, and West) is home to countries with different dietary preferences, natural endowments, and income levels. Japan is a rich, industrialised, densely populated, protein-saturated country with a highly protected farm sector; Myanmar, a small, poor, agrarian country newly emerging as a pulse exporter; and Turkey, a fairly large country at the top end of the developing country scale, which emerged as a major chickpea producer and exporter in the 1980s. India, with a per capita income of US$ 320 in 1994 and largely producing for its own population, shows a slowly increasing daily per capita intake between 49 g in 1970 and 52 g in 1995.[12] A largely vegetarian country, India had 33 g of daily protein from cereals in 1970, as against 35 g in 1995, and 10 g and 8 g from pulses, respectively. Dispelling some notions about the strictly vegetarian Indian diet, animal protein stood at 6 g in 1970 and overtook pulses for the first time in 1980, and has been consistently above pulse protein since 1988, reaching 9 g in 1995 (Figure 6).

China, with a per capita income of US$530 in 1994 was the world's second largest net exporter of pulses in 1995. It shows a strong growth in per capita protein intake, a very large increase in animal protein, and pulse protein declining from 5 g to 1 g per day (Figure 7).

Latin America Brazil had a 1994 per-capita-income of US$ 2970. It belongs to a group of meat and pulse-consumers and shows little volatility in its relatively high protein intake, which rose from 52 g in 1970 to 65 g in 1995 (Figure 8). Cereals accounted for 36% of the daily protein intake in 1970-72, but by 1995 this share had fallen to 31%. (Actual consumption rose, however.) Pulses contributed 13 g in 1970-72 and have since declined to 10 g in 1995. Animal products contributed 20 g in 1970-72 and the share has since gone up to 35 g in 1995.

Africa Nigeria is a large pulse growing country with a per capita income of only $260. The protein intake has been low and highly volatile (Figure 9). Animal protein has seen no expansion since 1985. Cereals have contributed most to the total, while pulses have fluctuated widely. It appears that pulses have since 1985 served to replace a share of animal products in the diet.

North America The U.S.A. are an excellent example of a protein-saturated exporter (Figure 10). Per capita income is a very high US$ 25,880. Pulses are produced for export as well as home consumption, driven by several pulse-eating ethnic groups. Only in the last year, has protein intake from pulses risen in the USA, from 2 to 3 g/day per capita[13].

4.2   Supply vs. Demand Constraint

Why has per capita consumption of pulses declined in developing countries? Is there a lack of supply so that demand can not be met (supply constraint), or are consumers not interested in purchasing more pulses (demand constraint)? India may serve again as an example. Radhakrishna and Ravi (1990) and Kumar et al. (1996) have estimated expenditure and price elasticities of demand for various food commodities, i.e., they have shown how much of a change in consumption will occur with a one percent increase in the price of that commodity (own price elasticity) or a one percent increase in total household expenses (expenditure elasticity). In a demand-constrained pulse economy, we would expect very low or negative expenditure elasticities, and likewise low price elasticities - i.e., an inelastic demand, which will not react to income or price changes.

---

[12] Note that the peaks and valleys are due to whole figures in the database.

[13] Values are rounded off to the nearest whole number in the FAO data for the USA.

Figure 3. Protein Consumption from Various Sources, 1970-94
World

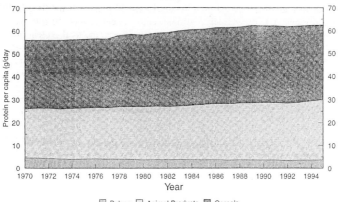

Source: FAO

Figure 4. Protein Consumption from Various Sources, 1970-94
Developed Countries

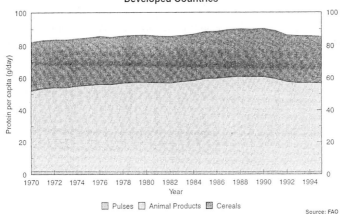

Source: FAO

Figure 5. Protein Consumption from Various Sources, 1970-94
Developing Countries

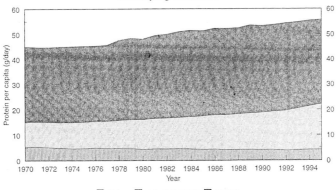

Source: FAO

Figure 6. Protein Consumption from Various Sources, 1970-94
India

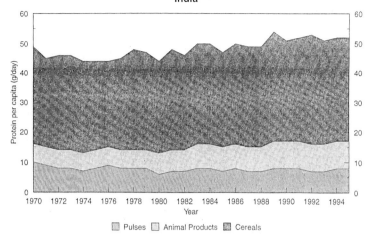

Pulses ☐ Animal Products ▨ Cereals

Figure 7. Protein Consumption from Various Sources, 1970-94
China

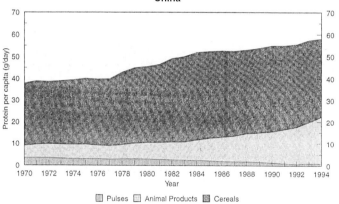

Pulses ☐ Animal Products ▨ Cereals

Figure 8. Protein Consumption from Various Sources, 1970-94
Brazil

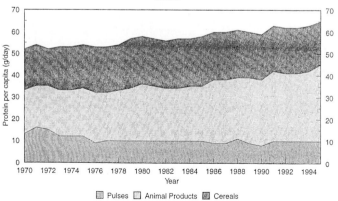

Pulses ☐ Animal Products ▨ Cereals

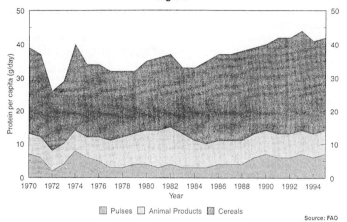

Figure 9. Protein Consumption from Various Sources, 1970-94
Nigeria

Source: FAO

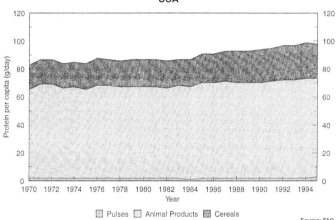

Figure 10. Protein Consumption from Various Sources, 1970-94
USA

Source: FAO

Elasticities In both studies, expenditure elasticities of demand for pulses are higher than for rice and wheat, but lower than for milk and meat, indicating that as incomes go up, Indian consumers buy relatively more pulses than cereals, but less pulses than animal products. (Table 13) Price elasticities of demand for pulses are higher than for the cereals in both studies. For meat and milk, the outcome is not quite clear, as Kumar et al. (1994) estimate lower price elasticities of demand for pulses, while Radhakrishna and Ravi (1990) give higher elasticities for pulses. In either case, a one percent increase in pulse price leads to a 0.5% to 1.1% reduction in pulse consumption, which points very clearly to a supply constraint.

Using their estimated elasticities and assuming a 5% p.a. income growth and a 1.9% population growth rate, Kumar et al. (1994) have estimated growth in demand for pulses in India at 3.3% p.a. to the year 2010, which is comparable to a 1.1% increase in per capita consumption. Past trends do not bear out this prediction. Figure 11 shows a steady decline in consumption in India since 1970 from 16.3 to 13.3 kg per capita per annum. This is curious, as real incomes in India rose throughout the period. The only reasonable explanation for the reduction in consumption is the high growth in prices of pulses observed during this period.

Figure 11. Total Pulse Consumption in India, 1970 to 1995

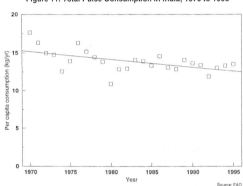

Source: FAO

Pulse Prices Too High Table 14 brings together changes in production, per capita availability, and real prices for selected pulses as well as "all pulses", wheat, rice, and milk. Pulses as a group registered a higher growth rate in price than wheat, rice, and even milk. Falling real (deflated) prices for wheat and rice attest to the impressive gains in productivity through technical change in these crops over the period during which per capita availability of wheat increased by 50%, and that of rice by 17%. At the same time, real prices of wheat and rice fell by 29% and 16%, respectively. Technological improvements in the dairy sector tell the same story: Gains in productivity through improved breeds and management resulted in an 85% increase in milk availability per capita in India over the 25-year period, while real prices rose by a mere 4%. By comparison, "all pulse" prices increased by almost 30%.

Table 14. Production, Per Capita Availability, and Price Indices of Selected Foods in India, 1970-72 and 1994-96

	Production ('000 MT)		Growth	Per capita availability (kg/yr)		Growth	Real-price indices (1970 = 100)		Growth
	Triennium Average			Triennium Average			Triennium Average		
	70-72	94-96	(%)	70-72	94-96	(%)	70-72	91-93	(%)
Wheat	23445	62742	167.6	44.6	67.2	50.8	94.8	67.2	-29.1
Rice	62269	120338	93.3	110.3	129.6	17.4	99.0	82.6	-16.4
Milk	21373	60948	185.0	38.9	72.0	85.1	108.0	111.8	3.5
Chickpeas	5275	5730	8.6	9.3	6.2	-33.1	101.3	141.3	39.5
Pigeonpeas	1803	3000	66.4	3.2	3.2	1.6	99.8	135.9	36.2
Lentils	390	792	103.1	0.7	0.9	35.2	120.0	138.4	15.4
Mungbean	595	1459*	145.3	1.1	1.7*	60.1	112.2	129.0	14.9
Blackgram	601	1675*	178.5	1.1	1.9*	82.2	126.6	115.0	-9.2
All Pulses	11446	15143	32.3	20.1	16.8	-16.6	106.0	136.6	28.9

* Based on 1990-92 triennium average

As for differences between pulse crops, chickpea and pigeonpea, which account for well over half the pulse production in India, have registered real price increases of 40% and 36% since 1970, more than any other food group. This dampenend any positive effects from rising incomes. Surprisingly, this rise in price had little effect on production. Throughout the 70s and 80s, the profitability of oilseeds and cereals rose relative to pulses; the former due largely to rising prices (Ashok Gulati, personal communication, 1997), and the latter by lowering per-unit cost of production through irrigation and adoption of improved cultivars. More than anything else, the lack of a technological breakthrough in pulses and slow growth in productivity (yields increased by a mere 0.6% per annum between 1970 and 1995) led to the shortfall in production in India. Because of import restrictions and the generally higher level of prices paid for imported pulses (world market price plus tariffs ranging between 35% and 20%, more recently 10%), per capita availability fell steadily at the rate of 0.7% p.a., or 16.6% over the period.

These findings clearly support a supply-constrained situation in India with respect to pulses. Demand does not appear to be limiting, as implied by the high expenditure elasticity of demand and high (negative) price elasticity of demand for most pulse crops. Population and income growth ensure a healthy long-term demand for this commodity in India, assuming prices can be kept down.

Can Supply Keep Up? In order to simply maintain per capita availability of pulses in India, production would have to increase at the rate of population growth, i.e., slightly less than 2% p.a. Production increases may occur from increases in area or yield. In the past, the area under pulses has remained stagnant while yields have increased by only 0.6% p.a. Yields in India currently average 610 kg/ha. Assuming no expansion in area, an increase of 1.9% p.a. to the year 2010 would require an additional 245 kg/ha, raising yields to 855 kg/ha. To achieve the required 3.3% p.a. aggregate growth in consumption estimated by Kumar et al. (1996), yields would have to increase to nearly 1000 kg/ha. Only a significant breakthrough in research and its rapid adoption would permit such an advance in India. New technologies exist, but much needs to be done to assess their potential and in identifying on-farm constraints that may limit their uptake.

The above scenario assumes that pulses will continue to be grown primarily on the land now allocated to them, i.e., the more marginal areas. A shift in government policy, resulting in a change in their price, relative to cereals and/ or oilseeds, could alter the situation dramatically. But as shown above, higher pulse prices have a major adverse effect on domestic demand. This brings out the major challenge for pulse research: How to close the gap between existing supply and unmet demand? In other words, what can be done to increase supply at a rate that would make pulses "cheaper"? Oram and Agcaoili (1994) suggested five possible solutions, the most promising being to increase productivity through technological change. They pointed out that "current pulse yields in most countries are both low and highly variable", which leads us to an analysis of yield variability. In an earlier paper, Muehlbauer et al (1996) indicated that technological progress in germplasm development for pulses was promising to alleviate multiple stresses in some pulses, thus making them more adaptable to adverse growing conditions. Later in this paper we discuss the possibility of closing the gap through imports.

4.3   Yield Variability

Productivity is defined as output relative to input use, without consideration of cost. However, output is most often measured relative to that input factor which is in shortest supply -- typically land. Thus, crop productivity is often measured as yield. As sustainability concerns are broadening the interpretation of legume productivity (see Byerlee and White, 1997), other measurements of productivity may need to be defined. This leaves unaffected the extreme output variability noted by Agostini and Khan (1988, p.464-465) and Oram and Agcaoili (1994, p.35-36). We undertake a closer look at this phenomenon.

4.3.1   Yield Variation between Regions

To understand pulse production and outputs, one must consider the basic systems in which they are grown. This also provides an indication of the factors in short supply.

Figure 12 shows the yield levels in each of the major growing regions of the world. Yields in Europe and North America are triple and double the world average of 810 kg/ha. Adding CIS and Oceania, the developed countries achieve yields of 1640 kg/ha as opposed to 670 kg/ha for the developing countries (Africa, Asia, and Latin America and the Caribbean). Which factors account for these differences?

Figure 12. Total Pulse Yields, Regional Averages, 1994-96

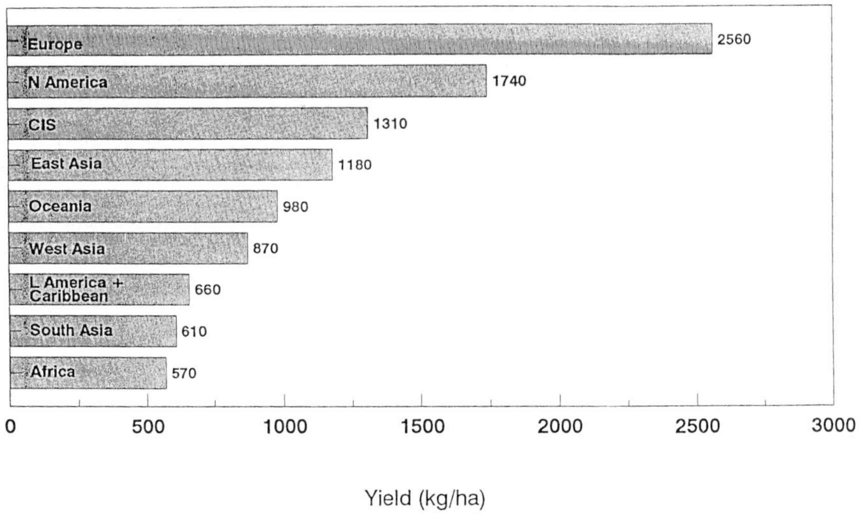

Yield (kg/ha)

Climate/Soil: Most of the developed countries are in the temperate zone, where lower temperatures and longer days during summer provide an environment conducive to high yields. Evapo-transpiration is lower, and moisture stress is not as frequent in these environments. In the tropics or sub-tropics where most developing countries are located, double cropping is a common practice in systems which include pulses, which tends to reduce the individual yields of each crop. -- There is a low rainfall in much of the tropical rainfed area and terminal heat/drought stress and poor, highly degraded soils occur, particularly in Africa. Short-duration pulse varieties would be one answer to the challenge faced by producers in developing countries (Kumar et al., 1996).

Subsistence: Agricultural production continues to have a subsistence objective in many of the developing countries, as opposed to the highly commercial production of developed countries, especially those who do not protect their farmers from world competition. Where farmers produce primarily for the market, they adjust factors of production in response to relative prices, i.e., productivity for all factors would be higher in such systems. Where employment opportunities exist outside of agriculture, labor productivity will be much higher than in subsistence agriculture, and all limiting factors will be used to produce as much as possible for the market. Farmers in a subsistence situation will exert less effort in raising yields if the meeting of household production targets, which change little from year to year, is the objective.

By-Products: Producers in developing countries often have a need for non-grain by-products (Kelley et al., 1996; Kelley et al., 1993). For pulses, this includes fodder (e.g., cowpea, pigeonpea, lentil) or fencing (pigeonpea). Grain yield would then be only one of several objectives, and lower (grain) yields would be the opportunity cost of meeting multiple objectives.

Level of Development/ Infrastructure: The general level of development determines the support for agriculture indirectly in terms of infrastructure, and more directly as income support. Better infrastructure provides easier access to inputs such as seed, fertilizer, pesticides, mechanization, and credit. It also means more highly developed markets, better transport, and information services. All this translates into higher yields not only for pulses, but all crops. In addition, research investments -- public and private -- are higher in developed countries (Gowda et al., 1997). Where economies mature, agriculture releases labor into higher-paying occupations, which puts pressure on governments to provide "parity" for the remaining agricultural population, leading to agricultural support policies. These are in place in many developed countries (e.g., EC and other European countries, Japan) or were until recently (e.g., USA, New Zealand,

CIS), which can have a direct effect on pulse yields[14] (e.g., dry peas in France). Agriculture in developing countries, on the other hand, is often net-taxed (Gulati and Sharma, 1995). Where policies have been favorable to agriculture, cereal crops such as wheat and rice have benefited often at the expense of pulses and coarse cereals (Gulati and Sharma, 1997).

4.3.2 Yield Variation Between Pulse Crops

Average global yields for specific pulse crops are shown in Figure 13. Dry peas, 80% of which are produced in developed countries, register the highest average yields at 1.6 t/ha. More than any other pulse crop, dry peas, whether grown in developed or developing countries, appear to have comparatively higher yields. This is illustrated in Table 15 where dry pea yields are higher than "total pulses" yields in virtually every region. In West Asia, for example, yields of dry peas are almost 2 t/ha, compared to the region's average total pulse yield of 0.9 t/ha. This points to a *crop-specific effect* underlying variations between regions which may partly result from "more sophisticated research attention" afforded certain (usually non-tropical) pulses (C Johansen, personal communication, 1997).

Figure 13. Average Global Yields for Specific Pulse Crops, 1994-96

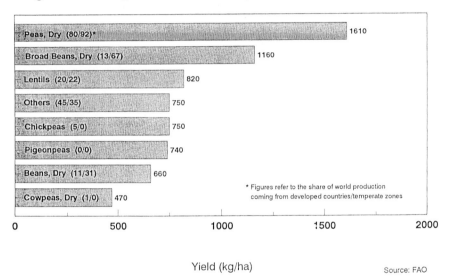

Yield (kg/ha)

Source: FAO

Broad beans are the next highest yielding with 1.2 t/ha most of which is produced in East Asia mainly China. Much of the crop area is in the temperate zone. A similar type crop-specific effect is also evident for broad beans, but more difficult to tease out because of the obvious regional (temperate climate) effect. Yield levels of lentil, chickpea, pigeonpea, and dry beans are roughly similar -- 820 to 660 kg/ha, and the vast majority of these is produced in developing countries. Cowpea, grown almost exclusively in Africa, a particularly disadvantaged region in the warm tropics, has the lowest yields at 470 kg/ha.

---

[14] If there is no upper limit to total payments, producers will increase non-land inputs to achieve higher output. In the case of set-aside requirements, it is usually the poorest land which is set aside, thus leading to an increase of average productivity across the remaining cropped land.

As already discussed in Section 4.3.1, differences in pulse crop yields at the global level can be explained in part by regional effects[15]. Nevertheless, there also appear to be crop-specific effects, as the data in Table 15 illustrate for dry peas and broad beans.

Table 15. Average yields (kg/ha) of Major Pulses by Region for the period 1994-96.

	World	Africa	Asia			L Amer + Carib	CIS	Europe	N Amer	Oceania
			East Asia	West Asia	South Asia					
Beans,	660	700	1,030	1,340	430	640	1,140	900	1,810	640
Peas,	1,610	680	1,540	1,950	830	860	1,350	4,030	1,950	1,100
Chickpeas	750	560	660	740	760	1,220	-	580	-	970
Pigeonpeas	740	730	630	-	750	730	-	-	-	-
Broad Beans	1,160	1,230	1,730	1,860	-	710	-	1,820	610	1,100
Lentils	820	600	1,150	920	880	750	930	670	1,260	1,050
Cowpeas	470	460	690	970	-	710	-	2,890	900	400
Others	750	480	990	820	510	880	1,190	1,840	-	950
Tot. Pulses	**810**	**570**	**1,180**	**870**	**610**	**660**	**1,320**	**2,560**	**1,740**	**980**

4.3.3   Yield Variations within Regions over Time

As geography and type of pulse crop influence yield levels, so does development over time as expressed in growth rates. More than the other two measures, yield growth reflects variable inputs. Of greatest interest to scientists are the medium- and long-term growth rates which reflect technological progress and/ or rates of return on inputs. Table 16 shows yield growth rates of total pulses for the period 1980-1996.

Table 16. Yield Growth Rates (% per annum) of 'Total Pulses', 1980-96 and 1990-96, by Region

Region	1980-96	1990-96
Africa	0.12	1.09
East Asia	0.26	0.07
West Asia	-1.03	0.21
South Asia	1.68	1.83
L. Amer +Carib	1.42	1.83
CIS	1.76	-2.91
Europe	6.73	-0.39
N. America	0.78	-0.70
Oceania	-0.06	-0.14
World	**1.26**	**-0.78**
Developing	0.72	0.35
Developed	3.05	-1.20

Overall, developed countries achieved the highest growth in yields of 3.1% per annum, compared to less than 1% per annum for the developing countries. Europe had an impressive 6.7% annual growth rate due mainly to strong support policies, but made possible by higher inputs. Yields stagnated in Africa, East Asia, and Oceania, and shrank in West Asia. North America had a slight growth, while South Asia and Latin

---

[15] For example, average yields for dry beans are low for the developing country regions (Africa, South Asia, Latin America/Caribbean), moderate for East Asia (of which much is temperate), and high for North America. The case of lentils is similar, though not as dramatic. For chickpea, where the major producers are located in tropical climates under low rainfall conditions, Oceania (primarily Australia, which is a developed country) has yields roughly 30% higher than South and West Asia, pointing to a developed country (infrastructure) effect. Pigeonpea yields are low relative to other pulse crops and they are grown exclusively in developing countries.

America and the Caribbean increased their yields by about 1.5% per annum. The factors accounting for the inter-regional differences are related to[16]:

a) level of development/ infrastructure (see 4.3.1. above)

b) crop-specific effects (see 4.3.2. above)

c) changes in cropping pattern: Pulse crops are being pushed into more marginal lands and this fact in sometimes reflected in declining, static, or slow-growing yields, e.g., chickpea in India (Kelley and Parthasarathy Rao, 1994) or sorghum in many parts of Africa (ICRISAT/FAO, 1997). Alternatively, as crop area expands into better endowed regions due to price changes, average yields can grow without technological change, which would be visible in a high correlation between yield increase and area increases.

d) Statistical anomaly: Area reductions (increases) in countries where pulse yields are above (below) the regional average will have a depressing (inflating) effect on yield growth rates even without any changes in the other countries' yields.

e) Base-yield level: All other factors being equal, growth rates will tend to be higher the lower the base level. Of course, low base yields are highly correlated with the developing region effect, and hence this effect may be difficult to tease out.

4.3.4. Yield Stability

Variability in yield is a major source of production instability and merits close investigation. In this analysis the coefficient of variation (CV) in yields is used as a measure of yield variability. The CV around the trend[17] was calculated for major pulse crops for each of the world regions using yield data from 1980 to 1996. Only the CVs for pulses which have at least 2% of the world share in any region are listed (Table 17).

Table 17. Pulse Yield Variability (as CV %) between 1980 and 1996 for Major Pulse Crops, by Region (1980-96)

	World	Africa	Asia			L Amer +Carib	CIS	Europe	N Amer	Oceania
			East Asia	West Asia	South Asia					
Beans	2.9	3.3	7.8	**	6.4	8.1			6.5	
Peas	10.8		9.8				19.6	8.3	10.3	17.3
Chickpeas	6.8			7.7	8.6	-			-	24.8
Pigeonpeas	6.1			-	7.0	-		-	-	-
Broad Beans	9.8	9.7	10.7		-		-	11.4		37.4
Lentils	6.4			13.4	5.7				19.3	
Cowpeas,	10.7	11.3					-			
Other	7.9	6.6	2.7		4.2			12.8	-	14.0
Tot. Pulses	4.0	4.1	6.9	7.6	4.6	7.1	18.3	12.2	6.2	14.4

*CV (coefficient of variation) was calculated after detrending the yield data using linear estimation.
*Blank cells indicate the region's share was less than 2% of world production.

At the global level, dry peas, cowpeas and broad beans have the largest CV values (10.8%, 10.7%, and 9.8%, respectively), indicating that these crops exhibit the greatest yield variability. The source of this instability can be ascertained from the table where yields of dry peas in CIS are shown to be extremely variable (CV = 19.5%), as they are in Oceania (CV = 17.3%). They are more moderate in Europe (8.3%) and N. America (10.3%). The high CV in CIS probably reflects the seemingly wild swings in production--due to area and yield changes--throughout the 80s and first half of the 90s, but particularly the latter. High CVs in Oceania are generally characteristic of that region due to great weather variability. Additionally, Oceania is, with respect to pulse geographic area, relatively small, compared to other regions (South Asia, Africa, etc.),

---

[16] It is beyond the scope of this paper to attempt to econometrically model differences in current yield levels, yield growth rates, or yield variability over time for pulses. We simply indicate those factors thought to be most relevant.

[17] That is, the standard error of regression divided by the mean, i.e., a de-trended CV. This is almost identical to the Cuddy-Della Valle index (Singh and Byerlee, 1989).

and there is less scope for smoothing out the fluctuations in yield from one sub-region to another, as for example in South Asia, or even within India.

The high CV for cowpeas suggests that year to year yield variability on this low input crop is probably strongly influenced by climatic events, rainfall mainly, in Africa. Note as well the relatively high CV for broad beans in Africa (9.7%), another low input pulse crop of which only a small amount is grown, mainly in Egypt. Dry beans have the lowest CV of any pulse crop, 2.9%. However, this is very likely due to the fact that "dry beans" includes numerous species and year to year variability in one species is averaged out by the others.

Yield stability is influenced by factors such as climate, irrigation, degree of commercialization, political and economic instability, and geographic coverage. Table 18 illustrates a few of these points. For instance, the CV of yields for chickpea and pigeonpea are higher in Rajasthan, an arid region, compared to Madhya Pradesh, a higher rainfall region. The CVs at the all-India level are lower compared to the individual states as India has a large geographical coverage, so that the variations in yield tend to average out.

Table 18. Yield Variability, (as coefficients of variation) for Selected Crops in India and Australia[*]

	Chickpea	Pigeonpea	Wheat
All-India	7.7	8.5	3.4
Madhya Pradesh(Semi-arid)	7.4	16.8	6.1
Rajasthan (Arid)	13.9	45.2	12.9
Australia	25.0		18.0

[*]CV (coefficient of variation) calculated after detrending the yield data. Data for chickpea and wheat: India, M.P. and Rajasthan, 1980-94. Pigeonpea: India and M.P., 1980-94; Rajasthan, 1980-92. Data for Australia for wheat from 1980-96 and for chickpea, 1983-96.

Source: Government of India, Agricultural Statistics at a Glance, 1994

The more market-oriented a farm economy, the more responsive its output will be to product and input prices. This may lead to large year-to-year fluctuations compared to production in countries where the crop is grown largely for subsistence. Figure 14 illustrates the case of India and Australia. Chickpea production in Australia roughly follows the trends in market price, while chickpea production in India appears almost unrelated to price. Thus the CVs in Table 18, which are considerably higher for Australia than for India, may also be explained by level of commercialization. Climatic variability in Australia probably contributes significantly as well, as does the narrow range of crop geographic area.

Finally, political, social, and economic instability probably play a role in variability in yields, i.e., high CVs, as characterized by the CIS since the early 1990s.

5. POINTS TO PONDER WHEN HAZARDING PREDICTIONS

The world market for pulses has been strongly influenced by unexpected developments over the past 15 to 25 years. These range from advances in competing crops such as Green Revolution cereals and soybean, and competing foods such as milk, meat, and vegetables, over government policies as varied as soyameal replacement in the EC (Gent, 1994), withdrawal of rail rate subsidies in Canada (Thompson, 1997; McVicar et al., 1997), and change of whole political systems (break-up of the former USSR and an export regime for Myanmar, formerly Burma), to technological progress in pulse production such as fallow replacement in Turkey (Acikgoz et al, 1994) and new methods of cultivar selection (Muehlbauer et al., 1996), to advances in pulse utilization (such as the semi-prepared pulse-based convenience foods and extruded snack foods now appearing on supermarket shelves in Indian cities), to name just a few. Byerlee and White (1998) discuss some of the more complex issues in pulse production systems.

The reversible (policies) as well as irreversible (technological progress) changes in the pulse economy have had a marked impact on market volumes and players, as discussed in Section 3 above. Many of the changes were unpredictable, and we believe they are likely to occur again in other forms in the future. Scientists concerned with world food issues have been waiting for technological breakthroughs in pulse production. The question of if, when and how there will be a "Green Revolution in pulses" continues to be discussed (Muehlbauer and Kaiser, 1994.) Therefore, we are unlikely to produce correct medium-term predictions for any market details. Only five years after Agostini and Khan's paper (1988), Oram and

Agcaoili (1994) dismissed FAO predictions, and any predictions based on production trends during the late 1980s did not hold for the early 1990s.

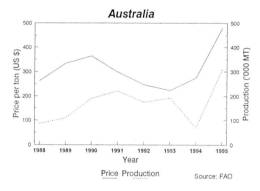

Figure 14. Chickpea Production and Prices

What we can discuss are the factors which will influence pulse production and markets. Oram and Agcaoili (1994) provided a list to which we have added other factors (Table 19).

Table 19. Factors Affecting Pulse Demand and Supply

		Demand	Supply
Social	Population growth	+	
	Migration/Tourism	+	
	Urbanization/Convenience foods	?	
	Concern for the poor	?	?
Dietary	Health Concerns	+/-	
Economic	Currency convertibility		+
	Import substitution	-	-
	Free Trade	+	-/+
Agricultural/			
Environmental	Sustainable systems		+
	Reduced chemicals		+/-
	High cost of production	-	-
	Yield Variability		-
Infrastructural	Research investments		+
	New Uses	+	
	Transportation	+	+
	Intelligence/Communication	+	+

Source: Adapted from Oram and Agcaoili, 1994

Social factors:

* Population growth, especially in India and China, will not have as strong an impact on demand as assumed earlier given the decline in per-capita consumption of pulses in those nations. Even though consumption of animal protein has gone up in both cases, the growth rate for feed use of pulses in their respective world region have been relatively low: 1.7% for India, and 0.3% for East Asia.

* Migration and tourism, have caused ethnic foods to become available all over the world. These activities are also introducing pulse-based dishes from South Asia and Latin America to affluent populations in developed countries. This is illustrated by the fact that Trinidad and Tobago, with a large population of ethnic Indians, boasts the world's highest pigeonpea yields of 1,620 kg/ha. The U.S.A., which is home to many immigrants from Latin America and South Asia, and also to a fast food chain called *Taco Bell* which offers Latin American bean dishes, has just seen an increase in the per capita intake of protein from pulses from the long-term level of 2 grams per day to 3 grams per day[18]. It would not surprise us to see a fast food chain emerge which offers Indian food.

* Urbanization, while reducing the importance of traditional pulse dishes, contributes to acceptance of non-traditional pulse preparations which arise out of imports, inter-regional trade, and new cultivars. It also gives rise to a demand for convenience foods, especially since dry pulses require long cooking times. In India, "Ready-to-mix" pulse preparations are fast entering supermarkets; in the U.S., pulses are sold frozen (Oram and Agcaoili, 1994); and Nestle Lanka is using mung beans in infant meals (Kana, 1997).

* A concern for the poor is behind much of the aid to developing countries, and is behind aid targeted specifically to small farmers in poorly endowed environments. This could influence investments in pulse research increasing production as well as alternative rural development which would take the poorest soils out of production.

Dietary factors:

• Health concerns in protein-saturated countries are not as yet showing up as an increase in pulse consumption. However, there has been a decline in consumption of all protein, corresponding to the decline in animal protein, in Germany, which has an affluent and health-conscious population and many "new" vegetarians (Figure 15).

Figure 15. Protein Consumption from Various Sources, 1970-94 Germany

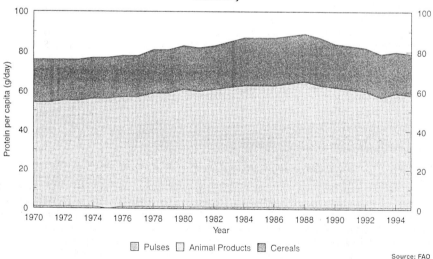

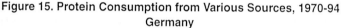

Source: FAO

---

[18] The change occurred in 1995 (FAOSTAT, 1997). Many countries report only whole numbers in this statistic, which leads to distortions.

* Reduced meat consumption can be expected to lower the demand for feed pulses; a trend which is concealed by the agricultural policies promoting feed pulse utilization in Europe, which has the world's highest growth rate for feed use.

Economic/ political factors:
* The alternative to producing domestic pulse requirements is to meet them through imports. Trade has played a minor role in many developing countries in the past, but is likely to become more important if domestic policies on trade and exchange rates are changed. Pakistan, for example, has opened its markets to pulse imports at the expense of domestic production (Kyi et al., 1997). How readily consumers take to imported commodities will depend on their quality and price.

Exchange rates are extremely important to agriculture. Myanmar's export expansion followed a devaluation of the kyat (Kyi, 1997). In India, full convertibility of the rupee would benefit domestic pulse producers who would not only gain from higher prices as imports became more expensive, but also improve their competitive position relative to oilseeds.[19]

* Import substitution in the EC was the driving force behind the pulse expansion in Europe discussed above. A reversal will further reduce the size of the world feed pulse economy, but will re-open markets for non-EC imports.

* The World Trade Agreement (Uruguay Round of GATT) "reduced and bound domestic agricultural subsidies" to lay open trade distortions and encourage a higher degree of free trade, which allows agricultural production to move to locations of comparative advantage and which decreases instability in the world markets. The U.S.A. made sweeping changes in their 1996 farm legislation; the EC passed one agricultural reform packet in 1992 and is discussing further changes to a system of direct income support (Thompson, 1997).

Agricultural/ environmental factors:
* Pulses are beneficial to cropping systems (Byerlee and White, 1998). As a result of concerns about sustainability, pulses may be made integral to international aid for agricultural development and also to new income support measures in the EEC.

* Penalties for high-input of fertilizers in Europe would reduce production costs of pulses there.

Infrastructural factors:
* Research investments in pulses have been relatively low, and more so in developing countries (Gowda et al, 1997). More funds could direct more research attention to technological breakthroughs in pulses, which would bring down their prices and increase supply and demand.

* "New Uses" would also benefit from research attention and could open up new markets for pulses as such programs have done elsewhere, e.g., biodegradeable plastics and ink from maize in the U.S.A.

* Transportation is crucial to market development and trade. Breakthroughs in this area would have a high impact on the pulse economy.

* Information and communication are important to efficient market development and trade. With increased trade in pulses and an integration of the markets for developed and developing countries, we should expect higher investments in market research and marketing efforts all over the world.

6. Summary and Conclusions

The expansion in the pulse economy of the 1980s and present stagnation are in fact a "bubble" largely based on only two regions, the USSR and the EC, where agricultural subsidies created a supply of dry peas which was utilized domestically as feed with the help of a second set of subsidies to processors (EC) (Hubertus Wolfgarten, personal communication) or as a direct subsidy to the livestock industry (USSR) (Thompson, 1997). With the break-up of the USSR and the reforms to the CAP, the bubble burst. We would assume the world pulse economy to return to a scenario of slow, but steady growth in production and consumption of pulses for food (mostly in developing countries, but also for select quality in developed countries) and feed (mostly in developed countries). Trade will play an increasingly important role in the pulse economy, and the sophisticated marketing techniques of developed country exporters can be expected to have an impact on developing country markets, particularly if coupled with improvements in infrastructure.

---

[19] Pulse producers switched over to soybean, rape/ mustard, sunflower and groundnut as a result of the 1986 Technology Oilseeds Mission set up to reduce imports of edible oils (Ashok Gulati, personal communication). We would expect edible oils to be imported if India would open up the markets.

Protein deficiency in developing-country diets is not directly related to declining per capita intake of pulse protein. We have shown that people in all but the poorest developing countries are consuming increasing amounts of protein from animal sources in addition to cereals. Even in India, animal products have overtaken pulses as a source of protein in the human diet. However, pulse consumption would increase considerably if their price came down.

A review of the supply/demand balance for India points to a supply-constrained situation for pulses, at least in the short to medium term. Declining consumption of pulses is related to slow growth in productivity, rendering pulses relatively less profitable than competing crops. This has resulted in production shortfalls and a subsequent rise in real prices, pushing pulses which exhibit price elasticities of demand between -0.5 and -1.1 further out of the diet.

Yield variability, a major problem in the pulse economy, is influenced by factors such as climate (especially rainfall), political, social, and economic instability, access to irrigation, degree of commercialization or price responsiveness, and geographical coverage. A coefficient of variation can be constructed for comparisons between pulses produced in different regions. It might also help in defining strategies to assure greater stability in world food supply.

International and interregional trade is becoming a major factor in the world pulse economy as markets are opening up, allowing production to begin to move to the relatively "best" location, promising higher welfare to all market participants. At the same time, concerns for sustainability are being built into the development and agriculture policy agenda. For pulses, this may imply a re-focusing on the more poorly endowed, rainfed environments and the need to re-define productivity of pulse systems in terms of a factor that may well prove to be scarcer than land: water.

Acknowledgements

We would like to thank C. Johansen for thought-provoking comments on an earlier draft of this paper, S. Valasayya for his valuable assistance with data processing, and Rama Laxmi and Shanti Priya for their unfailing and cheerful computer support.

References

Acikgoz, N., Karaca, M., Er, C. and Meyveci, K. 1994. In: *Expanding the Production and Use of Cool Season Food Legumes*. pp 388-98 (eds F.J. Muehlbauer and W.J. Kaiser). Kluwer Academic Publishers.

Agostini, B.B. and Khan, D. 1988. In: *World Crops; Cool Season Food Legumes*, pp 461-82 (ed R.J. Summerfield). Kluwer Academic Publishers.

Byerlee, D. and White, R. 1998. In *Proceedings IFLRC-III: Linking Research and Marketing Opportunities for the 21st Century*. (ed. R. Knight) Dordrecht: Kluwer Academic Publishers (in press).

Carrouee, B., Gent, G.P. and Summerfield, R.J. 1997. In *Proceedings IFLRC-III: Linking Research and Marketing Opportunities for the 21st Century*. (ed. R. Knight) Dordrecht: Kluwer Academic Publishers (in press).

FAO (Food and Agriculture Organization of the United Nations). 1997. FAOSTAT Production, Utilization and Trade Tapes.

ICRISAT and FAO. 1997., FAO Commodities and Trade Division and the Socio-Economics and Policy Division, International Crops Research Institute for the Semi-Arid Tropics (ICRISAT).

Gent, G.P. 1994. In: *Expanding the Production and Use of Cool Season Food Legumes*. pp 361-66 (eds F.J. Muehlbauer and W.J. Kaiser) Kluwer Academic Publishers.

Gowda, C.L.L., Ali, M., Erskine, W., Halila, H., Johansen, C., Kusmenoglu, I., Mahmoud, S.A., Malik, B.A., Meskine, M., Rahman, M.M., Sapkota, R.P. and Zong Xuxiao. 1998 In *Proceedings IFLRC-III: Linking Research and Marketing Opportunities for the 21st Century*. (ed. R. Knight) Dordrecht: Kluwer Academic Publishers (in press).

Gulati, A. and Sharma A. 1997. *Agricultural Trade Liberalization and Efficiency Gains - An Analysis of Major Indian Crops*. NCAER/ICRISAT study. Unpublished Report.

Gulati, A. and Sharma, A. 1995. *Subsidy Syndrome in Indian Agriculture. Economic and Political Weekly, September 30*.

Kana, N.L. 1997. *Pulses Trade Study. Paper presented at the Regional Workshop on Market Prospects of Upland Crops in Asia*, in Bogor, Indonesia, 25-28 February.

Kelley, T.G., P. Parthasarathy Rao, E. Weltzein R. and M.L. Purohit 1996. *Experimental Agriculture*: 32 (2): 161-172.

Kelley, T.G. and P. Parthasarathy Rao 1996. In *Adaptation of Chickpea in the West Asia and North Africa Region* pp. 239-254 (eds N.P. Saxena, M.C. Saxena, C Johansen, S.M. Virmani, and H. Harris.). ICRISAT/ICARDA, 1996.

Kelley, T.G. and P. Parthasarathy Rao 1994. *Economic and Political Weekly*: 29 (26): 89-100.

Kelley, T.G., Parthasarathy Rao, P. and Walker, T.S. 1993. In *Social Science Research for Agricultural Technology Development: Spatial and Temporal Dimensions*. pp. 88-105 (ed Dvorak K.) CABI, London.

Kumar, P. 1996. *Market Prospects for Upland Crops in India*. Working Paper No. 20. The CGPRT Centre, November 1996.

Kumar, J., Sethi, S.C., Johansen, C., Kelley, T.G., Rahman, M.M., and Van Rheenen, H.A. 1996. *Indian Journal Dryland Agriculture Research & Devolopment.* 11(1), 28-32.

Kyi, H. 1996. *Production, Consumption and Marketing of Selected Pulses in Myanmar.* Unpublished Report, Yangon, July 22.

Kyi, H., Mruthyunjaya, Khan, N.A., Liyanapathirana, R. and Bottema, J.W.T. 1997. *Market Prospects for Pulses in South Asia: International and Domestic Trade.* Working Paper 27, The CGPRT Centre, Bogor, Indonesia.

McVicar, R., Slinkard, R.E., Vandenberg, A. and Clancey, B. 1998. In *Proceedings IFLRC-III: Linking Research and Marketing Opportunities for the 21st Century.* (ed. R. Knight) Dordrecht: Kluwer Academic Publishers (in press).

Mueller, R.A.E., Parthasarathy Rao, P., and Subba Rao, K.V. 1990. In *The Pigeonpea* (ed Nene, Y.L.), UK: CAB International.

Muehlbauer, F.J., Johansen, C., Singh, L., and Kelley, T.G. 1996. *Crop Improvement - Emerging Trends in Pulses.* Paper presented at the 2nd International Crop Science Conference in New Delhi, November.

Oram, P.A. and Agcaoili, M. 1994. In: *Expanding the Production and Use of Cool Season Food Legumes.* pp 3-49 (eds F.J. Muehlbauer and W.J. Kaiser). Kluwer Academic Publishers.

Pachico, D. 1993. In Trends in CIAT Commodities, 1993. Working Paper No. 128, July. CIAT.

Pala, M., Armstrong, E., and Johansen, C. 1998. In *Proceedings IFLRC-III: Linking Research and Marketing Opportunities for the 21st Century.* (ed. R. Knight) Dordrecht: Kluwer Academic Publishers (in press).

Parthasarathy Rao, P. and Von Oppen, M. 1987. pp. 54-63 In *Food Legume Improvement for Asian Farming Systems:* Proceedings of an International Workshop, 1-5 September, 1986, Khon Kaen, Thailand (Wallis, E.S. and Byth, D.E. eds) Canberra, Australia: ACIAR

Radhakrishna, R. and Ravi, C. 1990. *Food Demand Projections for India.* Hyderabad, A.P., India: Centre for Economic and Social Studies.

Roberts, E.H. 1994. In: *Expanding the Production and Use of Cool Season Food Legumes.* pp 983-8 (eds F.J. Muehlbauer and W.J. Kaiser). Kluwer Academic Publishers.

Singh, U., Williams, P.C. and D.S. Petterson. 1998 In *Proceedings IFLRC-III: Linking Research and Marketing Opportunities for the 21st Century.* (ed. R. Knight) Dordrecht: Kluwer Academic Publishers (in press).Singh, A.J., and Byerlee, D. 1989. Indian Journal of Agricultural Economics

Thompson, R.L. 1997. *Presidential Address,* XXIII International Conference of Agricultural Economists, Sacramento, California, USA, 10 August.

Walker, T.S. 1987. In *Food Legume Improvement for Asian Farming Systems:* pp. 208-10 Proceedings of an International Workshop, Khon Kaen, Thailand (eds Wallis, E.S. and Byth, D.E.) Canberra, Australia: ACIAR.

World Bank. 1997. *World Development Indicators.* The World Bank, Washington, D.C.

Agricultural Systems Intensification and Diversification through Food Legumes: Technological and Policy Options

Derek Byerlee[1] and Robert White[2]

1 Rural Development Dept., World Bank, Washington, USA; 2 Dept. of Agricultural Economics, Michigan State University.

Abstract

Food legumes provide an important opportunity to contribute to world food supplies in a sustainable way, through intensification and diversification of agricultural systems in the developing world and by providing a major source of nutrition for the world's poor. However, over the past two decades per capita production and consumption of food legumes has steadily fallen, in contrast to the rapid growth of cereals, soybeans, and livestock products. Yields of food legumes, especially those grown in the cool season, have stagnated as crop area has often been forced into marginal areas and no yield breakthroughs have occurred. Meanwhile, there are serious problems of sustainability of cereal-based systems in the favored areas. This is due, in part, to an increased dominance of monocropping with displacement of legumes. Legume production is also risky. A high variation in yield is compounded by seasonal and annual price fluctuations in both domestic and international markets, which are often thin and fragmented.

Recent examples of the intensification and diversification of agricultural systems through the introduction of legumes in Turkey, Pakistan, and West Africa are discussed. These success stories highlight key issues in increasing food legume supplies, especially the identification of appropriate varieties for specific niches, reducing risks through tolerance to biotic and abiotic stresses, greater emphasis on consumer preferences for grain type, and seeking opportunities to lower costs, especially labor costs. Increased emphasis should also be given to identifying niches in more favored environments which offer potential for expanding the legume area, especially through double and triple cropping. Whether for favored or marginal areas, targeted extension campaigns will be an important element of any strategy. Finally, there is potential to improve markets and reduce price fluctuations through liberalization of trade and investment in market information, infrastructure, public awareness programs, and research on convenience foods.

INTRODUCTION

The world continues to be challenged by the need to produce sufficient food to meet the growth in population and incomes. The current best projection is that grain supply for developing countries will have to increase by at least 50% by the year, 2020 (Rosegrant et al., 1997). At the same time, traditional sources of growth in production, through expansion of land area, developing irrigation, and applying Green Revolution-type technologies are being exhausted. Indeed, annual growth in world grain production has slowed sharply from 2.7% in 1967-76 to 1.3% in the past decade.

Cereal cropping covers close to half of the world's cultivated land area but the system is under threat due to increased reliance on monocropping, and the mismanagement of irrigation and agri-chemicals. Yet intensification of cereal systems is critical to meeting world food requirements in the next century. Such systems need to be diversified in order to enhance their sustainability and

31

R. Knight (ed.), Linking Research and Marketing Opportunities for Pulses in the 21st Century, 31–46.
© *2000 Kluwer Academic Publishers. Printed in the Netherlands.*

protect the natural resource base. The introduction, or in many cases, re-introduction, of legumes into cereal systems will be central to increasing productivity and sustainability.

Food legumes complement cereals in production and consumption. On the production side, legumes in rotational cropping, inter-cropping and alley cropping, can be a source of nitrogen, which is ecologically sustainable and economically viable. Rotation with legumes also raises crop responsiveness to fertilizer, increases organic matter, reduces leaching losses, and helps control diseases and pests. Their restorative powers were extolled as early as the first century BC when the poet Virgil recommended:

...as the season changes you will sow the golden emmer on land whence you have earlier taken off the prolific bean with its rattling pods, or the tender growth of the vetch, and the fragile stems of the bitter lupine and its crackling foliage

Cereal-based systems as diverse as the rice-wheat irrigated systems of the Indo-Gangetic plains and the maize systems of southern Africa have been adversely affected by a reduction in the area under legumes that has occurred with cereal monocropping. In the Punjab of Pakistan, where legumes have been largely displaced, evidence suggests that cereal yield growth has slowed and factor productivity has declined in the past two decades, suggesting serious problems of resource degradation (Figure 1) (Ali, 1997).[1] Likewise, maize yields in Africa have increased very slowly despite the application of Green Revolution-type technologies (Byerlee and Eicher, 1997).

At the same time, there are examples of the successful intensification of cereal systems through the introduction of legumes, especially in the rainfed areas of West Asia and Australia, and some irrigated systems of Asia. Understanding the technological and policy factors underlying these cases is critical to enhancing productivity and sustainability

On the consumption side, food legumes are a source of protein and calories, especially for the one billion poor people, who subsist on a daily wage of less than one dollar. Expanding supplies is a potentially effective means of addressing nutritional deficits in developing countries (Deshpande, 1992). Pulses usually provide the cheapest source of protein to the poor, and sometimes energy (Figure 2). Even so daily protein consumption is often below the minimum recommended intake. Even for the rich of the world, pulses are enjoying a revival as a component of a balanced diet, following health concerns about the consumption of livestock products.

This paper reviews trends in pulses production and consumption in developing countries, emphasising the technology and policy factors affecting supply and demand. We have analyzed data by regions and for key countries and reviewed hundreds of studies on food legumes. Our approach is to consider pulses as part of a larger system that includes competing crops and sources of energy and protein. For example we refer to soybeans, an increasingly important pulse in crop rotations and human diets, as well as a major feed source. We make little effort to distinguish between cool and warm-season pulses as trends in supply and demand have been similar and most factors influence both types of pulses.

GLOBAL TRENDS IN THE PRODUCTION AND CONSUMPTION OF PULSES

Kelley, Parthasarathy Rao and Kelley, in these proceedings provide a picture of the global performance of food legumes over the past 15 years. In this section, we highlight a few trends, and compare the performance of food legumes in developing countries with competing crops.

---

[1] Byerlee and Hussain (1992) found that yields of wheat sown after a rotation legume crop were significantly higher that in continuously cropped fields.

Looking especially at the developing world, the following trends emerge consistently

The growth rates in production over the past 20 years has been rapid in the industrialized countries of North America, Europe, and Oceania, and slow in developing regions, but with substantial regional variation. Except in Europe, an expansion in area has been the main source of growth in these countries as they diversify cropping patterns and exploit export market opportunities.

In developing countries, the area and yield performance of warm-season legumes has been better than for cool-season legumes (Figure 3).

By contrast soybean production has expanded rapidly in developing countries through increases in area and yield. Soybeans are nearly always grown in rotation with a cereal crop (Figure 3).

Yield growth for pulses has been particularly slow in nearly all regions, averaging only 0.5 % annually over the past 20 years in developing regions The major exception is Europe where field peas have enjoyed a remarkable transformation in two decades encouraged by incentives under the Common Agricultural Policy. Yield growth has been particularly slow in Latin America and West Asia. The slow growth in yields for food legumes contrasts with relatively rapid growth in yields of competing crops, notably cereals and soybeans.

There are differences between developed and developing countries in the yields of legumes and competing crops. In developing countries, yield performance in cereals and soybeans has outpaced that in developed countries. However, for pulses, the picture is reversed. Their yield performance in developed countries has been very favorable compared to that in developing countries, and to yield growth in wheat and soybeans (Figure 4).

As a result of these trends, the yield gap between developed and developing countries in soybeans, and especially cereals, is rapidly narrowing while the gap for pulses is large and widening (Figure 5).

Bean, pigeon pea and chickpea yields of between 0.5 and 0.75 t/ha have grown by less than one percent annually since 1967. While cowpea yields achieved an annual growth rate of 2%, their absolute yields remain very low at less than 0.4 t/ha.

Yields of pulses overall are more variable than for cereals, although within specific countries, they are no more variable than for cereals grown under rainfed conditions (Figure 6).

The above trends have led to a declining per capita production of pulses in most regions (Figure 7). Increasing imports, especially to South Asia and Latin America, have only partly compensated for these deficits. Per capita consumption of food legumes has generally declined. Their role as a source of protein has diminished as per capita consumption of cereals and especially livestock products has increased. Even in a predominantly vegetarian country such as India, livestock products now provide more protein than pulses (Figure 8).

In sum, the picture is not encouraging. On the production side, yield growth has been particularly disappointing. Meanwhile soybean, a relatively new crop in the developing world, has expanded rapidly, generally in a cereal-soybean system. On the consumption side, pulses now provide only a small, but still important share, of total protein supplies, behind cereals and livestock products.

34

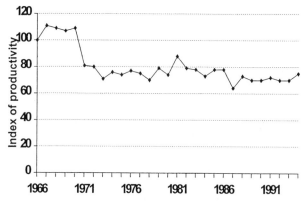

Figure 1: Index of Yields and Total Factor Productivity, Rice-Wheat System, Pakistan (Source: Ali 1997)

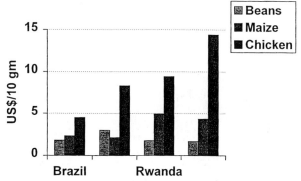

Figure 2: Cost per 10 g of Protein from Pulses, Cereals and Meat (Source: Pachico, 1993)

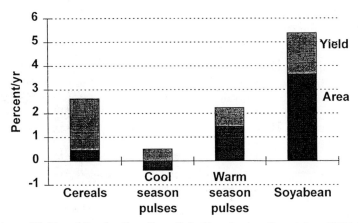

Figure 3: Area, Yield and Production Growth in Developing Countries, 1977-96 (Calculated from FAOSTAT)

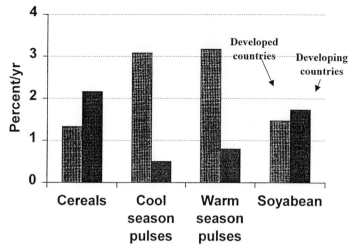

Figure 4: Annual Average Growth of Yields, 1977-96 (Calculated from FAOSTAT)

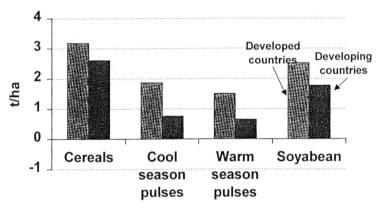

Figure 5: Annual Average Yields, 1977-96 (Calculated from FAOSTAT)

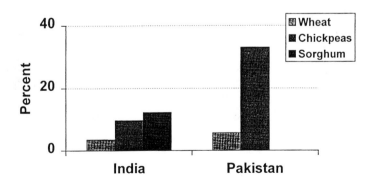

Figure 6: Coefficient of Variation of Yields around Trend, 1976-96

Factors Underlying the Decline in Pulses

A policy issue is whether the declining role of food legumes is a supply or demand problem. The observed decline in per capita consumption could indicate pulses are "inferior" foods whose consumption declines with increasing incomes. In fact, there is considerable evidence that an increase in consumption of pulses is moderately responsive to increasing incomes. Income elasticities for food legumes (the percentage increase in demand for a one percent increase in income) range from 0.3 to over 1.0, similar to those for cereals such as rice and wheat (Figure 9)., These are lower of course than for livestock products, which are a substitute source of protein, so that the *share* of pulses in expenditure patterns is likely to fall with increasing income (but not *absolute* consumption levels). The shift from rural to urban areas, where consumption of food legumes is lower, will further accelerate these trends.

The decline in per capita consumption of pulses reflects the rise in *real* consumer prices (Figure 10). This has been due to a lack of any significant technological breakthroughs, combined with consumers' relatively high responsiveness to the price of pulses—a 1% increase in consumer prices typically leads to a 1% decrease in consumption (Figure 11). With increasing real prices, consumers switch to other foods, such as cereals (whose real prices have been declining for three decades), livestock products, and vegetables. Data from Pakistan shows that with changing prices, there is significant substitution of pulses with these other foods (Ali and Ullah, 1997).

Rising real prices has resulted form the increasing costs of production due to rising costs of labor and land, and a lack of technical change. Meanwhile technical changes have occurred in competing crops. Without a reduction in the cost of pulses there is little likelihood of reversing their decline in human diets. Even with lower production costs and higher returns than other crops, the release of varieties incompatible with consumer preferences have often discouraged production. Without a yield break-through, it is unlikely pulses will be competitive as feed for livestock, given the progress in soybeans and their widespread availability.

On the supply side, the story for food legumes is generally consistent across countries and regions. Rapid technical change in cereal production has displaced food legumes from the favored areas toward marginal areas unsuitable for cereals. For example, prior to 1971, five northern states in India with good rainfall and irrigation accounted for over 70 percent of the chickpea area (Kelley and Parthasarathy, 1996). However, by 1989, the northern area had decreased to be equal to the more marginal southern zones. In such zones, pulses are produced under erratic rainfall and poor soil conditions. Intercropping and crop rotations have also given way to monocropping of both cereals and legumes in their respective production areas. For example, intercropping of beans with maize in Mexico has fallen from 50% of the bean area to just 15% in two decades, encouraged by labor-saving technologies in maize (mechanization and herbicide use) (World Bank, 1990). The spread of irrigation, especially in Asia, and the lack of suitable pulse technology for irrigated areas, has accelerated this trend.

Government policies have also played a part. Subsidies, have encouraged the use of nitrogen (N) fertilizer, for the higher-return cereal and cash crops. Governments, to achieve food security, have focused on cereals in research, extension, subsidized irrigation, marketing and price support policies.

The movement to marginal areas has increased the risks of production and has been a disincentive for farmers to invest in improved technologies. Yields of pulses in most countries are more variable than yields of cereals and the detrended coefficient of variation of chickpea yields between 1971 and 1989 was 19%, compared to 8% for wheat (Kelley and Parthasarathy Rao 1994).

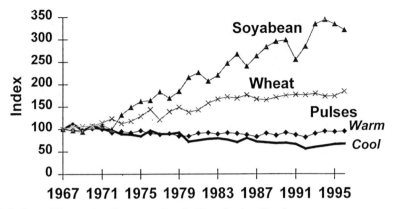

Figure 7: Index of Percapita Consumption in Developing Regions
(Calculated from FAOSTAT)

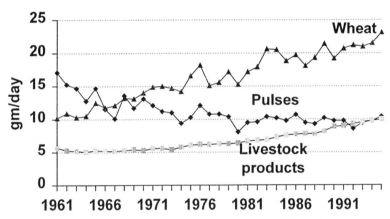

Figure 8: Percapita Supply of Protein from Pulses, Meat, and Wheat in India
(Calculated from FAOSTAT)

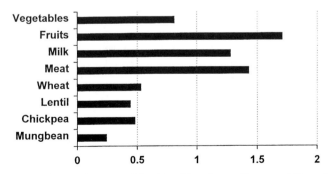

Figure 9: Income Elasticity of Demand for Food Products (Source: Ali et al., 1996)

38

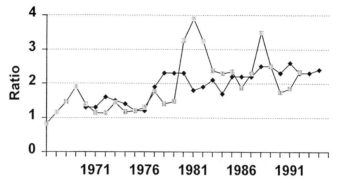

Figure 10: Ratio of Producer Price of Chickpea to Wheat, India and Pakistan

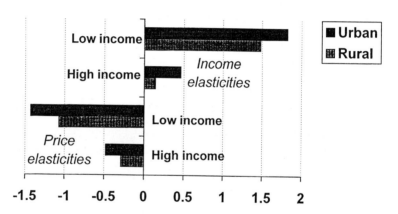

Figure11: Income and Price Elasticities of Demand for Chickpeas in India (Kelley and Rao, 1996)

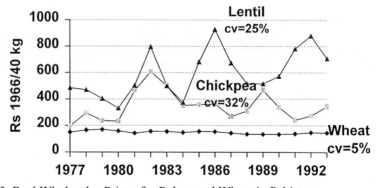

Figure 12: Real Wholesales Prices for Pulses and Wheat in Pakistan

These problems are particularly severe with cool-season pulses, due to their direct competition with wheat, which has shown spectacular gains over the past thirty years. As a result, these pulses have declined relative to warm-season pulses and cereals. Stagnant production and increasing demand have led to an increase in imports of food legumes to many developing countries.

Cool season pulses have done poorly relative to competing legumes, especially soybeans. The area of soybeans has expanded in all regions over the past 20 years, often in association with cereals. In India where pulse production has stagnated, soybeans are a spectacular success in the state of Madya Pradesh, with the use of previously fallowed land, strong extension efforts, high price supports and investments in processing (Wallis, 1977).

The thin and fragmented markets for pulses are also a problem. This is especially so for production in marginal areas. With poor infrastructure in these areas, localized consumer preferences, and variable production, market prices are unstable acting as a further disincentive (Figure 12). This contrasts with the relative price stability of competing crops such as wheat, where price support policies are the norm, or oilseed legumes which have the advantage of multiple uses for food, feed, and oil and an expanding demand.

Thus there are many agronomic and institutional factors constraining the production of cool season food pulses. Because they are grown in diverse and variable agroecological areas it is not easy to transfer uniform technological packages to raise production. Institutional obstacles to effective extension and technology transfer further complicate efforts to implement technologies.

SUCCESSFUL INTENSIFICATION AND DIVERSIFICATION

Despite the overall picture of a relative decline in the developing world, there are examples of the successful expansion of cool- and warm-season legumes in cereal-based systems.

Fallow substitution in rainfed wheat systems of Turkey

The substitution of fallow with chickpea and lentil crops is well documented for the medium rainfall wheat areas of the Anatolian plateau of Turkey (Duratan, et al., 1990; Wallis, 1997). The Government implemented a project to develop appropriate varieties and management practices for legumes, and launched an extension program to transfer the technology to farmers. The program also provided credit and inputs to expand production. The transition to pulses and their widespread adoption was facilitated by a policy that paid farmers above world prices. High returns were also partly due to the low level of farm investment required for intensification.

The fallow area in Turkey fell by 37% between 1979 and 1992 while wheat production expanded 20% and barley 35% (Wallis 1997). Nonetheless, the program depended on payment of high prices, including subsidies on exports, as well as inputs. The fall in the food legume area in the 1990s may indicate the program was not sustainable in light of the fiscal stringencies of the 1990s.

Fallow substitution in irrigated systems of Pakistan

A remarkable success has taken place in several irrigated districts of Pakistan over the past five years, where short-season mung beans have substituted for fallow in the less intensive irrigated wheat-based systems of the Punjab. This occurred despite the decline in per capita production of pulses (especially cool-season pulses) since the Green Revolution from 9.5 kg a year in 1970 to 3.4 kg a year in 1993.

As elsewhere in South Asia, pulses in Pakistan were displaced to marginal lands during the 1960s and 1970s by high-yielding cereal varieties in a policy environment favorable to cereals. However, in the 1990s, successful research, technological and policy advances have stimulated mungbean production in three districts. The key to success was the new very-short season varieties (60 days to maturity) with good disease resistance, a preferred grain type, and a short, erect stature that was suitable to mechanical harvesting (Ali et al. 1997). These varieties were rapidly adopted, displacing the traditional long-season variety, and almost tripling the area sown.

The average farmer yield was 45 percent higher than for local varieties. In addition, the larger seed and shiny seed coats were preferred by consumers, leading to higher prices (Table 1). Although farmers applied inputs to boost yields, these were offset by reduced labor costs resulting from lower weed growth. Net returns were three to four times higher than for local varieties (Ali et al. 1997). In addition, farmers applied 40% less nitrogen to wheat following mungbeans, while achieving 6% higher wheat yields (Table 2). Overall the program is estimated to have generated benefits of US$21 million in just three districts with 30% of the benefits resulting from higher returns in the following wheat crop.

Table 1 Benefits for Wheat of Rotation with Mungbeans, Pakistan

Rotation	Wheat yield (t/ha)	N applied (kg/ha)	Cost (Rs/ha)	Return (Rs/ha)
Fallow-wheat	2.65	119	8,209	2,731
Mungbean-wheat	2.80	84	6,524	5,036

Source: Ali et al., 1997

Table 2. Comparison of Local and Improved Mungbeans, Pakistan

Variety	On farm yield (kg/ha)	Days to maturity	Price grain (Rs/kg)	Net returns (Rs/ha)
Local	579	90	894	1215
Improved	900	58	988	5109

Source: Ali et al., 1997

Pigeon peas in India

A similar story holds for the early-maturing disease-resistant pigeon peas in rainfed cereal systems of southern India. Faced with rapidly spreading wilt, a high proportion of farmers adopted the new varieties. Adoption in northern Karnataka increased from 5% in 1987 to 60% in 1993 leading to average on-farm yield increases of 57%. Adoption rates in some villages reached 100% (Bantilan and Joshi. 1996). Bantilan and Joshi estimated there was an internal rate of return to the research investment of 62%. Using earlier-maturing pigeonpea cultivars meant there was a reduced risk of delays in the sowing of the following cereal crop (Wallis et al., 1988).

Cowpeas in West Africa

In West Africa, traditional intercropping of cowpeas with cereals was displaced by monocropping with the introduction of improved maize production (Byerlee and Eicher, 1997). Over the past ten years, however, improved cowpea technologies have led to a rapid expansion usually in a maize-legume rotation. In Ghana, new cowpea varieties and a strong extension effort have led to adoption rates of over 70 percent by 1994 (Denkyi et al., 1995). Most farmers also adopted insecticides and row planting; however, problems with insecticide use have stimulated research into integrated pest management.

In irrigated areas of northern Nigeria, cowpeas are a profitable crop in the cool season, competing favorably with wheat in rotation with coarse grain cereals (Inaizumi, et al., 1987). Production of cowpeas in the cool season has the advantage of reducing pest damage and hence costs. This example illustrates that in the tropics and subtropics, food legumes provide flexibility, often fitting into unexpected niches in cropping patterns. In addition, the distinction between cool-season and warm-season legumes no longer applies.

MAJOR CHALLENGES TO EXPANDING LEGUME PRODUCTION

Technological challenges

The development of technologies and their transfer are the key to increased production. of food legumes. In the past, research and extension have often ignored these crops in favour of cereals to improve food security. However, since about the mid-1970s, there has been substantial investment in pulse research and some of it has led to high returns (e.g., Bantilan and Joshi, 1996). Oram and Agcoaili (1992) estimated that 8% of crop scientists in developing countries in the 1980s worked on food legumes implying a research intensity (percent of output value invested in research) of about 0.5% compared to about 0.2% for cereals. Likewise in India, Mruthyunjaya and Ranjita (1995) found that pulse research was well supported relative to other crops. These data and the increasingly constrained budgets for public sector research suggest that budgets for pulses might decline rather than increase in the future.

One of the arguments for investing in food legumes is the benefit they provide to the total cropping system. The Pakistan experience described above indicated these can be substantial and may add up to 50 % to benefits computed for pulse production alone. Likewise in India wheat yields following legumes were as high as 50% greater than wheat yields following sorghum (Singh and Singh 1991). More work is needed along these lines if we are to make a case for increased investment in research on pulses.

For now, the issue is how to utilize research resources more effectively. One decision to be made is on the relative balance between favored and marginal areas. The shift to marginal areas has understandably encouraged research into their problems. The international agricultural research centers (notably ICRISAT, ICARDA, and CIAT) direct most of their food legume research to less favored areas. However, the future potential may be in the more favored areas, from which the pulses have been displaced, and where monocropping of cereals is no longer sustainable. This question has implications for the type of food legume to grow and whether warm-season legumes will be favored over cool-season legumes.

The technological issues can be classified into: appropriate varieties for ecological niches, consumers and end users, risk management, and reduced production costs.

Varieties for specific niches

The development of improved varieties seems to us to be the key to a technical break through in most areas. Without them it is unlikely farmers will invest in inputs and labor. At the same time, experience suggests varietal development and adoption, is a daunting task. In food legumes, more than for cereals, there are many factors to "get right" if new varieties are to be accepted including maturity, resistance to biotic stresses, improved drought, heat or cold tolerance, erectness to facilitate mechanisation, and grain type to fit consumer preferences.

The common ingredient in all of the success stories described above, was the development of varieties to fit specific niches in cropping patterns, especially ones with early maturity. In favored areas, the pulses compete with higher valued cereal and vegetable crops and farmers give priority to sowing and harvesting these crops. In addition, even where there is a gap in the cropping pattern to grow legumes, the gap is often made shorter by a lack of labor and of irrigation water. At the same time in marginal areas, earlier maturity is often needed to escape drought or heat stress.[2] Thus early maturity is often more important than yields in a farmers' acceptance of new varieties. Nonetheless, researchers need to be alert to opportunities to develop high-yielding varieties that will provide sufficient returns to allow substitution of cereals by legumes. This suggests that an important priority is to characterise existing and *potential* cropping patterns in terms of *both* agroclimatic and socioeconomic variables.

Consumer preferences

A pervasive factor in a farmer's acceptance of new varieties is his appreciation of consumer preferences. In many cases, a variety has been developed which has grain characteristics that lead to reduced market prices which negate any yield advantage (e.g., Tutwiler, 1995). Consumer preferences for grain type in food legumes are more variable between regions than for cereals. Table 3 shows how consumer preferences for bean types varied in Mexico. In marginal areas, the quantity and quality of crop residues as a livestock feed often is also important (Capper, 1990).

Table 3. Percent Consumption of Bean Types by Region, Mexico

Class	N. E.	N. W.	Centre	South
Black	1	0	11	90
Azufrado	5	98	14	2
Bayo	25	1	9	6
Pinto	45	0	2	0
Flor de Mayo	11	0	34	0
Flor de Junio	0	1	21	0
Others	13	1	9	2

Source: Castellanos, n.d.

Capacity for nitrogen fixation.

Another tradeoff in varietal development is the relative priority given to yield and nitrogen fixation. In southern Africa, for example, scientists have emphasised soybean varieties that give high yields but have little capacity to self-nodulate. Farmers however, have adopted the not-recommended self-nodulating varieties because they provide high soil enrichment (Kumwenda et al., 1997). Of course, N fixation can be enhanced through inoculation but this has rarely succeeded under small-holder conditions, despite considerable efforts to transfer the technology.

Reduction in risk

Varietal development will also be the main ingredient in reducing yield risks. Earlier maturity is important, as is increased resistance to biotic and abiotic stresses. In some cases, there are important interactions to be considered. In WANA, changing the date of sowing of chick pea

---

[2] In India, advances in chickpea cultivar development in India have produced new cultivars which are better adapted regions where chickpea production has expanded in recent years (Kelley and Parthasarathy Rao, 1996).

from the spring to the more favorable winter season stabilizes yield but requires improved disease resistance (Tutwiler, 1995).

Reducing production costs

In middle income countries such as Mexico and those in WANA, where farm wages average $3-7 a day, weeding and harvesting by hand account for up to 70% of labor costs (Haddad and Snobar 1990). In addition, if pulses are to fit into intensive cropping patterns, the timely completion of these operations is critical. Varieties suitable for mechanical harvesting will be an important part of the solution to labor constraints.

In favored areas, the use of pesticides adds to the production and environmental costs. The development of integrated pest management practices is needed to reduce these costs and risks.

Technology Transfer

The success of cereal production and other crops, especially soybeans has often resulted from campaigns that have integrated production technology, input supply, market support, and extension information. The rapid expansion in soybean production has been attributed to investments in research, mounting wide-scale extension programs, supporting producer prices, and encouraging industrialists to develop processing plants and export markets. Similar efforts on cool-season food legumes are needed. The advantages of incorporating legumes into crop rotations need to be widely demonstrated to farmers and policy makers.

Food legume production is often constrained by poor seed multiplication systems. In countless surveys in Latin America and Africa, farmers identify poor seed availability as the constraint limiting their adoption of new cultivars (e.g., Janssen 1989; Viana Ruano et al., 1996). Seed availability and viability are especially important for food legumes, which in many areas are minor crops and do not attract private sector participation in seed production. In addition, because of the large seed size, farmers have difficulty storing the seeds during the humid warm season and this compounds the problem of availability. In the absence of formal distribution systems, non-governmental organizations and local farmer associations have sometimes been effective in seed distribution.

Marketing and price policy

Price risk is often high for food legume producers. The question then is what policy options are available to reduce risks, especially for small farmers? In the past, some governments have implemented price support programs through governmental marketing agencies. While this has been less common than for cereals, there have been instances in Turkey and Mexico. The question is, should governments use this policy more widely?

There are several reasons why this may not be the most efficient option. The first is that a high support price to producers must be paid either by consumers, to the detriment of the poor who are consumers of food legumes, or from tax revenues, a strategy which is unsustainable, given the pressure on governments to reduce budget deficits. In the long run, improvements in infrastructure, especially roads, grain storage facilities, and market information are the most efficient ways of reducing price instability and governments have an important role in facilitating these improvements. Secondly, legumes in the past have been neglected relative to other crops. This has led to indirect disincentives since input and price subsidies have been for competing crops, such as cereals, or for sources of nitrogen (chemical fertilizers). During the 1990s, most countries have been phasing out such subsidies and few continue to subsidize nitrogenous fertilizers. In some European countries, there is a tax on its use. This more level playing field should provide incentives for expanding legume production.

Markets can also be improved through promotion of legumes and by informing consumers of the nutritional and health benefits. For example, a major campaign in Nigeria has led to a change in consumer perspectives of cowpea, from a food legume for the poor, to one that is now widely consumed by higher income households. With rapid urbanization in the developing world, research is needed on quick and easy preparations for food legumes. In countries with high incomes, the consumption of pulses is increasing with public awareness of their role in healthy diets, the development of new "easy-to-use" food preparations, and the small but growing number of vegetarians for whom pulses are important to achieving a balanced diet. Some of the cost of the research and promotion campaigns could be met from a small levy on legume output, controlled by farmers' associations as is done in Australia.

Finally, if the cost of production can be sharply reduced, grain legumes may have value and a demand in animal feeding, which will be the major source of growth in grain consumption in the developing world over the next few years. This may also have implications for the breeding of grain that has a nutrient content suitable for animal diets.

CONCLUSIONS

This overview has highlighted the world-wide decline in the role of food legumes in production and consumption. The key question is whether this is part of a long-term trend and if it is whether our attention should shift to more profitable and dynamic crops? In our opinion, there still are critical roles for food legumes that merit their priority in future strategies for food security and environmental preservation. This is especially so on the production side, where the trend to monocropping of cereals has to be reversed if productivity is to be sustained to meet the challenge of the world's demand for food. Rotations must be more diversified with food legumes playing an integral part in these rotations. Other legumes, especially soybeans and forage plants, will also play an increasing role in these systems.

Policy makers and farmers in the developing world must be made aware that monocropping, especially of cereals, is unsustainable. Scientists, farmers and policy makers will need to find niches within existing systems for legume crops, or substituting them for existing crops, including cereals. In this paper we have discussed success stories that have involved imaginative leadership.

Within this framework, there will always be options. New improved varieties have enabled changes to be made in sowing dates and in areas of adaptation. The traditional distinction between cool-season and warm-season legumes is no longer relevant in the tropics and subtropics, where so-called warm-season legumes are sometimes grown in the cool season (e.g., cowpeas in northern Nigeria) and vice versa (e.g., chickpeas in southern India). In any event, both types of legumes can play a role in the diversification of cropping systems, One important issue is the relative emphasis that should be placed on food legumes for favored and marginal areas. While food legumes are increasingly produced in marginal areas, the greatest potential is likely to be in the more favored areas in rotation with cereals or other cash crops to ensure sustainability

The development of new varieties is a key component in the promotion of legume crops. Early maturing varieties are essential in most systems, and maturity rather than yield is often a major factor in a farmer's decision to adopt a new variety. Besides maturity, yield stability is needed and will be obtained by breeding for resistance to biotic and abiotic stresses.

Scientific research will be a key to the expansion of legume crops. The wide diversity of actual and potential environments and the relatively large number of species complicates this research effort. The benefits of food legumes to succeeding crops in a rotation need to be highlighted when seeking resources for research.

Policy issues will be important and in many cases will involve the removal of policy distortions such as support prices, input subsidies, and special programs to promote cereals and soybeans that have occurred in the last two decades. Where appropriate technologies are available, it is appropriate to consider raising the general awareness of the role of legumes, and to target seed and extension programs.

Food legumes have had an important but declining role in human nutrition as a source of low-cost protein. Expanding their production and reducing their price is a means of enhancing that role. The development of alternative food preparations, especially for urban populations, and increasing the awareness of their nutritional value could help increase their role in human diets. Cost has to be considered as one component among the complex requirements of a nutritional diet. Finally any breakthroughs that result in a sharp reduction in prices will open up a role food in animal feeding. The expansion of the livestock industry with growing incomes in the developing world provides a virtually unlimited market for feed that can be exploited by producers in industrialized and developing countries.

References

Ali, M., 1997. *"Technological Change and Resource Productivity in Pakistan's Agriculture"*, Draft paper, Rural Development Dept., World Bank, Washington, DC.

Ali, M., and A. Ullah. 1996. *"Supply, Demand and Policy Environments for Pulses in Pakistan."* Draft paper, Shanhua, Taiwan:Asian Vegetable Research and Development Center.

Ali, M., I.A. Malik, H. M. Sabir and B. Ahmad. 1997. *Technical Bulletin No. 24.* Shanhua, Taiwan:Asian Vegetable Research and Development Center.

Bantilan, M.C.S. and P.K. Joshi. 1996. *Impact Series No. 1.* Andhra Pradesh, India:ICRISAT.

Byerlee, D. and C.K. Eicher, 1997. *Africa's Emerging Maize-based Revolution.* Boulder, Colorado: Lynnw Reinner.

Byerlee, D., and T. Husain, eds., 1992. *Farming Systems of Pakistan: Diagnosing Priorities for Agricultural Research.* Vanguard Publishing, Lahore.

Capper, B.S. 1990. In *The Role of Legumes in the Farming Systems of the Mediterranean Areas* (eds A.E. Osman, M.H. Ibrahim, and M.A. Jones). Kluwer Academic Publishers.

Castellanos, J., n.d., Campo Experimental El Bajio, Instituto Nacional de Investigacion Forestal, Agricultura y Pecuaria, Celaya, Mexico.

Denkyi, A.A., V.M. Anchirinah, A.I. Apau, B. Asafo-Adjei, M.A. Hossain and F. Ansere-Bioh. 1995. Ghana:Crops Research Institute, Kumasi.

Deshpande, S.S. 1992. *Critical Reviews in Food Science and Nutrition.* 32(4):333-363.

Durutan, N., K. Meyveci, M. Karaca, M. Avci and H. Eyuboglu. 1990. In *The Role of Legumes in the Farming Systems of the Mediterranean Areas* (eds A.E. Osman, M.H. Ibrahim, and M.A. Jones). Kluwer Academic Publishers.

Haddad, N.I. and B.A. Snobar. 1990. In *The Role of Legumes in the Farming Systems of the Mediterranean Areas* (eds A.E. Osman, M.H. Ibrahim, and M.A. Jones). Kluwer Academic Publishers.

Inaizumi, H., Adesina, A.A., and B. B. Singh, 1997. Paper presented at the 23rd International Conference of Agricultural Economists, Sacramento, USA, August 10-16.

Janssen, W., C. A. L. Gonzalez, and E. Lopez Salinas. 1989. *"La Adopcion de la Variedad Negro Huasteco 81 en las Huastecas de Mexico."* Cali, Colombia: Centro Internacional de Agricultura Tropical.

Kelley, T.G. and P. Parthasarathy Rao. 1994. *Economic and Political Weekly.* June 25:89-100.

Kelley, T.G. and P. Parthasarathy Rao. 1996. In: *Adaptation of Chickpea in the West Asia and North Africa Region* (eds N.P. Saxena, M.C. Saxena, C. Johansen, S.M. Virmani and H. Harris). Andhra Pradesh, India:ICRISAT.

Kumwenda, J.D.T., S.R. Waddington, S.S. Snapp, R.B. Jones, and M.J. Blackie, 1997 In: *Africa's Emerging Maize-based Revolution* (eds Byerlee, D. and C.K. Eicher). Boulder, Colorado: Lynne Rienner

Mruthyunjaya, P. Ranjitha, and S. Selvarajan, 1995, *Congruency Analysis of Resource Allocation in Indian Agricultural Research System,* Divisional Report, Division of Agricultural Economics, Indian Agricultural Research Institute, New Delhi.

Oram, P. and M. Agcaoili. 1994. In: *Expanding the Production and Use of Cool Season Food Legumes.* pp 3-49 (eds F.J. Muehlbauer and W.J. Kaiser). Kluwer Academic Publishers..

Pachico, Douglas. 1993. In *Trends in CIAT Commodities, 1993.* Working Paper No. 128, July. CIAT.

Pardey, P.G., J. Roseboom, and J.R. Anderson. 1991. eds. *Agricultural Research Policy: International Quantitative Perspectives.* Cambridge University Press, Cambridge.

Rahman, M.M. and R.N. Mallick. 1988. In: *World Crops; Cool Season Food Legumes,* pp 227-34 (ed R.J. Summerfield). Kluwer Academic Publishers.

Rosegrant, M., M. Sombilla, R.V. Gerpacio, and C. Ringler, 1997, Paper presented at the Illinois World Food and Sustainable Agriculture Program Conference, May 28, Urbana-Champaign, Illinios.

Russell, J.S., D.F. Beech, and P.N. Jones, *Food Policy,* May, 1989.

Singh, R., and K. Singh. 1991. *Indian Journal of Agricutlural Sciences.* 61(10):709-714.

Sperling, L., U. Scheidegger, and R. Buruchara. 1996 *Network Paper No. 60.* January. London, England: Overseas Development Institute.

Tutwiler, R. N. 1995. *ICARDA Soil Science Papers.* No. 4. Aleppo, Syria: ICARDA.

Viana Ruano, A. and J. A. Martinez. 1996Guatemala City: Instituto de Ciencia y Tecnologia Agricolas.

Wallis, E.S., R.F. Woolcock, and D.E. Byth. 1988. Paper No. 15. Bogor, Indonesia: Coarse Grains, Pulses, Roots and Tubers Centre.

Wallis, J.A.N. 1997. *World Bank Technical Paper. No. 364,* Washington, DC:World Bank.

White, K. D. 1970. *Agricultural History.* 44(2):281-90.

World Bank. 1990. Mexico: *Agricultural Technology Review.* Unpublished paper, World Bank, Washington.

Trends in support for research and development of cool season food legumes in the developing countries[1]

C.L.L. Gowda[1], M. Ali[2], W. Erskine[3], H. Halila[4], C. Johansen[1], I. Kusmenoglu[5], S.A. Mahmoud[6], B.A. Malik[7], M. Meskine[8], M.M. Rahman[9], R.P.Sapkota[10], and X. X. Zong[11]

1 ICRISAT, Patancheru, A P 502324 , India; 2 Indian Inst of Pulses Research, Kalyanpur, Kanpur 208024, Uttar Pradesh, India; 3 ICARDA, PO Box 5466, Aleppo, Syria; 4 Food Legumes Laboratory, INRAT, Avenue Hedi Karray, 2049 Ariana Tunisia. ; 5 Central Res Inst Field Crops, PO Box 226, Ulus, Ankara, Turkey; 6 Agriculture Research Centre Field Crops Research Institute Giza, Egypt; 7 NARC, PO Box NARC, Islamabad, 45500, Pakistan; 8 INRA/CRRA, Meknes B.P. 578 Morocco; 9 Pulses Research Centre, BARI, Joydebpur, Bangladesh; 10 Nepal Agricultural Research Council, Khumaltar, Lalitpur, P.O. Box No.5459, Nepal; 11 Institute of Crop Germplasm Resources, 30 Bai Shi Qiao Road,Beijing 100081, PR China

Abstract

In addition to their role in human nutrition, food legumes are an integral part of farming systems world wide. Their role in diversifying cropping systems and in maintaining soil fertility to sustain agricultural production is being realized increasingly among scientists and policy makers in most developing countries. Current (1996) world production is around 57 million tonnes. The population in developing countries is expected to be 6.06 billion by 2010, and the demand for food legumes is expected to be around 110.65 million tonnes. This poses a challenge to scientists and policy makers to meet this demand. The growth trends during 1990-94 for area, production, and productivity globally were negative, with few exceptions. Many countries in Asia need to increase production by at least 50% by 2010, and double it by 2020, to meet the needs of the growing population.

The current research and development (R&D) thrusts, in developing countries, are geared towards increased production, but with varied success. An increased reliance on plant breeding and extensive cultivation of legumes in marginal areas has lead to over-exploitation of the limited genetic resources (breeding for adaptation to harsh conditions thus losing genes for high yield). Some national governments (eg Turkey and India) have programs to increase production. The early successes have plateaued, and shifts in direction are needed.

Research infrastructure, staff, and funding for agricultural research are inadequate in most developing countries. Compared to 3.29% of agricultural gross domestic product (GDP) invested in R&D by the developed countries, the developing countries were spending on average only 0.39% in the late 1980s. This has declined further in the 1990s. The major proportion (50 to 75%) of the R&D funds in the developing countries is allocated to staple cereals, and only a small portion of the remaining budget is available for legumes. Although there are specialized research institutes or programs for major cereals, food legumes are lumped together and hence research efforts are scattered and superficial when compared with cereals.

The following strategies are suggested to strengthen support for food legume research.

Integrated cropping systems management (variety + agronomic practices + crop rotations) to bridge the yield gap in different agroclimatic conditions.

Initiate strategic research to breach yield ceilings, and to develop cultivars that can produce high and stable yields in better-endowed environments and thus compete with cereals.

Strengthen research collaboration within and among national programs and with the international agricultural research centers.

Increased role of regional, networks and working groups to enhance technical co-operation among developing countries (TCDC).

Increase the collaboration between public and private sectors and exploit their comparative advantages to achieve mutual goals.

Create Food Legume Councils (that include farmers, traders, and exporters) which support R&D by levying taxes or cesses on commodities and value-added products.

R. Knight (ed.), Linking Research and Marketing Opportunities for Pulses in the 21st Century, 47–58.
© 2000 Kluwer Academic Publishers. Printed in the Netherlands.

INTRODUCTION

Food legumes (pulses) play a role in human nutrition and more recently as animal feed, in the developing world. They contain minerals and vitamins essential for a balanced diet in humans. In many developing countries food legumes provide the necessary protein and amino acids (in predominantly vegetarian India, Bangladesh, Nepal, Myanmar and Sri Lanka) and supplement the protein diet of people in other countries. Since 1980-82 per capita consumption has declined by 6 % in developing countries where relative pulse prices have gone up and consumption of animal protein (eg milk) has increased. The importance of legumes as animal feed is increasing. The compound growth rate for feed use during 1980-95 was 7.97% compared to 1.5% growth for food use during the same period (Kelly et al., 1997). As an integral part of farming systems, food legumes, in rotation with cereals and tuber crops, assist in maintaining soil fertility and the sustainability of production systems (Rego et al., 1996). Owing to higher prices in comparison with cereals, food legumes are increasingly being grown to supplement farmers' incomes. The major food legumes grown in developing countries are: dry bean (*Phaseolus vulgaris*), faba bean *(Vicia faba)*, dry pea *(Pisum sativum)*, chickpea *(Cicer arietinum)*, lentil *(Lens culinaris)*, mung bean *(Vigna radiata)*, black gram (Vigna mungo) pigeonpea *(Cajanus cajan)* and Lathyrus (*Lathyrus sativus)*. Oil crops such as groundnut (*Arachis hypogaea*) and soybean (*Glycine max*) are food legumes but are not discussed in this paper. The terms food legume and pulse are used synonymously in this paper.

Cool season food legumes (faba bean, chickpea, lentil, and pea) contribute almost 60% of total world pulse production and 40% of the area (Oram and Agcaoili, 1994). Chickpea and lentil are produced predominantly in developing countries and dry peas in developed countries, while faba bean production is more evenly distributed. Overall yields in developing countries are only about half of those of developed countries (Oram and Agcaoili, 1994).

The world's population in 1994 was around 5.5 billion. It is expected to double by the year 2050. Most of this growth is expected to occur in developing countries in Asia, Africa, and Latin America (Swaminathan, 1995). This implies that food legume production needs to doubled or triple to meet the needs of the human population alone, not accounting for the demand for animal feed. The need to meet nutritional requirements (protein, amino acid, vitamins, minerals, etc) will be much greater in countries where people are likely to suffer deficiencies from predominantly cereal-based diets (James, 1997). The population in developing countries is 4.6 billion in 1997, and is estimated to be 6.06 billion in 2010 (assuming an average 2% growth rate). Current (1996) production of food legumes is around 57 million tonnes. Per capita consumption at a moderate rate of 50 g per day (18 kg per annum) will need at least 110.65 million t by 2010. This is the challenge facing food legume scientists, policy makers, and national governments and requires a meticulously planned approach. This calls for an R & D strategy developed by scientists, supportive funding support from research administrators and the political will of governments.

GROWTH TRENDS AND PROJECTIONS FOR FOOD LEGUMES

During the period (1990-'94) global growth rates for area, production and yield for the pulses under review were negative, although chickpea showed a marginal growth, and lentil recorded a 2 to 3% growth. Africa had highly negative growth rates for area and production, although yields showed positive trends for all crops, except chickpea. In Asian, positive growth trends were observed for area and production, but were negative for yield. Growth trends of selected countries in Asia and Africa are given in Table 1.

In Asia, growth rates for area are positive for all countries except Nepal, while production has shown negative trends in China, Nepal, and Pakistan. Yield growth rates were positive only in Bangladesh and India. Egypt, Morocco, Tunisia, and Turkey recorded negative growth rates for both area and production, with the exception of Tunisia that showed positive production growth. However, growth rates for yield are positive for all countries, except Morocco.

Current and projected demands for pulses in selected countries in Asia and Africa are given in Table 1. In many countries, production has to be increased by about 50% by year 2010 and to double by the year 2020 to meet the projected demands. Based on available data, it is expected Nepal and Myanmar will have surpluses and the remaining countries deficits.

Table 1. Current and projected demand (million tonnes) and growth rates (1990-94) for all food legumes in selected developing countries of Asia and Africa[1]

Country	Demand		Growth rates (1990-94) (%)		
	Current	Projected (Year)	Area	Production	Yield
Bangladesh	0.56	0.70 (2010)	0.50	0.66	0.16
China	8.00	15.00 (2020)	0.84	-2.84	-3.64
India	18.00	25.90 (2010)	0.36	1.76	1.62
Myanmar	NA[2]	NA	27.46	22.92	-3.58
Nepal	0.19	0.33 (2010)	-1.46	-1.54	-0.20
Pakistan	0.87	1.21 (2010)	0.64	-7.02	-7.64
Egypt	0.54	0.68 (2010)	-2.74	-15.92	8.82
Morocco	0.27	0.52 (2000)	-11.92	-12.94	-0.96
Tunisia	0.05	0.07 (2000)	-7.10	1.36	9.00
Turkey	1.95	3.00 (2010)	-4.76	-4.52	0.28

1 Sources : FAO Year Books (1990-94) for area, production and yield. Current and projected demands estimated / supplied by co-authors. 2 Data not available.

CURRENT RESEARCH IN FOOD LEGUMES

Current research in food legumes in most developing countries is geared towards increasing production to meet domestic demand and/or export. Activities include a) developing high yielding varieties with resistance to diseases and pests, and b) agronomic practices to increase productivity. Because of the demand, food legumes are being grown in marginal lands as better lands are devoted to high yielding stable and remunerative crops such as cereals. This has necessitated breeding for tolerance to poor soil fertility, drought, and salinity. In some areas there is a build-up of pests and diseases due to crop intensification. For example, intensification of chickpea and lentil cultivation in some areas in Turkey has resulted in epidemics of soil-borne and leaf diseases, mostly *Ascochyta* blight. This has necessitated breeding for resistance and for research on integrated pest management. Delayed or early sowing, to fit into cropping systems, has lead to the breeding of varieties adapted to the changed sowing dates, and for early maturing varieties. All these activities were more of `fire-fighting' efforts to meet immediate needs. Few countries have invested in long-term improvement programs to improve yields and the stability of production.

Realising the need to maintain soil fertility and the sustainability of agricultural production, natural resource management research should receive priority in many countries. Now both agronomic management and varietal improvement are receiving equal emphasis in many developing countries.

SPECIAL PROGRAMS AND POLICIES TO INCREASE LEGUME PRODUCTION

Faced with deficits in the supply of pulses, many countries (especially in Asia) resorted to imports to meet demand in the 1980's. Some countries such as Turkey and Australia have increased their production for export. In Turkey, for example, the production of chickpea and lentil was increased by expanding to marginal areas, utilizing fallow lands, and including legumes in cereal-cereal rotations. The government also provided policy incentives to increase production and exports. As a result the chickpea and lentil area increased from 3% of the total cropped area in late 1970 to 11% in late 1980s (Anonymous, 1990). On the other hand, importing countries initiated `Special Programs' to increase production. For example, the Indian National Agricultural Research Systems (NARS) initiated the `Technology Mission on Pulses' in 1991 to co-ordinate the efforts of different agencies and boost production of chickpea, lentil, pigeonpea, mungbean, and black gram. Various `Micro-missions' operate under the "Technology Mission" to deal with crop production; post-harvest technology; input and resource support to farmers; and price support, storage, processing, and marketing. The mission has been reasonably successful, and has been extended to the 9th five-year plan period. The Crop Diversification Program (funded by Canadian and Dutch projects) in Bangladesh (1990-'95) was aimed at increasing the production of some non-rice crops to increase food supplies. A project to increase production of chickpea and other pulses funded by the Pakistan government during 1994-96 has been highly successful. Chickpea production increased from 0.41 million t in 1993-94, to 0.68 million t in 1995-96. Because of this increased production, there have been no imports of *desi* chickpea during 1995-96. The project has been extended to cover other legumes in the 9th five-year plan. If the developing countries have to meet the projected demands for food legumes by the year 2010, such special programs are essential to accelerate production.

RESEARCH INFRASTRUCTURE, INSTITUTES, STAFF, FUNDING

After the `Green Revolution' of the 1960s and 1970s, there was concern among the national programs of many developing countries regarding the need for diversification from cereal mono-cultures. Many national programs initiated crop improvement programs on food legumes in the late 70s and early 80s. New institutes, divisions, or projects were started. Multi-disciplinary teams were either appointed or identified however, funding of R & D has been meagre (Table 2). Investment by developed (high-income) countries showed continued growth, with an average 3.29% of agricultural gross domestic product (GDP) invested in R&D by the late 1990s, with Japan reporting 3.36% in 1992, and Australia 3.54%. However, corresponding figures for developing (low-income) countries was approximately 0.39%. Agricultural research expenditures relative to total government funding have declined over time. Latest available figures of government funding in low-income Asian countries (Bangladesh, China, India, Pakistan) range from 0.25 to 0.52% (Pardey et al., 1997). Current actual funding in selected countries, along with the number of institutions and staff working on food legumes research is given in Table 3. Compared to the total human resources in each country, resources allocated to food legumes are small. The proportion of full time researchers on legumes in China is 0.3%, in Bangladesh 1.5%, and in India 6.8%. Data are not available for other countries, but the number is likely to be <3% in most developing countries.

Table 2. Investments in agricultural research and development (expressed as percentage of national agricultural GDP)

Country	1971-75	1976-80	1981-85	1986-90	Latest Year
Bangladesh	0.13	0.16	0.25	0.26	0.25[b]
China	0.40	0.48	0.41	0.38	0.43[c]
India	0.21	0.33	0.38	0.48	0.52[a]
Indonesia	0.13	0.21	0.26	0.27	0.27[a]
Pakistan	0.39	0.52	0.58	0.59	0.47[b]
Sri Lanka	0.40	0.53	0.50	0.37	0.36[c]
Low-income	0.27	0.37	0.39	0.40	0.39[a]
Malaysia	0.51	0.85	1.04	1.08	1.06[b]
South Korea	0.27	0.26	0.36	0.39	0.56[c]
Taiwan	1.14	1.70	2.34	3.03	4.65[b]
Thailand	0.73	0.65	0.89	0.94	1.40[b]
Middle-income	0.60	0.65	0.89	0.94	1.34[b]
Australia	2.56	2.93	3.51	3.11	3.54[b]
Japan	1.97	2.24	2.81	3.03	3.36[b]
High-income	2.06	2.33	2.92	3.04	3.29[b]
Total	**0.48**	**0.58**	**0.60**	**0.59**	**0.58[a]**

a 1990 figure b 1992 figure c 1993 figure   Source: Pardey et al., 1997

Despite the already low level of funding for agricultural research (including pulses), many national governments are imposing further cuts in research budgets. In a few countries, such as China, research institutes are being asked to generate their own funds and link up with the private sector to attract grants; while in Morocco the semi-public institutes are already generating 8-12% of their budget from internal income. Most of the national programs receive substantial contributions from bilateral or multilateral donors to support agricultural R&D. If this funding is withdrawn, it will severely affect the research in many countries.

Table 3. Research institutions, staff, and funding (US$m) for food legumes research in selected countries.

Country	No. of Institutions[1]	No. of Staff[2] (Approx.)	Total Budget[3] (for legumes)	Total Agri. Res. Budget[4]
Bangladesh	6	25	5.3 (1990-95)[5]	132.8
China	29	150	1.5 (1986-96)	1867.6
India	37	345	10.0 (1992-97)	1561.8
Myanmar	2	35	NA [6]	NA
Nepal	4	30	0.3 (1985-95)	NA
Pakistan	9	102	0.4 (1989-98)	198.3
Egypt	1	50	0.08[7] (1997)	60.0
Tunisia	2	8	0.43[8] (1996)	NA
Turkey	6	15	0.15 (1997)	4.7
Morocco	6	25	3.27[9] (1993-96)	26.9

1. Consists of institutes working on food legumes
2. Not necessarily full time staff, but indicates total person years. Based on information supplied by co-authors
3. Approximate amount allocated / spent for legumes. Figures are estimates by the co-authors representing the country
4. Source: Pardey et al., (1997). For comparison, local currency units are converted to US dollars
5. Local budget figures not available. Given figures are from a donor project
6. Data not available
7. Excluding salaries and donor supported project funds
8. Excluding donor supported project funds
9 Including external project funds, excluding salaries. Total agricultural research budget (of Institut National de la Recherche Agronomique) excluding external support project funds.

INTERNATIONAL AGRICULTURAL RESEARCH SYSTEM

Five International Agricultural Research Centers (IARCs) have food legumes as mandate crops:
Asian Vegetable Research and Development Centre (AVRDC) - Mungbean, soybean
Centro International de Agricultura Tropical (CIAT) - Phaseolus beans
International Center for Agricultural Research in the Dry Areas (ICARDA) - Lentil, faba bean and a regional mandate for chickpea (with ICRISAT)
International Crops Research Institute for the Semi-Arid Tropics (ICRISAT)-, Chickpea pigeonpea, and groundnut
International Institute of Tropical Agriculture (IITA) - Cowpea

With the exception of the bean program at CIAT, the emphasis at these centers on food legumes (in terms of research funds, staff, and resources) is less than for cereals or tuber crops. A comparative study of fund allocation during 1985-95 to cereals, legumes, and resource management programs at ICARDA and ICRISAT is presented in Table 4. At ICARDA, the four legumes received 25.4 % of the funds compared with 29.5% for two cereals and 45% for resource management. At ICRISAT, core funds (excluding special projects) for two cereals were higher (29.3%) than for three legumes (28.3%) during 1985-94. These comparisons indicate that the cereal crops garnered more funds because of the perceived need to produce more staple food. We cannot possibly hope to reverse the trend, but could influence a more equitable share of R&D support to legumes. With the recent funding cuts experienced by the Consultative Group on International Agricultural Research (CGIAR) it is likely that there will be further reductions in funds to legume research. Because of the reduced resources, and other reasons, the IARCs will now concentrate more on basic and strategic research, and are expected to work with their NARS partners in applied and adaptive research. IARCs will conduct research that will provide `international public goods' for use by the national programs. They will also not develop any finished products or technologies, but will supply enhanced germplasm, intermediate products, and components of technology. The NARS would use these intermediate products (segregating materials, populations, etc) to develop varieties or a complete set of technologies. All this change puts the onus on the NARS for the applied/adaptive research and development oriented programs.

Table 4. Funding (thousand US$) for research programs at ICRISAT and ICARDA, 1985-95.

Program	1985	1986	1987	1988	1989	1990	1991	1992	1993	1994	1995	Total
ICRISAT												
Cereals (2)	3969	4080	4330	4426	5443	5727	5943	4963	5000	4501	NA[1]	40333
Legumes (3)	2870	3520	4076	3660	3948	3850	4248	3920	3886	4892	NA	38870
Res. management	2236	2393	2986	2932	3263	3686	3852	3747	3685	3586	NA	32366
Tech. transfer	1666	1986	2011	2717	3500	3511	3943	2828	2065	1814	NA	26041
ICARDA												
Cereals (2)	NA[2]	2352	2467	2568	2806	2729	2825	2700	2970	3055	3083	27555
Legumes (4)	NA	2091	2559	2579	2596	2523	2577	2017	2214	2276	2298	23730
Res. management	NA	3431	3528	3917	4075	3913	3976	3833	4059	5147	6215	42094

1 Data for programs is not available as programs were re-organized in to research divisions.   2 Data are not reported for 1985.

Source: Finance Division, ICRISAT, Patancheru, India &   W. Erskine, ICARDA, Aleppo, Syria.

FUTURE RESEARCH TO INCREASE FOOD LEGUME PRODUCTION

Future food supplies are to come from increased productivity both by harnessing the genetic potential of crops and agronomic management to increase yields (Islam, 1995). This will need sustained research to develop technologies that can overcome the ever-changing constraints to production, and to contribute to sustainability. To achieve this, national governments must invest more in R & D in the next decade. Previous investments in food legumes are not encouraging and national governments must now act to save their countries from food shortages.

Although reliable data on funds and personnel involved in cereal research *vis-à-vis* pulses are not easily available, the lopsidedness in favour of cereals is evident in all countries. Major cereals such as rice and wheat receive almost 50 to 75% of the national research funds, while all the other crops (> 25) share the remaining 25-50% resources. There are individual national research institutes for rice, and wheat (eg, the Bangladesh Rice Research Institute, Directorate of Wheat Research in India, etc). However, three or more pulses are grouped together in research programs or institutes (e.g., Pulses Research Center, Bangladesh; Food Legumes Programs in Egypt, Morocco, Turkey, Tunisia). This change itself is a welcome development in a few countries where legumes were either clustered with oilseeds or as secondary crops, or as other field crops. However, more recently individual programs and projects are being initiated, such as the bifurcation of chickpea and pigeonpea as separate co-ordinated programs in India. It is imperative that other national programs emphasise the role of legumes in systems, and establish R&D programs.

Legumes are largely cultivated as rainfed crops, or under residual moisture conditions in post-rainy seasons. Through years of natural selection under low-moisture, low-fertility, and allied harsh conditions, they have been adapted to survive and produce minimal acceptable yields. The advent of high yielding cultivars of rice, wheat, maize, sorghum and soybean and the expansion of their cultivation to the better and more productive areas, has pushed the `poor-cousin' legumes to marginal and less fertile lands. These factors have resulted in legumes being non-responsive to inputs (irrigation and fertilizers). The excessive growth of legumes under good fertility and moisture conditions leads to increased disease, lodging and lower yields. Good progress has been made with peas in Europe, changing the ideotype (e.g. reduced leaf area) to make the crop responsive to inputs and to mechanised cultivation. Limited progress has been made in faba bean, lentil and chickpea to develop input responsive ideotypes to fit into changed cropping systems and production needs.

Increasing production can result from a) expanding the area, including irrigation and b) increasing yield. However availability of land is limited. Swaminathan (1990) has estimated that by the year 2000, per capita land availability will be 0.1 ha in China and 0.11 ha in India. A similar situation is likely in much of Asia, but not to the same extent in Africa or Latin America. Land resources can be extended to some extent by utilising fallow lands (as in Bangladesh, China, India, and Turkey) and increasing cropping intensity (intercropping, etc). Other possibilities are: a) introducing short-duration pulses to fit into cropping sequences, b) substituting low-profit cereal crops with high-value pulses, and c) shifting pulse cultivation from marginal areas to limited irrigation areas where yields can be increased 2-3 times with one or two irrigations and d) developing value added products and the use of legumes by industry to increase demand and remuneration to farmers.

VARIETIES, CROP MANAGEMENT AND CROP PRODUCTION.

Improving yields is a major component of increasing the production of pulses. As a result of efforts by IARC and NARS scientists, improved varieties have been developed. Many are high-yielding, and combine resistance/tolerance to pests or diseases. However, the farmer's realised yields are one-fourth the yields on experiment stations. Bridging this gap, by improved management (nutrition, soil and water conditions, crop husbandry) and protection against biotic and abiotic stresses are essential to achieve increased production (Plucknett, 1995). There is need for more strategic agronomic research, to improve yields and sustainability. Research needs to be focused on both genetic and management improvements when attempting to close the gap between potential and realised yields.

BRIDGING THE YIELD GAP

A review of literature indicates the technology exists (variety and agronomic management) for increasing production of most pulse crops. For example, in groundnut the highest dry pod yields recorded (the potential yield) on large plots range from 10 t ha^{-1} in Zimbabwe to 11.2 t ha^{-1} in China (Johansen and Nageswara Rao, 1996). At ICRISAT Patancheru in peninsular India (tropical environments) yields up to 7 t ha^{-1} on small plots and up to 5 t ha^{-1} on large plots have been reported. However, the realized average national yield in many developed countries (with the exception of China) is around 1 t ha^{-1} pod yield. As apparent in Figure 1, the yield gap (the difference between the farmer's realised yield and potential yield) is large, and is usually greater than 5 t ha^{-1} in most developing countries (Johansen and Nageshwara Rao, 1996). Such yield gaps exist in other legumes, implying there is a considerable scope for increasing yields through appropriate management and by identifying and alleviating the constraints to higher productivity.

BREACHING THE YIELD CEILING

In addition to bridging the yield gap, there is a need to raise the yield plateau or breach the yield ceiling to enable higher yields in the better environments. This can be achieved in two ways. One is by modifying the effects of limiting environments by management. For example, by using a polythene mulch to alleviate low temperatures in temperate regions. This has been used to increase groundnut yields in China and South Korea and is being evaluated in north and north-eastern India for spring season groundnut. The second is genetic alteration of the plant to tolerate stress, or to make better use of the ambient environmental conditions (Johansen and Nageshwara Rao, 1996).

FUTURE TRENDS IN RESEARCH COLLABORATION

Since-World War II there has been an increased exchange of knowledge of advances in agriculture among countries. Prior to this, most countries developed and used technologies in relative isolation (Plucknett, 1995). Today, the `global agricultural research system' is in place, and many countries have benefited. The system consists of the NARS of developing countries, IARCs, and research institutes in developed and developing countries. The interaction includes bilateral and multilateral agreements, contract research, consortia, and networks. This has lead to collaborative research that has helped many NARS overcome production constraints. A major constraint to plant breeding/crop improvement in the developing countries is the reduction in public funds for research. Many NARS depend on IARCs for breeding material and improved varieties (Duvick, 1995). Because of the reduction in funding to the IARCs they are reducing their applied/adaptive research. Instead of releasing improved varieties they are developing intermediate products, and enhanced germplasm. Some programs have been cut and others are likely to be pruned in the near future. Although commercial (private sector) plant breeding is getting established in a few countries, it

54

cannot replace public sector NARS and IARC plant breeding, especially for self pollinated crops. Many small-scale private seed firms will still depend on public-domain breeding for advanced material and finished varieties. Continued and increasing support of publicly funded plant breeding is essential for continued yield gains in the developing countries (Duvick, 1995).

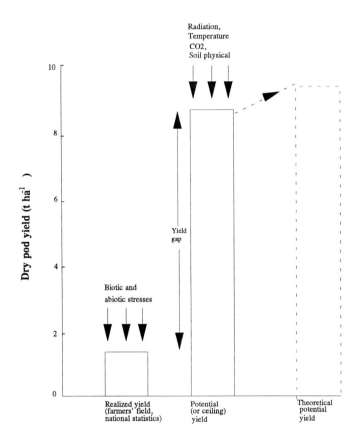

Figure 1. Representation of realized and potential yields and their relationships. (Source: Johansen and Nageshwar Rao, 1996)

TCDC APPROACH FOR RESEARCH AND DEVELOPMENT COOPERATION

The concept of Technical Co-operation among Developing Countries (TCDC) was approved at the United Nations Conference held in Buenos Aires in 1978. It recognises the role of developing countries in promoting and implementing TCDC but also the responsibility of the international community for TCDC. The main objective is to promote the individual and collective self-reliance of developing countries through the sharing of experience, knowledge, and technology. The South-South co-operation is seen as complementary to North-South co-operation. TCDC is initiated, managed, and principally financed by partner countries but is facilitated by the FAO of the United Nations (FAO, 1992). TCDC encompasses a range of social and economic activities according to the needs and capabilities of co-operating countries. Some of the activities are:

Information exchange, using directories or inventories of capabilities and needs.

Technology transfer using experts from one country to assist/facilitate R&D in other countries.

Group training, using faculties and facilities within developing countries.

Exchange of equipment among countries.

`Twinning', where a comparatively mature country/institution assists in developing another institution by providing technical assistance.

Networks, where a group of institutions/countries come together for technical co-operation to address common problems.

REGIONAL FORA AND RESEARCH COLLABORATION

As an extension of research collaboration, regional fora are becoming important in the R&D. Examples are the Southern Africa Development Council (SADC), the Association for Strengthening Agricultural Research in Eastern and Central Africa (ASARECA), the Association of Agricultural Research Institutes in the Near East and North Africa (AARINENA), and the Asia Pacific Association of Agricultural Research Institutions (APAARI). These fora aim to exchange research technology and materials among member countries and to strengthen the weaker NARS in the region. These fora also represent the region in various `Global Forums' to ensure their needs are addressed and to influence policies and the allocation of resources for R&D. APAARI for example, is playing a role among its member countries in setting priorities for regional research, liaising with regional and international research organisations and donor groups to achieve equitable technology and information exchanges in the Asia-Pacific region. APAARI also plans to help the member NARS by supporting research collaboration directly or through regional networks. In future, the regional fora are likely to play a greater role in food legumes R&D.

COLLABORATIVE RESEARCH NETWORKS

Networks are groups of individuals or institutions linked together to collaborate on common problems, and to use existing resources more effectively (Faris, 1991). The reduction in funds for agricultural R&D is motivating scientists and institutions to work together and use their resources more effectively.

Networks, such as the Cereals and Legumes Asia Network (CLAN) appear to be crop related, but include natural resource management research (NRM) in the appropriate production systems. On the other hand, the Crop and Resource Management Network (CREMNET) looks at NRM issues in the rice-based cropping systems. The main advantages of the networks are that research is planned and executed by members, and the results are shared. Examples are the exchange of germplasm and breeding material, production technologies, information, training to improve research capabilities, and the exchange of scientists to help the weaker NARS.

RESEARCH WORKING GROUPS OR CONSORTIA

Individual laboratories are unable to undertake comprehensive research due to a scarcity of funds facilities, expertise and staff. Working Groups (WG) are defined as groups of scientists with a common interest in finding a solution to a high priority regional problem (Gowda et al., 1996). International Working Groups facilitate co-operative research by bringing together expertise from developed and developing countries, international research centers, universities, private sectors, and non-governmental organizations, to achieve the critical mass needed to achieve objectives.

The concept is not new, and scientists around the world have been pooling resources, and sharing research agenda and results. What is new is that it is now more structured, has set objectives and a focus, and the `Working Group' Co-ordinator provides leadership in the research effort. The Working Groups also provide a forum for its members to meet every 2 to 3 years to review the research and plan for the future.

COLLABORATION BETWEEN THE PUBLIC AND PRIVATE SECTORS

The challenge posed by the global food situation requires the limited resources to be used effectively to develop sustainable systems. There is a need for collaboration between the public and private sectors to ensure production is doubled or tripled in the next 50 years (James, 1997). The private sectors' investment in R&D, especially in the seed sector, biotechnology, post-harvest, food processing, and agricultural implements is large and is expected to increase at a faster rate than public sector investments (Anderson et al., 1994).

Governments in developing countries must build partnerships with the private sector to use the comparative advantages of the public and private sectors to achieve national goals. Governments can use policy incentives to encourage private sector involvement in joint ventures, and donors can facilitate implementation of such programs (Anderson et al., 1994)

Many governments in developing and developed countries are encouraging the participation of the private sector in areas where it has a comparative advantage such as the seed industry and biotechnology. A challenge is to find ways to transfer appropriate technologies, which are often proprietary, from the private sector to the public sector. Current private sector investments in agriculture R&D are about $11 billion in developed countries and $2 billion in developing countries; compared to $8.5 billion and $8.8 billion, respectively, by the public sector. In most instances the sectors are spending their funds independently and there would be a synergy if the same amount were invested in a co-ordinated manner (James, 1997). Even then the private sector will be targeting those who can pay for their products. There will still be the need for public sector research to develop technologies relevant to poor farmers.

The establishment of the "Private Sector Committee of CGIAR" is an example of such co-ordination. The CGIAR institutes are expected to work with the private sectors in developed and developing countries to conduct research and to ensure that the CGIAR's research reflects the goals of its research partners (James, 1997).

PARTICIPATORY RESEARCH METHODS

The generation of new technology and adoption of existing technologies needs capable on-station and on-farm research scientists, a good extension service, and a committed government. Many NARS follow a top-down approach to technology transfer that is not suitable for the varied agroecological conditions where pulses are grown in the developing countries. Participatory technology development, including plant breeding, on-farm research, indigenous knowledge, and empowerment of poor farmers, should be disseminated at all levels – from undergraduate curricula to upgrading of senior research and extension staff. This requires a greater investment in education to produce a new generation of scientists and for the retraining of staff.

FUTURE TRENDS IN FUNDING FOR FOOD LEGUMES RESEARCH

Traditionally, agriculture research in most developing countries has been funded by the government. As discussed earlier, this has declined over the years, especially during the 1980's and 1990's. Apart from lobbying governments for increased funds, scientists have to look beyond government funds. The following are suggested:

Linking public and private sector efforts for their mutual benefit (as mentioned above).

Involving non-government organizations. In Turkey, for example, the Mediterranean Exporters Union supported projects on disease management in lentil. Involving the private seed sector and biotechnology groups would be welcome but it is likely public sector research in developing countries will continue to be the primary source of technologies for poor farmers.

Creating "Food Legume Councils" that include farmers, traders, and exporters to support research. Such councils exist in Australia, Canada, USA and some European countries. The councils could, in association with government agencies, levy a tax or cess on the commodities and any value-added products. This will create awareness among the stakeholders (farmers, traders, etc.) in research planning, technology generation, and dissemination. The feeling of `ownership' of the research process is a big asset.

Highlighting the role of legumes in nitrogen fixation, sustainability of agriculture, providing a balanced diet for humans, and feed for livestock and poultry. This has to be a planned public relation exercise by advertising to the general public and by convincing policy makers of the long-term benefits of food legume R&D.

Influencing national policies to support production and marketing. These policies should include the inputs (seed, fertiliser, pesticides, etc.), support price, and export policies to ensure profitability to farmers.

The IFLRC should provide a forum for linkage between research leaders, policy makers, research institutions, development agencies, NGOs, and donor agencies. This would create a better understanding of the research needs by the policy makers. It could also lead to a co-ordinated effort among funding bodies and research groups to avoid competition and duplication of R&D efforts.

CONCLUSIONS

An increasing world population will create a huge demand for food legumes for human consumption and animal feed. However, the global trends for legumes (1990-94) in the area cultivated and production have been negative, except in Asia. The challenge to scientists is to reverse this trend. Investment in research in developing countries is meagre and is declining rapidly. The IARCs are also facing funding constraints, and are shifting their emphasis to strategic research, and the development of intermediate products. The burden of the applied and adaptive research will be on the developing countries.

Future research on pulses should emphasise both varietal improvement and agronomic management, including strategic research on crop ideotypes, bridging the yield gap, and breaching the yield ceiling. Developing countries should collaborate with other countries, and with regional and international institutes to access information, material, and technology. Regional fora, networks, and consortia will play a critical role in technology exchange.

Partnerships between the public and private sectors will be essential to harness their comparative advantages and to be cost effective. The creation of "Food Legume Councils" that levy taxes or a cess on commodities to support R&D should be pursued. Scientists should lobby for support, and influence national policies on R&D. IFLRC should evolve into a forum for linkage and dialogue between scientists, policy makers, the private sector and the donor community.

References:

Anderson, J.R. , Pardey, P.G. and Roseboom, J. 1994. *Agricultural Economics* 10:107-123.

Anonymous, 1990. *The summary of agricultural statistics.* Publication no.8, Printing Division, State Institute of Statistics, Ankara, Turkey.

Duvick, D.N. 1995. . In: *Population and food in the early twenty-first century: meeting future food needs of an increasing population* (ed Islam, N.). Washington, D.C., USA: International Food Policy Research Institute.

FAO 1992. *Handbook on TCDC.* Rome, Italy: Food and Agriculture Organization.

Faris, D.G. 1991. *Agricultural research networks as development tools: Views of a network coordinator.* Ottawa, Canada: International Development Research Centre; and ICRISAT, Patancheru, 502 324, Andhra Pradesh, India:.

Gowda, C.L.L., Ramakrishna, A., Reddy, D.V.R., and Renard, C. 1996. In: *Groundnut Virus Diseases in the Asia-Pacific Region:* Fourth Meeting of the International Working Group, 12-14 Mar 1995, Khon Kaen, Thailand and ICRISAT Patancheru, 502 324, Andhra Pradesh, India:

Islam, N. 1995.. In: *Population and food in the early twenty-first century: meeting future food demands of an increasing population.* (ed Islam, N.). Washington, D.C., USA: International Food Policy Research Institute.

James, C. 1997. *Issues in Agriculture*, No.9. Washington, D.C. USA: Consultative Group on International Agricultural Research, 48pp.

Johansen, C. and Nageshwara Rao, R.C. 1996. In: *Achieving high groundnut yields*: Proceedings of an international workshop, 24-29 Aug 1995, Laxi City, Shandong, China (Gowda, CL.L., Nigam, S.N., Johansen, C. and Renard, C. eds.) ICRISAT Patancheru 502 324, Andhra Pradesh, India:

Kelly, T.G., Parthasarathy Rao, P. and Grisko-Kelley, H. 1998. *Proceedings IFLRC-III: Linking Research and Marketing Opportunities for the 21st Century.* (ed. R. Knight) Dordrecht: Kluwer Academic Publishers (in press).

Oram, P.A. and Agcaoili, M. 1994. In: *Expanding the Production and Use of Cool Season Food Legumes* (eds Muehlbauer, F.J. and Kaiser, W.J.). Kluwer Academic, The Netherlands.

58

Pardey, P.G. and Alston, J.M. 1995. *Revamping agricultural R&D*. Brief 24, 2020 Vision of IFPRI Washington, D.C.; USA: International Food Policy Research Institute.

Pardey, P.G., Roseboom, J. and Fan, S. 1997. In: *Financing Agricultural Research: A Sourcebook* (eds Tabor, S.R., Janssen, W. and Bruneau, H.). The Hague, Netherlands: International Service for National Agricultural Research.

Plucknett, D.C. 1995. In*: Population and food in the early twenty-first century: meeting future food demands of an increasing population* (ed. Islam, N.). Washington, D.C., USA: International Food Policy Research Institute.

Rego, T.J., Nageshwara Rao, V. and Kumar Rao, J.V.D.K. 1996. *Presented at the Workshop on Residual Effects of Legumes in Rice-Wheat Cropping Systems of the Indo-Gangetic Plains,* 26-28 Aug 1996, ICRISAT Asia Center, Patancheru 502 324, Andhra Pradesh, India.

Swaminathan, M.S. 1990. *Sir John Crawford Memorial Lecture,* Consultative Group on International Agricultural Research, Washington, D.C., USA.

Swaminathan, M.S. 1995. *Issues in Agriculture,* No.7, Washington, D.C., USA: Consultative Group International Agricultural Research.

Trends in Support for Research and Development of Cool Season Food Legumes in the Developed Countries

RM. Gareau [1] , F Muel [2] , and JV. Lovett [3]

1 USA Dry Pea & Lentil Council, 5071 Hwy. 8 West, Moscow, Idaho, 83843-4023, USA; 2 UNIP, 12 avenue George V, 75008 Paris, France; 3 Grains Research and Development Corp., Kingston, ACT 2604 Australia

Abstract

This paper compares trends in financial support for research in cool season pulse crops in Australia, Canada, France, and the U.S.A. While the research objectives and priorities tend to be similar among the countries, there are significant differences in the amounts of public funding for research. France, Australia, and Canada have had increased funding for grain legumes research while in the U.S.A. it has declined. Differences are seen in government support programs, grower assessments and private industry support.

INTRODUCTION

Developed countries differ in the amount and type of support that is dedicated to the research and development of cool season food legumes. Comparisons were made by examining the sources and amounts of funds directed toward pulse crops, the number of scientific person years, and the areas of research using the countries of Australia, France, Canada, and the United States of America as examples for the developed world.

Using the best available current estimates, this paper summarizes and compares the trends and direction of the pulse crop research investment in Australia, Europe and North America. These estimates are further broken down into the number of scientific person years involved in seven research areas: 1) Crop Adaptation; 2) Genetics and Breeding; 3) Crop Management; 4) Crop Protection; 5) Uses and Quality; 6) Equipment and Machinery; 7) Post-Harvest Storage. Estimates of public research funding from all countries examined were multiplied by a factor of three, to represent all associated inputs into research programs.

Countries such as USA, that have a longer history of dry pea, lentil, and chickpea production, have witnessed a decline in the number of scientists and research programs devoted to these crops. Conversely, countries such as Canada and Australia, with relatively young cool season food legume industries, are in a period of expansion of funding and research personnel coinciding with their expanding production.

Overall, while publicly-funded grain legume research programs are experiencing reductions in their budgets, the countries with robust pulse crop industries have been able to overcome these obstacles by tapping into private industry and grower funded sources.

AUSTRALIA

Australia has had a rapid expansion in pulse crop production and the amount of research investment has mirrored its growth. Lupins (*Lupinus angustifolius* and *L. albus*) and field peas (*Pisum sativum*) dominate the pulse legumes in Australia, but it is expected that chickpeas (*Cicer arietinum*) and faba bean (*Vicia faba*) will account for most of that country's pulse crop expansion. Other cool season pulses of commercial significance are lentil (*Lens culinaris*) and vetch (*Vicia spp.*)

Australian funding figures in Table 1 and personnel in Table 2 are based on reports from the Grains Research and Development Corporation (GRDC) and the result of a survey made in the early 1990's that collated the funds and staff dedicated to the Australian pulse industry and trends in resources allocated to research. The GRDC has been central to influencing research directions in the Australian pulse crop

59

R. Knight (ed.), Linking Research and Marketing Opportunities for Pulses in the 21st Century, 59–66.

industry, since it is the manager and distributor of funds received through a joint grower and government pool. In Australia, all growers pay a levy based on the gross value of production at the farm gate, that is matched on a 1:1 basis by the federal Government. The GRDC coordinates and disperses the research funds from this pool across the grain industry, of which pulse crops have now taken greater prominence. Support from GRDC for pulses has historically exceeded their current crop value, indicating a recognition of the potential value and emerging importance of pulse crops in Australian farming systems. The amount of funding directed into pulse crops is about 14% of total funds received. Interestingly, about 10% of the Australia's cropland area is in pulse crop production.

Table 1. GRDC Grain Legume Research Program funding in Australia from 1986 to 1991 in US $

	1986-1987	1987-1988	1988-1989	1989-1990	1990-1991
Crop Improvement	107,514	315,699	423,423	446,261	725,842
Crop Production	94,652		190,603	304,334	323,415
Crop Protection		156,926	295,970	305,093	435,679
Farming Systems		94,223	29,926	48,033	48,340
Storage & Handling	18,250		23,593	23,689	52,878
Research & Marketing	30,357	117,336	144,663	75,334	195,699
Communication & Training	14,768	10,716	56,187	64,295	150,099
Total	265,540	694,901	1,164,365	1,267,038	1,931,952

Table 2. Research staff and funds devoted to crop improvement in Australia in 1991-1992.

Crop	Professional Scientists	Total Staff	GRDC Funds $'000 (US)	Total Funds $'000 (US)
Lupins	10	28	409	1,242
Faba beans	1	5	36	83
Field Peas	6	18	193	804
Chickpeas	4	16	69	607
other pulses*	6	25	157	1,169
Multicrop pulses	1	2	31	49
Total pulses	28	94	894	3,956
Total grains	221	670	8,100	28,081
% of Total crop funding	13	14	11	14

* other pulses include peanuts, navy beans, mung beans, pigeon peas, and cow peas

In 1992, Clements et al. reported that pulse crop research received approximately 14% of total grain research funding for 1991-1992. In addition, another 4% was estimated to be allocated to multicrop research projects. In Australia, there has been a positive correlation between the amount of pulse crop research funding and the number of scientists involved in the research (table 2).

The resources allocated to pulse crops by the GRDC has increased significantly in more recent years (Figure 1) It is estimated these funds represent about one-third of the total industry funding for pulses (Lovett. 1992). While crop improvement has been and remains the primary focus, recently there has been more emphasis placed on market research and programs aimed at matching supply to market requirements, identification, development, and testing of quality standards, and new product development.

The Australian pulse crop industry offers the potential to be a valuable option for farmers in terms of both agronomic and economic benefits. The industry has experienced rapid growth over the last 20 years but with this, a host of new problems have emerged. Even in the lupin industry, which was seen as mature, the potential for new problems to emerge and significantly set back the industry are evident, such as the outbreak of Anthracnose in 1996.

Over the last 10-15 years, resources (funds and staff) allocated to pulses have increased significantly, corresponding to the increase in the number of crop species and varieties but it is now recognized that the research effort needs to be more focused on a few key pulse crops and a few core problems that restrict yield and marketability. As such, the industry's peak body, Pulse Australia is working with the GRDC to identify the crops and issues to be addressed in research and market development programs. Whilst this is likely to focus on cool season pulses such as lupins, field peas, faba beans, chickpeas and lentils, efforts will also be maintained on a limited range of warm season pulses such as mung beans.

$ US

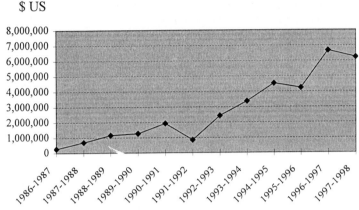

Figure 1. Investment by GRDC Australia in pulse research

FRANCE

France produces about 75% of the pulse production of the European Union. Pulse crops gained prominence in France in the late 1970's following the USA soybean embargo of 1973. At that time, the European Community decided to support pulse crop production in an attempt to reduce their deficit and dependency on imported protein.

Field pea is the main pulse crop in France and 90% of the production is used by the feed industries. Peas are the most competitive pulse crop when compared with other major crops in Europe, such as winter wheat, winter rapeseed, maize, sunflower, and secondary cereals, due to the high yield potential and good yield stability of peas.

The rise in pulse production in France coincides with the research efforts of UNIP (Union Nationale Interprofessionelle des plantes riches en Proteines). UNIP was created in 1976 to gather together economic partners with an interest in pulse crops and to set up a marketing organization. UNIP is partly funded by a technical fee, paid by farmers, of 0.45 Francs per quintal (100 kg) of pulses produced (approximately 1%). The technical service of UNIP strives to connect all scientists worldwide in order to get better efficiency and value from research projects involving pulse crops. One of the main partners of UNIP is INRA (Institut National de la Recherche Agronomique), funded by the French Ministries of Agriculture and Research. UNIP allocates research funds to INRA research projects involved with pulse crops, which help cover costs of supplies, occasional staff, and graduate student stipends. Part of the recent increase in the number of scientists involved with pulse crops at INRA is due to the recent INRA Council decision to make pea one of their model plants for study.

The other main economic partners represented in UNIP by these professional organizations are: FOP, an association of oilseed and protein crop growers; ITCF, a technical institute working mainly for growers and emphasizing cereal crops now receiving one-third of their funding from UNIP for pulse crop research; FNAMS, similar to ITCF, but focused on seed production; AMSOL, an association of seed manufacturers, pulse seed marketing, pricing and quality, who serve also as lobbyists for breeding companies; FFCAT/INAC, a group of cooperatives and private storage associations, dealing with market organization, market prices, and trade markets for all crops; SYNCOPAC/SNIAA, cooperatives and foodstuff company associations, feed uses; FNLS, dry legume food industry representatives, niche markets.

Table 3. Comparison of the number of Scientific Person Years in France devoted to pulse crop research in 1980 and 1997.

Research Area	1980		1997	
	Public	Private	Public	Private
Crop Adaptation				
Genetics and Breeding	5	18	15	14
Crop Management	1	2	10	10
Crop Protection	2	2	7	5
Use and Quality	4	2	10	5
Equipment and Machinery	0	1	0	2
Post-harvest Storage	0	1	0	1
Total	12	26	42	37
Grand Total	38		79	

UNIP also helped in setting up and managing a group of private pea breeders (GSP) to conduct joint research with geneticists from INRA. GSP is composed of seven companies involved mainly with disease resistance in collaboration with INRA. Funding of GSP is derived from member fees, and subventions from AMSOL, the Ministry of Agriculture, and UNIP. There is also a push towards increasing cooperation between UNIP and pulse crop research activities outside of France. The AEP (European Association of Research in Grain Legumes) was set up through the efforts of UNIP in an attempt to improve relations between scientists and to initiate new research throughout Europe and the world.

The only other European countries with organizations similar to UNIP in France are the U. with PGRO and Germany with UFOP. Both PGRO and UFOP come close to what UNIP is doing, however, their potential to support research is much lower. There are no other existing organizations similar to UNIP in the other European countries.

Table 4. Resources allocated to research on pulse crops in France (estimates in 1,000 US $).

Research Inputs	1980	1997	1997 self financing	Subventions (grants)
ITCF/FNAMS - applied research	413	3,300	1,980	1320
Private Breeding programs	825	3,300	2,970	330
Public research: ENSA*	83	660	495	165
Universities	83	165	83	83
INRA	191	6,930	5,445	1,485
Total	1,595	14,355	10,973	3,383
UNIP	165			2,228
Ministry of Agriculture/Research	413			825
European Union	0			330
Total of subvention	578			3,383

ENSA Ecole Nationale Superieure Agronomique

NORTH AMERICA

Both Canada and the USA have developed their cool season food legume industries around research programs funded jointly by grower organizations and government sponsored research programs. The past decade has seen the two countries reverse their roles, with Canada becoming the major producer of cool season food legumes in North America. This trend is also reflected in the amount of funding directed to the research programs in each country.

63

Canada

Pulse crop research programs in Canada receive funding through government programs and grower-financed boards established in the provinces of Saskatchewan, Alberta and Manitoba. Growers pay a check-off levy of 0.5% of the gross value of all pulses sold, resulting in a pool of money used to finance and administer the needs of their pulse crop industry. Government support for pulse crop research is provided via the federal Agriculture and Agri-Food Canada (field crops branch), and the Provincial Departments of Agriculture in Alberta, Saskatchewan and Manitoba.

Pea and lentil areas have risen dramatically in Western Canada during the past twenty years, to the present level of nearly 3,000,000 acres (1.21 m ha). Major efforts in pea and lentil breeding programs, as well as in crop management have resulted in Canada quickly becoming an important player on the world pulse production scene. Other contributing factors have been the recent reductions in wheat support and rail subsidies, which now make pulse crops a competitive alternative crop which fits well with current rotations.

The amount of funding provided from Canada's federal government has been declining since 1995 (Figure 2). This trend has been more than compensated for by large increases form provincial governments and private industry.

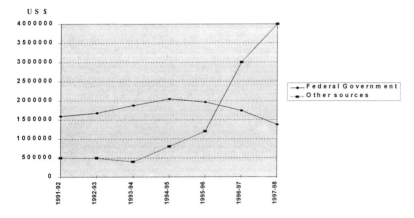

Figure 2. Canada's funding sources for pulse crop research during 1991 to 1998.

Table 5. Percentage of Scientific Person Years involved in each pulse crop research area per pulse producing province in Canada during 1992 to 1997.

Research Area	Saskatchewan	Manitoba	Alberta
Crop Adaptation	25	15	10
Genetics and Breeding	25	30	50
Crop Management	20	20	10
Crop Protection	20	15	10
Use and Quality	10	10	10
Equipment and Machinery	0	10	5
Post-harvest Storage	0	0	5
Total	100	100	100
Scientific Person Years *	32	44	12

*Actual number of Scientific Person Years

USA

Funding and personnel for research on cool season grain legumes in the USA (tables 6 & 7) is primarily through the Federal Government programs of the USDA (Unites States Department of Agriculture) and grower assessments administered by the USA Dry Pea and Lentil Council. The Agriculture Research Service (ARS) arm of the USDA contributes directly to the Grain Legume Genetics Physiology Research Unit based at Washington State University in Pullman, Washington. This program is augmented by grower-allocated funds provided by dry pea, lentil and chickpea industry members and collected by the Washington and Idaho State Commissions. The Idaho growers are currently paying an assessment of 2% of net sales value, and the Washington growers are paying an assessment of 1% of net sales value. Further supplementing grain legume research work is STEEP (Solutions To Environmental and Economic Problems), another USDA-ARS program which is estimated to contribute about one-third of its funds towards pulse crops. USDA-CSREES provides additional research funding through the Cool Season Food Legume (CSFL) Research Program. CSFL receives federal, state and private funding which serves to enhance grain legume projects in the USA.

Table 6. Research funding (per funding source) for cool season food legumes in the USA from 1976 to 1998 (US$).

	USDA-ARS	CSFL	STEEP	USADPLC	Total
1976-77				44,836	44,836
1977-78				35,254	35,254
1978-79				25,799	25,799
1979-80				35,601	35,601
1980-81	150,000			57,680	207,680
1981-82	160,000			55,540	215,540
1982-83	170,000			77,268	247,268
1983-84	180,000			69,855	249,855
1984-85	190,000			101,707	291,707
1985-86	200,000			66,220	266,220
1986-87	210,000			60,372	270,372
1987-88	220,000			50,900	270,900
1988-89	226,000			54,951	280,951
1989-90	361,700			94.515	361,794
1990-91	356,700			86,600	443,300
1991-92	361,400	355,500	300,000	96,815	1,113,715
1992-93	372,000	366,358	300,000	97,561	1,135,919
1993-94	372,000	365,391	300,000	92,452	1,129,843
1994-95	376,100	343,652	300,000	169,820	1,189,572
1995-96	376,100	96,757	300,000	188,784	1,174,064
1996-97	372,500	309,180	150,000	142,000	973,680
1997-98	369,600	307,485	150,000	150,000	977,085
Total	5,024,100	2,144,323	1,500,000	1,760,110	10,940,956

Table 7. Percentage of Scientific Person Years involved in each pulse crop research area per funding source in the U.S.A. during 1992 to 1997.

ResearchArea	USDA-ARS	CSFL & STEEP	USADPLC
Adaptation	15	0	10
Breeding	45	10	65
Management	5	50	5
Protection	20	25	10
Use & Quality	10	10	10
Equipment	0	5	0
Storage	5	0	0
Total	100	100	100
Scientific Person Years *	15	12	9

* Actual Scientific Person Years

The pulse crop research community in the USA has experienced a decline in amount of funding over the past ten years. There is potential for future increases in pea and lentil production in the states of North Dakota and Montana, which may result in a larger grower-based funding pool for research programs in the future.

CONCLUSIONS

While the four countries examined have similar sources for pulse crop research funds, the differences in size of their funding pools are worth noting. France has a well-organized and extensive network of funding agencies which has allowed their growing scientific community to flourish. Australia's system of pooling grower levies from all crops has benefited research activities directed towards a young pulse crop industry. In fact, the amount of research funds that Australia's pulse crops receives is greater in proportion to the amount received by cereal crop research programs. Canada's federal government pulse crop research funding has been declining, however, increased production areas resulting in more grower assessments coupled with provincial government and private industry support has more than made up for this shortfall. Of the countries examined, only the U.S.A. has experienced dwindling pulse research funding over the past decade. As recently as 1991-1992, the U.S.A. had been leading all countries in amounts of funding for research projects involved with peas, lentils and chickpeas. Recent cuts in federal programs that had supported cool season grain legume research projects in the past have placed the U.S.A. in the position of playing 'catch-up' to the rest of the developed world (Figure 3).

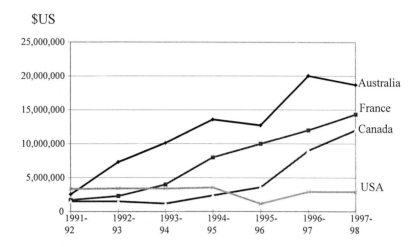

$US

Figure 3. Comparison of research funding dedicated to cool season food legumes in Australia, France, Canada and the USA.

References
Agriculture and Agrifood Canada, Research Branch. Ottawa, Ontario, Canada.
Alberta Pulse Growers Commission. Leduc, Alberta, Canada.
Carrouee, B., G.P. Gent, and R.J. Summerfield. 1998. *In Linking research and market opportunities for the 21st century,*(ed R Knight) Kluwer Academic Press.(in press)
Clements, R.J., A.A. Rosielle, and R.D. Hilton. 1992. *National Review of Crop Improvement in the Australian Grains Industry,* A report to the Board of the Grains Research and Development Corporation.
Hamblin, J. 1996. *The Future R and D Outlook* - Paper presented to the Australian Pulse Industry Workshop - Meeting the Market Challenge of the Indian Ocean Region, Agribusiness Conference, Fremantle.
Manitoba Pulse Growers Association Inc. Carman, Manitoba, Canada.
Saskatchewan Pulse Crop Development Board. Saskatoon, Saskatchewan, Canada.
USA Dry Pea and Lentil Council. Moscow, Idaho, USA.
United States Department of Agriculture - Agriculture Research Service.

Regional Reviews, Region 1: North America

F.J. Muehlbauer[1] and A.E. Slinkard[2]

1 U.S. Department of Agriculture, Agricultural Research Service, Washington State University, Pullman WA 99164-6434; 2 Crop Development Center, University of Saskatchewan, Saskatoon, Saskatchewan S7N 0W0, Canada.

Abstract

Cool season food legumes including dry pea, lentil, and chickpea are important components of farming systems in the dry-land regions of the western United States, most notably the Palouse region of eastern Washington and northern Idaho, and the provinces of western Canada, particularly Saskatchewan, Alberta and Manitoba. Production has increased dramatically in western Canada over the past five years and also production increases have taken place in North Dakota and South Dakota. California is a major U.S. producer of chickpeas. Current production in the U.S. stands at about 75,000 tonnes of lentils, 150,000 tonnes of peas and 16,000 tonnes of chickpeas, while production in Canada stands at about 420,000 tonnes of lentils, and 1,250,000 tonnes of peas. Production of faba bean, lupin and grasspea is minimal in the U.S. and Canada. Major constraints to production include Ascochyta blight of lentil and chickpea; and root rots, wilts, powdery mildew, Mycosphaerella blight and viruses of pea. Viruses are problematic on chickpea, lentil and peas in the US in years when aphid-vector populations increase early in the growing season. Current breeding programs are concentrated on the development of cultivars that will alleviate constraints to production while meeting market demands. As a result of the success of the breeding programs, cultivars in use have changed considerably over the past 10 years, a trend that is likely to continue. Most of the U.S. food legume crop is exported as food while the Canadian crop is exported as both food and feed. Current trends point to increased production in Canada and in North Dakota and South Dakota. The trend to replace summer fallowing in the western United States and Canada with annual cropping has provided considerable additional land area for expanded production of cool season food legumes.

INTRODUCTION

The Palouse region of the U.S. Pacific Northwest has been the traditional area of production of the food legumes in North America from the 1920s to the 1970s. Canada, while producing minor amounts of cool season food legumes up to 1970, has since greatly increased production and has surpassed the Palouse in area sown, amounts produced and amounts exported. The extensive land areas in the western provinces of Canada and the competitiveness of the food legumes compared to other crops in the rotations have been overriding reasons for the continued expanding production in Canada. During this period, production in the Palouse region has remained constant.

The value of legumes in the predominantly cereal-based rotations has been a major factor in maintaining production in the Palouse. Expansion of cool season food legumes in the U.S. has taken place in the states of North Dakota and South Dakota, while the introduction of winter chickpea has been a factor for increased production in California.

A major factor in favour of expanded production of food legumes in the U.S. is the trend to replace summer-fallow in areas that receive sufficient rainfall to permit annual cropping. Competitiveness with alternative crops, usually spring wheat, continues to be a major factor in the expanded production of legumes in the western provinces of Canada. Current objectives of breeding programs are to increase yields and improve product quality.

R. Knight (ed.), Linking Research and Marketing Opportunities for Pulses in the 21st Century, 67-70.

CURRENT STATUS OF THE COOL SEASON FOOD LEGUMES IN NORTH AMERICA

The most important region for cool season food legumes in North America is the three provinces of western Canada, where nearly 2 million tonnes are produced on about 900,000 hectares annually (Saskatchewan Agriculture and Food, 1996) (Tables 1&2). That region is projected to expand production by over 20% by the year 2000. Current production areas of the U.S. are located in the Palouse region of eastern Washington state and northern Idaho where an average of 200,000 tonnes is produced on about 150,000 hectares annually (U.S.A. Dry Pea and Lentil Council, 1996). The expanding area of production in North Dakota and South Dakota, which is contiguous with the vast production areas of western Canada, is projected to produce 100,000 tonnes per annum by the year 2000. Fluctuations in production in North America are mainly the result of weather variations and market forces.

Table 1. Area sown and production of dry peas in North America, 1987-1996.

Year	USA 1000 ha	USA 1000 MT	Canada 1000 ha	Canada 1000 MT	USA and Canada 1000 ha	USA and Canada 1000 MT
1987	82	152	237	415	319	567
1988	75	171	271	320	346	491
1989	72	185	150	234	72	419
1990	68	108	124	264	192	372
1991	82	175	199	410	281	585
1992	72	116	273	505	345	621
1993	67	153	506	970	573	1123
1994	52	98	696	1441	748	1539
1995	80	200	820	1455	900	1655
1996	75	95	579	1245	654	1340
10-year Average	72	145	386	726	458	871

Table 2. Area sown and production of lentils in North America, 1987-1996.

Year	USA 1000 ha	USA 1000 Mt	Canada 1000 ha	Canada 1000 MT	USA and Canada 1000 ha	USA and Canada 1000 MT
1987	62	77	218	286	280	363
1988	29	38	136	58	165	96
1989	38	49	103	96	141	145
1990	46	41	134	213	180	254
1991	52	76	238	343	290	419
1992	53	69	279	348	332	417
1993	63	93	373	349	436	442
1994	90	96	399	450	489	546
1995	70	99	334	432	404	531
1996	57	52	318	417	375	469
10 year Average	56	69	253.2	299.2	309.2	368.2

Production in western Canada has increased greatly in recent years because of the need for crop diversification and the need to provide farmers with additional cash income. Cool season food legumes were obvious choices, based primarily on the benefits to be derived from legumes in the rotation, value added processing of the food legumes, and the emphasis on sustainable agricultural systems. The situation in the U.S. is somewhat different. Legumes produced in the Palouse region are valued as "break" crops in primarily cereal-based rotations and are credited with providing a measure of control of cereal diseases. In addition, legumes provide opportunities for the control of weeds that infest cereal crops and they also produce at least part of their nitrogen needs through symbiosis with *Rhizobium* sp.

Canada produces mostly yellow-cotyledon dry peas that are exported and used either for food or animal feeding. However, interest is growing in Canada in producing green-cotyledon dry peas and currently the green type represents over 20% of the dry pea crop. Good quality green-cotyledon peas are more difficult to produce because of bleaching of the seeds during wet periods at maturity but before harvest. While the Canadian crop is primarily yellow peas, production in the Palouse region has concentrated on the green type. Since the harvesting period in the Palouse region is usually dry, bleaching of the peas is not often a problem.

Besides the extensive production of dry edible peas in the U.S. and Canada, a large pea processing industry relies on the western U.S. for good quality seed peas. It is estimated that 10,000 hectares of seed peas of processing types are produced in the states of Washington and Idaho annually. Production of processing peas in the states of Washington, Oregon, Idaho, Wisconsin, Minnesota and other states, and Canada as well, is estimated at nearly 100,000 hectares.

Lentil production in Canada is currently over 300,000 hectares while the area sown in the U.S. has fluctuated from 29,000 to 90,000 hectares over the past 10 years.

Chickpea production has expanded in the U.S. to an estimated 8,000 hectares in the Palouse region, with a similar area in California. Minor production is taking place in other states. In Canada, interest in the production of chickpea has occurred recently and estimates place production at 8,000 hectares with a projection 25,000 hectares by the year 2000. Nearly all of the production in the U.S. and Canada is Kabuli type, but about 20% of the production in Canada is Desi type

Other cool season food legume crops such as faba bean, grasspea and lupin are not widely grown in North America.

UTILIZATION OF COOL SEASON FOOD LEGUMES

Most of the Canadian production of 1.2 million tonnes of dry peas is targeted for the feed pea market in Europe, with the remaining amount being exported mostly to South America or India. Lentils grown in western Canada are mostly the 'Laird' cultivar, which has gained a large market share because of its large seed size and attractive seed-coat coloration. 'Eston' is also an important cultivar in Canada. In the U.S., the most commonly grown cultivar is 'Brewer'; however recent releases of large seeded cultivars are expected to eventually replace 'Brewer'. 'Redchief' a large red cotyledon type, is also important in the U.S. About 70% of U.S. production of lentil is exported as food mostly to South America and India with smaller amounts to other countries. Of the 30% of the dry pea crop utilized domestically in the U.S., nearly all is decorticated, split and used as food. Similarly for the lentil crop, most of the U.S and Canadian production is exported. In the U.S. and Canada, chickpeas are almost entirely reconstituted and canned for use in salad bars.

PRODUCTION CONSTRAINTS

Major pea diseases in western Canada include Mycosphaerella blight caused by *Mycosphaerella pinodes*, a disease that is favored by cool wet summers. None of the currently available cultivars in Canada is resistant. In the U.S., the major disease problems for dry peas are root rots caused by *Aphanomyces eutieches*, *Fusarium solani* f. sp. *pisi* and *Pythium ultimum*, wilts caused by races of *Fusarium oxysporum* f. sp. *pisi*, powdery mildew caused by *Erysiphe polygoni* and viruses (Slinkard *et al.*, 1994). The insect problems of dry peas include the pea weevil (*Bruchus pisorum*) and the pea leaf weevil (*Sitona lineatus*) and several aphid species in the US but these insects present no problems in Canada (Slinkard *et al.*, 1994).

Major diseases of the lentil crop in Canada and the U.S. include Ascochyta blight caused by *Ascochyta lentis*, and certain aphid transmitted viruses. Root rots and wilts have not been a problem (Slinkard *et al.*, 1994).

Ascochyta blight of chickpea, caused by *Ascochyta rabiei*, is the most serious disease problem of chickpea in North America (Kaiser *et al.* 1995). In the Palouse region of the U.S. the disease is being controlled through the use of resistant cultivars; however in California the disease is still a problem. Resistant cultivars will be widely available in Canada in 1998. Also, Fusarium wilt, caused by *Fusarium oxysporum f. sp. ciceris*, is a serious problem in California but is controlled through resistant cultivars. Aphid transmitted viruses have caused devastation in the Palouse region and in California.

GOALS AND TRENDS

The trend is for increased production of the cool season food legumes in the western provinces of Canada and for continued production of these crops in the Palouse region of the U.S. Other areas where production is likely to increase include areas of North Dakota and South Dakota, which are contiguous with the western provinces of Canada and have similar climates and production systems. Chickpea production has increased substantially in the Palouse region and indications are the trend will continue. Evaluations of other cool season food legumes are underway and potential exists for production of lupin, faba bean and grasspea however these crops are a long way from substantial production in North America.

Demand for cool season food legumes is projected to increase over the next 10 years (Oram and Agcaoili, 1994) and it appears that the extensive areas available in North America will be used to meet some of the demand. The trends in breeding programs are for increased yields and improved produce quality. Also, the breeding programs have begun to emphasize specialty types to meet the demands of niche markets.

References

Oram, P.A. and M. Agacoili. 1994. In: *Expanding the Production and Use of Cool Season Food Legumes.* pp. 3-49 (eds F.J. Muehlbauer and W.J. Kaiser), Kluwer Academic Publishers, Dordrecht, The Netherlands.

Kaiser, W.J., F.J. Muehlbauer and R.M. Hannan.1994. In: *Expanding the Production and Use of Cool Season Food Legumes.* .pp. 849-858 (eds F.J. Muehlbauer and W.J. Kaiser), Kluwer Academic Publishers, Dordrecht, The Netherlands.

Slinkard, A.E., G. Bascur and G. Hernandez-Bravo. 1994. In: *Expanding the Production and Use of Cool Season Food Legumes.* pp. 195-203 (eds F.J. Muehlbauer and W.J. Kaiser), Kluwer Academic Publishers, Dordrecht, The Netherlands.

Saskatchewan Agriculture and Food. 1996. Specialty Crop Report. Saskatoon, Saskatchewan Canada. 16p.

U.S.A. Dry Pea and Lentil Council. 1996. Directory of U.S. Suppliers and Industry Information. U.S.A. Dry Pea and Lentil Council. Moscow, Idaho. 16p.

Region 2: South America

J.U Tay[1], E Peñaloza[2] and JA Morales[3]

1. Instituto de Investigaciones Agropecuarias (INIA), Centro Regional Quilamapu. Casilla 426, Chillán, Chile.; 2. Instituto de Investigaciones Agropecuarias (INIA). Centro Regional Quilamapu. Casilla 58-D, Temuco, Chile.; 3. Instituto Nacional de Investigaciones Forestales y Agropecuarias (INIFAP), Campo Experimental Costa de Hermosillo. Apartado Postal 1031, CP 83000, Hermosillo, Sonora, México.

Abstract

Cool season food legumes are grown in most Latin American countries. Lentils, peas, chickpeas and lupins are among the most important food legumes in these countries, where they are part of the population´s diet as an inexpensive source of protein. These legumes are produced mostly by small farmers who use traditional methods of cultivation and obtain only low yields. As requirements for a more competitive agriculture are increasing in Latin America, some countries have begun to mechanize the production of crops such as lentil, chickpea and lupin, thus diverting their culture to the more technically advanced farmers. The most important biotic stresses experienced in production are those caused by fungal pathogens, while abiotic constraints vary and are closely related to the local environments where the crops are grown.

INTRODUCTION

Cool season food legumes in Latin America comprise about 2% of the total area devoted to these pulses in the world. However, they play an important role in the diet of the population. This figure increases to more than 12% when dry beans (*Phaseolus vulgaris* L) are included. At present, dry bean is by far the most important food legume produced by the region with Brazil being the highest producer. Dry beans are grown in more than 95% of the area sown with legumes in Latin America and represent 30% of the world's total production of this legume. Cool season food legumes have a socio-economic significance in the region as they can be established in areas not suited for other crops and are the only source of proteins for people living on small farms. Areas of cultivation in the region range from almost sea level to up 3000 m. altitude and from tropical to subtropical and Mediterranean climates. Although they are widely distributed through the region there are geographical concentrations of production such as dry bean in Brazil, chickpeas in Mexico, peas in Colombia, lentils in Argentina and lupins in Chile.

THE CURRENT SITUATION

The Distribution and productivity of cool season food legumes

The cool season food legumes, chickpeas, peas, lentils, faba beans and lupins are grown in many Latin American countries. The area has shown a steady decline during the last decade from 536, 000 ha in 1987 to 482,000 ha in 1996. An important feature of the distribution of the species in Latin America is that the main production of each species occurs in a different country. Any change in the area in each of these countries will be reflected in the current area sown with legumes in the region. Thus our discussion will be concentrated on analysing each species and how its production has been affected by changes that have occurred in the highest producing country.

R. Knight (ed.), Linking Research and Marketing Opportunities for Pulses in the 21st Century, 71–77.

Lentils

Six countries in the region grow lentils with Argentina being at present the largest producer, followed by Colombia and Chile. The present situation is far different from that observed at the beginning of the decade, when 44% of the total area in the region, devoted to lentil, was in Chile (Table 1). The area in the region sown to lentil has shown a sharp decline during the last decade from 107,000 ha in 1987 to 75,000 ha in 1996, a tendency that can be explained mainly by a reduction in the area in Chile. According to FAO figures, the region's share of the world's area of lentil has decreased from 3.19% in 1987 to 2.0% in 1996. This situation has changed production of this commodity to such extent that the region has begun to import large quantities of lentils in order to meet domestic requirements.

Table 1. Area, yield and production of lentil in the four main producing countries in the region for the period 1987 to 1996

Country	1987	1988	1989	1990	1991	1992	1993	1994	1995	1996
Area ('000ha)										
Chile	47	33	15	14	15	19	13	10	11	11
Argentina	16	18	22	25	27	25	24	27	27	28
Colombia	17	17	17	17	17	17	17	17	17	17
Mexico	20	17	15	20	13	13	14	15	15	11
Tot Region	107	92	77	84	81	83	77	78	79	75
Tot World	3179	3253	3173	3171	3303	3240	3465	3340	3351	3384
Yield (kg/ha)										
Chile	520	614	531	589	799	832	733	869	721	900
Argentina	774	833	1136	880	956	960	1046	963	963	1000
Colombia	353	353	353	353	400	412	412	412	412	412
Mexico	1000	1000	1000	1000	919	932	893	867	867	909
Av. Region	635	684	753	726	765	771	779	769	745	786
Av. World	818	824	637	789	807	791	810	824	839	828
Production ('000 MT)										
Chile	25	20	8	8	12	16	10	9	8	10
Argentina	12	15	25	22	26	24	25	26	26	28
Colombia	6	6	6	6	7	7	7	7	7	7
Mexico	20	17	18	20	12	12	13	13	13	10
Tot Region	68	63	58	61	62	64	60	60	59	54
Tot World	2600	2681	2020	2501	2665	2564	2806	2834	2813	2800

Yields of lentils in the region are below the world average, reflecting both their cultivation in marginal and dry lands in Chile, and in other unfavourable environments as in Colombia, where altitude and diseases are major constraints to production. Only the stable and higher yields of Argentina and Mexico are having a beneficial effect on the average yields of the region.

Chickpeas

Chickpea in the region represents 1.18% of the total area of this crop cultivated in the world. Its production is concentrated mainly in Mexico, a country that contributes 75% of the area of the crop in the region (Table 2). Thus any change in Mexico's cultivated area will be reflected in the regional area sown with this legume. Chickpea has shown an outstanding development in Mexico, with the desi type being produced for feed and kabuli-type mainly for the export market. Besides Mexico, chickpeas are grown in Colombia, Chile and Argentina.

The average yield for the region was 1,236 kg/ha in 1996, much higher than the world average of 734 kg/ha, with no significant changes during the last decade. Because of the importance of Mexico as a contributor to production, the average seed yield that characterises the region is strongly influenced by Mexico. Mexico has the highest yield in the world (1,500 kg/ha), a reflection of the irrigated conditions in which the crop is grown in the country.

Table 2. Area, yield and production of chickpea in the three main producing countries in the region for the period 1987-1996

Country	1987	1988	1989	1990	1991	1992	1993	1994	1995	1996
Area ('000 ha)										
Mexico	150	140	120	150	122	122	77	54	79	105
Colombia	23	23	23	23	23	23	21	21	21	21
Chile	14	14	8	9	12	13	11	9	9	10
Tot Region	194	184	156	187	163	164	114	89	105	141
Tot World	9937	8720	9743	9732	11464	9052	10115	9992	11151	11924
Yield (kg/ha)										
Mexico	1200	1179	1000	1133	1599	1598	1389	1896	1470	1443
Colombia	483	478	478	478	478	436	476	476	476	476
Chile	761	536	531	695	722	1447	1003	1187	1161	600
Av Region	1072	1027	897	1027	1349	1402	1166	1449	1494	1236
Av World	691	691	723	679	706	736	667	706	722	734
Production ('000 MT)										
Mexico	180	165	120	170	195	195	107	103	116	152
Colombia	11	11	11	11	11	11	10	10	10	10
Chile	11	8	4	6	9	19	11	10	11	6
Tot Region	208	189	140	192	220	230	133	129	136	173
Tot World	6870	6028	7043	6605	8089	6660	6751	7055	8016	8748

Dry peas

Dry peas are grown in almost all Latin-American countries and the species is the only cool season food legume to show an increase in area through the decade (Table 3).

Table. 3 The area, yield and production of dry peas in the four main producing countries of the region for the period 1987-1996

Country	1987	1988	1989	1990	1991	1992	1993	1994	1995	1996
Area ('000 ha)										
Colombia	57	57	57	57	60	60	55	55	55	55
Peru	13	20	19	20	20	21	20	23	28	28
Argentina	9	9	25	15	16	20	26	27	27	27
Ecuador	15	15	13	7	14	13	11	12	8	12
Tot Region	123	127	141	127	137	141	141	143	144	147
Tot World	9671	9981	9591	9233	8694	7806	8081	7991	7625	7328
Yield (kg/ha)										
Colombia	614	614	614	614	726	733	727	727	727	727
Peru	818	899	872	900	900	905	858	861	914	914
Argentina	1172	1250	960	1333	1290	1500	1538	1057	1057	1057
Ecuador	244	260	500	522	315	226	634	339	401	410
Av. Region	682	708	773	811	810	872	936	845	875	850
Av. World	1625	1558	1671	1872	1493	1824	1864	1834	1513	1551
Production ('000 MT)										
Colombia	35	35	35	35	44	44	40	40	40	40
Peru	11	18	16	18	18	19	18	20	25	25
Argentina	10	11	24	20	20	30	40	28	28	28
Ecuador	4	3	9	3	4	3	7	4	3	5
Tot Region	84	90	109	103	111	123	132	121	126	125
Tot World	15720	15487	16023	17286	12983	14239	15063	14654	11535	11364

At present, dry peas occupy 25% of the area sown to cool season legumes in the region and 2.0% in the world. The main producers are Colombia, Peru, Argentina and Ecuador, which together account for 83% of the area devoted to dry peas in the region. This figure, however, does not include green peas, which occupy about 50% of the area sown to dry peas. The production of green peas for canning or freezing is increasing in most countries in the region. In addition, green peas are sown and consumed by most of the small farmers of Latin-America.

The average yield in the region (850 kg/ha) is well below the world average of 1,551 kg/ha, due to the fact that its production is mainly concentrated in small farms, with the exception of Argentina.

Faba bean

The area sown to faba beans in Latin America is about 4.3% of the total area of the crop in the world (Table 4). It is cultivated in Bolivia, Ecuador, Guatemala, Mexico and Peru, with Mexico and Peru being the largest producers. Although FAO figures indicate Brazil as the main faba bean producer in the region, there are only small areas sown to this species in the southern part of the State of Río Grande do Sul. According to Nascimento and Giordano (1990), cited by Bascur (1993) the figures given by FAO as faba beans in Brazil should correspond to cowpea (*Vigna unguiculata* L Walp.) and lima bean or pallar (*Phaseolus lunatus* L) both known by the local name of "faba". The *Vicia faba* faba bean is an important cash crop for small farmers of the highland areas of Bolivia, Peru and Ecuador, where at an altitude above 2500 m few other crops can be grown. The low average yield that characterizes faba beans in the region (1,075 kg/ha) is below the world average (1,174 kg/ha), a situation that is explained because it is produced by small farmers and because it is grown mainly for personal consumption. Only in Mexico and Peru are yields near or at the level of average world yields

Table 4. The area, yield and production of faba beans in the three main producing countries in the region for the period 1987-1996

Country	1987	1988	1989	1990	1991	1992	1993	1994	1995	1996
Area ('000 ha)										
Mexico	30	32	32	31	23	23	23	23	23	24
Guatemala	23	23	18	17	18	18	18	19	19	19
Peru	16	20	20	21	25	20	20	22	24	24
Tot Region	112	119	117	114	107	102	105	116	118	119
Tot World	3072	3050	3065	3015	2280	2903	2702	2788	2759	2760
Yield (kg/ha)										
Mexico	1333	1281	1281	1323	1122	1130	1130	1152	1152	1167
Guatemala	453	453	664	647	683	683	685	682	682	682
Peru	1052	1171	1010	1024	1144	1000	1218	1186	1159	1159
Av Region	1035	1967	1042	1072	1056	1049	1142	1068	1059	1075
Av World	1349	1308	1263	1381	1418	1280	1392	1313	1227	1174
Production ('000 MT)										
Mexico	40	41	41	41	26	26	26	27	27	28
Guatemala	10	10	12	11	12	12	12	13	13	13
Peru	16	24	21	21	28	20	25	26	28	28
Tot Region	144	155	158	159	143	129	144	149	150	153
Tot World	4309	4146	4028	4337	4239	3833	3889	3778	3499	3356

Lupin

Lupin is an ancient crop of the Andean highlands of Peru and Ecuador. The genus is represented by *Lupinus mutabilis* (called tarwi or chocho in Peru, and chocho or lupino in Ecuador), which is used mainly for human consumption. Lupin is also an important crop in Chile where the cultivated land is shared between the species: *L. albus* and *L. angustifolius* introduced from Germany and Australia, respectively. Lupin is the other grain legume that has shown an increase in the area sown in the region with the main contribution coming from Chile. In contrast to the current situation with lupin in the Andean highlands of Peru and Ecuador, lupins are being grown as a commercial crop in Chile either for export (bitter *L. albus*) or for feeding in the domestic market (sweet *L. angustifolius*).

The average yield for the region is 1,020 kg/ha, but with Chile outstanding with an average yield of 1,800 kg/ha (INE, 1996), (Table 5). The introduction of new cultivars as well as the investment in research is expected to have an impact on lupin production in the near future in Chile. It may compensate for the reduction in the area of cool season legumes that has occurred in the country during the last decade.

Table 5 Area, production and yield of lupin. 1995 - 1996

Country	Area (ha)	Production(MT)	Yield (kg/ha)
Chile	21,320	38,376	1,800
Ecuador	4,980	1,230	259
Peru	5,520	5,619	1,020
Total	31,820	45,225	1,020

Source: Chile, INE, 1996 . Ecuador, Peralta, E. personal communication. Perú, Minag - OIA.1997.

TRENDS IN PRODUCTION.

The production of cool season food legumes in the region has closely followed changes in the sown area as no large changes in yield have been observed during the last decade.

Lentil production in the region comprises about 2% of the total world production (Table 1). Compared to 1987, its production has decreased 12.45% mainly due to a 75% reduction in Chilean lentils

Chickpea production in the region also represents about 2% of the world production. It shows large variations between years as a consequence of variations in Mexico, the highest producer. In Mexico, the area sown under irrigation either increases or decreases according to the water availability as well as with the prices of other crops such as corn, wheat, rice and vegetables. As an example, 122,000 ha were sown during 1990-1992 in Mexico, but this was reduced to 74,419 ha in the 1993-1995 period (Gomez 1995, unpublished; INEGI, 1995).

Dry pea production in the region is about 1% of the world's production (Table 3). The region's production has increased during the last decade. As the highest producer, Colombia's contribution has increased, not due to an increase in the cultivated area but in yield. At present, peas are occupying areas in this country often sown to wheat, because peas are more profitable to grow (Checa, personal communication). Some countries such Argentina, Chile and Venezuela have increased the production of peas but as green peas for frozen food.

Faba bean production has not shown major variation in the region, as it is grown mainly by small farmers with a high proportion consumed by the farmer and his family. At present, faba bean is being developed in Chile both as a rainfed and as an irrigated crop for the external market with a high productivity and income.

Regarding lupins, there are good prospects for increases in production in the near future in Chile. Local and international markets are being developed for sweet and bitter lupins, respectively. No changes are expected to occur with this species in Ecuador or Peru where lupins are grown mainly for home consumption.

TRENDS AND PATTERN OF CONSUMPTION

The per capita supply and domestic utilization of cool season food legumes shows a strong variation from country to country. Although FAO data separate dry bean and dry pea from the total of pulses, it is interesting to note that pulses in some countries include cowpeas, pigeon peas (*Cajanus cajan* L Millsp.) or lima beans. With the exception of Argentina and Bolivia, dry bean is the main food legume consumed in Latin American countries. Brazil shows the highest consumption of dry beans (16.1 kg/year) whereas other pulses represent only 0.4 kg/year (Table 6). Peas, lentils, chickpeas and faba bean are part of the traditional diet of most of the countries in the region whereas human consumption of lupins is important only in Peru, Bolivia and Ecuador.

Table 6 Per capita consumption. (kg/year) of pulses in Latin American countries

Country	1987-89	1989-91	1992-94	Country	1987-89	1989-91	1992-94
Argentina	1.7	1.0	1.0	El Salvador	8.6	10.6	11.4
Bolivia	3.5	3.1	3.0	Guatemala	11.9	11.8	10.9
Brazil	15.8	14.7	16.8	Honduras	9.8	11.2	9.6
Chile	2.9	2.9	3.3	Nicaragua	17.0	18.1	15.7
Colombia	6.3	6.7	8.0	Panama	5.5	5.4	5.4
Costa Rica	10.3	11.9	10.1	Paraguay	9.9	8.8	8.5
Cuba	13.6	12.2	13.6	Peru	4.8	4.6	4.8
Rep. Dominica	13.0	11.9	10.2	Uruguay	2.3	2.4	2.7
Ecuador	3.4	3.6	4.0	Venezuela	5.5	5.7	6.0

Source. FAO, 1996.

Among non-traditional legumes, cowpeas are consumed in Brazil, Colombia, El Salvador, Peru and Venezuela, pigeopeas in El Salvador and Venezuela, whereas pallar (*Phaseolus lunatus*) is used mostly in Brazil, Bolivia, Peru and El Salvador.

REGIONAL TRADE IN PULSES

Because of the observed reduction in cultivated area and the fact that it is consumed in almost all countries of the region, lentils are being imported to Brazil, Colombia, Ecuador, Peru and Venezuela. The main exporters of this commodity within the region are Chile and Argentina, whereas Canada and USA have also become main exporters to the region during recent years.

Considering eight selected countries (Table 7) the total production of pulses during the period 1992 - 1994 was of 5.07 million MT. In the same period, 386,000 MT were exported and imports exceeded 549,000 MT, equivalent to 7.6% and 10.8%, respectively of the total production. In 1990, Brazil imported 55,473 MT of dry beans (US$ 41,768,000) from Argentina, Bolivia, Canada, Chile and USA, and 3,605 MT of lentils (US$ 2,252,000) from Argentina, Canada, Chile and Turkey. With regard to chickpeas, Mexico exported 88,236 MT in 1995 (US$ 72,859,000), with Spain, USA, Italy, Venezuela and Brazil the main buyers (Inegi, 1995). In the same year Mexico imported 43,556 MT of pulses (US$ 20,062,000), mainly dry bean and on a minor scale lentil and pea, from USA, Argentina and Canada.

Table 7. Production, exports and imports (MT) of pulses of selected Latin America countries. Averages 1992 to 1994.

Countries	Production	Exports	Imports
Argentina	263,000	229,000	2,000
Bolivia	29,000	6,000	3,000
Brazil	2,908,000	-	164,000
Chile	111,000	55,000	7,000
Colombia	185,000	9,000	13,000
Cuba	25,000	-	88,000
Mexico	1,506,000	87,000	42,000
Venezuela	44,000	-	88,000
Total	5,071,000	386,000	549,000

Source. FAO, 1996.

Canada is the main exporter of cool season food legume to the region. As an average for the period 1992-1994, Canada exported 255,435 MT of lentil (C$ 110,946,000) to ten countries in the region and 495,494 MT of peas (C$ 114,525,000) to twelve countries (Canadian Pulses Report, 1995).

In order to meet their internal demands, many Latin American countries have had to import dry peas, especially split and whole grain for canning, from Canada and USA and to import seeds from Europe.

FUTURE OUTLOOK

Since their introduction following the Spanish conquest, cool season food legumes have played an important role in the diet of Latin American people. The present figures of per capita consumption show minor changes indicating that most of the countries have begun to import legumes in order to meet the needs of domestic consumption. There has been a decrease in production during the last decade as a consequence of a reduction in the area sown. One clear example is the Chilean experience in which the policy of opening markets has been reflected in a strong reduction in the area of legumes due to the lower prices of the imported commodities. The situation is expected to affect other countries in the near future as they open their markets to international trade.

The integration of some countries into market treaties such as MERCOSUR or the North American Free Trade Area (NAFTA) may have an impact on agriculture, especially on traditional crops. As an example, a recent study by agricultural economists concluded that the entry of Chile into the two treaties would reduce traditional crop production significantly. MERCOSUR would have negative effects on Chilean agriculture overall, while NAFTA would benefit exporters, especially of processed products. The entry into force of NAFTA has considerably influenced agriculture in Mexico. Trade liberalisation under the agreement has resulted in a 17% increase in agricultural imports from, and 7% increase in exports to, the USA in 1994. In Argentina, all forms of subsidies, market interventions and export taxes were eliminated. Only minor support measures to agricultural production existed in the country.

Changes in the above scenario will depend on governments adopting measures in favour of traditional agriculture or regulations to protect small farmers, as is the case of some countries. If no changes occur, then the Chilean case will be an example of how traditional legumes will be produced only for personal consumption by small farmers whereas commercial production will disappear or will be replaced by non-traditional crops such as lupins. Alternatively, countries should concentrate on producing legumes for which they have some advantages (such as chickpeas in Mexico) and become exporters within the region.

It seems clear that countries in the region will begin to specialize in the production of those legumes with which they can compete in international markets. With this scenario, it is expected that legumes produced for self-consumption on small farms will continue at their present level, whereas those legumes produced as commodities for sale and export will have to meet the requirements of the open market.

Acknowledgements

We are pleased to recognize the information given in the preparation of this paper by the Ing. Gar. Eduardo Peralta and Angel Murillo, of INIAP of Ecuador; Ing. Agr. José Villamil and Daniel Macías de Uruguay; Ing. Agr. Rosa María Gómez Garza of INIFAP of Mexico and Ing. Agr. Oscar Checa Coral of Colombia.

References

Anuario Estadístico del Comercio Exterior de México. 1995. *Instituto Nacional de Estadísticas, Geografía e Informática* (INEGI).

Bascur, B.g. 1993. *La lenteja y el Haba en América Latina: su importancia, factores limitantes e investigación.* Rapporte de Estudio Especial. ICARDA

Canadian Pulses Report.1995. *Annual Review 1994 - 1995.* Agriculture Canada, Market and Industry Services Branch. Canada.

Estadísticas Agropecuarias 1996. *Año Agrícola 1996.* Instituto Nacional de Estadísticas (INE)Chile.

Food and Agriculture Organization of United NATIONS. 1995. Production Yearbook, VOL. 49. Rome, Italy: FAO.

Food and Agriculture Organizations of United NATIONS. 1996. Hojas de Balance de Alimentos. ROME, Italy: FAO.

FAO. 1996. *The state of food and agriculture 1995.* FAO Agricultural Series N° 28. Rome

Gómez, G.R., Avilés, G. m., y PÉREZ, v .J. 1996. *Manejo del cultivo de Garbanzo Blanco en el centro de Sinaloa.* INIFAP, Fundación Produce, Sinaloa. México.

IICA - BID - PROCIANDINO. 1989. *Investigación para la producción de haba, lenteja, arveja y garbanzo en la Subregión Andina.* (Eds. Hernández-Bravo, G. Y Ramakrisna, B.)

MINAG - OIA. 1997. *Siembras de los principales cultivo de programación nacional.* Ministerio de Agricultura, Perú.

Production and uses of grain legumes in the European Union

B. Carrouée[1], G.P. Gent[2] and R.J. Summerfield[3]

1 UNIP,(Union Nationale Interprofessionnelle des plantes riches en proteines),12 avenue Georges V, 75008, Paris, France; 2 PGRO (Processors and Growers Organisation), The Research Station, Great North Rd., Thornhaugh, E8 6HJ Peterborough,UK; 3 University of Reading, Department of Agriculture, Early Gate, PO Box 236, Reading,RG6 6AT, Berkshire, UK

INTRODUCTION

Europe, with Russia included, comprises 36 countries and in 1997 had approximately 720 million inhabitants. The region covers extremely diverse climates, soils, and eating habits and produces a wide range of grain legumes. This paper will focus on the 15 countries (the EU 15) of the western part of Europe which form the European Union (EU) a group with a Common Agricultural Policy (CAP). It represents 370 million inhabitants and 3.2 million km². Before Austria, Sweden and Finland joined in 1995 there was the EU 12.

Only grain legumes harvested at maturity for their dry seeds are considered, and not the immature seeds and pods used as vegetables (eg garden peas and green beans) or as forage. We will focus on the six main Euro-Mediterranean grain legumes, the "cool season food legumes", ie pea, faba bean, lupin, chickpea, lentil and vetches.

Current situation of grain legumes in the EU

Eight grain legume crops were each grown on more than 40,000 ha in 1996 in the EU (Table 1).

Table 1. European (EU 12) grain legumes in 1995/1996. Area in '000 ha, yield in t/ha other quantities in '000 t (from EU sources for the resources, EUROSTAT for imports-exports and UNIP estimates for uses)

	Dry pea	Faba bean	Sweet lupins (1)	Chick pea (4)	Lentil (4).	Vetches (2) (4)	Total	Soya bean	Dry bean (3)
Area	893	227	59	131	41	189	*1 540*	305	247
Yield	4.10	2.42	2.04	0.23	0.37	0.38	*2.89*	2.86	0.59
Production	3 660	548	120	30	15	72	*4 445*	697 (5)	146
Imports	1 160	88	295	105	185	0 ?	*1 833*	12 355 * (5)	425 *
(% resources)	24	14	71	78	93	-	*29*	95	74
RESOURCES	4 820	636	415	135	200	72	*6 278*	13 052 (5)	571
Feed uses	4 345	414	403	7	10	68	*5 247*	12 813 (5)	29
(%)	90	65	97	5	5	94	*84*	98	5
Fooduses	200	150	0	118	181	4	*653*	150 (5)	506
(%)	4	23	-	87	91	6	*10*	1	89
Seed uses	245	55	12	(4)	(4)	(4)	*312*	53 (5)	11
(%)	5	9	3	-	-	-	*5*	0,4	2
Exports	30	17	0	10	9	0 ?	*66*	36 (5)	25
(%)	0,6	3	-	7	4	-	*1*	0,3	4
Ave. price ex farm 95/96	114	115	114	-	-	-	-	-	-
(% of Soft Wheat price) 96/97	133	137	133	-	-	-	-	-	-

(1) Sweet cultivars of *Lupinus albus, L. luteus* and *L. angustifolius*.   (2) mainly *Vicia sativa* and *V. ervilla*
ainly Common bean (*Phaseolus vulgaris*) and all other *Phaseolus* and *Vigna* species (source FAO for area and production)
eas and production devoted to seed multiplication (the 3 crops depend on 2 different European Price support schemes depending on
er they are grown for seeds or for other uses) (5) without Soya bean seeds are converted into deoiled meal with the coefficient 0,8

R. Knight (ed.), Linking Research and Marketing Opportunities for Pulses in the 21st Century, 79–97.
© 2000 *Kluwer Academic Publishers. Printed in the Netherlands.*

Table 2. Contribution of the 6 major producing countries of the EU 15 to the production of pulses in 1996
(Source : EU Commission except for chickpea, lentil and vetches in Spain, national sources)

CROPS	Dry pea	Faba bean	Sweet lupins	Chickpea	Lentil	Vetches	Total
AREA 1000 ha							
France	539	10	2	0,6	4,2	0	556
%	*57*	*5*	*3*	*0,4*	*9*	*-*	*32*
United Kingdom	71	105	0	0	0	0	176
%	*8*	*50*	*-*	*-*	*-*	*-*	*10*
Germany	87	21	42	0	0	0	150
%	*9*	*10*	*63*	*-*	*-*	*-*	*9*
Spain	97	16	17	148	40	308	626
%	*10*	*8*	*25*	*97*	*87*	*98*	*36*
Denmark	70	0	0	0	0	0	70
%	*7*	*-*	*-*	*-*	*-*	*-*	*4*
Italy	11	51	3	0,6	0,7	1,8	68
%	*1*	*24*	*4*	*0,4*	*2*	*0,6*	*4*
total EU 15	**943**	**211**	**67**	**152**	**46**	**313**	**1 732**
YIELD (t/ha)							
France	4.83	3.88	2.60	1.75	1.52	-	4.77
United Kingdom	3.90	3.40	-	-	-	-	3.60
Germany	3.36	3.28	3.07	-	-	-	3.27
Spain	0.92	0.96	0.80	0.58	0.69	0.62	0.69
Denmark	3.90	-	-	-	-	-	3.90
Italy	3.39	1.46	1.70	-	-	-	1.72
EU 15	4.00	2.73	2.34	0.65	0.78	0.64	2.79
PROD. 1000 t							
France	2 601	39	6	1	6.4	0	2 653
%	*69*	*7*	*4*	*1*	*18*	*-*	*55*
United Kingdom	277	357	0	0	0	0	634
%	*7*	*62*	*-*	*-*	*-*	*-*	*13*
Germany	292	69	129	0	0	0	490
%	*8*	*12*	*82*	*-*	*-*	*-*	*10*
Spain	89	15	14	96	28	193	435
%	*2*	*3*	*9*	*97*	*78*	*96*	*9*
Denmark	273	0	0	0	0	0	273
%	*7*	*-*	*-*	*-*	*-*	*-*	*6*
Italy	38	74	5	-	-	-	117
%	*1*	*13*	*3*	*-*	*-*	*-*	*2*
total EU 15	**3 770**	**577**	**157**	**99**	**36**	**201**	**4 840**

•   **Dry pea** (*Pisum sativum*) is the major species in Europe being grown from Spain to Sweden, although France, provides nearly two-thirds of the European production (Table 2). Peas produce the highest yields, slightly more than faba bean, and far above all other grain legumes[1]. More than 90% of pea production is used for animal feed, mainly for pigs and to a lesser extent poultry.

[1] : Comparisions of yields have to be done with care ; in particular, they must never be done directly with the European mean since the weight of the different countries having different yield potential varies between crops. Even if two crops are grown in the same part of a country, there may be a bias : for instance, the average yield of faba bean in France is more than 20% less than dry pea, mainly because peas are grown on fertile soils, unlike faba beans which are grown on heavy stony soils. In the same soil and climate the difference would be 10%.

Although pea was domesticated in Europe 5,000 years ago (Smartt, 1990) it has only a recent history as a field crop. Before the increase during the eighties (Figure 1) dry pea production was targeted at two specific foods, dehulled split peas and whole "marrowfat" peas.

European production is entirely of white-flowered low-tannin varieties, with the exception of some colour flowered[2] varieties used for pigeon feed or export. To date, most varieties are spring types, sown at the end of the winter.

Soyabean (*Glycine max*) is the second grain legume crop in terms of EU production. Its cultivation is limited to specific regions with warm summers and high rainfall or irrigation such as the Pô valley of Italy. Although it is possible to grow early maturing varieties up to the north of France and in England, in practice, the yields are too small to be competitive with local crops. Production is mainly for animal feed in the form of de-oiled meal, or as extruded whole seed. European production is small relative to the huge amount of imported soyabean (5% in terms of the de-oiled product).

Faba bean (*Vicia faba*) ranks third in production among the European grain legumes. It is a more traditional dry seed field crop than pea. It has been used for decades for human food in the south of Europe ("broad bean") and for cattle or pigeon feed ("horse bean" and "tick bean") in most EU countries[3]. It is grown in almost all countries, except those of northern Europe where it is too late maturing. Its reputation for adaptability comes from its very large tolerance of poor seed-beds and its ease of sowing and harvesting. The major producers are the United Kingdom, Germany and Italy. The faba bean area in France has declined (Figure 1). Most current varieties have coloured-flowers and the seeds contain tannin. Even in the UK where tannin-free varieties are grown on significant areas, there is not a specific market for low-tannin seeds, although their nutritional value is better. In the UK, Italy and Spain, most faba beans are autumn-sown (Gent 1997; Abbate *et al.* 1997), unlike France and Germany where they are spring-sown.

Vetches (*Vicia spp.*) are fourth in production in the EU but were second in 1996 in terms of area, excluding seeds for multiplication. They are largely a Spanish crop and Spain produces 98% of the EU production! The reason is that in the driest areas of Europe, vetches yield as much as the other grain legumes. They have the smallest seeds among pulses (40 to 80 g per1000 seeds, as compared to 200 to 300 g per 1000 seeds for pea and 500 to 600 g for faba bean). This enables them to have the smallest production costs. Two main species are grown for seed purposes, *Vicia sativa* and *Vicia ervilla*[4]. Vetches were traditionally grown in Spain for animal feed some decades ago. They experienced a prolonged decline until recently when they expanded rapidly after the setting up of a support scheme in 1989 (figure 2). However, they remain a traditional cheap grain fodder for dry lands, and little research and development work has been done on the crop in Europe.

Chickpea (*Cicer arietinum*) and **Lentil** (*Lens culinaris*) are crops with similar patterns of production and use. Both are used almost exclusively for food and are mainly imported. Their production has fallen for decades in the Mediterranean countries but has rallied recently, for the same reason as for vetches in Spain, whereas France and Italy, whose climate is suitable, remain far behind (Figure 2). The current varieties are spring types, although these are sometimes sown in the autumn in the south of Spain and Italy. The chickpeas are of the "kabuli type" (large white seeds) which are mostly grown and consumed in Europe. For lentil there is a preference for "green lentils". Recently, geographical "origin denominations", (i.e. specific varieties with a specific crop management in specific areas) have been developed for lentil and chickpea in France and Spain (De la Rosa, 1997). However, until now, research and development work, as well as markets, have been limited in Europe. Chickpea is better adapted to southern regions. Lentil can be grown everywhere in Europe but is limited by its low profitability as compared to other northern crops and by its susceptibility to *Botrytis* in wet climates.

---

[2] : For pea and faba bean, all white-flowered varieties have low-tannin seeds and all coloured flowered varieties have high tannin seeds.

[3] : In. Italy, Greece, Portugal, this traditional crop has declined for several decades (Monti *et al* .1991; Podimatas and Karamanos 1994; Dordio 1997).

[4] : Those two species and several other vetches (e.g.*Vicia villosa*) are also grown as green fodder or as plant cover on set-aside.

82

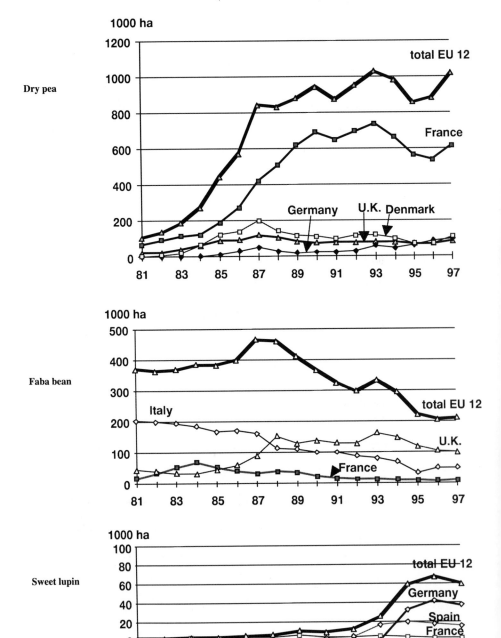

Figure 1 :
Pea, faba bean and sweet lupin
areas in the EU
(Source : EU Commission)

Figure 2 :
Chickpea, lentil and vetches areas in the EU
(excluding areas in seed multiplication)
(Source : EU Commission)

Chickpea

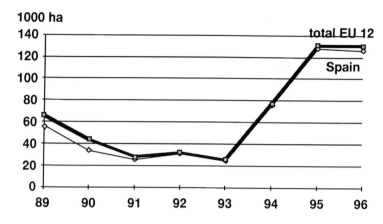

Lentil

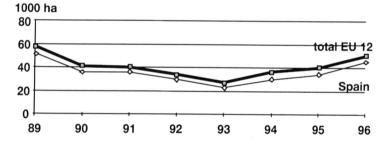

Vetches

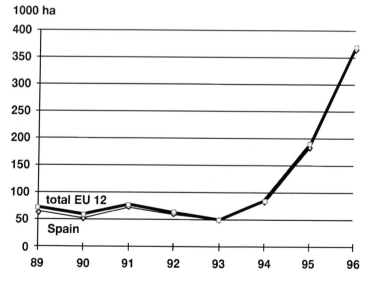

Dry beans form a complex group, dominated by common bean (*Phaseolus vulgaris*) but with other "American" or "Asian" beans (*Phaseolus lunatus, P. coccineus, Vigna unguiculata, V. radiata, V. mungo and V. adzuki*). Combined they represent the major group of pulses consumed by humans in the EU. Most are imported, mainly from America. They do not benefit from any support scheme in the EU, unlike the previous species which explains why no development has occurred (Gent, 1994). Moreover, accurate statistical data for them are difficult to obtain, because of the absence of any support scheme and because of the different uses of the crop (depending on the stage of harvest : green pods, immature seeds or dry seed, and on the colour and shape of the seeds, which results in various and confusing classifications). Hence it is difficult to describe the current situation of dry beans in the EU. It is nevertheless clear that production has steeply declined in southern countries for decades, and has been offset by increased imports.

Sweet Lupins form the most recent group of grain legumes cropped in the EU. Their expansion occurred following the development of non-bitter varieties and by the setting up of a support scheme in 1984. Three species are included in the scheme, *Lupinus albus, L. luteus* and *L. angustifolius*, but only the first two ones have increased and only to a moderate extent, in Germany, Spain and France. Three traits limit their cultivation :
- poor adaptation to calcareous soils ;
- excessive lateness of the current varieties of *Lupinus albus*, (the highest yielding type in Europe), which prevents it being grown in the northern countries ;
- difficulty of controlling the major disease *Colletotrichum gloeosporioïdes* which causes serious concerns in Germany and Austria.

Sweet lupin seeds are used almost entirely for animal feed and large quantities of *L. angustifolius* seeds are imported from Australia. However, lupins have benefited from active research work during recent years in Europe

The eight groups of grain legume are usually classified into different categories according to their regulation status, to their main uses, and to their origin (Table 3). None of these classifications groups the crops in the same way leading to frequent misunderstandings between people using different classifications.

Table 3. Classification of the different grain legume crops in the EU

	Dry pea	Faba bean	Sweet lupins	Vetches	Lentil	Chickpea	Dry beans	Soyabean
Origin	Europeo-Mediterranean						America	Asia
usual category	cool season food legumes						warm season food legumes	
	Pulses							oilseed
Regulatory support scheme	Protein crops			Grain legumes			No support	Oilseed crops
Main use	Animal feed (whole seed)				Human food			Feed (deoiled meal)
Main source of supply	European production		Both imports and European production				Imports	

It is important to note that, in spite of recent developments, grain legume are still a small part of arable crops in the EU (Table 4), as compared to countries such as Australia, USA, Canada or India. Only 2.5% of the "arable crops", benefiting from a common support scheme in the EU, are grain legume crops (2% for pea, faba bean and lupins and 0.5% for soyabean) whereas set-aside represented five times more area in 1995! There are significant differences between countries. In France, the UK and Denmark, pulses represent approximately 4% of arable crops compared to 1% in Germany and Italy. In comparison, soils and rotational constraints mean that grain legumes should be grown on up to 20% of arable lands.

Table 4. Share of Protein crops in the arable crops of EU 15 in 1995 (Source EU Commission)

	Arable crops area (including set aside) **(5)** 1000 ha	Protein crops(1) % of arable crops	Oilseed crops (2) (of which soyabean)	Cereals (3)	Set-aside (4)
France	13 550	4.2	14 (0.8)	69	11
United Kingdom	4 449	4.3	10 (0)	69	12
Germany	9 994	1.2	10 (0)	74	11
Spain	9 102	1.1	13 (0)	69	16
Denmark	2 017	3.9	8 (0)	76	11
Italy	5 060	0.8	10 (3.6)	76	13
Total EU 15	51 444	2.3	11 (0.6)	72	12

(1) :Dry Pea, Faba bean and Sweet Lupins.

(2) :Soyabean, Sunflower and Oilseed Rape (including areas on set aside for non-food uses).

(3) :Wheat, Maize, Barley, Oats, Triticale.

(4) :mainly compulsory set-aside devoted to the balance of cereal market.

(5) :Arable crops as defined here are the group of crops on the same support scheme, which comprises mainly the 4 former groups and a few other crops taken into account in the set-aside obligation.

Analysis of demands and quality requirements

Animal feed

In recent decades in Europe there has been an increase in the demand for Materials Rich in Protein (MRP) (Coléou, 1996). MRP includes all feed materials containing >15% crude protein. They are used to balance animal diets based on cereals, which contain insufficient protein (Carrouée and Coléou, 1996). In addition, for pigs and poultry, the amino acid profile of the MRP has to be rich in lysine, methionine plus cystine, tryptophan and threonine, otherwise it is necessary to add expensive industrial amino acids to the diet.

From 1973 the European consumption of MRP (when converting all MRP in tons of protein according to their protein content) has almost tripled (Figure 3). It now represents the equivalent of 59 kg of protein per inhabitant of the EU.

Figure 3 :
Trends in production and consumption
of Materials Rich in Proteins in the EU 12

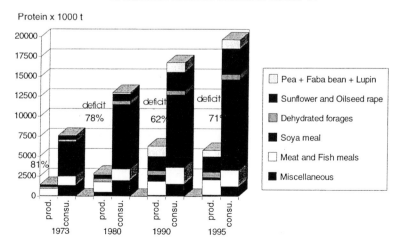

86

The change is not explained by an increase in the human population, which has been slow, but by significant changes in eating habits, notably:
- an increase in meat consumption, mainly pork, chicken and to a lesser extent beef (Coléou , 1995). Consumption of pork rose from 28 kg per capita per year in 1971 to 41 kg in 1995 and chicken consumption doubled (Table 5), but with significant differences between countries (Table 6)
a trend to produce diets more concentrated in protein in order to reduce the growth period of animals and their total feed intake (which is only 2 kg per kg of chicken and 3 kg per kg of pork) and to favour the accumulation of meat protein rather than fat.

Table 5. Meat consumption (kg/capita/year) in the EU 12 (Sources : OFIVAL - EUROSTAT)

kg per capita per year	1971	1985	1995
Pork	30	37	41
Chicken	10	16	20
Beef	24	23	20
Total :	62	76	81

Table 6. National averages of pork, chicken and beef consumption per capita in 1995 (Sources : OFIVAL - EUROSTAT)

kg/year	Pork	Chicken	Beef
Denmark	63	15	19
Germany	55	13	16
Spain	52	24	13
Netherlands	45	22	20
Belgium	49	22	21
Ireland	37	29	16
Portugal	37	24	18
France	36	23	28
Italy	34	19	25
Greece	23	19	21
United Kingdom	22	22	15
EU 12	40	20	20

In future, the consumption of MRP in the EU will probably stabilize or even ease slightly for two reasons
- meat consumption has probably reached a maximum and the European population is now more or less stable ;
- the necessity to reduce nitrogen losses, which leads to a need to formulate diets with the required minimum amount of essential digestible amino acids and with the minimum of non-essential or non-digestible amino acids.

The major question concerning future trends of MRP consumption in the EU is the potential for an increase in exports of pork and chicken onto the world market.

The EU is heavily dependent on imports for its MRP needs. Imported soya meal represented 71% of the requirements in 1995. This proportion was 81% in 1973 and only 62% in 1990 due to the increase in home production of oilseed and protein crops. However, imports have recently climbed because consumption has increased whereas EU production has tended to decline (Figure 3). In 1995, proteins from pea and faba bean represented only 5% of the MRP consumption as compared to 53% for the protein from soya meal. Protein crops represent only the third group of European resources in MRP, after rapeseed and sunflower, and meat and fishmeals (Table 7).

Table 7. Balance sheet of Materials Rich in Proteins in the EU 15 in 1995/1996 (Source : UNIP, 1997 provisional balance)

Raw material	PRODUCTION (1 000 t)	PRODUCTION % Crude protein	PRODUCTION protein (1 000 t)	Raw material		CONSUMPTION (1 000 t)	CONSUMPTION % Crude protein	CONSUMPTION protein (1 000 t)	Ratio Prod. Cons. (%)
Soyabean	907	38	345	Soyabean	seed	1 629	37	603	-
					meal	23 300	47	10 951	3
Sunflower	3 360	16	538	Sunflower	seed	434	16	69	-
					meal	4 440	30	1 332	38
Oilseed Rape	7 776	20	1 555	Oilseed Rape	seed	923	20	185	-
					meal	5 230	35	1 831	77
Pea + Faba bean + Lupins	3 650	22	803	Pea + Faba bean + Lupins		5 100	22	1 122	72
Meat meal + Fish meal	3 369	65 - 55	1 910	Meat meal + Fish meal		3 519	65 - 55	2 097	91
Dehydrated forages	4 065	16	650	Dehydrated forages		3 954	16	633	103
Miscellaneous*	2 471	-	562	Miscellaneous*		13 726	-	2 964	19
Total	25 598	-	6 363	Total		62 255	-	21 786	29

* Linseed, Groundnut, Coprah, Corn Gluten Feed

This limited contribution reflects the availability of pulses, both in Europe and worldwide. Three factors are important for the animal industry :

1. *A high nutritional value* determined by high digestible energy, high digestible amino acid content and the absence of antinutritional factors. Feed peas (ie peas with round seeds, free of tannin and with low antitrypsic activity) are valuable for that purpose (Bastianelli *et al.* 1995). Their main deficiency is a low concentration of tryptophan and of methionine plus cystine which may require the addition of expensive industrial amino acids when pea is included at a high level, for instance with maïze or tapioca (Table 8). Chickpea or tannin-free faba bean have a similar nutritional value to pea and could also be valuable for pigs and poultry with a very different composition. Lupins are more suitable for ruminants. Little is known about the nutritional values of lentil and vetches, and no zero-tannin varieties have been bred. This means they are less digestible than pea for pigs and poultry.

Table 8 Nutritional value of feed pea* compared to feed wheat and soyabean meal (Sources ITCF, UNIP, PGRO, 1996)

	Feed pea*	Wheat	Soya 48
Crude protein (g/kg as fed)	206	113	457
Lysine (g/kg as fed)	15.4	3.2	29.2
(% protein)	*7.5*	*2.8*	*6.4*
Methionine + cystine (g/kg)	5.2	4.6	13.7
(% protein)	*2.5*	*4*	*3*
Tryptophan (g/kg as fed)	1.7	1.3	5.9
(% protein)	*0.8*	*1.2*	*1.3*
Protein digestibility for poultry (pea pelleted)	85%	88%	92%
ENERGY (MJ/kg as fed)			
Net Energy for pig	9.8	10.6	7.9
Metabolisable Energy for poultry	11.1	12.4	9.7
Metabolisable Energy for ruminants	11.6	11.8	11.6

* feed pea : type of pea grown in the EU 12 for animal feed with round seed, tannin free and with a low antitrypsic activity.

2. *A competitive price.* For a given nutritional value, a raw material has to be cheaper than other materials to be included in a diet. A feature of animal feeds is that the materials may substitute for each other. Thus the price of feed pulses depends relatively little upon their supply but directly upon the market price of two major feedstuffs, cereals and soya meal (Huard and Lapierre., 1995).

3. *A guaranteed supply.* The animal feed industry requires a regular and abundant supply of raw materials. A factory typically requires 100 to 200 t per month for each of the major raw materials. A break in supply will favour substitute products. This explains why feed pulse imports in the EU have been strengthened by the increase in Europe-grown pulses. From 1982 to 1993, European production of "protein crops" increased by a factor of 5 and imports by 9 (Figure 4).

Figure 4:
Trends in dry pea, faba bean and sweet lupin imports into the EU 15
(Sources: E.U. and EUROSTAT)
Left ordinate and right ordinate of the graph for imports and production respectively.

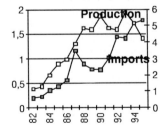

The demand from the feed industry would reach a maximum if supply were not a limiting factor. This maximum can be estimated from the situation in France, the only EU country where local production of dry pea is sufficient to meet the needs of the French animal feed industry along with exports. Since 1990, the average rates of inclusion have been 25% for pig compound feeds and 10% for poultry (Carrouée and Huard, 1995). Such rates, if applied to the European animal feed industry and farm-made pig feedstuffs would indicate a demand of 17 million t in 1994 (Table 9). However, there are different national reference standards for nutritional values (Jensen, 1996). At present, most non-French animal feed industries do not use such high inclusion rates, either because the price calculated from their own standards is not attractive, because they worry about risks of bad performances with high rates or because sufficient amounts of pea are not available in their region.

Table. 9 Potential utilization of feed pea and faba bean in pigs and poultry feedstuffs in the EU 12 in 1994 (based on inclusion rates of 25% for pigs and 10% for poultry) (Source : UNIP 1997)

	Feed Industry (1000 t)	Farm made feedstuffs (1000 t)
France	2 440	400
Netherlands	2 280	0
Germany	1 980	2 000
Spain	1 900	600
Denmark	920	400
United Kingdom	1 020	300
Total EU 12	13 400	4 200
Total :	17 6 Million t	

Human food

The characteristics of this outlet for pulses are very different from those for animal feed!

1 *Limited volume.* While consumption of meat per capita in EU has reached record levels, consumption of grain legumes has fallen to a low level. Consumption is estimated as 3.5 kg per capita per year in 1995, compared to 7 kg as the world average and 14 kg for India (Oram and Agcaoili, 1994). Consumption in Europe has declined from the beginning of the century and now seems stable. This decline has been more dramatic in the southern countries, for instance, in Italy where it was estimated at 14 kg per capita per year in 1901-1905 (mainly faba bean and common bean) but has declined to 4 kg per capita per year in 1976-1978 (Monti *et al.*, 1991).

In future, a limited recovery is likely due to health requirements for a diet low in fat and cholesterol and rich in fibre. This could promote a limited shift from meat to pulses. However pulses would not be the only candidates. There is a parallel tendency to consume new grains such as quinoa (*Chenopodium quinoa*) and to prefer foods ready-to-eat or ones that are quickly cooked.

2. *Segmented market,* with premium opportunities in niche markets. Contrary to the feed outlet, food pulses can hardly be substituted by others foods. Price and quality standards are not related to nutritional value but to appearance (colour, size, shape, absence of stains and of bruchid holes), response during processing (eg ability to soak and cook rapidly) and taste are important.

Prices of food varieties are generally higher than feed varieties of the same species, especially when they yield poorly or are difficult to grow. But they are much more susceptible to the level of supply in the marketplace, and thus are likely to change radically from one year to the next.

In the EU, niche markets are most organised in the United Kingdom (Gent, 1996) (Table 10). In comparison, France has only a few domestic outlets in the non-feed industry, requiring approximately 7,000 t for dehulled split green pea, which is the only traditional food use of pea in France. A few thousand tonnes of yellow pea are processed into micronised hulls for delicatessen or yoghurts and into extruded dehulled meal for fish feedstuff. 4,000 t of faba bean flour are incorporated in wheat flour, with the aim of whitening the crumb and 1-2,000 t of ultrafine dehulled sweet lupin flour are used to replace egg yolks in cakes (UNIP, 1994; Leterme and Fenart, 1997).

Table 10. U.K. pulse markets (Sources : British Edible Pulses Association)

Use	Special requirements	Varieties	Likely value (£/t)	Estimated requirement	U.K. production 1995
Animal feed compounding	Market for high tannin faba bean and the samples that fail to meet standards for premium markets	All	138	"unlimited"	230 000
Tannin-free animal feed for pig and poultry rations	White pea & tannin-free faba bean	All white pea plus Caspar & Vasco bean	140	"unlimited"	372 000
Animal feed micronising	Green-seeded varieties used almost exclusively. Secondary use for marrowfats and seed peas which do not meet quality standards	Solara, Hampton, Lantra, Elen	157	50 000	35 000
Export standard faba bean	Beans must be clean and sound and free from Bruchid damage. Pale skinned varieties used.	Alfred, Victor, Scirocco	-	35 000	10 000
Canning marrowfat pea	Samples must be clean, sound and with only low levels of moth damage.	Maro, Guido, Bunting, Progreta	162	24 000	24 000
Packet marrowfat pea (including fish shop trade & frozen mushy peas)	As above, but must also retain green colour.	As above, plus Princess	162	15 000	15 000
Export quality marrowfat pea	Premium samples with excellent colour, large size and freedom from serious blemishes.	Maro, Progreta, Princess, Guido	172	8 000	4 000
Canning small blues	Samples must be clean, sound and pass cooking tests.	Orb, Froidure, Arena	155	6 000	4 000
Pigeon Trade	Small size brown, white and green-seeded peas and beans.	Maple pea, Orb, Maris Bead	165 155 N/A	7 000 3 000 4 000	3 000 3 000 2 000

3. *Opportunities for export*. Until 1993 it was difficult to export European pulses due to the support schemes. Now there are no obstacles, and traders provide food pulses to Mediterranean countries where they are a major component of the human diet.

c. Non-food and non-feed uses

Examples of current non-food and non-feed uses of grain legumes in the EU include:

protein concentrates of faba bean to make coatings for paper in Italy ;

starch extracts of rich amylose wrinkled pea to produce biodegradable plastics in a pilot plant in Germany;

alkaloid extracts of bitter lupin to make crop protection products in a pilot plant in Germany.

Theoretically, there is a wide range of uses in industrial outlets. However, as for the animal feed industry, the non-food sector requires cheap and abundant raw materials, and in addition, precise quality standards.

Factors leading to changes in production

The Common Agricultural Policy (CAP)

If demand is determining the current and potential use of grain legumes, the EU production is tightly determined by CAP regulations. Regulations for the different crops are complex (Table 11), but their impact on production is simple. Since some crops benefit from a support, either through an area compensatory payment or through a custom protection, other crops, which do not have that support, are mainly imported (eg common bean). Consequently the development of each crop is connected with the setting up of a specific support scheme, with a few years of delay (compare Table 11 with Figures 1 and 2).

Table 11. A summary of support schemes of the main crops in the EU

Crops		First support scheme		Present support scheme
Cereals (Soft and Durum Wheat, Barley, Maïze, ...)	1958	Intervention price (guaranted price approximately twice that of feed cereal prices in the world market) High customs taxes to protect European production	1993	Same basic scheme but : Lower intervention prices (approximately 1,6 higher than feed cereals in the world market) Compensatory payments per hectare to farmers Compulsory set-aside dependent on the EU cereals stocks
Oilseed crops a. Oilseed Rape, sunflower	1966	High minimum prices guaranted to farmers ; subsidies paid to the users for European oilseeds no customs barrier to imports	1992	No more guaranted prices -> directly linked to the world market Compensatory payments paid to farmers
b. Soyabean	1974	High minimum prices guaranted to farmers ; subsidies paid to the users for European oilseeds no customs barrier to imports		Maximum guaranted ceiling of about 4,9 Mha after the "Blair House agreement" with the USA, plus a part of oilseed meal produced on non food set aside Set-aside obligation linked to cereals stocks
Protein crops a. Dry pea, faba bean b. Sweet lupin	1978 1984	idem oilseeds idem	1993	idem oilseeds except the absence of Maximum Guaranteed Ceiling
Other grain legumes Chickpea, lentil, vetches	1989	Area aid per hectare paid to farmers, at the same level throughout EU	1995	idem 1989 but : a higher area aid Maximum Guaranteed Ceiling at 400 000 ha no set-aside obligation

For example, pea production soared from 1981 to 1987, following the setting up of the first protein crop scheme in 1978. In 1988, a "Maximum Guaranteed Quantity" (MGQ) was fixed at a low level for protein crops (3.5 million t as compared to 160 million t for cereals) and provoked a sudden change. There were also poor yields of pea crops in northern Europe, in 1987, particularly Denmark, due to rainy weather and a difficult harvest. The MGQ prevented the extension of protein crops up to 1992. Then a new regulation scheme was set up in 1993. But from 1993 to 1995, the compensatory payments for cereals increased, and remained fixed for protein crops, leading to a downturn in the pea crop in many countries.

Three main aspects need to be emphasised:

cereals benefit from a special status, with strong price protection, involving high taxes to prevent imports and variable "restitution subsidies" to enable exports, whereas oilseed and protein crops are directly in competition with the world market (Carrouée and Coléou, 1996). This explains why the latter need a higher compensatory payment to maintain their competitiveness vis-à-vis cereals.

compensatory payments vary according to the regional average yields in cereals, oilseed and protein crops but not for the so-called "Other grain legumes". Consequently pea, faba bean and lupin tend to be grown in high yielding regions, whereas chickpea, lentil and vetches which have the same subsidy in all regions, are grown only in the low yielding areas of southern Europe.

protein crops are the only ones without any kind of ceiling: their cultivation could be expanded without entailing either set-aside increases (unlike cereals) or a decrease in the area payment (as occurs for oilseed crops and "other grain legumes"). Thus the possibility for increasing protein crops is only limited by the economic competitiveness and other agronomic constraints. However, it is likely there will be a reform of the CAP in the near future which will modify again the relative standing of all arable crops in the EU.

Others factors

European regulations have a strong impact on the production of grain legumes but are not the only factor:
The production cost per tonne, has an obvious impact. It is mainly the yield potential which determines pulse production costs per tonne. Differences concerning variable costs are limited. All legumes have relatively low costs for fertilisers, thanks to their ability to fix nitrogen, but most have large seeds which implies high seeding rates and high seed multiplication costs. The exceptions are lentil and vetches, with their sm ll seeds. It was the high yield potential of pea which enabled its production to increase quickly and to meet the demand for cheap MRP for animal feed.

Agronomic constraints are other obvious factors. The requirements of soyabean for warm temperatures and ample rain or irrigation and the susceptibility of lupins to calcareous soils (which are widespread in Europe) have already been mentioned. None of the grain legumes is well adapted to all soils and climates in Europe, compared with the broader adaptation of wheat or barley for instance.

Concerning the pea crop, with a growth habit which is adapted to almost all climates, particular limiting edaphic factors exist with the current varieties. They are:
- a susceptibility to lodging which makes harvesting difficult on stony or clay soils;
- its susceptibility to some soil diseases, particularly *Aphanomyces euteiches* which can prevent pea cultivation in some fields or require farmers to wait several years between crops.

The **farming systems** and the skill of the farmer play a significant role. For instance, the small farms with a dominant animal husbandry, numerous in southern Germany are not well adapted to produce grain legumes, in contrast to the large farms of eastern Germany (Pahl, 1997). In addition, under present regulations, the specific compensatory payment to oilseed and protein crops is paid only to farmers who are subject to set-aside. Small farms, with less than about 20 ha of arable crops, can choose not to have set-aside, which leads them to grow only cereals and forage crops.

Another farming system unfavourable to grain legume cropping is found in regions with large areas of sugar beet, potatoes or vegetable crops, such as in Belgium. There is no need for break crops such as the protein crops.

Finally, the impact of national research and development organizations must be mentioned. In the EU, each country has its own professional structures for agriculture. It is clear that the presence of specialised organizations in some countries explains the development of legume crops. The UNIP in France and the PGRO in the United Kingdom, which are private organizations, gather levies from farmers, merchants or processors to support R & D work. They also provide market information and organize agreements between producers and users regarding quality standards (Delplancke, 1995). In contrast, there is no equivalent organization in the other European countries, nor at the EU level, except in the last few years when a small structure was set up for research co-ordination on grain legumes - the European Association for Grain Legume Research (A.E.P) based in Paris.

The role of public research institutes is also significant. In France the National Institute of Agronomic Research (INRA) has carried out research on protein crops for animal feed from as early as 1974, four years before the protein crops scheme was set up.

RETROSPECT AND PROSPECTS

From the above it is evident
1. the potential use of pulses in animal feed is more than three times greater than current supply
2. although smaller than for feed pulses, the European demand for pulses as food, mainly common bean, lentil and chickpea, is much greater than present production
3. the compulsory set-aside, set up in the EU to maintain high prices for cereals, has accounted for the equivalent of two to six times the area of grain legumes for the last four years ;
4. soil and rotational constraints are not a limiting factor in most EU regions. Thus it is clear that both demand and land availability could enable the further development of grain, mainly the protein crops of which extension has limited since 1988 by a low "Maximum Guaranteed Quantity" and since 1993 by high set-aside obligations.

In the coming years, the key point will be the evolution of the Common Agricultural Policy. The reform of 1993 was intended to be a transition. Debates have already begun. There are two main options, whose consequences will be very different for grain legumes :

to maintain high protected prices for cereals. This would involve an obligation to maintain a high level of compulsory set-aside, since the amount of subsidized exports is now limited by international agreements with the EU. That option would also limit or push back all other arable crops unless they still benefit from higher compensatory payments than the cereals

or to use the whole potential of crop production of the EU in ending the set-aside obligation. That would require a lowering of the cereal price protection to enable exports without subsidies. In that case, all crops would receive the same support; area payments would be intended to support agricultural incomes and employment, and not a given type of production. Only market demands and production costs would dictate the balance between different crops.

Several European governments and the European Commission are backing the second option. The difficulty is that it would imply either income losses for the farmers or a budget increase to offset price cuts.

This possible uncoupling of production and agricultural support would certainly benefit those crops which are presently under-supported. It could perhaps enable common bean development or an extension of lentil or chickpea, elsewhere than in Spain, at the expense of imported food pulses.

But major competition would occur between those crops in the current "arable crops" scheme, cereals, oil crops and protein crops. A low intervention price for cereals along with a free entrance of feed cereals from the world market into the E.U. would certainly lead to a drop in the prices of feed cereals and near-cereals-raw-materials (of which feed pulses). On the contrary, milling wheat, malting barley and other specific cereals should withstand less supported prices much better, since their world price is higher than the price of feed cereals such as maïze (cf Table 12). Feed pulses such as feed pea would probably lose part of their competitiveness vis-à-vis milling wheat for instance, but not vis-à-vis feed cereals such as maïze or feed barley.

Table 12. Relative prices of maïze, wheat and soya bean meal in the USA and in the EU

US $/t		94/95	95/96	96/97	Average
USA	Soyabean meal	183	243	274	233
(Chicago)	Maïze	92	144	117	117
	Ratio S.meal/maïze	**1.99**	**1.69**	**2.34**	**2.00**
	Wheat	137	183	149	156
	Ratio wheat / maïze	**1.49**	**1.27**	**1.27**	**1.34**
E.U.	Soyabean meal	214	264	306	261
(FOB)*	Maïze	194	209	169	191
	Ratio S.meal/maïze	**1.1**	**1.26**	**1.81**	**1.37**
	Wheat	181	190	168	180
	Ratio wheat / maïze	**0.93**	**0.91**	**0.99**	**0.94**

* prices quoted in European ports (Rouen for wheat, Bordeaux for maïze, Lorient for soya bean meal)

94

The further development of feed pulses in the EU and the possibility of meeting the potential demand will depend mainly on two factors:
1. changes in relative prices of feed cereals and soyabean meal. The absolute values are extremely variable and difficult to forecast (Lacampagne, 1996). But it is likely that the world price of soya meal will remain about twice as expensive as the world price of maïze (Table 12), or even become relatively more expensive. Indeed the demand for MRP in developing countries is increasing quickly due to the rapid rise in meat consumption (Figure 5), mainly poultry which requires high concentrations of protein in the diet;
2. changes in the production costs of the main feed materials. It is worth examining their potential future trends :
   - Nutritional output per hectare of crops in the most fertile regions of Europe is similar to those of crops in the most fertile regions of American. A pea/wheat/oilseed rape/wheat rotation in Picardie (a region of the Parisian Bassin) produces as much as a soyabean/maïze/soyabean/maïze rotation in Iowa (a region of the "Corn Belt" in the USA) in terms of energy, protein and essential amino-acids for pigs and poultry feed, and significantly more of edible oil (Table 13). This combination of raw materials (wheat, pea and oilseed rape meal) has certainly the best ability among European crops to withstand increased competition from imported products in the future.

Figure 5 :
Global development of meat consumption
In 1989, per capita meat availibility was 13 g protein/day in China
as compared to 60 g protein/day in developed countries
(Source : J. Coléou, 1996)

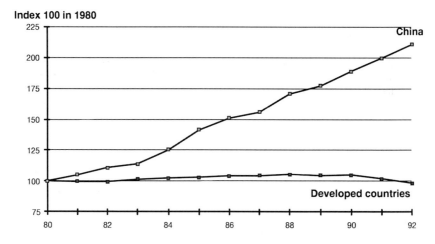

Table 13. Total production per hectare of feedstuff and of oil of a four year rotation in the yighest yielding regions of the EU and of the USA

Region			Picardie (UE)		Iowa (USA)	
Typical four year rotation and yields (t/ha)			Spring pea	5.5	Soya bean	2.9
			Winter wheat	8.5	Maïze	8.6
			Winter O.S. Rape	3.8	Soya bean	2.9
			Winter wheat	5.5	Maïze	8.6
Nutritional output	- Net Energy (pigs)	$(10^6$ J/ha)	248		220	
for pigs	- ME (cockerels)	$(10^6$ J/ha)	285		278	
and poultry	- Crude protein	(kg/ha)	3 800		3 750	
(for four years)	- Lysine	(kg/ha)	178		182	
	- Methionine + Cystine	(kg/ha)	138		131	
	- Tryptophane	(kg/ha)	40		41	
Edible Oil output (for four years) (kg/ha)			1 600		960	

- Variable costs (seeds, fertilizers, and chemicals) in the EU have been dramatically reduced during the last decade. Figure 6 gives the example of pea and wheat in France, where the reduction has been particularly rapid for pea. A further but slower reduction could be reached with more appropriate choices in plant densities, weeding and disease control strategies (Leveau and Carrouée, 1997). Concerning mechanization costs, significant progress is being made through improvements in resistance to lodging which reduces repair costs of harvesters.
- **Yields.** From 1981 to 1990, pea yields improved significantly but since then progress has been slow (Figure 7). For a sixteen year period (in the case of France), the increase was about 60 kg/ha/year for pea as compared to 120 kg/ha/year for wheat. Therefore the gap between the crops has increased.

The relative yield of wheat after wheat is significantly less than the yield of wheat after a break-crop, by approximately 10% on average. But this figure could change. New seed dressings are being tested which reduce eyespot (a cereal root disease due to *Gaeumanomyces graminis*). If effective they could make continuous wheat more profitable and perhaps more competitive than breakcrops such as pea or rapeseed.

In contrast, the soil born disease of peas *Aphanomyces euteiches* is increasing and could jeopardize crops in some EU regions (Carrouée et al., 1995).

In conclusion, whereas food pulses could benefit from a reform of the CAP which would give the same support to all crops, the case of protein crops is more complex. Dry pea, faba bean and sweet lupin have a strong potential for further expansion subject to one major condition : yield improvement!. A breakthrough is required for faba bean and lupin, and to a lesser extent pea which already has been steadily improved (Knott , 1996). Recently, for these species, research efforts are targeting reliable winter varieties, since winter sowings may enable higher and more stable yields than spring sowings (Ney and Duc, 1997 ; Milford and Huyghe, 1997 ; Erskine and Malhotra, 1997).

Further improvement of the resistance to lodging in pea is still necessary, to make this crop possible in any stony or clay soil and to reduce mechanization costs. The new concern due to soil born diseases requires specific attention and research to find either genetic or chemical solutions.

These challenges are difficult. But these crops are still relatively recent, and wide genetic resources can be exploited further. The increasing research and development work, which is currently being undertaken, justifies a realistic optimism.

Figure 6.
Changes in variable costs for wheat and pea crops in France
(seeds + fertilizers + chemicals)
(Sources : ITCF-UNIP)

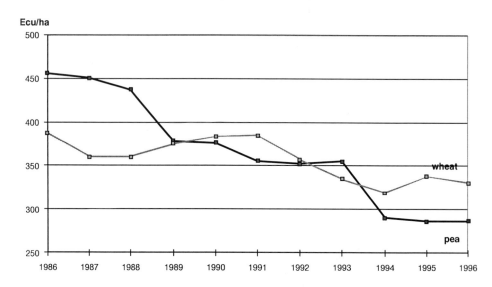

Figure 7.
Yield progress of pea and wheat in France
compared to maïze and soyabean in the USA

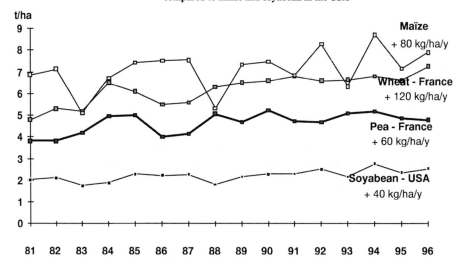

Acknowledgement.
We thank J.P. Lacampagne (UNIP) for his help with data computation.

References

Abbate V., Cavallaro V., Crino P. and Ranalli P. 1997. In *Problems and Prospects for Winter Sowing of Grain Legumes in Europe*, 3-4 December 1996, Dijon (FR), pp19-26. Paris:AEP.

Bastianelli D., Carrouée B., Grosjean F., Peyronnet C., Revol N. and Weiss Ph. 1995 . *Peas, utilisation in animal feeding* (Eds B.Carrouée and F.Gatel). Paris : UNIP-ITCF, 99 pp.

Carrouée B. and Huard M. 1995 . In *2nd European Conference on Grain Legumes*, 9-13 July, Copenhagen, pp 8-9. Paris:AEP.

Carrouée B., Verger S., Wicker E., Seguin B. 1995. *Perspectives Agricoles* 207: 55-59

Carrouée B. and Coléou J. 1996 .. *Perspectives Agricoles* 219:16-22

Coléou J. 1995. *Grain Legumes* 9: 22

Coléou J. 1996. *Perspectives Agricoles* 209: 8-15

De la Rosa 1997. *Grain Legumes* 16: 24-25

Delplancke D. 1995. *Compte-rendu de l'académie d'agriculture* 5: 35-48

Dordio A.M. 1997. In *Problems and Prospects for Winter Sowing of Grain Legumes in Europe*, 3-4 December 1996, Dijon (FR), pp15-18. Paris:AEP.

Erskine W. and Malhotra R.S. 1997. In *Problems and Prospects for Winter Sowing of Grain Legumes in Europe*, 3-4 December 1996, Dijon (FR), pp 43-50. Paris:AEP.

Gent G. 1994. *Grain Legumes* 5:14

Gent G. 1996. Pea usage in UK. *Grain Legumes* 14: 23-24

Gent G. 1997. In *Problems and Prospects for Winter Sowing of Grain Legumes in Europe*, 3-4 December 1996, Dijon (FR), pp 7-9. Paris:AEP.

Huard M. and Lapierre O. 1995. In *2nd European Conference on Grain Legumes*, 9-13 July, Copenhagen, pp 10-11. Paris:AEP.

Jensen H.A. 1996. *Grain Legumes* 12: 24

Knott C.M. 1996. *Plant varieties and seeds* 9: 167-180

Lacampagne J.P. 1996. *Grain Legumes* 13: 22-23

Leterme P. and Fenart F. 1997. New oulets in the food industry. *Grain Legumes* 15: 24

Leveau C. and Carrouée B. 1997. *Perspectives Agricoles* 220: 41-43

Milford G.F. and Huyghe C. 1997. In *Problems and Prospects for Winter Sowing of Grain Legumes in Europe*, 3-4 December 1996, Dijon (FR), pp 59-64. Paris:AEP.

Monti L.M., de Pace C., Scarascia Mugnozza G.T. 1991. In *Options Méditerranéennes*, série A (Eds J. I. Cubero and M. C. Saxena), 10:143-151. Paris:CIHEAM.

Ney B. and Duc G. 1997. In *Problems and Prospects for Winter Sowing of Grain Legumes in Europe*, 3-4 December 1996, Dijon (FR), pp 35-42. Paris:AEP.

Oram and Agcaoili M. 1994. In *Expanding the production and use of cool season food legumes* pp 3-49 (ed. F.J. Muehlbauer and W.J. Kaiser), Kluwer Academic Publishers,

Pahl H. 1997. *Grain Legumes* 16: 22-23

Podimatas C. and Karamanos A.J. 1991. In *Options Méditerranéennes*, série A (Eds J. I. Cubero and M. C. Saxena), 10:139-142. Paris:CIHEAM.

Smartt J. 1990. *Grain Legumes , evolution and genetic resources*. University Press, Cambridge, UK. 379 pp.

I.T.C.F., U.N.I.P., P.G.R.O. 1996. *Feed peas, a quality raw material for animal feeding* (ed. UNIP, Paris)

U.N.I.P. 1994 *Statistiques Plantes Riches en Proteines* (Ed. UNIP, Paris)

U.N.I.P. 1997 *Statistiques Plantes Riches en Proteines* (Ed. UNIP, Paris)

Region 4: Sub-Sahara Africa

G. Bejiga and Y. Degago

Debre Zeit Agricultural Research Center, P. O. Box 32, Debre Zeit , Ethiopia

Abstract

Food legumes are the major source of protein and an important component of farming systems in sub-Sahara Africa. The area and production of these legumes has increased by 4.82% and 9.97% respectively during the period 1993 to 1995. The major cool-season legumes are faba bean, chickpea, fieldpea and lentil. Lathyrus is also important in Ethiopia. These crops occupy 13.36% of the total food legume area and contribute about 18.56% of the total production of pulses in Africa. More than 65% of the area and 88% of the production of faba bean is from the Nile Valley countries of Egypt, Eritrea, Ethiopia and Sudan. For chickpea, more than 70% of the area and about 47% of the production in sub-Sahara comes from Ethiopia. Ethiopia also contributes more than 66% of the area and about 60% of the production of lentil.

Efforts have been made to improve these crops in many African countries, but more progress has been made in the Nile valley countries. Egypt, Ethiopia and Sudan have released 22 faba bean, 11 chickpea, 12 lentil and 6 field pea varieties. Kenya has released 1 chickpea variety. Agronomic packages, which advise on sowing dates, seeding rates, time and rates of application of fertilizers, fungicides, herbicides, insecticides and frequency of irrigation have been developed. Wilt, root rot, viruses, powdery mildew, rust, chocolate spot and aphids were identified as the major constraints of faba bean, chickpea, field pea and lentil production in sub-Sahara Africa. The Nile Valley and Red Sea Regional Networks on Wilt/root rots and Aphid and viruses coordinated by the International Center for Agricultural Research in the Dry Areas (ICARDA) have enhanced the identification of major diseases and sources of resistance.

INTRODUCTION

Consumption

Cool-season food legumes are important food crops of sub-Sahara Africa particularly in the Nile Valley countries: Egypt, Eriteria, Ethiopia and Sudan. Faba bean, chickpea, lentil and field pea provide a considerable portion of the diet of the people. They are major sources of low cost protein for the population of the Nile Valley who can not afford animal products. Consumption has also been rising due to urbanization, population growth, changing consumption habits and the rising prices of meat (Bejiga and Anbessa, 1994a, 1994b; Ahmed 1996). Increasing the production of these legumes would alleviate problems of malnutrition in these countries (Solh, 1995).

The crops are consumed in different forms such as soup, snacks and sauces. In Ethiopia, their consumption is extremely high during the two fasting months when orthodox Christians do not take meat, eggs or other animal products. An infant and child food " faffa" which consists of 57% wheat, 10% chickpea, 18% de-fatted soy, 5% skim milk powder, 9% sugar, iodized salt, and 1% vitamins and minerals (Backlander, 1987) is manufactured in Ethiopia. A malted weaning food consisting of 60% germinated barley, 30% germinated chickpea, 5% skim milk powder and 5% sugar was found to provide a good balance of nutrients (Yetneberk and Wondimu, 1994). After threshing and separating the seeds, the straw is used as cattle feed. Pulse haulms are considered to be of high quality in Ethiopia and are believed to have 5 to 8 percent protein whereas tef straw is equivalent to medium quality hay. The residues of other cereal crops are only of poor to fair quality (Mengistu, 1991).

R. Knight (ed.), Linking Research and Marketing Opportunities for Pulses in the 21st Century, 99–105.
© *2000 Kluwer Academic Publishers. Printed in the Netherlands.*

CURRENT AREA AND PRODUCTION

Ccording to FAO (1995) cool season food legumes occupy 13.36% of the total food legume area and account for about 18.56% of the total pulse production in sub Sahara Africa

The total area and production of food legumes have increased considerably in Ethiopia, Eritrea, Malawi and Sudan while only small increases have been observed in Kenya, Tanzania and Zaire (Republic of Congo). For instance between 1993 and 1995 the total area of pulses in Ethiopia increased from 979,00 ha to 1,242,00 ha and production from 872,00 to 1,108,000. However both area and production of faba bean have remained constant in all the major producing countries of Sub-Sahara Africa except Sudan (FAO, 1995). Ethiopia and Egypt are the major faba bean producers in the region and thus constitute together 60-70% of the area and 74-82% of the production of faba bean in Africa. Large increases in the area and production of chickpea and field peas have only been observed in Ethiopia. Between 1993 and 1995 the area of chickpea increased from 150,000 ha to 180,000 ha and production from 119,000 to 126,000 MT. Similarly, over the same period the area of field peas increased from 126,000 ha to 170,000 and production from 109,000 MT to 150,000 MT. Similar trends have been observed in the area and production of lentils. The major factors responsible for these increases in Ethiopia were changes in Government policies on land holdings (from communal farms under a socialist structure to private holdings) and the introduction of free markets (removal of fixed prices for crops). Currently, cool season food legumes particularly lentil, chickpea and field pea fetch very high prices; higher than most cereals. Moreover the pulses have become an important component of the farming systems due to increased prices of fertilisers. These crops will continue to be important as human food and feed for animals in Sub-Sahara Africa.

Chickpea is important in Tanzania and Uganda. Chickpea does not compete for land with most other crops in Tanzania, as these crops are cultivated during the rainy season (Nov/Dec-Feb/Mar) whereas chickpea is grown in March/April-July/August after the harvest of maize. The remaining area is left fallow (Pundir et al., 1996). Hence, there is a great potential to increase chickpea cultivation in Tanzania by replacing the fallow.

The production of field pea is mainly concentrated in Ethiopia, the Republic of Congo (Zaire), Tanzania and Burundi (FAO, 1995). Both its area and production are increasing in Ethiopia and Tanzania.

COOL-SEASON FOOD LEGUMES IN FARMING SYSTEMS

These crops have been used for centuries in rotations in most of sub-Sahara Africa and have helped to sustain the soil by nitrogen fixation where commercial fertilizers are not largely used. The crops have played a key role in sustaining soil fertility, productivity and environmental health mainly in the Nile Valley countries where they occupy large areas. With the present soaring prices of commercial fertilizers and concerns for environmental health, cool season food legumes will continue to play a major role. A review by Doran et al. (1996) on soil health and sustainability, showed the adverse effect of agricultural inputs on animal and human health due to contamination of soils, ground water, streams, and other surface water supplies which caused the development of cancer and other debilitating diseases. Reports of Clancy (1986) and Culliney et al (1992) cited by Doran et al (1996) showed that nitrate in drinking water can cause the potentially fatal methemoglobinemia or blue baby syndrome. It can also have more insidious carcinogenic effects if transformed in the body to nitrosamines a chemical which can be formed from the reaction of nitrate with atrazine herbicides. Such excess accumulation of nitrates as the result of inefficient nitrogen management (Doran et al., 1996) is well recognized. Biological nitrogen fixation could be exploited to replace inorganic N-fertilizers (Beyene, 1988). Several reports, reviewed by Doran et al (1996) showed legume-based systems improve total C, N, and P, available N, P and K, CEC, microbial biomass and respiration, organic matter content, greater water stable aggregates and reduce nitrate leaching compared to conventional systems, while yields were almost equal. Presently, the massive application of chemical fertilizers is drawing the ire of environmentalists and concerned consumers. Hence, the demand for organic farm products is increasing. These facts show that food legumes will be valuable as healthy foods. The crops are environmentally friendly and they sustain the productivity of soils.

PRODUCTION CONSTRAINTS

Surveys have been made in the Nile Valley region: Egypt, Ethiopia and Sudan to determine the biotic and abiotic factors that limit the production of these crops. Diseases and some insects are the most limiting factors and have received attention recently. The importance of some diseases has resulted in the formation of Regional Networks on (1) Wilt/root rots of food legumes (2) Aphids and Viruses and (3) Botrytis in which scientists from Ethiopia, Egypt, Sudan, Eritrea, Yemen, ICARDA and ICRISAT are involved. These networks are financed by the Netherlands government through ICARDA and coordinated by the ICARDA-Cairo office. Tunisia collaborates with the team working on wilt/root rots of chickpea.

These joint efforts have helped regional scientists exchange information, identify diseases and prioritize resistance breeding based on their economic importance. Sources of resistance have been identified for major diseases such as *Fusarium* wilt and root rots of chickpea, lentil and faba bean. The major diseases that are recorded on each crop are discussed below.

Biotic Stresses

Disease

Faba bean: Surveys made in these countries show that Chocolate spot (*Botrytis fabae* Sard), Fusarium wilt, root rots (foot rot, black root rot, collar rot, dry root rot etc.) rust and different viruses are important. Gorfu and Beshir (1994a) listed all the pathogens known to cause diseases of faba bean in Ethiopia. The important foliar diseases were, chocolate spot (*Botrytis fabae* Sard.), rust (*Uromyces viciae-fabae* (Pres.) Schr) Ascochyta blight (*Ascochyta fabae* Speg,) and powdery mildew. Of the soil-borne diseases, black root rot *Fusarium solani* (Mart.) Appel and Sacc is economically important in Ethiopia. A review by Ali (1996) indicated *Fusarium solani f. sp. fabae, Fusarium oxysporum, Fusarium moniliforme var. intermedium, F.acuminatum, Pythium sp., Rhizoctonia solani* and *Macrophomina phaseolina* were associated with wilt and root rot diseases of faba bean in Sudan. In Egypt, the survey results of 1993/94 showed that *R. solani, F. solani* (two isolates), *F. semitecum, Fusarium sp., M. phaseolina, Verticillum sp., Cephalosphorium sp,* and *F. oxysporum* were associated with faba bean (Doreiah et al., 1994). Abu Zeid (1996) reported that about 42-60%, 12-31% and 6-30% of the isolates were *Fusarium* spp., *Rhizoctonia solani* and *Macrophomina phaseolina* in the Assiut governorate of Egypt. These results indicate that *Fusarium oxysporum, Fusarium solani, Rhizoctonia solani, Macrophomina phaseolina* and *Sclerotium rolfsii* are the major soil-borne pathogens on faba bean in the region. Chocolate spot (*Botrytis fabae* Sard.) is the most important foliar disease.

Chickpea: Several diseases were recorded on chickpea in Ethiopia (Hulluka and Taddese, 1994. However, the most economically important ones are Fusarium wilt (*Fusarium oxysporum*), dry root rot (*Rhizoctonia bataticola*), collar rot (*Sclerotium rolfsii*) and wet root rot (*Rhizoctonia solani*) (Beniwal et al., 1992; Bejiga and Eshete,1996). Ascochyta blight (*Ascochyta rabiei*) is often observed on early sown crops at the Research Stations. Stunt virus is an important disease, while rust is often found on chickpea planted in the off-season. The review made by Faki et al (1996) also revealed that wilt (*Fusarium oxysporum*) and root rots (*Rhizoctonia bataticola* and *Rhizoctonia solani*) are the most important soil borne-diseases in Sudan. Powdery mildew (*Oidiopsis taurica)*was also recorded. The survey reported by Ali (1995) also showed that *Fusarium oxysporum* and *Rhizoctonia bataticola* are important on chickpea in Sudan. He indicated the variability among the populations of *F. oxysporum f.sp.ciceri* and the effect of irrigation interval on disease incidence. Chickpeas sown in early October and irrigated at 3-weeks intervals were more highly affected than plants sown late (Dec.) and irrigated at 1-week intervals. A survey of newly reclaimed areas of Nubaria in Egypt showed that Wilt/root rot and Sclerotinia stem rot were important in chickpea. Similar work has been undertaken in Assiut governorate and six pathogens were isolated from collected samples and three were still unidentified. The major isolates were *Sclerotinia sclerotium, Macrophomina phaseolina, Fusarium spp, Alternaria spp., Rhizoctonia solani, Cephalosporium* spp. Among these isolates *Sclerotinia sclerotium, Fusarium spp.* and *Rhizoctonia solani* were economically important in Egypt. Khattab and El-Sherbeeny (1996) also indicated that root rots, wilt, stem rot and Ascochyta blight cause considerable yield loss of which wet root rot (*Rhizoctonia solani*), dry root rot (*Rhizoctonia bataticola*), *Fusarium* wilt and verticillium wilt (*Verticillium* spp) are the most important in upper Egypt.

Lentil :Rust (*Uromyces fabae*) is the major disease in Ethiopia followed by collar rot and wilt caused by *Sclerotium rolfsii* and *Fusarium oxysporum* and other *Fusarium* spp. were also important in Sudan, whereas *Rhizoctonia solani,Fusarium oxysporum* and other *Fusarium* spp. are important in Egypt.

Field pea: Although it is difficult to get information from all countries in the region, about 20 diseases were recorded in Ethiopia, among which the following are important powdery mildew *(Erysiphe polygoni* DC), leaf, stem and pod spots/blight *(Ascochyta pisi)*, blotch *(Septoria pisi)*, stem lesions *(Phoma medicagini)* and downy mildew (*Peronospora pisi)* (Gorfu and Tesfaye 1994) .

Insects

The major insects affecting cool-season food legumes are African bollworm (*Helicoverpa armigera),* bruchid (*Callosbruchus chinensis)* and aphids (*Aphis craccivora, Aphis fabae)* (Ali and Tibebu,1994; Mohamed and Bushara, 1996) . These insects are often found on these crops, but aphids (*Aphis craccivora*) threaten lentil and field pea in Ethiopia and are an important pest on faba beans in Egypt and Sudan. Efforts are being made, through the Nile Valley Regional Network on Aphids and Viruses, to identify sources of resistance or tolerance. Aphids transmit different viruses and the Network intends to develop an integrated management of both aphid (vector) and virus.

Abiotic stresses

All cool-season food legumes are cultivated as rainfed crops in the region, except in Egypt and Sudan where they are grown under irrigation. Hence, these crops are the integral part of dryland farming and are always affected by drought. This is particularly true for chickpea and lentil which are planted on vertisols at the end of the rainy season (Bejiga and Anbessa, 1994; Pundir et al, 1996) . Since they are grown on residual soil moisture, they often suffer from terminal drought. These crops are also affected by frost in the Ethiopian highlands (Bejiga and Anbessa, 1994a, 1994b). Waterlogging is a major constraint both in irrigated and rainfed agriculture. Screening of some lentil accessions in Ethiopia showed that lines such as ACC-36115, ACC-215711, ACC-36140, ACC-215348, NEL-944, Gudo and FLIP-89-63L had relatively good tolerance to waterlogging. Similarly, some chickpea accessions such as ICC-4918, ICC-12795, ICC-7539, ICCL-890025/85-DZ/5-1, ICC-12855, ICC-13896, ICCL-84202, ICC-14131, ICC-13829, ICCL-89225, ICC-14158 and ICC-12635 showed better tolerance in pot screening and they are being further evaluated. Recently, The Ethiopian National Chickpea and Lentil Improvement Program and the Centre for Legumes in Mediterranean Agriculture (CLIMA), in Western Australia have started collaborative work to identify sources of resistance to waterlogging

RESEARCH EFFORTS

In the Nile Valley the national research efforts of Ethiopia, Sudan and Egypt are strong and well organised. The Nile Valley Regional programs of ICARDA have played a role in training young scientists and providing laboratory and other research facilities. The Nile Valley Regional program has been efficient in coordinating the national programs of the three countries to support and complement each other. It has also attracted other regional countries and recently Eritrea and Yemen have joined to work in collaboration with the three national programs. The Nile Valley Regional program has become the Nile Valley and Red Sea Regional program.

Problems that were common to all countries were identified and 'Regional Networks' formed. The networks that are efficiently operating are concerned with

1 Wilt and root-rot of food legumes,
2 Botrytis,
3 Aphids and viruses,
4 Socio-economic factors.

The main objective of the networks is to make efficient use of the limited resources, facilities and trained manpower available in the region. Each network consists of scientists from Ethiopia, Eritrea, Egypt Sudan and Yemen, ICARDA and ICRISAT (chickpea). The Tunisian national program also collaborates with 'wilt and root rots' network particularly on chickpeas. An annual Coordinating Meeting is held to review and discuss the results, and plan for the next season. We acknowledge the Netherlands Government for financing all the networks through the ICARDA Cairo office.

GENETIC RESOURCES

Most of the Sub-Saharan countries have their own Genetic Resource Centres. The Biodiversity Institute of Ethiopia collects germplasm every year from the regions that were not previously covered. In addition targeted germplasm collections from specific regions and for specific traits are made in collaboration with breeders. Several chickpea accessions were collected from Ethiopia in collaboration with ICRISAT. Similarly ICRISAT collects chickpeas jointly with the Tanzanian Gene bank. Germplasm from the region is also found at ICARDA

CULTIVAR DEVELOPMENT

Although information is scanty in countries such as Republic of Congo (Zaire), Tanzania, Uganda and others, a large number of faba bean (Table 1), chickpea (Table 2) and lentil (Table 3) cultivars have been released in Ethiopia, Egypt and Sudan. Eight field pea cultivars namely, Mohandefer, Tegegnech, Marcos, Hassabe, Adi, Milky, Holetta, Sefinesh and Adet-1 were released in Ethiopia (Ethiopian Seed Industry personal communication). Two *Lathyrus* cultivars that gave high yields over 3-4 seasons were identified but could not be released for cultivation due to Lathyrism, a disease that often occurs in drought hit regions where *Lathyrus sativa* is harvested.

Table 1. Released faba bean cultivars

Ethiopia	Egypt	Sudan
CSDK20	Giza 3	Selaim
NC58	Giza 402	SM-L
Bulga 70	Giza 429	Basabeer
Tesfa	Giza 461	Hudeiba 93
Mesay	Giza 643	Shambat 104
	Giza 674	Shambat 75
	Giza 714	Shambat 616
	Giza 716	
	Giza 717	
	Giza Blanca	

Table 2. Released chickpea cultivars.

Ethiopia	Egypt	Sudan	Kenya
DZ-10-4	Giza195	Shendi-1	ICCL-83110
DZ-10-11	Giza 531	Jebel Marra-1	
Dubie			
Mariye			
Worku			
Akaki			

Table 3. Released lentil cultivars

Ethiopia	Egypt	Sudan
EL-142	Giza	Selaim
R-186	Precoz	Rubatab1
Chalew	Giza 370	Aribo 1
Chekol		
Gudo		
Adaa		

MARKETING

In most of the Sub-Saharan countries domestic demand is high and thus local prices are competing with prices on international markets. A recent statistical report indicates that Ethiopia does not export these crops. This may be due to the high domestic demand particularly from drought-hit areas and to the increased prices of animal products. The Ethiopia government considers these legumes to be high value crops and thus efforts are being made to promote the products and their marketing. There is a great potential for Ethiopia to recapture its traditional export position. In Sudan all the faba beans, chickpeas, and lentils produced are consumed domestically. Ahmed (1996) indicated that Sudan's annual consumption of lentil has been estimated to be about 15,000 tonnes but more than 16,000 tonnes were imported in 1994. To satisfy domestic demand of faba beans, some imports are made by the Sudanese Government in some years. Egypt is now self-sufficient particularly in faba bean and chickpea, mainly due to the increased production of faba bean and less domestic consumption of chickpea. Generally an increase in population increases domestic demand and affects export markets.

FUTURE PROSPECTS

The high rate of population growth in sub-Sahara Africa is increasing the demand for pulses. This population increase is likely to lead to a severe protein deficiency in the region, as the majority of the people will not be able to afford animal products. Hence, cool-season food legumes will play a role in alleviating malnutrition particular in countries such as Ethiopia, Egypt, Eritrea, Sudan, Tanzania, Burundi, Republic of Congo (Zaire). Since pulses are also considered as a healthy food in the western world, there is also scope for export markets. This has been realized in many countries and the governments are now investing in these crops to satisfy both domestic demand and export markets. Efforts need to be made to increase the crop's productivity by breeding for resistance to the major diseases, insect pests, drought, excess moisture and to be responsive to inputs (eg. irrigation). These crops also need to be competitive with other crops to be attractive to producers.

References
Abu-Zeid, N. 1996 *Survey of root rot/wilt diseases distribution in some governorates of Egypt* (Unpuplished)
Ahmed, T.A. 1996 In *Production and Improvement of Cool-Season Food Legumes in the Sudan* pp 7-14 (Eds S. H. Salih, O. A. Ageeb, M. C. Saxena and M. B. Solh) ICARDA/ARC, Aleppo, Syria.
Ali, M.E.K. 1995 In *The Summary of the 1993/1994 Results and Workplan of 1994/95*, Wilt /Root Rot Network, NVRSRP/ICARDA
Ali, M.E.K. 1996 In *Production and Improvement of Cool-Season Food Legumes in Sudan* pp 153-167 (Eds S. H. Salih, O. A. Ageeb, M. C. Saxena and M. B. Solh) ICARDA/ARC, Aleppo, Syria.
Ali, K. and Habtewold,T. 1994 In *Cool-Season Food Legumes of Ethiopia* pp 367-396 (Eds A. Tilaye, G. Bejiga, M. C. Saxena and M. B. Solh) ICARDA/IAR, Aleppo, Syria.
Backlander, C.1987 *FAFFA saves life*, SIDA, Addis Ababa, Ethiopia.
Bejiga, G.and Anbessa 1994a In *Cool-Season Food Legumes of Ethiopia* pp 138-160 (Eds A. Tilaye, G. Bejiga, M. C. Saxena and M. B. Solh) ICARDA/IAR, Aleppo, Syria.
Bejiga, G.and Anbessa 1994a In *Cool-Season Food Legumes of Ethiopia* pp 162-182 (Eds A. Tilaye, G. Bejiga, M. C. Saxena and M. B. Solh) ICARDA/IAR, Aleppo, Syria.
Bejiga, G and Eshete M. 1996 In *Adaptation of chickpea in the West Asia and North Africa region* pp 137-153 (Eds N. P. Saxena, M. C. Saxena, C, Johansen, S. M. Virmani and Harris) ICRISAT/ICARDA.
Beniwal, S. P. S., Ahmed, S. and Gorfu, D 1992 *Tropical Pest Management* 38:48-51.
Beshir, T. 1996 In *Summary of the 1995/96 Results and Workplan of 1996/97*, Wilt/Root Rot Network, NVRSRP/ICARDA.
Beyene, D 1988 In *Nitrogen fixation by legumes in Mediterranean Agriculture* pp 73-78 (Eds D. P. Beck and L. A. Materon) the Netherlands.
Clancy, K. L. 1986 *American Journal of Alternative Agriculture* 1: 11-18.
Culliney, T.W., Pimentel, D. and Pimentel, M.M. 1992 *Ecosystems Environ.*1 297- 320
Doran, J. W. , Sarrntonio, M and Liebig, M. A.1996 *Advances in Agronomy* 56: 1-54.

Doreiah, E. S. 1994 *Survey of wilt /root rot diseases of food legumes in Egypt.* NVRP/ICARDA Coordination Meeting, Cairo, Egypt. (Unpublished)

FAO 1995 *Production Yearbook*, 49: 31-36.

Faki, H., Mohamed, A.I.S. and Ali, M.E.K. 1996 In *Adaptation of chickpea in the West Asia and North Africa region,* pp 155-169 (Eds N. P .Saxena, M. C. Saxena, C, Johansen, S. M. Virmani and Harris) ICRISAT/ICARDA.

Gorfu, D. and Beshir, T. 1994a In *Cool-Season Food Legumes of Ethiopia* pp 328-345 (Eds A. Tilaye, G. Bejiga, M.C. Saxena and M.B. Solh) ICARDA/IAR, Aleppo, Syria.

Gorfu, D. and Beshir,T. 1994b In *Cool-Season Food Legumes of Ethiopia*, pp 317-327 (Eds A. Tilaye, G. Bejiga, M.C. Saxena and M.B. Solh) ICARDA/IAR Aleppo, Syria.

Gorfu, D. 1994 In *Cool-Season Food Legumes of Ethiopia* pp 315-316 (Eds A. Tilaye, G. Bejiga, M.C. Saxena and M.B. Solh) ICARDA/IAR, Aleppo, Syria.

Hamdi, A., Hussanein, A.M. and El-Garhy, A.M. 1995 *Survey of fungal diseases of lentil in North Egypt.* NVRP/ICARDA Coordination Meeting ICARDA, Aleppo, Syria. (Unpublished)

Hulluka, M and Taddse,N. 1994 In *Cool-Season Food Legumes of Ethiopia* pp 346-365 (Eds A. Tilaye, G. Bejiga, M.C. Saxena and M.B. Solh) ICARDA/IAR, Aleppo, Syria.

Kattaab, A.M and El-Sherbeeny M.H 1996 In *Adaptation of chickpea in the West Asia and North Africa region* pp 127-136 (Eds N.P.Saxena, M.C.Saxena, C, Johansen, S.M.Virmani and Harris) ICRISAT/ICARDA.

Mengistu, A. 1991 In *Legume genetic resources for Semi-arid temperate environments*, pp 50-64 (Eds A.Smith and L.Robertson), ICARDA, Aleppo,Syria.

Pundir, R.P.S., Nyange, L.R and Sambai , L.M. 1996 *International. Chickpea and Pigeon pea Newsletter* 3: 11-12.

Solh, M.B. 1996 *A Model for Technology Transfer*, NVRP/ICARDA.

Yetneberk, S. and Wondimu, A. 1994 In *Cool-Season Food Legumes of Ethiopia* pp 346-365 (Eds A. Tilaye, G. Bejiga, M.C. Saxena and M.B. Solh) ICARDA/IAR, Aleppo, Syria.

Region 5: Near East

M.H. Halila[1], N.I. Haddad[2], B. Sakr[3], and I. Küsmenoglu[4]

1 Food Legumes Laboratory, INRAT, Avenue Hedi Karray, 2049 Ariana, Tunisia;. 2 ICARDA, PO Box 950764, Amman 11195, Jordan.; 3 Food Legumes Laboratory, CRRA, PO Box 589, Settat, Morocco.; 4 CRIFC, P.O. Box 226, Ulus, Ankara, Turkey.

Abstract

This review examines developments in the cool season food legumes since IFLRC II in 1992. The area, production and yield have generally decreased in the Maghreb and increased in West Asia. Climatic conditions during the period and changes in supporting policy programs seem to be the reason for these developments. Consumption tends to be stable but projections are that demands will increase and supplies will be limited, resulting in an increase in imports to most of the countries. Turkey will remain the main producer and exporter in the region. Research has made substantial progress but its impact is still to be seen or quantified.

INTRODUCTION

This review covers Region 5 as described by the International Steering committee of IFLRC III and consists of the countries of West Asia and North Africa (WANA). Afghanistan and Ethiopia, although belonging to the WANA region, are not discussed as no information was available from the former, and the latter is included in Region 4 (Sub-Sahara Africa).

Among the cool season food legumes (CSFL), faba bean, chickpea, lentil and to a lesser extent pea will be considered because of:

1. Their importance among the CSFL species grown and produced in the region.
2. The availability of published agricultural statistics in different countries of Region 5.

It is important to mention the statistics used are those published officially by different ministries of the countries and by FAO (1995 and 1996). Figures from these sources often match, although discrepancies may sometimes occur. In general these statistics are published based on national administrative levels masking variation within a country, and among countries, for geography and weather conditions.

In this paper we have attempted to examine the changes and progress (or regression) that have occurred in Region 5 since IFLRC II. For this, and to make the presentation clearer, two periods have been arbitrarily used; the first one from 1989 until 1991 and the second from 1992 onwards.

TRENDS IN AREA

Maghreb and Egypt

The total area of the three main CSFL, faba bean, chickpea and lentil have in general declined in the Maghreb (-13%) (Table 1). This reduction is even higher when the pea area is included. These crops lost ground mainly in Morocco (-25%) followed by Tunisia (-11%) and Algeria (-7%). The area of faba bean, which is the most important CSFL in the Maghreb, suffered an overall reduction of 16% to which Morocco contributed the biggest share (-33%). The area under chickpea also declined particularly in Tunisia where the reduction was 25% and was almost linear with time.

R. Knight (ed.), Linking Research and Marketing Opportunities for Pulses in the 21st Century, 107–114.

Table 1. Change (+/- in %) in area ('000 ha) of the main food legume crops grown in selected countries of West Asia and North Africa

Country	Faba bean			Chickpea			Lentil			Total 1			Pea		
	89-91[1]	92-96	+/-	89-91	92-96	+/-	89-91	92-96	+/-	89-91	92-96	+/-	89-91	92-96	+/-
Egypt	129	144	+11	6	7	+16	7	6	-15	142	157	+10	NA	1	-
Morocco [3]	205	138	-33	75	68	-9	54	44	-19	334	250	-25	61	36	-41
Algeria	56	49	-12	40	41	-	3	1.7	-43	99	92	-7	13	11	-15
Tunisia	42	44	+5	35	26	-25	4	2.6	-35	81	72	-11	7	7	-
Maghreb & Egypt	432	375	-13	156	142	-9	68	54	-20	656	571	-13	79	53	-31
Syria	8	7	-12	49	75	+53	131	113	-14	188	195	+4	NA	0.7	-
Iran	NA	NA	-	387	673	+74	128	213	+66	515	886	+72	NA	NA	-
Turkey	39	31	-20	849	782	-8	858	672	-22	1746	1484	-15	2	2	-
West Asia [2]	41	38	-20	1285	1530	+19	1117	998	-11	2439	2565	+5	-	-	-

(1) FAO production year book 1995, (2) Data for 92-96 are from FAO spreadsheet 92-96, (3) Data for 92-96 are from the Ministries of Agriculture of Morocco, Algeria and Tunisia.

In Egypt, the area of CSFL increased by 10% due mainly to an increase in the faba bean area. However, lentil area decreased by 15%.

West Asia

In West Asia, Turkey, Syria and Iran are the main CSFL producing countries. As a whole, the area devoted to CSFL in these countries has remained relatively stable (+5%). Taken individually, the area increased substantially in Iran (+72%) and to a lesser extent in Syria (+4%) (Table 1).

Turkey continues to be the dominant food legume producer in the region, particularly of lentil and chickpea. However, here the CSFL area has decreased by 15%, due mainly to the reduction in the lentil area (-22%) (Table 1).

In Iraq, Lebanon, Libya, Jordan and Cyprus, CSFL are relatively minor crops despite an important increase in their area. This increase is in the chickpea and faba bean areas in Iraq (data not shown).

TRENDS IN PRODUCTION AND PRODUCTIVITY

Food legumes are produced and used mainly as dry pulses. In some areas, however, an important share of the production is consumed as a green vegetable. This is quite common in the Maghreb countries and Egypt and concerns mostly faba bean (large or *equina* seed types) and pea. It is not always evident whether this type of production is accounted for in the published statistics.

Maghreb and Egypt

Production of pulses has decreased in the Maghreb virtually for all crops (Table 2). The only exception was faba bean production in Tunisia where it has increased by 13% leading to an overall unchanged level of production level of pulses in the country. The most dramatic reduction happened in Morocco (-55%) (Table 2).

Looking at the individual crops, it is apparent faba beans performed poorly. The production of this crop had a reduction of 9.1% a year in Morocco, 3.3% in Algeria and was the main cause of the total pulse reduction in the Maghreb region. For the other crops, it is striking to note their negative performance in Morocco where chickpea and lentil production were reduced by almost half and pea by 74% (Table 2).

Table 2. .Production ('000t) of the main food legume crops grown in selected countries of West Asia and North Africa

Country	Faba bean			Chickpea			Lentil			Total 1			Pea		
	89-91[1]	92-96	+/-	89-91	92-96	+/-	89-91	92-96	+/-	89-91	92-96	+/-	89-91	92-96	+/-
Egypt	339	402	+18	11	12	+10	13	9	-30	363	423	+16	4	3	-25
Morocco [3]	168	75	-55	56	29	-48	36	21	-42	260	125	-52	57	15	-74
Algeria	25	20	-20	20	19	-5	1	0.5	-50	46	40	-15	3	2.5	-17
Tunisia	30	34	+13	20	17	-15	2	1.2	-40	52	52	-	5	4	-20
Maghreb & Egypt	562	633	+12	107	77	-28	52	31.7	-38	721	640	-11	69	24.5	-65
Syria	10	14	+40	26	53	+103	74	113	+52	110	180	+63	NA	0.8	-
Iran	NA	NA	-	161	324	+101	60	131	+118	221	455	+105	NA	NA	-
Turkey [2]	73	58	-20	799	724	-9	669	634	-5	1541	1417	-8	5	4	-20
West Asia [2]	83	72	-13	986	1101	+12	803	878	+9	1872	2052	+10	-	-	-

(1) FAO production year book 1995, (2) Data for 92-96 are from FAO spreadsheet 92-96, (3) Data for 92-96 are from the Ministries of Agriculture of Morocco, Algeria and Tunisia

With the exception of a few cases, the negative growth in pulse production was the result of reductions in both area and yield.

Although several reasons (*Orobanche* infestation, low prices, high production costs) may have been responsible for the reduction in the CSFL area and production in North Africa, climatic factors are considered to be the main reason for the decline. Severe droughts were experienced in three out of the last five agricultural seasons. The droughts of 1994 and 1995 were the most damaging in all the three Maghreb countries.

In Egypt, faba bean production increased only by 18% despite an 11% increase in the area of this crop. Biotic stresses like viral diseases, such as necrotic yellows, may have been responsible. This particular virus was already the cause of the drastic drop in national faba bean production in 1991-92 in Egypt (ICARDA, 1993).

West Asia

The growth pattern in pulses in this sub-region varied between countries. In Turkey, a modest reduction was observed (-8%) and stemmed mainly from a decline in faba bean production. The reduction in productivity (5%) seems to have been buffered by an increase the yield of lentils (+19%) thus nullifying the 22% decrease in lentil area in the corresponding period (Table 1)

Total production doubled in Iran while Syria had an increase of 63% (Table 2). The doubling in Iran originated mainly from an increase in the area of chickpea and lentil. This increase was associated with a reasonable improvement in the yield, particularly of lentils (Table 3).

Table 3. Yield (kg/ha) of the main food legume crops grown in selected countries of West Asia and North Africa

Country	Faba bean			Chickpea			Lentil			Pea		
	89-91[1]	92-96	+/-	89-91	92-96	+/-	89-91	92-96	+/-	89-91	92-96	+/-
Egypt	2618	2857	+10	1851	1782	-4	1922	1514	-21	NA	2733	-
Morocco [3]	841	550	-35	752	420	-44	681	480	-30	921	440	-52
Algeria	430	410	-5	490	450	-8	360	320	-11	212	210	-
Tunisia	710	780	+10	589	660	12	447	460	+3	708	575	-19
Maghreb	660	580	-12	610	510	-16	496	420	-15	613	408	-34
Syria	1338	1950	+45	512	706	+38	593	989	+67	-	1272	-
Iran	NA	NA	-	417	482	+15	464	617	+32	NA	NA	-
Turkey [2]	1864	1513	-19	938	925	-2	799	948	+19	2498	2480	-
West Asia [2]	1601	1731	+23	622	704	+13	618	851	+38	-	1876	-

(1) FAO production year book 1995, (2) Data for 92-96 are from FAO spreadsheet 92-96, (3) Data for 92-96 are from the Ministries of Agriculture of Morocco, Algeria and Tunisia.

CONSUMPTION

Unfortunately, consumption is not well documented in many countries. Usually consumption figures are generated from surveys, which are conducted at different times in individual countries. Updated consumption data are therefore scarce.

In Turkey, per capita consumption of 3 kg/year for lentil (Bayaner *et al*, 1995) and 2.5 kg/year for chickpea (Kusmenoglu and Bayaner, 1995) are reported. In the Maghreb, per capita dry seed consumption seems to be interestingly similar for Morocco (Driuochi, 1992) and Algeria (Zaghouan, 1995) with 6.4 and 7 kg/year, respectively, whereas in Tunisia, per capita consumption is only 3.2 kg/year (Khaldi, personal communication). For the latter, per capita consumption of "green" food legumes, although declining, is also important (5.8 kg/year). Given an average moisture content of 50% for these green products, we can safely conclude that the per capita consumption of pulses in Tunisia would be about 6 kg/year.

On the other hand, in certain countries, food legumes mainly faba bean are also partly used for animal feeding. For example, 19% of faba bean production in Morocco is used to supplement livestock feeding and this should have increased at an annual rate of 2.2% (Driouchi, 1992). The figures for similar use in Algeria and Tunisia are 4% and 5%, respectively (Chaffai, 1994).

Previous analyses of the food legume sector had predicted a steady increase in pulse consumption (Oram and Belaid, 1989).

IMPORTS AND EXPORTS

Import/export data are fragmentary and, in some countries, aggregated in one single commodity labelled "legumes".

Regional trade of pulses is dominated by Turkey and by two commodities, lentils and chickpeas. For this country, lentil exports from 1989-91 to 1992-95 progressed by 33% despite a decline in production. Occasionally, this was achieved by importing lentils to meet Turkey's export commitments. In contrast, chickpea exports decreased by 15% for the same period. Turkey exports pulses to more than 50 countries (Bayaner *et al.*, 1995). The main factors that have affected its strong presence in world trade are:

- The implementation in the eighties of a "Fallow replacement program" which resulted in an increase in crop area of 2.3 millions ha of which 77% was allocated to lentil and chickpea (Kusmenoglu and Bayaner, 1995).
- The introduction by the Turkish Grain Board (TMO) of a subsidy program for pulses (Bayaner et al, 1996).

Competition in the world market from Canada and Australia reduced Turkey's share of world exports. Internal prices of lentil and chickpea have declined and farmers have shifted to other crops such as sunflower and melon.

In the Maghreb, Morocco, which in the past used to be an exporter of pulses to Europe, has become a net importer since 1992. In Tunisia, the food legume trade fluctuates and is closely influenced by production each season, which is governed by climatic conditions. The trends are however, towards increasing imports, particularly of lentil. Algeria, in contrast, has been importing food legumes for a long time and this trend is increasing especially of chickpeas, which represent the biggest share of the imports (Kelly and Parthasarathy, 1996).

In West Asia, Iraq, Jordan, Lebanon and Iran were importers of pulses during the 1989-91 period and imported variable quantities of chickpea. For the same period, Syria was a modest exporter (Kelly and Parthasarathy, 1996).

GOVERNMENT POLICIES AND FUTURE PROSPECTS

In the seventies and the eighties, pulse production was supported by many governments through price support schemes and policy incentives such as subsidies on inputs (seeds, herbicides, etc.). However, at no time has the extent of this support been at the levels provided to the cereals. In the last few years the support has started to decrease. In Turkey, the Government has been gradually reducing the subsidy and the guaranteed purchasing program was abolished in 1994 (Bayaner et al., 1995).

The prospects for food legume production and exports from Turkey fluctuate between optimistic and pessimistic, particularly for lentil (Bayaner et al, 1995). From an optimistic perspective (Kusmenoglu and Bayaner, 1995), it is expected that by year 2000, there will be an increase in area of chickpea and lentil (this is not supported by present trends, however). Production and domestic consumption have been projected to be 2,655,000 t and 2,194,000 t, respectively, resulting in an available surplus for exports of 500,000 t. When making these projections, an important assumption was that national policy would resume its support of the production of CSFL crops.

In the Maghreb, the economic liberalisation program is being implemented at different speeds. The speculation is that pulses will be among the main losers in this new situation. They have already experienced serious competition from other crops for land occupation. For example, in certain regions of Tunisia and Morocco, chickpea is battling sunflower, for a place in the cereal-based rotation systems. For chickpea, most countries in WANA will fall into a deficit position by year 2000. Imports will rise to meet the increasing demand (Kelly and Parthasarathy, 1996).

BIOTIC AND ABIOTIC CONSTRAINTS

Biotic and abiotic stresses facing the production of CSFLs in the region have been highlighted (Halila and Beniwal, 1994; Kusmenoglu and Haddad, 1994) and still remained basically the same (Beniwal at al., 1996; Kusmenoglu and Bayaner, 1995).

Among biotic constraints, necrotic yellows has become the most serious faba bean virus disease in Egypt (ICARDA, 1993, 1994) and also has been frequently found in the Maghreb (Ouffroukh, personal communication). Surveys have shown that Fusarium wilt is becoming one of the dominant chickpea diseases in Syria (ICARDA, 1993) and North Africa (Halila, 1995; Bouznad et al, 1996). Orobanche is an established problem for faba bean in Morocco where it has even became a threat to lentils and peas (Sakr, personal observation).

A breakdown of resistance to Ascochyta blight in chickpea is reported to have occurred during the last five years. This breakdown was observed in Syria in cultivar ILC 482 and was attributed to the emergence of new pathotypes of the fungus (Malhotra et al., 1996).

Among insects, there has been an increase of Lygus bug damage on red lentil in Turkey (Kusmenoglu, personal observation). Leaf miner caused severe damage in certain areas of North Africa (Halila, personal observation).

Drought was the most important and frequently reported abiotic stress during the last five years. Morocco experienced severe droughts in 1992, 1993 and again in 1995. Drought hit Algeria in 1994, and Tunisia observed a century record drought the same year. Similarly, Turkey had a relatively dry season in 1994. As mentioned earlier, these weather events have contributed to the decline in the area and production of CSFLs in the region.

RESEARCH ADVANCES AND HIGHLIGHTS

New cultivars in the region

Research, conducted since 1991, on genetic improvement and germplasm enhancement, have resulted in the release of a relatively high number of promising lines in many countries of the region (Table 4). In most cases, these lines were developed from collaborative research between ICARDA and national programs (ICARDA, 1995a). The lines combine one or more desirable traits such as yield, resistance to biotic stresses, suitability for mechanised farming and quality. In addition, improved agronomic packages were developed for the majority of CSFL crops. One good example of these advances is the winter or early spring sowing of chickpea. Despite the substantial increase in yield generated by the winter/early spring sowing, which has been clearly demonstrated in all countries where it has been tried, adoption by farmers of this technology has been slow. This may have been due to an insufficient availability of seeds of the improved varieties and lack of intensive and continuous efforts on technology transfer. Tutwiler (1996) studied the constraints to winter sowing of chickpea in Morocco and demonstrated that socio-economic factors largely determined the extent of adoption of new technology and therefore should be at the front of research on technology transfer.

Table 4. Food legume varieties released, since 1991, by national programs in the Near East Region (ICARDA, 1995)

Country	Chickpea	Lentil	Faba bean
Turkey			
1991	Akcin (87AK7115) -1991	Sazak 91 (ILL 854)	
1992	Aydin 92 (82-259C) - 1992		
1992	Menemen 92 (85-14C) - 1992		Hama 1
1992	Izmir 92 (85-60C) - 1992		
1994	Damla (85-7C) - 1994		
1994	Azizie (84-15C) - 1994	Baraka (ILL 5582)	
Syria	Ghab 3 (82-150C)	Toula (86-2L)	
1991	ILC 482		Reina bianca
Iran	ILC 3279		Giza 461
1995	84-48C		
Iraq	Rafidain (ILC 482)		
1991	Dijla (ILC 3279)		
1991		Essafsaf (78S26002)	
1994	Balella (85-50C)		
Lebanon			
1993	ILC 195		
1995	Rizki (83-48C)		
Egypt	Douyet (84-92C)		
1992	Farihane (84-79C)		
1993	Moubarak (84-145C)		
Morocco	Zahor (84-182C)		
1992	84-79C		
1992	84-92C		
1995	ILC 484		
1995			
1995			
Algeria			
1991			
1991			
Libya			
1991			
1993			

In Turkey, research has led to the development of winter red lentils for central Anatolia.

Biotechnology and new research tools

The use of molecular techniques and biotechnology tools have expanded considerably in the region, at ICARDA (ICARDA, 1995b) and in national programs (Morjane *et al.*, 1994; Bossis *et al.*, 1996; Kusmenoglu, 1995). The techniques are applied to almost all crops and concentrated in the development of marker-assisted selection and characterisation and identification of fungal pathogens and nematodes. Application of molecular biology techniques to the improvement of CSFLs in the region is expected to expand due to training at ICARDA and other institutions of an active and interested generation of young scientists.

COLLABORATIVE RESEARCH WORK

The volume and quality of research in the region has improved considerably. Here, we present a sample in the area of collaborative research.

ICARDA, which is the only International Centre working on CSFL in the region, has strengthened its relations with national programs through bilateral and regional collaborative research projects and through support to specific food legume research networks. The Regional Networks in the Nile Valley Program, and particularly the Wilt/Root-Rots Network and the Aphids and Viruses Network, are good examples of such collaboration.

In addition, collaboration among various national programs has been enhanced through Networks. Networks on diseases of chickpea in the Maghreb were developed within a UNDP-funded project on the diseases of cereals and food legumes. The project has been extended for two years. Furthermore a Maghreb Faba Bean Research Network, funded by GTZ and national programs of Morocco, Algeria and Tunisia, was initiated in the last five years and will continue to operate till the year 2000. Regional breeding trials on lentil and chickpea have been conducted in the Maghreb.

CONCLUDING REMARKS

The above review covers some of the most important developments in CSFLs in the WANA region. The review cannot be exhaustive as certain crops have been omitted or insufficiently covered. In future, peas will deserve more attention. This crop, although not highly important in the region at present, will receive more attention from policy makers in many countries of the region. It is a "future crop" in rainfed-farming systems mainly as a protein source in animal feeding.

We believe socio-economic studies have not been sufficiently included in the research agenda on the CSFL crops in national programs, which tend to be more oriented towards crop improvement. The statements made by Oram and Agcaoili (1994) in IFLRC II regarding these issues remain valid and topical.

Acknowledgements

We appreciate the contributions of scientists from the region who are not co-authors. The authors thank particularly Dr S. P. S. Beniwal, Coordinator of the Highlands Regional Program, for reviewing the manuscript and for his useful comments.

References

Bouznad, Z.. Maatougui, M. H. and Labdi, M. 1996. In.: *Proceedings of the Regional Symposium on Cereal and Food Legume diseases,* pp. 13 -19 (Eds. B. Ezzahiri, A. Lyamani, A. Farih & M. El Yamani).

Bayaner, A., Uzunlu, V. and Kusmenoglu, I. 1995. In: *Autumn-sowing of lentil in the highlands of West Asia and North Africa,* pp. 4 -14. (eds. D. H. Keatinge and I. Kusmenoglu) CRIFC. Turkey.

Bossis, M., Abbad, F., Caubel, G and Esquibet, M. 1997. In: *Colloques "GRAM"* de l'INRA France. (In press).

Beniwal, S. P. S., Haware, M. P. and Reddy, M. V. 1996. In: *Adaptation of chickpea in the West Asia and North Africa Region,* pp. 189 - 198. (eds. N. P. Saxena, M. C. Saxena, C. Johansen, S. M. Virmani and H. Harris). ICRISAT: India and ICARDA: Syria.

Chaffai, A. 1994. Etude sectorielle. GTZ/REMAFEVE. Laboratoire des léguminueses, INRAT, Ariana, Tunisia. pp. 36.

Driouichi, A. 1992. In: *Le secteur des légumineuses alimentaires.* Editions Actes, Rabat: Morocco.

Food And Agriculture Organization of the United Nations. 1995. *FAO Production Yearbook.* Rome, Italy: FAO.

Food And Agriculture Organization of the United Nations. 1996. *FAO Spreadsheet 92 - 96.* Rome, Italy: FAO.

Halila, H. and Beniwal, S. P. S. 1994. In: *Expanding the production and use of Cool Season Food Legumes,* pp. 951 - 955 (eds. F. J. Muehlbauer and W. J. Kaiser). Kluwer Academic Publishers.

Halila, M. H. 1995. Projet RAB 007/91 *Sur la surveillance des maladies des céréales et des légumineuses alimentaires dans les pays du Maghreb.* PNUD (Maroc, Tunisie et Algérie).

International Center for Agricultural Research in the Dry Areas (ICARDA). 1993. *Annual Report for 1993*. Aleppo Syria:

International Center for Agricultural Research in the Dry Areas (ICARDA). 1994. *Annual Report for 1994*. Aleppo Syria:

International Center for Agricultural Research in the Dry Areas (ICARDA). 1995a. *Annual Report for 1995*. Aleppo Syria:

International Center for Agricultural Research in the Dry Areas (ICARDA). 1995b. *Legume Program: Annual Report for 1995*. Aleppo Syria:

Kelley, T. G. and Parthasarathy Rao, P. 1996. In: *Adaptation of chickpea in the West Asia and North Africa Region*, pp. 239-254 (eds. N. P. Saxena, M. C. Saxena, C. Johansen, S. M. Virmani and H. Harris). ICRISAT: India and ICARDA: Syria.

Kusmenoglu, I. and Haddad, N.I. 1994. In: *Expanding the production and use of Cool Season Food Legumes*, pp. 956-958 (eds. F. J. Muehlbauer and W. J. Kaiser). Kluwer Academic Publishers.

Kusmenoglu, I. 1995. Ph.D. thesis, Selcuk Unversitesi, Konya, Turkey. 191 pp. (In Turkish)

Kusmenoglu, I. and Bayaner, A. 1995. *2nd European Conference on Grain Legumes*. Copenhagen, Denmark.

Malhotra, R. S., Singh, K. B., van Rheenen, H. A. and Pala, M. 1996. In: *Adaptation of chickpea in the West Asia and North Africa Region*, pp. 217 - 232. (eds. N. P. Saxena, M. C. Saxena, C. Johansen, S. M. Virmani and H. Harris). ICRISAT: India and ICARDA: Syria.

Morjane, H., Geistlinger, M., Harrabi, M., Weising, K. and Kahl, G. 1994. *Current Genetics* 26, 191-197.

Oram, P. A. and Belaid, A. 1990. *Legumes in farming systems*. A joint ICARDA/IFPRI Report. Aleppo, Syria: ICARDA.

Oram, P. A. Agcaoili, M. 1994. In: *Expanding the production and use of Cool Season Food Legumes*, pp. 3 - 49 (eds. F. J. Muehlbauer and W. J. Kaiser). Kluwer Academic Publishers.

Tutwiler, N. 1996. In: *Adaptation of chickpea in the West Asia and North Africa Region*, pp. 233 - 237 (eds. N. P. Saxena, M. C. Saxena, C. Johansen, S. M. Virmani and H. Harris). ICRISAT: India and ICARDA: Syria.

Zaghouan, O. 1995. In: *Autumn-sowing of lentil in the highlands of West Asia and North Africa*, pp. 35 - 40. (eds. D. H. Keatinge and I. Kusmenoglu) CRIFC. Turkey.

Region 6: Asia -Pacific: Meeting the Challenge

J.B. Brouwer[1], B Sharma[2], B.A. Malik[3] and G.D. Hill[4]

1 Victorian Institute for Dryland Agriculture, Private Bag 260, Horsham, Victoria 3402, Australia; 2 Division of Genetics, Indian Agricultural Research Institute, Delhi 110012, India; 3 National Agricultural Research Center, Park Road, Islamabad, Pakistan; 4 Department of Plant Science, Lincoln University, P.O. Box 84, Lincoln, Canterbury, New Zealand

Abstract

The Asia-Pacific region represents the most diverse pulse growing area in the world. This review concentrates on the countries of Pakistan, India, Bangladesh, Nepal, Myanmar, the Peoples Republic of China, Australia and New Zealand. The major cool-season pulses, chickpeas, peas, lentils and faba beans, all play an important role here, although not to the same extent everywhere. In addition, grasspea and lupin are found in some farming systems. The region includes economies, where production either has a focus on domestic consumption, is primarily export-oriented or where there is a balance in import-export activities. The agroecological differences are vast within and between the Asian-Pacific countries and pulse production methods vary from mono-cropping to inter- and mixed cropping. Trends in productivity are therefore difficult to detect. Biotic and abiotic constraints are worth considering, as both remain very important factors. Improved production technology is becoming available, but its on-farm application is still lagging as socio-economic factors often influence its adoption.

INTRODUCTION

The Asia-Pacific region is vast geographically, stretching from 50^0N to 45^0 S and 60^0 to 150^0 E. It is extremely diverse in terms of cultures, population densities and agroecological environments, which cover tropical to temperate climates. This review concentrates on the countries of Pakistan, India, Bangladesh, Nepal, Myanmar, the Peoples Republic of China, Australia and New Zealand, where cool season pulses commonly occur but where they differ in their distribution pattern.

The region accounts for 49% of the world's total pulse area. This includes both cool season (*rabi*) and summer (*kharif*) crops. The area sown to all pulses in the region of approximately 34 million ha has not shown a significant increase (only 0.2%) over the last decade (Paroda, 1995), while the annual growth rate in the remaining parts of the world has been 1.5%. The annual growth rate of the region's production of pulses was only 1.4% compared to 3.4% elsewhere.

Pulses are the major source of protein in human diets in Asia, where people obtain only 20% of their protein intake from animal sources (Paroda, 1995). Pulses play a significant role both in human nutrition and in the sustainability of agricultural systems. Yet, pulse production has increased at a much slower pace over the last two decades compared to cereals and oilseed crops. The annual growth rate for pulse production has been low at 1.4% with the gap widening between the different food crops as indicated by a pulse:cereal ratio of 1:14 in 1960 moving to 1:32 by 1990 (Paroda, 1995). The availability of pulse grain in the region has also declined per head of population from 1971 (10.3 kg per annum) to 1996 (8.2 kg). The region's population was estimated at more than 3 billion in 1996, including 1.2 billion in both China and in the Indian sub-continent. This number is expected to double in the next 50 years.

R. Knight (ed.), Linking Research and Marketing Opportunities for Pulses in the 21st Century, 115–129.
© 2000 *Kluwer Academic Publishers. Printed in the Netherlands.*

PRODUCTION OF COOL SEASON PULSE CROPS

Chickpeas

Despite significant gains in world pulse production over the last 15 years with a 1.9 % annual growth rate during the two decades 1971 - 1991, chickpea production has grown only slowly (Kelley and Parthasarathy Rao, 1996) at 0.3% per year. Chickpea yields rose by only 0.08 t/ha per annum worldwide, while the area was virtually unchanged during that period. The Asia-Pacific region contributes the greatest proportion of the world production of chickpeas (6.7 million t or 80%), with India being the dominant producer at 86% within the region and 69% worldwide (Table 1, average 1994-96). Production increased in India by 17% from an average of 4.9 million t over 1982-84 to 5.7 million t over 1994-96 (Table 1). Thus, the region's trend in chickpea production reflects largely the situation in India. The total chickpea area in the Asia-Pacific region remained static at 8.7 million ha over the last 15 years. The area in South Asia decreased over the same period by 215,000 ha but Australia compensated by adding 221,000 ha over 1994-96.

TABLE 1. Area ('000ha), production ('000t) and yield (t ha^{-1}) of cool-season pulses in the Asia-Pacific region as three year averages 1994-1996

	Chickpea			Lentils			Field Pea(dry)		
	Area	Production	Yield	Area	Production	Yield	Area	Production	Yield
Bangladesh	85	62	0.73	207	169	0.82	40	25	0.63
India	7206	5730	0.80	1224	792	0.65	620	561	0.91
Myanmar	134	86	0.64	3	1	0.29	37	24	0.66
Nepal	22	14	0.64	174	110	0.63			
Pakistan	1066	536	0.50	59	30	0.52	137	75	0.54
Thailand									
SOUTH ASIA	8514	6428	0.76	1667	1102	0.67	833	685	0.82
CHINA	1.5	2.7	1.78	95	112	1.18	722	1142	1.58
Australia	221	219	0.99	11	19	1.81	387	389	1.00
New Zealand				4	6	1.50	19	60	3.16
OCEANIA	221	219	0.99	15	25	1.67	406	449	1.11
ASIA-PACIFIC	8736	6649	0.76	1777	1239	0.69	1961	2275	1.16
% of world	79	80	101	52	44	83	28	19	67
WORLD	11019	8287	0.75	3398	2818	0.83	7044	12248	1.74

	Faba Bean (dry)			Lupins					
	Area	Production	Yield	Area	Production	Yield			
China	1087	1883	1.73						
Australia	86	97	1.13	1260	1187	0.94			
OCEANIA	86	97	1.13	1260	1187	0.94			
ASIA-PACIFIC	1173	1981	1.68	1260	1187	0.94			
% of world	51	58	113	89	89	100			
WORLD	2294	3421	1.49	1421	1339	0.94			

Source: FAOSTAT Datebase 1997. New Zealand lentil data: A.C. Russell

Chickpea yields in South Asia fluctuate dramatically. However, positive but non-significant growth rates were evident (Kelley and Parthasarathy Rao, 1996). The three-year average yield (1994-

96) in India of 0.80 t/ha (Table 1) showed an encouraging increase of 12% compared to the average yield over 1991-1993 of 0.71 t/ha (FAOSTAT Database 1997).

Since India's production is such a dominant influence (Table 1), it is interesting to examine the factors driving its production. Between 1971/73 and 1988/89 the area declined by 1.7 million ha in the traditional chickpea growing northern states of Bihar, Haryana, Punjab, Rajasthan and Uttar Pradesh (Kelley and Parthasarathy Rao, 1996). However, sowings increased in the central and southern states, including Andhra Pradesh, Gujarat, Karnataka, Madhya Pradesh, Maharastra and Orissa. While in 1971, 70% of the crop was located in the northern region, this geographical shift added approximately 880,000 ha to the central and southern region. As a result, the Indian chickpea area is now equally distributed between the northern and central-southern states. The decline in area in the northern states is expected to continue because of the replacement of chickpeas by more profitable, post-rainy season crops.

Lentils

The Asia-Pacific region produces 52% of the world's lentils from 44% of the world lentil area (Table 1). *Microsperma* cultivars are the predominant types in South Asia, where most (89%) of the Asia-Pacific region's lentils are produced (Table 1). The South Asian growing areas can be divided agroclimatically into the North western Plains (Punjab Province of Pakistan and Indian states of Haryana, Punjab, western Uttar Pradesh; 10% of lentil area), North eastern Plains (parts of Myanmar, Bangladesh, eastern Uttar Pradesh, Bihar and West Bengal; 35-40%), Central Highlands/Plateaux (Madhya Pradesh, Uttar Pradesh, and Maharastra; 35-40%) and the *Terai* region (near the Himalayan foothills of Pakistan, India and Nepal; 10%) (Bahl and Sharma, 1993).

During the last 15 years there has been an accelerated growth in lentil production in South Asia, from an annual average of 745,000 t, in the period 1982-84, to 1,102,000 t between 1994-96 (Table 1). Yield has also showed a remarkable upward trend. Within the region, Bangladesh achieved the highest average yield in 1996 at 0.82 t/ha. India showed the largest relative change in yield (+57%) over the last 15 years (average 1982/84 vs 1994/96). The average lentil yield in South Asia increased over the same period from 0.54 to 0.66 t/ha (+22%) compared to average world yields which went from 0.68 to 0.83 t/ha (also +22 %) (FAOSTAT Data Base 1997).

Within India, the cultivation of lentils is confined to two major zones. In Central India (Madhya Pradesh and Jhansi Division of Uttar Pradesh) lentils are grown almost exclusively as rainfed crops. Bold seeded cultivars (>3 g/100 seeds) are preferred. The other major area is the sub-Himalayan plains of northern India where the yield potential is higher and opportunities exist for irrigation. Small seeded (1.8-2.2 g/100 seeds) types are common here (Bahl and Sharma, 1993; Sharma, 1995). Approximately 90% of area and production of lentils is provided by three states, Uttar Pradesh, Madhya Pradesh and Bihar. Nepal has increased its production in recent years to 174 000 t (1994/96) largely by expansion of the area sown.

China and Oceania (Australia and New Zealand) are small lentil producers by comparison with an average of 112,000 t and 25,000 t for 1994/96 respectively (Table 1). However, yields are higher than on the Indian subcontinent (1.18 and 1.67 t/ha for China and Oceania respectively). Production is also increasing with a steady expansion of the area in China to 95,000 ha (1996) and in Australia to 40,000 ha (1997) (Materne and Brouwer, 1996).

Peas

The Asia-Pacific region supplies 19% of the world's dry pea or field pea grain from 28% of the world's pea growing area (Table 1). Within the region, China is the largest producer. Its average yields of 1.73 t/ha are the highest after New Zealand's 3.19 t/ha. After an initial expansion of the field pea area in Australia during the 1980s (Walgott and Rees, 1995; Siddique and Sykes, 1997) the area has remained at 350,000 – 400,000 ha in 1997. New, higher yielding cultivars are expected to increase the competitiveness of the field pea with alternative crops (Hamblin *et al.*, 1998). Field peas rank first among the pulses in yielding ability in India. Highest yields are being obtained with this crop in Rajasthan using irrigation (1.34 t/ha), but Uttar Pradesh is the main production centre with 59% of the area and 79% of Indian production (Sharma, 1995).

In addition to the dry pea crop, the region produces 23% (1.1 million t) of the fresh green pea harvested in the world. China is the largest producer (15% or 0.8 million t).

Faba beans

The major producer of dry faba beans is China with a current production of approximately 1.88 million t from 1.1 million ha (Table 1) or 55% of the world's total crop. Ninety percent of the Chinese crop is grown along the Yangtze River. However, three autumn sown and four spring sown regions can be recognised (Lang *et al*, 1993a, b). Although yields have more than doubled in China from 0.77 t/ha in the 1950s to 1.73 t/ha in the mid-1990s, the total area has declined since the 1960s because of the change in cropping system from double cropping (rice and a winter crop) to triple cropping (two crops of rice + a winter crop). The faba bean life cycle is too long to fit into this new regime and it is being replaced by rape seed or barley. (Lang *et al*, 1993a).

The Peoples Republic of China produces approximately 3% (25,000 t) of the world's total green broad bean harvest of 0.9 million t. A further 7% (71,000 t) is harvested in Taiwan.

Grasspea

Grasspea or khesari (*Lathyrus sativus* L) has declined in area because of adverse publicity and administrative restrictions related to its content of the neurotoxin B-N oxalyl-L- B- , diaminopropionic acid (ODAP), which can cause a paralysis of the lower limbs in humans. The crop still occupies 32% (239,000 ha) of the total pulse area in Bangladesh (Sarwar, 1995), where approximately 174,000 t (5 year average 1987/91) is produced. In India it is mainly grown in Madhya Pradesh (563,000 ha) but a sizeable area is also cultivated in Bihar (242,000 ha). Grasspea is a hardy crop, giving satisfactory yields when grown under zero-tillage and rainfed conditions (*utera*) in a relay system with rice (Sharma, 1995).

Lupin

Australia is the only country in the region with a substantial production of lupins, mostly of the narrow-leafed species (*Lupinus angustifolius*) (Table 1). Its history and development are comprehensively covered by Hamblin *et al.* (1998).

CONSUMPTION

The Asia-Pacific region in its broadest sense, has a population of over 3 billion people. Nearly half of them live in the Indian subcontinent where about 20% of the people are vegetarians and pulses are the major source of protein in the diet (Paroda, 1995). According to FAO and WHO, one gram of protein is required per day for each kg of body weight to maintain proper growth. Daily consumption of pulses in China and Nepal was reported to be at 15 g (Gai and Jin, 1995) and 26 g per capita (Neupane, 1995) respectively. In India the daily intake has declined over the last four decades (1951-1991) from 60 g to 40 g (Paroda, 1995). This is mainly due to the stagnant production coupled with increases in the human population.

In South Asia, the per capita availability of chickpeas has declined from 6.9 g/day (av. 1975-1977) to 4.8 g/day (av. 1989-1991), a 30% decline over 15 years (Ryan, 1995). This follows closely the trend in India where the stagnant pulse production and negligible imports (until recently) have combined with a 2.1% annual growth rate in population (Ryan, 1995). The decline in availability of pulses in India by about 1.2% per year since 1970 has been caused largely by the 32% decline in availability of chickpea per capita from 24 g/day to 16 g/day (annual growth rate -2.5%) (Kelley and

Parthasarathy Rao, 1996). The real price of chickpeas rose by 1.9% per year throughout the 20-year period (1970-1990) compared to 0.8% for lentils. This significant price increase for chickpeas was caused largely by the decline in its production and per capita availability. As a result, consumers have turned to other pulses and other foods. As their incomes goes up, consumers spend a proportionally higher share of their income on pulses other than chickpeas. In addition, consumers buy proportionally less chickpeas than other pulses for equivalent price increases.

Daily consumption of pulses varies with cultural and religious custom. In South Asia, consumption, especially of chickpeas, increases during the Muslim holy fasting month of Ramadan and the Hindu festivals such as Holy, Deepawali and Dassara. The form in which they are consumed varies between countries (Table 2). Pulses are used mainly as human food in South Asia, and are consumed as whole grain, dehulled and split (dhal) or as flour. In China, faba beans are used for human food and animal feed, while in Australia the pulses are mainly used as stockfeed.

Faba beans lend themselves well for use as fodder and feed. However in some countries, there is a lack of local industries, i.e. livestock, feed and food processing facilities, to take advantage of such value-adding opportunities. This absence of infrastructure is seen as disincentive to the cultivation of the crop in China (Lang *et al.*, 1993b. On the Indian sub-continent, the grasspea, clouded by its anti-nutritional reputation for human consumption, is seen as having potential mainly as animal feed.

Table. 2 Forms of utilization as human food among different pulses in South Asia and China.

	UTILIZATION IN SOUTH ASIA
CHICKPEA	
Kabuli	Whole, boiled and green grains minced with beef and other meat
Desi	Mainly as dhal or splits (40%), whole (10%) and flour or besan (50%)
PEAS	
Dun types (yellow cotyledon)	Mainly as flour, some dhal
White (yellow cotyledon)	Mainly as flour, some dhal, some whole (roasted)
LATHYRUS	Mainly as flour (mixed with peas), some as dhal
LENTILS (RED)	Whole (10%), remaining dhal (some only dehulled, not split)
	UTILIZATION IN CHINA
FABA BEAN	Fried as snacks, fermented into sauces and pastes, noodles etc from flour

TRADE

The Asia-Pacific region appears to be meeting most of its demands for pulses from within the region (Paroda, 1995). Yet, some countries are more import oriented, eg India, Pakistan, Sri Lanka and Thailand, whereas exports from China, Nepal, Australia and New Zealand far outstrip their imports (Table 3).

Table 3 Average (1993-1995) pulse imports and exports (t) within the Asia-Pacific region and the world.

		Average imports/exports (t)					
		Chickpeas	Lentils	Dry Peas	Dry Faba Beans	Lupins	Totals
Bangladesh	Imports	35,449	2,139	9,203			46,790
	Exports						0
India	Imports	86,103	37,723	157,830	33		281,688
	Exports	1,576	15,843	634			18,054
Myanmar	Imports						0
	Exports						0
Nepal	Imports			4,198			4,198
	Exports		20,059	0			20,059
Pakistan	Imports	98,054	16,789	37,124	37		152,004
	Exports		675				675
Sri Lanka	Imports	6,882	49,676	6,780			63,338
	Exports		2	7			9
Thailand	Imports		45	2,620	1,735		4,400
	Exports	6	240	8	61		315
SOUTH ASIA	Imports	226,487	106,371	217,755	1,806		552,419
	Exports	1,582	36,820	649	61		39,112
							0
CHINA	Imports	93	1,067	22,370	2,131		25,661
	Exports	577	55,092	30,309	311,596		397,574
							0
Australia	Imports	127	1,568	2,386	59	0	4,141
	Exports	161,907	447	274,195	67,831	608,544	1,112,925
New Zealand	Imports	46	12	356	1		416
	Exports	130	1,886	44,881	5		46,902
OCEANIA	Imports	174	1,580	2742	61		4,557
	Exports	162,037	2,333	319,076	67,836	608,544	1,159,827
							0
ASIA-PACIFIC REGION	Imports	226,753	109,018	242,867	3,998		582,637
	Exports	164,196	94245	350,034	379,493	608,544	1,596,513
WORLD	Imports	467,045	659,608	2,874,250	582,199	110,964	4,694,066
	Exports	442,778	738,702	3,236,160	607,383	609417	5,634,440

Source: FAOSTAT Database 1997

World trade in chickpeas is less than 0.5 million t annually (Kelley and Parthasarathy Rao, 1996). This is only 6.5% of total chickpea production. However, there is an increasing trend in world trade in chickpeas. Since 1975-77 the market has trebled and exports of chickpeas are likely to increase as production shifts to areas of greater comparative advantage as represented by the export-oriented countries of Australia and Canada (Ryan, 1995). Production in South Asia is not expected to rise from its 1989 level of 5.6 million t, against a demand for chickpeas rising 33%, well above the predicted supply level. The combination of population and income growth in South Asia with the relatively high income elasticity of demand for chickpeas, suggests the demand will continue (Kelley and Parthasarathy Rao, 1996). This is despite a gradual shift to other pulses and to meat products. High price elasticity will sustain a high level of chickpea consumption in South Asia when prices drop as a result of improved productivity. Higher productivity will have an impact in two ways, by increasing the profitability to growers and by benefiting consumers through lower prices (Ryan, 1995).

Increased export opportunities have led to an expansion of the lentil area in Nepal. This was facilitated by a decline in grasspea and chickpea production and an increase in the number of growers of lentil crops. The hills' production is mainly for home consumption, with the *terai* supplying both local customers and exports, mainly to India. The quantity exported is difficult to determine because of the open border between the two countries but is expected to increase beyond the 1993 level of almost 49,000 t. Lentils are very important to the Nepalese economy. In 1986/87 they contributed 45% of total earnings from agricultural commodities (Neupane, 1995). No lentil exports have been reported by any other country of the Indian sub-continent. In China, exports climbed from nil in the 1980s to 75,000 t in 1994. Both Australia and New Zealand began exporting lentils only a few years ago. Australia's exports are expected to increase with an expansion of the crop area to an estimated 40,000 ha in 1997.

The failure of the Pakistani chickpea crop in 1992/93 because of severe frosts, the replacement of lentils by cash crops such as sugar cane with the introduction of irrigation facilities in the *barani* areas of Punjab, and the increasing human population have all created a need for increased imports of pulses (Bashir and Malik, 1995).

MARKET INTERVENTION AND PRICE SUPPORT

In developing countries such as India, Pakistan and Bangladesh, governments can control market prices and the supply of pulses to some extent. This includes manipulation of the price paid to growers. The Pakistan Agricultural Storage and Supply Organisation (PASCO), for example, can buy pulse grain from farmers and keep it in storage. In addition, the Public Utility Store Corporations of Pakistan can procure local produce and supplement shortfalls with imports. These measures give price support to farmers and price protection for consumers. The Pakistan Government has a price support policy for chickpeas, determined by the costs associated with crop production. This policy is reviewed annually.

The prices of individual pulses fluctuate sharply during the year as observed at the wholesale market in Delhi, India (Figure 1). Such price changes can have a marked influence on a farmer's decision on whether to sow a pulse crop and which one to sow. Yet, direct government intervention in the marketing and import of pulses in the Indian sub-continent is often limited to public announcements on support prices and arrangements for procurement of grain stocks to stabilise prices. It is likely that different markets for pulses will evolve in South Asia for the middle to high income class who are prepared to pay more for high quality in terms of grain appearance, cooking time and taste. Packaged products are also gaining acceptance. The existing marketing procedures will continue for low income groups, to whom price rather than quality is the determinant factor.

DOMESTIC MARKETING CONSTRAINTS

The marketing price structure often constrains the adoption of pulses as profitable crops. A case for government intervention in developing countries can be made, where the widening gap between farm gate price and retail price to consumers becomes a disincentive to growers. Traders operating at different levels, play a dominant role in the marketing of the grain. For example, the profit share between growers and traders in Bangladesh in 1985 was estimated at 26% and 74% respectively (Sarwar, 1995). The average costs of marketing lentils in Bangladesh was estimated at 30% of the cost to end-users (Bakr, 1993). Post-harvest losses during storage and processing can significantly affect marketing arrangements and cost/price structures (Bakr, 1993).

ECONOMIC RETURNS FROM PULSES

An analysis of the data for yield, area and real prices, for the years 1970 to 1989, for wheat and chickpeas in India, showed the high growth rate in yield of wheat of 3.1% per annum more than offset its decline in price of −2.6% per annum (Kelley and Parthasarathy Rao, 1996). This translated into a 1.4% linear increase in the area sown to wheat. In contrast, there was a decline in the chickpea area despite a strong increase in price. The main reason was that chickpea yields had lagged behind those of wheat.

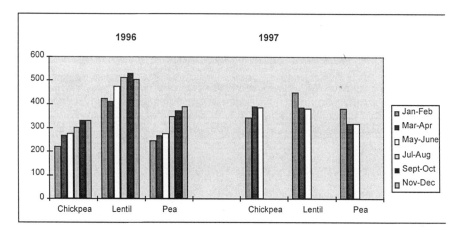

Figure 1. Fluctuations in wholesale prices of pulses in $US/t at Delhi, India during 1996 and 1997 ($US 1 = Rs 35)

The use of early maturing faba beans in China as green manure has added to their economic value. Green faba bean seed is used as a vegetable while the stems and leaves are used as green manure. In Zhejiang and Shanghai provinces 30-50% of the faba bean area is utilized for green manure, especially in the rice and cotton growing areas (Lang *et al*, 1993b). Seed size can also determine usage. Large and medium seed sizes are mostly destined for human consumption, while small seeded types are used as both food and feed in the spring-sown region and as green manure in the autumn-sown region.

The association of pulse growing in both China and South Asia with marginal land and lower profit margins compared with cash crops such as sugar cane and with cereals such as wheat and rice, is often a disincentive for growers to invest in production inputs. Yet, some studies indicate favourable benefits to costs in Pakistan and Bangladesh (Table 4).

Table 4 Benefits to costs of pulse and cereal production in Pakistan and Bangladesh

	BENEFITS TO COSTS	
CROP	PAKISTAN	BANGLADESH
Chickpea	1.83[a]	2.63[c]
Lentil	2.46[a]	2.53[c]
Lathyrus	1.60[a]	2.65[c]
Dry pea	1.44[a]	
Maize (grain)	1.11[b]	2.42[c]
Maize (forage)	1.72[b]	
Wheat	1.17[b]	
Sorghum	1.26[b]	
Millet	1.37[b]	1.31[c]

Source: a Pulses program, NARC, Pakistan 1996; b Soil sciences research unit, NARC, Pakistan 1993; c Sarwar (1995)

ENVIRONMENT AND PRODUCTIVITY

Adaptation to environments and cropping systems

Pulses in the Asia-Pacific region are grown under widely diverse growing conditions, including edaphic and climatic variations and an array of cropping systems.

For instance, the lentil growing areas in South Asia can be divided agroclimatically into the North Western Plains, North Eastern Plains, Central Highlands/Plateaux and the *Terai* region (Bahl and Sharma, 1993). Lentils in the North Western Plains are small seeded cultivars of 130-140 days maturity, with a bushy growth habit, profuse branching and a shallow root system. They grow mainly on conserved soil moisture, but yield potential can be high where irrigation (1-2 applications) is available. Lentils of the North-Eastern Plains grow under rainfed conditions after rice or on uplands after jute or other crops. They also grow in low-lying areas after monsoon/flood waters have receded. In the Central Highlands /Plateaux lentil is monocropped in *kharif* fallows. These lentils are mostly bold-seeded (3-5 g/100 seeds) with an upright growth habit and few branches, a long tap root adapted to black vertisols and early maturing. In the subtropical *terai* region near the Himalayan foothills of Pakistan, India and Nepal, lentils usually receive adequate rainfall but humidity is high and tolerance is required to severe winter frost (Bahl and Sharma, 1993).

While the pulses are grown only as monocrops in New Zealand and Australia, in South Asia and China a plethora of permutations in monocropping, double cropping, intercropping and mixed cropping can be found. Predominant crop sequences in India with lentils are (Ali *et al.*, 1993; Bahl and Sharma, 1993): monocropping in Central India; double cropping (after *kharif*) occurs as rice with lentils in double or relay methods, as fallow/lentil or after *kharif* crops such as millet, cotton, maize; mixed cropping of lentil with barley (NW plains), mustard (NE plains) or linseed (Central Plains); intercropping is also practised with autumn planted sugar cane.

The major lentil cropping patterns in Bangladesh are (Miah and Rahman, 1993):paddy rice/fallow/lentil; jute/fallow/lentil; long season rice/lentil/fallow; deep-water rice/lentil or fallow (flooded areas)/lentil. Chickpeas are also intercropped with mixtures of mustard and linseed. Time of sowing is crucial in most crop sequences and yield performance is strongly influenced by sowing date in the different cropping systems as the winter season is often short (100-110 days in Bangladesh). Delayed sowing, in particular after long-season rice, incurs a severe yield penalty.

The cool-season pulses in Pakistan are winter-grown (October- March) and are concentrated in the rainfed (*barani*) areas on poorer soils. Chickpeas are grown on sandy soils in the Thal Desert of the Punjab and in the North West Frontier Province. They are also grown in the rice growing areas of Sind and Balochistan (Bashir and Malik, 1995). Early maturing cultivars are required to suit late planting in rice-based cropping systems.

Fifteen agroclimatic zones, derived from five major soil zones and three rainfall regimes, are recognised in Myanmar (Virmani, 1996). The chickpea crop of 134 000 ha (average 1994-96) which is approximately 30% of the total pulse area, is grown mainly as a relay crop or sequential crop (100-110 days duration) after rice in the lowlands (Naing, 1995). In the uplands, it is grown mostly on soils with a good waterholding capacity after an early short-duration crop of maize or summer pulse, or after fallow. Chickpeas are sometimes intercropped with wheat. In the Ayeyarwady River delta chickpeas are sown after flood waters recede.

In the Nepal *terai* lentil, chickpea and grasspea are mostly grown as rainfed crops in a wide range of agroclimatic conditions. As pulses have the reputation of possessing a deep root system which provides drought tolerance, they are seen as a risk-reduction option and are intermixed or relay cropped with cereals and oilseeds (Neupane, 1995). Cropping patterns vary widely between the *terai*, inner-*terai* and hills and valleys.

The Chinese faba bean area can be divided into two major regions: the southern autumn-sown and the northern spring-sown regions (Lang *et al*, 1993b). Faba beans are commonly grown in the Changjiang valley and parts of south-west China in a double-cropping system where they follow rice, corn, and sweet potato in the autumn. They are also intercropped with cotton, and linseed in rainfed fields. South of the Yangtze River where the human population density is high, land is limited but where environmental conditions favour cropping, triple-cropping is common and a winter crop, such as faba bean, follows two crops of rice. The trend towards more intensive land use has required shorter duration crops to fit in with two crops of rice. The growing period of faba bean is too long and, as in many other countries, its yield is not stable or high.

In the eastern part of north-west China, faba beans are grown as a single annual spring crop in rotation with wheat, oats or sometimes maize. It is generally grown only once in every 2 or 3 years (Lang et al., 1993b, Gai and Jin, 1995). In this area, yields are comparatively stable as flowering and pod set coincide with mild temperatures, long days and good light intensity favouring seed development. The practices of mixed-cropping with barley or rapeseed, intercropping with wheat and undercrop-sowing with cotton, corn or potato can also be found in different provinces. Increases in yield per unit area can be demonstrated in comparison with single crop stands (Lang et al., 1993b).

CONSTRAINTS TO PRODUCTION

A number of reasons have been given (Paroda, 1995; Gai and Jin, 1995; Singh, 1995) for the stagnant or declining trends in pulse production and productivity in some individual countries within the region. Most pulses (>90%) are grown as rainfed crops in marginal areas, which are often poor in resources and do not attract much support in research and development. As pulses constitute less than 3% of food grains in the Asia -Pacific region, they tend to be of a lower priority in government policies. Only in India do pulses account for a greater proportion at about 8% (Paroda, 1995). There is a multitude of different pulse crops, resulting in limited support for any one crop. Cultivars with a high yield potential are still lacking for each species. Cultivars with a greater tolerance to stress have been released but the harvest index of pulses in most areas is still only 15-20% compared to 25-50% among high yielding wheat and rice cultivars. Furthermore, there is a lack of early maturing cultivars to fit the late planting regime imposed by the preceding rice crops in many countries.

Adoption of new technology is lagging. Seed of good quality is in short supply and as a consequence, farmers do not replace their seed stocks often enough. Diseases and pests still exact a heavy toll and integrated pest management systems are not well developed. Pulses tend to show a lack of responsiveness to higher inputs, such as fertilizer (P), and present a risk because of their low yield stability. There is inadequate training in extension and technology transfer is not effective. In addition, fluctuating prices and the absence of local industries for food and feed processing restricts the attractiveness of pulse crops.

BIOTIC CONSTRAINTS

Diseases

Diseases and insects affect the productivity and quality of cool season pulses. Their low and unstable yields are often attributed to biotic stresses. Major emphasis needs to be placed on managing the chickpea diseases, Ascochyta blight (*Ascochyta rabiei*), Fusarium wilt (*Fusarium oxysporum* f.sp. *ciceri*), dry root rot (*Rhizoctonia bataticola*), Botrytis grey mould (*Botrytis cinerea*) and stunt virus in South Asia (Beniwal et al., 1996, Ryan, 1995). Fusarium wilt and wet root rot (*Rhizoctonia solani*) are important in Myanmar. Collar rot (*Sclerotium rolfsii*) is a problem on the Indian-subcontinent. In Australia, *Phytophthora megasperma* f.sp. *medicaginis* and luteoviruses can cause major losses in chickpeas in northern New South Wales and Queensland (Siddique and Sykes, 1997).

In lentils, Ascochyta blight (*Ascochyta lentis*), Fusarium wilt (*Fusarium oxysporum* f.sp. *lentis*), rust *(Uromyces fabae)* and Botrytis grey mould (*Botrytis cinerea*) are seen as major problems in Pakistan (Bashir and Malik, 1995). Control of Fusarium wilt, collar rot (*Sclerotium rolfsii*), rust (*Uromyces fabae*) and powdery mildew (*Erysiphe polygoni*) is a high priority in India (Agrawal et al.,

1993). Control of Fusarium wilt is also a high priority in Nepal (Neupane, 1995). The need for resistance to rust and resistance or tolerance to Stemphylium blight (*Stemphylium* sp.) in lentils in Bangladesh has been emphasised by Sarwar (1995).

Yield losses in the region have been reported in peas from powdery mildew (*Erysiphe pisi*), downy mildew (*Peronospora viciae*), Ascochyta blight complex (*Mycosphaerella pinodes*, *Phoma medicaginis* var. *pinodella* and *Ascochyta pisi*) and rust (*Uromyces* sp.) (Sharma, 1995, Hamblin *et al.*,1998). In New Zealand Aphanomyces root rot (*Aphanomyces* sp.) limits the growing of field peas in rotation to once every six years. Grass pea is able to withstand biotic stresses well but powdery mildew (*Uromyces* sp.) and downy mildew (*Peronospora viciae*) can still be observed at a late growth stage.

Rust (*Uromyces viciae-fabae*), chocolate spot (*Botrytis fabae*) and Ascochyta blight (*Ascochyta fabae*) are common diseases of faba beans wherever they are grown in the region, whereas zonate spot (*Cercospora zonata*) and root rot (*Fusarium solani*) are a problem in China (Lang *et al.*, 1993b).

Brown leaf spot (*Pleiochaeta setosa*) which causes both root rot and leaf spotting in lupins has been the major disease in Australia but more recently (Hamblin *et al.*, 1998) the outbreak of anthracnose (*Colletotrichum gloeosporioides*) has been of great concern to growers of both narrow-leafed and albus (*Lupinus albus*) lupins.

Insect pests

There is only limited information on the economic importance of insect damage in South Asia (Weigand, 1996). For the most important pest, the pod borer (*Helicoverpa armigera*), the information relates to pod damage but the associated yield losses are difficult to quantify. The main insect pests of faba beans in China are the faba bean beetle (*Bruchus rufimanus*), aphids and root nodule weevil (*Sitona amurensis*). In Australia, pod borer (*Helicoverpa* spp.) attacks all pulses while peaweevil (*Bruchus pisorum*) control is essential in most pea crops. Other *Bruchid* species cause major storage problems in most South Asian countries.

The lack of early maturing cultivars, where pulse crops follow rice, (Johansen *et al.*, 1996) is also seen as a constraint to production, while anti-nutritional factors limit the cultivation of grasspea.

ABIOTIC CONSTRAINTS

Ryan (1995) estimated that economic losses from abiotic constraints in chickpea production exceed losses caused by biotic constraints. Drought was a major abiotic constraint in South Asia, where terminal drought was the prime manifestation. Its severity increased sharply towards the lower latitudes (Johansen *et al.*, 1996; Virmani, 1996). However, waterlogging is of particular importance in Bangladesh. In the subtropics of South Asia soil water levels, which are close to field capacity, can cause excessive vegetative growth resulting in lodging and the spread of foliar disease (Johansen *et al.*, 1996).

Frost caused major damage to the Pakistani chickpea crop in 1992/93. In Oceania and South Asia freezing temperatures are not so frequent as to cause plant death. Night frosts or extremely hot days (>30^0 C) do occur in Australia during spring when pea crops are particularly vulnerable to flower, pod and ovule abortion (Siddique and Sykes, 1997). Autumn sowing of peas can be risky in New Zealand because of late spring frosts at flowering time. Sub-optimal temperatures (0-10^0C) during the early reproductive phase can prevent, or delay, pod set in chickpeas (Johansen *et al.*, 1996). To alleviate the problem, the vegetative phase could be extended, but this causes the reproductive phase to be postponed into conditions which are more favourable for insect pests (*Helicoverpa*) and foliar diseases in South Asia. It also exposes the crop to terminal heat and drought stress towards maturity in most areas.

NEW TECHNOLOGY

Government initiatives in research, development and extension services

Australia and New Zealand

Pulse research and development were, and still are, led by government agencies in New Zealand and Australia (Hamblin *et al*., 1998; Walgott and Rees, 1995). More recently, private organisations such as grower groups, seed companies and food processors, have recognized opportunities to take on research into crop management, cultivar evaluation, processing quality and market development. In South Asia and China this trend towards private interests is not as discernible and government initiatives continue to provide the overriding stimulus.

India

In India the Directorate of Pulses Research was upgraded in 1993 into the Indian Institute of Pulses Research (IIPR) under the direction of the Indian Council of Agricultural Research (ICAR). IIPR conducts and monitors all pulse research across seven divisions or disciplines but is mainly focused on basic research. Nationally, co-ordinated projects contribute towards the regional development of cultivars and of technologies for production and plant protection (Singh, 1995). All India Coordinated Research Projects, relevant to pulses, are spread across different agroclimatic zones and provide improved technology in the form of superior cultivars and crop management (Singh, 1995).

Pakistan

A systematic approach to improvement of pulse production in Pakistan has brought together activities at the national (NARC, NIAB and NIFA) and provincial level (Punjab, NWFP, Sind and Balochistan) with the main focus on the rainfed areas (Bashir and Malik, 1995). However, responsiveness to irrigation, rhizobium inoculation and other inputs are also targeted. The impact of improved cultivars of chickpeas and lentils and improved management techniques is expected to be noticeable within the next few years resulting in increased productivity levels (Bashir and Malik, 1995).

Nepal

The National Agricultural Council (NARC) was made an autonomous body in 1991 with the Grain Legumes Research Program (GLRP) responsible for cultivar development, plant protection, agronomic and nutritional research and post-harvest technology (Neupane, 1995). Interaction has been strengthened with ICARDA, ICRISAT and more recently in a collaborative project sponsored by the Australian Centre for International Agricultural Research (ACIAR). This links lentil research in Nepal with similar research at ICARDA, in Pakistan and in Australia.

Bangladesh

Under the Crop Diversification Program (CDP) of the government of Bangladesh the Pulses Research Center (PRC) at the Bangladesh Agricultural Research Institute (BARI) has the objectives of developing cultivars of seven crops, including grasspea, lentil, chickpea and field pea, generating improved management packages for different agroecological zones and post-harvest technologies for grain storage (Sarwar, 1995). Inter-institutional linkages exist between PRC and BINA (Bangladesh Institute for Nuclear Agriculture), BAU (Bangladesh Agricultural University), IPSA (Institute of Post-Graduate Studies in Agriculture) and Chittagong University. The On-Farm Research Division (OFRD) of BARI and the Directorate of Agriculture Extension are responsible for promoting the adoption of local technology by farmers.

China

In the Peoples Republic of China the Institute of Plant Genetic Resources of the Chinese Academy of Agricultural Sciences is responsible for organising the national research program covering all pulse crops (Gai and Jin, 1995). This work is in collaboration with provincial and local academies, agricultural institutes and universities. Extension services covering cultivar replacement, seed distribution, advice on cropping systems and pesticide use are provided mainly through government advisory stations. Many new cultivars of both faba bean and field pea have been released in different provinces by collaborating institutions (Gai and Jin, 1995). High experimental yields of faba bean of 9.7 t/ha in Qinghai province have been obtained with new cultivars (Lang *et al.*, 1993b). The Shaoxing Institute of Agricultural Sciences in Zhejiang has achieved a good adoption by farmers of a management system for faba bean undercropped with late rice (Lang *et al.*, 1993b).

International linkages

Regional networks have been established such as the Cereals and Legumes Asia Network (CLAN) and Food Legumes and Coarse Grains Network (FLCGNET), which bring together scientists, administrators and institutions committed to solving problem in Asia. An example is the Working Group on Botrytis Grey Mould of Chickpeas (Faris, 1993). Crop management research and genetic material generated by ICARDA and ICRISAT have contributed to pulse improvement in the region, with many cultivars having been released following major inputs from these two organisations.

FUTURE

Supply and demand

By the year 2000 China's total requirement for pulse grain is estimated to be 22.5 million t, double its current production (Gai and Jin, 1995). The availability of cereals and pulses in India may reach approximately 460 and 70 g per capita per day respectively by that time. This means a total demand for pulses in India of 24.7 million t by the end of the century (Singh, 1995). The availability of pulses in Pakistan has stagnated at 16g per day. The country's pulse production would have to increase three-fold if it were to provide the two-fold per capita availability required by the projected population growth (Bashir and Malik, 1995). Similar predictions can be made for Bangladesh (Sarwar, 1995).

Productivity

Ryan and co-workers (Ryan, 1995) have provided estimates of potential gains to be made in alleviating region-specific biotic and abiotic constraints to productivity from the application of either genetic crop improvement or improved crop management techniques. These estimates can serve as useful guidelines in setting priorities for research and extension programs.

Some of the traits required to increase chickpea productivity in South Asia are an ability to escape the terminal drought stress by early maturity, without incurring a yield penalty, to set pods at sub-optimal temperatures in sub-tropical winters and to be tolerant to heat at the pod-filling and seedling stage (Johansen *et al.*, 1996). For instance, the availability of a cold-tolerant chickpea that could mature by the end of February in India, Pakistan, Bangladesh and Nepal, may be a way to avoid foliar disease as the low temperatures ($15\text{-}20^0$ C) prevailing at that time would not favour epiphytotics (Beniwal *et al.*, 1996).

Insect control by synthetic chemicals is common in Australia but is not a viable option in many areas of South Asia and China, where bio-insecticides such as the neem seed extract may offer a better solution (Weigand, 1996).

There is ample scope for increasing farmer yields in the region. In Bangladesh there is a large discrepancy between farmer's yields using current practices (0.5 – 0.7 t/ha) and improved technology (1.0 – 1.6 t/ha). If improved technology were applied to the present area of production, an estimated 879,000 t could be achieved, which is 49% of the demand of 1.81 million t expected in Bangladesh by the year 2000 (Sarwar, 1995). "Front Line Demonstrations", a scheme launched by ICAR in India during 1991-1992, has indicated yield advances averaging 73%, would occur if new cultivars replaced local types of *rabi* (chickpeas, lentils, peas) pulses in farmer's fields (Singh, 1995). The technology appears to be available, but adoption is lagging, often due to socio-economic reasons and insufficient funds for the extension services.

Area expansion

There is scope for expanding the chickpea production area. In India, an additional 3 million ha could be used for late-sown chickpeas in the north-eastern plains in rice fallows (Singh, 1995). Likewise, opportunities exist in Myanmar to expand the chickpea area by including pulses, to a greater extent, in the cereal-based systems in the higher and medium rainfall zones and in Lower Myanmar if cultivars were available with tolerance to high temperatures and acid soils (Naing, 1995; Virmani, 1996). In addition, pulses could grown in a parts of the high *barind* tract in Bangladesh, where 160,000 ha remain fallow in winter (Sarwar, 1995).

STRATEGIES TO MEET THE CHALLENGE

The challenge is to ensure adequate nutrition for the region's population, while providing sufficient profitability to sustain the region's farming systems. The following strategies are pertinent (Gai and Jin, 1995; Singh, 1995; Ryan, 1995).
–   Provide incentives such as loans to allow the purchase of inputs in the developing countries of Asia. The wide gap between farm-gate and consumer price also needs to be addressed.
–   Develop national strategies aimed at strengthening R&D.
–   Promote the use of pulses to change narrow based rotations (eg rice-fallow, rice-wheat, intercropping with sugar cane or cotton) and provide greater sustainability.
–   Promote the adoption of improved technology, including higher inputs and mechanisation.
–   Provide incentives to establish processing of pulses at a local level to stimulate domestic and international demand and increase economic value.

References
Agrawal, S.C., Singh, K. and Lal, S.S. 1993. In: *Lentil in South Asia, Proceedings of the seminar on lentil in South Asia,* 11-15 March 1991, New Delhi, India, 147-167 (Eds. W. Erskine and M.C. Saxena), ICARDA.
Ali, M., Saraf, C.S., Singh, P.P., Rewari, R.B. and Ahlawat, I.P.S. 1993. In: *Lentil in South Asia, Proceedings of the seminar on lentil in South Asia,* 11-15 March 1991, New Delhi, India, 103-127 (Eds. W. Erskine and M.C. Saxena), ICARDA.
Bahl, P.N., Lal, S. and Sharma, B.M. 1993. In: *Lentil in South Asia, Proceedings of the seminar on lentil in South Asia,* 11-15 March 1991, New Delhi, India, 1-10 (Eds. W. Erskine and M.C. Saxena), ICARDA.
Bakr, M.A. 1993. In: *Lentil in South Asia, Proceedings of the seminar on lentil in South Asia.* 11-15 March 1991, New Delhi, India, 195-205 (Eds. W. Erskine and M.C. Saxena), ICARDA.
Bashir, M. and Malik, B.A. 1995. In: *Production of pulse crops in Asia and the Pacific region,* 169-190 (Eds. S.K. Sinha and R.S. Paroda), RAPA/FAO Publication No. 1995/8, Bangkok.
Beniwal, S.P.S., Haware, M.P. and Reddy, M.V. 1996 In: *Adaptation of chickpea in the West Asia and North African region* 189-198 (Eds. N.P.Saxena, M.C. Saxena, C. Johansen, S.M. Virmani and H. Harris), ICRISAT/ICARDA.
Faris, D.G. 1993. In: *Lentil in South Asia, Proceedings of the seminar on lentil in South Asia,* 11-15 March 1991, New Delhi, India, 220-231 (Eds. W. Erskine and M.C. Saxena), ICARDA.
Gai, J. and Jin, J. 1995. In: *Production of pulse crops in Asia and the Pacific region,* 93-100 (Eds. S.K. Sinha and R.S. Paroda). RAPA/FAO Publication No. 1995/8, Bangkok.
Hamblin, J., Hawthorne, W.A. and Perry, M. and 1998 *Regional Review: The Australian scene. In: Proceedings International Food Legume Research Conference,* (Ed. R. Knight), Kluwer Academic Publishers.
Johansen, C., Saxena, N.P. and Saxena, M.C. 1996. In: *Adaptation of chickpea in the West Asia and North Africa region,* 181-188 (Eds. N.P.Saxena, M.C. Saxena, C.Johansen, S.M. Virmani and H. Harris) ICRISAT/ICARDA.
Kelley, T.G. and Parthasarathy Rao, P. 1996. In: *Adaptation of chickpea in the West Asia and North Africa region,* 239-254 (Eds. N.P.Saxena, M.C. Saxena, C.Johansen, S.M. Virmani and H. Harris) ICRISAT/ICARDA.
Lang, Li-juan, Zheng, Zhuo-jie and Hu, Jia-peng 1993a. In: *Faba bean production and research in China. Proceedings of an international symposium,* 24-26 May 1989, Hangzhou, China, 1-18 (Eds. M.C. Saxena, S. Weigand and Lang, Li-juan). ICARDA.

Lang, Li-juan, Yu Zhao-hai, Zheng Zhao-jie, Xu Ming-shi and Ying Han-qing 1993b. *Faba bean in China: state-of-the-art review,* ICARDA, 144pp.

Materne, M.A. and Brouwer, J.B. 1996. In: *New Crops, New Products. Proceedings of the First Australian New Crops Conference,* University of Queensland, Gatton College, 8-11 July 1996, 45-52 (Eds. B.C. Imrie *et al.*).

Miah, A.A. and Rahman, M.M. 1993. In: *Lentil in South Asia, Proceedings of the seminar on lentil in South Asia,* 11-15 March 1991, New Delhi, India, 128-138 (Eds. W Erskine and M.C. Saxena), ICARDA.

Naing, T. 1995. In: *Production of pulse crops in Asia and the Pacific region,* 141-145 (Eds. S.K. Sinha and R.S. Paroda). RAPA/FAO Publication No. 1995/8, Bangkok.

Neupane, R.K. 1995. In: *Production of pulse crops in Asia and the Pacific region,* 147-162 (Eds. S.K. Sinha and R.S. Paroda). RAPA/FAO Publication No. 1995/8, Bangkok.

Paroda, R.S. 1995. In: *Production of pulse crops in Asia and the Pacific region*, 1-13 (Eds. S.K. Sinha and R.S. Paroda). RAPA/FAO Publication No. 1995/8, Bangkok.

Ryan, J.G. 1995. In: *Production of pulse crops in Asia and the Pacific region*, 225-248 (Eds. S.K. Sinha and R.S. Paroda). RAPA/FAO Publication No. 1995/8, Bangkok.

Sarwar, Ch.D.M. 1995. In: *Production of pulse crops in Asia and the Pacific region*, 77-91 (Eds. S.K. Sinha and R.S. Paroda). RAPA/FAO Publication No. 1995/8, Bangkok.

Sharma, B. 1995. In: *Sustaining crop and animal productivity. The challenge of the decade.* 101-130 (Ed. D.L. Debb), Associated Publishing Co., New Delhi.

Siddique, K.H.M. and Sykes, J. 1997. *Australian Journal of Experimental Agriculture* 37: 103-11.

Singh, D.P. 1995. Pulses in India. In: *Production of pulse crops in Asia and the Pacific region,* 101-112 (Eds. S.K. Sinha and R.S. Paroda). RAPA/FAO Publication No. 1995/8, Bangkok.

Virmani, S.M. 1996. In: *Adaptation of chickpea in the West Asia and North Africa region,* 13-19 (Eds. N.P. Saxena, M.C. Saxena, C. Johansen, S.M. Virmani and H. Harris), ICRISAT/ICARDA.

Walgott, J. and Rees, R.O. 1995. In: *Production of pulse crops in Asia and the Pacific region,* 61-76 (Eds. S.K. Sinha and R.S. Paroda). RAPA/FAO Publication No. 1995/8, Bangkok.

Weigand, S. 1996. In: *Adaptation of chickpea in the West Asia and North Africa region,* 199-205 (Eds. N.P. Saxena, M.C. Saxena, C. Johansen, S.M. Virmani and H. Harris), ICRISAT/ICARDA.

Regional reviews: The Australian Scene

John Hamblin,[1,] Wayne Hawthorne,[2] and Michael Perry,[1,3]

1. The Cooperative Research Centre for Legumes in Mediterranean Agriculture, University of Western Australia, Nedlands, Western Australia 6907 ; 2. South Australian Research and Development Institute, P.O. Box 618, Naracoorte South Australia 5271 ; 3. Agriculture Western Australia, Baron-Hay Court, South Perth, Western Australia 6151

Abstract

This paper examines the changing situation of pulse production in Australia over the last 30 years and projections to the year 2005. During the 30 years there has been a rapid change in the industry. In 1967 pulse production was 30,000 tonnes whilst in 1996 some 2,300,000 tonnes were produced. It is anticipated that between 3,500,000 and 4,000,000 tonnes will be produced in 2005. For most of the last 30 years *Lupinus angustifolius* has been the most important pulse. Australia is the major producer and the only country to export lupins into world markets. Over the last decade we have seen expansion in production of peas, chickpeas, faba beans and lentils as well as lupins. The production of these alternative pulses is now some 1,000,000 tonnes and should double by 2005. We predict Australia will be the largest pulse exporter in the world by year 2005.

INTRODUCTION

If this conference had been convened thirty years ago it is unlikely we would have met in Australia. At that time cool season pulses were not important in this country, as the total area was some 30,000 ha of which virtually all were peas grown in the states of South Australia and Victoria (Hamblin 1987). By chance 1967 was the year of the release of Uniwhite, the first sweet (low alkaloid) narrow-leafed lupin (*L. angustifolius*) variety bred in Western Australia by Dr John Gladstones the father of the Australian lupin industry (Gladstones 1994). Twenty years ago in 1977 the area of cool season pulses was about 130,000 ha of which some 90,000 ha (70%) was lupins (Hamblin 1987, Siddique and Sykes 1997). In a decade, lupins had become the dominant pulse in Australia.

Ten years later in 1987 and only a decade ago the area of cool season pulses was 1.5 million ha of which just over 1.0 million ha was lupins (67%). Lupins and the other cool season pulses had increased 10 fold in a decade. Two crops, new to Australia, had become established and significant areas of chickpeas and faba beans were grown. Commercial chickpea production commenced in 1979 with the identification of a suitable, specific strain of inoculum (Corbin et al 1977), the opening of Indian food markets, and the release of the first adapted desi varieties Tyson in the north (Beech and Brinsmead 1980) and Dooen in the south (Anon 1987). The release of the variety Opal (Anon 1980) enabled a small kabuli industry to develop. Faba beans became commercially significant after the identification of their potential (Laurence 1979) and the release of the first locally adapted variety, Fiord, in 1980 (Knight 1994). Large seeded or broad beans became a smaller but significant industry in high rainfall areas of south-eastern Australia from about 1986 when the locally adapted variety Aquadulce was grown commercially by farmers, rather than just being grown in gardens (Hawthorne 1995a).

Last year we had about 2 million ha of pulses and produced some 2.3 million tonnes. Lupins are still dominant at 1.414 million tonnes (62%), but each decade their share of the total production has fallen. In the last decade commercial lentil production has become significant with the release of ICARDA derived, locally adapted red lentil varieties Cobber (Brouwer 1995a), Digger (Brouwer 1995b) and Aldinga (Ali 1995) in 1993 and 1994.

Thus, although Adelaide, Australia in 1967 would have been an unlikely venue for this conference, it is highly appropriate that it is the site for the meeting in 1997. It is also appropriate because as we look to the future we expect the area of cool-season pulses in Australia will continue to grow and that by 2005 AD we will grow more than 3.6 million tonnes (lupins 56%).

In thirty years Australia has gone from being a non-player to being the second largest exporter in the world. If we achieve our 2005 target, then Australia will be by far the major exporter of pulses in the world. By then we will be exporting 3 species of lupins (*Lupinus angustifolius,* the major species, *L. albus,*

131

R. Knight (ed.), Linking Research and Marketing Opportunities for Pulses in the 21st Century, 131–142.

currently under threat from anthracnose, and *L luteus,* released to farmers in 1997) as well as significant quantities of dry peas, chickpeas, faba beans and lentils. These figures do not include possible production from other Vicia (*V. narbonensis, V. sativa*), Lathyrus (*L. sativus, L. cicera*) or from the rough seeded lupins (*L. atlanticus, L.pilosus*), all of which will probably have a small but locally significant role by 2005. Research is being undertaken on these species at several locations.

Past, present and future production

Australia produces its cool season pulses from three main areas (Figure 1). The south-west area, with large areas of acidic, sandy soils with a Mediterranean environment of winter rainfall, and is exclusively in the state of Western Australia. The south-eastern area is predominantly alkaline, heavy to sandy soils and a Mediterranean environment of winter rainfall, and includes parts of the states of South Australia, Victoria, the southern areas of New South Wales and Tasmania. The north-eastern area is predominantly alkaline, heavy soils with a semi tropical environment, hence minimal winter rainfall. It includes the northern areas of New South Wales and the southern areas of Queensland.

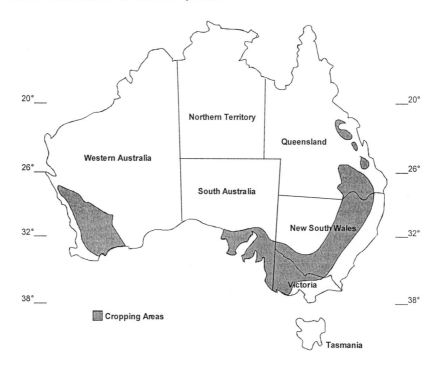

Figure 1. The major pulse cropping areas of Australia

There is however a wide range of soils, rainfalls and environments within each major area. The change in total area sown to pulses and the area sown to lupins over the last 30 years and the projection to 2005 from the Australian Pulse Industry Strategic Plan (Anon 1995 a and b) are shown in Figure 2. This clearly demonstrates the importance of lupins. If we achieve the 2005 objectives, we will need to double production from peas, lentils, chickpeas and faba beans and have a significant (50%) increase in lupin production.

Production in1985, currently and the projection for 2005 in Australia is given in Table 1 by crop and state. It is apparent that the area has expanded dramatically in the last decade, from some 830,000 tonnes in 1985 to nearly 2.3 million tonnes in 1995, an increase of 176% in 11 years.

133

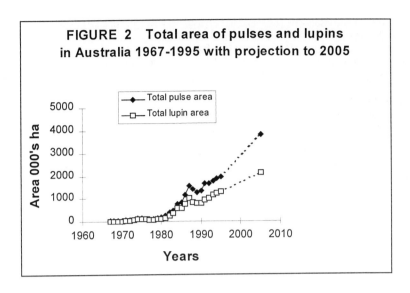

Table 1. The production ('000t) of cool season pulses grown in Australia in 1995, the production in 1985, the percentage increase from 1985 - 1995, two estimates of predicted tonnage in 2005 and the percentage increase expected between 1995 and 2005 (Data from ABARE and Anon 1995 a and b).

Tonnage in 1995			Crop			
State	Lupins	Chick peas	Faba beans	Dry peas	Lentils	Total
New South Wales	110	40	21	23	0.3	148.3
Victoria	45	170	40	205	35	375
Queensland	0	8	0	0	0	40
Western Australia	1200	25	24	28	0.3	1210.3
Southern Australia	72	15	34	208	2.5	292.5
Australian Total	**1417**	**258**	**119**	**464**	**38.1**	**2296.1**
Tonnage in 1985	**588**	**26**	**19**	**210**	**0**	**844**
% increase 1985-95	241	992	626	221	infinite	272
Estimate for 2005[1]	**2150**	**490**	**250**	**900**	**20**	**3810**
% increase 1995-2005	52	90	110	94	-48!	84
Estimate for 2005[2]	**1498**	**756**	**407**	**560**	**127**	**3348**

1 (Anon 1995) 2 Hamblin (pers.comm)

In 1994 the pulse industry initiated a Pulse Strategic Plan developed by the Meyer Strategy Group for the Grains Council of Australia and the Grains Research and Development Corporation (Anon 1995 a and b). It predicted that over the next 11 years from 1995 to 2005, there would be a further increase to about 3.8 million tonnes of cool season pulses. This is a highly significant increase, but in percentage terms will be less than occurred in the previous two decades. There are some sectors of the grains industry which doubt whether the prediction for year 2005 is achievable. In a recent re-evaluation, our current estimate for 2005 is some 3.6 million tonnes. Both this target and the strategic plan target highlight confidence in a controlled expansion of the industry which has set high quality and production targets.

Besides the Australian Pulse Industry Strategic Plan (Anon. 1995b), overviews have been produced recently for pulses as a whole (Richards 1995, Siddique and Sykes 1997) and for individual species; lupins (Grain Pool of WA 1995, Perry et al. 1994), field peas (Smith and Mahoney 1995), faba beans (Hawthorne 1995a), chickpeas (Knights 1995) and lentils (Hawthorne 1995b, Materne and Brouwer 1997).

Two states, Victoria and Western Australia, are currently the major producers of pulses (Table 1) and this is likely to continue. Because lupins are adapted to sandy and acidic soils, where other pulses perform poorly, they will continue to be the species with the greatest production. Significant production of chickpeas, faba beans and dry peas will occur on suitable heavier alkaline soil types. The pulse industry strategic plan in 1994 (Anon 1995 a and b) expected that lentils would reach a total production of 20,000 tonnes by 2005, however this was achieved in 1995. It is likely that lentil production will be significant by 2005 (we estimate127,000 tonnes), but probably will be less than that of other cool season species.

Pulse crops in Australia have expanded to environments outside of their initial introduction. Chickpea commenced as a north-eastern, warm season crop, but is now also widely grown in southern, Mediterranean climates. Faba beans have done the opposite and spread north. Lupins have also spread from the acid sands of the west to heavy but acidic soils in eastern Australia. Hence a range of pulse crops are grown across Australia which differ in their specific adaptation to different soil types, rainfall and market demands. This is illustrated in Figures 3 & 7 which show where pulses are produced in Australia. Considerable effort has gone into expanding the adaptation of pulse crops in Australia.

CONSUMPTION

Pulse consumption as food in Australia is estimated at only 6 g/head per day, despite there being a large immigrant population of southern European, middle-eastern and Asian origin (FAO 1994, Figure 8). This is half the level consumed in the United Kingdom and a fifth of Indian consumption and is approximately 40,000 tonnes (1.7% of production). It is an objective of the Cooperative Research Centre for Legumes in Mediterranean Agriculture (CLIMA) to increase this amount by 50% to 60,000 tonnes in 2005. However if all targets for 2005 are met, internal consumption of pulses for food will still have fallen by some 0.2% of total production. Pulses for animal feed is a more significant use of local production. This occurs on farms, mainly as sheep and cattle feed supplements at critical periods of the year, and as components of rations in intensive animal industries. The relationship between production, exports and residual (domestic use) is shown in Table 2.

Table 2. Average Australian production, exports and the residual (000 t) for 1989-1993 (Data from ABARE)

Year	1989	1990	1991	1992	1993
Production	1620	1339	1353	1857	2041
Exports	867	792	730	930	1447
Residual	753	547	623	927	594

If human food consumption of pulses in Australia is 40,000 t/annum then on average over the last 5 years some 650,000 tonnes of pulses have been used as seed or feed on farms, or sold as animal feed to intensive livestock industries. As pulse production increases in areas where beef feed-lotting occurs, then local consumption of lower value and lower priced pulses will increase. This is unlikely to be important in Western Australia in the period to 2005 as feed-lotting will only be a small component of that state's animal production however increases are likely to occur in Queensland and New South Wales.

The actual local consumption in different regions will depend on the types of pulses grown, their price competitiveness for food or feed and the value of alternative activities on farm.

IMPORTS

Even if all the cool season pulses consumed as food in Australia were imported, it would be only a small proportion of production (1.7% see above). ABARE (1996) gives no values for pulse imports into Australia. Although surprising, it means that imports are extremely low and are likely to decline as a wider range of species and varieties are grown locally. With new varieties, improved agronomic practices and quality assurance a wider range of quality products will become available locally.

EXPORTS

Table 2 shows exports from Australia during the period 1989 to 1993, averaged 952,000 t/annum (ABARE 1996a). This has increased over the last three years to an average of 1,185,000t (ABARE 1996b). Over the period 1989-93, Australia ranked second to France (average exports 1,183,000 t / annum) and China third, with 762,000t.

If predictions are correct, it is likely Australia will produce 3.6 million tonnes of cool season pulses by 2005, locally consume about 1 million tonnes (food 60,000t, feed and seed 940,000t) and export some 2.6 million tonnes. Australia's exports will then be significantly more than the combined exports of the next two most important nations, likely to be China and France.

FACTORS INFLUENCING PULSE PRODUCTION IN AUSTRALIA FROM NOW TO 2005.

The factors influencing pulse production and how to overcome them were outlined in the pulse industry strategic plan (Anon. 1995a and b) and some of the difficulties, faced by farmers trying to produce what the markets want, were outlined by Hawthorne (1993). Over the next 10 years, the factors likely to affect pulse production include price, varieties adapted to new areas and management systems that lead to improved yield, stability and quality. An improved understanding of the role of pulses in rotation will provide a better estimate of their impact on the yield of following crops. Also needed is an efficient system of technology transfer between scientists, farmers and marketeers. Information on the market requirements of different countries will be required, along with better marketing of the crop. Focused research and development aimed at meeting these requirements will be driven by farmers. Funding will come from State Governments and the farmers themselves through the levy system operated by the Grains Research and Development Corporation on their behalf.

Price

There is no price support system for Australian pulses, farmers take the price the market is prepared to pay. Price will be a determinant of future production, both the absolute price and the price relative to other farming options such as other crops and livestock. In areas where several pulses can be grown, their ease of management, yield and relative price will determine which are grown.

Varieties

Breeding of many pulses in Australia is a relatively new activity. The oldest continuous breeding program is that on lupins in Western Australia which started in 1954 (Gladstones 1994). Apart from lupins, most pulse production in Australia commenced with the release of overseas varieties. Subsequent progress has been with local breeding programs using local and introduced material. Breeding of all species except lupins is much more recent and only now is a continuous series of locally bred varieties appearing.

There is an increasing level of national co-operation and co-ordination within the Australian breeding and evaluation programs. This has been driven by several factors. The pulse industry although significant, will never be as important as wheat. Many of the pulse species need further research to adapt them to the range of environments in which they could be grown. No one state organisation can provide varieties for the full range of species for their state from their own resources as costs are too great (Brinsmead et al. 1991,

Hawthorne 1997). At the same time farmer funded support for research and development has insisted that there be more interstate co-operation to reduce duplication and maximise return on the dollars invested. These factors have lead to improved co-operation with different state departments taking responsibility for the breeding of different species and co-ordinating testing across all states. Recently Hamblin (pers. comm.) has estimated that approximately $18-20,000,000 per year are being invested in pulse improvement in Australia. The benefits from this investment are just beginning to show.

The key issues for the breeding programs are reliability of production; an ability to harvest the crops in mechanised systems, good disease resistance to avoid yield and quality losses and a range of development patterns to ensure varieties are adapted to a range of climatic conditions. Adaptation to soil type is a lower priority as it is usually possible to find an alternative pulse species which is adapted to the major soils types found in Australia.

There are good skills in Australia for the genetic transformation of pulses. Of the seven species being discussed we have the technology to transform all but *Lupinus albus*. This work is referred to in other papers. Once transformation is available for a species, the key issues are access to needed components of the intellectual property involved and dealing with regulatory issues in Australia and overseas. If the entry costs are too high, useful genes may be incorporated into pulses, but because of low production levels and high costs to have the material cleared for markets, the material is not used. This could be particularly unfortunate if genes become available for pest and disease control.

Pests and diseases

Most pulse diseases have been in Australia for a considerable time. They include the black spot complex, downy and powdery mildew and bacterial blight on peas, Ascochyta on lentils, chocolate spot and Ascochyta on faba beans, Phytophthora and grey mould on chickpeas, and brown leaf spot, Plieocheata root rot and Phomopsis on lupins. Cucumber mosaic virus and bean yellow mosaic virus are wide spread in lupins, and luteoviruses are important on chickpeas. Major and wide spread pests include Heliothus and red legged earth mites. Pea weevil is a major pest of peas, particularly in regions where the area has increased. Major pests and diseases of pulses in Australia and an indication of their importance are given in Table 3. The listing is not all inclusive. Some pests and diseases can be locally important but not nationaly. Recently *Ascochyta rabiei* has been confirmed on chickpeas in South Australia and Victoria. It is likely that the disease has been present for a while, but has not been devastating over wide areas. Delay in its confirmation may have been due to confusion with phoma blight which has been present for some years. Screening of current varieties has shown there is significant resistance in some of our commercial chickpea varieties. There has been a quarantine ban on material from eastern Australia entering Western Australia, which is likely to slow down the rate of flow of germplasm from the national breeding program in Tamworth in New South Wales into Western Australia whilst the situation is being assessed.

In the spring of 1996, an epidemic of anthracnose (*Colletotricum gloeosporoides*) was found in the northern agricultural zone of Western Australia initially on *Lupinus albus*. The *L. albus* variety grown is Kiev Mutant. It is extremely susceptible to this disease. Initial efforts to eliminate the disease have not been possible in Western Australia, but a much smaller, unrelated outbreak in *L. angustifolius* on the Eyre Peninsula in South Australia may be eradicated. The sources of the epidemics are unknown, but may be ornamental Russell lupins where the disease was identified a few years ago in a nursery in the Dandenong Ranges in Victoria. In 1966 the disease was found on garden plants in New South Wales, Victoria and Western Australia. Thanks to a rapid response by Agriculture WA scientists, CLIMA and GRDC, in cooperation with New Zealand scientists, it was possible to screen material over summer (1996/97) in New Zealand and significant levels of resistance to anthracnose were found in two recent releases of *L. angustifolius* and in wild material. There is a campaign under way to reduce the risk of infection by ensuring good hygiene in terms of clean seed, seed dressings, crop rotation and destruction of sources of infection. There are restrictions on sending material from the national lupin program in Western Australia to eastern Australia. The anthracnose experience is of particular significance to crop scientists. A disease may occur on garden plants and although identified, its significance for agriculture may not be appreciated.

In the last few years in northern New South Wales rust has been found to be severe on faba beans and luteo viruses on chickpeas. In all cases disease resistance is a key component of control complemented by crop hygiene.

Table 3. The main pests and disease of pulse crops

Pests / diseases	Lupins Narrow lf	Lupins white	Lupins yellow	Dry peas	Faba beans	Chick peas	lentils
Insects							
Redlegged earth mite[1]	x	x	xxx	xx	xx	ns	ns
Cutworms or brown pasture loopers	x	x	x	x	x	ns	x
Pea weevil	ns	ns	ns	xx	ns	ns	ns
Heliothis / Helicoverpa	xx	xxx	xxx	xxx	xxx	xxx	xxx
Lucerne flea	ns	ns	ns	ns	x	ns	xx
Aphids (feeding damage)	x	x	xxx	ns	x	ns	x
Virus diseases							
Bean yellow mosaic virus	xx	xx	xx	x	x	ns	x
Cucumber mosaic virus	xx	ns	xx	ns	ns	xx	xx
Lueteo viruses	ns	ns	ns	ns	xxx	xxx	xx
Pea seed-borne mosaic virus	ns	ns	ns	x	x	ns	xx
Alfalfa mosaic virus	ns	ns	ns	ns	ns	xx	xx
Fungal diseases							
Brown leaf spot	xxx	xxx	x	ns	ns	ns	ns
Pleiocheata root rot	xxx	xxx	ns	ns	ns	ns	ns
Rust	ns	ns	ns	ns	xx	ns	ns
Anthracnose	xxx	xxx	xxx	ns	ns	ns	ns
Ascochyta	ns	ns	ns	xxx	xxx	xxx	xxx
Powdery mildew	ns	ns	ns	xx	ns	ns	ns
Downy mildew	ns	ns	ns	xx	ns	ns	ns
Botrytis	x	x	x	ns	xxx	xx	ns
Sclerotinia	x	x	x	ns	x	xx	xx
Phomopsis	xx	x	xx	ns	ns	ns	ns

[1]xxx= can be a major problem   xx= can be a problem   x= can be a minor problem

ns= either the problem is not significant, it does not affect this species or the problem is not yet present in Australia or we do not have sufficient knowledge of the problem.

The information.in this table was provided by M Sweetingham, R Jones, D Hardie and J Ridsdill-Smith.

With insect pests, control is based on insecticidal sprays, however considerable effort is going into improving plant resistance. With concerns about the over use of chemicals and with the development of quality assurance systems for markets, there is an increasing emphasis on plant resistance. An example is the use of an alpha amylose inhibitor gene from Phaseolus being inserted into peas to control the pea weevil (Schroeder et al., 1998). It is likely this type of approach will increase in the future.

Improved agronomic practices

Apart from variety adaptation, which is determined by the breeders, the key to successful pulse production is agronomic packages for the different species and production areas. These packages, unlike varieties, have to be developed locally. The resources and enthusiasm for this activity vary across the nation. The experience with lupins and more recently with other pulses, suggests that unless there is a local champion for a species, it is difficult to convince extension advisers or farm consultants to accept a new species for which there is little local knowledge. Farmers and their advisers need support from scientists who have the technical knowledge to provide confidence in the suitability of the pulse species, varieties and production systems. If this support is not available the risk of failure is increased and farmers are unlikely to take up the crop. The importance of new, locally developed, proven agronomic packages to the development of lupins as a new crop is discussed by Nelson and Hawthorne(1998).

Nett returns to growers

While the decision on whether to grow a pulse, and which one, is influenced by agronomic and marketing factors, the final decision is driven by the anticipated dollar return to the grower. This may include the gross

138

margin (price by yield less costs) in the year of production and the carry over effects in subsequent crop and livestock operations. A whole farm economic analysis would be ideal, using a computer model like MIDAS (Model of Integrated Dryland Agricultural Systems) developed for lupins in WA (Kingwell and Pannell 1987). Some indicative values of simple gross margins for one region and year are given in Table 4, appreciating that yields, prices and costs vary considerably between areas and years.

Table. 4 Summary of gross margins(A$) for 1997 in the south east region of South Australia

	CROPPING ENTERPRISES			
	G.M $/hectare	Yield (t/ha)	Costs ($/ha)	Price ($/t)*
Wheat (ASW)	194	3.0	238	144
Biscuit Wheat	217	2.7	253	174
Malting Barley	226	2.8	240	167
Feed Barley	104	3.0	216	107
Oats	69	3.0	204	91
Triticale	162	3.0	231	131
Oaten Hay	283	5.0	267	110
Peas	224	1.8	226	250
Beans (broad)	325	1.8	215	300
Faba Beans	292	2.0	228	260
Chick Peas	246	1.5	234	320
Vetch	194	1.5	181	250
Lupins	146	1.6	174	200
Lentils	330	1.2	210	450
Canola	346	1.7	284	360
Safflower	141	1.0	159	300
	LIVESTOCK ENTERPRISES			
	G.M $/hectare	G.M $/animal		G.M $/d.s.e
Merino Ewe Flock	77	30.18		15.46
Merino Wether Flock	39	8.58		7.80
Prime Lamb Flock	108	32.68		21.51
Beef Cattle Herd	28	90.04		5.64

Yield trends for the different pulse species

Long term national yield trends integrate the effects of seasons, including pests and diseases, varieties and agronomic management practices used by farmers. For many of the species of interest there are insufficient data to be conclusive. Apart for lupins, national yields per hectare have not changed significantly (Table 5). Lupin yields have increased at about 2% per year since at least 1978. If we had projected back further it is likely we would find an even higher rate of improvement as the yields of lupins in the period 1970 - 1977 were extremely low because many farmers were growing poorly adapted, long season varieties, with limited agronomic knowledge, during a period of drought.

Table 5. Yield trends in farmer's fields for cool season pulses in Australia

Species	Yield trend over time	Significance	Years
Lupins	Y = 0.018 x - 34.80	< 0.05	1978 - 1995
Peas	Y = 0.005 x - 9.46	ns	1980 - 1995
Chickpeas	Y = -0.012 x + 25.77	ns	1983 - 1995
Faba beans	Y = -0.029 x + 59.03	ns	1983 - 1995
Lentils	Y = 0.050 x - 98.93	ns	1990 - 1995

During the periods assessed for the non lupin crops, areas have increased considerably, including expansion into new, often lower rainfall or lower yielding environments. Also there have been few variety releases during the period. Until recently the faba bean industry was based on one variety. The chickpea releases for the northern areas required Phytophthora resistance rather than yield increases, but the disease has not occurred in recent drier years. Field pea releases have been targeted at improved quality, rather than yield. In some areas, where there has been a long history of pea crops in an intense rotation, yields are perceived to be declining, with black spot, downy mildew, sulphonyl urea residues and their nutritional interactions the suspected causes. Lentil releases were the basis of the new industry and so the industry is too recent to assess. As well, new yield limiting factors have emerged with each crop as they expand into new areas or as production intensifies. Examples are viruses and fungal pathogens in chickpeas, rust in northern faba beans and vetch, downy and powdery mildew in peas, and now anthracnose in lupins. All become new breeding and crop management objectives to maintain yields and stability.

The lack of change in yields, other than in lupins, is due to several causes, including the more recent start to the breeding effort, a lower investment in the agronomy of the species, a series of poor years in north-eastern Australia in the 1990's and less experience amongst growers. These should improve in the next decade and yields increase. Since 1987 there has been more investment in pulse Research and Development (R & D) with a greater contribution from farmer levies. Prior to 1987 only lupins in Western Australia were paying a consistent levy (Hamblin 1987). Now all pulses are part of the levy system collected in all states of Australia and administered by the Grains Research and Development Corporation (GRDC).

Industry Coordination

In the early days of the pulse industry in Australia there was limited national co-ordination and co-operation. From 1987 to 1995 the pulse industry was guided by the National Grain Legume Consultative Committee (NGLCC), a voluntary group representing the major organisations in the industry. The NGLCC set national receival and export standards and organised visits aimed at understanding overseas markets. It paved the way for the Pulse Industry Strategic Plan (Anon 1995a), and was superseded by the formation of Pulse Australia in 1995. The formation of Pulse Australia as the peak industry body is seen as a major step to achieving the year 2005 production targets (Anon. 1995a). Its goals include: improved industry co-ordination, leadership and planning; developing new markets and expanding market share in existing domestic and international markets.

Market requirements and sales opportunities

The marketing arrangements for lupins exported from Western Australia (WA) have assisted the development of lupins in that state and nationally. The system involves the compulsory acquisition and marketing of lupins by the Grain Pool of WA. The system does not exist outside of Western Australia, nor for any other pulse in Australia. There is much debate on the relative merits of a single marketing organisation such as the Grain Pool of WA versus unrestricted marketing by many organisations. The benefits of a single marketeer appear to be more apparent when developing a new crop, but may diminish once it is well established on farms and in the international market place. In the early stages of pulse development in eastern Australia, co-operatives were formed by growers to market pulses, eg SA Pea Growers Co-operative, Lupin Growers Association, Grain Legume Co-operative. Most were superseded by the entry of the Australian Wheat Board and the Australian Barley Board into the market place for bulk commodities. Pool payments, cash sales and forward contracts are all marketing options for growers offered by the boards, co-operatives and, more recently by private buyers.. Some grower co-operatives still exist and concentrate on producing for speciality, niche markets, eg PeaCo, Australian Pulse Co-operative .

There is an increasing realisation in Australia that marketing success and maximising returns to farmers requires that we meet market needs, rather than produce poorly targeted, low priced, bulk commodities for world markets. Considerable effort is being put into defining the market requirements of countries and producing to those specifications as well as targeting specific markets. Pulse Australia is developing a system of Quality Assurance to meet market specifications.

Rotational effects

The benefits of growing pulses in rotations in Australia has been clearly demonstrated for lupins (Reeves et al., 1984; Rowlands et al., 1988; Bayliss and Hamblin 1986), and for the other pulse crops (Heenan et al., 1984; King 1984; Marcellos 1984; Rovira and Venn 1985; Rovira 1986; Evans et al., 1991; Schultz 1995). They include the control of root diseases (Rovira and Venn 1985, Wilson and Hamblin 1990) and weeds that affect other phases of the rotation, particularly grasses difficult to control in the cereal phase (Cheam et al. 1987). The nitrogen balance of the soil is improved, along with effects on soil structure and other nutrients, for example potassium (Rowland et al. 1986, 1988). These effects are often difficult to separate, but can be substantial leading to 50% cereal yield improvements.

These benefits were particularly important in the uptake of pulses by Australian farmers because cereals remain the main cropping enterprise. Areas with sand-plain soils in Western Australia were dominated by either continuous cereals or with poor grassy pastures in rotation, both of which led to a build up of root diseases and nitrogen deficiency problems. Introducing lupins as a phase in the rotation removed these constraints and enabled cereals to better utilise the limited water available during the growing season. Over the past decade this has led to an improvement in wheat yields of 50 kg/ha/year.

As the range of rotational options has increased in the winter-cropping areas of southern Australia, we have seen the traditional wheat/sheep (ie wheat/pasture) rotation become diversified. Rotations can now be continuos cropping, and include a range of pulses and oilseeds as well as wheat, barley and oats. There are also different pasture options available where pastures remain in the rotation, including legume pastures sown or sprayed to be free of grasses, green manuring with legumes, or phased pasture systems where the rotation has several years of cropping followed by several years of pasture based on new pasture species which are cheap and easy to harvest and therefore can be re-sown after the end of a cropping phase.

New industrial uses and processes

Despite considerable interest in Australia we have not been successful in adding value to our products although they have been shown to be good for processing,. The sitaution is likely to continue, however there are small niche markets such as Chik Nuts, a roasted chickpea snack, and Quick Pulse which packages partially processed pulses which reduces preparation times. With increasing production it is likely the situation will continue and the majority of our products will be exported in bulk. We hope to improve pulse quality and increase product identification overseas so that Australian pulses are recognised as consistent, high quality products that fetch a premium in the market place.

The role of Research and Development (R & D)

There has been a major investment in pulse R & D in Australia over an extended period. The incentives of the late 1960's and early 1970's were driven by the search for alternative enterprises during a period of low wool and cereal prices. There was also a perceived shortage in the world of protein, which could only be met from increased pulse production. Wheat quotas were imposed in Australia at that time to overcome excess production. While these triggers to R & D investment did not persist, the momentum for pulse crop R & D continued. An illustration is the investment in lupin technology, going back over 40 years. Lupin R & D led to the development of improved varieties; better production systems including sowing and harvesting methods, better fertiliser and weed control recommendations, as well as methods of managing pest and diseases. The role of lupins in crop rotations and animal diets was also assessed. A farm extension program was developed and efforts made to have lupins recognised in international markets. R & D has also been essential for other pulses. Each crop has been developed uniquely, but with a similar general methodology. For example the first lentil varieties were released in conjunction with a commercial partner to ensure growers, marketeers and advisors were all initially involved and committed to developing the new crop (Materne and Brouwer 1997).

Initially pulse development was not co-ordinated in Australia as the country has separate states, with separate governments and agricultural departments. Pulse work was funded by the state departments or as a small part of a levy system collected on wheat and barley production by the Wheat Research Council and the Barley Research Council. Since 1987 there has been an increased investment in pulse R & D and a greater contribution to this from nationally collected farmer levies. Now all pulses are part of the national levy system collected on all grains in all states of Australia, and along with the federal government contribution, administered by the Grains Research and Development Corporation (GRDC).

We estimate the current annual investment in pulse R & D in Australia to be between $26-30 million dollars. To this must be added contributions (not fully known) by organisations such as the Grain Pool of WA, the Federal Government through the CRC program and chemical companies through their work on pesticide and fertiliser recommendations.

Farmers and others engaged in the industry are great innovators and have helped develop improved production systems. This is seen in activities such as improving weed management, developing minimum tillage, sowing practices and improving mechanical harvesting, handling and storage.

CONCLUSIONS

Australia's success in expanding from nothing to being a major force in the pulse markets of the world in 30 years has been based on a long term vision and investment decisions taken by many individuals and organisations. Timing has also been important and probably fortuitous as we have often had the right product available at a time when there was a major need for it.

With this experience it is likely that Australia will be the major pulse exporter in the world within the next decade. This will be achieved by improved communication between scientists working at the molecular level at one end of the information chain through to cooks using the products to feed their families at the other. We are becoming increasing aware of the importance of these links in the chain and the key role of information flowing up and down the chain. We look forward with confidence to being able to report on reaching our objectives at the fifth IFLRC conference.

References

ABARE 1996a. *Australian Commodity Statistics 1996*, Australian Bureau of Agricultural and Resource Economics, Canberra. 195-203.

ABARE 1996b. *Australian Crop Report*, 3rd Dec 1996 number 97 Australian Bureau of Agricultural and Resource Economics, Canberra. 21.

Ali, S.M. 1995. *Australian Journal of Experimental Agriculture* 35: 557-558.

Anonymous 1980. *Journal of the Australian Institute of Agricultural Science* 46: 263.

Anonymous 1987. *Journal of the Australian Institute of Agricultural Science* 53: 128.

Anonymous 1995a. "*Inventing the future. Australian Grain Legumes Industry Strategic Plan*" volume 1, Consultants Report. (Grains council of Australia).

Anonymous 1995b. *Inventing the future. Australian Grain Legumes Industry StrategicPlan*". volume 2, Research Report. (Grains council of Australia).

Barber, A., Peake, B. Rendell, D. and Potter, T. 1997. *Gross margins Mid and Upper South East districts*. Primary industries South Australia.

Bayliss, J.M. and Hamblin, J. 1986. *Proceedings of the 4th International Lupin Conference*, Geraldton. 161-172.

Beech, D.F. and Brinsmead, R.B. 1980. *Journal of the Australian Institute of Agricultural Science* 46: 127-129.

Brinsmead, R.B., Brouwer, J.B., Hawthorne, W.A., Holmes, J.H.G., Knights, E.J. and Walton, G.H. 1991. In *'Proceedings of Grains 2000 Conference, Canberra.'*

Brouwer, J.B. 1995a. *Australian Journal of Experimental Agriculture* 35: 115.

Brouwer, J.B. 1995b. *Australian Journal of Experimental Agriculture* 35: 116.

Corbin, E.J., Brockwell, J. and Gault, R.R. 1977. *Australian Journal of Experimental Agriculture and Animal Husbandry* 17: 126-134.

Cheam, A.H., Ralph, C., Hamblin, J., and Nelson, P. 1987. *8th Australian Weed Conference*, Sydney.

Evans, J., Fettell, N.A., Coventry, D.R., O'Connor, G.E., Walsgot, D.N., Mahoney, J. and Armstrong, E.L. 1991. *Australian Journal of Experimental Agriculture and Animal Husbandry* 42: 31-430.

FAO 1993. *FAO production year book*, Volume 47.

Gladstones, J.S. 1994 *Proceedings of the 1st. Australian Technical Symposium*, (Ed Dracup and Palta) Department of Agriculture, Western Australia, 1-38.

Grain Pool of W.A. 1995. In *Australian Grains: a complete reference book on the grain industry* pp 460-465 (Ed Bob Coombs) Morescope Publishing, Camberwell, Australia.

Hamblin, J 1987. In *Proceedings of 4th Australian Agronomy Conference*, La Trobe University, Australian Agronomy Society, 65-81.

Hawthorne, W.A. 1993. In *Proceedings of 7th Australian Agronomy Conference*, Adelaide: Australian Agronomy Society, 64 - 67.

Hawthorne, W.A. 1995a. In *Australian Grains: a complete reference book on the grain industry.* pp 446-452 (Ed Bob Coombs) Morescope Publishing, Camberwell, Australia.

Hawthorne, W.A. 1995b.. In *Australian Grains: a complete reference book on the grain industry.* pp 466-470 (Ed Bob Coombs) Morescope Publishing, Camberwell, Australia Hawthorne, W.A. 1997. *Proceedings of the First Australian New Crops Conference* 2: 37-44.

Heenan, D.P., Taylor, A.C., Cullis, B.R., and Lill, W.J. 1994. *Australian Journal of Agricultural Research* 45: 93-117.

King, P.M. 1984. *Australian Journal of Experimental Agriculture*, 24: 555-564.

Kingwell, R.S. and Pannell, D.J. 1987. *MIDAS, A bioeconomic Model of a Dryland farming System.* Pudoc, Wageningen.

Knight, R. 1994. In *Research and development prospects for faba bean* pp14-15 (Ed C. Piggin and S. Lack) ACIAR technical Publication No 29.

Knights, E.J. 1995. In *Australian Grains: a complete reference book on the grain industry.* pp 436-441 (Ed Bob Coombs) Morescope Publishing, Camberwell, Australia

Laurence, R.C. N. 1979. *Australian Journal of Experimental Agriculture and Animal Husbandry* 19: 495-503.

Marcellos, H. 1984. *Journal of the Australian Institute of Agricultural Science* 50: 111-113.

Materne, M and Brouwer, J.B 1997. *Proceedings of the First Australian New Crops Conference* 2: 45-52.

Perry, M.W., Garlinge, J, Landers, K, Marcellos, H, Mock, I.T., Walker, B and Hawthorne, W.A. 1994 *Proceedings of the 1st. Australian Technical Symposium,* (Ed Dracup and Palta) Department of Agriculture, Western Australia, 1-38.

Reeves, T.G., Ellington, A. and Brooke,H.D. 1984. *Australian Journal of Experimental Agriculture and Animal Husbandry* 24: 595-600.

Richards, R. 1995. In *Australian Grains: a complete reference book on the grain industry.* pp 429-431 (Ed Bob Coombs) Morescope Publishing, Camberwell, Australia

Rovira, A.D. and Venn, N.R. 1985. In *Ecology and Management of Soil borne Plant Pathogens* pp 255-258 (Eds C.A. Parker, A.D. Rovira, K.J.Moore, P.T.Wong and J.F.Kollmorgen.) American Phytopathological Society:St Paull, MN, USA.

Rovira, A.D. 1986. *Phytopathology* 76: 669-673.

Rowland I.C., Mason, M.G. and Hamblin J. 1986. *Proceedings of the 4th International Lupin Conference* Geraldton. 96-111.

Rowland I.C., Mason, M.G. and Hamblin J. 1988. *Australian Journal of Experimental Agriculture*: 28, 91-97.

Schultz, J.E. 1995. *Australian Journal of Experimental Agriculture*, 35: 865-876.

Siddique, K.M.H and Sykes J. 1997. *Australian Journal of Experimental Agriculture*: 37: 103-111.

Silsbury, J.H. 1990. *Australian Journal of Experimental Agriculture* 30: 645-649.

Smith, I and Mahoney, J. 1995. In *Australian Grains: a complete reference book on the grain industry.* pp 453-459 (Ed Bob Coombs) Morescope Publishing, Camberwell, Australia

Wilson, J. and Hamblin, J. 1990. *Australian Journal of Agricultural Science* 41: 619-631.

World and Regional Trade: Quantity versus Quality

R. O. Rees[1] R. Richards[2] and F. Faris[1]

1 Primary Industries South Australia,GPO Box 1671, Adelaide 5001, South Australia; 2 Bowman Richards & Associates, 275 Alfred St. N Sydney 2060 NSW Australia

Abstract

Since IFLRC1 was held in Spokane in 1986 aggregate world production of pulses has changed only marginally, rising by 7 per cent. However, there have been significant shifts between winter and summer pulses and between varieties over this time. The proportion of the summer pulses produced has increased with total dry bean production increasing by 29 per cent to 18.6 million tonnes, due mainly to increases in India, Brazil, Myanmar, Mexico and China. Production of the main winter pulse, dry peas, has fallen over the period by 4 million tonnes due mainly to reduced production in Russia and China. This is despite pea production almost doubling in Europe to 6.4 million tonnes. Production of chickpeas has increased by 2 million tonnes, due mainly to Indian growers responding to relatively profitable prices and increasing production and improvements in Iran and Australia. Of note is the relatively poor performance of Turkey, a major exporter of chickpeas and lentils, where production has been either static in the case of chickpeas and falling in the case of lentils.

Trade in pulses increased by around 70 per cent in the period 1987 to 1996 with the two major increases being in dry peas, up 2 million tonnes to 3.7 million tonnes, mainly for use in stockfeed and dry beans, up 600 kt, mainly to South America. The proportion of bulk shipments of average quality increased relative to container loads. Trade in other cool season food legumes was characterised by volatility but no substantial increase was registered for any of the other key food legumes.

The two biggest challenges facing pulse growers world wide is quality assurance and food safety. Systems are being designed in Australia to meet customer driven specifications in order to build greater credibility with customers. They are designed to underpin the present system of QA grain in bulk handling facilities and at port terminals. The Australian system developed by Pulse Australia has its own HAACP (Hazard Analysis of Critical Control Points) based quality system, which addresses food safety and product quality issues. The feature of the system is that it has created a world first by integrating quality parameters into the whole farm system based on best management practices. The ultimate goal as the system develops is to provide QA for cooking time, taste and flavour.

INTRODUCTION

Jean Jacques Pernet of Agrogen (Pernet 1988), when speaking about international trade in dry beans described it as an *obsolete concept*. He saw the trade at that time (1988) as international charity full of tripartite agreements, food aids, tea funds, Export Enhancement Programs, GSM credits, Public Law 480, Biceps and other US Government programs. Whilst trade in cool season legumes is not as charitable, nevertheless it is dominated by international feed prices and the purchasing power of, in particular, the countries of the Indian subcontinent.

The paper will address world trends in area, yield and production of pulses since the first International Food Legume Research Conference (IFLRC 1), the changing face of consumption (particularly in India), forecast demand, world trade trends, the emerging supermarket trade and developments in quality assurance. Trade in pulses increased by around 74 % between IFLRC 1, held in 1986, to 8 million tonnes traded in 1996 (FAO 1988, 1997), the volume of which seems relatively insignificant alongside 95 million tonnes of wheat and 88 million tonnes of coarse grains.

As our Chief Executive (Mutton, 1997) said at the trade launch opening for this conference, trade in pulses is viewed in many different ways by delegates to this conference;

R. Knight (ed.), Linking Research and Marketing Opportunities for Pulses in the 21st Century, 143–153.
© *2000 Kluwer Academic Publishers. Printed in the Netherlands.*

Trade reflects a failure of supply to meet demand.

High prices have led to a shift in consumption from pulses to cereals, animal products and fruit and vegetables.

Relatively high import prices for pulses on the Indian subcontinent have led to a significant expansion of production, particularly in Canada, Australia and Myanmar.

Prices and incomes are critical for increased consumption.

Pulses are a cheap source of protein and energy for livestock development.

Imports are used to reduce inflated domestic prices and offset hoarding.

Dry peas have experienced the largest growth in trade, increasing by 138 %, and account for the largest volume of pulse exports (Table 1). Dry beans, the second largest traded pulse, should not be ignored. They are an integral part of a basket of pulses demanded by many countries including India, Pakistan and Spain. The positive influence that mung bean has had on subsequent wheat yields in Pakistan has recently become apparent (Redden, pers. comm.). South and Central America for the most part are consumers and producers of black beans and pinto beans but if prices of these basic types rise too high then some degree of substitution occurs, generally with lentils, peas or other alternate bean varieties.

Trade in pulses is viewed in many different ways, although at its most basic level, trade essentially reflects a failure of supply to meet demand. With respect to pulses, in recent times high prices have led to a shift in consumption from pulses to cereals, animal products, and fruit and vegetables. Thus, prices and incomes are critical for increased consumption. Pulses provide a cheap source of protein and energy for livestock development. Imports are used to reduce inflated domestic prices and offset hoarding.

Table 1: World Pulse Exports (kt)- 1987 vs 1996

	1987	1996	Change (1987 to 1996)
Lentils	456	676	+48.4 %
Faba beans	340	516	+51.8 %
Chickpeas	322	333	+3.6 %
Dry peas	1,543	3,673	+138.0 %
Dry beans	1,924	2,504	+30.2 %
Total pulses	4,844	7,967	+64.5 %

Source: FAO Yearbook, 1988, and FAO Trade Data, 1997.

Why has trade almost doubled since the IFLRC 1 in 1986? To answer the question it is necessary to critically evaluate the pulse industry and to ascertain whether supply issues exist and/or whether demand is increasing.

WORLD PULSE SUPPLY

Pulse Area

The area sown to pulse crops is shown in Table 2. Of particular interest is the 5 % increase in total dry bean area, particularly in India and Myanmar. The increased production of pigeon pea and black gram from these two sources, as a result of increases in area sown, has influenced prices for cool season food legumes.

Table 2: Average areas of world pulses('000 ha) and percentage change.

	1984-86	1989-91	1994-95	Change (1984-86 to 1994-95)
Lentils	2,600	3,156	3,396	+30.6 %
Faba beans	3,231	3,163	2,864	-11.4 %
Chickpeas	9,874	11,078	10,544	+6.8 %
Dry peas	9,089	9,374	7,808	-14.1 %
Dry beans	25,861	25,649	27,189	+5.1 %
Total pulses	69,846	68,982	66,137	+5.6 %

Source: FAO Production Yearbooks, various.

The total area sown to lentils increased by 31 %, with India up 23 %, and Canada by 290 %. Bangladesh also had a significant increase in the area sown to lentils.

The strong volatility in the dry pea area may be attributed to the increased interest in stockfeed peas. Initially Australia, and then Canada, planted large areas of peas primarily for stockfeed before discovering a viable market niche for food use in the Indian subcontinent. The Canadian area rose 700 % over the period, by 217 % in France and 137 % in Australia.

Overall, the area sown to faba beans by all major producers has declined. Several exporting countries are interested to see whether China can maintain its relatively high area which has enabled it to dominate world trade.

While the overall area planted to chickpeas from 1984-86 to 1994-95 rose by 7 %, the average area planted in India fell by 5 % (Table 3). In Turkey, the major producer, chickpea area rose by 85 % over the period but since 1994-95 when price subsidies were removed the area sown has fallen sharply. There have also been significant increases in area sown in Australia (470%) and Iran (719%). In the 1997/98 season Australia planted some 250,000 ha to chickpeas, primarily of the desi type.

Table 3: Average area ('000ha) of chickpeas by country and the percentage change.

	1984-86	1989-91	1994-95	Change (1984-86 to 1994-95)
India	7,243	6,897	6,881	-5.0 %
Pakistan	1,011	1,023	1,055	+4.4 %
Turkey	415	850	765	+84.5 %
Iran	75	120	615	+719.3 %
Australia	32	110	183	+470.3 %
World	9,874	10,078	10,544	+6.8 %

Source: FAO Production Yearbooks, various.

Pulse Production

Although the total area sown to pulses increased by only five % during the period, production of dry beans expanded by 24 %, while dry pea and faba bean production fell sharply (Table 4). The general decline in faba bean production can principally be accredited to a lowering of production incentives in the European Union, the reduction in the livestock herd in the Former Soviet Union and the pressures placed on Chinese faba bean growers to move into alternative, more profitable crops.

Table 4: Average world production of pulses (kt) and percentage change

	1984-86	1989-91	1994-95	Change (1984-86 to 1994-95)
Lentils	1,890	2,302	2,805	+48.4 %
Faba beans	4,269	4,296	3,400	-20.4 %
Chickpeas	6,891	7,116	8,887	+29.0 %
Dry peas	13,140	16,542	11,083	-15.7 %
Dry beans	14,802	16,105	18,371	+24.1 %
Total pulses	51,694	58,030	56,517	+9.3 %

Source: FAO Production Yearbooks, various.

Table 5 clearly illustrates which countries have moved to fill the gap left by India and Turkey. There has been a significant increase in chickpea production by Iran and Australia.

Table 5: Average production of chickpeas (kt) by country

	1984-86	1989-91	1994-95	Change(1984-86 to 1994-95)
India	4,998	4,847	5,155	3 %
Turkey	445	801	690	55 %
Pakistan	543	534	485	-11 %
Iran	53	49	320	507 %
Australia	14	131	294	1,947 %
World	6,891	7,116	8,887	29 %

Source: FAO Production Yearbooks, various.

Indian pulse production rose to 14.5 million tonnes in 1996-97, up 13 % on the 1995-96 level. There is some evidence of an increase in the area sown to arhar or pigeon pea, which is grown in rotation with wheat.

Despite the fall in aggregate pea production to 11 million tonnes over the 1994-95 period, production of dry pea in France in 1997-98 surpassed 3 million tonnes as growers experienced a near record yield of 5.1 tonnes per hectare.

Pulse Yields

A disturbing trend is the failure of the major chickpea producers to make much headway in improving yields since IFLRC 1, except for a small increase by India (Table 6).

Table 6: Trends in yield (kg/ha) for chickpeas by country and percentage change.

	1984-86	1989-91	1994-95	Change (1984-86 to 1994-95)
India	689	702	750	+8.8 %
Turkey	1,058	940	902	-14.8 %
Pakistan	537	521	459	-14.5 %
Iran	702	417	520	-26.0 %
Australia	1,113	1,194	829	-25.6 %
World	696	706	714	+2.5 %

Source: FAO Production Yearbooks, various.

However, taking into account prices received and yields achieved in Australia, the rewards for planting chickpeas relative to wheat are either equal to, or in most years exceed, the returns to wheat (Figure 1). Given the increased returns, and the Indian subcontinent market's preference for chickpeas, it seems somewhat surprising that there has not been a larger adoption of chickpeas in similar climates. The potential rewards accrued from growing lentils could be even greater.

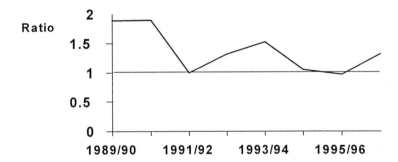

Figure 1: Australian Chickpea/Wheat Price by Yield Ratio

The basic axiom is that if prices are rising in real terms then investment will be attracted to that commodity. In countries such as India, it is difficult to find the balance which allows the market to ration investment without interference, while at the same time allowing the Government to keep prices from rising beyond the means of the poorer population through the domestic import of bulk shipments of dry peas, black gram and chickpeas.

DEMAND FOR PULSES IN INDIA

The basic demand determinants for most food items are population, income and changes in taste preferences.

As the Indian economy continues to grow at around 6-8 % a year, it represents a typical situation where the combined pressures of population and income growth place a strain on available supplies. Based on a 3 % annual rate of growth in pulse consumption in India and Pakistan and a 1 % annual rate of growth in their pulse production (Gordon, et al., 1997), the deficit between annual production and consumption could increase to over 6 million tonnes in the medium term (Table 7). However, the solution lies in considering not only the protein component of the diet, but also the level of calories, vitamins and minerals, which are the more important criterion.

Table 7. Forecast pulse demand for India

	Actual 1996 ('000)	Forecast* 2011 ('000)
Production	13,190	15,313
Consumption	13,578	22,032
Short-fall	388	6,719

Source: 1996 figures; de Boer & Pandey, 1997. Forecast figures; refer footnote.

Nevertheless, as this deficit and consequently pulse import demand grows, it is possible that the resultant increases in world prices will generate faster growth in domestic production than the projected 1 %. Further investment in pulse production on the Indian subcontinent will only occur if there is minimum interference in the domestic market by Government. Until this occurs, it is not possible to develop the full economies of scale and investment in infrastructure required to have an efficient distribution system and adequate volumes of production. As income levels rise, there is generally a shift away from pulses as a staple food source, toward superior classes of food and services. Such a transformation is currently apparent in China, with per capita consumption of wheat for food purposes falling while wheat for livestock feed is beginning to rise. In addition, when coupled with the general fall in birth rates as income rises, this places less pressure on available supplies.

The Changing Face Of Consumption In India And Other Rapidly Developing Economies

During the next 40 years India's population is expected to increase by 50 %. The resulting growth in demand for mass market products, such as packaged wheat flour, biscuits, poultry and liquid milk, will be staggering. It is predicted that by 2005, the overall market for value added foods will treble from US$21 billion to more than US$65 billion (de Boer & Pandey, 1997).

Despite having 75 % more arable land than China, India produces 30 % less agricultural crops and yields are 25 to 40 % of world-best levels (Chong, 1997).

India's changing consumption patterns are driven by both rising incomes and liberalisation of the Indian economy, which offers its population a greater choice of products and has in turn caused greater experimentation with food.

An analysis of the development of eating patterns across 20 countries shows that they go through a distinct evolutionary process (Table 8). During the initial stage, emphasis is on obtaining basic foods for survival; cereals, fats, oils, pulses, fruit and vegetables. Following this stage, when purchasing power reaches US$1,000 per head, the subsistence diet is supplemented with foods such as milk and dairy products, meat, fish, poultry and eggs. The premium stage comes into play at an income level of US$5,000, as consumers move to increased processing, better quality and more exotic ingredients. It has been argued, that because of the diversity of the Indian economy, all three evolutionary stages are present at once (Chong, 1997; de Boer & Pandey, 1997).

---

* Key Assumptions:

India's population will be 1,259,000,000 by 2011 (ACIL, 1996 cited in Agriculture Western Australia, Pulses and Oilseeds Program Strategic Plan, 1996).

Consumption per head will be 17.5 kg per head (ACIL, 1996 cited in Agriculture Western Australia, Pulses and Oilseeds Program Strategic Plan, 1996).

Production will increase at an average of 1% per annum (Gordon, et al., 1997).

Excludes possible exports from India.

148

Table 8: Relationship between income and spending characteristics

Annual Income ($US)	Spending Characteristics
0-1,000	Heavy concentration on basic foodstuffs Little discretionary spending
1,000-2,000	Some consumer product expenditure More people eating out Limited product range in supermarkets
2,000-3,000	Wide range of consumer products in supermarkets Noticeable recreation/leisure spending Rising expenditure on consumer durables Purchase of small cars and motorcycles for personal use
3,000-5,000	Diversification of alcoholic drink consumption Leisure spending diversifies to include travel/holidays Consumer durable spending widens to include non-basic durables (videos/hi-fi) Private health expenditure rises Car purchase increases
5,000-10,000	Expenditure on eating out rises Basic food stuffs replaced by frozen processed meals Luxury beverage market emerges Leisure spending includes overseas holidays/luxury goods Financial services sector emerges
10,000 & over	Financial services Luxury goods sector Home entertaining

Source: S G Warburg Securities reported in *Asia Pacific Food Industry*, Jan 1993.

GLOBAL FORECAST FOR THE NEXT FIVE YEARS

According to the World Bank, the Big Five - China, Indonesia, India, Russia and Brazil - will see their share of world trade increase to 1.5 times that of the European Union by 2020, compared with barely a third today. The significance of this statement is that they are all major pulse players.

Following the GATT round, the support prices for cereal grains and pulses were lowered significantly. The use of cereal grains and their incorporation rates in compound feed rations was expected to increase following the lowering of prices. As dry peas are regarded as a contributor to the total energy of the ration and thus compete with cereals, they will be under some competitive pressure from cereals and incorporation rates may fall (Gordon, et al., 1997).

The emerging demand is in southern Europe through Spain, Portugal and Italy where there is strong demand for peas, lupins and faba beans. Factors favouring competition are the distance that intra-European Union peas have to travel, exporters are able to use smaller ports in the Mediterranean, there is growing prosperity in the region, timing of southern hemisphere shipments coincide with peak European demand, and late Canadian crops freeze (Gordon, et al., 1997).

The general movement toward freer global trade has led to a relaxation of self sufficiency policies in some Asian countries. In China, for example, the Government has moved from a relatively strict interpretation of these policies to one of acceptance of limited self sufficiency - importing 5 % of the annual grain consumption (Gordon, et al., 1997).

Globally, the relatively strong growth in demand and the changes in Government policies are projected to result in the prices of vegetable oils, food pulses and malting barley increasing in real terms over the medium term. However, reflecting weaker demand and productivity improvements, the real prices of wheat, rice and feed protein (pulses and oilseed meal) are projected to decline in the outlook period to 2001 (Gordon, et al., 1997).

Total oilseed production is projected to increase by 15 % over the next five years. For soybean, the increase is expected to occur mainly in the United States, Brazil, Argentina, China, and India. Production of oilseeds in China and India is expected to expand in response to favourable prices, increasing domestic demand for meal[1] and oil, as well as Government policies that promote the processing sector (Gordon, et al., 1997).

TRADE

The top six exporting countries have increased their market share from 50 % of world trade in 1986 to 67 % in 1994 (Table 9). Since that time smaller faba bean production in China has emerged, record pea areas in the European Union have been planted, near record pea yields occurred in France this year, and Canadian pulse production has increased to around 1.8 million tonnes.

Table 9: Top six pulse exporters' share of World Trade

	1986		1994	
	Tonnes	% of World's Exports	Tonnes	% of World's Exports
Canada	265,000	5.3 %	1,068,000	13.7 %
USA	575,000	11.5 %	599,000	7.7 %
China	436,000	8.8 %	1,462,000	18.8 %
Myanmar	150,000	3.0 %	500,000	6.4 %
France	532,000	10.7 %	898,000	11.5 %
Australia	506,000	10.2 %	674,000	8.7 %
Total	2,464,000	49.5 %	5,201,000	66.7 %

Source: FAO Trade Yearbooks, various.

In contrast to the exporting countries, the top six importing countries have lost their relative share from 54 % to 48 % in 1994 (Table 10). Of the six, the only food market of importance is India. Whilst the European countries do import food pulses, the majority of these imports are for feed purposes.

Table 10: Top six pulse importers' share of World Trade

	1987		1994	
	Tonnes	% of World's Imports	Tonnes	% of World's Imports
Netherlands	851,000	16.4 %	726,000	9.8 %
Germany	689,000	13.3 %	558,000	7.6 %
India	475,000	9.2 %	555,000	7.5 %
Belgium-Luxembourg	359,000	6.9 %	520,000	7.0 %
Italy	317,000	6.1 %	419,000	5.7 %
Spain	95,000	1.8 %	748,000	10.1 %
Total	2,785,000	53.7 %	3,526,000	47.7 %

Source: FAO Trade Yearbooks, various.

The major chickpea importers have been consistently Pakistan, India and Bangladesh (Table 11).

---

[1] India is expected to produce 4.5 million tonnes of soybean in 1997-98 but still export 2.75 million tonnes of soybean meal rather than feed the expanding livestock industry (USDA, 1997, *Oilseeds; World Markets and Trade,* October).

Table 11: Top Five Chickpea Importers (tonnes)

	Average 1993-1995	High	Imports from Australia 1996/97
Pakistan	98,054	180,069	0
India	86,103	150,181	236,268
Bangladesh	35,449	52,346	58,771
Algeria	17,535	20,717	0
Saudi Arabia	15,521	17,000	1,377

Source: FAO Trade Yearbook, 1996, and ABARE, Australian Commodity Statistics, 1997.

The major faba bean exporters (Table 12) are China (see Asia) and Australia (see Oceania), as food to Egypt, Saudi Arabia and the United Arab Emirates. Trade in Western Europe is mainly internal for stockfeed use.

Table 12: Faba Bean Exports (tonnes)

	1995	1996
Asia	439,938	246,897
Western Europe	153,846	185,108
Oceania	83,916	58,820
India	0	0
World	740,209	515,730

Source: FAO Trade Data, 1997.

The two major exporters of peas are Canada and Australia (Tables 13 and 14). Canada focuses primarily on the European stockfeed market and in recent years on food to India. Australia, on the other hand, focuses mainly on food markets and the domestic feed market.

Table 13: Whole and split pea exports from Canada (tonnes)

	Aug 1996 - May 1997
Spain	209,913
Belgium	133,276
India	72,875
United States	36,234
Colombia	28,523

Source: Statistics Canada, 1997.

Table 14: Whole and split pea exports from Australia (tonnes)

	Jul 1996 - Jun 1997
Bangladesh	146,101
India	69,084
China	36,243
United Arab Emir.	20,875
Malaysia	12,083

Source: Australian Bureau of Statistics, 1997.

The demand for food pulses, predominantly kabuli chickpeas, lentils and faba beans is expected to increase in Middle Eastern countries, especially Turkey. The area sown to chickpeas in Turkey has been on a downward trend in recent years, a direct result of the removal of price support for chickpeas in 1994-95.

The United States has lost market share in traditional commercial export markets in South America, Spain and India, while Canadian and Australian exports to those markets have risen. United States exporters have attempted to defend market shares by creating product niches, based on quality differences. Yet despite this, price is still the major factor in the dry pea and lentil trade.

A major feature of the WANA (West Asia North Africa) region is the availability of free trade zones in Bahrain and Dubai. These two ports are strategically located to re-export Australian pulses, in a processed or semi processed form, to more than a billion people in the immediate region. Apart from the Gulf countries, particularly Iran and Kuwait which did not have bulk receival facilities due to war damage, the ports are easily able to service countries on the Red sea including Egypt, Ethiopia and Sudan, whilst the

Indian subcontinent, the largest food legume consuming area, is within easy reach (Rees, 1992). In Dubai, the population is expected to increase by 40 % by the year 2000, most of these being immigrants from the subcontinent.

Trade in chickpeas by Turkey is being strongly influenced at present by barter trade to Iraq in exchange for oil on the food for oil scheme. Turkish processors have also indicated that they believe the country is more likely to become a net importer of green lentils from Canada and red lentils from Australia because of greater relative production efficiency in those two countries compared to Turkey.

This raises the issue of quality standards generally. If Canada exports bulk lentils then the value adding and packaging is left with the importing country. This effectively reduces the opportunities for cleaned, graded, split and bagged pulses in containers from Canada, United States of America, Turkey and Australia. This is also the case with Chinese and Australian faba bean, and all country pea exports to the Indian subcontinent. Quality becomes subsumed by the economies associated with bulk shipments. Effectively countries are left with relatively minimal export sales to speciality niche markets of high quality pulses.

Evidence for this trend is reflected in the fall in sales of field peas and lentils grown in the Palouse and elsewhere in the United States, which have fallen by 37 % on last year. This is due mainly to stiff price competition from bulk Canadian sales. In addition there is reduced PL480 food aid demand due to cut backs in the US aid budget.

SUPERMARKET TRADE

There are a number of sessions in the conference which will examine developments for food legumes in the supermarket trade for a range of countries. In recent years income growth in many countries has had a major impact on the consumer habits on populations in these countries. These sessions will identify those products with better quality characteristics, faster preparation time, pre-cooked preparations, snack foods, and refrigerated products. Similarly, different tastes for different cultures will be identified and the resultant products may become household names in many countries in a relatively short time.

The goal of the Australian food industry is to increase its position as a reliable supplier of quality products, to develop easy or quick cooking food legumes, prepared or total meal products, split or low level processed products, and as feed ingredients for the poultry and pig industries.

The distribution and promotion of pulses in Europe and other developed countries is old fashioned. New products can be launched; the consumers have money and they are open to new ideas, especially when diet are concerned. In essence, the developed world needs to more rigorously examine its own backyard and aim for investment in food products suitable for the time. This is where growth and value adding will be found for sizeable populations with incomes willing to purchase innovative, practical, easy to cook, or ready prepared meals (Schluter, 1997).

QUALITY

So where does quality fit into the cut-throat business of the international pulse trade?

A quality product or service will need to meet the customer's specification in terms of conformity, consistency and cost. McDonald's is perhaps the best recognised example of a quality product, ie. predictable, uniform and consistent.

The driving forces behind the adoption of quality assurance is improving Australia's competitiveness and maximising market access. A recent study for the Agrifood Council of Australia found that the perceptions Australian firms of their delivery of quality and reputation, exceeded their Asian customer's perceptions of product quality, service and value.

The growth in quality systems and practices has been rapid in Australia, with the many different quality systems that have emerged creating considerable confusion. The development in quality systems and practices is shown in (Figure 2).

152

Figure 2. : The Development of Quality Systems in Australia

The focus of industry schemes until recently has been on 'through chain' quality assurance, ie. quality assurance starting at the farm gate and continuing through to the end consumer. Many industry schemes have followed, which has meant considerable confusion and cost to the farmer's business. The current Australian Government, through its Supermarket to Asia scheme, is looking to develop a single on-farm system.

Pulse Australia has developed its own HACCP based quality system, which addresses food safety and product quality issues. As the industry matures, cooking parameters and eating quality assurance will become increasingly important. The diagram below highlights the progression in quality assurance, for which the Australian pulse industry is striving.

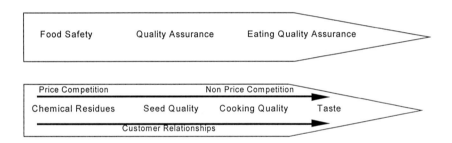

Figure 3: The Elements of Quality Assurance

First essential requirement of a quality system is to meet food safety needs

As markets mature, opportunity to differentiate product on assurance of product quality factors and consumer or eating quality factors

As move to these consumer quality assurance systems, opportunity to move away from price competition

For the pulse industry this means as initial focus on chemical residues and other food safety issues, moving (quickly in some markets) to a focus on cooking and taste factors

International community awareness of environmental and food health and safety issues is an increasingly important influence on food (and feed) consumption patterns. Concerns about chemical residues in food and animal feed, biological risks of aflotoxin in grains; salmonella and mad cow disease from meats and the sustainability and environmental impact of production systems is influencing demand, especially in premium markets. The development of genetically engineered plants and animals is already being practised in protein, oilseed and animal product markets.

The Pulse Australia quality assurance program is aimed at improving the industry's competitiveness. Adoption of the system will signal to customers that the industry is proactive and committed to meeting their needs. This should result in stronger customer support for Australian pulses, preferred supplier status and improved profitability for industry participants. In particular, Pulse Australia believes it provides the basis for market development, targeting the high value, discerning markets.

Development of the domestic market for pulse food products will provide a strong incentive for pulse growers to meet strict quality standards. This approach to quality will provide Australia with a solid basis to launch its products into the emerging supermarket business.

CONCLUSION

Removal of the supply side blockages is a key issue for many pulse consuming countries. There is strong demand potential as incomes rise and investment in pulse production leads to lower prices. For the Big Five exporters the major challenge is to encourage investment in value added products suitable for their domestic supermarket shelves which are quick and easy to use and in tune with today's cultural imperatives; ease of preparation, taste, length of cooking time and products which are interesting.

References

Agriculture Western Australia 1996. "Pulses and Oilseeds Program Strategic Plan, 1997-2002", 21 November.

Chong, F. 1997. "India Opens Toothsome Markets for Technology", *The Australian*, 3 July.

De Boer, K., and Pandey, A. 1997. "India's Sleeping Giant: Food" *The McKinsey Quarterly*, No. 1.

FAO 1996. *Trade Yearbook*, Foreign Agriculture Organisation of the United Nations, Rome Italy.

Gordon, S., Hardcastle, S., Hooper, S., Krieg, A., Martin, P., Podbury, T., and Foster, M. 1997. "Outlook for Grains, Oilseeds and Pulses", *Outlook 97 Conference Proceedings*, ABARE.

Mutton, D. 1997. Official Exhibition Opening Address, delivered to the *International Food Legume Research Conference III*, Adelaide, 21 September.

Pernet, Jean-Jacques 1988. *Michigan Dry Bean Digest*, Vol.12, No. 4 (Summer).

Rees, R. 1992 "Oilseeds and Grain Legumes Outlook", *Outlook 92 Conference Proceedings*, ABARE.

Richards, R. 1997. "Australian Pulses: The Opportunities for India", *Outlook 97 Conference Proceedings*, ABARE.

Schluter, V. 1997. Delivered to the *International Food Legume Research Conference III*, Adelaide, 25 September.

S G Warburg Securities 1993. *Asia Pacific Food Industry*, January.

PROCESSING AND GRAIN QUALITY TO MEET MARKET DEMANDS

U. Singh[1] P.C. Williams[2] and D.S. Petterson[3]

1 International Crops Research Institute for the Semi-Arid Tropics (ICRISAT), Patancheru, Andhra Pradesh 502 324, India.; 2 Grain Research Laboratory, Canadian Grain Commission, Winnipeg, Canada.; 3. National Pulse Quality Project, Agriculture Western Australia, Perth, Australia.

Abstract

Considerable resources have been directed towards improving the nutritional quality of cool season food legumes with respect to protein content and amino acid pattern and to reducing the content of antinutritional factors such as trypsin inhibitors and haemaglutinins. Less attention has been paid to the processing and grain quality factors that affect the utilisation of these legumes. Two important market considerations are the dhal yield and consumer acceptance of the product. These are influenced by the size, shape, colour and, chemical composition of the grain, by storage conditions and any pre-treatment before use. The cooking time, texture, water absorption and dispersibility of solids are determinants of quality of these legumes as food. Under adverse storage conditions, the legumes can develop hard-to-cook defects, depending on genotype and cultural practices. Nutritional quality needs to be considered in terms of protein digestibility, antinutritional factors, availability of carbohydrates and content of essential micronutrients such as vitamin A, iron, copper and zinc. Available technologies such as roasting, steaming, germination, fermentation, and extrusion cooking, and protein isolation/concentration play a role in determining the product quality. These topics are reviewed and future research needs are suggested in the paper.

INTRODUCTION

Nearly 80% of proteins and 90% of calories consumed by humans in developing countries are supplied by plant products. The cool season food legumes are increasingly important in human nutrition, because they are less expensive sources of proteins than the animal proteins, especially in developing countries. In addition, they are valuable sources of carbohydrates, minerals and vitamins. Their protein quality for food is low, however when mixed with cereals the total diet quality can exceed 70% of casein or lactalbumin. In several developing countries, mixing cereals and legumes to provide a good quality diet was in practice long before people understood the nutritional importance of the practice. The world production of cool season food legumes has increased during the past few decades, mostly due to an increase in the area under cultivation. World production of both dry peas and lentils almost doubled in 1980-90 (FAO 1990). However, in recent years per capita availability of pulses has declined in some regions of Asia, mainly due to increases in population. The daily per capita consumption of pulses were reported to be 43 g for Pakistan, 35.7 g for India, 23 g for Indonesia, 42 g for China, and 25-50 g for the Middle East (Singh and Singh 1992).

The majority of pulses are consumed in the regions where they are grown. Those entering international markets do so in the raw or processed forms. Although Asian countries produce nearly 55% of the world's pulse crops, the demand will continue to increase because of the expanding population, particularly on the Indian subcontinent. Consumers in these countries are increasingly aware of the factors that affect the nutritive value of the food they consume. This paper reviews research on processing and grain quality and the needs for further research on end-uses and market requirements.

THE FOOD LEGUMES

Dry beans (*Phaseolus vulgaris*) can be broadly divided into culinary beans, weighing from about 250 mg (e.g. Great Northern, red kidney) to 600 mg (Borlotti, Cannelloni), and navy beans, 120-220 mg. Dry beans are used in a diversity of soup, salad and vegetable dishes. Immature green beans are also harvested green for vegetable use.

R. Knight (ed.), Linking Research and Marketing Opportunities for Pulses in the 21st Century, 155–166.
© *2000 Kluwer Academic Publishers. Printed in the Netherlands.*

Field peas (*Pisum sativum*) traditionally, are harvested to be eaten fresh or field dried before storage and consumption. A significant proportion is also harvested as green, immature seeds for commercial freezing. There are three main types used for human consumption: 1) vining peas for canning fresh ('garden peas'), quick freezing, or artificial drying; 2) threshed dry peas which may be split after dehulling; and 3) pulling peas sold as fresh peas in pods.

Chickpea (*Cicer arietinum* L.), Bengal gram or garbanzo bean is second to dry beans in area grown and third in production to dry beans and dry peas. Kabuli seeds are generally large and light coloured while desi seeds range from yellow to black. They are generally smaller, and have a rougher surface. Some new desi types have 'large' seeds, approaching the size of kabuli seeds.

Chickpeas have a very low content of antinutritional factors and tend to be richer in calcium and phosphorus than most other pulses. Kabuli types are mostly used as a cooked vegetable and sometimes as a roasted snack. Desi types are used in soups, as dhal and made into flour, besan, for a variety of bread, soup and vegetable uses. Chickpea/cereal blends are used to make a variety of fermented, deep-fried, toasted, baked and puffed traditional foods.

Faba beans (*Vicia faba*) provide significant levels of protein and other nutrients to be important in the diets of developed and subsistence populations. They are grown on nearly four million hectares of land in about 50 countries with total production of 4.7 million tonnes (Anderson et al. 1994).

Lentil (*Lens culinaris*) is used in mujadurra, a lentil and rice dish, and lentil soup. Dhal curry is a common dish in the Indian sub-continent whereas in Western societies lentils are most commonly used in soups. Other uses are in blends with cereals, snack foods, invalid and weaning foods.

Lupin (*Lupinus* spp). Small quantities of the lupini bean (large seeded bitter *L. albus*), and the Albus lupin (medium-sized sweet *L albus*) are consumed as a snack food in Mediterranean countries. Grain of *L. angustifolius* can be used for making tempe, an Indonesian fermented food.

PRIMARY PROCESSING

Dehulling

The testa of pulses, with its high content of cellulose and hemicellulose, is often indigestible, and sometimes has a bitter taste due to the presence of tannins. In view of this, primary processing improves palatability and reduces cooking time. Proper dehulling is essential for marketing and human consumption. Two steps are involved 1) removing the seed coat and 2) splitting the cotyledons.

Methods of dehulling

Dehulling methods include small-scale processing by households in villages or large-scale processing by commercial mills in urban areas. For small-scale processing, wooden pestles and mortar, and stone chakki (quern) are used for dehulling chickpea, field pea and lentil in several Asian countries. For large-scale processing, carborundum rollers and discs are employed by the commercial units. These large-scale processing operations are located in cities in the Indian subcontinent, Australia and Canada.

Attrition-type dehullers and roller mills are particularly suitable for dehulling and splitting legume grains with loose seed coats; whereas abrasive-type dehullers are suitable for dehulling grains with more tightly adhering seed coats. The use of an emery-coated roller is a common practice in commercial dhal mills in India (Singh 1995). The emery-coating, also called carborundum, is made of silicon carbide and used for abrasive or refractory action. Some millers use a carborundum roller for both dehusking and splitting, while others use a roller and disc sheller alternatively for this purpose to reduce dehulling losses. The disc sheller is generally used for wet processing and works on the principle of attrition, which removes the husk and splits the cotyledons simultaneously resulting in excessive breakage of cotyledons. The disc shellers are generally used for dehusking rice.

A comparison of dehulling yields, achieved by commercial processors in India and Australia, is given in Table 1. Dhal yields of chickpea and field pea were higher in India than in Australia, indicating that losses in terms of broken chips, powder, and husk are lower in India. The seed characteristics and agronomic factors that would affect these losses, from crops grown in different countries, need to be examined because they influence the market price of the commodity. In Australia, legume decorticators are typically used for dehulling. The outer layers of the cotyledons are invariably scarified by the abrasive action of the dehulling process resulting in nutrient losses (Singh et al. 1992) of protein, calcium, and iron which are in the powder fraction (Table 2). These losses could be reduced with better methods and by selecting more suitable genotypes. This would improve the market price of processed products and their nutritive value

Table 1. Processing yields (decorticated dry split cotyledons) of different legumes obtained by commercial processing mill

	Seed coat %[a]	Split yield (%)[b]		Theoretical Yield
		Australia	India	
Chickpea	14.2	70.0	80.0	85.8
Faba bean	18.5	60.0	-	81.5
Field pea	10.0	72.0	78.0	90.0
Lentil	8.4	-	73.0	91.6

a   Average of values from different sources.
b   Average of values from several processing mills in Australia and India.

Table 2. Effect of dehulling on chemical constituents (g/100 g sample) of dhal and powder fractions of chickpea cv Annigeri.[a]

Dehulling time	Dhal					Powder				
(min)	Protein (%)	Sugar (%)	Starch (%)	Calcium	Iron	Protein (%)	Sugar (%)	Starch (%)	Calcium	Iron
0	18.6	6.8	56.2	43.7	5.7	-	-	-	-	-
2	18.0	6.5	57.8	39.5	5.0	23.6	12.1	48.0	85.0	12.0
4	17.5	6.3	57.8	38.0	4.8	21.8	10.5	50.3	65.5	10.5
8	17.5	6.0	58.0	35.5	4.3	19.8	9.5	52.0	45.0	8.5
12	16.4	6.1	60.8	35.0	3.8	18.9	8.6	55.4	45.0	7.0
SE±	0.18	0.21	0.31	0.78	0.21	0.21	0.13	0.51	1.90	0.32

a   For each treatment, results are averages of two samples obtained as in Table 1 and analysed separately.   Results are expressed on a moisture-free basis.   Source: Singh et al. 1992.

FACTORS THAT AFFECT QUANTITATIVE AND QUALITATIVE DEHULLING LOSSES

Conditioning

The loosening of testas by preconditioning, generally results in a higher dhal yield. Soaking in sodium bicarbonate followed by oven-drying may produce the highest dhal yield from chickpea and field pea (Table 3). This pre-treatment loosens the seed coat and reduces the cooking time of dhal. Soaking in water (2-14 hr) followed by sun-drying is a common practice before dehulling small quantities in India. Dehulling is made easier by prolonged soaking in water, but the dhal remains uncooked and tough even after prolonged boiling. Preheating the seeds at a higher temperature helps loosen the seed coat and increases the dhal yield, but excessive heating may result in breakage of cotyledons and reduce the dhal yield.   Therefore, the preconditioning treatments needs to be carefully monitored and standardised to increase the dhal yield. Some interactions between preconditioning treatments and legume genotypes may be expected.   Newly developed high yielding genotypes should be examined for their preconditioning effects on dhal yield.

158

Table 3. Effect of preconditioning treatments on dhal/split yields of chickpea and field pea [a]

Treatment	Dhal/split yields	
	Chickpea[a]	Field pea[b]
Control	71.1	79.5
Water	73.6	81.4
Sodium chloride (1.0%)	75.2	82.0
Sodium bicarbonate (1.0%)	76.1	82.8
Edible oil (0.5 w/w)	-	81.7
Preheating	-	81.0

a Source: Iyer and Singh (1997).   b Source: Black et al. (1998).

Seed characteristics

Seed characteristics affect the dehulling efficiency in terms of increasing the dhal yield and reducing nutrient losses. Genotypes with lower seed coat contents give higher dhal yields. Other factors are seed size, shape and hardness. Small seeds are generally difficult to dehull. Greater than 75% of the variability in dehulling efficiency could be accounted for by grain hardness and the resistance to splitting into the individual cotyledons (Reichert et al. 1984). This would suggest that losses in terms of the broken and powder fraction are greater, if grains are too hard, requiring more abrasive force during dehulling. If grains are too soft, they tend to break, resulting in more losses in terms of the broken fraction and powder.

Williams et al. (1993) reported that dehulling and splitting of lentil was easier with large seeds, thin testas, short storage periods and correct wetting (conditioning) practices. Very angular seeds lose excessive amounts, because the dehulling process attacks sharper edges. There is a loss also from flat seeds. Generally, rounder seeds split more readily than flatter seeds. Variability in dehulling is influenced by genotype. Results from a study on 24 field pea genotypes indicated a large variability in dhal/split yield (71.1-85.7%) when evaluated by the laboratory type Satake mill (Black et al. 1998). Similar variations have been reported for chickpea (67.7-84.8%) (ICRISAT 1987) and lentil (62.1-80.2%) (Erskine et al. 1991). However, the role of agroclimatic conditions on dhal yield has not been clearly defined. Faba beans usually are not dehulled but are consumed as whole seeds after soaking and cooking, or as bean flour.

SECONDARY PROCESSING AND FOOD USES

Secondary processing to make food products or value-added snack items includes soaking, boiling, frying, roasting, steaming, canning, baking, germination, fermentation and extrusion.. All these processes improve appearance, texture and cooking quality. Some important products, are listed in Table 4.

Pulses are often soaked in water as a pre-treatment to reduce their cooking time. Soaking also completely or partially reduces the content of undesirable substances such as phytate, trypsin inhibitors, tannins, and flatulence-forming carbohydrates. Boiling is an indispensable process and most commonly used to cook the pulses and their products. Cooking time is therefore important in terms of market requirements. Large differences exist among pulse grains (Table 6). According to Williams and Singh (1987) cooking time is a fairly heritable characteristic, in that if genotypes differing widely in cooking time are grown at different locations, the differences between genotypes persist. Cooking time is positively and significantly correlated with seed size in chickpea (ICRISAT 1987), lentil (Erksine et al. 1985) and field pea (Black et al. 1998). Further, there seems to be no correlation between the cooking time of whole seeds and the dhal components in chickpea (ICRISAT 1987). This indicates that the time taken to cook is greatly affected by the nature of the seed coat. The cooking time is greatly reduced by dehulling these legumes (Table 5). The cooking time of chickpea seed is reduced by soaking overnight to the extent that most of the differences between cultivars are eliminated (Williams et al 1989). Erskine et al. (1985) reported that the genotypic variance for cooking time in lentil was significantly greater than any genotype-location interaction. Genotypes that require less time to cook therfore should be developed. Abas El Faki et al. (1984) reported that cooking improved the protein digestibility of faba bean and chickpea due to inactivation of trypsin inhibitors. Cooking also improves carbohydrate digestibility, possibly by changing the content and chain length of the amylose components.

Table 5 Variability in cooking time of cool season food legumes.

Legume	Component	No. of samples	Cooking time (min.) Range	Mean	Reference
Chickpea	Whole seed (Kabuli)	303	30-296	118.0	(Williams et al. 1989)
	Whole seed (Desi)	125	52-98	78.5	(ICRISAT, 1987)
	Dhal (Desi)	125	26-46	32.0	(ICRISAT, 1987)
Faba bean	Whole seed	16	171-274	220.0	(Singh et al. 1988)
Field pea	Whole seed	24	79.0-150	101.6	(Black et al. 1996)
	Dhal	24	19.0-45.0	29.1	(Black et al. 1996)
Lentil	Whole seed	25	29.9-45.0	33.2	(Erskine et al. 1985)

Soaking and frying could be helpful in expanding the snack-food market for pulses. The usual process is to soak the field peas or chickpeas overnight, and then to fry them in hot oil. Sometimes peas are coated with other materials such as rice flour before frying to provide different flavours.

Germination reduces antinutritional factors such as protease inhibitors and oligosaccharides in legumes (Singh 1984). It also helps develop desirable flavours for certain products. Germinated and boiled faba beans make a very good sauce (Li 1987). Snack items can be prepared by roasting and there are clear cut differences in the roasting quality of chickpea genotypes (Table 6). Although India is the largest producer of chickpea in the world, there is only a limited number of genotypes that are suited for making roasted products such as 'phutana'. The market demand for 'phutana' is very great and the development of genotypes for this purpose would pay rich dividends.

Table 6 Roasting quality of chickpea desi and kabuli genotypes, Hisar 1985/86

Genotype	Sensory evaluation[a] Colour	Texture	Flavour	Taste	General acceptability
G 130	3.8	3.3	3.3	3.2	3.5
ICCV 6 (ICCC 32)	1.7	2.8	2.8	2.5	2.3
SE±	0.18	0.12	0.09	0.13	0.11

a   Five point hedonic scale.

Fermented chickpea products are popular in the Indian sub-continent but fermented products of other cool season pulses are less popular. However the use of these pulses to produce tempe, natto, dhokla, kiyit injera and other fermented foods is becoming popular in South-East Asia. Extrusion is a relatively new process involving both pressure and heat. This process has a wide application in making value-added products, particularly from dry peas. Peas are ground into fine flour that is passed through an extruder under pressure to create different sizes and shapes. The extruded shapes may be fried, seasoned, and packaged.

160

Table 4. Some important and potential food preparations of cool season food legumes around the world

Method	Chickpea Food	Chickpea Country	Faba bean Food	Faba bean Country	Field pea Food	Field pea Country	Lentil Food	Lentil Country
Boiling	Dhal-curry	India, Bangladesh	Kit wot	Ethiopia	Soup	Australia, India, Central Asia, Middle East	Soup	Mediterranean region
	Homos	Syria, Jordan, Lebanon, Egypt, Nepal, Pakistan	MedamisFS	Egypt	Veg. Curry	India, Pakistan, Europe, Middle East	Koshary	Egypt
Frying	Lablebi (whole seed)	Jordan, Tunisia, Turkey	Salty faba beans	Lebanon, China	Flour coated peas	Egypt, China	Dhal curry	Indian subcontinent
	Cocido	Spain	Lablebi	Middle East, Egypt			Nifro	Ethiopia
	Seviya, pakoda	Indian subcontinent	Falafel	Egypt, Syria				
	Falafel	Syria, Jordan, Lebanon, Egypt, Turkey	Orchid bean	China				
Roasting	Phutana (dry seed)	India, Pakistan	Nabet soup	Egypt, Syria, Jordan, China			Kollo	Ethiopia
	Badami safra	Syria, Egypt, Lebanon	Tempe	Indonesia				
	Segared nuts	Turkey, Lebanon	Sauce	Japan				
Germination	Sprouts	India, Nepal	Snacks (sugar coated)	Indonesia				
Fermentation	Tempe, natto, dhokla, kiyit injera	India, Japan, Ethiopia	Shiro wot	Ethiopia, Indonesia				
Baking	Bread/roti	Turkey, India, Pakistan, Turkey, Burma	Fool Mekalley	Egypt	Kitta	Ethiopia, Australia		
Extrusion	Noodle	India, Nepal, Pakistan						
Green immature seeds	Veg. Curry	India, Pakistan			Veg. Curry	North America, Europe, Middle East		

A greater emphasis must be placed on marketing new products from legumes. For example, quick-cooking dhal, also called instant dhal has a good market potential. Even though dehulling and splitting into dhal reduces the cooking time, the cooking process is time and energy consuming. Special soaking solutions containing inorganic salts have been used for quick-cooking of legumes (Rockland et al. 1979). The cooking time of field peas could be reduced by soaking in sodium carbonate or bicarbonate solutions (Black et al. 1998). However, such treatment adversely affect the quality. Pectinase enzyme treatment significantly decreased the cooking time as compared to salt solutions and it also improved the acceptability of the product (Singh and Rao 1995). The legumes can be used in various foods including noodles, breads, snacks, soups, stews, and purees. Legume flour can be blended up to 20-30% in sausages and beverages with acceptable results. Chickpea flour can be used in tortilla and unleavened bread. Protein concentrates by pin-milling and protein isolates by precipitation of these legumes can be prepared containing approximately 50-60% and 80-90% protein, respectively. They should have a place in second generation functional foods, with and without blending with cereals.

STORAGE

Storage conditions affect the processing and grain quality of pulses which are stored at farmer, trader, or government levels in various structures. Quantitative and qualitative losses occur during storage. The result from interactions among the storage conditions, physiological changes during ageing, and the activity of storage micro-organisms and insects. The pulse beetle (bruchid) is an important pest of chickpea, lentil and field pea. Sometimes, the insect begins its infestation in the field and is carried into the store with the grain. The dehulling yields are considerably reduced by insect infestation as the bruchid causes physical damage to the seed. The greater susceptibility of large-seeded cultivars of soybean to field deterioration was used to explain their poor storability in comparison to small seeded cultivars (Verma and Gupta 1975).

Cultivars differ in their susceptibility to bruchid attack. However, seed size, colour, and texture have not been related to bruchid preferences, but seed-coat thickness may have an influence.

Storage under conditions of high temperature and humidity may render the grain susceptible to the hard-to-cook (HTC) defect. Beans with this defect require long cooking times, they are less acceptable to the consumer and are of lower nutritive value. According to Vindiola et al. (1986), the mechanisms causing HTC could be 1) limited hydration of intercellular protein; 2) pectin insolubilisation in the middle lamella by Ca and/or Mg; and 3) cross-linking of phenolics (lignification) and/or protein in the middle lamella of the cell wall. None of these phenomena has been systematically studied in the cool season food legumes. The accumulation of uric acid in pulses stored for 4-5 months generally make them unacceptable to the consumers. Protein quality is also adversely affected when pulses are stored for long periods.

Fungi and bacteria are also harmful to stored pulses. Chickpeas can become contaminated with mycoflora and mycotoxins (Nahdi et al. 1982). Aflatoxin B1 and B2 were present in 50% of the bean samples tested in Columbia (Usha et al. 1991). The low germination of pea seeds following commercial storage has been attributed to *Aspergillus* infection (Fields and King 1962). Storage effects on some quality parameters such as colour and hydration capacity of faba bean, lentils, and field peas have been noticed (U Singh unpublished). As the storage period increased, the time taken for cooking also increased irrespective of the type of pulse stored and the treatment given (Vimala and Pushpamma 1985). The loss of protein solubility and digestibility, development of off-flavours, and oxidative rancidity as a result of storage, need to be studied.

FUNCTIONAL PROPERTIES AND THEIR APPLICATION

Functional properties are assuming significance in view of the utilisation of legumes in cereal-based composite flours. Functionality can be defined as the set of properties of a protein or protein ingredient that contributes to the desired flavour, texture, and nutritive value of a food. Functional characteristics of defatted flours and concentrates are provided both by the proteins of the seed and by the carbohydrates (Martinez 1979). A better understanding of the functional properties is essential if future genotypes are to be utilised for the development of value-added food products. In this context, chickpea and field pea proteins are expected to have wider applications. Such functional properties as water absorption, oil absorption, emulsification capacity, flour solubility, swelling capacity, gelation capacity, gel consistency (gel spread), nitrogen solubility index (NSI) and paste-viscosity are considered important.

Because of the increasing costs of cooking oil and from a health point of view, the preparation of deep-fried products, using a minimum of oil is desirable. Chickpea proteins greatly reduce oil absorption by deep-fried products (Singh et al. 1993). According to this study, the oil absorption of the product ('Seviya') differed significantly among genotypes and this could be due to differences in the chemical and physical nature of the proteins. Emulsification has been reported to influence the product stability and large differences exist among chickpea genotypes. At ICRISAT, we have observed that the emulsion capacity of desi genotypes was twice that of the kabuli genotypes. However, the emulsion capacity appears to be lower than that of soy flour and winged bean. More importantly, the emulsion capacity of chickpea flour was 2-3 times higher than that of wheat flour (Iyer and Singh 1997). This trait could have a beneficial effect in wheat-chickpea composite.

The nature and type of protein and starch influence the gelling ability of cereal and legume flours. The gel consistency of cereal flour could be decreased by supplementation with legume flours, particularly chickpea and field pea. Unlike wheat starches, no break down of peak viscosity is noticed in chickpea, faba bean, field pea, and lentil. There were large differences in peak viscosity and gelatinisation temperature of pulses as calculated from the Brabender viscoamylograph (Table 7). Generally, cereals contain fragile swollen starches, which first swell and then break down under the continuous stirring of the viscoamylograph. This characteristic is not noticed in cool season legumes (Table 7), indicating that starch granules of these legumes have greater stability against mechanical shear than those of the cereal starches. Generally, lentil, faba bean and field pea starches have a high water binding capacity (92.4 to 98.0%), being comparable to that of wheat starch (Bhatty 1988). These legumes also show restricted swelling power that result in a type C Brabender curve (i.e. no pasting peak at 95°C). This property is typical of high amylose corn starches. Such starches may have only limited uses in food applications. But they may be modified, for example pea starch, by acetylation and phosphatisation. This destroys the heat and acid stability of the starch granules and makes them functionally similar to corn starch derivatives (Comer and Fry 1978). The addition of chickpea flour to wheat flour may not adversely affect such dough properties, but would enhance product stability due to the higher emulsion capacity of chickpea flour. These are some examples of the improved functional properties of cereal-legume composite flours (Iyer and Singh 1997).

Table 7. Some functional properties of cool season food legume flours[a]

Legume	Water absorption[a] g/g	Swelling capacity[b] g/g		Gelatinisation temp. (°C)	Peak viscosity(BU)
		65°C	95°C		
Chickpea	1.12	2.30	6.14	72.0	340
Faba bean	1.20	1.75	3.80	82.5	80
Field pea	0.83	2.03	4.15	70.5	115
Lentil	1.04	2.46	8.24	73.5	525

a   Mixing the flour with water at 30°C for 30 min.

b   Heating the sample in a block heater at 65°C and 95°C respectively.

NUTRITIONAL COMPOSITION

There is a voluminous literature on the nutritional composition of pulses, including the cool season pulses. They are good sources of carbohydrates and proteins, which together constitute about 70-80% of the total dry seed weight. Large variations have been reported in the protein content of chickpea, field pea, lentil, and faba bean. Legume seeds contain a considerable amount of non-protein nitrogen (NPN). In one study, chickpea NPN varied between 5.84 and 16.48% of meal nitrogen (Williams and Singh 1987). Any large variation in NPN could affect the estimated true protein of the sample. Agronomic and environmental factors affect the protein content. Salinity may significantly reduce the seed size and protein content in pulses.

Chickpea seed crude protein content ranges between 12.6-30.5% (Williams and Singh 1987). Starch is the principal carbohydrate constituent and generally ranges between 40-50%. The starch contains 20-30% amylose and the remainder is amylopectin (Singh 1985). The bioavailability of carbohydrates is important in terms of calorific value. Unfortunately, the concentration of unavailable carbohydrates in chickpea is the highest among the commonly consumed Indian pulses. Chickpea can add a considerable amount of fat to a consumer's diet. Among the cool -season legumes, chickpea contains the highest fat content ranging between 4.0 and 10.0%. Chickpea is also a good source of minerals and trace elements. Faba beans contain considerable amounts of protein (25-30%), carbohydrates (60-65%), ash (3-4%), and fat (2-3%). Traditionally, beans are utilised by soaking and cooking in the house or consumed as commercially processed canned beans. Generally, beans require a long cooking time to achieve satisfactory palatability and to improve the digestibility of proteins and carbohydrates. Peas contain 25-35% protein, 40-50% starch, 5-10% soluble sugars, 2-4% ash and 4-10% crude fibre, depending on the varieties and growing environments.

The protein content of lentil ranges between 19.5 and 36.4% and about 90% of the protein is present in the cotyledons (Bhatty and Christison 1984). Lentil is lower than most pulses in antinutritional factors such as haemagglutinins, oligosaccharides and favogens. Starch is the major component of lentil carbohydrates and may vary from 35 to 53% (Reddy et al. 1984). Lentil seeds contain < 1.0% lipids. Because of low lipid content, the gross energy of lentil is similar to that of wheat and barley. The low levels of antinutritional factors, together with a higher protein level and a shorter cooking time than other common pulses, make lentil very suitable for human consumption.

The nutritional quality of these legumes is greatly influenced by processing practices. Germination improves amino acid availability, increases the availability of vitamins, and decreases concentrations of phytic acid and trypsin inhibitors. Fermentation improves nutritional quality by solubilising proteins, inactivating antinutritional compounds, and increasing water-soluble vitamins. David and Verma (1981) reported improved nutritional quality of faba bean 'tempe' over non-fermented faba bean. Bhatty and Christison (1984) evaluated the nutritional quality of pea, faba bean, and lentil as meal, protein concentrates and protein isolates and found the products as digestible as casein protein.

ANTI-NUTRITIONAL AND TOXIC FACTORS

Many pulses are known to contain anti-nutritional factors (ANFs). However, recent epidemiological work suggests that some of these compounds may even have beneficial effects for humans. For example; phytate, some tannins and trypsin inhibitors also have an anti-oxidant effect that may help prevent the onset of cancer (McIntosh and Topping, 1998, these proceedings).

Alkaloids: The lupin alkaloids are usually bicyclic (e.g. lupinine), tricyclic (e.g.angustifoline) or tetracyclic (e.g. sparteine) derivatives of quinolizidine. An important exception is gramine, which is found in some cultivars of *L. luteus*. The major alkaloids of commercial cultivars are: *L. albus*: lupanine, 13-hydroxylupanine, sparteine (some European cultivars); *L. angustifolius*: lupanine, 13-hydroxylupanine, angustifoline; *L. mutabilis*: lupanine, sparteine, 13-hydroxylupanine; *L.luteus*: gramine (some cultivars only), lupanine, cytosine.

Modern cultivars of 'sweet' *L. angustifolius* typically contain < 200 mg alkaloids/kg (Harris and Jago 1984) in contrast to 'bitter' wild and green manuring types which may contain 5-40 g alkaloid/kg depending on the species and growing conditions. The typical alkaloid profile of 'sweet' *L. angustifolius* is: lupanine (42-59%), 13-hydroxylupanine (24-45%), angustifoline (7-15%), α-isolupanine (1-1.5%) and traces of other alkaloids (< 1% total). There is an effect of growing environment on seed alkaloid content, and new cultivars are lower in alkaloid than the cultivars they replaced.

Wink et al., (1995) reviewed the alkaloids in lupins and the changes that have occurred with breeding low-alkaloid cultivars. For example, albine is not present in cultivars of *L.albus* grown in Australia, and gramine is present in new cultivars of *L. luteus* being developed in Poland and Australia. These compounds have a low toxicity, LD50 values around 2000 mg/kg. Their pharmacological effects include blocking ganglionic transmission, decreasing cardiac contractility and contracting uterine smooth muscle (Mazur et al,. 1966).

Phytate: Phytic acid is an inositol hexaphosphate which forms complexes with minerals and protein, interfering with their availability. Together with the lesser substituted homologues and their collective salts these compounds are referred to as phytate. Phytate can form insoluble complexes with divalent cations, particularly calcium and zinc, rendering them less available for absorption and utilisation. The net effect of phytates depends on the overall composition of the food or diet, particularly the protein content and characteristics and the total mineral content. The amount of phytate in lupins does not seem to be of major concern. For example, Petterson et al. (1994) found the absorption of Zn from a range of lupin (*L. angustifolius*) based foods to be overall higher than from comparable soy products. They concluded that lupin milk could be an attractive alternative to soy milk for infant formulae (Petterson et al. 1994). The phytate content of lupin seed can be lowered by practices such as germination (Dagnia et al. 1992) and fermentation (Fudiyansyah et al. 1995).

Polyphenols: Polyphenols are loosely or interchangeably termed as tannins. Practically all tannins are polyphenols, but all polyphenols are not tannins. This is particularly true for cool season food legumes which contain significant amounts of polyphenols in their seed coats/husk and small amounts of tannins. Polyphenols of chickpea and pigeonpea reduce the activities of protein and carbohydrate hydrolysing enzymes (Singh 1984).

Tannins: Tannins are compounds of plant origin with molecular weights ranging from about 500 to 2,000 daltons, and with one to two phenolic hydroxyl groups per 100 daltons. This enables them to form cross-linkages between proteins and other macromolecules (Griffiths, 1991). The two sub-groups are hydrolysable tannins, which typically have central glucose core with the hydroxyl groups being wholly or partly esterified with gallic acid or hexahydrodiphenic acid, and condensed or ion-hydrolysable tannins which are higher oligomers of flavan-3-ols with varying degrees of substitution. In lupins, as with other grain legumes, the polyphenols are concentrated in the seed coats and dehulling minimises any effects. The concentration of condensed tannins, those most responsible for protein-binding, is so low in lupins that it is unlikely to affect human nutrition. Concentrations in *L. albus* and *L. angustifolius* are typically < 0.01% (Petterson and Mackintosh, 1994).

Saponins: Saponins can produce a soapy foam even at low concentrations, and are well known for their ability to lyse red blood cells *in vitro*. Saponins are glycosides with the non-sugar moiety being known as a sapogenin, which may be a steroid or a triterpenoid compound. Saponins are generally harmless to humans, and some may be beneficial by lowering blood cholesterol levels (Fenwick et al. 1991). The saponins of liquorice and quinoa used in the confectionary industry and are among the active ingredients of ginseng and other health foods and tonics (Fenwick et al 1991).

Protease inhibitors: The cool season pulses may contain protease inhibitors. Variations exist in trypsin and chymotrypsin activities of desi and kabuli genotypes of chickpea (Williams and Singh 1987). Lentils contain small concentrations of trypsin (Bhatty, 1988). Trypsin inhibitor activity is very low in lupins, ranging from < 0.01 to 0.28 mg/g in *L. angustifolius* and from 0.1 to 0.2 mg/g in *L. albus* (Petterson and Mackintosh, 1994). Chymotrypsin inhibitor activity was reported at < 0.01 to 0.59 mg/g in *L. angustifolius*.

Lectins: The conventional assay procedure agglutination of at least some of a wide range of different types of red blood cells have not detected any sign of lectin activity in *L. albus* or *L. angustifolius* (B.N. Greirson and D.S. Petterson, 1990, Perth, unpublished results; Liener, 1989). A slight activity can be induced if the red cells are specially treated (A. Pustzai, Aberdeen, 1995, personal communication), however this is not thought significant. Duranti et al., (1995) isolated conglutin γ from *L. albus* seed and showed it was capable of binding to various N-glycosylated proteins. Immunological homology with two bean lectins suggests there could be some lectin activity by this protein, which is almost certainly the activity observed by Pustzai (cited above).

Oligosaccharides: Chickpea is well known for its content of oligosaccharides. Field peas, lentils and faba bean contain lesser amounts. The oligosaccharides of pulses belong to the raffinose family. Raffinose has one galactose moiety linked to a sucrose molecule through an α 1,4 bond, stachyose has an α 1,4 linkage to the galactose unit of raffinose, and so on through verbascose and ajugose. These compounds cannot be metabolised by monogastrics and they pass through to the colon where bacteria break them down to produce carbon dioxide, methane and hydrogen. This causes abdominal discomfort, cramps and flatulence leading to a reduced interest in consuming these pulses in many societies. However the oligosaccharides have a role in osmotic regulation in the gastrointestinal tract which may be beneficial in maintaining flora that help prevent the onset of colonic cancer.

Neurotoxins: Prolonged consumption of *Lathyrus sativus* causes neurolathyrism in humans due to the ODAP present in all parts of the plant. It is highly desirable to develop genotypes free of this toxic compound.

FACTORS INVOLVED IN MARKETING OF COOL SEASON FOOD LEGUMES

In an era where population increase threatens to outstrip food production the dogmas that 'hungry people will eat anything' and 'if you feed the people they will not fight each other' are not necessarily true. People do have taste and texture preferences, and in most countries, including 'developing' countries, provision of foods of inferior quality results in waste. Some of the wasted food is used to feed domestic animals but this is an inefficient use of the funds allocated to the alleviation of hunger.

Food safety is becoming an important factor in marketing. Customers are demanding assurance that grains are either free, or meet specifications on pesticide, herbicide or fumigant residues, mycotoxins and other substances. Monitoring for chemical residues is expensive, and costs usually have to be built into the purchase price. Certification under ISO 9000 is becoming important, and some agencies will not accept analytical data from laboratories unless they have attained the necessary certification.

Scientific institutions should devote more attention to the organoleptic evaluation of foods. Members of taste panels need not be scientists; and other people capable of identifying differences in aroma, taste and texture. The most important attribute of a taste panel member is consistency of evaluation.Institutions, need to establish taste panels, to identify the most important characteristics of foods.

FUTURE CONSIDERATIONS

Genotypes and processing practices influence the nutritional quality and utilisation of pulses. Genotypic variation exists in nutritional factors, cooking time/quality and processability of these legumes. Processability is affected by the chemical and physical nature of the seed coat and by cotyledon hardness, which are influenced by the environment. The heritability of these characteristics needs to be established. Although large-scale pulse processing methods appear to be adequate for dehulling, further studies of preconditioning treatments are needed. To meet the consumer demands in various markets, quick cooking products of these legumes could be developed by combining suitable genotypes with pre-treatments that improve dehulling and efficiency and reduce cooking time. In particular, improving the cookability of whole seeds of cool season food legumes would reduce the need for alternative processing such as sprouting, germination, fermentation or dehulling. The development of quality standards and suitable genotypes for specific consumer needs should be emphasised. The flour, protein concentrate (air-classified fraction), and protein isolate are the three major components of pulses that could be used in value-added food products and more efforts are needed to enhance the use of these fractions in food products. Keeping in mind the new market requirements in pulse importing countries, the development of varieties of these legumes for specific end-uses must be emphasised.

References

Abas El Faki, H., Venkataraman, L.V. and Desikachar, H.S.R. 1984. *Qualitas Plantarum* 34:127-133.

Anderson, J.C., Idowu, A.O., Singh, U. and Singh, B. 1994. *Pl. Fd. Hum. Nutr.* 45: 371-379.

Bhatty, R.S. 1988. *Canadian Institute of Food Science and Technology Journal J.* 21: 144-160.

Bhatty, R.S. and Christison, G.I. 1984. *Qualitas Plantarum* 34: 41-51.

Black, R.G., Singh, U. and Cheryl, S. 1998. *Journal of the Science of Food and Agriculture* (In press).

Comer, F.W. and Fry, M.K. 1978. *Cereal Chem*istry 55: 818-824.

Dagnia, SG, Petterson, DS, Bell, RR and Flanagan, FV 1992. *Journal of the Science of Food and Agriculture* .60: 419-423.

David, I.M. and Verma, J. 1981. *Journal of Food Technolnology* 16: 39-50.

Duranti, M, Gius, C and Scarafoni, A. 1995. *Journal of Experimental Botany* 46: 725-728

Erskine, W., Williams, P.C. and Nakkoul, H. 1985. *Field Crops Research* 12: 153-161.

Erskine, W., Williams, P.C. and Nakkoul, H. 1991. *Journal of the Science of Food and Agriculture* 57: 85-92.

166

Fenwick, GR, Price, KR, Tsukamoto, C and Okuba K 1991. Saponins In *Toxic substances in Crop Plants* Ch 12 (eds JPF D'Mello, CM Duffus and JH Duffus) The Royal Society of Chemistry. Cambridge.

Fields, R.W. and King, T.H. 1962. *Phytopathology* 52: 336-339.

Fudiyansyah, N., Petterson, DS, Bell RR and Fairbrother, AH 1995. *International Journal of Food Science and Technology* 30: 297-306

Griffiths, DW 1991. Condensed tannins. In *Toxic substances in Crop Plants* (eds JPF D'Mello, CM Duffus and JH Duffus) The Royal Society of Chemistry. Cambridge.

Harris, DJ and Jago, J.,1984. Report of the Government Chemical Laboratories, Perth Australia pp 12

ICRISAT 1987. Annual Report. International Crops Research Institute for the Semi-Arid Tropics. pp. 157-159.

Iyer, L. and Singh, U. 1997. *Food Australia* 49: 27-31.

Keeler, RF 1973. *Teratology* 7: 23-30

Li, Q.L. 1987. In: *Techniques of Grain Food Processing*. pp. 294-299, Beijing, China, Food Publishing House of China.

Liener, IE 1989. In *Recent advances of research in anti nutritional factors in legume seeds* pp 6-13 (ed J Huisman, TFB van der Poel and IE Liener) Pudoc Wageningen

Martinez, W.H. 1979. *J. Amer. Oil Chem. Soc.* 56: 280-283.

Mazur, M, Polakowski, P and Szadowska, A 1996. *Acta Physiologia Polonica* 17 299-309

McIntosh G. and Topping, DL 1998 In *Proceedings IFLRC-III: Linking Research and Marketing Opportunities for the 21st Century.* (ed R. Knight) Dordrecht: Kluwer Academic Publishers (in press).

Nahdi, S., Nusrath, M., Batool, H. and Nagamani, V. 1982. *Indian Journal of Botany* 5: 1196-199.

Petterson, DS and Mackintosh, JB 1994. *The chemical composition and nutritive value of Australian grain legumes* Grains Research and Development Corporation, Canberra

Petterson, DS, Sandstrom, B and Ceberblad, A. 1994 *British Journal of Nutrition* 72: 865-71

Reddy, N.R., Pierson, M.D., Sathe, S.K. and Salunkhe, D.K. 1984. *Food Chemistry* 13: 25-41.

Reichert, R.D., Oomah, B.D. and Young, C.G. 1984. *Journal of Food Science* 49: 267-272.

Rockland, L.B., Zaragosa, E.M. and Oracca-Tetteh, R. 1979. *Journal of Food Science* 44: 1004-1007.

Singh, K.B., Nakkoul, H. and Williams, P.C. 1988. *Journal of the Science of Food and Agriculture* 44: 135-142.

Singh, U. and Seetha, R. 1983. *Journal of Food Science* 58: 853-855.

Singh, U. 1984. *Nutrition Reports International* 29: 745-753.

Singh, U. 1985 *Qualitas Plantarum* 35: 339-351

Singh, U., Rao, P.V. and Seetha, R. 1992. *J. Fd. Comp. Anal.* 5: 69-76.

Singh, U. and Singh, B. 1992. *Economic Botany* 46: 310-312.

Singh, U. 1995. *Journal of Food Science and Technology* 32: 81-93.

Singh, U. and Rao 1995. *Journal of Food Science and Technology* 32: 122-125.

Verma, R.S. and Gupta, P.C. 1975. *Seed Research* 3: 39-44.

Vimala, V. and Pushpamma, P. 1985. *Journal of Food Science and Technology* 22: 327-329.

Vindiola, O.L., Seils, P.A. and Hoseney, R.C. 1986. *Cereal Foods World* 31: 538-552.

Williams, P.C., Nakkoul, H. and Singh, K.B. 1983. *Journal of the Science of Food and Agriculture* 34: 492-496.

Williams, P.C. and Singh, U. 1987. In *The Chickpea,* pp. 329-356 (eds M.C.Saxena and K.B. Singh) CAB International, Wallingford, Oxon, UK.

Williams, P.C., Erskine, W. and Singh, U. 1993. *Lens Newsletter* 20: 3-13.

Wink, M, Meissner, C and Witte, L., 1995. *Phytochemistry* 98: 139-53

Produce quality: bulk and niche market opportunities for food and feed.

T. D. McGreevy,

USA Dry Pea and Lentil Council ,5071 Hwy. 8 West, Moscow, Idaho, 83843-4023, USA

Abstract

The world's population is predicted to increase from its current 5.8 billion to over 8 billion by the year 2030. The emerging markets of today will require high quality food and feed legumes to meet the growing demand for protein and dietary fiber. The transition to a global economy will create new wealth and new opportunities for legumes that can be easily processed into consumer-friendly products. As the economy of the world grows, so too will the demand for animal protein. Niche markets for high quality legumes for animal consumption will continue to expand as the livestock industry increases its need for inexpensive protein supplements with superior quality. As the world population and economy expands, the demand for a variety of high quality legume products for human and animal consumption will increase accordingly.

WORLD POPULATION

In November 1996, delegates from around the world gathered at the World Food Summit to discuss global food supplies and the programs needed to feed a growing population. The World Bank predicts that the current world population of roughly 5.8 billion people will increase to 8.7 billion people by the year 2030. Thirty-five years from now the world will have 3.0 billion more people to feed and shelter. The majority of this population increase will occur in the Indian Sub-Continent, Asia, Africa and South America. The rapid rise in world population will create increased demand for high-protein legumes for human and animal consumption in the next half century.

WORLD LEGUME PRODUCTION & CONSUMPTION

The Food & Agriculture Organisation (FAO) estimated world legume production (peas, lentils, and beans) at just over 58 million metric tons in 1994. Current estimates by the FAO put world legume production at about 60 million metric tons. During the 1960's and 1970's, legume production stood at around 40 million metric tons and grew slowly at about 4% per year. Production picked up to a yearly 5.5% during the 1980's as the EU, Turkey, Canada and Australia made a concerted effort to expand the use of legumes in their cereal crop rotations. Collecting information on carry-over stocks in some of the largest legume producing countries is difficult. However, the FAO reports that production and consumption of legumes closely follow each other.

POTENTIAL WORLD DEMAND FOR LEGUMES

If the world's population reaches 8.7 billion people and the consumption of legumes stays constant on a per capita basis, production of legumes will need to exceed 90 million metric tons by the year 2030. World legume production will need to increase by at least 30 million metric tons in the next 35 years in order to accommodate the increasing demand.

R. Knight (ed.), Linking Research and Marketing Opportunities for Pulses in the 21st Century, 167-171.
© *2000 Kluwer Academic Publishers. Printed in the Netherlands.*

Table 1 Countries of Major Significance in the World Legume Market for the year 1994.

COUNTRY	Pulse Production MT	Pulse Imports MT	Pulse Exports MT	Apparent Consumption
World	58,647,000	7,384,164	7,793,434	58,237,730
India	14,536,000	554,888	59,497	15,031,391
China	6,078,000	72,190	1,462,004	4,688,186
Brazil	3,307,000	258,384	382	3,565002
Russian Fed.	3,000,000	0	48,000	2,952,000
France	3,450,000	168,654	898,255	2,720,399
Ukraine	2,636,000			2,636,000
Nigeria	1,750,000	40	0	1,750,040
Mexico	1,702,000	83,072	164,135	1,620,937
Turkey	1,752,000	13,155	411,672	1,353,483
U.S.A.	1,552,000	89,523	599,307	1,042,216
Canada	2,028,000	31,317	1,067,609	991,708
Spain	211,000	748,411	7,494	951,917
Pakistan	621,000	184,197	794	804,403
Germany	204,000	557,756	15,531	746,225

FAO Trade and Production Yearbooks

GLOBAL PERSPECTIVE

The long-term outlook for legumes appears to be promising, given population growth and current consumption patterns. The question the international research community faces in the next 35 years is how to reach 90 million metric tons of production with a fixed or declining amount of arable land. The research community also faces the challenge of developing new varieties that meet higher quality standards and consumer acceptance. The movement toward a global economy is providing new opportunities for high-quality legume products. In the long-term consumer demand looks bright, but in the short term the competitive market for dry peas, lentils and beans is cut-throat.

Focus on Quality

To survive in today's market you must produce a quality product at a competitive price. As buyers become more sophisticated, I predict that quality standards will increase for both food and feed uses. Legume buyers have their own set of quality standards and specifications for colour, size, appearance and cooking characteristics. These variations make it difficult for plant breeders who may produce a variety that works for one market but is less desirable in another.

Legume buyers around the world share an appreciation for a clean, wholesome product that meets the requirements of the contract. In the United States, all legumes that are exported must have a Certificate of Inspection from the United States Department of Agriculture Federal Grain Inspection Service. The grading standards in tables 2-4 have evolved over the past 35 years to reflect the increasing quality demands placed on our farmers from our international customers. Farmers in the United States are interested in filling the high-quality niche markets around the world, and these quality standards reflect their commitment to producing a quality product. Several of our customers demand an even higher degree of quality than is represented by these grading standards.

Table 2a Grading standards of the U.S. Department of Agriculture for whole peas (%)

	Weevil Damage	Heat Damage	Other Damage	Other Classes	Bleach
US No 1	0.3	0.2	1.0	0.3	1.5
US No 2	0.8	0.5	1.5	0.8	3.0
US No 3	1.5	1.0	2.0	1.5	5.0

Table 2b. Grading standards of the U.S. Department of Agriculture for whole peas (%)

	Split Peas	Shrivelled Peas	Cracked Seedcoats	Foreign Material	Minimum for Colour
US No 1	0.5	2.0	5.0	0.1	good
US No 2	1.0	4.0	7.0	0.2	good
US No 3	1.5	8.0	9.0	0.5	fair

Table. 3a. Grading standards of the U.S. Department of Agriculture for split peas (%)

	Weevil Damage	Heat Damage	Other Damage	Contrasting Peas	Whole Peas
US No 1	0.5	0.2	1.0	0.3	0.5
US No 2	1.0	0.5	1.5	0.8	1.0
US No 3	1.5	1.0	2.0	1.5	2.0

Table 3b. Grading standards of the U.S. Department of Agriculture for split peas (%)

	White Caps	Bleached Peas	Foreign Material	Minimum for Colour
US No 1	0.5	2.0	5.0	good
US No 2	1.0	4.0	7.0	fair
US No 3	1.5	8.0	9.0	poor

Table. 4. Grading standards of the U.S. Department of Agriculture for lentils (%)

	Weevil Damage	Heat Damage	Foreign Material	Stones	Skinned	Minimum for colour
US No 1	0.3	0.2	0.2	0.1	4.0	good
US No 2	0.8	0.5	0.5	0.2	7.0	fair
US No 3	0.8	1.0	0.5	0.2	10.0	fair

Market Opportunities

The remainder of my comments will focus on the market opportunities for the various classes of peas, lentils and chickpeas produced by growers in the U.S.A. The following two markets as well as the surrounding countries are the largest legume producers in the world, and are likely to be the largest importers in the future.

India

India has a population of 940 million people. By the year 2050, India is expected to grow to 1.8 billion people. Eighty percent (80%) of the population is vegetarian. India is the largest producer of legumes at 12 to 14 million metric tons. The country is also one of the largest importers of pulses at over 500,000 metric tons. Projected legume consumption is estimated to be 28 million metric tons per year. India has an emerging middle class of 200 million people who demand quality food products. The demand for high-quality peas, lentils and chickpeas in India, Pakistan, Bangladesh, and Sri Lanka will continue to grow.

China

China, with its 1.2 billion people, is one of the largest legume producers in the world. They produce between 4.0 to 6.0 million metric tons of peas, beans, and lentils annually. Traditionally, China has been a net exporter of legumes, but that trend is slowly changing as their population grows at a rate of slightly over 1.0% per year. Due to a short crop, China imported over 193,000 metric tons of dry peas and beans in 1996. The imports were dominated by dry peas, with purchases over 122,000 metric tons. Eighty percent of the dry pea imports went to animal feed with the remaining 20% purchased by snack and bean paste manufacturers. As the population and economy grows, so will the demand for peas and other legumes for bulk livestock feed and for high edible quality.

Whole Green Peas
Specifications: Light green seed coat, dark green cotyledon, 6.0-8.0 mm in size
Primary End Users: Consumers, Canners, Food Manufacturers (Snackfoods, Soups, etc.)
Quality Demands: Dark green colour, Uniform size and Cooking quality, Fresh Pea Taste
Biggest Market: India- 940 million people in 1996. Projected to have 1.8 billion in 2050. Primary Markets: Indian Subcontinent, Asia, Europe, Central and South America.

Green Split Peas
Specifications: Dark green cotyledon
Primary End Users: Consumers, Food Manufacturers (Soups, etc.)
Quality Demands: Dark Green Colour (eye appeal), Uniform Size, and Good Cooking Quality
Primary Markets: Americas, Europe

Marrowfat Green Peas
Specifications: Large Seed 8.0 to 10 mm with a light green seed coat and dark green cotyledon
Primary End Users: Consumers, Canners, Food Manufacturers (Snackfoods)
Quality Demands: Consistent dark green colour with a large uniform size
Primary Markets: Asia, Europe

Whole Yellow Peas
Specifications: Deep yellow Cotyledon, large uniform size 6.0 to 9.0 mm
Primary End Users: Livestock Feed and Human Consumption
Quality Demands: Livestock producers want higher protein content above 25% and a consistent nutritional profile in order to better formulate their livestock rations. Human consumers want uniform seed size, colour and good cooking quality.
Primary Markets: European livestock feeders remain the single biggest market for yellow peas in the world. Europe consumes between 400,000 to 500,000 MT of field peas each year with the majority of those peas of the yellow variety. China has emerged as a large buyer of field peas, primarily yellow, as they purchased over 100,000 MT last year. Human consumption of whole yellow peas is concentrated in the Middle East, Indian subcontinent, and Cuba.

Yellow Split Peas
Specifications: Deep yellow cotyledon
Primary End Users: Human Consumption, Food Manufacturers
Quality Demands: Dark Yellow Colour (eye appeal), Uniform Size, and Good Cooking Quality
Primary Markets: Yellow splits are a high-quality niche market spread throughout the Indian Subcontinent, Middle East, and parts of the Americas.

Austrian Winter Peas
Specifications: Mottled dark green/brown seed coat with yellow cotyledon.
Primary End Users: Green Manure, Forage Crop, Bean Paste, and Bird Seed
Quality Demands: Consistent size, good germination and good plant growth
Primary Markets: Small niche market for producers interested in green manure crops or forage for livestock. There is a small market ion Asia for use as bird seed.

Large Green Lentils
Specifications: Large bright green, non-mottled seed coat with yellow cotyledon 6.0-8.0 mm in size
Primary End Users: Consumers, Food Manufacturers (soups, etc)
Quality Demands: Large consistent size above 7.0 mm, consistent bright green colour, and good cooking quality
Primary Markets: Europe, Mediterranean, Central and South America, Africa and the U.A.E.

Medium Green Lentils
Specifications: Medium sized 5.0-6.0 mm with light green seed coat and yellow cotyledon
Primary End Users: Consumers, Food manufacturers (soups, etc)
Quality Demands: Consistent size and light green colour, good cooking quality
Primary Markets: Mediterranean, The Americas, Arab and African Countries.

Small Green Lentils
Specifications: Small seed size 3.5 to 5.0 mm seed size with light green seed coat and yellow cotyledon.
Primary End Users: Consumers, Food Manufacturers, and Canners
Quality Demands: Consistent size and light green colour, good cooking quality
Primary Markets: Europe (Greece), Central and South America, Africa and the U.A.E.

Red Lentils (Small seeded)
Specifications: Small seeded 3.5-5.0 mm with a reddish-brown seed coat and red cotyledon.
Primary End Users: Consumers, Food Manufacturers
Quality Demands: Consistent size with bright red cotyledon and good cooking quality
Primary Markets: Red lentils are most popular in the Indian sub-continent, Africa, the U.A.E., and Europe. Most of the red lentils are decorticated (or split) for use in curries and dals in these countries.

Red Lentils (Large seeded)
Specifications: Large seed size 5.0-7.0 mm with red cotyledon
Primary End Users: Consumers
Quality Demands: Consistent large size with bright red cotyledon
Primary Markets: Niche market for quality conscience end users in the Americas, Indian sub-continent, and Middle East.

"Pardina" Small Seed Brown Lentils
Specifications: Small seeded 3.5-5.0 mm Speckled greyish brown seed coat with yellow cotyledon.
Primary End Users: Consumers, canners, and food manufacturers
Quality Demands: Consistent Colour, High Quality, and Good Cooking characteristics.
Primary Markets: Spain.

Kabuli Chickpeas (Large Seeded)
Specifications: Large Seeded 7.0-10.0 mm, creamy-white seed coat; golden yellow cotyledon
Primary End Users: Consumers, canners, food manufacturers
Quality Demands: Large seed size, consistent colour, and good cooking characteristics.
Primary Markets: Europe, Mediterranean, Africa and the U.A.E., Americas, Asia, Indian Sub-continent. Large milky white chickpeas are also popular in the same markets as the Kabuli type blond chickpeas.

Desi Chickpeas (small seeded)
Specifications: Small Seeded 6.0-9.0 mm with brownish seed coat.
Primary End Users: Consumers
Quality Demands: Consistent size and colour, good cooking characteristics
Primary Markets: Indian sub-continent, Africa and the U.A.E., Europe.

CONCLUSION

As the population increases substantially over the next 30 years, so too will the demand for high-protein crops such as legumes. Peas, lentils, and chickpeas enjoy a wide, global acceptance, and niche markets for legumes will grow along with the population as consumers in various cultures seek legumes that fit their varied tastes and traditions. The world research community, in tandem with growers, face the task of developing legume crops that can accommodate this growing population and still meet the varied characteristic needs of a diverse, worldwide market.

Produce quality of food legumes: Genotype (G), environment (E) and (GxE) considerations

T. Nleya', A. Vandenberg', G. Araganosa[2], T. Warkentin[3], F.J. Muehlbauer[4] and A.E. Slinkard'

1 Crop Development Centre, University of Saskatchewan, 51 Campus Drive, Saskatoon, SK, Canada. S7N5A8; 2 Department of Applied Microbiology and Food Science, University of Saskatchewan, 5 1 CampusDrive, Saskatoon, SK, Canada S7N 5A8; 3 Agriculture and Agri-Food Canada, Morden Research Centre, Unit 100-101 Route 100, Morden, MB, Canada. R6M 1Y5; 4 USDA-ARS, Johnson Hall, Washington State University, Pullman, WA 99164, USA

Abstract

Food legume crops are grown in a wide range of environments on every continent. Investigations of the quality of pulses for human consumption can be classified as nutritional or culinary quality characters. Nutritional quality is usually described in terms of protein concentration, protein quality and the type and concentration of antinutritional factors. The marketplace in general does not determine the price of a pulse on its nutritional quality. In contrast, culinary quality is the major price-determining factor. Culinary traits include seed size, seed shape, seed coat colour, cotyledon colour, uniformity and purity. If pulses are used in industrial food manufacturing, characters such as thermal processing, splitting, or flour milling and attributes such as hydration coefficient, cooking time, texture, viscosity and flavour are important. Studies of the effects of genotype (G), environment (E), and GxE interaction on nutritional and culinary quality of chickpea, lentil, faba bean, pea and common bean indicate that many quality traits have not been thoroughly investigated. An analysis of variance can be used to distinguish between crossover and non-crossover GxE interactions. This type of analysis provides information on the strategies that will be most effective for the improvement of quality traits of food legumes.

INTRODUCTION

Food legumes are an important component of human diets, and in some regions such as South Asia, are a major source of dietary protein. In many countries, they are attracting more interest as a result of the increased awareness of the benefits of using food legumes as a protein (replacing animal protein) and complex carbohydrate (fibre) source in human nutrition. The nutritional quality of food legumes has been a focus of research as part of a global effort to improve the human food supply. Culinary quality is a focal point for marketplace negotiation of the value of food legumes and breeders are increasingly interested in improving the level and stability of quality over a range of environments.

Produce quality depends on the genetic constitution of the cultivar and on the environmental conditions experienced during the growing season. It may be further influenced by storage conditions (environment) that modify culinary quality. Knowledge of the type of gene action involved and the degree to which the genes are influenced by environmental conditions is fundamental to the development of breeding strategies to improve quality.

The large body of literature on antinutritional factors in food legumes and their effects on livestock feeding is not extensively reviewed in this paper. The focus is on the influence of genotype and environment on the quality traits that influence human nutrition and acceptance in food markets. In this paper produce quality is defined as those characteristics that affect (i) nutritional quality and (ii) culinary quality related to marketability such as physical appearance and cooking, manufacturing and thermal processing characteristics. Principal aspects of nutritional quality include protein concentration, protein quality (amino acid composition and digestibility) and antinutritional factors (Table 1). Culinary quality includes physical and chemical properties of the dry and cooked grain that influence consumer acceptance and processor

R. Knight (ed.), Linking Research and Marketing Opportunities for Pulses in the 21st Century, 173–182.

standards (Hosfield, 1991) (Table 1). For example, various consumers have different preferences in colour, seed size, texture, appearance and ease of cooking. Food legume packagers and manufacturers are influenced by consumer preferences, but they also seek commercial lots that are uniform for physical and for processing characteristics. For instance, thermal processors require legumes that hydrate uniformly and leave no sludge at the bottom of the can while maintaining acceptable drained weight and texture after cooking (Hosfield, 1991, Williams et al., 1994). In this paper, the discussion will focus on chickpea *(Cicer arietinum)*, faba bean *(Vicia faba)*, lentil *(Lens culinaris)* and field pea *(Pisum sativum)*.

Table 1. Main components of quality for chickpea, faba bean, lentil and pea

Nutritional quality factors	Culinary quality factors
Protein concentration	Seed characteristics
	-appearance
Protein quality	-size, shape, uniformity
- digestibility	- colour
- amino acid composition	-damage (cracking, splitting)
Antinutritional factors	Thermal processing
- tannins	- water absorption
- protease inhibitors	- cooking time
-glycosides	- texture
- flatus factors	-granulation
	-viscosity
	- flavour

GXE INTERACTION

Plant traits are expressed through a phenotype, which consists of genotypic (G) and environmental (E) components. The E component can be either predictable or unpredictable. Predictable E factors such as harvest management, storage temperature, storage humidity, handling procedures and manufacturing processes occur systematically and can be controlled relatively easily. Unpredictable E conditions, usually associated with climatic factors, can mask the G component. Depending on the environment and the agronomic regime, E components are described in terms of location (L), year (Y) or season (S) effects. If unpredictable E effects are large, significant genotype x environment (GxE) interactions occur as a result of differential response among genotypes. The GxE interactions reduce the correlation between genotypic and phenotypic values and complicate therefore the design and conduct of breeding programs.

The changes in genotypic response may be in the form of rank (qualitative/crossover) or changes in response may be a matter of scale (quantitative/non crossover) (Baker, 1988a; Romagosa and Fox, 1993). Quantitative GxE interactions result from differential responses of some genotypes compared with others in terms of scale. Genotypes with superior traits will also be superior in other environments. Quantitative interactions do not cause major problems to breeding programs because sampling across environments is predictable when no change in rank occurs. Qualitative or crossover interactions, however, complicate the selection and decision making process because of the changes in rank across environments. This makes it difficult to identify superior cultivars. GxE crossover interactions for quality traits imply that the same genotypes grown in different environments differ in nutritional or culinary quality.

Baker (1988b) discussed GxE interaction in terms of differential responses to environmental stresses. Resistance, adaptability and stability may refer to differential responses to changes in environmental stress. When changes in ranking of genotypes occur, the interactions may be due to differential response to two or more limiting factors. The GxE interaction may also be due to genotypic differences in response to environmental stresses that differ in pattern from one environment to another. For example, if the environmental stress is disease to which some cultivars are resistant, obvious crossover GxE interactions will occur if the disease is severe in some environments and not in others. Likewise a late season rain in one environment, but not in another, will result in GxE for yield. If the genotypes of indeterminate food legumes differ in maturity, they consequently differ in the ability to respond to late season rainfall.

GXE INTERACTION AND FOOD LEGUME NUTRITIONAL QUALITY

The range in protein concentration for four major food legumes is shown in Table 2. Chickpea has the narrowest range in protein concentration, while the range in the other three legumes is wide. Protein digestibility is low in faba bean (59%) and high in lentil, chickpea and pea (71-94%) (Williams andNakkoul, 1983). Digestibility is reduced by the presence of anti-nutritional factors such as protease inhibitors and tannins in the seed, which are generally high in faba bean (Williams et al., 1994).

Table 2. Protein concentration in seeds of four cool season food legumes

Crop	Form	Protein concentration (N x 6.25 as mg/kg)
Desi chickpea	Whole uncooked	172-220[x]
Kabuli chickpea	Whole uncooked	181-240[x]
Faba bean	Whole uncooked	186-378[y]
Lentil	Whole uncooked	195-355[z]
Lentil	Whole decorticated uncooked	195-325[z]
Lentil	Whole uncooked	205 - 314[z]
Lentil	Whole decorticated uncooked	225- 290[z]
Pea	Whole uncooked	155 - 397[y]

x Bhatty (1997, unpublished)   y Williams et al. (1994).   z Bhatty (1995a).

Food legumes are rich in lysine (LYS) and poor in methionine (MET) and therefore are complementary supplements to cereals that are rich in MET, but poor in LYS. In comparison with the FAO reference egg protein, food legumes are deficient in most essential amino acids with the exception of arginine, LYS and leucine (Williams et al., 1994).

Chickpea

In addition to genetic make up, environmental factors such as location (L), year (Y), and season (S) influence protein concentration of chickpea seeds (Dodd and Pushpamma, 1980, Dahiya et al., 1982, Awasthi and Abidi, 1987, Dahiya et al., 1982, Hulse, 1994). Significant GxL and GxS interactions occurred for protein concentration among chickpea cultivars (Dahiya et al., 1982, Dodd and Pushpamma, 1980, Singh et al., 1990). Location effects on protein concentration of chickpea seed may be more pronounced than G effects (Dodd and Pushpamma, 1980) or S effects (Singh et al., 1990). However, chickpea cultivars that maintain high protein concentration over different seasons and different locations are available (Dahiya et al., 1982, Singh et al., 1990).

The levels of MET and tryptophan vary in chickpea (Awasthi and Abidi, 1987) and are influenced by L effects (Dodd and Pushpamma, 1980), but no interaction for amino acid composition of chickpea seed has been reported.

Variation in the level of antinutritional factors such as tannins and alpha-amylase inhibitors, and oligosaccharides has been reported (Chavan et al., 1989). The extent to which GxE interactions occur for the content of specific amino acids or antinutritional factors has not been reported.

Faba bean

The protein concentration in faba bean seed is influenced by G, Y, L, yield and seed position within the plant (Chavan et al., 1989, Griffiths and Lawes, 1978, Sjodin, 1982). The great part of the variation in protein concentration among faba bean genotypes is due to heritable factors (Griffiths and Lawes, 1978, Sjodin 1982). Significant GxY (Picard, 1977) and GxL interaction (Bond, 1977) effects for protein concentration in faba bean seed may occur. However, both reports (Picard, 1977, Bond,1977) concluded that the variation due to GxE interaction was far smaller than the variation due to G and that no important changes in ranking of genotypes occurred in different environments.

Considerable variation exists in amino acid composition of faba bean cultivars (Bond, 1977, Chavan et al., 1989). Faba bean has variable concentrations of antinutritional factors such protease inhibitors, tannins, lectins, flatus factors and beta-glycosides, such as vicine and convicine (Chavan et al., 1989, Hussein, 1982, Pitz et al., 1981). Considerable variation for protease inhibitors and tannins exist among faba bean Genotypes (Hussein, 1982) but GxE interaction studies have not been reported.

Lentil

The protein concentration of lentil was strongly influenced by Y effects, but the GxE interaction for protein concentration was not significant (Bhatty, 1984). Erskine et al. (1985), working with 24 lentil genotypes grown in three environments in Syria and Lebanon, observed non-significant GxE interaction effects for protein concentration. The broad-sense heritability for protein concentration was 0.71, indicating strong genetic control of protein concentration.

Pea

Ali-Khan and Young (1973) studied the variation in protein concentration of 19 pea cultivars grown for a single year at four locations and 10 pea cultivars grown for three years at a single location in western Canada. They reported significant effects on protein concentration due to G, Y, L and GxL interaction.

A more recent analysis of GxE effects on protein content of field pea, using data for recently developed cultivars (Grande, Carneval, Keoma and Radley) for two years at three locations in western Canada, showed almost half of the total variation for protein content was due to G effects (Report of the 1995 and 1996 Field Pea Co-operative Trials). Variation in protein concentration due to L contributed about 25% of total variation, whereas variation due to Y was much less. The GxY interaction for protein concentration was significant. The average protein concentration among cultivars varied from 218 to 244 mg/kg, among years from 228 to 237 mg/kg and among locations from 221 to 238 mg/kg. The level of trypsin inhibitors in pea has been a concern to pea breeding programs in Europe where this trait is used to discriminate among new cultivars. To date this has not been an issue of concern in the food pea market. Recent research in New Zealand (A Russell, personal communication) suggests that the number of isoforms and concentration of trypsin inhibitors in food pea cultivars is highly variable.

DISCUSSION

An important issue is how much notice is taken of protein in the marketplace for human food. Cereal and oilseed crops are usually processed prior to consumption as refined food products, and the marketplace segregates produce on the basis of protein or oil quantity and quality. Increasingly, processors are restricting purchases to specific cultivars based on quality characteristics. For food legumes, neither protein quantity nor quality is recognized as a pricing factor, unless the issue of the presence or the level of antinutritional factors comes into play as a regulatory issue. Current examples are toxins found in the seed of grass pea *(Lathyrus sativus)* or common vetch *(Vicia sativa)*. Even in feed pea markets, protein concentration and quality are not marketing issues. Traders focus on foreign matter content and the comparative value of feed pea relative to the dominant protein and energy sources (soybean meal and corn).

Some food manufacturers express interest in the chemistry of food legumes, either to comply with food labelling regulations or to gain market share. In value-added markets, much of the interest relates to nutritional information on specific compounds such as soluble fibre level, or concentration of folic acid, tannin, or other components. The quest for increased market share, based on unique food chemistry and the development of unique value-added products, may result in differential pricing for various chemical quality attributes in the future.

As food legumes become more important in food processing and manufacturing, levels of protein, fibre and oil will become factors in the pricing and marketing structure. The ability to overcome environmental fluctuation in nutritional quality by influencing levels of protein quality, protein quantity, fibre, and oil through conventional breeding or biotechnological applications has improved dramatically in recent years.

The question still remains as to whether or not application of these technologies represents a sound investment in research and development for specific food legume crops.

GXE INTERACTION AND FOOD LEGUME CULINARY QUALITY

Culinary quality attributes are the primary marketing factors for food legumes. Food legumes as a group are sold on the basis of seed characteristics or processing characteristics (Table 1). The limits of culinary quality are determined by the genotype and the agronomic environment of the crop, but can be manipulated through post-harvest handling and storage environments and by processing.

Chickpea

A significant GxE interaction for seed weight and cooking time was reported for kabuli chickpea (Singh et al., 1990). Table 3 summarizes the analysis of the effects of G, L and GxL on canning quality traits for six kabuli chickpea cultivars grown at six locations in western Canada in 1996. GxL effects were significant for all traits except percent washed drained weight, texture and appearance. The light/dark colour score (L score, Hunter colorimeter) showed a GxE interaction for the uncooked dry seeds but not for the cooked seeds.

Table 3. Mean squares from the analysis of variance, means and ranges for different canning quality traits for six kabuli chickpea cultivars grown at six locations in western Canada in 1996

			Mean squares		
Trait	Location(L)	Cultivar(C)	LxC	Mean	Range
Dry Seed					
100-seed wt (g)	329.7 **	1152.3 **	12.2 *	43.7	23.7-63.3
HC^u (Ratio)	0.040 **	0.014 **	0.003 **	2.10	1.97- 2.25
Surface colour^v					
L	78.7**	29.5**	4.5**	50.11	42.9-56.4
a$_L$	4.6**	4.7**	0.4**	7.9	5.6-9.5
b$_L$	3.2**	3.5**	0.3*	15.3	14.2- 18.2
Cooked seed					
WDWT^w (g)	53.5**	103.5**	17.8**	186.3	178.7- 198.5
PWDWT^x (%)	7.0*	12.2**	2.9	58.8	51.1-68.1
Texture^y (g/kg)	3494.9 **	1167.8 **	162.1	164.8	115.7- 196.9
Surface colour^v					
L	6.8**	11.4**	1.2	48.1	42.9- 50.6
a$_L$	14.9 **	5.4 **	0.4 **	9.3	6.8 - 11.8
b$_L$	4.3 **	4.7 **	0.5 **	19.2	17.2-21.1
Appearance^z	4.3 **	1.5 **	0.3	4.3	3.0 - 5.0

u- Hydration coefficient. v- Hunter colorimeter scores. w - Washed drained weight.
x- Percent washed drained weight. y - As measured with a Kramer shear press.
z- Scale of 1 poor to 5 good. *, ** Significant at the 0.05 and 0.01 levels, respectively

Dhal, one of the most important forms of chickpea consumption, is produced by tempering the seed with water or steam followed by removal of the seed coat (decortication) and splitting the cotyledons. The extent to which GxE interactions affect the recovery rate during this process is unknown but reduced seed size and quality can be a problem.

Faba bean

Wide variation was reported for cooking characteristics in 93 faba bean samples from commercial fields in Egypt (Shehata, 1982). Variation in seed weight, swelling coefficient, hydration coefficient, softness, granulation and cooked bean colour were reported (Table 4). Significant E effects were observed for these cooking characteristics but the importance of the GxE interaction was not reported. GxL interaction effects cooking characteristics but the importance of the GxE interaction was not reported. GxL interaction effects for seed size, hydration coefficient and hard seeds were significant for faba bean cultivars grown in a number of locations in Sudan (Saxena and Stewart, 1983).

Table 4. Range of variability in cooking characteristics of 93 faba bean samples from commercial fields in Egypt in 1979 and 1980 (Source: Shehata 1982)

Cooking characteristic	Range
Dry seeds	
Weight of 1000 seeds (g)	430- 926
Hydration coefficient (after 8 hours)	1.61-1.95
Swelling coefficient (after 8 hours)	1.72-2.19
Stewed beans	
Hydration coefficient	2.20-2.91
Colour score (1 poor to 5 good)	2.1-4.0
Softness score (1 soft to 10 hard)	5-10
Granulation score (out of 10)	0.0-9.1
Penetrometer reading (kg/cm2)	17.6-80.3

Lentil

The cooking quality of lentil has been extensively studied (Bhatty, 1984, 1995; Bhatty et al., 1983, 1984; Erskine et al., 1985). Bhatty et al. (1983) reported significant GxL interaction effects for cooking quality (texture, measured using shear force) for two lentil cultivars grown at 12 locations (L). G had a small effect on texture of cooked lentil compared to L. The influence of E on cooked lentil quality was further investigated by Bhatty et al. (1984) who reported GxL and GxY interaction effects for texture. Bhatty (1984) concluded that available soil phosphorus (P) levels and its effect on phytic acid (PA) concentration in the seed lead to these GxE effects on cooked lentil texture. Reduction in lentil cooking time with increased levels of seed PA and soil P was later confirmed by Bhatty (1995b). Lentil seed samples requiring less cooking time (less shear force or reduced texture) had the highest concentration of PA and P. Significant effects of L and GxL on cooking time due to seed PA concentration and seed P concentration were reported.

No significant GxE interaction effects for cooking time or seed weight were observed by Erskine et al. (1985) working with 24 lentil genotypes grown at three environments in Syria and Lebanon. Broad-sense heritability estimates for cooking time and seed weight were 0.82 and 0.98, respectively, suggesting strong genetic control of these traits.

Diameter of lentil seed affects the grading of large-seeded (macrosperma) lentil and dehulling and splitting of small-seeded (microsperma) lentil. Macrosperma lentil marketing is based on seed size and cultivars with the largest seeds command the highest prices, provided other quality attributes such as uniformity and colour are optimum (Williams et al., 1993). Within the microsperma range, large seeds are preferred because they are easier to split and dehull. Seed size, though a varietal characteristic, is strongly influenced by L and E (Erskine et al., 1985) and significant GxE interaction may occur (Erskine et al., 1991). Erskine et al. (1991) also looked at the effect of E on dehulling and concluded that the influence of G was greater than that of either growing L or S. Rounder lentil seeds are easier to decorticate and split, compared to flatter seeds. Seed shape is under strong genetic control with little influence of E (Williams et al., 1993).

Pea

Factors influencing the cooking quality of field pea have been extensively studied (Chernick and Chernick, 1963, Halstead and Gfeller, 1964, Mattson et al., 1951). Chernick and Chemick (1963) concluded that cooking quality of pea (measured by weight ratio of puree to hard-to-cook material) is under genetic control. However, Halstead and Gfeller (1964) reported significant GxE interaction effects for cooking quality (texture of puree) for two pea cultivars grown in the greenhouse. Mattson et al. (1951) reported significant GxE interaction effects for cooking time of field pea, and seed phytin content was associated with cooking quality. Possible factors of influence were P and K fertility.

Chemick and Chemick (1963) suggested that cooking quality of peas is improved by a combination of factors promoting complete maturity of seed such as late harvest, large seeds and seeds from the lower pods. Drake and Muehlbauer (1985) reported cultivar differences for hydration ratio in canned green pea.

Food quality data from 1995 and 1996 in western Canada for four pea cultivars grown at three locations were analysed to determine the level of GxE interaction. Colour of puree was a cultivar characteristic with little influence from growing conditions (Table 5). Granulation of puree was strongly influenced by L effects (Table 5). The Y x L x G interaction for granulation of puree was also significant, indicating that granulation scores for individual cultivars were inconsistent across environments. Puree viscosity was a cultivar characteristic though L effects may influence this trait (Table 5). Unpublished investigations (Crop Development Centre, University of Saskatchewan) of GxE effects on seed shape, seed dimpling, and tendency to retain green seed coat for yellow cotyledon pea suggest that these factors show significant non-crossover GxE interaction.

Table 5. Mean squares from the analysis or variance for color, granulation and viscosity of puree for four pea cultivars grown in two years at three locations in western Canada.

Source of variation	d.f.	Colour	Granulation	Viscosity
Year (Y)	1	0.845	0.011	12.25
Location (L)	2	0.365	2.128*	41.40
Y x L	2	0.121	0.130	11.42
Rep in Y x L	12	0.059	0.105	1.90
Cultivar (C)	3	1.156*	0.206	92.63*
Y x C	3	0.041	0.073	4.14
L x C	6	0.075	0.114	27.56*
Y x L x C	6	0.083	0.125*	5.65*
Error	36	0.048	0.044	1.72
Total	71			

For green cotyledon pea, seed colour intensity and consistency are important quality traits because bleached product loses value. G and E effects are important considerations for pea breeding programs. Analysis of bleaching scores for seven green cotyledon pea cultivars at 13 locations in 1995 and 12 locations in 1996, showed that in both years, about 57% of the total sums of squares was due to G effects compared to 20% due to both GxL and L effects (Table 6).

Table 6. Analysis of variance for bleaching scores of seven pea cultivars grown in 13 locations in 1995 and five cultivars grown in 12 locations in 1996 in western Canada

Source of variation	d.f.	S.S.	M.S.
		1995	
Location (L)	12	12.12	1.01**
Rep in L	26	2.43	0.09
Cultivar (C)	6	61.25	10.21 **
LxC	72	30.39	0.28 **
Error	156	10.57	0.07
Total	272	106.77	
		1996	
Location (L)	11	12.07	1.10**
Rep in L	24	3.67	0.15
Cultivar (C)	4	68.30	17.07**
LxC	44	24.77	0.56**
Error	96	11.33	0.12
Total	179	120.14	

DISCUSSION

Table 7 summarizes the status of GxE interaction studies for quality traits in chickpea, faba bean, lentil, pea, and common bean *(Phaseolus vulgaris* L.). Investigations of this nature are more numerous for common bean (Ghaderi et al, 1984, Hosfield et al , 1984, Shellie and Hosfield, 1991), and include extensive analyses of canning traits. Thermal processing quality traits such as clumping, splitting, and washed drained weight of bean are strongly influenced by GxE interactions (Hosfield et al., 1984). As consumption of food legumes increases, demand for ready-to-eat and canned food legume products also increases. The high quality standards for canned products suggest that the importance of genotype and environmental effects on canning quality of food legumes must be studied more thoroughly.

Table 7. Summary of GE interactions reported for major quality traits of chickpea, faba bean, lentil, pea and dry bean

Trait	Chickpea[v]	Fababean[w]	Lentil[x]	Pea[y]	Dry bean[z]
Nutritional Quality					
Protein content	*	*	n.s	*	*
Anti-nutritional factors	n/a	*	n/a	n/a	*
Thermal Processing					
Cooking time	*	*	*	*	*
Texture/granulation	n.s	n/a	*	*	*
Viscosity	n/a	n/a	n/a	*	n/a
Hydration coefficient	*	*	n/a	n/a	*
Water absorption	n/a	n/a	n/a	*	*
Colour	*	n/a	n/a	n.s	*
Seed characteristics					
Size	*	*	*	*	*
Shape	n/a	n/a	n/a	n.s	n/a
Bleaching/colour	*	n/a	n/a	*	*

*Significant GE interaction reported.   n.s GE interaction not significant.   n/a Information on GE not available.

Sources for significant GE interaction are v (Dahiya et al. 1982, Dodd and Pushpamma 1980, Singh et al. 1990) w (Bond 1977, Picard 1977, Pitz et al. 1981, Saxena and Stewart 1983) x (Bhatty et al. 1983, 1984 and Erskine et al.1991) y (Ali-Khan and Young 1973, Halstead and Gfeller 1964, Mattson et al. 1951) z (Ghaderi et al. 1984, Hosfield et al. 1984 and Shellie and Hosfield 1991)

IMPLICATIONS OF GXE EFFECTS ON QUALITY FOR CROP IMPROVEMENT

If crop improvement efforts are increasingly directed to produce quality, the presence and nature of GxE interactions must be investigated for the major quality traits that affect marketability. Scientists involved in crop improvement must develop market awareness to determine which quality traits have highest priority. Experiments can be specifically designed at relatively low cost to determine the presence and type of GxE quality interactions. Detection of non-crossover interactions for a particular trait implies that long-term breeding strategies are required to improve the trait. The identification of crossover interactions implies that (i) the trait may have high heritability and (ii) the set of cultivars may show regional adaptation.

Consideration of adaptation and growth habit is critical to understanding how GxE interactions can be managed in crop improvement programs. In some environments, an indeterminate growth habit may lead to more variability in quality because the reproductive period is spread over a wider range of environmental conditions. Some conflicting reports on GxE interaction may be more easily interpreted in the context of the overall climate. For cooking quality (texture) in lentil, where GxE interaction was high in western Canada but low in Syria, the overriding consideration may in fact be that the Canadian studies were conducted in a northern temperate climate in which seed filling and maturity occurs as temperatures decrease. The physiological development of the quality profile may be entirely different in the Mediterranean climate in which seed filling occurs as temperatures increase.

Table 7 indicates that gaps remain in our knowledge of GxE interactions particularly on seed shape and colour characteristics. In addition to seed size, primarily identified by seed diameter, the GxE influences on uniformity of seed diameter has not been thoroughly investigated. Seed cleaning and processing companies express preferences for some cultivars based on uniformity of seed size because the cost saving during seed cleaning and processing. The extent to which uniformity is under genetic or environmental control is not well established. Genetic improvement in any trait can be achieved if sufficient variation exists for the trait in question and heritability for the trait is high. The above review indicates that in most cases where a large number of genotypes were studied for a particular trait, genetic variation was found. When niche markets are considered, for instance, marrowfat pea for food processing, or food legumes for sprouting, very little information is available on GxE interaction effects on quality.

Nutritional quality and culinary quality of food legumes are subject to variation caused by environmental factors. Significant GxE interaction exists for most quality traits. The extent to which breeders may address GxE interaction for a particular trait in a particular food legume depends on the economic importance of the trait in question, the crop, the type (non-crossover or crossover) and source of GxE interaction (year or location) and the relationship between the trait in question and other quality traits and/or yield.

Quantitative GxE interactions will not interfere with the identification and selection of genotypes with superior traits of interest and, therefore, are of minor consequence to plant breeders. If rank order of genotypes for a particular trait changes from one environment to another, GxE interactions should concern breeders. They will need to use the appropriate statistical analytical tools to understand the nature of GxE interactions (Romagosa and Fox, 1993; Baker, 1996). Few of the reports reviewed in this paper differentiated between crossover and non-crossover interactions.

Food legume pricing is based currently on visual aspects rather than on "hidden" quality factors. GxE interaction effects for visual traits such as seed diameter, seed shape, seed colour and uniformity are of primary importance to the breeder geared for export markets. For example, some kabuli chickpea production is geared for the snack market and large seed size and uniformity in seed size are the major criteria for grading. Since seed size of the legumes can be affected by environmental conditions (Table 7), breeders need to exercise strict standards to ensure quality at the lowest price for the export market. While "hidden" qualities such as protein concentration or protein quality do not determine prices for food legumes, breeders need to ensure that environmental variation does not reduce protein concentration or quality to unacceptable levels since legumes are an important source of protein in human nutrition.

The crop in question and the end-use of the product will also determine the importance of GxE interaction. For example, GxE interaction for canning quality traits is comparatively important for kabuli chickpea because the canned product is a major form of consumption in many markets.

The cause of the GxE interaction for a particular trait must be considered by plant breeders.. Using the analysis of variance, the GxE interaction term may be partitioned into components attributed to GxL, GxY and GxLxY. When the GxL interaction component comprises a major proportion of the total sums of squares, the problem may be easier to deal with than when GxY interaction is predominant. This is because variation due to location may be handled by subdividing the production area of the crop into uniform subregions and growing genotypes adapted within each region (Romagosa and Fox 1993). In certain instances, subdivision of the production area may be unnecessary, especially if the source of the GxE interaction at particular locations can be identified and corrected. For example, application of specific fertilizers may eliminate the GxE effects caused by fertility differences among locations. If, on the other hand, GxY and GxYxL interaction predominates, breeders may need to test genotypes more widely, sampling both spatial and temporal variation.

References
Ali-Khan, S. T. and Young, C. G. 1973. *Canadian Journal of Plant Science* 57: 85-92.
Awasthi, C. P. and Abidi, A. B. 1987. *Indian Journal of Agricultural Chemistry* 20(2): 163-70.
Baker, R. J. 1988a. *ISI Atlas of Science:Animal and Plant Science.* 1-4.
Baker, R. J. 1988b. In: *Proceedings of the 2nd International Symposium on Quantitative Genetics,* pp 492-504 (eds. B.S. Wier, E.J. Eisen, M.M. Goodman and G. Namkoong). Sunderland: Sinauer Associates, Inc.
Baker, R. J. 1996. In: *Proceedings of the 5th International Oat Conference* and *7th International Barley Genetics Symposium,* pp. 235-239 (eds. A. E. Slinkard, G. Scoles and B. Rossnagel).University of Saskatchewan: University Extension Press.
Bhatty, R. S. 1984. *Journal of Agricultural and Food Chemistry* 32(2): 1161-1166.
Bhatty, R. S. 1995a. *Cereal Foods World* 40(5): 387-392.
Bhatty, R. S. 1995b. *Journal of the Science of Food and Agriculture* 68: 489- 496.
Bhatty, R. S., Nielsen, M. A. and Slinkard, A. E. 1983. *Canadian Institute of Food Science and Technology Journal* 16(2): 104-110.
Bhatty, R. S., Nielsen, M. A. and Slinkard, A. E. 1984. *Canadian Journal of Plant Science* 64: 17-24.
Bond, D. A. 1977. In: *Protein Quality Leguminous Crops.* Luxembourg: Commission of European Communities.

Chavan, J. K., Kadam, S. S. and Salunkhe, D. K. *1989.* In: *CRC Handbook of World Legumes,* pp. 247-288 (eds. D. K. Salunkhe and S.S. Kadam). Boca Raton, Florida:CRC Press.

Chavan, J. K., Kute, L. S. and Kadam, S. S. 1989. In: *CRC Handbook of World Food Legumes,* pp. 223-245 (eds. D.K. Salunkhe and S.S. Kadam). Boca Raton, Florida:CRC Press.

Chemick, A. and Chemick, B. A. 1963. *Canadian Journal of Plant Science* 43: 175-183.

Dahiya, B. S., Kapoor, A. C., Solanki, I. S. and Waldia, R. S. 1982. *Experimental Agriculture* 18: 289-292.

Dodd, N. K. and Pushpamma, P. 1980. *Indian Journal ofAgriculturalScience* 50(2):139-44.

Erskine, W., Williams, P.C. and Nakkoul, H. 1985. *Field Crops Research* 12: 153-161.

Erskine, W., Williams, P.C. and Nakkoul, H. 1991. *Journal of the Science of Agriculture* 57: *85-92.*

Furedi, J. 1970. In: *Protein Growth by Plant Breeding,* pp. 99-128 (eds. A.Balint).Budapest:Akademiai Kiado.

Ghaderi, A., Hosfield, G. L., Adams, M.W. and Uebersax, M. A. 1984. *Journal of American Society of Horticultural Science* 109(1) 85-90.

Griffiths, D. W. and Lawes, D. A. 1978. *Euphytica* 27: 487-495.

Halstead, R. L. and Gfeller, F. 1964. *Canadian Journal of Plant Science* 44: 221-228.

Hosfield. G. L. 1991. *Food Technology* 45: 98, 100-103.

Hosfield, G. L., Uebersax, M. A. and Isleib, T. G. 1984. *Journal of American Society of Horticultural Science.* 109(2): 182-189.

Hulse, J. H. 1994. In: *Expanding the Production and Use of Cool Season Food Legumes,* pp. 77-97 (eds. F.J. Muehlebauer and W.J. Kaiser). Dordrecht: Kluwer Academic Publishers.

Hussein, L. 1982. In: *Faba Bean Improvement,* pp. 333-341 (eds. G. Hawtin and C. Webb). The Hague: Martinus Nijhoff Publishers.

Mattson, G. L., Akerberg, E., Eriksson, E., Koutler-Andersson, E. and Vahtras, K. 1951. *Acta Agriculturae Scandinavica* : 40-61.

Picard, J. 1982. In: *Protein Quantity for Leguminous Crops.* Luxembourg: Commission of European Communities. 153-161.

Pitz, W. J., Sosulski, F. W. and Rowland, G. G. 1981. *Journal of the Science of Food and Agriculture* 32: 9-16.

Romagosa, I. and Fox, P. N. 1993. In: *Plant Breeding Principles and Prospects,* pp. 373-390 (eds. M.D. Hayward, N.O. Bosemark and I. Romagosa). London: Chapman and Hall.

Saxena, M. C. and Stewart, R. A. 1983. *Faba Bean in the Nile Valley.* The Hague: Martinus Nijhoff Publishers.

Shehata, A. M. E, 1982. In: *Faba Bean Improvement,* pp. 354-362 (eds. G. Hawtin and C.Webb). The Hague: Martinus Nijhoff Publishers.

Shellie, K. C. and Hosfield, G. L. 1991. *Journal of American Society of Horticultural Science* 116(4):732-736.

Singh, K. B., Williams, P. C. and Nakkoul, H. 1990. *Journal of the Science of Food and Agriculture* 53: 429-441.

Sjodin, J. 1982. In: *Faba Bean Improvement,* pp. 319-331 (eds. G. Hawtin and C. Webb).The Hague: Martinus Nijhoff Publishers.

Williams, P. C., Bhatty, R. S., Deshpande, S. S., Hussein, L. A. and Savage, G. P. 1994. In: *Expanding the Production and the Use of Cool Season Food* Legumes, pp. 113-125 (eds. F. J.Muehlbauer and W. J. Kaiser). Dordrecht: Kluwer Academic Publishers.

Williams, P. C., Erskine, W. and Singh, U. 1993. *LENS Newsletter* 20(1): 3-13.

Williams, P. C. and Nakkoul, H. 1993. In: *Proceedings of the International Workshop on Faba Beans, Kabuli Chickpeas and Lentils in the 1980s,* pp. 245-256 (eds. M. C. Saxena and S. Varma). Aleppo, Syria: The International Centre for Agricultural Research in the Dry Areas.

Breeding for yield: The direct approach

A.E. Slinkard[1], M.B. Solh[2] and A. Vandenberg[1]

1 Crop Development Centre, University of Saskatchewan, Saskatoon SK S7N 5A8 Canada; 2 Nile Valley Research Project, ICARDA, P.O. Box 2416, Cairo, Egypt.

Abstract

Yield, a trait of low heritability, is conditioned by many genes with small additive effects in self-pollinated crops and is a difficult trait to improve. Thus, more efficient methods of breeding for yield are required and increased selection pressure must be placed on yield. The Recombinant-Derived Family method uses early generation selection for yield in $F_{2.4}$, $F_{2.5}$ and $F_{2.6}$ families to eliminate inferior crosses and inferior F_2-derived families. Regional and state yield tests are conducted on the best $F_{2.7}$ and $F_{2.8}$ families. A homogeneity phase for reducing phenotypic variability and increasing pre-Breeder and Breeder seed of selected $F_{6.7}$, $F_{6.8}$ and $F_{6.9}$ families is conducted concurrently with regional and state testing of $F_{2.7}$ and $F_{2.8}$ families. Results of the F_2–derived family yield tests determine which F_6-derived family will be released as the improved cultivar. The Recombinant-Derived Family method also results in the release of the improved cultivar one to five years earlier than from standard breeding methods, further emphasizing its greater efficiency.

INTRODUCTION

A plant breeding program should be designed to isolate, increase and release a superior cultivar as quickly and efficiently as possible. The release of a superior cultivar one or two years earlier than normal greatly increases the value of that cultivar to society since the benefits accrue in current funds, rather than in discounted funds in one or two years time. Likewise, a more efficient breeding program will result in the release of one or more superior cultivars or even an outstanding cultivar at less cost, or in a shorter time, than normal. Efficiency in a breeding program is usually measured in the number of years or generations between the time the cross is made to the time a superior cultivar from that cross is released. A more precise measure of efficiency is genetic gain per year or per unit cost.

Breeding objectives normally involve improving yield, disease resistance and quality. In most food legumes quality is not a major issue in that crosses are made between parents with acceptable quality and the selections should have acceptable quality. Thus, normally all that is required is to double-check the most promising lines prior to increase and release to avoid the "out-liers". Niche market quality attributes will require extra evaluation. In many food legumes, a major component of quality relates to visual attributes of the seed such as large size, round shape, a specific cotyledon color, a white seed coat or even a specific type of mottling on the seed coat. In most cases, other than size or colour, small differences in quality do not result in a premium price or a discount, at least not yet.

Most plant diseases cause varying degrees of loss in yield and quality and partial resistance is often adequate to minimize disease losses. The exception is a devastating disease, such as *Ascochyta rabiei* [Pass.] Labr. on chickpea (*Cicer arietinum* L.). Evaluation of the most promising lines for yield in the area of commercial production will result in the selection of genotypes with an acceptable, although marginal, level of disease resistance. However, the breeder must continue to upgrade the level of resistance in the breeding program. Thus, the primary objective of the breeding program

R. Knight (ed.), Linking Research and Marketing Opportunities for Pulses in the 21st Century, 183–190.
© *2000 Kluwer Academic Publishers. Printed in the Netherlands.*

must be to improve yield. This is reinforced by the fact that the food legume producer is paid for yield and yield only, assuming that quality is reasonable and acceptable.

If yield is the primary objective, the breeding program must be modified to increase selection pressure for yield. This can be done by early generation testing for yield in a manner that results in the accumulation of data on selected F_2-derived families. Since this method is most appropriate for self-pollinated crops (pea *Pisum sativum* L., lentil *Lens culinaris* Medikus, chickpea, etc.), discussion will be limited to its use in self-pollinated crops.

Food legumes differ from cereals in several respects that have a profound effect on the efficiency of a breeding program. The major difference is that cereals have a determinate flowering habit, whereas most food legumes have an indeterminate habit, i.e., they often have ripe pods and open flowers on the same plant. Indeterminate genotypes of a food legume have a higher yield potential then determinate genotypes, since they often are able to respond to a timely rain unless the plants have irreversibly senesced. Genotype by environment (G x E) interactions for yield often occur when determinate and indeterminate genotypes of a crop are tested over a number of environments.

Genotypes that differ in disease reaction also frequently show G x E interactions for yield when they are evaluated over a number of environments. Other cases of G x E interactions for yield cannot be readily explained. A crossover G x E interaction involves a significant change in rank of the genotypes (Baker, 1988), whereas a non-crossover G x E does not. Crossover G x E interactions require evaluation over additional environments in an effort to explain the cause. Often, the effect of crossover G x E interaction can be minimized by selecting one or two genotypes for one subset of environments and another one or two genotypes for another subset of environments.

A second major difference between food legumes and cereals is the "reproductive ratio", the average number of seeds per plant in commercial fields. Food legumes produce large seeds and the reproductive ratio is usually between 10 and 30. Wheat (*Triticum aestivum* L.) typically has a reproductive ratio between 50 and 100, whereas canola (*Brassica napus* L.) has a reproductive ratio between 200 and 600. Food legumes present special problems in that single plant selections often must be increased for two generations before enough seed is available for replicated yield trials at several locations. Thus, selection for yield in food legumes is often delayed at least one year longer than in cereals.

A third difference between food legumes and cereals relates to the ability of food legumes to symbiotically fix much of their nitrogen requirement. Nitrogen fixation is an energy intensive process and between 4 and 10 g C are required per g N_2 reduced (Schubert, 1982). Thus, the N_2-fixing food legume plant will have a reduced yield relative to a cereal plant in a specific environment.

A fourth difference between food legumes and cereals relates to the higher protein concentration in the food legume seed relative to the cereal seed, e.g., many food legumes have a seed protein concentration of 18 to 28%, relative to 9 to 14% for many cereals. Since plants require more energy to produce one gram of protein than one gram of carbohydrates, the yield of food legumes rarely will equal that of cereals in a specific environment. Since extra energy is required for the production of high protein seed in food legumes, it will be more difficult to attain a 10% increase in the genetic potential for seed yield in a food legume than in a cereal. The net result is that a "yield breakthrough", such as occurred in semidwarf wheat (*Triticum aestivum* L.), semidwarf rice (*Oryza sativa* L.), hybrid corn (*Zea mays* L.) and hybrid sorghum (*Sorghum bicolor*) will never occur in the food legumes. Efficiency of plant breeding must be increased in all crops, but it is most critical in food legumes in view of their low reproductive ratio, low seed yield, high photosynthetic energy requirements for N_2 fixation and protein synthesis. There is a continuing need for legume proteins to complement the essential amino acid profile of cereal proteins in human diets. This review suggests various means of increasing the efficiency of a plant breeding program, including the use of a modified breeding method.

STANDARD BREEDING METHODS FOR SELF-POLLINATED CROPS

The standard breeding methods are the 1) pedigree, 2) backcross, 3) bulk, and 4) various modifications of the bulk method, such as the mass pedigree (Harrington, 1937) and the single seed descent method (Grafius, 1965; Brim, 1966). In addition, many breeders have developed modifications of these methods as their breeding program matured. Recently, the doubled haploid method has been developed, based on the *Hordeum bulbosum* technique in barley (Kasha and Kao, 1970) or on anther culture in some other crops (Clapman, 1973). The single seed descent and the doubled haploid method are more efficient than the other methods. However, they both suffer from the absence of yield data prior to development of homozygosity, which must be followed by at least one generation of increase before yield testing is initiated. Unfortunately, food legumes are not amenable to anther culture at the present time and so the doubled haploid method will not be discussed further.

All breeding methods are effective, provided adequate genetic variability is present for the traits under consideration and the breeder can successfully separate the genetic from the environmental variability. Standard breeding methods are not very efficient when selecting for traits of low heritability, such as yield, in that selection is delayed several generations longer than deemed necessary by some breeders.

Several facts relating to yield and breeding for yield in self-pollinated crops are widely accepted by breeders and can be considered conventional wisdom:

Yield is conditioned by many (hundreds) genes with small additive effects.

Yield is a trait of low heritability.

Yield of an individual plant provides a very poor estimate of the yield of that genotype.

Yield of spaced plants provides a poor estimate of yield in a competitive stand.

Yield of an individual plot provides a poor estimate of the yield of that genotype in a crop.

Yields from a replicated yield trial in one environment provide only a preliminary estimate of the yield of those genotypes, relative to the yield of two or more check cultivars.

Yields from replicated yield trials in several environments (at least four), using incomplete block designs, provide the first meaningful estimate of yield of these genotypes, relative to the yield of two or more check cultivars.

Yields from replicated yield trials in many environments (at least 15), using incomplete block designs, provide a realistic estimate of the yield of these genotypes, relative to the yield of two or more check cultivars.

YIELD TESTING IN EARLY SEGREGATING GENERATIONS

The effectiveness of early generation testing (EGT) for yield has been evaluated in two ways. Yield trials of F_2 bulk populations were indicative of the potential of the different crosses (Harlan *et al.*, 1940; Harrington 1940, Immer 1941). However, Kalton (1948), Atkins and Murphy (1949) and Fowler and Heyne (1955) reported that yield trials of F_2 bulk populations were of little value in predicting the F_2 population that contained the highest yielding lines. Other researchers have selected plants in F_2 or F_3 (pedigree method) and evaluated their yield potential in replicated trials in succeeding generations. The general conclusion has been that individual plant selection in F_2 or F_3 is effective for highly heritable traits, such as plant height or maturity. However, individual plant selection for yield in early generations generally has not been successful (Fiyzat and Atkins, 1953; Briggs and Shebeski, 1971; Knott and Kumar, 1975).

The pedigree method was used in many of these early tests and individual plants were selected within the best F_3 lines, increased in F_4 and grown in a replicated yield trial in F_5. The F_5 yield was poorly correlated with yield of the original F_2 plant, no yield results were available until F_5, and usually only in one environment. Results like these have convinced many plant breeders to avoid use of the pedigree method in its strictest sense.

However, these failures should not be used to downplay the importance of EGT. Specifically, EGT should be used to identify F_2 plants with superior progenies, using the F_2 progeny method of Lupton and Whitehouse (1957). Frey (1954) was the first to propose a plant breeding method whereby selection was based on yield of F_2-derived families in the F_4 and subsequent generations. Individual plant selections would be made within the highest yielding F_2-derived families in F_6 to F_{10}, increased, evaluated for yield for three or more years and the best one released. This method has the advantage of EGT (F_4, F_5, F_6) for yield, but then the homozygous selections in F_6 to F_{10} must be increased and tested again before release. This version of the F_2-derived family method is only marginally more efficient than the standard breeding methods. However, Frey (1954) did comment on the apparent uniformity of plant type within F_2-derived F_3 ($F_{2:3}$) families and $F_{3:4}$ families.

Lupton and Whitehouse (1957) were more specific and proposed selecting individual plants from the $F_{2:6}$ families that were highest yielding in F_4, F_5 and F_6, increasing them in F_7 and then evaluating these selected F_6-derived progenies in F_8, F_9 and F_{10} with the best F_{10} selections being evaluated in regional or state trials in F_{11} and later generations. This is a two-phase selection scheme where the first selection phase is based on yield of F_2-derived lines in F_4, F_5 and F_6, and the second selection phase is based on yield of F_6-derived selections in F_8 to F_n. The scheme was designed to provide reasonable homogeneity in the resulting improved cultivar.

Boerma and Cooper (1975a) also used EGT to isolate high yielding lines. $F_{2:3}$ lines were grown in unreplicated single row plots, and selected $F_{2:4}$ and $F_{2:5}$ families were yield tested in three replications at two sites. Boerma and Cooper (1975b,c) yield tested $F_{5:6}$ lines within selected $F_{2:5}$ families and concluded that selection within the heterogeneous $F_{2:5}$ families could result in improved yield. Cooper (1990) further refined his EGT procedure: $F_{2:4}$ – 1 to 30 lines per selected cross; $F_{4:5}$ – 10 to 50 lines per selected F_2 family; $F_{4:6}$ – 1 to 50 lines per selected F_2 family. The key points of this approach were 1) to yield test $F_{2:3}$ families to identify best crosses, 2) yield test $F_{2:4}$ families to identify the best F_2 plants in the best crosses and 3) yield test $F_{4:5}$ and $F_{4:6}$ to identify the best pure lines remaining. Cooper (1990) used unreplicated plots up to F_6, rationalizing that family selection, using lines within a family (or cross) as a form of replication, partially offsets the disadvantage of unreplicated yield data. Replicated yield testing was concentrated on the high yielding lines in the F_7 and later generations. This procedure increases efficiency by reducing the land and labor requirements per cross, but it is not efficient in terms of the number of generations between making the cross and releasing Breeder's seed of an improved cultivar.

Singh and Smithson (1986) evaluated 46 chickpea (*Cicer arietinum* L.) crosses by selecting 232 F_2 plants and testing them as $F_{2:3}$, $F_{2:4}$ and $F_{2:5}$ families. The best five were evaluated as $F_{2:6}$ families in replicated yield trials in India. These $F_{2:6}$ families ranked 1, 4, 7, 17 and 54 (wilt susceptible) in yield in a 63-entry test, suggesting that the F_2-derived family method was effective in isolating high yielding and widely adapted families. They suggest multilocation yield testing in F_4 and F should be used$_5$ to avoid strong selection for a specific environment. These $F_{2:5}$ families are heterogeneous, resulting in yield stability. For the breeding of self-pollinated crops, Singh and Smithson (1986) also saw the F_2-derived family method as an effective way for the International Agricultural Research Centers to provide selected heterogeneous material to national programs for further selection.

Muehlbauer and Slinkard (1985) and Slinkard (1993) proposed a further modification of the F_2-derived family method that would result in the release of Breeder seed in the F_8 or F_9, one or more generations earlier than the standard breeding methods. Slinkard (1993) increased the efficiency of this scheme by omitting the testing of the F_6-derived families. Thus, he initiated regional or state yield trials using $F_{2:7}$ and $F_{2:8}$ families with release in F_8 or F_9. However, he concurrently made about 200 individual plant selections at random from each of the most promising $F_{2:6}$ families in an effort to provide reasonable uniformity for phenotypic traits, such as height, maturity, seed size, color, etc. (homogeneity phase). Each $F_{6:7}$ selection was grown as a microplot (short row) and all atypical or visibly segregating rows were discarded. Then each $F_{6:8}$ selection was grown as a miniplot (long row), either on site or in an off-season increase. Again, any atypical or segregating rows were discarded. The $F_{6:9}$ selections were either increased again or bulked and released as Breeder seed of the improved cultivar, depending upon seed size and the reproductive ratio.

Singh (1994) proposed the use of "gamete selection" for simultaneous improvement of multiple traits. He assumed that multiple-parent crosses are required for simultaneous improvement of multiple traits and that EGT is required for the selection of improved cultivars. Intercrosses are made between several parents that carry desirable dominant or co-dominant alleles. These heterozygous plants serve as the heterogametic male parent in crosses with several homozygous elite female parents. The resulting progeny are F_1-derived families and are increased for one or two generations. The $F_{1:3}$ or $F_{1:4}$ families are grown in replicated yield trials in areas characterized by different production stresses, where selection is for combinations of these multiple traits and EGT is practiced on these F_1-derived families.

Singh's "gamete selection" is merely a further modification of the F_2-derived family method, except that in this case the F_1 is the first segregating generation, whereas in hybrids between homozygous parents, the F_2 is the first segregating population. It is proposed that the name of the F_2 progeny method, as first proposed by Lupton and Whitehouse, and modified by Slinkard (1993) and others and the gamete selection scheme by Singh (1994), be changed to the "recombinant-derived family" (RDF) method. In each case, the unit of selection was a recombinant plant derived from the first segregating generation and it was evaluated on a family basis, with emphasis on EGT.

ADVANTAGES OF THE RECOMBINANT-DERIVED FAMILY METHOD

Selection of the best RDF capitalizes on the genetic variation among RDF, but apparently fails to exploit variation within the selected RDF, since the F_6-derived Breeder seed is based on random selections within the selected RDF. Elimination of off-type and segregating selections among $F_{6:7}$, and $F_{6:8}$ lines results in a heterogeneous cultivar that is relatively uniform in phenotypic appearance. Heterogeneous cultivars generally have wider adaptation and greater tolerance to various stresses than homogeneous cultivars.

A second advantage of the RDF method is that low yielding crosses and families within crosses are rapidly eliminated from the testing program. This helps overcome the primary limitation of most plant breeding programs - the inability to adequately test all selections for yield. Regardless of what people may say, plant breeding is still a numbers game and the sooner low yielding selections are eliminated, the more efficient is the breeding program. This allows the breeder to evaluate more crosses, further increasing the probability of developing an improved cultivar.

A third advantage of the RDF method is that Breeder seed of an improved cultivar can be released in F_8 or F_9 relative to F_{10} to F_{15} for many other methods, further improving the efficiency of the breeding process.

DISADVANTAGES OF THE RECOMBINANT-DERIVED FAMILY METHODS

The primary disadvantage of the RDF method is that the resulting heterogeneous cultivars may not be acceptable under "plant variety protection" legislation since they cannot be precisely described in a court of law. However, this is not a major concern in many public breeding programs and certainly not in most developing countries. In many parts of the world the priority is to develop an improved cultivar and to provide it to the grower as quickly as possible.

Cultivars developed by the RDF method are heterogeneous and for this reason may not be acceptable to some seed certification agencies. However, an agency's primary concern is with uniformity of phenotype as a guard against genetic admixture. Cultivars developed by the RDF method are phenotypically uniform within relatively narrow limits and are mostly acceptable to seed certification agencies.

OFF-SEASON INCREASES

The efficiency of a breeding program can be improved by growing more than one generation a year in off-season increases either in a greenhouse or in other areas of the world that have suitable growing conditions in the off-season.[4] For example, the University of Saskatchewan has grown four generations per year of bean (*Phaseolus vulgaris* L.) in Colombia.

Any generation in which yield is not evaluated can be grown in an off-season increase. Thus, the crosses can be made and the F_1, F_2 and F_3 can all be grown as off-season increases, if deemed necessary. In addition, $F_{6:7}$, $F_{6:8}$ and $F_{6:9}$ (Breeder seed) selections can be screened for phenotypic uniformity and increased in an off-season site. This makes Breeder seed somewhat expensive, but the economic benefit of having an improved cultivar available one year earlier, more than justifies the cost. The problem often is how can these benefits be reflected in increased funding for the breeding program to pay for the off-season increase of Breeder seed.

PHILOSOPHICAL CONSIDERATIONS

Parent selection is critical. A general guideline is to cross high yielding (adapted), but somewhat unrelated parents. Assuming additive gene effects for yield, the mean yields of the resulting segregating populations should equal the mean yield of the parents and the presence of complementary genes for yield in some crosses should result in some higher yielding RDF.

Since yield is so difficult to breed for and producers get paid for total production, it is critical that increased yield be the primary objective of a breeding program. To emphasize this, the authors state their breeding program has three objectives: 1) yield, 2) more yield and 3) still more yield. Thus, strong selection pressure must be placed on yield at every opportunity, starting in the $F_{2:3}$, if enough seed can be produced on the F_2 plant for a two-replicate test in F_3. The $F_{2:5}$ and $F_{2:6}$ yield tests must be conducted at several locations in the area of production in order to avoid extremely narrow adaptation.

Another philosophical consideration relates to the question of using many crosses with few selections per cross or few crosses with many selections per cross. For example, Hurd (1969) suggested growing 25,000 F_2 plants per cross and selecting the best 1000 or so for increase in F_3 and yield testing in F_4. This implies that the breeder can successfully select for yield based on the top 4% of the F_2 plants. However, earlier, it was concluded that selection for yield on an individual plant basis was ineffective. In many instances an F_2 population of 200 plants will provide a realistic sample of the yield potential of a cross. If a small F_2 population (200) is used, the breeder can evaluate more crosses and concentrate on the most promising ones, realizing that fewer than 5% of the crosses ever result in an improved cultivar. The quicker the plant breeder can discard the remaining 95% of the crosses, the more efficient the breeding program. Thus, Simmonds (1989) concluded that the frequency of high yielding genotypes is fairly high in plant breeding populations, but that selection efficiency is too low to fully exploit them. If this is true, then plant breeders must devise methods for increasing selection efficiency for yield. This means abandoning such superficial things as phenoptypic selection for yield and selecting for yield components and ideotypes (after Donald, 1968), which at best are yield neutral in their effect.

Yield should be evaluated under above average growing conditions so that genetic differences in yield of the various lines can be expressed. The breeder should not select for yield in a trial where the average yield is lower than the average commercial yield in an adjacent field. For example, a salt-tolerant cultivar of barley (*Hordeum vulgare* L.) may have a higher yield than an improved cultivar in a saline site. However, the field is not uniformly saline and an improved cultivar will result in higher total production from the field than the salt-tolerant cultivar which has a lower yield in the more favorable parts of the field.

DISCUSSION

It is imperative that the efficiency of breeding for increased yield be improved. This is more critical for food legumes than for cereals because of their low reproductive ratio, low seed yield, high photosynthetic energy requirements for N_2 fixation and protein synthesis and the increasing need for legume proteins to complement the essential amino acid profile of cereal proteins in human diets. Efficiency can be expressed as genetic gain per year. Thus, something as simple as releasing an improved cultivar one year earlier than normal can be considered an increase in efficiency. An earlier than normal release of an improved cultivar also increases the benefits to society of that new cultivar.

To increase the efficiency of breeding for yield means that the selection pressure for yield must be increased. EGT for yield in F_4, F_5 and F_6 will eliminate low yielding crosses and low yielding progenies within the high yielding crosses. The standard procedure is to select plants from the best F_6 progenies, test these relatively homogeneous F_6 progenies for yield and then release the best one as an improved cultivar (if it is truly improved) in F_{10} to F_{14}.

The Recombinant-Derived Family method is a more efficient modification of the above procedure that involves regional and state testing of $F_{2:7}$ and $F_{2:8}$ families as a means of determining the highest yielding family. Concurrent with the $F_{2:7}$ and $F_{2:8}$ yield tests, 200 plants are selected at random from each of the most promising $F_{2:6}$ families, increased and off-type progenies eliminated in $F_{6:7}$ family microplots and $F_{6:8}$ family miniplots. Subsequently, based on the $F_{2:6}$, $F_{2:7}$ and $F_{2:8}$ yield test results, progeny lines the best $F_{6:8}$ family are bulked for release as Breeder seed in F_9. This is one to five generations (years) earlier than with standard plant breeding methods.

Efficiency, measured as the number of years between making crosses and releasing a cultivar from those crosses, can be increased by using off-season increases of all generations except those in which a yield test is required. In the RDF method, yield testing is done with the $F_{2:4}$, $F_{2:5}$, $F_{2:6}$, $F_{2:7}$ and $F_{2:8}$ families, whereas off-season increases can be made in F_1, F_2, $F_{2:3}$, $F_{6:7}$ and $F_{6:8}$. Thus, an improved cultivar can be released in a minimum of six years (five years of yield testing) after making the cross using the RDF method.

References

Atkins, R.G. and Murphy, H.C. 1949 *Agronomy Journal* 41: 41-45.

Baker, R.J. 1988. *Canadian Journal of Plant Science.* 68: 405-410.

Boerma, H.R. and Cooper, R.L. 1975a. *Crop Science.* 15: 225-229.

Boerma, H.R. and Cooper, R.L. 1975b. *Crop Science.* 15: 300-302.

Boerma, H.R. and Cooper, R.L. 1975c. *Crop Science.* 15: 313-315.

Briggs, K.G. and Shebeski, L.H. 1971. *Euphytica* 20: 453-463.

Brim, C.A. 1966. *Zeitschrift fur Pflanzenzuchtung* 65: 285-292.

Clapman, D. 1973. *Zeitschrift fur Pflanzenzuchtung.* 69: 142-155

Cooper, R.L. 1990. *Crop Science.* 30: 417-419.

Donald, C.M. *Euphytica* 1968. 17: 385-403.

Fiyzat, Y. and Atkins, R.E. 1953. *Agronomy Journal* 45: 414-416.

Fowler, W.L. and Heyne, E.G. 1955. *Agronomy Journal* 47: 430-434.

Frey, K.J. 1954. *Agronomy Journal* 46: 541-544.

Grafius, J.E. 1965. *Crop Science.* 5: 377.

Harlan, H.V., Martin, M.L. and Stevens, H. 1940. *USDA Technical Bulletin* 720.

Harrington, J.B. 1937. *Journal of American Society of Agronomy* 29: 379-384.

Harrington, J.B. 1940. *Canadian Journal of Research* 18: 578-584.

Hurd, E.A. 1969. *Euphytica* 18: 217-226.

Immer, F.R. 1941. *Journal of American Society of Agronomy* 33: 200-206.

Kalton, R.R. 1948. *Iowa Agricultural Experiment Station Research Bulletin* 358: 699-732.

Kasha, K.J. and Kao, K.N. 1970. *Nature* 225: 874-876.

Knott, D.R. and Kumar, J. 1975. *Crop Science.* 15: 295-299.

Lupton, F.G.H. and Whitehouse, R.N.H. 1957. *Euphytica* 6: 169-184.

Muehlbauer, F.J. and Slinkard, A.E. 1985. In Proc. *Intern. Workshop on Faba Beans, Kabuli Chickpeas, and Lentils in the 1980's* pp 351-363 (Eds. M.C. Saxena and S. Varma). Aleppo, Syria:ICARDA.

Schubert, K.R. 1982. *American Society Plant Physiology* Workshop Summaries – I. Rockville, MD, 31 pp.

Simmonds, N.W. 1989. *Biological Reviews* 64: 341-365.

Singh, O. and Smithson, J.B. 1986. *International Chickpea Newsletter* 15: 2-4.

Singh, S.P. 1994. *Crop Science*. 34: 352-355.

Slinkard, A.E. 1993. In *Breeding for Stress Tolerance in Cool Season Food Legumes*. pp 429-438 (Eds. K.B. Singh and M.C. Saxena). New York:Wiley-Sayce.

Breeding for increased biomass and persistent crop residues in cool-season food legumes.

W. Erskine[1], I. Kusmenoglu[2], F. J. Muehlbauer[2] and R. J. Summerfield[3]

1 International Center for Agricultural Research in Dry Areas (ICARDA), P.O. Box 5466, Aleppo, Syria; 2 United States Dept. of Agriculture-Agriculture Research Service, 303W Johnson Hall, Washington State University, Pullman, WA 99164-6434, USA; 3 The University of Reading, Department of Agriculture, Earley Gate, P.O. Box 236, Reading, RG6 6AT, UK

Abstract

The impetus to breed for increased biomass and persistent crop residues in food legumes comes from the value of the straw, particularly of lentil *(Lens culinaris)* for livestock feeding in the Middle East, and from the inadequate quantity and quality of crop residues, especially of pea *(Pisum sativum)* and lentil, for controlling erosion in the Palouse region of U.S.A. Genetic variation for biomass is broad within germplasm collections and in breeding material of food legumes and is sufficiently heritable to allow responses to selection. To date, selection for seed yield in food legumes has resulted in an indirect response in increased biomass. Crosses of chickpea *(Cicer arietinum)* with *C. echinospermum and C. reticulatum* have given segregants that are transgressive for biomass; whereas crosses of lentil with *L. culinaris ssp. orientalis* have not given equivalent transgressive segregants. Quantum increases in biomass in chickpea and lentil have been achieved by altering the crop management, particularly early sowing, combined with selection for new environmental conditions. In conclusion, there is scope for rapidly increasing biomass in the cool-season food legumes; however, the quality and persistence of crop residues is largely unexplored.

INTRODUCTION

Cool-season food legumes are grown primarily for their seed. As a by-product, their crop residues (straw), comprising pod walls, dead leaves, branches and sometimes roots, are variously used. In developed countries, food legumes are generally harvested by combine harvester and the 'trash' is left on the soil surface for grazing or soil-incorporation. In developing countries, food legume residues are mostly fed to livestock with some faba bean *(Vicia faba* L.) and chickpea straw also used as fuel. In North Africa and West Asia, the estimated six million tonnes of food legume straw can supply only 1% of the feed requirements of the ruminant livestock population, compared with cereal straws which supply 30% (Capper, 1990). Nevertheless, the economic value of food legume residues varies widely between species, farming systems and, even within a system, also over time depending on the relative price and availability of alternative feeds. The feeding of lentil residues to sheep is particularly important in Syria in winter (Nordblom and Halimeh, 1982; Capper, 1990). The straws of food legumes contain more crude protein than those of cereals. Lentil and faba bean straws also have greater metabolizable energy values than cereal straws, whereas chickpea and pea have smaller values (Table 1). The dehulling of food legume seed for *dhal* or splits produces another by-product - the seed testa, which is fed to livestock.

In the North-westem USA, farmers have found the quantity of organic matter returned to the soil after a legume harvest to be inadequate to control soil erosion from rainfall, particularly in the autumn (Frazier *et al.,* 1983), which limits the popularity of legumes in the crop rotation. Additionally, the stems of pea and lentil are brittle and broken into small particles by harvesting machinery. These fragments are easily blown away and, even where they remain on the soil surface, may well afford little protection against the impact of rain drops which disturb the soil surface and cause erosion. Hence, both the quantity and quality of straw produced are important traits in determining how much organic matter is returned to the soil at seed harvest. The U. S. National Resources Conservation Service has set minimum requirements for crop residues (USDA Soil Conservation Service, 1992). In Western Australia wind erosion is also a recognised problem in pea stubbles grazed by sheep (Russell, 1993). The approach taken in Australia to reduce soil erosion in pea has focused on crop husbandry rather than plant breeding and on the retention of cereal stubble followed by rolling before seeding the peas. Elsewhere, the widespread use of semi-leafless peas has reduced straw yield compared to conventional leaf types mainly due to improved assimilate partitioning to fruits (Snoad, 1985; Zain *et aL,* 1983).

191

R. Knight (ed.), Linking Research and Marketing Opportunities for Pulses in the 21st Century, 191–197.

Table 1. Compiled information on laboratory analyses and indicators of the feed value of food legume straws, cereal straws, hays and concentrate feeds; from review of Capper (1990)

Feed	Laboratory analysis			Digestibility [1]	Indicators of feed value	
	Crude protein %	Crude fibre %	Phosphorus %		ME[2] (MJ/kg)	Voluntary consumption [3]
Chickpea straw	6.0	44.4	0.12	39.0	5.9	35
Lentil straw	5.8	37.1	0.07	51.9	7.9	55
Faba bean straw	5.7	33.7	na	50.0	7.4	55
Pea straw	8.4	39.5	0.11	43.0	6.5	40
Barley straw	4.2	35.6	0.16	45.0	6.8	40
Wheat straw	3.8	38.9	0.34	39.0	5.9	30
Alfalfa hay	16.7	35.0	0.22	51.0	7.7	60
Oat hay	7.9	37.6	0.22	50.0	7.8	55
Vetch/Barley hay	11.2	27.4	na	57.2	8.6	68
Barley grain	10.0	5.0	0.44	86.0	13.1	90-100
Wheat bran	16.7	8.3	1.00	61.0	10.1	50
Cooonseed cake	26.7	24.4	1.22	51.0	8.7	40
Soyabean meal	50.0	6.1	0.67	79.0	12.3	90-100

1 Digestibility g digestible dry matter /100 g dry matter   2 Metabolisable energy value megajoules Kg[-1] dry matter
3 Voluntary consumption g dry matter/kg liveweight [0.75] / day.   na, Data unavailable

Interest in the residues of lentil as a feed in the Middle East and of lentil and pea in soil erosion control in the USA has fueled breeding efforts toward biomass (i.e. seed plus residue) improvement. However, these endeavors have distinctly different straw quality aims. In the Middle East, focus is on the maintenance of digestibility, metabolizable energy value and voluntary consumption for small ruminants; whereas in the USA the aim is to increase the lignin content to decrease straw break-down during threshing. In the other cool-season food legumes, the economic value of crop residues is insufficient to justify breeding for biomass improvement *per se*.

In this discussion, a *caveat* on the measurement of food legume biomass is essential. There are differences among- and within- species in the extent and timing of the shedding of leaves and leaflets. For example, most genotypes of lentil begin to shed imparipinnate leaflets between physiological maturity and 100% fruit maturity, whereas a small minority of genotypes such as BARI Masur 3 (ex-Bangladesh) retain green leaflets after 100% fruit maturity. Consequently, the timing of harvest is critical to the weight of leaflets recorded in a biomass sample. Additionally, it is often unclear whether biomass measurements result from harvest at the soil surface and so comprise solely above-ground dry matter, or whether biomass comes from harvest by hand-pulling and also contains some root material. Comparisons among studies should be made with great care.

GENETIC VARIATION FOR BIOMASS AND PERSISTENT CROP RESIDUES

Data on seed and biomass yields from germplasm collections of cultivated chickpea (kabuli), faba bean, lentil and dry pea grown at ICARDA, Syria are given in Table 2. The irrigated faba bean gave the highest yields of seed and straw with a mean biomass > 10 t ha[-1] (Singh *et al.,* 1983; Erskine and Witcombe, 1984; ICARDA, 1991; Robertson and El-Sherbeeny, 1993). The smallest mean yields of seed and straw came from pea with spring-sown kabuli chickpea and lentil occupying intermediate positions. There is considerable genetic variation within all of these species for both seed and straw yield.

Data from yield trials of lentil and pea confirm the considerable genetic variation in the respective yields of biomass, seed and straw and that this variation is heritable. For example, pea biomass in trials over three locations and seasons in the USA ranged from genotype means of 3,365 to 8,032 kg ha[-1] (McPhee and Muehlbauer, 1997). There were also major seasonal differences. In several studies on lentil, sufficient genetic variation in straw yield has also been found to warrant its selection (Erskine, 1983; Hamdi *et al.,* 1991; Kusmenoglu and Muehlbauer, 1997 a, b; McPhee *et al.,* 1997). The broad-sense heritability for straw yield was $h_{bs} = 0.85$ with 24 lines in three environments and $h_{bs} = 0.95$ with 34 diverse lines over 10 environments. Biomass yield is controlled by a combination of additive and non-additive gene effects

Table 2. Compiled information on variation in germplasm collections of cultivated food legumes for yields of biomass, seed, straw (kg ha[-1]), and harvest index (%).

Parameter	Chickpea[1] (kabuli)	Faba bean[2]	Lentil[3]	Pea[4]
Biomass				
Mean + SE	$2,041 \pm 10.8$	$10,963 \pm 122.0$	4,219	$1,167 \pm 31.2$
Minimum	349	0	78	0
Maximum	5,333	25,960	10,383	2,758
Seed				
Mean $\pm$ SE	987 ± 5.4	$3,574 \pm 49.9$	$1,287 \pm 8.5$	238 ± 11.7
Minimum	73	0	10	0
Maximum	2,924	7,510	3,258	1,000
Straw				
Mean $\pm$ SE	1,054	7,389	$2,932 \pm 20.9$	928 ± 25.0
Minimum	na	na	63	0
Maximum	na	na	7,983	2,279
Harvest index	0.49	0.32	0.31	0.20
No. accessions	3,263	840	3,586	351

1 Singh *et al.* 1983   2 Robertson and El-Sherbeeny 1993   3 Erskine and Witcombe 1984
4 ICARDA 1991   * Data unavailable

(Hamdi, 1987). In lentil there is a strong and positive correlation between seed yield and straw yield in all but very wet environments. Although both seed and straw yields are high in such wet environments, the indeterminate growth habit of the crop causes the correlation between these traits to be negative.

POTENTIAL OF WILD SPECIES AND RELATED GENERA

Among the annual species of the genus *Cicer,* some accessions of *C. reticulatum and C. echinospermum,* both of which are crossable with *C arietinum,* overlapped the cultigen in biomass production (Robertson *et al.,* 1995). Crosses between chickpea and these species were made at ICARDA and pedigree selection was used until progeny were bulked in the F_5 generation. A yield trial of 22 selected F_7 lines (10 lines from cultivated x *C reticulatum* and the others from cultivated x *C echinospermum)* was undertaken at ICARDA in 1995 (Singh and Ocampo, 1997). One line from the *C reticulatum* cross and eight lines from the *C echinospermum* cross yielded significantly more biomass than the respective most productive parent.

In the genus *Lens,* the wild species in general produce smaller plants than the cultivated lentil. However, some accessions of *L. culinaris ssp. orientalis,* which is the closest among wild *Lens* taxa to the cultigen, overlapped with the cultigen for biomass productivity (Robertson and Erskine, 1997). Crosses of wild *(L. culinaris ssp. orientalis)* x cultivated lentil were made in a line x tester mating design with three diverse cultivated lines and seven parents of *L. culinaris ssp. orientalis* (ICARDA, 1995). Biomass yield varied among the cultivated parents firom 151 to 343 g m^{-2} and among the wild parents from 4 to 98 g m^{-2}. Heterosis above the better parent was only present in crosses with a poor yielding cultivated lentil line and was not found in crosses with the currently recommended cultivar. Given this heterosis (deviations from the model of the mid-parent mean) for biomass, specific combining ability (SCA) was of greater importance than general combining ability, indicating the importance of non-additive gene effects in the genetic control of biomass in the wild x cultivated crosses.

Given that the more distant *Lens* species have less biomass than *L. culinaris ssp. orientalis,* it may be worth considering wider intergeneric crosses of lentil with different *Vicia* spp., such as *V sativa,* to increase biomass yield. Efforts at wide-crossing lentil with various *Vicia* spp. with a suitable chromosome complement of 2n = 14 and a superior biomass are justified, but ovule rescue techniques will be needed. Looking beyond the primary and secondary gene pools, the techniques currently used to produce transgenic plants allow the transfer of very few genes. But, as biomass yield is controlled by many genes of individual small effect, the scope of this approach is limited, unless a *gigas* gene is discovered.

VARIATION FROM MANAGEMENT PRACTICES AND THEIR INTERACTION WITH GENETIC VARIATION

Early sowing offers exciting prospects to increase biomass production in areas receiving winter rainfall. For example, compared to a spring-sown evaluation of kabuli chickpea which gave average yields of 2,041 kg biomass ha[-1] and 987 kg seed (Table 2), winter-sown germplasm (6221 accessions) gave double the yield with a mean biomass of 5,170 kg ha[-1] and 2,450 kg seed ha[-1] under lowland Mediterranean conditions where the crop is traditionally spring-sown (Singh et al., 1991). These spring and winter evaluations were made in different seasons and so the comparison is gross. Nevertheless, comparisons of winter versus spring sowing within the same growing season and location also revealed increases in the yields of biomass (59%) and seed (70%) averaged over three sites and ten seasons (Singh et al., 1997). There is, however, a need for a change in genotype to accompany the change in sowing date. Winter sowing is only feasible using cultivars with cold tolerance and resistance to Ascochyta blight caused by Ascochyta rabiei.

Lentils are winter-sown in the lowland Mediterranean region; but above approximately 850 m elevation, where the winter is colder, c. 400,000 ha of lentil are spring-sown in Iran and Turkey. A mean yield advantage of winter over spring lentil of >50 % has been recorded in the highlands, indicating that winter-sown lentil has a significantly greater yield potential than the spring sown (Sakar et al., 1988). However, the extension of early winter/autumn sowing to elevations above ca 850 m will require cultivars with increased winter hardiness. There is a similar interest in replacing spring with winter sowing in the Pacific NW of the USA. Early sowing is without doubt a promising route to a quantum increase in lentil biomass and seed yield in both the highlands of W. Asia and the Pacific NW of the USA (Muehlbauer, 1997; Erskine and Malhotra, 1997).

In contrast to the boost in biomass from early sowing, changes in crop management may also influence biomass negatively. For example, machine harvesting with a cutter bar, as compared to hand pulling, leaves a stubble in the field and a reduced harvest of biomass and seed. In lentil, the losses from cutting with a double knife cutter bar at ground level, or at 5 or 10 cm above ground level, compared to hand-pulling, were 10, 15 and 20% in seed yield and 16, 40 and 45% in straw yield, respectively (Silim et al., 1989). However, although that part of the plant below the harvest level (some pods, stem and roots) is a harvest loss, it is by no means a complete loss to the system as it may be grazed or incorporated into the soil. The stubble contained on average an additional 2.7, 7.7 and 8.3 kg N ha[-1] from harvests at ground level, 5 cm and 10 cm above ground level in comparison to the hand harvested plots.

ASSOCIATIONS OF BIOMASS WITH OTHER TRAITS

In food legumes, with an indeterminate flowering habit, an increase in the size of the vegetative frame of the crop will usually result in an increased number of potential flowering nodes and hence potential yield. Conversely, selection for increased seed yield has often resulted in an indirect increase in biomass because a positive correlation between seed and straw yields is the norm. In lentil there is a strong and positive correlation between seed yield and straw yield in all but very wet environments (Erskine, 1983; Hamdi et al., 1991; Kusmenoglu and Muehlbauer, 1997a, b).

Biomass and straw yield are labour intensive to measure; so cheaper alternative indirect selection criteria have been sought. Kusmenoglu and Muehlbauer (1997a) have recommended the use of plant height as an indirect criterion for biomass in lentil. Kusmenoglu and Muehlbauer (1997b) indicated that seed and straw yield can be simultaneously increased by selecting genotypes of lentil that have short vegetative and generative growth periods, large rates of crop, seed and straw growth, and high seed partitioning coefficients. Visual scores of seedling vigour, correlated with measurements of seedling dry weight, in the Mediterranean environment have also been correlated with final biomass in lentil (Silim et al., 1991, 1993).

There is little information on the quality of food legume straw in terms of erosion control or ruminant digestibility. Consequently, there is also no information on the relation between straw quality and quantity. There is, however, an indication that differences in straw quality (digestibility) among genotypes are related more to differences in the partition of dry matter between various plant parts than to substantial variation among genotypes in the relative digestibility of the individual components (Erskine et al., 1990).

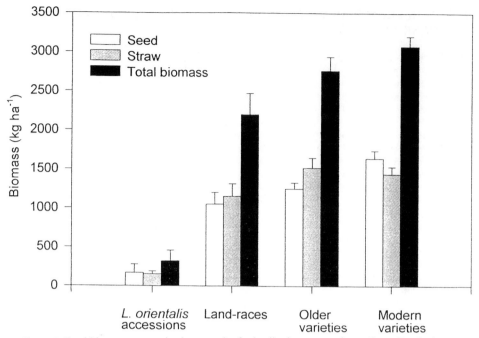

Figure 1. Total biomass at reproductive maturity for lentils along a genetic gradient. Standard errors are represented by vertical bars (Whitehead *et aL* , 1997).

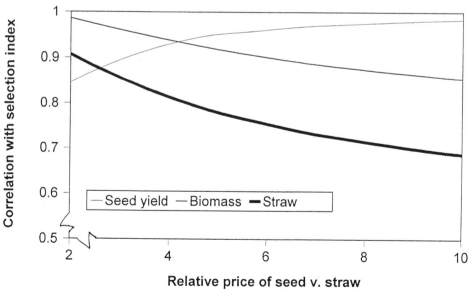

Figure 2. The correlations between a selection index based on economic value of seed and straw and the yields of seed, residue and biomass for different relative prices of seed and straw using data from a germplasm collection of 2277 accessions of lentil (Erskine and Witcombe, 1984).

PROGRESS AND PROSPECTS IN BREEDING

On a theoretical basis, simultaneous selection for seed and straw, using a selection index based on the economic value of the seed and straw, is more efficient than individual selection for either seed, straw or biomass yield alone (Falconer, 1989). However, if it is impractical to use such an index. At what relative value of seed v. straw does it become more efficient to select for seed rather than biomass? To answer this question, the correlations between an economic selection index and the yields of seed, straw and biomass were modelled for various different relative price scenarios of seed and straw using data from a germplasm collection of 2,277 accessions (Erskine and Witcombe, 1984) (Figure 1). It is clear that when straw price is relatively expensive, (> 0.25 of the seed price) selection is more profitably based on biomass, but when the price of seeds exceeds that of straw by more than four times, then selection for seed yield alone is more efficient than selection for biomass alone in terms of economic return. Selection for straw yield is always less efficient than selection for biomass yield.

Response to selection for seed yield has been analyzed by the study of a genetic gradient comprised of *L. culinaris ssp. orientalis,* land races and ancient and modern cultivars (Whitehead *et al.,* 1997). The preliminary findings are of a gradual increase in seed yield moving along the selection gradient toward modern varieties (Figure 2). There was a parallel increase in biomass yield along the gradient, with an indication that modern cultivars have the greatest harvest index because of a mobilization of non-seed biomass to fruits. As selection for yield has intensified, stems have acquired additional lignified tissue in isolated bundles at the extremities of stems (i.e. within the ridges). This contrasts to the arrangement in the wild progenitor of supportive tissue concentrated at the centre of the stem (Whitehead *et al.,* 1997).

Another, more general, way of looking at the response to selection for seed and straw yields is to compare the land races of food legumes with those of forage legumes. Land races of food legumes are the products of selection for seed yield. They may be considered 'wild' for straw yield, which resulted from a correlated response to selection to seed yield, and their harvest index is typically above 0.3 (Table 2). In contrast, land races of forage legumes have been selected for forage yield. They may be considered 'wild' for seed yield, resulting from a correlated response to selection to forage yield. The harvest indices of forage legume crops are often at or below 0.1 (Lorenzetti, 1981).

In conclusion, there is substantial scope for rapidly increasing biomass in the cool-season food legumes, however, quality aspects of the persistence of crop residues are largely unexplored.

References

Capper, B.S. 1990 In *The role of legumes in the farming systems of the Mediterranean Areas*, pp 151-162. (Eds A.E. Osman, H. Ibrahim and M. Jones). Aleppo: ICARDA.

Erskine, W. 1983 *Field Crops Research* 7: 115-121.

Erskine, W. and Malhotra, R.S. 1997. In *Proceedings of Workshop on Problems and Prospects forWinter Sowing of Grain Legumes in Europe.* (In press)

Erskine, W., Rihawe, S. and Capper, B.S. 1990 *Animal Feed Science and Technology* 28: 61-69.

Erskine, W. and Witcombe, J.R. 1984 *Lentil Gerrnplasm Catalog.* Aleppo: International Center for Agricultural Research 'in the Dry Areas (ICARDA).

Falconer, D.S. 1989 *Introduction to quantitative genetics.* Harlow: Longman.

Frazier, B.E., McCool, D.K. and Engle, C.F. 1983 *Journal of Soil and Water Conservation* 38: 70-74.

Hamdi, A.H.I. 1987 Ph.D. thesis, University of Durham, U.K.

Hamdi, A., Erskine, W. and Gates, P. 1991 *Euphytica* 51: 109-116.

ICARDA 1991 *Food Legume Improvement Program*: Annual Report 1990. Aleppo: International Center for Agricultural Research in the Dry Areas (ICARDA).

ICARDA 1995 *Germplasm Program Legumes*: Annual Report for 1994. Aleppo: International Center for Agricultural Research in the Dry Areas (ICARDA).

Kusmenoglu, I. and Muehlbauer, F.J. 1997a *Crop Science* (In press)

Kusmenoglu, 1. and Muehlbauer, F.J. 1997b *Crop Science* (In press)

Lorenzetti, F. 1981 *Report of the Meeting of the Fodder Crops Section of EUCARPIA.* 57-74. Ghent:

McPhee, K.E. and Muehlbauer, F.J. 1997 *Journal of Production Agriculture.* (In press)

McPhee, K.E., S.C. Spaeth and F.J. Muehlbauer. 1997 *Journal of Production Agriculture* (In press) Muehlbauer, F.J. 1997 In *Proceedings of Workshop on Problems and Prospects for Winter Sowing of Grain Legumes in Europe.* (In press)

Nordblom, T. and Halimeh, H. 1982 *LENS Newsletter* 9: 8- 1 0.

Robertson, L.D. and Erskine W. 1997 In *Biodiversity in Trust.* Cambridge: Cambridge University Press.(In press)

Robertson, L.D. and El-Sherbeeny, M. 1993 *Faba bean germplasm catalog: Pure line collection.* Aleppo: International Center for Agricultural Research in the Dry Areas (ICARDA).

Robertson, L.D., Singh, K.B. and Ocampo, B. 1995 *A catalog of annual wild Cicer species.* Aleppo:International Center for Agricultural Research in the Dry Areas (ICARDA).

Russell, J. 1993 *Western Australia Journal of Agriculture* 34: 113-117.

Sakar, D., Durutan, N. and Meyvecl, K. 1988 In *World Crops: Cool Season Food Legumes.* pp 137-146 (ed. R-J. Summerfield) Dordrecht: Kluwer Academic Publishers.

Silim, S.N., Saxena, M.C. and Erskine, W. 1989 *Field Crops Research* 21: 49-58.

Silim, S.N., Saxena, M.C. and Erskine, W. 1991 *Experimental Agriculture* 27: 145-154.

Silim, S.N., Saxena, M.C. and Erskine, W. 1993 *Experimental Agriculture* 29: 9-19

Singh, K.B. and Ocampo, B. 1997 *Theoretical and Applied Genetics* (In press)

Singh, K.B., Malhotra, R.S., Saxena, M.C. and Bejiga, G. 1997 *Crop Science* (In press)

Singh, K.B., Malhotra, R.S. and Witcombe, J.R- 1983 *Kabuli chickpea germplasm catalog.* Aleppo: International Center for Agricultural Research in the Dry Areas (ICARDA).

Singh, K.B., Holly, L. and Bejiga, G. 1991 *A catalog of kabuli chickpea germplasm.* Aleppo: International Center for Agricultural Research in the Dry Areas (ICARDA).

Singh, K., Malhotra, R., Saxena, M. and Bejlga, G. 1997 *Crop Science* (In press)

Snoad, B. 1985 In *The pea crop.* Pp 391-412 (eds P.D. Hebblethwaite, M.C. Heath and T.C.K. Dawkins)London, Butterworths.

USDA Soil Conservation Service 1992 *Crop residue management specification sheet No. 1.* Practice 344fFSA crop residue use. USDA Soil Conservation Service, Spokane, Washington.

Whitehead, S.J., Summerfield, R.J. and Muehlbauer, F.J. 1997 Production and structure of biomass in lentil. *Field Crops Research* (In preparation)

Zain, Z.M. Gallagher, J.N. and White, J.G.H. 1983 *Proceedings of Agronomy Society* 13: 95-102.

Direct and indirect influences of morphological variations on diseases, yield and quality.

A. Porta-Puglia[1], T.W. Bretag[2], J.B. Brouwer[2], M.P. Haware[3] and S.A. Khalil[4].

1. Istituto Sperimentale per la Patologia Vegetale, Rome, Italy; 2. Victorian Institute for Dryland Agriculture, Horsham, Victoria, Australia; 3. International Crop Research Institute for the Semi-Arid Tropics, Patancheru, India; 4. Field Crop Research Institute, Giza, Egypt.

Abstract

Interest in morphological variation in food legume species is increasing as plant breeders search for new variants to satisfy the adaptation requirements from new or changing environments or the needs of new end-users. Examination of evolutionary pathways often provides leads in understanding morphological or physiological variation, which may offer opportunities for exploitation in plant breeding. Variation has direct and indirect effects on yield stability and quality through several parameters acting within the plant and the crop.

Traits affecting the development of the crop canopy or the seed, including for example photosynthate repartitions, can have an impact on yield, quality and diseases. Yet the information available is often incomplete for practical use or is very environment specific. Examples are given of the potential utilisation of genetic diversity conserved in different geographic areas as are available in lentils (pilosae types) and chickpeas (kabuli-desi introgression). The concept of quality in pulses is often dominated by morphological traits and the appearance of the seed. There are also instances where the morphological traits affect nutritional and processing quality, (e.g., the novel alleles at the loci controlling both seed shape and starch composition in pea or the gene for zero tannin in lentil).

Where prospects are still remote for developing cultivars with high levels of resistance to important diseases, more emphasis needs to be put on other components of integrated disease management. Some plant characteristics, such as growth habit and canopy structure (modulated by sowing date, plant density, etc.), can contribute to control of diseases. However, experiments have shown that an increase in disease incidence due to increased plant density can be compensated for by a yield increase as is the case with chocolate spot and rust in faba bean. Of interest also are morphological traits, which can slow penetration by the pathogen, enabling the plant to deploy post-infection physiological mechanisms of resistance.

Increased attention to these complex interactions through international multi-disciplinary cooperation has contributed, and could further contribute, to progress in breeding and disease management. This will result in an improvement of the yield potential, yield stability and quality of these crops.

INTRODUCTION

The morphology of plants of a species is the result of the interaction between its genome and the environment during evolution, including the action of man who has effected changes in legumes during domestication. Modification has affected growth habit, suppressed seed

R. Knight (ed.), Linking Research and Marketing Opportunities for Pulses in the 21st Century, 199–220.

dispersal mechanisms and seed dormancy, changed breeding systems (from exogamy to autogamy) and ploidy levels, etc. (Smartt, 1990). Many components of morphological variation are spatially and temporally variable. These concepts are discussed in two reviews dealing with the canopy (Norman and Campbell, 1989) and root systems (Smucker, 1993). Plant architecture and root form and functions in food legumes are reviewed by Heath et al., (1994) and Gregory et al. (1994), respectively. Other aspects studied include the morphology and physiology of stomata, which regulate gas exchange between the plant and atmosphere (Weyers and Lawson, 1997). The level of complexity increases as we proceed from individual organs to plants and to communities in the field, each level being determined by genotype × environment interactions. This complexity is the reason for sometimes-conflicting reports.

Most plant breeders have a model on which they base improvements of cultivars in their target environments. Sometimes these models have been formulated as ideotypes such as proposed by Berry and by Hedley and Ambrose (Davies et al., 1985) for peas. Testing models is costly and can be misleading as ideotypes tend to be developed for adaptation to specific environments. Yet, genetic variation in morphological traits has attracted interest in improving yields of pulses. Such variation is expressed in the vegetative and reproductive stages and in all plant structures. The interest is not only in morphological traits, which contribute to yield. A large number of mutants have been used to improve other features such as plant height or standing ability required for mechanical harvesting.

Other variants are useful in enhancing the quality of the grain. They can increase disease resistance or reduce disease incidence by acting on factors affecting disease development. New approaches likely to have an impact are techniques such as "positional cloning" which clarifies the role of genes beginning with the identification of genetic loci responsible for phenotype expression (cf. Caetano-Anollés, 1997).

There is a great potential for exploitation, by breeding, of morphological variation in grain legumes and their wild relatives. This paper reviews aspects related to morphology in the broadest sense of the field crop and individual plants. Reviews exist for many aspects not discussed here, (Summerfield and Roberts 1985; Saxena and Singh 1987, Muehlbauer and Kaiser 1994, Ranalli and Graham 1997).

PEAS

Effect of phenotype and crop density on the severity of Ascochyta blight in peas.

The severity of Ascochyta blight in field peas (*Pisum sativum* L.) varies between crops and years (Bretag 1991). Plant density and phenotype can influence the microclimate within a crop (Ali et al., 1978; Hedley and Ambrose, 1981) and also effect the severity of the disease.

It is often supposed that Ascochyta blight will be more severe in crops sown at a high seeding rate as this leads to denser canopies and the microclimate in the canopy will be more favourable for the disease. Many varieties, currently grown in Australia (eg. Dun, Dundale, Derrimut) are tall indeterminate, trailing types that form a dense canopy (Gent, 1988). It is important to know whether shorter varieties with a semi-leafless habit are less affected by Ascochyta blight. As varieties vary in their responses to temperature and photoperiod, they mature at different times (Berry and Aitken, 1979). The Australian varieties have a wide range of maturities and it is unclear whether this affects their disease resistance. Previous studies have indicated that early maturing varieties may be more susceptible than late varieties (Hare and Walker, 1944).

By monitoring disease progress during the growing season on a wide range of varieties, with contrasting phenotypes and maturity classes, it should be possible to determine the extent to which phenotype and maturity influence Ascochyta blight. Studies were undertaken in Australia on conventional pea varieties sown at different rates to determine whether differences in height, leaf habit and maturity affected susceptibility to Ascochyta blight (Bretag, 1991; Bretag and Brouwer, 1995).

A wide range of plant densities was tested, Buckley: 11 to 171, Dinkum: 8 to 105, Dun: 9 to 116, Maitland: 9 to 114 and Whero: 6 to 109 plants m^{-2}, respectively. While there were significant differences between varieties in the severity of disease and yield losses, the effect of plant density was similar for all varieties. Severity was lowest at a seeding rate of 25 kg ha^{-1} and increased at higher rates however above 100 kg ha^{-1} there was no further increase in severity. Plant density had a greater effect on yield than on disease severity. As density increased, there was an increase in yield which outweighed the decrease caused by disease. The lower the plant density the greater were the yield losses.

There were large differences between varieties in plant height, days to first flower, node of first flower on the main branch and the total number of nodes formed. Early in the season, the rate of formation of new leaves was similar on all varieties and differences in plant height were associated with variation in internode length. As varieties changed from the vegetative to the reproductive stage, there was a reduction in the rate of leaf formation. Consequently, the early maturing varieties produced fewer leaves than late varieties. For each variety tested, plant growth, development and disease progress were similar in plots sown at either 40 or 120 kg/ha of seed. Comparison of disease progress showed that while there were differences in the levels of disease at a given time, for each variety, disease progress followed a similar pattern to plant growth. Logistic functions successfully fitted the data. The correlation coefficients showed that the models fitted the observed data with a high level of accuracy.

There were large differences between varieties in the shapes of their disease progress curves, but there was no effect of differences in seeding rate. At maturity, there were no differences between varieties in the amount of disease on the first five internodes of the main branch, but large differences occurred on the internodes 6 to 10. There were few symptoms above the tenth node on any variety. The leaves and stems of early maturing varieties (eg. Buckley) were generally more severely affected than those of late varieties (eg. Mackay). Severity did not appear to be associated with varietal differences in plant height. Likewise, there was little evidence that severity was lower on the leafless (Filby) and semi-leafless (Dinkum, Maitland and Mega) varieties. In addition, the range in severity was similar for all phenotypes despite large differences in canopy structure. The differences between varieties with the same leaf type were as great or greater than the differences between varieties having different leaf types. The late maturing varieties often appeared to have lower disease levels at maturity than the early varieties. This may occur because many of the leaves formed at the end of the growing season on the late varieties escape infection. Conditions are more favourable for infection early in the growing season. As a result, the lower leaves usually become infected, when there are high levels of air-borne inoculum and conditions are ideal for infection. In contrast, little disease develops on the upper leaves produced late in the season when there are low levels of inoculum and conditions are less favourable for infection. Irrespective of plant density and canopy structure, all plants usually become infected by Ascochyta blight early in the growing season. At that time, there are small differences between varieties in their canopy structure and plant height. During the early stages of growth, there is sufficient moisture for infection to occur and plant density appears to have little effect on the amount of disease on each plant. It follows that the varieties best suited to their regions, irrespective of their plant type should be used by growers and sown at their optimum seeding rates for maximum grain yield.

Cold and drought tolerance

Genes for cold tolerance in peas are linked to *ilo* characters, pigmented hilum and pigmented seed coat (Monti, Frusciante and Romano 1993). Reports on the drought tolerance of the *afila* genotypes differ (Monti, Frusciante and Romano 1993, Gonzalez-Lauck, 1990), but leafless cultivars have advantages over conventional cultivars on heavier soils and under wet conditions (Jackson, 1985).

Vegetative plant parts

Several studies have indicated that major genes controlling morphological traits (leaflet status, stipule size, internode length, seed shape and bracts) can be associated with variation in yield per plant (Berry, 1981; Gonzalez-Lauck, 1990; Cousin, 1997). It is often difficult to resolve whether these genes have a direct causal effect (pleiotropy) or are linked with other unidentified genes causing differences in yield. The utility of these major genes can be assessed in a range of genetic backgrounds by crossing the parental lines or by using isogenic lines. Some backgrounds may favour differentially specific gene or gene combinations.

Thirty-three genes are known to modify the size or form of pea leaves (Davies et al., 1985). Of these *af*, the *afila* gene, controlling the semi-leafless trait has been used most readily by breeders and has transformed the crop in many countries. The gene *af* confers advantages over traditional types through better standing ability from interplant support and more uniform seed ripening. However, the effect of *af* on yield has varied in different environments. Its effect is not always positive and normal leaf types have often given greater yields (Berry, 1981, Goldman and Gritton, 1992). It is possible that the structural improvement associated with the *af* allele could outweigh the potential yield disadvantage in certain environments. The lower yield of *af* plants may suggest a reduction in source capacity is limiting yield.

A characteristic of the *af* gene is its reproducibility across environments and genetic backgrounds. However, its relationship with yield can be altered by genetic background (Lanfond et al., 1981) and in some cases *af* types yield more only in certain environments (Pate and Armstrong, 1996), especially when the mutant-leaf gene is incorporated into more favourable genetic backgrounds (Gonzalez-Lauck, 1990). However, in most studies, plant densities for the mutant and normal varieties were constant, which makes the comparisons biased as the mutants have a reduced leaf area per plant. Mutant types should be sown at their optimal population density to assess their potential yield (Hedley and Ambrose, 1985).

Another gene which regulates leaf area is *St*. The recessive *st* allele causes the stipule area to be reduced by nearly half. The interest in using this gene in combination with *af* (*af af st st* as leafless peas) has waned. The decrease in average seed size in *st* compared to *St* plants (Berry, 1981) suggests there is a source capacity limitation to yield. However, Blixt (Berry, 1981), found significantly fewer seeds per pod and a lower setting rate in *st* lines. In contrast, he observed no difference in seed size and little difference in number of pods per node.

Wehner and Gritton (1981) evaluated near-isogenic lines of all possible homozygous combinations of the genes *af*, *st* and *tl* (leaflets in the place of tendrils) in eight different backgrounds of commercial processing (green) varieties at a constant plant density. All mutant genotypes tended to yield the same as the normal leaf varieties except for *af tl* and *af tl st* which yielded much less. Snoad (1974) evaluated 15 different experimental lines with the *af* and *st* gene combinations and found the dry seed yield was similar to the normal leaf genotype for *af* or *st* individually. If both genes were together, in the *af st* combination, yield decreased significantly. With near-isogenic genotypes, Snoad (1981) determined that semi-leafless *af* peas yielded less even though they had a higher harvest index. Cardi and

Monti (1985) grew dry peas on trellises and reported that *st* had a negative effect on yield and its components. This was especially true in combination with the *af* gene.

Little agronomic advantage over conventional leaf types for green pea production could be demonstrated from plants carrying the *tac* gene (tendriled acacia) in combination with *af* in comparisons of near-isogenic sets in three genetic backgrounds at two plant densities (Goldman and Gritton, 1992; Goldman et al., 1992). The expression of the gene varied significantly with genetic background. Where source capacity is limiting, the complementary dominant *Br* and *Bra* genes which produce leafy bracts on the inflorescences, could usefully contribute to yield. The expression of the trait is variable and depends on the environment as well as the background genotype. Berry (1981) found that plants with bracts were on average much higher yielding than those without bracts. The difference was probably due to an association with apex longevity. On average there were about five more reproductive nodes on the main stem of the plants with bracts than on those without them. Plants with bracts also had more reproductive nodes on branches.

Growth development

Generally, the leaf mutant genes can be demonstrated to have differential effects on dry matter accumulation, root growth (Berry, 1981), light interception (Hedley and Ambrose, 1985), photosynthesis (Gonzalez-Lauck, 1990), and semi-leafless plants show an intermediate growth rate compared to the normal and leafless peas (Hedley and Ambrose, 1985). Associations with final yield are more difficult to prove, probably because "only a small part of production capacity is used" in peas (Pate, 1975), although most studies show a good relationship between light interception and grain yield (Gonzalez-Lauck, 1990). There is no difference in stem, petiole, rachis and tendril anatomy for conventional or *afila* peas (Gonzalez-Lauck, 1990). Tendrils have 50% less stomata than the mean of the upper and lower leaf surfaces of both normal and semi-leafless varieties. The chloroplasts in tendrils of semi-leafless and normal genotypes are of the same size and distribution. However, genes *sa1, sa2, sa3,* can be used to vary the density of the stomata.

Growth habit and stems

In *Pisum sativum* 27 mutant genes are known (Davies et al.1985) to affect stems. Breeders have mostly used the dwarfing gene *le*. Berry and Blixt agreed that tall plants (*Le*) had a yield advantage over dwarf (*le*) plants due to more pods per node, which is reflected in a higher pod number and hence seed number (Berry, 1981). Plants possessing the dominant *Le*, were of greater vigour (Berry, 1981) and this was reflected in the higher yields, principally through greater seed numbers.

Stem fasciation, in particular that controlled by the genes *fa* and *fas*, has also been proposed (Berry, 1981) as a trait to improve growth habit and the mutual support of plants. On fasciated plants the leaves occur all around the stem instead of in one plane, thus increasing tendril contact between plants at the reproductive nodes. The spatial distribution of seed yield should also be improved as a false umbel of pods is produced at the top of the plant rather than at axillary inflorescences along the stem. The key is the flowering gene *Hr* which as a recessive allele *hr* brings about rapid apical senescence after commencement of flowering. In contrast, *Hr* delays apical senescence. Thus, fasciation and the gene *hr* are both necessary to produce the cluster of pods, if this is a desirable trait for adaptation to specific environments or cropping systems where uniformity of maturation is essential.

Increased stem lignification and strength associated with the *rms* gene may also have significance (Erskine et al., 1988).

Armstrong and Pate (1994) attributed the poor growth of semi-leafless and tare-leaved types, compared to conventional types grown in Western Australia, to their lower vigour which resulted in poorer ground cover, reduced photosynthetic area and lower photosynthate production. Floyd and Murfet (1986) found an interaction between dwarf and flowering genes in their effect on branching habit in that lines which had shorter internodes and were photoperiod sensitive also tended to have more basal laterals than taller or photoperiod neutral lines. Non-competitive plants with an unbranched habit have been suggested for the UK (Hedley and Ambrose, 1985). However, there is trend now among European breeders to select for types with basal branching because of the high seed cost of sowing required to achieve the high plant density necessary when sowing single-stemmed plants (P. Lonnet, personal communication).

Walton (1990) in Western Australia found that tall cultivars had a higher number of aerial branches whereas dwarf cultivars produced a higher number of basal branches. The proportion of basal to aerial branches varied for cultivars but together they contributed over 50% of the seed yield in low rainfall environments. Non-branching plants (especially tall in peas) allow greater control over the relationship between vegetative and reproductive growth by adjustment via plant density. However, contributions to grain yield from aerial branches tend to increase where environments are more favourable (Armstrong and Pate, 1994; Walton, 1990).

Root morphology

Seven root and shoot characters of peas were studied to assess genetic variability and to measure broad sense heritability. Estimates of heritability exceeding 50% were noted for the number of laterals, length of shoot and weight of shoot and root in *P. sativum* and *P. fulvum* (Ali-Khan and Snoad, 1977). The seed weight was correlated with the length of the longest lateral and weight of the root.

Reproductive structures

Some of the factors, which control the number of flowers initiated in each raceme, have been identified (Davies et al., 1985; Hardwick 1988). The number can vary from one to seventeen and is controlled by two duplicate recessive genes. Most cultivars have one or two flowers per node. Multi-podded lines usually set fewer seeds per fruit and similar numbers of seeds per node can be found on single-podded types with a high ovule fertility (Davies et al., 1985). Consequently, there is less interest in the 'multi-podded' habit in peas than in chickpeas or faba beans.

Walton (1990) suggested that for the drier areas of Western Australia pea breeders should select for increased seed number per pod in early flowering, medium to tall plants, to improve the total seed weight from the first three reproductive nodes on the main stem. A highly determinate flowering pattern has also been advocated in Europe (Walton, 1990), but some degree of indeterminacy would be beneficial in seasons where finishing rains permitted the filling of later-formed seeds (Pate and Armstrong, 1996).

Five major genes have been shown to control in pod shape. One gene, *bt*, which determines pod apex shape, has an effect, either by linkage or pleiotropy, on ovule number (Berry, 1981), for in all segregating populations, in several studies, blunt pods (*Bt*) were associated with low ovule number and pointed pods (*bt*) with high number. Ovules do not abort at random with respect to position within the pod. There is evidence of within-fruit positional variation in *Pisum* as well as *Lupinus* (Hardwick, 1988).

The photosynthesis by pods is important in seed filling and the optimum positioning of pods in the canopy of leafless peas may thus be equally important (Erskine et al., 1988). There is considerable variation in size, shape, wall thickness and growth rates of pods but associations with yield are not known. One might expect a less efficient conversion of assimilates by genotypes with thick-walls, in view of the greater investment in non-metabolizable dry matter of walls and possibly poorer photosynthetic conservation of respired carbon (Pate and Armstrong, 1996).

Relatively high proportions of seed mass occur as the seed coat (14-29%) in small-seeded primitive forms (*Pisum humile* and *Pisum elatius*) as compared to the 8-9% typical of large-seeded modern cultivars of *P. sativum*. This should be considered when using 'wild germplasm' for genetic improvement as it is likely to impair the conversion of assimilates into cotyledonary reserves. Pod dehiscence is a problem under dry hot conditions at harvest. Three genes reduce seed losses from pod shattering. The first two (*p* and *v*) suppress the production of sclerenchymatous cells and lignification in the pod wall (giving edible pods), resulting in a wall which closely envelops the seeds at maturity. The other gene (*def*) increases funicle strength and the attachment of the seed to the pod, even after dehiscence (Erskine et al., 1988). The funicle part, which may remain attached to the seed, even after threshing, may affect the marketability of the grain.

Effect of seeds on growth or yield

About 50 genes are known to influence seed shape, seed size, and colour of the testa, but most of these control the patterns and colour of the testa. Large seed usually produce a more vigorous seedling which favours the rapid establishment of a canopy. Seed size is highly heritable and simple to breed for. Interactions between size and number of seeds can be significant but have only minor effect on yield per plant (Berry, 1981). Both components are largely independent and crosses between lines of widely different seed size are likely to produce families with much higher yields than either parent.

Lodging

The performance of the pea as a crop plant is affected by the weakness of lower stems, a procumbent habit and a tendency for potentially productive source leaves to become buried by later-developing shoot parts (Pate and Armstrong, 1996). Lodging not only reduces light conversion to dry matter (Heath and Hebblethwaite, 1987) but affects yield and quality (Gonzalez-Lauck, 1990; Pate and Armstrong, 1996). Both conventional leaf (*Af*) and semi-leafless types (*af*) can lodge completely by maturity. The main difference is that the semi-leafless trait delays lodging. Although leaf type influences lodging, there are factors such as basal stem stiffness, intensity of inter-winding and differences in top growth with environment that have a major impact on standing ability.

Yield components and harvest index

Negative correlations are often found between the components of yield and compensation occurs. Correlations vary with plant density. Developmental homeostasis occurs through the abortion of buds, flowers, immature pods, ovules and seeds within pods (Hardwick, 1988). Selecting too intensely for one yield component can disrupt the balance between components. Hedley and Ambrose (1985) pointed to the association of large seed size with high variance for plant size and for partitioning of dry matter into grain within the pea crop as this leads to lower harvest indices among the plant population. They suggested selection for large embryonic axis and relatively small seed size, which would give increased growth rate early in the season without the problem of seed abortion, normally associated with large seeds or high planting density. In general, leafless and semi-leafless peas have higher harvest index values than conventional peas (Erskine et al., 1988; Gonzalez-Lauck, 1990). Pate and Armstrong (1996) suggested that yield improvement could be achieved from increased synchrony of seed maturation through multi-podded nodes and near-simultaneous flowering of neighbouring nodes. An increase in pods per node, using the genes *fn* and *fna*, can increase sink size without changing the duration of flowering period. Berry (1981) however, did not detect a corresponding increase in source capacity and multi-podded genotypes had generally poor seed setting rates.

Quality

In morphological terms, texture, size and colour define quality. Requirements for these attributes differ with traditions and end-uses of the grain. Dark pigmented testas contain tannins which can lessen digestibility and the availability of proteins when the grain is used as stockfeed.

Quality can be influenced by the lack of standing ability at harvest with significant percentages of the grain being classed as 'waste and stain' (Davies et al., 1985). A reduced 'blonding' problem was observed in shelled peas from semi-leafless plants (Wehner and Gritton, 1981). The difference between smooth-seeded and wrinkled-seeded cultivars has determined their usage as dry field (combining) peas or as green processing (vining) peas respectively. The genetic mechanism for smooth versus wrinkled seed has been shown to have a pleiotropic effect on grain yield, 100-grain weight and protein content (Shia and Slinkard, 1977). Wrinkled seeds, as governed by the complementary *r* and *rb* (*rugosus*) loci, have a reduced starch content. Some of the alleles at the *r* and *rb* loci change the amylose proportion of the starch and another changes the shape of the starch granules. Additional mutations have been generated, which are not complementary to *r* and *rb* such as *rug-3*, *rug-4* and *rug-5* but also are associated with wrinkled seed. The *rug-3* alleles have been shown to give rise to mature seeds with virtually no starch or, when starch is present, it contains no amylose but only amylopectin (Harrison and Wang, 1995). Starches of such different compositions and structures are of interest in processing for food and industrial purposes.

CHICKPEA

Agronomic and physiological factors influencing Botrytis gray mold of chickpea

Botrytis gray mold (BGM) disease of chickpea (*Cicer arietinum* L.) caused by *Botrytis cinerea* Pers. ex Fr. is common in areas where humidity and temperatures are high. Prospects of developing cultivars with high levels of resistance seem unlikely in the immediate future. Emphasis has to be put on chemical, biological, and agronomic components of integrated disease management, which are effective in augmenting the low levels of resistance. Management factors (eg. sowing date, row spacing, and plant density) and plant characters (eg. growth habit and open canopy structure) have been studied as a means of reducing BGM incidence. Relative humidity (RH), duration of leaf wetness (Butler 1993) and temperature of the air in canopies seem to be important. Direct effects of temperature have been found on the epidemiology of the disease (Rewal and Grewal 1989), with possible indirect effects on duration of leaf wetness. A flow of air could modify humidity and temperature in canopies, and change leaf wetness and disease incidence. There is no detailed published work on such effects. Light (photoperiod) seems to affect the reproduction of the pathogen (Rewal and Grewal 1989), and foggy (Mahmood et al., 1989) and cloudy conditions seem to favour disease development. In the subtropics of South Asia where BGM is important, residual soil moisture from the rainy-season, and winter rains, combined with mild winter temperatures result in profuse growth if the crop is sown soon after the rainy-season (in October or early November). High seed rates and close row spacings contribute to the development of dense canopies. These conditions are conducive to the development of BGM.

Late sowing of chickpea, in which BGM incidence is low, seems to create a mismatch between the phase of the crop most vulnerable to the disease and climatic conditions favourable to disease development. For example, at Gurdaspur in Punjab State, a location where epidemics occurred during 1980-81 and 1981-82, sowing in October exposes the crop to increasing RH and falling temperatures during the period the crop is progressing towards flowering. Sowing in late November or early December results in this phase coinciding with periods of falling RH and increasing temperatures.

A similar effect of climatic conditions on Ascochyta blight (AB) is seen in Mediterranean environments in the West Asia and North Africa (WANA) region. Low AB incidence on spring (March/April) sown, compared with winter (November/December) sown crops seem to be due to the unfavourable climatic conditions for AB during spring. Although yields of spring sown chickpea are almost half those of winter sown crops (Saxena 1984), spring sowing developed as a traditional practise because cultivars resistant to the disease and cold were not available in the past. These traits are essential if chickpea is to be grown as a winter crop in WANA.

In Australia it has been observed that early sowing causes excessive vegetative growth and greater BGM incidence, and late sowing causes reduced dry matter production, both conditions reducing yield, (Brinsmead 1992; Haware and McDonald, 1993). Effects of sowing date on phenology, shoot mass, yield and harvest index are also commonly observed in India. At Hisar, a lower shoot mass and higher harvest index from late sowing shows that vegetative growth is inhibited in the late-sown crop. Crop growth is related to temperature, terminal heat and a fall in RH. It is presumed that excessive vegetative growth and dense canopies, following early sowing, causes microclimate conditions favourable to disease incidence. The reverse is expected to be true from late sowing. However, data are not sufficient to support these supposed effects.

A wider row spacing at a constant plant density reduces disease incidence (Reddy et al., 1993), perhaps because of improved ventilation in the canopy. The advantage of sowing paired rows, compared with uniform row spacing (ICRISAT 1995), should combine the positive effects of wide spacing, and increasing plant densities, but this needs to be confirmed. There are few studies on the effects of sowing geometry, row spacing, and row orientation, with respect to wind direction and effects on airflow through crop canopies, leaf wetness, and disease incidence. Information on these aspects would be useful to improve management of the disease.

Genotypes with an erect compact growth habit have less BGM than genotypes with a bushy spreading habit (Reddy et al., 1993). Again the effect is attributed to the differences in microclimate conditions however, no data are available to confirm this conclusion.

A large variation in the number of leaves or nodes per plant exists within chickpea germplasm. The variation in leaf number is generally associated with differences in seed size, large-seeded types having fewer leaves. For example, K 850 (28 g/100 seeds) has fewer nodes and therefore, leaves per plant (129 nodes plant[-1]) than Annigeri (145 nodes plant[-1]; 18 g /100 seeds), when grown at ICRISAT Asia Center.

K 850, a genotype with a semi-compact growth habit, has less disease than the bushy types (M.P. Haware, personal communication). However, large-seeded types have large pinnule size. Accessions with fewer pinnules (ICC 5680) or narrow pinnules (ICC 14330) are available. Both traits could be useful to reduce the pinnule size of large-seeded commercial varieties, provided the two traits are not closely linked.

Some farming practices, which reduce vegetative growth, also may influence disease incidence. Detopping (nipping) of young shoots and the top portion of branches with three to four leaves, is a practice where growth is profuse, as in some parts of India (Saxena and Yadav, 1975) and Bangladesh. The shoots are used as a vegetable ("Shag"), in the Barind tract of Bangladesh. Light grazing by goats or sheep is practised in some regions of India and Pakistan. The effect of these practices on disease is not well known.

Effects of host morphology and physiology on Ascochyta blight

In many plants, pre-infectional physical barriers such as cuticle, epidermis or other structures may prevent or slow down penetration by the pathogen, (Akai and Fukutomi,

1980; Campbell et al., 1980). In order to find the relation between morphology and resistance to *Ascochyta rabiei*, four kabuli cultivars with different morphological and agronomical characteristics and various reactions to *A. rabiei* (Pass.) Labr. [teleomorph: *Didymella rabiei* (Kovachevski) v. Arx] were studied (Venora and Porta-Puglia, 1993). The cultivars showed differences in epidermal and sub-epidermal structures of the stem in terms of cell wall thickness and cell volume. The cell walls of the epidermis were thicker in the resistant cultivars. In cross sections studied by image analysis, 'Sultano' had the thickest epidermal cell walls and the largest mean area of the parenchymal cell of the outer layer. This cultivar was one of the most resistant to Ascochyta blight among accessions tested in Italy (Saccardo et al., 1987; Calcagno et al., 1992). Since the epidermal layer is also the site of storage of the isoflavones biochanin A and formononetin, precursors of phytoalexins in *C. arietinum* (Barz et al., 1993), we could assume that the protection provided by a thick wall can be completed by chemical mechanisms of resistance. Furthermore, the resistance in cv Califfo seems to be associated with both wall thickness of the epidermal cells and size of the first parenchyma cell layer. The susceptible cvs Principe and Calia do not differ significantly in cell wall thickness of the epidermal layer, while the cell area of the first parenchyma layer is significantly different between the two cultivars.

In another investigation (Angelini et al., 1993), the first and fourth internodes of the stem of fifteen-day-old plants of cv Calia and Sultano were analysed histologicaly. 'Sultano' had a higher number of xylem cells and xylem parenchyma cells produced by the vascular cambium in the interfascicular regions, when compared with 'Calia'. Although these results need to be confirmed with more varieties, they may be indicative of the defence mechanisms acting against penetration by *A. rabiei*. Whether *A. rabiei* penetrates by mechanical force or with the aid of cell-wall degrading enzymes, a thick epidermis is desirable in that it retards penetration and gives time for the plant to deploy defence mechanisms triggered by the penetration. Among these mechanisms, peroxidase (POD) and diamine oxidase (DAO) can play an important role (Angelini et al., 1990). Structural differences between 'Sultano' and 'Calia', as well as wall autofluorescence and lignosuberized depositions during fungal infection were investigated. Non-inoculated fourth internodes of the resistant 'Sultano' showed a greater POD and DAO activity than the susceptible 'Calia'. After inoculation, both enzyme activities were significantly higher in 'Sultano'. A crucial difference between the cultivars is that invasion of pith parenchyma cells in the fourth internodes was never observed in 'Sultano' while massive lysis of these cells occurred in 'Calia'. The infection areas in both cultivars were surrounded by a barrier made by lignosuberization of cell walls of cortical parenchyma. It is noteworthy that lignosuberized barriers appeared to a greater extent in the resistant ' Sultano'. Histochemical POD and DAO activities were detected in the barriers. Similar findings were observed in infected first internodes. These results suggest that both the structure of xylem tissue and the higher enzymatic activities of DAO and POD after inoculation may play a role in the complex process of to *A. rabiei*.

Vegetative plant parts

Various mutant leaf forms for shape and positional arrangement are known. They are controlled by single recessive genes (Muehlbauer and Singh, 1987; Erskine et al., 1988). Benefits from the simple leaf mutant on yield still need to be established.

Growth habit and stems

Variation in growth habit and branching is used by breeders to improve erectness and suitability for mechanized harvesting. Growth habit, which can be prostrate to erect, is controlled by dominant Mendelian genes, *Hg* for erectness over prostrate (*hg*) and *Br* for basal branching over "umbrella" type branching (*br*) (Muehlbauer and Singh, 1987). There is quantitative genetic variation in the degree of branching (Erskine et al., 1988). In Western

Australia, when Siddique and Sedgley (Erskine et al., 1988) eliminated the later formed basal laterals by debranching, no effect was observed on above-ground biological yield but harvest index increased by 13% and yield from 1.35 t ha^{-1} to 1.87 t ha^{-1}

Tall plants are easier to harvest mechanically. Variation for height is considerable in kabuli germplasm, with a range from 15-50 cm and a mean of 29.7 cm (Erskine et al., 1988). Tall genotypes have an increased internode length and node number and a more compact habit because of their narrower branching angles, but they have a similar yield potential to winter-planted conventional types in West-Asia. There is a tendency for tall kabuli accessions to have small seeds, with a seed type intermediate between desi and kabuli, but this negative correlation can be broken.

According to Sedgley et al., (1990), an erect habit and few branches is the most suitable plant habit for high input systems and good weed control. This type, when sown at high densities, reduces water depletion in winter resulting in a higher biological yield and harvest index.

Kumar et al. (1996) compared the conventional bushy type and the tall, erect type which are both in cultivation, with intermediate types developed at New Delhi. The traits included yield, number of branches, seeds per pod, pod number, seed weight and plant height. The tall types recorded the highest values for the traits which had the lowest values in conventional types and vice-versa. They concluded desirable genes with positive contributions to productivity may occur in both tall and bushy types. The intermediate plant types had high mean values for most yield components indicating that they possess better combinations of desirable traits inherited from both the bushy and the tall types.

Root morphology

Genetic differences in the rooting patterns of seedling have been recorded in chickpea (Erskine et al., 1988) with tolerance to drought being found in small seeded cultivars with long roots and a root length/plant height or root weight/biomass ratio of 2 (Calcagno and Gallo, 1993). Selection for large root size gave a greater degree of drought resistance measured as the decrease in yield with progressively increasing drought (Saxena and Johansen, 1996).

Reproductive plant parts

Although one flower or one pedicel per peduncle is normal, double flowered genotypes are quite common. A single recessive gene (sfl) controls the difference. The proportion of double podded nodes on double-pedicel genotypes varies over environments but when well expressed, it confers a yield advantage of 6-11% (Erskine et al., 1988). Pods and seeds from double-podded mutants are generally smaller. A recessive pod disposition gene (pdfr), which places the pods above the leaves rather than in the normal pendant position (Muehlbauer and Singh, 1987) is of interest with the view to improving pod photosynthesis. However, when pods of normal cultivars were reorientated and exposed to the sun, in each of three locations, no yield advantage could be demonstrated (Erskine et al., 1988). Pod dehiscence is not a problem (Erskine et al., 1988) but genetic variation exists for resistance (E.J. Knights, personal communication). Pod drop is a problem for mechanical harvesting during the hot summers in areas of south-eastern Australia but little information is available on its inheritance.

Yield-associated traits and harvest index

Progress in increasing harvest index by selecting for determinant flowering is expected to be slow as the heritability of determinancy is poor and characterised by a predominance of non-additive gene action (Erskine et al., 1988). Pod number per plant and seed size have the greatest direct effect on seed yield (Muehlbauer and Singh, 1987). Branching per plant and pod number per plant are consistently correlated with grain yield in segregating populations. Increased branching tends to increase flower and pod numbers, with positive effects on yield. Branch and pod numbers are the most important components of yield and most often used for selection. Yet, no correlation of pod number with seed yield was found when 3,269 kabuli germplasm accessions were evaluated at ICARDA (Muehlbauer and Singh, 1987). Strong correlations have also been reported between seed yield and biological yield and to a lesser extent with plant height (Muehlbauer and Singh, 1987). Indirect effects on yield are also expected from non-random associations found between five morphological traits (growth habit, canopy width, days to flowering, plant height and seed weight) and plant responses to cold stress, Ascochyta blight and leaf miner infestation (Jana, 1995). A multivariate analysis of 6,400 kabuli accessions indicated that these associations developed in the process of adaptation to diverse agroclimatic conditions in five continents.

Evolution and effects of seed on yield and quality

The intra-specific classification of *Cicer arietinum* into macrosperma and microsperma types based on seed and other morphological characters indicates a divergent geographic distribution. The microsperma group is found throughout the geographic distribution of the species but is scarcer in the Mediterranean region where the macrosperma group predominates. The distinction into desi, kabuli and "pea" or "intermediate" types is a parallel system more commonly used by traders (Hawtin and Singh, 1980). It is based on seed shape, size and colour, and takes into account geographic origin and use as food. The kabulis originated from the more primitive desi by selection primarily for seed size and suitability for human consumption. White flowered phenotypes became more acceptable as a result of the correlated response, commonly observed in legumes, where white flowered genotypes show low or zero tannin and a reduction in other anti-nutritional factors. Useful traits can be found in both kabulis and desis and can be transferred by intercrossing (Hawtin and Singh, 1980). The kabuli group has a greater range in seed size, tends to have more primary branches, greater cold tolerance, a more upright and, in some cases, taller growth habit and greater resistance to Fe chlorosis. Desis tend to have a bushier growth habit, more seeds per pod, more pods per plant and greater tolerance to drought and heat. Traits such as double-podding and resistance to wilt and salinity were also identified in desi backgrounds. As the gene pools have been separated for many years, genes for traits related to yield may differ between the groups. Reports of transgressive segregation for growth habit, number of branches, plant height, seed size, pod number or yield from kabuli × desi crosses are therefore not surprising (Bahl, 1980; Hawtin and Singh, 1980). Other associations relate to more rapid seedling growth of kabulis or to a higher percentage of field emergence in desis with a pigmented seed coat compared to non-pigmented kabulis (Auld et al., 1988). In both groups an effect of larger seed size can be demonstrated on germination, emergence and seedling (Roy et al., 1994; Saxena, 1987; Smith et al., 1987). Biderbost et al. (1980), who compared four seed-size categories of one cultivar for germination rate, total germination percentage and seed yield, did not find an effect of seed size on grain yield. However, others (Carter and Bretag, 1997; Eser et al., 1991; Vadivelu and Ramkhrishnan, 1983) have obtained higher yields from using the larger seed within cultivars. Murray and Auld (1987) reduced sowing costs by 30%, without sacrificing grain yield or 100-grain weight, when they used seed 14% smaller than the largest size seed in seed lots of the kabuli cv UC-5.

There is a strong financial incentive for growers of kabuli cultivars to sell their largest grain and keep the smallest for sowing as there can be a large price differential on grain size. In Australia, growers have noticed that grain size and size distribution vary from year to year and want to know how to produce high yields with large grain. Seeds of cv Kaniva kabuli were graded into different size lots (6,7,8 and 9 mm screen sizes, the equivalent of 19.5, 29.2, 38.1 and 43.2 g/100 seeds) and sown at different densities (Carter and Bretag, 1997).

Crop establishment was poorest from the smallest seed. At least twice the number of 6 mm size seeds were required to achieve similar plant populations to those obtained with 8 mm or 9 mm size seeds. Highest yields were obtained from the largest seeds (8 mm and 9 mm). Increased seeding rates resulted in large increases in grain yield (0.92 to 1.83 t ha^{-1}) but a small reduction in grain size (8.6 to 8.1 mm or 39.2 to 33.6 g/100 seeds). Reducing seeding rates can result in small increases in grain size but is likely to cause a large reduction in yield. Thus, growers will not save money by using small seeds (<8 mm) for sowing, as the large seeds are more likely to result in higher yields.

Quality

There is strong consumer preference for specific seed sizes in different countries. A difference between consumers in their preferences occurs particularly for kabuli types but preferences also vary among consumers of desi types. Larger seed are often cooked whole, rather than split into dhal. In kabuli the small-seeded class is <25 g/100 seeds, medium size 25-40 g and large-seeded >40 g (Singh, 1987). In some countries such as Spain, consumers will not accept any size less than 60 g/100 seeds and in Canada anything less than 52 g/100 seeds. Seed size for canning should be around 50 g/100 seeds.

Seed size is positively correlated with longer cooking time and hydration capacity (Williams and Singh, 1987) but not with seed coat thickness (Gil and Cubero, 1993; Williams and Singh, 1987), which indicates that development of larger desi types with a thin coat would be feasible. Coat thickness is linked with flower colour and exhibits monogenic inheritance with the thin kabuli seed coat being recessive (Gil and Cubero, 1993). It affects dhal recovery in processing desi grain.

Seed size uniformity is important as a high proportion of small seeds lowers the value of a sample. Larger seeds process more easily into foods and have a lower husk content (Williams and Singh, 1987). Seed shape also influences processing. Rounder seeds are easier to mill, roast or coat with sugar. Deeply convoluted seeds carry more dirt and seeds with a pronounced beak tend to lose more on decortication. Decortication loss is a significant factor in processing and market value and is determined by seed shape and size, thickness of the hull, hardness of the seed and the decortication process. The larger the seed, the smaller the proportion removed by decortication. However in kabulis, which are usually consumed without decortication, consumer preference is for a seed with deep surface folds.

Both positive (Muehlbauer and Singh, 1987) and negative (De Haro and Moreno, 1992) correlations have been found between seed size and protein content. Oil content in both kabuli and desi types were positively correlated with seed size, while positive correlations were found for seed size with stearic acid content in kabulis and with oleic acid content in desis (De Haro and Moreno, 1992).

Preferences for colour of the seed can vary considerably. In the kabuli type, preference is for beige or light cream-white colour. Among desis the preference varies between countries and even within country. Some preferred colours are yellow, brown, black, green and pink. Numerous genes influence colour and several have pleiotropic effects on flower, stem and leaf colouration. Their interactions cause a wide range of seed phenotypes, which are of interest to breeders satisfying diverse market requirements (Muehlbauer and Singh, 1987). Gene *gr*, for example, causes green cotyledons, green testa and bluish-green leaves and three dominant genes (*Rs, Rsa* and *Rsb*) control rough seed surface.

FABA BEAN (*VICIA FABA* L.)

Resistance to chocolate spot and mode of inheritance

Chocolate spot (*Botrytis fabae* Sard.) is a major constraint to faba bean production in many areas of the world. In Egypt, chocolate spot and rust [*Uromyces fabae* (Grev.) DeBy ex Fuckel] seriously affect production in the Nile Delta region (Bernier et al., 1984), with losses from both diseases of up to 50% (Ibrahim et al., 1979) or more (Mohamed, 1982). A collaborative program between ICARDA and the National Agricultural Research System in Egypt, identified sources of resistance to these diseases (Khalil, et al., 1984). Large differences were found by Abo-El-Zahab et al.(1994) among genotypes across seasons and generations in their reaction to chocolate spot. Crosses involving the resistant parents ILB 438 and ILB 938 gave the highest levels of resistance. These ICARDA lines, along with BLP 710, BLP 1179 and BLP 1196 have been used in breeding programs in several countries (Duc, 1997). Both general and specific combining ability occur indicating that additive and dominance components affect resistance, with the additive variance being more important. The above lines are recommended as parents in any crosses if the additive × additive genetic variances are to be exploited.

Utilisation of genetic resources

Khalil et al (1993c) evaluated 21 F5 lines from six crosses. Trials were conducted during three years in two North Delta locations in fields and pots. Plants in the pots were inoculated with a highly virulent isolate of *B. fabae* and revealed large differences. Eight lines were selected for their chocolate spot resistance and yielding ability. The overall disease score was 36%, 34% and 44% lower than for the check cultivars Giza 2, Giza 3 and Giza 402, respectively.

The average 100-seed weight of the resistant lines was greater (18.4%) than that of the check cultivars. The dark or light green seed coat colour was associated with highly or moderately resistant recombinants, derived from crosses involving ILB 938 or ILB 438. Seed weight and seed-coat colour were used as indicators or gene markers during the breeder seed production to maintain resistance and seed quality.

Six new cultivars have been developed from ICARDA , Egyptian, Mediterranean and European material: Giza 461 (Khalil et al., 1994), Giza 714, Giza 717 (Khalil et al., 1993a), Giza 716, Giza 643 and Gizablanca (Khalil et al., 1996). All have a higher yield potential than the check varieties and were resistant to rust and chocolate spot. The new varieties are also resistant to lodging and produce seeds with improved commercial quality. They were released to the farmers in the North Delta and Nubaria areas.

Integrated approaches

To investigate the possibilities for an integrated approach to disease management, experiments were carried out in Egypt over three years (Khalil et al., 1993b) with the recommended cultivar Giza 3 (medium seed size) and the accession Reina blanca (large seed size) sown at two densities (16 and 33 plants m^{-2}) with and without chemical control of disease. 'Reina blanca' was more resistant to chocolate spot and rust than 'Giza 3'. The severity of both diseases was increased at the higher seeding rate. The average yield of 'Reina blanca' was 45% higher than that of 'Giza 3'. The higher plant density gave higher yields. Giza 3' was taller and produced more pods and seeds per plant than 'Reina blanca'. However, the latter gave more yield and heavier seeds (31.1% and 83.30% increase respectively). Higher plant density increased plant height, while lower density increased number of pods, number of seeds per plant and yield per plant. Chemical control with mancozeb increased yield by an average of 92.8%, together with plant height, number of pods, seed per plant and 100-seed weight.

Studies on cultivars, resistant and susceptible to chocolate spot, indicated that resistance was associated with the accumulation of phytoalexins in plant tissues. The highest concentration was found in the resistant accession ILB 938, followed by the new cv Giza 461, while the lowest was in the susceptible 'Giza 402' (Omar et al., 1992).

Plant architecture and yield

Natural variability and induced mutation are available and are used when breeding for reduced lodging and abortion of reproductive organs. Indeterminate cultivars are tall, but short-strawed cultivars are available. In determinate types ("topless") the *ti* gene is present (Sjödin, 1971). A *ti-s* gene, controlling the "semi-determinate" growth habit was also described (Frauen and Brimo, 1983). In semi-determinate winter types developed by Le Guen and Duc (1992), meristems senesce after 10-12 flowering nodes, whereas conventional types may reach 25 nodes. Frauen and Sass (1989) have identified a "stiff-straw" character, which is of simple, monogenic inheritance.

Effects of plant architecture on crop performance and productivity is discussed by Heath et al. (1994), Bond (1987) and Duc (1997). Determinate types produce less yield than indeterminate types, because of their less efficient plant architecture and physiological disadvantages in assimilate transfer from stem to pod (Pilbeam et al., 1989).

In spring sown faba beans the optimum plant densities for determinate types are higher than for the indeterminate types (Pilbeam et al., 1990). No difference was observed between short strawed and indeterminates (Cleal, 1991). By contrast, plant morphology does not affect the optimum density in winter sown cultivars (Pilbeam et al., 1991).

The trait of independent pod vascular supply has been used to reduce reproductive losses. The high pod number per node and the high number of seeds per pod are generally considered to be positive traits, aimed at concentrating yield on a few nodes (Duc, 1997).

Relationship between flower colour and yield

S.A. Khalil (unpublished data, 1995) crossed a tannin-free accession, Triple White, which has a white wing petal (WF), with five other genotypes all with black spotted wing petals (BS). The F1 plants showed complete dominance of BS over WF. However, the F2 segregated into BS, WF and yellow spotted wing petal colours (YS). The BS plants gave the highest estimated yields, the YS plants the lowest and WF plants had intermediate yields. These findings indicate the relation between quantitative and qualitative characters. Undesirable characters, such as smoken seeds (Ibrahim, 1963), are genetically linked with white flowers. Other examples are yellow flowers being linked with pod dehiscence, pod shedding and shattering and to susceptibility to chocolate spot and rust, all of which have an effect on yield (Mohamed, 1982).

LENTIL

Vegetative plant parts

Several atypical leaf types are known in lentils (*Lens culinaris* Medik.) but they do not appear to have the potential to improve yield (Erskine et al., 1988). However, the gene *tnl* which governs the presence of tendrils at the end of the leaves may be useful for canopy formation and the standing ability of the crop, making it more suitable for mechanized harvesting (Muehlbauer et al., 1995).

Growth habit and stems

Lentil plants have a freely branching growth habit and poorly defined stems. A single gene *Gh* controls growth habit, with *ghgh* giving prostrate and *GhGh* giving erect plants (Muehlbauer et al., 1995). Erect lentils are easier to harvest than lodged plants and lodging increases losses of both seed and straw. The tendency for plants to lodge is related to stem diameter and probably stem lignification (Erskine et al., 1988). Variability in stem height is continuous and pods can be placed in the range 3 -30 cm from the base, allowing for selection for suitability for mechanized harvesting, which requires a clearance of about 12 cm. Advantageous and positive correlations have been found between plant height and yield in lentil. An increased height often ensures a large vegetative frame on which many reproductive nodes can develop. However, deleterious correlations have been reported between plant height and both time of maturity and tendency to lodge (Erskine et al., 1988). Growth habit can also affect the severity of diseases (W. Erskine and T.W. Bretag, personal communication) as infection by *Sclerotinia sclerotiorum* (Lib.) de Bary is often enhanced when the plant tops touch the ground after lodging.

Reproductive structures

Reproductive nodes commonly carry one to three flowered racemes per peduncle. A single gene determines the difference between double and triple flowered genotypes but up to seven flowers per peduncle have been recorded for plants grown in a glasshouse. The effect of the multi-podded habit on yield needs further investigation (Erskine et al., 1988, Muehlbauer et al., 1995). Pod length was found to have a direct positive effect on yield (Muehlbauer et al., 1995). Problems with pod drop account for 65% of losses in mechanical harvesting under dry conditions, and pod dehiscence contributes to the remainder. Excellent resistance to pod dehiscence has been found; a single recessive gene pi controls pod indehiscence. The wild progenitor, *L. culinaris ssp. orientalis* has *PiPi* and cultivated *L. culinaris spp. culinaris* has *pipi*. Variability for shattering occurs in the cultivated type even in the presence of *pi*, which indicates the presence of modifier genes. Genetic variation for pod drop may be limited (Erskine et al., 1988; Muehlbauer et al., 1995).

Path coefficient analyses and positive correlations of grain yield with a range of morphological and physiological traits have been discussed by Verma et al. (1993) and Muehlbauer et al. (1995). Breeding for increased harvest index has been advocated in lentils (Erskine et al., 1988). In Middle Eastern countries however, lentil straw is still a valuable commodity and requires a large biological yield rather than a high harvest index. Prolonged leaflet retention may also contribute to a larger forage yield.

Effects of seed size on yield

Seed size is highly heritable (Erskine et al., 1988; Muehlbauer et al., 1995) and has been studied for its relation with yield. Correlations of seed size with yield differ between studies, with positive and negative values having been found (Singh and Singh, 1969; Verma et al., 1993; Muehlbauer et al., 1995).

Seed size was positively correlated with pod size and leaf size among bold seeded genotypes. Correlations between pod size and leaf size were also demonstrated, leading to the suggestion leaf size may be a fairly reliable selection index for seed size in segregating populations, even at the seedling stage (Sharma et al., 1993). However, seeds per pod is negatively correlated with 1000 grain weight, pod size and leaf area in bold-seeded lentils in India (Sharma et al., 1993).

Some of the contradictory associations reported for seed size are worth considering in the context of varying environments and diverse gene pools. Of interest are the studies of Erskine (1996), who examined adaptation to temperature and rainfall in Syria, Turkey and

Lebanon among accessions and breeding lines of two seed-size groups represented by large-seeded, yellow cotyledon lentils and by small-seeded red cotyledon lentils (seed mass < and > 4.5 g/100 seeds or seed diameter < and > 6 mm respectively). West Asia historically has been the only lentil growing area in the world where both types are largely grown by farmers.

The large seeded material had a longer reproductive growth period than the small-seeded group (2.8 days longer). An extended period is considered to be required to fill the greater seed mass per pod among the yellow-cotyledon types, which also produces taller plants and more straw. The germplasm with large seeds was less susceptible to winter cold. The group also had an advantage in yield over the small-seeded types in average temperatures <10 °C, with the converse being true at higher temperatures.

Allied to the higher cold tolerance of the large-seeded group was their response to cool conditions throughout the season. They showed an advantage in yield at the two wetter sites, with the cool and wet seasons allowing a protracted period for vegetative and reproductive growth, whereas the small-seeded group was better adapted to dry environments. To cope with dry warm environments, drought escape through early flowering and early maturity is important (Erskine, 1996). Selection among the large-seeded lentils should be for early flowering, as an extended reproductive period may be required by all large-seeded lentils.

The environment, acting as an evolutionary force creating and preserving genetic variation useful for lentil improvement, should not be underestimated (Erskine et al., 1989). The obvious discrimination is in seed size (Muehlbauer et al., 1995) as macrosperma types are common in the Mediterranean basin and Western Hemisphere, while microsperma are dominant on the Indian subcontinent.

Promising results were obtained for yield improvement in Bangladesh by introducing genes for larger seed size (>2.5 g/100 seeds) from macrosperma into a locally-adapted microsperma variety (1.3-1.5g/100) (Sarker et al., 1992). The lentils in South Asia are of a specific ecotype (*pilosae*), characterized by the pubescent trait and by their short plant height and small seed. They are thought to have been derived from a small founding population as they show a lack of variability compared to lentils in other regions (Erskine et al., 1996). The lack of variability is seen as a bottle-neck to genetic improvement and much progress is expected to come from hybridization with "exotic" germplasm.

Quality

Quality factors such as seed size and seed thickness affect the processing and cooking quality and seem to be strongly heritable. Cooking time is more related to seed size (r=0.92) than to environment (Muehlbauer et al., 1995). Seed coat colour, which is important in determining market potential is controlled by several genes and pleiotropic action is likely for epicotyl and flower colour, with parallels in *Cicer* and *Pisum* (Muehlbauer et al., 1995). Five alleles have been identified for the *Scp* gene which controls the trait of spotted seed coat; the nonspotted genotype is *scp scp*. Background colour of the seed coat is controlled by two genes. Dominant *Ggc* determines grey ground colour, while the dominant *Tgc* produces a tan ground colour. When both are present (*Ggc Tgc*) a brown seed coat is produced. The double recessive (*ggc tgc*) has a green seed. A recessive gene *tan* for zero tannin in the coat is useful in reducing darkening of the seed with age (Muehlbauer et al., 1995), but the coat is thinner and more fragile, making the seed more susceptible to rot. Two loci are involved in determining the colour of lentil cotyledon, with dominant red colour (*Yc-*) epistatic to both yellow (*I-*) and green (*ii*) (Muehlbauer et al., 1995).

A small but positive correlation (r=0.26) has been reported between protein content and seed weight (Sharma et al., 1993) and between seed size and protein per seed (r=0.31) suggesting that breeding for larger grain does not necessarily cause a decline in protein content.

LUPINS

Plant architecture

In lupins (*Lupinus* spp.) pod set and growth on the main shoot, thickening of the main stem and vigorous growth of the first-order lateral branches all occur at the same time (Gladstones, 1994). This is likely to generate competition for resources. In addition, the branches grow above the main shoot and shade its leaves and inflorescence. The competitive effects of these branches have been demonstrated by cutting them off at an early stage. The main shoot then sets more pods and produces a greater yield. As often happens, more pods are set than will fill seeds. As a result of this perceived wasteful process, genotypes of narrow-leaf lupins (*Lupinus angustifolius* L.) with less branching have been developed and have given promising yields in the short-season environments of Western Australia (Hamblin et al., 1986). However, interest in reduced-branching lost favour among breeders for a number of years because of problems associated with high alkaloid levels, insufficient *Phomopsis* resistance, and small seed size (Gladstones, 1994). Reduction in branching and the late production of new leaves were also considered to enhance susceptibility to brown leaf spot (*Pleiochaeta setosa* [Kirchn.] Hughes). A single incompletely dominant gene, *Det,* controls reduced branching. In plants carrying *Det* or similar genes, enhanced pod set appears to be achieved indirectly through suppression of competing branch growth. The *Det* genotypes lack the capacity for new leaf development after initial setting regardless of when it occurs. Therefore, new leaves may be able to function later in the season, when older leaves have senesced and/or fallen down. It is this part which makes reduced-branching lupins so vulnerable to brown leaf spot as it causes premature loss of older leaves. More recent comparisons have shown restricted-branching types have the potential for higher yields than and normal types of narrow-leafed lupins and *albus* lupins when compared in otherwise similar genetic backgrounds (Dracup and Kirby, 1996).

CONCLUDING REMARKS

There is a scarcity of genetic resistance to the many foliar diseases of the cool-season pulses. Morphological characters, which can create microenvironments less favourable to the development of epidemics, can have significant effects on yield and quality.

The economic value of the variation in morphological traits available in germplasm of the pulses still remains largely unexploited. A reason for this is that the underlying physiological factors interact and are complex to understand. Their study requires an interdisciplinary and global approach. Collaboration between diverse disciplines to widen our knowledge is a recent trend not yet fully exploited.

Computer models, which can simulate plant development and growth (Jeuffroy and Ney, 1997) will enhance our understanding of crop physiology and the effect of plant morphology on yield, disease resistance and quality. They will help in dealing appropriately with the complex interrelationships.

When using morphological variation, it is important to be mindful of the specificity of the cropping system and of the environment in which this variation will be exploited. Strategies for the genetic improvement of low input systems in tropical areas should be different from those aimed at high input European systems. This applies particularly where the option exists for disease control by means of morphological traits rather than the application of chemicals.

Techniques developed recently, such as positional cloning, can help overcome limitations in genetic improvement, where traits depend on complex interactions between genes, gene products and the environment.

The expression of many morphological traits may vary with the genetic background of the individual plant. New traits may also upset the balance among other factors which have evolved in the species and still influence performance. Genetic improvement aimed at achieving an "equilibrium" are, therefore more likely to succeed than those where a reputedly "positive" trait is over-emphasised.

International cooperation has, and should continue to have, an important role in collecting, characterising and maintaining germplasm collections. Finally, a large variability also exists among wild relatives of several legume species. Joint international and national efforts are needed to save this variability for present and future exploitation.

References

Abo-El Zahab, A.A., Khalil, S.A, Abou Zeid, N.M, El-Hinnawy H.H and El-Hady, M.M. 1994. *Proceedings 6th Conf. Agronomy, Al-Azhar University, Cairo, Egypt, Sept. 1994*, Vol. 2, pp. 675-691.

Akai, S. and Fukutomi, M. 1980. In: *Plant Disease. An Advanced Treatise*, Vol. 5, pp 139-159 (eds. J.G. Horsfall and E.B. Cowling). New York: Academic Press.

Ali, S.M., Nitschke, L.F. and Dube, A.J. 1978. *Pea breeding programme.* South Australian Department of Agriculture and Fisheries, Technote 3/78: 1-4.

Ali-Khan, S.T. and Snoad, B. 1977. *Annals of Applied Biology.* 85:1313-136.

Angelini, R., Manes, F., Federico, R., 1990. *Planta* 182: 89-96.

Angelini, R., Bragaloni, M., Federico, R., Infantino, A. and Porta-Puglia, A. 1993. *Journal of Plant Physiology*, 142: 704-709.

Armstrong, E.L. and Pate, J.S. 1994. *Australian Journal of Agricultural Research* 45:1347-1362.

Auld, D.L., Bettis, B.L., Crock, J.E. and Kephart, K.D. 1988. *Agronomy Journal* 80:909-914.

Bahl, P.N. 1980. 75-80. *Proceedings International Workshop on Chickpea Improvement*, 28 February-2 March, 1979, pp. 51-60. (eds. J.M. Green, Y.L. Nene and J.B. Smithson). Hyderabad, India: ICRISAT.

Barz, W., Bless, W., Gunia, W., Höhl, B., Mackenbrock, U., Meyer, D., Tenhaken, R. and Vogelsang, R. 1993.. In: *Breeding for stress tolerance in cool-season food legumes* pp. 193-210 (eds. K.B Singh. and M.C. Saxena). Chichester, U.K: John Wiley & Sons.

Bernier, C.C., Hanounik, S.B., Hussein, M.M. and Mohammed, H.A. 1984. Inf. Bull. No. 3, ICARDA, Aleppo, Syria, pp. 40.

Berry, G.J. 1981 PhD Thesis, University of Melbourne, pp. 421.

Berry, G.J. and Aitken, Y. 1979. *Australian Journal of Plant Physiology* 6:573-587.

Biderbost, E., Peretti, D. and Errasti, J. 1980. *Revista de Ciencias Agropecuarias* 1: 39-58.

Bond, D.A., 1987. *Plant Breeding* 99: 1-26.

Bretag,T.W. 1991.. La Trobe University, PhD Thesis.

Bretag, T.W. and Brouwer, J.B 1995. *Proceedings of the Second European Conference on Grain Legumes*, p.92, 9-13 July, Copenhagen, Denmark: AEP.

Brinsmead, R.B. 1992. *Program and abstracts. 2nd International Food Legume Research Conference*, 12-16 April, Cairo, Egypt. Cairo, p. 64.

Butler, D.R. 1993.:*Summary Proceedings of the Second Working Group Meeting*, 14-17 Mar 1993, Rampur, Nepal, pp. 7-9 (eds. M.P Haware, C.L.L. Gowda and D. McDonald). Patancheru, India: ICRISAT.

Caetano-Anollés, G. 1997. *Field Crops Research* 53: 47-68.

Calcagno, F., Crinò, P. and Saccardo, F. 1992. *Program and Abstract 2nd International Food Legume Research Conference, Cairo, Egypt,* p. 40.

Calcagno, F. and Gallo, G. 1993. In: *Breeding for Stress Tolerance in Cool-season Food Legumes*, pp. 293-309 (eds. K.B. Singh and M.C. Saxena). Chichester, U.K.: John Wiley & Sons.

Campbell, C.L., Huang, J.-S. and Payne, G.A., 1980. In: *Plant Disease. An Advanced Treatise,* Vol. 5, pp 103-120 (eds. J.G. Horsfall, and E.B. Cowling). New York: Academic Press.

Cardi, T. and Monti, L.M. 1985. *Pisum Newsletter* 17:6.

Carter, J.M. and Bretag, T.W. 1997. *Proceedings First Australian New Crops Conference*, pp. 9-13. University of Queensland, Gatton College, 8-11 July, 1996.

Cleal, R.A.E. 1991. In *Production and Protection of Legumes,* pp. 89-94 (eds. R.J. Froud-Williams et al.) Wellesbourne, Warwick, U.K.: The Association of Applied Biologists, Horticulture Research International.

Cousin, R. 1997. *Field Crops Research* 53: 111-130.

Davies, DR, Berry, G.J., Heath, M.C. and Dawkins, T.C.K. 1985. In: *Grain Legume Crops*, pp. 147-198 (eds. R.J. Summerfield and E.H. Roberts). London: Collins.

De Haro, A. and Moreno, M.T. 1992. *Proceedings of First European Conference on Grain Legumes*, 1-3 June 1992, Angers, France, pp. 95-96.

Dracup, M. and Kirby, E.J.M. 1996. *Lupin development guide*, Nedlands, WA: University of Western Australia Press, 97pp.

Duc, G. 1997. *Field Crops Research* 53: 99-109.

Erskine, W. 1996. *Journal of Agicultural Science, Cambridge*, 126: 335-341.

Erskine, W., Nassib, A.M. and Telaye, A. 1988. In: *World crops: Cool season food legumes*, pp. 17-125 (ed. R.J. Summerfield). Dordrecht: Kluwer Academic Publishers,.

Erskine, W., Adham, Y. and Holly, L. 1989. *Euphytica* 43: 97-103

Erskine, W., Chandra, S., Chaudry, M., Malik, I.A., Sarker, A., Sharma, B., Tufail, M. and Tyagi, M.C. 1996. *Diversity* 12: 64.

Eser, D., Ukur, A. and Adak, M.S. 1991. *International Chickpea Newsletter* 25:13-15.

Floyd, R.S. and Murfet, I.C. 1986. *Pisum Newsletter* 18:12-15.

Frauen, M. and Brimo M. 1983. *Zetschrift für Pflanzenzüchtung* 91: 261-263.

Frauen, M. and Sass, O. 1989. *XII Eucarpia Congress*, 13-8, p. 15.

Gent, G.P.1988. *The PGRO Pea Growing Handbook.* Thornhaugh, Petersborough, U.K.: Processors and Growers Research Organisation, 264 pp.

Gil, J. and Cubero, J.I. 1993. *Plant Breeding* 111, 257-260.

Gladstones, J.S. 1994. *Proceedings of the First Australian Lupin Technical Symposi*um, Pert, Western Australia 17-21 October, p. 1-38.

Goldman, I.L. and Gritton, E.T. 1992. *Crop Science* 32:851-855.

Goldman, I.L., Gritton, E.T. and Flannery, P.J. 1992 *Crop Science* 32:855-861.

Gonzalez-Lauck, V.W. 1990. PhD Dissertation (Agronomy), University of Wisconsin, Madison, USA, 142 pp.

Gregory, P.J., Saxena, N.P., Arihara, J. and Ito, O. 1994. In: *Expanding the Production and Use of Cool Season Food Legumes,* pp.809-820 (eds. F.J. Muehlbauer and W.J. Kaiser). Dordrecht: Kluwer Academic Publishers.

Hamblin, J., Delane, R.J, Bishop, A. and Gladstones, J.S. 1986. *Australian Journal of Agricultural Research*. 37: 611-620.

Hardwick, R.C. 1988. In: *World Crops: Cool Season Legumes,* pp. 886-896 (ed. R.J. Summerfield). Dordrecht: Kluwer Academic Publishers.

Hare, W.W. and Walker, J.C. 1944. *Wisconsin Agricultural Experimental Station, Research Bulletin* No. 150: 1-31.

Harrison and Wang, 1995. *Proceedings of the Second European Conference on Grain Legumes,* p. 395, 9-13 July, Copenhagen, Denmark: AEP.

Haware, M.P. and McDonald, D. 1993. In: *Recent advances in research on Borytis gray mold of chickpea:* Summary proceedings of the Second Working Group Meeting, 14-17 March, Rampur, Nepal), pp. 3-6 (eds. M.P Haware, C.L.L Gowda and D. McDonald). Patancheru, India: ICRISAT.

Hawtin, G.C. and Singh, K.B. 1980. *Proceedings International Workshop on Chickpea Improvement*, pp. 51-60 (eds. J.M. Green, Y.L. Nene and J.B. Smithson), 28 February-2 March, 1979. Hyderabad, India: ICRISAT.

Heath, M.C. and Hebblethwaite, P.D., 1987. *Annals of Applied Biology* 110: 413-420.

Heath, M.C., Pilbeam, C.J., McKentie, B.A. and Hebblethwaite, P.D., 1994. In: *Expanding the Production and Use of Cool Season Food Legumes* (eds. F.J. Muehlbauer and W.J. Kaiser), pp. 771-790. Dordrecht: Kluwer Academic Publishers.

Hedley, C.L. and Ambrose, M.J. 1981. *Advances in Agronomy* 34: 225-277.

Hedley, C.L. andAmbrose, M.J. 1985. In: *The pea crop. A basis for improvement*, p.95-104 (eds. P.D. Hebblethwaite, M.C. Heath and T.C.K. Dawkins). London: Butterworths.

Ibrahim, A. A. 1963. Faculty of Agriculture, Cairo University. PhD Thesis

Ibrahim, A.A., Nassib, A.M and El-Sherbeeny, M. 1979. In: *Food Legume Improvement and Development,* ICARDA-IDRC. Ottawa: IDRC Pub. 126e.

ICRISAT 1995. *ICRISAT Asia Region Annual report* 1994. Patancheru, India: ICRISAT. 224 pp. (Semi-formal publication).

Jackson, M.B. 1985. In: *The pea crop. A basis for improvement,* 163-172 (eds. P.D. Hebblethwaite, M.C. Heath and T.C.K. Dawkins). London: Butterworths.

Jana, S. 1995. *Proceedings of the Second European Conference on Grain Legumes,* p. 175, 9-13 July, Copenhagen, Denmark: AEP,

Jeuffroy, M-H. andNey B., 1997. *Field Crops Research,* 53: 3-16.

Khalil, S.A., Nassib, A.M., Mohamed, H.A and Habib, W.F. 1984.*Proceedings, Seminar EEC programme,* 3-9 April, pp.80-94 (eds. G.P. Chapman and A.A.P. Tarwal), The Hague, The Netherlands: Martinus Nijhoff.

Khalil, S.A, Amer, M.I.A., El-Hadi, M.M., Dissouky R.F. and El-Baraee, M.A. 1993a *Egyptian Journal of Applied Science, Zagazig University,* 8: 218-232.

Khalil, S.A., Dissouki, R.F., Amer, M.I., El-Hady, M.M and. Hassan, M.W. 1993b. *Journal of Agricultural Sciences, Mansoura University,* 18:1306-1314.

Khalil, S.A., El-Hady, M.M, Dissouki, R.F., Amer, M.I. and Omar, S.A. 1993c. *Journal of Agricultural Sciences, Mansoura University,* 18. 1315-1328.

Khalil, S.A., Saber, H.A., El-Hady, M.M., Amer, M.I., Omar, M.A., Mahmoud, S. A. and Abou-Zeid, N.M., 1994. *International Symposium on Pulses Research,* April 2-6, 1994, New Delhi, India, Abstract P1.6.11, p 147-148.

Khalil, S.A., Saber, H.A., El-Hady, M.M., Amer, M.I., Omar, M.A., Mahmoud, S. A. and Abou-Zeid, N.M. 1996. *Premier seminaire du réseau Maghrebin de recherche sur fève (Remafeve),* 24-27 May 1995, Rabat, Morroco, pp. 47-54.

Kumar, J., Yadav, S.S. and Singh, R. 1996. In: *Abstracts: 2nd International Crop Science Congress,* p. 232, New Delhi, National Academy of Agricultural Sciences, India and Indian Council of Agricultural Research.

Lanfond, G., Ali-Khan, S.T. and Evans, L.E. 1981. *Canadian Journal of Plant Science* 61:463-465.

Le Guen, J. and Duc, G. 1992. In: *Amélioration des Espèces Végétales Cultivées,* pp.189-203 (eds. A. Gallais and H.Bannerot). Paris, INRA.

Mahmood, M., Sinha, B.K., and Kumar, S. 1989. *Journal of Research—Rajendra Agricultural University* 7(1-2): 39-43.

Mohamed, H.A. 1982. In: *Faba bean improvement. Proceeding of the International faba bean conference,* 7-11 March 1981, Cairo, Egypt, pp. 213-225.

Monti, L., Frusciante, L. and Romano, R. 1993 peas. In: *Breeding for stress tolerance in cool-season food legumes* pp. 63-73 (eds. K.B. Singh and M.C. Saxena),. Chicester, U.K.: John Wiley & Sons.

Muehlbauer, F.J., Kaiser, W.J.(eds.), 1994. *Expanding the Production and Use of Cool Season Food Legumes.* Dordrecht: Kluwer Academic Publisher.991 pp.

Muehlbauer, F.J., Kaiser, W.J., Clement, S.L. and Summerfield, R.J. 1995. *Advances in Agronomy* 54:283-332.

Muehlbauer, F.J. and Singh, K.B. 1987. In: *The chickpea,* pp. 99-125 (eds. M.C. Saxena and K.B. Singh) Wallingford, Oxon U.K.: CAB International.

Murray, G.A. and Auld, D.L. 1987. *Journal of Applied Seed Production* 5:10-17.

Norman, J.M., and Campbell, G.S., 1989. In: *Plant Physiological Ecology,* pp.301-325 (eds R.W. Pearcy, J.R. Ehleringer, H.A. Mooney, and P.W. Rundel). London and New York: Chapman and Hall.

Omar, S.A., Khalil, S.A. and El-Hady, M.M., 1992. *Egyptian Journal of Applied Sciences.,* 7(1), 42-47

Pate, J.S. 1975. In: *Crop physiology, some case histories* pp.191-224 (ed. L.T. Evans), Cambridge University Press.

Pate, J.S. and Armstrong, E.L. 1996. In: *Photoassimilate distribution in plants and crops. Source-sink relationships* pp. 625-642 (eds. E. Zamski and A.A. Schaffer). Marcel Dekker Inc.

Pilbeam, C.J., Hebblethwaite, P.D. and Clark A.S. 1989. *Field Crops Research,* 21: 203-214.

Pilbeam, C.J, Duc, G. and Hebblethwaite, P.D. 1990. *Journal of Agricultural Science, Cambridge.* 114:19-33.

Pilbeam, C.J., Hebblethwaite, P.D., Ricketts, H.E. and Nyongesa, T.E. 1991. *Journal of Agricultural Science, Cambridge* 116: 375-383.

220

Ranalli, P. and Graham, P.H.(eds), 1997. *Field Crop Research*, 53: 1-218.

Roy, A., Paul, S.R. and Sarma, R.N. 1994. *Annals of Agricultural Research* 15:383-384.

Reddy, M.V., Ghanekar, A.M., Nene, Y.L., Haware, M.P., Tripathi, H.S. and Rathi, Y.P.S. 1993. *Indian Journal of Plant Protection* 21: 112-113.

Rewal, N. and Grewal, J.S. 1989. *Indian Phytopathology* 42: 79-83.

Saccardo, F., Crinò, P., Vitale, P., Calcagno, F. and Porta-Puglia, A. 1987. *Terra e Sole*, 42: 82-88.

Sarker, A., Rahman, A. and Rahman, M. 1992. *Proceedings of First European Conference on Grain Legumes,* 1-3 June, Angers, France, 91-92.

Saxena, M.C. 1984. In *World crops: production, utilization and description* pp 123-139. (eds. M.C. Saxena, and K.B. Singh) The Hague, The Netherlands: Martinus Nijhoff / Dr. W. Junk Publishers.

Saxena, M.C. 1987. In: *The chickpea*, pp. 207-232 (eds. M.C. Saxena and K.B. Singh), Wallingford, Oxon U.K.: CAB International.

Saxena, M.C. and Singh, K.B. (eds.), 1987. *The chickpea,* Wallingford, Oxon, U.K.: CAB International. 409 pp.

Saxena, M.C. and Yadav, D.S. 1975. *Proceedings of the International Workshop on Grain Legumes,* 13-16 January, pp. 31-61 Patancheru, India: ICRISAT.

Saxena, N.P. and Johansen, C. 1996. *Abstracts: 2nd International Crop Science Congress*, November 17-24, p. 232. New Delhi, National Academy of Agricultural Sciences, India and Indian Council of Agricultural Research.

Sedgley, R.H., Siddique, K.H. and Walton, G.H., 1990. *Proceedings of the Second International Workshop on Chickpea Improvement*, pp. 87-92. (eds. H.A. van Reenen, M.C. Saxena, B.J. Walby and S.D. Hall). Patancheru, India: ICRISAT.

Sharma, B., Tyagi, M.C. and Asthana, A.N. 1993. *Proceedings of the seminar on lentil in South Asia,* 11-15 March 1991, pp. 22-38 (eds. W. Erskine and M.C. Saxena). New Delhi: ICARDA.

Shia, G. and Slinkard, A.E. 1977. *Crop Science* 17:183-184.

Sjödin, J. 1971. *Hereditas* 67: 155-180.

Singh, K.B. 1987. In: *The chickpea* , pp. 127-162 (eds. M.C. Saxena and K.B. Singh), Wallingford,Oxon, U.K.: CAB International.

Singh, K.B. and Singh, S.I. 1969. *Indian Journal of Agricultural Science* 39:737-741.

Singh, U., Kumar, J., Jambunathan, R and Smithson, J.B. 1980. *International Chickpea Newsletter* 3:18.

Smartt, J. 1990. *Plant Breeding Abstracts*, 60: 725-731.

Smith, C.W., Wiesner, L.E., Lockerman, R.H. and Frisbee, C. 1987. *Applied Agricultural Research* 2: 342-344.

Smucker, A.J.M., 1993. *Annual Review of Phytopathology* 31: 191-216.

Snoad, B. 1974. *Euphytica* 23: 257-265.

Snoad, B. 1981. *Scientia Horticulturae* 14: 9-18.

Summerfield, R.J. and Roberts, E.H. (eds.) 1985. *Grain Legume Crops*. London: Collins,. 859 pp.

Vadivelu, K.K. and Ramkrishnan, V. 1983. *Seed Research* 11: 177-181.

Venora, G. and Porta-Puglia, A. 1993. *Petria*, 3: 177-182.

Verma, M.M., Singh, I.S. and Brar, J.S. 1993. *Proceedings of the seminar on lentil in South Asia,* 11-15 March 1991, pp. 39-57 (eds. W. Erskine and M.C. Saxena). New Delhi: ICARDA.

Walton, G.H. 1990. *Australian Journal of Agricultural Research* 42:79-94.

Wehner, T.C. and Gritton, E.T. 1981. *American Society for Horticultural Science* 106:272-278.

Weyers, J.D.B. and Lawson, T. 1997. *Advances in Botanical Research* 26: 317-352.

Williams, P.C. and Singh, U. 1987. In: *The chickpea*, pp. 329-356 (eds. M.C. Saxena and K.B. Singh), Wallingford, Oxon, U.K.: CAB International.

Market Demands and Research Opportunities: addressing the supply / demand gap for pulses

J V Lovett[1] and G P Gent[2]

1 Grains Research and Development Corporation, PO Box E6, Kingston ACT 2604, Australia ; 2 Processors & Growers Research Organisation, The Research Station, Great North Rd, Thornhaugh, Peterborough PE8 6HJ, United Kingdom.

Abstract

The market demands for pulses are diverse, ranging from an economical source of protein, to convenience and heath foods, to stockfeed. A common trend amongst all markets is that consumers are becoming increasingly sophisticated in their demands. Market growth is occurring in both feed and food markets and there is likely to be a widening of the supply/demand gap unless productivity can be improved. Research opportunities to enable suppliers to meet this growing market demand relate to both productivity and quality and the balance of investment in these activities varies between suppliers. For the sub-continent, a net importer of pulses, the focus is clearly on productivity. For Europe, the focus, for both the stockfeed and well-established food processing industry, is primarily on quality attributes. For Australia, the focus is both on productivity, to ensure pulses are part of the farmer's cropping program and required supply volumes can be met, and on quality attributes to better meet customer requirements and gain access to higher value markets. In the short term, a focus on productivity factors may provide greatest benefit to the industry, as long as this is not at the expense of quality attributes.

1.   INTRODUCTION

The past century has seen a dramatic increase in the world's population, and an increase in per capita food consumption. But food availability and consumption are not evenly distributed and the gap between food-rich and food-poor countries continues to widen. There have been numerous claims that world demand will soon outpace the growth of supply, largely due to population and income growth in countries such as China and India. Already the demand / supply imbalance is emerging in relation to pulses.

Both developing and developed countries see the pulse industry as one of enormous opportunity with a strong demand for both food and feed. The supply of pulses to date has fluctuated significantly and has failed to match the growth in demand. However, it can be expected that the level and stability of pulse production will improve due to increases world-wide in the allocation of resources to pulse crop improvement, the improvement of grower experience and management skills and the application of new techniques such as biotechnology. This offers significant opportunities for all major producers of pulses, but particularly for exporting nations such as Australia, North America, Turkey and Burma.

The demand for pulses is diverse and influenced by a range of factors. In developing countries they have been used traditionally as a staple part of the diet and to provide an economical source of protein. Pulses are particularly important in regions such as the Indian sub-continent where much of the population is vegetarian. In developed or westernised countries, the use of pulses varies significantly with regional food preferences. Whilst they are a commonly consumed product in Mediterranean Europe and parts of the US, in countries such as Australia pulses are only used in small quantities and awareness has only increased recently in response to health concerns. However, as production has increased in developed countries, the focus and demand for pulses has switched to their use in the feed industry. Feed now forms the predominant use in Europe and Australia and is becoming increasingly important in North America.

Despite the variation in the use of pulses, a common trend in all markets – of food and feed – is that consumers are becoming increasingly sophisticated in their demands.

World trade in pulses has been steadily increasing over the past ten years. The proportion of pulses traded, to that produced, rose from a stagnant 3 percent over many years to about 6 percent in 1980, 8 percent in 1985 and is currently about 11 percent. This is comparable to the level of cereals; i.e. 15-20 percent (Oram 1992).

R. Knight (ed.), Linking Research and Marketing Opportunities for Pulses in the 21st Century, 221–233.

Growing demand generates a number of opportunities for exporters of pulses while also challenging importing countries to increase their self-sufficiency. Attention has been focused on the research sector to assist in improving both the productivity and quality of pulses. The allocation of research resources for the two areas varies between nations.

For the Indian sub-continent, a net importer of pulses, the focus is clearly on building productivity. Whereas for Europe, the focus is primarily on quality attributes for both the stockfeed and the well-established food processing industry. The focus for Australia is both on productivity – to ensure pulses are part of the farmers' cropping program and required supply volumes can be met – and lately on quality attributes, to ensure that products meet customer requirements and gain access to higher value markets.

In the short term, a focus on productivity factors may provide greatest benefit to the pulse industry, as long as it is not at the expense of quality attributes.

Section 2 of this paper examines the demand / supply situation for pulses. Section 3 examines research opportunities to address the supply / demand imbalance and finally, some concluding remarks are made. This paper takes a global perspective of the pulse industry, however it uses the situation in the European Union, Australia and India as examples to illustrate key points.

2.   THE DEMAND / SUPPLY SITUATION FOR PULSES

2.1   Trends in Demand for Pulses

In the late 1990s demand for pulses is increasing for both food and feed uses. As a consequence, a number of opportunities for exporting nations has arisen over the past three decades and is driving growth in the pulse industries of Europe, Australia and Canada.

World consumption of pulses grew from around 40 million tonnes in the mid 1980s to over 50 million tonnes in the early 1990s (Table 1). Consumption grew by only 2.5 percent per annum between the four-year period ending 1987 and that ending 1991, but grew by 18 percent over the next four-year period. Whilst population growth has underpinned continued expansion in total usage, per capita consumption of pulses in traditional markets such as India is declining. India, and other nations, continue to consume protein below levels recommended by the Food and Agricultural Organization (FAO).

Table 1. Changes in Total Pulse Consumption ('000 tonnes), between 1984-87 and 1992-95

		World	North America	Oceania	Europe
Food	1984-87	29,182	800	64	1,410
	1988-91	30,770	967	71	1,502
	1992-95	32,504	1,214	240	1,662
Feed	1984-87	6,887	25	440	3,529
	1988-91	9,256	83	493	5,887
	1992-95	14,522	245	700	6,051
Seed	1984-87	3,035	88	106	358
	1988-91	3,098	102	116	401
	1992-95	3,662	174	141	351
Other	1984-87	99	0	0	86
	1988-91	152	0	0	119
	1992-95	196	0	0	143
Total	1984-87	39,204	913	610	5,383
	1988-91	43,277	1,152	679	7,909
	1992-95	50,885	1,633	1,082	8,207

Source FAO 1997

The major growth has been in consumption for feed use, with this increasing by 34 and 56 percent respectively for the two periods, compared to only 5 percent for food purposes. This growth has been greatest in Europe and Oceania, where this market segment represents over 65 percent of total pulses used.

Imports have also continued to grow at an average annual rate of 5 percent (Table 2). Import growth has been strongest for pea – annual average growth of 9.26 % – reflecting the strong growth in feed usage, as well as an increasing demand from developing nations.

Table 2 World Pulse Imports 1986-1995, by commodity ('000 tonnes)

Year	Beans (dry & broad)	Peas, dry	Chick peas	Lentils	Others	Total
1986	2,375	1,387	219	421	449	4,850
1987	1,892	2,202	405	431	546	5,475
1988	2,019	2,252	492	714	500	5,977
1989	2,031	2,287	351	468	428	5,564
1990	2,319	2,824	427	519	571	6,660
1991	2,262	2,680	457	451	396	6,246
1992	2,357	2,514	456	563	740,	6,630
1993	2,050	2,562	632	566	756	6,565
1994	2,671	2,724	438	754	814	7,400
1995	2,513	3,324	331	630	628	7,427
Ann Average Increase (%)	1.01	9.26	5.64	8.18	7.95	5.04

Source: FAO 1997

2.1.1 Population increase is driving demand in developing countries

Across many developing countries, total pulse consumption is growing in line with population, although not always maintaining per capita consumption. For many countries, production is not keeping pace with demand and per capita consumption is declining as imports are not affordable. Demand in developing countries is highly influenced by internal crop production and an ability to pay.

The demand / supply gap for pulses in India is widening every year as pulse production struggles to keep pace with India's population growth; i.e. pulse production has increased by only 30% since 1961, whilst population has doubled. The situation has been compounded by farmers switching to cereal and oilseed production, which have more attractive support measures. The Indian Government has tried to manage the level of pulse imports in the past by monitoring import duties; however pulse imports are now allowed under the Open General Licence. After taking into account annual imports of around half a million tonnes, the net supply shortfall is estimated at 3 million tonnes (Pulse Australia, 1997). Pulse imports into India have declined in the five years to 1995, compared to the five year period to 1990 (Table 3). This reflects the constraints of world availability and prices rather than demand.

Table 3. Pulse Imports into India, 1986-1995 ('000 tonnes)

Commodity	Average Five Years 1986-1990	Average Five Years 1991-1995	Percentage Change
Beans, dry	218	74	-195.34
Peas, dry	221	139	-58.42
Chick peas	138	87	-59.04
Lentils	61	24	-151.50
Others	82	163	49.94
Total	719	487	-47.51

Source: FAO 1997

The markets of developing countries are changing and becoming more segmented as wealth increases unevenly throughout the economy. The Indian market can now be classified into two broad segments; namely, the higher income "wealthy" middle class and the others. The higher income segment is quality conscious and has a focus on cleanliness, size, packaging, information, shelf life, taste and texture, cooking time and colour. They can afford to pay for quality products and a packaged product is more common. For the majority of the population however, price is the major determinant of the quantity and quality of pulses consumed.

2.1.2 Demand for food pulses is increasing slowly in developed countries

The use of pulses for food in developed countries needs to be considered in terms of traditional and non-traditional consumers of pulses.

Some of the traditional consumers of pulses are shown in Table 4. In the countries of Mediterranean Europe, there is a long history of pulses in the diet. Consequently, there is high consumer awareness and knowledge of pulses–including how to use them. However, consumption of processed foods, particularly vegetables, is either static or declining in Europe due to competition from fresh products.

Table 4. Main Premium Food Importing Countries

Region	Country	Product
Mediterranean Europe	Spain, Italy, France, Greece, Turkey, Syria	Beans, Lentils, Chick peas, Lupins
North Africa	Morocco, Algeria, Tunisia, Libya	Beans, Lentils, Chick peas, Lupins, Yellow / Green Peas
Red Sea	Egypt, Sudan, Ethiopia, Jordan, Saudi Arabia, Yemen	Beans, Lentils, Chick peas, Lupins, Yellow / Green Peas
Persian Gulf	Oman, UAE, Kuwait, Iraq	Beans, Chick pea, Lentil, Dry Peas

Source:Gadd 1996

The European Union (EU) has a well-developed and technically innovative food processing industry with food legumes providing important elements in factory operations. In the UK, frozen vining (garden) peas and Phaseolus beans – in the form of navy or baked beans – are key products and provide a healthy, nutritious and convenient food suited to modern lifestyles. The UK currently imports around 100,000 tonnes of Phaseolus beans. In Scandinavia frozen peas are important, whilst a range of other food legumes, including canned peas and green and flageolet beans, are produced in France. Numerous smaller industries are established in other EU countries.

Awareness and use of pulses in countries that are not traditional consumers, such as Australia, is being driven by the increasing health consciousness of consumers and by the spread of Mediterranean, Middle East, African and Asian cuisines. In contrast to Europe, Australia has a very low usage of pulses for food – around 30,000 tonnes – and little processing or value adding based on pulses.

An increasing share of the food dollar in non-traditional pulse markets is being spent on food consumed outside the home; this has further implications for product development. The availability of pulses in a form to meet convenience and quality demands will be an important factor influencing the acceptance of pulses.

2.1.3 Strong growth in feed demand expected to continue

The use of pulses for feed has increased substantially, such that stockfeed now is the major market for pulses in the EU and Australia. This growth is now slowing but remains significant. In Europe this trend is already apparent for the period from 1992 to 1995. Feed use has also grown strongly in North America – increasing by 232% between 1988-91 and 1984-87 and by 195% between 1995-95 and 1988-91 – however, this was from a very low base.

The demand for feed pulses is growing in line with the intensive livestock industries. It is driven largely by the more developed Asian nations who are increasing their protein consumption from meat and dairy sources. Anti-nutritional factors are a key issue in this market, along with available energy and protein contents.

The stockfeed market is possibly the most developed market for pulses and is highly sophisticated in its ingredient usage. Pulses need to be competitive on a price / quality basis. Higher prices can be obtained from niche markets, however these markets require special varieties and qualities.

In the EU peas, and to a lesser extent faba beans, have to compete with imported soybeans in terms of price, quality, availability and trading security. This often means that production is based on relatively low prices. Peas have become well established for this use in France and Denmark, but in most other EU countries (including the UK) soy products remain the first choice of animal feed compounders. Australia is a major supplier to this market.

The Australian stockfeed sector is similarly a major user of pulses, with the best estimate of current usage being in the order of 750,000 tonnes a year (Meyers Strategy Group, 1995). In contrast to its counterparts elsewhere in the world, the Australian industry is highly sophisticated and utilises a wide range of ingredients in its rations. Pulses have become an increasingly popular feed source, with field peas the most widely used pulse in the intensive livestock industries. However, lupins are rapidly becoming recognised as a superior food source as production on the east coast expands and subsequent incorporation into ration occurs.

In line with increasing feed use, imports of pulses into the EU have increased significantly, primarily peas and lupins (Table 5). In total, imports have increased at an average annual rate of 7 percent and are influenced as much by the domestic support policies and local production levels as by market demand.

Table 5. European Union Pulse Imports, by commodity ('000 tonnes)

Year	Beans, dry	Beans, broad	Peas, dry	Chick peas	Lentils	Others	Total
1986	412	334	1,124	48	154	18	2,090
1987	473	474	1,721	52	143	27	2,887
1988	500	418	1,717	100	180	33	2,948
1989	396	663	1,621	102	119	32	2,932
1990	443	579	2,069	96	145	26	3,357
1991	515	612	2,154	120	164	7	3,572
1992	479	520	2,074	130	188	11	3,402
1993	454	440	2,071	115	2191	11	3,310
1994	443	473	2,178	129	2325	12	3,469
1995	471	341	2,734	109	193	13	3,862
Annual Av Increase (%)	1.96	3.05	10.59	11.54	4.01	6.14	6.93

Source: FAO 1997

2.1.4   Understanding market requirements

The pulse industry worldwide generally lacks data and analyses on current and projected trade patterns and current and future changes in supply / demand. For some markets, pulse products are well established and requirements are understood by both the buyer and supplier. The challenge for pulse suppliers is to obtain a better understanding of market requirements /consumer preferences in those markets which are less developed. In addition there is a need to quantify the demand in developing countries and to assess the *potential* demand in industrialised countries. For the two consumer groups of pulses discussed in section 2.1.2, namely traditional and non-traditional consumers, key market requirements driving demand are:

quality, including (colour, size, flavour, processability / cooking), food safety and consistency / reliability;

product availability and convenience;

diversity in terms of varieties and uses; and

market support i.e. promotion, recipes etc.

Within Europe, the specialty markets are most developed in the UK and are summarised in Table 6. These markets can be very demanding in terms of the size, shape, colour and cooking properties of the produce and often present a challenge to plant breeders who have to retain the quality attributes of old varieties and yet improve agronomic performance.

2.2   Trends in Supply of Pulses

World production of pulses has increased slightly over the last ten years but is not matching demand. The average annual growth in pulse production has been less than one percent in this time period (Table 7). Chickpeas have shown the strongest growth at an average of 3.5 percent a year, which is comparable to an increase in consumption also at around 3.5 percent.

226

Table 6 Market for Pulses in the UK and Crop Production (tonnes)

Use	Special requirements	Requirement	1996 Production
Animal feed compounding	Market for high tannin beans and other samples that fail to meet standards for premium markets	600,000	240,000
Tannin-free animal feed for pig and poultry rations	White peas and tannin free beans	500,000	340,000
Animal feed micronising	Green-seeded varieties used almost exclusively. Secondary use for marrow-fats and seed peas which do not meet quality standards	55,000	45,000
Export standard beans	Beans must be clean and sound and free from Bruchid damage. Pale skinned varieties used.	20,000	12,000
Canning marrowfat peas	Samples must be clean, sound and with only low levels of moth damage	24,000	22,000
Packet marrowfat peas (including fish shop trade and frozen mushy peas)	As above, but must retain green colour	15,000	15,000
Export quality marrowfat peas	Premium samples with excellent colour, large size and freedom from blemishes	8-10,000	6,000
Canning small blues	Samples must be clean, sound and pass cooking tests	4,000	4,000
Pigeon trade	Small size brown, white and green seeded peas and beans	Maple peas 7,000 Small blues 3,000 Maris bead 4,000	2,000 3,000 1,500

Source: Gent 1997

Pulse production is concentrated in a few countries, with the top six producers accounting for 80 percent or more of total production for most commodities. Developing countries produce over half the volume of the world's pulses with the vast majority consumed domestically. Because of the unreliability of pulse production in developing countries, total world pulse production tends to fluctuate significantly and productivity is low.

Table 7. World Production of Major Pulses ('000 tonnes)

Year	Beans, faba & broad	Peas, dry	Chick peas	Lentils	Others	Total
1987	18,507	14,866	6,884	2,698	10,225	53,179
1988	20,474	15,129	5,840	2,672	10,670	54,785
1989	18,705	15,148	7,151	2,089	11,279	54,371
1990	20,964	16,608	6,776	2,569	11,167	58,084
1991	19,102	12,231	8,094	2,633	11,151	53,211
1992	17,332	13,321	6,734	2,604	10,853	50,843
1993	19,503	14,736	6,767	2,802	11,905	55,714
1994	21,464	14,579	7,087	2,351	11,382	57,361
1995	21,361	11,221	8,866	2,849	12,010	56,260
1996	22,170	10,945	8,908	2,819	11,932	56,774
Annual Av Increase (%)	2.18	-2.18	3.55	1.32	1.64	0.78

Source: FAO 1997

The yield differences between developed and developing countries reflects the improved technology in developed countries. For example, the average yield of pulses in India is 0.58 t/ha compared to 1.06 t/ha in Australia, 3.5 - 4.0 t/ha in the United Kingdom and up to 5 t/ha in France.

In contrast to production, exports of pulses (Table 8) are predominantly from developed countries. There has been an expansion in the production of all pulses in Australia, pea production in the EU and peas and lentils in North America as these countries seek to capture export markets.

A number of factors influence a farmer's decision to grow pulses as well as influencing the resulting net return or profit as compared to other crop options. The economics of growing pulses is determined by both prevailing market conditions and the yield achieved, while major influences on pulse production worldwide are farmers' rotations and practices, government policies and new techniques, varieties and practices.

Table 8. Pulse Exports by Commodity ('000 tonnes)

Commodity	Average exports 1986-1990	Average exports 1990-1995	% Change
Beans (dry and broad)	2,185	2,771	26.82
Peas, dry	2,148	2,952	37.39
Chick peas	457	468	2.42
Lentils	574	641	11.70
Others	329	318	-3.38
Total	5,693	7,150	25.59

Source: FAO 1997

2.2.1 Whole farm systems

The focus on whole farm systems and the move to incorporate alternative crops in farm rotations has played a key role in driving increased production of pulses. For example, in Australia production of pulses has increased dramatically over the last 20 years as both economic, but particularly agronomic benefits were recognised. During the 1980s, pulses benefited from the low cereal prices and export opportunities that emerged. This is reflected in Figure 1.

The adoption of pulses into the rotation has had a number of economic and non economic benefits for farmers. For example, as highlighted by Robertson (1997), the development and adoption of lupin / wheat production package in the Western Australian sandplains over the past decade has:
  doubled annual wheat yields;
  increased farming profitability (land prices have increased from A$450 to A$600 a ha over seven years);
  dramatically reduced wind erosion and
  reduced groundwater recharge.

Today in Australia's sustainable farming systems, there is a wide range of cool and warm season pulse crops grown in rotation with cereals, oilseeds and pasture. Despite the growth over the last few decades, pulses still comprise only around 10 percent of the cropping area in Australia. It is estimated that this could be increased significantly, possibly to reach 16 percent, as there are large areas of soil types suitable for a range of pulse crops and new, better adapted pulse varieties are becoming available (Siddique and Sykes 1996).

228

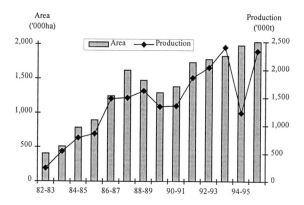

Figure 1: Area and production of pulses in Australia, 1982-83 to 1995-96 (ABARE 1990 and ABARE 1996)

In recent years there has been strong competition from alternative crops such as canola. The expansion of the pulse industry is dependent on the development of more reliable and higher yielding varieties. The industry's strategic plan forecasts that the Australian pulse industry could achieve production of 4 million tonnes per annum by 2005 from a base of 2.3 million tonnes in 1995-96. To achieve the target, yield stability and potential are key priorities.

2.2.2   Disease, pest and agronomic management

There are strong implications for trade in addressing agronomic issues in pulse production. Disease resistance and agronomic management are seen as the key elements through which global pulse productivity will be improved. New techniques such as biotechnology are likely to play a key role in achieving this.

Disease, pest and agronomic issues, and the resulting research opportunities they present, are addressed further in section 3.1

2.2.3   Government policy

In both developed and developing countries, government policies play an important role in influencing the level of pulses planted.

Production of all major arable crops in the EU is underpinned by area payments. In the UK, payments range from about $US400 per hectare for cereals to $US800 per hectare for linseed. Peas, faba beans and sweet lupins are supported with a payment of about $US600 per hectare, whilst lentils, vetches and chickpeas are supported at a lower level. No aid is payable for navy and other Phaseolus beans.

A production decline of pulses in the EU during 1995 and 1996 was largely due to an imbalance in aid towards cereals and favourable world cereal prices. However both trading and political developments are correcting this for 1997 and, hopefully, for future years. Further changes to the EU support arrangements are expected in the near future and whilst the impact is unknown, it is speculated that cereal and pulse profitability will be slightly enhanced, with oilseeds declining. Currently production of pulses in the EU is about 5 million tonnes per annum. However, this figure could double if soy could be replaced in animal feeding rations.

In India, the area planted to pulses has also been influenced by the level of support for, not only pulses, but also cereals and oilseeds. The total area under pulse cultivation has varied between 20-24 million hectares over the past few decades, with annual production of between 9 and 14 million tonnes. Chickpea and pigeon pea account for 45 percent of total pulse output. Whilst the Indian government has had ambitious targets for growth in pulse production, these have not been met, Table 9. This has been due largely to more attractive support programs and greater technical support for cereals and oilseeds and thus farmers have switched area to these crops.

Australian pulse growers operate in an open economy with little or no domestic price support or protection, so that the incomes of these farmers are linked closely to movements in international prices and exchange rates. The Australian pulse industry is a major exporter, with approximately 70 percent of its annual production sold on international markets, though it is a relatively small producer of pulse crops by world standards. India is the major market for Australian chickpeas, lentils and dry peas for human consumption where rising income and population growth is expected to lead to continued strong growth in pulse consumption. Development in the international trading environment are, therefore, particularly important to Australian pulse exports where the fortunes of their export-oriented industry depends heavily on world prices and access to international markets.

Table 9 Targeted and Actual Production of Pulses in India ('000 tonnes)

Year	Target	Actual	% Difference
1985-86	13,500	13,360	-1
1986-87	14,000	11,710	-16
1987-88	14,500	10,960	-24
1988-89	13,300	13,850	4
1989-90	14,750	12,860	-13
1990-91	15,000	14,260	-5
1991-92	15,000	12,020	-20
1992-93	14,500	12,820	-12
1993-94	15,500	13,100	-15
1994-95	15,500	(est)14,330	-8

Source: Pulse Australia

2.2.4　New technology

There has been considerable investment in research over the past decades to increase pulse productivity as pulses tend to be seen globally as high-risk crops with variable yields. There is high volatility in price and few risk management tools available to growers. Research has been successful to varying degrees across the different nations. It is expected that over the remainder of this decade and into the next, the continued investment in pulse research and extension will start to see greater gains in productivity and quality, which will be enhanced by continued improvement in grower skills and management practices.

Australia is currently amongst the leaders in the field of biotechnology, although Europe now has a strong focus in this area and should make significant progress with peas in the coming years. The Australian focus is on improving seed quality, pest and disease resistance and modifying plant growth. The first material from these programs is likely to be released in the next 3-5 years.

New techniques and varieties have had a significant impact on productivity and profitability of pulses and is illustrated in the following examples.

The growth of the lupin industry in Australia in the last 30 years is fundamentally due to the domestication and introduction of new lupin species and subsequent improvements in yield, disease resistance and quality. The development of new varieties took approximately 25 years to complete and was largely due to the work of one researcher, Dr John Gladstones, Agriculture WA. Not only did the research program result in a new crop, but it also created the opportunity for a new farming system, wheat / lupin production, as described in section 2.2.1.

Afila (semi-leafless) varieties of peas in Europe, having boosted productivity by at least 25 percent, now account for 70 percent of EU pea production and have allowed the expansion of areas suitable for pulse production.

2.3　Summary of the Supply / Demand Balance

It is evident from the trends highlighted in the preceding sections that there is a widening gap between demand and supply. The gap is a result of growth in the demand for pulses for both food and for feed use.

230

It is expected that the supply / demand relationship will become more dynamic as the imbalance worsens. This gap is reflected in the increase in imports (or potential import demand) by developing countries for food and developed countries for feed.

3. RESEARCH OPPORTUNITIES TO ADDRESS THE SUPPLY / DEMAND GAP

There are significant opportunities emerging for the pulse industry from the growth in demand. The challenge is to increase supply on two levels; namely increasing absolute quantity and by better matching products to markets.

Research opportunities exist in relation to:

production potential or yield (variety potential, disease / pest / weed control and agronomic practice);

quality – food (quality parameters / product form)

– feed (anti-nutritional factors and quality parameters); and

value adding pulses and developing new uses.

A balance is required between research opportunities because even though increasing production or yield in many cases will provide economic benefit for producers, the greatest long term strategic benefit may come from matching products to markets; i.e. addressing quality parameters.

For Australia the challenges are perhaps greater than for either the EU or India. Whilst only a small producer in world terms, Australia is a significant exporter. Despite the rapid growth in pulse production throughout the 1980s, pulses have not delivered on the potential initially projected. They have played an important role in farming systems, but have not secured a place in rotations on economics alone. This is due to the high-risk perception of the crop, failure to reach yield potential and lack of market development. As such, the Australian industry has allocated resources across a wide diversity of species and characteristics in order to gain production and market advantages. Whilst emphasis has been on crop improvement, this has focused on both production and quality issues. Biotechnology is seen as a key element that could greatly assist the future growth of the pulse industry.

There has been a significant increase in the number of crop species and varieties grown in Australia and it is now being recognised by the industry that research efforts need to be more focused, and concentrated on a few crops and a few core problems that restrict yield and marketability. As such, the industry's peak body, Pulse Australia, is working with the Grains Research and Development Corporation (GRDC) to identify the core crops and issues to be addressed in research and market development programs. Whilst this is likely to focus on cool season pulses such as lupins, field peas, faba beans, chick peas and lentils, efforts on a limited range of warm season pulses such as mung beans will be maintained.

3.1 Improving Production

Disease, pest and agronomic management remain key problems for all nations in respect to maximising yield potential of pulses. The three regions reviewed in this paper show varying levels of development within the pulse industry. However, all three have a focus on building productivity.

3.1.1 European Union

Within the EU, pulse production is stable and the industry is well established. There is ongoing investment in production techniques to retain and improve agronomic ability; specific areas being harvesting practices and pest and disease control. Key areas of concern are pests that spoil produce, such as pea moth, *Cydia nigricana* – a long established pest of human consumption and seed peas – and the bean bruchid beetle which has inhibited development of trade into the Middle East.

The EU industry is benefiting from a successful breeding programme on spring and winter peas, spring beans and, to a lesser extent, winter beans, winter lupins and navy beans. Such R&D is vulnerable because of the high costs of variety development and the relatively small, but diverse, areas of temperate pulses compared to major crops such as maize, soy and rice. There is, however, economic sense in breeding for agronomic factors such as yield, plant type, timeliness or disease resistance because these are readily appreciated by farmers.

A further challenge to plant breeders is emerging with global warming. For most of northern Europe this should have a beneficial effect on pea production, but it could further depress the position of spring sown faba beans. Dry Phaseolus beans – which can be grown in selected areas of southern UK and other southerly areas of Europe – would also benefit from global warming, but without area aid there is no prospect of significant production. Currently in the UK, global warming is an accepted meteorological phenomenon but there is uncertainty about the rate it will affect temperature and rainfall.

3.1.2   India

The key priority of the Indian research program is to improve both yield potential and the consistency of production. Indian pulse production, in contrast to the EU, is highly variable and yields remain well below those of developed countries at 0.6-0.8 t/ha. Factors identified by Pulse Australia (1997) which constrain production include:

dependence on dryland farming, with more than 90 percent of pulses cultivated on marginal lands under rainfed, non irrigated conditions;

inability of farmers to afford high cost inputs such as fertilisers, rhizobia, cultures and pesticides;

the shift towards remunerative crops such as rice, wheat and other cash crops which have high yields per unit area;

higher susceptibility of pulses to crop diseases and pest infestation than cereals;

pulses being usually grown in poorer soils;

wide fluctuations in prices from year to year, whilst cereals and other cash crops tend to have more price stability; and

preference for growing cereals as these are India's major staple food and form the basic diet for farmers and their cattle.

A significant approach in the past to lift India's pulse yields has been breeding for disease resistance. India suffers significant disease problems due to tight rotations. Generally 2-4 crops per year are grown resulting in little opportunity to break disease or pest cycles. There has been considerable progress made in the area of disease resistance, however yield responses are still proving difficult to achieve. A possible reason for this is that varieties have been selected over time for both low rainfall and fertiliser environments. To further increase yield potential, the research efforts are now switching to identifying varieties that are more tolerant of a range of conditions.

The other more recent initiative that has been undertaken in India, to increase production, is the search for pulse varieties that can fit into new farming systems, as there are few other areas into which pulses could expand. There has been a shift in some areas of rice and wheat production, creating an opportunity for short duration pulse varieties to fit into these systems; for example, short duration chickpea and lentil varieties are being developed to follow rice and short duration pigeon peas to follow wheat.

3.1.3   Australia

Australia, to some degree, reflects elements of both the EU and Indian situations. As indicated earlier, pulse production has undergone a revolution in Australia in the last 20 years. Hamblin (1996) noted that this has resulted in two key features emerging, namely:

an increase in the range of species grown and

significant increases in the area sown of most of these species.

The industry's strategic plan established a target of 4 million tonnes of pulses by 2005. This would require both a significant increase in average yields and an increase in the share of the cropping area devoted to pulses. In order to improve the production potential, Australian research efforts have been focused on:

improving varieties which both perform better in terms of yield potential and stability;

improving identification, management and resistance to diseases, pests and weeds. The challenge is to identify likely pest and disease problems before they are encountered. Weed control is crucial to the agronomic package. The focus has been to channel R&D into problems likely to be significant over large areas;

species development, which involves moving traditional species such as chickpeas, dry peas, faba beans, lentils and narrow leaf lupins into new areas and

development of species which are less familiar on world markets e.g. Lathyrus, vetch and various new lupins;

improved coordination and cooperation between breeders to ensure new varieties are adapted to new and existing areas of production in Australia. Similarly, agronomic packages need to be tailored to local environments, with this being addressed through programs such as TOPCROP, which works with grower groups to establish agronomic and economic benchmarks and

improved crop management practices to improve the yield and quality of pulses.

3.2   Improving Quality

As markets become more discerning and as exporters and local producers seek higher value markets, the focus on quality will become increasingly important. Quality will not only be focused on visual quality parameters such as size, colour and shape, but also on cooking properties. These latter properties directly affect suitability for particular markets. Generally, quality factors are more difficult to incorporate into varieties and market requirements can change during the ten year development period of a new variety.

For the EU, the emphasis and research effort is on improving quality attributes including reducing / removing anti-nutritional factors. Whilst in general, requirements in the food processing sector are well known and met by existing products, for emerging niche markets, there needs to be a focus on quality parameters such as cooking properties.

Australia also has a strong emphasis in its crop improvement program on better understanding and improving the quality parameters of varieties to ensure they better meet market needs. For India, the focus is primarily on yield, with little attention on quality parameters at this stage.

3.3   Value Adding and New Uses

As the pulse industry grows, the use and markets for pulses will become more varied. New uses and value-adding opportunities are currently being explored through both research and commercial institutions within countries such as Australia, Europe and North America. This includes:

convenience products such as pre-cooked, ready to use pulses e.g. "Quickpulse" a product newly available in Australia;

micronised techniques e.g. in Europe a technique used to produce peas that have been heated and flaked for use in a range of pet and animal feeds, with this market returning a premium over compounding values;

fresh products such as sprouts, dips;

snack foods, in particular, for Asian markets;

canned products for both traditional markets such as the Middle East and non traditional markets;

ingredient manufacture to produce flour, starch and protein, eg, in Europe, pea flour is used by people with a gluten allergy;

use of peas as a "filler" in prepared meats rather than soy and

non-food products, eg biodegradable plastics and pharmaceuticals, such as antibodies and vaccines, in pea seeds.

4.   CONCLUSION

This paper provides evidence of the widening supply / demand gap that is occurring in the world pulse market. Whilst it is important to understand and predict the demand side factors, the critical constraint in the short term is productivity. There is considerable diversity in yields achieved throughout the world and most producing nations have yet to develop high yielding adaptable varieties. Stability of supply is imperative.

However, there must be a greater focus on understanding markets and the diversity of end uses as markets become increasingly demanding. If the industry neglects this focus, whilst targeting production, then a result may be large volumes of product unsuited to markets. This would mean the product would need to be utilised in low value segments such as developing countries seeking low cost protein sources.

In the future, a focus on value-adding of pulses and new uses will be important in accessing both traditional and non-traditional markets. This will be influenced by income growth, particularly in Asia, and increasing health consciousness of consumers.

Despite this, the greatest profitability for growers will be through improved productivity and thus, research opportunities exist to improve yields and reduce constraints to production. Over time, markets will become increasingly specialised and quality conscious and the world pulse industry must position itself to capture this opportunity through a focus on quality parameters and new uses / processing of pulses.

References

ABARE 1996. *Commodity Statistical Bulletin*, 364pp, AGPS Canberra

ABARE 1990. *Commodity Statistical Bulletin*, 271pp, AGPS Canberra

FAO 1997. FAOSTAT Database [online], <URL:http://apps.fao.org/lim500/nph-wrap.pl?CBD.Crops
AndProducts&Domain=CBD>, accessed May 1997

Gadd, B. 1996., *Paper presented to the Australian Pulse Industry Workshop* – Meeting the Market Challenge of the Indian Ocean Region, Agribusiness Conference, Fremantle

Gent, G. P. 1997. *Market Demands and Research Opportunities – Europe*, Paper prepared on behalf of Processors and Growers Research Organisation, England

Hamblin, J. 1996. *The Future: R&D Outlook, paper presented to Agribusiness Opportunities in the Indian Ocean* Rim Workshop, Agribusiness 1996 Conference, Perth

Meyers Strategy Group 1995. *Feed Grains Study Volume 1*, commissioned by Grains Research and Development Corporation, Dairy Research and Development Corporation, Pig Research and Development Corporation, Chicken Meat Research and Development Corporation and the Egg Industry Research and Development Council, April

Oram, J. 1992. *Current Status and Future Trends in Supply and Demand of Cool-Season Food Legumes*, International Food Policy Research Institute, Washington DC

Pulse Australia 1997. Ne*w Horizons, Report on New Horizons* Trade Mission to India, November 1996, 26pp

Robertson, G. 1997. *In Outlook 97: Proceedings of the National Agriculture and Resources Outlook Conference* vol. 2 Agriculture 75-79 (Eds ABARE)

Siddique, K. H. M. and Sykes, J. 1996. *Pulse Production in Australia: Past, Present and Future*, Paper presented to the Australian Pulse Industry Workshop – Meeting the Market Challenge of the Indian Ocean Region, Agribusiness Conference, Fremantle

ACKNOWLEDGMENTS The authors would like to thank the following people for their assistance in preparing this paper: Ms Trishia Bancroft, Ms Maureen Cribb, Ms Rosemary Richards, Dr KHM Siddique and Ms Emily Whitten

The impact of production from the Indian sub-continent on world trade in pulses

Kothari Soya Ltd, Bombay, India

THE INDIAN PULSE SCENARIO

India is the world's largest producer and consumer of pulses. The Indian population today stands close to 950 million people and the growth rate is 2.5 % a year. A large segment of the population is vegetarian and pulses are the best source of protein for them. Today the requirement for pulses is about 17 million tonnes but production is 13 to 14 million tonnes leaving a gap of 3 to 4 million tonnes. This large gap between demand and supply is due to stagnant production, drought or a poor-crop as happened this year (1997) for Tyson chickpeas. India has imported about 200 thousand tonnes of Tyson chickpeas from Australia this calendar year and expects another 70 thousand tonnes to come from Australia between September and December 1997. Demand for pulses in India is also sometimes affected by their high price, as consumption is price sensitive. If the price of a particular pulse is high, its consumption falls and is supplemented by vegetables or other lower priced pulses.

The production of pulses in India is normally between 12-14 million tonnes every year. The area under pulses fluctuates in a narrow range between 22-24 m hectares with the yield varying between 483 kg and 610 kg per hectare. Despite India being the largest producer, with a share of 35-36% in the world in terms of area and 27-28% in terms of production, the people do not get enough pulses to eat. Coupled with the population growth, the per capita net availability has fallen by half from 69 g per day in 1960-61 to 37 g in 1995 as against the minimum requirement of 80 g as suggested by WHO & FAO. It is also lower than the minimum level of 47 g prescribed by the Indian Council of Medical Research.

Although India accounts for the largest share of the world's output of pulses, production in India is not encouraging when compared to that of cereals and oilseeds. Pulse crops accounted for only 14.6% of the total area cropped in 1991 as compared to 61.1% shared by cereals and millets. In spite of Government emphasis on improving production through special programs, the area under pulses has not increased significantly. The key constraint to raising productivity is that pulse crop are largely grown in small and marginal lands and in rain-fed areas. The irrigated area is only 9% of the total area and there has been no breakthrough in developing improved varieties with high seed yields. The production of pulses in the last 25 years has risen only marginally. Production has been stagnating between 10-14 million tonnes. In 1960-61 it was 12.7 million tonnes, declining to 10.6 million tonnes in 1980-81 and it was just 13.2 million tonnes in 1995-96. Production has not kept pace with the increased population.

India produces over a dozen pulse crops. Out of these, gram (chickpeas) and Toor (pigeon peas) account for over 50% of the total output. The other important pulses are mung beans, black matpe (*Vigna umbellata*), lentils, cow peas and moth beans (*V. aconitifolia*).

In view of the shortage of pulses and the special focus needed for their development, the National Pulses Development Project, which is being implemented in the 25 States and Union Territories, was brought under the purview of the Technology Mission on Oilseeds in 1990. One option to increase production is to increase the area under pulses. But since pulses are grown by small and marginal farmers, who have limited land, there is not much scope for increasing the area. So the only way is to increase productivity levels. The productivity of dry land and rain-fed crops like pulses, has been low in comparison with that of developing countries. The following table gives an idea of the area and production in the last 5 years

YEAR	AREA (Million Hectares)	PRODUCTION(MillionTonnes)
1991-92	25.5	12.0
1992-93	22.4	12.8
1993-94	22.2	13.3
1994-95	23.0	14.1
1995-96	23.9	13.2

R. Knight (ed.), Linking Research and Marketing Opportunities for Pulses in the 21st Century, 235–236.
© *2000 Kluwer Academic Publishers. Printed in the Netherlands.*

To plug the gap between demand and supply, India has to import pulses but imports are hardly sufficient. Imports during the last 5 years have averaged about 500 thousand tonnes annually. The prominent countries exporting pulses to India are Myanmar, Australia, Canada, Turkey, Iran and Syria. Many countries see India as a major and very important market for their trade in pulses. Myanmar exports almost its entire production of toor (pigeon peas) only to India. Large quantities of black matpe and green mung beans from Myanmar also move to India. Large quantities of Tyson chickpeas and Dun peas from Australia are being imported into India every year. Canada exports annually almost 100 thousand tonnes of its edible peas i.e. green peas and yellow peas to India.

Imports of pulses are tabulated below: -

YEAR	IMPORTS (Tonnes)
1991-92	450,000
1992-93	400,000
1993-94	628,000
1994-95	550,000
1995-96	450,000

Since India is the largest consumer of pulses in the world, the fluctuations in its production have a wide effect on the world trade in pulses. Although India exports some of the pulses against the special licenses given by the government, it remains a net importer of pulses to plug the gap between demand and supply. Due to drought or bad crops sometimes the country has to import large quantities, which leads to price hikes internationally.

CONCLUSION

Despite India being the largest producer, it will continue to import pulses to meet the gap between demand and supply. By virtue of its immense production and consumption, India has a major role to play in the international pulse market. The large Indian population, with its traditional food habit of emphasizing pulses as the main protein source, ensures that India will remain a major player in the world's pulse trade, for times to come. With the Government giving pulses their rightful due and extending all the help to increasing production, a day may come when India will emerge as a major exporter, instead of the net importer it is at present.

Impact of Turkish Marketing Policies on World Trade in Pulses

A. BAYANER[1] and V. UZUNLU[2]

1 Agricultural Economics Research Institute, P.O Box 34, Bakanliklar 06100, Ankara, Turkey.; 2 Central Research Institute for Field Crops, P.O Box 226, Ulus 06042 Ankara, Turkey

Abstract

In Turkey, chickpea and lentil have been the two most widely cultivated pulses for many years and they are an important component of the Turkish diet. A major change in production has occurred since 1982 when fallow areas began to be replaced with pulse crops, supported by a generous price policy. Since then, Turkey has become a major exporter of these two pulses and has had a significant impact on world trade. In this study, the impact of these changes is examined together with the potential implications of Turkey's export prospects on world markets. A strong integration between Turkish pulse export prices and world prices is evident. However, although Turkey is still one of the major exporter of pulses, recent changes in the marketing policies will most probably affect Turkey's role in world trade.

INTRODUCTION

There has been a considerable development in the Turkish economy since the establishment of the Turkish Republic in 1923. That development has also been reflected in pulse production. Pulses are second to cereals with 11 % of the total cultivated area. Lentil and chickpea are the two most widely cultivated pulses in Turkey. The value of pulse production accounts for about 3.4 percent of total agricultural production and 6.7 percent of field crop production in Turkey. Of the 3.4 percent, 70 percent is from lentil and chickpea production (SIS, 1994). The terms pulses and lentil-chickpea will be used interchangeably in the text since more than three-quarters of the overall food legume areas are occupied by lentil and chickpea (SIS, 1995).

A moderate increase in the area sown to pulses had been observed until 1980's. Then the production increased more than five fold. Although there had been a large increase in both area and production of pulses since 1930's, the increase in yield has not been as great. Pulse production in Turkey is subject to harsh weather conditions, and climate is a limiting factor to productivity. From year to year, yields have ranged from 0.5 t/ha in the bad years of the 1930's to 1.3 t/ha in a better year in 1953 (SIS, 1995).

Pulses have been grown in rotation with cereals in the transitional zones of Central Anatolia since late 1970's and used mainly for human food or sometimes as a green manure. They play an important role in the national diet. Pulses are consumed heavily in the rural areas and in poorer areas around large cities. Their consumption is increasing faster than the population increase. As high inflation reduces the real incomes of the lower middle class and poor, these groups increasingly turn to pulses as a low cost protein alternative to meat. Pulses also provide an excellent complement to cereals and an important component of dry land farming systems in warmer regions. All these factors combined with the favourable price support of the Turkish Grain Board (TMO) made the research and extension project, known as the Utilisation of Fallow Areas (UFA), a success in Turkey. UFA was directed at replacing fallow in crop sequences with chickpea and lentil crops mainly in Central Anatolia where the cultivated area is predominantly devoted to cereals and the annual rainfall exceeds 410 mm. The project was later extended nationally (Keatinge, 1994). By 1988, fallow areas in Turkey were reduced by 37 percent (3 million hectares) and correspondingly, field crop areas were increased by million hectares, of which 43 percent and 34 percent were lentil and chickpea, respectively (Bayaner et al, 1993; Kusmenoglu and Bayaner, 1995). A major increase in the production of pulses occurred after 1982 from the impetus of the project. The increase in the production of pulses in the Near East is essentially a reflection of the outstandingly successful Turkish campaign to expand food and feed legumes in Anatolia (Oram and Agcaoili, 1992).

R. Knight (ed.), Linking Research and Marketing Opportunities for Pulses in the 21st Century, 237–242.

The UFA project placed a heavy emphasis on technology transfer and additionally farmers realized that pulses were potentially important cash crops. Consequently, pulse production increased substantially over a short period of time. The total area sown to pulses is given in Figure 1. In the figure, the total area is partitioned into its components of which the lentil and chickpea areas are the major contributors. The area of lentil increased 4 times from 255,000 ha in 1981 to its peak level of 997,000 ha in 1989. Likewise, the area of chickpea increased 4.5 times from 200,000 ha in 1981 to the peak level of 890,000 ha in 1990. The production of lentil increased 300% and chickpea 400% in the same period. A further large increase in production has been projected in the 7th Five Year State Plan (Anonymous, 1994).

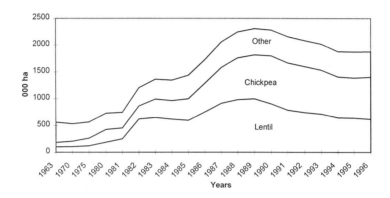

Figure 1. The total area of pulses in Turkey ('000 ha) and the component areas devoted to chickpeas and lentils and other pulses.

Other factors contributed to the increased production of pulses in Turkey. Pulse production along with that of other crops was supported by the Government of Turkey. Policy incentives for pulses had a great impact. In addition, returns were high in a short period. Major incentives were the attractive world prices and the government's recommendation to plant winter-hardy pulses on fallow areas. The devaluation of the Turkish Lira was also timely: export expanded from 0.1 million t in 1980 to 1.1 million t in 1988 (SIS, 1992; Anonymous, 1997). Tax refunds in exported goods were also a further incentive for the production of pulses.

Within the economic stabilization and liberalization programs of the 1980's, Turkey made a great stride in opening its borders to imports and reducing controls on exports. Quantitative restrictions were replaced by a system of tariffs and duties. All trading monopolies and quotas were gradually phased out. As a result, the trading regime has become more transparent and the flow of relevant information has improved. This combined with the boosted pulse production made Turkey an important pulse exporter, playing an important role in the world export markets and helping to establish the world price. India is the other principle producer, however India has never been a factor in the pulse export market primarily because domestic demand exceeds production (Blain, 1988). Commercial lentil production in Canada has now increased significantly, exploiting the success of an effective research program, and as a result Canada as also become a major exporter of pulses, including lentils.

In Turkey chickpea is grown in all areas, whereas, lentil production is localised mainly in Central Anatolia and Southeast Anatolia accounting for 70 percent of the total production (Kusmenoglu and Bayaner, 1995). However, Tekinel et al (1990) stated that the cropping pattern in Southeast Anatolia would be likely to change dramatically in the near future to the disadvantage of lentil due to the Southeast Anatolia Project (GAP). Approximately a 90 percent decrease is expected in the cultivated area and production of lentil in the GAP region. This implies a 60 percent reduction in total national production.

About 70 percent of the 50 million tons of world production of pulses is produced in developing countries under marginal conditions. Production in many countries fluctuates from year to year, affecting the volume and the price of the world trade. Pulses in Turkey are an important dietary component and it is estimated that the total domestic demand of lentil and chickpea is around 700,000 tons a year. Therefore, much of the pulse production is intended to satisfy the import demand from more than 50 countries, mainly developing countries. The marketing ratios of lentil and chickpea are 75 percent and 76 percent, respectively (SIS, 1995).

NATIONAL PULSE POLICIES

Farmers in Turkey believe the government declares the price and buys almost every kilogram of the products not consumed or sold in the local market. This reduces the farmer's market risks to as low a level as possible, although their production risks remain. Farmers feel secure because they know they will earn from as much as they can produce and sell.

The Turkish Grain Board (TMO) was established to support purchases of wheat, coarse grains, and some other commodities under its responsibility, to sell these on the domestic market, to augment domestic supplies with imports and to export surpluses, when they were authorized. Occasionally the TMO has been assigned to purchase pulses at prices determined by the Ministry of Agriculture and Rural Affairs. The main function of the TMO as defined in Article 4 is:

"to prevent a decrease in domestic cereals below the normal level for producers and/or and abnormal increase in prices to the detriment of consumers, to take measures when necessary to carry out duties involving pulses ... in accordance with a government decision..."

A subsidy program was introduced nationally for pulse production. The Government of Turkey supplied free inputs to farmers who participated in the UFA project to make it more attractive. In addition, the Bank of Agriculture provided the farmers with credit, five to six months in advance and at a low interest rate. A strong emphasis was placed on this program with the implementation of the UFA. The main functions of the TMO are to prevent a decrease and/or an abnormal increase in domestic prices. Occasionally the TMO has been assigned to purchase pulses at prices which are determined annually by the Ministry of Agriculture and Rural Affairs (MARA). In declaring the prices the MARA takes the price parity into account. The price parity of green lentil to wheat was around 2.6, red lentil 2.9, and chickpea 2.3 in the years TMO has purchased pulses. However, these parities have declined to around 2.0 on average in the last couple of years.

TMO started purchasing pulses in 1977. The amount as well as the price, has varied from year to year since then. During the 1990's the pulse purchases by the TMO has decreased. This, combined with the decision that the Government of Turkey excludes pulses, leaving only three commodities in the subsidy program has had an important impact on the production of pulses in Turkey. The area devoted to pulses has gradually declined from 1.75 million hectares in 1988 to 1.4 million hectares in 1994, and the production from 1.8 million tons to 1.25 million tons in the same period. The future is not very promising either.

In keeping with the general trend toward privatization, the Government of Turkey (GOT) has eliminated price support for a number of agricultural commodities, including pulses. The GOT continues to support farmers by subsidizing production loans and fertilisers. The Turkish Agricultural Bank provides farmers with loans at interest rates about half the cost of the ongoing commercial rate. Fifty percent of the cost of fertilisers is also subsidised.

TMO and GUNEYDOGUBIRLIK, a quasi government organization representing cooperatives, used to buy a good percentage of the total lentil production and either sell it at subsidized prices or keep it stored for several years. As a result of the government's move toward privatisation, TMO and GUNEYDOGUBIRLIK stocks have essentially been liquidated. As a result, Turkish lentil stocks are at very low levels relative to historical trends. Although progressing slowly, the Government of Turkey is trying to establish commodity exchanges to regulate the domestic market.

TRADE

Turkey's share in total pulse exports in the world varies from year to year depending on the production. For example, Turkey's share in the world's export market was as high as 22.5 % in 1988. In the same year, the share of lentil in world exports was 69 %, and of chickpea 80 %. Turkey's share in total world trade in pulses has been declining in recent years mainly due to the low level of production, which has been a result of lack of well organized farmers unions, combined with inadequate government policies. The share of lentil has declined to be only 22 % and chickpea 37 % in 1995. The share of these two crops in the Turkey's total pulse export was 94 %.

One major problem encountered in international pulse markets is variation in quality, leading to price fluctuations. Product quality fluctuates considerably due to relatively unsophisticated crop husbandry practices. To preserve market share, Turkey has to ensure customers receive a continuity of supply of consistent quality and at relatively stable prices. Recently, national seed cleaning technology has been improved. Quality is steadily improving thus helping Turkey to become a reliable exporter of pulses. Turkey will retain market share as long as it produces enough to export after the needs of national consumption are met.

Exports have been reduced in the last years due to reduced supply and increasing domestic demand. The West Asia and North African countries, together with India, Iraq, Sri Lanka, England, Pakistan, Saudi Arabia, Bangladesh, Iran, Spain and Italy are traditional importers of Turkish pulses. France and Egypt are other important potential markets. Along with unprocessed chickpea, Turkey has also exported a significant amount of roasted chickpea.

Exports by the TMO has been affected directly by the quantity it has purchased, which in turn has been affected by the production and price support. The TMO's exports reached the highest level in 1988 due to its high level of purchases in 1987.

ANALYSIS OF MARKET INTEGRATION

Data on the pulse export prices in Turkey are published by the Export Promotion Center, TMO, and Exporters Unions. World trade prices are published by FAO. Data used in this analysis were obtained from Akdeniz Exporters Unions of Turkey (Anonymous, 1997).

It was hypothesised that the impact of Turkish marketing policies on world trade could be explained by the relationship between the Turkish export price and the world prices. In other words, there existed a long-run relationship between world prices and Turkish export prices.

The presence of long-run relationship between economic data series can be investigated by different methods. Cointegration can be regarded as the empirical counterpart of the theoretical notion of a long-run or equilibrium relationship. It will indicate if markets are integrated, meaning that prices in the long-run are tied and do not drift far apart. Tests for cointegration provide evidence for the existence of such relationships Hallam et al., 1992; Sanjuan and Gil, 1996).

In the bivariate case, the cointegration regression is:

$Pit = \alpha + \beta\, Pjt + \upsilon$

where:

α = constant

β = slope coefficient

Pit = price in country i in time t

Pjt = price in country j (j=i)

υ=error term

Cointegration indicates that prices in Turkey and the rest of the world are linked in the long-run, implying they are integrated. The degree of integration depends on the magnitude of the β coefficients. The closer to β is to 1, the greater the degree of integration. Integration between markets is perfect when β is 1. It means price variations in one market are reflected perfectly in the other in the long-run. Both prices vary in the same proportion and a single or representative price exists (Sanjuan and Gil, 1996).

Widely used tests are the cointegration regression Durbin-Watson (CRDW) test of Sargan and Bhargava (1983), and the Dickey-Fuller (DF) and augmented Dickey-Fuller (ADF) tests (Dickey and Fuller, 1979 and 1981). Test for cointegration involves testing whether linear combinations of the variables are themselves integrated to the same order as the individual variables.

Tests for the integration of Turkish lentil and chickpea export prices with world prices and for quantities exported were carried out. The statistical results are not reported here but revealed that all the series are integrated of the order one I(1), fulfilling the first condition of cointegration. The second step estimates the cointegration regressions. The results are given in Tables 1 and 2 for lentil and chickpea, respectively.

Table 1. Cointegration Regressions and Associated Tests for the Models (Lentil)

WLPt = 9.66 - .691 TLPt
$R^2 = 0.82$ DW = 1.04   DF = -2.57(-4.11) ADF = -1.92(-4.22)
TLPt = 12.63 - 1.198 WLPt
$R^2 = 0.82$ DW = 1.06   DF = -2.03(-4.11) ADF = -1.72(-4.22)
TLQt = 21.89 - 1.576 TLPt
$R^2 = 0.69$ DW = 1.91   DF = -3.54(-4.11) ADF = -.91(-4.22)

Estimated coefficients are statistically significant at 0.05 level.
Variables are in natural logarithms.
WLP = World traded lentil price,
TLP = Turkish lentil export price,
TLQ = Turkish exported lentil quantity.

Table 2 Cointegration Regressions and Associated Tests for the Models (Chickpea)

WCPt = 10.21 - 0.786 TCPt
$R^2 = 0.92$ DW = 1.67   DF = -3.17(-4.11) ADF = -2.30(-4.22)
TCPt = 12.44 - 1.171 WCPt
$R^2 = 0.92$ DW = 1.39   DF = -2.75(-4.11) ADF = -1.88(-4.22)
TCQt = 19.34 - 1.149 TCPt
$R^2 = 0.55$ DW = 1.94   DF = -2.75(-4.11) ADF = -1.62(-4.22)

Estimated coefficients are statistically significant at 0.05 level.
Variables are in natural logarithms.
WCP = World traded chickpea price,
TCP = Turkish chickpea export price,
TCQ = Turkish exported chickpea quantity.

Failure to reject the hypothesis of no cointegration between pairs of variables need not necessarily imply that no long-run relationship exists between them. Rather it may be that such relationship involves more than two variables (Hallam et al. 1992). Here, only two variable cases (world price versus Turkish prices and quantities traded versus prices) were examined. The results would be more precise if a longer time series of data were available.

The DF and ADF tests cannot reject the hypothesis of no cointegration. Only in the case of CRDW test, hypothesis of no cointegration is rejected. Failure to reject no cointegration may reflect an under-parameterisation (Hendry, 1986) trough the omission of further I(1) variables.

It appears that the strict form of a Law of One Price (LOP) does not hold (Sanjuan and Gil, 1996). However, a strong integration is evident from the analysis that the Turkish pulse markets are integrated with the world pulse market although cointegration test results provided little support for the models. Cointegration is only supported by the CRDW statistic, ie. only CRDW test rejects the hypothesis of no cointegration in both directions. Inconclusive results may be misleading. There must be some other factors affecting the movement in the world and Turkish pulse prices. A more general model including variables concerning the national and international policies, should be specified for such relationships.

CONCLUSION

Contrary to theory, producers in Turkey respond to promises from the government more than they do to prices, although prices are good explanatory variables in the pulse supply response (Bayaner, 1996). Because of the reduced government support to pulses (mainly price support), since the early 1990's, together with the high labour cost and limited availability of labour, pulse production has steadily declined thus losing market share. The TMO has a regulatory role in the supply of cops in support schemes. When the TMO does not purchase, middlemen take this role. They stockpile pulses at a relatively cheap price and this forces rapid price increases. Countries like Canada, USA, China, and Australia are becoming leaders in pulse trading in the world. No export subsidies are applied to pulses. Price parity of legumes to cereals in last five or six years fallen to around 2.0 form 2.7. Real prices also have been decreasing in the last several years. With the inconsistencies in pulse production, Turkey will most likely lose its share and reliability in the world's export markets.

Turkey therefore needs to take some serious measures to hold its position in international markets. To continue to earn from the export of pulses, high quality and standardized production should again be promoted and supported. An active producer organization would be worth establishing, to protect producers from severe free-market conditions and to promote an increase in quality of product. The Ministry of Agriculture and Rural Affairs has stated recently that TMO will procure pulses in 1997. This is expected to be a relief to producers. This can promote the production of pulses in the coming year. Since the TMO had removed itself as a purchaser of pulses, domestic marketing and trade had been handled by the private sector. Processed food and value added products have a high marketing margin. Thus, the development of a processed pulse industry, principally for export, should also be supported nationally. In addition, Turkey has to carry out marketing research on pulses to more precisely identify the needs of customers in international markets.

Bayaner (1996) indicated that pulse crops respond to prices. In that respect, it would be possible to control the supply of pulses by manipulating the price. Since the pulses are excluded from the support scheme, and since Turkey is obliged to obey multilateral agreements such as GATT and the Custom Union with EU, it is more likely that Turkey will have to give up some of its export earnings from pulses. Turkey must decide whether to include pulses in the support scheme to keep her international market share and export earnings.

Results indicate the Turkish pulse market is integrated to some extent with the world market. Therefore, whatever policy action the Government of Turkey takes affects the world market somehow since Turkey is still an important producer and exporter of pulses. Although recent changes have resulted in a downward trend in pulse production and export earnings, Turkey can still boost her pulse production with some conventional policy measures.

References

Anonymous, 1994. *Seventh Five Year Development Plan.* Turkish Republic, State Planning Organization. Ankara.

Anonymous, 1997. Akdeniz Exporters Unions Annual Reports. Mersin.

Bayaner, A., Uzunlu, V., Keatinge, J.D.H., and Tutwiler, R. 1993. *Highland Regional Program.* A Collaborative Research Project between TARM and ICARDA. Central Research Institute for Field Crops. Ankara.

Bayaner, A. 1996. Post Doctorate Study Report. Department of Agricultural Economics and Management. Reading University.

Blain, H.L. 1988. In: *World Crops; Cool Season Food Legumes,* pp 501-511 (ed R.J. Summerfield). Kluwer Academic Publishers.

Dickey, D.A. and Fuller, W.A. 1979. *Journal of American Statistical Association,* 74(366): 427-431.

Dickey, D.A. and Fuller, W.A. 1981. *Econometrica,* 49(4):1057-1072.

Hallam, D., Machado, F. and Rapsomanikis, G. 1992. *Journal of Agricultural Economics,* 43(1): 29-37.

Hendry, D.F. 1986. Oxford Bulletin of Economics and Statistics 48 201-212.

Keatinge, J.D.H. 1994.. *Journal of Economics, Agriculture and Environment* (MEDIT) 5, 19-23.

Kusmenoglu, I. and Bayaner, Ahmet. 1995 In. *Improving Production and Utilization of Grain Legumes.* Proceedings of Second European Conference on Grain Legumes. AEP. 9-13 July, Copenhagen-Denmark.

Oram, P.A. and Agcaoili, M. 1994. In: *Expanding the Production and Use of Cool Season Food Legumes.* pp 3-49 (eds F.J. Muehlbauer and W.J. Kaiser). Kluwer Academic Publishers.

Sargan, J.D. and Bhargave, A. 1983. *Econometrica,* 51(1): 153-174.

Sanjuan, A.I. and Gil, J.M. 1996. Presented at the EAAE Conference. 3-7 September 1996. Edinburgh.

SIS, 1992. *Foreign Trade Statistics,* Turkish Republic, Prime Ministry, State Institute of Statistics, No: 1615. Ankara.

SIS, 1994. *Agricultural Structure and Production,* Turkish Republic, Prime Ministry, State Institute of Statistics, Ankara.

SIS, 1995. *Statistical Indicators, 1923-1995.* Turkish Republic, Prime Ministry, State Institute of Statistics, No: 1883. Ankara.

Tekinel, O., Dinc, E., Erkan, O., Cevik, B., Tuzcu, O., and Saglamtimur, T. 1990. *Cukurfova University Faculty of Agriculture GAP Agricultural Research and Development Project Final Report.* GAP Publication, No: 33. Adana.

Trends in pulse crop diversification in western Canada and effects on world trade

R. McVicar[1], A. E. Slinkard[2], A. Vandenberg[2] and B. Clancey[3]

1 Saskatchewan Agriculture and Food, 125 - 3085 Albert St., Regina, Saskatchewan S4S OB1 Canada; 2 Crop Development Centre, University of Saskatchewan, Saskatoon, Saskatchewan S7N 5A8 Canada; 3 Stat Publishing Canada, Box 8110-361, Blaine, WA. 98230 USA.

Abstract

Canada has become a major producer and exporter of pea (*Pisum sativum L.*) and lentil *(Lens culinaris* Medikus). Production is concentrated in western Canada (Saskatchewan, Alberta and Manitoba), where the grains industry has undergone a significant change. Crop diversification, depressed prices for traditional grains, along with the recent end of transportation subsidies and production support programs are factors in this change. Successes in pulse crop research, commercial production, producer group organization, and export markets has encouraged the investigation and adoption of new crops. In-country primary processing has expanded rapidly to match increases in pulse crop production and provide a consistent high quality product. Adapted *Rhizobium* inoculant products have been developed that assist in maintaining relatively low input costs in pulse crop production. Pulse crop research is a high priority for funding agencies and producer groups. Diversification into chickpea (*Cicer arietinum* L.) and dry bean (*Phaseolus vulgaris* .L) is poised to enter large-scale commercial production.

INTRODUCTION

Traditionally, Canada has been an exporter of hard red spring wheat *(Triticum aestivum L.)*, durum wheat (*T. durum L.*), canola (*Brassica napus, B. rapa*) and barley (*Hordeum vulgare L.*). In recent years Canada has become a major producer and exporter of pea (*Pisum sativum L.*) and lentil (*Lens culinaris* Medikus*)*. This production is concentrated in western Canada (Saskatchewan, Alberta and Manitoba), where the grains industry has undergone significant change in the past 10 years. The most important of these changes have been crop diversification (Table 1) and the potential for production of higher value crops such as pulse crops. More recent changes include the end of transportation subsidies and production support programs which favoured the production of wheat. Pulse crop producer organizations are playing a more active role in providing research funding, guiding research objectives, and identifying barriers to export trade such as tariffs and duties.

Table 1. Area seeded to speciality crops* in western Canada, 1981-1997 (Saskatchewan Agriculture and Food, 1982-1996).

Year	Hectares	Year	Hectares
1981	416,000	1990	734,000
1982	415,000	1991	749,000
1983	350,000	1992	868,000
1984	450,000	1993	1,300,000
1985	438,000	1994	1,730,000
1986	588,000	1995	1,625,000
1987	753,000	1996	1,435,000
1988	800,000	1997 estimate	1,690,000
1989	715,000		

Specialty crops include: pea (*Pisum sativum*), lentil (*Lens culinaris*), canaryseed (*Phalaris canariensis*), sunflower (*Helianthus annuus*), mustard (*Sinapis alba* and *Brassica juncea*), coriander (*Coriandrum sativum*), caraway (*Carum carvi*), buckwheat (*Fagopyrum esculentum*), dry bean (*Phaseolus vulgaris*), chickpea (*Cicer arietinum*), fababean (*Vicia faba*), and safflower (*Carthamus tinctorius*)

R. Knight (ed.), Linking Research and Marketing Opportunities for Pulses in the 21st Century, 243–249.
© *2000 Kluwer Academic Publishers. Printed in the Netherlands.*

244

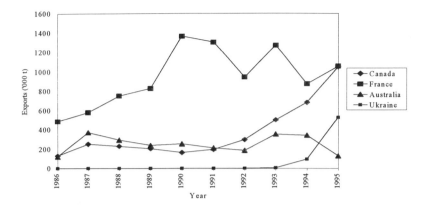

Figure 1 Pea exports by major exporting countries (FAO 1997). No data are available for China or Russia.

DIVERSIFICATION OF CROPS GROWN IN CANADA

In the past, producers have spent much energy protesting against the low price of wheat and seeking deficiency payments from government when wheat prices were low. However, producers have changed from the practice of growing traditional crops, such as wheat regardless of market demand or price, to the position where they are now able to respond to market signals by producing crops that the market needs. In doing so, producers have gained experience in the production of many "non-traditional" crops, such as pea and lentil.

EXPANSION OF PULSE CROP PRODUCTION

Pulse crop production has led this trend in crop diversification in western Canada over the past 20 years, and has risen to production levels of over 1.65 million tonnes in 1996. Canada is now a major producer and exporter of pea (Figure 1) and yellow cotyledon lentil (Figure 2).

As expertise in pea production was developing in Canada, the European soybean embargo had a major impact on the market. Europe developed alternatives to soybean meal as sources of protein for animal feed, and Canada began making significant bulk shipments of feed pea to Europe in 1986. Expansion of Canadian pea production occurred rapidly and as more uses for the crop were identified, and the currency exchange rates favoured exports from Canada, the price paid to producers increased. Before 1986, Canada was only a small player in the pea market, supplying mainly domestic human consumption, and was not significantly affected by world market factors. Canada has now reached the level of pea production where such things as global weather patterns and world production of feedstuffs greatly affect our industry, and our pea production levels in turn affect world markets. The greatest potential for growth in pea markets is seen in the livestock sector; mainly swine and poultry especially in Canada and in Pacific Rim countries. Interest in Pacific Rim countries is increasing in the Canadian pea and canola-meal blended product as it provides a consistent, balanced protein source, with no foreign material.

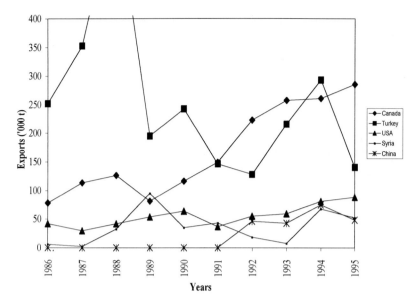

Figure 2 Lentil exports by the major exporting countries. (FAO 1997). The exceptionally high value for Turkey in 1988 was 606,000 tonnes.

Development of the Laird lentil was the single most important factor contributing to the rapid expansion of the lentil industry in Canada. Laird lentil is well adapted to most of the lentil-producing region of western Canada, and has met the requirements of the large-sized yellow-cotyledon lentil market.

As drought conditions swept across the lentil producing areas of northern Africa in the 70's and 80's, a shift in the origin of supply and an increase in prices occurred in the lentil market. The large-sized lentil was a premium-priced niche market, filled mostly by Turkey. The Laird lentil allowed Canada to offer large volumes of consistent quality, large-sized lentil and thereby enter this premium market. Lentil production has increased to the point that Canada is a major supplier in this market, while yellow cotyledon lentil production in Turkey has declined. Production of yellow cotyledon lentil in Canada has levelled off in recent years at 400,000 - 450,000 tonnes.

Slinkard and Blain (1988) reported on the hectares of dry pea and lentil grown in western Canada from 1949 to 1986. In 1986, pea and lentil were grown on 258,000 ha in western Canada. Pulse production has continued to expand in Canada since then, surpassing 1.2 million hectares of pea, lentil, dry bean, and chickpea in 1997, resulting in the production of over 1.9 million tonnes of pea and lentil in 1997 (Table 2).

World demand for pulse crops is expected to expand. As the Indian subcontinent experiences an increase in wealth, caloric intake will increase. If their capacity to produce pulse crops remains at current levels, imports are likely to increase, and pulse crops and soybean are the most likely candidates, placing increased demand on world supplies. However, the Canadian pulse industry believes there is more certainty in the expansion of pulse exports into livestock feed markets than into human consumption markets.

Table 2. Annual production ('000 tonnes) of pulse crops in Canada, 1987-1997 (Statistics Canada, 1997)

Year	Pea	Lentil	Dry Bean	Chickpea	Total
1987	415	286	110	0	811
1988	320	59	59	0	438
1989	234	96	71	0	401
1990	264	213	100	0	577
1991	410	343	113	0	866
1992	505	349	53	0	907
1993	970	349	131	0	1,450
1994	1,441	450	171	0	2,062
1995	1,455	432	118	0	2,005
1996	1,245	417	110	5	1,777
1997est	1,500	410	120	20	2,050

Shift in Area of Pea Production

Before 1980, pea production was centred in southern Manitoba. Long-vined, normal- leafed, late-maturing cultivars, suited to southern Manitoba conditions, dominated production and were sold into the small domestic food pea market. With the expansion of dry pea into other regions of western Canada, cultivars with shorter growing season requirements were introduced, mainly from northern Europe. By 1997, over 62 pea cultivars were registered in western Canada (Saskatchewan Agriculture and Food, 1997). Semi-leafless pea cultivars with improved lodging resistance and harvestability, as well as improved herbicides, have encouraged more producers to grow pea. Pea production is now centred in Saskatchewan and Alberta. This provides a larger, more stable base for dry pea production, resulting in a large consistent supply to export markets.

Increased Threat of Mycosphaerella Blight

The continued expansion of dry pea production in western Canada has led to an increased level of *Mycosphaerella pinodes* infection. Cool, moist summer conditions have prevailed in western Canada for the past seven years and this has favoured pea production and the subsequent spread of this disease. This constraint to pea production is being addressed by pathologists and breeders.

Reduction in Transportation Subsidies

For more than 90 years, the cost of transporting bulk grains from the producing areas of western Canada to port were regulated by Federal legislation. Costs to producers were subsidised. This benefit commonly known as the "Crow Rate" was eliminated in 1995 and producers must now pay the full cost of shipping grains to port. The result is that producers and grain companies are now looking for high value, low volume crops. They are also increasing the amount of primary processing within the area of production, including removal of foreign material before shipping the product to port in order to reduce freight costs. Because pulse crops usually produce lower volumes of higher value product and are cleaned in-country, Canadian producers will continue to produce pulse crops for export to world markets.

Expansion of In-Country Value-Added Processing

Dry pea, lentil and bean production has supported the development of a thriving pulse crop industry infrastructure capable of primary processing of the crops. Over 300 small to medium-sized facilities in western Canada clean, bag, handle and transport pea and lentil (Slinkard et al., 1992). The primary processing facilities are being consolidated, as small on-farm cleaning plants are being replaced by larger, more efficient plants on main rail lines. The rapid expansion of the special crops processing industry has resulted in many much-needed new enterprises and increased employment in rural western Canada. It has also allowed Canada to develop a reasonable reputation for quality. Canadian trading companies have been able to provide a consistent supply of quality product to world markets.

Secondary processing, such as four successful dry pea splitting facilities and two pea starch and pea fibre facilities are in operation. Other diversification efforts include development of several snack food products based on pulses and infrared processing for rapid cooking. Several facilities are currently under construction, which will be capable of dehulling and splitting both red cotyledon lentil and desi chickpea. The development of red lentil and chickpea splitting capacity at the same time commercial production of these two crops is expanding will allow Canada to further diversify its pulse crop presence in export markets.

Within the past two years, three feed mills have been constructed in western Canada to produce a pea and canola-meal blend as an excellent replacement to soybean meal in swine rations. The recent emphasis on doubling the swine herd in western Canada will provide further opportunities for the expansion of processing.

Expansion of in-country processing has encouraged additional producers to grow speciality crops and, thus, the speciality crop industry has become more integrated, linking production, processing and marketing.

Development of the Legume Inoculant Industry

Prior to the development of the pulse crop industry in western Canada, all commercial legume inoculant products were imported from the U.S.A. A thriving inoculant industry now exists in western Canada with three Canadian companies involved in research, production and marketing of inoculants. These companies have selected *Rhizobium* strains and developed products best adapted to western Canadian conditions and crop seeding systems. These products have improved the nitrogen-fixing capacity of pulse crops and contributed to the stabilisation of pulse crop yields in western Canada. Commercial producers have widely adopted the use of *Rhizobium* inoculant products. This has reduced the use of nitrogen fertiliser in pulse crop production and is seen as a significant benefit in reducing input costs and improving the competitiveness of Canadian producers. Current research is investigating the suitability of soil-applied granular inoculants for pulse crop production. Recently, thirty-eight strains of chickpea *Rhizobium* were evaluated for their ability to promote growth and enhance nitrogen accumulation in kabuli and desi types. A highly effective *Rhizobium* strain was identified for chickpea production in western Canada (Walley et al., 1996). Chickpea *Rhizobium* strain BCF32 is well adapted and was marketed on an introductory basis in 1997.

Adoption of Direct Seeding / Reduced Tillage Technology

Producers in western Canada are rapidly adopting minimum disturbance crop production. This has been facilitated by a greatly reduced price for glyphosate, a widely used herbicide for pre-seeding weed control and chemical fallowing, and the discovery that direct-seeded pea and lentil outyield pea and lentil seeded into land tilled before seeding. The soil moisture loss during and after tillage was the cause of reduced yields. Excellent machinery for direct seeding is now manufactured in Canada. This seeding equipment is capable of safely handling large-seeded pulses such as chickpea, dry bean and dry pea. Direct seeding/minimum disturbance technology is facilitating the spread of pulse crop production into the vast expanse of lower rainfall areas. Millions of hectares, previously under cereal crop/fallow production, are being fallowed less frequently. Producers are lengthening their crop rotations, and are recognising that extended rotations should include a pulse crop to improve soil structure, soil fertility and protein levels in the succeeding cereal crop. Producers are selecting lentil, pea and chickpea as the best-adapted pulse crops for their farms. The result is a more sustainable crop production system in dry areas and a more consistent supply of pea and lentil. It is anticipated that western Canada will also become a major producer and exporter of dry bean and chickpea in the next 10 years.

Increased Use in the Domestic Feed Industry

To maintain a competitive advantage, Canadian livestock producers are searching for cost-effective sources of protein and energy in rations. The Canadian pulse industry has identified the domestic feed market as crucial for their industry's expansion. This is leading to an increased use of pulse crops, especially feed pea in swine rations.

248

Current swine rations in western Canada include a significant portion of soybean meal imported from the USA. (Agriculture Canada, 1994). Replacement of imported soybean meal with domestic feed pea, as a source of protein supplement, would require approximately 300,000 tonnes of pea per year. A large domestic feed market would reduce the reliance of western Canadian pea production on the European feed market, and reduce the impact of large Canadian pea crops on world markets. The swine industry in western Canada is expanding rapidly and a doubling of the herd is expected within the next 5 years.

Increased Pulse Crop Research

Success in pulse crop development, such as the Laird lentil, has given producers, traders, and researchers in Canada, the confidence to look to further diversification within pulse crops. Crops previously grown on small areas, or not at all, are now receiving attention from government funding agencies, producer organisations and scientists. Agriculture and Agri-Food Canada has a pea breeder at Morden, Manitoba and a bean breeder at Lethbridge, Alberta. Alberta Agriculture has a pea breeder at Edmonton, Alberta and a pea and lentil breeder works at the University of Saskatchewan in Saskatoon. Recent changes in the support system for pulse crop breeding and the introduction of new cultivars into field-scale production in Saskatchewan have allowed for the hiring of a second pulse crop breeder at the Crop Development Centre in Saskatoon. The primary focus of this plant breeder position is in dry bean and chickpea development. Additional pulse crop research projects are underway in western Canada with the following objectives:
- Disease management systems in pulse crops
- Pulse crop transformation systems
- Molecular marker assisted selection systems
- Crop quality research in pulse crops
- Producer demonstration and applied research where breeding lines designated for first year comparative trials will be tested in large-scale plots
- Early maturing dry bean with resistance to *Pseudomonas syringae* (bacterial blight)
- Development of early maturing, extra-large kabuli chickpea and early maturing, small desi chickpea
- Development of a bleach resistant green cotyledon pea
- *Ascochyta fabae f.sp. lentis* resistant yellow cotyledon lentil
- *Mycosphaerella pinodes* tolerance in dry pea
- Development of an adapted red lentil with improved market acceptance
- Development of adapted *Rhizobium* inoculant products for chickpea and dry bean
- Effect of fungicidal seed treatments
- Fertility in pulse crops
- Direct seeding of pulse crops in cereal stubble
- Weed control of pulse crops
- Water use relations in pulse crops
- Harvestability of pea
- Salt tolerance of pulse crops
- Feeding quality of pea and chickpea
- Canning quality of pulse crops
- Anti-nutritional factors in feeding pulse crops to livestock
- Drying, decorticating and splitting technology for pulse crops
- Anthracnose of lentil and bean
- Variability in *Mycosphaerella pinodes*
- Control of *Erisyphe polygoni* (powdery mildew) of pea

Research played a major role in the early development of pulse crops, enabling Canada to grow from a residual supplier to a dominant exporter in some classes of pulses. Research will continue to play a significant role in pulse crop production and, consequently, exports of pulse crops from Canada.

Relative Absence of Insect Pests in Pulse Crops

Weevils (*Sitona spp.*), bruchids (*Bruchus spp.*) and many other insect pests cannot survive the severe winter cold in western Canada and thus are not a problem. Grasshoppers do not prefer pulse crops, but they will eat the flower buds of lentil and insecticide application is required in some areas. The relative absence of insect pests allows for lower costs of production, fewer storage problems, and higher quality of pulses destined for export markets.

Increased Desire and Ability to Grow New Crops

The "frontier spirit" on the part of scientists, producers and trading companies working in speciality crops, including pulse crops, has maintained the desire to look for potential new crops from around the world and investigate their adaptation in Canada. This attitude was reinforced during the late 1980's and early 1990's when many of the traditional grains produced under dryland agriculture in western Canada did not provide enough return to sustain a farm business. Pulse crop yields under dryland production often equal yields of wheat, and have provided equal or superior returns (Slinkard et al., 1992). The large land base of many farms provides the area needed to try a number of new crops on a small scale with little risk. The land base also provides the opportunity to produce low cost crops on a large scale with relatively lower yield requirements due to economies of scale. For western Canada, the large land base reduces the risk of a country-wide crop failure and helps stabilise the development of new crops.

CONCLUSION

Pulse crops could reasonably account for 20% of the land seeded to annual crops in western Canada, and help develop and maintain an agronomically sound crop rotation. The pulse crop industry in Canada recently supported this prediction by identifying the potential for an annual production of 4 million hectares (1.2 million ha in 1997) of pulse crops (Pulse Canada, 1997). Producers who have included pulse crops in their rotations are reporting superior cereal crop yields and protein levels compared to continuous cereal or cereal/fallow sequences. With the potential for temporary reductions in livestock feed imports in Europe, the need to develop further value-added processing and greater pulse crop usage in the domestic feed industry in Canada is crucial to the expansion of pulses. The long term trend toward lower prices for traditional grains, such as wheat, reduces the opportunity for producers to sustain their farm by growing only traditional grains. These facts continue to support the trend to diversification of crops and increased production of pulse crops in western Canada.

References

Agriculture Canada, 1994. *Oilseeds Sector Profile*, January 1994. Soymeal Imports by Province. pp. 15. Agriculture and Agri-food Canada, Markets and Industry Services Branch, Ottawa

FAO, 1997. FAOSTAT Agriculture Statistics Database. Internet address: www.fao.org/waicent/agricul.htm. Rome, Italy.

Pulse Canada, 1997. *Business Plan*. 1997. Pulse Canada, Winnipeg, Manitoba.

Saskatchewan Agriculture and Food, 1982-1996 *Specialty Crop Report*, Saskatchewan Agriculture and Food, Regina, Saskatchewan.

Saskatchewan Agriculture and Food, 1997. *Dry Pea Varieties*, 4 pp. Saskatchewan Agriculture and Food, Regina, Saskatchewan.

Slinkard, A.E. and Blain, H.L. 1988. In: *World Crops: Cool Season Food Legumes*, pp. 1059-1063 (ed R.J. Summerfield). Dordrecht: Kluwer Academic Publishers.

Slinkard, A.E., van Kessel, C., Feindel, D.E., Ali-Khan, S.T. and Park, R. 1992. In: *Expanding the Production and Use of Cool Season Food Legumes*, pp.877-889 (ed. F.J.Muehlbauer and W.J. Kaiser). Kluwer Academic Publishers.

Statistics Canada, 1997. Annual Production ('000 Tonnes) of Pulse Crops in Canada, 1987-1997 Ottawa, Ontario.

Walley, F.L., Hnatowich, G.L., Stephens, S. and van Kessel, C. 1996. *Pulse Crop Research Workshop*, Calgary, Alberta. November 1996.

Relations between demand and production of edible pulses for food and feed in China

ZUO Mengxiao
Deputy Director, Grain and Oil Crops Division, Department of Agriculture, Ministry of Agriculture, People's Republic of China

Abstract

China has a long history of pulse cultivation. The country is the centre of origin of soybean and red bean and a secondary centre of some other pulses. In recent years the area sown to edible pulses has exceeded 11 million ha, 72% of which were devoted to soybeans. Soybean production in the last four years has been between 13.3 and 16.0 million metric tonnes. Approximately 6 million tonnes of these soybeans are used for oil extraction. Improvements in the production of soybeans have fallen behind the improvements in wheat and rice. Production of soybeans and the other pulses varies widely with market conditions. Of these other pulses, broad beans have occupied about 1 million ha and produced about 1.9 million tonnes which is 45 % of the production of these pulses. Recently pea production has amounted to about 780,000 tonnes. Edible pulses have been a traditional export commodity, as grain or as processed products. Soybean will continue to be an important crop in China. It is believed there are additional and extensive areas of north-east China which could be used for soybean production. It is forecast soybean production will reach 17.3 million tonnes annually by the year 2000 and 19.0 million tonnes by 2010.

1.   EDIBLE PULSES IN CHINA

China has a great variety and rich resources of edible pulses, which include soybean (*Glycine max*), broad bean (*Vicia faba* L), pea (*Pisum sativum*), mung bean (*Vigna radiata*), red bean (*Vigna angularis*), cowpea (*Vigna unguiculata*), common bean (*Phaseolus vulgaris*), kidney bean (*Phaseolus vulgaris*), lentil (*Lens culinaris*) and other minor pulses. China is not only the centre of origin of soybean and red bean, but also one of the centres of origin of mung bean, and the secondary centre of origin of common bean and cowpea. China has a long history of pulse cultivation. The Chinese have cultivated soybean for over 5,000 years, red bean, mung bean, broad bean and pea for over 2,000 years, and the other pulses for several centuries or a millennium.

In 1995, the area sown to edible pulses in China was 11.23 million ha, accounting for 10.2% of the total area sown to grain crops. The production of pulses was 17.87 million metric tonnes (MT) or 3.8% of the total grain production; and the average yield of pulses was 1.59 t/ha. The area sown to soybean was 8.13 million ha, or 72.3% of the total pulse area; and soybean production was 13.50 million MT, or 75.5% of the total pulse production, with an average yield of 1.66 t/ha. There were only small changes in these values between 1995 and 1996. In 1996, China's edible pulse production was 17.90 million MT of which 13.22 million MT were soybeans with a yield of 1.77 t/ha

1.1   Soybean

Soybean is a major grain crop in China, with the area sown and production ranking fourth after rice, wheat and corn. China has had a long experience of soybean cultivation and has developed many varieties that are widely distributed in the major agricultural areas of the country. Historically and over a long period of time, soybean production in China ranked first in the world. The area sown in 1936 reached 8.93 million ha with a total production of 11.30 million MT or over 90% of the world's total. Later, the imperialist invasions of China and other factors seriously destroyed this production. Production recovered fairly rapidly after the founding of the People's Republic in 1949 (Table1). The area sown reached 12.75 million ha and total production 10.05 million MT in 1957, both were very large increases from 1949. After 1958, due to the nationwide tight grain supply situation, the area of high-yielding crops such as corn and sweet potato was expanded at the cost of

251

R. Knight (ed.), Linking Research and Marketing Opportunities for Pulses in the 21st Century, 251–256.

soybean, resulting in a decline in soybean production. By 1977, China's soybean area had fallen to 6.84 million ha with a production of 7.26 million MT.

After the reforms of 1978, the state mobilized the farmers' enthusiasm for growing soybean by greatly raising its procurement price. This, together with the large-scale extension of high-yielding cultivation practices, led to a rapid recovery and expansion of soybean production. In 1987, the area sown recovered to 8.44 million ha and production reached 12.18 million MT. The production fluctuated for five consecutive years after 1988. China's soybean area and production in 1992 were 7.22 million ha and 10.30 million MT respectively, about 15% less than in 1987 for both. These fluctuations caused a tight supply situation and pushed up the soybean market price by a large amount. As a result, the year 1993 saw unprecedented developments of the soybean crop, with the area reaching 9.45 million ha and production 15.31 million MT. Despite a slight fall in area in 1994, both the total production and yield reached a record level of 16.00 million MT and 1,734 kg/ha. Soybean production remained at a relatively high level in 1995 and 1996.

At present, the main areas are Northeast China, the lower and middle reaches of the Yellow River, the Huaihe River basins and the Yangtze River basin while Heilongjiang is the largest producing province, where the area and production account for more than 30% of the national total. Other provinces, or autonomous regions, such as Henan, Hebei, Jilin, Inner Mongolia, Jiangsu and Sichuan are also major producers in China.

Table 1: Area, production and yield of soybean

Year	Area (million ha)	Production (million t)	Yield (kg/ha)
1949	8.32	5.09	611
1952	11.68	9.52	815
1957	12.75	10.05	788
1962	9.50	6.51	685
1967	8.50	8.27	972
1972	7.58	6.45	851
1977	6.84	7.26	1060
1982	8.42	9.03	1073
1987	8.44	12.18	1443
1988	8.12	11.65	1434
1989	8.06	10.23	1269
1990	7.56	11.00	1455
1991	7.04	9.71	1380
1992	7.22	10.30	1427
1993	9.45	15.31	1619
1994	9.22	16.00	1735
1995	8.13	13.50	1661

Source: China agriculture statistics

The growth of soybean production in China has been relatively slow when compared with the major grain crops. In the 46 years between 1949 and 1995, China's total grain production increased by 312 % at an annual growth rate of 3.13%, of which, wheat production increased by 641% and rice by 281% whereas soybean increased by only 165% at an annual rate of 2.15%. The main reason for this slow growth has been the low and unstable yields of soybean. In 1952, the national yield was 815 kg/ha, which was 35% of rice yield and 61% of corn yield but exceeded the wheat yield of 735 kg/ha. In 1995, the average yield reached 1,661 kg/ha, equivalent to only 28% of rice, 34% of corn and 47% of wheat. Apparently, the gap between the yield of soybean and the three major crops has widened.

1.2 Other Pulses

In China, pulses other than soybean are generally referred to as sundry pulses. Their production fluctuates widely under the influence of the market. The area sown to them and their production are unstable. Compared with the area in 1991, the area sown in 1995 was 3.11 million ha, up by 46.4% and production was 4.37 million MT, up by 56.1%(Table 2). The average yield was 1,406 kg/ha, up by 6.5%. Among these pulses, the broad bean area was 1.083 million ha with a production of 1.97 million MT, accounting for 45.1% of the total production of sundry pulses. In 1995, mung bean and red bean (Table 2) accounted for 14.1% and 7.4% of the total production of sundry pulses. The production of pea, kidney bean, black bean, cowpea and fresh soybean was 784,000 MT, 269,000 MT, 204,000 MT, 43,000 MT and 35,000 MT respectively. These crops are mainly distributed in Sichuan, Yunnan, Jiangsu, Hubei, Hebei, Inner Mongolia, Shanxi, Henan and Jilin provinces/regions, whose production accounts for about two thirds of the national total and each of them produces over 100,000 MT.

Table 2: The production of sundry pulses (minus soybean). Area in million ha and production in million MT

Year	Total Area	Total Prod.	Broadbean Area	Broadbean Prod.	Mung bean Area	Mung bean Prod.	Red bean Area	Red bean Prod.
1957	-	-	1.498	1.300	1.639	0.811	0.378	0.295
1986	-	-	0.885	1.560	0.547	0.500	0.183	0.220
1993	2.922	4.200	0.861	1.810	0.943	0.729	0.256	0.265
1995	3.106	4.370	1.083	1.970	1.294	0.618	0.346	0.323
1996	3.073	4.680	1.068	1.990	1.179	0.582	0.336	0.323

Source: China agriculture statistics

In recent years, along with the improving living standards of the people, the demand for vegetable protein has been growing. The pulses and their processed products have won more and more favour among consumers, leading to increased demand. The minor pulses have a high nutritive value. They are not only a traditional food grain but are also considered a modern health food. Meanwhile, the minor pulses are also China's traditional export products for foreign exchange earnings, with a strong export demand.

2. CHARACTERISTICS OF EDIBLE PULSES

Edible pulses have a high content of protein, generally between 20% and 40%, which is 2-4 times higher than in graminaceous crops. For example, the protein content of soybean is about 40%, pea 23%, broad bean 25% and red bean 22%. Edible pulses have not only a high protein content but also a wide variety of amino acids to meet the requirements of the human body. The Chinese people have always treasured and loved edible pulses, which is evidenced by a popular saying in the past that "green vegetables and bean curd keep people well". At the same time, the pulses are also an important source of nutrition for livestock and poultry. The seeds, the by-products after processing, and stems and leaves are all good animal feeds.

Farmers recognize that growing edible pulses is an effective way to increase soil fertility. Part of the nitrogen, fixed by the rhizobia, is used by the plants for their own growth, the rest remains in the soil and improves soil fertility before a cereal crop. According to tests, one hectare of soybean can fix 50-75 kg of N a year, broad bean 50-140 kg, cowpea 75-220 kg, pea 75-80 kg, common bean 45-90 kg and mung bean 30 kg. If the stalks and leaves are ploughed under the soil, they increase the organic matter and N, P, K in the soil as well as improving the soil's physical structure.

Edible pulses and their products are traditional export commodities. Each year, China exports soybeans, broad beans, peas, mung beans, red beans, lentils, common beans, cowpeas and chickpeas (Table 3). Processed products include bean cakes, bean vermicelli and canned beans. Soybean is among China's three main export farm products (tea, silk and soybean). In 1929, China exported 1.7 million MT of soybean, which accounted for 85-90% of the world's total export at that time. After the founding of New China, our average annual export in the 50's was above 1 million MT and reached 1.77 million MT in 1959. Later due to stagnation and decline in production, it dropped gradually to 0.13 million MT by 1977. Following the reforms, China began to restore its soybean exports, which reached 1.14 million MT in 1985. During 1988-91 period, it was around 1 million MT. In recent years, exports have declined again to around hundreds of thousands MT.

Minor pulses are also among China's traditional export commodities. Many of them are special, famous and good quality products, very popular domestically and abroad. According to statistics provided by the State General Administration of Customs, in 1995 China exported 375,000 MT of soybean valued at US$99.67 m, 1,041,000 MT of dried beans valued at US$359 m and 78,000 MT of fresh or frozen beans valued at US$69.88 m.

Table 3: The export of edible pulses (1,000 t)

Year	1992	1993	1994	1995
Soybean	658.2	373.2	831.8	375.1
Kidney bean	335.3	256.4	491.4	375.1
Broadbean	367.6	274.3	427.1	233.3
Mungbean	53.4	71.3	159.1	183.8
Red bean	52.0	73.9	103.3	72.6
Lentil	46.0	42.8	74.5	47.9
Pea	64.3	24.7	42.3	23.7

Source: China customs statistics

Edible pulses have a wide adaptability, occupying a special position in crop farming. China has a huge territory with a diverse climate and topography. On hilly land and infertile dryland, which account for a big percentage of the total area in China, it is difficult to grow high-yielding crops like rice, corn and wheat, but the land is suitable for growing edible pulses. Edible pulses have more flexibility and complementarity when it comes to geographic and seasonal cropping arrangements. Many have a short growing season, which makes them good catch-crops and disaster-relief crops. For example, mung bean and red bean have a growing season of 70-90 days and early-maturing pea 80 days. If these pulses can be quickly sown following disasters, good harvests can be expected. Hence the name "disaster-relief crops". Some pulses are good choices for multiple cropping in places where temperatures are too low for two crops a year. Others can endure drought, infertility and shade when grown in fragmented areas like farm yards, field margins and ditch banks, thus filling gaps in crop farming production and helping to increase food production.

3.   DEMAND AND PROSPECTS FOR EDIBLE PULSES IN CHINA

China has basically solved the problem of feeding and clothing its population. In 1996, our per capita grain availability exceeded 400 kg, close to the world average. However, the diet still features a high proportion of cereals with a low nutritional level. In terms of total food consumption, only a small proportion is high quality protein food and the people consume inadequate amounts of animal-based food and pulses. Edible pulses, particularly soybean, are a special group of products that have an important bearing on our national economy and the people's livelihood. As the Chinese move from a "subsistence" to a more "comfortable" life style and the Chinese diet from "calorie-focused" to "nutrient-focused", the demand for pulses will keep growing because of their role as high-protein food, high-protein feed and as an export commodity. Soybean is still the most important pulse in China, for oil and for bean product processing. In 1985, only 2.1 million MT of soybean were used to extract oil which was 20% of the total production. In 1995, around 6 million MT of soybean were used to extract oil, which was 45% of the total production. On the other hand, the consumption of processed soybean products grew by a big margin. Only a small percentage of total soybean production was used as animal feed.

In 1993, the State Council issued the Outline for the Reform and Development of China's Food Structure in the 1990s. It suggested that by the year 2000, the average daily protein intake of the Chinese people would be 72 g, one third of which would be high quality protein and that per capita consumption of pulses (mainly soybean) would be 8 kg yearly. As the population grows and consumption improves, the demand for pulses, particularly soybean, will keep growing. It is estimated that by the year 2000 (Table 4), the total demand for pulses will be 22 million MT of which 10.4 million MT will be for consumption by urban and rural residents, 8 million MT for oil extraction, 0.8 million MT for animal feed and 1.27 million MT for seed. By the year 2010, the total demand for pulses would be 24 million MT, including 11.5 million MT for consumption by urban and rural residents, 9 million MT for oil extraction, 0.9 million MT for animal feed and 1.12 million MT for seed.

Table 4: The forecast consumption of edible pulses (1,000 t) based on statistics given below the table.

Year	1995	2000	2010
Total consumption	17,880	22,000	24,000
Food	8,000	10,400	11,500
Oil extraction	6,000	8,000	9,000
Feed	720	800	900
Stock	350	200	200
Export	1,110	800	800
Seed	1,200	1,270	1,120
Loss	500	530	480
Consumption per capita(kg)	15.0	16.9	17.1

(1) Population:1.3 billions in 2000, 1.4 billion in 2010

(2) Extracting oil: 46% in extracting oil in 2000, 57% in 2010

(3) Seed: 100kg/ha in 2000, 90kg/ha in 2010

4. PRODUCTION PROJECTION FOR EDIBLE PULSES IN CHINA

In the long term, because of increasing consumption of edible pulses in the domestic market, which causes an unfavorable supply and demand balance, it is not practical to expect China to become once again a major exporter of soybeans. China's policy for production is to expand the area, improve yields, become self-sufficient domestically and participate in international trade. Recently, China has reaped good harvest for several consecutive years and the market price for rice, wheat and corn have all dropped drastically. However the price for soybean remains high, making soybean farming very profitable. In some places, there has been a "soybean boom" where farmers have cut down on their corn area, converted to soybean cultivation and increased inputs. This has brought about a new development phase for soybean production in China. In 1997, China's soybean area is expected to increase by 300,000 ha over the previous year, up by 4%. By the year 2000 (Table 5), the total pulse area in China is expected to reach 12.7 million ha, producing 22.4 million MT of pulses, including a soybean area of 9.4 million ha producing 17.3 million MT at an average yield of 1,840 kg/ha. By the year 2010, the total pulse area will be 12.4 million ha, producing 24 million MT, which will include 9.2 million ha of soybeans producing 19 million MT at an average yield of 2,065 kg/ha

Table 5: The present and forecast production of sundry pulses and of soybean. Area in millions ha and production millions tonnes.

Year	Sundry pulses		Soybean		
	Area	Production	Area	Production	Yield (kg/ha)
1995	3.106	4.369	8.126	13.502	1,661
2000	3.300	4.600	9.400	17.300	1,840
2010	3.200	5.000	9.200	19.000	2,065

The ups and downs in China's edible pulse production are mainly caused by variation in the cropping area and low average yields. In 1994, the world's average soybean yield was 2,182 kg/ha but country averages were 3,606 kg/ha for Italy, 2,883 kg/ha for Egypt, 2,815 kg/ha for the United States and only 1,735 kg/ha in China. Since China's soybean average yield was only 80% of the world average, there is a big potential for improvement. If China can restore its soybean area to 9.2 million ha and bring the average yield up to the 1994 world average, total soybean production will be expected to reach 20.1 million MT, sufficient to meet demand in the coming few years (table 6).

Table 6: Previous changes and forecast changes in the yield of soybean.

Years	Interval (years)	Yield (kg/ha)	increase
1949-1952	3	611-815	10.10%
1949-1995	46	611-1661	2.00%
1987-1995	8	1442-1661	1.78%
1995-2000	5	1661-1840	2.07%
2000-2010	10	1840-2065	1.16%

To realize the production targets for China's edible pulses, we need to take the following measures:

1 Stabilize and expand the area of edible pulses. Northeast China is a region with vast land resources suitable for soybean. It is estimated there are 2.7 million ha of exploitable land available, offering a huge potential for further pulse production. Other effective ways of expanding the area include intercropping, relay cropping, mixed cropping, multiple cropping and bund cropping.

2 We should popularize high-yielding and high-quality new varieties. In recent years, China has developed a number of high-yielding soybean varieties with yields of 3,000-3,700, some even 4,500 kg/ha. These new varieties have created favorable conditions for improving China's soybean yields.

3 We should popularize applicable yield-increasing technology. Science and technology is playing an inadequate role in edible pulse production in China and this has been the most important obstacle to increased production. Through years of efforts, agricultural research and extension organizations have accumulated experience regarding cultivation, fertilization, irrigation and integrated pest management of edible pulses. When they are widely used by farmers, China's production will surely progress.

4 We should improve production conditions with multi-level inputs by the national government, the local governments and individual farmers. We will set up national-level soybean producing bases in order to stabilize soybean production, meet market demands and improve economic benefits.

Opportunities for improved adaptation via further domestication.

G Ladizinsky[1] and J Smartt[2]

1 The Hebrew University of Jerusalem, Faculty of Agriculture, PO Box 12, Rehovot 76100 Isarel; 2 Department of Biological Sciences, University of Southampton, Bassett Crescent East, Southampton, SO16 7XP, UK

Abstract

The concept of improved adaptation via further domestication is based on the premise that wild legumes may contain useful diversity not present in the existing food legumes. It is a diversity which cannot be transferred to the food legumes by conventional breeding. For a wild legume to be considered for domestication it has to comply with three criteria: 1 A relatively high harvest index 2 A reasonable prospect of producing a wholesome product, free of toxins and unpalatable substances and 3 An ability to exploit environments in which conventional food legumes cannot produce economic yields. Adaptation in this regard includes the ability not only to cope with environmental stresses but also to offer a better product. At the moment, wild species can be domesticated as food legumes by: 1 Selecting for some key characters, such as soft seed coat and non-shattering pods 2 Creating diversity in key characters by mutagenesis and 3 Transferring the domestication syndrome from cultivated relatives. Wider adaptation may be achieved by domesticating some perennial chickpea and *Lathyrus* species, after study of their potential as food legumes.

INTRODUCTION

In evolutionary terms, adaptation is the process by which organisms become fitted to their environments. In wild populations, adaptation is the total response of the individuals to aerial, edaphic and biotic factors. The same factors play a role in crop plant adaptation. In addition, crop adaptation includes the response to human selection of specific attributes, which may be negatively adapted and limiting to survival in wild populations.

Domestication is a process by which wild plants are modified morphologically, genetically and physiologically to fit the environment in which they are grown, and to better satisfy human needs. In some crop plants, the beginning of domestication can be identified with the modification of a single trait: non-brittle rachis in cereals, lack of seed dormancy in legumes, and toxin-free genotypes in many crops. Domestication is an ongoing process of improving the adaptation of crop plants to their environment and to human demands.

Natural, and particularly human selection have been the major augmentor of diversity in many crop plants. A large part of that diversity has been preserved by conscious or unconscious acts of the grower, and is usually not expressed in the crop's progenitors. The food legumes, which are the subject of this conference, are of southwestern Asiatic and Mediterranean origin. For thousands of years these legumes have been selected for better adaptation to the growing conditions in these geographical areas and to the specific demands of their growers. Diffusion out of their nuclear area of domestication has exposed them to different ecological conditions, resulting in the selection of genotypes which can cope and yield in the new environments. In the new territories, the food legumes have not lost their basic growing requirements, namely, a cool wet period during the vegetative phase followed by a hotter, dry period during flowering and seed set. Despite the marked diversity in their responses to different growing conditions, the cool season food legumes have remained crops of Mediterranean or Mediterranean-like climates. They have adapted somewhat to temperate zones. They have not established themselves in the humid tropics, where vegetative growth

257

R. Knight (ed.), Linking Research and Marketing Opportunities for Pulses in the 21st Century, 257–263.
© *2000 Kluwer Academic Publishers. Printed in the Netherlands.*

may be possible, but successful flowering and fruiting is prevented by wet conditions and high humidity and temperature.

In the last few decades, modern plant breeding has widened the adaptation of many crop plants. This has been achieved by manipulating the crop's responses to its physical and biotic environment, including pest and disease resistance. When attempting to widen the adaptation of legume crops by domesticating other wild legumes, it is important to bear in mind that crop plants generally exhibit much wider adaptation than their wild progenitors. The crop plants may possess the necessary diversity. Selection among cultivated germplasm for a set of ecological conditions may be more successful than screening the populations of their wild progenitors. When the desired diversity for adaptation to marginal environments is lacking in the crop's germplasm, the breeder may choose to proceed in one of two possible ways: 1 To select over a long period within the crop's germplasm with only a slow rate of success, as adaptation may be under complex genetic control. 2 To look for crop relatives which have the appropriate ecological adaptation, and to attempt to establish the domesticated syndrome in the wild plants.

The situation in lupin may serve as an example. The cultivated lupin species, are mostly adapted to acidic soils with poor performance in calcareous soils. The wild species *Lupinus pilosus* grows on calcareous soil with a higher pH. Domesticating this large seeded species might expand the lupin's growing area.

Domestication of wild species with adaptation to marginal environment involves genetic manipulation. But it also has economic implications. There are three major economic reasons for domesticating wild species:

1 To produce entirely new products (eg evening primrose oil) or substitute products to replace those from sources no longer available (eg jojoba replacing whale products)

2. To reduce costs and enhance local self-sufficiency

3. To increase crop diversity and improve the sustainability of indigenous farming systems. Domesticating wild species for any of the above purposes must furnish an economically competitive crop, at least locally where traditional crops fail. Even when a wild legume possesses the necessary adaptability, its domestication is not a simple process. The alteration of several traits is a prerequisite for making it a seed crop. Among these are the elimination of pod-shattering and of seed dormancy as a result of the hard seed coat, typical of wild legumes. In seed legumes, pod indehiscence is almost invariably controlled by a single recessive gene (Table 1). Although much less information is available on the genetics of seed dormancy in legumes, the limited data suggest its control varies from a single to a few genes (Forbes and Walls, 1968; Donnelly, et al. 1972; Ladizinsky. 1985; Verma and Ram, 1987).

Table 1. Genetics of pod indehiscence in several legumes

Crop	Genetics	Reference
Chickpea *Cicer arietinum*	1 recessive gene	Kazan et al. (1993)
Cowpea *Vigna unguiculata*	1 recessive gene	Rawat (1975)
Lentil *Lens culinaris*	1 recessive gene	Ladizinsky (1979)
Lupins		
L. luteus	1 recessive gene	Sengbusch & Zimmernann (1937)
L angustifolius	1 recessive gene	S&Z(1937),Gladstones (1967)
L. digitatus	1 recessive gene	Gladstones (1967)
Pea *Pisum sativum*	1 recessive gene	Waines (1975)
Soybean *Glycine max*	1-2 additive genes	Tsuchiya (1987)

Until genetic engineering techniques enable the domestication of wild legumes there are three possible methods:

1. Selecting for certain key traits in wild populations. By growing over a million wild plants, Sengbusch and Zimmermann (1937) selected mutants of *Lupinus luteus* with indehiscent pods, as did Gladstones (1967) in *L. digitatus.* Independently, these authors

selected pod indehiscent mutants in *L. angustifolius*. Since pod indehiscence is controlled essentially by a single recessive gene, it is reasonable to expect the same control in other wild legumes, as predicted by Vavilov's law of homologous series.

2. Induced mutation in key characters. Since some of the key characters of domesticated legumes are controlled by single genes, the necessary variation may be produced by mutagenesis.

3. Transferring the domestication syndrome by hybridization. This mode of domestication may be relevant to species of the crop's secondary gene pool. The common way of exploiting these species in breeding is by a single or several backcrosses of the partially fertile interspecific F_1 hybrid and its derivatives, to the domesticated parent. However, when the genes controlling the characters in question are located on non homologous chromosome segments, it may be easier to exploit the wild relative by transferring the relatively few genes controlling the domestication syndrome from the crop to its wild relative. This approach was recently used to domesticate the wild tetraploid oats *Avena magna* and *A. murphyi* by transferring non-shattering seed, glabrous and yellow lemmas and lack of awns to them from the cultivated hexaploid oat (Ladizinsky 1995). These wild oats were domesticated because of the high protein content of their seeds, approximately 30% more than in common oat cultivars.

To what extent then can domesticating wild species improve the adaptation of food legumes to stressful environments? Buddenhagen and Richards (1988) concluded that the dry matter production of drought-resistant natural vegetation is no greater, and is often less, with the same water supply, than that of a mesophytic crop. On the other hand, marked differences in drought tolerance is evident among the food legumes, with grasspea being much more tolerant than other legumes. These authors also believed that gradual improvement might be mediated through a better matching of crop phenology with the environment. Undoubtedly, the wild relatives of the food legumes will continue to be an important source of diversity when selecting for a response to biotic stresses, but this alone barely justifies their domestication.

It must also be borne in mind that a newly domesticated food legume will only survive if it is able to compete economically with the common cultivars of the traditional crop. In other words, in the habitat to which it is better adapted than the traditional crop, it must produce reasonable economic yields. Alternatively, it may possess other characteristics, which are lacking in the traditional crop. It is obvious then, that newly domesticated wild legumes may become crops in limited geographical areas under specific conditions.

PLANT MATERIAL FOR DOMESTICATION

A wild legume species adapted to a unique environment may be considered for domestication as a food plant if it possesses:

1 A relatively high harvest index and the potential to produce economic yields

2 Either a lack of substances which are toxic or unpalatable, or substances which can be removed by breeding or processing

3. The ability to adapt to habitats where other legume crops have failed.

Of the many wild annual legume species of the Middle East only a handful have been domesticated for food. Is this a proof that the others are inappropriate for domestication? Perhaps, but alternatively it may suggest that when the domesticated forms of several wild legumes were established, there was little or no interest in further domestications.

Of the five food legumes, the faba bean, pea, lentil and grasspea are members of the tribe Vicieae and chickpea is classified in the monogeneric tribe Cicereae. These tribes, and more specifically the genera to which the food legumes belong, are the natural place to look for species containing attributes which make them candidates for domestication.

The genus *Vicia*

Faba bean *(Vicia faba* L*)* is the only food legume in this large genus. Morphologically and cytogenetically it holds a unique position in the genus, and its wild progenitor has not been discovered. Furthermore, no other species of *Vicia* is a member of the faba bean's secondary gene pool. The genus *Vicia* contains several crop plants used for forage and fodder. Their seeds are bitter and cannot be considered as food for humans. Members of section Faba are unique because of their relatively large leaflets, pods and seeds and may be attractive for domestication, but their seeds are as bitter as those of other *Vicia* species. Further research is needed to determine whether non-bitter genotypes exist in the wild species of section Faba, or whether they can be induced by mutagenesis.

The genus *Cicer*

Chickpea, *Cicer arietinum* L, is currently a crop in countries of all continents. It grows under diverse climatic and soil conditions and at different latitudes and altitudes. Consequently, it possesses considerable diversity in terms of adaptation. This is in contrast to its wild progenitor which, geographically and ecologically, is much more restricted.

Besides the cultivated chickpea, the genus *Cicer* contains eight annual and 32 perennial wild species (Van der Maesen, 1987). In addition to the chickpea wild progenitor, *C. arietinum* subsp. *reticulatum,* another two wild species, *C. echinospermum* and C. *bijugum* have relatively large seeds and in this respect may be considered for domestication. Both grow on basaltic soil in northern Syria and Iraq and southern Turkey, where chickpea is a major crop. It can be concluded that the adaptation of these two wild chickpeas to the local conditions is shared by the local chickpea land races. Other annual chickpea species have much smaller seeds and they all grow in areas where chickpea is extensively grown. None of the wild annual species has shown sufficient adaptation to stressful environments to justify their domestication.

The perennial species have been grouped in three sections: Acanthocicer, Chamecicer and Polycicer (Van der Maesen 1987). Relative to the annual chickpeas, the perennial species are much less known and species' boundaries have not been tested by breeding experiments. The morphology of some of them is rather similar and they may present geographical variants of one polymorphic species. Some of the perennial species show interesting characteristics: besides their shrubby habit, they have relatively large seeds, are confined to stony habitats at high altitudes, 1500-4000 m, and may be common locally. Moreover, they can withstand the cold conditions of high altitudes and grow in areas where arable land is scant or totally lacking. Domesticating some of these perennial species presents the following potential advantages: 1 Planting may be needed only once in several years, 2 They may provide some income from land where annual crops cannot be grown and which is otherwise useless. Growing perennial chickpea may have some advantages in the highlands of central Asia in terms of local impact where these species are native plants and their green pods and seeds are often used. It is only in the second phase, with more breeding efforts that these domesticated perennials may be adapted to arable land as well. The species, which could be considered for domestication are *C. anatolicum* and *C. songoricum* of section Polycicer and their related species. Species of section Acanthocicer are problematical because of their spiny plant parts: stipules in C. *acantophyllum* and *C. macracanthum,* and rachis in *C. incanum.* Other species of this section are even more spiny.

As already mentioned, little is known about the biology and genetics of the perennial chickpeas, including those which may be considered for domestication. Only further study will show whether domesticating perennial species is a viable economic option.

The genus *Lathvrus*

Grasspea, *Lathyrus sativus* L, is the most heat- and drought-tolerant of the food legumes. Unfortunately, its seeds contain a non-protein amino acid toxin which is the causative agent of lathyrism, a crippling neurological disorder of the lower limbs. The nomenclature of the toxin is β-*N*-oxalyl-α,β diaminopropionic acid, β-*N*-oxalylamino-L-alanine , *l-3*-oxalylamino propionic acid also known as ODAP, BOAA and OAP, respectively. The lathyrism agent has been detected in another 22 of the 30 examined species (Bell, 1973). Besides *L. sativus,* several other *Lathyrus* species are occasionally cultivated for their seeds or for forage and fodder (Kearney and Smartt, 1995). Eight of these belong to section Lathyrus and five are members of four different sections. This large number of partially or locally cultivated species suggests that more species of the genus *Lathyrus* are amenable to cultivation and domestication. However, the primary reason for domesticating additional species may be not so much for improving environmental adaptability, as for establishing a toxin-free stock. Selection for low toxin content is a major breeding objective in *L. sativus.* Responses have been encouraging but the final solution may come from new domestication.

The genus *Lens*

The cultivated lentil, *Lens culinaris* Medikus, is an important crop in southwest Asia, the Indian subcontinent, East Africa, the Mediterranean region and to a lesser extent Europe and the Americas. The wild progenitor of lentil, *L. culinaris* subspecies *orientalis,* is native to southwest and central Asia in the range of latitude 32°-43° N, longitude 28°-72° E. It is usually confined to poor habitats with stony, shallow soil. It tolerates soils of different origin such as calcareous, basalt, metamorphic and igneous rocks, at altitudes up to 1800 m. Adaptation to this wide range of environments is shared by the cultivated lentil which exhibits adaptation to even more varied ecological conditions, including territories and climatic conditions in which the wild progenitor does not grow.

In the last few years, the genus *Lens* has been enriched by three more species. Besides the well-known *L. nigricans* and *L. ervoides, L. odemensis* (Ladizinsky, 1986) and *L. tomentosus* (Ladizinsky, 1997) have been described, and the validity of separating *L. lamottei (= L. tenorei)* (Czefranova, 1971) from *L. nigricans* has been confirmed (Van Oss et al. 1997). The distribution range of *L. odemensis* and *L. tomentosus* is in the Middle East and in habitats where subsp. *orientalis* grows, while *L. lamottei* is confined to the western Mediterranean region. Throughout the distribution range of the wild lentil species the domesticated lentil is a common crop and the potential for improving the crop's adaptability via new domestication is rather slim.

Exceptions may be some populations of *L. ervoides* from East Africa. This species grows in the Middle East and southern Europe, but has also been reported near Addis Ababa, Ethiopia, and on the slopes of Mt. Muhavura, in the southwest corner of Uganda, on the other side of the equator. Unlike other wild lentils, including *L. ervoides* in the Mediterranean region, which mature under long days, those of East Africa mature under a short day photoperiod. Lentil is a major crop in Ethiopia where day length is neutral. Lentil is not a crop in Uganda and definitely not in the Kisoro area where *L. ervoides* grows naturally. Domesticating the Ugandan *L. ervoides* as a mean of introducing the crop to Uganda is obviously inappropriate because it would fail to compete with beans, maize, potatoes, banana and taro all of which are well adapted and productive in that region. Alternatively, for developing of a crop in Uganda, the Ethiopian lentil germplasm which is pre-adapted to the ecological conditions of Uganda may be tried. *Lens ervoides* has an important disadvantage as a candidate for domestication in its small seeds. On the other hand, the domestication syndrome of the cultivated lentil can be transferred to *L. ervoides* through hybridization and embryo culture (Ladizinsky et al. 1985).

The genus *Pisum*

Pea, *Pisum sativum* L originated in the Middle East but today its main growing areas lie outside that region. Traditionally, the pea has been grown as a garden plant for its green pods and immature seeds. However, in modern agriculture it is grown for its dry seeds and for canning. Two wild peas, *P. humile* and *P. elatius,* are members of the pea primary gene pool. They are closely related to the cultivated pea, although chromosomal rearrangements may be apparent in specific accessions. Genuine populations of these wild peas grow in the Mediterranean vegetation zone of the Middle East. *Pisum humile* exhibits a preference for the more open, herbaceous niches in the region whereas *P. elatius* is characterized by a climbing habit in maquis vegetation, also occurring sporadically in southern Europe. If these wild peas contain the diversity in adaptation, which is lacking in the cultivated pea, it could be exploited by conventional breeding techniques.

The wild pea *P. fulvum,* on the other hand, is classified in the secondary gene pool of pea. It can only be crossed with the cultigen unilaterally as a pollen parent. In the other cross direction, the embryo dies shortly after fertilization. The rare seeds obtained in this latter cross have failed to germinate. Young hybrids between *P. sativum* and *P. fulvum* often have twisted and partially chlorotic leaves, as if affected by a virus. These symptoms gradually disappear towards the upper leaves. The hybrids are heterozygous for a reciprocal translocation and are partially sterile.

Pisum fulvum is native to the Middle East and its distribution partially overlaps that of *P. elatius* and *P. humile,* occasionally growing by their side. Of the three wild peas, *P. fulvum* is apparently the most drought and heat tolerant. This is indicated by its distribution in southern Israel, where annual rainfall is no more than 300 mm. Reports of *P. humile* from that region refer to feral forms of *P. sativum* which are always confined to roadsides and the edges of cultivation. To what extent heat and drought tolerance is unique to *P. fulvum* is not clear, but if it is unique, domesticating this wild pea might provide new germplasm with some other interesting characters, such as pod shape and underground pods. As already indicated, the syndrome of domesticated characters can be easily transferred to *P fulvum* from *P. sativum.*

CONCLUSIONS

The food legumes have always been crops of Mediterranean or Mediterranean-like climates and are not adapted to humid tropical conditions. Wider adaptation to stressful environments within their natural growing area, as well as pest and disease resistance, may be sought within the existing germplasm of the individual crops, and in their close wild relatives. Domesticating wild species for the sake of improving adaptation is a last resort and may have a local effect, particularly in places where none of the traditional food legumes can produce economic yields. The most attractive candidates for domestication are several perennial wild chickpea species, but further study is needed to determine their prospects as food legumes before embarking on such a project. Wild *Lathyrus* species may be a source of a crop, free of the lathyrism toxin.

Domesticating wild species as food legumes can be achieved in a much shorter time than that required for the traditional crops. Domestication of species which are in the tertiary gene pool of any of the food legumes can be achieved by selecting mutants in the wild populations for some key characters which are required for a domesticated seed crop. Alternatively, these traits may be obtained via mutagenesis. Wild members of the secondary gene pool may be domesticated by gene transfer from their respective domesticated relatives.

References
Bell, E.A. 1973. In: *Toxicants Occurring Naturally in Foods.* pp. 153-169, National Academy Science Wash. D.C.

Buddenhagen, I.W.and Richards, R.A. 1988. In: *World Crops: Cool Season Food Legumes,* pp.81-95 (ed. R.J. Summerfield), Kluwer Academic Publishers.

Czefranova, Z. 1971. Review of species in the genus *Lens* Mill. (in Russian).Nov. *Siestemat Vysshikh Rastenii.* 8: 184-191.

Donnelly, E.D., Watson, J.E. and McGuire, A. 1972.. *Journal of Heredity* 63: 361-365.

Forbes, I. and Walls, H.D. 1968. *Crop Science.* 8:195-197.

Gladstones, J. S. 1967. *Australian Journal of Experimantal Agriculture and Animal Husbandry* 24: 360-366.

Kazan, K., Muehlbauer F.J., Weeden, N.F. and Ladizinsky, G. 1993. *Theoretical and Applied Genetics* 86: 417-26.

Kearney, J. and Smartt, J. 1995. In:*Evolution of Crop Plants.* pp. 266-270 (eds J. Smartt and N.W. Simmonds). Longman Sci. Tech.

Ladizinsky, G. 1985. *Euphytica* 34: 539-543.

Ladizinsky, G. 1979. *Journal of Heredity* 70: 135-137.

Ladizinsky, G. 1986. A *new Lens species from the Middle East.* Noytes Roy.Bot. Gard.

Ladizinsky, G. 1995. *Theoretical and Applied Genetics* 91: 639-646.

Ladizinsky, G. 1997. A new species of *Lens* from south east Turkey. *Botanical Journal of the Linnean Society.*

Ladizinsky, G., Cohen, D. and Muehlbauer, F.J. 1985. *Theoretical and Applied Genetics* 70: 97-101.

Rawal, K.M. 1975. *Euphytica* 24:699-707.

Sengbusch, R. and Zimmermann, K. 1937. *Zücter* 9: 57-65.

Tsuchiya, T. 1987. *Japanese Agricultural Research Quarterly* 21: 166-175.

Van der Maesen, L.J.G. 1987. In: *The Chickpea,* pp. 11-34 (ed M.C. Saxena and K.B. Singh) CAB International.

Van Oss, H., Aron, Y., and Ladizinsky, G. 1997. *Theoretical and Applied Genetics* 94: 452-457.

Verma, V.D. and Ram, H.H. 1987. *Journal of Agricultural Science* 108: 305-310.

Waines, J.G. 1975. *Bulletin Torrey Botanic Club* 102: 385-395.

Using models to assess the value of traits of food legumes from a cropping systems perspective

M.J. Robertson[1], P. S. Carberry[1], G. C. Wright[2] and D. P. Singh[3]

[1]Agricultural Production Systems Research Unit, CSIRO Tropical Agriculture, Toowoomba, Queensland, Australia; 2 Queensland Department of Primary Industries, Kingaroy, Queensland, Australia; 3 Department of Genetics and Plant Breeding, Govind Ballabh Pant University of Agriculture and Technology, Pantnagar, Uttar Pradesh, India

Abstract

Plant breeders spend considerable time, effort and expense in conceptualizing and selecting for improved plant types. Given the climatic variability and diverse management practices in different regionsto which crops are exposed, different plant types may be needed for different agro-ecological zones. By capturing physiological understanding in a predictive framework, crop modelling offers the potential to interpret and predict the performance of individual genotypes in different environments, thus offering a possible decision support role in plant breeding. We propose that, when considering the adaptation of food legumes to the environments in which they are growing, the impact on the whole system should evaluated, including the N economy of the cropping system and impacts on the productivity of associated crops. We present some case studies of the use of static analytical models and simulation to evaluate traits in food legumes, from a cropping systems perspective. In one case study we show that the yield difference between determinate and indeterminate soybean types in this semi-arid environment will depend upon interactions between seasonal conditions, flowering date and starting soil water, which models can assist in analysing. Conducting analyses over the climatic record also allows any yield advantage to be assessed against the impact on the riskiness of production. In another study, we highlight the interaction between agronomic management (such as row spacing) and the performance of cowpea genotypes differing in height and leaf posture when grown as intercrops with maize. In a third case study we address the question of the impact of chickpea genotypes differing in potential N fixation on system performance of a chickpea-wheat rotation under dryland conditions. The results show the trade-off between the gains or losses in chickpea and wheat yields by introducing chickpea with different traits into the rotation.

In summary, the case studies are intended to demonstrate that breeding objectives often need to incorporate the impact of altered plant traits beyond the yield of the targeted crop and the interaction of management and climate on the differential performance of genotypes. This is particularly the case for food legume crops which are typically grown as important components of crop rotations.

INTRODUCTION

Plant breeders spend considerable time, effort and expense in conceptualising and selecting for improved plant types. Given the climatic variability and diverse management practices in different regionsto which crops are exposed, different plant types may be needed for different agro-ecological zones. By capturing physiological understanding in a predictive framework, crop modelling offers the potential to interpret and predict the performance of individual genotypes in different environments, thus offering a possible decision support role in plant breeding.

Several reviews have considered the potential role that physiology and simulation modelling could have in assisting plant breeding (e.g. Shorter et al. 1991; Lawn and Imrie, 1991; Hammer et al. 1996). Models can be used in both a static and a dynamic simulation sense. In the static approach, the performance of a genotype can be analysed in terms of functional components of seed yield. For example, Wright et al. (1996) used a simple physiological model to improve understanding of the basis of G x E interactions in groundnut under drought conditions, where grain yield was analysed in terms of water transpired, transpiration efficiency and harvest index. Hamdi et al. (1992) used a regression approach to analyse the yield response of lentil to water supply.

R. Knight (ed.), Linking Research and Marketing Opportunities for Pulses in the 21st Century, 265–278.

Simulation models can be also be used to analyse plant performance and to analyse responses of probe genotypes for improved interpretation of multi-environment trials. When combined with optimisation techniques, models have been used to optimise combinations of genotype and management over the target environment domain. For example, Hammer *et al*. (1996) used this approach to determine optimal maturity and plant density for sunflower grown in a variable rainfall environment.

By far the most common application for crop simulation models is in assessing the value of specifictraits for selection of improved plant genotypes. When coupled with long-term climatic data, crop simulation offers the opportunity to analyse the value of specific crop traits at different locations inclimatically-variable environments. A number of criticisms of using models to evaluate crop traits have been raised. These relate to the fact that there is insufficient understanding of the mode of action of the traits; no accountis made of genetic or phenotypic correlations among traits (Jackson *et al*. 1996); promising traits that are identified are often difficult to screen in large segregating populations (Hunt, 1993); and they provide a substantial increase in the number of selection criteria for which breeders have to potentially select. It is likely that these limitations have led to studies that focus upon traits that are easily observable; are often the target of selection in breeding programs; and are able to be parameterised in, and predicted by, crop models with a high degree of confidence. Phenology is the most important aspect of crop adaptation and yield determination, and can be accurately simulated by crop models. The range in genetic variation for phenology within a breeding population is often better characterised than for other traits. The consequences of different phenology on grain yield have been assessed using a crop simulation approach for a number of crops including peanut (Bailey and Boisvert, 1989), sorghum (Jordan *et al*. 1983, Hammer and Vanderlip, 1989, Muchow and Carberry, 1993), rice (O'Toole and Jones, 1987), and sunflower (Hammer *et al*., 1996).

Only a few crop simulation studies have examined the impact on grain yield of traits other than phenology. For example, Hammer and Vanderlip (1989) simulated the impact of differences in radiation-use efficiency of old and new sorghum cultivars; Jordan *et al*. (1983) and Jones and Zur (1984) examined the value of osmotic adjustment and deep rooting traits. Muchow and Carberry (1993) assessed the impact of lower and higher radiation-use efficiency and transpiration efficiency on growth and yield of sorghum, maize and kenaf grown in a semi-arid environment; and Palanisamy *et al* (1993) evaluated the impact of genotypic differences in dry matter partitioning on yield of rice.

Food legumes are often grown as part of a multi-crop system (rotations, intercrops, relay crops), often with a non-legume crop forming the dominant part of the system. Legumes are grown not only because they provide a high protein food source, and offer a disease break in crop rotations, but because they can contribute to the nitrogen (N) economy of the cropping system through N fixation. We propose that when considering the adaptation of food legumes to differing environments, an assessment of their impact on the N economy of the cropping system and productivity of associated crops must be included. Attempts to use models to evaluate crop traits have, until now, concentrated upon the impact on the crop under consideration. This paper represents a new perspective in this area by considering the impact upon the crop within its system. Hence, in this paper we present some case studies on the use of static and simulation models to evaluate putative traits in food legumes, and consider not only the impact on yield of the food legume, but also the yield of the associated species in the cropping system.

STATIC ANALYTICAL MODELS

Description of the crop models

A number of crop analytical models have been proposed to 'dissect' yield into a small number of independent physiological components, which effectively integrate numerous complex processes into fewer 'biologically meaningful' parameters. In a practical breeding program a great deal can be achieved by working with these integrated parameters rather than with yield alone (Williams, 1992). Such simple modelling/analytical approaches can greatly assist in evaluating genotypic adaptation in stress environments, and hence improve the efficiency of genetic enhancement in food legume breeding programs.

A useful conceptual framework for analysing yield (Y) variation in food legumes is provided by the relationship:-

$$Y = TDM \times HI \qquad (1)$$

where TDM is total above-ground dry matter and HI is harvest index. Silim *et al*. (1993) examined variation in Y of various lentil lines under drought in terms of TDM and HI. For a modest investment in collection of TDM at maturity, in addition to seed yield, the breeder can immediately determine whether yield of specific genotypes is being limited by factors associated with (a) a low HI rather than TDM, or (b) low TDM rather than effects on HI. In empirical breeding programs, one can then ask 'How often are genotypes with high levels of either the TDM or HI characteristic (but not together) rejected in early generations as a result of selection on seed yield alone?' Such basic knowledge can immediately assist the breeder in identifying parents with high levels of both characteristics for future crossing purposes. However, in assessing the importance of TDM in food legumes it is important to recognise that leaf abscission before the harvest at maturity reduces the reliability of TDM values.

The components of Equation (1) can be further partitioned into functional components that describe more detailed physiological processes responsible for variation in TDM and HI. A number of these analytical frameworks have been proposed and include the crop growth model developed by Duncan et al. (1978) in which seed yield (Y) is considered to be a function of the crop growth rate (C), the duration of reproductive growth (Dr), and the proportion (p) of C partitioned to yield:

$$Y = C \times Dr \times p \qquad (2)$$

Monteith (1977) expressed Y as a product of the amount of radiation intercepted by the crop (RI), the efficiency of conversion of radiation into dry matter production (RUE), and the partitioning of dry matter into the reproductive component (HI), such that:

$$Y = RI \times RUE \times HI \qquad (3)$$

In water-limited environments, Passioura (1977) considered Y to be determined by the relationship:-

$$Y = T \times WUE \times HI \qquad (4)$$

where T is the water transpired by the crop and WUE is the efficiency of use of water in producing TDM.

Each sub-component of these relationships represents an integrated function of a number of developmental, morphological, physiological and biochemical attributes(Hardwick, 1988). For instance, in Equation (2), C is an integrated measure of the 'source' capacity of a crop, and can be further evaluated through the effects of RI and RUE in Equation (3). Any characteristic thought to be beneficial for adaptation to specific stress conditions can then be evaluated in terms of its functional relationship and strength of correlation to one of the yield components.

The key to utilising these analytical relationships in crop improvement is critically dependent on an ability to obtain reliable measurements on each attribute for the large numbers of genotypes within large scale breeding programs (Cooper et al., 1993). Until recently, relatively sophisticated and resource-intensive techniques have been required to estimate these model parameters, allowing only a few genotypes to be analysed within such frameworks (e.g. Mathews et al., 1988; Wright et al., 1991). These difficulties have recently been addressed following some novel and pragmatic approaches that allow quantification of parameters on large numbers of food legume genotypes (Williams and Saxena, 1991; Greenberg et al., 1992; Ntare, 1992; Williams, 1992, Redden and Wright, 1993; Wright et al, 1996; Turner et al., 1997). The approach relies on estimation of parameters using biomass sampling, phenological observations and even 'reverse engineering' of analytical models to derive functional component parameters, which can be obtained simply for very little extra investment in data collection.

Case study

In a recent evaluation of a range of bean genotypes (both common bean, *Phaseolus vulgaris* and lima bean, *P. lanatus*) using the water model approach outlined in equation 4 to analyse yield variation among contrasting genotypes (Wright and Redden, 1995; Wright and Redden, unpublished data), it was found that high yield under both stressed and well-watered conditions was achieved with a combination of drought component characteristics (Table 1). The highest yielding line was Bridgeton, a lima bean, which was shown to have high WUE, T (amount of water transpired) and HI compared to the other common bean genotypes. A strong negative association between WUE and T was observed in the common bean genotypes (Figure 1), in agreement with the study by White et al. (1990). Interestingly though, the lima bean genotype was observed to be an outlier in this relationship (Figure 1, circled), thus indicating that high levels of T and WUE are possible within a single *Phaseolus* genotype. Clearly, this analysis can allow the bean breeder to :-

be aware that in *P. vulgaris*, WUE and T are negatively associated, and concurrent selection for both characters will be necessary

identify genotypes with high levels of specific characters directly associated with yield, which could be used as potential parents in future crossing programs

use simply-measured physiological characters as screens for T, WUE and HI in early generations

Table 1. TDM (kg/ha), pod yield (kg/ha) and the physiological components of yield (T (mm), WUE (g/kg) and HI) estimated from the "water model" (equation 4X) for 10 *Phaseolus* genotypes grown in a field experiment at Kingaroy, Qld, Australia. IRR = irrigated environment, RF = rainfed environment.

Genotype	TDM		Pod Yield		Harvest Index		Pred WUE		Pred T	
	IRR	RF	IRR	RF	IRR	RF	IRR	RF	IRR	RF
C40	3930	2585	1321	471	33.6	18.2	3.41	3.09	115	84
Bridgeton	6516	4905	1992	1133	30.6	23.1	3.83	3.72	170	132
Mex 685B	4201	2895	1334	622	31.8	21.5	3.43	3.39	122	85
ICA21477B	5492	3419	1885	636	34.3	18.6	3.23	3.20	170	107
Narino11	4917	3086	1968	670	40.0	21.7	3.03	3.35	162	92
Sirius	5600	3040	1822	535	32.5	17.6	3.30	3.48	170	87
Rainbird	4655	3778	1509	599	32.4	15.9	3.06	3.07	152	123
Turrialba	5988	3122	1662	549	27.8	17.6	3.08	2.91	195	107
Acc 1280	5508	4287	1834	847	33.3	19.8	2.87	2.90	192	148
Acc 54	4659	3735	1293	611	27.8	16.4	3.25	3.42	143	109

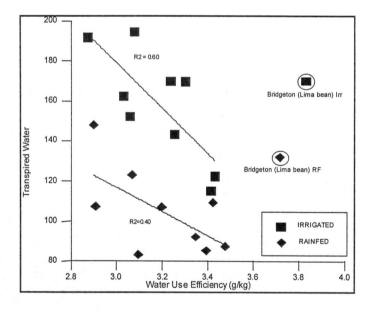

Figure 1. Relationship between transpired water (mm) and water-use efficiency (g/kg), estimated from the "water model" (equation 4) for 10 *Phaseolus* genotypes in the field.

DYNAMIC SIMULATION MODELS

Description of the crop models

Previously, efforts on modelling crop production have centred on the use of individual crop models that respond to climate, water, nitrogen and management. However, they do not deal with other important features of the cropping system such as crop sequences, intercropping, crop residue management, and feedback between the crop and soil fertility status. The software system, APSIM (Agricultural Production Systems sIMulator), (McCown *et al.*, 1996) represents a new mode of simulating cropping systems. It allows models of crop and pasture production, residue decomposition, soil water and nutrient flow, and erosion to be readily configured to simulate various production systems, including crop sequences and intercropping, and soil and crop management, to be dynamically simulated using conditional rules. A key concept of APSIM is that the soil provides a central focus; crops, seasons and managers come and go, finding the soil in one state and leaving it in another. In this way APSIM is ideally suited to examine crop production issues where feedback between the crop and soil is of primary concern. APSIM has also been designed to deal with intercropping systems, and has been validated with maize / cowpea (Carberry *et al.*, 1996a) and maize / stylosanthes systems (Carberry *et al.* 1996b).

Crop growth simulation modules in APSIM have been developed and validated for a number of crops, including the food legumes soybean, cowpea and chickpea (Carberry, 1996). Cereal and legume models were based, respectively, on the framework of CERES-Maize (Jones and Kiniry 1986) and the soybean model of Sinclair (1986). The models have been calibrated to the genotypes and environments of northern Australia (Birch *et al.* 1990) and semi-arid east Africa (Keating *et al.* 1991). The models have evolved to incorporate several innovative routines, which are described in detail elsewhere (Carberry and Abrecht 1991). In the legume models, alternative routines were developed to simulate leaf area development in branching crops, nitrogen fixation and light interception as influenced by row spacing. The chickpea model has been used to analyse the productivity of wheat-chickpea rotations (Probert *et al.* 1998).

All crops models use a similar framework. They simulate crop development, growth, yield and nitrogen accumulation in response to temperature, photoperiod, soil water and nitrogen supply. Daily maximum and minimum temperature, solar radiation and rainfall are climatic inputs. Crop phenology is divided into phases, the duration of each based on daily temperature and photoperiod. The duration from sowing to flowering is simulated as four phases: sowing to emergence; emergence to the end of the basic vegetative(or juvenile) period, which is a photoperiod-insensitive phase; a photoperiod-induced stage which depends upon the cultivar's photoperiod sensitivity and which ends at floral initiation, and a floral development period which ends at flowering. Leaf area development is described using functions for the appearance, expansion and senescence of leaves. Leaf senescence is a function of age, light competition, drought and frost and has resultant impacts on leaf area, dry matter and plant nitrogen. Potential above-ground biomass production is predicted from leaf area index, a radiation extinction coefficient and the crop's radiation use efficiency. Actual daily biomass increase is calculated from the minimum of two potential crop growth rates, one determined by the intercepted radiation, limited by temperature and nitrogen stresses, and the other by soil water supply. This group of functions also grows grain and partitions and retranslocates carbon between leaf, stem and grain. The demand, uptake and retranslocation of nitrogen is also simulated, as is whole plant death due to stress. The crop has a defined minimum, critical and maximum N concentration for each plant part. Demand for nitrogen in each part attempts to maintain nitrogen at the critical (non stressed) level. Nitrogen demand on any day is the sum of the demands from the pre-existing biomass of each part required to reach critical N content, plus the N required to maintain critical N concentrations in that day's potentially assimilated biomass. If nitrogen demand cannot be satisfied by mass flow then it is supplied by diffusion. If both mass flow and diffusion supplies can't satisfy demand then nitrogen is sought from N fixation. The potential daily rate of nitrogen fixation is a function of crop biomass (i.e. the size of the crop), discounted for soil water stress.

In the legume models, the demand for assimilate, for grain yield accumulation is defined by the linear increase with time of the harvest index. The ability to meet the grain yield demand is determined by the rate of biomass accumulation and the retranslocation of dry matter that had accumulated in the leaves and stems before the start of pod-filling. If assimilate supply is in excess of grain growth requirements, the excess is used for new leaf and stem growth. The rate of increase in linear harvest index with time is a genotypic parameter. Indeterminate genotypes will tend to have a low rate of harvest index increase over a longer pod-filling duration, as seen in pigeonpea (R. Ranganathan, ICRISAT Centre, Hyderabad, India, personal communication). The consequence of the lower harvest index increase is that indeterminate types will accumulate pod yield less quickly and leaf and stem dry matter more quickly during pod-filling than determinate types (Bushby and Lawn, 1992). The incorporation of routines to simulate growth and partitioning of biomass in determinate and indeterminate genotypes is a novel feature of the food legume modules.

SIMULATION ANALYSES

APSIM was run in three case studies to highlight different features of the adaptation of legumes to cropping systems, in response to variation in crop traits.

Case study 1: Indeterminacy and phenology in soybean

The availability of water is a serious constraint to crop productivity in the semi-arid tropics. In designing plant types for water-limited environments, approaches may involve better matching of crop phenology to expected water supply to improve crop yield potential to exploit the good seasons, or the incorporation of traits that confer drought resistance to minimise risk in the poor seasons. Dryland soybean in the semi-arid tropics is exposed to high variability in the amount and temporal distribution of rainfall, both among and within seasons, which makes the evaluation of alternative plant types difficult.

This case study addresses the question of whether there is an advantage for developmental plasticity (i.e. indeterminate growth) in dryland soybean in the semi-arid tropics, as suggested by Ludlow and Muchow (1990). Turk and Hall (1980) observed differences in harvest dates as large as 21 days for cowpeas that were sown at the same time but were grown under limited or abundant water supply. Indeterminate growth will confer advantages in those years when water deficits are relieved late in the season, and the crop can take advantage of improved water supply. Villalobos-Rodriguez and Shibles (1985) reported that the relative effects of water deficit during reproductive growth were greater in determinate than in indeterminate tropical soybeans. Saxena *et al* (1993) suggested selecting for indeterminacy in chickpea in early flowering backgrounds, thus permitting plants to flower and set pods early but also continue growing, flowering and podding if the season extends. However, indeterminate growth is often associated with a low rate of partitioning to pods during pod-filling, which may penalise yield in seasons where terminal water deficits are experienced. Lawn (1982) and Sinclair *et al*. (1987) found that the developmental plasticity of cowpea and mungbean contributed to their superior performance compared to soybean in water-limited environments. Phenology can therefore interact with growth habit to determine crop duration, and hence influence the probability of escaping drought through early maturity.

Flowering date was modified in the soybean model by changing the duration of the basic vegetative phase. In the absence of any quantitative information on rate of pod-filling in indeterminate types, an arbitrary reduction was used. As such this represents a "best guess" and needs to be confirmed with experimental studies. Genotypes with indeterminate growth during pod-filling were simulated by reducing partitioning to reproductive growth, achieved by decreasing the rate of linear increase in harvest index from 0.011 to $0.007d^{-1}$ and increasing the duration of pod-filling from 660 to 800 ^{o}Cd. The duration of pod fill was lengthened in the indeterminate types to ensure that both types would attain the same final harvest index as the determinate types and hence have similar yield potential.

The soybean model was run under rainfed conditions for 38 seasons (1951 to 1988) using climatic data from Emerald (latitude 23.4 ^{o} S), in northern Australia. Simulations treated each year independently; with all parameters being re-initialised at sowing. At sowing, soil water was initialised either to 50 or 100% of plant available water and soil mineral nitrogen to 20 kg ha^{-1}. Sowings were conducted on the first day of the months of September, October, November, December, January, February and March. Soybean cv. Davis was used in model validation and is taken as the standard for these analyses. The impact of changed phenology on crop production was assessed by altering crop parameters as described above.

Figure 2 shows the pattern of stress development and grain yield accumulation for determinate and indeterminate types sown in contrasting seasons. In 1955, favourable water supply late in the season allowed continued grain yield accumulation, which resulted in the indeterminate type out-yielding the determinate type. In contrast, severe terminal water deficit in 1957 curtailed any growth in the indeterminate type which resulted in little yield advantage due to indeterminacy.

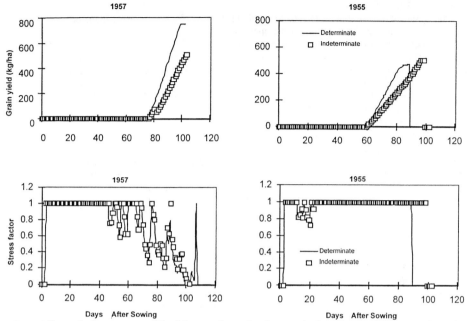

Figure 2. Example time-courses of the 0-1 stress factor for photosynthesis and grain yield versus days after sowing for determinate and indeterminate genotypes of soybean in the contrasting seasons of 1955 and 1957. For the stress factor, 0 is maximum and 1 is no water stress.

For a given flowering date, indeterminate types, on average, had a higher yield when the season started with a full profile as a consequence of a potentially longer grain-filling period (Figure 3). There was a diminishing returns response to delayed flowering date. On the other hand, when the season started with a 50% full profile, determinate types were superior, particularly for early flowering dates, because they escaped drought. There was no response in yield for flowering dates later than 55 days.

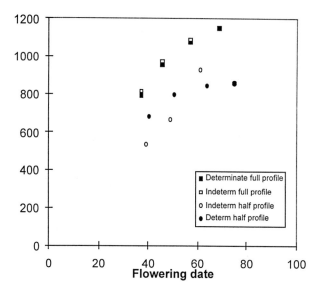

Figure 3. Simulated response of soybean grain yield to flowering date for indeterminate and determinate genotypes, with either half or full profile of starting soil water. Each point represents one genotype and is the mean across sowing dates and years.

Figure 4 shows the trade-off between yield and its variability. There was a linear response for both indeterminate and determinate types when starting on a full profile, indicating that higher yield comes with a cost of more season-to-season variability. When starting with a 50% full profile, therewas a wide range of variability for the same grain yield, indicating that it is possible to choose a genotype which minimises risk for a given yield level.

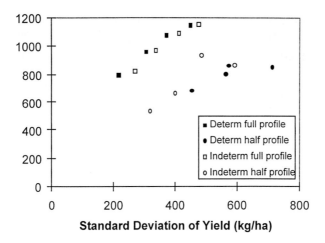

Figure 4.: Relationship between mean grain yield and standard deviation of grain yield. Each point represents one genotype and is the mean across sowing dates and years.

In conclusion, this case study shows that the yield difference between determinate and indeterminate types in this semi-arid environment will depend upon interactions between seasonal conditions, flowering date and starting soil water, which models can assist in analysing. Conducting analyses over the climatic record also allows a yield increase due to trait modification to be assessed against the impact of the trait on the riskiness of production.

Case study 2: Cowpea canopy morphology in maize / cowpea intercropping

Intercropping is widely practised by farmers in low-input systems in the tropics. Its advantages over sole cropping have been attributed to improved biological efficiency of resource use by intercrops. The complexity of intercropping systems is often highlighted as a major reason for the difficulty in attributing cause and effect to experimental results (Fukai, 1993). Simulation modelling can make a significant contribution to the quantitative evaluation of the processes involved in an intercrop system (Keating and Carberry, 1993). Of particular interest to the plant breeder is the design of plant types that maximise productivity and economic returns from the competing mixture. Competition for radiation in mixtures will be influenced by genotype characteristics such as morphology (height and canopy structure), leaf area development and phenology, relative to its competitor (Davis and Woolley, 1993). These factors will also interact with agronomic considerations such as row configuration, plant population density and sowing time (Midmore, 1993).

This case study addresses the question of the impact of different morphologies of cowpea, when grown in an intercrop with maize under low-input production systems. This issue has been addressed experimentally by Davis and Garcia (1987) who showed that climbing beans versus bush beans varied significantly in their ability to compete with maize in an intercrop situation. Varieties differed in their distribution of dry matter among branches and the main stem, and this was related to their yield performance when intercropped with maize. In this situation, the relative yields of the component crops would be expected to vary with factors such as row configuration, plant population density and sowing time. Davis and Woolley (1993) reviewed some of the techniques available to improve the efficiency of selection and testing of intercropped species. However, these authors did not address the potential for simulation modelling to evaluate putative plant types, particularly in different cropping systems. In this case study we use a crop modelling approach to address the issue of the trade-off between the loss in maize yield in the intercrop relative to the sole crop, against the gain in cowpea yield for different cowpea genotypes.

In APSIM, the arbitrator module enables two or more crops to be intercropped by simulating the competition for light by the respective canopies, and competition for water and nitrogen uptake. The component crops are sown as if sole crops. Competition for above-ground resources (i.e. light) is simulated by calculating the light intercepted by each crop component using the crop height, LAI profile, and extinction coefficient provided to it by each crop. Competition below ground is simulated by allowing the order of extraction of water and N between crops on alternate days (Carberry *et al.*, 1996a).

For this exercise, morphology in cowpea was modified in the cowpea model by changing the maximum attainable height of the canopy (80 vs. 160 cm), thus mimicking bush versus climbing types, and the radiation extinction coefficient (0.47 vs. 0.60) to mimic types with erect versus prostrate leaf posture. Results were evaluated in terms of the impact on both maize and cowpea grain yields. Carberry *et al.* (1996a) outlined a method for analysing the economic dimension of this question. They calculated the value of cowpea grain required to offset the loss in maize grain, where the nominal price for cowpea grain was expressed as maize equivalent units.

The maize and cowpea models were run under rainfed conditions for 31 years (1957 to 1988) using climatic data at Katumani, (latitude 1.5° S), in semi-arid eastern Kenya. Simulations treated each year independently; all parameters were re-initialised at sowing. On this date soil water was initialised at the lower limit of plant available water and soil mineral nitrogen was initialised to 30 kg ha^{-1}. APSIM was configured to sow Katumani composite B maize during the short and long rainy seasons, with no fertiliser application, which was consistent with local practice. A sowing density of 4 maize and 5 cowpea plants m^{-2} was used, and all crop residues were removed at harvest.

Figure 5 shows the mean maize yield, expressed as the reduction from the sole crop situation plotted against the cowpea grain yield attained with different morpho-types in narrow and wide maize rows. In narrow (50cm) maize rows the cowpea yield responded best to increases in maximum height, with little response to an increase in the extinction coefficient of the cowpea. However, there was an interaction, where the biggest response to increasing height was at the larger extinction coefficient. In contrast, in the wide maize rows, the biggest increase in cowpea yield was obtained when the extinction coefficient was increased, with a much smaller response to increasing height.

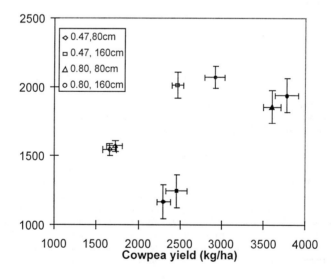

Figure 5. Relationship between mean decrease in maize yield versus cowpea yield for 4 cowpea genotypes differing in extinction coefficient (0.47, 0.80) and maximum height (80, 160cm), grown in 50cm maize rows spacing (hollow symbols) and 100cm row spacing (filled symbols). Error bars are the standard error of the mean.

This case study highlights the complex interactions between agronomic management (such as row spacing) and the performance of different genotypes in competing mixtures. Models can help analyse some of these interactions by quantifying the likely trade-offs between increasing the yield of one species at the expense of the other.

Case study 3: Chickpea biomass in chickpea / wheat rotation

Legume crops are often considered as rotational or break crops to the preferred cereals in the production systems of Australia, producing comparable economic returns as well as contributing to the nitrogen economy of the system. There has been a range of experimental studies that have tried to estimate the benefits of food legumes in cropping rotations in northern Australia (Doughton *et al.* 1993, Dalal *et al.*, 1994). APSIM provides the capability to undertake similar experimentation through simulation (Probert*et al.* 1998).

In water-limited environments, grain yield and the levels of biomass and nitrogen produced by a legume crop will be closely linked to the amount of water available to the crop. Higher legume productivity will require higher water use, one consequence of which will often be less residual water available for, and in turn lower yield prospects for, a following cereal crop. Breeding for crop traits that improve the productivity of a legume crop will thus impact on the performance of the whole system, possibly negatively, by reducing the amount of stored soil water available to the following cereal crop, or positively in increasing the amount of available nitrogen through transfer of more root and surface residues. Hence, there will be trade-offs between legume traits and the relative productivity of the food legume and cereal crop.

This case study uses a simulation modelling approach to address the question of the impact of different chickpea genotypes on system performance of a chickpea-wheat rotation under dryland conditions. Of interest in this analysis is the trade-off between the gains or losses in chickpea and wheat yields by introducing chickpea with different traits into the rotation. Relative to a standard genotype, two hypothetical chickpea genotypes are considered, both of which contain traits aimed at increasing chickpea residual and root biomass in order to improve their nitrogen contribution to the system. The first trait reduced crop harvest index (HI), by delaying chickpea maturity date and producing a crop of similar yield potential to the standard genotype but with higher biomass production - achieved via increasing growth duration by 20% commensurate with a similar reduction in grain growth rate. The second trait increased the transpiration efficiency (TE) of chickpea by 20% (from 0.005 to 0.006 Pa), resulting in more biomass produced per amount of transpired water.

The simulation scenario is similar to that reported by Probert *et al.* (1998). The APSIM-Chickpea and APSIM-NWheat modules were run under rainfed conditions for 25 years (1971 to 1995) using climatic data for Brookstead (latitude 27.6° S), in the semi-arid subtropics of north-eastern Australia. At this site, thesoil could potentially store 260mm available soil water and could mineralise approximately 55kg N ha^{-1} yr^{-1} in a continuous wheat cropping system. Simulations were run continuously for the 25 years in order to quantify the effects of crop sequence on system performance. A wheat-chickpea rotation (including both phases) was compared with continuous wheat with five rates of applied fertilizer nitrogen (0, 40, 80, 120, 160 kgN ha^{-1}). Both wheat and chickpea were sown zero-till once 25mm rainfall was received over 5 days between 1st May and 31st July each year. Wheat cultivar Hartog was sown at 110 plants m^{-2} and chickpea cultivar Amethyst (used in model validations and taken as the standard for these analyses) was sown at 35 plants m^{-2}.

The simulated response of wheat to applied nitrogen fertilizer was as expected with average yield increases diminishing as rates increased from 0 to 160kg N ha^{-1} (Figure 6). On this low fertility soil, the nitrogen contribution of the standard chickpea genotype to following wheat crops was about 100kg N ha^{-1}. Changing the harvest index of the chickpea genotype produced more chickpea biomass and N fixation with little change in grain yield, but also resulted in a reduction in the average yield of wheat grown in rotation. This later maturing chickpea produced more biomass and N but also used more soil water which had a negative impact on the subsequent wheat crops. In contrast, the chickpea genotype that produced more biomass via improved water use efficiency (increased kg biomass per mm transpired water) resulted in increases in both average chickpea and wheat yields. The improvement in simulated wheat yields was due to the increased nitrogen contribution from the bigger biomass chickpea crops.

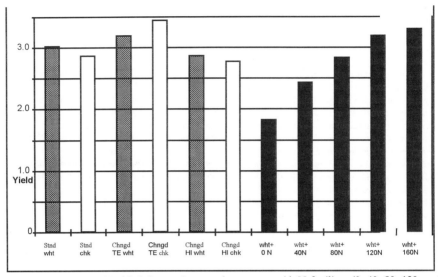

Figure 6: Average grain yields (t/ha) for continuous wheat grown with N fertilizer (0, 40, 80, 120, or 160 kg/ha) and wheat and chickpea yields from rotations using chickpea genotypes that were standard or with changed TE or HI. In the graph wheat, chickpea, standard and changed are abbreviated to wht, chk, Stnd and Chngd respectively.

This case study was intended to demonstrate that breeding objectives often need to incorporate the impact of altered genotypic traits beyond the yield of the targeted crop. This is particularly the case for food legume crops which are typically grown as important components of crop rotations.

CONCLUDING REMARKS

This paper has illustrated, with examples, the use of static and dynamic simulation models to assist plant breeding. The unique advantage that modelling can offer is its ability to interpret and predict the performance of individual genotypes in different environments.

The static, or analytical, model framework can allow the breeder to effectively measure and exploit the components that relate directly to seed yield. The challenge remains to effectively incorporate these yield component traits as selection criteria in large scale breeding programs, by developing simple screening methodologies and better understanding the genetics of the component traits. Static models also allow an assessment of any negative associations between yield components, which need to be identified and understood before relevant breeding strategies are developed.

The advent of dynamic simulation models, such as APSIM, which can account for cropping systems issues, can make the use of crop models more applicable to food legumes. They allow an assessment of the N economy of the cropping system and hence a further evaluation of the impact on productivity of associated crops. Also, they can allow exploration of the ineractions between climate, management and genotypic performance.

References

Bailey, E. and Boisvert, R. N. 1989. *Australian Journal Agricultural Economics* 33: 153-169.

Birch, C. J.; Carberry, P. S.; Muchow, R. C.; McCown, R. L. and Hargreaves, J. N. G. 1990. *Field Crops Research* 24: 87-104.

Bushby, H. V. A. and Lawn, R. J. 1992. *Australian Journal Agricultural Research* 43: 1609-1628.

Carberry, P. S. 1996. Final Report to the Grains Research and Development Corporation, Project CSC9, 33pp.

Carberry, P. S. and Abrecht, D. G. 1991. In: *Climatic risk in crop production: Models and management in the semi-arid tropics and sub-tropics*, pp. 157-182. (eds RC Muchow and JA Bellamy) CAB International, Wallingford.

Carberry, P. S.; Adiku, S. G. K.; McCown, R. L. and Keating, B. A. 1996a. In: *Dynamics of Roots and Nitrogen in Cropping Systems of the Semi-Arid Tropics*, pp. 637-648. (eds O Ito, C Johansen, J J Adu-Gyamfi, K Katayama, J V D K Kumar Rao, and TJ Rego) Japan International Research Centre for Agricultural Sciences.

Carberry, P. S.; McCown, R. L.; Muchow, R. C.; Dimes, J. P.; Probert, M. E.; Poulton, P. L. and Dalgliesh, N. P. 1996b. *Australian Journal Experimental Agriculture* 36: 1037-48.

Cooper, M.; DeLacy, I. H. and Eisemann, R. L. 1993. In: *Focused Plant Improvement - Towards Responsible and Sustainable Agriculture.* Vol. 2, pp. 116-131, (eds B. C. Imrie and J B Hacker) Proceedings of the Tenth Australian Plant Breeding Conference, Australian Plant Breeding Society, Canberra.

Dalal, R. C.; Strong, W. M.; Weston, E. J.; Cahill, M. J.; Cooper, J. E.; Lehane, K. J.; King, A. J. and Gaffney, J. 1994. *Transactions. 15th International Congress Soil Science* 5a:62-74.

Davis, J. H. C. and Garcia, S. 1987. *Field Crops Research* 16:105-116.

Davis, J. H. C and Woolley, J. N. 1993. *Field Crops Research* 34:407-430.

Doughton, J. A.; Vallis, I. and Saffigna, P. G. 1993. Nitrogen fixation in chickpea. *Australian Journal Agricultural Research* 44:1403-13.

Duncan, W. G.; McCloud, D. E.; McGraw, R. L. and Boote, K. L. 1978. Physiological aspects of peanut yield improvement. *Crop Science 18,* 1015-1020.

Fukai, S. 1993. *Field Crops Research* 34:239-467.

Greenberg, D. C.; Williams, J. H.; and Ndunguru, B. J. 1992. *Annals Applied Biology* 120, 557-566.

Hamdi, A.; Erskine, W. and Gates, P. 1992. *Crop Science* 32: 987-990.

Hammer, G. L.; Butler, D. G.; Muchow, R. C. and Meinke, H. 1996. In: *Plant Adaptation and Crop Improvement,* pp. 419-442. (eds M Cooper and GL Hammer) CAB International, Wallingford, UK.

Hammer, G. L. and Vanderlip, R. L. 1989. *Crop Science* 29: 385-391.

Hardwick, R. C. 1988. In: *World Crops: Cool Season Food Legumes*, pp. 885-96 (ed R. J Summerfield) . Kluwer Academic Publishers, Dordrecht.

Hunt , L. A. 1993. In: *Systems Approaches for Agricultural Development*, pp. 3-18 (Eds.FWT Penning de Vries et al.) Kluwer Academic Publishers, The Netherlands.

Jackson, P. A.; Robertson, M. J.; Cooper, M and Hammer, G. L. 1996. *Field Crops Research* 49: 11-37.

Jones, C. A. and Kiniry, J. R. (Eds). 1986. *CERES-Maize: a simulation model of maize growth and development.* Texas A & M University Press, College Station. 194 pp.

Jones, J. W. and Zur, B. 1984. Irrigation Science 5: 251-264

Jordan, W. R.; Dugas, W. A. and Shouse, P. J. 1983. Agriculture and Water Management 7:281-299.

Keating, B. A. and Carberry, P. S. 1993. *Field Crops Research* 34:273-301.

Keating, B. A.; Godwin, D. C. and Watiki, J. M. 1991. In: *Climatic risk in crop production: Models and management in the semi-arid tropics and sub-tropics*, pp. 329-358. (Eds. RC Muchow and J. A Bellamy) CAB International, Wallingford..

Lawn, R. J. 1982. *Australian Journal Agricultural Research* 33: 481-496.

Lawn, R. J. and Imrie, B. C. 1991. *Field Crops Research* 26: 113-139.

Ludlow, M. M. and Muchow, R. C. 1990. *Advances in Agronomy* 43: 107-153.

Mathews, R. B.; Harris, D.; Rao, R. C. N.; Williams, J. H.; and Wadia, K. D. R. 1988. *Experimental Agriculture* 24: 191-202.

McCown, R. L.; Hammer, G. L.; Hargreaves, J. N. G.; Holzworth, D. P. and Freebairn, D. M. 1996 *Agricultural Systems* 50: 255-71.

Midmore, D. J. 1993. *Field Crops Research* 34: 357-380.

Monteith, J. L. 1977. *Philosophical Transactions Royal Society* London. Ser. B. 281: 277-294.

Muchow, R. C. and Carberry, P. S. 1993. In: *Systems Approaches for Agricultural Development*, pp 37-62 (eds F.WT Penning de Vries et al.),. Kluwer Academic Publishers, The Netherlands.

Ntare, B. R. 1992. *Euphytica* 59: 27-32.

O'Toole, J. C. and Jones, C. A. 1987. In: *Weather and Rice.* Proceedings of the International Workshop on the Impact of Weather Parameters on Growth and Yield of Rice, pp. 255-269. IRRI, Los Banos, Phillipines.

Palanisamy, S.; Penning de Vries, F. W. T.; Mohandass, S.; Thiyagarajan, T. M. and Kareem, A. A. 1993. In: *Systems Approaches for Agricultural Development,* pp. 63-76. (eds F. W. T. Penning de Vries et al.), Kluwer Academic Publishers, The Netherlands.

Passioura, J. B. 1977. *Journal of the Australian Institute Agricultural Science* 43: 117-120.

Probert, M. E.; Carberry, P. S.; McCown, R. L. and Turpin, J. E. 1998. *Australian Journal Agricultural Research* (in press).

Redden, R. J. and Wright, G. C. 1993. In: *Farming - from Paddock to Plate.* pp. 88-91. (eds G. K McDonald and W. D Bellotti) Proceedings of the 7th Australian Agronomy Conference Australian Society of Agronomy, Parkville.

Saxena, N. P.; Johansen, C.; Saxena, M. C. and Silim, S. N. 1993. In *Breeding for Stress Tolerance in Cool-Season Legumes,* pp. 245-270. (eds K. B Singh and M. C Saxena) Wiley-ICARDA Co-publication, U.K.

Shorter, R.; Lawn, R. J. and Hammer, G. L. 1991. *Experimental Agriculture* 27: 155-175.

Silim, S.N., Saxena, M. C. and Erskine, W. 1993. *Experimental Agriculture* 29: 9-19.

Sinclair, T. R. 1986. Water and nitrogen limitations in soybean grain production. *Field Crops Res.* 15: 125-141.

Sinclair, T. R.; Muchow, R. C.; Ludlow, M. M.; Leach, G .J.; Lawn, R. J. and Foale, M. A. 1987. *Field Crops Res.* 17: 121-140.

Turk, K. J. and Hall, A. E. 1980. *Agronomy Journal* 72: 413-420.

Turner, N. C.; Wright, G. C. and Siddique, K. H. M. 1997. In *Management of Agricultural Drought: Agronomic and Genetic Options.* (ed NP Saxena) ICRISAT Publication (in press).

Villalobos-Rodriguez, E. and Shibles, R. 1985. *Field Crops Res.* 10: 269-281.

White, J.W.; Castillo, J.A. and Ehleringer, J. R. 1990. *Australian Journal Plant Physiology* 17: 189-198.

Williams, J. H. 1992. In: *Groundnut - a Global Perspective:* pp. 345-52. (ed SN Nigam) Proceedings of an International Workshop, ICRISAT, Patancheru, India.

Williams, J. H. and Saxena, N. P. 1991. *Annals Applied Biology* 119: 105-112.

Wright, G. C. and Redden, R. J. 1995. *ACIAR Food Legumes Newsletter* No. 23, 4-5.

Wright, G. C.; Hubick, K. T. and Farquhar, G. D. 1991. *Australian Journal Agricultural Research* 42: 453-470.

Wright, G.C.; Rao, R. C. N. and Basu, M. S. 1996. In: *Plant Adaptation and Crop Improvement,* pp. 365-381. (ed M Cooper and G. L Hammer) CAB International, Wallingford.

Autumn sowing of lentil in the Mediterranean highlands: Lessons for chickpea

J.D.H. Keatinge [1], R.J. Summerfield [1], I. Kusmenoglu [2] and M.H. Halila [3]

1 Department of Agriculture, The University of Reading, Earley Gate, P.O. Box 236 Reading RG6 2AT, UK; 2 Central Research Institute for Field Crops, P.O. Box 226, Ulus, Ankara, Turkey.; 3 Food Legumes Laboratory, INRAT, Avenue Hedi Karray, 2049 Ariana, Tunisia.

Abstract

Options exist for substantial increases in the area sown and yield of chickpea in the Mediterranean highlands as a result of an advancement of sowing date. To realise this potential requires genotypes with enhanced disease resistance and improved winter hardiness. Recent experience with lentil illustrates how genotypes suited to early sowing can be most readily and efficiently identified through a clear understanding of the variability in photothermal sensitivity of flowering available in the world germplasm pool. This understanding, coupled with simple agro-meteorological modelling of the risks of damage from disease and cold, can identify suitable genotypes for testing much more quickly and efficiently than using conventional breeding and traditional agronomic testing strategies.

INTRODUCTION

The range of crops grown successfully under dryland conditions in the semi-arid Mediterranean highland areas of west Asia and north Africa is limited, and often confined to cereals. Alternative crops such as chickpea (*Cicer arietinum*) and lentil (*Lens culinaris*) are often intolerant of the severe winter cold and disease problems associated with sowing in late autumn or in early spring. The continuous production of cereals, often without fertiliser input, is an unsustainable and environmentally degrading system; alternatively, the need for fallow years is a very inefficient use of land in an environment where rainfall receipt typically exceeds 300 mm per annum (Guler and Karaca, 1988). When food legumes are grown in these highland areas, late spring sowing is often the traditional practice to avoid problems associated with disease and lack of winter-hardiness in local landraces (Kusmenoglu and Aydin, 1995). Sowing in late spring frequently implies a very poor potential yield because of the then restricted growing season which is terminated abruptly by severe heat and terminal drought in early summer. However, given the availability of germplasm with adequate winter hardiness, with appropriate crop duration and with disease resistance, suggests that the use of late-autumn sowing could substantially increase seed yields compared with traditional practices, as shown in lowland Mediterranean areas (Pala and Mazid, 1992; Singh and Saxena, 1996). The thrust of current breeding efforts for the highlands is therefore to stimulate crop diversification, and especially of food legumes, by breeding winter-hardy material with good yield potential which is capable of maturing sufficiently early to avoid severe terminal drought.

This paper uses Anatolia in Turkey as the principal case study region and describes research on diverse lentil genotypes in controlled and field environments. It explores how phenological factors and a better understanding of the climatic risks associated with late-autumn or early-spring sowing in the highlands can help guide and improve the efficiency in breeding programs devoted to lentil and chickpea. It examines as well the additional risks associated with earlier sowing and the implications of this change in agronomic strategy for potential yield increases over a range of Mediterranean highland environments. This study also attempts to provide guidelines for the selection of winter-hardy and Ascochyta blight (*Ascochyta rabiei*)

279

R. Knight (ed.), Linking Research and Marketing Opportunities for Pulses in the 21st Century, 279–288.
© *2000 Kluwer Academic Publishers. Printed in the Netherlands.*

resistant parents for breeding programs. The expectation is that these outputs could lead to substantial savings in national agricultural research services (NARS) expenditure on breeding and agronomy trials (Keatinge *et al.*, 1995; 1996). Appropriate cultivars might then be selected or developed locally more efficiently and over a shorter time scale than the typical duration of 8-10 years.

Meyveci *et al.* (1993) have recently demonstrated in a study over several years at various sites in central Anatolia, that an advance in chickpea sowing date of up to 60 d (from mid-May to mid-March) is possible when using germplasm lines which, though cold susceptible, are partially resistant to Ascochyta blight. Early sowing increased seed yield by more than fifty percent. Just how much earlier sowing dates could be without incurring substantial and unacceptable risks from cold and disease is not known for the next generation of winter-hardy and blight resistant germplasm.

Given these potential radical changes in traditional agronomy for chickpea cropping in Anatolia, it is important to recognise that the primary factor in the adaptation of food legumes to different environments is an appropriate crop phenology (Erskine *et al.*, 1989). Phenology in chickpea is modulated primarily by temperature and photoperiod (Roberts *et al.*, 1985). Substantial variation in the responses of flowering to both temperature and photoperiod is present between and within food legume species. A simple model has been developed in collaboration with ICRISAT and ICARDA to describe the flowering behaviour of genotypes:

$$1/f = a + bT + cP \qquad \text{(Equation 1)}$$

where f is the time (days) from sowing to first flowering, T and P are the respective values of mean temperature (°C) and mean photoperiod inclusive of civil twilight (hd^{-1}) during the same period and a, b and c are genotypic constants. The model has been validated in the field over a wide range of environments for each of chickpea (Ellis *et al.*, 1994) and lentil (Erskine *et al.*, 1994), in both cases including substantial diversity in photothermal responsiveness (e.g. Figure 1; Erskine *et al.*, 1990). Previous modelling work has included chickpea ecotypes partially adapted to the Anatolian highlands such as ILC 482 and ILC 195 -2 (Summerfield *et al.*, 1987). Both of these older generation lines have also been extensively tested in the field with early spring-planting in central and western Anatolia and have now been released as cultivars in Turkey (Meyveci *et al.* 1993): ILC 482 (Guney Sarisi) for winter sowing in south-eastern Anatolia and ILC 195-2 for early-spring sowing in western Anatolia.

Is winter-hardy germplasm available for autumn sowing? For lentil, Erskine *et al.*, (1981) identified 238 out of 3592 accessions as potentially cold tolerant following one autumn sowing in the severe environment of Haymana in central Anatolia. For nine of these lines, the photothermal constants required for Equation 1, were determined as part of the study by Erskine *et al.* (1990) and are shown in Figure 2 (extracted from Figure 1). For the transitional zones on the fringes of the Anatolian plateau Kayi-91 and Sazak-91 are suitable for autumn-sowing at lower elevations. Evidence that more truly cold tolerant material is forthcoming is now available (Aydin *et al.*, 1997). For chickpea, Malhotra and Singh (1990) have reported that cold tolerance is controlled by at least five sets of genes but that heritability is generally high. Yet, with currently available genotypes, autumn sowing of chickpea in severe environments such as central or eastern Anatolia is not immediately feasible. However, given the availability of the ILC 482 radiation-induced mutant ILC 861 claimed by Haq and Singh (1994) to be super winter-hardy, as well as lines with very much improved winter-hardiness and Ascochyta resistance (e.g. ILC 8262), the potential range of chickpea adaptation could extend into much colder environments. This possibility is well worth investigating beginning at the modelling level before commitment to expensive field testing. It is recognised, of course, that in specific instances lethal temperatures can occur. However, this is not necessarily the central problem for early sowing in the environments of concern. Here the problem is more general and involves winter-hardiness traits and cumulative effects of exposure to cold temperatures at the seedling stage and which are ultimately related to poor stands and low yields.

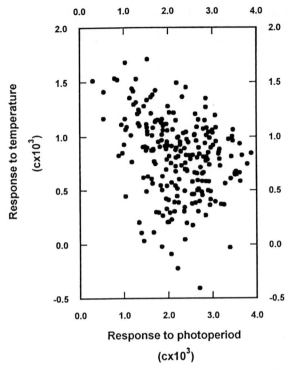

Figure 1. Sensitivity of rate of progress from sowing towards flowering (Equation 1) to temperature (b) and photoperiod (c) of 230 cultivated lentil accessions (Σ) and one wild lentil accession from Turkey ($\Box$). Source Erskine *et al.* (1990).

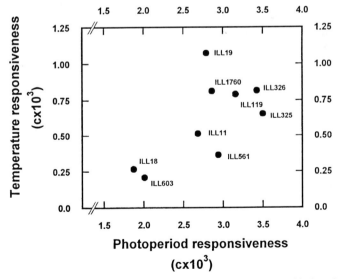

Figure 2. Temperature- and photoperiod-response constants (*b* and *c* of Equation 1) of nine cold-tolerant lentil genotypes for rate of progress from germination towards first flowering (1/*f*). Source Keatinge *et al.* (1996).

For Ascochyta blight there is limited quantitative information linking blight incidence to meteorological variables (Kaiser, 1992; Trapero-Casas et al., 1996). The temperature limits for blight development range from 5°-30°C. Humid air and cool temperatures (5° - 10°C) enhance the formation of the fruiting bodies of the pathogen (Trapero-Casas and Kaiser, 1992). Moreover, occurrence of rain, rather than its amount per se, seems to be an important factor in the development of the disease (Trapero-Casas et al., 1996). Such thermo-moisture regimes are in the normal spring range throughout the highlands of Turkey and North Africa even with conventional sowing dates. It is also clear that durations of leaf wetness (>10 h in each diurnal cycle) is a crucial factor and so an assessment of the increased risk of consecutive raindays with early sowing will necessarily be a target for initial evaluation. Although a theoretical model predicting Ascochyta risk for various geographical zones and growth seasons has been developed (Dickman, 1992), no comprehensive weather-driven model of phenology/disease interaction is yet available for Mediterranean food legume crops. There is a clear need to combine published information, evidence of historical blight events for which there is associated climatic data available, and substantial anecdotal information based on west Asian and north African experience and to pool this knowledge into meteorologically based probability statements for risk assessment.

RECENT DEVELOPMENTS WITH GERMPLASM EVALUATION FOR AUTUMN SOWING OF LENTIL IN THE ANATOLIAN HIGHLANDS

Eyupoglu et al. (1995) examined the potential for autumn sowing of lentil in the central Anatolian highlands with existing cultivars such as Firat 87 for lack of any better alternative at that time as Firat-87 was formally released for winter sowing only in south-eastern Anatolia. During the period 1991-93 productivity was poor (< 1 t ha^{-1}) in most years whether the crop was sown in autumn or in spring. Yet, spring sowing certainly gave less consistent yields. Damage to autumn-sown crops was particularly severe when snow cover was insufficient to provide adequate protection from very cold air temperatures and desiccating winds. They also indicated that inadequate weed control could be an additional severe constraint to the introduction of autumn sowing. Weeds are a common problem through the west Asia and north African highlands (e.g. Halila, 1995). However, Eyupoglu et al. (1995) suggested that the prospects for stable, higher yielding cultivars would make the practice of autumn-sowing much more acceptable to farmers.

In support of these agronomic findings, recent field measurements of winter-hardiness in a range of new cold-tolerant lentil germplasm sown at Konya and Haymana in central Anatolia (Kusmenoglu and Aydin, 1995) have revealed positive correlations in cold tolerance scores between sites, which suggests that genotypic differences rather than environmental differences are the causative agent for variability in hardiness. They also observed substantial variability in cold tolerance, plant height and dates of flowering and maturity in a collection of 152 landraces.

Erskine and Muehlbauer (1995) have warned that winter survival of lentil often requires tolerance to factors other than cold including frost-heaving, water-logging and resistance to root pathogens, Ascochyta blight and other aggressive foliar and root diseases (Beniwal et al., 1995). They warn that a myopic preoccupation with cold-tolerance per se should be avoided. They also point out that the annually distributed ICARDA Cold Tolerance Nursery has been successful in identifying species with good adaptability to extreme highland conditions such as ILL 5865 now released as Shiraz 96 for the highland areas of Pakistan (Asghar Ali et al., 1991). Avoidance of disease risk is also a primary motivating factor for farmers to adopt later sowing even in substantially less-severe lowland environments (Halila, 1995).

Keatinge et al. (1995) obtained daily climatic data for the uninterrupted period 1950-1989 [maximum and minimum air temperature (°C), daily precipitation total (mm) and daily snow depth (cm)] from the Turkish Meteorological Service for three locations viz. Çorum (40° 33'N 34° 58'E, elevation 798m), Sivas (39° 45'N 37° 01'E, 1285m) and Erzurum (39° 55'N 41° 16'E, 1869m; Figure 3). These sites were chosen for duration and quality of unbroken data run and because they represent the three main agro-ecological subdivisions of central and eastern Anatolia as defined by Guler et al. (1990).

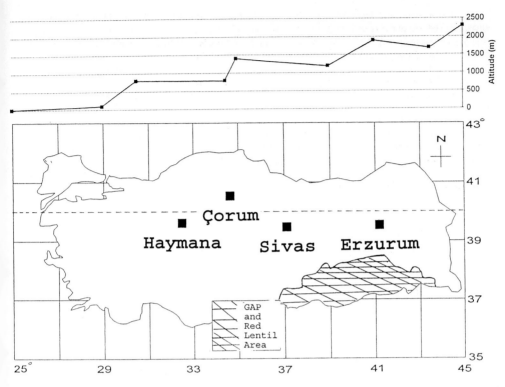

Figure 3. Site locations, and (cross-hatched) the SE Anatolian Project (GAP) irrigation command area and area of current red lentil production, and a representational E-W cross-section of the elevation (m) profile of Turkey at 40°N (dashed line). Source Keatinge *et al.* (1996).

From these computerised data sets, the probabilities and associated dates of specific climatic events at each location were calculated using programs written in Fortran (Prospero Fortran, 1989). The probability of a specific climatic event was taken to be the direct proportion (%) of years when that event occurred during the 40-year period. Based on a knowledge of crop physiology and agronomy, and of local farming practices for lentil cultivation in Turkey, Keatinge *et al.* (1995) defined specific meteorological events and their associated probabilities of occurrence at representative locations in Anatolia. The events defined were: the window of opportunity for safe sowing, germination and emergence in autumn, and, the window of opportunity for safe flowering (and eventual seed maturity) in spring. Examining these definitions at the three target locations exposed the potential narrowness of the window of opportunity for safe sowing in autumn at Sivas, and the comparatively early start to the potential cropping season at Erzurum (Figure 4).

Keatinge *et al.* (1996) have subsequently suggested that it is possible to derive important information for the selection of winter-sown lentils for the Anatolian highlands from the output of a photothermal phenological model of flowering (Summerfield *et al.*, 1985). Evidence that unresponsive accessions (i.e. comparatively photothermally-insensitive genotypes) are likely to be severely damaged or killed across the Anatolian highlands is presented (Figures 2 and 4) irrespective of having been rated as cold tolerant by Erskine *et al.* (1981). Likewise, date of sowing experiments at several locations were suggested to be essentially redundant for these phenological traits because it was only photothermally insensitive genotypes which showed a substantial response to sowing date in late autumn and these were unable to survive the severe winter conditions (Figures 5 and 6). Given the poor adaptation of photothermally-insensitive lines to winter-sowing, we suggest that this class of "cold tolerant" material can now be excluded from the breeding programme for highland areas.

Ellis and Hong (1995) supplemented this recommendation in suggesting that all genotypes with a base temperature for germination > 1.5 °C can also be excluded for consideration for field trials with little risk of excluding cold-tolerant germplasm. These recommendations are of substantial importance in that they imply that a majority of the world lentil collection with a photoperiod response constant, c, below values of 0.003 (from Equation 1) can be ignored (Figure 1; Erskine *et al.*, 1990). Therefore the only likely sources of truly winter-hardy material will probably be of Turkish or Iranian origin (Erskine *et al.*, 1994). In addition, there are the further implications for finding truly cold-tolerant chickpeas given that Summerfield *et al.* (1987) and Ellis *et al.* (1994) have reported most c values are smaller than 0.003. In contrast, Jana and Singh (1993) have indicated the availability of potentially useful variability in germplasm congruent upon geographic dispersion from the centre of origin in Anatolia (van der Maesen , 1972).

Keatinge *et al.* (1995) also estimated the first safe flowering date in spring for their target sites at different elevations to be 8 May for Çorum, 7 May for Sivas and 6 June for Erzurum. With the introduction of photothermally-sensitive accessions such as ILL 119, ILL 326 and ILL 325, the model output (Figure 2) implies a comfortable fit within the "*safe windows of opportunity*" for germination and first flowering,

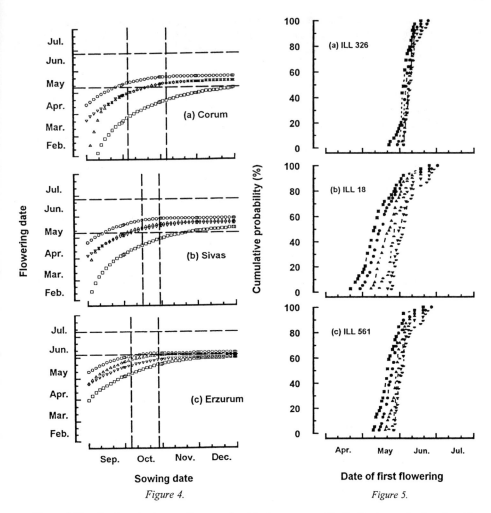

Figure 4. Mean first flowering date over 40 years predicted for four accessions of lentil with winter sowing dates from 1 Sept to 31 Dec at Çorum, Sivas and Erzurum. The two broken vertical lines represent the safe window of opportunity for germination for winter sowings and the two broken horizontal lines represent the safe window of opportunity for first flowering in spring (squares ILL 18, triangles up ILL 19, triangles down ILL 561 and circles ILL 326). Source Keatinge *et al.* (1996).

Figure 5. Cumulative probability of first flowering date for accessions ILL 326, ILL 18 and ILL 561 at Erzurum for selected dates of sowing: 15 Sept (squares), 1 Oct (circles), 15 Oct (triangles up), 30 Oct (half-circles up), 15 Nov (triangles down), 30 Nov (half-circles down). Source Keatinge *et al.* (1996).

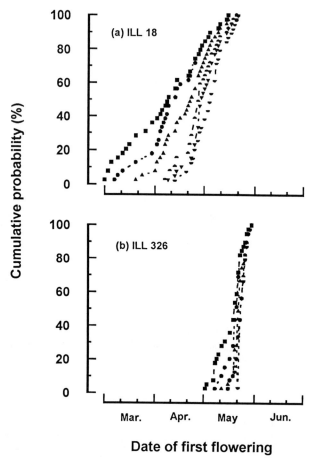

Figure 6. Cumulative probability of first flowering date for accessions ILL 18 and ILL 326 predicted for selected dates of sowing at Çorum: 15 Sept (squares), 1 Oct (circles), 15 Oct (triangles up), 30 Oct (half-circles up), 15 Nov (triangles down), 30 Nov (half-circles down). Source Keatinge *et al*. (1996).

except at Erzurum. It suggests first flowering dates approximately 14 d earlier at all locations than would be expected, based on field work at the Central Research Institute for Field Crops, Ankara, with their existing genotypes (Figure 4). A reproductive period either advanced or extended by 14 d in such a Mediterranean environment should improve yield potential (Silim *et al.*, 1993) without incurring risk from cold damage. This type of photothermally-sensitive material would therefore seem to be a germplasm group from which rapid progress might well be made. Where environments are less severe than in central and eastern Anatolian highlands, such as on the fringes of the Anatolian plateau, germplasm with intermediary photothermal sensitivity (such as ILL 19 and ILL 561) could be more productive. Niche environments demanding greater or less temperature responsiveness could then be served more effectively, given the variation for this characteristic in the world germplasm of both chickpea and lentil.

CONCLUSIONS

Lessons can be learned for chickpea from the experiences of seeking lentil germplasm suitable for autumn sowing in the Mediterranean highlands. A more complete analysis of photothermal sensitivity of flowering in a wide range of germplasm might well facilitate the rapid identification of genotypes with a phenology more suitable for germination in autumn, for winter survival and for flowering in the safe period following the last frost in spring and before the onset of terminal summer drought. In addition, further research will be required to relate the probability of an economically damaging outbreak of Ascochyta blight with standard weather station data. Without modelling such a relation it will be very difficult to determine whether autumn sowing or very early spring sowing is likely to be the more risky strategy and whether the increased risk of an earlier sowing date can be compensated for fully, or only partially, by germplasm with improved disease resistance. Such research would help to reduce the need for large-scale chickpea germplasm screening trials undertaken at many locations with several dates of sowing, which is a very costly and time-consuming exercise. Experience of working with the Turkish Ministry of Agriculture and Rural Affairs on autumn sowing of lentil has shown that this expense can be largely avoided (Keatinge *et al.*, 1996).

Acknowledgements
This document is an output from a project partially funded by the Overseas Development Administration (ODA) Plant Sciences Programme, managed by the Centre for Arid Zone Studies, University of Wales, Bangor, for the benefit of developing countries. The views expressed are not necessarily those of the ODA. The authors acknowledge the considerable help of the Turkish Meteorological Office for access to archives.

References
Asghar Ali, Keatinge, J.D.H., Roidar Khan, B and Sarfraz Ahmad 1991. *Journal of Agricultural Science, Cambridge* 117: 347-353.
Aydin, N., Kusmenoglu, I., Aydogan, A., Erskine, W. and Beniwal, S.P.S. 1997. In *Proceedings of the IFLRC III* (Ed. R. Knight) Dordrecht:Kluwer Academic(in press).
Beniwal, S.P.S., Kaiser, W.J. and Dalkiran, H. 1995. In *Autumn-sowing of lentil in the highlands of west Asia and north Africa*, pp 84-94 (Eds J.D.H. Keatinge and I. Kusmenoglu) Ankara:CRIFC.
Dickman, M. 1992. *Plant Disease Reporter* 79: 409-412.
Ellis, R.H., Lawn, R.J., Summerfield, R.J., Qi, A., Roberts, R.H., Chay, P.M., Brouwer, J.B., Rose, J.L., Yeates, S.J. and Sandover, S. 1994. *Experimental Agriculture* 30: 271-282.
Ellis, R.H. and Hong, T.D. 1995. In *Autumn-sowing of lentil in the highlands of west Asia and north Africa*, pp 95-105 (Eds J.D.H. Keatinge and I. Kusmenoglu) Ankara:CRIFC.
Erskine, W., Meyveci, K. and Izgin, N. 1981. *LENS Newsletter* 8: 5-9.
Erskine, W., Adham, Y. and Holly, L. 1989. *Euphytica* 43: 97-103.
Erskine, W, Ellis, R.H., Summerfield, R.J., Roberts, E.H. and Hussain, A. 1990. *Theoretical and Applied Genetics* 80: 193-199.
Erskine, W., Hussain, A., Tahir, M., Baksh, A., Ellis, R.H., Summerfield, R.J. and Roberts, E.H. 1994. *Theoretical and Applied Genetics* 88: 423-428.
Erskine, W. and Muehlbauer, F.J. (1995). In *Autumn-sowing of lentil in the highlands of west Asia and north Africa*, pp 51-62 (Eds J.D.H. Keatinge and I. Kusmenoglu) Ankara:CRIFC.
Eyupoglu, H., Meyveci, K., Karagullu, E., Isik, M. and Orhan, A. 1995. In *Autumn-sowing of lentil in the highlands of west Asia and north Africa*, pp 72-83 (Eds J.D.H. Keatinge and I. Kusmenoglu) Ankara:CRIFC.

288

Guler, M. and Karaca, M. 1988. In *Winter cereals and food legumes in mountainous areas*, pp 41-49 (Eds J.P. Srivastava, M.C. Saxena, S. Varma and M. Tahir). Aleppo, Syria:ICARDA

Guler, M., Karaca, M. and Durutan, N. (1990). *Turkiye Tarimsal Iklim Bolgeleri*. Ankara: Tarla Bitkileri Merkez Arastirma Enstitusu, 87 pp.

Halila, M.H. 1995. In *Autumn-sowing of lentil in the highlands of west Asia and north Africa*, pp 20-28 (Eds J.D.H. Keatinge and I. Kusmenoglu) Ankara:CRIFC.

Haq, M.A. and Singh, K.B. 1994.. *Mutation Breeding Newsletter* 41: 6-7.

Jana, S. and Singh, K.B. 1993 *Crop Science* 33: 626-632.

Kaiser, W.J. 1992. In *Disease Resistance Breeding in Chickpea*, pp 117-134 (Eds K.B. Singh and M.C. Saxena) Aleppo:ICARDA.

Keatinge, J.D.H., Aiming Qi, Kusmenoglu, I., Ellis, R.H., Summerfield R.J., Erskine, W. and Beniwal, S.P.S.1995. *Agricultural and Forest Meteorology* 74: 251-263.

Keatinge, J.D.H. Aiming Qi, Kusmenoglu, I., Ellis, R.H., Summerfield R.J., Erskine, W. and Beniwal, S.P.S. 1996. *Agricultural and Forest Meteorology* 78: 53-65.

Kusmenoglu, I and Aydin N. 1995. In *Autumn-sowing of lentil in the highlands of west Asia and north Africa*, pp 63-71 (Eds J.D.H. Keatinge and I. Kusmenoglu) Ankara:CRIFC.

Malhotra, R.S. and Singh, K.B. 1990. *Journal of Genetics and Breeding* 44: 227-230.

Meyveci, K., Eyupoglu, H. and Karagullu, E. 1993. *Ankara:Tarla Bitkileri Merkez Arastirma Enstitusu* 39pp.

Pala, M. and Mazid, A. 1992. *Experimental Agriculture* 28: 175-184.

Prospero *Fortran 1989. Version 2.1 User Manual, Parts 1-3.* London:Prospero Software.

Roberts, E.H., Hadley, P. and Summerfield, R.J. 1985. *Annals of Botany* 55: 881-892.

Silim, S.N., Saxena, M.C. and Erskine, W. 1993. *Experimental Agriculture* 29: 9-19.

Singh, K.B. and Saxena, M.C. 1996. *Winter chickpea in Mediterranean-type environments. A Technical Bulletin.* Aleppo, Syria:ICARDA 38pp.

Summerfield, R.J., Roberts, E.H., Erskine, W., and Ellis, R.H. 1985. *Annals of Botany* 56:659-671.

Summerfield, R.J., Roberts, E.H. and Hadley, P. 1987. In *Proceedings of a Consultant's Workshop on the Adaptation of Chickpea and Pigeonpea for Tolerance to Physical Stresses,* Hyderabad, India :ICRISAT pp. 33-48.

Trapero-Casas, A. and Kaiser, W. 1992. *Phytopathology* 82:1261-1266.

Trapero-Casas, A., Navas-Cortes, J.A. and Jimenez-Diaz, R.M. 1996. *European Journal of Plant Pathology* 102:237-245

van der Maesen, L.J.G. 1972. *A monograph of the genus, with special reference to the chickpea (Cicer arietinum L.), and its ecology and cultivation.* Wageningen:Veenman and Zonen.

Adaptation of chickpea (*Cicer arietinum* L.) and faba bean (*Vicia faba* L.) to Australia

K.H.M. Siddique[1], R.B. Brinsmead[2], R. Knight[3], E.J. Knights[4], J.G. Paull[3] and I.A. Rose[5]

1 Agriculture Western Australia, Locked Bag No 4, Bentley Delivery Centre, WA 6983, and Centre for Legumes in Mediterranean Agriculture, University of Western Australia, Nedlands WA 6009; 2 Hermitage Research Station, Queensland Department of Primary Industries, Warwick, Qld 4370; 3 Department of Plant Science, Waite Campus, University of Adelaide, PMB1, Glen Osmond, SA 5064; 4 The Tamworth Centre for Crop Improvement, New South Wales Agriculture, RMB 944, Tamworth, NSW 2340; 5 Australian Cotton Research Institute, New South Wales Agriculture, PMB, Myall Vale, Narrabri, NSW 2390

Key words: Adaptation, chickpea, faba bean, breeding, yield, germplasm, phenology, pulses, diseases

Abstract

In the last 20 years chickpea and faba bean have become major pulse crops in Australia. They are grown during winter over a latitudinal range which extends from 10-40° S for chickpea, and 20-40° S for faba bean. In low latitudes these crops grow mainly on water stored in the soil from summer rainfall with supplementary irrigation in some areas, while in the higher latitudes, they are grown in Mediterranean type environments relying solely on winter rainfall. Farmers recognise the benefit of including these pulses in rotation with cereals and the interest in these crops is continuing to expand, notably faba bean in subtropical north-eastern areas and of both pulses in Western Australia. Breeding programs are continuing to develop cultivars adapted to the wide range of latitudes and with disease resistance. Some germplasm introduced from overseas has required minimal selection before being released, while other genotypes have been poorly adapted and genetic changes are required. Faba bean produces highest yields when sown as early as possible and when diseases are managed by genetic resistance or fungicide application. Chickpea is less affected by delayed sowing and some crops are sown in mid-winter to avoid radiation frosts during early spring. Studies have indicated significant differences between strains of rhizobia in terms of crop growth and the development of acid tolerant strains have been critical for the production of faba bean on low pH soils. A high proportion of both pulses is exported; chickpea to the Indian sub-continent and faba bean to the Middle-East. To meet the requirements of consumers, the effects of the environment on quality are being studied and efforts made to overcome deficiencies through breeding.

INTRODUCTION

Australian pulse production has increased rapidly over the past three decades as a result of the appreciation by farmers of the financial returns from growing these crops and the role of pulses in sustainable crop rotations (Rees *et al.*, 1994). The area sown to pulses has increased from almost nothing in 1965 to 2 million ha in 1996, or about 10% of the area cropped (Siddique and Sykes, 1997). In this period chickpea (*Cicer arietinum* L.) and faba bean (*Vicia faba* L.) have been integrated into farming systems covering a broad agroclimatic and latitudinal (10°-40° S) range. Chickpea extends from the tropical Ord River Irrigation Area (ORIA) in Western Australia (WA) to the mainly rainfed subtropical environments of Queensland (Qld) and northern New South Wales (NSW), and the Mediterranean type environments of WA, South Australia (SA) and Victoria (Vic). Faba bean is grown in the Mediterranean type environments of WA, SA, Vic and southern NSW, and extends into the subtropical environment of northern NSW and southern Qld.

Both pulses have been introduced into a variety of regions that differ in rainfall distribution and are grown as a winter or spring-sown crop. In southern Australia, they typically follow one to three years of winter cereals. Alternatively, they can follow canola (*Brassica napus*) or other winter pulses. In the northern

R. Knight (ed.), Linking Research and Marketing Opportunities for Pulses in the 21st Century, 289–303.
© *2000 Kluwer Academic Publishers. Printed in the Netherlands.*

areas there are similar rotations based on winter cereals, but the summer-dominant rainfall also enables opportunistic double-cropping immediately following a summer cereal crop or cotton.

In the Ord River Irrigation Area of northern WA, irrigated chickpea is grown during the dry winter in rotation with melons, sorghum, maize, cotton or sugarcane. Faba bean is also produced with supplementary irrigation in both southern NSW, where it is grown in rotation with summer crops such as maize, and in northern NSW where it is grown following cotton.

In all of these rotations chickpea and faba bean have helped alleviate problems caused by continuous cereal cropping such as fungal root diseases, the depletion of soil organic matter and nitrogen levels, degradation of soil structure (Hamblin and Kyneur, 1993) and herbicide resistance in weeds. Reports of the nitrogen fixed by chickpea and faba bean crops have ranged from 29-85 kg/ha (Herridge *et al.*, 1995) and 0-99 kg/ha (Schwenke and Herridge, 1997). Faba bean has a greater potential for dry matter production and nitrogen fixation than chickpea. In the cotton-based rotations of northern NSW, dry matter production of faba bean often exceeds 10 t/ha and nitrogen fixation has ranged from 10-300 kg/ha, with residual nitrogen for following crops of 0-200 kg/ha (Rochester *et al.*, personal communication). Including pulses in a rotation increases the yield and protein content of the following cereal crop. Cereals following chickpea or faba bean produce 15-103% more seed yield than cereals following cereals (Jessop and Mahoney, 1985; Strong *et al.*, 1986; Marcellos, 1984).

CULTIVAR DEVELOPMENT

Chickpea

Chickpea was first tested in Australia in the 1890s (Valder, 1893), but a comprehensive evaluation did not commence until 1972 when a diverse range of germplasm was introduced and tested at many locations throughout Australia. This testing culminated in the release of the Indian desi cultivar C235 (renamed 'Tyson') in Qld in 1978. Tyson proved to be widely adapted and its seed was acceptable to emerging export markets in the Indian sub-continent (Beech and Brinsmead, 1980). It was subsequently grown in other regions of Australia where it remained the only desi cultivar available for nearly a decade.

The first cultivar developed from an Australian hybridisation program was the desi cultivar Amethyst, which was released in NSW in 1987 (Table 1). Amethyst was well suited to mechanised harvesting because of its tall stature and lodging resistance. These features helped facilitate an extension of chickpea cropping into low rainfall areas of north-eastern Australia. Subsequent breeding efforts in this region centred on disease control. Resistance to Phytophthora root rot became a major breeding objective after the disease emerged as an important production constraint in Qld and NSW in the early 1980s. Field resistance to the disease (Brinsmead *et al.*, 1985) was developed in desi cultivars Barwon and Norwin, released in 1991 and 1992, respectively. During the 1990s viruses replaced Phytophthora as the major disease problem in some parts of the north-eastern region. A desi landrace selection from Iran possessing field resistance to virus, was released in NSW in 1997 under the name Gully.

New desi cultivars for the south-eastern and western regions have been released by local breeding and evaluation programs. Hitherto, these programs relied on introduced germplasm and breeding lines as the source of improved local adaptation. Significant yield increments were realised with the release of Dooen (1986), Desavic (1993) and Lasseter (1996) in south-eastern Australia (Vic and SA) and with the early maturing ICRISAT-derived cultivars Sona and Heera released in WA in 1997 (Table 1).

Cultivar development in kabuli chickpea has relied mainly on introductions: five of the six cultivars released originated from overseas collections or breeding programs. The first cultivar (Opal) was released in NSW in 1980 but was not widely grown due to its comparatively small seed (35 g/100 seed) and late maturity. Significant expansion of the rainfed kabuli area did not occur until the release of the early maturing, large-seeded cultivars Garnet and Kaniva. The large-seeded cultivar Bumper (50 g/100 seed) was subsequently released in 1997 for all rainfed production areas. Irrigated production of kabuli chickpea in the ORIA commenced with the release of the very large-seeded Mexican cultivar Macarena (55 g/100 seed) in 1983.

Table 1. Yield improvement from new desi chickpea cultivars in various chickpea producing regions in Australia

Cultivar	Year of release	NSW	QLD	VIC	SA	WA
		Yield as % of Tyson				
Tyson	1978	100	100	100	100	100
Amethyst	1987	110	105			
8805-78H*		121	107			
8813-74H*			112			
Dooen	1986			108		
Desavic	1993			111	112	
Lasseter	1996			116		
86130-5*				125		
8511-14*					113	
Heera	1997					117
Sona	1997					117
Tyson Mean yield (kg/ha)		1305	1709	1242	1486	1272

* Advanced breeding lines not commercially released.

Faba bean

Faba bean improvement in Australia commenced in 1972 in Adelaide, SA. Germplasm was introduced from many countries and it soon became evident that the best-adapted and highest yielding material originated from the Mediterranean region with a similar environment to southern Australia. Of less immediate value was material from high altitudes in the Mediterranean region, temperate European environments or irrigated areas of Egypt, Sudan, Iraq or Iran.

Germplasm from the high altitude regions of South America, although not well adapted to most of Australia because of its late flowering, has genes for resistance to the diseases chocolate spot (*Botrytis fabae*) and rust (*Uromyces fabae*). For these reasons, selections from this material have found a place in areas of Australia with long growing seasons or irrigation. Recent collecting, undertaken by the International Centre for Agricultural Research on Dry Areas (ICARDA) Syria and the national agricultural research services of the South American countries, attempted to locate early flowering genotypes. Germplasm from Ethiopia has good potential for the subtropical areas of Australia (M. English, personal communication), but has only produced average yields in areas with a Mediterranean type climate.

There are seven faba bean cultivars grown in Australia (Table 2). A factor in the release of the first Australian bred cultivar, Fiord, was its small seed size, which enables farmers to sow and harvest the crop with machines used for cereal crops. Soon after the release of Fiord it was noticed there was variation among plants of the cultivar in resistance to Ascochyta blight (*Ascochyta fabae*). The cultivar Ascot was bred by mass selection from within Fiord and progeny testing, and has a high level of resistance to this disease. Within Fiord there was also variation in plant habit. The cultivar Barkool was selected for bearing more pods at high nodes and a partially determinant growth habit. Fiord, together with its two derivatives Ascot and Barkool, are at present the dominant varieties grown in Australia. Fiesta is a recent release with moderate resistance to Ascochyta and chocolate spot and good seed quality, originating from Spain. Of the cultivars originating from South America, Icarus and Rossa were released for their chocolate spot resistance and Rossa may have a special market because of its red seed coat and be of value as a green manure crop in vineyards.

Table 2. Faba bean cultivars released in Australia

Variety	Origin	Date of release	Seed weight (g/100 seeds)	Characteristics
Aquadulce [1]	?	?	150	Introduced-natural adaptation
Fiord	Naxos, Greece	1980	45	High yield
Ascot	Naxos, Greece	1992	45	Selected from Fiord (Ascochyta resistant)
Barkool	Naxos, Greece	1994	40	Selected from Fiord (growth habit)
Icarus	Ecuador	1992	85	Chocolate spot resistant
Fiesta	Spain	1997	60	High stable yield
Rossa	Ecuador	1997	55	Red seed, rust resistant

1 The Aquadulce in Australia is of unrecorded origin and has been derived probably from material of that name imported twenty to thirty years ago. Some natural and farmer selection has occurred.

MATCHING PHENOLOGY TO ENVIRONMENT

Matching the growth cycle of a crop to the environment is the single most important factor affecting crop adaptation in Australia.

Climatic features of four locations, representing the major growing regions of chickpea and faba bean, are presented in Figure 1. Emerald (Qld) and Moree (NSW) represent subtropical environments of Australia where the two pulses are grown during the cool winter season on water stored in the soil from summer rainfall. In these regions date of sowing is frequently determined by the date of harvest of the summer crop and by land preparation procedures. In the Mediterranean type environment of southern Australia, (Merredin, WA and Horsham, Vic) sowing of chickpea and faba bean is determined by the timing of the first autumn rains. This is usually coincident with falling air temperatures during May. The growing season ends soon after the last effective spring rains in the period of October to November when the crops experience high temperatures and moisture stress. Although Fig. 1 shows some rain during summer at Merredin and Horsham, this moisture is generally ineffective because of high temperatures and evaporative demand at this time. Summer crops are rarely grown in these regions. Changes in daylength are not given in Fig 1 but they closely reflect the annual cycle of temperature. Harvest dates range from late October at Merredin to December and January at Horsham.

The balance between vegetative and reproductive growth, as determined by time of first flower, is particularly important. Early flowering is an advantage in enabling a long pod-filling period before the onset of drought or high temperatures, and this is particularly important for cropping areas in WA, NSW and Qld. However, winter and early spring frosts can cause serious yield losses in early flowering crops in many regions.

Chickpea

Chickpea is a long day plant and genotypes differ in their response to temperature and photoperiod. Drought escape, through early flowering accounts for the superior adaptation of the chickpea cultivars Lasseter in the short season region of Vic, and Heera and Sona to the dry parts of WA (<400 mm p.a.). Enhanced cold tolerance in the early reproductive phase increases pod set of chickpea in these environments (Siddique and Sedgley, 1986). Genotypic variation for cold tolerance has been reported (Siddique *et al.*, 1994) and *in vitro* pollen selection techniques and molecular markers are currently being developed for this trait (Lawlor *et al.*, 1988).

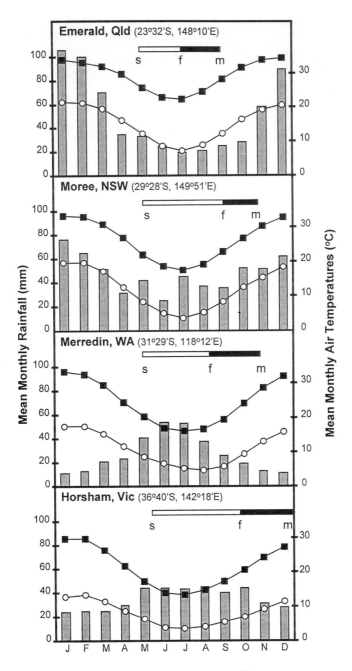

Figure 1. Mean monthly rainfall (vertical bars) and maximum (■) and minimum temperatures (□) for four sites in Australia's cropping regions. Also indicated is the average duration from sowing (s) to first flower (f) and first flower to maturity (m) for chickpea (cv. Tyson).

The north-eastern region of Australia, (represented by Moree and Emerald in Fig. 1), has summer dominant rainfall and the growing season is shorter than in the south, but severe radiation frosts from August to early October often negate the benefit of early flowering, especially for chickpea. Sowing chickpea is usually delayed until early winter to prevent the onset of flowering during the period of maximum frost risk. Mid flowering chickpea cultivars (e.g. Tyson, Amethyst, 8805 78H and 8813 74H) have generally given the highest and most stable yields in these environments (Table 1).

In the short season environments of WA and central Qld, chickpea produced the largest yields from mid April to late May sowings (Siddique and Sedgley, 1986; Brinsmead, 1992). Yields of chickpea have been less sensitive to sowing time in the medium rainfall areas of SA, Vic, NSW and southern Qld (Pye 1987; Brinsmead 1992). In these environments sowing at any date between April and late June does not have a major effect on yield.

Faba bean

Faba bean is also a long day plant and genotypes differ in their photoperiodic and temperature requirements. Sowing date can strongly affect the length of the vegetative period and it can show interactions with genotype. When sown early in SA (e.g. early May) at latitude 35°S, daylength and temperature are adequate for the requirements of some genotypes and these flower very early, 41 days after emergence (Fig 2). However, for those genotypes where their requirement is not met, flowering is delayed until daylength and temperature have risen in spring and they flower 76 days after emergence. When sowing is delayed after the shortest day of the year (e.g. in July), this differential response does not occur and the variation between genotypes in days-to-flower is far less. The yield benefit of early sowing and flowering is not always achieved in faba bean because of frost damage, low bee activity in mid-winter and a high incidence of foliar disease resulting from a large canopy development.

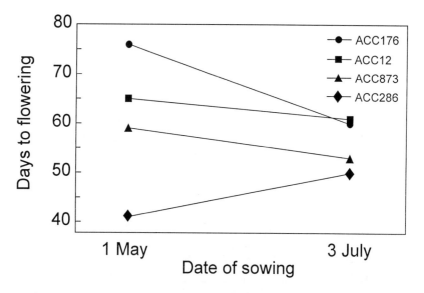

Figure 2. Effect of date of sowing on the number of days from emergence to flowering of faba bean accessions in South Australia.(After Adisarwanto 1988)

Delayed sowing has been shown to significantly reduce yields with current faba bean cultivars, as there is less biomass and limited podding sites. Studies on faba bean have shown a marked yield reduction with late sowing (Baldwin, 1980; Marcellos and Constable, 1986; Adisarwanto and Knight, 1997; Loss and

Siddique, 1997). Experiments in subtropical environments indicate the earliest flowering genotypes are generally the highest yielding and even when irrigation is used to remove the effects of the late season drought stress, yields correlate strongly with the time of flowering.

CROP ESTABLISHMENT

Chickpea shows high yield plasticity over a range of plant densities, but the economic optimum plant density is 40-45 plants/m^2 for desi types and 30 plants/m^2 for kabuli (Siddique *et al.* 1984; Beech and Leach, 1988; Leach and Beech, 1988; Beech and Leach, 1989; Brinsmead *et al.*, 1996). Most commercial crops are established at these plant densities. This is achieved by using germination-tested seed which is treated with fungicide if seed borne disease is suspected. Early studies concluded that conventional, narrow row spacing (18 cm) enhanced crop radiation interception, thereby increasing biomass accumulation and weed suppression (Siddique *et al.*, 1984; Beech and Leach, 1989). However, direct seeding into undisturbed cereal stubble, a technique which increases chickpea yield by 11% compared to establishment in stubble mulched cultivated seedbeds (Felton and Marcellos, 1997), normally requires a wide row spacing. Studies in the north-eastern wheatbelt show that extending row spacing from 18-25 cm to 50 cm has no effect at moderate to high yield levels, while in low yielding situations (<1.5 t/ha) no reduction is observed at 100 cm row spacing (Brinsmead *et al.*, 1996; Felton and Marcellos, 1997).

The large yield advantage of early sowing of faba bean in SA is evident from Figure 3. It is also apparent that when sowing early it is necessary to lower density to achieve maximal yields. The need to change sowing rate and plant density depending on the date of sowing is not commonly practised by farmers who sow to achieve a density of 30 plants/m^2 irrespective of sowing date. The results in Figure 3 were obtained in a trial in which foliar diseases were absent. These diseases are usually more prevalent on early sown crops and need to be controlled by fungicidal sprays. . Most farmers in SA prefer to suffer the yield penalty of sowing in June and avoid the cost of disease control. Part of the objective when breeding Ascochyta resistant cultivars such as Ascot is to take advantage of the high yields that result from early sowing without high levels of disease infection.

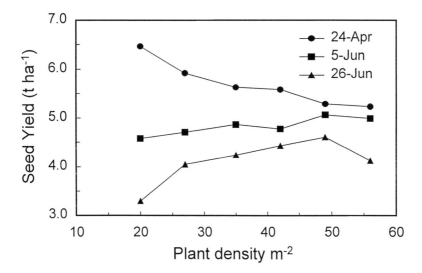

Figure 3. Effect of plant density and sowing date on yield of faba bean cv Fiord in South Australia.(After Adisarwanto and Knight 1997)

WEED MANAGEMENT

Chickpea and faba bean differ markedly in their early vegetative growth and their competitiveness with weeds (Siddique et al., 1993, Thomson and Siddique, 1997). Chickpea with its slow growth, particularly in cool southern Australia, is vulnerable to weed competition whereas faba bean is more vigorous and competitive. Both chickpea and faba bean tolerate seeding depths up to 8 cm without yield penalty (Siddique and Loss, 1997) and deep sowing minimises any adverse effects of pre-emergent herbicides. Genotypic variation in tolerance of chickpea to pre-emergent herbicides indicates scope for adjusting application rates according to cultivar (Bretag and Brouwer, 1995b). The post-emergent herbicide pyridate has extended options for managing broad-leaf weeds, but it does not control the dominant cruciferous weeds. The grass weeds Avena fatua and Lolium rigidum are alternative hosts for cereal diseases and the control of these weeds in pulse crops with post-emergent herbicides is a factor in the rotational benefits attributed to pulses.

FUNGAL DISEASES

As already mentioned, fungal diseases have had a major impact on the establishment and expansion of the Australian chickpea and faba bean industries. Both crops were relatively free of disease when first grown in Australia, but as the area increased the incidence and severity of diseases limited further expansion. For example, in SA the area sown to faba beans increased from 2,000 ha in 1982 to 40,000 ha in 1987, but decreased to 20,000 ha in 1989, largely as a result of increased incidence of foliar diseases. Production in SA has remained at this level since 1989. Breeding disease resistant cultivars is a priority in the programs for both crops.

Chickpeas

Phytophthora root rot (*Phytophthora medicaginis*) is the major root disease of chickpea in northern NSW and southern QLD, although relatively unimportant elsewhere in the world. Since its identification in 1979 (Vock et al., 1980) Phytophthora has caused yield losses up to 20% in some districts (M. Schwinghamer, personal communication). Genetic resistance is the most effective control option (Bretag and Mebalds, 1987), and can be complemented by field selection, crop rotation and seed dressings. Field resistance was identified in the chickpea accession CPI 56564 and resistance screening techniques developed (Brinsmead et al., 1985). The resistance from CPI 56564 was subsequently incorporated into the cultivars Barwon and Norwin. These cultivars have only moderate resistance with yields about 50% higher than for susceptible cultivars under disease conducive conditions. The wild species Cicer echinospermum has been identified as a superior source of resistance (Schwinghamer, personal communication).

Pythium seed rot and damping off, incited by *Pythium* spp. (Cother, 1977), is a less serious soil-borne disease. However, kabuli chickpea is highly susceptible to pythium in wet soils, because of their thin seedcoats and low concentrations of the fungistatic phenolic compounds (Knights and Mailer, 1989). Fungicide treatment of seed provides an effective control of seed borne diseases including Pythium. Damage to rhizobia by fungicide and dispersal of Botrytis grey mould (BGM, *Botrytis cinerea*) in inoculum applied on the seed are avoided by separate application of fungicide and rhizobia shortly prior to sowing (T. Bretag, personal communication).

The most serious foliar disease in southern Australia is BGM. Sclerotinia (*Sclerotinia sclerotiorum* and *S. trifoliorum*), Phoma blight (*Phoma medicaginis* var *pinodella*) and Ascochyta blight (*Ascochyta rabiei* asexual stage) can also cause serious sporadic damage. Foliar diseases cause economic loss through reduced yields and seed quality. The use of lodging resistant cultivars (e.g. Amethyst) and management practices such as low seeding rates and wide row spacings reduce disease severity through improved canopy aeration. Some useful resistance to BGM has been demonstrated in glasshouse tests (Bretag and Brouwer, 1995a). Seedling soft rot caused by seed transmission of BGM results in establishment losses up to 75% (Burgess et al., 1997) and in Vic the seedling phase of BGM is considered more serious than the foliar phase (T. Bretag, personal communication).

Faba bean

The foliar diseases, chocolate spot (*Botrytis fabae*), Ascochyta blight (*Ascochyta fabae*) and rust (*Uromyces fabae*) cause economic losses in Australia. The importance of each varies on a regional basis and seasonal conditions. The diseases are managed by a combination of agronomic practices, including the grazing and incorporation of the pulse stubble, extended rotations, time of sowing, sowing rate and application of fungicides. Genetic resistance has been identified for all diseases and the development of resistant cultivars has a high priority in breeding programs. Much of the disease resistant germplasm used in the programs was obtained from ICARDA.

Chocolate spot affects crops in all regions of Australia and is most severe under conditions of early sowing and high biomass production together with high humidity during the warm spring period. This disease usually can be controlled by fungicide application, but in some seasons the disease develops into an aggressive stage and yield loss and seed discoloration occur even with repeated applications. In the Australian breeding material there is a range of resistance to chocolate spot. Fiord, which until recently was the only cultivar in Australia, is very susceptible to chocolate spot. Botrytis resistant germplasm, mostly originating from the Andean region of South America, has been obtained from ICARDA and the resistant cultivar Icarus was selected from BPL710 and released in 1992.

Ascochyta blight is a winter disease of the cool, southern faba bean growing regions. Genes conferring resistance occur in many populations, especially those from the Mediterranean region, however most of these populations are heterogeneous. Single plant selections have been made within a number of populations to develop homozygous resistant lines. The Ascochyta blight resistant cultivar Ascot was selected from Fiord and released in 1992. Segregation in the F_2 of Ascot crossed with Icarus, which is susceptible to Ascochyta, suggested the resistance of Ascot was conferred by a single recessive gene (Ramsey *et al.*, 1995). However, Lawsawadsiri (1992) reported resistance in the population ILB752 was dominant and the nature of inheritance of resistance to the disease differed between populations.

Rust is primarily a serious disease of the warm temperate and subtropical districts, particularly northern NSW and southern Qld where in most years it is the main fungal pathogen. It sometimes occurs following late spring rains in the cool southern regions. Field trials conducted at Narrabri NSW in 1995 and 1966 have identified several rust resistant accessions, especially the ICARDA line ILB3025 which originated from Ecuador. The cultivar Rossa originates from this line but was released because of other attributes. Several accessions from Ecuador, including the cultivar Icarus, are also rust resistant. Single plant selection and progeny testing has been used to successfully establish rust resistant lines from many heterogeneous populations of diverse geographic origins. In total 36 accessions and 54 single plant progenies originating in 16 countries have been identified with rust resistance, relative to the cultivar Fiord. Hybrid populations have been established and are being studied to determine the genetic control of rust resistance in ILB3025.

Viruses

Major losses in chickpea and faba bean from virus diseases first occurred in northern NSW in 1992, where the outbreak was confined initially to the Liverpool Plains causing crops failure. Yield reductions were less in subsequent epidemics, but the affected area expanded to include most of northern NSW. Four aphid borne viruses (subterranean clover redleaf virus, bean leafroll virus, alfalfa mosaic virus and cucumber mosaic virus) have been implicated in the disease syndrome (M. Schwinghamer, personal communication). A moderately resistant chickpea accession (T1315) is being released as the cultivar Gully for virus affected areas and yields of 1.0 t/ha have been obtained under high disease pressure where yields of other commercial cultivars have ranged from 0.1-0.3 t/ha (Knights, unpublished data).

The faba bean cultivars Fiord and its derivatives are extremely susceptible to the luteoviruses and little progress has been made in the search for resistant genotypes. In addition to breeding, management practices that deter aphids such as stubble retention from a previous crop, rapid canopy closure (narrow rows and high plant population) and cereal borders, are being examined.

Nematodes

Two root lesion nematodes (*Pratylenchus thornei* and *P. neglectus*) are important pests in northern NSW, southern Qld, Vic and SA (Vanstone *et al.*, 1993). Both nematode species have low multiplication rates on faba bean, but chickpea crops generate high soil populations that may affect subsequent crops of susceptible cereal cultivars. The only current nematode control measures are using a non-host crop (e.g. canola) or resistant species or cultivars to precede and follow susceptible crops such as chickpea.

The oat race of stem nematode (*Ditylenchus dipsaci*) is found in South Australia. It infests faba bean where it can multiply to high numbers and is a problem where susceptible oat cultivars are included in the rotation. The race of stem nematode in Australia is different to that found in Europe and stem nematode resistant accessions introduced from ICARDA are susceptible to the Australian race (M. Scurrah, personal communication).

Insect pests

Pod borers (*Helicoverpa punctigera* and *H. armigera*) are the only insect causing direct economic damage to chickpea (Titmarsh, 1987) and faba bean in all regions. Although both pulses are vulnerable to larval attack in the seedling stage, temperatures are generally too low for moth activity until flowering commences in spring. From that time, crops are monitored until threshold larval numbers are reached. Large yield losses are almost completely avoided by one or two insecticide sprays usually applied by aeroplane. However, the costs of the application, insecticide resistance and environmental considerations are demanding a more integrated pest management approach involving biological control and host resistance.

Aphids can infest crops in the warm months of both autumn and spring in the northern region and faba bean sown in autumn can therefore be attacked on two separate occasions. The role of aphids in transmitting viruses means that control in either period will be critical to the success of the crop. Aphids tend to be sporadic problems in southern areas depending upon seasonal conditions. Other insect pests of faba bean include red legged earthmite (*Halotydeus desmetor*) and lucerne flea (*Sminthurus viridis*) and these require chemical control in some seasons. Chickpea is resistant to both these insect pests (Thackray *et al.*, 1997).

Pollination of faba beans

As faba bean is partially cross-pollinated, the degree of fertilization will depend on the autofertility of the genotype and the presence of effective pollinators. In Australia, the honeybee is the only pollen vector for faba bean and bumblebees, a major vector in Europe, do not occur in Australia (Stoddard, 1991). Bee visitation and pollination of the cultivar Fiord has been studied in southern Australia (Stoddard, 1991) and northern NSW (Marcellos and Perryman, 1988). In neither case was there evidence of yield being limited by ineffective pollination, with the mean frequencies of pollination or fertilization over many crops being 80% and 97%, respectively. The conclusion that there are sufficient feral honeybees to effect pollination is based on the autofertility and phenology of Fiord. This might not be the case for genotypes with a low autofertility, which would rely to a greater extent on insect-mediated pollination. However, as the Australian breeding trials are conducted on farms and rely on feral populations of honeybees for pollination, it is unlikely that a cultivar would be selected and released whose pollination requirements were not being met.

SOILS

Chickpea was the first winter pulse grown on a significant scale in north-eastern Australia, a region where vertisols (grey clays and black earths) are common. A factor in the crop's adoption in this region was its preference for neutral-alkaline soils (Jessop and Mahoney, 1982). Nonetheless, crop failures were first experienced on the poorly-drained vertisols, due to root rot, waterlogging and sodicity. Waterlogging is now recognised as a major cause of production loss on fine textured soils in Qld and NSW (Schwinghamer, 1994; Cowie *et al.*, 1996) and shallow duplex soils in WA (Siddique *et al.*, 1993) and losses are now contained by confining chickpea to the well-drained soils.

Chickpea has demonstrated a yield advantage on the deep grey clays of the Victorian Wimmera and the slightly acid to neutral clay-loams of WA when compared with more coarse-textured and acid soils. Faba bean is generally adapted to a similar range of soils as chickpea and performs best on neutral to alkaline loams or clay-loams. Well-drained soils are preferred, but faba bean tolerates transient waterlogging better than most pulses, particularly after emergence.

NUTRITION

Chickpea shows little or no response to applied phosphorus (P), even in soils where the P status is low. There was no significant response to P applied to vertisols in northern NSW (Cole and Lonergan, unpublished data), or at several sites in WA, except on virgin soils (Bolland *et al.*, 1997). On the other hand, faba bean is responsive to soil applied P and yield increases of 20 to 25% have been reported at soil P levels considered adequate for legume pastures (Lewis and Hawthorne 1996, Bolland *et al.*, 1997).

Faba bean is also responsive to zinc, but only if the P nutrition is adequate (Lewis and Hawthorne, 1996). Susceptible genotypes of both pulses suffer iron chlorosis on high pH soils, but sources of iron efficient genotypes are available. The chickpea cultivar Tyson originally contained 25% of iron inefficient plants and a new iron efficient cultivar was subsequently developed and released as "Reselected Tyson". Iron chlorosis occurs in the south-east of SA where a broad bean industry, based on the iron-efficient cultivar Aquadulce, is located. High subsoil concentrations of boron reduce cereal yields in parts of the low rainfall regions of southern Australia (Nable and Paull, 1991), and are a potential constraint to other crops in these areas. Most Australian cultivars of chickpea are highly sensitive to boron under controlled environmental conditions (Bagheri *et al.*, 1993), but Australian faba bean cultivars are largely tolerant (J. Paull, unpublished data).

NITROGEN FIXATION

The identification of appropriate commercial strains of rhizobia has been essential for the successful cultivation of chickpea and faba bean in Australia. Strains specific to chickpea were not present in Australian soils when chickpea was first introduced. Five strains were evaluated by Corbin *et al.* (1977) and one (CC1192) was developed into the commercial inoculant Group N. Inoculation of seed was at first mandatory for all "new" fields, and in many areas is still recommended irrespective of cropping history, particularly where soils are slightly acid in pH. Crops are generally well nodulated and Schwenke *et al.*, (1997) did not find any poorly nodulated crops in a recent survey of northern NSW.

Faba bean is inoculated by the rhizobium strain (Group E) also used for field pea, lentil and vetch. This strain persists in the soil from one pulse crop to the next in neutral to alkaline soils (Carter *et al.* 1995), but seed of each crop needs to be inoculated when sown in acid soils. The current commercial inoculant (NA533 syn SU303) was released following studies by Carter *et al.* (1994) who demonstrated a yield advantage over the previous inoculum, particularly on acidic soils. The release of the new inoculant has enabled faba bean cultivation to extend to soils previously considered too acidic (Siddique *et al.* 1993, Loss and Siddique 1997).

HARVESTABILITY

All chickpea and faba bean crops in Australia are machine harvested. Significant harvest losses occur when the crops are short, especially with the chickpea cultivar Tyson and the faba bean cultivars Fiord and Ascot when grown in low rainfall conditions. All chickpea cultivars released since Tyson are taller, and are less prone to harvesting losses. However, conditions favouring high biomass sometimes result in harvest losses due to lodging. This applies particularly to northern NSW and southern Qld where the effects of competition, mechanical damage and foliar disease have reduced the yields of lodging-susceptible chickpea genotypes by up to 74% (Knights, unpublished data). Resistance to lodging has been achieved in the cultivar Amethyst which has a recessive gene for erectness. This gene confers increased stem strength, better resistance to bending, and longer internodes which increase plant height by 25% and the height of the lowest pod by 30% (Knights 1989).

There is considerable genetic variation among Australian faba bean breeding material in pod height and many genotypes set their lowest pod higher than Fiord. Although there is a positive relation between days to commencement of flowering and the height of the lowest pod there is also variation within each flowering class allowing selection for high pods and early flowering. As harvesting often occurs under very hot conditions, dehiscent pods can lead to large seed losses. In the faba bean breeding program, genotypes with dehiscent pods are excluded.

MARKETS AND GRAIN QUALITY

The most important factor contributing to the expansion of chickpea production in Australia was the opening up of export markets to the Indian sub-continent (ISC) in the mid 1980s. The ISC is now Australia's largest market for chickpeas. As Australian consumption accounts for only 1% of production, growth of the industry will continue to rely on export markets (Siddique and Sykes, 1997). Bulk consignments of desi chickpea are mainly shipped to the ISC ports in time to exploit demand before the local harvest in March/April (Siddique, 1993). The small-seeded desi cultivars Tyson and Amethyst were well suited to the low value, bulk (splitting) market. However, an increasing emphasis on seed size and colour has been reflected in the new cultivars Semsen, Lasseter, Heera, Sona and Gully. These have large, light-coloured seeds and are sold in high value niche markets.

Increased seed size has also been a breeding objective in kabuli chickpea. The cultivar Macarena produces very large seeds (60 g/100 seeds) in irrigated crops (McNeil, 1988), and the most recent kabuli cultivar Bumper produces mainly premium, 9-10 mm diameter (50-55 g/100 seeds) grades under rainfed conditions (Fig. 4). The importance of appearance in seed quality has focussed attention on the damage that can occur during harvesting and grain handling. The problem is often compounded in sub-tropical regions where pre-harvest rainfall causes weathering of the seed. Cultivars have been developed with weathering resistance (Knights, 1993), but the character is correlated with low splitting recovery.

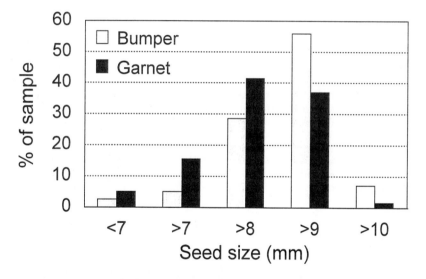

Figure 4. Comparison of seed size distribution for two kabuli chickpea cultivars in New South Wales

Markets for faba bean have been developed in the Middle-East, Europe and South-east Asia where the beans are sold whole, split, or canned. Seed size, an ability to be split, and testa colour are relevant to specific markets and consumer preferences. All these characters are affected by environmental conditions. The light buff colour of the seed coat preferred by many markets is sometimes not attained because of adverse weather before harvest and changes to an unacceptable dark brown colour is a problem when beans are stored for several months. The breeding program is selecting against this progressive seed coat discolouration and is also developing low tannin types.

The green colour of the testa of Icarus, while acceptable to some faba bean consumers, is unimportant if the bean is split but can make it unsuitable for canning as whole seed. Now that faba bean production in the northern subtropical region is expanding, the interaction between grain quality and the environment is increasingly evident. Mean seed weights for any one cultivar are generally less in the north than in the southern regions. There is a need to develop genotypes with large seed weights to contribute to national marketing strategies. Although the potential exists to breed for quality traits they must be assigned a low priority compared to disease resistance and yield in faba bean.

FUTURE IMPROVEMENTS TO ADAPTATION

The current national average yields of chickpea and faba bean of about 1.0 and 1.3 t/ha could be increased to 1.4 and 1.6 t/ha, respectively, with additional cultivar improvements and better crop management. Genetic improvement of chickpea in the last two decades has increased yield potential in the various regions by 7-25% (Table 1). However, biotic constraints still prevent the realisation of yield potential even in the most recent cultivars of both crops. All chickpea cultivars are at least moderately susceptible to BGM and root lesion nematode, and there is scope for further gains in virus and phytophthora resistance.

Good sources of resistance to the three major foliar fungal pathogens of faba bean have been identified. Several cultivars with resistance to a single disease have been released but cultivars with combined resistances to the diseases prevalent in specific regions are still required. Identification of sources of virus resistance and development of virus resistant cultivars adapted to the sub-tropical region is a high priority for both crops. In addition a better understanding of the pathotypes and epidemiology of the diseases present in Australia is required to develop effective long-term breeding and management strategies.

Yields can be improved by increasing water use efficiency (WUE), which is lower for chickpea than for other cool season pulses such as field pea and faba bean (Siddique and Sedgely, 1985 and 1987; Thomas and Fukai, 1995; McDonald, 1995; Loss *et al.*, 1997). This can be attributed in part to low biomass production in chickpea (Siddique *et al.*, 1993; Thomson and Siddique, 1997). A more rapid biomass accumulation would improve WUE, particularly in Mediterranean type environments where 60% of the water used by chickpea crops can be lost through soil evaporation (Siddique and Sedgley, 1987).

The cultivation of faba bean and chickpea in Australia is still in an expansion phase and the full potential of these crops in the various regions is still unknown. The optimum phenology pattern has not been determined for all potential regions of production and it will be necessary to evaluate germplasm from various overseas environments to identify the most appropriate types. One of the major management issues facing faba bean production in all regions of Australia is the relation between time of sowing, yield potential and incidence of disease. Faba bean has consistently demonstrated a positive yield response to early sowing across a range of environments, although at a higher risk of foliar disease. Disease resistant cultivars would not only improve yield through reduced losses due to diseases, but would also enable early sowing to maximise the yield potential.

Acknowledgments

We acknowledge the Grains Research and Development Corporation of Australia (GRDC) for their financial support of faba bean and chickpea improvement and industry development in Australia. We also thank Dr Stephen Loss and Kerry Regan for their comments on the manuscript.

References

Adisarwanto, T. 1988 Ph D Thesis, University of Adelaide, South Australia.

Adisarwanto, T. and Knight, R. 1997 *Australian Journal of Agricultural Research* 48: 1161-68

Baldwin, B.J. 1980 *FABIS* 2: 39.

Bagheri, A., Paull J.G. and Rathjen A.J. 1993 *Proceedings Tenth Australian Plant Breeding Conference* 72-3.

Beech, D.F. and Brinsmead, R.B. 1980 *Journal of the Australian Institute of Agricultural Science* 46: 127-129.

Beech, D.F., and Leach, G.J. 1988 *Australian Journal of Experimental Agriculture* 28: 367-376.

Beech, D.F., and Leach, G.J. 1989 *Australian Journal of Experimental Agriculture* 29: 241-246.

Bolland, M.D., Siddique, K.H.M., Loss, S.P. and Baker, M.J. 1997 *Nutrient Cycling in Agroecosystems* (in press).

Bretag, T. and Brouwer, J. 1995a In *Proceedings of the 2nd European Conference on Grain Legumes*, Copenhagen, Denmark, 98. AEP, Paris.

Bretag, T. and Brouwer, J. 1995b In *Proceedings of the 2nd European Conference on Grain Legumes*, Copenhagen, Denmark, 133. AEP, Paris.

Bretag, T.W. and Mebalds, M.I. 1987 *Australian Journal of Experimental Agriculture* 27:141-148.

Brinsmead, R.B., Rettke, M.L., Irwin, J.A.G., Ryley, M.J. and Langdon, P.W. 1985 *Plant Disease* 69: 504-506.

Brinsmead, R.B. 1992 In *Proceedings of the 6th Australian Agronomy Conference*, Armidale, Australia, 244-246 (eds K.J. Hutchinson and P.J. Vickery). Australian Society of Agronomy.

Brinsmead, R.B., Thompson, P.R. and Martin, W.D. 1996 In *Proceedings of the 8th Australian Agronomy Conference*, Toowoomba, Australia 627 (ed M. Asghar). Australian Society of Agronomy.

Burgess, D.R., Bretag, T. and Keane, P.J. 1997 *Australian Journal of Experimental Agriculture* 37: 223-229.

Carter, J.M.,Gardner,W.K. and Gibson A.H. 1994 *Australian Journal of Agricultural Research* 45: 612-23.

Carter, J.M.,Tieman, J.S. and Gibson, A.H.1995 Soil *Biology and Biochemistry* 27: 617-23

Corbin, E.J., Brockwell, J. and Gault, R.R. 1977 *Australian Journal of Experimental Agriculture and Animal Husbandry* 17: 126-34.

Cother, E.J. 1977 *Seed Science and Technology* 5:593-597.

Cowie, A.L., Jessop, R.S. and MacLeod, D.A. 1996 *Plant and Soil* 183: 97-103.

Felton, W. and Marcellos, H. 1997 *Australian Grain* (in press).

Hamblin, A. and Kyneur, G. 1993 *Trends in Wheat Yields and Soil Fertility in Australia*. Australian Government Publishing Service: Canberra.

Herridge, D.F., Marcellos, H., Felton, W.L., Turner, G.L. and Peoples, M.B. 1995 *Soil Biology and Biochemistry* 27: 545-551.

Jessop, R.S. and Mahoney, J. 1982 *Australian Journal of Soil Research* 20:265-268.

Jessop, R.S. and Mahoney, J. 1985 *Journal of Agricultural Science, Cambridge* 105: 231-236.

Knights, E.J. 1989 M.Sc.Agric. Thesis, University of Sydney, Australia.

Knights, E.J. 1993 *International Chickpea Newsletter* 29: 25-27.

Knights, E.J. and Mailer, R.J. 1989 *Journal of Agricultural Science, Cambridge*. 113: 325-330.

Lawler, H.J., Siddique, K.H.M., Sedgley, R.H., and Thurling, N. 1988 *Acta Horticulturae* (in press).

Lawsawadsiri, S. 1992 Ph D Thesis, University of Adelaide, South Australia.

Lewis, D. C. and Hawthorne, W.A. 1996 *Australian Journal Experimental Agriculture* 36:479-84.

Leach, G.J., and Beech, D.F. 1988 *Australian Journal of Experimental Agriculture* 28: 377-383.

Loss, S.P. and Siddique K.H.M. 1997 *Field Crops Research* 52: 17-20.

Loss, S.P., Siddique, K.H.M. and Tennant, D. 1997 *Field Crops Research* 54: 153-62.

Marcellos, H. 1984 *Journal of the Australian Institute of Agricultural Science* 50:111-113.

Marcellos, H. and Constable, G. A. 1986 *Australian Journal of Experimental Agriculture* 26: 443-6

Marcellos, H. and Perryman, T. 1988 *Australian Journal of Agricultural Research* 39: 579-87.

McDonald, G.K. 1995 *Australian Journal of Experimental Agriculture* 35: 515-523.

McNeil, D.L. 1988 *International Chickpea Newsletter* 18: 34-35.

Nable, R. and Paull, J.G. 1991 *Current Topics in Plant Biochemistry and Physiology* 10: 257-273.

Pye, D.L. 1987 In *Proceedings of the Australian Chickpea Workshop*, Warwick, Queensland, 27-29 October 1987, 58-62 (eds R.B. Brinsmead and E.J. Knights) Australian Institute of Agricultural Science.

Ramsey M., Knight, R. and Paull, J. 1995 In *Proceedings of the 2nd European Conference on Grain Legumes*, Copenhagen, Denmark, 164-5. AEP, Paris.

Rees, R.O., Brouwer, J.B., Mahoney, J.E., Walton, G.H., Brinsmead, R.B., Knights, E.J. and Beech, D.F. 1994 In: *Expanding the Production and Use of Cool Season Food Legumes*. pp 412-425 (eds F.J. Muehlbauer and W.J. Kaiser). Kluwer Academic Publishers.

Schwenke, G.D. and Herridge, D. 1997 *Australian Grain* 7: 15-16.

Schwenke, G.D., Herridge, D.F., Peoples, M.B. and Turner, G.L. 1997 In *Managing legume nitrogen fixation in cropping systems of Asia*. Proceedings International Workshop, ICRISAT, Hyderabad, India, 20-24 August 1996, 00-00 (eds O.P. Rupela, C. Johansen and D.F. Herridge).(in press)

Schwinghamer, M.W. 1994 *Grower guide to identification of chickpea diseases in northern NSW*. NSW Agriculture/Grains Research and Development Corporation. NSW Agriculture Communications Unit, Dubbo, NSW, Australia.

Siddique, K.H.M. 1993 Cooperative Centre for Legumes in Mediterranean Agriculture, Occasional Publication No. 5. University of WA.

Siddique, K.H.M., Walton, G.H. and Seymour, M. 1993 *Australian Journal of Experimental Agriculture* 33: 915-922.

Siddique, K.H.M. and Loss, S.P. 1997 *Australian Journal of Experimental Agriculture* (submitted).

Siddique, K.H.M. and Sedgley, R.H. 1985 *Field Crops Research* 12: 251-269.

Siddique, K.H.M. and Sedgley, R.H. 1986 *Australian Journal of Agricultural Research* 37: 245-261.

Siddique, K.H.M. and Sedgley, R.H. 1987 *Australian Journal of Agricultural Research* 37:599-610.

Siddique, K.H.M. and Sykes, J. 1997 *Australian Journal of Experimental Agriculture* 37: 103-111.

Siddique, K.H.M., Sedgeley, R.H., Davies, C.L. and Punyavirocha, S. 1994 In: *Abstracts of the International Synposium on Pulses Research*, April 2-6, 1994, New Dehli, India, 85.

Siddique, K.H.M., Sedgley, R.H. and Marshall, C. 1984 *Field Crops Research* 9: 193-203.

Stoddard, F.L. 1991 *Australian Journal of Agricultural Research* 42: 1173-8.

Strong, W.M., Harbison, J., Nielson, R.G.H., Hall, B.D. and Best, E.K. 1986 *Australian Journal of Experimental Agriculture* 26:353-359.

Thackray, D.J., Ridsill-Smith, T.J. and Gillespie, D.J. 1997 *Plant Protection Quarterly* 12: 141-144.

Thomas and Fukai, S. 1995 *Australian Journal of Agricultural Research* 46: 35-48.

Thomson, B.D. and Siddique, K.H.M. 1997 *Field Crops Research* 54: 189-99.

Titmarsh, I.J. 1987 In *Proceedings Australian Chickpea Workshop*, Warwick, Queensland, 27-29 October 1987. 117-22 (eds R.B. Brinsmead and E.J. Knights). Australian Institute of Agricultural Science.

Valder, G. 1893 *Agricultural Gazette*, NSW 4: 916-917.

Vanstone, V.A., Nicol, J.M., and Taylor, S.P. 1993 In: *Proceedings of the Pratylenchus Workshop*, 9th Biennial Australian Plant Pathology Society Conference, Hobart, 8-9 July 1993 (eds V.A. Vanstone, S.P. Taylor and J.M. Nicol).

Vock, N.T., Langdon, P.W. and Pegg, K.G. 1980 *Australasian Plant Pathology* 9: 117.

Innovations in processing and marketing-Lessons from the vertical integration of the U.S. soybean industry

Lockwood Marine
Marine Associates, Fort Wayne, Indiana, USA

INTRODUCTION

The cultivation and processing of the soybean has an interesting history. It is generally agreed that soybeans were first cultivated in China. They were probably consumed as a pulse crop and later developed into a variety of food products, including tofu, miso, tempeh, soymilks and, of course, soy sauces. The extraction of oil for use in direct consumption and cooking other food products is known to have occurred in China and other Asian countries for hundred of years. However, the oilseed processing industry in the United States and Western Europe is young, dating from about 1910.

EARLY DEVELOPMENT OF THE U.S. SOYBEAN INDUSTRY

Soybeans were first introduced to the United States around 1900 and used as a forage and green manure crop. The high protein content of the plants and their ability to fix nitrogen from the air made them an attractive forage crop for the depleted soil areas of the south and south Atlantic states of the United States.

The first soybeans were crushed for oil and meal in the United States in 1911, using soybeans imported from Manchuria and processed at a facility in Seattle, Washington. During World War I the shortage of fats and oils led to an expanded processing of imported and domestically grown material, primarily at cottonseed processing plants in the southeast and southern parts of the U.S. Soybean production spread from the south and southeast into the mid-western part of the United States and the first plant, built specifically for processing soybeans, was completed in 1922 by the A.E. Staley Company of Decatur, Illinois. This plant was followed by others in the 1920's and early 1930's including Funk in 1924, Goodrich in 1926, American Milling in 1927, Shellabarger and Archer, Daniels, Midland in 1929, Central Soya in 1934 and Spencer Kellogg in 1935. During this period the major problem was to acquire sufficient soybeans to use the existing capacity since production did not equal processing capacity in the U.S. until 1954. From the earliest days of the U.S. industry, there has been cooperation between producers, marketers, and processors to encourage production, look for new uses for products and to develop new markets. Industry teamwork began as early as 1928 when a contract was developed for the production of 50,000 acres (22,240 ha) of soybeans at a minimum price of $1.35 per bushel (1 bushel of soybeans weighs 27.2 kg)

It was during the period 1935 to 1945 that the U.S. soybean industry established itself as a leader in the fats and oils industry. In 1934, there were 19 processing plants, which crushed a total quantity of 9 million bushels. By 1945 the shortages of imported fats and the need to support the world-wide war effort led to an industry composed of 160 plants with a capacity to crush 189 million bushels per year. The U.S. industry continued to expand rapidly to the present time, with 1997 production estimated at over 2.7 billion bushels (approx. 73.4 million MT) and a processing volume of almost 1.5 billion bushels. From humble beginnings only 75 years ago, the U.S. industry has grown into a major factor in world markets with an annual sales value over $18 billion per year.

Throughout this development period, there have been three factors responsible for the growth (1) innovations in production, (2) innovations in processing and development of new products, and (3) joint industry market development activities. I will discuss each of these factors in detail, as I believe there are lessons for the worldwide pulse industry. Throughout the growth of the industry, a unique situation has been the degree of cooperation between agricultural producers, marketers, and processors. I know of no other

305

R. Knight (ed.), Linking Research and Marketing Opportunities for Pulses in the 21st Century, 305–309.
© *2000 Kluwer Academic Publishers. Printed in the Netherlands.*

major crop or producing and processing industry where industry-wide cooperation has been as important as in the U.S. soybean industry.

SOYBEAN PRODUCTION DEVELOPMENTS - RESEARCH

As I have mentioned, early varieties were introduced to the United States from breeding stock produced in China. From those early days, representatives of seed companies and processing companies have worked together to develop improved varieties with better yields, resistance to disease and improved physical characteristics. Credit for much of the success in variety development, and indeed for much of the success of U.S. agriculture, goes to the Land Grant Universities which sponsored research in production, screened and bred varieties for improved yields, and developed incomes. Today seed companies and private breeders do a larger share of the research on varieties, but the land-grant institutions continue to be a force in soybean improvement in the United States. In the 1950's, a joint producer-processor organisation, the National Soybean Crop Improvement Council, was formed to promote production. This organisation sponsored yield contests, improved production practices, and regularly brought together the leading breeders and crop production specialists from the major land-grant universities throughout the soybean belt. Advancements have included the development of high-yielding, disease resistant varieties, the development of weed and pest control chemicals, development of tillage methods including reduced tillage and no-till options, and an improved composition of the soybean.

In recent years a major source of funding for research and production has come from the United Soybean Board, a national agency responsible for managing the funds collected from the soybean producer check-off program. The national check-off involves collecting one-half of one percent of the value of the soybeans, at the first point of sale. This money is ultimately used by the Board to support research in improving production methods, improving soybean products, market development activities throughout the world and improvement of producer and product user communications.

INNOVATIONS IN PROCESSING AND NEW PRODUCTS

Innovations in processing and development of new soybean based products have contributed significantly to the growth of the industry. Advancements include the adoption of continuous solvent extraction plants in the 1930's and early 1940's; the adoption of a technique to eliminate an anti-nutritional factor known as trypsin inhibitor activity; the adoption of energy efficient processing machinery; and the development of process control equipment, and pollution control facilities. The development of procedures for desolventizing and toasting soybean meal improved the nutritional value of the product and contributed to the worldwide growth in poultry and livestock production over the last forty years. These and other innovations have made the U.S. a world leader in processing costs and in the production and marketing of soybean products.

The U.S. processing industry began development in the late 1940's and early 1950's of industrial and food products derived from soybean and its component products. The list includes all of the products for direct human consumption such as tofu and soymilk, as well as products for use as ingredients in food products including soyflour, soy protein concentrate, isolated soy proteins, and fibre from soybean hulls. In addition to the standard food, fat and oil products, there is a range of products for other uses, such as soybean lecithin, which is natures most plentiful natural emulsifier, special biodegradable lubricants, soybean oil based printing inks, paints, glues, and of course raw materials for producing specialty fatty acids through fractionation. I will discuss various protein food products in more detail in a later section.

MARKET DEVELOPMENT

The U.S. soybean industry has long been involved in market development and promotion of soybeans and soybean products. The first national Soybean Growers Association was formed in 1920 and became the American Soybean Association in 1929. The National Soybean Processors was formed in 1930 to promote production, the interests of processors and the marketing of products on a worldwide basis. I have already mentioned the Soybean Crop Improvement Council which was a joint producer/processor organisation to stimulate production. At the end of WW II, producers and processors began to look to export markets as a source of a major part of their future growth. In the late 1940's representatives of the American Soybean Association and sales people of the processing industry travelled to foreign countries to search for new markets. Japan and Germany were two of the earliest customers. In 1956 processors and producers went together to form the Soybean Council of America with the objective of promoting soybean products in forty countries. The Council established international soybean offices and in 1960 signed a global contract with the Foreign Agricultural Service of the United States Department of Agriculture, which made federal government funds available for the promotion of exports of soybeans and soybean products. At about the same time, the U.S. Congress passed Public Law 480, designed to promote the export of U.S. agricultural products particularly to food deficient developing countries. This program was useful in introducing large quantities of soybean oil to consumers in developing countries in South America, Eastern Europe, the Mediterranean region, Africa and the Asian Sub-continent. The millions of tons of oil exported, as well as soybean based products distributed under the Food for Peace Program made millions of consumers around the world aware that the products were nutritionally sound, economical and readily useable in existing diets and food preparation systems. These programs also contributed to worldwide agricultural development since most of the commodities were sold in local currency subsequently given back to the respective governments for use in agricultural projects. Likewise the international offices of the American Soybean Association, with early support from the processors and continuing support by the Foreign Agricultural Service and, more recently, the United Soybean Board, have been instrumental in promoting the use of soybeans and soybean products throughout the world. Feeding trials to show producers the value of soybean meal in livestock and poultry rations, food promotions of products using soybean oil, and numerous other market development and promotion programs have greatly expanded the world market for U.S. soybeans. During the 1950's and 1960's, when there were world-wide surpluses of food and feed grains, cotton and other basic agricultural commodities, the growth in demand for soybeans enabled the U.S. farmer to shift a significant area to soybeans to meet the growing world demand. As a result, soybeans and soybean products have become the U.S. second largest crop and its largest export earner in the agricultural field.

IMPORTANCE OF TRADITIONAL MARKETS

One advantage of the soybean is that it produces co-products which are in growing demand in world markets. Despite all the attention often given to new food and industrial products from soybeans, the basic products of oil and meal drive the fortunes of the industry. Because it is used in food products, the oil always has been the higher valued fraction in terms of price per pound. However, since soybeans contain approximately 80% protein meal, it is the world-wide growth and demand for meat products and the need for a quality protein to use in the livestock and poultry rations, that has led to much of the growth of the soybean industry. Certainly the demand for protein has been and will continue to be the key driver in the world export markets. In order for the U.S. industry to grow and prosper, it will need to exploit fully its competitive advantages. Competing producers of oilseeds, fats and oils are certainly very viable and provide strong competition. This includes countries like Brazil and Argentina who produce soybeans and sunflower seed; countries in Europe producing rapeseed and sunflower seed and the Asian countries of Malaysia and Indonesia who are major and growing producers of palmoil for export. However, the U.S. industry will remain a world-class supplier and be able to prosper if it exploits its advantages. These are large amounts of fertile land available for shifting into oilseeds, one of the world's most favourable temperate growing climates, the world's most efficient marketing and transportation network, an outstanding base of production, marketing and technology information for producers, marketers and plant breeders, as well as the world's largest and most cost efficient processing industry.

CHALLENGES OF THE FOOD PRODUCT MARKETS

Soybeans have been eaten directly or processed into various types of food for thousands of years. Use of soy based products has occurred, and still occurs, primarily in the Asian countries, of China, Japan, Korea and, to a lesser extent, India. Having consumed these products for generations, the populations have developed a taste for the particular flavour of soybeans and its products. The popularity of the products continues to exist and large quantities of soybeans are used for food throughout Asia. However, the first uses of soybeans in food products in the United States and Western Europe occurred much later. The first significant efforts were made during WW II when soy flour was used as extenders in meat products such as sausages, as protein additions to bread and other baked goods and as an alternate for milk products in infant formulas for children allergic to dairy products. Unfortunately the products produced in the 1940's were not of good quality, had poor flavour and had the problem, common to members of the pulse family, of producing flatulence. As a result, products have had to fight the stigma of flavour and flatulence ever since. However, the improved flavour of soy protein concentrates and isolates today has encouraged their use in a variety of products in rapidly growing quantities. As a result of dietary concerns about excessive meat consumption, the traditional soy based products such as tofu are enjoying increased popularity in North America and Western Europe. Only a few of the processing companies around the world are involved in the development and marketing of speciality products from soybeans including the protein concentrates and isolates. With the exception of a few major integrated processing operations, we have yet to see a major food producer in the North American-Western European market attempt to develop soy based food products. Certainly the Asian markets are large and growing and probably offer opportunities for far-sighted companies in the production of food products. Over the long term, food products based on vegetable proteins, whether from soybeans, pulses, or other crops has significant potential in developing countries. Vegetable-based proteins can have a cost advantage in meeting the protein needs of the lower income segments of the population. However, in the developed countries of North America and Europe, vegetable protein products will have a struggle to replace proteins from the favourite sources meat, milk, and eggs produced by animal industries. Clearly there are developing niche markets which hold significant promise, not only for the soybean industry, but also for segments of the worldwide pulse industry. An example is the growing demand for a wide variety of humus products, which are produced and sold in delicatessens and major supermarkets in the United States. Today one can buy humus in a variety of flavours, whereas a few years ago it was available only in a limited number of ethnic restaurants. I suspect that in the United States most of the humus is consumed as a flavourful and nutritious snack, rather than as a main meal course. Regardless of how it is used, it is an example of what product development and smart marketing of age-old products can do in a changing market.

CHANGES IN INDUSTRY STRUCTURE

Historically the U.S. soybean industry has consisted of three rather distinct segments. They were farmer producers, marketer-processing firms, and end-product producers and marketers such as the food and animal feed companies. To this day, the production of soybeans is largely in the hands of private family farms. There is a limited amount of contract production of specialty food soybeans and soybeans bred to have special oil characteristics. Some individual entrepreneurs and groups of farmers have found niche markets within the specialty food area. An example would be a group of farmers in Kansas or Nebraska who grow special varieties uniquely suited to the production of tofu and, after special cleaning, ship them in containers to buyers in Japan, either trading companies or tofu producing companies. There is one major processing and refining company in the United States which is owned by a group of regional cooperatives and, ultimately by farmer producers. However, this company deals with their farmer-owners in much the same way corporate integrated processor-refiners deal with individual farmers.

Through most of its history, the U.S. processing industry has been composed of individual entrepreneurs or corporations which may have operated grain collection facilities as well as processing plants and, in some cases, refineries. However, the few major food companies, such as Ralson-Purina and Proctor and Gamble, who once spanned the distance between raw soybeans and consumer food products, have now virtually exited the processing and refining segments of the industry.

On a historical basis, the development of new food products and marketing them through supermarkets has been done by the multinational food companies such as Proctor and Gamble, Nestle, and Unilever. This situation is expected to continue.

Over the past twenty years, there has been a consolidation within the processing, refining and marketing segments of the U.S. soybean industry. Today the U.S. processing industry is composed of only fifteen firms with two new entrants due to come on line within the next twelve months. The top eight processing firms also control a very significant percentage of the facilities for the collection, transportation, and exporting of soybeans as well as the oil refining capacity. These companies buy soybeans from producers or local elevators, process the soybeans into meal and oil, refine the majority of the oil into edible products, sell the meal to feed companies and the oil, lecithin, and protein products to food companies who use them in food manufacturing. Today there is little integration between the processing-refining segment and the major food producing companies, although there is cooperation in designing ingredients to meet specific purposes, in changing or improving the quality of products and, in some cases, contractual arrangements to provide supplies on an annual basis. The processing refining companies use their experience and skills in assembling large quantities of raw material, processing it at a low cost and moving high quality ingredients to the next level of users. The feed and pet food companies and major food producing companies exploit their experience and skills in food chemistry, new product development, brand management, and retail marketing and advertising. This structure, where each segment does the portion of the chain at which it is most efficient, seems to be working very well.

U.S. MISTAKES - LESSONS FOR THE PULSE INDUSTRY

The U.S. soybean industry has shown tremendous growth and, by and large, has been a profitable enterprise. However, the industry has made mistakes, which have opened the door to competitors and resulted in the size of the industry and the rate of growth over the last twenty years being substantially less than it might have been. The two principal areas where the industry could have improved its fortunes include, exploiting its natural advantages in a better fashion and better or more complete cooperation between segments of the industry. The industry would be larger and more profitable today if it had not been for the restrictive agricultural policies of the 1980's and early 1990's. The tight supply management policies of the U.S. Government, which virtually forced farmers to plant program crops such as corn, wheat, rice and cotton, in order to receive income supports, prevented land from being shifted into soybeans which were enjoying good growth and world wide demand. These policies, coupled with the two major U.S. embargoes on agricultural products, opened the door and created a boost to the growth of the oilseed industries in our two major competitors, Brazil and Argentina. The embargoes were geo-political matters on which the US industry probably could have had little influence. But the industry could have worked harder to modify or eliminate the restrictive government policies which caused the U.S. industry to stagnate after 1980 and presented a major share of the world market to our good competitors to the south. The second major mistake of the U.S. soybean industry was a failure to consistently work together to promote the production and worldwide marketing of soybeans and soybean products. During most of the history of the U.S. industry, cooperation between growers, marketers and processors has been very good. However, there have been periods when petty jealousies and perceived differences in objectives have kept the industry from reaching its full potential. I feel that it is critical that all segments of an industry realise that they are in the business together and that no one segment can prosper unless the total segment prospers.

Opportunities for Pulses in Foods and Non-Food Industrial Processes by Fractionation

Z.Czuchajowska[1]

1 Department of Food Science and Human Nutrition, Washington State University, Pullman, Washington 99164-6376, U.S.A. in cooperation with IMPACT.

Abstract

An important way to increase utilization of legumes in the food and non-food industry is through the wet fractionation process. The main components of legumes, (starch, protein and fiber), utilized separately, could increase the value of legumes by broadening their application. The complexity of the wet fractionation process could be simplified by selecting a material which has desirable properties. Research of the microstructure of legume seeds, distribution of main components within the cotyledon, dry milling and the relationship between them could be very helpful in this selection. The well-defined properties of legume main components separated in the wet process are necessary for their proper utilization.

INTRODUCTION

Legumes have traditionally been consumed as whole seeds or as a ground flour after dehulling. Legume flour is utilized in a variety of traditional products, depending on the part of the world or country (Aykroyd and Doughty, 1982; Chavan et al., 1989; Singh and Seetha, 1993). Cereal grain products such as bread, muffins and especially noodles are frequently fortified by legume flour in order to increase their nutritional value (Deshpande et al., 1983; Duszkiewicz-Reinhard et al., 1988; McWatters, 1990; Bahnassey et al., 1986; Hung and Nithianandan, 1993; Woodin, 1981; Lu and Al Yasser, 1986; Chompreeda et al., 1987; Chompreeda et al., 1988; Taha et al., 1992; Bahnassey and Khan, 1986), since legumes contain almost two times more proteins and minerals, and three times more dietary fiber than wheat flour (Sathe et al., 1984). In addition, legume proteins are rich in lysine, an important essential amino acid, the presence of which is limited in cereal grains (Müller, 1983).

At present there are at least two reasons, which attract the attention of researchers to legume main components (protein, starch and fiber) separated by the process of wet fractionation. First, there is strong public interest in natural, unmodified sources of ingredients. The second reason is the impetus to increase the value of legumes by broadening their utilization.

Fractionation of legumes into main components provides a great opportunity for reaching these goals, and may represent in the future the only option of legume utilization in food and non-food products. The present article, based on selected literature and the author's own experience, is focused on the fractionation process of legumes. Specifically, the article discusses the importance of physical and chemical properties of the material selected for the wet process, the effect of the microstructure of seeds as related to dry milling performance, flour particle size, chemical composition and the thermal behaviour of the legume prime and tailings starches.

R. Knight (ed.), Linking Research and Marketing Opportunities for Pulses in the 21st Century, 311–321.

FRACTIONATION OF LEGUMES

One of the most powerful and effective ways to increase the value of legumes is through their efficient fractionation into starch, protein and fiber, and utilization of these components in the food and non-food industries. Legumes in general are known for their high protein content (15-40%) and, as such, have been the subject of protein extraction studies (Colonna et al., 1980; Desphande and Damodaran, 1990; Swanson, 1990).

The starch content of legumes ranges from 35-60% of the dry weight of beans (Rockland and Melzler, 1967; Sosulski and Youngs, 1979; Desphande and Damodaran, 1990). Legume starches have been identified as a new and attractive food ingredient (Gregory and Bilidt, 1990; Beck and Kevin, 1995; Gujska et al., 1994).

The problem with the utilization of legume starches is that the processes for separating starch and protein components from the seed are often lengthy, laborious and costly. This is due to a highly hydratable fiber from the legume cotyledon cell walls, and the strong adherence of large amounts of insoluble proteins to starch granules (Schoch and Maywald, 1968). Techniques for the isolation of legume starches were originally reported by Kawamura et al. (1955) and Kawamura and Tada (1958). Their method involved treatment with a 0.2% NaOH solution, washing with water, and dehydration with ethanol and water (Desphande and Damodaran, 1990). The procedures most widely used today are three methods developed by Schoch and Maywald (1968), who established optimum conditions for starch isolation from legumes. In the first method (for mung beans, garbanzo beans and dehulled split yellow peas) pure starch was obtained by steeping legume seeds in warm water (in the presence of toluene to prevent fermentation), wet grinding and repeated screening. The first part of the second method (for lentils, lima beans and white navy beans) was similar to the first method, and was followed by resuspension in a 0.2% NaOH solution (to dissolve most of the protein) and several tabling steps. The third method (for wrinkled seeded peas) requires exhaustive alkaline steeping and washings of the isolated starch. The above procedures or their modifications were used, among others, by Haase et al. (1987), Honingfort (1988) and Hoover et al. (1991). Olsen (1978) described a batch and continuous ultrafiltration method for isolation of faba bean protein isolates. The process can be used to obtain pure (0.5% protein) pea starch as a by-product.

Some methods have been developed that combine the initial dry separation of protein and starch with a wet fractionation process. A final starch washing and purification step was included to increase starch purity. Vose (1977) pin milled and air-classified smooth-seeded peas and obtained a starch-rich flour fraction. This flour was then suspended in water (1:10 flour to water ratio), screened to remove the fiber and repeatedly washed and centrifuged to yield a starch with less than 0.4% protein. The starch, however, was highly damaged by the pin milling. Comer and Fry (1978) and Colonna et al. (1980) used a similar method as Vose (1977), except that the flour was soaked for 30-40 minutes prior to fractionation. These wet methods increased starch purity, yet problems still existed with the length of the procedure, damage to the starch and excessive water use.

Czuchajowska and Pomeranz (1994) developed a method of legume fractionation superior to existing methods (Fig. 1.). This technique for isolation of pure starches and protein concentrates combines elements of both dry and wet fractionation procedures. The method involves an initial selection of milled legume flours, followed by a simple and fast wet fractionation procedure that does not require presoaking or chemicals, and that needs much less water than the commonly used method by Schoch and Maywald (1968). This method was applied to garbanzo beans (*Cicer arietinum L.*), with some limited consideration given to other legumes. The yield of prime starch was 41.3% and the content of protein, ash and free lipids was 0.42%, 0.09% and 0.20%, respectively. For tailings starch the yield was 12.7%, content of protein 8.1%, ash 1.33% and free lipids 0.58%. The patented method of Czuchajowska and Pomeranz (1994) was further modified (Otto et al., 1997a) by reducing and recycling the amount of water used in the process, without

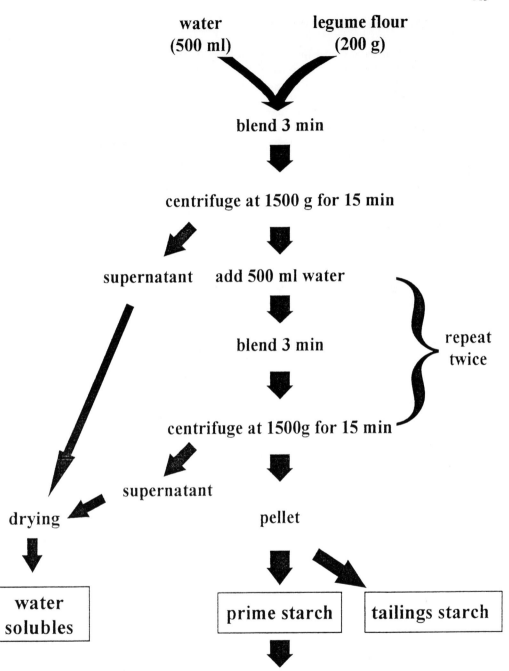

Figure 1   Schematic of wet fractionation process of legumes (Otto et al., 1997a).

affecting yield and purity of isolated fractions of garbanzo bean flour. This modified wet fractionation method was successfully extended to smooth pea (*Pisum sativum*) varieties. High purity of prime starches obtained from garbanzo bean and smooth pea cultivars Latah and SS Alaska by reducing and recycling water methods was indicated by low protein (0.33-0.45%) and ash (0.01-0.17%) content (Table 1).

Table 1. Yield, protein and ash contents of prime starch fractionated by the reduced water and recycled water methods[1] (Otto et al, 1997a)

Prime Starch	Yield (%)		Protein (%)		Ash (%)	
	Reduced[3]	Recycled	Reduced	Recycled	Reduced	Recycled
Garbanzo Bean	46.4a	46.7a	0.41a	0.42a	0.17a	0.17a
Pea						
cv Latah	37.1a	33.6a	0.33a	0.41a	0.01b	0.06a
cv SS Alaska	42.2a	41.1a	0.35a	0.45a	0.08a	0.04b

1 Results expressed on a dry weight basis.   2 Protein = N x 6.25

3 Water to flour ratio 2:1   4 Water to flour ratio 2:1, water from 3rd wash of previous batch was recycled for 1st wash.

Under modified conditions, however, the authors had difficulty in fractionation of wrinkled pea. This was probably due to its highly hydratable fiber component (Schoch and Maywald, 1968). Therefore, contrary to smooth pea, wrinkled pea showed a low yield of prime starch of only 16.1%, but a high yield of 35.8% of tailings starch. Also, the protein and ash contents of prime starches were higher than for smooth pea starches. This lower purity of starch granules could be due to the composite shape of wrinkled starch granules, which causes a greater adherence of the protein matrix onto the starch surface (Colonna et al., 1980; unpublished data of author's laboratory). Based on the results of Czuchajowska and Pomeranz (1994) and Otto et al. (1997a), an attempt to define the desirable properties of materials selected for the wet fractionation process can be made.

-     The starch-rich material originating from the central part of the legume's cotyledon is lower in ash, fiber and lipids, which is preferable to the protein-rich, especially fiber-rich, outer part of the cotyledon (Table 2).

-     In order to hydrate the material quickly, one has to recognize the importance of physical properties. The small particles, mostly below 63μm, but with low damage of starch, are important to the wet process.

-     Another crucial feature of selected material is low water holding capacity, preferably below 1.0. Water must work as a medium to wash out the solubles, instead of being absorbed by highly hydratable flour.

Table 2. Characteristics of flour fractions[1] separated from garbanzo beans and peas (Otto et al, 1997a)

Sample	Protein[2] (%)	Free Lipids (%)	Ash (%)	Starch (%)	Total Fiber(%)
Garbanzo Bean					
<86μm	17.97	5.49	3.03	55.8	13.6
>86μm	22.17	6.62	3.70	46.5	19.9
cv Latah					
(1+2+3)B+1R[3]	29.16	0.65	3.22	53.6	10.4
2R+3R[4]	31.17	0.76	3.39	51.0	16.4
cv SS Alaska					
(1+2+3)B+1R	23.24	1.18	2.89	56.8	13.1
2R+3R	26.73	1.51	3.27	49.2	19.4
cv Scout					
(1+2+3)B+1R	28.39	2.03	3.31	47.3	19.3
2R+3R	29.59	2.27	3.40	45.9	22.9

1 Results expressed on a dry wt. basis   2 Protein = N x 6.25   3 Blend of the three break and first reduction flours obtained from inner and intermediate part of cotyledon.   4 A blend of the 2nd and 3rd reduction flours obtained from outer part of cotyledon.

PROPERTIES OF LEGUME FLOUR FOR THE WET FRACTIONATION PROCESS

Fractionation studies of legumes as reviewed above have shown the complexity of the wet fractionation process. The lengthy procedures, however, could be simplified by selecting legume flour having desirable physical and chemical properties, as has been shown by recent studies (Czuchajowska and Pomeranz, 1994; Otto et al., 1997a). The selected garbanzo bean and pea flours from dry milling, originating from the central part of the cotyledon, did not require presoaking or application of chemicals in the wet process. The initial selection of flour fractions from the dry stone milling process took advantage of the milling differences between the starch-rich central part of the garbanzo bean cotyledon, and the protein and fiber-rich cotyledon periphery.

A study of 17 smooth and wrinkled pea varieties investigated by Kosson et al. (1994b) also showed that during dry roller milling the outer and inner parts of these legumes milled differently. Break legume flour originated from the softest (central) part of the cotyledon, while reduction stream flours originated from the harder (outer) part of the cotyledon. These results inspire research on legume seeds focusing on seed hardness, microstructure and particle size of dry milled flour.

Most of the studies on structure regarding handling and milling have been carried out on cereal grains, advancing their technology of dry and wet milling processes (Stenvert and Kingswood, 1977; Pomeranz et al., 1986; Pomeranz et al., 1985; Pomeranz and Czuchajowska, 1985; Martin et al., 1987; Glenn and Johnston, 1992). Kosson et al. (1994a, b), by studying hardness of twelve smooth and four wrinkled pea cultivars, reported a highly statistically significant correlation between hardness tests in whole seeds (compression and abrasion) and pea flour (particle size index).

Contrary to cereal grain seeds, the study of legume seed microstructure has been focused primarily on the textural defects of seeds that have decreased the market value of legumes (Sete-Dedeh and Stanley, 1979; Agbo et al., 1987; Joseph et al., 1993; Liu et al., 1993). Textural defects like hard-to-cook or hard shell create barriers to legume consumption because they prolong cooking time (Stanley and Aguilera, 1985). So far, little is known about the relationships between the microstructure of legume seeds and their milling performance. However, the rapidly growing interest in utilization of legume flours promises research focusing on these relationships.

The microstructural study of garbanzo beans and peas using scanning electron microscopy (SEM) has revealed large structural differences between the inner and outer layers of the cotyledon (Otto et al., 1997b). SEM pictures of smooth pea cv. Latah (Fig. 2) show the outer and inner parts of the cotyledon. The outer part of the cotyledon shows a compact structure. Cells are elongated and tightly packed. In the inner part of the cotyledon, cells are loosely packed, with large intercellular spaces. Garbanzo beans, smooth pea cv. SS Alaska and wrinkled pea cv. Scout in general showed a similar pattern of cotyledon structure to cv. Latah. However, some cultivar differences in size of cells were observed. The large differences in the microscopic structure between the inner and outer layer of cotyledon in four investigated samples suggest that these two layers would produce different characteristics of flours during milling.

Kosson et al. (1994b) determined the chemical composition of the inner and outer layers of pea cotyledon separated by gradual abrasion and reported that they were significantly different (Table 3). They also reported that the chemical composition of the inner layer of the cotyledon was similar to that of a blend of the three break and the first reduction flours obtained from roller milling of peas. Flour fractions from the inner layer of the cotyledon of garbanzo beans and pea cultivars were lower in protein, lipids, ash and fiber, and contained more starch than those from the outer layer of the cotyledon (Otto et al., 1997b).

Besides large significant differences in chemical composition between flours originating from inner and outer parts of peas and garbanzo beans, the SEM revealed large differences in flour particles originating from these two layers (Fig. 3). Particles in the coarse fraction of garbanzo

316

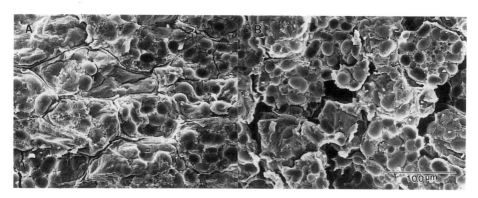

Figure 2   Scanning electron micrographs of smooth pea cv. Latah; outer (A) and inner (B) part of the cotyledon (Otto et al. 1997b).

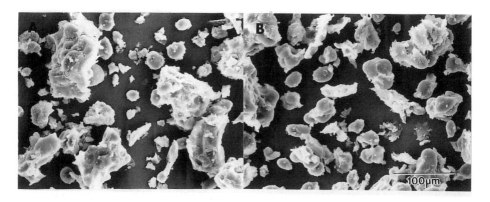

Figure 3   Scanning electron micrographs of flours of smooth pea cv. Latah; their particles originating from the outer layer (A) and the inner layer (B) of the cotyledon (Otto et al., 1997b).

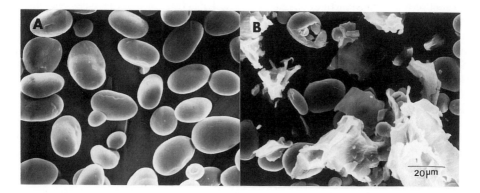

Figure 4   Scanning electron micrographs of prime starch (A) and tailings starch (B) isolated by the wet process from smooth pea cv. Latah (Otto et al., 1997b).

beans are much larger and show a more compact structure than those in the fine fraction. Starch granules in the coarse flour fraction of garbanzo beans are buried in the protein matrix. Flour particles from the central part of the cotyledon are generally much smaller than those of the coarse flour fraction and contain many free starch granules. Therefore, the study using SEM confirmed a good relationship between the structure of intact legume cotyledons and the size and structure of flour particles from the milling process, indicating that microstructure of seeds could be helpful for selecting suitable legume varieties for the milling and fractionation process.

Table 3. Free lipid and protein distribution1 of pea cotyledon fractions from abrasion studies (Kosson et al, 1994b)

part of cotyledon[2]	% as is			
	cv SS Alaska (smooth green)		cv Latah (smooth yellow)	
(% by wt)	lipid	protein	lipid	protein
10	1.80a	26.4a	1.00a	33.2a
20	1.32b	23.3b	0.70b	31.1b
35	1.22c	20.9c	0.61c	28.1c
residual 35	1.00d	18.7d	0.45d	25.5d

1 Results followed by the same letter within a column are not significantly different (p<0.05)   2 Layers of cotyledon starting from outside to inside of seed

LEGUME STARCHES

Only the highly efficient wet fractionation processing of legumes into their main components can be economically meaningful. It extends the utilization of legumes and also is beneficial to food and non-food products into which the legume components are incorporated. The possibility of utilization of legume main components instead of whole seeds or flours requires research focusing on, first, a simple, fast, economical technique of wet fractionation (as indicated in the first two sections of this article) and, second, on well-defined properties of fractions originating from the wet process.

During the wet separating process of legumes, three fractions are yielded: water solubles, tailings starch and prime starch. This section of the article will focus on legume starches because of the need to expand their utilization. It is crucial to define the differences between prime and tailings starches. Prime starch is a high purity starch as indicated by low protein content (normally <0.5%) and low ash content. High yield of prime starch is the goal of a wet process (Schoch and Maywald, 1968; Czuchajowska and Pomeranz, 1994; Otto et al., 1997a). Yield of prime starch depends on raw material and purification steps during the wet process. Tailings starch, contrary to prime starch, contains considerable amounts of protein, ash and fiber. As evaluated in the author's laboratory, protein, lipid and ash content of tailings starches were significantly higher than those of prime starches in garbanzo beans and peas, ranging from 4.6%-10.2%, from 0.1%-0.4% and from 1.24%-2.1%, respectively.

SEM pictures of prime starch and tailings starch originating from smooth pea cv. Latah showed differences between these starches (Fig. 4). The surfaces of prime starch granules are clean, without any protein residue, and have an elliptical shape with occasional shallow indentations. The SEM picture shows that the tailings starch retains small and broken starch granules, along with fibrous cell wall materials and protein residues. Therefore, tailings starch is a mixture of starch granules, cell wall materials and insoluble protein.

Legume starches are consistently higher in amylose than cereal grain starches (Beck and Kevin, 1995). Garbanzo beans and smooth-seeded peas have 30%-45% amylose (Schoch and Maywald, 1968; Vose, 1977; Colonna et al., 1980; Chavan et al., 1989), while wrinkled-seeded peas can have from 60% to over 80% of linear starch fraction (Schoch and Maywald, 1968; Vose, 1977; Colonna and Mercier, 1984; Stute, 1990). The amount of amylose within a starch granule influences starch gelatinization temperature, viscosity and stability of starch pastes, clarity of the starch pastes and gels, stability and mechanical properties of the gels, and gel retrogradation (Whistler and Daniel, 1985; Russell, 1987; Leloup et al., 1991). Therefore, high amylose starches are expected to add certain functional properties to products to which they are applied.

Differential scanning calorimetry (DSC) is a method of thermal analysis that can be used to follow the endothermic processes of starch gelatinization and retrogradation in aqueous systems. For example, the process of gelatinization occurs at about 65-70°C for garbanzo bean and smooth pea starches (Schoch and Maywald, 1968; Lineback and Ke, 1975; Biliaderis et al., 1979; Biliaderis et al., 1980). Wrinkled pea starch, being higher in amylose content, requires temperatures in excess of 100°C for gelatinization to occur (Sterling, 1978; Biliaderis et al., 1979). The temperature range over which the peak appears and the size of the peak (transition enthalpy) corresponds to the gelatinization characteristics of a particular starch (Kosson et al., 1994a; Paredes-LÛpez et al., 1994). The beginning of the peak corresponds to the start of birefringence loss (Hoseney, 1986). Kosson et al. (1994a) reported that thermogram peaks which occurred over a higher temperature range corresponded with pea starches having higher gelatinization temperatures and amylose contents. If other components are present along with the starch, such as protein or fiber (as with tailings starch), the peak will occur at a higher temperature, over a wider temperature range, and will have a lower enthalpy (Szczodrak and Pomeranz, 1991).

DSC thermograms are also useful for showing starch retrogradation characteristics. This is typically accomplished by re-scanning a gelatinized starch sample after cooling and storage. Melting endotherms at around 60°C is attributed to the slow recrystallization of amylopectin (Eerlingen et al., 1994; Paredes-Lûpez et al., 1994). Amylose, if present in sufficient quantity, will exhibit a melting endotherm at the much higher temperature of 150°C (Eerlingen et al., 1994; Paredes-Lûpez et al., 1994; Czuchajowska et al., 1991). This is attributed to the longer chain sections that can be involved in amylose crystallite formation, compared to the shorter sections of branched amylopectin.

The linear starch molecule recrystallization process not only causes gel firming (retrogradation), but also eventual gel shrinkage and syneresis (Sterling, 1978). Gel storage temperature and time will influence the extent to which these changes occur. The gel will also become more opaque as retrogradation continues, due to the reflection of light from the increasing number of molecular junction zones (Craig et al., 1989). Therefore, because garbanzo bean and pea starches are high in amylose, they can be expected to form firm gels with a high degree of opaqueness. Paste or gel clarity is an important characteristic to consider when formulating food products. For instance, starch used in a fruit pie filling should be transparent, but an opaque appearance is necessary for applications such as salad dressing.

Starches, when incorporated into food products, impart certain textural attributes, depending on the thickening and gelling properties of that starch. Textural characteristics vary with each type of starch. Textural properties of 8% prime and tailings starch gels were evaluated (Fig. 5) (unpublished data obtained in author's laboratory). Hardness of gels stored at 4°C followed the same pattern as those stored at 22°C. However, due to facilitated retrogradation at 4°C, the hardness of gels increased by 3.8 N for garbanzo beans, 5.3 N for smooth pea cv. Latah, 8.4 N for cv. SS Alaska and 4.7 N for autoclaved starch of wrinkled pea cv. Scout. A strong positive correlation was obtained between amylose content and hardness of gel. Correlation coefficients were r=0.965 for gels stored at 22°C and r=0.967 for gels stored at 4°C. The strong relationship between amylose content of starch and hardness of starch gel provides useful information for selecting starches for specific uses.

Once the texture profile of a starch is defined, it can be matched with specific products requiring those textural attributes. The unique properties of legume starches as discussed above are necessary for many food products, including sausage and paté-type meat products, gluten-free oriental noodles, canned and retorted products, extruded snacks, products with crisp coatings and confections (Vose, 1977; Orford et al., 1987; Leloup et al., 1991; Beck and Kevin, 1995). Garbanzo bean and pea starches are consistently high in amylose (Beck and Kevin, 1995), and high amylose starches are finding use in the production of enzyme-resistant starch (Kevin, 1995).

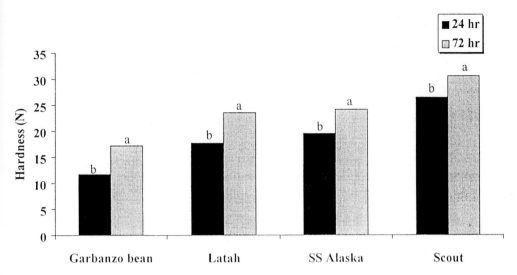

Figure 5   Hardness of 8% (dry basis) prime starch gels from garbanzo beans, pea cultivars (cv. Latah, SS Alaska) and wrinkled pea cv. Scout. Gels for each legume starch were prepared by heating to 93°C for 10   min. except cv. Scout, which was prepared by autoclaving at 126°C for 1hr. Gels were stored at 4°C for 24 and 72 hrs. The different letters marked on the top of each bar indicate significant differences at the 5% level (unpublished results of the author).

Enzyme-resistant starch is a recrystallized starch which is resistant to attack by amylolytic enzymes, and is used as a dietary fiber additive (Berry, 1986; Sievert and Pomeranz, 1989).

Pea and garbanzo bean starches are also well suited to many non-food uses, such as in adhesives and paper production (Vose, 1977). Vose reported that, due to its stability in dilute-caustic solution, decreased amounts of pea starch can be used in adhesive formulations and still maintain the desired functionalities.

Starch is an excellent raw material used to modify food texture and consistency. Therefore, information on starch properties in the starch-water system, such as thermal behavior, rheological properties of paste and thickening and gelling properties is important to improve the texture of food products containing starch. Improper texture of food products not only creates problems during handling, but also lowers consumer acceptance. Once characteristics of a starch are defined, it is easy to find suitable application for the starch in food and non-food products.

References

Agbo, G. N., Hosfield, G. L., Uebersax, M. A. and Klomparens, K. 1987. *Food Microstructure* 6:91-101.

Aykroyd, W. R. and Doughty, Y. 1982. Rome: FAO/UN.

Bahnassey, Y. and Khan, K. 1986. *Cereal Chemistry* 63:216-219.

Beck, H. and Kevin, K. 1995. *Food Processing* 2:58.

Berry, C. S. 1986. *Journal of Cereal Science* 4:301-314.

Biliaderis, C. G., Grant, D. R. and Vose, J. R. 1979. *Cereal Chemistry* 56:475-480.

Biliaderis, C. G., Maurice, T. J. and Vose, J.R. 1980. *Journal of Food Science* 45:1669-1674, 1680.

Chavan, J. K., Kadam, S. S. and Salunkhe, D. K. 1989. CRC *Critical Reviews in Food Science and Nutrition* 25:107-158.

Chompreeda, P., Resurreccion, A. V. A., Hung, Y. C. and Beuchat, L. R. 1987. *Journal of Food Science* 52:1740-1741.

Chompreeda, P., Resurreccion, A. V. A., Hung, Y. C. and Beuchat, L. R. 1988. *International Journal of Food Science and Technology* 23:555-563.

Colonna, P., Gallant, D. and Mercier, C. 1980. *Journal of Food Science* 45:1629-1636.

Colonna, P. and Mercier, C. 1984. *Carbohydrate Research* 126:233-247.

Comer, F. W. and Fry, M. K. 1978. *Cereal Chemistry* 55:818-829.

Craig, S. A. S., Maningat, C. C., Seib, P. A. and Hoseney, R. C. 1989. *Cereal Chemistry* 66:173-182.

Czuchajowska, Z. Sievert, D. and Pomeranz, Y. 1991. *Cereal Chemistry* 68:537-542.

Czuchajowska, Z. and Pomeranz, Y. 1994. *U.S. Patent* 5,364,471.

Deshpande, S. S., Rangnekar, P. D., Sathe, S. K. and Salunkhe, D. K. 1983. *Journal of Food Science* 48:1659-1662.

Desphande, S. S. and Damodaran, S. 1990. In *Advances in Cereal Science and Technology* pp. 147-241. (Ed Y. Pomeranz). St. Paul:AACC.

Duszkiewicz-Reinhard, W., Khan, K., Dick, J. W. and Holm, Y. 1988. *Cereal Chemistry* 65:278-281.

Eerlingen, R. C., Jacobs, H. and Delcour, J. A. 1994. *Cereal Chemistry* 71:351-355.

Glenn, G. M., and Johnston, R. K. 1992. *Food Structure* 11:187-199.

Gregory, D. and Bilidt, H. 1990. In *Gums and Stabilizers for the Food Industry* pp 63-67. (Eds G. O. Phillips, P. A. Williams and W. D. J. Wedlock). Oxford: IRL.

Gujska, E., Reinhard, W. D. and Khan, K. 1994. *Journal of Food Science* 59(3):634-637.

Haase, N. U., Kempf, W., Tegge, G. and Dîheur, U. 1987. *Die Stärke* 39:416-421.

Honingfort, T. April, 1988. Masters of Science Thesis. Berlin: University of Berlin, Institute of Food Technology, Department of Cereal Technology.

Hoover, R., Rorke, S. C. and Martin, A. M. 1991. *Journal of Food Biochemistry* 15:117-136.

Hoseney, R. C. 1986. In *Principles of Cereal Science and Technology*: A General Reference on Cereal Foods. St. Paul:AACC.

Hung, T. and Nithianandan, V. 1993. *ASEAN Food Journal* 8:26-31.

Joseph, E., Crites, S. G. and Swanson, B. 1993. *Food Structure* 12:155-162.

Kawamura, S., Tuboi, Y. and Huzii, T. 1955. *Studies on legume starches.* I. Technical Bulletin Kagawa Agricultural College 7:87-90.

Kawamura, S. and Tada, M. 1958. *Chemical Abstracts* 52:20428h.

Kevin, K. 1995. *Food Processing* 56(1):14.

Kosson, R., Czuchajowska, Z. and Pomeranz, Y. 1994a. *Journal of Agriculture and Food Chemistry* 42(1):91-95.

Kosson, R., Czuchajowska, Z. and Pomeranz, Y. 1994b. *Journal of Agriculture and Food Chemistry* 42:96-99.

Leloup, V. M., Colonna, P. and Buleon, A. 1991. *Journal of Cereal Science* 13: 1-13.

Lineback, D. R. and Ke, C. H. 1975. *Cereal Chemistry* 52:334-347.

Liu, K. Hung, Y. - C. and Phillips, R . D. 1993. *Food Structure* 12:51-58.

Lu, J. Y. and Al Jasser, M. 1986. *Journal of Food Processing and Preservation* 10:177-187.

McWatters, K. H. 1990. *Journal of the American Oil Chemists Society* 67:272-275.

Martin, C. R., Convers, H. H., Czuchajowska, Z., Lai, F. S. and Pomeranz, Y. 1987. *Applied Engineering in Agriculture* 3(1):104-113.

Müller, H. P. 1983. In: *Seed Proteins* p 309. (Eds W. Gottschalk and H.P. Müller) Hague: Martinus Nijhoff/Dr W, Junk.

Olsen, H. S. 1978. *Lebensmittal-Wissenschaft Technologie* 11(2):57-64.

Orford, P. D., Ring, S. G., Carroll, V., Miles, M. J. and Morris, V. J. 1987. *Journal of Science Food and Agriculture* 2:169-177.

Otto, T., Baik, B. - K. and Czuchajowska, Z. 1997a. *Cereal Chemistry* 74:141-146.

Otto, T., Baik, B. - K. and Czuchajowska, Z. 1997b. *Cereal Chemistry* (in press).

Paredes-Lûpez, O., Bello-Pèrez, L. A. and Lûpez, M. G. 1994. *Food Chemistry* 50:411-417.

Pomeranz, Y. and Czuchajowska, Z. 1985. *Food Microstructure* 4:213-219.

Pomeranz, Y., Czuchajowska, Z., Martin, C. R. and Lai, F. S. 1985. *Cereal Chemistry* 2: 108-112.

Pomeranz, Y., Martin, C. R., Traylor, D. D. and Lai, F. S. 1986. *Cereal Chemistry* 61:147-150.

Rockland, L. B. and Melzler, E. A. 1967. *Food Technology* 21:344-349.

Russell, P. L. 1987. *Journal of Cereal Science* 6:133-145.

Sathe, S. K., Deshpande, S. S. and Salunkhe, D. K. 1984. CRC *Critical Reviews in Food Science and Nutrition* 20:1-46.

Schoch, T. J. and Maywald, E. C. 1968. *Cereal Chemistry.* 45:564-573.

Sete-Dedeh, S. and Stanley, D. W. 1979. *Food Technology* 33(10):77-83.

Sievert, D. and Pomeranz, Y. 1989. *Cereal Chemistry* 66:342-347.

Singh, U. and Seetha, R. 1993. *Journal of Food Science* 58:853-855.

Sosulski, F. and Youngs, C. G. 1979. *Journal of the American Oil Chemists Society* 56:292-295.

Stanley, D. W. and Aguilera, J. M. 1985. *Journal of Food Biochemistry* 9:277-323.

Stenvert, N. L. and Kingswood, K. 1977. *Journal Science Food and Agriculture* 28:11-16.

Sterling, C. 1978. *Journal of Texture Studies* 9:225-255.

Stute, R. 1990. *Starch/Staerke* 42:178-184.

Swanson, B. G. 1990. *Journal of the American Oil Chemists Society* 67(5):276-280.

Szczodrak, J. and Pomeranz, Y. 1991. *Cereal Chemistry* 68:589-596.

Taha, S.A., ¡CS, E. and Sagi, F. 1992. *Acta Alimentaria* 21:153-162.

Vose, J. R. 1977. *Cereal Chemistry* 54:1141-1151.

Whistler, R. L. and Daniel, J. R. 1985. In *Food Chemistry*, 2nd ed. Pp 69-138 (Ed O. Fennema). New York: Marcel Dekker, Inc.

Woodin, G. B. 1981. *Food Development* 15:20-21, 24-25.

The Role of Legumes in Sustainable Cereal Production in Rainfed Areas

M. Pala[1], E. Armstrong[2] and C. Johansen[3]

1 ICARDA, P. O. Box 5466, Aleppo, Syria; 2 NSW Agriculture, Agricultural Research Institute, Wagga Wagga, NSW 2650, Australia; 3 ICRISAT, Patancheru P. O., Andra Pradesh 502 324 India

Abstract

Cool season food legumes (CSFL) are minor crops compared to cereals, but they are important in farming systems, in human and animal nutrition and as a source of biological N. They protect the soil from erosion, add organic matter, fix nitrogen and spare soil mineral N as well as helping control cereal diseases. They provide more flexible weed control options. Two major factors may lead to the increased use of cool season food legumes in cereal based cropping systems. The first is the demand for grain with a high protein content and other nutritional factors, for human and animal consumption. The second is the realisation of their importance as a "break" crop in continuous non-legume cropping systems.

The contribution of pulses to increased cereal production will come from a better understanding of rotations and production packages, better information on the residual effects of legumes, development of simulation models to predict these effects and the enhancement of biological N_2 fixation. The pulses themselves require continued support to develop better disease resistance, higher yields, better fashioned cultivars for specific cropping environments, technological packages and extension support and ,more sophisticated local and international marketing strategies. Cereal-legume rotations enable the use of different herbicides in the respective crops and reduce the risk of herbicide resistance in weeds.

INTRODUCTION

Sustainable agriculture has become a concern all over the world in the last decade because of the increasing degradation of natural resources, low commodity prices leading to low-input systems and concern about food quality and the welfare of rural life. Sustainable agriculture systems are designed to use existing soil nutrient and water cycles, and naturally occurring energy flows to produce food and feed that is nutritious and harmless to human and animal health. In practice, such systems should rely on crop rotations, crop residues, animal manure, legumes, green manure, off-farm organic wastes, mechanical cultivation, and mineral bearing rocks to maintain soil fertility and productivity, and on natural biological and cultural controls for insects, weeds, and other pests (MacRae et al., 1990).

Cool season food legumes (CSFL) are minor crops, compared with cereals, but are important for human and animal nutrition and as a source of biological N. The ability to fix N_2 enables them to grow on low-N soils and to produce seed high in protein. They provide nutritionally rich crop residues for animal feed; and play a key role in maintaining the productive capacity of soils with their N fixation (Hamblin, 1987; Beck and Materon, 1988; Osman et al. 1990) and by breaking disease and pest cycles (Papendick et al. 1988). Legumes ideally should comprise 30-50% of the cropland (Parr et al., 1983).

RECENT TRENDS IN PULSE PRODUCTION

Although the production of pulses in WANA and South Asia has increased over time (Table 1) supply has fallen short of demand, resulting in high prices and the need to import about 1.3 million MT (FAO, 1994a). As can be seen from Table 1, the area of all pulses in the region of WANA, South Asia, including Pakistan, and Oceania (Australia and New Zealand) covers about 50% of the world area under pulses, and

R. Knight (ed.), Linking Research and Marketing Opportunities for Pulses in the 21st Century, 323–334.
© 2000 Kluwer Academic Publishers. Printed in the Netherlands.

40% of world production of pulses. However for CSFLs, this proportion increases significantly to 60% of total world area. The area of pulses and CSFLs worldwide increased by 12% and 10% respectively between 1979-81 and 1989-91, and production by a remarkable 38% and 45% respectively. Since then, there has been little change to CSFLs in the world. Regional increases in area and production of these crops are higher than that of the world figures since 1979-81 (Table 1). However, irrespective of their contribution to the farming systems, all pulses including the CSFLs have not assumed their rightful place. Their area and production has increased very little relative to cereals since 1979-81(Table 2).

Table 1 Regional distribution of area and production of cereals, pulses, CSFLs and wheat with barley

Cereals	(Area, 1000 ha))			(Production, * 1000 MT))		
Region	1979-81	1989-91	1993-95	1979-81	1989-91	1993-95
N. Africa	20640	22971	24712	23812	33641	35370
W.Asia	31029	31127	35667	44688	55125	64327
S.Asia	128981	128978	126508	182135	252603	270053
Oceania	16197	12995	13957	21974	22207	23655
Region total	196847	196071	200844	272609	363576	393404
World	717137	708039	695198	1573337	1901136	1915903
% of world	27.4	27.7	28.9	17.3	19.1	20.5

Wheat +Barley	(Area, * 1000 ha)			(Production, * 1000 MT)		
N. Africa	10803	12383	11293	10216	17574	15599
W.Asia	27246	31251	32556	38133	47468	55339
S.Asia	32077	34153	35451	48753	71059	79571
Oceania	14134	11063	11906	18310	18055	19614
Region total	84260	88850	91206	115412	154156	170123
World	316135	301591	291989	591333	729118	701633
% of world	26.7	29.5	31.2	19.5	21.1	24.2

Pulses	(Area, * 1000 ha)			(Production, * 1000 MT)		
N. Africa	1801	1825	1983	1723	1766	1789
W.Asia	1452	3257	3293	1439	2640	2889
S.Asia	25583	26260	26768	11928	14876	15636
Oceania	235	1522	1975	259	1596	2089
Region total	29071	32864	34020	15349	20878	22403
World	60754	68285	68802	40628	55949	56888
% of world	47.9	48.1	49.4	37.8	37.3	39.4

CSFL	(Area, * 1000 ha)			(Production, * 1000 MT)		
N. Africa	1421	1290	1231	1417	1347	1161
W.Asia	867	2524	2583	838	1974	2199
S.Asia	10487	10315	10338	6067	7123	7177
Oceania	96	615	661	147	698	756
Region total	12871	14744	14813	8469	11142	11293
World	23089	25467	24529	19818	28770	27564
% of world	55.7	57.9	60.4	42.7	38.7	41

Source FAO Production Year Book 1994 1995

There has been a significant increase in the areas under pulses in Australia and Turkey, over the last two decades (FAO, 1994b; FAO, 1995). While the area in Turkey decreased slightly over the last 5 years, in Australia, the area has increased from almost nothing in the mid 1970s to about 2 million hectares now (Siddique and Sykes, 1997)

Table 2. Area and production of pulses and CSFL as percentage of cereals, and CSFL as percentage of pulses and wheat+barley

Region	Pulses area (% of cereals)			Pulses production (% of cereals)		
	1979-81	1989-91	1993-95	1979-81	1989-91	1993-95
N. Africa	8.7	7.9	8.0	7.2	5.2	5.1
W.Asia	4.7	10.5	9.2	3.2	4.8	4.5
S.Asia	19.8	20.4	21.2	6.5	5.9	5.8
Oceania	1.5	11.7	14.2	1.2	7.2	8.8
Region total	14.8	16.8	16.9	5.6	5.7	5.7
World	8.5	9.6	9.9	2.6	2.9	3.0
	CSFL areas (% of cereals)			CSFL production (% of cereals)		
N. Africa	6.9	5.6	5.0	6.0	4.0	3.3
W.Asia	2.8	8.1	7.2	1.9	3.6	3.4
S.Asia	8.1	8.0	8.2	3.3	2.8	2.7
Oceania	0.6	4.7	4.7	0.7	3.1	3.2
Region total	6.5	7.5	7.4	3.1	3.1	2.9
World	3.2	3.6	3.5	1.3	1.5	1.4
	CSFL areas (% of pulses)			CSFL production (% of pulses)		
N. Africa	78.9	70.7	62.1	82.2	76.3	64.9
W.Asia	59.7	77.5	78.4	58.2	74.8	76.1
S.Asia	41	39.3	38.6	50.9	47.9	45.9
Oceania	40.9	40.4	33.5	56.8	43.7	36.2
Region total	44.3	44.9	43.5	55.2	53.4	50.4
World	38	37.3	35.7	48.8	51.4	48.5
	CSFL areas(% of wheat+barley)			CSFL production (% of wheat+barley)		
N. Africa	13.2	10.4	10.9	13.9	7.7	7.4
W.Asia	3.2	8.1	7.9	2.2	4.2	4
S.Asia	32.7	30.2	29.2	12.4	10	9
Oceania	0.7	5.6	5.5	0.8	3.9	3.9
Region total	15.3	16.6	16.2	7.3	7.2	6.6
World	7.3	8.4	8.4	3.4	3.9	3.9

PROBLEMS RESULTING FROM CONTINUOUS CEREAL PRODUCTION

There is increasing concern about the deterioration of crop/livestock systems because of the pressure put on them by the ever-rising demands for food and feed. Continuous cereal systems are increasing in parallel with the increasing demands for food from humans and feed by animals in the regions of WANA and South Asia (Harris et. al., 1991; Jones, 1993; Harris, 1994; Paroda et al., 1994).

Cereal-fallow or continuous cereal cropping are the common rotations in the WANA region, but the inclusion of legumes has been accompanied by many benefits. For example, the organic matter content of the soil has been increased under wheat-legume systems compared to continuous wheat or wheat-fallow systems (FRMP, 1993; Masri, 1996). The inclusion of legumes in the crop rotation in the WANA region has alleviated problems created by replacing a 'fallow-cereal' rotation with a 'continuous cereal', such as the build-up of noxious weeds, pests and pathogens, and an accumulation of allelopathic compounds. For example, continuous cereal production leads to yield decline accompanied by cereal cyst nematode (*Heterodera avena*), soil-inhabiting fungi such as *Cochliobolus sativus* syn. *Helminthosporium sativum,* take-all diseases (*Gaeumannomyces graminis* var. *tritici*) and wheat ground beetle (*Zabrus tenebroides*) (Saxena et al., 1991, Harris, 1994).

The high input rice-wheat cropping systems in the Indo-Gangetic Plain of India are reaching productivity limits, and the edaphic resource base is under threat of degradation (Paroda et al., 1994). Evidence for this includes the plateauing of rice and wheat yields in regions of high productivity, declining organic matter and productivity of soils, increasing salinity, and build-up of pests, diseases and weeds (Hobbs and Morris,

1996). Many of the maladies associated with continuous rice-wheat cropping are yet to be precisely diagnosed (e.g. soil chemical, physical and biological factors).

Similarly in Australia, farming practices prior to the 1970s traditionally centred around wheat/sheep enterprises in which cereals were rotated with legume-based pastures in a ley farming system (Greenland 1971). Typically, 3-5 years of a pasture containing subterranean clover, medic, lucerne and grasses was followed by a crop of oats, then by several years of wheat to conclude with a crop of wheat or barley undersown with pasture. Such a long cereal cropping phase obviously became unsustainable in terms of the supply of nutrients such as N and by the perpetuation of cereal diseases such as take-all (*G. graminis*), yellow leaf spot (*Pyrenophora tritici-repentis*) and crown rot (*Fusarium graminearum*). This dilemma was ultimately bought to a head in 1969 when Australian authorities introduced wheat quotas to prevent a potential overproduction of wheat. Farmers were then quick to look for profitable alternative crops to fill the void. This resulted in a new era of farming in Australia whereby broad-leaf crops became accepted and more widely grown in rotation with cereals in a more sustainable and profitable manner (Farrington, 1974; Gladstones, 1975; Hamblin, 1987; and Robson 1994).

In the wet environmental conditions of the United Kingdom large amounts of take-all (*G. graminis*) in wheat-wheat and wheat-barley rotations have also caused significant yield losses (Prew et al. 1985;McEwen et al. 1989).

Soil erosion and the loss of soil organic matter and essential nutrients are increasing problems in all three regions. Legume cultivation to protect against erosion, add soil organic matter, fix atmospheric nitrogen, to spare soil mineral N, to eliminate cereal diseases, and to provide more flexible weed control options, offers a means of maintaining production and environmental resources in the face of ever-increasing crop intensification in less favourable areas.

ADVANTAGES OF LEGUMES IN CEREAL ROTATIONS

A. Increased Crop Yield

Increased cereal yields may well be one of the most practical justifications for having legumes in crop rotations (Prew et al. 1985; Armstrong, 1986a, 1986b; Saxena, 1988; McEwen et al., 1989; Mason and Rowland, 1990; Silsbury, 1990; Evans *et al.*, 1991; Karaca et al. 1991; Mitchell et al., 1991; Reeves, 1991; Hamblin *et al.*, 1993: Martin and Felton, 1993; Heenan *et al.*, 1994; Rowland *et al.*, 1994; Karlen et al., 1994; Harris, 1995; Saraf et al., 1997; and Yadav et al., 1997). The predominant consensus of all these reports is that crop rotation increases cereal yields and rotational profits, and allows for more sustained system of production. The following case studies highlight many of these points.

Karaca et al (1991) reported that wheat following lentil, chickpea and vetch provided yields that were nearer to those from wheat following fallow, and much better that wheat following sunflower and safflower, in a two-year crop rotation in highland areas of Turkey. Wheat after wheat was the most yield reducing sequence. Wheat yield following chickpea, lentil, vetch and fallow increased 68, 69, 75, and 80% respectively over the yield of wheat after wheat, respectively. Badaruddin and Meyer (1994) indicated that pulses had a positive effect on the subsequent wheat crop suggesting that they should be considered in higher moisture areas of the northern Great Plains to maintain crop productivity.

In an area of Syria with a Mediterranean-type climate, Harris (1995) obtained data over a seven-year period from 1985/86 to 1991/92 for a two-course rotation of wheat following either wheat, medic, chickpea, lentil, vetch, melon or fallow. Relative to the yield of wheat following wheat, the percentage increases in wheat yields followed these other crops were 39, 46, 82, 84, 119 and 126%, respectively. The highest yields (2.26 t/ha) were obtained from a wheat-fallow rotation, however this result must be credited to a two- rather than a one-year period, while the lowest yield (1.00 t/ha) was from wheat after wheat. The experiment is still running and the most recent results show very similar trends (Figure 1). Saxena (1988) reported that the grain yield of wheat following lentil, faba bean and dry pea was significantly higher than that of wheat following wheat under rainfall of about 340 mm in a Mediterranean-type climate. Large yield increases in the yield of barley following medic, vetch, faba bean and chickpea compared to continuous barley, and barley after oat or ryegrass were also reported in Cyprus (Papastylianou, 1988). Barley-vetch rotations are reported to be more productive than continuous barley or fallow-barley systems in northern Syria (Harris et al., 1991) and Cyprus (Papastylianou, 1993).

Results from a series of experiments on the typical Australian wheat rotation system, conducted at five sites, (Armstrong 1986a and 1986b) showed that wheat yields after lupin, pea or faba bean were 50-55% higher than wheat yields after wheat and gross margins were 137-153% higher in situations where no fertiliser was used (Figure 2). In Australia, a farmer's choice of the pulse to grow should be based on issues of local adaptation, farm infrastructure and marketing opportunities as the beneficial effects of the different pulses on the following wheat crop appear to be similar.

A. Wheat yield

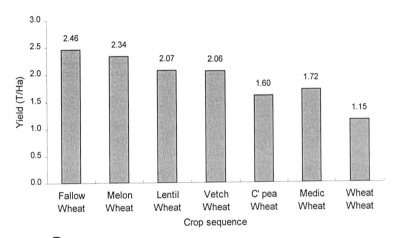

B. Rainfall

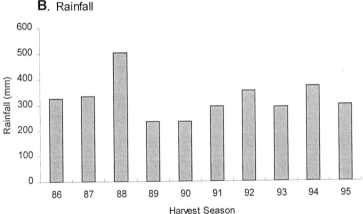

Figure 1. Mean grain yield of a wheat crop following various pulses in a two-course rotation study over a 10 year period (1986-95) at Tel Hadya, ICARDA Syria and (B.) annual rainfall incident over the same period.

B. Additions of Fixed Nitrogen

Nitrogen input is the single most limiting soil fertility factor constraining productivity and water-use efficiency of crops in dry areas. In N-deficient environments, the soil N-budget will be improved or at least the amount of N removed from the soil decreased (soil mineral N sparing) when well-managed nodulated legumes are grown in rotation with cereals. This has been well documented across many regions, including WANA (Beck and Materon, 1988; Saxena, 1988; Keatinge et al., 1988; Beck et al., 1991; Afandi et al., 1995), South Asia (Saraf et al., 1997; Yadav et al., 1997), and Australia (Evans et al., 1989; Armstrong et al., 1994; Herridge et al., 1994; Unkovich et al., 1997).

Build-up of inorganic soil-N following pulses is a direct consequence of both their N-fixation and their ability to spare soil mineral N during growth, and these amounts often equate to or exceed those of wheat-fallow rotations (Papastylianou and Jones 1989; Papastylianou 1993; Badaruddin and Meyer 1994; Rupela et al., 1995). The fertiliser replacement values (ie. the fertiliser N equivalent required by continuous cereal to equal the yield of cereal after legume with zero N) of various pulses have been reported to vary between 30 to 60 kg N/ha (McEwen et al., 1989; Papastylianou 1993; Saraf et al.1997).

However, N-fixation is affected by many factors, including legume biomass, efficiency of N_2 fixation, soil mineral-N, rhizobium strain and inoculation, water regime and soil and crop management. For example, root nodulation and N-fixation substantially decrease as both plant-available soil-N increases (Bergerson et al., 1989; Brockwell et al., 1989; Beck et al., 1991) and as water deficits become more severe (Kirda et al. 1989). N-fixation in legumes is generally proportional to biomass production, and as such, fixation commonly varies between 25-150 kg N/ha depending on environmental conditions and the inherent ability of the species/genotype to produce biomass (Beck and Materon, 1988; Evans et al., 1989; Beck et al., 1991; Saxena et al., 1993; Armstrong et al., 1994; Shwenke et al., 1997; Unkovich et al., 1997). In southern Australia, field pea and lupin crops are likely to fix around 18-20 kg N for each additional tonne of shoot DM produced per hectare (Evans et al., 1989; Armstrong et al., 1994). Therefore, given comparable levels of biomass and N_2 fixation efficiencies, chickpea; lentil, faba bean, pea, grass pea and chickpea should fix similar amounts of N and have comparable effects on subsequent crops (Papendick et al., 1988; Saraf et al., 1997; Yadav et al., 1997). In budgeting for the final residual N benefit of pulse crops, account has to be taken of the N exported from the system as either forage, hay, straw or grain.

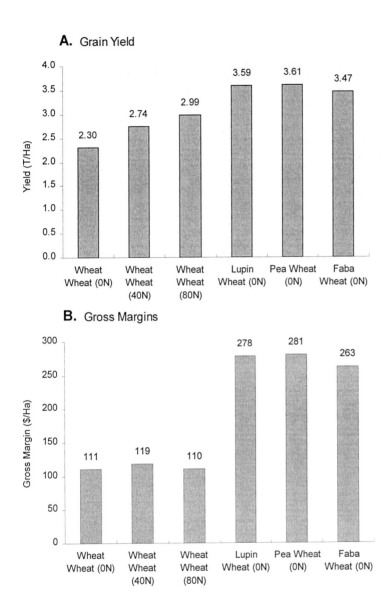

A. Grain Yield

B. Gross Margins

Figure 2 Yields (A.) and gross margins (B.) of wheat following crops of wheat, lupin, pea and faba bean. Trials were located at five sites in the Central West of NSW between 1984-86 and results here are presented as means across the sites. The continuous wheat plots were further divided in year 2 into 3 nitrogen treatments. Wheat on-farm prices per tonne of $130 and production costs per tonne of $188 were used in these calculations. N was costed at $1/kg and its spreading at $10/ha.

N$_2$ fixation efficiency can increase in cereal/legume intercrops as N$_2$ fixation of the legume component is stimulated by absorption of free soil mineral N by the cereal (Adu-Gyamfi et al. 1996). However, transfer of fixed N from the legume to the cereal in intercropping systems is more circumspect and undergoing closer scrutiny (Chalk 1996; Chalk and Smith 1997).

C. Reduction in cereal diseases and pests

The vastly superior performance of wheat after pulses compared to N-fertilised wheat after wheat reported by Armstrong (1986a and 1986b) suggests factors other than N nutrition can severely limit wheat productivity. In these case studies, the poor yields suffered in continuous wheat plots were due to the disease take-all (*G. graminis*), resulting in large proportions of unfilled spikelets (white heads). Only one season of a grass-free pulse crop was sufficient to break this disease cycle, allowing maximum expression of yield in the following wheat. This work also showed the use of fertiliser N on wheat may in some instances be unprofitable, particularly when factors such as disease, moisture or weeds are limiting performance. Therefore, despite fertiliser N increasing wheat yields by up to 30% in some instances, gross margins were largely unaffected due to extra costs of the fertiliser (Figure 2). Similar observations have been made at ICARDA Syria (Figure 3), where N applications to continuous cereal was more inefficient than N application to wheat following fallow or pulses. Well maintained bare fallows are as effective as pulses in breaking cereal disease cycles, however the land remains unproductive during this process, soils are more exposed to erosion and extra management and expenses incurred in eliminating weeds (Robson, 1990).

Unpublished work, by the main author, has shown that under conditions favouring severe infestations of wheat ground beetle (*Zabrus tenebroides*), yields of continuous wheat are reduced by up to four-fold compared to wheat after food or forage legumes.

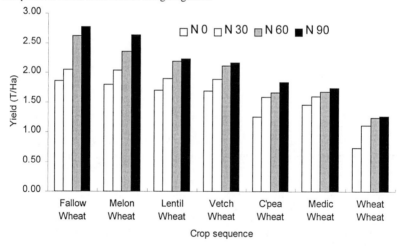

Figure 3 Mean grain yield of a wheat crop, given four rates of N and following various pulses in a two-course rotation study over a 10 year period (1986-95) at Tel Hadya, ICARDA Syria.

D Water-use efficiency (WUE)

Water scarcity limits agricultural productivity in the rainfed areas of the world. This is often important in Mediterranean environments because of the temporal and spatial variability in rainfall. Any attempts to increase production must focus on a more efficient use of water.

With ever increasing demands for food and land-use, long fallow periods are no longer possible in most parts of the world since WUE and productivity of one crop spread over a two year cycle inevitably compromises efficiency. This is particularly true in the WANA region where efficiency of moisture conservation in fallows can be low due to both low rainfall and shallow soils (Durutan et al., 1989; Harris et al., 1991; Harris 1995). Continuous cereal production is not an efficient alternative since these systems experience some of the lowest WUE and yields recorded (Karaca et al. 1991). Rather, replacing a fallow with either a leguminous crop, forage or pasture enhances the overall efficient use of water and nitrogen in the system, and further, legumes such as lentil or vetch provide greater efficiencies since they mature up to one month earlier than alternatives such as chickpea wheat or medic. As a result, additional residual water is left for use by the subsequent cereal (Harris et al., 1991; Papastylianou, 1993; Harris, 1994).

E. Other benefits to cereal production

Soil Quality effects

Longer-term implications from using legumes in rotation centre around improvements to soil quality such as its structure (water-stable aggregates and reduced bulk density), water infiltration and storage, resistance to erosion, total N, organic C and pH (Arihara et al. 1991; Coventry and Slattery, 1991; Chan *et al.*, 1992; Heenan and Chan, 1992; Chan and Heenan, 1993, 1996a, 1996b; Harris 1995; Heenan and Taylor,1995; Heenan *et al*, 1995; Masri, 1996; Helyar *et al.*, 1997; and Saraf et al., 1997;). Reliance on fertiliser N rather than biologically fixed N can lead to greater rates of soil pH decline and to increases of soil Al and Mn to toxic levels (Heenan and Taylor,1995). Since measurement of these effects require studies over long-term periods, many issues are still currently under investigation. All ultimately lead to a more favourable environment for root growth, allowing more efficient exploitation of water and nutrients and consequently, positive effects on long-term crop production.

Increased availability of other nutrients

Some species of legumes such as field pea have an ability to form root symbiotic associations with vesicular-arbuscular mycorrhizal fungi. This has been shown to facilitate both the uptake of N and P (Martensson and Rydberg, 1994) and assist interplant transfer of N (Frey and Schuepp, 1993). Others such as albus lupins have the ability to produce dense clusters of rootlets, or proteoid roots, and these have been associated with more efficient uptake of nutrients such as P (Gardner et al. 1981). Deeper rooting of some tap-rooted legumes such as lupins and chickpea can assist with recycling of nutrients from the deeper soil layers. The acidic root exudates of chickpea have been shown to increase the availability of P in some soils, particularly those where calcium phosphates dominate (Arihara et al. 1991). All the above mechanisms allow to varying degrees an increase in the availability of P to the cropping system. Healthy nutrient balances of legumes differ inherently from those of cereals and this in itself allows more efficient utilisation of limited major and minor nutrient resources across the system (Saraf et al., 1997).

POTENTIAL FOR IMPROVING THE CONTRIBUTION OF PULSES TO CEREAL PRODUCTION

Two factors favour an increased use of CSFLs and other grain legumes in the cereal-based systems across WANA, South Asia, and Oceania regions of the world. The first is the increase in demand for grain high in protein and nutrients for human and animal consumption. The second is the vulnerability of non-leguminous cropping systems and the need to use the ameliorative effects of legumes. There are increasing possibilities for reducing the risks associated with cultivation of grain legumes by incorporating increased host plant resistance to biotic stresses and by adopting better techniques for integrated pest, diseases and weed management. The development of shorter duration types of CSFLs presents the possibility of escaping various end of season stresses and for fitting these crops into novel cropping niches (Malhotra et al., 1996; Kumar et al. 1996). Some of these issues have been discussed earlier (Pala et al., 1994).

However, other specific research and development needs include:
- the improved dissemination of information and the demonstration of improved legume production technologies which are currently available, but not yet widely adopted,
- the breeding of more robust erect varieties with high stable yields, better disease resistance, improved quality and easier to harvest,
- research on the residual non-N effects of CSFLs,
- development of simulation models to predict the residual effects of legumes and demonstrate effectively their beneficial consequences on sustainability,
- improving the efficiency of N_2 fixation of legumes and their residual N benefits, by genetic changes to the crop, reducing N harvest indices and improving their amelioration effects on soil,
- research on the economics of legumes in rotations and their contributions to the sustainability of cereal-based systems,
- breeding cultivars for specific cropping systems; e.g. better matching crop duration and adaptation to specific cropping windows,
- overcoming the herbicide resistance in weeds that results from the continual use of 'dim' and 'fop' type grass herbicides in pulse crops,
- re-examining the possibilities of cereal-legume intercropping,

The expansion of CSFL production will not be without problems and there is an urgent need for governments to maintain research and development activities. There is also a need for more sophisticated local and overseas marketing strategies, better management packages designed to produce high quality pulses destined for premium markets and for more technical and agronomic support to growers and industry (this has already been successfully applied in Turkey and Australia).

Addressing many of these potential areas will undoubtedly lead to more assured future of pulses in sustainable cereal/crop/livestock systems.

References

Adu-Gyamfi, J.J., Katayama, K., Gayatri Devi, Rao, T.P. and Ito, O. 1996. In *Dynamics of Roots and Nitrogen in Cropping Systems of the Semi-Arid Tropics* pp. 493-506 (eds O. Ito, C. Johansen, J.J. Gayamfi, K. Katayama, J.V.D.K Kumar Rao and T.J. Rego). JIRCAS International Agriculture Series No.3. Tsukaba, Japan.

Afandi, F., Trabulsi, N. and Saxena, M.C. 1995. *Biological nitrogen fixation by cool season food and feed legumes: impact on wheat productivity in Syria and Lebanon.* Regional Soil Fertility Workshop, 19-23 November, 1995, Aleppo, Syria.

Arihara, J., Ae, N., and Okade, K. 1991. In *Problems nutrition of grain legumes in the semi-arid tropics*, pp. 183-194, (C. Johansen, K.K. Leer, and K.L. Sahrawat, eds.), ICRISAT, Patancheru, India.

Armstrong, E. L. 1986a. In: *Proceedings of Symposium on N cycling in Agricultural Systems of Temperate Australia, Agricultural Research Institute, Wagga.* 14-16 July, 1986. Aust. Soc. Soil Sci. (Ed. P.E. Bacon, J. Evans, R.R. Storrier and A.C. Taylor)

Armstrong, E. L. 1986b. In *Proceeding of the Fourth International Lupin Conference*, Geraldton, Western Australia. 15-22 August, 1986. International Lupin Association & Western Australia

Armstrong, E. L., Pate, J. S., and Unkovich, M. J. 1994. *Australian Journal of Plant Physiology* 21, 533-49.

Badaruddin, M., and Meyer, D.W. 1994. *Crop Science Society of America* 34: 1304-1309.

Beck, D.P., and L.A. Materon (eds.). 1988. *Nitrogen Fixation by Legumes in Mediterranean Agriculture.* Proceedings of the Workshop, 14-17 April, 1986, Aleppo, Syria: ICARDA.

Beck, D.P., J. Wery, M.C. Saxena, and A. Ayadi. 1991. *Agronomy Journal* 83: 334-341.

Bergersen, F.J., Brockwell, J., Gault, R.R., Morthorpe, L.J., Peoples, M.B., and Turner, G.L. 1989. *Australian Journal of Agricultural Research* 40: 763-780.

Brockwell, J., Gault, R.R., Morthrope, L.J., Peoples, M.B., Turner, G.L., and Bergersen, F.J. 1989. *Australian Journal of Agricultural Research* 40: 753-762.

Chalk, P.M. 1996. In *Dynamics of Roots and Nitrogen in Cropping Systems of the Semi-Arid Tropics* (eds O. Ito, C. Johansen, J.J. Gayamfi, K. Katayama, J.V.D.K Kumar Rao and T.J. Rego). JIRCAS International Agriculture Series No.3. Tsukaba, Japan.

Chalk, P.M. and Smith, C.J., 1997. *Biology and Fertility of Soils* 24: 239-42

Chan, K. Y. and Heenan, D. P. 1993. *Australian Journal of Agricultural Research* 44: 1971-84.

Chan, K. Y. and Heenan, D. P. 1996a. *Australian Journal of Experimental Agriculture* 36: 539-43.

Chan, K. Y. and Heenan, D. P. 1996b. *Soil and Tillage Research* 37: 113-25.

Chan, K. Y., Roberts, W. P., and Heenan, D. P. 1992. *Australian Journal of Soil Research* 30: 71-83.

Coventry, D. R. and Slattery, W. J. 1991. *Australian Journal of Agricultural Research* 42: 391-97.

Durutan, N., Pala, M., Karaca, M., and Yesilsoy, M.S. 1989. In: *Proceedings of Soil, Water, and Crop/Livestock Management Systems for Rainfed Agriculture in the Near East Region,* pp. 60-77 (eds.C.E. Whitman, J.F. Parr, R.I. Papendick, and R.E. Meyer) January 1986 Amman Jordan USDA, Washington DC, USA

Evans, J., Fettell, N. A., Coventry, D. R., Oconnor, G. E., Walsgott, D. N., Mahoney, J., and Armstrong, E. L. 1991. *Australian Journal of Agricultural Research* 42, 31-43.

Evans, J., O'Connor, G. E., Turner, G. L., Coventry, D. R., Fettell, N., Mahoney, J., Armstrong, E. L., and Walsgott, D. N. 1989. *Australian Journal of Agricultural Research* 40: 791-805.

FAO. 1994a. Food and Agriculture Organization, Trade Yearbook, Vol. 48, Rome, Italy.

FAO. 1994b. Food and Agriculture Organization, Production Yearbook, Vol. 48, Rome, Italy.

FAO. 1995. Food and Agriculture Organization, Production Yearbook, Vol. 49, Rome, Italy.

Farrington, P. 1974. *The Journal of the Australian Institute of Agricultural Science* 99-108.

Frey and Schuepp, 1993. *Soil Biology and Biochemistry* 25: 651-658.

FRMP. 1993. *Farm Resource Management Program Annual Report for 1992.* pp. 85-91, ICARDA, Aleppo, Syria.

Gardner et al. 1981. *Plant and Soil* 60: 143-147.

Gladstones, J. S. 1975. *The Journal of the Australian Institute of Agricultural Science* 227-240.

Greenland, D. J. 1971. *Soil Fertility* 34: 237-257.

Hamblin, J. 1987. In: *Proceedings of 4th Australian Agronomy Conference*, 24-27 August, 1987. La Trobe University, Melbourne, Victoria. pp. 1-16.

Hamblin, J., Delane, R, Bishop, A., and Adam, G. 1993. *Australian Journal of Agricultural Research* 44: 645-59.

Harris, H.C. 1994. *Aspects of Applied Biology* 38: 165-172.

Harris, H.C. 1995. *Advances in Soil Science* pp. 447-469.

Harris, H.C., Osman, A.E., Cooper, P.J.M., and Jones, M.J. 1991. In: *Proceedings of Soil and Crop Management for Improved Water Use Efficiency in Rainfed Areas,* pp. 237-250 (eds H.C. Harris, P.J.M. Cooper, and M. Pala) Ankara, Turkey, ICARDA, Syria.

Heenan, D. P. and Chan, K. Y. 1992. *Australian Journal of Soil Research* 30: 977-988.

Heenan, D. P. and Taylor, A. C. 1995. *Soil Use and Management* 11: 4-9.

Heenan, D. P., McGhie, W. J., Thompson, F. M., and Chan, K. Y. 1995. *Australian Journal of Experimental Agriculture* 35: 877-84.

Heenan, D. P., Taylor, A. C., Cullis, B. R., and Lill, W. J. 1994. *Australian Journal of Agricultural Research* 45: 93-117.

Helyar, K. R., Cullis, B. R., Furniss, K., Kohn, G. D., and Taylor, A. C. 1997. *Australian Journal of Agricultural Research* 48:

Herridge, D.F., O.P. Rupela, R. Serraj, and D.P. Beck. 1994. In: *Expanding the Production and Use of Cool Season Food Legumes* pp. 472-492 (eds F.J. Muehlbauer and W.J. Kaiser) Dordrecht: Kluwer Academic Publishers.

Hobbs, P., and Morris, M. 1996. *NRG Paper,* 96-101. CIMMYT, Mexico, D.F.

Jones, M.J. 1993. In: *Proceedings of The Agrometeorology of Rainfed Barley-based Farming Systems* pp.129-144 (eds M.Jones, G. Mathys, and D. Rijks) 1989, Tunis.

Karaca, M., M. Guler, N. Durutan, K. Meyveci, M. Avci, H. Eyuboglu, and A. Avcin. 1991. In: *Proceedings of Soil and Crop Management for Improved Water Use Efficiency in Rainfed Areas*, pp. 251-259 (eds H.C. Harris, P.J.M. Cooper, and M. Pala) ICARDA, Syria

Karlen, D.L., G.E. Varvel, D.G. Bullock, and R.M. Cruse. 1994. *Advances in Agronomy* Vol. 53: 1-45.

Keatinge, J.D.H., N. Chapanian, and M.C. Saxena. 1988. *Journal of Agricultural Science, Cambridge.* 110: 651-659.

Kirda, C., Danso, S.K.A., and Zapata, F. 1989. *Plant Soil* 108: 87-92.

Kumar, J., Sethi, S.C., Johansen, C., Kelley, T.G., Rahman, M.M., and van Rheenen, H.A. 1996. *Indian Journal of Dryland Agricultural Research and Development* 11: 28-32.

MacRae, R.J., S. B. Hill, G.R. Mehuys, and J. Henning. 1990. *Advances in Agronomy* Vol. 43: 155-198.

Malhotra, R.S., K.B. Singh, H.A. van Rheenen, and M. Pala. 1996. In: *Adaptation of Chickpea in the West Asia and North Africa Region* pp.217-232 (eds N.P. Saxena, M.C. Saxena, C. Johansen, S.M. Virmani, and H. Harris) ICRISAT/ICARDA.

Martensson and Rydberg, 1994. *Swedish Journal of Agricultural Research* 24: 13-19.

Martin, R. J. and Felton, W. L. 1993. *Australian Journal of Experimental Agriculture* 33: 159-165.

Mason, M. G. and Rowland, I. C. 1990. *Australian Journal of Experimental Agriculture* 30: 231-236.

Masri, Z. 1996. Ph.D thesis, KSAU University, Russia. 325 pp.

McEwen, J, R.J. Darby, M.V. Hewitt, and D.P. Yeoman. 1989. *Journal of Agricultural Science, Cambridge* 115: 209-219.

Mitchell, C.C., R.L. Westerman, J.R. Brown, and T.R. Peck. 1991. *Agronomy Journal* 83: 24-29.

Osman, A.E., Ibrahim, M.H. and Jones, M.A., eds. 1990. *The Role of Legumes in the Farming Systems of the Mediterranean Areas.* Proceedings of a Workshop sponsored by UNDP/ICARDA, 20-24 June, 1988, Tunis, Tunisia.

Pala, M. Saxena, M.C., Papastylianou, and Jaradat, A.A. 1994. In *Expanding the Production and Use of Cool Season Food Legumes.* pp 130-143 (eds F.J. Muehlbauer and W.J. Kaiser) Dordrecht Kluwer Academic Publishers.

Papastylianou, I. and M. Jones. 1989. In: *Proceedings of Challanges in Dryland Agriculture: A Global Perspective*, pp. 822-825 (Eds. P.W. Unger, T.V. Sneed, W.J. Jordan, R. Jensen), August 15-19, 1988. Amarillo/Bushland, Texas, USA.

Papastylianou, I. 1988. In: *Proceedings of Nitrogen Fixation by Legumes in Mediterranean Agriculture*, pp.55-64 (D.P. Beck and L.A. Materon, eds.), Martinus Nijhoff Publishers, Dordrecht, The Netherlands.

Papastylianou, I. 1993. *European Journal of Agronomy* 2 (2): 119-129.

Papendick, R.I., Chowdhury, S.L., and Johansen, C. 1988. In: *Proceedings of World Crops: Cool season food legumes*, pp.237-255 (ed R.J. Summerfield). Kluwer Academic Publishers.

Paroda, R.S., Woodhead, T., and Singh, R.B. 1994. *Sustainability of rice-wheat production systems in Asia.* Regional Office for Asia and the Pasific (RAPA) publication: 1994/11, FAO/UNDP, Bangkok, Thailand, 209 pp.

Parr, , J.F., R.I. Papendick, and I.G. Youngberg. 1983. *Agroecosystems* 8: 183-201.

Prew, R.D., B.M. Church, A.M. Dewar, J. Lacey,N. Magan, A. Penny, R.T. Plumb, N.GillianThorne, A.D. Todd & T.D. Williams. 1985. *Journal of Agricultural Science, Cambridge* 104: 135-162.

Reeves, T.G. 1991. In: *Proceedings of Soil and Crop Management for Improved Water Use Efficiency in rainfed Areas*, pp. 274-285, (H.C. Harris, P.J.M. Cooper, and M. Pala, eds.) 15-19 May, 1989, Ankara, Turkey, ICARDA, Aleppo, Syria.

Robson, A. D. 1994. In: *Australian Grains - A Complete Reference Book on the Grains Industry.* pp. 288-90 (Ed. B.Coombs) Morescope Publishing Pty Ltd., Camberwell Victoria,.

Robson, A.D. 1990. In: *Proceedings of The Role of Legumes in the Farming Systems of the Mediterranean Areas* pp. 217-236 (eds A.E. Osman, M.H. Ibrahim, and M.A. Jones) Kluwer Academic Publishers, Dordrecht. The Netherlands

Rowland, I. C., Mason, M. G., Pritchard, I. A., and French, R. J. 1994. *Australian Journal of Experimental Agriculture* 34: 641-646.

Rupela, O.P., Wani, S.P., Danso, S.K.A., and Johansen, C. 1995. *Journal of Soil Biology and Ecology* 15: 127-134.

Saraf, C.S. Rupels, O.P., Hedge, D.M., Yadav, R.I., Shivakumar, B.G., Bhattarai, S., Razzague, M.A. and Sattar, M.A. 1997. Paper presented at the workshop on '*Residual effects of Legumes in Rice-Wheat Cropping Systems of the Indo-Gangetic Plain*' held at ICRISAT Asia Center. Patancheru, India.

Saxena, M.C. 1988. In: *Proceedings of Nitrogen Fixation by Legumes in Mediterranean Agriculture*, pp. 11-24 (eds D.P. Beck and L.A. Materon), Martinus Nijhoff Publishers, Dordrecht, The Netherlands.

Saxena, M.C., Moneim, A.A., and Beck, D. 1991.In *Proceedings of the Twenty-fourth Meeting of the Board of Trustees*, 30-31 May, 1990, ICARDA, Aleppo, Syria, pp. 72-75.

Saxena, N.P., Johansen, C., Saxena, M.C., and Silim, S.N. 1993. In *Breeding for stress tolerance in cool season food legumes*, pp. 245-270, (K.B. Singh, and M.C. Saxena, eds.), John Wiley & Sons and Sayce Publishing, St. Leonards, UK., and ICARDA, Aleppo

Shwenke, G.D., Herridge, D.F., Peoples, M.B., and Turner, G.L. 1997. In *Managing Legume Nitrogen Fixation in Cropping Systems of Asia* (eds OP Rupela, C. Johansen and DF Herridge) ICRISAT, Asia Center, Patancheru, India

Siddique, K. H. M. and Sykes, J. 1997. *Australian Journal of Experimental Agriculture* 37: 103-111.

Silsbury, J. H. 1990. *Australian Journal of Experimental Agriculture* 30: 645-649.

Unkovich, M. J., Pate, J. S., and Sanford, P. 1997. *Australian Journal of Agricultural Research:* 48, 267-293.

Yadav, R.L., Dwivedi, B.S., Gangwar, K.S. and Prasad, K. 1997. In *Residual effects of legumes in rice and wheat cropping systems of the Indo-Gangetic plain* (ed Kumar Rao, JVDK, Johansen, C.and Rego TJ) New Delhi, India:Oxford and IBH publishing.

Nitrogen Nutrition of Legume Crops and Interactions with Water

P.J. Gregory[1], J. Wery[2], D.F. Herridge[3], W. Bowden[4], N. Fettell[5]

1 Department of Soil Science, The University of Reading, Whiteknights, PO Box 233, Reading, RG6 6DW, U.K.; 2 ENSAM - INRA UFR d'Agronomie et Bioclimatologie, 2 Place Viala, 34060 Montpellier, Cedex 1, France.; 3 NSW Agriculture, PMB 944, Tamworth, NSW 2340, Australia.; 4 W.A. Department of Agriculture, Baron Hay Court, South Perth, WA 6151, Australia; 5 NSW Agriculture, PO Box 300, Condobolin, NSW 2877, Australia.

Keywords: nitrogen fixation, nitrogen/water interaction, legumes, rotations, N balance
Running title: Nitrogen/water interactions

Abstract

Accumulation of nitrogen by legumes occurs by uptake of soil NO_3^-, which dominates in young plants, and by fixation of N_2. The total quantity of N accumulated is highly correlated with shoot mass showing that N concentration is conservative (about 22 g/kg for a range of legumes). Rates of nitrate uptake and N_2 fixation are influenced by the availability of soil water with N_2 fixation more sensitive to mild water deficits than growth and nitrogen assimilation. The amount of N_2 fixed by grain legumes varies widely and is a consequence of both soil nitrate supply and growth. Relations between these quantities, when combined with an estimate of crop offtake in grain, can be used to provide a simple N balance facilitating improved N management in crop rotations. The N dynamics of cereal/legume rotations are complex but the residual benefits of the fixed N_2 have been widely demonstrated. Results from experiments in places such as Syria, France and Australia suggest that grain legumes should be considered as crops able to produce N-rich grains and straw in the absence of N fertilizer without detrimental effects on soil N. Nevertheless, the sustainable production of high-yielding cereals is only possible when N fertilizer is applied in the cereal phase of the rotation.

INTRODUCTION

The interaction of water and nutrients is a central feature of crop production in most rainfed systems. For nitrogen, in particular, the interactions are complex because not only the processes affecting chemical transport (e.g. diffusion) depend on water, but the biological processes affecting availability (e.g. mineralization) are also strongly dependent on water. The interaction is seen practically in the seasonally-dependent response of dryland crops such as wheat and barley to N fertiliser (Jones and Wahbi 1992), and the additional yields of many crops when water and N fertiliser are applied (Cooke 1986). In legume crops, nitrogen absorption and accumulation is reduced in water-limited conditions and the interaction of water and nitrogen is complicated further by the effects of water on the N_2-fixing system. Water contributes to the substantial variation in both growth and amount of N_2 fixed by leguminous crops in different seasons.

The inclusion of legumes in crop rotations has numerous benefits including breaks of pests and diseases, increasing N content of soils, and enhancing subsequent cereal yields. In Australia, where inputs of N fertiliser to crops are small, it has been estimated that 96% of the N in shoot biomass of cereals comes from non-fertiliser sources (Vallis, 1990). This means that correct management of legume crops and pastures is an important component of achieving sustainable cereal and protein yields. Hamblin and Kyneur (1993) concluded that contributory factors to the declining cereal protein yields and associated decline in soil fertility in Australia were inadequate legume pastures, too small a proportion of grain legumes in the rotation, and inadequate inputs of N fertiliser. Similarly, in West Asia and North Africa, rotations that include legumes in place of fallow have been shown to assist intensification of the cropping cycle because of

R. Knight (ed.), Linking Research and Marketing Opportunities for Pulses in the 21st Century, 335–346.

the additional N available (Keatinge and Chapanian, 1991). Because of the large seasonal effects of water on N_2 fixation and grain yield, the growth of the subsequent crops is affected not only by the rainfall in that season but also by the carry over effect on N balance of the legume crop.

In this paper we summarise the current knowledge about the accumulation of N by legumes and the effects of water deficit on growth and N_2 fixation. Recent attempts to predict the amount of fixed N carried over to subsequent crops are reviewed and the opportunities for further improvements described. Finally, the rotational benefits of grain legumes to cereal crops are explored and N balances of cereal/legume rotations summarised.

NITROGEN ACCUMULATION BY LEGUMES

Grain legumes acquire nitrogen by both uptake from the soil and fixation by *Rhizobium* and *Bradyrhizobium* in nodules. The balance of these processes changes during the life of the crop so that typically, as seed reserves are depleted during early growth, NO_3^- is taken up from the soil and reduced (mainly in the leaves), but as the plant develops so N_2 fixation becomes increasingly important. For example, Fig. 1 shows that nitrate reductase activity (a measure of the rate of N assimilation) in a chickpea crop was high initially but declined to almost nothing by the start of flowering (because of depletion of soil NO_3^-) whereas the rate of acetylene reduction (an indicator of the rate of N_2 fixation) was low until the start of flowering and peaked at the start of grain filling (Wery *et al.*, 1988). Many studies have shown this pattern and established that field-grown legumes will fix negligible amounts of atmospheric N_2 if the soil supply of available N is greater than or equal to plant demand. For example, if N fertilisers are applied or mineralised soil N is enhanced by rain or irrigation, then N assimilation is increased and N_2 fixation reduced. Conversely, large amounts of N (>200 kg N/ha) can be fixed by high biomass crops if plant demand for N exceeds supply and there are adequate populations of effective rhizobia in the soil. During the development of the crop, the rate of apparent N fixation (C_2H_2 reduction) increases with the rate of dry matter production (Fig. 1) which is, in turn, dependent on the production and expansion of leaves to intercept radiation and the conversion of the light energy intercepted to dry matter. At some point during grain filling, the rate of N_2 fixation decreases possibly because the grain competes more effectively than the nodules for the limited supplies of carbohydrate and because of the general senescence of the plant leading to cessation of N accumulation and the remobilization of N from the older leaves and stems.

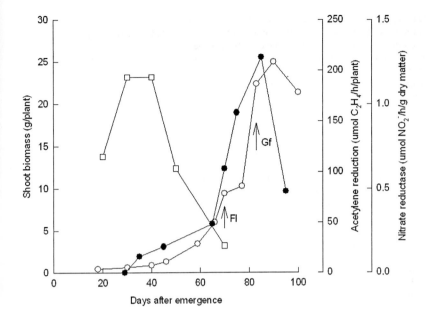

Figure 1 Changes in shoot biomass (O), apparent N_2 fixation (● measured using the acetylene reduction technique) and nitrate reductase activity (□) for a field-grown crop of chickpea (from Wery et al., 1988).

Generally, the amount of N accumulated by the crop is highly correlated with shoot biomass showing that N concentration is almost constant (Fig. 2). Under a wide range of growing conditions in south-eastern Australia, France and Syria, N concentration at maturity in lupin (*Lupinus angustifolius*), pea (*Pisum sativum*), lentil (*Lens culinaris*), faba bean (*Vicia faba*), and chickpea (*Cicer arietinum*) crops was conservative at 22 g/kg dry matter (r^2 0.92). Similarly, analysis of 10 lentil crops and 22 crops of chickpea grown at Tel Hadya, Syria over a period of 17 years showed highly significant linear relations between shoot N content and shoot dry matter although the concentration of N was higher in lentil (20.6 g/kg) than chickpea (16.9 g/kg; Pilbeam et al., 1997). This indicates that the partitioning of N between the stems/leaves and grain is such that the interspecific variability of grain N concentration is counterbalanced by the variability in straw N concentration and harvest index of the dry matter. However, the proportion of the N accumulated derived from the atmosphere by fixation varies considerably between species and between cropping situations. In the studies underlying Fig. 2, estimated fixation ranged from 20 to 97% in Australia, 44 to 92% in France, and 8 to 81% in Syria.

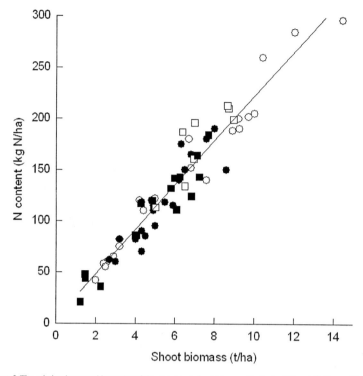

Figure 2 The relation between N content of the shoot and shoot biomass for lupin (○) and field pea (●) crops grown in Australia, and several food legume crops (see text for details) grown in France (□) and Syria (■). The line is the linear regression of all data.

WATER, N AVAILABILITY AND N₂ FIXATION

Water affects most of the processes determining the availability of N in soils to crops. Among the many transformations of N occurring in soils, and the numerous pathways of potential loss (see Fillery 1992 for a description of legume-based pastures), the mineralization of organic N to nitrate is critical. This biologically-mediated process is characterised by dependence on temperature and soil water and typically the rate increases to a maximum as soil water content increase close to field capacity (Gregory et al., 1997). In some soils with low organic matter content, the rate of mineralization is low and largely unaffected by water content presumably because the rate is substrate limited.

Two major issues require resolution in rainfed grain legume systems. First, rainfed systems are usually characterised by a marked break between the dry and wet seasons leading to the rapid mineralization at the start of the rainy season if temperatures are warm. This is particularly pronounced in the humid tropics (Birch, 1960; Warren et al., 1997) but may also occur in winter rainfall zones. In the tropics, the considerable amount of NO_3^- produced before a crop has established is frequently lost beneath the crop root zone from whence it may be leached to water courses or adsorbed by kaolinitic clays (Wong et al., 1990). For the cool season grain legumes, the magnitude of leaching losses early in the season is largely unknown but given that this is the only supply of N during early growth, it is important to know its fate. Second, legumes are used to improve fertility and there is a need to quantify better the mineralization characteristics of legume residues and the inputs of below-ground N. Recent studies (e.g. Russell and Fillery, 1996; McNeil et al., 1997) suggest that below-ground inputs of N from lupin and pasture legumes are underestimated if only N in the macro-roots is recovered, and that the mineralization of below-ground biomass (including fine roots, rhizodeposition and fine-root turnover) may contribute significantly to subsequent wheat crops.

Besides affecting the availability of N in the soil, water also affects the growth of the leaf canopy. Young, expanding leaves are affected by even short periods of water deficit because both cell division and cell expansion are inhibited. For example, in peas the expansion of leaf area can be described in relation to the fraction of transpirable soil water remaining in the soil (Lecoeur et al., 1996). The reduced canopy size coupled with smaller stomatal conductance under conditions of more severe water deficit, limits the carbon supply to the nodules. Similarly, the reduction in growth rate of the expanding leaves and branches limits the N demand of the shoot which can have a negative feedback effect on nodule activity. The combination of these two effects is a close relation between the rates of shoot growth and N_2 fixation.

Generally, N_2 fixation is reduced more by mild water deficits, than growth and nitrogen assimilation. For field-grown chickpea in France, Wery et al. (1988) determined that nitrate assimilation (nitrate reductase activity) integrated over the season in non-irrigated crops was very similar to that of irrigated controls whereas the N_2 fixation (integrated acetylene reductase activity) in non-irrigated crops was only 70% that of irrigated controls. Similar results have been found by Obaton et al. (1982) with soyabean. Although some physiological explanations have been advanced for the different relative susceptibility of the two processes to drought, the most dominant factor, at least for cool season grain legumes, lies in the relative timings of the NO_3^- assimilation and N_2 fixation processes. Water deficits are most common during grain filling (especially in Mediterranean climates) where nitrate assimilation has ceased but N_2 fixation is occurring. The net result is that grain yield is often less affected by water deficit than the amount of N_2 fixed. In a 2 year experiment on spring-sown chickpea, rainfed crops yielded 76% (1989) and 87% (1990) of the irrigated crops but the amount of N_2 fixed was only 40% (1989) and 56% (1990) that of the irrigated crops. (Table 1).

Table 1. Effect of terminal water deficit on grain yield, shoot biomass and N_2 fixation for spring-sown chickpea crops grown on a deep soil in southern France. The same letter after a given measure indicates no significant difference (P = 0.05) for that year with the Newman-Keuls test (Wery, 1996).

Year	Water deficit	Grain Yield (t/ha)	Total shoot biomass(t/ha)	N_2(% total N fixed)	Fixation (kg N/ha)
1989	Irrigated	3.78 a	11.81 a	61.6 a	133.5 a
	Rainfed	2.87 b	7.18 b	44.3 b	53.5 b
1990	Irrigated	2.73 a	9.42 a	64.9 a	108 a
	Rainfed	2.37 a	7.44 b	47.3 b	60 b

Because grain yield is relatively less affected by water deficits than the amount of N_2 fixed, the N balance resulting from the growth of legume crops can be substantially affected by the incidence and severity of water shortage. Using a simple N balance (N_2 fixed and N from seeds - N exported with grain) Beck et al. (1991) and Wery (1996) found that in Mediterranean regions, rainfed legume crops had a negative balance in 22 out of 24 cases but irrigated crops of pea and chickpea had positive balances in 9 out of 10 cases. Moderate water deficits and soils with moderate nitrogen contents gave balances closer to zero.

NITROGEN FIXATION - QUANTITIES AND PREDICTION

Previous reviews of nitrogen fixation by grain legumes have reflected the variable quantities fixed, reflecting the genetic capacities of the different species to fix N_2, the environmental constraints (including water) to those capacities, and the effects of cultural practices on both (Herridge et al., 1994; Peoples et al., 1995). Given the considerable range of interactions possible, it is unsurprising that both the amount of N_2 fixed and the percentage of N in a crop derived from the atmosphere should vary widely within a single crop species (Table 2). Similarly, the more limited review of annual legumes in Australian Mediterranean agriculture by Unkovich et al. (1997) while demonstrating the superiority of lupin (L. angustifolius) in the amounts of N fixed (mean 165 kg/ha) relative to field pea (83) and chickpea (70), found a considerable range for each species (30 - 283 for lupin, 26 - 183 for field pea and 43 - 124 for chickpea). Unkovich et al. (1997) also showed the considerable range of values for the proportion of N derived from the atmosphere (29 - 97% in the case of lupin).

Table 2. Range of values for the % of N in a crop derived from the atmosphere and the quantity of N_2 fixed. Data from Evans *et al.*, 1989; Herridge *et al.*, 1994; and Peoples *et al.*, 1995.

Species	% N in crop from atmosphere	N_2 fixed (kg ha^{-1})
Cool-season food legumes		
chickpea	8 - 82	3 - 141
lentil	39 - 87	10 - 192
pea	23 - 83	17 - 330
faba bean	59 - 92	78 - 330
lupin	29 - 97	32 - 288
Warm-season food legumes		
pigeonpea	10 - 81	7 - 235
cowpea	32 - 89	9 - 201
common bean	0 - 73	0 - 125
groundnut	22 - 92	37 - 206

Many of the experimental results published come from experimental plots rather than farmers' fields. Plots are often more fertile and yields are often larger because of more intensive husbandry so that estimates of N_2 fixation may be greater or smaller than those pertaining on farms, depending on the balance between soil nitrate supply and crop growth. Recent surveys in farmers' fields show that N_2 fixation was generally 70-80% of N in the shoot biomass except for chickpea and faba bean in Australia and mung bean (*Vigna radiata*) in Pakistan (Table 3). In this survey, there was no marked effect of species and the extremes (86% for lupin in Australia and 28% for chickpea also in Australia) most likely reflected soil nitrate and environmental conditions as well as the response of individual species to these factors. These values for N_2 fixation are clearly greater than the often quoted 50% value determined for N_2 fixation by warm-season food legumes such as soyabean (Rennie *et al.*, 1992).

Table 3. A summary of nitrogen fixation measurements made in farmers' fields.

Species	Country and region	Number of fields	% N fixed	N_2 fixed (kg/ha)
Cool-season food legumes				
Lentil	Pakistan - NWFP	40	78	47
	Nepal - Terai	15	77	-
Chickpea	Syria, Lebanon	27	72	-
	Pakistan - Punjab	83	75	-
	Nepal - Terai	13	80	-
Faba bean	Syria, Lebanon	19	60	-
	Australia - NSW	21	28	22
Lupin	Australia - NSW dryland	24	52	69
	Australia - NSW irrigated	39	75	150
	Australia - WA	6	86	222
Warm-season food legumes				
Mung bean	Pakistan - NWFP	40	47	28
	Pakistan - Punjab	12	77	-
Pigeonpea	Nepal - Terai	4	70	-

Sources: Unkovich *et al.*, 1994; Ali *et al.*, 1997; Aslam *et al.*, 1997; Maskey *et al.*, 1997; Schwenke *et al.*, 1997; Shah *et al.*, 1997; Rochester, Afandi, Peoples and Herridge, unpublished.

As explained previously, the regulation of N_2 fixation occurs through the balancing of soil nitrate supply and biomass increases which may change depending on the crop and season. Data from surveys in the North West Frontier Province (NWFP) of Pakistan provide some insight into this (Shah *et al.*, 1997). A total of 40 fields used for cool season lentil, and 40 fields used for the warm-season mung bean and mash bean (*Vigna mungo*), were sampled during the 1993/94 winter and 1994 summer for biomass, grain yield and N_2 fixation. The %N derived from the atmosphere (fixed) was different for the two types of crop. With lentil, the mean was 78% (range 55-100%) with the majority in the range 60-90% whereas with mung and mash bean, the mean was 47% (range 0-100%) with a slight concentration around 50%. In this environment, N mineralization would occur most rapidly during the summer when soils are moist and warm so that the smaller percentage of N_2 fixed by mung and mash bean is probably associated with the release of NO_3^- from organic matter. The total amount of N_2 fixed was <50 kg/ha for most crops ranging from 16 to 100 kg N/ha for lentil (mean 47) and 0-55 kg N/ha for mung and mash bean (mean 28). Correlation matrices show that grain yield and shoot biomass were the major determinants of N_2 fixation in lentil with the %N fixed generally high and uniform. Conversely, total N_2 fixed in mung and mash bean was highly correlated with the %N fixed but not with shoot biomass; this reflects the wide range of values for %N fixed and the probable inputs of soil nitrate.

It follows that the prediction of N_2 fixation by grain legumes requires relationships between N_2 fixation and crop yield, and N_2 fixation and soil nitrate. When combined with an estimate of crop offtake, these could be used to provide farmers with information on N balance of fields that should facilitate improved management decisions about the legume and subsequent cereal (or other crop) phases of the rotation. Essentially the following are required:

1) An estimate of total biomass. This can be obtained from grain yield and harvest index although the highly variable harvest index of some legumes (e.g. *L. angustifolius*) may complicate this.

2) An estimate of N in the biomass. This is reasonably robust because N concentration in biomass is fairly conservative (Fig. 2).

3) Estimate the percentage of N obtained by fixation. This could be estimated if a relation between %N fixed and mineral N at sowing (i.e. a soil test) could be determined; the relation would also need to take into account total N uptake because the %N fixed increases as total plant N (and dry matter) increases. Alternatively soil N supply could be estimated from a non-fixing crop with similar growing cycles and growing conditions.

4) Calculate the N removed in legume grain. Again, variation in N concentration in grain is small.

These functions may be separate and stand-alone or incorporated into paper-based decision aids for N management (e.g. NITROGEN IN '96 for northern NSW, Australia by Martin *et al.*, 1996), or into more complex models of growth and soil processes (e.g. HOWWET by Freebairn *et al.*, 1994). As a first step in model development, Herridge *et al* (1997) used regression analysis to relate the total N_2 fixed and the % N derived from the atmosphere to soil nitrate at sowing (to 0.9 m depth), total shoot N, and grain yield. The functions were derived from 9 field experiments with chickpea in the northern grain belts of Australia at 5 sites over 5 seasons involving different amounts of soil N as treatments. Soil nitrate ranged from 19 to 158 kg N/ha at sowing and crop N (shoot N x 1.3) at harvest from 61 to 194 kg N/ha. The total N_2 fixed ranged from 0 to 124 kg N/ha and the % of N fixed was 0-79%. Data from the 9 experiments were then subjected to simple and multivariate regression using %N fixed and total N_2 fixed as dependent variables and soil nitrate, shoot N and grain yield as independent variables. With the simple regressions, %N fixed was most strongly correlated with soil nitrate (r^2 0.59, P<0.001) and grain yield (r^2 0.62, P<0.001), than with shoot N (r^2 0.26, P<0.01). Multivariate regression analysis using soil nitrate plus shoot N, and soil nitrate plus grain yield improved the regression coefficients to 0.74 and 0.82 (Table 4). For total N_2 fixed, only soil nitrate plus grain yield could be used as independent variables and the regression coefficient (0.68) was not quite as strong as that for %N fixed.

Table 4.Regression functions describing the relations between nitrogen fixation of chickpea and soil nitrate (0 - 0.9 m depth at sowing), shoot N content, and grain yield.

Variables	Regression Function	Regression Coefficient (r^2)
% N fixed		
soil nitrate (z)	% N fixed = 93 - 0.97z + 0.003z^2	0.59
soil nitrate (z)	% N fixed = 12 - 0.54z + 0.0002z^2 + 1.12w - 0.004w^2	0.74
shoot N (w)		
soil nitrate (z)	% N fixed = -52 - 0.89z + 0.003z^2 + 145x - 34.8x^2	0.82
grain yield (x)		
N_2 fixed (kg/ha)		
soil nitrate (z)	N_2 fixed = - 128 - 0.97z + 0.004z^2 + 235x - 55.9x^2	0.68
grain yield (x)		

While trying to maximise grain yield, farmers would prefer to grow legumes that have a positive N balance i.e., fix more N than is harvested in grain and other products. In the 9 experiments of Herridge et al. (1997), N balance ranged from -67 to +61 kg/ha so that farmers must manage the system to ensure a positive N balance. Using the regression functions in Table 4 with the knowledge that chickpea grain contains 3.5% N, it is possible to demonstrate the conditions under which a positive N balance, will result. For example, in a soil with low nitrate content at sowing (40 kg N/ha), grain yields of 1 and 2 t/ha change the balance from -16 to +16 kg N/ha but for a soil with high nitrate content (120 kg N/ha), crops with the same yield will both have negative balances (-35 and -10 kg N/ha). Thus to sustain soil N, chickpea must be grown in soils low in N at sowing. With low-yielding crops, N_2 fixation is constrained by the small biomass but with high-yielding crops, substantial amounts of N are removed in the grain.

In some countries, grain legumes are grown not only for their grain but the straw is also marketed and its N content must be accounted for. For example, in dry regions where forage production is low (e.g. lentil in Syria or cowpea and groundnut in Senegal), grain legume straw is sold as forage and gives economic returns equivalent to those for grain. The use of legumes as dual-purpose crops is generally restricted to those species with high N concentration in the straw (12.6 g/kg in lentil) while those with low N concentration (6 g/kg for chickpea) are generally used only for grain. Beck et al. (1991) demonstrated that the export of straw had a limited effect on the N balance in chickpea (-29.7 kg N/ha for the dual-purpose crop compared with -13.8 kg N/ha for the grain crop) but a major effect on lentil (-44.0 kg N/ha when dual-purpose but +18.0 kg N/ha for grain only). Thus the common use of legumes as both grain and forage crops in dry regions has major impact on the N balance, and the improvement of soil N fertility depends largely on the development of forage legumes together with the ploughing of grain legume straw.

ROTATIONAL BENEFITS OF GRAIN LEGUMES

Yields of cereals grown after legumes are almost always increased compared with cereals grown after cereals especially in rainfed cropping systems with low chemical inputs (Evans et al. 1991, Australia; Keatinge et al. 1988, Syria). The benefit varies seasonally depending on the weather both during the legume season and the cereal season. For example, Gregory (1997) in Western Australia found that wheat yields in 1993 were increased by an average of 39% following a range of legumes in 1992 and by 135% in a separate study in the next season (1994) because of different amounts and patterns of rainfall; grain N was increased in the first season by legumes (2.07% compared to 1.95% with continuous cereals) but not in the second season (2.33% irrespective of previous crops). In 1994, wheat yields were 3.7 t/ha after lupin, 2.6 t/ha after pea, 2.4 t/ha after chickpea and 2.1 t/ha after faba bean. These correlated well with the residual N returned to the soil after the legume crops (Asseng et al 1998). On some sites there are also non-nitrogen benefits (particularly reductions in cereal root diseases) but these are site-specific and difficult to separate from nitrogen benefits. However, provision of N fertilizer alone rarely gives cereal yield equal to those obtainable in rotation with cereals (Rowland et al. 1994).

The N dynamics of legume-cereal rotations is not well researched but are important if the rotations are to be managed for optimum yields and in an attempt to sustain cereal production. The additional N available to the cereal can come from 3 sources:

"Spared" N resulting from less NO_3^- uptake by the legume crop than by a cereal (see Unkovich *et al.* 1997 for a discussion of this source).

N from root exudates and decomposition of roots and nodules.

N from the decomposition of above-ground crop residues.

The quantity of legume residues is a dominant feature in predicting N carry over to cereals (e.g. the NPDECIDE model of Bowden and Burgess, 1992) and has a marked influence on the amount of mineral N in the soil at the start of the cereal season. Clearly, the N concentration of the residues will also affect the rate at which N is mineralised and once mineralised this N will be susceptible to other pathways of loss such as leaching. Compared with inorganic sources of N, one of the advantages of residue N may be its smaller susceptibility to leaching as the intensity of rainfall increases. Table 5 shows results from a series of experiments conducted on deep, yellow sands (field capacity about 75 mm/m) in Western Australia. The results show that the effectiveness of organic sources relative to inorganic N supplied at sowing as ammonium nitrate increases both as the proportion of N in the organic material increases and as the intensity of leaching increases. This, and other studies in columns containing yellow loamy sands (e.g. Diggle and Bowden, 1991), shows that although leaching may be smaller with organic additions than inorganic fertilizers, crop growth may be inhibited because of slow mineralisation and although N may be leached, the roots may catch up with this N.

Table 5. Effect of leaching intensity on the effectiveness at anthesis of residue organic nitrogen sources applied 2-4 weeks before sowing relative to ammonium nitrate applied at sowing. The mixtures shown were made from de-hulled lupin seed (6.1% N) and ground wheat straw (0.5% N).

Source	%N	Site - mm leaching rain		
		Lancelin (208)	Badgingarra (174)	Meckering (104)
Clover seedlings	6.9	4.3	1.6	1.1
Mixture	6.0	2.1	1.6	1.0
Lupin leaves	3.8	1.8	1.2	0.7
Mixture	3.0	2.1	1.1	0.8
Mixture	1.0	0.0	0.1	0.2

Management of the legume crop will also influence the N dynamics and residual benefit. For example, earlier sowing of peas increased biomass N and the % of N fixed resulting in a large N benefit (O'Connor *et al.*, 1993). Similar conclusions, at least on the apparent N balance of the legume crop, were drawn in a comparison of winter- and spring-sown chickpea in Mediterranean regions (Beck *et al.*, 1991). Depending on the residue applied and the pattern of weather, the benefits of the residual N may not be apparent in the first season after the legume crop. Fettell (unpublished) found no response of wheat to the sowing time of peas in the first season but there were responses in the second and third seasons. Overall, early sowing of peas increased wheat yield from 5.77 to 6.29 t/ha (sum of 3 years) and protein from 9.7 to 10.1%.

Several recent studies have examined the sustainability of 1:1 grain legume : cereal rotations. Armstrong *et al.* (1997) studied 4 grain legume : wheat rotations and a barley : wheat rotation for 1 cycle in south-eastern Australia. Only the *Lupinus angustifolius* : wheat rotation achieved a positive N balance (10 kg N/ha) and that meant that the wheat had to be highly efficient in obtaining N from the lupin. All other rotations (*Lupinus albus*, field pea and chickpea) had net N deficits (-6 to -80 kg N/ha) with field pea (-80) and chickpea (-74) rotations depleting soil N more than the barley : wheat rotation (-67 kg N/ha). These results suggest that 1:1 rotations are rarely able to achieve sustainable N sufficiency.

A longer 1:1 legume : cereal and cereal : cereal study was initiated in 1989 in northern NSW Australia to examine the propositions that first, chickpea could fix sufficient N_2 to maintain soil N; and second, that chickpea could provide sufficient N to the following wheat crop for high levels of productivity (Herridge *et al.* 1997). Selected treatments from 6 experiments (combinations of 3 sites and 3 seasons) were used to provide the data. The shoot biomass of the chickpea crops ranged from 4.8 to 6.4 t/ha (average 5.8 t/ha) and shoot N was 84 to 113 kg/ha (average 102 kg/ha). Grain yields and grain N content varied by more than 2 fold from 1.2 to 2.8 t/ha and 45 to 105 kg/ha respectively. The calculated total N_2 fixed ranged from 4 to 116 kg/ha (average 73 kg/ha) and the % N fixed from 4 to 79% (average 57%). In the same 6 experiments, the amount of NO_3-N in the soil prior to harvest was 41 kg N/ha (range 13 to 63 kg N/ha) beneath chickpea and 26 kg N/ha (7 to 43 kg N/ha) beneath wheat. Soil nitrate increased during the summer fallow at all sites to an average 89 kg N/ha for chickpea and 56 kg N/ha after wheat (Fig 3). Thus, an additional 18 kg N/ha

344

was mineralised during the fallow after chickpea. The distribution of pre-harvest NO_3^- was almost identical in the upper 60 cm but there was greater depletion by wheat from 60 to 120 cm (Fig 3). By the end of the summer fallow, NO_3^- was more evenly distributed and concentration has increased to 80 cm depth. Although the N content of the chickpea residues (average 60 kg N/ha) was always greater than that of wheat (average 17 kg N/ha) and more soil NO_3^- accumulated during the fallow after chickpea than wheat, the calculated mineralization rates were insufficient to sustain an N-sufficient crop. On average, chickpea fixed sufficient N (73 kg/ha) to offset N harvested in chickpea grain (71 kg/ha) but insufficient to compensate for N harvested in the wheat grain (65 - 130 kg/ha). Thus soil N can only be maintained through inputs of fertilizer N in the cereal phase.

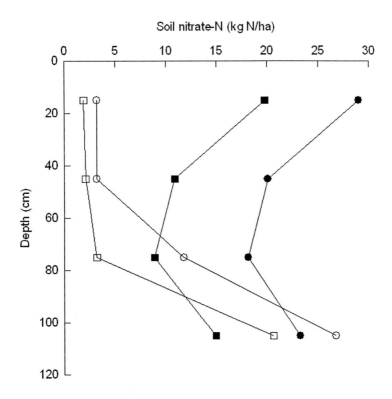

Figure 3 Profiles of nitrate-N in the soil beneath crops of chickpea (circles) and wheat (squares) at pre-harvest (open symbols) and post-harvest (closed symbols). Original data from Herridge et al., 1997.

CONCLUSIONS

Considerable progress has been made in determining the principal interactions between soil water availability and the accumulation of soil NO_3^--N and N_2 fixation by grain legume crops. It is clear that the amount of N_2 fixed is a consequence of both the supply of soil nitrate and the biomass production of the crop; both of these are influenced by many factors including the availability of water.

Simple models to describe the residual benefits of grain legume crops to subsequent cereal crops are being developed and increasingly adopted by farmer groups. Results from experiments in places such as Syria, France and Australia suggest that grain legumes should be considered as crops able to produce N-rich grains and straw in the absence of N fertilizer without detrimental effects on soil N. Improvement of soil fertility and the achievement of high yields by subsequent cereals can only be obtained through inputs of fertilizer N in the cereal phase.

References

Ali, S., Yasmin, K., Mustaq, N., Mann, M.I., Peoples, M.B. and Herridge, D.F. 1997. In *Managing Legume Nitrogen Fixation in Cropping Systems of Asia:* Proceedings of an International Workshop eds O.P. Rupela, C. Johansen and D.F. Herridge). ICRISAT: Patancheru, India.

Armstrong, E.L., Heenan, D.P., Pate, J.S. and Unkovich, M.J. 1997. *Australian Journal of Agricultural Research* 48: 39-47.

Aslam, M., Peoples, M.B. and Herridge, D.F. 1997. In *Managing Legume Nitrogen Fixation in Cropping Systems of Asia:* Proceedings of an International Workshop eds O.P. Rupela, C. Johansen and D.F. Herridge). ICRISAT: Patancheru, India.

Asseng, S., Fillery, I.R.P. and Gregory, P.J. 1998. *Australian Journal of Experimental Agriculture* (In press).

Beck, D.P., Wery, J., Saxena, M.C. and Ayadi, A. 1991. *Agronomy Journal* 83: 334-341.

Birch, H.F. 1960. *Plant and Soil* 12: 81-96.

Bowden, J.W. and Burgess, S.J. 1992. In *Transfer of Biologically Fixed Nitrogen to Wheat*, pp. 77-84 (ed. J.F. Angus). Canberra: Grains Research and Development Corporation.

Cooke, G.W. 1986. *Philosophical Transactions of the Royal Society* London A316: 331-346.

Diggle, A.J. and Bowden, J.W. 1991. *Australian Journal of Agricultural Research* 42: 1053-1064.

Evans, J., Fettell, N.A., Coventry, D.R., O'Connor, G.E., Walsgott, D.N., Mahoney, J. and Armstrong, E.L. 1991. *Australian Journal of Agricultural Research* 42: 31-43.

Evans, J., O'Connor, G.E., Turner, G.L., Coventry, D.R., Fettell, N., Mahoney, J., Armstrong, E.L. and Walsgott, D.N. 1989. *Australian Journal of Agricultural Research* 40: 791-805.

Fillery, I.R.P. 1992. In *Transfer of Biologically Fixed Nitrogen to Wheat*, pp. 5-13 (ed. J.F. Angus). Canberra: Grains Research and Development Corporation.

Freebairn, D.M., Hamilton, A.H., Cox, P.G. and Holzworth, D. 1994. HOWWET? A Computer Program. Toowoomba, Queensland: APSRU, DPI-CSIRO.

Gregory, P.J. 1997. *Australian Journal of Agricultural Research* (in press).

Gregory, P.J., Simmonds, L.P. and Warren, G.P. 1997. *Philosophical Transactions of the Royal Society* London B 352: 987-996.

Hamblin, A. and Kyneur, G. 1993. *Trends in wheat yields and soil fertility in Australia.* Canberra: Australian Government Publishing Service.

Herridge, D.F., Marcellos, H., Felton, W.L., Turner, G.L. and Peoples, M.B. 1997. *Australian Journal of Agricultural Research* (in press).

Herridge, D.F., Rupela, O.P., Serraj, R. and Beck, D.P. 1994. In *Expanding the Production and Use of Cool Season Food Legumes,* pp. 472-492 (eds F.J. Muehlbauer and W.J. Kaiser). Dordrecht: Kluwer Academic Publishers.

Jones, M.J. and Wahbi, A. 1992. *Experimental Agriculture* 28: 63-87.

Keatinge, J.D.H. and Chapanian, N. 1991. *Journal of Agronomy and Crop Science* 167: 61-69.

Keatinge, J.D.H., Chapanian, N. and Saxena, M.C. 1988. *Journal of Agricultural Science, Cambridge* 110: 651-659.

Lecoeur, J., Wery, J. and Sinclair, T.R. 1996. *Agronomy Journal* 88: 467-472.

Martin, R.J., Marcellos, H., Verrel, A. and Herridge, D.F. 1996. In *Decisions about Nitrogen Fertiliser for High Yield and Quality Wheat.* 13pp. GRDC/NSW Agriculture: Australia.

Maskey, S.L., Bhattarai, S., Peoples, M.B. and Herridge, D.F. 1997. In *Managing Legume Nitrogen Fixation in Cropping Systems of Asia:* Proceedings of an International Workshop eds O.P. Rupela, C. Johansen and D.F. Herridge). ICRISAT: Patancheru, India.

McNeil, A.M., Zhu, C. and Fillery, I.R.P. 1997. *Australian Journal of Agricultural Research* 48: 295-304.

Obaton, M., Miquel, M., Robin, P., Conejero, G., Domenach, A.M. and Bardin, R. 1982. *Compte Rendus d'Academie de Science,* Paris, Serie III, 294: 1007-1012.

O'Connor, G.E., Evans, J., Fettell, N.A., Bamforth, I., Stuchberry, J., Heenan, D.P. and Chalk, P.M. 1993. *Australian Journal of Agricultural Research* 44: 151-163.

Peoples, M.B., Herridge, D.F. and Ladha, J.K. 1995. *Plant and Soil* 174: 3-28.

Pilbeam, C.J., Wood, M. and Jones, M.J. 1997. *Experimental Agriculture* 33: 139-148.

Rennie, R.J., Dubetz, S., Bole, J.B. and Muendel, H.H. 1992. *Agronomy Journal* 74: 725-730.

Rowland, I.C., Mason, M.G., Pritchard, I.A. and French, R.J. 1994. *Australian Journal of Experimental Agriculture* 34: 641-646.

Russell, C.A. and Fillery I.R.P. 1996. *Australian Journal of Agricultural Research* 47: 1047-1059.

Schwenke, G.D., Peoples, M.B., Turner, G.L. and Herridge, D.F. 1997. In *Managing Legume Nitrogen Fixation in Cropping Systems of Asia:* Proceedings of an International Workshop eds O.P. Rupela, C. Johansen and D.F. Herridge). ICRISAT: Patancheru, India.

Shah, Z., Shah, S., Herridge, D.F. and Peoples, M.B. 1997. In *Managing Legume Nitrogen Fixation in Cropping Systems of Asia,* pp. (in press) (eds O.P. Rupela, C. Johansen and D.F. Herridge). Patancheru: International Crops Research Institute for the Semi-Arid Tropics.

Unkovich, M.J., Pate, J.S. and Sanford, P. 1997. *Australian Journal of Agricultural Research* 48: 267-293.

Unkovich, M.J., Pate, J.S., Sanford, P. and Armstrong, E.L. 1994. *Australian Journal of Agricultural Research* 45: 119-132.

Vallis, I. 1990. *Journal of the Australian Institute of Agricultural Science* 3: 19-23.

Warren, G.P., Atwal, S.S. and Irungu, J W. 1997. *Experimental Agriculture* (in press).

Wery, J. 1996. In *Memoire d'Habilitation a Diriger des Recherches en Physiologie Vegetale*. Universite Montpellier, France.

Wery, J., Deschamps, M. and Leger-Cresson, N. 1988. In *Nitrogen Fixation by Legumes in Mediterranean Agriculture,* pp. 287-301 (eds D.P. Beck and L.A. Materon). Dordrecht: Martinus Nijhoff.

Wong, M.T.F., Hughes, R. and Rowell, D.L. 1990. *Journal of Soil Science* 41: 655-663.

EFFECTS OF REDUCED TILLAGE ON FOOD LEGUME PRODUCTIVITY

H. Marcellos [1], R. C. Dalal [2] and W. L. Felton [1]

1 New South Wales Agriculture, Tamworth Centre for Crop Improvement, RMB 944, Tamworth. New South Wales, 2340, Australia.;
2 Queensland Department of Natural Resources, Queensland Wheat Research Institute, Toowoomba, Queensland, 4350, Australia.

Abstract

The trend toward eliminating tillage before sowing food legumes directly into crop residues has been driven by the need to reverse soil degradation. It has been stimulated by the accrual of benefits such as improved timeliness of sowing, less labour and machinery input, effective weed control using herbicides, less erosion and runoff, increased soil water storage and improved soil physical and organic fertility.

The productivity of food legumes is not reduced by eliminating tillage, and in many instances, is higher due to improvements in some site-specific limiting factor(s). The nitrogen dynamics of food legumes tend to be improved under no-tillage, with small increases being reported in nitrogen fixation and the proportion of plant nitrogen derived from fixation. The beneficial effects of no-tillage in some environments, has been due to the improved establishment of food legumes and in others to a reduction in pest and disease incidence. In rain-fed agriculture, benefits of no-tillage have been due largely to improved soil water storage and availability to food legumes.

INTRODUCTION

It was said in the early 1800's that "tillage is beneficial to all sorts of land" (Tull cited by Elmore 1990). The claim is no doubt justified on land where specific problems exist as for example land where soil strength needs to be reduced, drainage or water infiltration improved, cycles of insect pests and diseases disrupted, or the mineralisation of soil nitrogen increased. Tilling soil has been however primarily necessary to eliminate weeds, and prepare a seedbed into which seed and fertiliser could be placed.

Tillage refers to any of many methods for disturbing soil (mouldboard, disc or chisel ploughing; tine or disc cultivation; tine or disc planting; hand hoeing), depth of cultivation, and timing of operations (Table 1). Many variations in tillage practice, combined with treatment of crop residue after harvest are possible giving rise to many choices in fallow management. In less developed countries (LDC's), above-ground crop residues are highly valued for fuel or animal feed, and so are gathered and not incorporated into the soil.

Given that tillage continues to be a cornerstone of agriculture for most farmers, in most parts of the World, "why is there interest in reducing it? ". The pressure to reduce, or eliminate tillage has arisen relatively recently, to better sustain the cultivation of crops in countries with cereal-fallow systems, and in which continuous cropping has involved tilled fields, low fertiliser inputs, and incorporation of crop residues. These practices have proven unsustainable, invariably leading to loss of soil through erosion, and a decline in soil organic matter (Dalal and Mayer 1986).

Attempts to remedy problems arising from frequent tillage and removal of crop residues have stimulated research to develop cropping systems in which soil disturbance other than that required to sow the crop is either eliminated or minimised (reduced), and the residues of previous crops are retained. These are typically in regions where agriculture has become highly mechanised and inputs of manual labour have decreased, and broad-spectrum, non-residual herbicides, like glyphosate, have become available to control weeds.

R. Knight (ed.), Linking Research and Marketing Opportunities for Pulses in the 21st Century, 347–353.
© *2000 Kluwer Academic Publishers. Printed in the Netherlands.*

Table 1. Terms used to describe tillage practices.

Tillage	Crop residue[1]	Abbrev.[2]	Crop residue management	Frequency of tillage[3]	Tillage method[4]
Conventional	Removed	CT	Burned, feed	3 - 7	MP, DP, CP
Conventional	Retained	CT	Mulched	3 - 7	MP, DP, CP
Minimum	Removed	MT	Burned, feed	1 - 3	DP, CP
Minimum	Retained	MT	Mulched	1 - 3	DP, CP
No-till (Zero-till)	Removed	NT, ZT	Burned, feed	0	Nil
No-till (Zero-till)	Retained	NT, ZT	Mulched	0	Nil
Direct-drill	Retained	DD	Grazed	0	Nil

1 Fate of above-ground residues of previous crop. 2 Abbreviation. 3. Number of times field cultivated during the fallow. 4. Mouldboard plough (MP); Disc plough (DP), Chisel or tined implements, including hand-held hoe (CP).

A new lexicon has arisen, with terms such as "conservation farming" and "conservation tillage" appearing. Conservation tillage is defined as those practices which leave >30% residue cover on the soil surface after sowing (CTIC 1995) and is regulated in the United States by legislation. In Europe burning winter cereal residues after harvest is prohibited to reduce atmospheric pollution. The greatest application of reduced and no-tillage has been in countries like the United States, Canada, Australia, and Brazil and in Europe. As practices, they have been assessed for semi-arid, temperate regions (Rasmussen and Collins 1991), the semi-arid tropics (Anderson and Muir 1996) and temperate Australia (Martin *et al.* 1995). Their adoption by farmers has depended on demonstration that positive benefits such as erosion control, conservation of soil organic fertility, effective weed control, reduced labour and input costs, and improved timeliness of mechanical operations can be expected. Yield benefits are desirable and achievable when they improve site-specific variables like the amount of stored soil water, or alleviate a stress.

Many farmers in less developed countries (LDCs) rely on their own labour to cultivate small farms in which winter and cereal grain, and food legume and vegetable crops are inter-, double- or triple-cropped. Traditional support for tillage is strong in these countries. In Ethiopia, (Ghizaw and Molla 1995) one of the causes of low yield in faba bean and field pea is held to be inadequate seedbed preparation (Ghizaw and Molla 1995). Seeding chickpea by broadcasting is common in West Asia (Singh and Saxena 1996), and requires a cultivated seedbed for seed and fertiliser to be incorporated into the soil; in many other areas, such as terraced highlands, farmers do not have access to tillage equipment.

Farmers in many LDCs therefore must produce food from land, which may be already degraded, and will be at the mercy of whatever occurs in the way of drought, flood, pests and disease. It is in these areas where policy intervention and support from governments will be needed to introduce new practices aimed at restoring and sustaining soils and their productivity.

THE BENEFITS OF CONSERVATION TILLAGE IN CEREAL SYSTEMS.

Research with cereals has shown that eliminating or reducing tillage during the fallow may have significant effects on grain yield and quality, operating through effects on:

    soil structure, organic fertility and nitrate-N,
    infiltration and conservation of soil water,
    reduction in soil erosion,
    cooler seedbed temperature,
    higher levels of soil biota,
    slower crop biomass accumulation,
    sometimes higher disease incidence of cereal diseases such as yellow leaf spot (*Pyrenophora* sp.), and crown rot (*Fusarium graminearum*) for example.

Tillage itself may not to have a large effect on yield, as was the case for wheat in eastern Australia; the retention of cereal residue was the major factor underlying 10% less productivity under continuous no-tillage (Felton *et al.* 1995).

Even with the reduction or elimination of tillage prior to seeding, and retention of stubble from the previous crop (mulching), soil organic matter is likely to decline in cereal systems of semi-arid regions, albeit at a slower rate (Rasmussen and Collins 1991). This will be affected by the local climate, use of fertilisers, and the relative amounts of C and N in the system or removed in grain. The introduction of pasture leys, and forage and food legumes to break up cereal mono-culture may also have positive effects on soil organic fertility in the long term (Dalal et al. 1995) or in the short term (Herridge et.al. 1994: Horn et al. 1996a: Dalal et al. 1997).

TILLAGE AND PRODUCTIVITY

There has been a general reduction in the frequency of tillage, during the fallow, in countries like the United States, in Europe, Brazil and Australia with extensive cereal-fallow based systems. There has also been a trend towards the retention of crop residues and the greater use of chisel cultivators. Under these circumstances, food legumes like chickpea, faba bean, field pea, lupin, and others are likely to be grown using similar approaches, and have proven well suited to no-tillage.

In this paper, we show how eliminating or consider how reducing tillage prior to sowing, with or without retention of prior crop residues, has been reported may affect firstly to increase the productivity of food legumes, and benefit some also of the underlying processes. We will also discuss how food legumes can be successfully planted using no-tillage.

The literature abounds with information on the influence of conservation tillage on the productivity of cereals and soybeans. Data for food legumes are scarce especially from the LDCs. Productivity benefits should be measured in terms of both marginal changes in legume grain yield directly attributable to reducing or eliminating tillage, and also collateral benefits to the system such as improved nitrogen dynamics, impact on biotic and abiotic factors affecting legume production, improved soil structure and improved weed and pest management.

Effects of tillage on soybean productivity

Although not a mandate crop for the International Food Legume Research Conference III, data for soybean serve to demonstrate the benefits that could be gained for other legumes by adoption of no-tillage. Research in the mid-western and southern United .States. has shown the existence off differences in productivity between tillage practices, and attributed them to site-specific management or environmental factors. No "tillage x cultivar", or "tillage x sowing date x cultivar" interactions have been observed (Elmore 1990). Soybean yields in the United States are usually similar, or slightly higher in comparisons of tilled with no-tilled systems (Lindemann et al. 1982; Dick and van Doren 1985; Bharati et al. 1986; Elmore 1987; Webber et al. 1987; Edwards et al.1988). In dry years, no-till soybean may yield more than tilled (Desborough 1984; Webber et al. 1987); poorly drained, fine-textured soils may require tillage to produce satisfactory soybean yield (Van Doren and Reicosky 1987); poor weed control (Lindemann et al. 1982), herbicide injury, and disease may reduce no-till soybean yields relative to yields of tilled treatments.

The evidence for soybean therefore is that productivity under no-till should be equivalent to, and in some cases higher than under conventional tillage if the practice improves some site factor (Wagger and Denton 1989). Greater conservation of water, long term reduction in soil erosion, saving of time at sowing and the capacity to plant earlier leading to higher yields are site factors which are advantaged by no-tillage.

Effects of tillage on yields of cool-season food legumes

Eliminating tillage when growing cool season food legumes generally results in similar or mostly higher yields and no evidence for generally lower yields (Table 2). The variation reported in grain yield increases from no-till could have been due to differences in crop residue and weed management during the preceding fallow period, to more timely sowing and/or better crop establishment.

Soil type or condition may influence the suitability of the system for no-till. For example, lupin yields on sandy soils in Western Australia were sometimes increased in the range 5-15%, by deep tillage (Jarvis 1986). Lupin yields were increased by deep tillage of fine-textured soils in southern Australia, probably due to improvements in drainage (Ellington 1986).

Table 2. The influence of no-tillage (NT) compared with tillage (CT) on grain yield of food legumes

Food legume	N[1]	Yield NT[2]	Yield CT[3]	Yield change (%)	Reference
Chickpea	3	1065	910	17	Dalal et al. 1997
	1	2416	2285	6	Doughton et al. 1993
	1	1940	1290	50	Horn et al. 1996
	16	2040	1830	11	Felton & Marcellos 1997
Faba bean	5	5150	4450	16	Rizk et al. 1991
	6	2840	2350	21	Felton & Marcellos 1997
Lupin	12	1100	1100	0	Ellington et al.1979
Lentil	-	861	979	-12 (ns[4])	Miah & Rahman 1993
Cowpea	3	1020	870	17	Herridge & Holland 1992
Mung bean	2	1500	1390	8	Herridge & Holland 1992
Soybean	3	1560	1430	9	Herridge & Holland 1992
	4	1470	1190	24	Herridge & Holland 1992
	4	2590	2600	ns	Hughes & Herridge 1989

1 Number of site-years of data, 2 No-tillage; 3 Conventional tillage; 4 Not significant.

Yields of cool-season food legumes therefore tend to be higher under no-till under most conditions. The practice may be recommended, particularly when considered together with the other benefits, which accrue in terms of soil conservation and soil fertility.

INFLUENCE OF REDUCED TILLAGE ON SYSTEM VARIABLES

Nitrogen fixation.

Total N_2 fixation by a species of food legumes (Nfix) is directly related to two variables, crop biomass production, and the tolerance of its N_2 fixation mechanism to soil nitrate (Peoples*et al.* 1995). Management practices, which affect these, will influence Nfix, and the percentage of crop biomass N derived from fixation (Pfix). Tillage normally accelerates the mineralisation of soil organic matter, leading to higher levels of soil nitrate-N in the profile at the time the legume is seeded. This should result in lower values for Pfix by the food legume. If more soil water is stored, and soil nitrate-N levels are lower following no- cereal fallows, both Nfix and Pfix should be higher.

Tillage practice will influence Nfix and Pfix according to its effect on accumulation of biomass N. No-tillage has increased Nfix in soybean by as much as 52 kg/ha (Table 3), and in other assessments, Pfix tended to be higher under no-till by a few percent in some cases, and slightly less in others (Herridge and Holland 1992).

Table 3. N budgets for soybean grown with cultivation or no-tillage [1]

Tillage	Nodule mass	Crop N	Seed N	Fixed N	N balance
	(mg/plant)	*(kg N/ha)*	*(kg N/ha)*	*(kg N/ha)*	*(kg N/ha)*
Cultivated	86	245	150	180	30
No-tillage	139	264	152	232	80

1 From Hughes and Herridge (1989).

Nfix has been measured in chickpea (Table 4) following either fallow or grain sorghum where sorghum residue was either removed, incorporated or retained on the surface with NT (Doughton *et al.* 1993). Tillage had no effect on total Nfix, and leaving stubble on the soil surface with no-till produced as much fixed N_2 as conventional till where stubble was incorporated. It was concluded that no-tillage, as a technique for preventing rainfall run-off and soil erosion did not disadvantage Nfix or yields in chickpeas and could therefore be recommended to farmers. In another study with chickpea over two years, Nfix was low (15-32 kg N/ha), and not significantly affected by tillage (Horn *et al.* 1996b).

Therefore, Nfix by food legumes should not be adversely affected, and under most conditions it is likely to be increased by no-till.

Table 4. . N_2 parameters for chickpea grown with no-tillage or cultivation.

Variable	NT	CT	N	Reference
Nfix	9	7	4	Dalal *et al.* 1997
	103	97	1	Doughton *et al.* 1993
Pfix	30	21	2	Dalal *et al.* 1997
	90	85	1	Doughton *et al.* 1993

SOIL CONSERVATION

The axiom "no soil, no crop" is often ignored, but must be given special consideration when food legumes are grown in extensive, highly mechanised cropping systems. Little above-ground crop residue is left following harvest of food legumes, and if not removed for feed or fuel, will break down more readily than cereal straw due to a lower C:N ratio. Inadequate soil cover during the fallow increases the risk of rainfall run-off, and erosion of soil. Soil water recharge may be impeded due to lack of soil roughness in some soils like non-cracking clays, and tillage may be required after the legume.

Food legumes are usually sown at lower populations than cereals (approx. 30 m^{-2}), and may accumulate biomass more slowly. Soil may therefore be inadequately protected against erosion during the legume crop's early growth. An option is to plant food legumes directly into cereal residue following a no-till fallow. No significant yield reduction in chickpea and faba bean was associated with widening the between-row spacing to as much as 75 cm to facilitate sowing no-till to preserve cereal residue cover between rows (Felton and Marcellos 1996). The cereal residue persisted following seeding, and continued to provide soil protection even after the food legume crop was harvested.

SEEDING AND ESTABLISHMENT OF FOOD LEGUMES

Crop establishment following seeding may be improved by no-till into crop residue, due to a range of factors. Press wheels are often used on no-till seeders and these can improve seed-soil contact, resulting in better crop establishment. Evaporation losses immediately after sowing are substantially reduced, due to surface cover and minimum soil disturbance in no-till, thus creating better conditions for germination and crop establishment.

In sandy soils, no-till also provides other benefits. Direct drilling lupins into wheat residues on sandy, coastal soils in Western Australia is an important requirement for establishing the crop, as the cereal residue protects seedlings from sand blast (Nelson *et al.*1983), reduces water run-off and increases rainfall infiltration. Establishment of soybean, mung bean, cowpea and pigeon pea under no-till was generally equal to or better than under conservation tillage; it was poorer in only one very dry season (Herridge and Holland 1992).

DISEASE INCIDENCE

Food legumes are susceptible to many diseases which cause considerable losses in yield and quality of grain.

Ascochyta blight (*Ascochyta fabae*) and chocolate spot (*Botrytis fabae*) are foliar diseases of autumn-sown faba beans in many regions. One approach to reducing disease pressure in the Mediterranean environment of South Australia, is to sow in wide rows to delay canopy closure and the onset of micro-environmental conditions which favour disease. No-tillage is more feasible when a wide row spacing is used.

Viruses are important diseases of food legumes world-wide. Bean yellow mosaic potyvirus (BYMV) is serious in lupin (*Lupinus angustifolius*) in Western Australia (Jones 1994), whereas bean leaf roll luteovirus (BLRV), subterranean clover red-leaf (SCRLV), alfalfa mosaic (AMV) and cucumber mosaic (CMV) viruses have caused serious disease in food legumes in northern grains region of New South Wales (Schwinghamer, M., Moore, K. J., Southwell, R, Bambach, R. and Marcellos, 1997 unpublished poster paper). Experiments to monitor effects of fallow management of winter barley straw have shown that Barley Yellow Dwarf Virus (BVDV) infection may be higher following conventional cultivation (43 - 57% infected plants)), and least in direct-drilled plots (3%), possibly due to effects on survival of arthropods which predate the aphid vectors of this disease (Anon 1991). These data for barley indicate the potential for similar effects in food legumes. The addition of cereal straw mulch to lupin plots immediately after seeding reduced the infection rate of BYMV from 20% to 5% in one experiment, but had little effect in a second experiment (Jones 1994). Potential benefits exist in terms of reduced virus infection, from seeding lupins (and possibly other food legumes) directly into cereal stubble.

Using a disc plough to control weeds, to reduce stubble or incorporate herbicide before seeding can dramatically increase the incidence of root rot of lupins (*Pleiochaeta setosa*) in Western Australia (Sweetingham 1990, 1992). The disc plough inverts a band of surface soil, concentrating spores at depth where they may more readily infect roots. Direct drilling is therefore recommended for lupins.

INCIDENCE OF SOIL BIOTA

The influence of tillage on soil biota associated with food legumes does not appear to have been well investigated, despite its well-known effects on soil physical and biological properties. Some biota are important in nutrient cycling, and others are pests. It may well be that reducing tillage has important effects on soil biota which interact with the productivity of food legumes.

The biomass of earthworms in the top 10 cm of soil under direct-drilling in southern Australia was found to be more than twice that of conventional cultivation (Haines and Uren 1990).

On the other hand, cultivation to a depth of 5-10 cm after harvesting cotton is the best way to kill over-wintering pupae of *Heliothis armigera* without having to use chemicals (Slack-Smith *et al.* 1997) This will reduce the insect pressure on food legumes like faba bean sown in rotation following cotton.

Increases in populations of the seed-corn maggot (*Delia platura* Meigen) affecting corn and soybeans are associated with tillage systems involving substantial soil disturbance (Hammond 1997); no-tillageNTno-till systems do not enhance populations of this organism.

There are likely to be other cases where tillage has an effect on soil-borne biota.

SUMMARY AND CONCLUSIONS

The trend toward eliminating tillage during the fallow period before sowing cereal and legume crops directly into crop residues with a minimum of soil disturbance has been driven by the need to reverse soil degradation. It has been stimulated by the recognition that other benefits such as improved timeliness of sowing, less labour and machinery inputs, better weed control using herbicides, reduced soil erosion and runoff, increased soil water storage and improved soil physical and organic fertility may result. It has much to commend its adoption in LDCs where the need to maintain productivity is paramount.

The productivity of food legumes is not reduced by eliminating tillage, and in many instances, is higher due to improvements in some site-specific limiting factor. The nitrogen dynamics of food legumes tend to be improved under no-till, with small increases in Nfix and Pfix being reported. In some environments, the establishment of food legumes is improved by sowing no-till, in others, the effects of pests and diseases are mitigated by eliminating tillage.

References

Anderson , C. A. and Muir, L L. 1996. *Australian Journal of Agricultural Research* 36: 915-1080.

Anon. 1991. Soil cultivations and their effect on virus control. *The Agronomist* (BASF). No.3. 10-16.

Bharati, M. P., Whigham, D. K. and Voss, R. D. 1986. *Agronomy Journal* 78: 947-950.

CTIC. 1995. *Conservation Technology Information Center* 12: 1-6.

Dalal, R. C. and Mayer, R. J. 1986. *Australian Journal of Soil Research* 24: 265-279.

Dalal, R. C., Strong, W. M., Weston, E. J., Cooper, J. E., Lehane, K. J., King, A. J. and Chicken, C. J. 1995. *Australian Journal of Experimental Agriculture* 35: 903-913.

Dalal, R. C., Strong, W. M., Doughton, J. A., Weston, E. J., Cooper, J.E. and McNamara, G. J. 1997. *Australian Journal of Experimental Agriculture 37*: (In press).

Desborough, P. J. 1984. p. 89. In: *World Soybean research Conference III, Abstracts* (ed R. Shibles). Iowa State University Press, Ames: Iowa.

Dick, W. A. and van Doren, D. M. 1985. *Agronomy Journal 77*: 459-465.

Doughton, J.A., Vallis, I., and Saffigna, P.G. (1993). *Australian Journal of Agricultural Research 44*: 1403-1413.

Edwards, J. H., Thurlow, D. L. and Eason, J. T. 1988. *Agronomy Journal 80*: 76-80.

Ellington, A. 1986. In: A Review of Deep Tillage Research in Western Australia.. 100-127. (Ed. M. A. Perry).

Ellington, A., Reeves, T. G., Boundy, K. A. and Brooke, H. D. 1979. *49th ANZAAS Congress* Auckland New Zealand.

Elmore, R. W. 1987. *Agronomy Journal 79*: 114-119.

Elmore, R. W. 1990. *Agronomy Journal 82*: 69-73.

Felton, W. L. and Marcellos, H. 1996. *Proceedings 8th Australian Agronomy Conference*. Toowoomba.30 January - 2 February 1996..251-253.

Felton, W. L. and Marcellos, H. 1997. In: *Australian Grain* 7: Northern Focus i - iv.

Felton, W. L., Marcellos, H. and Martin, R. J. 1995. *Australian Journal of Experimental Agriculture.* 35, 915-921.

Ghizaw, A. and Molla, A. 1995. In: *Cool-season Food Legumes of Ethiopia.* . 205 (Eds. A. Telaye, G. Bejiga, M. C. Saxena and M. B. Solh) Aleppo, Syria: ICARDA.

Hammond, R. B. 1997. *Crop Protection* 16: 221-225.

Haines, P. J. and Uren, N. C. 1990. *Australian Journal of Experimental Agriculture.* 30: 365-371.

Herridge, D. F. and Holland, J. F. 1992. *Australian Journal of Agricultural Research* 43: 105-122.

Herridge, D.F, Marcellos, H., Felton, W.L, Turner, G.L, and Peoples, M.B. (1994). *Soil Biol.Biochem.* 27, 545-551.

Horn, C. P., Birch, C. J., Dalal, R. C., Birch, C. J. and Doughton, J. A. 1996a. *Australian Journal of Experimental Agriculture* 36: 695-700.

Horn, C. P., Birch, C. J., Dalal, R. C., Birch, C. J. and Doughton, J. A. 1996b. *Australian Journal of Experimental Agriculture* 36: 701-706.

Hughes, M. and Herridge, D. F. 1989. *Australian Journal of Experimental Agriculture* 29: 671-677.

Jarvis, R. J. 1986. In: A Review of Deep Tillage Research in Western Australia. 40-51. (Ed. M. A. Perry).

Jones, R. A. C. 1994. *Annals Applied Biology.* 124: 45-58.

Lindemann, W. C., Randall, G. W. and Ham, G. E. 1982. *Agronomy Journal* 74: 1067-1070.

Martin, R. J., Grace, P. R., Anderson, C. A. and Fegent, J. C. 1995. *Australian Journal of Experimental Agriculture* 35: 857-928.

Miah, A. A. and Rahman, M. M. 1993. In: Lentil in South Asia. 128-138. Aleppo, Syria: ICARDA.

Nelson, P., Hamblin, J., and Williams, A. 1983. Western Australian Department of Agriculture Bulletin No. 4079.

Peoples, M. B., Ladha, J. K. and Herridge, D. F. 1995. *Plant and Soil* 174: 83-101.

Rizk, M. A., Hussein, A. H. A. and El-Borai, M. 1991. *Egyptian Journal of Agricultural Research.* 70; 559- 573.

Rasmussen, P. E. and Collins, H. P. 1991. *Advances in Agronomy* 45: 94-135.

Singh, K. B. and Saxena, M. C. 1996. In: *Winter Chickpea in Mediterranean-Type Environments.* 20 Technical Bulletin. Aleppo, Syria: ICARDA.

Slack-Smith, P., Pyke, B., and Schoenfisch, M.1997. Australian Cotton Research Institute Newsletter.Research Extension Education Program. No. 3.

Sweetingham, M. W. 1990. *Journal of Agriculture, Western Australia.*1990. 31: 5-13.

Sweetingham, M. W. 1992. *Journal of Agriculture, Western Australia.*1990. 33: 6-7.

van Doren, D. M. and Reicosky, D. C. 1987. In: *Soybean Improvement, Production and Uses.*. 391-428. (ed J. R. Wilcox).

Wagger, M. G., and Denton, H. P. 1989. *Agronomy Journal* 81: 493-498.

Webber, C. L., Gebhardt, M. R. and Kerr, H. D. 1987. *Agronomy Journal* 79: 952-956.

TOWARDS THE MORE EFFICIENT USE OF WATER AND NUTRIENTS IN FOOD LEGUME CROPPING

Masood Ali[1], R. Dahan[2], J.P. Mishra[1] and N.P. Saxena[3]

1 Indian Institute of Pulses Research, Kanpur 208024 India 2 INRA, Settat, Morocco 3 ICRISAT, Patancheru, India

Abstract

Nutrient imbalance and soil moisture stress are the major abiotic constraints limiting productivity of cool season food legumes. These constraints are more pronounced in the semi-arid tropics and sub-tropics which are the principal production zones of chickpea, lentil and faba bean. The legumes are generally grown on residual moisture as a mono crop and consequently face drought especially during the reproductive phase. In recent years, chickpea, lentil, peas and faba bean have been grown in some areas with an irrigated/assured water supply under intensive cropping to sustain cereal based systems.

An increased water supply favourably influences productivity in dry environments. Faba bean, French beans and peas show a relatively better response to irrigation. The pod initiation stage is considered most critical with respect to moisture stress. Excessive moisture often has a negative effect on podding and seed yield. Eighty to ninety percent of the nitrogen requirements of leguminous crops is met from N_2 fixation hence a dose of 15-25 kg N ha^{-1} has been recommended. However, in new cropping systems like rice-chickpea, higher doses of 30-40 kg N ha^{-1} are beneficial. Phosphorus deficiency is wide spread and good responses occur to 20 to 80 kg P_2O_5 ha^{-1}, depending on the nutrient status of soil, cropping systems and moisture availability. Response to potassium application is localized. The use of 20-30 kg S ha^{-1} and some of the micronutrients such as Zn, B, Mo and Fe have improved productivity. Band placement of phosphatic fertilizers and use of bio-fertilizers has enhanced the efficiency of applied as well as native P. Foliar applications of some micronutrients have been effective in correcting deficiencies. Water use efficiency has been improved with some management practices such as changed sowing time, balanced nutrition, mulching and tillage.

INTRODUCTION

Food legumes are an important and cheap source of protein for human and animal nutrition in developing countries. Their importance as a builder and restorer of soil fertility has long been recognized in the semi-arid tropics and subtropics and they continue to be components of subsistance cropping in Asia, Africa, Oceania and South America. Cool season food legumes viz. chickpea, pea, lentil, faba bean French bean, and grasspea together share 40% of total food legume area (Pala et al. 1991).

The productivity of cool season food legumes is often constrained by moisture scarcity and nutrient imbalance especially in the semi-arid tropics and subtropics. In some areas, such as the highlands of Ethiopia, lowland areas of south east Asia, coastal areas around the Mediterranean and water harvesting catchment areas, the highlands in West Asia and North Africa (WANA), ephemeral water logging limits productivity of these legumes (Saxena et al. 1994). In most parts of the world, food legumes are generally grown in marginal and submarginal lands which are impoverished of plant nutrients and their yield potential is not realized. The ability to fix atmospheric nitrogen enables legumes to meet a large proportion of their N-requirement provided moisture and nutrient status of the soil are favourable for the host plant and the rhizobia. Soil moisture is the most scarce input in the semi-arid tropics and management practices to improve its use is the key to an enhanced production. In this paper we review the responses of cool season food legumes to water supply and plant nutrients, interactive effects, biofertilizers, beneficial effects of legumes in cropping systems and management practices to improve the efficiency of these inputs.

355

R. Knight (ed.), Linking Research and Marketing Opportunities for Pulses in the 21st Century, 355–368.
© 2000 *Kluwer Academic Publishers. Printed in the Netherlands.*

FOOD LEGUMES IN CROPPING SYSTEMS

The ability of legumes to fix nitrogen, improve soil health and perform better than many other crops under harsh climatic and edaphic conditions has made them important components of subsistence cropping in the semi-arid tropics for many centuries. The beneficial effect on succeeding cereals has been well established. Kacemi (1992) reported that in Morocco, wheat grown after legumes had higher grain yields and water use efficiency (Table 1). The yield of faba bean was 2.95 t ha^{-1} as against 1.08 t ha^{-1} under continuous wheat. Chickpea and lentil also showed a beneficial effect. The evapotranspiration (ET) under different cropping systems did not vary significantly but the water use efficiency (WUE) after legumes and fallow was better. Meena and Ali (1985) working on rice-based cropping systems observed that on sandy loam soils of Kanpur (India) lentil, fieldpea, chickpea and kidney bean improved the productivity of rice 0.4 to 0.8 t ha^{-1} as compared to wheat-rice system with the kidney bean showing the most favourable effect. Similar observations were made by Ahlawat et al. (1981). The effects have been quantified in terms of nitrogen equivalent and in most studies were 25-30 kg N ha^{-1}. The benefits are not due solely to N fixation but to increased nutrient availability, reduced incidence of disease, increased mycorrhizal colonization, etc. (Wani and Lee 1995).

Table 1. Grain yield of wheat (t ha^{-1}), water use and water use efficiency as affected by previous crop (Morocco)1991 Source : Kacemi (1992)

Rotation	Wheat grain yield	ET[1]	WUEG[2]	WUEDM[3]
Wheat-Wheat	1.08	245	4.4	21.7
Wheat-fallow	2.48	244	10.2	27.1
Wheat-corn	1.81	244	7.4	25.5
Wheat-chickpea	1.95	247	7.9	24.7
Wheat-fababean	2.09	243	8.6	29.6
Wheat-lentil	1.43	238	6.0	26.3
LSD(5%)	0.40		2.1	5.1

1 ET = evapotranspiration   2 WUEG = Wheat grain yield water use efficiency

3 WUEDM = wheat dry matter yield water use efficiency

The legumes are grown in cropping systems as a sole crop and as an intercrop both under mono as well as sequential cropping systems depending upon the availability of soil moisture and domestic needs. In the rainfed areas of south and central Asia, North Africa and the Mediterranean these crops are largely grown on conserved soil moisture as a monocrop. In the rainfed areas of India, Bangladesh, Nepal and Pakistan, intercropping of chickpea and lentil with mustard, linseed and barley is popular. In peninsula India, chickpea/safflower is also an important intercropping system. The practice of intercropping is primarily for diversified and stable production. In the subhumid central Alberta, barley intercropped with field pea is profitable. This system also returned more nitrogen to the soil than a monocrop of barley (Izaurralde 1990).

In recent years, new varieties and attractive prices have led to the cultivation of legumes in irrigated areas in intensive systems. For example, French bean (kidney bean), has been introduced as a new winter crop in the north-east plains of India in a maize-cowpea-mungbean system (Ali and Lal 1991). Chickpea and Lentil are also grown under double cropping with rice in eastern India, Bangladesh and Nepal. On uplands of Pakistan and India maize-chickpea, maize-peas and maize-lentil are important rotations whereas in the semi arid tropics of Morocco, faba bean-wheat, chickpea-wheat and lentil-wheat are important. In south-east Asia, relay cropping of grasspea and lentil with short duration rice is practised on low land to use the residual moisture. In the Mediterranean region of Central Asia, winter-sown lentils and chickpeas were distinctly superior to spring-sown crops, which often face terminal drought.

NUTRIENT MANAGEMENT

Plancquaert (1991) showed that under good conditions, a crop of faba bean producing 6.5 t ha^{-1} may remove 405 kg N, 102 kg P$_2$O$_5$ and 258 kg K$_2$O ha^{-1}. Since the soil of the major pulse producing regions are impoverished of plant nutrients, an adequate and balanced fertilization is necessary to boost productivity. The ability to fix N_2 enables leguminous crops to meet a large proportion of their nitrogen requirements. Huber et al. (1987) at Zurich found that faba bean fixed on an average 90% of its N-requirement. A well-managed crop of chickpea may fix up to 270 kg N ha^{-1} (Nutman 1969). Since food legumes are harvested as grain, the amount of N in the crop residue and left on the soil does not replenish the N removed by the crop. It is imperative therefore to enhance nitrogen fixation or resort to N-fertilization in the already poor soils of the semi-arid tropics and subtropics (Buresh and Datta, (1991).

Inoculation and Nitrogen

Inoculating seed with an appropriate rhizobial strain increases nodulation, nitrogenase activity and yield (Whiting 1985, El Khadir 1991, Reddy 1992, Srivastava 1993 and Haque and Haq 1994). In the All India Co-ordinated Pulse Improvement Project (AICPIP), seed inoculation enhanced the productivity of chickpea and lentil by 10-15% in different parts of India (Chandra and Ali 1986). In the semi-arid regions of Morocco inoculation increased the seed yield of faba bean by 80% over the check (Table 2).

Table 2 Effect of N application (120 kg N ha^{-1}) and rhizobium inoculation on chickpea nodulation and grain yield, Morocco 1990-91

Treatment	Number of nodule/plant	Nodulation index	Grain yield (t ha^{-1})	% increase
N check	5.0	2.0	1.36	74
Untreated	2.2	1.5	0.78	-
Inoculated	23.5	4.0	1.40	80

Source : El Khodir (1991)

Strains of rhizobia are not equally effective and it is necessary to use specific strains to enhance fixation. Tekalign (1994) reported on the rhizobial strain specificity of faba bean and lentil in Ethiopia.

Legume species not only differ in their nodulation but cultivars within species also differ significantly, suggesting host factors are determinants of assimilation. Significant interactions of rhizobial strain x genotypes have been observed for chickpea by Patel et al. (1986) in central India. They observed that in genotype BG 209, strain KG 31 was most effective whereas in Dohad Yellow, F 75 and in Chaffa, H 45 were the most effective strains (Table 3). Similar interactions have been reported for peas and faba beans under Australian conditions (Herdina and Silsbury 1989).

In view of biological nitrogen fixation, a starter dose of 15-30 kg N ha^{-1} has been recommended for normal sown rainfed legumes. The French bean crop on the north-east plains of India, however, needed 100-120 kg N ha^{-1} to realize its yield potential, as the native rhizobial strains were not effective (Fig. 1). Similarly in rice-based cropping systems on low lands, late-sown chickpea generally responds well up to 40 kg N ha^{-1} (Ali and Mishra 1996). Under irrigated conditions, peas and faba bean also show good responses to increased nitrogen level up to 40 kg N ha^{-1}.

Table 3. Effect of Rhizobium inoculation on the grain yield (t ha^{-1}) of four genotypes of Kabuli chickpea.

Rhizobium strain	Genotypes				% increase over control
	BG 209	Dohad Yellow	JG 315	Chaffa	
F 6	3.02	3.10	3.49	2.93	49
Ca 181	2.78	3.02	3.49	2.62	42
KG 31	3.17	2.54	4.24	1.89	41
F 75	3.10	3.17	4.44	2.54	58
H 45	2.54	2.94	3.41	2.41	41
Uninoculated control	1.91	1.91	2.62	1.98	-

LSD(P=0.05) Genotypes(G) 0.34;   Rh. strain(S) 0.14;   G x S 0.29

Source : Patel et al. (1986)

358

a. North east plain zone

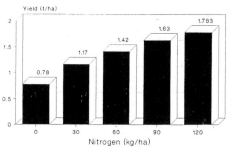

Fig 1: Effect of N levels on yield of frenchbean

b. Central zone

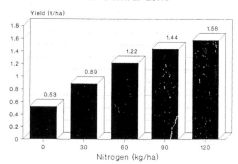

Phosphorus Effect of Rhizobium inoculation on the grain yield (t ha^{-1})of four genotypes of Kabuli chickpea.

For most crops, P is the second most critical plant nutrient but for legumes it is the first. Besides metabolic functions, P has a role in root proliferation, nodule development and biological nitrogen fixation. Phosphorus deficiency is wide spread in South Asia and Africa. Tandon et al. (1992), found that out of 371 districts in India, 150 were low in P status while only 17 showed K deficiency. Khanom and Islam (1984) reported that on calcareous brown flood soils of Bangladesh an application of 60 kg P_2O_5 ha^{-1} significantly improved chickpea yields. On grey flood plain soils of Jamalpur (Bangladesh), a response was observed up to 90 kg P_2O_5 ha^{-1} (Islam 1989). In the WANA region, lentil showed a good response to 48 kg P_2O_5 ha^{-1} (Singh and Saxena 1986). Under Pakistan conditions, Malik et al. (1991) found that lentil responded up to 40 kg P_2O_5 ha^{-1}. On a sandy loam soil of Delhi (India), an application of 60 kg P_2O_5 ha^{-1} significantly increased yield, nodulation and nitrogen assimilation in peas (Kasturi 1995). In the semi-arid regions of Morocco, the yield of winter chickpea increased by 10-48% with application of 60 kg P_2O_5 ha^{-1} depending upon moisture availability and P status of soil. Yield responses were higher on soils testing low in P (EL Khadir, 1991).

Potassium

The K status of soils of the semi-arid tropics and subtropics is generally high, however, in coarse textured and acidic soils which do not contain illite clay minerals, a deficiency of K have been observed. Response to K application has been recorded in laterite, red, coastal and deltic alluvial soils of India (Khera et al. 1990). Results of AICPIP trials on lentil showed that in the north-east plains of India, an application of 20 kg K_2O ha^{-1} increased mean yield by 219 kg ha^{-1} but in the north-west plains, there was no response (Ali et al. 1993). Responses to a high dose of K have been reported from Egypt. El Fouly et al. (1989) observed that faba bean responded to 120 kg K_2O ha^{-1}. From 8 experiments, the mean increase in yield was 9% over the control when 120 kg K_2O ha^{-1} was applied. This is an exceptionally high dose and limited to a specific location.

Secondary nutrients and micronutrients

Among the secondary nutrients, sulphur appears to be most important. In general, the S uptake by legumes is almost double that of cereals for each unit of grain production. Further, sulphur containing amino acids like methionine, cystein and cystine are low in legumes and therefore an adequate sulphur fertilization is important. Sulphur deficiency has been reported for light textured soils in India. In the multilocation studies of AICPIP, applications of 20 kg S ha^{-1} increased yield of chickpea and lentil by 0.3-0.6 t ha^{-1} in northern India (Fig. 2). The light textured soils of the north-west plains showed greater responses. Applications of S beyond 20 kg S ha^{-1} did not prove beneficial (Ali and Singh 1995). However, on sulphur deficient soils of Ludhiana (north-west plains of India), Aulakh and Pasricha (1986) reported that application of 40 kg S ha^{-1} increased yields of lentil by 27% over the control. Among micronutrients, responses to Zn, Mo, B and Fe have been reported from different parts of India. In the calcareous clay loam soils of the foot hills of Shiwalike range, wide spread deficiencies of Zn have been observed. Among the crops, lentil has been found most sensitive to Zn deficiency. Soil applications of 12.5-15.0 kg ZnSO$_4$ ha^{-1} or foliar spray of 0.5 kg ZnSO$_4$ ha^{-1} twice (15 and 45 days after sowing) corrected Zn deficiency and increased yield. (Gangwar and Singh 1986). Different responses of genotypes to nutrient stress have been observed (Sakal et al. 1982). On sandy loam calcareous soils of north Bihar (India) which are deficient in iron and boron, lentil genotype L 4076 performed better than Precoz sel. Seed treatment with aqueous solutions of ferrous sulphate increased the yield of Precoz sel by 33.8% whereas L 4076 did not respond (Sinha 1988). In another study, variety DPL 77-2 was found tolerant to Zn deficiency and PL 406 to Fe stress (Singh et al. 1984). Responses of chickpea to foliar spray of 2% FeSO$_4$ on calcareous soils of southern India have also been observed (Perur and Mithyantha 1985). These results suggest a breeding program aimed at transferring genes from tolerant land races/cultivars may help alleviate micronutrient disorders.

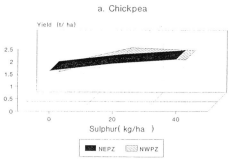
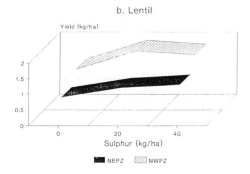

Fig 2.Response of chickpea and lentil to sulphur fertilization

Molybdenum being an essential component of the nitrogenase enzyme, has a role to play in N_2- fixation. Several workers have reported significant increase in yields of pea (Srivastava and Ahlawat 1995), lentil (Sharma and Chahal 1983), chickpea (Pal 1986) and faba bean (Brhada 1986) with applications of 1.0-1.5 kg sodium molybdate ha^{-1}. Ali and Mishra (1997) at Kanpur, India found foliar sprays of 0.01% ammonium molybdate increased yields of Kabuli chickpea by 0.38 t ha^{-1} (Table 4). A response to boron was also observed. Foliar application of 0.2% borax enhanced grain yield by 0.59 t ha^{-1}.

Table 4 Effect of foliar spray of Molybdenum (0.01% Ammo. Molybdate) and Boron (0.2% borax) on seed yield (t ha^{-1})of Kabuli Chickpea genotypes (Kanpur, India).

Foliar Nutrition	Genotypes				Mean
	BG 267	BG 1003	HK 89-96	L550	
Water spray	1.26	1.06	1.04	1.36	1.18
Molybdenum spray	1.66	1.59	1.48	1.52	1.56
Boron spray	1.76	1.64	1.73	1.94	1.77
Mean	1.56	1.43	1.42	1.61	
	Genotype	Foliar spray	Interaction		
SE mean	0.07	0.04	0.09		
LSD (P=0. 05)	0.15	0.13	NS		

Source : Ali and Mishra (1997)

The genotypes did not show different responses. Similar results have been obtained in other countries (Khanom and Islam 1984).

Improving nutrient use efficiency

Mineral nutrient stresses are a wide spread constraint to food legume production in the semi-arid tropics and sub-tropics. Water being a solvent and carrier of mineral nutrients, plays a role in fertilizer use efficiency. In dry areas, the amount of fertilizers to be applied is dictated by water availability in addition to edaphic factors. For example, irrigated chickpea in north India responded favourable up to 80 kg P_2O_5 ha^{-1} whereas under rainfed conditions the response was limited to 30-40 kg P_2O_5 ha^{-1} (Ali 1994).

Method of nutrient application

The methods of application and the mobilization and fixation dynamics of the soil influence the efficiency of applied nutrients. Band placement of phosphatic fertilizers is more efficient than broadcasting and mixing due to the low mobility and fixation of P-ions. Studies at ICRISAT showed that in vertisols of south India, deep banding (15 cm) of 10 kg P ha^{-1} was as effective as mixing 20 kg P ha^{-1} in rainfed chickpea (Arihara et al. 1991). Under rainfed conditions, when dry top-soil layers do not permit top dressing of basal applications of fertilizers, foliar sprays may supply small amount of macronutrients. Several studies under AICPIP revealed that under dryland conditions foliar sprays of 2% urea and di-ammonium phosphate enhanced productivity of chickpea (Ali 1994). Foliar nutrition of micronutrients corrects nutritional disorders rapidly. Gangwar and Singh (1986) working at Pantnagar (India) reported that on calcareous clay loam soils, foliar sprays of 0.50% $ZnSO_4$ mixed with 0.25% lime at 15 and 45 days after emergence of lentil were more efficient in correcting Zn deficiency than a basal application of 10 kg $ZnSO_4$ ha^{-1}

Soil micro-organisms :

Rhizobium spp, *Bacillus* spp. and Vesicular Arbuscular mycorrhizae (VAM) are involved in the nutrition of food legumes. The proliferation and colonization of these micro-organisms, the growth of the host plant and the rhizospheric environment (nutrient status, moisture, aeration etc.) are interdependent. In a study conducted at Ludhiana (India), seed inoculation of lentil coupled with an application of 20 kg P_2O_5 ha^{-1} produced yields comparable with 40 kg P_2O_5 ha^{-1} without inoculation (Dhingra et al. 1988). It is commonly accepted that soil micro-organisms can stimulate the mycelial growth of mycorrhizal fungi, which may increase the availability of phosphorus (Bieleski 1973).

The role of phosphate-solublizing bacteria (PSB) in P nutrition is well recognized. Their ability to enhance P availability of less expensive citrate soluble phosphatic fertilizers, like rock phosphate, is a boon to poor farmers of semi-arid regions. Rathore et al. (1992) working on vertisols of central India found that inoculation of lentil seeds with PSB (*Bacillus megatericum*) increased yields by 16% over the control. At Delhi, seed inoculation with *Bacillus polymixa* increased N and P uptake by peas. Dual inoculation with *Rhizobium* and PSB was most effective (Srivastava and Ahlawat 1995).

The symbiotic association between plant roots and fungal mycelia i.e. mycorrhiza has received attention in recent years. VAM enhances plant growth by improved mineral nutrition particularly of P, and water uptake, due to the network of hyphae originating from the mycorrhizal roots. Rao et al. (1986) observed high N and P concentrations in the shoots of chickpea due to *Glomus fasciculatum* inoculation. Reddy (1992) working on the efficacy of biofertilizers on nodulation and yield of lentil found that *Rhizobium* and VAM both improved nodulation and yield (Table 5). Thus, it is clear that VAM improves not only the P nutrition of the legumes but also that of rhizobial bacteroides and the efficiency of N_2 fixation.

Table 5. Effect of bio-fertilizers on root nodulation (80 DAS) and grain yield of lentil (New Delhi, India)

Bio-fertilizers	No. of nodules per plant	Wt.of nodules (mg/plant^{-1})	Grain yield (t ha^{-1}) 1989-90	1990-91
Control	17.5	4.2	0.84	0.80
Rhizobium	26.1	5.4	1.03	0.98
VAM fungi	20.5	6.3	1.09	1.01
Rhizobium +VAM fungi	29.8	7.2	1.19	1.14
LSD. (P=0.05)			0.04	0.05

Source : Reddy (1992)

Positive interactions

An understanding of the positive interactions among mineral nutrients, soil micro-organisms and organic manures may improve the efficiency of applied fertilizers. Enhancing the growth and colonization of *Rhizobium* spp. and reducing fixation/leaching losses of mineral nutrients due to organic manuring and residue management are well recognized. Prasad et al. (1990) working on a rice-lentil system at I.A.R.I., New Delhi reported that resulting from the incorporation of lentil residues, 46-89 kg ha^{-1} of primary nutrients (15-36 kg N, 2.5-10.0 kg P_2O_5 and 28.5-43 kg K_2O) were saved and thereby the soil nutrient balance improved.

An application of molybdenum improves nodulation, fixation and yield as Mo is a constituent of the enzyme nitrogenase. Mohandas (1985) reported that soaking seeds of French bean in 2 ppm sodium molybdate for 1 hour increased nodulation, N uptake and yield. A similar synergistic effect of 0.5 kg Mo ha^{-1} was noticed on peas by Srivastava and Ahlawat (1995). Pal (1986) observed that in the presence of P, the efficiency of applied Mo improved. Phosphorus application improves the N_2-fixation in legumes and uptake of several other nutrients due to its active role in root proliferation. The P x S interaction is generally positive at low to medium levels of applied P but at higher level of P, the effect is antagonistic (Nayak and Dwivedi 1990). Significant interactions of P and B in lentil have also been reported (Singh and Singh 1983).

WATER MANAGEMENT

Crops grown on conserved soil moisture, often face moisture stress during growth. In the Mediterranean region, crops sown in spring also face stress. In some regions, like northern Europe and the central and eastern highlands of Ethiopia, excessive soil moisture may adversely affect productivity. The problem of excess moisture is also experienced on deep vertisols of south India as well as coastal regions where the rainfall is bimodal. However, the extent of the moisture scarcity problem is far greater than moisture excess.

Responses to irrigation

Cool season food legumes generally respond to limited irrigation. Studies carried out at ICARDA, Tel Hadya (Syria) on chickpea showed a linear response to moisture for seed yield and total biomass (Saxena and Saxena 1991). The yield from plants with adequate moisture was three times that of the rainfed treatment. Rizk and Hassan (1991) reported that the average seed yield of lentil in Sharkia Governerate of Egypt increased from 1.36 t ha^{-1} without irrigation to 2.01 t ha^{-1} with irrigation at 20 and 50 DAS. Studies in agro-ecological zones of India revealed that chickpea, lentil, peas and French bean responded to irrigation in the central and southern zone. French bean was the most responsive followed by field pea. Late sown chickpea showed a better response to irrigation than a normal sown crop (Ali 1994). Beneficial effects of supplemental irrigation in spring and winter season chickpea in Syria (Saxena, 1992), and winter sown lentil in Bangladesh (Hassan and Rahman 1987) have been reported. Srivastava and Srivastava (1992) working in the plateau region of north Bihar (India) found that the irrigation was the most important input contributing 45-50% towards productivity. Responses have also been reported from Europe. In Poland, irrigation increased the productivity of faba bean and peas by 47.4% and 20.9% respectively over the controls (Borowezak and Szukala 1991).

When scheduling irrigation, various approaches have been advocated, such as crop day, phenological stage, day interval and cumulative pan evaporation. The reproductive phase particularly pod development is the most critical with respect to moisture stress and irrigation at this stage has given higher yields. Mohamad et al. (1988) in Sudan observed that the omission of one irrigation at the reproductive phase of faba beans reduced pod development and seed yield whereas missing an irrigation during the vegetative phase did not have a negative effect (Table 6). The initial moisture profile and soil type influences the requirement for irrigation. For example, on light soils of Kanpur (India), irrigation at maximum branching was as important as at pod development (Ali 1991). Kasturi (1995) working on peas at Delhi (India) observed that moisture stress in the vegetative stage was most detrimental for nitrogenase activity and nodule production. Tiwari and Tripathi (1995) working at Raipur (India) reported that chickpea grown in rice fallows needed irrigation at branching and at podding. Similarly in lentil, a 51% increase in yield was recorded with irrigations at branching and podding (Rathore 1992). Dhar and Singh (1995) used cumulative pan evaporation data to schedule irrigation of chickpea on a silty clay loam soil of Pantnagar (India). They found that irrigation scheduled at IW/CPE ratio of 0.6 gave a higher yield and WUE than other treatments (Table 7). The yield response was higher during the dry year of 1989. Studies on scheduling irrigation on the basis of day interval in the Sudan showed that a 7-days interval was the most productive for faba bean. In another study, Mohamad et al. (1988) reported that irrigation at a 15-day interval was most economical and efficient in faba bean. Studies in French bean in the north-east plains and central zone of India showed that the 25-day crop stage was most critical for irrigation (Ali and Lal 1991). The mean increase in productivity from irrigation at this stage was 2.1 t ha^{-1} in NEPZ and 0.4 t ha^{-1} in the central zone. The crop responded to 3-4 irrigations (Fig 3). In Egypt, the 30-day crop stage was found to be most critical for irrigation of lentil (Attia 1988).

363

Table. 6 Effect of water stress during stages of plant growth on grain yield (t ha⁻¹) of faba bean at 3 locations in Morocco

	Yield		
Treatments	Hudeiba	Shambat	Wad Madani
1. Irrigation every day	2.57	1.85	0.97
2. Irrigation every 15 day	1.92	1.89	0.64
3. As (i) but 3rd irrigation missing			
4. As (i) but 7th irrigation missing	1.38	2.02	0.68
5. As (i) but 9th irrigation missing	2.02	1.96	0.80
6. As (i) but 3rd and 7th irrigation missing	1.55	1.85	0.59
7. As (i) but 7th and 9th irrigation missing	1.39	1.65	0.56
8. As (i) but 3rd and 9th irrigation missing	1.98	1.54	0.64
9. As (i) but the 3rd, 7th and 9th irrigation missing	1.37	1.65	0.51
Mean	1.86	1.99	0.71
SE	0.08	0.11	0.06

Mohamad et. al. (1988)

Table 7. Effect of irrigation on seed yield (t ha⁻¹) of lentil and water use efficiency (kg grain ha⁻¹- cm) (Pantnagar, India).

Irrigation at IW/CPE	Seed yield		Water use efficiency	
	1989-90	1990-91	1989-90	1990-91
0.4	1.85	1.08	66.3	51.8
0.6	2.02	1.46	74.2	59.0
0.8	1.92	1.31	51.6	39.6
No irrigation	1.93	0.96	68.5	43.5
LSD(P=0.05)	0.33	0.34		

Source : Dhar and Singh (1995)

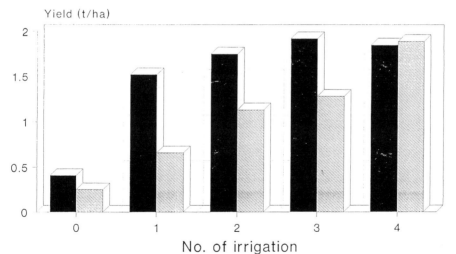

Fig 3. Response of frenchbean to irrigation

Excessive moisture

Excessive moisture, from poor drainage leads to crop damage from the anaerobic conditions, accumulation of toxic acids, reduction in mineral nutrient availability, inhibition of nitrogen fixation and increased susceptibility to root diseases. In the Ethiopian highlands, the problem is acute. Sowing on broadbed and furrows considerably increased the yield and economic return of chickpea, lentil and faba bean. Field observations in Egypt have also revealed that broadbed and furrow systems improved the productivity of irrigated lentil on the heavy soil of the region by preventing the transient water logging which is common feature in basin irrigation (Saxena et al. 1994).

In some crop species genotypic differences exist in response to anoxia. Alcalde and Summerfield (1994) subjected pot grown plants of two genotypes of lentil to water logging for 2,4 and 6 days after 15 days of emergence. The genotype Pant L 406 was better nodulated and developed more extensive root system than variety Precoz under water logging conditions. Genotypic variation with respect to anoxia needs to be exploited in breeding programs.

Efficient use of water

In-situ conservation of rainwater and the efficient use of conserved as well as irrigation water are keys to enhanced productivity. Various practices such as improved tillage, choice of appropriate crop and variety, sowing time/season, use of balanced and adequate plant nutrients, weed management, use of mulches, anti-transpirants etc. influence productivity and water use efficiency. In dryland areas, where the top layer of soil may be dry at the beginning of the season, a technique to place seeds in the moist zone to establish a good plant stand is important. A low water use efficiency in dryland areas is commonly associated with low yields on account of a low plant population. In the drier regions of central India, deep sowing of chickpea (10-12 cm) is practised to ensure contact of seed with moist soil.

Shifting the cropping season, according to evaporative demand of a region, is another way of improving the use of available water. An example is the winter cultivation of chickpea in the WANA region. Chickpeas are normally sown in spring on the residual soil moisture supply, which decreases with the growth of the crop. The crop faces drought in its late vegetative or reproductive stages because of increasing evaporative demand. Shifting the cropping season from spring to winter and growing an appropriate variety has led to increase in yield and WUE (Table 8). The mean increase in productivity was 87%. The effect was most pronounced at locations where seasonal rainfall was low.

Table. 8 Chickpea grain yield (t ha^{-1}) for winter and spring sowing at four locations in Morocco 1988-1991.

Location	Av. seasonal rainfall (mm)	Av. yield winter	spring	% difference
Douyet	451	2.93	1.79	64
Merchouch	388	1.81	0.99	83
Jemma Shaim	305	2.56	1.26	103
Jemaa Riah	277	1.20	0.52	131
Mean		2.13	1.14	87

Water Use efficiency

An adequate and balanced nutrition improves WUE, primarily due to a greater seed yield. However, potassium being involved in stomatal regulation helps reduce transpiration losses and consequently increases crop yield and WUE. Results of field experiments at Kanpur (India) showed that foliar sprays of 2% KCl at flowering increased grain yield and WUE of chickpea. WUE, with foliar sprays of KCl was 16-17 kg grain ha^{-1} -

mm water as compared to 14-15 kg grain ha^{-1} -mm water in the control (Ali 1985). Abd-Alla and Wahab (1995) working on water stressed faba beans found that applying KCl at 150 mg kg^{-1} soil, helped alleviate moisture stress and enhanced N fixation. The use of `Jalshakti', a water absorbing polymer is also promising in increasing moisture availability. At Kanpur (India), soil incorporation of Jalshakti at 12 kg ha^{-1} or furrow placement at 2 kg ha^{-1} at sowing enhanced yield of chickpea by 12% over the control Ali (1994) (Table 9).

Table 9 . Effect of Jalshakti on yield (t ha^{-1}) of chickpea (Kanpur, India)

Jalshakti	Grain yield
Soil incorporation 8 kg ha^{-1}	2.77
Soil incorporation 12 kg ha^{-1}	2.70
Furrow placement 1 kg ha^{-1}	2.72
Furrow placement 2 kg ha^{-1}	2.95
Harrowing (dust/mulch)	2.89
Control	2.63
LSD (P=0.05)	

Source: Ali (1994)

The effect of phosphorus in increasing water use and root proliferation is well recognized. Venkateswarlu and Ahlawat (1993) working on phosphorus x irrigation interactions found that the response of applied P in the presence of irrigation, was much higher than under unirrigated conditions. The yield of lentil with the application of 70 kg P$_2$O$_5$ ha^{-1} was 1.55 t ha^{-1} when irrigation was scheduled at 0.6 IW/CPE ratio whereas at low moisture regime (0.35 IW/CPE ratio) it was 1.43 t ha^{-1} and under unirrigated conditions 0.75 t ha^{-1} (Table 10). Khera et al. (1991) working on P nutrition under irrigated and rainfed conditions of southern India reported that on vertisols an application of 52 kg P$_2$O$_5$ ha^{-1} under irrigated conditions increased yields of chickpea by 1.68 t ha^{-1} whereas in the absence of P, the yield increment was only 1.19 t ha^{-1}

Table 10 Effect of soil moisture and phosphorus on seed yield (t ha^{-1}) of lentil (New Delhi, India)

Soil moisture (IW/CPE ratio)	Phosphorus (kg P$_2$O$_5$ ha^{-1})		
	0	35	75
No post-sowing irrigation	0	0.64	0.75
Irrigation at 0.35 IW/CPE	0	1.17	1.43
Irrigation at 0.60 IW/CPE	0	1.30	1.55
LSD (P=0.05)		0.04	

PBI = Phospho bacterial inoculation   Source:Venkateswarlu and Ahlawat (1993)

Increased WUE due to application of surface mulches, anti-transpirants and light reflectants have been advocated. In Syria, Saxena et al. (1992) observed increases in chickpea yields with dust mulching when the seasonal rainfall was 372 mm. Mandal and Mahapatra (1990) working on lentil/barley intercropping in the tropics of India found applying a straw mulch at 7.0 t ha^{-1} improved the yield of lentil by 11-17%. Applications of a light reflectant (Kaolin) at 100% flowering in chickpea increased yields in Syria whereas the effect on faba bean was not significant (Saxena et al. 1992). Lal et al. (1994) found a foliar spray of a 6% Kaolin suspension 80 DAS increased the relative water content of leaves and WUE in lentil.

CONCLUDING REMARKS

The importance of factors such as nutrition and water supply to the growth of cool season food legumes has been discussed above for the cropping systems of the semi-arid tropics, subtropics, and the Mediterranean regions. It was suggested the realized yields were far below potential productivity due to several production constraints.

However, detailed information is lacking and requires the attention of research scientists on the means of improving management practices which have a bearing on WUE in different production systems. Advancing the cropping season from spring to winter in the WANA region is an example of efficient crop management. Variability in the land races/genotypes with respect to their tolerance to drought has been observed and should be exploited. In the regions where excess moisture is constraining productivity, appropriate bed management techniques need to be developed.

Imbalance and an inadequate availability of plant nutrients in most regions often limits productivity. Among plant nutrients, phosphorus is most important, and good response to 30-40 kg P_2O_5 ha^{-1} in rainfed areas and up to 80 kg ha^{-1} in irrigated areas have been observed. The escalation in the price of P-fertilizers, in recent years, particularly in India, calls for efforts to improve the efficacy of P-fertilizers. Limited research has shown that band placement and use of PSB help improve the efficiency of P-fertilizers. In south Asia, sulphur deficiency has been observed widely. Applications of 20-40 kg S ha^{-1} are beneficial particularly on light textured soils having a low S status. Crop response studies have shown the beneficial effect of fertilization with micronutrients such as zinc, molybdenum, boron and iron under certain edaphic conditions. Genotypic variation with respect to zinc and iron stress has been observed. It is imperative to identify areas with deficiencies of micronutrients, to find out critical limits and to improve the efficiency of applied nutrients. Screening of germplasm to find genotypes tolerant to these deficiencies should be an important part of future research.

Most of the information on the use of nutrients in food legumes is for the component crops rather than cropping system. Since nutrient mobilization, the fixation dynamics of a soil, water availability and the nature of the preceding crops, influence the efficiency of applied nutrients, comprehensive information on fertilizer management for the whole cropping system needs to be generated. More emphasis should be given to integrated nutrient management especially for the new cropping systems in irrigated areas. Information on nutrient x water interactions are limited. A better understanding of these interactions would help improve the productivity of legumes under the existing constraints.

References

Abd-Alla MH, Wahab AMA 1995. *Journal of Plant Nutrition* 18: 1391-1402.

Ageeb OAA, Ali MA, Salih FA 1989. *FABIS* Newsletter. 24: 8-10.

Ahlawat IPS, Singh A, Saraf CS 1981. *Experimental. Agriculture* 17: 57-62.

Alkalde JA, Summerfield RJ 1994. *LENS* Newsletter 21 (1) : 22-29.

Ali M 1985. *Legumes Research.* 8 (1): 7-11.

Ali M 1991. *Annual Report,* Directorate of Pulses Research, ICAR, Kanpur. pp. 11-12.

Ali M 1994. In *25 years of Research on Pulse in India.* pp 19-22 (Ali, M et al. eds.) International Symp. on Pulses Research, 2-6 April 1994, New Delhi, India.

Ali M, Ahlawat IPS, Mishra JP 1994. In *Proc. of Integrated input management for efficient crop production.* pp 101-112 Indian Society of Agronomy, New Delhi, India.

Ali M, Singh KK 1995. *Technical. Bulletin.* Indian Institute of Pulses Research, Kanpur, India pp. 17

Ali M, Lal S 1991. *Technical. Bulletin* No. 3, Directorate of Pulses Research ICAR, Kanpur, India pp 12.

Ali M, Mishra JP 1996. *Indian Farming.* 46 (8): 67-71.

Ali M, Mishra JP 1997. (unpublished).

Ali M, Saraf CS, Singh PP, Rewari RB, Ahlawat IPS 1993. In *Lentil in South Asia* pp 103-127 (ed Erskine, W. and Saxena, MC) 11-15 Mar 1991, New Delhi, India.

Arihara J, Ae N, Okada K 1991, In Proc symp *Phosphorus Nutrition of Grain legumes in the semi-Arid Tropics.* pp 157-166 ICRISAT, Patancheru.

Attia AN 1988. *Journal of Agricultural Sciences* 13 (4 A): 497-503.

Aulakh MS, Pasricha NS 1986. *Fertiliser News* 31 (9): 31-35.

Bieleski RL 1973. *Annual Review Plant Physiology.* 24: 225-52.

Borowczak F, Szukala J 1991. 15 *Krajowe Sympozjum Nawadniania Roslin.* Bygoszcz (Poland)

Brahada F 1986. In *Proc of Workshop on Biological Nitrogen Fixation on Mediterranean type Agriculture,* pp 241-249 Aleppo Syria 14-17 April 1986.

Buresh RJ, De Datta SK 1991. *Advances in Agronomy.* 45:1-59.

Chandra S, Ali M 1986. *Technical Bulletin* No.1, Directorate of Pulses Research Kanpur, India.

Dhar S, Singh, NP 1995 *Indian Journal of Agronomy.* 40: 620-25.

Dhingra K K, Sekhon H S, Sandhu PS, Bhandari SC 1988. *Journal of Agricultural Science* (Camb.) 110: 141-44.

El Fouly, Fowzi AFA, El Baz FK 1989

El Khadir M 1991. In *Sixth Morocco/ICARDA Food legume Coordination Meeting*, October. 1991, INRA-Rabat, Morocco.

Gangawar KS, Singh NP 1986. *LENS* Newletter. 13 (1) 20-22.

Hassan AA, Rahman MA 1987. *Thai Journal of Agricultural Science* 20:277-281.

Haque MM, Haq MF 1994. *LENS* Newsletter 21(2): 29-30.

Herdina, Silsbury, JH 1989. *Australian Journal of Agriculture Research* 40:991-100

Huber R, Keller ER, Schwendimann F 1987. *FABIS* Newletter 17: 14-20.

Islam MS 1989. In *Advances in Pulses Research in Bangladesh*. pp 51-59 Proc of the 2nd National Workshop on Pulses 6-9 June 1989, Joydalpur, Bangladesh.

Izaurralde RC, Juma NS, Gill WB 1990. *Agronomy Journal* 82: 295-301

Kacemi M 1992. Ph.D. Diss. Colorado State University, Fort Collins, Colorado, USA.

Kasturi K 1995. Ph.D. Thesis, IARI, New Delhi, India.

Khanom D, Islam MS 1984. *Journal of Soil Science* 20:117-121.

Khera MS, Deshmukh VN, Talukdar MC 1990. In *Soil Fertility and Fertilizer Use* pp 69-93 Vol. IV Nutrient Management and Supply System for Sustaining Agriculture in 1990s (ed Virendra Kumar et al.). IFFCO New Delhi.

Lal B, Kaushik SK, Gautam C 1994.*Indian J. Agronomy* 39: 241-45.

Malik MA, Tanvir A, Hayee MA, Ali A 1991. *Journal Agricultural Research* (Pakistan) 29:333-38.

Mandal BK, Mahapatra SK 1990. *Agronomy Journal* 22:1066-1068.

Meena NL, Ali M 1985. *Annual Report,* Directorate of Pulses Research ICAR) 1985, pp. 14.

Mohamad GE, Salih FA, Ageeb AA 1988. *FABIS* Newsletter 22:17-19.

Mohandas S 1985. *Plant and Soil*. 86: 283-85.

Nayak GS, Dwivedi AK 1990. *FABIS* Newsletter.27:13-15.

Nutman PS 1969. *Proceedings of the Royal Society* 172:417-437.

Pal AK 1986. *Environmental Ecology*. 4:642-647.

Pala M, Saxena MC, Papastylianou I, Jaradat AA 1994. In *Expanding the Production and Use of Cool Season Food Legumes* pp 130-43. (ed Muehlbauer, F.J. and Kaiser, W.J.) Kluwer Academic Publishers London.

Patel S, Thakkar NP, Chaudhari SM, Shah RM 1986. *International chickpea* Newsletter. 14:22-24.

Parihar SS, Tripathi RS 1989. *Experimental. Agriculture* 25:394-355.

Perur NG, Mithyantha MS 1985. In *Soils of India and their management* pp 177-207, (ed Alexander T.M.) 2nd ed. New Delhi, India; Fertilizer Associate of India.

Plancquaert Ph 1991. In *IFA World Fertilizer Use Mannual* pp. 161-167 (ed W. Whichmann.). International Fertilizer Industry Association, Paris.

Rao NSS, Tilak KVBR, Singh CS 1986. *Plant and Soil*. 95 (2):351-359.

Rathore RS, Khandwe R, Khandwe N, Singh PP 1982). *LENS* Newsletter. 19 (1) : 17-19.

Reddy K 1992. Ph.D. Thesis IARI, New Delhi 1992.

Rizk MA, Hassan MS 1991. In *Nile Valley Regional programme on cool season food legumes and cereals, 1990-91*. Annual Report Egypt pp. 53-54, Cairo, Egypt-ICARDA.

Saxena MC, Gizaw A, Rik MA, Ali M 1994. In *Expanding the Production and Use of Cool Season Food Legumes* pp 633-41 (eds F.J. Muehlbauer,F.J. and Kaiser, W.J..). Kluwer Academic Publisher, London.

Saxena MC, Salim SN, Singh KB 1990. *Journal of Agricultural Science* (Camb) 114:285-293.

Saxena MC, Saxena NP 1991 *Legume Programme Annual Report ICARDA*, Aleppo, Syria pp 90-91.

Sharma PK, Chahal VPS 1983. *Journal of Research of Punjab Agriculture University*. 204 563-566.

Singh NP, Saxena MC 1986. *LENS* Newsletter . 132 : 27-28.

Sinha RP 1988. *LENS* Newsletter. 152: 9-10.

Srivastava GP, Srivastava VC 1992 *Indian Journal of Agronomy* 36: 143-147.

Singh BP, Singh B 1983. *Journal Indian Society of Soil Science* 38:769-771

Srivastava TK 1993. Ph.D. Thesis IARI, New Delhi 1993.

Srivastava TK, Ahlawat IPS 1995 *Indian Journal of Agronomy* 40: 630-635.

Tandon HLS 1992.. In *Fertilizers, Organic Manures, Recyclable Wastes and Bio-fertilizer* pp 12-35 (ed. H.L.S. Tandon FDCO, New Delhi.

Tekalign M 1994. In *Cool Season Food Legumes of Ethiopia* pp 293-311 (eds Geletu, B. et al.) Proc of National Cool Season Food Legumes Conference Addis Ababa, 16-20 Dec., 1993.

Tiwari OP, Tripathi RS 1995. *Indian Journal of Agronomy* 40:513-517.

Venkateswarlu U, Ahlawat IPS 1993. *Indian Journal of Agronomy* 38 : 236-243.

Whiting AL 1985. *Agronomy Journal* 17: 474-49.

Wani SP, Lee KK 1995 In *Plant microbe interaction in sustainable agriculture* pp 62-88. (eds. Bhal R.K., Khurana, A.L. and Dogra RC.) Bio science publishers, Hisar, India.

Lathyrus sativus: a crop for harsh environments

M.A. Malek, Ali Afzal, M.M. Rahman and ABM Salahuddin

Bangladesh Agricultural Research Institute, Joydebpur, Gazipur-1701, Bangladesh.

INTRODUCTION

Grass pea (*Lathyrus sativus* L.), known as `Khesari` in Bangladesh, is a protein-rich legume crop grown in a range of harsh conditions from very dry to low lying, water logged, flooded land. The crop is exclusively cultivated in India, Bangladesh, Burma, Nepal, and Pakistan in cold winter months under rainfed agriculture. It is also grown to a small extent in the Middle Eastern countries, Southern Europe, and parts of Africa and South America. In Bangladesh, grass pea is cultivated in an area of 239,343 ha with a production of 174,245 tonnes, the mean yield being 728 kg/ha. Among the pulses, it occupies the highest area (33%) and production (34%).

Grass pea is the hardiest of the pulse crops because it can tolerate flooding, drought, and moderate soil salinity. This tolerance of stress has made it a popular pulse crop for food and for cattle feed among the poor farmers of Bangladesh. It is commonly cultivated as a relay crop in wet rice fields without any inputs or care.

ENVIRONMENTAL VARIATION IN AREAS SOWN TO GRASS PEA.

Grass pea is grown in adverse agricultural conditions in Bangladesh. These range from low lying rice fields where flood waters remains stagnant for long periods to very dry and saline conditions. When the flood waters starts to recede, the heavy clay soil looses moisture so quickly that the sowing of winter crops is difficult. The harsh conditions under which grass pea is grown in Bangladesh may be classified into three distinct environments.

Environment I

This environment is experienced in the central part of Bangladesh and is characterized by heavy rainfall during the monsoon season (June-August) and inundation by floodwater every year. The water remains stagnant for a long period, then starts receding very quickly and the calcareous dark gray/brown soil looses its moisture rapidly after the rice harvest making sowing of winter crops extremely difficult. The farmers have no alternative other than to relay crop with grasspea under such conditions. Even grasspea suffers from water stress as the winter season advances and the yield is reduced in consequence.

Environment II

This environment is experienced in northern parts of Bangladesh where there are very low temperatures and no rainfall during the winter season. The gray terrace soil becomes very hard and cracks easily and no winter crop is grown during November to April. The farmers are forced to relay crop grass pea with rice or to fallow the land.

R. Knight (ed.), Linking Research and Marketing Opportunities for Pulses in the 21st Century, 369–373.

370

Environment III

This environment occurs in the entire coastal belt facing the Bay of Bengal in the south of Bangladesh. The soil is a Gangetic tidal flood plain/gray flood plain saline phase. The land is exposed to salinity making it impossible to grow other winter crops profitably. Again the farmers either relay sow grass pea with rice or keep their land fallow.

GERMPLASM COLLECTION AND EVALUATION

Until 1982, there were a few collections of grasspea germplasm but they were either damaged or lost due to a lack of proper storage. No passport data are available. A collecting programme was initiated in 1992 by the Bangladesh Agricultural Research Institute (BARI) in collaboration with the Directorate of Agriculture Extension (DAE). 2078 accessions of grass pea were collected from areas with environments II and III. The germplasm has been stored at the Pulses Research Centre (PRC) and Genetic Resources Centre (GRC) of BARI. In 1995, a collecting mission collected sixty-two accessions from areas with environmentsI & III under a BARI/CLIMA/ICARDA collaborative project with the title "Collection and Conservation of Bangladeshi land races of Lentil and Lathyrus". The passport data sheets of all the collections are available at PRC/GRC, BARI. Half of the material of each accession was sent to ICARDA.

In 1993, 1994, and 1995, the materials collected from environments I, II and III were evaluated at BARI. Each entry was grown in 1.5 m row plot with a row spacing of 50 cm. Days to 50 percent flowering, days to maturity, pod length, seeds per pod and 1000-seed weight were recorded. The ODAP content of the germplasm was also analysed (Table 1, 2 & 3).

Environment I

The germplasm from 27 districts of Bangladesh with environment I conditions, consisting of 748 accessions, were evaluated at BARI. The results for various characters are given in Table 1.

Table 1. Results for accessions from environment I.

Character	Range	Mean	SD	CV(%)
Days to 50% flowering	50-83	70	11.01	15.2
Days to maturity	102-125	114	5.62	4.1
Pod length (cm)	2.9-4.1	3.0	0.17	4.9
Seeds/pod (No.)	3.2-5.8	4.6	0.42	9.9
ODAP (%)	0.0857-0.2307	0.1239	0.0466	29.9

Environment II

The germplasm (748 accessions) from environment II areas were similarly evaluated at BARI (Table 2).

Table 2. Results for accessions from environment-II.

Character	Range	Mean	SD	CV(%)
Days to 50% flowering	55-88	71	4.96	7.0
Days to maturity	105-119	113	3.81	3.4
Pod length (cm)	2.5-3.3	2.8	0.15	5.5
Seeds/pod (No.)	2.3-4.9	3.4	0.41	11.9
ODAP (%)	0.0817-0.2209	0.1512	0.0455	30.1

Environment III

The results for 535 accessions from Environment .III (CV) are given in Table 3.

Table 3.Results for accessions from environment III.

Character	Range	Mean	SD	CV(%)
Days to 50% flowering	43-82	69	10.09	14.6
Days to maturity	99-122	114	4.92	4.3
Pod length (cm)	2.8-3.6	3.2	0.15	4.8
Seeds/pod (No.)	3.4-5.7	4.4	0.35	8.0
ODAP (%)	0.0791-0.2315	0.1621	0.0545	28.2

Germplasm from all three environments showed a wide range of variation in days to 50% flowering, seeds/pod and ODAP content. The highest CV was observed in ODAP content and the lowest in days to maturity.

The germplasm from the three environments were similar for days to 50 percent flowering and days to maturity. The entries from environment II had a smaller pod length and smaller seed size. The mean ODAP content was less in environment II material.

Table 4. Summary of the mean values for germplasm from three environments.

Character	Env.-I	Env.-II	Env.-III
Days to 50% flowering	70	71	69
Days to maturity	114	113	114
Pod length (cm)	3.0	2.6	3.2
Seeds/pod (No.)	4.6	3.4	4.4
ODAP (%)	0.1239	0.1512	0.1621

The characterisation of the collected germplasm is under way with help from CLIMA but will take some time after which the materials will be conserved in GRC of BARI.

VARIETAL IMPROVEMENT

BARI and BINA (Bangladesh Institute of Nuclear Agriculture) have the mandate to carry out research on grass pea. BARI uses conventional breeding techniques whereas BINA applies nuclear techniques in developing varieties. BARI has released two varieties BARI-Khesari-1 and BARI-Khesari-2 having high yields and low ODAP content compared with local checks (Table 5). The breeder's seed of these varieties have been given to Bangladesh Agricultural Development Corporation (BADC) and Non-Government Organisations (NGO) for multiplication. Since outcrossing occurs in grass pea, these seed multiplying agencies have been advised to grow the crops in isolation. To maintain purity of seeds the farmers will be supplied with seed every year (a particular variety for a particular locality).

Table 5. Average performance and ODAP content (%) of the BARI released varieties of grass pea.

Variety	Days to maturity	Plant height (cm)	1000-seed weight (g)	Yield (kg/ha)	ODAP content
BARI-Khesari-1	115	70	64	1720	0.06
BARI-Khesari-2	115	70	68	1727	0.14
Local Check	112	60	40	1616	0.53

Seven entries were tested in the three environments for the final year in 1995-96. The entries were sown in 8 x 4 m plots at a row distance of 40 cm. There were 3 replicates in a RCBD design. The results are given in table 6

The entries 104/11-2 and 114/26-1 have been identified as high yielding and with a low ODAP content relative to the local check. They will be proposed for registration.

Table 6.Average performance of the entries in different locations.

Line	Days to 50% flowering	Days to maturity	1000-seed weight	Yield (Kg/ha)				Mean	ODAP content(%)
				Joy	Isd	Rah	Jes		
112/14-1	85	114	51.0	1105	911	651	1655	1105	0.272
112/15-1	84	116	61.0	1735	1200	612	1625	1293	0.280
112/7-2	85	116	64.6	1239	1132	612	1481	1119	0.285
114/26-1	84	115	64.4	1664	1718	573	1712	1417	0.178
114/6-1	87	116	52.5	942	1393	495	1643	1118	0.275
110/3-1	91	112	62.0	908	1171	814	1436	1082	0.233
104/11-1	86	116	60.7	1638	1661	547	1384	1308	0.100
Local check	90	119	60.3	1378	1390	468	1251	1122	0.533
F. test	*	**	**	**	**	**	NS	-	-
CV(%)	2.95	0.78	3.43	2.60	21.37	9.25	21.1		
LSD(0.05)	4.406	1.566	3.519	60.1	500.7	137.9			

CROPPING SYSTEMS & AGRONOMIC PRACTICES

No agronomic research on grass pea has been conducted in Bangladesh. Grass pea is grown popularly as a relay crop in low-lying areas in broadcast aman rice fields. Grass pea seed at a rate of 30-40 kg/ha is broadcast in the rice crop in wet soil from mid-October to the end of November, 4-5 weeks before the rice harvest. Growing of grass pea as an intercrop with sugarcane in some upland areas has gained popularity only recently. Some rhizobial strains, compatible with grass pea nodulation and leading to increased yields, have been identified by BARI, BINA & BAU (Bangladesh Agricultural University)

CONSUMPTION AND MARKETING

Grass pea is mainly consumed as `dal` with rice. Boiled pods and roasted seeds are also consumed by the villagers. Soaking overnight and decanting the water has been found to make a `dal` free from ODAP and safe for consumption. But when Khesari powder (besan) is used for making pakoras, chapatis, dalpuri etc. the danger of lathyrism remains. To be on the safe side, low toxin varieties should be used for commercial production.

The majority of the farmers are poor and cannot keep their produce for long. During harvest time the price is at a minimum when the farmers sell their crops. The price goes up as the season advances and the poor farmers are deprived of this high price. The price should be fixed before harvest and Government. agencies should be responsible for procurement as they are for rice and wheat.

RESEARCH PRIORITIES

The priorities are
Characterization and conservation of the collected germplasm.
Breeding for short duration, salinity tolerant, disease resistant (downy and powdery mildew) varieties suitable for relay cropping and intercropping.
Breeding for low toxin varieties with better nutritional quality and consumers' acceptance.

CONCLUSIONS

After the harvest of low-land rainfed rice in Bangladesh, large areas remains fallow during the winter season where grass pea cultivation could be expanded. Farmers grow this low input crop because of its versatility and ability to grow well in water logged, drought and saline conditions. The low toxin, bold seeded varieties developed by BARI should reach the farmers without delay. Seed multiplication programmes should be strengthened to ensure the supply of high quality seeds to farmers. The price of Khesari should be fixed before harvest and Government agencies should be responsible for its procurement as they are for wheat and rice. Low toxin, bold-seeded and high yielding varieties suitable for relay cropping and dual purposes should be bred for commercial production.

Relevant literature

Bangladesh Bureau of Statistics. *Statistical Yearbook of Bangladesh*, 1995.

Bangladesh Bureau of Statistics. *Yearbook of Agricultural Statistics of Bangladesh*, 1995.

Bangladesh Agricultural Research Institute. *Annual Reports on Pulses Breeding. Pulses Research Centre.* 1987 to 1995.

Haque, R.M., Hussain and L.Labein. 1993. *Lathyrus* Newsletter. Third World Medical Research Foundation, New York, USA and London, U.K.

Sarwar, C.D.M. and M.A. Malek. 1995. *Collecting grasspea germplasm in Bangladesh* (submitted for publication in IPGRI Newsletter for Asia, The Pacific and Oceania).

Elias, S.M. 1991.. *In Advances in Pulses Research in Bangladesh*: p. 169-175 Proceedings of the Second National Workshop on Pulses, 6-8 June 1989, Joydebpur, Bangladesh and ICRISAT, Patancheru 502 324, A.P., India.

Quader, M. 1985. Ph.D. thesis, Division of Genetics, Indian Agricultural Research Institute, New Delhi 110 012, India.

Yadav. C.R. 1995. *Genetic Evaluation and Varietal Improvement of Grasspea in Nepal.* Proc. of Regional Workshop Raipur India 27-29 December, 1995.

Sarwar, C.D.M., M.A. Malek, A. Sarker and M.S. Hassan. 1995. *Genetic Resources of Grasspea (Lathyrus sativus L.) in Bangladesh.* Proc. of Regional Workshop Raipur India 27-29 December, 1995.

Malek, M.A., C.D.M. Sarwar, A. Sarker and M.S. Hassan. 1995. *Status of Grasspea Research and Future Strategy in Bangladesh.* Proc. of Regional Workshop Raipur, India 27-29 December 1995.

When and where will vetches have an impact as grain legumes?

C M Francis[1], D Enneking[1], A M Abd El Moneim[2]
1 Centre for Legumes in Mediterranean Agriculture, University of Western Australia, Nedlands,W.A.6009, Australia.; 2 International Centre for Agricultural Research in Dry Areas, Aleppo, Syria.

Abstract

The diversity of the genus *Vicia* L. (vetches) provides a wealth of domesticated crop plants for Mediterranean and temperate agriculture. The levels of domestication (soft seededness, high grain yield, high harvest index, +/- reduced shattering, +/- high biomass) found in several *Vicia* spp make them attractive as grain and forage crops for dryland farming. Species such as *V. sativa* L., *V. narbonensis* L., *V. articulata* Hornem. and *V. pannonica* are candidates for further domestication towards grain legume status. Vetches have an advantage over most of the food legumes in being better suited to hay production and most are thus dual purpose crops. They will have an increasing role as animal protein becomes scarcer and more expensive. Vetches already occupy special niches in farming systems because of their wide adaptation and ability to grow where food legumes are not suited. These include extremes of winter cold and very dry conditions. They may also be considered as alternative leguminous break crops where diseases and pests may otherwise limit the production of traditional food legumes. Despite these inherent advantages, their potential role as food legumes in human diets is largely unexplored. Traditional barriers need to be overcome and breeders will have to ensure that their product is free of toxins. The current program at ICARDA is one of the few in the world aimed at enhancing the value of vetches by improving quality, harvest index and harvestability, while Australian scientists are screening *V. sativa, V. narbonensis* and *V. ervilia* collections for genotypes with higher seed production, reduced toxicity and improved palatability. The relative lack of breeding effort in itself is a barrier to domestication.

INTRODUCTION

Vetch cultivation in Mediterranean agriculture is currently going through a renaissance, which could well lead to a revolution in current farming. Vetches, being multi-purpose crops, allow for either fodder conservation or immediate cash returns through hay or grain production, while at the same time providing a green manure and grazing option. They differ currently from food legumes largely in terms of end use. This distinction however has become blurred and with adequate research, several species could assume a higher profile as alternative food legumes.

FAO (1987) reported that in 1985 there was 1.3 million hectares sown to vetches globally for a total production of 2.2 million tonnes of seed. There is no evidence that the area has decreased. The demand for meat and milk is strong and increasing and in the West Asia - North Africa region alone, the current number of small ruminants is estimated to be well over 340 million (ICARDA, 1993). A shortage of animal feed is inflicting a heavy burden on the rangelands, which are deteriorating due to the rapidly growing livestock and human population. Severe feed and food deficits have led to the replacement of the fallow barley rotation in the dry areas. New oil seeds or grain legumes permit a diversification of the cropping rotations and allows different options for weed and disease control.

Although over 20 species have been tried as a food, there is today no significant use of vetches as a food despite their widespread use as forage legumes. Of the 160 Vicia species, apart from *V. faba*, only *V. ervilia, V. sativa, V. pannonica, V. villosa* and *V. benghalensis* are widespread in the world and are sown as hay crops in mixtures with oats or barley. Several, like narbon vetch, *V. narbonensis* and *V. ervilia* have the morpho-agronomic attributes of grain crops but most have anti-dietary or toxic factors which preclude their use as food legumes. The grains with their high protein are well suited as supplements for ruminant production, although the amount which may be included in a diet and the effect of such diets on end-product quality require further

R. Knight (ed.), Linking Research and Marketing Opportunities for Pulses in the 21st Century, 375–384.
© *2000 Kluwer Academic Publishers. Printed in the Netherlands.*

delineation. Development of vetches as feeds for mono-gastric animals and aquaculture is still in its early stages.

The current program at ICARDA is one of the few in the world aimed at enhancing the value of vetches by improving quality, harvest index and harvestability, while Australian scientists are screening *V. sativa, V. narbonensis* and *V. ervilia* collections for genotypes with higher seed production, reduced toxicity and improved palatability.

VETCH GERMPLASM

The concept of replacing bare fallow fields with leguminous crops (Carter, 1978) has given new impetus to research on legumes, including vetches. A wealth of germplasm of species such as *V. sativa, V. villosa, V. narbonensis, V. ervilia, V. pannonica, Lathyrus sativus, L. cicera, L. ciliolatus* and *L. ochrus* has been collected by ICARDA (Robertson et al., 1996). As well as the contribution of ICARDA's collecting program, many accessions have been introduced from IPK Gatersleben and Instituto de Germplasmo Bari. ICARDA's collection of *Vicia* spp is probably the world's largest standing at nearly 5500. Since the inception of the program there have been large-scale evaluations of the species for agronomic traits and for their reaction to biotic and abiotic stresses (Saxena et al., 1993). Valuable variation has been recorded in these attributes, but many accessions show the typical wild growth habit and their domestication will depend on further enhancement and breeding. The recent collaboration between the Vavilov Institute, St. Petersburg and ICARDA provides a vital link to further enhance the genetic conservation of this agriculturally important genus.

1. REQUIREMENTS FOR DOMESTICATION

The requirements for domestication were summarised by Bailey (1952) when he listed the essential characteristics of a successful grain legume (Table 1). Most of these are relevant to areas with a predominantly winter rainfall and relatively short seasons. We have added requirements for soft seededness in species to be regularly cropped and for improved seed quality through low seed toxin levels.

Table 1. Desirable traits for a grain legume in southern Australia (after Bailey 1952)

1. Ability to be harvested by conventional machinery
2. Growth period of 6 months or less
3. Tolerance to pests and diseases
4. Tolerance to adverse conditions (cold and drought)
5. Grain and fodder yields at least equal to alternatives like peas
6. Large seed size, approximately the size of a pea
7. A residual soil cover after summer grazing
8. Uniform maturity and shatter-resistant pods for mechanical harvesting
9. Soft seededness
10. Seed quality and non toxicity to monogastrics (including humans)

Progress towards domestication

Of the 10 most cultivated vetch species there is variation in their potential for success as grain legumes. Their immediate potential depends on their content of toxins, harvestability, seed yield, harvest index, pod shattering and of course environmental adaptation. The present state of progress in some of these features is summarised in table 2. The ideal grain legume would have an erect growth habit, high harvest index and a high seed yield of fairly large, toxin free seeds. High seed protein would be desirable as well as an amino acid profile with a higher content of S containing amino acids. It is obvious from the table that most of the now cultivated species require further improvement. Narbon bean and bitter vetch (*V. ervilia*) appear well placed whilst considerable progress is being made in the development of crop types of *V. sativa*. Pod shattering, a typical wild habit characteristic, while rare for *V. ervilia*, is a major constraint in the use of many species as grain crops, but lines of *V. sativa* and more recently *V. villosa* with non-shattering pods and good agronomic characteristics have been selected (Abd El Moneim and Saxena, 1995). These improvements could lead to vetch replacing fallow in rotations, a change which would be of good economic value in North Africa and West Asia.

Table 2. Vicia spp-Summary of their grain legume characteristics

Vicia species	Adaptation rainfall, winter temp	Relative Seed yield	Seed size	Hard seed	Seed protein %	Seed Toxins	Growth habit	Reduced Shattering
articulata Hornem.	+/- 300 mm, mild	***	***	*	22 -26	low -mod	vine	available
benghalensis	>400 mm, mild	**	**	soft	28 -34	high	vine	yes
ervilia	>300 mm, cold	***	**	soft	21 -25	low -mod	erect	yes
faba	>350 mm. mild	*****	**** *	soft	21-24	low -high[1]	erect	yes
hybrida	>350 mm, cold	**	*	***	22-24	low -mod	vine	No
narbonensis	<300mm, mod cold	****	****	*	21 -30	low ?[2]	erect	yes
monantha	+/- 300 mm, mild	**	**	**	22 -26	mod - high	vine	No
pannonica	>350 mm, very cold	**	**	*	24 -26	low	vine	available
sativa	>300 mm, mod cold	****	**	**	24 -32	mod -high	vine [3]	available
villosa	>350 mm, very cold	*	*	****	28 -34	high	vine	available

1 Favism affects genetically susceptible individuals (see Enneking and Wink, this conference).

2 Erect type available (Moneim pers comm)

3 Further research needed

Soft seededness is available and selection for improved seed size is feasible in most species. Whilst the relative seed yields in the table are equivocal and are generally lower than faba bean (or pea) yields, most at least equal or exceed the yield of lentil or chick peas. Specific environmental adaptation will also give them advantages over the traditional food legumes. Hungarian vetch and Bitter vetch for example are often the legume crops of choice in very cold winter rainfall localities where traditional pulses cannot tolerate the extreme cold. On shallow limestone soils, bitter vetch is a species of choice in northern Morocco whilst on sandy acid soils, species like *V. benghalensis* and *V. articulata* will usually out perform food legumes. *V narbonensis* is the best adapted *Vicia* species on heavier alkaline soils of the drier margins of the Australian cereal belt (ca 250 mm/annum). Similar results have been obtained in the ICARDA mandate region.

Biotic stresses may also preclude the consistent production of traditional grain legumes. Where food legumes like chick pea and faba bean are ravaged by *Orobanche crenata,* resistant varieties from species like *Vicia villosa* will prove useful as would some of the *Lathyrus* spp like *Lathyrus ochrus*. Inter and intra-specific variation in *Vicia* species for resistance to major diseases is widespread and durable sources of resistance have been found to *Ascochyta* blight (*Ascochtya pisi* f.sp. *viciae*), downy mildew (*Peronospera viciae*), powdery mildew (*Erysiphi pisi* f.sp *viciae*), *Botrytis* blight (*Botrytis cinerea*), cyst nematode (*Heterodera ciceri*) and root knot nematode (*Meloidogyne artiellia*) (Abd El Moneim and Saxena, 1995).

THE PAST UTILISATION OF *VICIA* SPECIES AS GRAIN LEGUMES

The historical background to the use of *Vicia* species as grain legumes dates back to antiquity, with finds of large hoards of *V. ervilia* dating to 7000 BC, and the earliest finds of *V. faba* to 5000 BC. Since ancient times, some 20 species of *Vicia* have been utilised in agriculture, but most are no longer in cultivation. Table 3 lists *Vicia* species known to have been cultivated and their use in agriculture (Enneking, 1994, 1995; and references therein). Low seed production of locally well adapted species often hindered their further development (Lechner, 1959). Some species are also of medicinal (*V. faba, V. ervilia*), biochemical diagnostic (*V. graminea,* anti-N-lectin) and ornamental value.

Table 3. Previously cultivated *Vicia* species and their use as grain (G) or forage (F).

Species	Synonym	Common name	Use
V. articulata Hornem.	*V. monanthos* (L.) Desf.	One-flowered vetch	G, F
V. benghalensis L.	*V. atropurpurea* Desf	Purple vetch	F
V. monantha Retz	*V. calcarata* Desf. (Dem)	Bard vetch	G
V. ciliatula Lipsky	*V. ciliata* Lipsky		
V. cracca L.	*V. tenuifolia* Gren&Godr	Tufted vetch	F
ssp.*tenuifolia* Gaud	*V. tenuifolia* Roth		F
V. ervilia Willd	*Ervum ervilia* L.	Bitter vetch	G, F
V. faba L.	*Faba vulgaris* Moench	Broad bean	G
V. fulgens Battand.		Scarlet vetch	
V. graminea Smith	*V.selloi* Vogel		F, G
V. hirsuta(L.) Gray		Hairy tare	F
V. johannis Taman.			G
V. narbonensis L.		Narbonne vetch	G, F
V. pannonica Crantz		Hungarian vetch	G, F
V. peregrina		Broad Pod vetch	F
V. pisiformis L.		Pale flowered vetch	G
V. sativa L. ssp. *sativa*		Common vetch	G, F
ssp. *amphicarpa*		Subterranean vetch	F
V. tetrasperma(L.) Schreb		Smooth vetch	F
V. unijuga A.Br.	*Orobus lathyroides* L.	Two leaved vetch	F
V. villosa Roth		Hairy Vetch	F
ssp. *varia* (Host) Corb	ssp. *dasycarpa* (Ten)Cav	Woolly pod vetch	F

Vicia species as food grains

   Although the current world consumption of vetches is very small, they have been used frequently as food during famines. Unorthodox foods are often consumed in times of famine (Salih et al., 1992). This extends to biblical times and may help explain the large hoards of *V. ervilia* found in some of the early Neolithic settlements documented by Zohary and Hopf (1988). Kunkel (1984) has listed the species, occasionally used for human consumption (Table 4). Several authors have mentioned that species such as *V. sativa*, *V. ervilia*, *V. narbonensis* and *V articulata* have served as food (Enneking, 1994, 1995). The practice of soaking, leaching and fermenting seeds would probably have evolved to make them more edible. *V. ervilia* is ground, soaked and used in soups in the Rif Mountains of Morocco. Leaching of bitterness is easily monitored by taste.

   V. sativa in the absence of the cyanogenic glycoside vicianine and despite the favism toxin vicine is quite palatable. Similar to beta-ODAP, beta-cyanoalanine is a glutamate analogue and can act as a flavour enhancer and phagostimulant. Combined with an attractive red cotyledon the cultivar Blanchfleur gained popularity amongst the grain trade as a lentil mimic (Tate and Enneking, 1992). [See also Enneking & Wink, this conference]. The human consumption of a white tare which was also called the Canadian lentil or Napoleon pea (*V. sativa alba*) has been reported (Wilson, 1852; Birnbaum and Werner, 1882). It had white or cream coloured seeds, with an apparently milder taste than other cultivars and was used to adulterate wheat flour by 10% in France. With its dwarf habit it produced a greater quantity of seeds than the other varieties of *V. sativa*. It was cultivated far more extensively in France and Canada than in Britain, chiefly for the sake of its seeds. Prjanischnikow (1930) reported that during the famine in 1921 the farmers in the region of Moscow include *V. sativa* flour at a rate of 25-50 % with rye flour for use as bread.

Table 4. Vicia species occasionally used for human consumption

Species	Seedtoxins *	Locality	Parts utilised, comments
V. articulata Hornem.	CAN	Medit. Region	Seeds used like lentils (black lentil)
V. cracca L.	GEC,CAN	Eurasia	Young shoots used as a pot-herb, leaves also used as tea, seeds used as food (Hedrick)
V. ervilia (L.) Willd	CAN	Mediterranean.	Seeds eaten in soups
V. nigricans Hook.			
ssp. *gigantea* (Hook)		California	Seeds edible
V. hirsuta (L.) S. F. Gray	CAN	Eurasia, N Africa	Weedy, young leaves and shoots eaten (boiled?), seeds cooked or roasted (Tanaka)
V. monantha Retz	CAN	Mediterranean.	Seeds used in soups
V. narbonensis L.	GEC	S. Europe	"A vegetable" (Tanaka), Hedrick: seeds eaten
V. noena Boiss.		Asia Minor	Seeds edible
V. bakeri Ali		Himalayan	Cultivated, as above? (Hedrick)
V. pisiformis L.		Europe, cult.	Seeds used like lentils
V. sativa L.	BCNA	Eurasia, cult.	Seeds ground into flour used in soups and bread, young shoots a pot-herb ; leaves as tea
V. sepium L.	BCNA	Eurasia	Seeds used as a food (Hedrick)
V. villosa Roth	GEC, CAN	Eurasia, cult.	" A vegetable" (Tanaka)

* CAN = canavanine, BCNA = beta cyano alanine and derivatives, GEC = gamma glutamyl - S ethenyl -cysteine

In Eastern Turkey, near Dyarbakir, *V. narbonensis* is apparently eaten as a pulse, after it has been boiled with some salt (Enneking, pers. com., Ergani village, 1991). A white-seeded variety of *V. sativa*, is used for human consumption in the province of Ratcha in the Caucasus (Hanelt, pers. comm. 1992).

Seed of *V. monantha* and/or *V. articulata* has been used for human consumption in Spain and was known as black lentil. It is however considered inferior in taste to lentils was only rarely found in markets (Fruwirth, 1921; Hegi and Gams, 1924). *V. articulata* locally known as 'lentis nigra' is still cultivated in South American countries such as Ecuador (L Robertson and R Reid, unpublished observations) where it had been introduced from Spain. Clearly, while there is ample evidence for the human consumption of *Vicia* species, other than *V. faba*, this has been limited and there is little evidence for their use as staple foods.

Vicia species as feed grains

The vetches have a major advantage in being dual purpose crops with a value both as hay and seed. Although the use of vetch seed for human consumption has been limited, its use for animal feed is well established in some regions. *V. ervilia, V. narbonensis, V. sativa, V. articulata* and *V. pannonica* are all used for this purpose. Production is greatest in countries of the former Soviet Union, the Anatolian plateau and eastern Europe, where there is a need to conserve fodder or grain to meet shortages over the freezing winter months.

a) *V. sativa* (Common vetch)

Whilst best adapted to neutral to alkaline loamy soils, it is because of its wide adaptation that it has received most attention by breeders (Lechner, 1959). For mediterranean climates the species is the subject of plant improvement programs at ICARDA, the Aegean Agricultural Research Institute in Izmir, and in South Australia. ICARDA has produced high seed yielding, non-shattering types and more recently a type with an erect habit of growth (Abd El Moneim and Saxena, 1995). All programs are selecting for the dual purpose of forage and grain - but basically for ruminant production. Major aims are to improve harvest index and reduce pod shattering at harvest thus strengthening the crop's dual purpose. High seed yield, reduced shattering, disease and pest resistance and an erect type of growth, coupled with white seeds, as achieved within the ICARDA program, can well change common vetch from a predominantly hay species to one where the grain is the primary product. A zero tannin, white-seeded variety was recently developed in Albacete, Spain. Virtually all research has centred on *V. sativa ssp sativa* but for a grain legume the relatively large seeded *ssp macrocarpa* needs investigation (Maxted, 1995).

The possible toxic effects of the amino acid, beta-cyanoalanine, the favism toxin vicine and the cyanogenic glycoside vicianine found in *V. sativa* grain need to be kept in mind.

For ruminants there is little or no experimental evidence of any adverse effects from the feeding of *V. sativa* seeds (Pandey and Pal, 1960). In Cyprus, Koumas and Economides (1987) were able to show feed intake and conversion rates of lambs and kids were similar when soybean meal was replaced partly or completely with common vetch or faba bean and there were no detrimental effects on the health of the animals. Valentine and Bartsch (1996) compared *V. sativa* seed meal with lupins as protein sources for high quality barley/silage dairy feed and found no adverse effects at 4kg/head/day, the lupin ration resulting in higher milk yields (p<0.001) while vetch fed cows gained 0.2 kg more weight/day over 49 days during early lactation.

For monogastrics the situation with respect to toxicity is far less promising and levels above 20 percent of the ration should not be exceeded for fattening pigs, and 10% for piglets (Piccioni, 1970). Similarly for poultry there are numerous reports of toxicity resulting in mortality rates of as much as 90 percent with seed at 40% of the diet (Harper and Arscott, 1962) and sharp reductions in intake level and body weight gains of chickens following feeding of split red vetch (Enneking, 1994, 1995).

In Australia progress has been made with the aid of the ICARDA collection, toward selection for reduced toxin levels. Varieties with beta cyano alanine levels around 20% of the average for the species have been located amongst natural populations (J. Rathjen, M. Tate and L. Robertson, unpublished data). Further reductions can be expected by recombinant breeding.

b) *V. ervilia* (Bitter vetch)

This species has promise as a low input grain legume well adapted to shallow calcareous loamy soils. In such an environment it is very common in the Rif mountains of Morocco (Francis et al., 1994; Enneking et al., 1995) It is a domesticated species with an upright growth habit, relatively non shattering pods, soft seeds and a fairly high harvest index. Improvement in seed size, plant height and lodging resistance are desirable. Bitter vetch is a very old crop plant and probably has been selected for a lower toxicity during the last 10,000 years. Compared to other *Vicia* species the content of canavanine in *V. ervilia* seed is low (Garcia and Ferrando, 1989), however, the biochemical nature of its bitterness and its pharmacological properties remain to be investigated.

Bitter vetch seed can be fed to ruminants in large quantities and has been a favoured grain for milking cows in Spain and Morocco. Soaking of the grain prior to being fed to ruminants has the purpose of making it more palatable and digestible.

Recently (Enneking, unpublished data) has selected very low canavanine lines of bitter vetch. Some have almost round seeds and an attractive red kernel suited to splitting. The end product, like the vegetative plant, itself is a lentil mimic. It has a relatively high content of methionine for a grain legume and is certainly a species with promise both as a feed and food grain.

c) *V. narbonensis* (Narbon bean)

Narbon bean has many of the characteristics of a grain legume in terms of its upright growth habit, large seeds and reduced shattering. Its performance relative to faba beans in dry areas has led to breeding and selection programs both at ICARDA and in Australia. Resistance to major grain legume diseases, *Ascochyta* blight, *Botrytis* blight, downy and powdery mildew, and nematodes exists amongst narbon varieties.

The species contains very high levels (1-3 %) of γ-L-glutamyl - S-ethenyl -L- cysteine (GEC) which imparts a sulphurous flavour to the beans and greatly limits their palatability. No acute toxicity has been reported with the species but it may be that the unpalatability limits intake and prevents the ingestion of sufficiently high concentrations of GEC. Feeding trials have given mixed results even with wool growth when one would have considered the very high sulphur content might have been beneficial (Allden and Geytenbeek, 1980). Jacques et al. (1994) with Merino wethers noted superior animal performance for the narbon over pea supplements at low allocations of grain (1% of live weight/day). The peas offered to the sheep were eaten within half a day, whereas narbon grain was consumed more slowly over two days. The lower palatability of narbon beans compared to peas, may result in a more even consumption over time and hence their better utilisation as a grain supplement.

In the European literature, grain of *V. narbonensis* has been identified as a suitable ruminant feed, especially for cattle (Mateo Box, 1961; van der Veen, 1960). Thus the crop holds particular promise for increased beef production, while for milk production there may be some off-flavour problems. It is important to assess the flavour of meat produced from narbon fed animals.

For non ruminants Davies (1987) found that narbon grain (line RL 140001) depressed feed intake when fed as a major component (35%) of pig diets. Mateo-Box (1961) noted *V. narbonensis* to be distasteful to poultry. This was partly confirmed by Eason et al. (1990) who tested the nutritive value of *V. narbonensis* line RL 140004 in the diets of day old broiler chickens. In this study, a 10% inclusion of narbon seed in the diet reduced intake 4% and live weight 2.6% as compared to soya bean meals. However in the same trial, feeding peas at the same level resulted in a 2.7% reduction in live weight. There is thus potential for utilisation in poultry diets, if lines with low levels of anti-nutritional factors can be selected.

Breeding and selection in Victoria, Australia has produced lines with substantially lower GEC levels (Castleman and Mock, unpublished data) and reduced shattering. Best adapted to alkaline heavy soils, yields of the best lines (Siddique et al., 1997) indicate that in dry areas the species may well be a significant grain legume of the future.

d) *V. articulata* (One flowered vetch, Algarroba lentil)

V articulata Hornem. is closely related to, and often confused with *V. monantha* Retz. The illustration of *V. monantha* in Duke (1981) for example, is clearly *V. articulata*. *V. articulata,* is best known in Spain (Barulina, 1930; Fisher, 1938; Lopez Bellido, 1994). It is adapted to non calcareous and acid sandy soils and could be grown as an alternative to lupins. Its bushy growth habit closely resembles a lentil, but its vigour and superior plant height are advantageous in terms of its grain legume potential.

Fruwirth (1921) considered this crop to be most important as a green fodder, however besides the valuable straw, the cracked grain in quantities of up to 1/3 of the ration, was considered useful for the fattening of animals. Mixed with other feeds it has been fed to dairy cattle and horses.

An early flowering selection from the US cultivar Lafayette was released in Australia (Bailey, 1952; Herbage register 1968). In Spain, the area once sown to these species, exceeded that of lentils (Barulina, 1930). It is still commonly found in southern Spain but mostly as a contaminant in other vetch crops or as a weed in cereals. Large areas are sown in Ecuador (Reid and Robertson, unpubl.) as a multi-purpose crop, including its use as a "black lentil" for local consumption. It is favoured because of it greater vigour, relatively large seed size and seed yield as compared to lentil. Maxted and Sabanci (private communication) report the species is still cultivated in South Eastern Turkey.

Relatively large seed size and reliable seed set are good features for a grain legume. The level of pod shattering, though normally less than common vetch, and seed canavanine levels are problems which limit its potential value as a grain for monograstric consumption in the short term. Scope for selection does exist. Two *V. articulata* lines tested in Australia contained 0.1-0.2% canavanine (Enneking and Francis, unpublished data).

e) *V. pannonica* (Hungarian vetch)

Hungarian vetch is native to eastern Europe and Caucasus and is well adapted to severe winter cold. The cultivation of this species is expanding rapidly in Turkey, replacing less productive *Vicia spp.* (Sabanci pers. comm.). Adapted to heavy soils it tolerates poorly drained soils better than other vetches (Duke 1981). The species has distinct promise as a grain crop as relatively non shattering lines are available. Similar to *V. villosa* it is an outbreeding species where fertilisation and fruit set can be increased by visits of pollinators (Zhang & Mosjidis 1995). Like all vetches, its main uses have been for hay and green manure. In Moldavia, Avedeni (1989), found lines with high cold tolerance, high biomass, seed weight, seed yield and protein content. ICARDA is focussing on the selection of genotypes with higher harvest index, reduced shattering and resistance to *Ascochyta* blight. Progress has been rapid and several lines have produced seed and biological yields equal to the best *V. sativa* lines with seed yields of 1372 kg/ha recorded at Tel Hadya, Syria.

The species is extremely low in known vetch toxins (canavanine, beta cyano-alanine) in two samples examined. Bell & Tirimanna (1965) found low levels of VA3, recently identified as GEC (Enneking et al., 1998) in the seeds of this species but the seed is certainly worth testing by monogastric bioassay to assess its performance. Coupled with its resistance to extreme cold it is a potentially a very valuable feed, if not a food grain, and deserving of a greater breeding effort.

f) *V. villosa* (Winter or Hairy vetches)

Hairy and winter or woolly pod (sometimes described as *ssp dasycarpa*) vetches are second in the world in area of vetch cultivation and are prized for their herbage, hay yields, resistance to pests and diseases. *V villosa* is highly resistant to *Orobanche crenata* and many varieties also have resistance to major disease like *Ascochyta* and nematodes. The tolerance to the biotic stresses and abiotic stress, especially frost, ensure reliable dry matter production. Seed yields are however poor relative to the other *Vicia* spp mentioned here and can be improved by pollinators, as it is a species with a substantial degree of outcrossing (Zhang and Mosjidis 1995).

The seeds contain canavanine in high quantities (> 2.0 percent) and the pods shatter freely. The high content of toxins has been observed to have dramatic effects on the intake of pigs at inclusion levels as low a 4%.

Enneking et al. (1993) showed that intake was reduced by 75% with 8% inclusion rates of cv. Namoi vetch seed in the diet. Likewise the toxicity to poultry of diets comprising mainly *V. villosa* seeds is well documented (Arscott and Harper 1964). Although mortality has been observed in cattle grazing green crops, the species is valued widely as a fodder for ruminants. Current knowledge suggests that the herbage becomes toxic during seed formation (Enneking, 1994, 1995).

With its vigorous and reliable dry matter production, disease and cold tolerance, and ability to grow on a wide range of soils, including sandy acid soils, it is worthy of serious attempts to select for the grain legume characteristics - reduced shattering, larger seed and low toxicity. Such aims are likely to require long term efforts and the major place of the species will probably remain as a forage rather than a grain legume.

g) *V. benghalensis* (purple vetch)

This species has promise in temperate Mediterranean climates as a grain species as it has a low level of shattering and can easily be harvested mechanically. In the US its seed yield has been similar to common, Hungarian and monantha vetches and it has the added ability to grow in acid as well as alkaline soils (Duke 1981). In Western Australia a relatively early selection cv Barloo is in production but its seed yield have been only half that of *V sativa* cv Blanchfleur. Nevertheless, the great improvement in seed production of the early selection over the 'parent' vetch (cv. Popany) suggests further gains are achievable.

Like hairy vetch, the species has a very high content of canavanine - often > 2 %. A similar picture in terms of adverse effects on non ruminants can therefore be expected with purple vetch. This species can also be toxic to grazing animals at seed set. Particular care should be taken with earlier flowering cultivars where the toxic stage is reached earlier in the season. If selection for reduced canavanine content can be achieved, then the species has considerable potential as a feed grain.

h) Other Species

In the eastern Mediterranean region, on alkaline soils, *V. lutea* (yellow vetch) and *V. hybrida* (hybrid vetch) are occasionally dominant in farmer's fields and used as volunteer hay crops. Widespread in the regions where they occur, little effort has been made to domesticate them. Both species shatter freely and are vine types. Hybrid vetch contains both cyanoalanine acids and vicine. Yellow vetch has vicine but appears to be a relatively low toxin species, however, further screening of this species for antinutritional factors by bioassay is required before firm conclusions can be drawn.

CONCLUSION

Vetches are widely distributed ecogeographically and have been cultivated over a long period of time. Most are not well domesticated in terms of plant habit for use as a grain legume. They have considerable potential for producing high seed yields, to be relatively free from pests and diseases and some are adapted to adverse conditions such as severe winter cold. Their main role has been as dual purpose species suitable for hay and increasingly for grain production. Those species that are crop types, like bitter vetch and narbon vetch, have anti-nutritional factors which need to be corrected before the species could be widely used as feed or food grains. Hungarian vetch has almost no known vetch toxins and must have potential as a food legume in areas with cold climates.

It is likely however the main use of vetch grain will be for animal feed, especially ruminants. The rapid increase in human population in West Asia and North Africa accentuates the demand for higher protein in the diets. The grain legume resource base is currently inadequate and increased production is urgent (Moneim and Saxena 1995). Vetches are likely therefore to be of increasing value as dual purpose crops largely for animal production and a valuable legumes in areas where current food legume are not adapted or where animal industries are the most profitable. Their prospects as food legumes must be further investigated. Several species like *V. ervilia*, *V. sativa*, *V. narbonensis* and *V. pannonica* require far more breeding and research support than is currently directed to them.

References

Abd El Moneim, A M., Saxena M.C. 1995. *Diversity* 11: 120 -21

Abd El Moneim, A.M. 1992. *Journal of Agronomy and Crop Science* 169: 347-53

Abd El Moneim, A.M. 1993. *Z. Pflanzenzucht*, 110: 168-171

Abd El Moneim, A.M., Cocks, P.S., and Swedan, Y. 1988. *Journal of Agricultural Science, Cambridge* 111: 295-301.

Abd El Moneim, A.M., Khair, M.A., and Cocks, P.S. 1990b. *Journal of Agronomy and Crop Science* 164: 34-41.

Abd El Moneim, A.M., Khair, M.A., and Rihawi, S. 1990a. *Journal of Agronomy and Crop Science* 164: 85-92.

Allden, W.G. and Geytenbeek, P.E. 1980. *Proc Aust Soc Anim Prod* 13: 249-252.

Arscott, G.H. and Harper, J.A. 1964. *Poultry Science* 43: 271-73.

Avadeni, L.P. 1989 Nauchno Tekhnicheskii Byulleten' Vsesoyuznogo Ordena Lenina i Ordena Druzhby-Narodov-Nauchno-Issledovatel'Skogo-Instituta-Rastenievodstva-Imeni-N.I.-Vavilova, 190: 21-23 [Plant-Breeding-Abstracts 1991 061-03532; 7G Seed-Abstracts 1991 (014-02767)

Bailey, E.T. 1952.. CSIRO Division of Plant Industries Technical Paper No. 1:

Bailey, E.T. 1965.. *Australian Journal of Experimental Agriculture and Animal Husbandry* 5: 466-469.

Barulina, H. 1930. *Bulletin of Applied Botany, Genetics & Plant Breeding,* Leningrad (Supplement) 40: 265-304.

Bell, E.A. and Tirimanna, A.S.L. 1965. *Biochemical Journal* 97: 104-111.

Birnbaum, K., and Werner, E. 1882. Thiel's Landwirt-schaftliches Konversationslexikon, 762-765. Leipzig: Thiel.

Carter, E.D. 1978. *Legumes in Farming Systems of the Near East and North African Region.* Report to ICARDA

Davies, R. L. 1987. *In Grain legumes for low rainfall areas.* Final Report D. Georg. Adelaide: South Australian Department of Agriculture.

Duke, J. A. 1981. *Handbook of legumes of world economic impor*tance. New York: Plenum Press.

Eason, P., Johnson, R., and Castleman, G. 1990. *Australian Journal of Agricultural Research.* 41: 3, 565-571; 18 Ref.,

Eason, P.J., Johnson, R.J., and Castleman, G. 1987. Proc. Nutr. Society Australia 12: 119.

Enneking, D. 1994, 1995 *The toxicity of Vicia species and their utilisation as grain legumes* Ph. D. thesis, University of Adelaide, South Australia. Occasional Publication 6. Nedlands, Western Australia: Centre for Legumes in Mediterranean Agriculture (CLIMA).

Enneking, D., Giles, L.C., Tate, M.E., and Davies, R.L. 1993. *Journal of the Science of Food and Agriculture,* 61: 315-325

Enneking, D., Lahlou, A., Noutfia, A., and others, 1995. *Al Awamia* 89;. 141-148

Enneking, D.; Delaere, I. M., and Tate, M. E. 1998. *Phytochemistry,* accepted.

Ergun, A., Colpan, I., Kutsal, O., and Yalcin, S. 1986. Doga, Veterinerlik Ve Hayvancilik. 1986. 10: 2, 144-152; 29 Ref.,

Ergun, A., Yalcin, S., Colpan, I., Dikicioglu, T., and Yildiz, S. 1987. Veteriner Fakultesi Dergisi, Ankara Universitesi. 1987. 34: 3, 449-466; 28 Ref.,

Fischer, A. 1937. *Züchter* 9: 286-288.

Fischer, A. 1938. *Züchter* 10: 51-56.

Francis, C.M., Bounejmate, M., and Robertson, L.D. 1994. *Al Awamia* 84: 17-27.

Fruwirth, C. 192. *Handbuch des Hülsenfruchterbaues* . Berlin: Paul Parey.

Gabriel and Gottwald 1887. *Journal Für Landwirtschaft* 35: 239-247.

Garcia, A.M. and Ferrando, I. 1989. *Annales De Bromatologia,* 4: 167-175

Harper, J.A. and Arscott, G.H. 1962. *Poultry Science* 41:

Jacques, S., Dixon, R.M., and Holmes, J.H.G. 1991. *Proc. Australian Society Animal Production* 19: 249.

Jacques, S., Dixon, R.M., and Holmes, J.H.G. 1994. *Small Ruminant Research* 15: 39-43

Koumas, A. and Economides, S. 1987. Technical Bulletin, Agricultural Research Institute, Cyprus, 88:

Kunkel, G. 1984. *Plants for human consumption.* Koenigstein: Koeltz Scientific books.

Lechner, L. 1959. In *Handbuch der Pflanzenzüchtung* Bd. IV. Züchtung der Futterpflanzen, 52-95 Eds H. Kappert and W. Rudorf. Berlin: Paul Parey.

Mateo-Box, J. M. 1961. Leguminosas de grano. Barcelona: Ed. Salvat.

Pandey, J.N. and Pal, A.K. 1960. *Indian Journal of Physiology and Allied Sciences* 20: 87-90.

Piccioni, M. Diccionario de Alimentacion Animal, 819 pp. Zaragoza: Ed. Agribia [An Italian edition is published by Ed Agricole].

Prjanischnikow, D. N. 1930. Spezieller Pflanzenbau. Der Anbau der landwirtschaftlichen Kulturpflanzen, 364 Ed E Tamm. Berlin: Springer.

Robertson, L.D., Singh, K.B., Erskine, W., and Abd El Moneim, A.M. 1996. *Genetic Resources & Crop Evolution* 43: 447-460

Salih, O.M., Nour, A.M., and Harper, D.B. 1992. *Journal of the Science of Food and Agriculture* 58: 417-24.

Saxena, M. C., Abd El Moneim, A. M., and Ratinam, M. 1993. Proceedings of the Vicia/Lathyrus workshop, Perth, Western Australia 22-23.9.1992, 1-7 Eds J. R. Garlinge and M. W. Perry. Perth, Western Australia: Co-operative Research Center for Legumes in Mediterranean Agriculture (CLIMA).

Siddique, K.H.M. and Loss, S.P. 1996. *Australian Journal of Experimental Agriculture* 36: 587-593

Siddique, K.H.M., Loss, S.P., and Enneking, D. 1996.. *Australian Journal of Experimental Agriculture* 36: 53-62.

Tate, M.E. and Enneking, D. 1992. *Nature* 359: 357-8.

Valentine, S. C. and Bartsch, B. D. 1996. *Australian Journal of Experimental Agriculture*, 36 (6) : 633-636

Van der Veen, J.P.H. 1960. *Journal of the British Grassland Society* 15: 137-144.

Wilson, J. M. 1852. The Rural Encyclopedia or a General Dictionary of Agriculture Vol. IV, Q-Z, 581-588 . Edinburgh.

Zhang, X. and Mosjidis, J.A. 1995. *Crop Science* 35: 1200-1202

Zohary, D. and Hopf, M. 1973. *Science* 182: 887-894.

Zohary, D., and Hopf, M. 1988. *Domestication of plants in the old world* . Oxford: Clarendon.

Impact of biotechnological interventions on productivity and product quality

HJ.Jacobsen[1] and G. Ramsay[2]

1 Lehrgebiet Molekulargenetik, Universitat Hannover, Herrenhduserstr.2, D-30419 Hannover,Germany; 2 Scottish Crop Research Institute, Invergowrie , GB-Dundee DD2 SDA, Scotland

INTRODUCTION

Biotechnology is a term used for a variety of different approaches to modern crop improvement programs, ranging from apparently simple (but nevertheless successful techniques) like
(a) micropropagation for mass production of healthy seed- and planting material,
(b) embryo-rescue protocols for saving the products of wide crosses between related species or genotypes from the same gene pool that are difficult to hybridize,
to the somewhat more sophisticated techniques like gene isolation and transfer.

The basic principle of genetic engineering is the transfer of distinct genes directly targeted to add, delete or modify specific characters in a plant. The rationale is that a specific gene, which is engineered under the control of a useful promoter, is introduced into and expressed in the plant genome and acts as a plant gene. These genes theoretically can be of any origin (plant, bacterial, fungal or even human). The advantage of gene transfer is that only a defined genetic entity or entities are transferred, with the consequence that the host genome is only marginally modified, in contrast to other approaches like, for instance, wide crosses or somatic hybridizations, where in principle, two different genomes can interact

The use of DNA markers (RFLPs, DNA-fingerprint patterns, RAPDs) has enabled the development of marker-assisted breeding as a new tool for the breeder. It occurred as a logical spin-off from the successful application of markers in human genetics and its introduction to applied plant sciences through plant molecular biologists. Environmentally and developmentally neutral, markers will help breeders follow the inheritance of desired traits in progenies much more reliably than is possible today.

The application of these markers in breeding programs will not only become the basis for more rational selection systems, but will also be indispensable for the identification of relevant genes in the near future, provided highly probe saturated genetic maps are available for most of the crops ("positional cloning" or "map-based-cloning").

WHY BIOTECHNOLOGY?

Through conventional plant breeding, many crops in the developed countries now have a high productivity (yield, yield stability, quality), but there are still many breeding objectives to be attained. These, in particular, are increased resistance or tolerance to biotic and abiotic stresses. While for some objectives, classical breeding offers solutions (which necessarily are time and labour consuming), others are hardly or never achievable, in particular when the gene pool of a crop is limited. In this situation, foreign genes have to be introduced, either by wide crosses, somatic hybridization or the transfer of specific genes.

Methods for gene transfer

To transfer genes, two principally different approaches have been developed:

R. Knight (ed.), Linking Research and Marketing Opportunities for Pulses in the 21st Century, 385–387.

(a) the vector-mediated indirect gene transfer by *Agrobacterium tumefaciens*,

(b) different vector independent, direct transfer methods (protoplast transformation, particle bombardment).

Surprisingly most of the known protocols applied in grain legumes make use of the Agrobacterium mediated gene transfer, preferably for transforming easy-to-regenerate meristems.

WHERE ARE WE TODAY?

The application of biotechnological methods (including genetic engineering) to the improvement of grain legumes (and other important crop plants) is still in its infancy. Despite the fact that the first transgenic plant (tobacco) was developed almost two decades ago, we are still far away from using the potential of biotechnological methods as a set of routinely applicable tools for crop improvement as they are available in conventional breeding. The lesson to be learnt is that the time frame has to be expanded considerably. This is particularly relevant to the food grain legumes, which are the focus of the IFLRC.

Although transformation in principle is possible, substantial efforts are necessary before we will have a genotype-neutral and robust transformation system for all the species and lines. It needs to be remembered, all biotechnological approaches are components of so-called "pre-breeding". In other words, biotechnology contributes to the enhancement of genetic variability and provides new and novel breeding lines which have to be tested for their respective performance in conventional breeding programs. The important feature is that genetic engineering allows the use of foreign gene pools, which are inaccessible by conventional means.

TRANSGENIC GRAIN LEGUMES

Transgenic soybeans, resistant to herbicides are on the market (and have evoked consumer protests in Europe). Convincingly transgenic peas were first reported in Australia (Schroeder et al. 1993) as well as *Vicia narbonensis* with a putatively improved amino acid composition, which were developed in Europe and are now being explored in Australia. Peas, with resistance to the pea weevil, have been evaluated in the field in Australia as well as peas with an enhanced grain sulfur level. In both cases, herbicide resistance and a scoreable marker are also expressed (htt p://www.dist.gov.au/science/gcmac/pis~book/pr59.htm and pr6l.htm). In the USA, artificially sweet peas have been claimed by a company and more transgenic soybeans are likely to come, but what else? Phaseolus bean species still are difficult to transform, as is *Vicia faba* and the other pulses. Promising prototype systems have been reported, but in many cases there was no apparent commercial follow-up. Constraints and perspectives for biotechnological interventions are being discussed, based on the experience from the EU-funded projects, TRANSLEG (which focused on the development of grain legume transformation protocols) and PRELEG, a shared-cost project, which was launched after the breeders set some priorities. The main objective of PRELEG is to test different transgenesis-based approaches to resistance to fungal diseases (a "proof-of-principle"-project).

From the perspective of European breeders, the poor yield stability of grain legumes is the major limiting factor to high productivity. The factors negatively affecting stability are given in table 1 (TRANSLEG Annual Report 1995).

Table 1: Priority of objectives identified by the breeders as affecting yield stability

Objective	Factor	Genetics/knowledge
1. diseases	a) *Micosphorea pinodes*	not clear
	b) *Botrytis*	not clear
	c) root rot complex	not clear
	-*Fusarium*	
	-*Aphanomyces*	
	-*Phoma*	
	d) viruses - –	
	PsbMV & PEV	system of resistance not clear
	e) downy mildew	
	f) bacterial blight	
2) standing ability	a) control of stem elongation	
	b) stem architecture	
	-stem stiffness -branching	
3) engineering of flowering		breeding programmes available
4) product quality *		

*Product quality was found difficult to determine.

It was concluded more basic research on quality was needed to provide the information required by the breeder. For the disease-related problems solutions are available or likely to come from conventional breeding and genetic engineering. Plant architecture, control of flowering and product quality were issues which were heavily debated, because different solutions are required in different regions and climates.

In principle, one can conclude that transgenic grain legumes offer the possibility for improvement in productivity and product quality: This may be illustrated by examples:

If a transgene is found, which confers horizontally resistance to fungal diseases, it is logical to expect that grain legume products will be produced in future less contaminated by mycotoxins or fungicide residues. Grain legumes, which express a cell wall degrading enzyme under the control of a germination-specific promoter, may have better processing qualities for human or animal nutrition. While better storage should be feasible with pulse seeds which are resistant to weevils or bruchids, it is questioned whether an enhanced level of sulfur-containing amino acids will necessarily improve nutrition. Nutrition is never based on a single product and cereals can compensate for deficiencies in grain legumes and vice versa.

References

Schroeder, H.E., Schotz, A.H., Wardley-Richardson, T., Spencer, D. and Higgins, T.J.V. 1993 *Plant Physiology* 101:751-757.

Gene Technology for Improved Weed, Insect, and Disease Control and for Seed Protein Quality.

H.E. Schroeder[1], J.E. Barton[2], L.M. Tabe[1], L. Molvig[1], J.E. Grant[3], M. Jones[4] and T.J.V. Higgins[1]

1. CSIRO Division of Plant Industry, GPO Box1600, Canberra ACT 2601, Australia; 2. The University of Western Australia, Nedlands, WA 6907, Australia; 3. New Zealand Institute for Crop and Food Research Limited, Private Bag 4704, Christchurch, New Zealand; 4. Western Australian State Agricultural Biotechnology Centre, School of Biological and Environmental Sciences, Murdoch University, WA 6150, Australia

Key words: Gene Technology, Lupins, Peas, Chickpeas, Lentils, Herbicide Tolerance, Pea Weevil Resistance, Viral Diseases, High MethionineRunning title Gene Technology for Cool Season Grain Legumes

Abstract

The cool season legumes, narrow-leaf lupin, yellow lupin, pea, chickpea and lentil have been shown to be amenable to genetic transformation using recombinant DNA and tissue culture procedures. Transformation methods for faba bean and grasspea are under development. Transgenic lupins and peas containing genes conferring herbicide tolerance, insect resistance, virus resistance or nutritional enhancement of seed protein have been evaluated in field trials in Australia and New Zealand during the last two years. Commercial development of some of these plants is expected within the next 3-4 years.

INTRODUCTION

Genetic technology is a powerful tool for introducing new traits into plants. Genes encoding desirable traits can be derived from any organism, and can be introduced into a plant singly or in known combinations, thereby preserving the pre-existing characteristics of elite cultivars. Gene transfer techniques are currently being used to develop crops that are resistant to herbicides, pests or diseases or with improved quality characteristics. With recent advances in regeneration of transformed plants, the potential now exists to transform almost any crop plant with a foreign gene, which is then inherited in a stable manner.

Successful use of gene technology requires that the gene for the new trait be identified, isolated and then reconstructed for expression in the relevant organ of the new host. In addition gene transfer procedures and appropriate tissue culture methods must be developed for each target species to regenerate fertile, transgenic plants. The most commonly used procedure for introducing foreign genes into dicotyledonous plants makes use of a disarmed version of the pathogenic bacterium *Agrobacterium tumefaciens* as a gene vector. Chemical or physical methods such as polyethylene glycol treatment, microprojectile bombardment, electroporation and micro-injection are also used for DNA transfer into plant cells (Walden and Wingender, 1995, Chowira *et al.* 1996). The latter methods have primarily been used in the transformation of monocotyledonous plants, although *A. tumefaciens*-mediated transformation of some cereals has also been achieved (Hiei *et al.* 1994). Where possible, DNA transfer via *A. tumefaciens* seems preferable because few gene copies are transferred by *Agrobacterium*; giving rise to transgenic plants which show more stable inheritance of transgenes in subsequent generations. The presence of many introduced gene copies in transgenic plants has reportedly been associated with gene silencing or inactivation in subsequent generations (Finnegan and McElroy, 1994).

In general, grain legumes have been difficult to transform and regenerate compared to other dicotyledonous plants. Tobacco was first transformed in 1983 (Zambryski *et al.* 1983) and the first insect resistance gene was transferred into a crop plant in 1987 (Fishhoff *et al.* 1987). The first stable transformation of a large-seeded legume (soybean) was achieved in 1988 (McCabe *et al.* 1988, Hinchee *et al.* 1988) but over five years passed before stable transformation of peas (*Pisum sativum*) (Schroeder *et al.* 1993), chick pea (*Cicer arietinum*) (Fontana *et al.* 1993) narrow leaf lupin (*Lupinus angustifolius*) (Atkins *et al.* 1994, Molvig *et al.* 1994, Somsap *et al.* 1995), and yellow lupin (Li *et al.* 1996) were reported. The

R. Knight (ed.), Linking Research and Marketing Opportunities for Pulses in the 21st Century, 389–396.
© *2000 Kluwer Academic Publishers. Printed in the Netherlands.*

390

reasons for this are many, but among the most important are, little research on the topic prior to late 1980s, poorly defined regeneration systems and the lack of effective selectable marker genes.

Here we will discuss progress on the genetic engineering of cool season grain legumes identified at the Second International Food Legume Research Conference, namely peas, faba beans (*Vicia faba*), chick pea, lentils (*Lens culinaris*) and grasspea (*Lathyrus sativus*). In addition we add the narrow leaf lupin, the major cool season grain legume in Australia (Siddique and Sykes 1997) and the high-protein, yellow lupin (*Lupinus luteus*), a potential new broad acre crop for Western Australia (Cowling, 1994).

STATUS OF TRANSFORMATION

All plant genetic engineering procedures involve firstly the introduction of the foreign gene into an isolated fragment of the target plant (the explant) and the regeneration through tissue culture of a viable, fertile, transformed plant from that explant. Grain legumes in general are still difficult to transform reproducibly, as evidenced by the relatively low transformation frequencies reported (for review, see Atkins and Smith 1997a). Reproducible, stable transformation of peas and the two lupin species was achieved when the *bar* gene, conferring resistance to the herbicide glufosinate, was used as the selectable marker (Table 1). The *bar* gene has also been used successfully in lentil and chickpea transformation (Barton and Molvig, unpublished) and is being used to develop gene transfer methods for faba beans and grasspea (Barton, Grant, unpublished). In an important advance, Grant *et al.* (unpublished) have transformed peas using the *nptII* gene, which confers resistance to the antibiotic kanamycin as a selectable marker. Chowira *et al.* (1996) have recently reported the generation of transgenic grain legumes via *in planta* electroporation, without a selectable marker. However, this method gave rise to a poor rate of transmission of transgene activity to subsequent generations (Chowira *et al.* 1996).

The most common explant sources in grain legume transformation are immature or germinating seeds. In the case of peas, slices of the immature embryonic axis (Schroeder *et al.* 1993), the cotyledons of immature seeds (Grant *et al.* 1995), or lateral cotyledonary meristems from germinating seeds (Bean *et al.* 1997) have been used. Slices of the immature embryonic axis (Molvig *et al.* 1994, 1997), or the shoot apex of the germinating embryonic axis (Atkins *et al.* 1994, Pigeaire *et al.* 1997, Li et al. 1996, pers com,1997, Hoffmann et al. 1996) have been used successfully for lupins. The lateral cotyledonary meristem from germinating chickpea seed (Molvig, unpublished) or the shoot apex of germinating embryonic axes of chickpea and lentil (Barton, unpublished) have been successful. Regeneration of stably transformed transgenic plants from all five species is by organogenesis. In spite of concerted efforts in many laboratories only peas, lupins and chickpeas have so far been stably transformed with genes that have potential for crop improvement (Table 1). Some of these genes have been stably expressed and inherited for up to six generations.

Table 1. Summary of traits transferred to cool season grain legumes via genetic engineering techniques

Species	Trait	Reference
Pisum sativum	Weevil resistance	Shade *et al.* 1994
		Schroeder *et al.* 1995
	Herbicide tolerance + increased seed methionine	Schroeder *et al.* 1994
	AMV resistance	Grant *et al.* unpublished
	PSbMV resistance*	
Lupinus angustifolius	Herbicide tolerance	Pigeaire *et al.* 1997
	Increased seed methionine	Molvig *et al.* 1997
	BYMV resistance*+ Herbicide tolerance	Atkins *et al.* unpublished
Lupinus luteus	BYMV resistance	Jones *et al.* unpublished
Cicer arietinum		
(Kabuli)	Herbicide tolerance*	Barton *et al.* unpublished
(desi)	Herbicide tolerance* + increased seed methionine	Molvig *et al.* unpublished
Lens culinaris	Herbicide tolerance*	Barton *et al.* unpublished

*expression of trait not yet confirmed in glasshouse or field trials

Herbicide tolerance

Weed control in modern agriculture depends largely on the use of chemical herbicides. In some situations herbicides, which are selectively toxic to either monocots or dicots, can be used to preferentially kill weeds of one type while leaving a crop of the other type unaffected. Genetic engineering procedures have been used to confer tolerance to herbicides such as glyphosate, glufosinate, bromoxynil, 2,4-D, sulfonylureas and atrazine on crop and pasture plants, thus allowing non-selective herbicides to be used selectively on weeds in a transgenic crop. More than 30% of all field trials of transgenic plants have involved a herbicide resistance trait (Dale 1996). An additional feature of herbicide tolerance genes is their potential as screenable markers in breeding programs involving transgenic plants. Because of their dominant phenotype and their tight linkage to other genes transferred in the same T-DNA, herbicide tolerance genes can be used to track genes whose activity is less easily detectable, such as those conferring insect or virus resistance.

The *bar* gene, from a soil bacterium, *Streptomyces hygroscopicus* (Thompson *et al.* 1987), encodes the enzyme phosphinothricin acetyltransferase (PAT) which confers resistance to glufosinate ammonium (active constituent of the herbicides, Basta^{TM} and Liberty^{TM}). The *bar* gene has been widely used as a selectable marker (DeBlock *et al.* 1987) and has been invaluable in the development of transformation technology for grain legumes. It has been inserted either alone or in combination with other genes into lupins, peas, chickpeas and lentils (Table 1).

Field trials of transgenic, glufosinate tolerant lupins and peas, derived from commercial cultivars, were conducted in Australia in 1995, 1996 and 1997. Because glufosinate is not yet registered for broad acre farming of pulses, field application rates of the herbicide were predicted from glasshouse studies. Spray results showed that transgenic lupins and peas containing the *bar* gene tolerated application rates well in excess of those required to kill non-transgenic parental lines (Fig 1). If glufosinate is registered for use on lupins, the commercial release of a Basta tolerant cultivar of narrow leaf lupin could be expected in 2001.

Fig 1.Plants of transgenic pea lines (Laura 1 and Laura 2) carrying the BAR gene tolerated an application rate of 800 g/ha glufosinate. Plants of the non-transgenic parental line, Laura, were killed by the herbicide. Photographed 20 days after spray application.

Transgenic   Laura   Transgenic
Laura 1   Laura 2

INSECT RESISTANCE

The first transgenic plants with resistance to insects contained genes for insecticidal proteins called δ-endotoxins from the soil microorganism, *Bacillus thuringiensis* (Bt). Bt-protected cotton, potato and corn were introduced into the market place in 1996 (Estruch *et al.* 1997). *B. thuringiensis* has been a rich source of insecticidal genes (Kociel *et al.* 1993) and, until recently, all field tested insecticidal plants had made use of the genes for Bt δ-endotoxins to control insect pests. A number of other genes encoding insecticidal proteins such as proteinase inhibitors (Hilder *et al.* 1987, Ryan 1990), α-amylase inhibitor (Huesing *et al.* 1991, Ishimoto and Kitamura 1989), chitinases (Ding 1995) and lectins (Chrispeels and Raikhel 1991) have been proposed as potential insect resistance genes.

The development of two Bruchid beetle pests of stored grain, *Callosobruchus maculatus* (cowpea weevil) and *C. chinensis* (azuki bean weevil) was inhibited by bean α-amylase inhibitor (αAI) in an artificial diet (Ishimoto and Kitamura 1989). A modified α-amylase inhibitor gene *(αai)* from the common bean (*Phaseolus vulgaris*) was transferred to peas (Schroeder *et al.* 1995). The levels of αAI in seeds of the transgenic peas ranged up to 1.27% (w/w) of seed meal. Seeds expressing αAI were assayed for resistance to infestation with either cowpea or azuki bean weevils (Shade *et al.* 1994). A level of 0.15% (w/w) of αAI killed all azuki bean weevils, while a higher level of 0.77% (w/w) of αAI was required to confer resistance to cowpea weevil.

A related Bruchid, the pea weevil (*Bruchus pisorum*), is the major pest of pea crops in Australia (Hardie *et al.* 1995, Ali *et al.* 1994). There is no natural resistance to pea weevil in *P. sativum* although some resistance is found in *P. fulvum* (Hardie *et al.* 1995). In glasshouse bioassays, transgenic pea seeds containing αAI (0.77% w/w) caused total inhibition of pea weevil development (Schroeder *et al.* 1995). In order to test for any anti-nutritional effects of αAI in animals, flour of αAI transgenic pea seeds was compared with control pea flour in rat feeding experiments by the late Dr Eggum at the National Institute for Animal Science in Denmark. No differences were found between true digestibility, biological value or digestible energy of diets containing flour of the transgenic line or of the control parent line.

A pea line homozygous for the *αai* gene (5th generation) was tested for resistance to natural challenge with pea weevil under field conditions in 1996. Equal initial infestation of control and transgenic seed of approximately 70 % was recorded. However, in control plants, adult pea weevils emerged from 99.6 % of infested seeds while in αAI transgenic seeds 0.25 % of infested seeds produced an adult pea weevil (Fig 2). Thus the *αai* gene shows great potential for weevil control in peas.

Fig.2. Level of damage due to pea weevil in seeds of a non-trangenic pea line, and a transgenic pea line carrying the α-amylase inhibitor gene.

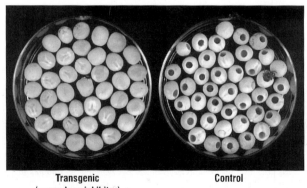

Transgenic   Control
(α-amylase inhibitor)

NUTRITIONAL ENHANCEMENT

Pulses are a high quality source of protein and energy in formulated animal feeds, where their amino acid profile complements that of cereals (Tabe *et al.* 1993). Despite a relatively high content of lysine, grain legumes, including peas, lupins and chickpeas, are deficient in the sulfur amino acids methionine and cysteine when compared to animal nutritional requirements (Waddel 1958). A number of genes have been identified which encode proteins with unusually high contents of methionine and cysteine. Amongst them is a gene from Brazil nut (*Bertholletia excelsa* H.B.K) which codes for a 2S albumin containing 18% methionine plus 8% cysteine (Altenbach *et al.* 1992), another from sunflower (*Helianthus annuus*) codes for a sunflower seed albumin (SSA) which contains 16% methionine plus 8% cysteine (Kortt *et al.* 1991). Genes encoding the Brazil nut protein have been transferred into soybeans (Townsend and Thomas 1994), narbon bean (Saalbach *et al.* 1994, 1995) and the common bean (Aragao *et al.* 1996) with the aim of improving the nutritive value of the pulse seeds. However, the use of the Brazil nut protein for nutritional enhancement of seed protein is now in question, at least for human use, because of its recently-demonstrated allergenicity (Nordlee *et al.* 1996).

A gene encoding the sunflower seed albumin, SSA, has been transferred to lupin, pea and chickpea (Table 1). This resulted in seed specific expression and the SSA protein accumulated to levels of 2-5% of extractable protein in the mature seeds of transgenic plants. One transgenic line of lupins expressing SSA at approximately 5% of extractable seed protein, showed an increase of 90% in methionine content compared with the non-transgenic parent line. Rat feeding trials demonstrated that the transgenic lupins had significantly increased nutritive value compared to control lupins, as indicated by increases in net protein utilisation, biological value and live weight gains of rats fed the transgenic seed (Molvig *et al.* 1997). Transgenic lupin seed will be used in a poultry feeding trial during 1997.

It has previously been demonstrated that wool growth in sheep can be significantly increased by supplementation with methionine, either by post-ruminal infusion or by feeding rumen-protected methionine. Since SSA has been found to be rumen stable, it is expected that the SSA-containing lupins will provide an effective sulfur amino acid supplement for sheep in summer, to maintain the quality of wool when pasture is poor. Commercial release of a transgenic lupin line with enhanced sulfur amino acid content is expected in 2002.

In peas, efforts to increase the sulfur amino acid content by traditional breeding were unsuccessful because the levels of the two major seed protein fractions, which contain sulfur amino acids, were negatively correlated (Schroeder 1982, Schroeder and Brown 1984). This would make the pea an obvious target for the modification of seed protein quality by genetic engineering. In seeds of transgenic peas and transgenic chickpeas containing a chimeric *ssa* gene, the SSA protein accumulated to between 2 and 5% of extractable protein, accompanied by an increase in seed content of methionine and cysteine. These transgenic seeds are currently being characterised.

FUNGAL RESISTANCE

Gene transfer strategies aimed at providing generalised resistance to fungal infection have been limited by our knowledge of host-pathogen interactions and by the availability of suitable genes to control infection. Approaches to fungal control through gene transfer include attempts to limit fungal invasion using expression of genes encoding enzymes such as chitinases, glucanases and peroxidases or a polygalacturonase-inhibiting protein (PGIG). Other approaches aim to enhance the ability of the plant to produce natural fungal control agents such as phytoalexins, specific pathogenesis-related (PR) proteins, or anti-microbial peptides (Shah *et al.* 1995).

In cool season grain legume crops there is a great need for the introduction of resistance to combat the effects of fungal pathogens of both roots and leaves. The most serious and widespread fungal pathogens of grain legumes in Australia include the Ascochyta complex (A. *pisi* and A. *pinodes* in peas, A. *faba* in faba beans, A. *rabiei* in chickpeas), the Fusarium wilts (*Fusarium oxysporum, F.solani, F sp pisi* in peas and faba beans), *Botrytis cinerea* in chickpeas and lupins, and *Rhizoctonia species, Colletotrichum gleosporoides/acutatum (anthracnose)* and *Pleiochaeta setosa* in lupins. A recent, serious outbreak of anthracnose in lupins in Western Australia has served to focus local attention on fungal diseases. Strategies being adopted (Barton, unpublished) include the transfer of genes for specific isoforms of peroxidases from

Stylosanthus humilus which are produced upon fungal infection (Harrison 1995). Genes encoding phenylalanine ammonia lyase (PAL) or antimicrobial proteins will also be assessed for the ability to provide generalised resistance to fungal diseases including *C. gleosporoides* in lupins (Barton, unpublished).

In peas, attention is focussed on resistance to Ascochyta blight, the major fungal disease of peas in Australia (Wroth, 1995). Genes encoding an anti-fungal peptide from radish (Terras *et al.* 1995), or a phytoalexin-synthesising enzyme from bean (Hain *et al.* 1993), have been transferred to peas, and transgenic plants are being evaluated for resistance to *Mycosphaerella pinodes*, the major causative agent of Ascochyta blight (K. Bateman and H. Schroeder, unpublished).

VIRAL RESISTANCE

Transformation strategies that have been shown to limit the establishment of viruses in other plants have been adopted for the development of virus resistant legumes (Di et al. 1996). In *L. angustifolius*, cucumber mosaic virus (CMV) and bean yellow mosaic virus (BYMV) are the major causes of viral disease (Jones and McLean 1989, Mathews *et al.* 1994) whereas BYMV is the major viral pathogen of L. luteus since natural resistance to CMV has been identified (Jones and Latham 1996). CMV coat protein and defective replicase genes were transferred to tobacco and tested for their ability to protect transgenic plants against Western Australian isolates of CMV subgroup 2 viruses from lupins and CMV subgroup 1 from banana and subterranean clover. Whilst it was effective against the subgroup 1 isolates of CMV 1, these genes were less effective against lupin isolates of CMV (Singh *et al.* 1997). A range of constructs derived from genes of the coat protein, viral replicase (Nib) and protease/Vpg (Nia) from aggressive strains of BYMV isolated from lupins have been transformed into *L.angustifolius* (Somsap *et al.* 1996) and *L. luteus* (Li *et al.* in preparation). Transgenic *L. angustifolius* and *L.luteus* lines (T2-T4 generations), containing BYMV-derived genes, are being tested in glasshouse and screenhouse trials for possible resistance to BYMV (Jones, unpublished).

The major viral pathogens of peas include alfalfa mosaic virus (AMV) and pea seed-borne mosaic virus (PSbMV). A gene encoding the coat protein of AMV has been successfully introduced into a number of pea cultivars and the relationship between transgene expression and virus resistance is currently being examined. Translatable and untranslatable genes encoding the coat protein of PSbMV genes have been introduced into peas and clonally propagated R_0 plants from many independently derived transgenic lines are being screened for virus resistance.

CONCLUSION

The application of gene technology for the improvement of cool season legumes looks promising. Small-scale field trial results for lupins with herbicide tolerance and improved methionine levels in seed protein, as well as peas with resistance to pea weevil, show that the transgenic plants perform well with respect to expression of the new genes. Further field studies are required to evaluate these plants for agronomic characters in a range of environments.

References

Ali, S.M., Sharma, B. and Ambrose, M.J. 1994 *Euphytica* 73:115-126.

Altenbach, S.B., Kuo, C.C., Staraci, L.C, Pearson, K.W., Wainwright, A.,Georgescu, A. and Townsend, J. 1992 *Plant Molecular Biology* 18:235-245.

Aragao, F.J.L., Barros, L.M.G., Brasileiro, A.C.M., Ribeiro, S.G., Smith, F.D., Sanford, J.C., Faria, J.C. and Rech, E.L. 1996 *Theoretical and Applied Genetics* 93:142-150.

Atkins, C.A., Smith, P.M.C. and Pigeaire, A. 1994 In *Proceedings of the First Australian Lupin Technical Symposium*, (Eds M Dracup and J Palta). Department of Agriculture Western Australia pp.119-122.

Atkins, C.A., Smith, P.M.C., Gupta, S., Jones, M.J.K. and Caligari, P.D.S. 1997(b) In *Lupins their Products and Usage*. (Eds Gladstone, J. Atkins, C.A. and Hamblin, J.) CABI, Wallingford, UK. (in press).

Atkins, C.A. and Smith, P.M.C. 1997a In *Biological Fixation of Nitrogen for Ecology and Sustainable Agriculture* pp.283-304. (Eds A. Legocki, H. Botag and A. Puehler) Springer Berlin Heidelberg

Bean, S.J., Gooding, P.S., Mullineaux, P.M. and Davies, D.R. 1997 *Plant Cell Reports* 16 (in press.)

Chowira, G.M., Akella, V., Fuerst, P.E. and Lurquin, P.F. 1996 *Molecular Biotechnology* 5:85-96.

Chrispeels, M.J. and Raikhel, N.V. 1991 Plant Cell 3:1-19.

Cowling, W.A. 1994 In *Proceedings of the First Australian Lupin Technical Symposium*, (Eds M. Dracup and J. Palta) Department of Agriculture Western Australia pp.119-139.

Dale, P.J. 1995 *Trends in Biotechnology* 13:398-402.

DeBlock, M., Bottermann, J., Vandewiele, M., Dockx, J., Thoen, C., Gossele, U. 1987 *EMBO Journal* 6:539-548.

Di, R., Purcell,V., Collins, G.B.and Ghabrial, A.S. 1996 *Plant Cell Reports* 15:746-750.

Ding, X. 1995 PhD Dissertation, Kansas State University, Manhattan.

Estruch, J.J., Carzzi, N.B., Desai, N., Duck, N.B., Warren, G.W. and Koziel, M.G. 1997 *Nature Biotechnology* 15, 137-141.

Finnegan, J. and McElroy, D. 1994 *BioTechnology* 12:883-888

Fishhoff, D.A., Bowdish, K.S., Perlak, F.J., Marrone, P.G., Mc Cormick, S.M., Niedermaeyer, J.G., Dean, D.A., Kusano-Kretzmer, K., Mayer, E.J., Rochester, D.E., Roger, S.G., Fraley, R.T. 1987 *BioTechnology* 5:807-813.

Fontana, G.S., Santini, L., Careho, S., Frugis, G., Mariotti, D. 1993 *Plant Cell Reports* 12:194-198.

Grant, J.E., Cooper, P.A., McAra, A.E., Frew, T.J. 1995 *Plant Cell Reports* 15:254-258.

Hain, R,. Reif, H.J., Krause, E., Langebartels, R., Kindl, H., Vornam, B., Wiese, W., Schmelzer, E., Schrier, P.H., Stocker, R.H. and Stenzel, K. 1993 *Nature* 361:153-156.

Hardie, D.C., Baker, G.J. and Marshall, D.R. 1995 *Euphytica* 84:155-161.

Harrison, S., Curtis, M.D., McIntyre, C.L., MacLean, D. and Manners, J.M. 1995 *Molecular Microbe Interactions* 8: 398-406.

Hiei, Y., Ohta, S., Komari, T. and Kumashiro, T. 1994 *Plant Journal* 6:271-282.

Hilder, V., Gatehouse, A., Sheerman, S., Barker, R. and Boulter, D. 1987 *Nature* 330:160-163.

Hinchee, M.A.V., Connor-Ward, D.V., Newell, C.A., McDonald, R.E., Sato, S.J., Gasser, C.S., Fishhoff, D.A., Re, D.B., Fraley, R.T. and Horsch, R.B. 1988 *Bio Technology* 6:915-922.

Hoffman, K., Somsap, V., Li, H. and Jones, M.G.K. 1996 *Factors influencing the Agrobacterium-mediated transformation of Lupinus angustifolius.* 7th Annual Combined Biological Sciences Meeting, Perth (16 August), p.5.

Huesing, J.E., Shade, R.E., Chrispeels, M.J. and Murdock, L.L. 1991 *Plant Physiology* 96:993-996.

Ishimoto, M. and Kitamura, K. 1989 *Applied Entomology and Zoology* 24:281-286.

Jones, R.A.C. and Latham, L.J. 1996 Natural resistance to cucumber mosaic virus in lupin species. *Annals of Applied Biology* 129:523-542.

Jones, R.A.C, and McLean, G.D. 1989 *Annals of Applied Biology* 114:609-637.

Kim, J.W. and Minamikawa, T. 1996 *Plant Science* 117:131-138.

Kociel, M.G., Carozzi, N.B., Currier, T C., Warren, G.W. and Evola, S.V. 1993 *Biotechnology and Genetic Engineering Review* 11:171-228.

Kortt, A.A., Caldwell, J.B., Lilley, G.G. and Higgins, T.J.V. 1991 *European Journal of Biochemistry* 195:329-334.

Li, H., Somsap, V., Wylie, S.J., Hoffman, K. and Jones, M.G.K. 1996 *Genetic transformation of yellow lupin (Lupinus luteus).* 7th Annual Combined Biological Sciences Meeting, Perth (Aug.1996) p.83.

Mathews, A., Wylie, S., Jones, R.A.C. and Jones, M.G.K. 1994 *Proceedings of the First Australian Lupin Technical Symposium*, (Eds M Dracup and J Palta) Department of Agriculture Western Australia pp.119-122.

McCabe, D.E., Swain, W.F., Martinell, B.J. and Christou, P. 1988 *Bio Technology* 6 August.

Molvig, L. 1994 In *Proceedings of the First Australian Lupin Technical Symposium*, (Eds M. Dracup and J. Palta). Department Agricultureof Western Australia pp.119-122.

Molvig, L., Tabe, L.M., Eggum, B.O., Moore, A.E., Craig, S., Spencer, D. and Higgins, T.J.V. 1997 *Proceedings National Academy Science*, USA. (in press).

Nordlee, M.S., Julie, A., Steve, L., Taylor, P.H.D., Jeffry, A., Townsend, B.S., Laurie, A., Thomas, B.S. and Bush, R.K. 1996 *New England Journal of Medicine* 334:688-692.

Petterson, D.S. and Mackintosh, J.B. 1994 *The Chemical composition and nutritive value of Australian grian legumes.* Grains R and D Corporation, Canberra.

Pigeaire, A., Abernethy, D., Smith, P.M.C., Simpson, K, Fletcher, N., Lu, C-Y., Atkins, C.A. and Cornish, E. 1997 *Molecular Breeding.* (in press).

Russel, D.R. 1993 *Plant Cell Reports* 12:165-169.

Ryan, C. 1990 *Annual Review of Phytopathology* 28:425-449.

Saalbach, I., Pickardt, T., Machemehl, F., Saalbach, G., Schieder, O., Muenz, K. 1994 *Molecular and General Genetics* 242: 226-236.

Saalbach, I., Pickardt, T., Waddel, D.R., Hillmer, S., Schieder, O. and Muntz, K. 1995 *Euphytica* 85:181-192.

Schroeder, H.E. 1982 *Journal of Science and Agriculture* 33:623-633.

Schroeder, H.E. and Brown, A.D.H. 1984 *Theoretical and Applied Genetics* 68:101-107.

Schroeder, H.E., Gollasch, S., Moore, A., Tabe, L.M., Craig, S., Hardie, D.C., Chrispeels, M.J., Spencer, D. and Higgins, T.J.V. 1995 *Plant Physiology* 107:1233-1239.

Schroeder, H.E., Gollasch, S., Tabe, L.M. and Higgins, T.J.V. 1994 *Pisum Genetics* 26:1-5

Schroeder, H.E., Schotz, A.H., Wardley-Richardson, T., Spencer, D. and Higgins, T.J.V. 1993 *Plant Physiology* 101: 751-57.

Shade, R.E., Schroeder, H.E., Pueyo, J.J., Tabe, L.M., Murdock, L.L., Higgins, T.J.V. and Chrispeels, M.J. 1994 *BioTechnology* 12:793-796.

Shah, D.M., Rommens, C.M.T. and Beachy, R.N. 1995 *Trends in Biotechnology* 13:363-368

Siddique, K.H.M. and Sykes, J. 1997 *Australian Journal of Experimental Agriculture* 37:103-111.

Singh, Z., Jones, M.G.K. and Jones, R.A.C. 1998 *Australian Journal of Experimental Agriculture* (in press).

Somsap, V., Li H., Li, D., Hoffman, K., Mathews, A. and Jones, M.G.K. 1995 *Engineering lupins for virus resistance*. 35th Annual Meeting, Australian Society of Plant Physiology and New Zealand Society of Plant Physiology, Sydney, p.25.

Tabe, L.M., Higgins. C.M., McNabb, W.C., and Higgins, T.J.V. 1993 *Genetica* 90:181-200.

Terras, F.R.G., Eggermont, K., Kovaleva, V., Raikhel, N.V., Osborn, R.W., Kester, A., Rees, S.B. Torrekens, S., Van Leuven, F., Vanderleyden, J., Cammue, B.P.A. and Broekaert, W.F. 1995 *Plant Cell* 7: 573-88.

Thompson, C.J., Movva, N.R., Tichard, R., Crameris, R., Davies, J.E., Lauwereys, M. 1987 *EMBO Journal*. 6:2519-28.

Townsend, J.A. and Thomas, L.A. 1994 *Journal Cell Biochemistry* Supplement 18A:78.

Waddel, J. 1958 In *Processed Plant Foodstuffs*, (Ed Altschul, A.M.). Academic Press Inc New York pp.307-351.

Walden, R. and Wingender, R. 1995 *Trends in Biotechnology* 12: 883-888.

Wroth, J.M. 1995 PhD Dissertation, Faculty of Agriculture, University of Western Australia.

Zambryski, P., Joos, H., Genetyello, C., Leemands, J. Van Montagu, M. and Schell, J. 1983 *EMBO J*. 2:2143-2150.

How similar are the genomes of the cool season food legumes?

N.F. Weeden[1], T.H.N. Ellis[2], G.M. Timmerman-Vaughan[3], C.J. Simon[4], A.M. Torres[5], and B. Wolko[6]

1 Department of Horticultural Sciences, Cornell University, Geneva, NY 14456 USA; 2 John Innes Institute, Norwich NR4 7UH, England; 3 Crop and Foods CRI, Lincoln, New Zealand; 4 USDA, 59 Johnson Hall, Washington State University, Pullman, WA 99164 USA; 5 C.I.D.A., Departamento de Mejora y Agronomia, Aptdo. 4240, 14080 Cordoba, Spain; 6 Instytut Genetyki Roslin, ul. Strzeszynska 34, 60-479 Poznan, Poland

Abstract:

The genomes of pea, field-pea, lentil, faba bean, chickpea and lupin are compared in relation to size, level of polymorphism and arrangement of genes on the respective linkage maps. Lupin is quite distinct, as its distant phylogenetic relationship would indicate. Chickpea possesses a significantly smaller genome than the members of the Viceae and appears to display less polymorphism at the molecular level. However, examination of the linkage maps available for pea, lentil, faba bean and chickpea indicate that many loci linked in pea also exhibit linkage in the other pulse crops. The extent of conservation of linkage arrangements may be as much as 40% of the genome, suggesting that gene locations in one pulse crop are often easily transferable to the other crops and permitting a generalized linkage map to be constructed for these pulse crops.

INTRODUCTION

The exciting finding that linkage relationships are highly conserved within the grass family (Whitkus et al., 1992; Ahn et al., 1993; Devos et al., 1994; Yu et al., 1995) has stimulated researchers to consider the possibility of using linkage maps developed in one crop to predict linkages in related crops. Some evidence for linkage conservation has also been generated in the Solanaceae, particularly between potato and tomato (Bonierbale et al., 1988; Tanksley et al., 1992), and in the Brassicaceae (McGrath and Quiros, 1991; Cheung et al., 1997)

Although the family Fabaceae is a much older and a more diverse than the Poaceae, it is likely that genome structure has been conserved among members of the same tribe or even across wider taxonomic distances. In the initial paper comparing linkage groups in pea and lentil (both members of the tribe Viceae), about 40% of the loci appeared to exhibit similar linkage relationships (Weeden et al., 1992). Similarly, Boutin et al. (1995) determined that *Phaseolus* and *Vigna*, sister genera in the Phaseoleae, displayed a very similar arrangement of genes in linkage groups. When these researchers extended their comparison to soybean (a species in a closely related tribe, the Glycineae) linkage conservation was much less evident. Comparisons between the Viceae and Phaseoleae have revealed only a few possible cases of conserved linkages, and these are very short (Garvin and Weeden, 1994; Timmerman-Vaughan et al, 1996).

Linkage maps have now been developed in chickpea (Gaur and Slinkard, 1990; Simon and Muehlbauer, 1997), faba bean (Torres et al. 1993a; Torres et al., 1995; Satovic et al., 1996), sweet clover (Weeden and Cargnoni, 1994), diploid alfalfa (Brummer et al., 1993; Kiss et al. 1993; Kidwell et al., 1993), peanut (Halward et al., 1993), lupin (Wolko and Weeden, 1995; Brien et al., 1997). In addition, many new markers have been mapped in pea, soybean, and common bean. The intent of this paper is to analyze much of this information, particularly that available for pea, lentil, faba bean, chickpea, and lupin in order to evaluate the possibility of constructing a generalized linkage map for the pulse crops and identify markers that will be particularly useful for comparative mapping in legumes.

R. Knight (ed.), Linking Research and Marketing Opportunities for Pulses in the 21st Century, 397–410.
© 2000 *Kluwer Academic Publishers. Printed in the Netherlands.*

GENERAL PROPERTIES OF THE GENOMES OF COOL SEASON FOOD LEGUMES

The five cool season legume crops to be examined include four placed in the Viceae (pea, lentil, faba bean) or the closely related tribe Cicereae (chickpea). The fifth (lupin) has been assigned to the Genisteae based on morphological and karyological characters (Polhill, 1976; Bisby, 1981). The Viceae and Cicereae are thought to have been derived from a galegeate ancestor and belong to that section of the Papilionoideae that lack the inverted repeat in the chloroplast DNA (Palmer et al., 1988). In contrast, the lineage leading to *Lupinus* is believed to have diverged from the primitive Sophoreae early in the evolution of the Papilionoideae (Figure 1). *Lupinus*, is unusual for some morphological characters (e.g. the digitate leaves), its peculiar root nodules (Corby, 1981), in the serological reactivity of its proteins (Cristofolini and Feoli-Chiapella, 1984; Cristofolini, 1989) and the quinolizidine alkaloid content (Wink, 1993). For these reasons it has been segregated into the mongeneric subtribe Lupininae (Bisby, 1981). The distinctness of *Lupinus* was also confirmed by molecular studies of inverted repeat sequences (Palmer et al., 1987, 1988) and by comparison of chloroplast and nuclear gene sequences (Kass and Wink, 1995a, b).

The wide phylogenetic distance between the initial four crops and lupin suggest that the latter will be very different from the others in terms of genome structure. Indeed, if we examine a list of features generally used to characterize a genome, we observe that lupin is distinctive in several ways (table 1). It is the only clearly polyploid crop of the five, with n ranging mostly between 20 and 26. Lupin species also express a considerable number of duplicate loci for relatively conserved isozymes (Wolko and Weeden, 1990). In contrast, the other pulse crops have uniformly low haploid chromosome counts, with no evidence for changes in n within a genus (Goldblatt, 1981), and have little evidence for duplication of the genome. Lupin and chickpea have significantly smaller DNA content per haploid complement than the members of the Viceae. This last observation could indicate that comparisons between chickpea and the Viceae may prove unrewarding. However, as will become evident, chickpea is highly comparable to the pea and lentil with regard to linkage arrangements. Extrapolation to lupin still remains of questionable value.

Table 1. Comparison of the genomes of the pulse crops

Character	Field Pea	Lentil	Grass pea	Faba bean	Chickpea	Lupin
Chromosome no. (n)	7	7	7	6	8	16 to 26
DNA content (pg/n)	4.9	4.6	4.8	11.9	1.0	0.6 to 0.9
Marker evidence for duplication of genetic material	45S ribosomal	none	none	none	aldolase, esterase	polyploid
Molecular marker diversity within species	high	moderate	few data	moderate	low	low
Known translocations	many	several	no data	few (induced)	few	few

THE PEA LINKAGE MAP

The most recent version of the pea linkage map represents a composite of many crosses. Mapping of morphological and physiological mutants continues to be performed in several laboratories, particularly in Novosibirsk, Hobart, Poznan, Norwich, and Geneva (NY). Molecular marker maps have been developed in several laboratories, usually with a heavy emphasis on DNA markers (Weeden and Wolko, 1990; Ellis et al. 1992, Dirlewanger et al. 1994; Timmerman-Vaughan et al., 1996). Unfortunately, the selection of DNA markers used, differed between laboratories making it difficult to compare maps initially. Agreement on a set of standard markers (Weeden et al., 1993a) and exchange of probes and sequence information has permitted homologous linkage groups to be identified and has led to the composite map shown in Figure 2.

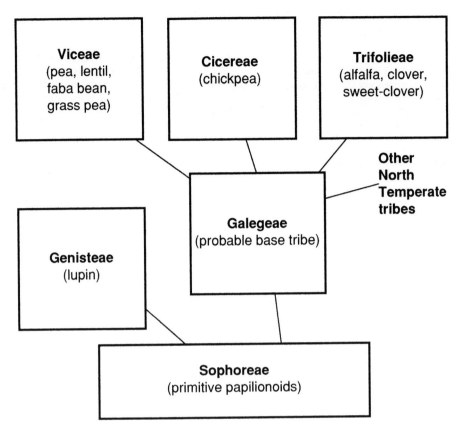

Figure 1. General phylogenetic relationships of the papilionoid tribes containing cool season food legume species.

Genetic Map of Garden Pea

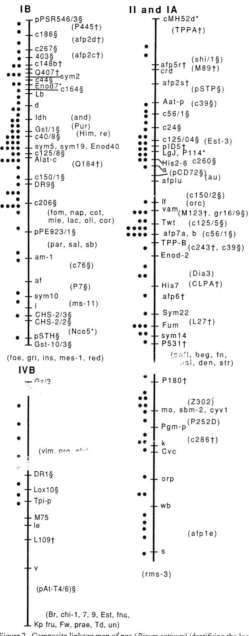

Figure 2. Composite linkage map of pea (*Pisum sativum*) identifying the locations of many of the standard and markers useful for comparative mapping studies. The number of lines (thickness of the chromosome) reflects the relative strength of the experimental data supporting the map in specific regions. The sources of the DNA markers have been designated by symbols (*= USA; ?= United Kingdom; §= New Zealand).

Genetic Map of Garden Pea

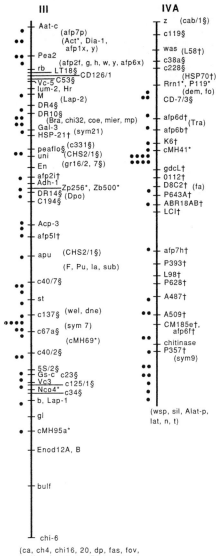

III

- Aat-c
- (afp7p)
- (Act*, Dia-1, afp1x, y)
- Pea2
- (afp2f, g, h, w, y, afp6x)
- rb LT18§ CD126/1
- Vc-5 C53§
- lum-2, Hr
- M (Lap-2)
- DR4§
- DR10§
- (Bra, chi32, coe, mier, mp)
- Gal-3
- HSP-21† (sym21)
- peaflo§ (c331§)
- uni (CHS2/1§)
- En (gr16/2, 7§)
- afp2i†
- Adh-1
- DR14§ Zp256*, Zb500*
- C194§ (Dpo)
- Acp-3
- afp5l†
- apu (CHS2/1§)
- (F, Pu, la, sub)
- c40/7§
- st
- c137§ (wel, dne)
- c67a§ (sym 7)
- (cMH69*)
- c40/2§
- 5S/2§
- Gs-c c23§
- Vc3 c125/1§
- Nco4* c34§
- b, Lap-1
- gl
- cMH95a*
- Enod12A, B
- bulf
- chi-6

(ca, ch4, chi16, 20, dp, fas, fov, fr, Kpa, l, och, op, Rf, sg-2, twp, wil, yp)

IVA

- z (cab/1§)
- c119§
- was (L58†)
- c38a§
- c228§ (HSP70†)
- Rrn1*, P119* (dem, fo)
- CD-7/3§
- afp6d† (Tra)
- afp6b†
- K6†
- cMH41*
- gdcL†
- 0112†
- D8C2† (fa)
- P643A†
- ABR18AB†
- LCl†
- afp7h†
- P393†
- L98†
- P628†
- A487†
- A509†
- CM185e†, afp6f†
- chitinase
- P357† (sym9)

(wsp, sil, Alat-p, lat, n, t)

Genetic Map of Garden Pea

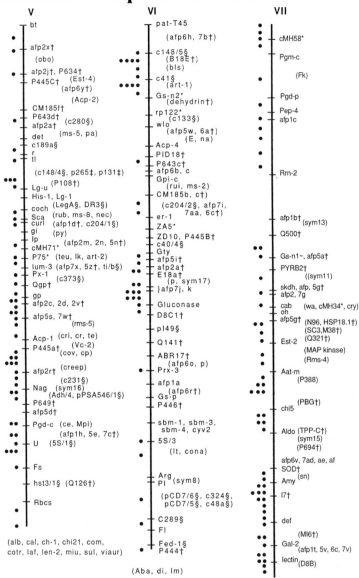

V

- bt
- afp2x†
- (obo)
- afp2j†, P634†
- P445C† (Est-4)
- (afp6y†)
- (Acp-2)
- CM185f†
- P643d†
- afp2a† (c280§)
- det (ms-5, pa)
- c189a§
- r
- tl
- (c148/4§, p265‡, p131‡)
- Lg-u (P108†)
- His-1, Lg-1
- coch (LegA§, DR3§)
- Sca (rub, ms-8, nec)
- curl (afp1d†, c204/1§)
- gi (py)
- lp
- cMH71* (afp2m, 2n, 5n†)
- P75* (teu, lk, art-2)
- lum-3 (afp7x, 5z†, ti/b§)
- Px-1 (c373§)
- Qgp†
- gp
- afp2c, 2d, 2v†
- afp5s, 7w† (rms-5)
- Acp-1 (cri, cr, te)
- P445a† (Vc-2) (cov, cp)
- afp2r† (creep)
- (c231§)
- Nag (sym16)
- (Adh/4, pPSA546/1§)
- P649†
- afp5d†
- Pgd-c (ce, Mpi)
- (afp1h, 5e, 7c†)
- U (5S/1§)
- Fs
- hst3/1§ (Q126†)
- Rbcs

(alb, cal, ch-1, chi21, com, cotr, laf, len-2, miu, sul, viaur)

VI

- pat-T45
- (afp6h, 7b†)
- c148/5§ (B18E†)
- (bls)
- c41§ (art-1)
- Gs-n2* (dehydrin†)
- rp122* (c133§)
- wlo (afp5w, 6a†)
- (E, na)
- Acp-4
- PID18†
- P643c†
- afp6b, c
- Gpi-c (rui, ms-2)
- CM185b, c†)
- (c204/2§, afp7i, 7aa, 6c†)
- er-1
- ZA5*
- ZD10, P445B†
- c40/4§
- Gty
- afp5i†
- afp2a†
- E18a† (p, sym17)
- }afp7j, k
- Gluconase
- D8C1†
- pl49§
- Q141†
- ABR17† (afp6o, p)
- Prx-3
- afp1a (afp6r†)
- Gs-p
- P446†
- sbm-1, sbm-3, sbm-4, cyv2
- 5S/3 (lt, cona)
- Arg
- PI (sym8)
- (pCD7/6§, c324§, pCD7/5§, c48a§)
- C289§
- FI
- Fed-1§
- P444†

(Aba, di, lm)

VII

- cMH58*
- Pgm-c
- (Fk)
- Pgd-p
- Pep-4
- afp1c
- Rrn-2
- afp1b† (sym13)
- Q500†
- Gs-n1~, afp5a†
- PYRB2† ((sym11)
- skdh, afp, 5g†
- afp2, 7g
- cab (wa, cMH34*, cry)
- oh
- afp5g† (N96, HSP18.1†)
- (SC3, M38†)
- Est-2 (Q321†)
- (MAP kinase)
- (Rms-4)
- Aat-m (P388)
- chi5 (PBG†)
- Aldo (TPP-C†)
- (sym15)
- (P694†)
- afp6v, 7ad, ae, af
- SOD†
- Amy (sn)
- l7†
- def
- (Ml6†)
- Gal-2 (afp1t, 5v, 6c, 7v)
- lectin (D8B)

This map is arranged in seven linkage groups, reflecting the progress made in the past five years at which time the 'classical' seven chromosomes of Lamprecht and Blixt (Blixt, 1972) had been broken into 10 to 12 linkage groups as a result of molecular marker data (Ellis et al., 1993; Weeden et al., 1993b). However even today there is debate about the structure of certain linkage groups. When comparing the pea map with those for other species, it is important to know which portions of the pea linkage map are well supported by experimental data and which have been identified as linked in only a few laboratories or in a limited number of crosses. The experimental support for the linkage arrangement is conveyed by the thickness of the line in that region of the map. As can be seen, several regions of the linkage map are only weakly supported by experimental evidence. In comparisons of this map with those of other species it is important to remember that the current pea linkage map has not been rigorously tested in all regions. Indeed, many translocations have been identified within the species (Lamm and Miravalle, 1959) and certain regions display either low marker density or high recombination frequency, making them particularly difficult to saturate with markers.

A description of each linkage group in pea may help identify these groups or portions of groups in other taxa. Linkage group IB (what remains of the classical linkage group I after transfer of the upper portion to linkage group II) has two important morphological markers on it: D, which directs the production of red pigment in leaf axils and occasionally on other parts of the stem, and I, which is Mendel's yellow/green cotyledon gene. This linkage group also possesses the major leghemoglobin cluster of genes as well as several other genes affecting nodule development (Kozik et al., 1996). The region between D and I is not well marked, with several genes near D and several near I but no easily scored marker in the middle. Thus, relatively few mapping experiments have verified that I and D are syntenic. There are no known translocation break points between D and I, but above D there appears to be at least one region that occasionally is involved in translocations (Temnykh et al., 1995). A translocation here may be the reason that the *lf -- a -- Aat-p* segment, now on the upper portion of LG II, was originally linked to IB (Blixt, 1972).

Group II contains a number of easily scored markers but also suffers from a lack of markers in the center region. The two end segments have been confirmed by many investigators, but the fusion of the two segments has been demonstrated only in three laboratories (Paruvangada et al., 1995; Gilpin et al., in press; Ellis, T.H.N. unpublished). The important markers for inter-genus comparisons include the plastid-specific aspartate aminotransferase (*Aat-p*), *LgJ*, the histone gene cluster *His2-6*, *Enod2*, *Pgm-p*, and convicilin (*Cvc*). An important translocation break point lies just below a (Temnykh et al., 1995).

The classical linkage group III appears to have withstood the rigors of molecular marker analysis. Ellis et al. (1992) provided strong evidence for synteny of *Vc-5* and *Vc-3*, markers at opposite ends of the linkage group. The linkage group has some excellent markers, including *Vc-5*, *M. uni* (=PeaFlo), *Adh-1*, *st*, *b*, and *Lap-1*. Cytogenetic investigations have placed the centromere between *st* and *b*. However, the middle of the linkage group has failed to materialize in several molecular mapping studies. To break the linkage group between *st* and *b* would be heresy as far as classical pea genetics is concerned, because if anything was demonstrated by repeated two point crosses, it was that *st* and *b* are linked and about 25 cM apart. In certain studies, mapping with molecular markers has linked the *st* region to markers on linkage group IV (Timmerman-Vaughan et al., 1996). In these instances, weak or no linkage is observed between the *st* and *b* regions, albeit these markers were not segregating in the populations. In other studies (Marx, 1987; Polans, 1993) serrated leaf margins, controlled by the gene *Td*, appeared to be linked to markers on LG III despite being originally placed on LG IV by Lamprecht (1945, 1948). In the study by Polans, *st* and *b* displayed the expected 25 cM recombination value, but *Td* exhibited linkage with both markers at a linkage intensity of about 22 cM. These results suggest a more generalized interaction between *Td* and the *st -- b* region than expected for a 3-point cross and may indicate a peculiar relationship between linkage groups III and IV.

Linkage group IV remains somewhat of an enigma in pea mapping. Cytogenetic evidence indicates that *le* (Mendel's dwarfing gene) is not on a chromosome with a nuclear organizer region (NOR) (Folkeson, 1985). However, Polans et al. (1986) found linkage between *Rrn1* and two morphological markers (*was* and *fa*) that had been placed on the same linkage group as *le*. Assuming that *Rrn1* is the site of an NOR, these two results are contradictory and suggest that LG IV is incorrect. The absence of strong data linking the

Rrn1 region with *le*, and the preliminary evidence of linkage between markers in the center regions of LG III and LG IV indicate that this is one region of the pea linkage map that requires further study. Several cDNA probes have become available relatively recently that should prove useful for defining this linkage group in other species.

In contrast to LG IV, LG V is the success story of the last five years. The classical map had this group split on two chromosomes (5 and 7). Several studies indicated that the two halves might be syntenic (Lamm, 1951; Marx, 1969; Folkeson, 1990; Weeden and Wolko, 1990), and since 1990 many studies, particularly by researchers in Novosibirsk, Poznan, and Tasmania, have confirmed and saturated the joining regions with morphological and molecular markers. The most important markers for comparative work include *r* (Pjam425), *Lg-1*, cMH71, *Px-1*, P649, *Pgd-c* and *Rbcs*.

Linkage Group VI appears to be another group that is marker poor near its center. In a number of mapping populations the linkage between the *wlo* region and *Prx-2* or *Pl* has not been able to be confirmed despite 200 to 300 markers being mapped in the cross. The linkage group has several excellent morphological markers (*wlo*, *na*, *Arg*, *Pl*) and some very important disease resistance genes (*er-1*, *sbm-1*, *cyv-2*). Presently, the linkage group is best defined by the genes *rpl22*, *wlo*, pID18, pI49, *Prx-3*, *Gs-p*, *Pl*, and *Fed-1*, although mutants homologous with *wlo* and *Pl* probably will not be readily found in other species.

LG VII is probably best marked by isozyme loci, with four above the ribosomal array, *Rrn2*, and seven below this region. The region immediately surrounding *Rrn2* is marker poor, and it is often difficult to observe linkage between the upper and lower portions of this linkage group. Important other genes on this group include a morphological marker, *oh*, and several DNA sequences (*Gs-n1*, the Cab gene cluster, P398, P694, and the lectin gene cluster).

We have discussed the pea linkage map at some length in order to establish a basis for comparison with the other pulse crops. However, pea is not centrally positioned among the pulses and possesses several features that distinguish it from the other species. Most importantly, it contains two 45S ribosomal arrays and two satellite chromosomes, whereas all other pulses, except relatives of chickpea (Galasso et al., 1996), possess only one such array. The second ribosomal array presumably was generated by some type of gene duplication event, possibly by a series of translocations as has been shown to lead to duplication of chromosomal segments in maize (Gopinath and Burnham, 1956). If such were the case, we would expect to find a series of duplicated genes adjacent to the ribosomal arrays. As yet, no evidence for parallel linkages of duplicated genes has been reported for LG IV and LG VII.

Pea also appears to be unusual among plants in that a general linkage map has been difficult to develop. Ellis, et al. (1992) emphasized that there were among-map inconsistencies in the mapping data they generated for several crosses. Their results have been confirmed in several laboratories working with a sufficient number of markers to develop complete linkage maps. As should have been evident from the discussion of the current linkage map, there are several regions that remain difficult to resolve in specific crosses. Translocations are common in the pea germplasm, but are usually easily identified by reduced fertility in the F_1 or F_2 generations. Most of the populations now being used for mapping appear to retain high fertility and thus are not suspected of having translocations affecting the linkage results.

LENTIL

Presently, lentil is the second best mapped of the pulse crops. The chromosome number and genome size for lentil are similar to pea, but the genus is believed to represent a different evolutionary line than that leading to *Pisum* and *Lathyrus*, the two lines thought to have diverged independently from an ancestral *Vicia* (Kupicha, 1977). Thus the lentil/pea comparison should be valuable for revealing the level of genome homology throughout the Viceae.

The most recent linkage map for lentil based on markers useful for inter-generic comparison is that of Simon et al. (1993). This map represents a consensus map based on several studies, but does not include all reported linkages in lentil, some of which are contradictory or are difficult to reconcile with the map presented by Simon et al. It appears that, as has been found in pea, certain crosses in lentil display linkages that are not reproducible in other crosses. The map produced by Simon et al. is about 640 cM long and consists of 12 linkage groups, suggesting the map is about half to two thirds complete. Most of the over 100 genetic markers shown on the map have homologues mapped in pea, permitting the direct comparison of the

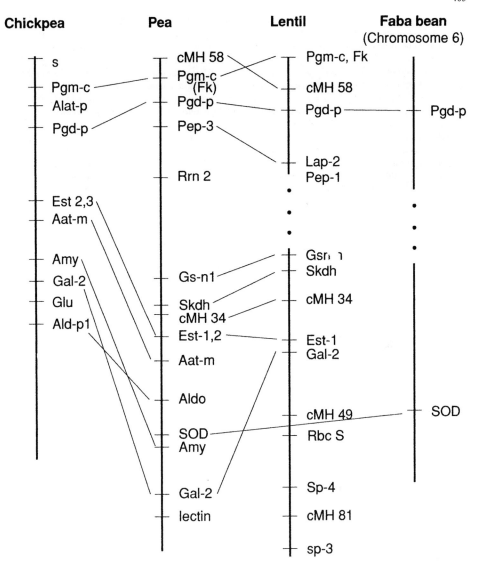

Figure 3. Comparison of linkage group VII in pea with homologous linkage groups in chickpea, lentil and faba bean.

arrangement of these genes in the two species. Weeden et al. (1992) estimated a 40% homology between the lentil and pea maps, and the revised map by Simon et al. (1993) reports a similar level of homology.

Several linkage groups identified in lentil can be immediately recognized as homologous to specific linkage groups of pea. One of the clearest parallels is that for pea linkage group VII where isozyme loci have been particularly useful for establishing homology. The upper portion of lentil VII is nearly identical to pea LG VII (Figure 3). The locus order for the lower section of VII is also very similar [*Skdh* -- cMH34 -- *Est-2* -- *Aldo* -- *Gal-2*]. Vaillancourt and Slinkard (1993) presented evidence for linkage between *Pgd-p* and *Skdh*, paralleling the pea map, except in lentil the single 45S ribosomal array is not situated between these two loci but rather lies on a different linkage group. Another segment of pea LG VII that does not appear to be present on lentil LG VII is that containing *Aat-m*, the gene encoding subunits of the mitochondrial aspartate aminotransferase. In several studies this gene displays linkage to *Me-2* and assorts independently of *Est-2* and *Aldo*. In one analysis (Tahir and Muehlbauer, 1994) the *Pgm-c* end of lentil LG VII displayed linkage with markers associated with pea LG III. This VII -- III connection has yet to be confirmed, and indeed does not appear to be present in other experiments. However, in pea the *Pgm-c* region has occasionally shown linkages to other supposed non-syntenic loci as well (Weeden unpublished, Ellis et al., 1992)

Comparison of pea and lentil LG III reveals several interesting aspects. A significant set of pea LG III markers also display linkage in lentil. The gene coding the cytosolic aspartate aminotransferase subunits (*Aat-c*) is linked to *Gal-1*, an actin sequence, and an acid phosphatase in both species. If these genes are all homologous, about half of LG III is conserved between pea and lentil. On a different lentil linkage group Havey and Muehlbauer (1989) placed two other markers (cMH69 and *Lap-1*) assigned to LGIII in pea. These two linkage groups in lentil may represent opposite ends of the same chromosome, for in pea the two groups of markers often assort independently.

The remaining linkage groups identified in lentil do not exhibit extensive homology with pea linkage groups. Four markers [*rpl22*, *Prx-3*, *Gpi-c*, and *Gs-p*] from LG VI of pea have each been placed on a separate short linkage group in lentil and appear to assort independently. The LG IV P119 -- cMH41 linkage appears in lentil, as was mentioned above, but no other LG IV markers have been mapped in lentil. In several crosses, markers located on pea LG IB and II have displayed linkage in lentil. However, the arrangement of loci clearly is different in the two species, and too few markers have been mapped in lentil to completely understand the changes involved. The 45S ribosomal array is linked to the major leghemoglobin cluster in lentil, while in pea the major leghemoglobin cluster assorts independently of either ribosomal array. *Aat-p* shows linkage with cMH52 and a seed protein gene in both species, but homology between the two seed protein loci has yet to be demonstrated.

FABA BEAN

Vicia faba belongs to what is considered the basal genus in the Viceae and thus information about the arrangement of genes on its genome should be very important for examining the evolution of the genome structure in the tribe. The faba bean possesses only six chromosomes a change from the $n = 7$ characterizing the other genera of the Viceae, but one (chromosome 1) is very large and possibly represents a fusion of two acrocentric chromosomes. The number of cytological tools available in faba bean for assigning genes and linkage groups to their respective chromosomes is limited to translocation stocks (Sjödin, 1971; Schubert and Rieger, 1991) and primary trisomics (Cabrera and Martín, 1989; Cabrera et al., 1989). To date, five of a possible six primary trisomics have been characterized by Martín and Barceló (1984), and these represent valuable tools for understanding linkage arrangements in *V. faba*.

The trisomic stocks in faba bean permit the isolation and examination of each linkage group independently, a very important capability judging from the difficulty researchers have had in pea and lentil in identifying complete linkage groups. The large size of the faba bean genome suggests that it may contain considerable amounts of repetitive DNA. Experiments involving a collection of deletion and duplication mutants in this species showed that duplication of genetic material was tolerated quite well, while deletions were not tolerated (Schubert et al. 1988). Thus, if the difference in genome size between faba bean and lentil or pea is the addition of non-coding transposon or retrotransposon DNA, the addition must have resulted from widely dispersed duplications that maintained the general arrangement/spacing of genes.

Unfortunately, relatively few markers useful for inter-generic comparisons have been placed on the faba bean linkage map, although a number of new isozyme markers have just been published (Satovic et al., 1996; Torres et al., in press). Only three chromosomes of faba bean can be matched with linkage groups in pea or lentil. Torres et al. (1993a) reported a linkage between *Prx-1* and *Acp-1*. In pea, *Px-1* is linked to *Acp-1*, *Nag*, and *Pgd-c* on LG V. The *Prx-1* -- *Nag* linkage also appears to be maintained in lentil (Muehlbauer et al., 1989), indicating a possible conserved linkage general to the Viceae. Nevertheless, in faba bean *Prx-1* has been assigned to chromosome 5 (Satovic et al., 1996), whereas the *Nag* -- *Pgd-c* group has been located on chromosome 1 (Torres et al., 1993b). These results suggest that the gene order in faba bean developed from a chromsome rearrangement occurring after faba bean diverged from the common ancestor to the three genera.

Chromosome 6 appears to be homologous with pea and lentil linkage groups VII. It possesses *Sod-1* on LG V of Satovic et al (1996) and *Pgd-p* on LG VI, both linkage groups being assigned to chromosome 6 through the use of the trisomic stocks. This result is particularly interesting because in pea *Pgd-p* and *Sod* also assort independently, and it has only been in a few experiments that the two loci have been identified as being on the same linkage group.

In a recent study in faba bean (A. M. Torres, unpublished), *Adh-2* and *Lap-1* have been assigned to the same linkage group (chromosome 2). This finding suggests yet another conservation of linkage that extends to the Cicereae. Both pea and chickpea have linkages between Adh and Lap loci (Weeden and Marx, 1987; Gaur and Slinkard, 1990). The homology of chromosome 2 of faba bean with LG III of pea is further supported by the placement of the vicilin gene sequences on this chromosome in faba bean (Macas et al., 1993). Two vicilin sequences (*Vc-3* and *Vc-5*) are located on LG III in pea. Thus, although the homology between the LAP, ADH and vicilin genes has yet to be demonstrated, the circumstantial evidence for identifying chromosome 2 in faba bean as homologous to LG III in pea is considerable. The extension to chickpea requires more confirmation.

CHICKPEA

Chickpea possesses a much smaller genome than any member of the Viceae and also contains eight chromosomes in its haploid complement. One might predict that its genome would exhibit a significant divergence in gene arrangement from what was observed in the Viceae. Lack of polymorphism for both isozyme and RFLP markers within *Cicer arietinum* has forced most of the initial mapping studies to use the interspecific cross *C. arietinum* x *C. reticulatum*. This cross does not appear to be troubled by chromosomal rearrangements, showing high fertility in the F1 and F2 generations, and therefore represents a particularly useful tool for general mapping. Gaur and Slinkard (1990) were able to arrange 26 isozyme loci and three segregating morphological traits into seven linkage groups (total length 257 cM). A similar study was performed by Kazan et al. (1993), and the combined data sets provide considerable material for comparing the chickpea linkage map with those of the Viceae.

Several cases of apparent linkage homology have been identified between chickpea and members of the Viceae. Chickpea LG II contains the loci *Aat-p* and *Pgm-p*, paralleling the same widely spaced loci on LG. II of pea. Linkage group VII is very similar to pea LG VII, with only the *Aldo* -- *Gal-2* segment inverted (Figure 3). Linkage group III in chickpea contains three loci that appear to correspond to three syntenic isozymes on pea LG III (*Lap-2* -- *Acp-3* -- *Adh-1*). Unfortunately, the homology between these loci is difficult to clearly demonstrate. *Lap-2* in chickpea could be homologous to either *Lap-1* or *Lap-2* in pea. Similarly there are multiple alcohol dehydrogenase and acid phosphatase isozymes in both pea and chickpea. DNA markers for specific single or low copy sequences are needed to provide definitive evidence for homology between these two linkage groups.

Gpi-c of chickpea linkage group VI is probably homologous to the *Gpi-c* on pea LG VI. However, this gene is known to have been duplicated early in the evolution of the Leguminosae (Weeden et al., 1989). Both *Cicer* and *Pisum*, along with other members of the Viceae and Trifolieae now express only one of the loci, but it is not impossible that different loci have been silenced in the two genera. *Fdh* has not been mapped in pea or other pulses, but the *Gpi-c* -- *Fdh* linkage also appears to be present in Phaseolus (Garvin et al., 1989), suggesting that this short linkage may be widely conserved in the Papilionoideae. Chickpea groups I, IV, and V are difficult to compare with pea due to lack of homologous marker loci.

More saturated linkage maps of the chickpea genome have been developed using DNA markers (Simon and Muehlbauer, 1997). Random amplified DNA polymorphisms (RAPDs) were particularly useful for generating a map because they exhibited a relatively high level of polymorphism. However, these markers do not usually have easily identified homologs in other legumes. Hence, the more saturated map is not particularly useful for intergeneric comparisons.

LUPIN

The maps that have been generated in lupin are for narrow-leaved lupin (*L. angustifolius*) and are primarily based on arbitrarily primed DNA markers (Wolko and Weeden, 1993; Brien, 1997). As was indicated in chickpea, such maps are of little utility for determining interspecific or intergeneric homologies because such markers are rarely conserved across such wide taxonomic distances. The map developed by Wolko and Weeden (1995) contained 14 isozyme loci, as well as the 105 RAPD markers. Of these, four of isozyme loci were placed in multilocus linkage groups. One of the groups contained the linked loci *Aat-1 -- Gpi-2 -- Pgd-2*. In the other pulses we have examined similar isozyme loci segregate independently, indicating that this linkage group is not conserved. However, a linkage reported by Marshall et al. (1974) between *Adh-1* and *Lap-2* is reminiscent of linkages found in both pea and chickpea. Although the polyploidy in *Lupinus* complicates the identification on homologous loci, the possibility that this small section of the linkage map has been maintained relatively intact in two very divergent branches of the Papilionoideae merits further investigation.

CONCLUSIONS

The comparison of linkage maps available among the cool season food legumes reveals a considerable amount of homology at least within the Viceae/Cicereae tribes, and opens the possibility of predicting gene location in several genera from mapping studies done on a single model species. However, it is also clear that several species have unique traits or tools associated with them that make them excellent candidates for model species. Pea possesses the most detailed linkage map and a large number of morphological mutations that are of interest to developmental biologists and physiologists. However, the large genome and possibly interactions between chromosomes have made a complete and consistent linkage map very difficult to develop in this species. Although faba bean has a very large genome, the availability of trisomics permits the identification of entire linkage groups, thereby circumventing the problems encountered in pea and lentil. At least as far as conserved linkage homology will permit, faba bean can be very useful for identifying on which linkage group a new fragment of DNA resides. Similarly, chickpea, with its smaller genome and relatively short linkage map is particularly useful for determining gene placement and gene order when polymorphism for that gene can be identified. Linkage maps in sweetclover, alfalfa, and clover also tend to be short and easy to generate. Thus, information currently being developed in the Trifolieae may be very useful for mapping not only in chickpea but also in the Viceae.

At present, relatively few among-map inconsistencies have been observed in lentil, faba bean, chickpea or field pea, possibly because relatively few maps have been generated within each of these latter crops. It may be associated with the large genomes of the Viceae, because linkage maps in alfalfa and clover are much shorter and more repeatable. Judging from linkage comparison with common bean, mung bean, peanut, and lupin, it does not appear that significant linkage homology extends much further than across closely related tribes. Unfortunately, this situation leaves lupin without an obvious partner to interact with. At least for the present, lupin researchers will not be able to take advantage of much of the mapping research done in other legume crops.

References

Ahn, S., J.A. Anderson, M.E. Sorrells, and S.D. Tanksley. 1993. *Molecular and General Genetics* 241:483-490.
Bisby, F.A. 1981. In: *Advances in Legume Systematics*.pp 409-425 (Eds. R.M. Polhill and P.H. Raven). Kew: Royal Botanic Gardens.
Blixt, S. 1972. *Agri Hort. Genetica* 30:1-293.

Bonierbale, M.W., R.L. Plaisted, and S.D. Tanksley. 1988. *Genetics* 120: 1095-1103.
Boutin S.R., N.D. Young, T.C. Olson, Z.H. Yu, R.C. Shoemaker, and C.E. Vallejos. 1995. *Genome* 38: 928-937
Brien, S.J., P.A. O'Brien, W.A. Cowling, R. H. Potter, M. Shankar, R.A.C. Jones, and M.G.K. Jones. 1997. *Plant & Animal Genome* V. Abstr.
Brummer, E.C., J.H. Bouton, and G. Kochert. 1993. In: *Genetics Maps* (Ed. S.J. O'Brien). Cold Spring Harbor Laboratory Press. pp. 6:82-83.
Cabrera, A. and A. Martín. 1989. *FABIS Newsletter* 24: 3-5.
Cabrera, A., J.I. Cubero, and A. Martín. 1989. *FABIS Newsletter* 23: 5-7.
Cheung, W.Y., G. Champagne, N. Hubert, B.S. Landry. 1997. *Theoretical and Applied Genetics* 94:569-582.
Corby, H.D.L. 1981. In: *Advances in Legume Systematics*. pp. 657-669 (Eds. R.M. Polhill and P.H. Raven). Kew: Royal Botanic Gardens..
Cristofolini, G. 1989. *Plant Systematics and Evolution* 166:265-278.
Cristofolini, G., and L. Feoli-Chiapella. 1984. *Webbia* 38:105-122.
Devos, K.M., S. Chao, Y. Li Q, M.C. Simonetti, and M.D. Gale M.D. 1994. *Genetics* 138:1287-1292.
Dirlewanger, E., P.G. Isaac, S. Ranade, M. Belajouza, R. Cousin, H. deVienne. 1994. *Theoretical and Applied Genetics* 88:17-27
Ellis, T.H.N., L. Turner, R.P. Hellens, D. Lee, C.L. Harker, C. Enard, C. Domoney, and D.R. Davies. 1992. *Genetics* 130:649-663.
Ellis, T.H.N., R.P. Hellens, L. Turner, D. Lee, C. Domoney, and T. Welham. 1993. *Pisum Genetics*, 25:5-12.
Folkeson, D. 1985. *Pisum Newsletter* 17:15-16.
Folkeson, D. 1990. Ph.D. Dissertation., University of Lund, Sweden
Galasso, I., M. Frediani, M. Maggiani, R. Cremonini, and D. Pignone. 1996. *Genome* 39:258-265.
Garvin, D.F., M.L. Roose, and J.G. Waines. 1989. *Journal of Heredity* 80:373-376.
Garvin, D.F. and Weeden, N.F. 1994. *Journal of Heredity* 85:273-278.
Gaur, P.M. and A.E. Slinkard. 1990. *Journal of Heredity* 81: 455-461.
Gilpin, B.J., J.A. McCallum, T.J. Frew, and G.M. Timmerman-Vaughan. *Theoretical and Applied Genetics* (in press).
Goldblatt, P. 1981 In *Advances in Legume Systematics*, pp 427-463 (Eds. R.M. Polhill and P.H. Raven). Kew: Royal Botanic Gardens.
Gopinath, D.M. and C.R. Burnham. 1956. *Genetics* 41:382-395.
Halward, T., H.T. Stalker, and G. Kochert. 1993. In: *Genetics Maps* (S.J. O'Brien, Ed.). Cold Spring Harbor Laboratory Press. pp. 6:86-87.
Kass, E. and M. Wink. 1993. *Proceedings of the 7th International Lupin Conference. Evora, Portugal*. pp. 267-270.
Kass, E. and M. Wink. 1995. *Botanica Acta* 108:149-162.
Kazan, K., Muehlbauer, F.J., Weeden, N.F., and Ladizinsky, G. 1993. *Theoretical and Applied Genetics* 86:417-426.
Kidwell, K.K., C.S. Echt, T.C. Osborn, and T.J. McCoy. 1993. In: *Genetics Maps* (S.J. O'Brien, Ed.). Cold Spring Harbor Laboratory Press. pp. 6:84-85.
Kiss, G.B., G. Csanadi, K. Kalman, P. Kalo, and L. Okresz. 1993. *Molecular and General Genetics* 238:129-137.
Kozik, A., M. Matvienko, B. Sheres, V.G. Paruvangada, T. Bisseling, A. Van Kammen, T.H.N. Ellis, T.A. LaRue, and N.F. Weeden. 1996. *Plant Molecular Biology* 31:149-156.
Kupicha, F.K. 1977. *Botanical Journal of the Linnean Society* 74:131-162.
Lamm, R. 1951. *Hereditas* 37:356-372.
Lamm, R. and J.R. Miravalle. 1959. *Hereditas* 45:417-440.
Lamprecht, H. 1945. *Hereditas* 31:347-382.
Lamprecht, H. 1948. *Agri Hort. Genet.* 6:10-48.
Macas, J., W. Weschke, H. Bäumlein, U. Pich, A. Houben, U. Wobus, and I. Schubert. 1993. *The Plant Journal* 3:883-886.
Marshall, D.R., P. Broue, and R.N. Oram. 1974. *Journal of Heredity* 65:198-203.
Martín, A. and P. Barceló, P. 1984. In: *Systems for cytogenetic analysis in Vicia faba L.* pp. 63-76 (Eds. G.P. Chapman and S.A. Tarawai). Martinus Nijhoff / Dr W. Junk Publishers, The Hague..
Marx, G.A. 1969. *Pisum Newsletter* 1:20-21.
Marx, G.A. 1987. *Pisum Newsletter* 19:38-39.
McGrath J.M. and C.F. Quiros. 1991. *Theoretical and Applied Genetics.* 82:668-673.
Muehlbauer, F.J., N.F. Weeden, and D.L. Hoffman. 1989. *Journal of Heredity* 80: 298-303.
Palmer, J.D., B. Osorio, J. Aldrich, and W.F. Thompson. 1987. *Current Genetics* 11:275-286.
Palmer, J.D., B. Osorio, and W.F. Thompson. 1988. *Current Genetics* 14:65-74.
Paruvangada, V.G., Weeden, N.F., Cargnoni, T., Yu, J. Gorel, F., Frew, T., Timmerman-Vaughan, G.M., and McCallum, J. 1995. *Pisum Genetics* 27:12-13.
Polans, N.O. 1993. *Pisum Genetics* 25:36-38.
Polans, N.O., N.F. Weeden, and W.F. Thompson. 1986. *Theoretical and Applied Genetics* 72:289-295.
Polhill, R.M. 1976. *Botanical Systematics* 1:143-380.
Satovic, Z., A.M. Torres, and J.I. Cubero. 1996. *Theoretical and Applied Genetics* 93:1130-1138.
Schubert, I. and R. Rieger. 1991. *FABIS Newsletter* 28/29:14-22.
Schubert, I., R. Rieger, and A. Michaelis. 1988. *Theoretical and Applied Genetics* 76:64-70.
Simon, C.J. and F.J. Muehlbauer. 1997. *Journal of Heredity* 88:115-119.
Simon, C.J., M. Tahir, and F.J. Muehlbauer. 1993. In: *Genetics Maps* (S.J. O'Brien, Ed.). Cold Spring Harbor Laboratory Press. pp. 6:96-100.
Sjödin, J. 1971. *Hereditas.* 68: 1-34.
Tahir, M., and F. J. Muehlbauer. 1994. *Journal of Heredity* 85:306-310.
Tanksley, S.D., M.W. Ganal, J.P. Prince, M.C. De Vicente, and M. W. Bonierbale. 1992. *Genetics*. 132: 1141-1160.
Temnykh, S.V., F.L. Gorel', V.A. Berdnikov, and N.F. Weeden. 1995. *Pisum Genetics* 27:23-25.
Timmerman-Vaughan, G.M., J.A. McCallum, T.J. Frew, N.F. Weeden, and A.C. Russell. 1996. *Theoretical and Applied Genetics* 93:431-439.
Torres, A.M., N.F. Weeden, and A. Martín. 1993a. *Theoretical and Applied Genetics* 85: 937-945.

410

Torres, A.M., T. Millán, S. Cobos, and J.I. Cubero. 1993b.. XXVIII *Jornadas Luso-Espanholas de Genética,* Faro, Portugal.
Torres, A. M., Z. Satovic, J. Canovas, S. Cobos, and J.I Cubero. 1995. *Theoretical and Applied Genetics* 91:783-789.
Torres A.M., M.C. Vaz Patto, Z. Satovic, and J.I. Cubero. (in press). *Journal of Heredity.*
Vaillancourt, R.E. and A.E. Slinkard. 1993. *Canadian Journal of Plant Science* 73:917- 926.
Weeden, N.F., J.J. Doyle and M. Lavin. 1989. *Evolution.* 43:1637-1651.
Weeden, N.F., and T. Cargnoni. 1994. *Plant Genome II,* San Diego. Abst.
Weeden, N.F. and G.A. Marx. 1987. *Journal of Heredity* 78:153-159.
Weeden, N.F., Muehlbauer, F.J., and Ladizinsky, G. 1992. *Journal of Heredity* 83:123- 129.
Weeden, N.F., Swiecicki, W.K., Timmerman, G.M., and Ambrose, M. 1993a. *Pisum Genetics* 23:13-14.
Weeden, N.F., Swiecicki, W.K., Timmerman, G.M., and Ambrose, M. 1993b. *Pisum Genetics,* cover.
Weeden, N.F. and B. Wolko. 1990. In: *Genetic Maps,* 5th edition, S.J. O'Brien (ed.) Cold Spring Harbor Laboratory Press, Cold Spring Harbor, NY pp.6.106-6.112.
Whitkus, R., J. Doebley, and M. Lee. 1992. *Genetics,* 132:1119-1130.
Wink, M. 1993. In: *Methods in Plant Biochemistry.* Pp 197-239 (Ed. P. Waterman) vol 8. London, Academic Press.
Wolko, B. and N.F. Weeden. 1990. *Genetica Polonica* 31:179-187.
Wolko, B. and N.F. Weeden. 1995. *Proceedings of the 7th International Lupin Conference. Evora, Portugal.* pp. 42-49.
Yu, G.X., A.L. Bush, and R.P. Wise. 1995. *Genome* 39:155-164

Wide crossing: opportunities and progress

B. Ocampo[1], C. Conicella[2] and J.P. Moss[3]

1 ICARDA, PO Box 5466, Aleppo, Syria; 2 CNR-IMOF, Research Institute for Vegetable and Ornamental Breeding, Via Università 133, Portici, Italy; 3 Ealachan Bhana, Clachan Seil, Oban, Argyll, PA 34 4TL, UK

Abstract

Cool season food legumes (CSFLs) are increasingly being grown in marginal areas, where yield and production stability are affected by biotic and abiotic stresses to which they lack adequate resistance. Wild relatives of CSFLs possess valuable genes for resistance/tolerance to several of these stresses. Recently wide hybridization has shown potential for the genetic improvement of CSFLs. Genes for resistance to cyst nematode, and tolerance to cold have been transferred to chickpea (*Cicer arietinum* L.) from its closest relatives. The transfer of tolerance to cold is also being pursued in lentil (*Lens culinaris* Medikus). Wide hybridization could play a role in improving the adaptation of CSFLs. The tertiary gene-pool (GP3) of the CSFLs includes desirable variation. Embryo-ovule culture could play a role in expanding the hybridization range of the CSFLs if this technique were refined. The superior agronomic phenotypes produced following wide hybridization in *Cicer* and *Lens*, underlines the importance of assessing the breeding value of wild accessions in hybrid combinations with their cultivated counterparts. Collaboration is needed at the international level, between national programs, gene banks and International Agricultural Research Centers, to tackle the complexity of wide crossing systematically. This paper reviews recent progress on wide hybridization in CSFLs.

INTRODUCTION

Cool season food legumes (CSFLs) are an important source of protein for people in developing regions. They are also valuable as feed and fodder for livestock and play a valuable role in crop rotations. Low yields and unstable production are major disadvantages of most CSFLs, which are therefore considered to be second-rate crops. They are grown under conditions of low inputs in spite of their recognized nutritional and agronomic importance. The fact that their cropping is increasingly being forced into marginal areas exposes them to constraints, causing a reduction in yield from their genetic potential and a further decline in their production stability. The control of pests and diseases by chemicals is seldom a possible solution in developing regions, because of their high costs, the lack of environmentally friendly chemicals and the lack of the physical facilities for their distribution. Breeding genetically enhanced cultivars appears to be an alternative, but there is inadequate variation in the cultivated germplasms. Wide hybridization provides an opportunity to improve the genetic variability of a disparate group of crops. Wild species may prove valuable when they have alleles for resistance, different from those in the cultigen. Encouraging reports about the value of germplasm of wild CSFLs, of successful introgressions, and of increased availability of seeds of wild taxa, have encouraged breeders, germplasm curators and geneticists to assess the breeding value of the wild species more systematically. The exploitation of this wealth is just beginning.

R. Knight (ed.), Linking Research and Marketing Opportunities for Pulses in the 21st Century, 411–419.
© 2000 Kluwer Academic Publishers. Printed in the Netherlands.

TAXONOMY

The CSFLs belong to the family Leguminoseae and subfamily Papilionaceae. Lentil, faba bean (*Vicia faba* L.), pea (*Pisum sativum* L.) and grasspea (*Lathyrus sativus* L.) are all members of the tribe Vicieae, whereas chickpea (*Cicer arietinum* L.) belongs to the monogeneric tribe Cicereae Alef. (Kupicha, 1975, 1977). The extent of the gene pools differs between the CSFLs, and presents different problems to breeders of these crops.

Cicer

The genus *Cicer* L. (2n = 16) comprises 43 species: 9 annuals, 33 perennials and 1 unspecified (van der Maesen, 1987). The annual *Cicer* species are *C. arietinum*, *C. bijugum* K.H. Rech., *C. chorassanicum* (Bge.) M. Pop, *C. cuneatum* Hochst. ex Rich., *C. echinospermum* P.H. Davis, *C. judaicum* Boiss., *C. pinnatifidum* Jaub. & Sp., *C. reticulatum* Ladiz., and *C. yamashitae* Kitamura. Phenetic, karyotypic, biochemical, and crossability studies indicate that *C. reticulatum* and *C. echinospermum* are the closest wild relatives of chickpea (Ladizinsky and Adler, 1976, Ahmad and Slinkard, 1992, Ocampo *et al.*, 1992, Labdi *et al.*, 1996, Tayyar and Waines, 1996, Singh and Ocampo, 1993). These two species and *C. arietinum* should be regarded as one biological species. *C. reticulatum* is considered to be the ancestor of chickpea (Ladizinsky and Adler, 1976). The remaining annual wild *Cicer* species belong to the GP3, because although most can be crossed with the cultigen, the gene flow is non-existent. A report claiming fertile offspring between the cultigen and members of the GP3 (Verma *et al.*, 1990) suggests that the sexual hybridization range of chickpea could still be an open question.

Lens

The genus *Lens* Miller (2n = 14) includes 4 annual species, *L. culinaris, L. odemensis* (Ladiz.), *L. ervoides* (Brign.) Grande and *L. nigricans* (M.Bieb.) Godron. *L. culinaris* encompasses two subspecies: *L. culinaris* ssp. *culinaris*, the cultivated type (hereafter *L. culinaris*), and *L. culinaris* ssp. *orientalis* (Boiss.) Hand.Mazz. (hereafter *L. orientalis*) (Ladizinsky, 1993). Genetic relationships, defined using molecular techniques, acknowledge *L. orientalis* as the progenitor of lentil, and *L. ervoides* and *L. nigricans* as the species most distantly related to the cultigen (for a review see Baum *et al.*, 1997). Cytological, hybridization and crossability studies suggest a common gene pool for all *Lens* species (Ladizinsky, 1993, Muehlbauer *et al.*, 1995, Ahmad *et al.*, 1995).

Vicia

The taxonomy of faba bean (2n = 12) is debated. One schools envisages three taxonomic ranks: (1) a generic (genus *Faba*, related to *Vicia*); (2) a subgeneric (subgenus *Faba* of the genus *Vicia*); and (3) a specific rank (faba bean as species in the genus *Vicia* section *Faba*) (Cubero, 1984). This species is the only member of the GP1, and it is genetically isolated from other taxa. The Narbonensis species complex, though morphologically the closest to faba bean (Cubero, 1984, Maxted *et al.*, 1991), shows substantial phylogenetic differences from this species (Chooi, 1971, Ladizinsky, 1975, Cubero, 1981, Raina and Rees, 1983, van de Ven *et al.*, 1993, Raina and Ogihara, 1994, 1995). Schäfer (1973) hypothesized the extinction of the progenitor of faba bean.

Pisum

Taxonomic interpretations for the genus *Pisum* L. are diverse (2n = 14). The most generally accepted considers two species (Davis, 1970): *P. sativum* and *P. fulvum* Sibth. & Sm. The species *P. sativum* encompasses two subspecies: *P. sativum* ssp. *sativum*, the cultivated type, and *P. sativum* ssp. *elatius*. Lamprecht (1966) suggested the monospecificity of this genus, as introgression between the two species is possible. The ancestry of the cultivated species is debated (Hoey *et al.*, 1996).

Lathyrus

Grasspea (2n = 14) belongs to genus *Lathyrus* L., section *Lathyrus*, along with 33 other annual and perennial taxa (Kupicha, 1983, Maxted and Goyder, 1988). The demarcation of the grasspea gene pools is difficult at present, as few of the possible interspecific crossings have been attempted.

MAJOR CONSTRAINTS TO THE PRODUCTION OF COOL SEASON FOOD LEGUMES

Yield, production stability and grain quality of CSFLs are adversely affected by an array of constraints. Diseases are the most important of the biotic stresses in the field. Drought is the major environmental stress as these legumes are commonly grown under receding soil moisture in rainfed agriculture.

The foliar diseases Ascochyta blight (caused by *Ascochyta rabiei*) and Fusarium wilt (caused by *Fusarium oxysporum* f.sp. *ciceri*) are the most serious biotic stresses of chickpea.

The most critical diseases of lentil are root rots (caused by *Pythium, Rizoctonia* and *Sclerotinia* spp.), rust (caused by *Uromices fabae*) and vascular wilt (caused by *F. oxysporum* f.sp. *lentis*). In the Mediterranean basin, the obligate weed parasite *Orobanche crenata* is becoming a serious problem.

Chocolate spot (caused by *Botrytis fabae* and *B. cinerea*), Ascochyta blight (caused by *A. fabae*), and rust (caused by *U. viciae-fabae*) are the major diseases of faba bean. In the Mediterranean basin, broomrape (*Orobanche* spp.) is dramatically reducing the crop area. When fields become infested, farmers have no other choice than to abandon the growing of faba bean. Aphids can be particularly harmful, directly injuring the crop (*Aphis fabae*) and by transmitting viruses (*A. craccivora* and *Acyrthosiphon pisum*).

Powdery mildew (caused by *Erysiphe pisi*) and Fusarium wilt (caused by *F. oxysporum* f.sp. *pisi*) are the most devastating diseases of pea.

Grasspea, although the most robust of the CSFLs, can be seriously damaged by Ascochyta blight (caused by *Ascochyta* spp.), stem blight (caused by *B. cinerea*) and powdery mildew (caused by *Peronospora* spp.).

USEFUL GENETIC VARIATION IN THE WILD GERMPLASM OF THE COOL SEASON FOOD LEGUMES

In light of the recognized narrow genetic variability present in CSFLs to counter constraints, breeders, germplasm curators and geneticists are paying increased attention to the breeding value of the wild forms of these crops. Sources of resistance, not available in the CSFLs, have been identified in their wild relatives. Many wild accessions have a combination of desirable traits.

The annual *Cicer* species are sources of genes conferring resistance to most important yield reducing and variation factors of chickpea. The wild taxa have higher levels of resistance to Ascochyta blight, Fusarium wilt, Botrytis gray mold (caused by *B. cinerea*) and leafminer (*Liriomyza cicerina*) than the cultivated chickpea (Singh *et al.*, 1982, Singh *et al.*, 1997). Sources of resistance to cyst nematode (*Heterodera ciceri*), lacking in chickpea, are available in *C. bijugum, C. pinnatifidum* and *C. reticulatum* (Di Vito *et al.*, 1996). Sources of resistance to root-knot nematode, *Meloidogine artiellia*, are present in C. *judaicum, C. pinnatifidum, C. chorassanicum* and *C. cuneatum* (Greco and Di Vito, 1993). High levels of cold tolerance are present in *C. reticulatum* and *C. bijugum* (Singh *et al.*, 1995). Unlike the cultigen, a few wild accessions possess sources of resistance to multiple stresses. Unfortunately, the most desirable genes are in *C. bijugum, C. judaicum* and *C. pinnatifidum*, and are not exploitable now as these species belong to the GP3.

Ladizinsky (1979) inferred that wild lentil germplasm possessed genetic variability for economic traits that are no longer available in the cultivated lentil. Higher levels of cold tolerance than that available in lentil are found in accessions of *L. orientalis* from the highlands of Turkey (Hamdi *et al.*, 1996). High levels of resistance to a Syrian isolate of Ascochyta blight (caused by *A. fabae* f.sp. *lentis*) have been found among accessions of ICARDA's (International Center for Agricultural Research in the Dry Areas) wild lentil germplasm (Bayaa *et al.*, 1994). Sources of resistance to vascular wilt are present in *L. orientalis* and *L. ervoides* (Bayaa *et al.*, 1995). Although lentil is reputed to stand drought better than most CSFLs, its yield can be jeopardised by low soil moisture.

Overall, the wilds forms have shown lower drought susceptibility indexes (variation of seed yield in response to variation in soil moisture) than the cultivated lentils (Hamdi and Erskine, 1996).

Vicia species show resistance to the aphids (*A. craccivora, A. fabae, A. pisum* and *Megoura viciae*) (Holt and Birch, 1984, Birch *et al.*, 1985). *V. narbonensis* L. possesses genes for resistance to Botrytis, Orobanche (*O. egyptiaca*), aphids, drought, and frost (Birch *et al.*, 1985). Several lines of *V. villosa* ssp. *dasycarpa* (Tan.) and a few of *V. sativa* L. possess resistance to *O crenata* (Linke *et al.*, 1993). One selection of *V. villosa* ssp. *dasycarpa* has genes conferring resistance to root-rot nematode (*M. artiellia*) (Abd El Moneim and Bellar, 1993). Sources of tolerance to frost are present in *V. johannis* Tam. (Birch *et al.*, 1985).

Sources of resistance to *Bruchus pisorum, Heterodera goettingiana* and *Pseudomonas pisi* have been identified in wild *Pisum* accessions (Di Vito and Perrino, 1978, Di Vito and Greco, 1986, Hardie, 1990, Ali *et al.*, 1994, Hardie *et al.*, 1995). *P. fulvum* and *P. sativum* ssp. *elatius* also have a high level of drought tolerance; the first presumably because it has a robust root system, and the second because its leaves are covered by a thick layer of wax (Ali *et al.*, 1994). Sources of cold tolerance and winter-hardiness have been identified in wild accessions of *Pisum* (Auld *et al.*, 1983).

L. ochrus (L.) DC possesses genes which confer high resistance to *O. crenata* (Linke *et al.*, 1993). Several accessions of this species combine this resistance with resistance to Ascochyta leaf spot, stem blight, Botrytis blight and downy mildew (ICARDA, 1995). *L. cicera* L. has a high level of tolerance to cold and drought (ICARDA, 1995). A high level of cold tolerance has also been found in *L. ochrus* (Robertson *et al.*, 1996). Both these species have rapid winter and spring growth (ICARDA, 1995). *L. cicera* showed an average content of β-N-oxalilamino-L-alanine (BOAA) four to five times lower than that of *L. sativus* L. and *L. ochrus* (Aletor *et al.*, 1994).

The known variability present in the wild germplasms of the CSFLs is believed to be an underestimation of their real variability. More valuable genes may be available in the future following further assessments of the genetic variation and from new collections. Actually, large collections are believed to hold rare desirable genes.

INTROGRESSION

During the last few years, wide hybridization has been used to transfer genes/alleles for qualitative and quantitative traits in chickpea and lentil.

The introgression into chickpea of resistance to cyst nematode from *C. reticulatum* (Di Vito *et al.*, 1996) is an example. Resistant selections are now being evaluated in yield trials at ICARDA's headquarters in north-west Syria. Other selections, with high levels of cold tolerance derived from *C. reticulatum* and *C. echinospermum* are now included in yield trials at ICARDA headquarters (ICARDA, 1995). These hybridisations have shown the potential of the technique to the improve agronomic characteristics of the crop (Singh and Ocampo, 1997). Selections have been bred which have phenotypes superior to those of the cultivated parent ILC 482 in field- experiments (Singh and Ocampo, 1997) (Fig. 1). These selections possess all the peculiar morphological and qualitative characteristics of the kabuli-type chickpea. Introgressions from *Cicer* species in the GP3 are highly important, as these species possess most desirable genes for improving the crop. Verma *et al.* (1990) claimed it was possible to cross chickpea with *C. bijugum, C. judaicum* and *C. pinnatifidum* but there have been no further reports on this work.

Ahmad *et al.* (1995) succeeded in crossing lentil with species that previously could only be crossed with it by means of *in-vitro* culture. This finding amplifies the value of the wild germplasm of lentil for the improvement of the crop, and is expected to result in a major use of wild *Lens* species in breeding programs. Segregating populations of a cross between *L. culinaris* and *L. orientalis* are being evaluated for response to cold (Hamdi *et al.*, 1996). The possibility of improving agronomic traits *via* wide hybridizations has also been assessed for lentil using *L. orientalis* (Abbo *et al.* 1992, ICARDA, 1995).

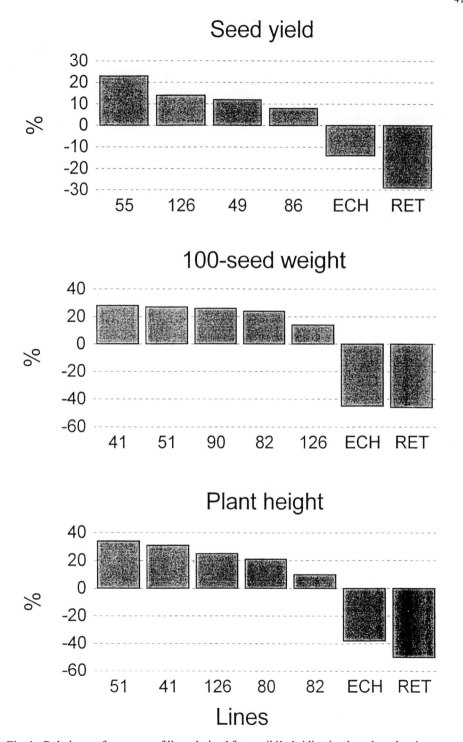

Fig. 1. Relative performances of lines derived from wild hybridization based on the phenotype of the cultivated parent chickpea ILC 482 (=0%). The data are the averages of three cropping seasons at Tel Hadya, Aleppo, Syria. RET = *C. reticulatum* ILWC 124; ECH = *C. echinospermum* ILWC 179.

In pea, using *P. fulvum* as the male parent makes interspecific hybridisation possible, whereas, reciprocal crosses abort because of unfavorable nuclear-cytoplasm interactions. The pollen fertility of the hybrids was very low, but increased after back-crossing with *P. sativum* (Errico *et al.*, 1996). By using secondary constrictions and FISH spots as cytological markers, pea lines with reconstructions in nucleolar chromosomes 5 and 7 were obtained by transferring the satellites from *P. fulvum* into *P. sativum* through interspecific hybridization, subsequent backcrosses and selfings. They showed an additional secondary constriction on chromosome 5 and a longer satellite on chromosome 7. The influence of modified chromosomes 5 and 7 were monitored for some morpho-physiological traits, but the additional rDNA loci did not correlate with any obvious change in the phenotype. However, modified chromosomes 5 and 7 were involved in segregation distortion events.

Faba bean can be crossed with a few species of the section *Faba*, but the hybrid embryos abort a few weeks after crossing (Pickersgill *et al.*, 1985, Lazaridou and Roupakias, 1993). The highest rates of fertilization have been achieved with *V. galilaea* and *V. bithynica* L. (Pickersgill *et al.*, 1985).

Crosses between grasspea and other taxa have not been successful and genetic improvement depends on the exploitation of landraces (Yunus and Jackson, 1991). F_1 hybrids were obtained between grasspea, and *L. amphicarpos, L. cicera, L. gorgoni* and *L. latifolius* but only the crosses with the first two species produced apparently normal plants and they were sterile (Khawaja, 1988, Yunus and Jackson, 1991).

Barriers to hybridisation and introgression

The barriers to wide hybridization in CSFLs are diverse. They were reported to occur mainly at the post-mating stage (Pickersgill *et al.*, 1985, Ahmad *et al.*, 1988, Yunus and Jackson, 1991, Ladizinsky, 1993).

Crosses between chickpea and its wild relatives included in the GP3 commonly produce pods with small and shrivelled seeds 2-5 mm long. Crosses between chickpea and *C. pinnatifidum* generate albino F_1s that die at the seedling stage. Similarly, non-viable, chlorophyll deficient hybrids have been reported in interspecific *Lens* crosses (Ladizinsky, 1993). Abbo and Ladizinsky (1994) attributed the F_1 sterility of *Lens* hybrids (*L. culinaris* H *L. ervoides* and *L. orientalis* H *L. culinaris*) to dominant gene action, while dominant and additive factors presumably played a role in following generations. Barriers to interspecific hybridization do not occur in *Pisum*, however, segregation distortion phenomena that involve single chromosomes, or chromosome segments, can limit the transfer of useful traits from *P. fulvum* to *P. sativum*. Two translocations involving chromosomes 1, 3, 5 and 7 distinguish *P. sativum* and *P. fulvum*. These translocations account for the decrease in male gamete viability (Errico *et al.*, 1991).

OPPORTUNITIES

Desirable traits in the GP1 are easily accessible by breeders; nevertheless the variation in existing collections is limited. More collections could be made, with an emphasis on the search for desirable genes, as was done with nematode resistance in *C. reticulatum* (Di Vito *et al.*, 1996). It is possible that many original collections from the wild were variable when collected and this would justify recollection when possible. Collecting from diverse sites in a species' distribution range would provide samples differing in their adapted allelic complexes. Variability in self-pollinated species is usually greater among populations, than within, populations (Crawford, 1990, Hamrick and Godt, 1990). The success of wide crosses often depends on the particular parents used and other factors. This underlines the importance of a better exploitation of genetic and non-genetic factors and their interaction. For instance, the success of wide hybridizations in cereals depends upon several factors including the accessions used in the crosses, the direction of the crosses, years, locations, and even research programs (Sharma, 1995). Crossing success may also depend on intra-accession variation (Sharma and Ohm, 1990). An example is the cross of *C. arietinum* H with *C. echinospermum*. Earlier attempts to make this cross failed (Ladizinsky and Adler, 1976, Ahmad, 1988) but fertile offspring were eventually produced at ICARDA (ICARDA 1989, 1990). Several cross combinations between CSFLs and their distant wild relatives engender aborted ovules, of variable dimensions, suggesting that barriers to gene flow might not be insurmountable. When this occurs, embryo/ovule culture may prove helpful to the production of fertile hybrids. The reliable use of this technique, which is successful for wide crosses in *Lens* (Cohen *et al.* 1984), needs to be refined for most CSFLs.

Molecular marker technology will increase the use of wild species in breeding. This technique will increase precision when assessing biodiversity and taxonomic relationships, and when screening segregating populations. The selection of superior lines, freed from undesirable genes of wild origins, is a demanding task that commonly discourages the use of wild species in breeding programs. But once tagged with molecular markers, genes can be efficiently screened in segregating populations, and transferred to targeted genotypes.

Wide hybridization helps in disclosing polymorphisms, which can be helpful when making genome maps, especially for crops such as chickpea that show little molecular polymorphism. High-density genome maps are being constructed for CSFLs (Gaur and Slinkard, 1990, Ellis *et al.*, 1992, Kazan *et al.*, 1993, Dirlewanger *et al.*, 1994. Millan *et al.*, 1995, and Baum *et al.*, 1997 for a review on lentil and faba bean). Pea breeding is already benefiting from these linkage maps. The well-established and novel technologies used in peas could be useful for the improvement of other CSFLs. In fact, preserved linkage associations among CSFLs (Muehlbauer *et al.* 1989, Kazan *et al.* 1993) provide an opportunity for easier mapping of less saturated linkage maps such as those of lentil and chickpea.

Gene transfer is a technique that aims to introgress definite DNA sequences from one living organism into the genome of another, irrespective of taxonomic differences between those organisms. Tens of species, including legumes, have been transformed with the introduction of several foreign genes (Uchimiya *et al.* 1993, Christou, 1994). Compared to other CSFLs, transformation of pea is at an advanced stage (Schroeder *et al.*, 1993). Pea cultivars resistant/tolerant to insect pests will probably be achieved in the near future by the introgression of insecticidal compounds, as has been done in other crops (Gatehouse *et al.*, 1991, Peferoen, 1992). Chickpea, lentil and faba bean have been reported to be susceptible to *Agrobacterium* spp. infection (Srinivasan *et al.*, 1988, Warkentin and McHughen, 1991, Ramsay and Kumar, 1991) however, frequencies of transformants and satisfactory rates of regeneration are still low. Protocols are too cumbersome for the routine use of this technique in CSFLs at present. Problems of integration and expression of the introduced gene, and social and legal impediments to the use of transgenic plants, are drawbacks that are usually absent when introgressions are made by sexual means. For the near future, conventional sexual hybridization, supported by refined *in-vitro* techniques will still be the leading approach in wide hybridization.

The production of unexpected superior variants for complex traits derived from wide hybridizations, as occurs in *Cicer* and in diverse groups of other crops (Cox *et al.*, 1995), illustrates the breeding value of wild genetic resources.

CONCLUSIONS

Wide hybridization is one means of broadening the genetic variation of cultigens. The success of crosses between chickpea and lentil, with their close relatives, has shown the potential for achieving the breeders' objectives of improving yield, yield-related traits, and adaptation to biotic and environmental stresses. The exploitation of wild species for the improvement of faba bean and grasspea by sexual hybridization is not possible at present because of karyotypic and genetic differences. The available evaluations of the wild germplasm of the CSFLs show just a little part of their potential for the genetic improvement of the crops. Assessment of the value of complex traits of the wild accessions must be made on the hybrid combinations with the cultivated counterpart. The future exploitation of the wild relatives depends on free interspecific gene flow, the exploitation of genetic and environmental variability and their interactions, the comprehensive and reliable evaluation of wild germplasm, and efficient use of *in-vitro* and biotechnological technologies. Collaboration between national programs and the International Agricultural Research Centers is needed to tackle systematically the introgressions from wild to cultivated taxa.

References

Abbo, S., Ladizinsky, G. and Weeden, N. F. 1992. *Euphytica* 58:259-266.
Abbo, S. and Ladizinsky, G. 1994. *Heredity* 72:193-200.
Abd El Moneim, A. M. and Bellar, M. 1993. *Nematologia Mediterranea* 21:67-70.
Ahmad, F. 1988 Ph. D. Thesis. University of Saskatchewan, Canada.
Ahmad, F., Slinkard, A. E. and Scoles, G. F. 1988. *Plant Breeding* 100:193-198.
Ahmad, F. and Slinkard, A. E. 1992. *Theoretical and Applied Genetics* 84:688-692.
Ahmad, M., Fautrier, A. G., McNeil, D. L., Burrit, D. J. and Hill, G. D. 1995. *Plant Breeding* 114:558-560.
Aletor, V. A., Abd El Moneim, A. M. and Goodchild, A. V. 1994. *Journal of the Science of Food and Agriculture* 65:143-151.
Ali, S. M., Sharma, B. and Ambrose, M. J. 1994 In *Expanding the Production and Use of Cool Season Food Legumes*,pp 540-558 (Eds F. J. Muehlbauer and W. J. Kaiser). Kluwer Academic Publishers, The Netherlands.
Auld, D. L., Ditterline, R. L., Murray, G. A. and Swensen, J. B. 1983 *Crop Science* 23:85-88.
Baum, M., Erskine, W. and Ramsay, G. 1997 In *Plant biotechnology and plant genetic resources for sustainability and productivity*, pp117-131 (Eds K. N. Watanabe and E. Pehu) R. G. Landes Company and Academic Press, Austin, Texas, U.S.A.
Bayaa, B., Erskine, W. and Hamdi, A. 1994. *Genetic Resources and Crop Evolution* 41:61-65.
Bayaa, B., Erskine, W. and Hamdi, A. 1995. *Genetic Resources and Crop Evolution* 42:231-5.
Birch, A. N. E., Tithecott, M. T. and Bisby, F. A. 1985. *Economic Botany* 39:177-190.
Chooi, W. Y. 1971. *Genetics* 68:213-230.
Christou, P. 1994. *Euphytica* 74:165-185.
Cohen, D., Ladizinsky, G., Ziv, M. and Muehlbauer, F. J. 1984. *Plant Cell Tissue Organ Culture* 3:343-347.
Cox, T. S., Sears, R. G. and Bequette, R. K. 1995. *Theoretical and Applied Genetics* 90:571-7.
Crawford, D. J. 1990. *Plant molecular systematics: macromolecular approaches*. Wiley-Inter Science Publ, New York.
Cubero, J. I. 1981. In *Faba bean improvement*, pp 91-108 (Eds G. Hawton and C. Webb) Martinus Nijhoff, The Hague, The Netherlands.
Cubero, J. I. 1984 In *Genetic resources and their exploitation-chickpeas, faba beans and lentils*, pp 131-144 (Eds J. R. Witcomb and W. Erskine) Martinus Nijhoff, The Hague, The Netherlands.
Davis, P. H. 1970. *Flora of Turkey*, Vol.3, 370-373. Edinburgh University Press, Edinburgh.
Dirlewanger, E., Isaac, P. G., Ranade, S., Balajouza, M., Cousin, R. and De Vienne, D. 1994. *Theoretical and Applied Genetics* 88:17-27.
Di Vito, M. and Greco, N. 1986. In *Cyst Nematodes*, pp 321-332 (Eds F. Lamberti and C. E. Taylor). New York, USA: Plenum Press.
Di Vito, M. and Perrino, P. 1978. *Nematologia Mediterranea* 6:113-118.
Di Vito, M., Singh, K. B., Greco, N. and Saxena, M. C. 1996. *Genetic Resources and Crop Evolution* 43:103-107.
Ellis, T. H. N., Turner, L., Hellens, R. P., Lee, D., Harker, C. L., Enard, C., Domony, C. and Davies, D. R. 1992. *Genetics* 130:649-663.
Errico A., Conicella, C. and Venora, G. 1991. *Genome* 34:105-108.
Errico A., Conicella, C., De Martino, T., Ercolano, R. and Monti, L. M. 1996 Journal of Genetics and Breeding 50:309-313
Gatehouse, J. A., Hilder, V. A. Gatehouse, A. M. R. 1991. In *Plant Genetic Engineering*, 105-135 (Ed. D. Grierson) Plan Biotechnology Series, vol. 1. Blackie & Son Ltd., London/Chapman and Hall, New York.
Gaur, P. M. and Slinkard, A. E. 1990. *Theoretical and Applied Genetics* 80:648-656.
Greco, N. and Di Vito, M. 1993. In *Breeding for stress tolerance in cool-season food legumes*, pp 157-166 (Eds K. B. Singh and M. C. Saxena). Chichester, UK: John Wiley & Sons.
Hamdi, A. and Erskine, W. 1996. *Euphytica* 91:173-179.
Hamdi, A., Küsmenolu, I. and Erskine, W. 1996. *Genetic Resources and Crop Evolution* 43:63-67.
Hamrick, J. L. and Godt, M. J. W. 1990 In *Plant population genetics, breeding and genetic resources*, pp 43-63 (Eds A. H. D. Brown, M. T. Clegg, A. L. Kahler and B. S. Weir) Sinauer Assoc Publ, Sunderland, Mass.
Hardie, D. 1990 In *Proceedings of National Pea Weevil Workshop*, pp 72-79 (Ed. A. M. Smith). Melbourne, Victoria, Australia: Department of Agriculture and Rural Affairs.
Hardie, D. C., Baker, G. J. and Marshall, D. R. 1995. *Euphytica* 84:155-161.
Hoey, B. K., Crowe, K. R., Jones, V. M. and Polans, N. O. 1996. *Theoretical and Applied Genetics* 92:92-100.
Holt, N. and Birch, N. 1984. *Annals of Applied Biology* 105:547-556.
ICARDA, 1989. Food Legume Improvement: Annual Report 1989. Aleppo, Syria: ICARDA.

ICARDA, 1990. Food Legume Improvement: Annual Report 1990. Aleppo, Syria: ICARDA.

ICARDA, 1995. Germplasm Program-Legume: Annual Report 1994. Aleppo, Syria: ICARDA.

Kazan, K., Muehlbauer, F. J., Weeden, N. F. and Ladizinsky, G. 1993. *Theoretical and Applied Genetics* 86:417-426.

Khawaja, H. I. T. 1988. *Euphytica* 37:69-75.

Kupicha, F. K. 1975. *Botanical Journal of the Linnean Society* 70:231-242.

Kupicha, F. K. 1977. *Botanical Journal of the Linnean Society* 74:131-162.

Kupicha, F. K. 1983. Notes from the Royal Botanical Garden Edinburgh 41:287-326.

Labdi, M., Robertson, L. D., Singh, K. B. and Charrier, A. 1996. *Euphytica* 88:181-188.

Ladizinsky, G. 1975. *Euphytica* 24:785-788.

Ladizinsky, G. 1979. *Euphytica* 28:179-187.

Ladizinsky, G., 1993. *Critical Reviews in Plant Sciences* 12:169-184.

Ladizinsky, G. and Adler, A. 1976. *Euphytica* 25:211-217.

Lamprecht, H. 1966. *Die Entstelung der Arten und Hoheren Kategorien.* Springer-Verlag. Vienna.

Lazaridou, T. B. and Roupakias, D. G. 1993. *Plant Breeding* 110:9-15.

Linke, K.-H., Abd El Moneim, A. M., Saxena, M. C. 1993. *Field Crops Research* 32:277-285.

Maxted, N. and Goyder, D. J. 1988. *Kew Bulletin* 43:711-714.

Maxted, N., Khattab, A. M. A. and Bisby, F. A. 1991. *Chronica Botanica* 10:435-465.

Millan, T., Hajj Moussa, E., Gil, J. and Moreno, M. T. 1995. 2nd AEP European Conference on Grain Legumes, 446-447, Copenhagen.

Muehlbauer, F. J., Weeden. N. F. and Hoffman, D. L. 1989. *Journal of Heredity* 80:298-303.

Muehlbauer, F. J., Kaiser, W. J., Clement, S. L. and Summerfield, R. J. 1995. *Advances in Agronomy* 54:283-332.

Ocampo, B., Venora, G., Errico, A., Singh, K. B. and Saccardo, F. 1992. *Journal of Genetics and Breeding* 46:229-240.

Peferoen, M. 1992 In *Plant Genetic Manipulation for Crop Protection*, pp 135-153 (Eds A. M. R. Gatehouse, V. A. Hilder and D. Boulter). Biotechnology in Agriculture Series No. 7, CAB International, Wallingford.

Pickersgill, B., Jones, J. K., Ramsay, G. and Stewart, H. 1985. In *Proceedings of the international workshop on faba beans, kabuli chickpeas, and lentils in the 1980s*, pp 57-70 (Eds M. C. Saxena and S. Varma). ICARDA, Aleppo, Syria.

Raina, S. N. and Rees, H. 1983. *Heredity* 51:335-346.

Raina, S. N. and Ogihara, Y. 1994. *Theoretical and Applied Genetics* 88:261-266.

Raina, S. N. and Ogihara, Y. 1995. *Theoretical and Applied Genetics* 90:477-486.

Ramsay, G. and Kumar, A. 1991. *Journal of Experimental Botany* 41:841-847.

Robertson L. D., Singh, K. B., Erskine, W. and Abd El Moneim, A. M. 1996. *Genetic Resources and Crop Evolution* 43:447-460.

Schäfer, H. I. 1973. *Kulturpflanzen* 21:211-273.

Schroeder, H. E., Schotz, A. H., Wardley-Richardson, T., Spencer, D. and Higgins, T. J. V. 1993. *Plant Physiology* 101:751-757.

Sharma, H. C. 1995. *Euphytica* 82:43-64.

Sharma, H. C. and Ohm, H. W. 1990. *Euphytica* 49:209-214.

Singh, G., Kapoor, S. and Singh, K. 1982. *International Chickpea Newsletter* 7:13-14.

Singh, K. B., Malhotra, R. S. and Saxena, M. C. 1995. *Crop Science* 35:1491-1497.

Singh, K. B. and Ocampo, B. 1993. Journal of Genetics and Breeding 47: 199-204.

Singh, K. B. and Ocampo, B. 1997. *Theoretical and Applied Genetics* 95:418-423.

Singh, K. B., Ocampo, B. and Robertson, L. D. 1997. *Genetic Resources and Crop Evolution* (in press).

Srinivasan, M., Gupta, N. and Chopra, V. L. 1988. *International Chickpea News*letter 19:2-3.

Tayyar, R. I. and Waines, J. G. 1996. *Theoretical and Applied Genetics* 92:245-254.

Uchimiya, H., Patena, L. F. and Brar, D. S. 1993. *International Crop Science* I, 633-639 *Crop Science Society of America*, Madison, WI, U.S.A.

van de Ven W. T. G., Duncan, N., Ramsay, G., Phillips, M., Powell, W. and Waugh, R. 1993. *Theoretical and Applied Genetics* 86:71-80.

van der Maesen, L. J. G. 1987. In *The Chickpea*, pp 11-34 (Eds M. C. Saxena and K. B. Singh). CAB International Publisher, UK

Verma, M. M., Sandhu, J. S., Brar, H. S. and Brar, J. S. 1990. *Crop Improvement* 17:179-181.

Warkentin, T. and McHughen, A. 1991. *Plant Cell Reports* 10:489-493.

Yunus, A. G. and Jackson, M.T. 1991. *Plant Breeding* 106:319-328.

Marker technology for plant breeding

M. Baum[1], N.F. Weeden[2], F.J. Muehlbauer[3], G. Kahl[4], S.M. Udupa[1], I. Eujay[1] F. Weigand[1], M. Harrabi[5], and Z. Bouznad[6]

1 *International Center for Agricultural Research in the Dry Areas (ICARDA), P.O. Box 5466, Aleppo, Syria ; 2 NYS Agricultural Experimental Station, Geneva, NY 14456, USA; 3 Grain Legume Genetics and Physiology Research, USDA-ARS, 303 Johnson Hall, Pullman, WA 99164-6434, USA; 4 Plant Molecular Biology Group, Department of Biology, Biozentrum, University of Frankfurt, D-60439, Frankfurt/Main, Germany;5.Institut National Agronomique de Tunisie (INAT), 43 Avenue Chales Nicolle, Tunis, Tunisia; 6 Institut Nationale Agronomique, Department de Botanique, El Harrach, Algiers, Algerie*

Abstract

Numerous molecular markers, linked with traits of agronomic importance in the food legumes, pea, chickpea and lentil, have been identified. Microsatellite markers are being developed and mapped in a collaborative effort between the International Center for Agricultural Research in the Dry Areas (ICARDA) and the University of Frankfurt to overcome the relatively low amount of information that can be derived from the widely used dominant markers in chickpea. Besides mapping and identifying host plant resistance, efforts are being made to characterize the pathogen populations. Once host plant resistance has been identified and mapped, it will be feasible to deploy the relevant resistance genes when shifts occur in the pathogen population. The technology for using these markers in marker-assisted selection (MAS) has been greatly improved. The ability to use MAS to pyramid genes will make this technology an essential tool for legume breeders.

1.   INTRODUCTION

Biochemical and molecular techniques have been used to identify and evaluate pathogen populations attacking legumes as well as to characterize and map the genes for resistance in the host plant. This knowledge is important when deploying resistance genes and formulating disease control strategies. The molecular characterization of the pathogen population will allow the deployment of suitable resistance genes in the host once the host's resistance genes are tagged.

2.   MOLECULAR TECHNOLOGY FOR THE POPULATION GENETICS OF PLANT PATHOGENS

Identification of *Aschochyta* species of legumes by RAPD analysis. Several species of fungi are responsible for blight on legumes. They are *Ascochyta pisi*, *Mycosphaerella pinodes*, and *Phoma medicaginis* var. *pinodella* on pea, *Ascochyta viciae* on vetch, *Ascochyta fabae* on faba bean, *Ascochyta lentis* on lentil and *Ascochyta rabiei* on chickpea. RAPD analysis with primers RAF-AE10, grouped these fungi into three groups: 1) *A. viciae, A. pisi,* and *A. lentis* 2) *A. rabiei* and 3)*M. pinodes* and *P.m. pinodella*. The primers, RAF-C18 and -C13 differentiated between *A. lentis* and the *A.viciae,* and *A. pisi* and *A. lentis*, respectively. The primer, RAF-E15 revealed 5 distinct groups: 1)*A. fabae*, 2) *A. pisi*, 3) *A. lentis*, 4) *A. rabiei* and 5) *M. pinodes* and *P.m. pinodella*. RAPD analysis also revealed that the three pathogens on pea can be distinguished easily. Furthermore, although *A. lentis* shows similarities with *A. fabae* and *A. viciae*, it is possible to differentiate between them.

R. Knight (ed.), Linking Research and Marketing Opportunities for Pulses in the 21st Century, 421–427.

Genotypic variation in *A. rabiei* isolates from Tunisia by DNA fingerprinting. DNA fingerprinting with simple repetitive sequences (SRS) was used to analyze genotypic variation among 156 isolates of *A. rabiei* obtained from five fields in the Beja region of Tunisia (Morjane et al. 1996). Hybridization of a set of synthetic oligonucleotides complementary to simple repetitive sequences to *Hin*fl or *Rsa*I restricted genomic DNA, revealed 17 genotypes, two of which, were common and represent 38.2% and 22.6% of the population, respectively. Hierarchical diversity analysis demonstrated that the majority of genetic variability was distributed within, rather than between, fields. Genetic variation estimated by Nei's diversity index varied between the extremes of 0.08 and 0.86 with a mean of 0.77. This study further showed that: 1) the genotypes were stable through five subcultures. 2) The genotypes were unevenly distributed across fields. 3) The genotypes occurred at different frequencies in different fields.

Population structure of *Ascochyta rabiei* in Syria. Conventionally, the population structure of a pathogen is determined by pathogenicity surveys (pathogenic variability) and the reaction to a set of differential cultivars. Such a study using three chickpea differential cultivars (ILC 1929, ILC 482 and ILC 3279) in Syria revealed the occurrence of three different pathotypes for *A. rabiei*. Additionally, a set of microsatellite and RAPD markers were also used (Udupa and Weigand 1997) to resolve pathoype diversity in this pathogen. All the pathogenicity surveys (1991-1995) using RAPD markers revealed the predominance of a single genotype (genotype-H and its phenotype is pathotype III) in all the chickpea growing regions, which was not detected during the earlier survey of 1982 (Reddy and Kabbabeh, 1985). The frequency of this genotype has increased in all the chickpea growing regions of Syria in recent years (33% in 1991 to over 80% in 1995). The genotype is highly aggressive and can kill the variety, Ghab 2 (ILC 3279) released for cultivation in Syria.

On the basis of the pathogenicity surveys (1991-1995), it is tempting to speculate on the evolutionary forces acting on the *A. rabiei* population in Syria. The increase in the aggressiveness, observed in the population in response to a change in host resistance in the field, could be due to mutation and selection exerted by the host population. A study of the plant pathogenic fungus, *Rhynchosporium secalis*, by Goodwin et al. (1994) showed that a high rate of mutation could occur for pathogenicity under natural conditions in response to the selection pressure exerted by a host's resistance. This kind of selection-induced genetic change is not due solely to random mutations but also to a process called "adaptive mutation" (Harris et al. 1994) described in bacteria and yeast. In addition to mutation and selection, migration/genetic drift affected genetic changes in the population. The most likely mechanism for migration is infected seeds. Since a single genotype was periodically sampled in a very large area, it can be concluded that only one clone exists. This observation further supports the view that under natural conditions the pathogen preferably reproduces asexually and sexual recombination plays a minor role in evolution of new pathotypes of *A. rabiei*. Furthermore, the increase in aggressiveness in the population over the years in response to resistance in the host-population suggests that selection is a strong force in establishing and maintaining genetic change in the population.

The pathogenicity surveys using DNA markers demonstrated there is a need to develop chickpea cultivars with stronger levels of resistance to the most predominant genotype (genotype-H) of the pathogen. New sources of resistance have been identified in chickpea. It could also be established that the level of resistance to the genotype-H could be improved through conventional pyramiding of resistance genes. Since the pathogen seems to migrate by means of infected seeds (as revealed by DNA markers), the use of disease free seeds for sowing and treating the seed with suitable fungicides can help in the management of the disease.

2. Molecular technology for genome mapping and gene tagging in lentils, chickpeas and peas

The aim when developing linkage maps in crops is to localize in the genome the position of important agronomic traits and to identify tightly linked markers to enable indirect selection by marker assisted selection. F_2 populations, or populations of random near recombinant inbred lines (RILs), are most commonly used for mapping.

Lentil

To maximize the polymorphism within the populations to be mapped, wide crosses are often made, frequently resulting in a distorted segregation. A test of distortion in the F_2 of potential mapping populations in later generations was carried out on two populations (Eujayl et al, 1997). One of the parents of each cross contained considerable amounts of wild germplasm of *Lens culinaris* ssp. *orientalis*. In cross 1, half of the RAPD markers exhibited segregation distortion, which was also found for isozyme and morphological loci. By contrast, cross 2 did not show distortion. Reduced recombination in interspecific crosses or chromosomal rearrangements may lead to segregation distortion and non-representative genetic distances and linkage relationships. Segregation distortion in interspecific crosses can be explained by small chromosomal rearrangements with little impact on chromosome pairing during meiosis and on fertility, linkage to an incompatibility locus or linkage to a lethal allele in gametes. Potential mapping populations can be analyzed for segregation distortion in the F_2 with a number of markers prior to the development of a comprehensive linkage map using later generations.

An F_2 population of the cross L92-013 (L962-15-8 (F) x ILL5588) was used to develop RILs by advancing F_2 individual plants to F_6 using Single Seed Descent (SSD) (Eujayl et al. 1997). In total 178 markers (89 RAPD, 79 AFLP, 7 RFLP and 3 morphological markers) were mapped in seven major linkage groups covering 1073 cM, with an average distance of 6.0 cM between adjacent markers. Using anchor RFLP markers we confirmed the synteny among lentil, pea and chickpea genomes (Tahir et al 1993). The morphological markers, pod indehiscense, seed coat spotting and flower color loci were mapped. Out of the total linked loci, 8.4 % showed segregation distortion. More than one quarter of the distorted loci were clustered in one linkage group. The loci detected by AFLP markers showed more segregation distortion than the RAPD markers. The AFLP and RAPD markers were intermingled and little clustering of AFLPs was observed. So far, only a loose linkage to fusarium wilt resistance and reaction to cold has been identified. However, the marker density of this map could be used for identification of markers linked to QTLs in this population.

Chickpea

Characterization of microsatellite markers and genome mapping. Microsatellites or simple sequence repeats (SSR) are tandemly arranged repetitions of 1 to 5 nucleotide units with unit numbers ranging from 2 to several hundreds or even thousands. They are ubiquitous and evenly distributed components of all eukaryotic genomes investigated so far. Microsatellites frequently change their length by replication slippage, unequal crossing-over or other less understood mechanisms, thereby generating a wealth of highly polymorphic alleles.

Using microsatellite-flanking sequences as primers, to amplify the microsatellites in between, is currently the most effective way of using microsatellite markers for genome mapping. The resulting locus-specific amplification products often exhibit considerable length differences, due to the variable number of tandem repeats (VNTR) within the microsatellite. These are called sequence-tagged microsatellite site markers (STMS). A high degree of polymorphism in these loci makes them suitable for genome mapping in chickpea. The crop is self-pollinating and exhibits only low levels of intraspecific variability for RFLPs and RAPDs (Udupa et al. 1993). STMS are single-locus, co-dominant markers with high polymorphic information content that have the potential for automatic, non-radioactive detection, and are reliable and easy to use.

A pilot study, in which several microsatellite motifs were hybridized against small-insert libraries, demonstrated that microsatellites are abundant in chickpea. As shown in Table 1, the three most abundant motifs, 1 tri- and 2 di-nucleotide repeats, account for more than 12,000 loci with an average genomic spacing of around 60 kb between two microsatellites assuming an equal distribution of loci within the genome. Because $(TAA)_n$ repeats were not only most abundant, but also contained the highest number of repeat units (which correlates with their polymorphism) further studies on the development of STMS markers focussed on the isolation of such loci and their conversion into STMS.

Table 1: Abundance, average genomic spacing and repeat unit numbers of three microsatellite motifs in the chickpea genome.

Motif	No. of loci in genome	Average distance between loci (kb)	Clones	No. repeat sequences
$(TAA)_n$	5,680	130	13	5-54
$(GA)_n$	3,780	195	13	9-32
$(CA)_n$	2,600	283	5	4-42
Total	12,060	61.2	31	

Using the same approach as in the pilot study, around 200,000 clones with insert sizes ranging from 350 to 750 bp were hybridized against $(TAA)_5$. A total of 600 (TAA)n microsatellite-containing clones were detected of which 242 were sequenced. 131 repeat units were perfect with repeat unit number ranging from 9 to 131.

For 184 microsatellite-flanking sequences, primers could be designed and used for the detection of microsatellite length polymorphisms in 6 relevant chickpea breeding cultivars and in *C. reticulatum* and *C. echinospermum.* Of these primer pairs, 112 generated clear patterns that were at least polymorphic between the species. In many cases all 8 tested accessions exhibited different allele lengths even on agarose or native polyacrylamide gels. In other studies many primer pairs produced bands from several annual species of the genus *Cicer* and from *C. anatolicum,* a perennial species thought to be closely related phylogenetically to chickpea. This indicates that microsatellite-flanking sequences are conserved within the genus *Cicer.* For some primer pairs 2 $(CT)_n$-, 2 $(GA)_n$ and several $(TAA)_n$ SSRs, amplified the respective loci from pea DNA, demonstrating the potential of STMS for use in synteny studies within legumes.

For the generation of an STMS marker map of the chickpea genome two populations of randomly recombinant inbred lines (RILs) were used that were obtained by advancing F_2 individual plants to F_6 using Single Seed Descent (SSD). One of the two RIL populations was derived from a wide cross between *C. reticulatum* and chickpea accession ICC 4958, whereas the other was derived from a narrow cross between a kabuli chickpea accession, C104, and desi accession WR 315. The latter population segregates for resistance to *Fusarium oxysporum* races 1 and 4, whereas the former segregates for several physiological and morphological traits.

In the wide cross, 82 STMS markers were mapped in 10 linkage groups (LGs), that at a LOD score of 3, cover 570 cM of the chickpea genome with eight markers being unlinked. One feature of this map is the dense clustering of microsatellite markers in distinct genomic regions. This phenomenon is most pronounced in the clusters around markers TA135 and CaSTMS28 in LG7 and between markers CaSTMS7 and TA5 in LG1. Several loci already located in the wide cross were also mapped in the narrow cross. In the narrow cross, 45 markers were mapped at a LOD score of 5 spanning around 210 cM in 9 LGs. Ten markers were unlinked. The position of markers is indicated in the map relative to their position in the wide cross map. At many loci both maps arecolinear, with slight changes in the relative position of closely linked markers. For example, the relative position of TA71, TR59 and TA5 on LG1 in the wide cross is changed in the narrow cross. In few cases markers map to very different positions or are closely linked in one but unlinked in the other cross. A possible reason for the discrepancies between the maps could be the high level of segregation distortion found at these loci.

In both populations skewed segregation of markers was observed. Unexpectedly, biased segregation was more pronounced in the narrow cross than in the wide cross (Table 2). In the wide cross a general tendency in favor of *C. reticulatum* alleles (68 %) was observed. In the wide cross 24 of the 82 markers (29%) showed strong segregation distortion as judged from χ^2 tests (p= 0.05). Of the loci, 23 exhibited a strong preference for the presence of the *C. reticulatum* allele. The majority of distorted loci were clustered at two distinct regions in LG1 and 7, each comprising 7 markers. It is at these loci, that discrepancies between narrow and wide cross maps are most pronounced. Some of these loci also segregate distortedly in the narrow cross. In this cross 73 % of the loci with biased segregation showed a bias towards the presence of the WR 315 allele, 22 % of markers were predominantly obtained from C104. In the narrow cross 19 out of 45 markers (42 %) displayed distorted segregation according to χ^2 tests with the WR 315 allele being dominant in 17 cases. In both populations, 4 % of markers showed perfect Mendelian inheritance.

Table 2: Summary of the segregation of STMS markers in near RILs from chickpea derived from a narrow and a wide cross. Numbers in brackets indicate the relation of loci that were preferentially obtained from one or the other parent.

Cross	Biased segregation %	Unbiased segregation %	Distorted segregation %
Narrow	96 (73 : 23)	4	42 (17 : 2)
Wide	96 (68 : 28)	4	29 (19 : 1)

Tagging genes for host plant resistance in chickpea. In chickpea, several important traits were characterized including resistance to fusarium wilt (races 1 and 4). Four sets of RILs were used for the analysis, one of which was from the wide cross *C. reticulatum* x FLIP 84-91C. This set of lines is about 10 times more polymorphic than the narrow crosses. The candidate markers were identified through selective genotyping of the RILs. In the case of fusarium wilt race 1, two markers (CS27 and UBC170) linked to a recessive gene for resistance to wilt were found (Mayer et al. 1997). This is the first report of a marker linked to wilt resistance in chickpea. It has about 7% recombination. One of the markers (CS27) was converted to ASAP (allele-specific associated primers). In the C-104 x WR315 set of RILs (100 lines) the linkage of resistance to race 4 to markers CS27 and UBC170 was determined. This is the first indication that the genes for resistance to races of wilt may be clustered in the chickpea genome (Tullu et al. 1997). Additionally, an inter-simple sequence repeat (ISSR) marker (UBC855) was also linked to the gene for resistance to race 4 in repulsion. The UBC855 marker is 0.6 cM from CS27 and closer to the gene for resistance (Ratnaparkhe et al. 1997, submitted to TAG). Both markers were mapped onto linkage group 6 in chickpea. In a different cross, CS27 was linked to one of the genes for resistance to race 4 at 10.7 cM. This was done using F3 families and some error is expected when compared to that from RILs. The result indicated that there is a gene for early versus late wilting irrespective of the gene for resistance that was analyzed.

Peas

Marker-assisted selection in peas. Marker-assisted selection in peas was initiated when the morphological markers *k*, *uni*, and *wlo* were found to be linked to the genes conferring resistance to pea common mosaic virus, pea enation mosaic virus, and pea seed borne mosaic virus, respectively (Marx and Provvidenti, 1979; Marx et al., 1986; Hagedorn and Gritton, 1971). The problem with using these morphological markers is that they are not usually desired in the final variety. Thus, after the resistance gene has been introgressed into the desired genetic background, back crossing and selection are necessary to break the linkage between the morphological marker and the gene.

Molecular markers offer an alternative that does not require a final cross to break specific linkages because such markers do not significantly affect plant phenotype. In addition, DNA techniques access the entire genome, facilitating the identification of markers closely linked to the target gene. Clearly, the more closely the marker is linked to the gene the more reliable marker-assisted selection becomes. It is possible to actually have a marker for the target gene itself, as is typical for transformation studies. However, for most of the interesting genes in cool season food legumes we have yet to reach that level of sophistication.

In pea there are now are at least 20 commercially important genes tagged with DNA markers (Table 3). Most genes conferring resistance to viral diseases have at least one DNA marker. *En* has at least four, with two of these having been converted to a type of marker that eliminates the need for electrophoresis of DNA fragments (Yu et al., 1995). One gene conferring resistance to powdery mildew has three markers (Timmerman et al. 1994, Dirlewanger et al. 1994), one of which also appears to be useful for marking a gene involved in the expression of tolerance to common root rot (Cargnoni et al, 1994). A region involved in conferring tolerance to Ascochyta was clearly identified in one study involving multiple markers (Dirlewanger et al. 1994), and QTLs for seed weight and color have been identified in other studies.

Table 3. Genes of interest to breeders with linked molecular markers in pea.

Gene	Function	Marker	Reference
En	Resistance to PRMV	Adh-1,	Weeden and Provvidenti, (1988)
		B500, P256	Yu et al. (1995)
		PeaFlo (=uni)	Marx et al. (1985)
mo	Resistance to BYMV	B302	Yu et al. (1996)
sbm-1	Resistance to PSbMV	G185	Timmerman et al. (1994)
cyv-2	Resistance to CYVV	G185	Timmerman et al. (1994)
er-1	Resistance to powdery mildew		
		OPD10	Timmerman et al. (1993)
		OPA5	Cargnoni et al. (1994)
Fw	Resistance to race 1 of Fusarium wilt	P236	Dirlewangeret al. (1994)
	Lack of anthocyanins	P254	Dirlewangeret al. (1994)
a	Lack of anthocyanins	PCD72	Weeden et al. (1996)
a-2	Leafless	OPA11	Weeden (unpublished)
af	Green cotyledon	CMH49	Weeden and Wolko (1990)
i	Wrinkled seed	NcoI	Weeden and Wolko (1990)
r	Wrinkled seed	Pjam425	Burton et al. (1995)
rb	Photo-period Insensitivity	Vicilin-5	Lee et al. (1988)
Sn	Aphanomyces tolerance		
	Resistance to ascochyta	Sod	Weeden et al. (1996)
QTL	blight	OPA5	Cargnoni et al. (1994)
2 QTLs		P236	Dirlewangeret al. (1994)
	Seed size	P227	Dirlewangeret al. (1994)
		P105	Dirlewangeret al. (1994)
QTL	Color retention	CM185	Timmerman-Vaughan et al. (1996)
		P9	Timmerman-Vaughan et al. (1996)
QTLs		Pjam425	McCallum et al. (1997)
		Lg-J	McCallum et al. (1997)

The actual implementation of marker-assisted selection in a breeding program has been greatly simplified and streamlined by recent advances in DNA technology. Most new DNA markers are either PCR-based or are being converted to this type of marker. RFLP technology requires considerably more DNA per assay than PCR systems. Particularly in crops with large genomes, such as pea, lentil, and faba bean, several leaves must be harvested for the DNA extraction step, precluding screening of very young plant material. For PCR techniques a leaf sample 5mm in diameter will suffice for 200 assays. Furthermore, the minute amounts of DNA needed for PCR analyses allows the extraction procedure to be simplified to such an extent that the processes can be automated, and inexpensive machines are now available that can process up to 1,000 samples per day.

To further simplify the screening procedure, allele-specific associated primers (ASAPs) (Gu et al., 1995) have been developed for several of the DNA markers in pea that permit the presence of the gene to be determined by a fluorescent assay directly after PCR (Yu et al., 1995, 1996). The approach basically uses primers that will amplify a major product only if the desired gene/marker is present. After amplification, ethidium bromide is added to each sample and the samples analyzed (either visually or by machine) for an intense fluorescence indicative of the presence of a large amount of double stranded DNA. If there is a marker that does not require an electrophoresis step, then the rate-determining factor for the number of samples processed per day to identify desirable genotypes is probably the number of PCR reactions. With two programmable thermo-cycling machines a laboratory could easily perform 200 to 500 PCR reactions per day. This should be sufficient to serve a modest to large breeding program at a cost per sample barely above that for the *Taq* polymerase, used in the amplification reaction.

With DNA markers being already available for many of the genes desired in a variety, and the process of tagging additional genes becoming relatively simple, it appears certain that marker-assisted selection will play an important role in future breeding programs. One problem that will have to be dealt with is how to maintain a sufficient number of plants beyond the marker-assisted selection step if the population is to be screened for five or more recessive genes. The expected frequency of the quintuple homozygotes in an F_2 population segregating for five recessive genes is 0.001, leaving a breeder very little to work with from a population of 1000 if only the quintuple recessive phenotype is selected. At least in the pulses, we should be able to maintain population size through the marker-assisted selection process by screening for the presence of the genes (homozygous and heterozygous genotypes), assuming that heterozygous plants could always be converted to the desired homozygous genotype by selfing. Under this scheme the screening for five genes would reduce a population of 1000 to approximately 240 plants, a number that would permit further selection on agronomic characters in the field.

References:

Burton, R.A., Bewley, J.D., Smith, A.M., Bhattaacharyya, M.K., Tatge, H., Ring, S., V. Bull, V., Hamilton, W.D. and Martin, C. 1995 *Plant Journal* 7: 3-15.

Cargoni, T.L., Weeden, N.F. and Gritton, E.T. 1994 *Pisum Genetics* 26: 11-12.

Dirlewanger, E., Iosaac, P.G., Ranade, S., Belajouza, M., Cousin, R. and deVienne, H. 1994 *Theoretical and Applied Genetics* 88: 17-27.

Eujayl, I., Baum, M., Erskine, W., Pehu, E. and Muehlbauer, F.J 1997 *Euphytica* 96: 405-412.

Eujayl, I., Baum, M., Erskine, W., Pehu, E. and Muehlbauer, F.J. 1995 In *Proceedings of second European Grain Legumes Conference*, Copenhagen, Denmark, 440-441.

Eujayl, I., Baum, M., Erskine, W., Powell, W. and Pehu, E. 1998 Submitted to *Theoretical and Applied Genetics*

Goodwin, S.B., Webster, R.K. and Allard, R.W. 1994 *Phytopathology* 84: 1047-1053.

Gu WK, Weeden NF, Zu J and. Wallace, D.H. 1995 *Theoretical and Applied Genetics* 91: 465-470.

Hagedorn, D.J. and Gritton, E.T. 1971 *Pisum Newsletter* 3:16.

Lee, D., Turner, L., Davies, D.R. and Ellis, T.H.N.1988 *Theoretical and Applied Genetics* 75: 362-365.

Harris, R.S. Longerich, S. and Rosenberg, S.M. 1994 *Science* 264: 258-260.

Marx, G.A. and Provvidenti, R. 1979 *Pisum Newsletter* 11:28-29.

Marx, G.A., Weeden, N.F.D. and Provvidenti, R. 1985 *Pisum Newsletter* 17: 57-60.

Mayer, M.S., Tullu, A., Simon, C.J., Kumar, J., Kaiser, W.J., JKraft .M. and Muehlbauer, F.J. 1997 *Crop Science* (in press).

McCallum, J.A. Timmerman-Vaughan, G., Frew, T.J. and Russell, A.C. 1997. *Journal of American Society of Horticultural Science* 122: 218-225.

Ratnaparkhe, M.B., Santra, D.K., Tullu, A. and Muehlbauer, F.J. 1997 Submitted to *Theoretical and Applied Genetics.*

Reddy, M.V. and Kabbabeh, S. 1985 *Phytopathologia Mediterranea,* 24: 265-266.

Timmerman-Vaughan, GM., McCallum, J.A., Frew, T.J., Weeden, N.F. and Russell, A.C. 1996. *Theoretical and Applied Genetics* 93: 431-439.

Tahir, M., Simon, C.J. and Muehlbauer, F.J. 1993 *LENS Newsletter* 20:3-9.

Timmerman, G.M., Frew, T.J., Weeden, N.F., Miller, A.L. and Jermyn, W.A.1993. *Theoretical and Applied Genetics* 85: 609-615.

Timmerman, G.M., Frew, T.J., Miller, A.L., Weeden, N.F. and Jermyn, W.A. 1994. *Theoretical and Applied Genetics* 88: 1050-1055.

Tullu, A., Muehlbauer, F.J., Simon, C.J., Mayer, M.S., Kumar, J. and Kraft, J.M.1997 Submitted to *Euphytica*

Udupa, S.M. and Weigand, F. 1997 In *DNA markers and breeding for resistance to ascochyta blight in chickpea,* pp 67-78 (Eds S.M. Udupa and F. Weigand). Aleppo: ICARDA.

Udupa, S.M., Sharma, A., Sharma, R.P and Pai, R.A. 1993 *Journal of Plant Biochemistry and Biotechnology* 2: 83-86.

Weeden, N.F. and Provvidenti, R. 1988 *Journal of Heredity* 79: 128-130.

Weeden, N.F., Swiecicki, W.K., Timmerman-Vaughan, G.M., Ellis, T.H.N., and Ambrose, M. 1996 *Pisum genetics* 28: 1-4.

Weeden, N.F. and Wolko, B. 1990. In *Genetic Maps*, 6.106-6.112 (Ed. S.J. O'Brien). Cold Spring Harbor: Cold Spring Habor Laboratory Press.

Yu J, Gu, WK, Provvidenti R., and Weeden N.F 1995. *Journal of American Society of Horticultural Science* 120: 730-33.

Yu, J., Gu, W.K., Weeden, N.F. and Provvidenti, R. 1996. *Pisum Genetics.* 28: 31-32.

Tissue Culture and Protoplast Fusion of Cool-Season Pulses: Pea (*Pisum sativum* L.) and Chickpea (*Cicer arietinum* L.)

Short Title: Tissue Culture and Protoplast Fusion of Pulses

E. C. K. Pang, J. S. Croser, K. T. Imberger, J. S. McCutchan and P. W. J. Taylor

Joint Centre for Crop Improvement, Institute of Land and Food Resources, The University of Melbourne, Parkville 3052, Victoria.

Abstract

This review discusses recent work on the *in vitro* regeneration of complete plants of pea (*Pisum sativum* L.) and chickpea (*Cicer arietinum* L.), anther culture in both species, and the feasibility of producing somatic hybrids between pea and *Lathyrus* sp. Effective protocols are available for regeneration *via* organogenesis and somatic embryogenesis in both species, though a few problems, such as root induction and the germination of somatic embryos are reported in a few cases. There is a paucity of information pertaining to anther culture of both species, and much work is necessary before this procedure may be used to produce pea and chickpea doubled-haploids on a routine basis. Somatic hybridisation between pea and other species has been attempted previously, but without success. Current attempts at producing pea-*Lathyrus* hybrids are in their early stages, and remains somewhat speculative.

Abbreviations: BAP - 6-benzyladenine; B5 - Gamborg's (1968) B5 Medium; 4-CPA - 4-chlorophenoxyacetic acid; 2,4-D - 2,4-dichlorophenoxy acetic acid; GA_3 -gibberellic acid; IAA - Indole acetic acid; IBA - indole-3-butyric acid; MS - Murashige and Skoog (1962) Medium; NAA - α-napthaleneacetic acid; 2,4,5-T - 2,4,5-trichlorophenoxyacetic acid; TDZ - thidiazuron (N-phenyl-N'-1,2,3-thidiazol-5'-ylurea); WH - White's (1943) Medium.

1. INTRODUCTION

In vitro techniques are important complements to the repertoire of tools available for modern plant improvement programs. The ability to regenerate seed-bearing plants from a variety of cell and tissue (explant) sources is crucial to the successful application of *in vitro* methods to the improvement of pea (*Pisum sativum* L.) and chickpea (*Cicer arietinum* L.). A reliable and efficient regeneration system is a prerequisite for the implementation of techniques which introduce new variation into existing crop cultivars, *e.g.* genetic transformation, somacloning and somatic hybridisation (protoplast fusion). Pulses have long been considered to be recalcitrant to cell and tissue culture, but progress has been made recently, particularly in the regeneration of complete plants from calli, immature cotyledons, embryonic axes and shoot meristems. However, pea and chickpea have remained difficult subjects in other areas, notably anther/microspore culture and the regeneration from protoplasts. The aim of this review is to evaluate the recent advancements in regeneration systems for pea and chickpea, and to discuss the rationale for, and the feasibility of producing somatic hybrids between pea and *Lathyrus* sp. For additional information on biotechnology, the reader is referred to reviews by Griga and Novak (1990) and Atlaf and Ahmad (1990).

2. REGENERATION SYSTEMS

2.1 Introduction and historical perspective

Regeneration systems developed for pea and chickpea may be divided into three categories,

R. Knight (ed.), Linking Research and Marketing Opportunities for Pulses in the 21st Century, 429–436.

Organogenesis from pre-existing meristems,
De novo organogenesis (from newly-formed meristematic zones in calli), and
Somatic embryogenesis

In pea, the induction of organogenesis from pre-existing meristems by meristem-tip or axillary bud culture was first demonstrated by Kartha *et al.* (1974). *De novo* organogenesis was first achieved by Hildebrandt *et al.* (1963) from callus cultures initiated from stem segments of pea. Somatic embryogenesis from callus (cell-suspension) cultures was first reported by Jacobsen and Kysely (1984), where torpedo-shaped embryos were formed, but the regeneration of complete plants was only achieved by Kysely *et al.* (1987) using callus derived from shoot apices of immature embryos. In chickpea, the production of complete plants by organogenesis was first demonstrated by Bajaj (1979), but the same result was not achieved *via* somatic embryogenesis until about ten years later (Rao and Chopra, 1989).

Recent efforts on plant regeneration have been mainly directed toward (a) the development of protocols for regenerating plants following *Agrobacterium*-mediated transformation of pea (*e.g.* Schroeder *et al.* 1993; Özcan *et al.* 1993) and chickpea (*e.g.* Kar *et at.* 1996; Adkins *et al.* 1995), or after transformation by electroporation or direct DNA uptake (*e.g.* Puonti Kaerlas *et al.* 1992) and (b) Rapid propagation (micropropagation) of desirable genotypes or early generation populations (*e.g.* Malik and Saxena 1992b; Kosturkova *et al.* 1997). Relatively modest efforts have been expended on producing regenerants from anthers or microspores of pea and chickpea (haploid, or doubled-haploid plant production), or the regeneration of plants from heterokaryons following somatic hybridisation between pea and various species. The following sections will examine the progress made in the development of regeneration systems for transformation studies and micropropagation of pea and chickpea, while the topics of anther culture and protoplast fusion will be discussed in subsequent sections.

2.2   Pea

Recently, a number of authors reported rapid rates of plant regeneration *via* organogenesis, with varying efficiencies (50-90%) of root induction (Table 1). Seeds, hypocotyls, or immature embryonic axes were usually plated on media containing a high cytokinin:auxin ratio (Table 1), though for cotyledon explants, a 10:1 ratio of NAA:BAP was found to produce the highest frequency of regenerants (Özcan *et al.* 1993). Rates of up to 20 shoots/explant in two weeks were reported (Sanago *et al.* 1996), and seeds may be collected from regenerated plants 9-11 weeks after initial culture, representing a significant improvement over other published procedures. The frequency of shoot induction was found by most authors to vary with genotype, and a previous report by Malmberg (1982) indicated that this trait is governed by a single recessive gene. For cotyledon explants, orientation was found to significantly affect shoot regeneration, as the highest rates were achieved from cotyledons plated with their distal ends in contact with the medium (Özcan *et al.* 1993). Rooting of regenerated shoots was achieved using basal MS or B5 media supplemented with NAA or IBA or both, with varying degrees of success (Table 1).

Somatic embryos were induced from cultured cotyledon and shoot apices at varying efficiencies (Table 1). In pea, the induction of somatic embryogenesis and the subsequent development of embryogenic structures may occur on the same (induction) medium, without the transfer onto a development medium (Bencheikh and Gallais, 1996a). The induction media used by the various authors usually consisted of MS (macro and micro salts) supplemented with B5 vitamins and a strong auxin (either 2,4-D, picloram or 4-CPA). Although very high embryogenic efficiencies were reported by all authors for some genotypes (Table 1), not all the somatic embryos produced germinated. Loiseau *et al.* (1995) reported, for example, that only 40% of the somatic embryos they obtained could be converted into plants. The reasons for this remains unclear, but may be associated with the high number of abnormal or poorly developed embryos (arrested development at the globular or torpedo stage) observed by most authors. Somatic embryogenesis was found to be highly genotype-dependent, and a recent study by Bencheikh and Gallais (1996b) indicated that this trait was under oligo- or poly-genic control, with a tendency for induction amenability to be governed by recessive genes.

Table 1. Summary of recent reports on *in vitro* morphogenesis in *P. sativum.*

Explant	Regeneration Type	Media	Final Product, and Efficiency	Authors
Seed	Organogenesis	MS + 50 μM BAP	Shoots, 4-8 shoots in 6 weeks.	Malik and Saxena (1992a)
Seed	Organogenesis	MS + 1-50 μM TDZ, Rooting on MS + 2.5 μM NAA	Complete plants, 9 shoots in one week. Rooting efficiency not mentioned.	Malik and Saxena (1992b)
Cotyledon	Organogenesis	MS + 2.2 μM BAP + 21.5 μM NAA, Rooting on 1/2 MS + 4.9 μM IBA	Complete plants, 12-14 shoots per explant after 5-6 weeks. 80-90% root formation.	Özcan *et al.* (1993)
Hypocotyl segments	Organogenesis	MS + (0.5-1.0) μM TDZ, Rooting on MS + NAA or IBA (1-2 μM)	Complete plants, 20 shoots per explant in 2 weeks. 50-60% root formation.	Sanago *et al.* (1996)
Immature embryonic axis	Organogenesis	MS + 10 mM BAP + 1.0 mM NAA Rooting on B5 + (5-25 mM IBA)	Complete plants, 3-4 shoots per explant in 3 weeks. 60% root formation.	Kosturkova *et al.* (1997)
Cotyledon	Somatic embryogenesis	MS + (27-215) μM NAA + (23-181) μM 2,4-D	Complete plants, 70-80% embryogenic efficiency, 4-9 embryos per explant. Germination rates of somatic embryos not mentioned.	Özcan *et al.* (1993)
Shoot apices	Somatic embryogenesis	MS + 4.5 μM Picloram OR 45 μM 4-CPA + 336 mM Fructose	Complete plants, 70-100% embryogenic efficiency, 40% germination of somatic embryos.	Loiseau *et al.* (1995)
Shoot apices	Somatic embryogenesis	MS + 2.0 mM Picloram	Somatic embryos. Max. 88% embryogenic efficiency.	Bencheikh and Gallais (1996a)

2.3   Chickpea

Efficient organogenesis from cultured explants of chickpea was recently demonstrated by a number of authors (Table 2), notably Kar *et al.* (1996) and Polisetty *et al.* (1997). The choice of explant appears to be a crucial factor in most procedures, with most authors (with *Agrobacterium*-mediated gene transfer in mind) favouring the use of immature cotyledons or embryonic axes (Table 2). A variety of cytokinins have been employed in the different experiments, but it is difficult to assess the efficacy of one over the other, given the differences in the explants and conditions used (Table 2). Some workers (Malik and Saxena 1992b; Shri and Davis, 1992; Adkins *et al.* 1995) experienced considerable difficulties in inducing roots from regenerated shoots, and where rooted plants were transferred to the glasshouse, a poor rate of survival was observed (Adkins *et al.* 1995). It has been observed that roots which develop at the base of callus-bearing regenerated chickpea shoots were often not physically attached to such shoots, or that the vascular connections between shoot and root(s) were poorly developed or altogether absent (Atlaf and Ahmad, 1990). Excessive vitrification of shoots (hyperhydricity) was also observed, especially when TDZ was employed (Malik and Saxena 1992b) and root formation was inhibited when shoots were subjected to prolonged culture on high BAP medium (Polisetty *et al.* 1997). There is a consensus that the frequency of organogenesis from callus is highly genotype-dependent, but no study is available to suggest whether such observed differences are heritable.

The regeneration of complete plants *via* somatic embryogenesis is a relatively recent development in chickpea. Consequently, there are few reports on the successful application of this procedure; the most recent of these are summarised in Table 2. The induction media usually comprised of MS supplemented with a strong auxin such as 2,4-D, 2,4,5-T or picloram. Some authors experienced difficulties in converting somatic embryos to plants (Barna and Wakhlu 1993), but the addition of zeatin to the maturation medium appeared to ameliorate the problem (Sagare *et al.* 1993).

Table 2. Summary of recent reports on *in vitro* morphogenesis in *C. arietinum*.

Explant	Regeneration Type	Media	Final Product, and Efficiency	Authors
Seed	Organogenesis	MS + (1-50) µM TDZ Rooting on MS + 2.5 µM NAA	Complete plants, 70 shoots per explant after 5 weeks. Stunted roots.	Malik and Saxena (1992b)
Immature cotyledons	Organogenesis	MS + 13.7 µM Zeatin + 0.2 µM IAA	Shoots, 64% explants forming cotyledon-like structures, 50% of these sprouted shoots within 20-40 days. Problems with root induction	Shri and Davis (1992)
Immature leaflets, shoot tips	Organogenesis from callus	MS + 10 µM NAA + 5.0 µM BAP (Callus induction), then MS + 10 µM BAP + 0.1 µM IBA Rooting on MS + 1.0 µM IBA	Complete plants, 1.8 shoots per callus after 4 weeks, 80% root formation.	Barna and Wakhlu (1994)
Meristem tips and cotyledonary nodes	Organogenesis	B5 + 4.4 µM BAP Rooting on WH + 2.5 µM IBA	Complete plants, 7-10 shoots per node in 3 weeks.68-72% root formation.	Brandt and Hess (1994)
Immature cotyledons, embryonic axis	Organogenesis	MS + 4.6 µM Zeatin + 0.2 µM IAA OR MS + 13.7 µM Zeatin + 0.2 µM IAA Rooting on MS + 11 µM NAA + 0.5 g/L activated charcoal	Complete plants,10-20 embryo-like structures per cotyledon explant in 2 weeks. 30% root formation.	Adkins *et al.* (1995)
Embryonic axis (Decapitated, De-radicled)	Organogenesis	MS + 4 X MS Micro Salts + 13 µM BAP + 0.005 µM NAA Rooting on MS + 0.25 µM IBA	Complete plants, 22 shoots per explant after 3 weeks, > 90% root formation.	Kar *et al.* (1996)
Seed (De-radicled)	Organogenesis	MS + 5.0 µM BAP Rooting on 1/4 MS +0.75% sucrose	Complete plants, Max. 14 shoots per explant after 30 days. >90% root formation.	Polisetty *et al.* (1997)
Immature leaflets	Somatic embryogenesis	MS + 25 µM 2,4-D Maturation on MS + 15 µM GA$_3$ + 1.0 µM IBA	Complete plants, 71% Embryogenic efficiency, 2% germination of somatic embryos.	Barna and Wakhlu (1993)
Immature cotyledons and embryonic axis	Somatic embryogenesis	MS + 11.7 µM 2,4,5-T Maturation 1/2 MS + 4.6 µM Zeatin	Complete plants, 50-70% embryogenic efficiency, 70% germination of somatic embryos.	Sagare *et al.* (1993)
Leaf Explants (13-d.o. seedlings)	Somatic embryogenesis	MS + 1.1 µM 2,4-D + 1.0 µM Picloram + 0.5 µM BAP Maturation on B5 + 1.1 µM BAP, then basal B5	Complete plants, 50% germination of somatic embryos.	Dineshkumar *et al.* (1995)

2.4   Factors Affecting the Success of Plant Regeneration

The recent literature suggests that a variety of factors contributed to the success of regeneration procedures, some of which have been outlined in previous reviews by Griga and Novak (1990) and Atlaf and Ahmad (1990). The most important of these factors appear to be the genotype(s) used in regeneration experiments, and by extension, the interaction between the genotypes and the media and cultural conditions used. The amenability of different genotypes for regeneration *via* organogenesis and somatic embryogenesis in pea has been shown to be heritable, and governed in the main, by recessive genes and additive effects

(Malmberg 1982; Bencheikh and Gallais 1996a,b). Genotypic effects are similarly reported for chickpea, but little is known of the heritability of regeneration capacity in this species. For both pea and chickpea, there is consensus that the choice of explant is extremely important, and in the instance of cotyledon explants, its orientation in the medium appears to significantly affect the rate of organogenesis. A few authors have reported difficulties in inducing root formation and subsequent poor transplantation survival of regenerants, indicating the need for further research in this area. Similarly, those engaged in producing plants via somatic embryogenesis have encountered problems with the germination of somatic embryos, highlighting the need to formulate more effective media and cultural conditions for (a) the production of higher proportions of cotyledon-type somatic embryos and (b) the maturation and germination of such embryos.

3.   ANTHER CULTURE IN PEA AND CHICKPEA

Anther or microspore culture is an attractive means of rapidly producing homozygous plants, and is being used successfully to produce pure-breeding lines from early segregating generations (F_1 or F_2) in e.g. canola (*Brassica napus*) for field evaluations. Such pure breeding lines are achieved in one generation, as opposed to the six or seven generations of inbreeding required via conventional means. Further, desirable recessive genes masked in early segregating generations are exposed and expressed in the doubled-haploid lines, allowing for their earlier selection. In genetic mapping studies, anther/microspore culture may be used to produce a set of pure breeding lines from the F_1, allowing for the mapping of quantitative traits tested in a number of environments. The current procedure for producing similar lines involves a time consuming advancement of random plants from the F_2 to the F_6-F_7 to produce recombinant inbred lines (R.I.Ls), which may however still contain residual heterozygosity.

Table. 3 Summary of reports on anther culture in *P. sativum* and *C. arietinum*

Explant	Regeneration Type	Media	Final Product, and Efficiency	Authors
Pisum sativum				
Anthers with microspores at uninucleate stage	N/A	MS + 23 µM IAA + 9.5 µM Kinetin	Callus and Embryo-Like Structures (ELS) 70% plated anthers produced callus. 0.3% "Pollen Embryos"	Gosal and Bajaj (1988)
Anthers with microspores at uninucleate stage	Organogenesis (Shoot Formation from Callus)	B5 + 4.5 µM 2,4-D + 0.75 mM Adenine + 292 mM Sucrose	Shoots, 10% plated anthers produced shoots (from callus)	Anderson (Unpublished)
Anthers with microspores at early-mid uninucleate stage	N/A	MS + 9 µM 2,4-D + 4 µM Picloram + 10 µM IAA + 292 mM Sucrose	Callus and ELS, 35% plated anthers produced callus	Imberger *et al.* (Unpublished)
Cicer arietinum				
Anthers with microspores at uninucleate and binucleate stages	Organogenesis (Root Formation from Callus)	MS + 9 µM 2,4-D + Coconut Milk (10% v/v) Regeneration: MS + 11 µM NAA + 0.1 µM BAP	Callus and roots	Khan and Ghosh (1983)
Anthers with microspores at uninucleate stage	N/A	MS + 23 µM IAA + 9.5 µM Kinetin	Callus and Embryo-Like Structures (ELS) 62% Plated Anthers Produced Callus, 0.5% "Pollen Embryos"	Gosal and Bajaj (1988)

In vitro morphogenesis of cultured anthers was first demonstrated in pea by Gosal and Bajaj (1988) and in chickpea by Khan and Ghosh (1983). These authors were not successful in producing complete plants from cultured anthers, but observed the formation of multicellular "pollen embryos" and roots from callus. Since then, very few published reports on anther/microspore culture in these two species have emerged, and even incorporating all known unpublished reports, there is a paucity of information on this topic (Table 3). All authors used anthers with microspores at the uninucleate stage as the explant material, and a variety of media were used. Given that none of these procedures were successful, it is presently difficult to assess the relative merits of the different growth regulators used. Recently Croser *et al.* (1997) were able to successfully induce somatic embryogenesis from anther-derived calli. However, the results are presently tentative, in that it is presently unclear as to the ploidy status of the regenerants, the source of the callus (microspore, or anther wall derived ?) and the (possible) beneficial effects of silver nitrate in the medium. It is clear that much research is needed before anther/microspore culture may be used on a routine basis to produce doubled-haploids in pea and chickpea.

4.   PROTOPLAST FUSION TO PRODUCE SOMATIC HYBRIDS BETWEEN PEA AND *LATHYRUS* SP.

The genus *Lathyrus* is grouped along with *Pisum*, *Lens* and *Vicia* in the tribe Vicieae. Members of this genus, particularly *L. sativus* (grasspea) and *L. cicera* are currently being investigated as potential new crops. From the perspective of transferring useful traits from *Lathyrus* to *Pisum*, and vice versa, strong crossing barriers prevent hybridisation between these two species, and *Lathyrus* may be properly regarded as comprising one component of the tertiary gene pool for pea (*P. sativum*). Previous assessments of *Lathyrus* revealed that 59 accessions of this genus, representing ten different species, were all highly resistant to the fungus responsible for the ascochyta-blight (*Mycosphaerella pinodes*) disease in pea (Pang and Brown 1993; Fig. 1). As effective resistance to ascochyta-blight has not yet been discovered within the primary and secondary (*P. fulvum*) gene pools for pea, it would be desirable (and perhaps necessary) to source such resistance from the tertiary gene pool. This resistance, if governed by one- or a small number of genes, may be transferred to pea *via* protoplast fusion (somatic hybridisation). This technique has been employed successfully, for example, to transfer blackleg (*Leptosphaeria maculans*) resistance from black mustard (*Brassica nigra*) to canola (Sjodin 1989; Sjodin and Glimelius 1988).

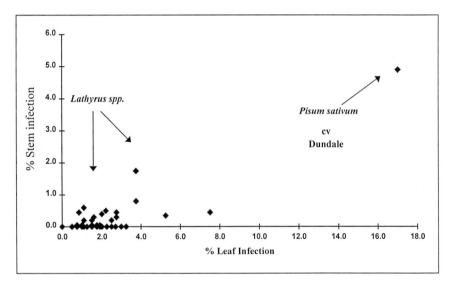

Figure 1. Plot of the severity of leaf- vs stem- infection for 59 accessions of *Lathyrus* spp. inoculated with pycnidiospores of *Mycosphaerella pinodes*.

The success of any protoplast fusion procedure relies heavily on the ability to effectively regenerate complete plants from the fusion products (hybrid cells or heterokaryons). This in turn depends on the availability of an efficient regeneration system from protoplasts for either parent. In pea, organogenesis and somatic embryogenesis from protoplast-derived calli was demonstrated by Lehminger-Mertens and Jacobsen (1989), Puonti-Kaerlas and Erikkson (1988), and more recently, by Böhmer *et al.* (1995). Unfortunately, most procedures are genotype-sensitive, and suffer from low reproducibility in different locations.

There are currently no reports of the successful production of somatic hybrids between pea and other species, but cell (heterokaryon) divisions were observed by Constabel *et al* (1975; 1976) following the fusion of pea and soybean (*Glycine max*) protoplasts. Research on the production of somatic hybrids between pea and *Lathyrus* is still in its early stages, and the current emphasis is on improving the frequency of cell divisions and minicalli formation from pea protoplasts (McCutchan *et al.* 1996). Future work will concentrate on developing an efficient nurse culture system, the creation of a ready source of *Lathyrus sativus* protoplasts (suspension culture) and optimisation of electrofusion protocols. This work presently remains speculative, but many of the essential criteria for its successful implementation, such as the availability of an effective regeneration system from pea protoplasts, and reliable electrofusion protocols, have been satisfied.

5. CONCLUSIONS

Effective protocols are available for regenerating pea and chickpea, *via* organogenesis or somatic embryogenesis from various somatic tissues. Genotype and explant type have consistently been identified to be the crucial factors determining the success and efficiency of regeneration protocols. Problems encountered in root induction and germination of somatic embryos are not universal, and pose no major hindrances to the regeneration of complete seed-bearing plants. Much work is required in developing techniques for producing doubled-haploids from anther culture of both species. Somatic hybridisation experiments to produce pea-*Lathyrus* hybrids are in their early stages.

References

Adkins, A. L., Godwin, I. D., and Adkins, S. W. 1995. *Australian Journal of Botany* 43: 491-497.

Atlaf, N., and Ahmad, M. S. 1990. In *Biotechnology in Agriculture and Forestry 10: Legumes and Oilseed Crops I* pp 100-113 (Ed. Y. P. S. Bajaj). Springer-Verlag, Berlin, Heidelberg.

Bajaj, Y. P. S. 1979. *Indian Journal of Experimental Biology* 17: 1405-1407.

Barna, K. S., and Wakhlu, A. K. 1993. *Plant Cell Reports* 12: 521-524.

Barna, K. S., and Wakhlu, A. K. 1994. *Plant Cell Reports* 13: 510-513.

Bencheikh, M., and Gallais, A. 1996a. *Euphytica* 90: 251-256.

Bencheikh, M., and Gallais, A. 1996b. *Euphytica* 90: 257-264.

Böhmer, P., Meyer, B., and Jacobsen H.-J. 1995. *Plant Cell Reports* 15: 26-29.

Brandt, E. B., and Hess, D. 1994. *In Vitro Cellular and Developmental Biology* 30P (1): 75-80.

Constabel, F., Dudits, D., Gamborg, O. L., and Kao, K. N. 1975. *Canadian Journal of Botany* 53: 2092-95.

Constabel, F., Weber, G., Kirkpatrick, J. W., and Pahl, K. 1976. *Zeitschrift fr Pflanzenphysiolische.* 79: 1-7.

Croser, J. S., Taylor, P. W. J., Pang, E. C. K., and Brouwer, J-B 1997. Proc. III International Food Legume Research Conference (IFLRC III), Adelaide 22-26 Sept. [Poster].

Dinesh-Kumar, V., Kirti, P. B., Sachan, J. K. S., and Chopra, V. L. 1994. *Plant Cell Reports* 13: 468-72.

Gamborg, O. L., Miller, R. A., and Ojima, K. 1968. *Exp. Cell Research* 50: 151-158.

Gosal, S., and Bajaj, Y. P. S. 1988. *SABRAO Journal* 20(1): 51-58.

Griga, M., and Novak, F. J. 1990. In *Biotechnology in Agriculture and Forestry 10: Legumes and Oilseed Crops I* pp 65-99 (Ed. Y. P. S. Bajaj).. Springer-Verlag, Berlin, Heidelberg.

Hildebrandt, A. C., Wilmar, J. C., Johns, H., Riker, A. J. 1963. *American Journal of Botany* 50:248-254.

Jacobsen, H. J., and Kysely, W. 1984. *Plant Cell, Tissue and Organ Culture* 3: 319-324.

Kar, S., Johnson, T. M., Nayak, P., and Sen, S. K. 1996. *Plant Cell Reports* 16: 32-37.

Kartha, K. K., Gamborg, O. L., and Constabel, F. 1974. *Zeitschrift fr Pflanzenphysiolische* 72:172-176.

Khan, S. K., and Ghosh, P. D. 1983. *Current Science* 52(18): 891-893.

Kosturkova, G., Mehandjiev, A., Dobreva, I., and Tzvetkova, V. 1997. *Plant Cell, Tissue and Organ Culture* 48: 139-142.

Kysely, W., Myers, J. R., Lazzeri, P. A., Collins, G. B., and Jacobsen, H. J. 1987. *Plant Cell Reports* 6: 305-308.

Lehminger-Mertens, R., and Jacobsen, H-J 1989. *In Vitro Cellular and Developmental Biology* 25(6): 571-574.

Loiseau, J., Marche, C., and Deunff, Y. L. 1995. Effects of cytokinins, carbohydrates and amino acids on somatic embryogenesis induction from shoot apices of pea. *Plant Cell, Tissue and Organ Culture* 41: 267-275.

Malik, K. A., and Saxena, P. K. 1992a. *Naturwissenschaften* 79: 136-137.

Malik, K. A., and Saxena, P. K. 1992b. *Australian Journal Plant Physiology* 19: 731-740.

Malmberg, R. L. 1982. *Pisum Newsletter* 14: 39-40.

McCutchan, J. S., Pang, E. C. K. and Taylor, P. W. J. 1996. International Association for Plant Tissue Culture (Australian Branch) Vth Meeting, Gatton, p 18.

Murashige, T., and Skoog, F. 1962. *Physiologia Plantarum* 15:473-497.

Özcan, S., Barghchi, M., Firek, S., and Draper, J. 1993. *Plant Cell, Tissue and Organ Culture* 34: 271-277.

Pang, E. C. K. and Brown, J. S. 1993. *Proc. Field Pea Breeding Wor*kshop, Oct. 1993, Horsham, Victoria.

Polisetty, R., Paul, V., Deveshwar, J. J., Khetarpal, S., Suresh, K., and Chandra R. 1997. *Plant Cell Reports* 16: 565-571.

Puonti-Kaerlas, J., and Eriksson, T. 1988. *Plant Cell Reports* 7 (4) 242-245

Puonti-Kaerlas, J., Ottosson, A., and Eriksson, T. 1992. *Plant Cell, Tissue and Organ Culture* 30: 141-148.

Rao, B. G., and Chopra, V. L. 1989. *Journal Plant Physiology.* 134: 637-638.

Sagare, A. P., Suhasini, K., and Krishnamurthy, K. V. 1993. *Plant Cell Reports* 12: 652-655.

Sanago, M. H. M., Shattuck, V. I., and Strommer, J. 1996. *Plant Cell, Tissue and Organ Culture* 45: 165-168.

Schroder H. E., Schotz, A. H., Wardley-Richardson, T., Spencer, D, and Higgins T. J. V. 1993. *Plant Physiology* 101:751-757.

Shri, P. V., and Davis, T. M. 1992. *Plant Cell, Tissue and Organ Culture* 28: 45-51.

Sjodin, C. 1989. Thesis, Dept. Plant Breeding, Swedish University of Agricultural Science.

Sjodin, C., and Glimelius, K. 1988. *Theoretical and Applied Genetics* 77: 651-656.

White, P. R. 1943. *A handbook of plant tissue culture*. J. Cattel, Lancaster.

Foliar Diseases of Cool Season Food Legumes and Their Control

W.J. Kaiser [1], M. D. Ramsey [2], K.M. Makkouk [3], T.W. Bretag [4], N. Açikgöz [5], J. Kumar [6] and F.W. Nutter, Jr [7]

1 USDA, ARS, Western Regional Plant Introduction Station, Washington State University, P.O. Box 646402, Pullman, Washington 99164-6402, USA; 2 South Australia Research & Development Institute, Field Crop Pathology Unit, Waite Research Precinct, Hartley Grove, Urrbrae, South Australia 5064, Australia; 3 ICARDA, P.O. Box 5466, Aleppo, Syria; 4 Victorian Institute for Dryland Agriculture, Private Bag 260, Horsham, Victoria 3401, Australia; 5 Aegean Agricultural Research Institute, P.O. Box 9, Menemen, Izmir 35661, Turkey; 6 ICRISAT, Patancheru, Andhra Pradesh 502 324, India; 7 Department of Pathology, Iowa State University, Ames, Iowa 50011, USA

Abstract

This paper reviews the most economically important foliar diseases of chickpea, faba bean, lentil, lupin and pea caused by fungi and viruses. Many of these pathogens are seed-borne which has aided their local and international spread and survival. Fungal foliar diseases caused by species of *Ascochyta, Botrytis, Colletotrichum, Didymella, Erysiphe, Mycosphaerella, Peronospora, Phoma, Pleiochaeta* and *Uromyces* will be discussed. Some 46 viruses naturally infect these five crops, but only information relating to the most important viral diseases will be presented. The importance of quarantine restrictions to prevent further spread of these pathogens will also be discussed. New and pertinent knowledge of the biology, epidemiology and control of these fungal and viral diseases, particularly that which has been published since the last IFLRC in 1992, will be reviewed.

INTRODUCTION

This paper reviews the foliar diseases of five globally important cool season food legumes: chickpea (*Cicer arietinum* L.), faba bean (*Vicia faba* L.), lentil (*Lens culinaris* Medik.), lupin (*Lupinus* spp.) and pea (*Pisum sativum* L.). All the crops, with the exception of lupin, appear to have originated in the Fertile Crescent of West Asia (Near East) (Smartt, 1984). Three of the major lupin crop species, *Lupinus albus* L. (white lupin), *L. angustifolius* L. (narrow-leafed or blue lupin) and *L. luteus* L. (yellow lupin) are of European origin, while *L. mutabilis* Sweet (tarwi) originated in the Andean highlands of South America (Pate et al., 1985). From their centers of origin, these crops and their associated pathogens have been transported to many countries over the centuries.

In the Proceedings of the first and second International Food Legume Research Conference (IFLRC) held in 1986 (Summerfield, 1988) and 1992 (Muehlbauer and Kaiser, 1994), there are reviews dealing with the important foliar diseases of chickpea, faba bean, lentil and pea. Lupins were not included. In this review, we will include new information on the etiology, biology, epidemiology and control of the economically important diseases of these crops that has been published since the last IFLRC.

ASCOCHYTA BLIGHT

Chickpea

Ascochyta rabiei (Pass.) Labrousse is the most important and devastating foliar disease of chickpea in many countries (Nene and Reddy, 1987), particularly when environmental conditions (cool, wet weather) favour disease development and spread. In the last 30 years the disease has been introduced, on infected seed, to Australia, Canada, southwestern Iran, and to Washington, Idaho and California in the USA (Kaiser, 1997). The situation in Australia is an interesting example of the problems associated with attempting to exclude the disease and problems of its proper identification.

437

R. Knight (ed.), Linking Research and Marketing Opportunities for Pulses in the 21st Century, 437–455.

438

A. *rabiei* was first observed in Australia in 1974 in trial plots on the Waite Campus, Adelaide (Cook and Dubé, 1989) but was successfully eradicated. The pathogen also was incorrectly reported in New South Wales (Cother, 1977); as these isolates were subsequently identified as *Mycosphaerella pinodes* (Berk. & Blox.) Vestergr. (J. Walker and E. Punithaningham, personal communication). Australia was considered free of *A. rabiei* (Chandreshankar and Culvenor, 1993) and maintained strict quarantine on seed imports.

In 1995, *A. rabiei* was identified at two sites in South Australia based on symptoms. Identity of these isolates was confirmed by RAPD PCR analyses and by mating to known isolates (Khan *et al.*, in press). The pathogen was probably present for several years but only caused sporadic problems. To date, only one mating type (MAT1-1) has been identified in Australia (Kaiser, 1997), and in the absence of the teleomorph (sexual state), the pathogen has probably survived at low levels in seed. Quarantine restrictions on seed should remain to prevent the introduction of the second mating type or possibly more virulent pathotypes.

Didymella rabiei (Kovachevski) Arx (syn. *Mycosphaerella rabiei* Kovachevski), the teleomorph of *A. rabiei*, was discovered by Kovachevski (1936) on over-wintered chickpea debris in Bulgaria. Subsequently, it has been reported from the former USSR, Greece, Hungary, Spain, Syria and the United States (Kaiser, 1997). In recent studies with naturally infected debris from 23 countries, the teleomorph developed on the debris from 13 countries, eight of which were new records (Kaiser, 1997). The fungus is heterothallic and two compatible mating types are required for fertile pseudothecia of *D. rabiei* to develop (Wilson and Kaiser, 1995).

Epidemiology

The fungus survives from one growing season to the next on seed, crop residues and volunteer seedlings which are infected. Infection of the crop may arise from seed-borne inoculum, or from rain-splashed conidia or wind-borne ascospores from infected debris. It has been shown that the teleomorph plays an important role in the epidemiology of the disease and is important therefore in attempts to control the disease in Spain (Trapero-Casas et al., 1995) and the United States (Trapero-Casas and Kaiser, 1992). The teleomorph may also be involved in increasing genetic diversity in the pathogen population, in the long-distance spread of air-borne ascospores which may serve as the primary inoculum for development of disease epidemics, and in the survival of the pathogen from one growing season to the next on infected debris.

Control

In the last 20 years, several national and international research centers have initiated breeding programs to develop Ascochyta blight-resistant cultivars. A number of blight-resistant cultivars have been released by the International Center for Agricultural Research in the Dry Areas (ICARDA) (Robertson et al., 1996), the International Crops Research Institute for the Semi-Arid Tropics (ICRISAT) (J. Kumar, personal communication), Turkey (Açikgöz, 1997), and the United States (Kaiser and Muehlbauer, 1994).

Controlling Ascochyta blight of chickpea may require several strategies rather than relying on blight-free seed or resistant cultivars. An integrated approach is being used in the U.S. Pacific Northwest (Kaiser et al., 1994). This involves sowing blight-free seed, applying seed-treatment and foliar fungicides, using a three-year crop rotation, managing infested debris, and planting blight-resistant cultivars.

Lentil

One of the most important and potentially devastating diseases of lentil worldwide is caused by *Ascochyta lentis* Vassiljevsky (syn. *A. fabae* f. sp. *lentis* Gossen et al.) (Nene et al., 1988). The fungus attacks all foliar parts of the plant and is seed-borne. Infection of commercial seed lots can often exceed 25% (Kaiser, 1992; Nasir and Bretag, 1997).

Movement of infected seed within and between countries is one reason for the widespread distribution of the fungus (Kaiser, 1997). Gene banks also may serve as important reservoirs of *A. lentis* (Kaiser and Hannan, 1986). The pathogen was still viable in seed stored for over 35 years at 4 C and 35% RH (Kaiser and Hannan, 1986).

Cool, wet weather is conducive to infection, disease development and spread. The fungus over-winters on infested debris and in seed. Isolates of *A. lentis* from different countries vary in their virulence to lentil (Ahmed et al., 1996b).

In 1992, *Didymella lentis*, the teleomorph of *A. lentis*, was found on over-wintered lentil debris from Genesee, Idaho, U.S.A. (Kaiser and Hellier, 1993; Kaiser et al., 1997). The fungus is heterothallic. Two compatible mating types are required for the teleomorph to develop (Kaiser and Hellier, 1993). Both mating types have been found in isolates of the pathogen from Algeria, Canada, Italy, Morocco, Pakistan, Spain, Syria, Turkey, the United States, and recently in Australia (Ahmed et al., 1996a; Kaiser, 1997). Until now, the teleomorph has only been found under field conditions in the U.S. Pacific Northwest. Further studies are needed to determine the distribution of the teleomorph and its importance in the spread, survival and epidemiology of Ascochyta blight.

Control

Most research on the control of Ascochyta blight of lentil has dealt with the use of seed treatment fungicides and resistant cultivars. Fungicides containing thiabendazole or carbendazim minimize the introduction of the pathogen on infected seed and have been effective in controlling seed-borne infection in several countries (Bretag, 1989a; Kaiser and Hannan, 1987).

Ascochyta blight-resistant lentil cultivars and germplasm lines have been identified and/or released by improvement programs in Australia, Canada, India, Pakistan, and Syria (Ahmed and Morrall, 1996; Andrahennadi et al., 1996; Nasir and Bretag, 1996; Robertson et al., 1996).

Faba bean

Ascochyta fabae Speg. is a major disease on faba beans worldwide. The pathogen infects the foliage, including stems and pods and is seed-borne. Seed staining caused by *A. fabae* infection also reduces the value of grain for human consumption. Seed infection is considered the most important source of disease in the United Kingdom and stringent seed standards are enforced to control the disease (Thomas and Sweet, 1990). Seedling growth is reduced significantly when greater than 25% of the seed coat is covered by *Ascochyta* lesions (Madeira *et al.*, 1992). However, survival of the teleomorph (*Didymella fabae* Jellis and Punith.) supports evidence that crop residues are also an important source of infection, particularly in autumn sown crops (Jellis and Punithalingam, 1991). *A. fabae* and *A. pisi* Lib. produce a toxin, ascochitine, in culture which has broad biological activity (Beed *et al.*, 1994a). The role of this toxin in determining virulence of *A. fabae* to faba beans was investigated; however, there was no relationship between the production of ascochitine by the fungus and the virulence of *A. fabae* of several isolates (Beed *et al.*, 1994b).

Control

In the United Kingdom, statutory testing of certified seed is used as a measure to control the spread of the disease (Thomas and Sweet, 1990). In Australia, however, crop residues are important sources of infection and a single application of the fungicide, mancozeb, early in the season is generally required to prevent yield loss from Ascochyta blight. Seed staining caused by Ascochyta blight can be reduced by applications of fungicides after flowering (M.D. Ramsey, unpublished data).

Resistance to Ascochyta blight has been identified in faba beans and the germplasm collection at ICARDA contains 24 resistant accessions (Robertson, 1995). The line 29H was highly resistant in France and has been used successfully in breeding programs to produce resistant cultivars (Maurin and Tivoli, 1992). In Australia, a population of the cultivar Fiord was subjected to three cycles of mass selection and one of progeny testing for resistance to the disease (R Knight personal communication). Selections were made and, thirty-one of them were significantly more resistant than the original population (Lawsawadsiri, 1994). A resistant cultivar Ascot was released which was a composite of the two most resistant selections. Resistance within Ascot is possibly due to a single recessive gene (Ramsey *et al.*, 1995). Other sources of resistance include a dominant gene identified in ILB752 and a recessive gene in resistant plants selected from NEB463 (Lawsawadsiri, 1994).

White flowered, low tannin lines were highly susceptible to *A. fabae* when grown from infected seed, and infection resulted in poor establishment and plant death (Fagbola and Jellis, 1994). However, the level of condensed tannins in seed does not influence the level of cultivar resistance to foliar infection by *A. fabae* (Helsper et al., 1994). Low tannin and Ascochyta resistance were easily combined in a crossing program. However, few progeny combined low vicine content with Ascochyta resistance indicating a genetic linkage between these characters (Jellis and Vassie, 1995).

Pea

Ascochyta blight of pea is a disease complex caused by three pathogens: *Ascochyta pisi*, which causes leaf, stem and pod spot, *Mycosphaerella pinodes* (anamorph: *Ascochyta pinodes* L.K. Jones) which causes leaf, stem and pod spot, and foot rot, and *Phoma medicaginis* Malbr. & Roum. var. *pinodella* (Jones) Boerema (syn. *Ascochyta pinodella* Jones), which causes leaf spot, stem lesions and foot rot. *M. pinodes* is the most prevalent and the most damaging of the three fungi and can be found in most pea growing regions of the world

In the field, the three pathogens often occur together and can be difficult to distinguish. They may also be difficult to separate by microscopic examination. Research is underway to compare and distinguish the species by isoenzyme analysis and RFLP of amplified rDNA spacers (Faris-Mokaiesh *et al.*, 1996), RAPD assay (Tohamy *et al.*, 1997), and by using monoclonal antibodies (Bowen *et al.*, 1996). This section will emphasize the disease incited by *M. pinodes*.

Yield Losses

There are many reports of serious crop losses caused by Ascochyta blight. However, most are qualitative rather than quantitative and do not provide accurate estimates of losses. The severity of Ascochyta blight varies between crops, years and regions and depends on environmental conditions. Moisture appears to be the most important environmental factor. It follows that in wet seasons large losses may occur, while in dry seasons crop losses will probably be small. The most reliable estimates of losses come from fungicide control experiments where the yields of peas, with and without disease, are compared. By controlling the disease, using fungicides, increases in yield of 15-75 % have been reported in field experiments (Bretag, 1989b; Bretag *et al.*, 1995a; Hwang *et al.*, 1991; Tivoli *et al.*, 1995; Warkentin *et al.*, 1996).

Studies in Australia (Bretag *et al.*, 1995a) have shown that individual pea cultivars have different levels of tolerance to *M. pinodes* and that there are large differences between cultivars in the yield loss at the same level of disease. Ascochyta blight caused substantial reductions in yield of all the pea cultivars grown in Australia but was usually most severe on the early maturing cultivars. Crop losses were much higher in early-sown crops than late-sown crops (Bretag, 1991; Heenan, 1994)..

Clulow *et al.* (1991c) and Wroth (1996) compared the host pathogen interactions between pea lines that were either resistant or susceptible to *M. pinodes.* They concluded that resistance was due to the failure of the fungus to penetrate the host and/or a hypersensitive reaction which restricted the fungus within the host. Nasir and Hoppe (1991) also found that the development and spread of the fungus was retarded in resistant plants.

Studies by Bretag (1991), Clulow *et al.* (1991a), Nasir and Hoppe (1991) Nasir et al. (1991), and Wroth (1996) have shown there is variation between individual isolates of *M. pinodes* in their virulence and aggressiveness, and different pathotypes of the fungus have been shown to exist. Individual isolates can be grouped together based on their reactions to different hosts. However, resistance to leaf or stem infection appears to be controlled by different mechanisms (Clulow et al., 1991b).

Epidemiology

The presence on seed, of fungi of the Ascochyta complex, can be detected using the agar plate technique and a range of media. Recent studies by Corbiere *et al.* (1994) found that malt agar medium + 2,4-D enabled quicker detection of *M. pinodes.* Faris *et al.* (1995) reported that Ascochyta blight fungi in pea seed could be detected and quantified using an ELISA technique. Seed surveys have demonstrated that the three fungi associated with the Ascochyta complex can all be seed-borne and survive in seed for more than five years (Czyzewska, 1993; Corbiere *et al.,* 1994; Bretag *et al.*, 1995b). The level of infestation varies between seed lots depending on the amount of disease present in the crop and the rainfall between flowering and maturity (Bretag *et al.,* 1995b).

Although seed transmission can occur, it appears to be of minor importance. In established pea growing regions, there was no correlation between the level of seed infestation by *M. pinodes* and the severity of Ascochyta blight; however, where the level of seed infection was high (>11%), there was a significant reduction in emergence, which resulted in lower grain yields (Bretag *et al.,* 1995b).

Under field conditions, the main source of primary inoculum is air-borne ascospores of *M. pinodes.* In Australia, ascospores also play an important role in the secondary spread of the disease (Bretag, 1991). Spore trapping studies have shown that high numbers of ascospores are released early in the season from the residues of previous crops but the number from this source declines rapidly during the growing season. However, as the disease progresses on the crop, ascospores are released in large numbers from senescent parts of infected plants. In France, Roger and Tivoli (1996) found that both pycnidiospores and ascospores are important in the secondary spread of disease. Pycnidiospores are rarely found in Australian pea crops and appear to be of minor importance in the in-field spread of disease (Bretag, 1991). The teleomorph of *P. medicaginis* var. *pinodella* was recently observed in culture, but has yet to be identified taxonomically (Bowen *et al.,* 1997). Information is lacking on whether the sexual state develops in nature and if it is important in disease development, spread and survival of the pathogen.

Control

Although there are many reports in the literature of resistance to the Ascochyta blight fungi, the genetics of resistance is poorly understood. Recent studies suggest it may be possible to determine the genetics of resistance in peas to *M. pinodes.*(Clulow *et al.* 1991c). However, genetic studies have always proven to be difficult because it is hard to distinguish between resistant and susceptible reactions. Ultimately, it may be possible to select molecular markers associated with the resistance genes.

Foliar sprays of chlorothalonil and benomyl have been effective in reducing the severity of Ascochyta blight and increasing the yield and seed weight of field pea (Bretag, 1991; Cors and Meeus, 1994; Gilet and Durand, 1996; Warkentin *et al.,* 1996).

In Australia, treatment of pea seed with thiabendazole plus thiram has been shown to reduce the severity of Ascochyta blight early in the season and increase grain yield by up to 20%. However, where the disease becomes severe late in the growing season the likelihood of better yields is low.

DOWNY MILDEW

Pea

Downy mildew (*Peronospora viciae* (Berk.) Casp.)) is a major disease of pea (*P. viciae* f.sp. *pisi*) and faba bean (*P. viciae* f.sp. *fabae*). Yield losses in peas can exceed 30% in Europe (Olofsson, 1966) and Australia (J. Davidson, personal communication). Infection of pea pods is common and reduces quality but not yield (Stegmark, 1994).

Oospores, which can survive for many years in soil, are the primary source of infection and infect pea seedlings though epicotyl and hypocotyl tissue but not roots (Stegmark, 1994). A protocol was developed for extracting and germinating oospores from plant tissue which facilitated their study (Vandergraag and Frinking, 1996b). *P. viciae* is homothallic and more oospores of *P. viciae* f.sp. *pisi* are produced in systemically colonized tissue at 20°C than at 10°C, although they are smaller at the higher temperature (Vandergraag and Frinking, 1996a). The fungus occurs as oospores and mycelium in seed; however, seed infection appears to be unimportant in the epidemiology of the pathogen (Stegmark, 1994). A specific rabbit antisera was developed for identifying infected seedlots by ELISA (Fegies *et al.,* 1992).

Sporangia are an alternative source of inoculum. They form extensively on surfaces of infected plants and are a means of spread over large distances (Stegmark, 1994).

Control

Acylalanine fungicide seed treatment is very effective for controlling the systemic infection of downy mildew in seedlings but does not prevent pod infection later in the season (Stegmark, 1994).

P. viciae is highly variable and there is considerable variation for resistance between pea cultivars (Olofsson, 1966; Stegmark, 1994; Yanase et al., 1995). Isolates of *P. viciae* from peas and faba beans are specific to their host species, although the infection efficiency of *P. viciae* f. sp. *fabae* on faba beans is much lower than for *P. viciae* f.sp. *pisi* on peas (Yanase et al., 1995). Both race and polygenic durable resistance have been identified to *P. viciae* f. sp. *pisi* in pea germplasm (Stegmark, 1994). Only one line (L1382) showed complete resistance in replicated trials using inoculation of pregerminated seed and this was related to characteristics of the seed coat (Stegmark, 1994). Durable resistance in the cultivar Dark Skin Perfection has been stable for over 30 years (Stegmark, 1994). A disease key for assessing downy mildew in peas on a 0-10 scale was developed by Falloon et al. (1995).

BOTRYTIS DISEASES

Chickpea

Gray mold caused by *Botrytis cinerea* Pers.: Fr. seriously retards production in most chickpea -growing regions of the world, particularly in south Asia. Production in southern Australia can also be affected. All cultivars are susceptible to gray mold however, there are large differences between cultivars in their susceptibility (Bretag and Brouwer, 1995).

Advances in research on Botrytis gray mold of chickpea were discussed at international meetings in 1991 and 1993 and summaries of the findings have been published by ICRISAT (Haware *et al.*, 1992; Haware *et al.*, 1993).

Studies on *B. cinerea* have generally focused on the foliar phase of infection, as epidemics are most common at flowering (Haware and McDonald, 1993). Joshi (1992) reported that gray mold causes an annual loss of 15% in production in Nepal. Infection of pods may lead to seed infection (Laha and Grewal, 1983). *B. cinerea* has been isolated from chickpea seed produced in Australia (Bretag and Mebalds, 1987). Burgess *et al.* (1997a) have shown that infection affects emergence and seed-to-seedling transmission can cause soft rot of seedlings. In 1994, seedling soft rot occurred in the Wimmera region of Victoria, Australia resulting in total crop failures).

Epidemiology

In many regions of the world, the relative importance of seed-borne inoculum and other sources of inoculum have yet to be determined. Rewal and Grewal (1989b) reported there were at least 5 different pathotypes of *B. cinerea* and further studies of pathogenic variability are required.

Seedlings grown from seed infested with *B. cinerea* often develop soft rot, and sporulation of the fungus can be seen at the base of the plant. The pathogen is able to survive in naturally infested seed from one season to the next. However, there is a rapid decline in viability of the fungus during storage, particularly at high temperatures (Singh and Tripathi, 1993). In most regions *B. cinerea* is thought to survive mainly on crop residues (Singh and Tripathi, 1992). *B. cinerea* has a wide host range and can survive on other crops and weeds. In glasshouse tests conducted by Rathi and Tripathi (1991), it was pathogenic on 8 different crop species and 21 weed species.

Butler (1993) provided a summary of how important microclimate is in the development of an epidemic. Moist conditions and moderate temperatures favor the disease (Mahmood *et al.*, 1989). Initial infection probably occurs in the lower parts of the plant early in the season and spreads to the upper leaves later in the season.

Control

Despite extensive screening of chickpea germplasm, lines with high levels of resistance to gray mold have not been found although there are many lines with moderate resistance. Rathi and Tripathi (1993) tested 8,500 accessions and identified 22 lines with useful resistance. Many chickpea lines with field tolerance produce good yields despite a high degree of flower infection (Shahu and Say, 1988; Dewan, 1993).

Late sowing of chickpeas can sometimes reduce the severity of gray mold but is likely to reduce yields in normal years (Kharki *et al.*, 1989). Chickpea cultivars suitable for late sowing need to be developed. Wide spacing of plants has been shown to reduce the incidence and severity of gray mold (Reddy *et al.*, 1988; Reddy *et al.*, 1993). It has also been suggested that plants with a more upright growth habit may be able to escape the disease (Bakr *et al.*, 1993).

Mahmood *et al.* (1989) showed that gray mold could be effectively controlled by seed treatment with captan followed by 3 sprays of carbendazim. Foliar sprays of vinclozolin have also been effective (Reddy *et al.*, 1993). Singh and Kaur (1990) showed that seed-borne infection (>94%) was checked by seed treatment with triadimefon, carbendazim + thiram, mancozeb or triadimenol. Seed treatment, followed by one spray of mancozeb, thiram, thiabendazole or triadimefon, 50 days after sowing or at the appearance of symptoms gave complete control of both primary and secondary infection. Mukherjee and Haware (1993) screened fungi isolated from the rhizosphere of chickpeas for activity against *B. cinerea*. An isolate of *Trichoderma viride* Pers.: Fr. was shown to reduce the severity of gray mold under glasshouse conditions. Burgess *et al.* (1997b) reported that *Gliocladium roseum* Bainier was also highly antagonistic to *B. cinerea* and as a seed treatment, was as effective as the fungicide thiram

At present, high levels of disease resistance are not available in chickpea cultivars. Therefore, an integrated disease program using a combination of moderately resistant cultivars, strategic use of chemicals and improved cultural practices appears to be the best way to minimize crop losses caused by gray mold.

Faba bean

Faba beans are infected by both *Botrytis fabae* Sardiña and *B. cinerea* Pers.: Fr. However, only *B. fabae* is a major cause of chocolate spot in the field (Harrison, 1988). Chocolate spot causes severe epiphytotics which are determined by weather conditions, inoculum density, and the age of plants. Epidemics generally occur on autumn-sown crops during late winter and spring and are exacerbated by damage to leaves and humid conditions (Harrison, 1988). The teleomorph of *B. fabae* has not been found (Harrison, 1988).

Control

Disease control strategies are based on modifying crop management, fungicides and cultivar resistance. Recent work with antagonistic fungi and bacteria indicates some promise for biological control in the future.

A single spray of benomyl and chlorothalonil at late flowering controlled both rust and chocolate spot in the UK. In many seasons a single spray is unlikely to be adequate to control these diseases (Palmer and Stevens, 1993). Mancozeb, dichlorofluanid and tebuconazole were the most effective fungicides for preventing yield losses caused by rust and chocolate spot outbreaks in New South Wales. Application of mancozeb early and during flowering provided an effective and economical increase in grain yield in 1990 and 1992 (Marcellos *et al.*, 1995). Pyrimethanil is a promising new fungicide which reduces the amount of host cell lysis during necrotophic growth of *B. fabae* by inhibiting the formation of cell wall degrading enzymes by the pathogen (Daniels and Lucas, 1995). Seed treatment with the plant regulator, ethephon induced resistance to both rust and chocolate spot diseases (Salem *et al.*, 1992).

Resistance to fungicides is an increasing problem in controlling diseases caused by *Botrytis* spp. (Palmer and Stevens, 1993). This has lead researchers to investigate alternative methods of control, including biological control strategies. *Penicillium chrysogenum* Thom inhibited growth of *B. fabae* in vitro and on detached leaves (Jackson et al., 1994). *Penicillium brevicompactum* Dierckx and *Cladosporium cladosporioides* (Fresen.) G.A. De Vries inhibited radial growth and caused hyphal browning of *B. fabae* in vitro. Isolates of both fungi significantly reduced chocolate spot infection when applied to plants with *B. fabae* (Jackson *et al.*, 1997). Prior infection of detached leaves by *A. fabae* restricted development of chocolate spot, and this effect may be associated with antibiotic activity of Ascochitine (Madeira et al., 1993).

The severity of chocolate spot increases with plant density. This relationship also occurred in organic farming systems where faba beans were intercropped with wheat. Furthermore, disease development was independent of the amount of wheat in the mixture (Bulson *et al.*, 1997).

Resistance to chocolate spot varies between faba bean cultivars, but most accessions are moderately susceptible (Harrison, 1988). Some lines collected from South America, particularly from Ecuador, have moderate levels of resistance. The breeding line BPL938 was tolerant under field conditions; however, tolerance was not stable and broke down in testing under controlled light and temperature conditions and at high inoculum concentrations (Tivoli *et al.*, 1992). The cultivar Icarus, which is a reselection from BPL710, is moderately resistant. However, the level of resistance should be sufficient to reduce losses in the field (Ramsey et al., 1995). The germplasm collection at ICARDA includes 21 accessions which have been identified with some resistance to chocolate spot (Robertson, 1995).

POWDERY MILDEW

Pea

Powdery mildew of pea incited by *Erysiphe pisi* DC (syn. *E. polygoni* DC) occurs in most countries where peas are grown commercially or for personal consumption. The disease is usually more damaging in mid- to late-maturing crops when temperatures during the day are warm and nights are cool. On heavily infected plants, yield can be reduced significantly and the viability and quality of the seed adversely affected.

Symptoms of the disease usually develop on the upper surface of older leaves first and consist of whitish, powdery spots of varying sizes. These spots contain fungal mycelium and spores (conidia). On the under surface of affected areas, the tissues turn purplish-brown. With favorable environmental conditions (dew without rain), the fungus may eventually cover the entire surface of leaves, stems and pods. Towards the end of the growing season, the pathogen produces small, round, black sexual fruiting bodies (cleistothecia).

In addition to *Pisum*, the host range of this obligate parasite includes species of *Lupinus, Medicago* and *Vicia*. The pathogen shows marked host specialization, which has resulted in the identification of four formae speciales (Boerema and Verhoeven, 1979). The host range of *E. pisi* f. sp. *pisi* is confined to *Pisum* spp.

Under field conditions, the fungus survives overwinter as cleistothecia (sexual fruiting bodies) on infested plant debris and infected seed. The cleistothecia produce ascospores which can initiate infection on susceptible peas in late spring or early summer. Once primary infection has occurred, conidia (asexual spores) develop in the lesions. Both ascospores and conidia are windborne.

Control

Control measures aimed at reducing infection consist of field sanitation (destroying infected crop debris by burying or burning), crop rotation, and early planting. Powdery mildew can be controlled by spraying the foliage with chemicals, such as benomyl or sulfur (Hagedorn, 1984), and by planting resistant cultivars. Resistance is controlled by at least two recessive genes, designated *er* and *er-2* (Kraft and Kaiser, 1993). Presently, new pea cultivars are being released with resistance to powdery mildew in different countries (Hagedorn, 1984; Kraft and Kaiser, 1993).

RUST DISEASES

Lentil

Rust caused by *Uromyces fabae* (Grev.) Fuckel. (syn. *U. viciae-fabae* (Pers.) Schroet.) is one of the most important foliar diseases of lentil worldwide. The disease can be particularly damaging in countries, such as

Chile, Ecuador, Ethiopia, India, Morocco, and Pakistan (Erskine and Saxena, 1993). Early infection and environmental conditions that favor disease development (temperatures 20-22 C and wet weather) can result in a total crop failure. Some countries, like the United States, have regulations to prevent the introduction of rust on lentil seed from countries where the disease occurs.

Rust infects all aerial parts of the plant. Yellowish-white pycnia (spermagonia) and aecia develop on the under surface of leaves and on stems. Later in the growing season, brown uredia form on both surfaces of leaves and other aerial parts. Dark brown to black teliospores develop in the same sori as urediospores. The teliospores are the resting spores of the pathogen. *Uromyces fabae* is an autoecious rust which completes its complex life cycle on one host. The fungus infects *Lathyrus, Lens, Pisum* and *Vicia* spp.

The pathogen survives from one growing season to the next as teliospores on infested lentil debris. Infection of the new crop is by basidiospores. Prasada and Verma (1948) demonstrated that lentil seedlings became infected when infested debris was mixed with seeds. Aeciospores are important in the spread of the disease during the growing season.

Various methods have been used to control the disease, including crop rotation, field sanitation, seed-treatment, foliar fungicides (Nene et al., 1988), and host resistance (Erskine and Saxena, 1993; Khare, 1981). In recent years, ICARDA has identified new sources of resistance in lentil to rust. (Robertson et al., 1996). Some of the accessions also have resistance to one or more diseases. ICARDA has identified new sources of rust resistance by screening lentil germplasm in Ethiopia, Morocco, and Pakistan where rust epidemics occur frequently.

Faba Bean

Rust of faba bean is found wherever the crop is grown (Diekmann, 1994). Rust is usually more severe on late-sown crops in humid climates. In some countries, heavy infection later in the growing season can lead to premature defoliation which may have an adverse affect on yield (Gaunt, 1983).

Rust-colored, blister-like lesions develop on the leaves and stems. The pustules containing urediospores may be surrounded by a chlorotic halo. Later in the growing season, darker teliospores are produced on foliar tissues (Gaunt, 1983).

Rust of faba bean is incited by *Uromyces fabae* (Grev.) Fuckel., the same pathogen that causes rust on lentils, peas, *Lathyrus* spp. and *Vicia* spp. Although *U. fabae* infects species in several genera, the fungus exhibits some degree of host specialization. Some specialized pathogenic forms are restricted to one host, while others have a broader host range (Boerema and Verhoeven, 1979; Gaunt, 1983). Research in Canada has identified several races on faba beans (Hanounik et al., 1993).

The fungus has a complex life cycle producing five spore types; pycniospores, aeciospores, urediospores, teliospores and basidiospores. Teliospores are the primary overwintering structures and they may survive adverse conditions for up to two years (Gaunt, 1983). In the spring, teliospores germinate to produce basidiospores which serve as the primary inoculum source to initiate infection on susceptible cultivars. Urediospores are then produced on the infected plants and provide the secondary inoculum which is responsible for the rapid build-up and spread of the fungus within and between fields. During the growing season, the fungus is spread primarily by urediospores. The basidiospores and urediospores are windborne.

In Australia, rust is considered the major fungal disease in northern New South Wales, where winters are warmer and drier than in southern Australia (Marcellos *et al.*, 1995). Early infection determines the amount of yield lost. Epidemics, which occur after pod set, have no affect on yield components. Epidemics reduce seed size (Marcellos *et al.*, 1995), grain weight per stem, average grain weight and the number of pods per stem, but not the number of grains per pod (Sache and Zaddocks, 1994).

Rust is also the major disease of common vetch (*Vicia sativa* L.), which is a significant alternative legume for drier areas of southern Australia. Glasshouse testing has established that specific strains of rust infect vetch and these do not infect faba beans (M.D. Ramsey, unpublished data).

Control

The fungus survives primarily on infested crop debris, but spores can also be carried on the surface of seeds. Control of faba bean rust should include a two-year rotation between crops to prevent build-up and survival of inoculum. Infested crop residues should be disposed of by plowing or burning. Sources of resistance to rust on faba bean have been identified in ICARDA's germplasm collection, such as Sel. 82 Lat.

15563-1 and several pure line (BPL) accessions (Robertson et al., 1996). In Egypt, ICARDA germplasm has been used to develop rust-resistant cultivars, like Giza 461, that have resistance to rust and chocolate spot (Robertson et al., 1996). Similarly several lines with rust resistance have been identified in New South Wales (I. Rose, unpublished data). Resistance to vetch rust has been identified within *Vicia sativa* and other *Vicia* spp., including *V. benghalensis* L., *V. villosa* Roth, *V. atropurpurea* Desf. and *V. ervilia* (L.) Willd. (M.D. Ramsey, unpublished data).

Mancozeb, dichlorofluanid and tebuconazole were the most effective fungicides for controlling rust in faba beans in field trials conducted in New South Wales. Mancozeb applications at early and late flowering provided an economical increase in grain yield in seasons when rust was severe (Marcellos *et al.*, 1995). Resistance can also be induced by treating crops with the growth promotant ethephon (Salem *et al.*, 1992).

ANTHRACNOSE

Lentil

Anthracnose caused by *Colletotrichum truncatum* (Schwein.) Andrus & W.D. Moore is a relatively new foliar disease of lentil in North America where it was first discovered in Manitoba, Canada in 1987 (Morrall, 1988). Subsequently, it was reported infecting commercial lentils in North Dakota, U.S.A. in 1992 (Venette et al., 1994). Anthracnose is a potentially devastating disease of lentil. Most, if not all, lentil cultivars being grown commercially in Canada and the U.S. are susceptible to anthracnose. In some areas of Canada and North Dakota, anthracnose has had a drastic effect on yield and seed quality. The disease will undoubtedly be found in other lentil-producing countries. Since 1992, the senior author has observed the disease on lentils in Bulgaria and isolated the pathogen from infested lentil debris and seeds from Bulgaria, infested lentil debris from Pakistan, and seeds from New Zealand. Cool, wet weather favors disease development and spread.

Lentils are susceptible to infection by the anthracnose pathogen at all stages of growth. The fungus produces lesions on the leaflets, stem, pods, and seeds. On the leaflets, lesions are usually white to gray, but may eventually turn brown. Infected leaflets may abscise prematurely. At times, black fruiting bodies (acervuli) develop in the lesions and produce masses of asexual spores (conidia). Conidia are spread by rain splash to the upper canopy and to adjacent plants. Heavy infection often results in dieback and/or death of plants. Dark, resting structures, called microsclerotia, develop in infected tissues (Buchwaldt et al., 1996). The pathogen is seed-borne, but usually at low levels (2% or less) and it is uncertain how important seed infection is in the disease cycle, and to what extent seed to seedling transfer occurs in nature (Buchwaldt et al., 1996; Morrall, 1988).

The host range of *C. truncatum* includes many seed and forage legumes. Isolates of *C. truncatum* from lentil have a limited host range. Faba bean and vetch (*Vicia sativa*) were highly susceptible to infection, but lupin (*Lupinus* sp.), alfalfa (*Medicago sativa* L.), bean (*Phaseolus vulgaris* L.), soybean (*Glycine max* (L.) Merr.) and round-leafed mallow (*Malva pusilla* Smith) were not hosts. Peas were infected but the lesions did not increase in size (Buchwaldt et al., 1996).

In Canada, Buchwaldt et al. (1996) found that the fungus survived for at least 44 months on buried infested debris, but viability of the fungus declined rapidly after 12 months in infested debris on the soil surface. They demonstrated that inoculum of *C. truncatum* from an infested lentil crop could be wind dispersed at least 240 m from a combine harvester. Lentil debris and soil dispersed by high winds from diseased fields after harvest were infective to lentil.

The most effective and economical means of controlling lentil anthracnose is by the use of resistant cultivars. Sources of resistance have been identified in the U.S. lentil germplasm that was screened in Canada (Buchwaldt et al., 1992) and Bulgaria (W.J. Kaiser, unpublished data). Breeders in Canada and the U.S. are incorporating this resistance into commercially-acceptable cultivars.

Lupin

Anthracnose (*Colletotrichum gloeosporioides* (Penz.) Pens. & Sacc.) is the most serious disease of white lupin (*Lupinus albus*) in France, Chile and Brazil (Gondran *et al.*, 1994). It also limits production of lupins in North America (Paulitz *et al.*, 1995). In New Zealand it caused the decline of the tree lupin (*L. arboreus* Sims) in pine plantations (Molloy, 1991).

Australia is the major lupin producing country with over 1 million tonnes, mainly *L. angustifolius,* produced annually. Until 1994, anthracnose had not been recorded on lupin crops in the major growing areas. In that year, an outbreak occurred in Western Australia on *L. albus, L. angustifolius, L. luteus* and *L. mutabilis* Sweet following an accidental introduction on seed imported from Germany (Sweetingham, 1995). This outbreak was eradicated; however, it alerted authorities to the disease risk. There were previous records of anthracnose on lupins, including *L. consentinii* Guss. in Western Australia, *L. angustifolius* in Queensland and ornamental lupins (*L. polyphyllus* Lindl.) in gardens and nurseries in Victoria. However, none of these previous records had threatened commercial production (Sweetingham, 1995). In 1996, the disease reoccurred on 132 properties in crops of *L. albus* and *L. angustifolius* in Western Australia, and on *L. polyphyllus* in Victoria, NSW and Tasmania. A major eradication campaign was established under state plant health legislation and large areas of crop were destroyed in Western Australia and South Australia. Unfortunately, due to the scale of the outbreak in Western Australia, eradication was not possible, and restrictions now apply to seed and machinery movement between states and districts to prevent further spread of the pathogen.

Several *Colletotrichum* spp. have been recorded on lupins by various authors *C. acutatum* J.H. Simmonds was recorded on *L. polyphyllus* in UK (Reed et al., 1996). *L. angustifolius* was also highly susceptible to race 1 or 2 of *C. trifolii* Bain and Essary (alfalfa race) and *C. fragariae* A.N. Brooks (Welty, 1984). Furthermore, *C. truncatum* and *C. gloeosporioides* f. sp. *aeschynomeme* infected several species of lupin, including *L. albus* in pathogenicity tests (Weidemann *et al.*, 1988).

An international collection of *C. gloeosporioides* isolates from lupins was screened for vegetative compatibility groups (VCG) using nitrate deficient mutants. Isolates could be divided into two groups and those from Australia were identical to those found in Europe and Chile (H.A. Yang, personal communication). There is some confusion about the identity of *Colletotrichum* spp. infecting lupins and other hosts. Sreenisvasaprasad *et al.* (1994) tested isolates of *C. gloeosporioides* from lupins for DNA homology and concluded they were more correctly described as *C. acutatum*.

Control

Several strategies are being implemented to control anthracnose in Australia. Southeastern Australia remains largely free of the disease as the outbreak in South Australia was restricted and eradication will be attempted. The magnitude of the outbreak in Western Australia makes eradication impractical and quarantine restrictions have been imposed on the movement of lupin seed and machinery to the eastern states of Australia.

Field evaluation of *Lupinus* germplasm was conducted in trials established in New Zealand during 1996. These trials confirmed that high levels of resistance occur within *L. angustifolius* germplasm, including two recent cultivar releases, Kaylia and Wonga (M.W. Sweetingham, personal communication). An integrated management strategy has been developed for infected areas in Western Australia which includes sowing uninfected seed, treating seed with fungicides, crop rotation and paddock selection, controlling naturalized blue lupins (*L. consentinii*), cleaning machinery, and sowing resistant cultivars (Sweetingham, 1997).

PLEOICHAETA LEAF SPOT

Lupin

Brown spot (*Pleiochaeta setosa* (Kirchiner) Hughes) is one of the major diseases affecting lupins (*Lupinus angustifolius*) in Australia (Sweetingham, 1990). In Europe, the disease is only serious on autumn sown crops of *L. albus* (Gondran *et al.*, 1994). *P. setosa* infects all above ground parts of *L. angustifolius* resulting in premature leaf defoliation, stem girdling and death of young seedlings (Sweetingham, 1990). It also causes a severe root rot (Sweetingham, 1989).

Cold climatic conditions and unsuitable soil types increase the risk of disease. Lupins are most susceptible at the 0-4 leaf stage (Sweetingham, 1990). Disease risk is highest in cooler areas where slow growth prolongs exposure to infection, or in low rainfall areas where the growing season is short and crops are unable to compensate (Sweetingham, 1990). Similarly, lupins growing slowly on unsuitable or phosphorus deficient soils have an increased risk of infection (Sweetingham, 1990).

The primary inoculum for foliar infection is soilborne conidia produced on defoliated leaves in the previous growing seasons. Conidia survive for several years in soil and are splashed from the soil onto leaves of young plants by rain (Sweetingham, 1990; Sweetingham et al., 1993). Diseased leaves abscise within days of symptom expression and produce large quantities of thick walled conidia under the crop canopy leading to further infection and survival of the pathogen (Sweetingham, 1990).

Control

The population of spores in the soil is reduced by half each 12 months in the absence of host plants (Sweetingham, 1990). P. setosa has a narrow host range and a long crop rotation depletes soilborne inoculum (Sweetingham, 1992). Early sowing (late autumn) is recommended in high-risk areas to increase early crop growth (Sweetingham, 1990). High sowing rates and improving phosphorus nutrition produce thick lupin stands which reduce rain-splash within crops (Sweetingham, 1990).

Dicarboximide fungicides (iprodione and procymidone) applied to the seed protects seedlings against brown spot (Loughman and Sweetingham, 1991). Fungicides controlled infection on leaves 1 and 2 but gave little protection to leaves 3 or 4 (Sweetingham et al., 1993).

Retaining cereal stubble reduces disease by protecting seedlings from rain-splashed spores on the soil surface. There is a benefit from retaining as little as 0.5 tonnes residue/ha (about 10% ground cover) with an optimum at 2 t/ha (about 80% ground cover) (Sweetingham et al., 1993). The effects of fungicide seed treatment and cereal stubble are additive. Fungicides protect early growth while slow breakdown of stubble offers direct protection against rain-splash throughout the season (Sweetingham et al., 1993).

The comparative resistance of Lupinus spp. (L. luteus, L. angustifolius and L. albus) was compared in glasshouse and field trials. Large differences existed between species. Lines of L. luteus were most resistant. There was variation between L. angustifolius lines but none were as resistant as L. luteus; and L. albus cv. Kiev Mutant was the most susceptible line (Sweetingham et al., 1994).

The reaction of L. albus cv. Kiev Mutant, L. angustifolius cv. Yorrel and L. luteus cv. Motiv 369 to P. setosa was studied by Yang et al. (1996). Field trials showed that 30-40 times more conidia attached to the lower than to the upper leaf surfaces of all cultivars and most spores stuck onto leaf hairs. Eight times more conidia attached to the susceptible cv. Kiev Mutant than other cultivars. The majority of infections resulted from direct penetration through the cuticle via appressoria. Conidia germinated more readily and infection efficiency was highest on Kiev Mutant. A resistant L. angustifolius cultivar Myallie was released in 1996 (W. Cowling, personal communication).

VIRUS DISEASES

Chickpea, faba bean, lentil, lupin and pea are affected worldwide by more than 45 viruses. However, only some are of economic importance and others are of little significance as they occur sporadically. This paper will not review the literature on all viruses that affect these crops, but will focus on those of regional or global economic importance, with special emphasis on those identified more recently. For more information, the reader is referred to detailed reviews published earlier (Hagedorn, 1974; Cockbain, 1983; Bos et al., 1988; Jones and McLean, 1989).

Information on viruses that have economic significance in one or more countries is summarized in Table 1. Economic importance is often based on reported incidences and severity in locations, but precise data on the loss of yield are often missing (Bos, 1982).

An important group of viruses that affect the five legume crops is one that causes yellowing, chlorosis, reddening and stunting. Until the early 1990s, the symptoms were attributed mostly to bean leaf roll luteovirus. Recent data, however, suggest that many other luteoviruses such as beet western yellows virus (BWYV) (Bosque-Perez and Buddenhagen, 1990), legume yellows virus (LYV) (Duffus, 1979), soybean dwarf virus (SbDV) (Edwardson and Christie, 1991; Makkouk *et al.*, 1997) and chickpea luteovirus (CpLV) (Horn *et al.*, 1995, 1996) are involved. In addition, non-luteoviruses such as faba bean necrotic yellows virus (FBNYV) (Katul *et al.*, 1993) and chickpea chlorotic dwarf germinivirus (CCDV) (Horn et al., 1993) were found to produce similar symptoms in many countries. The common factor among these viruses is that they are phloem-limited and associated with degeneration or necrosis of phloem tissue. Their epidemiology is more or less similar as they are persistently transmitted by insects, they are not seed-borne and they have similar host ranges (Table 1).

Luteoviruses (BLRV, BWYV, LYV, SCRLV, SbDV) are serious diseases in many regions of the world that affect many legume species, including the cool season food legumes. The symptoms they induce are mostly yellowing and stunting of infected plants. These viruses are phloem-limited and transmitted by aphids such as *Aphis craccivora* (Koch) and *Acyrthosiphon pisum* (Harris) in a persistent manner. They are not transmitted mechanically or by seed. In countries with a Mediterranean climate there is a trend to change from spring sowing to winter sowing of chickpea, a practice which proved to increase yield by 50-100%, through better use of winter rains. Such a shift seems to influence the spread of luteoviruses (as well as other viruses of similar ecology), depending on the region. In California, the shift led to increased virus incidence (Bosque-Peres and Buddenhagen, 1990), whereas in the Mediterranean basin, it led to reduced virus incidence (K.M. Makkouk, unpublished).

Faba bean necrotic yellows virus (FBNYV), a novel circular single stranded (ss) DNA virus, was recently described by Katul *et al.*, (1993). It is widespread on faba bean, chickpea and lentil in a number of countries in West Asia such as Egypt, Jordan, Syria, Turkey and is sporadic in others such as Morocco, Algeria, Ethiopia and Pakistan. Symptoms induced by FBNYV in most species include chlorosis, leaf rolling and stunting. The virus is not mechanically or seed-transmitted, but could be transmitted persistently by the aphid species *Aphis craccivora* and *Acyrthosiphon pisum*. The virus is found to naturally infect summer legumes such as *Phaseolus vulgaris* and *Vigna unguiculata* (L.) Walp. suggesting that these crops can act as over-summering hosts of FBNYV (Franz *et al.*, 1995).

Table 1. Viruses reported to naturally infect chickpea, faba bean, lentil, lupin and pea, which are of regional or global economic importance.

CROP / Viruses	Manner of transmission			Geographical distribution				
	Sap	Seed	Vectors	America	Europe	Africa	Asia	Australasia
CHICKPEA								
Alfalfa mosaic alfamovirus	+	+	aphids NP[a]	+	+	+	+	+
Bean yellow mosaic potyvirus	+	-	aphids NP	+	+	+	+	+
Bean leaf roll luteovirus	-	-	aphids P	+	+	+	+	+
Beet western yellows luteovirus	-	-	aphids P	+	-	-	+	-
Chickpea chlorotic dwarf geminivirus	-	-	leafhoppers P	-	-	+	+	-
Chickpea luteovirus	-	-	aphids P	-	-	-	+	-
Cucumber mosaic cucumovirus	+	+	aphids NP	+	+	+	+	+
Faba bean necrotic yellows virus	-	-	aphids NP	-	-	+	+	-
Legume yellows luteovirus	-	-	aphids P	+	-	-	-	-
Pea enation mosaic enamovirus	+	-	aphids P	+	+	-	+	-
Pea seed-borne mosaic potyvirus	+	+	aphids NP	+	+	+	+	+
Pea streak carlavirus	+	-	aphids NP	+	+	-	-	-
Red clover vein mosaic carlavirus	+	-	aphids NP	+	-	-	-	-
FABA BEAN								
Alfalfa mosaic alfamovirus	+	-	aphids NP	+	+	+	+	+
Bean yellow mosaic potyvirus	+	+	aphids NP	+	+	+	+	+
Bean leaf roll luteovirus	-	-	aphids P	+	+	+	+	+
Beet western yellows luteovirus	-	-	aphids P	+	-	-	+	-
Broad bean mottle bromovirus	+	+	beetles	-	+	+	+	-
Broad bean stain comovirus	+	+	beetles	-	+	+	+	+
Broad bean true comovirus	+	+	beetles	-	+	+	+	-
Broad bean wilt fabavirus	+	-	aphids NP	+	+	+	+	-
Chickpea chlorotic dwarf geminivirus	-	-	leafhoppers P	-	-	+	+	-
Chickpea luteovirus	-	-	aphids P	-	-	-	+	-
Cucumber mosaic cucumovirus	+	+	aphids NP	+	+	+	+	+
Faba bean necrotic yellows virus	-	-	aphids P	-	-	+	+	-
Legume yellows luteovirus	-	-	aphids P	+	-	-	-	-

Pea early browning tobravirus	+	+	nematodes	-	+	+	-	-
Pea enation mosaic enamovirus	+	-	aphids P	+	+	-	+	-
Pea seed-borne mosaic potyvirus	+	+	aphids NP	+	+	+	+	+
Subterranean clover red leaf luteovirus	-	-	aphids P	-	-	-	-	+
LENTIL								
Alfalfa mosaic alfamovirus	+	+	aphids NP	+	+	+	+	+
Bean leaf roll luteovirus	-	-	aphids P	+	+	+	+	-
Bean yellow mosaic potyvirus	+	+	aphids NP	+	+	+	+	+
Broad bean stain comovirus	+	+	beetles	-	+	+	+	+
Chickpea chlorotic dwarf geminivirus	-	-	leafhoppers P	-	-	+	+	-
Chickpea luteovirus	-	-	aphids P	-	-	-	+	-
Cucumber mosaic cucumovirus	+	+	aphids NP	+	+	+	+	+
Faba bean necrotic yellows virus	-	-	aphids P	-	-	+	+	-
Pea enation mosaic enamovirus	+	-	aphids P	+	-	-	+	-
Pea seed-borne mosaic potyvirus	+	+	aphids NP	+	+	+	+	+
Pea streak carlavirus	+	-	aphids NP	+	+	-	-	-
Soybean dwarf luteovirus	-	-	aphids P	-	-	-	+	-
LUPIN								
Alfalfa mosaic alfamovirus	+	+	aphids NP	+	+	-	-	+
Bean yellow mosaic potyvirus	+	+	aphids NP	+	+	+	+	+
Clover yellow vein potyvirus	+	-	aphids NP	+	+	-	-	+
Cucumber mosaic cucumovirus	+	+	aphids NP	+	+	+	+	+
Soybean dwarf luteovirus	-	-	aphids P	+	-	-	+	+
PEA								
Alfalfa mosaic alfamovirus	+	-	aphids NP	+	+	+	+	+
Bean leaf roll luteovirus	-	-	aphids P	+	+	+	+	+
Bean yellow mosaic potyvirus	+	+	aphids NP	+	+	+	+	+
Beet western yellows luteovirus	-	-	aphids P	+	-	-	-	-
Broad bean stain comovirus	+	-	beetles	-	+	+	+	+
Broad bean wilt fabavirus	+	-	aphids NP	+	+	+	+	+
Cucumber mosaic cucumovirus	+	+	aphids NP	+	+	+	+	+
Legume yellows luteovirus	-	-	aphids P	+	-	-	-	-
Pea early browning tobravirus	+	+	nematodes	-	+	+	-	-
Pea enation mosaic enamovirus	+	-	aphids P	+	+	-	+	-
Pea seed-borne mosaic potyvirus	+	+	aphids NP	+	+	+	+	+
Pea streak carlavirus	+	-	aphids NP	+	+	-	-	-
Red clover vein mosaic carlavirus	+	-	aphids NP	+	+	-	-	-

a: P= persistent transmission; NP= non-persistent transmission

Chickpea chlorotic dwarf geminivirus (CCDV) produces symptoms in chickpea plants which include yellowing, stunting, phloem browning in the collar region, and in the case of desi types, leaf reddening (Horn et al., 1993). In faba bean it causes yellowing and stunting (Makkouk et al., 1995). CCDV is widely distributed in chickpea growing areas in India and Pakistan (Horn, 1994) and more recently in Sudan (Makkouk et al., 1995). The virus is persistently transmitted by the leafhopper Orosius orientalis, which makes the ecology of this disease slightly different from the diseases caused by luteoviruses or FBNYV.

Other groups of viruses that are of global importance are those which are non-persistently transmitted by aphids and are seed-borne. Viruses such as pea seed-borne mosaic virus, cucumber mosaic virus, alfalfa mosaic virus and bean yellow mosaic virus cause serious losses. These viruses are known to be seed-borne in two or more of the five legume crops dealt with in this paper but not in all five. It was not until recently that alfalfa mosaic virus and cucumber mosaic virus (CMV) were reported to be seed-borne in chickpea and lentil (Jones and Coutts, 1996), bean mottle virus (BMV) in chickpea (Erdiller and Akbab, 1996), and bean yellow mosaic virus (BYMV) in lentil (Kumari et al., 1994; Erdiller and Akbab, 1996). Such information is important in relation to seed movement as well as to the epidemiology and control of these viruses.

As there is no direct practical way of curing crops from virus infection, all current control strategies emphasize measures that prevent or reduce infection. Cultural practices, such as varying sowing date, roguing infected plants early in the season, using borders of plants which are not hosts to the virus, etc. have been effective in reducing virus incidence. Jones (1993) reported a 90% reduction in BYMV incidence in lupin fields when cereal borders or admixtures with cereals were used. Since around 50% of viruses affecting legumes are seed-borne (Bos et al., 1988), it is always recommended to use virus-free seed for planting, especially when the virus is also transmitted by active vectors.

Host resistance, when available is the most acceptable control. There is useful resistance in pea (Johnson and Hagedorn, 1958) and faba bean (Makkouk and Kumari, 1995) to BYMV, in lupin to CMV (Jones and Latham, 1997) and in lentils to FBNYV (K.M. Makkouk, unpublished data). However, host resistance for a number of important virus diseases of cool season food legumes has not been found and control is dependent on other options.

Experience over the last several decades has shown that viruses cannot be controlled by preventive measures alone. The complicated ecology of many of the viruses, which affect cool season food legume crops and other crops, calls for a strategy, which is long term, sustainable, economically acceptable to resource-poor farmers and friendly to the environment. This can only be accomplished by employing a number of complementary interventions in an integrated disease management approach.

FUTURE RESEARCH NEEDS

Information is lacking on the biology, epidemiology and control of the foliar diseases. For example, Bowen et al. (1997) produced the teleomorph of *Phoma medicaginis* var. *pinodella* in culture with an Australian isolate of the fungus. What is the significance of the sexual state in the spread and survival of the pathogen? Will the teleomorph be found under natural conditions? The teleomorph of *P. medicaginis* var. *pinodella* was found on over-wintered pea debris in Pullman, Washington, but it was not possible to produce the sexual state in culture when different single ascospore and conidial isolates of the pathogen were crossed in mating studies (W.J. Kaiser, unpublished data). Similar information is lacking on the significance of the teleomorphic states of *Ascochyta fabae* and *A. lentis* in the disease cycle.

Many of the foliar diseases are seed-borne (Diekmann, 1994; Kaiser, 1997; Mathur et al., 1988; Neergaard, 1979). Presently, the laboratory tests used frequently to detect the seed-borne pathogens are agar and blotter tests. These plating techniques are generally not applicable to large samples of seed, especially when the level of infection is less than 0.1%. More sensitive techniques are needed to detect and identify pathogens in large samples of seed. Research has begun on molecular and serological tests to differentiate between *Ascochyta* species of chickpea faba bean, lentil and pea (Bowen et al., 1996; Faris-Moraiesh et al., 1996; Kaiser et al., 1997; Tohamy et al., 1997), to detect *A. lentis* in lentil seed (Tranberg et al., 1997) and the '*Ascochyta* complex' in pea seed (Faris-Moraiesh et al., 1995). With additional research, it may be possible to utilize these techniques to detect low levels of infection of seed by seed-borne pathogens.

CONCLUDING REMARKS

Within the last 20-25 years, there has been an increase in the inter- and intra-national movement of the seed of crops, including the five under discussion. This poses serious quarantine problems related to the introduction of new and potentially damaging pathotypes or strains of indigenous pathogens, or compatible mating types of heterothallic fungi which could lead to the development of the teleomorph in nature. Anthracnose of lupin incited by *Colletotrichum gloeosporioides* is an excellent example of the recent inadvertent introduction of an important pathogen into Western Australia (Sweetingham et al., 1995). Sweetingham et al. (1995) state that anthracnose poses a serious threat to the Australian lupin industry and that quarantine regulations are needed to regulate the importation of lupin seed into Australia.

In recent years, there has been a dramatic increase in the establishment of gene banks in different countries (Kaiser, 1997). Most gene banks store seed at 4° to -18^oC and relative humidities of <50%. These temperatures prolong the longevity of seeds of different plant species and their seed-borne pathogens. Gene banks have served as reservoirs of the seed-borne *Ascochyta* pathogens of chickpea, faba bean, lentil and pea (Hagedorn, 1984; Kharbanda and Bernier, 1989; Kaiser and Hannan, 1986; Kaiser et al., 1989; W.J. Kaiser, unpublished data). Kaiser et al. (1989) demonstrated with lentil seed infected with *A. lentis* that the pathogen survived temperatures of 4° to -196° C for four years without loss of viability and pathogenicity. Seeds of chickpea, faba bean and pea infected with their respective *Ascochyta* pathogens probably would survive similarly in gene banks at low temperatures. Special precautions need to be taken to prevent the introduction of these and other pathogens on imported seeds of these cool season food legumes.

References

Açikgöz, N. 1997. *Anadolu J. AARI* 7:1-8.

Ahmed, S. and Morrall, R.A.A. 1996. *Canadian Journal of Plant Pathology* 18:362-369.

Ahmed, S., and Morrall, R.A.A. and Kaiser, W.J. 1996a. *Canadian Journal of Plant Pathology* 18:347-353.

Ahmed, S., and Morrall, R.A.A. and Sheard, J.W. 1996b. *Canadian Journal of Plant Pathology* 18:354-361.

Andrahennadi, C.P., Slinkard, A.E. and Vandenberg, A. 1996. *Lens* Newsletter 23:5-7.

Bakr, M. A.. Rahman, M. M., Ahmed, F. and Kumar, J. 1993. In: *Recent Advances in Research on Botrytis Gray Mold of Chickpea,* pp. 17-19 (eds. M.P. Haware, C.L.L. Gowda,and D. McDonald). Patancheru, India: ICRISAT.

Beed, F.D., Rodney, E.S. and Strange, R.N. 1994a. *Mycological Research.* 98:1060-1076.

Beed, F.D., Rodney, Strange, R.N., Onfroy, C. and Tivoli, B. 1994b. *Plant Pathology* 43:987-997.

Boerema, G.H. and Verhoeven, A.A. 1979. *Netherlands Journal Plant Pathology* 85:151-185.

Bos, L., Hampton, R.O., and Makkouk, K.M. 1988. In: *World Crops: Cool Season Food Legumes* pp: 591-615 (ed. R.J. Summerfield). Kluwer Academic Publishers.

Bos, L. 1982. *Crop Protection* 1:263-282.

Bosque-Perez, N.A. and Buddenhagen, I.W. 1990. *Plant Disease* 74:372-378.

Bowen, J.K., Lewis, B.G. and Matthews, P. 1997. *Mycological Research.* 101:80-84.

Bowen, J.K., Peart, J., Lewis, B.G., Cooper, C. and Matthews, P. 1996. *Plant Pathology* 45:393-406.

Bretag, T.W. 1989a. *Annals of Applied Biology* 114 (Suppl.), *Tests for Agrochemicals and Cultivars* 10:44-45.

Bretag, T.W. 1989b. *Annals of Applied Biology* 114 (Suppl.), *Tests for Agrochemicals and Cultivars* 10:156-157.

Bretag, T. W. 1991. La Trobe University, Australia. Ph.D. Thesis. 297 pp.

Bretag, T. W. and Brouwer, J.B. 1995. In: Proc. 2nd European Conference on Grain Legumes. Copenhagen, Denmark. pp. 98.

Bretag, T. W. and Mebalds, M. I. 1987. *Australian Journal Experimental Agriculture* 27:141-148.

Bretag, T. W., Keane, P. J. and Price, T. V. 1995a. *Australian Journal Experimental Agriculture* 35:531-536.

Bretag, T. W., Price, T. V. and Keane, P. J. 1995b. *Australian Journal Experimental Agriculture* 35:525-530.

Buchwaldt, L., Morrall, R.A.A., Bernier, C.C., and Vandenberg, B. 1992. (Abstr.) *Phytopathology* 82:1141.

Buchwaldt, L., Morrall, R.A.A., Chongo, G. and Bernier, C.C. 1996. *Phytopathology* 86:1193-1198.

Bulson, H.A.J., Snaydon, R.W. and Stopes, C.E. 1997. *Journal of Agricultural Science* 128:59-71.

Burgess, D. R., Bretag, T. W. and Keane, P. J. 1997a. *Australian Journal Experimental Agriculture* 37: 223-229.

Burgess, D. R., Bretag, T. W. and Keane, P. J. 1997b. *Plant Pathology* 46:298-305.

Butler, D.R. 1993. In: *Recent Advances in Research on Botrytis Gray Mold of Chickpea,* pp. 7-9 (eds. M.P. Haware, C.L.L. Gowda, and D. McDonald) Patancheru: India: ICRISAT.

Chandreshankar, M. and Culvenor, B. 1993. Bureau of Resource Sciences, Canberra, Australia. 1993.

Chen, M. H., Huang, J.W. and Yein, C. F. 1994. *Plant Pathology.* Bull. 3:133-139.

Clark, J.S.C. and Spencerphillips, P.T.N. 1993. *New Phytologist* 124:107-119.

Clark, J.S.C. and Spencerphillips, P.T.N. 1994. *Plant Pathology* 43:56-64.

Clulow, S. A., Lewis, B. G. and Matthews, P. 1992. *Plant Pathology* 41:362-369.

Clulow, S. A., Lewis, B. G. and Matthews, P. 1991a. *Journal Phytopathology* 131:322-332.

Clulow, S. A., Lewis, B. G., Parker, M. L. and Matthews, P. 1991b. *Mycological Research.* 95:817-820.

Clulow, S. A., Matthews, P. and Lewis, B. G. 1991c. *Euphytica* 58:183-189.

Cockbain, A.J. 1983. In: *The Faba bean (Vicia faba* L.), pp. 421-462 (ed. P.D. Hebblethwaite). London: Butterworths.

Cook, R.P. and Dubé, A.J. 1989. South Australian Department of Agriculture. 142 pp.

Corbiere, R., Gelie, B., Molinero, V., Spire, D. and Agarwal, V. K. 1994. *Seed Research* 22:26-30.

Cors, F. and Meeus, P. 1994. In: 46th International Symposium on Crop Protection, Gent, Belgium. Mededelingen Faculteit Landbouwkundige en Toegepaste Biologische Wetenschappen, Universiteit Gent 59:1081-1087.

Cother, E.J. 1977. *Plant Disease Report.* 61:736-740.

Csizmadia, L. 1995. *Horticultural Science* 27:57-61.

Czyzewska, S. 1993. Biuletyn Instytutu Hodowli i Aklimatyzacji Roslin 188:289-299.

Daniels, A. and Lucus, J.A. 1995. *Pesticide Science* 45:33-41.

Dewan, B. 1993. *International Chickpea* Newsletter: 29:31.

Diekmann, M. 1994. ICARDA/Danish Government Institute of Seed Pathology for Developing Countries: Aleppo, Syria. 56 pp.

Duffus, J.E. 1979. *Phytopathology* 69:217-221.

Edwardson, J.R., and Christie, R.G. 1991. In: *CRC Handbook of Viruses Infecting Legumes,* pp. 245-261. Boca Raton, Florida: CRC Press.

Erdiller, G. and Akbab, B. 1996. *Journal Turkish Phytopathology* 25:93-101.

Erkine, W. and Saxena, M.C. 1993. In: *Breeding for Stress Tolerance in Cool-Season Food Legumes,* pp. 51-62 (eds. K.B. Singh and M.C. Saxena). Chichester, U.K.: John Wiley & Sons.

Fagbola, O. and Jellis, G.J. 1994. *Annals Applied Biology* 15:124.

Falloon, R.E., Viljanenrollinson, S.L.H., Coles, G.D. and Poff, J.D. 1995. *New Zealand Journal Crop and Horticultural Science* 23:31-37.

Falloon, R.E. and Sutherland, P.W. 1996. *Mycologia* 88:473-483.

Faris-Mokaiesh, S., Boccara, M., Denis, J.-B., Derrien, A. and Spire, D. 1996. *Current Genetics.* 29: 182-190.

Faris-Mokaiesh, S., Corbiere, R., Lyons, N.F. and Spire, D. 1995. *Annals of Applied Biology* 127:441-455.

Fegies, N.C., Corbiere, R., Molinero, V., Bethenod, M.T., Bertrandy, J., Champion, R. and Spire, D. 1992. In: Proc. 1st European Conference on Grain Legumes. Angers, France. pp. 315-316.

Garry, G., Tivoli, B., Jeuffroy, M. H. and Citharel, J. 1996. *Plant Pathology* 45:769-777.

Garry, G., Tivoli, B., Jeuffroy, M. H., Ney, B. and Jeudy, C. 1995. In: Proc. 2nd European Conference on Grain Legumes, Copenhagen, Denmark. pp. 82-83.

Gaunt, R.E. 1983. In: *The Faba Bean (Vicia faba* L.), pp. 463-492 (ed. P.D. Hebblethwaite). London: Butterworths.

Gilet, A. and Durand, N. 1996. Cultivar Rueil *Malmaison* 403:36-37.

Gondran, J., Bournoville, R. and Duthion, C. 1994. INRA Editions, Versailles, France.

Hagedorn, D.J. (ed.). 1984. *Compendium of Pea Diseases.* St. Paul, MN: American Phytopathological Society. 57 pp.

Hagedorn, D.J. 1974. Monograph No. 9. St. Paul, MN: *American Phytopathological Society.* 47 pp.

Hanounik, S.B., Jellis, G.H. and Hussein, M.M. 1993. In: *Breeding for Stress Tolerance in Cool-Season Food Legumes,* pp. 97-106 (eds. K.B. Singh and M.C. Saxena). Chichester, U.K.:John Wiley & Sons.

Harrison, J.G. 1988. *Plant Pathology* 37:168-201.

Haware, M. P., Faris, D. G. and Gowda, C. L. L. (eds.) 1992. *Botrytis Gray Mold of Chickpea.*Patancheru, India: ICRISAT. 22 pp.

Haware, M. P., Gowda, C. L. L. and McDonald, D. (eds.) 1993. *Recent Advances in Research on Botrytis Gray Mold of Chickpea.* Patancheru, India: ICRISAT. 32 pp.

Haware, M. P. and McDonald, D. 1993. In: *Recent Advances in Research on Botrytis Gray Mold of Chickpea,* pp. 3-6 (eds. M.P. Haware, C.L.L. Gowda, and D. McDonald). Patancheru, India: ICRISAT.

Haware, M. P., Rao, J. N. and Pundir, R. P. S. 1992b. *International Chickpea* Newsletter 27:16-18.

Heenan, D. P. 1994. *Australian Journal Experimental Agriculture* 34:1137-1142.

Helsper, J.P.F.G., Vannorel, A., Burgermeyer, K. and Hoogendijk, J.M. 1994. *Journal Agricultural Science* 123:349-355.

Horn, N.M. 1994. Wageningen Agricultural University, the Netherlands. Ph.D. Thesis. 137 pp.

Horn, N.M., Reddy, S.V., van den Heuvel, J.F.J.M., and Reddy, D.V.R. 1996. *Plant Disease* 80:286-290.

Horn, N.M., Reddy, S.V., Roberts, I.M., and Reddy, D.V.R. 1993. *Annals of Applied Biology* 122:467-479.

Horn, N.M., Makkouk, K.M., Kumari, S.G., van den Heuvel, J.F.J.M., and Reddy, D.V.R. 1995. *Phytopathologia Mediterranea* 34:192-198.

Hwang, S. F., Lopetinskey, K. and Evans, I. R. 1991. *Canadian Plant Disease Survey* 71:169-172.

Jackson, A.J., Walters, D.R. and Marshall, G. 1994. *Mycological Research.* 98:1117-1126.

Jackson, A.J., Walters, D.R. and Marshall, G. 1997. *Biological Control* 8:97-106.

Jellis, G.J. and Punithalingam, E. 1991. *Plant Pathology* 40:150-157.

Jellis, G.J. and Vassie, V. 1995. In: Proc. 2nd European Conference on Grain Legumes. Copenhagen, Denmark. pp.88-89.

Johnson, K.W., and D.J. Hagedorn. 1958. *Phytopathology* 48:451-453.

Jones, R.A.C. 1993. *Annals of Applied Biology* 122:501-518.

Jones, R.A.C. and Coutts, B.A. 1996. *Annals of Applied Biology* 129:491-506.

Jones, R.A.C. and Latham, L.J. 1997. *Annals of Applied Biology* 129:523-542.

Jones, R.A.C. and McLean, G.D. 1989. *Annals of Applied Biology* 114:609-637.

Joshi, S. 1992. In:*Botrytis Gray Mold of Chickpea,* pp. 12-13. (eds. M.P. Haware, D.G. Faris and C.L.L. Gowda) Patancheru, India: ICRISAT.

Kaiser, W.J. 1997. *Canadian Journal of Plant Pathology* 19:215-224.

Kaiser, W.J. 1992. *Plant Disease* 76:605-610.

Kaiser, W.J. and Hannan, R.M. 1986. *Phytopathology* 76:355-360.

Kaiser, W.J. and Hannan, R.M. 1987. *Plant Disease* 71:58-62.

Kaiser, W.J. and Hellier, B.C. 1993. (Abstr.) *Phytopathology* 83:692.

Kaiser, W.J. and Muehlbauer, F.J. 1994. In: Proc. of a Symposium on *Breeding of Oil and Protein Crops,* pp. 56-59. Albena, Bulgaria: Eucarpia.

Kaiser, W.J., Muehlbauer, F.J. and Hannan, R.M. 1994. In: *Expanding the Production and Use of Cool Season Food Legumes,* pp. 849-858 (eds. F.J. Muehlbauer and W.J. Kaiser) Kluwer Academic Publishers.

Kaiser, W.J., Wang, B.-C. and Rogers, J.D. 1997. *Plant Disease* 81:809-816.

454

Karki, P. B., Tiwari, K. R.., Singh, O. and Bharati, M. P. 1989. *International Chickpea* Newsletter 21:21-23.

Katul, L., Vetten, H.J., Maiss, E., Makkouk, K.M., Lesemann, D.E., and Casper, R. 1993. *Annals of Applied Biology* 123:629-47.

Khan, M.S.A., Ramsey, M.D., Corbiere, R., Porta-Puglia, A., Bouznad, Z. and Scott, E.S. 1997. 3rd International Food Legume Conference. (Abstr.) (In press).

Kharbanda, P.D., and Bernier, C.C. 1980. *Canadian Journal of Plant Pathology* 2:139-142.

Khare, M.N. 1981. In: *Lentils*, pp. 163-172 (eds. C. Webb and G. Hawtin). Slough, U.K.: CAB/ICARDA.

Kovachevski, I.C. 1936. Ministry of Agriculture and Natural Domains, Plant Protection Institute,Sofia, Bulgaria. 80 pp.

Kraft, J.M. and Kaiser, W.J.Jr. 1993. In: *Breeding for Stress Tolerance in Cool-Season Food Legumes*, pp. 123-144 (eds. K.B. Singh and M.C. Saxena). Chichester, U.K.: John Wiley & Sons.

Kumari, S.G., Makkouk, K.M., and Ismail, I.D. 1994. *Lens* 21:42-44.

Laha, S. K. and Grewal, J. S. 1983. *Indian Phytopathology* 36:630-634.

Lawsawadsiri, S. 1994. University of Adelaide, Australia. Ph.D. Thesis. 140 pp.

Larsen, R.C., Kaiser, W.J. and Wyatt, S.W. 1995. (Abst.) *Phytopathology* 85:1138.

Loughman, R. and Sweetingham, M.W. 1991. *Australian Journal Experimental Agriculture* 31:493-498.

Madeira, A.C., Clark, S., Rossall, S. and McArthur, A.C. 1992. *FABIS* Newsletter 30:48-51.

Madeira, A.C., Fryett, K.P., Rossall, S. and Clark, J.A. 1993. *Mycological Research.* 97:1217-1222.

Madhu, M. and Bedi, P. S. 1992. *Indian J. Mycol. and Pl. Pathol.* 22:213-219.

Mahmood, M., Sinha, B. K. and Kumar, S. 1989. *Journal Research,* Rajendra Agric. Univ. 7:39-43.

Makkouk, K.M., Dafalla, G., Hussein, M., and Kumari, S.G. 1995. *Journal Phytopathology* 143:465-466.

Makkouk, K.M., Damsteegt, V., Johnstone, G.R., Katul, L., Lesemann, D.-E., Kumari, S.G. 1997. *Phytopathologia Mediterranea* (in press).

Makkouk, K.M., and S.G. Kumari. 1995. *Journal of Plant Disease and Plant Protection* 102:461-466.

Marcellos, H., Moore, K.J. and Nickandrow, A. 1995. *Australian Journal Experimental Agriculture* 35:97-102.

Mathur, S.B., Haware, M.P. and Hampton, R.O. 1988. *In: World Crops: Cool Season Food Legumes,* pp. 351-365 (ed. R.J. Summerfield). Dordrecht, the Netherlands: Kluwer Academic Publishers.

Maurin, N. and Tivoli, B. 1992. *Plant Pathology* 41:737-744.

McDonald, G. K. and Dean, G. 1996. *Australian Journal Experimental Agriculture* 36:219-222.

Molloy, B.P.J., Partridge, T.R. and Thomas, W.P. 1991. *New Zealand Journal Botany* 29:349-52.

Morrall, R.A.A. 1988. *Plant Disease* 72:994.

Muehlbauer, F.J. and Kaiser, W.J. (eds.) 1994. *Expanding the Production and Use of Cool Season Food Legumes.* Kluwer Academic Publishers. 991 pp.

Mukherjee, P. K. and Haware, M. P. 1993. *International Chickpea* Newsletter 28:14-15.

Mukherjee, P. K., Haware, M. P. and Jayanthi, S. 1995. *Indian Phytopathology* 48:141-149..

Nasir, M. and Bretag, T.W. 1996. *Lens* Newsletter 23:7-9.

Nasir, M. and Hoppe, H. H. 1991. *Journal of Plant Disease and Plant Protection* 98:619-626.

Nasir, M., Hoppe, H. H. and Ebrahim-Nesbat, F. 1992. *Plant Pathology* 41:187-194.

Neergaard, P. 1979. *Seed Pathology* Vol. 1. New York: John Wiley & Sons. 839 pp.

Nene, Y.L., Hanounik, S.B., Qureshi, S.H. and Sen, B. 1988. In: *World Crops: Cool Season Food Legumes,* pp. 577-589 (ed. R. J. Summerfield). Kluwer Academic Publishers.

Nene, Y.L. and Reddy, M.V. 1987. In: *The Chickpea,* pp. 233-270 (eds. M.C. Saxena and K.B. Singh). Wallingford, UK: CAB International.

Olofsson, J. 1996. *Plant Disease Report* 50:257-261.

Ondrej, M. 1994. *Rocenka Geneticke Zdroje Rastlin* 1994:19-23.

Palmer, G.M. and Stevens, D.B. 1993. *Annals Applied Biology* 122:28-29

Pate, J.S., Williams, W. and Farrington, P. 1985. In: *Grain Legume Crops,* pp. 699-746. (eds. R.J.Summerfield and E.H. Roberts). London: Collins.

Paulitz, T.C., Atlin, G. and Gray, A.B. 1995. *Plant Disease* 79:319.

Prasada, R. and Verma, U.N. 1948. *Indian Phytopathology* 1:142-146.

Rahul, C., Singh, I. S., Gupta, A. K. and Chaturvedi, R. 1995. *Legume Research* 18:1-4.

Ramsey, M.D., Knight, R. and Paull, J. 1995. In: Proc. 2nd European Conference on Grain Legumes.Copenhagen, Denmark. pp.164-165.

Rathi, Y. P. S. and Tripathi, H. S. 1991. *International Chickpea* Newsletter 24:37-38.

Rathi, Y. P. S. and Tripathi, H. S. 1993 In: *Recent Advances in Research on Botrytis Gray Mold of Chickpea* pp. 10-11 (eds. M.P. Haware, C.L.L. Gowda, and D. McDonald). Patancheru, India: ICRISAT

Reddy, M. V., Singh, O, Bharati, M. P., Sah, R. P. and Joshi, S. 1988. *International Chickpea Newsletter* 19:15.

Reddy, M. V., Ghanekar, A. M., Nene, Y. L., Haware, M. P., Tripathi, H. S. and Rathi, Y. P. S. 1993. *Indian Journal of Plant Protection* 21:112-113.

Reed, P.J., Dickens, J.S.W. and O'Neill, T.M. 1996. *Plant Pathology* 45:245-248.

Rewal, N. and Grewal, J. S. 1989a. *Euphytica* 44:61-63.

Rewal, N. and Grewal, J. S. 1989b. *Indian Phytopathology* 42:265-268.

Rewal, N. and Grewal, J. S. 1989c. *Indian Phytopathology* 42:79-83.

Robertson, L.D. 1995. *Grain Legumes* 8:25-26.

Robertson, L.D., Singh, K.B., Erskine, W. and Abd El Moneim, A.M. 1996. *Genetic Resources and Crop Evolution* 4:447-460.

Roger, C. and Tivoli, B. 1996. *Plant Pathology* 45:518-528.

Rybus-Zajac, M. and Kozlowska, M. 1996. *Acta Physiologia Plantarum* 18: 211-216.

Sache, I. and Zadoks, J.C. 1994. *Plant Pathology* 44:675-685.

Salem, D.E., Omar, S.A.M. and Aly, M.M. 1992. *FABIS* Newsletter 31:29-33.

Shahu, R. and Sah, D. N. 1988. *International Chickpea* Newsletter 18:13-15.

Singh, G and Kaur, L. 1990. *Plant Disease Res*earch 5:132-137.

Singh, M. P. and Tripathi, H. S. 1992. *Indian J. Mycol. Pl. Path.* 22:39-43.

Singh, M. P. and Tripathi, H. S. 1993. *Indian J. Mycol. Pl. Path.* 23:77-179.

Smartt, J. 1984. *Experimental Agriculture* 20:275-296.

Sreenivasaprasad, S., Mills, P.R. and Brown, A.E. 1994. *Mycological Research.* 98:186-188.

Stegmark, R. 1994. *Agronomie* 14:641-647.

Summerfield, R.J. (ed.). 1988. *World Crops: Cool Season Food Legumes.* Kluwer Academic Publishers. 1179 pp.

Sweetingham, M. 1990. *Journal of Agriculture, Western Australia* 31:5-13.

Sweetingham, M. 1997. *Agriculture Western Australia Farmnote.* No 24/97: 4 pp.

Sweetingham, M.W. 1989. (Abst.) *Australian Journal Agricultural Research* 40:781-789.

Sweetingham, M.W., Cowling, W.A., Buirchell, B.J. and Henderson, J. 1994. In: *Advances in Lupin Research* pp. 368-370 (eds. J.M. Neves Martins, and L. Beirao da Costa). Lisbon:ISA Press.

Sweetingham, M.W., Cowling, W.A., Buirchell, B.J., Brown, A.G.P. and Shivas, R.G. 1995. *Australian Plant Pathology* 24:271.

Sweetingham, M.W., Voight, J. and Lange, B.L. 1992. *Australian Grain* 2:32-33.

Sweetingham, M.W., Loughman, R. and Porritt, S.E. 1993. *Australian Journal of Experimental Agriculture* 33:469-473.

Thomas, J.E. and Sweet, J.B. 1990. National Institute of Agricultural Botany. Cambridge, UK. 51 pp.

Tivoli, B. and Bonavita, J. 1995. Proc. 2nd European Conference on Grain Legumes, Copenhagen, Denmark. pp. 72-73.

Tivoli, B., Lemarchand, E., Masson, E. and Moquet, M. 1995. Proc. 2nd European Conference on Grain Legumes, Copenhagen, Denmark. pp. 84-85.

Tohamy, A.M., Nour Eldin, H.A., Mohamed, Z.K., Gad El-Karin, G.A. and Madkour, M.A.1997. (Abstr.) *Phytopathology* 87 (6 Suppl.). S95.

Tranberg, J.N., Taylor, J. and Morrall, R.A.A. 1997. (Abstr.) *Canadian Journal of Plant Pathology* 19:117-118.

Trapero-Casas, A. and Kaiser, W.J. 1992. *Phytopathology* 82:1261-1266.

Trapero-Casas, A., Navas-Cortés, J.A. and Jiménez-Díaz, R.M. 1996. *European Journal of Plant Pathology* 102:237-245.

Van der Gaag, D.J., Frinking, H.D. and Geerds, C.F. 1993. *Netherlands Journal of Plant Pathology* 99:83-91.

Van der Gaag, D.J. 1994. *Mycologia* 86:454-457.

Van der Gaag, D.J. and Frinking, H.D. 1996a. *Journal of Phytopathology* 144:57-62.

Van der Gaag, D.J. and Frinking, H.D. 1996b. *Plant Pathology* 45:990-996.

Venette, J.R., Lamey, H.A. and Lamppa, R.L. 1994. *Plant Disease* 78:1216.

Warkentin, T. D., Rashid, K. Y. and Xue, A. G. 1996. *Canadian Journal of Plant Science.* 76:67-71.

Weidemann, G.J., TeBeest, D.O. and Cartwright, R.D. 1988. *Phytopathology* 78:986-990.

Welty, R.E. 1984. *Plant Disease* 68:142-144.

Wilson, A.D. and Kaiser, W.J. 1995. *Mycologia* 87:795-804.

Wroth, J. M. 1966. University of Western Australia, Australia. Ph.D. Thesis.

Yanase, Y., Thomas, J.E. and Camps, A. 1995. In: Proc. 2nd European Conference on Grain Legumes. Copenhagen, Denmark. pp. 99

Yang, H.A., Sweetingham, M.W. and Cowling, W.A. 1996. *Australian Journal Agricultural Research* 47:787-799.

Soilborne Diseases and their Control

J. M. Kraft [1] , M. P. Haware [2], H. Halila [3] , M. Sweetingham [4] and B. Bayaa [5]

1 U.S. Department of Agriculture, Agricultural Research Service, WSU-IAREC,, 24106 North Bunn Road, Prosser, Washington, 99350-9687 USA; ,2 ICRISAT, Patancheru P.O., Andhra Pradesh 502 324, INDIA; 3 Institut National de la Recherche, Agronomique de Tunisie, Avenue de 'Indepencance, 2034 Ariana, TUNISIA; 4 Agriculture Western Australia, 3 Baron Hay Court, South Perth, Western Australia 6151; 5 ICARDA, P.O. Box 5466, Aleppo, SYRIA

Abstract

Seed and seedling diseases, root rots, and wilts are caused by a number of soilborne fungi, all of which are facultative saprophytes and can survive in soil for long periods in the absence of a susceptible host. In general, these diseases are serious yield constraints where short rotations or monoculture of legume crops are the rule. Seedling diseases and root rots are enhanced by poor seed vigor, poor seedbed preparation, and other biotic and abiotic stresses which predispose the host plant. Control of these diseases requires an integrated approach of genetic resistance/tolerance, cultural practices, appropriate seed treatments, and high seed vigor. The most economical and durable control of Fusarium wilt is to grow resistant varieties. New races of a wilt pathogen have arisen due to increased selection pressure from growing resistant varieties in short rotations but have not outpaced the development of resistant cultivars.

INTRODUCTION

Cool season food legumes, including chickpea (*Cicer arietinum* L.), lentil (*Lens culinaris* Medik.), pea (*Pisum sativum* L.), and lupin (*Lupinus angustifolius* L.), are susceptible to a number of soilborne fungal diseases. Seed and seedling rots are caused primarily by *Pythium* spp. and *Rhizoctonia solani*. Full season root rots and root diseases are caused by *Aphanomyces euteiches, Fusarium solani, Sclerotium rolfsii, Rhizoctonia solani,* and *Pleiochaeta setosa*. Wilt of these crops is caused primarily by various forma specialis of *Fusarium oxysporum*.

SEEDLING DISEASE

Pythium Seed and Seedling Rot

Factors which delay emergence or result in uneven plant stands including: a) cold, wet soil, b) poor seed vigour, and c) herbicide injury, predispose developing plants to seedling disease. Diseases caused by *Pythium* spp. are referred to as seed rot, damping-off, or root rot. Damage due to *Pythium* is more prevalent and severe when soil moisture is high and soil temperatures are in the 10-15°C range. When conditions are favourable for the fungus but less for the host, *Pythium* species can become very pathogenic. However, as the plant matures, *Pythium* attack is usually focused on root tips. Infection takes place when zoospores produce germ tubes or hyphal elements form appressoria, and penetrate the host plant by means of infection pegs (Kraft *et al.*, 1967). According to Stasz *et al.* (1980), there are at least three kinds of genetic resistance in pea to Pythium seed and seedling rot: 1) seeds lose juvenile susceptibility within 48 hrs. after imbibition begins, thus decreasing the number of seeds rotted; 2) peas with round, not wrinkled, seed exude reduced amounts of substances stimulatory to *Pythium*; and 3) peas with pigmented seed coats are nearly immune to *Pythium* due to the presence of fungistatic, phenolic compounds.

Seed quality and vigour have a major influence on *Pythium* seed infection. Seed with poor vigour and/or mechanical damaged seed exude more water soluble and volatile exudates than do high vigour seeds and are

457

R. Knight (ed.), Linking Research and Marketing Opportunities for Pulses in the 21st Century, 457–466.
© 2000 Kluwer Academic Publishers. Printed in the Netherlands.

more stimulatory to *Pythium* germination and infection (Matthews, 1971). In the United Kingdom, seed lots of cool season legumes are categorized based on electroconductivity tests of imbibed seed. Only seeds with relatively low EC readings are recommended for planting in very cold, wet soil where *Pythium* attack is most likely.

Rhizoctonia Seedling Rot

Rhizoctonia solani can survive in field soils for extended periods and can attack legume seedlings when warm, moist conditions are prevalent. For seedling infection to occur, the sclerotium or hyphal fragment germinates and grows up to several millimeters through soil to form an infection cushion on the host surface to penetrate the seedling plant. *Rhizoctonia* invades both inter- and intracellulary, and seedlings become less susceptible with maturity. *Rhizoctonia* is warm temperature dependent, occurring most frequently and severely when surface soil temperatures are in the 24-30°C range and in sandy soils because they warm rapidly. Seedling damping off, caused by *R. solani*, can be especially severe when soil moisture and surface soil organic matter are high and the legume crop is direct drilled into fields with reduced or no tillage.

Seedling hypocotyl and epicotyl symptoms, caused by *R. solani*, appear as water-soaked lesions turning reddish-brown to brown (Flentje & Hagedorn, 1964). The growing tip of a seedling may die as it emerges from the soil. On older plants, reddish brown, sunken lesions may occur on the epicotyl or hypocotyl, sometimes girdling the entire plant, resulting in severe plant stunting or death. *Rhizoctonia solani* (AG4) prefers well-aerated areas at or near the soil surface, so plant parts at or near the soil surface are most vulnerable to attack.

There are no reported varieties of any cool season legumes resistant to *Rhizoctonia*, but pea cultivars with vigorous, thick stems, which emerge rapidly may escape serious damage (McCoy & Kraft, 1984). Rotation with cereals, clean tillage of fields prior to sowing, and fungicidal seed treatment chemicals, such as PCNB or Demosan, provide some protection against *R. solani* (Hagedorn, 1984). Biological seed treatments, such as *Trichoderma viride* and/or *Gliocladium virens*, have also shown promise in protecting against Rhizoctonia seed and seedling rot (Chet & Baker, 1981; Harman *et al.*, 1980).

ROOT ROT

Root rots of cool season legumes are caused by several different soilborne fungal pathogens that produce similar symptoms. These disease complexes can encompass the whole root system and/or extend short distances above the soil surface. All of the pathogens discussed in this paper can be disseminated by water, movement of infected plant debris, or infested soil carried on farm implements. In most cases, where soilborne diseases are severe, a short or no interval between legume crops in a given field is the rule. Crop rotation has historically been a primary means of soilborne disease control (Bruehl, 1986). However, research is often focused on making monoculture or short rotations possible through developing resistant varieties and on biological or chemical control. In the Columbia Basin of Central Washington, some farmers were growing seed and processing peas in a double cropping sequence with sweet corn for processing. These growers received a double income from a given field, but peas were being grown every year. The result has been the development of severe root rot in less than 5 years that in some instances has caused a total loss of the crop. In Western Australia, the population of *Pleiochaeta setosa* spores in soil declines by about 50% every 12 months. Consequently, the longer the break between legume crops the lower the spore numbers and disease risk (Sweetingham, 1996).

Rhizoctonia Root Rot of Lupine

In Western Australia, a wheat-lupin rotation is the preferred rotation. As a result, both Rhizoctonia bare patch (AG8) and a slow-growing binucleate *Rhizoctonia* spp. have become serious problems in lupin production (MacLeod & Sweetingham, 1997). Yellow lupins are resistant to the binucleate form. Affected or disease patches caused by *Rhizoctonia solani* (AG8) and the binucleate *Rhizoctonia* range from 0.3 to 12 m in diameter. Patches caused by the binucleate *Rhizoctonia* are usually not visible before 7 wk after sowing. This is in contrast to Rhizoctonia bare patch (*R. solani* AG8), which is usually visible in lupin crops 4 wk after sowing. Symptoms caused by *R. solani* (AG8) are usually spear-tipped lateral and tap roots, compared to sloughing of the tap root and pinched-off lateral roots caused by the binucleate *Rhizoctonia* spp.

Rhizoctonia bare patch is most severe in minimum tilled crops and cultivation reduces disease severity (Rovira, 1986). Rhizoctonia bare patch control increased with cultivation depth. However, cultivation does not control hypocotyl rot caused by other strains. Severe outbreaks have been observed following a long pasture phase, which suggests inoculum build-up under clover or annual medics. At sites with a history of severe hypocotyl rot, increased seeding rates are recommended to compensate for losses in stand density. No known resistance to Rhizoctonia bare patch or hypocotyl rot has been reported in lupins.

Pleiochaeta Root Rot and Brown Spot of Lupins

Pleiochaeta setosa (Kirchn.) Hughes is the cause of a brown spot of lupins and is also a virulent and widespread root pathogen of lupin seedlings in Western Australia (Sweetingham, 1989). Pleiochaeta root rot occurs only in fields previously cultivated with lupins where soilborne inoculum exists in sufficient quantities. The severity of Pleiochaeta root rot is greatly reduced as sowing depth increases, due to avoidance of concentrated inoculum on or near the soil surface. Inoculum of *P. setosa* originates from sporulation on lupin leaf litter. Pleiochaeta root rot is more severe where the surface 10-cm of soil dries out shortly after sowing.

Spore numbers are usually high in fields sown to lupins the previous season. Consequently, double cropping of lupins is not a recommended practice. Retaining cereal stubble mulch on the soil surface reduces rain-splash of spores onto the foliage of plants, thus reducing brown spot severity. Seed treatment with fungicides is recommended and can reduce brown spot losses up to 4 wk after sowing. In Western Australia, a sowing depth of 5 cm is recommended for lupin establishment in fields where high concentration of *Pleiochaeta* spores exist (Sweetingham, 1996). Shallower seeding rates lead to increased disease, and planting deeper than 7 cm results in reduced plant stands. The variety 'Myallie' is reported more resistant to Pleiochaeta brown spot than other varieties of narrow-leafed lupin and is recommended for planting in high inoculum fields. However, Myallie is not root rot resistant and cultivars with resistance to Pleiochaeta root rot should be released in the near future (Sweetingham, personal communication).

Aphanomyces Root Rot of Peas

Aphanomyces root rot of pea is the most destructive disease of peas worldwide, most often occurring in poorly drained fields with heavy-textured soils (Hagedorn, 1984). The pathogen, *Aphanomyces euteiches* Drechs., can also infect roots of other legumes such as alfalfa, clover, common bean, lentil, and faba bean (Papavizas & Ayers, 1974). Aphanomyces root rot was first described in 1925 in Wisconsin and has continued to be a serious problem there (Hagedorn, 1984; Papavizas & Ayers, 1974). Since 1985, Aphanomyces root rot has been observed in northern Idaho, in the Palouse and Blue Mountain areas in eastern Washington, and in the Columbia Basin of central Washington.

Where inoculum levels are high, yellowing and stunting are evident soon after emergence, but symptoms can occur at any age given high soil moisture and warm temperatures. Infected cortical tissue is straw coloured and darkens as secondary organisms colonize. When infected plants are pulled from the soil, a strand of vascular tissue is all that remains of the root system. Microscopic observation of infected, cortical tissue reveals typical, thick-walled oospores (25-30 µ diam.), which can survive in soil for years.

There are no economically feasible control practices for Aphanomyces root rot other than to avoid planting a susceptible legume in heavily infested fields (Kraft *et al.*, 1990). Long-term rotations with crops that delay or reduce inoculum build-up, coupled with field indexing to determine field root rot potential, is the only means of control (Hagedorn, 1984; Kraft *et al.*, 1990). Commercial varieties are now being developed that have measurable levels of resistance to avoid severe economic loss in fields moderately infested with *A. euteiches* (Kraft *et al.*, 1995). The development of commercial varieties with resistance to *A. euteiches* is due to public breeding efforts in the last 15 years that have produced resistant germplasm lines approaching a commercial type (Davis *et al.*, 1995; Gritton, 1990; Kraft, 1981, 1989, 1992). However, resistance is not sufficient to withstand adverse environmental conditions and/or increased inoculum levels (Kraft & Boge, 1996).

Green manure plowdown of oats (*Avena sativa* L.) and several species of crucifers have shown promise in lowering the inoculum potential of *A. euteiches* in Wisconsin and Minnesota (Fritz *et al.*, 1995;

Muehlchen et al., 1990). Saponins are produced in oat roots and tops, and they are considered inhibitive to Aphanomyces and Pythium zoospores (Maizel et al., 1963). Research in Minnesota (Fritz et al., 1995) has shown a significant decrease in Aphanomyces root rot severity where green oat residue was incorporated to a shallow depth in the fall prior to sowing peas the next spring on Aphanomyces infested ground. Chisel plowing usually incorporates crop residue into the top 10 cm (Wilkins & Kraft, 1988). Placing the oat residue in the top 10 cm, where saponins are available to reduce primary zoospore inoculum and consequent infection of seedling roots, is necessary for significant control (Fritz et al., 1995).

Since Aphanomyces root rot was first described in northern Idaho (Bowden et al., 1985), a modified paper towel baiting technique was developed to determine field inoculum levels of A. euteiches (Kraft et al., 1990). Because oospores are the survival structure of this pathogen and are buried in susceptible legume root debris, a baiting technique using wet-sieved organic matter was developed. Using this technique, Aphanomyces was readily detected in several areas and fields. Use of this procedure also revealed that infective oospores of A. euteiches were as deep as 60 cm in the soil profile and were present in areas with poor drainage in fields with low overall inoculum levels.

The fungicide Tachigaren (hymexazole) has shown promise as a seed treatment to reduce Aphanomyces root rot. Both greenhouse and laboratory studies demonstrated that 1.8 to 3.7 g a.i./kg seed plus 1.6 g a.i./kg seed Apron and 1/8 g a.i./kg seed Captan resulted in significant control. In further work at Prosser, the combination of resistant germplasm, fungicides, or biological seed treatments improved pea seedling stands, disease control, and seed yields (Kraft, 1982; Kraft & Papavizas, 1983; Kraft et al., 1995).

Control of Aphanomyces root rot will depend on a multi-faceted program of resistant/tolerant varieties, green manure plowdowns, seed treatment chemicals, and longer intervals between susceptible legume crops.

Fusarium Root Rot of Chickpeas and Peas

Fusarium root rot of pea and chickpea is caused by Fusarium solani (Mart.) Appel and Wr. f. sp. pisi (F.R. Jones) Snyd. & Hans. (Kraft et al., 1981; Bhatti & Kraft, 1992a). Symptoms on both pea and chickpea consist of yellowing of the basal foliage, stunted growth, and reddening of the vascular tissue below the soil line. The common site of seedling infection by Fusarium is the cotyledonary attachment area, below ground epicotyl, and upper taproot (Kraft & Roberts, 1967). Penetration of pea seedlings often occurs through stomates on the epicotyl (Bywater, 1959). Infection can then extend upward to the soil line and downward into the root zone. Initial symptoms on seedling roots consist of reddish-brown to blackish-brown streaks, which usually coalesce. A red discoloration of the vascular system can occur in the taproot but usually does not progress above the soil line.

Data on actual pea crop losses due to Fusarium root rot are scarce but yield losses up to 30% have been reported (Kraft & Berry, 1972; Raghavan et al., 1982). Halila and Strange (1996) observed Fusarium root rot in 3% of Tunisian chickpea fields. The formae specialis responsible for Fusarium root rot of chickpea in many places has not been determined. However, F. solani f. sp. pisi was shown to be a virulent pathogen of chickpea roots when soil temperatures were 30°C or above (Bhatti & Kraft, 1992a).

Yield reductions in pea due to Fusarium root rot are influenced by previous cropping history, soil temperature, moisture, compaction, aeration, acidity, and fertility (Kraft et al., 1981; Allmaras et al., 1988; Bhatti & Kraft, 1992b). Likewise, Fusarium root rot of chickpea is more severe in compacted than in loose soil and root growth of chickpea was inversely correlated with soil compaction (Bhatti & Kraft, 1992b). In fact, the degree of root infection and damage caused by F. solani f. sp. pisi is directly dependent on the stress level to which the plant is exposed. Any condition(s) which decrease root growth will increase Fusarium root rot severity (Allmaras et al., 1988; Kraft et al., 1981; Kraft & Wilkins, 1989).

Chlamydospores of F. solani f. sp. pisi germinate to produce pre-infection growth when stimulated by root and seed exudates (Cook & Flentje, 1967). Rhizosphere effects may extend no more than 2 mm from the root surface and chlamydospore mobility is nil. Exudation from healthy pea roots is greatest near the root tip and along the zone of maturation (Rovira, 1973). Fusarium chlamydospores require 6 to 10 h for germination, and growth toward a substrate would probably miss the root tip with resultant contact with the zone of maturation where exudation is reduced. Poor aeration and/or soil compaction can reduce root growth and induce lateral root branching closer to the root apex, thus enhancing the probability that the germinating chlamydospore will make contact with the root tip and the exudation zone. Fusarium solani f. sp. pisi typically produces initial disease symptoms in the region of cotyledonary attachment, epicotyl, and hypocotyl, which are stationary (peas are hypogeal in germination) due to seed exudation that stimulates chlamydospore germination. It has been our experience that only when the entire root system is invaded does *F. solani* f. sp. *pisi* cause serious disease losses.

In eastern Washington and northeastern Oregon, a definite tillage pan in all pea, wheat, or wheat-fallow sites was found, regardless of soil type and whether dryland or irrigated (Kraft & Allmaras, 1985). Typically, *F. solani* f. sp. *pisi* propagules were found throughout the upper 60 cm of soil, but their numbers were low in the tillage pan. The low numbers in the tillage pan and their presence below it could be related to impaired drainage from tillage pan compaction and the saprophytic survival of *F. solani* f. sp. *pisi* in the drier subsoil. In fields not cropped to peas for five or more years, *F. solani* f. sp. *pisi* was not detected in the plow layer but was recovered in the subsoil and there was a corresponding increase in yield. Long-term cultivation has apparently produced an environment beneath the plough layer that is favourable for survival of *F. solani* f. sp. *pisi*. In the absence of other stress factors, inoculum of *F. solani* f. sp. *pisi* deep in the soil profile has little detrimental effect on pea growth and development up to anthesis, when the upper 20 cm of the root system is not infected (Rush & Kraft, 1986). When inoculum levels of *F. solani* f. sp. *pisi* were significantly reduced by methyl bromide fumigation in the 0-20 cm depth and compaction was reduced due to tillage, there was a decrease in root disease and an increase in root growth and dry seed yields (Kraft & Wilkins, 1989).

Glyphosate is being used as an alternative to mechanical weed control in the winter wheat-green pea rotation of southeastern Washington and northeastern Oregon. Glyphosate was found to stimulate proliferation of *F. solani* f. sp. *pisi* in the rhizosphere of some common weeds sprayed with it (Kawate *et al*., 1997). Apparently, after exposure to glyphosate, nutrients were released in sufficient quantities for *F. solani* f. sp. *pisi* to increase in population numbers.

A breeding program to incorporate resistance to *F. solani* f. sp. *pisi* in pea has been ongoing at Prosser, WA, since 1967. High seed vigour is an important consideration in comparing one pea line with another for resistance to *Fusarium*. A line with poor seed vigour may appear susceptible to Fusarium root rot when in fact it is genetically resistant (Kraft, 1986). Because seed and seedling vigour is important in the development of Fusarium root rot, a seed soak test to screen peas for resistance is now used (Kraft & Kaiser, 1993). Seeds of test lines with high vigour are soaked overnight in a conidial suspension of *F. solani* f. sp. *pisi* adjusted to 1 x 10^6 per ml and planted in coarse-grade perlite. Inoculated lines are read after 14 days and scored on a 0-5 scale. Good progress has been made and will continue to be made in developing peas with acceptable horticultural traits and inheritable resistance to Fusarium root rot (Kraft & Kaiser, 1993). Because resistance is not of a high level, an integrated control approach is needed which includes cultural practices, maintenance of good seed vigour, and genetic resistance.

Sclerotium Rot of Cool Season Legumes

Sclerotium rolfsii Sacc. (teleomorph *Athelia rolfsii* (Curzi) Tu & Kimbrough, 1978) is a serious pathogen of chickpea and causes a collar rot of lentil. This fungus has a wide host range of nearly 500 species with Graminaceous species being less susceptible (Punja, 1985). This pathogen is widely distributed in warm climates and the disease is usually observed under wet warm conditions and in fields where chickpea and lentil are sown following a paddy rice crop. Sclerotia formed on undecomposed tissues in the field are capable of initiating infection and are the primary source of inoculum (Punja & Grogan, 1981). *Sclerotium rolfsii* can cause damping-off of seedlings, stem canker, and/or root rot. When chickpea or lentil seedlings are attacked, this pathogen is capable of invading all parts of the host and they die quickly. Usually the infection on more mature plants begins on the succulent stem as a dark-brown lesion just below the soil line. The first visible symptoms appear as yellowing or wilting of the lower leaves which progresses to the upper leaves. The fungus grows upward in the plant and downward into the roots. White mycelium is always present in infected tissues and grows over the soil to adjacent plants to start new infections. On all infected tissues the fungus produces numerous small round-shaped sclerotia which are brown in colour. The mature sclerotia are not connected with mycelial strands and have the size, shape, and color of mustard seed (Punja & Rahe, 1992). *Sclerotium rolfsii* kills and disintegrates tissues by secreting oxalic acid, and also pectinolytic, cellulolytic, and other enzymes. Once established, production of mycelium and sclerotia are very rapid, especially during periods of high moisture and temperature (30-35°C). This pathogen usually attacks plants near the soil line and sclerotia are capable of surviving for long periods. The only economic control consists of long-term rotations with non-susceptible hosts and deep plowing of sclerotia.

Collar rot of lentil, caused by *S. rolfsii,* can be reduced by altering the sowing date so the seedling stage does not coincide with high soil moisture and temperatures above 25°C (Agrawal *et al.*, 1975). Crop rotation is unlikely to be an effective method of control due to the wide host range and *S. rolfsii's* persistence on numerous types of host debris. Seedling mortality in lentil can be significantly reduced by treating seed with combinations of fungicides, such as thiram + pentachloronitrobenzene or thiram + carbendazim (Agrawal *et al.*, 1975). Mancozeb seed treatment has also been found to reduce collar rot of lentil (Singh *et al.*, 1985). No biological control of this pathogen has been achieved under field conditions with either chickpea or lentil. However, several resistant lentils have been identified (Khare *et al.*, 1979; Kannaiyan & Nene, 1976; Abu-Mohammad & Kumar, 1986).

FUSARIUM WILT OF CHICKPEA, PEA, AND LENTIL

Fusarium wilt of pea, lentil, and chickpea is caused by the formae specialis *pisi, lentis,* and *ciceri,* respectively of *F. oxysporum.* These pathogens are soil inhabitants that can survive indefinitely in soil and be seedborne.

Chickpea Wilt

Among the diseases reported on chickpea, wilt caused by *F. oxysporum* Schl. emnd. Snyd. Hans. f. sp. *ciceri* [Padwick] Snyd. & Hans. is one of the most important diseases in North Africa, South Asia, and Southern Europe (Nene & Reddy, 1987), some areas in the United States (Buddenhagen & Workneh, 1988) and causes up to 10% losses in yield. In Tunisia, wilt is present in 30-40% of chickpea fields (Halila & Strange, 1996). The disease is more prevalent in the lower latitudes (0-30 N) where the chickpea-growing season is relatively dry and warmer than in the higher latitudes (30-40 N). The pathogen is soilborne surviving in soil for more than 6 years in the absence of a susceptible host (Haware *et al.*, 1986; Haware & Nene, 1982b) and is also seedborne (Haware & Nene, 1978).

This pathogen exhibits physiologic specialization and seven races have been reported from India, Spain, and the USA. Of these seven races, designated 0-6, races 1, 2, 3, and 4 were reported in India (Haware & Nene, 1982a) and 0, 5, and 6 in Spain (Jiménez-Díaz *et al.*, 1989). In Tunisia, morphological and pathogenic variability studies determined that race 0 predominates (Halila & Strange, 1996). In Morocco, race 1 was found to be the predominant pathogen (El-Hadi, 1993). Research conducted in Tunisia showed that chickpea cultivars varied in wilt symptoms from very early wilting to very late wilting (Halila & Strange, 1996). Late wilting is thought to be a form of partial resistance, governed by more than one gene, and that complete resistance can be obtained by crossing late wilting parents (Upadhaya *et al.*, 1983). Considerable progress has been made in identifying wilt resistant sources and the development of wilt-resistant and high-yielding chickpea cultivars. Breeding programs at national and international centers have developed and released resistant cultivars; however, these cultivars have not maintained resistance across locations due to area specific races of the wilt pathogen (Infantino *et al.,* 1996). At the International Crops Research Institute for the Semi-Arid Tropics (ICRISAT), over 13,500 chickpea accessions from 40 countries have been evaluated for wilt resistance (Haware *et al.*, 1992) and 160 accessions were found resistant to race 1. Of these, 150 were "desi" types.

Zote *et al.* (1996) reported that wilt susceptible, moderately susceptible, and resistant cultivars of chickpea all supported multiplication of *F. o.* f. sp. *cicèri* in the rhizosphere. Apparently, it is not possible to eliminate the chickpea wilt pathogen from infested soil by growing resistant cultivars alone. There is a need to employ other management practices such as long-term crop rotation to reduce the inoculum density in soil.

Pea Wilt

The pea wilt pathogen, *Fusarium oxysporum* Schlecht f. sp. *pisi* (van Hall) Snyd. & Hans, is a soil inhabitant that can survive indefinitely and can be seedborne (Kraft, 1994). Typical symptoms of pea wilt are pale foliage, downward curling of leaves, and vascular discoloration both below and above ground (Hagedorn, 1984). Often the above- and below ground vascular system turns a light yellow to brick-red color, and the lower subterranean portion of the stem becomes larger than normal. Penetration often occurs through root tips (Nyvall & Haglund, 1972). The pathogen concentrates in the larger xylem elements and can reproduce in the rhizosphere of both resistant and susceptible cultivars (Charchar & Kraft, 1989).

Race 1 (common wilt) of pea, caused by *F. oxysporum* Schl. f. sp. *pisi* race 1 Snyd & Hans., was first reported in Wisconsin in 1924 and resistance was attributed to a single, dominant gene. Race 1, after the release of resistant cultivars, was not a problem in commercial production in the United States again until 1972 (Kraft *et al.*, 1974). Race 2 (near wilt) of pea was described in 1933 and is found in most pea growing areas of the world (Hagedorn, 1984). Resistance to race 2 is also controlled by a single dominant gene. In 1970, *F. oxysporum* f. sp. *pisi* race 5 (Haglund & Kraft, 1970) was described from western Washington which killed varieties resistant to races 1 and 2. In 1979, an additional race of *F. oxysporum* f. sp. *pisi* was described, which killed varieties resistant to races 1, 2, and 5. This strain was named race 6 (Haglund & Kraft, 1979). Resistance to races 5 and 6 are again attributed to separate, single, dominant genes.

The pathogenicity of races 1, 2, 5, and 6 of *F. oxysporum* f. sp. *pisi* can be distinguished by their reaction on differential pea varieties (Kraft, 1994). The disease reactions are based on a resistance response (no observable disease) and a susceptible reaction (dead or severely stunted, chlorotic plants). However, pathogenicity tests are subjective because they are influenced by temperature, host plant age, method of inoculation, etc. Research during the last 10 years has classified strains of *F. oxysporum* using vegetative compatibility groupings (VGC) by pairing nitrate nonutilizing (nit) mutants generated on a potassium chlorate medium. Strains, or races of *F. oxysporum,* can be further characterized based on fungus genetics along with host-pathogen interactions. Within *F. oxysporum* f. sp. *pisi*, 4 VGS's have been reported (Correll, 1991). Molecular techniques have revealed that races 1, 5, and 6 are closely related, and that race 2 is distinct (Coddington *et al.*, 1987; Kistler *et al.*, 1991). So far, all isolates of race 2 have exhibited a highly conserved banding pattern, whereas isolates of the other races exhibited much more variability with the primers used.

There has been disagreement in the literature on classifying races of *F. oxysporum* f. sp. *pisi* (Armstrong & Armstrong, 1974; Haglund, 1974; Kraft & Haglund, 1978). Currently, the accepted method of classifying any isolate for a race designation is by host-pathogen response. Because inoculation procedures, genetic homozygosity of host and pathogen, environmental conditions, and inoculum levels all influence the host-pathogen response, standardization of procedures is important for repeatability (Huebbeling, 1974; Kraft & Haglund, 1978).

The only economic control of pea wilt is to use resistant varieties, index prospective fields for presence of a given race, and avoid planting a susceptible variety in infested soil.

Lentil Wilt

Several species of *Fusarium* have been associated with wilted lentils (Khare *et al.*, 1979). However, the primary pathogen appears to be *Fusarium oxysporum* Schlecht ex. Fr. f. sp. *lentis* Vasudeva and Srinivasan (Chattopadhyay & Sengupta, 1967). The disease is widespread where lentil is grown. Lentil wilt has been reported in Argentina, Canada, Chile, Colombia, Czechoslovakia, Egypt, Ethiopia, France, Hungary, India, Jordan, Morocco, Nepal, Sudan, Syria, Turkey, Tunisia, Uruguay, the USA, and the former USSR. The host range is primarily lentil, however *Vicia montbretii* Fisch. & C.A. can be infected when artificially inoculated (Bayaa *et al.*, 1995). Although variability in fungal sensitivity to nutrition, fungicides, and temperature exists in strains of *F. oxysporum* f. sp. *lentis,* races of this pathogen have not been defined.

Symptoms appear as patches of infected plants in the field at either the seedling or adult plant stage. Seedling wilt is characterized by sudden drooping, followed by drying of leaves and death of seedlings. Wilt in mature plants can appear from the flowering to late pod filling stage and is also characterized by sudden drooping of top leaflets, dull green foliage color, and wilting of individual branches or the whole plant. The root system of either infected seedlings or mature plants appears healthy, with a slight reduction of lateral roots but with very little vascular discoloration.

In India, seedling wilt of lentil appears during November and/or in the adult plant stage during April-May. The pathogen can survive in soil more than five years due to chlamydospore formation. Lentil wilt is more severe in sandy loam soil than in clay soil and the mortality of infected lentil plants increases with soil pH up to 7.5. Wilt is most severe during warm weather (Izquierdo & Morse, 1975; Bayaa *et al.*, 1986; Agrawal *et al.*, 1993). Sowing date also affects wilt incidence. In India, delayed sowing reduced disease incidence but late sowing reduces yield (Kannaiyan & Nene, 1975). A rotation out of lentils for 4 to 5 years reduced inoculum density but did not completely eradicate the disease. In India, cultivation of paddy rice or sorghum during the rainy season reduced wilt incidence the next winter (Kannaiyan & Nene, 1979). Biological control of lentil wilt under field conditions has not been achieved.

Cultivars with resistance to lentil wilt have been released from Bulgaria, Lebanon, and India. Sources of resistance have been found by many authors in cultivated lentil (Kannaiyan & Nene, 1976; Bayaa & Erskine, 1990; Hamdi et al., 1991; Hossain et al., 1985; Khare et al., 1979; Tiwari & Singh, 1980) and made available through the Lentil International Fusarium Wilt Nursery. Resistance has also been found among wild lentil relatives *L. culinaris* ssp. *orientalis* and *L. nigricans* spp. *ervoides* (Bayaa et al., 1995).

CONCLUSION

Soilborne fungal diseases of cool season grain legumes are described as seed and seedling blights, root rots, and wilts. Seed and seedling diseases are caused primarily by *Pythium* spp. and *Rhizoctonia solani*. However, in Western Australia, Pleiochaeta seedling and root rot has become a serious yield constraint. The most important fungi causing root rots include *Aphanomyces euteiches, Fusarium solani, Rhizoctonia solani*, and *Sclerotium rolfsii*. Wilt is caused primarily by various host-specific forms of *Fusarium oxysporum*.

Such diseases as *Aphanomyces, Fusarium, Pythium, Pleiochaeta*, and Sclerotium rot are dramatically increased by short rotations, which facilitate significant increases in soil inoculum. Resistance to these diseases is usually not sufficient to withstand high inoculum levels, especially in concert with other biotic and abiotic stresses which predispose these legumes to root diseases. Consequently, a multifaceted approach is needed to control these diseases including host resistance/ tolerance, biological and chemical seed treatments, longer rotations between susceptible crops, and cultural practices which enhance rapid root growth.

High levels of genetic resistance exist for most races of Fusarium wilt, which attack chickpea, pea, and lentil. Resistance to various races of wilt has been stable for long periods. However, new races of the wilt pathogen have developed in areas where the crop has been grown in monoculture or short rotations and where resistance to an existent race(s) is extensively grown. This has certainly been the case for the pea wilt pathogen in the Skagit Valley of Western Washington.

References

Abu-Mohammad & U. Kumar, 1986. *Indian Journal of Phytopathology* 39: 93-95.

Agrawal, S.C., M.N. Khare & L.K. Joshi, 1975. *Proceedings of the Annual Conference of Madhya Pradesh Science Academy.*

Agrawal, S.C., K. Singh & S.S. Lal, 1993. In: W. Erskine & M.C. Saxena (eds.), *Proceedings of the Seminar on Lentil in South Asia.*

Allmaras, R.R., J.M. Kraft & D.E. Miller, 1988. *Annual Review of Phytopathology* 26: 219-243.

Armstrong, G.M. & J.K. Armstrong, 1974. *Phytopathology* 64: 849-857.

Bayaa, B. & W. Erskine, 1990. *Arab Journal of Plant Protection* 8: 30-33.

Bayaa, B., W. Erskine & A. Hamdi, 1995. *Genetic Resources and Crop Evolution* 42: 231-235.

Bayaa, B., W. Erskine & L. Khoury, 1986. *Arab Journal of Plant Protection* 4: 118-119.

Bhatti, M.A. & J.M. Kraft, 1992a. *Plant Disease* 76: 50-53.

Bhatti, M.A. & J.M. Kraft, 1992b. *Plant Disease* 76: 960-963.

Bowden, R.L., H.S. Fenwick, L.J. Smith & J.M. Kraft, 1985. *Plant Disease* 69: 451.

Bruehl, G.W., 1986. *Soilborne plant pathogens.* Macmillan Publishing Company, New York, 368 pp.

Buddenhagen, I. & F. Workneh, 1988. *Phytopathology* 78: 1563 (Abstr.)

Bywater, J., 1959. *Transactions of the British Mycological Society* 42: 201-212.

Charchar, M. & J.M. Kraft, 1989. *Canadian Journal of Plant Science* 69: 1335-1346.

Chattopadhyay, S.B. & P.K. Sengupta, 1967. *Indian Journal of Mycological Research* 5: 45-53.

Chet, I. & R. Baker, 1981. *Phytopathology* 71: 286-290.

Coddington, A., P.M. Matthews, C. Cullis & K.H. Smith, 1987. *Journal of Phytopathology* 118: 9-20.

Cook, R.J. & N.T. Flentje, 1967. *Phytopathology* 57: 178-182.

Correll, J.C., 1991. *Phytopathology* 81: 1061-1064.

Davis, D.D., V.A. Fritz, F.L. Pfleger, J.A. Percich & D. K. Malvick, 1995. *Horticultural Science* 30: 639-40.

El-Hadi, M., 1993. M.Sc. Dissertation, Dept. of Plant Pathology, Washington State University, Pullman, Washington, USA.

Flentje, N.T. & D.J. Hagedorn, 1964. *Phytopathology* 54: 788-791.

Fritz, V.A., R.R. Allmaras, F.L. Pfleger & D.W. Davis, 1995. *Plant and Soil* 171: 235-244.

465

Gritton, E.T., 1990. *Crop Science* 30: 1166-1167.

Hagedorn, D.J. (Ed.), 1984. *Compendium of pea diseases. The American Phytopathological Society*, 57 pp.

Haglund, A.A., 1974. *Pisum Newsletter* 6: 20-21.

Haglund, W.A. & J.M. Kraft, 1970.. *Phytopathology* 60: 1861-1862.

Haglund, W.A. & J.M. Kraft, 1979. *Phytopathology* 69: 818-820.

Halila, M.H. & R.N. Strange, 1996. *Phyto. Pathology* 35: 67-74.

Hamdi, A., S.A.M. Omar & M.L. Amer, 1991. *Egyptian Journal of Applied Sciences* 6: 18-29.

Harman, G.E., I. Chet & R. Baker, 1980. *Phytopathology* 70: 1167-1172.

Haware, M.P. & Y.L. Nene, 1978. *Phytopathology* 68: 1364-1367.

Haware, M.P. & Y.L. Nene, 1982a. *Plant Disease* 66: 809-810.

Haware, M.P. & Y.L. Nene, 1982b. *Plant Disease* 66: 250-251.

Haware, M. P., Y.L. Nene & M. Natarajan, 1986. In: Abstracts, *Seminar on Management of Soilborne Diseases of Crop Plants,* Tamil Nadu Agricultural University, Coimbatore, Tamil Nadu, India.

Haware, M.P., Y.L. Nene, R.P.S. Pundir & J. Narayana Rao, 1992. *Field Crops Research* 30: 147-154.

Hossain, M.A., A. Ayub & H.U. Ahmed, 1985. Bangladesh Agriculture Research Institute, Joydepur. Abstracts of papers of First National Plant Pathology Conference, Joydepur (Bangladesh), pp. 5-6.

Huebbeling, N., 1974. Mededelingen van de Facultfit Landbouwwe Tenschappen Rijksuniversitfit Gent. 29: 991-1000.

Infantino, A.A., Porta-Pugalia & K.B. Singh, 1996. *Plant Disease* 80: 42-44.

Izquierdo, J.A. & R. Morse, 1975. *LENS Newsletter* 2: 20-28.

Jiménez-Díaz, R.M., A. Trapero-Casas & J. Cabrera de la Colina, 1989. In: E.C. Tjamos & C. Beckman (Eds.), *Vascular Wilt Diseases of Plants,* pp. 515-520. Springer-Verlag, Berlin.

Kannaiyan, J. & Y.L. Nene, 1975. *Madras Agricultural Journal* 62: 240-242.

Kannaiyan, J. & Y.L. Nene, 1976. *Indian Journal of Agricultural Sciences* 46: 165-167.

Kannaiyan, J. & Y.L. Nene, 1979. *Indian Journal of Crop Protection* 7: 114-118.

Kawate, M.K., Susan Colwell, A.G. Ogg & J.M. Kraft, 1997. *Weed Science.* (in press)

Khare, M.N., S.C. Agrawal & A.C. Jain, 1979. Technical Bulletin, pp. 1-29, Jawaharlal Nehru Krishi Vishwa Vidyalay, Jabaalpur, Madhya Pradesh, India.

Kistler, H.C., E.A. Momol & U. Benny, 1991. *Phytopathology* 68: 331-336.

Kraft, J.M., 1981. *Crop Science* 21: 352-353.

Kraft, J., 1982. *Plant Disease* 66: 798-800.

Kraft, J.M., 1986. *Plant Disease* 70: 743-745.

Kraft, J.M., 1989. *Crop Science* 29: 494-495.

Kraft, J.M., 1992. *Crop Science* 32: 1076.

Kraft, J.M., 1994. *Agronomie* 14: 561-567.

Kraft, J.M. & R.R. Allmaras, 1985. In: C.A. Parker, A.D. Rovira, K.J. Moore, P.T.W. Wong & J.F. Kollmorgan (Eds.), *Ecology and Management of Soilborne Plant Pathogens,* pp. 203-205. Academic Press, New York, USA.

Kraft, J.M. & J.W. Berry, 1972. *Plant Disease Reporter* 56: 398-400.

Kraft, J.M. & W.L. Boge, 1996. *Plant Disease* 80: 1383-1386.

Kraft, J.M., D.W. Burke & W.A. Haglund, 1981. In: P.E. Nelson, T.A. Toussoun, and R.J. Cook (Eds.), *Fusarium: diseases, biology, and taxonomy,* pp. 142-156. Pennsylvania State University Press, University Park, Pennsylvania, USA.

Kraft, J.M., V.A. Coffman & T.J. Darnell, 1995. *Biological and Cultural Control of Plant Diseases* 10: 139.

Kraft, J.M., R.M. Endo & D.C. Erwin, 1967. *Phytopathology* 57: 86-90.

Kraft, J.M. & W.A. Haglund, 1978. *Phytopathology* 68: 273-275.

Kraft, J.M. & W.J. Kaiser, 1993. In: *Breeding for Stress Tolerance in Cool Season Food Legumes* (eds. K.B. Singh & M.C. Saxena), pp. 123-144. John Wiley & Sons, Chichester.

Kraft, J.M., J. Marcinkowska & F.J. Muehlbauer, 1990. *Plant Disease* 74: 716-718.

Kraft, J.M., F.J. Muehlbauer, R.J. Cook & F.M. Entemann, 1974. *Plant Disease* Reporter 58: 62-64.

Kraft, J.M. & G.C. Papavizas, 1983. *Plant Disease* 67: 1234-1237.

Kraft, J.M. & D.D. Roberts, 1967. *Phytopathology* 59: 149-152.

Kraft, J.M. & D.E. Wilkins, 1989. *Plant Disease* 73: 884-887.

MacLeod, W.J. & M.W. Sweetingham, 1997. *Australian Journal of Agricultural Research* 48: 21-30.

Maizel, J.V., H.J. Burkhardt & H.K. Mitchell, 1963. *Biochemistry* 3: 424-426.

Matthews, S., 1971. *Annals of Applied Biology* 68: 177-183.

McCoy, R.J. & J.M. Kraft, 1984. *Plant Disease* 68: 491-493.

Muehlchen, A.M., R.E. Rand & J.L. Parke, 1990. *Plant Disease* 74: 651-654.

466

Nene, Y.L. & M.V. Reddy, 1987. In: *The Chickpea,* pp. 233-270 (eds M.C. Saxena & K.B Singh), CAB International, Wallingford, Oxon, OX10 8DE, United Kingdom.

Nyvall, R.F. & W. A. Haglund, 1972. *Phytopathology* 62: 1419-1424.

Papavizas, G.C. & W.A. Ayers, 1974. *USDA Technical Bulletin* 1484, 158 pp.

Punja, Z.K., 1985. *Annual Review of Phytopathology* 23: 97-127.

Punja, Z.K. & R.G. Grogan, 1981. *Phytopathology* 71: 1099-1103.

Punja, Z.K. & J.E. Rahe, 1992. *Sclerotium.* In: *Methods for Research on Soilborne Phytopathogenic Fungi,* pp. 166-170. (eds L.L. Singleton, J.D. Mihail & C.M. Rush), APS Press, St. Paul, Minnesota, USA.

Raghavan, G.S.V., F. Taylor, B. Vigier, L. Gauthier & E. McKyes, 1982. *Canadian Agricultural Engineering* 24: 31-34.

Rovira, A.D., 1973. *Pesticide Science* 4: 361-366.

Rovira, A.D., 1986. *Phytopathology* 76: 669-673.

Rush, C.M. & J.M. Kraft, 1986. *Phytopathology* 76: 1325-1329.

Singh, S.N., S.K. Srivastava & S.C. Agrawal, 1985. *Indian Journal of Agricultural Sciences* 55: 284-286.

Stasz, T.E., G.E. Harman & G. A. Marx, 1980. *Phytopathology* 70: 730-733.

Sweetingham, M.W., 1989. *Australian Journal of Agricultural Research* 40: 781-789.

Sweetingham, M., 1996. *Farmnote* No. 5/96. Agriculture Western Australia.

Tiwari, A.S. & B.R. Singh, 1980. *LENS* Newsletter 7: 20-22.

Upadhaya, H.D., M.P. Smithson, M.P. Haware & J. Kumar, 1983. *Euphytica* 32: 447-452.

Wilkins, D.E. & J.M. Kraft, 1988. In: Proceedings of International Soil and Tillage Research Organization, Edinburgh, Scotland, pp. 927-932.

Zote, K.K., M.P. Haware, S. Jayanthi & J.N. Rao, 1996. *Phytopathologia Mediterranea* 35: 43-47.

Opportunities for integrated management of insect pests of grain legumes

S.L. Clement [1] , J.A. Wightman [2], D.C. Hardie [3] , P. Bailey [4], G. Baker [4] and G. McDonald [5]

1 U.S. Department of Agriculture, Agricultural Research Service,Regional Plant Introduction, Station, 59 Johnson Hall, P.O. Box 646402, Washington State University, Pullman, WA 99164-6402, USA; 2 Pest Management International,9115 Octaira Court, Springfield, VA 25153, USA; 3 Agriculture Western Australia, 3 Baron-Hay Court, South Perth, WA 6151, Australia; 4 South Australia Research and Development Institute ,GPO Box 397, Adelaide, SA 5001, Australia; 5 Agriculture Victoria, Institute for Integrated Agricultural Development, RMB 1145, Rutherglen, VIC 3685, Australia

Abstract

Architects of the integrated control strategy (now known as Integrated Pest Management [IPM]) for insect pests envisaged a biologically-based system in which chemical control was to be used only at times and places where natural pest control was inadequate. Another interpretation of IPM accepted by many pest managers puts pesticides at the center of pest management in crops. This paper reviews research achievements which are supportive of IPM (insect forecasting and modeling, field sampling and monitoring, economic thresholds) and pest management methods (conventional and biological insecticides, plant resistance, transgenic plants, biological control, among other methods) for the major arthropod pests of the cool season food legumes. Despite many examples of research outcomes that have been successfully applied, the number of established IPM programs is scarce. For the present, conventional insecticides will continue to play a key role in grain legume production because farmers have few economic alternatives. However, pressures on pulse farmers to implement IPM programs will increase because of modern society's fear of pesticide residues in food, however minute, and tightened regulations governing pesticide use. Declining budgets for grain legume research, add uncertainty to future progress in IPM in many parts of the world.

INTRODUCTION

Integrated Pest Management (IPM) for many entomologists and pest managers is "an ecologically based system in which all available techniques are evaluated and consolidated into a unified program to manage pest populations so that economic damage is avoided and adverse side effects on the environment are minimized" (Metcalf, 1996). The available IPM control and management tactics involve several methods, categorized as biological, cultural, mechanical, physical, chemical, genetic, and regulatory (Luckmann and Metcalf, 1975). As conceived and proposed by the architects of the integrated control strategy (now known as IPM), farmers would only need chemical control if natural pest control was inadequate (Stern *et al.*, 1959). Another interpretation of IPM, albeit one widely accepted today by pest managers, puts pesticides at the center of pest management in agroecosystems (Gardner, 1996). Insecticides are the pest control method of choice because they are easy to use, are usually effective, and are relatively inexpensive. Grain legumes (pulses) are second only to the cereals in importance as food for humans and livestock and comprise a wide range of genera and species. Within this group are important cool season food legume crops, *viz.* chickpea (*Cicer arietinum* L.), field pea (*Pisum sativum* L.), lentil (*Lens culinaris* Medikus), faba bean (*Vicia faba* L.), and grasspea (*Lathyrus sativus* L.) (Oram and Agcaoili, 1994). Other annual legumes such as grain lupins (*Lupinus albus* L., *L. augustifolius* L.) are good sources of dietary protein and fat for animals and humans (Todorov *et al.*, 1996).

Worldwide, grain legume crops are attacked by many species of pestiferous insects and mites. For example, more than 55 species of arthropods attack chickpea, lentil, field pea, and grain lupins (Singh,

R. Knight (ed.), Linking Research and Marketing Opportunities for Pulses in the 21st Century, 467–480.
© *2000 Kluwer Academic Publishers. Printed in the Netherlands.*

468

1985; Reed et al., 1987; Ferguson, 1994; Bhatnagar et al., 1995), and more than 20 species attack faba bean (van Emden et al., 1988). However, few of these species are major field pests (Weigand et al., 1994) (Table 1). Yield and quality of grain legume seeds are also affected by insects during storage (Bushara, 1988).

Table 1. Major pests of grain legumes, the plant parts that they damage, and their global distribution on the crops

Pests	Crops attacked[a]	Plant parts damaged[b]	Distribution[c]
Helicoverpa armigera Hübner[d]	C, Lu, P	V, Re	B, C, D
Helicoverpa punctigera (Wallengren)[d]	All crops	V, Re	D
Liriomyza cicerina (Rondani)[e]	C	V	B
Acyrthosiphon pisum (Harris)[f]	C, F, Le, P	V, Re	A, B, C
Aphis craccivora (Koch)[f]	All crops	V, Re	A, B, C, D
Aphis fabae Scopoli[f]	F	V	B, C
Myzus persicae (Sulzer)[f]	Lu	V	D
Lygus hesperus Knight[g]	Le	Re	A
Sitona crinitus Herbst[h]	Le	R, V	B
Sitona lineatus (L.)[h]	F, P	R, V	A, B
Bruchus pisorum L.[i]	P	Re	A, B, C, D
Halotydeus destructor Tucker[j]	F, Lu, P	V	D

[a] Crops: C=Chickpea; F=Faba bean; Le=Lentil; Lu=Lupins; P=Field pea.

[b] Plant parts: R=Root; V=Vegetative organs (stems, leaves); Re=Reproductive organs (flower, pod and/or seed damaged).

[c] Insect species recorded on crops in: A=Americas; B=Europe, Africa, W. Asia; C=S.E. Asia, including Indian subcontinent; D=Australia.

[d] Lepidoptera: Noctuidae; [e] Diptera: Agromyzidae; [f] Homoptera: Aphididae; [g] Heteroptera: Miridae; [h] Coleoptera: Curculionidae; [i] Coleoptera: Bruchidae; [j] Acarina: Penthaleidae.

This review identifies and discusses grain legumes, production zones, issues, and research areas overlooked or not considered in previous reviews on the management of grain legume pests (van Emden, 1978; Reed et al., 1987; van Emden et al., 1988; Pimbert, 1990; Weigand et al., 1994). It also addresses research areas developed since the IFLRC II in 1992 (Muehlbauer and Kaiser, 1994). The pests of grasspea are not addressed, as there is a dearth of information about them. Nor do we discuss storage pests (see Bushara, 1988). Because much of the work deals with insecticidal control of insects, we will not attempt a comprehensive review of the literature of the field. Instead, we use IPM supportive activities (insect forecasting and modeling, field sampling, economic thresholds) and pest management methods as a framework for our discussions. We end with comments on the prospects for IPM in grain legumes, since the present control strategy for most pests of these crops involves insecticides (Weigand et al., 1994; Ogg et al., in press).

MAJOR INSECT AND MITE PESTS

Most of the major pests of grain legumes are immigrants (Table 1). For example, the redlegged earth mite (Halotydeus destructor Tucker [Acarina: Penthaleidae]) in Australia is thought to have immigrated from South Africa (Qin, 1997). One pest that probably co-evolved with its host plant (P. sativum) is the pea seed weevil, Bruchus pisorum L. (Coleoptera: Bruchidae), a Near East insect that has accompanied or followed its host plant (pea) to Europe, the Americas, and Australia (Smith, 1990b; Clement, 1992; Clement et al., in press). Other pest problems are new-encounters in which native insects on non-economic hosts adapted to introduced grain legumes, e.g., the native budworm, Helicoverpa punctigera (Wallengren) (Lepidoptera: Noctuidae), a pest of chickpea, field pea, and lupins in Australia (Zalucki et al., 1986; White et al., 1996).

The seasonal activity, abundance and the related damage caused by crop pests differs greatly from place to place and year to year. This phenomenon applies to several pests of grain legumes. It is illustrated by the high year to year variability in the occurrence of the pea aphid, Acyrthosiphon pisum (Harris) (Homoptera: Aphididae), in eastern Washington and northern Idaho (Palouse region of USA)

(Clement *et al.*, 1991). In this region, grain legume crops are devastated by pea aphid-vectored pathogenic viruses and by aphid feeding damage during "outbreak years" (1983, 1990, 1996) (Fig. 1) (Hagedorn, 1984; Clement *et al.*, 1991; Klein *et al.*, 1991). As well, other major pests (Table 1) exhibit variability in occurrence and pest status, typified by 1) the variable pest status of the gram podborer, *Helicoverpa armigera* (Hübner) (Lepidoptera: Noctuidae), across chickpea production zones in India (Pimbert, 1990; Wightman *et al.*, 1995); 2) yearly fluctuations in the magnitude of early spring infestations of *H. punctigera* on grain legumes in Australia (Gregg *et al.*, 1995; Walden, 1992); and 3) higher lygus bug (primarily *Lygus hesperus* Knight [Heteroptera: Miridae]) populations with increased feeding on lentils in some years in northern Idaho (USA), resulting in chalky spot damage to seed (O'Keeffe *et al.*, 1991b; Summerfield *et al.*, 1994). By contrast, some pests like the chickpea leafminer, *Liriomyza cicerina* Rondani (Diptera: Agromyzidae), occur in several countries in high densities every year (Weigand *et al.*, 1994).

A number of supportive activities, *e.g.*, insect forecasting/modeling, field sampling/monitoring, and the establishment and use of economic thresholds, among other activities, are required to make IPM decisions. Much research has addressed field sampling, economic thresholds, chemical control and other management methods such as host-plant resistance and biological/cultural control (Table 2).

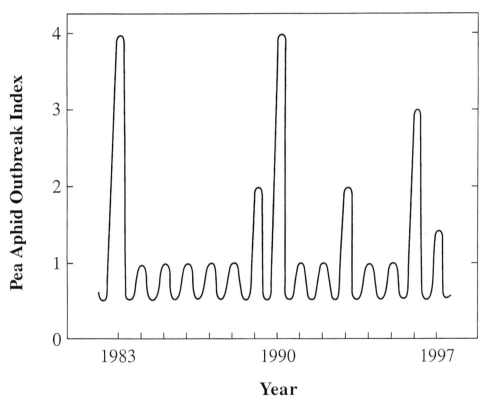

Figure 1. Schematic graph of year to year variability of pea aphid (*Acyrthosiphon pisum*) occurrence in eastern Washington and northern Idaho, USA. An outbreak index of 0-1 = pea aphid populations exceeding economic thresholds in 0-25% of the commercial pea fields, 1-2 = 26-50%, 2-3 = 51-75%, and 3-4 = 76-100% of the fields. Index derived from aphid suction trap counts, field sampling, incidence of pea aphid vectored viruses and virus epidemics in field peas, and farmer contacts.

Table 2 Activities supportive of integrated pest management and management methods for grain legume insects.

IPM Activities & Methods	Pests	Crops	Countries	Selected References
Forecasting / modeling insect dynamics	Helicoverpa armigera	Chickpea	Syria, Indian subcontinent	Reed et al. (1987), Gregg et al. (1995)
	Helicoverpa punctigera	Grain legumes	Australia	Dale et al. (1992), Gregg et al. (1995), Ridland et al. (1993a, b), McDonald and Bryant (1994), Miles and McDonald (1995), Walden (1995), Maelzer et al. (1996)
	Bruchus pisorum	Pea	Australia	Smith (1990c, 1992), Baker (1993)
	Acyrthosiphon pisum	Pea	USA, Sweden	Clement et al. (1991), Bommarco and Ekbom (1995)
Field sampling & monitoring	Helicoverpa armigera	Chickpea	India, Australia	Wightman et al. (1995), McIntyre and Titmarsh (1989)
	Helicoverpa punctigera	Pea, chickpea, faba bean, lupins	Australia	Michael et al. (1982), Hardie (1989), McIntyre and Titmarsh (1989), Ridland et al. (1993a,b), Parigi and McDonald (1995), Loss (1996), White et al. (1996)
	Bruchus pisorum	Pea	Australia, USA	Bailey and Baker (1988), Hardie (1989), Smith (1990b),O'Keeffe et al. (1992), Smith and Hepworth (1992)
	Sitona lineatus	Faba bean, pea	UK, USA, Denmark	O'Keeffe et al. (1991a), Ward and Morse (1995), Nielsen and Jensen (1993)
	Sitona crinitus	Lentil	Syria, Turkey	Kaya and Hincal (1987), ICARDA (1989)
	Acyrthosiphon pisum	Lentil, pea	Canada, USA	Maiteki and Lamb (1985), Schotzko and O'Keeffe (1989b), Soroka and MacKay (1990a,b), Homan et al. (1992).
	Aphis fabae	Faba bean	UK	Ward and Morse (1995)
	Liriomyza cicerina	Chickpea	Syria	ICADA (1990), Weigand and Pimbert (1993)
	Lygus hesperus	Lentil	USA	Schotzko and O'Keeffe (1986a,b; 1989a)
Economic thresholds & conventional insecticides	Helicoverpa armigera	Chickpea	India	Wightman et al. (1995)
	Helicoverpa punctigera	Pea, faba bean, lupins, chickpea	Australia	Michael et al. (1982), Bailey and Comery (1987), Ridland et al. (1993a), McDonald (1995, 1996), Loss (1996), White et al. (1996)
	Bruchus pisorum	Pea	Australia, USA	Smith et al. (1987), Hardie (1989), Horne and Bailey (1991), Michael et al. (1990), O'Keeffe et al. (1992)
	Sitona lineatus	Pea	USA	O'Keeffe et al. (1991a)
	Acyrthosiphon pisum	Lentil, pea	Canada,USA	Maiteki and Lamb (1985), Soroka and MacKay (1990b), Homan et al. (1992)
	Lygus hesperus	Lentil	USA	O'Keeffe et al. (1991b)
Management beyond conventional insecticides				
Biological insecticides	Helicoverpa armigera	Chickpea	India, Syria	Reed et al. (1987), Pimbert (1990), Weigand and Joubi (1994), Cowgill and Bhagwat(1996)
	Liriomyza cicerina	Chickpea	Syria	Weigand and Joubi (1994)
Conventional plant resistance	17 species, + Myzus persicae, Halotydeus destructor	Grain legumes, lupins	Worldwide	Reviewed by Clement et al. (1994, in press). Berlandier (1996), Thackray et al. (1997).
Transgenic plants	Helicoverpa punctigera	Grain legumes	Australia	Heath et al. (in press)
	Bruchus pisorum	Pea	Australia	Schroeder et al. (1995)
	Sitona lineatus	Pea	USA	Quinn and Bezdicek (1996)
Biological control	Helicoverpa armigera	Chickpea	India, Syria	Pawar et al. (1986), Reed et al. (1987), Pimbert (1990), Weigand et al. (1994), Murray et al. (1995)
	Helicoverpa punctigera	Pea	Australia	Murray and Rynne (1992), Ridland et al. (1993b)
	Bruchus pisorum	Pea	Australia, USA	Annis and O'Keeffe (1987), Baker (1990a,b)
	Acyrthosiphon pisum	Pea, lentil	USA	Schotzko and O'Keeffe (1989b), Karevia and Sahakian (1990)
	Liriomyza cicerina	Chickpea	Syria	Reed et al. (1987), Pimbert (1990), Weigand et al. (1994)
Cultural control	Helicoverpa armigera	Chickpea	India, Syria	Reed et al. (1987), Pimbert (1990), Weigand et al. (1994), Wightman et al. (1995)
	Bruchus pisorum	Pea	Australia	Baker (1990b), Smith (1990a)
	Sitona crinitus	Lentil	Syria	Weigand et al. (1994)
	Halotydeus destructor	Faba bean, lupins, pea	Australia	Michael et al. (1982), Loss (1996), McDonald et al. (1995)
Behavior modifying chemicals	Helicoverpa armigera	Chickpea	India	Rembold et al. (1990a,b), Rembold and Schroth (1993)
	Bruchus pisorum	Pea	USA	Clement et al. (1991)
	Sitona lineatus	Faba bean	UK	Blight and Wadhams (1987)

FORECASTING AND MODELING INSECT DYNAMICS

With the high variability in the occurrence of many grain legume pests across time and space, it is important that pest managers have the capability to forecast pest occurrence, abundance, population development, and damage potential. Timely forecasts give farmers sufficient lead time to make appropriate pest management decisions. The objective is to estimate pest development and movement over time and under various agrometerological conditions. This can be done at both the regional and the local level (Tummala et al., 1974). Australian researchers have analyzed the early season dynamics of the migrant H. punctigera from information on source breeding areas, seasonal light-trap catches of the moth, quantity of larval host plants, and weather variables (temperature, rain, synoptic trajectories) in an attempt to predict populations during late spring and early summer (Gregg et al., 1995; Maelzer et al., 1996). This research suggests it should be possible to make long-range forecasts of the movement of H. punctigera into southeastern Australia during early spring (Dale et al., 1992). Additionally, pheromone traps in southeastern Australia (Ridland et al., 1993a,b) and light traps in Western Australia (Walden, 1992, 1995) were evaluated to provide local forecasts of native budworm activity in grain legumes. Although pheromone trap catches indicate periods of major moth flights in southeastern Australia, particularly early in the season (Ridland et al., 1993a), peak catches are not always indicative of periods of peak egg-laying. Interpretation of pheromone trap catches is complicated by the stage of crop development, trap placement, and the prevailing weather conditions, as also identified in other Australian studies (e.g., Wilson and Morton, 1989). In addition, the failure to predict local damage from long-range forecasts and trap data may be attributed to lack of information on the host selection behavior of adults when alternative hosts occur in and around grain legume fields. Despite these limitations, pheromone traps are being used on a trial basis in a co-ordinated southeastern Australia campaign ("Budworm Watch") involving selected producer groups, pesticide dealers, and the Department of Agriculture to alert farmers of potential risk periods when 10-day running means of moth catches exceed locally derived thresholds (McDonald and Bryant, 1994; Miles and McDonald, 1995). A similar budworm alert system is operated by Agriculture Western Australia (D.C. Hardie, personal communication). It may be argued that pheromone traps play a larger role in maintaining the general interest and awareness of farmers in monitoring their crops than they do in actually providing pest alerts.

Pheromone traps at more than 60 locations across India, Pakistan, Bangladesh, and Sri Lanka were used to determine if catches of H. armigera moths could predict the need for insecticide applications on chickpea and other crops (Reed et al., 1987). Today, pheromone traps are used to better manage H. armigera in the Indian-subcontinent (J.A. Wightman, personal communication). Aphid suction traps hold promise for tracking the movement of pea aphids to the Palouse region of the USA and for alerting pea and lentil producers of potential outbreaks (Fig. 1) and the need to monitor their crops for damaging populations (Clement et al., 1991). By contrast, suction traps are not useful for tracking pea aphid migrants to pea fields in Sweden (Bommarco and Ekbom, 1995). Currently, aphid suction traps in the USA are not part of any organized IPM program to help farmers improve pea aphid management.

Modeling the development and population dynamics of grain legume pests for better management has received little attention. Only research on pea seed weevil (Smith 1990c, 1992; Baker, 1993; Smith and Ward, 1995) and lupin aphids (Thackray, 1996) in Australia and on pea aphid in Sweden (Bommarco and Ekbom, 1995) have addressed this area. Baker's (1993) study led to the development of a temperature-based model that predicts the commencement date of pea weevil invasion and the daily cumulative emergence of the weevil as invasion proceeds. This model is used in South Australia to limit control of B. pisorum to a single well-timed insecticide spray around the border (20-40 m) of the crop (Baker, 1993). Bommarco and Ekbom (1995) modeled the population development of A. pisum using accumulation of day-degrees to help pest managers in Sweden improve their forecasting of infestations in field peas. Finally, a model to forecast aphid outbreaks and virus epidemics in lupins in Western Australia is being developed (Thackray, 1996)

FIELD SAMPLING AND MONITORING

The control of insect infestations in a field should be based on estimates of the density of the insect (Schotzko and O'Keeffe, 1986a). Field sampling methods that maximize precision while minimizing sampling time and sampling costs (Gomez and Gomez, 1984) must be developed so pest managers can generate reliable estimates of insect density. Realizing the importance of sampling, entomologists have assessed a variety of methods, namely direct observation or counts, sweepnet and vacuum sampling, soil sampling, jarring or beating plants to dislodge insects, and trapping. Entomologists consider the insect species (including life stages) and crop plant (including growth stages) being sampled and other factors such as resources, cost, and size of the sampling area when selecting a sampling method.

Sampling procedures based on direct observations of insect-damaged plants and direct counts of insects have been assessed for: *H. armigera* on chickpea (counts of eggs and larvae) in India (Reed *et al.*, 1987; Wightman *et al.*, 1995); *L. cicerina* on chickpea (level of leafminer mining of leaflets) in Syria (Weigand and Pimbert, 1993); *B. pisorum* on field pea (egg counts on pods) in Australia (Smith and Hepworth, 1992); *S. lineatus* on faba bean (adult feeding on leaf margins and adult counts) in Denmark and England (Nielsen, 1990; Ward and Morse, 1995) and on pea (adult feeding on leaves) in the USA (O'Keeffe *et al.*, 1991a); *Sitona crinitus* Herbst. (Coleoptera: Curculionidae) on lentil (adult feeding on leaves) in Turkey (Kaya and Hincal, 1987); *Aphis fabae* Scopoli (Homoptera: Aphididae) on faba bean (direct counts of aphids on plants) in England (Ward and Morse, 1995); and *A. pisum* on field pea (direct counts of aphids on 20 cm of vine) in Canada (Soroka and Mackay, 1990a). In addition, monitoring pulse crops for *H. punctigera* eggs has been attempted in Australia and a theoretical sampling plan developed based on two levels of precision (Ridland *et al.*, 1993a).

Sweepnet sampling, the most frequently used method for making insect counts (Southwood, 1978), is widely used in grain legumes. Schotzko and O'Keeffe (1986a,b, 1989a) compared different sampling methods (sweepnet, vacuum, absolute sampling) and the influence of weather/time variables, insect spatial distribution, and sample size on sweepnet sampling of *L. hesperus* in lentil fields in the USA. In addition, Schotzko and O'Keeffe (1989b) compared sweepnet, vacuum, absolute sampling, and diel variation of sweepnet sampling estimates in lentils for *A. pisum* and insect predators (lady beetles, lacewings). The sweepnet is widely used for *H. punctigera* larvae in grain legumes in Australia (Ridland *et al.*, 1993a,b; Loss, 1996; White *et al.*, 1996), for *B. pisorum* in field pea in Australia (Bailey and Baker, 1988; Hardie, 1989; Smith, 1990b; Smith and Hepworth, 1992) and in the USA (O'Keeffe *et al.*, 1992), and for *A. pisum* in field pea in North America (Maiteki and Lamb, 1985; Homan *et al.*, 1992).

Soil sampling provided estimates of *S. crinitus* egg populations in lentil in Syria (ICARDA, 1989), of *S. lineatus* populations (all life stages except egg) in faba bean in Denmark (Nielsen, 1990), and is recommended for estimating *S. lineatus* adult populations in field pea in the USA (O'Keeffe *et al.*, 1991a). Additional sampling methods include shaking chickpea (McIntyre and Titmarsh, 1989), pea (Ridland *et al.*, 1993a,b), and lupin plants (Michael *et al.*, 1982) to dislodge *H. punctigera* larvae in Australia, the use of aggregation pheromone traps to monitor adult *S. lineatus* in grain legumes in Europe (Blight and Wadhams, 1987; Nielsen and Jensen, 1993; Biddle *et al.*, 1996), and water filled trays to collect *L. cicerina* larvae dropping from chickpea leaves in Syria (ICARDA, 1990).

The various limitations and shortcomings associated with insect counts, plant beating, and sweepnet sampling are recognized by entomologists. For example, counts of *H. punctigera* eggs on grain legumes can be a misleading indicator of ensuing crop damage because of *Trichogramma* egg parasitism (Ridland *et al.*, 1993b), possible losses of eggs to predators, and adverse weather (Sime, 1996). Also, the labor required to monitor *Helicoverpa* eggs is unlikely to be economically justifiable (McDonald, 1995). Sweepnet sampling is often the method of choice, but other methods may be more efficient and less variable. Plant beating was more accurate than net sweeping for monitoring larval *H. punctigera* in field peas in Australia, but excessive time requirements limit its usefulness (Ridland *et al.*, 1993a). Shortcomings aside, sweepnet sampling provides the only practical method that farmers and scouts in developed countries are likely to use in their pest management programs because it is not laborious and nets are inexpensive and readily available. On the other hand, direct counts of *Helicoverpa* larvae on chickpea plants in small fields can provide farmers in developing countries with estimates of damaging populations (J.A. Wightman, personal communication).

ECONOMIC THRESHOLDS AND CONVENTIONAL INSECTICIDES

The level of insect infestation or plant damage that triggers a management action to prevent an economic loss is referred to as the "economic threshold (ET)", whereas the lowest pest population density that will cause economic damage is the "economic-injury level (EIL)". Controls (usually insecticides) are applied to prevent an increasing pest population from reaching the EIL (Stern *et al.*, 1959; Pedigo, 1996). In several crops worldwide, modern insect pest management has brought about the substitution of fixed calendar insecticide sprays for timely and necessary sprays based on EILs and pest densities at ET levels. Parigi and McDonald (1995) provide an example from grain legumes where calendar sprays cannot be economically and environmentally justified for insect (*H. punctigera*) control.

Economic thresholds (sometimes called 'Action Thresholds') are available to time insecticide sprays against six pests of grain legumes (Table 2). The ET levels for *H. armigera* in India (Wightman *et al.*, 1995), *H. punctigera* in Australia (Ridland *et al.*, 1993a; McDonald, 1995, 1996), *B. pisorum* in Australia (Smith *et al.*, 1987; Horne and Bailey, 1991), and *A. pisum* in Canada (Maiteki and Lamb, 1985; Soroka and Mackay, 1990a,b) were derived from experimental data on pest density-yield loss relationships (EILs), with consideration of variables such as market price, crop stage, cultivar, and cost of insecticide treatment. While ETs in grain legumes are usually established via sweepnet sampling (Michael *et al.*, 1982; Maiteki and Lamb, 1985; Hardie, 1989; O'Keeffe *et al.*, 1991b; Homan *et al.*, 1992; O'Keeffe *et al.*, 1992; Ridland *et al.*, 1993a, b), other sampling methods (insect counts, soil sampling, observation of insect-damaged plants) are recommended in some instances (O'Keeffe *et al.*, 1991a; Wightman *et al.*, 1995). Information on specific ET levels for individual pests and crops and information on insecticides registered for use in grain legumes in individual countries is available in the papers cited in Table 2.

In the past decade, a significant change has been: the development of ETs to improve spray use and timing, the substitution of one class of insecticide for another (*e.g.*, synthetic pyrethroids for DDT in Australia) (Bailey and Comery, 1987); the loss of insecticide registrations to government regulation (*e.g.*, ethyl parathion in the USA) (Anonymous, 1992); the development of insecticide resistance (*H. armigera* in India) (Armes *et al.*, 1996); and more precise applications to reduce the quantity of insecticide used (*e.g.*, border spraying for *B. pisorum* control in Australia) (Smith *et al.*, 1987; Baker, 1990a; Horne and Bailey, 1991). To our knowledge, *H. armigera* and *Myzus persicae* (Sulzer) (Homoptera: Aphididae) are the only major pest of grain legumes to have developed resistance to synthetic insecticides (Berlandier and Dadour, 1992; Armes et al., 1996), although there is evidence of insecticide tolerance in *H. destructor* (Hoffmann *et al.*, in press). Insecticide research continues, exemplified by work on chemical control of *B. pisorum* (Michael *et al.*, 1990), seed treatments for *S. lineatus* control in faba beans (Ester and Jeuring, 1992), partial application of insecticide to faba bean crops to control *A. fabae* and *S. lineatus* (Ward and Morse, 1995), evaluation of insecticide mixtures for control of *H. armigera* on chickpea (Pal *et al.*, 1996), and compatibility studies on chemical control and host-plant resistance in chickpea (Cowgill and Bhagwat, 1996). The recent work of Bwye *et al.* (1997) on insecticidal control of aphid vectors and cucumber mosaic virus in lupin in Australia strengthens the view held by many entomologists that synthetic insecticides are seldom effective in controlling non-persistent aphid-transmitted viruses in grain legumes.

MANAGEMENT METHODS BEYOND CONVENTIONAL INSECTICIDES

Biological Insecticides

Conventional insecticides have been effectively eliminated for long-term control of *H. armigera* on chickpea and other crops in India because the insect has developed resistance to most of the classes of synthetic insecticide available (Armes *et al.*, 1996). This development and the possible adverse effects of insecticides on humans and the environment (Lateef, 1990) has stimulated interest in other control methods such as resistant genotypes and biological insecticides (Pimbert, 1990; Cowgill and Bhagwat, 1996; Wightman, unpubl. report). Entomologists responded to the need for new *Helicoverpa* controls by showing that nuclear polyhedrosis virus (NPV) sprays hold promise as a biological insecticide in India,

474

providing it can be made available in the form of a reliable product (ICRISAT, 1995a; Wightman, unpubl. report). However, uncertainties over the compatibility of NPV sprays with host-plant resistance have highlighted the need for more research on the compatibility of multiple control tactics (Cowgill and Bhagwat, 1996). As well, formulations of *Helicoverpa* NPV sprays show promise for native budworm control in Australia where they may also enhance the effectiveness of natural enemies (D.A.V. Murray, personal communication).

Other biological insecticides have been considered for insect control in chickpea, including preliminary research on neem seed extract for leafminer and gram podborer control (Weigand and Joubi, 1994) and formulations of *Bacillus thuringiensis* Berliner for *Helicoverpa* control (Reed *et al.*, 1987).

Conventional Plant Resistance

By 1992 and IFLRC II (Muehlbauer and Kaiser, 1994), plant resistance to at least 17 field and storage insect pests of cool season food legumes had been identified in national programs and at international agricultural research centers (ICARDA, ICRISAT) (Clement *et al.*, 1994). Unfortunately, this research has not reached farmers in the form of insect-resistant cultivars (Clement *et al.* in press). The recent registrations of chickpea germplasm lines with *L. cicerina* (Singh and Weigand, 1996) and *H. armigera* resistance (Singh *et al.*, 1997) give optimism that adapted chickpea cultivars with insect resistance will eventually be developed and released to farmers in the developing world. However, before the insect resistance in this material or in any other nonadapted germplasm can be used, plant scientists must overcome barriers to the development of cultivars with both disease and insect resistance (Clement *et al.*, 1994).

Entomologists continue to evaluate the world's germplasm stocks of grain legumes for insect resistance. There have been recent screenings of pea accessions for resistance to *A. pisum* in Sweden (Sandstrom, 1994) and to *B. pisorum* in Australia and the USA (Clement *et al.*, 1995; Hardie *et al.*, 1995), lentil genotypes for resistance to *H. armigera* in India (Ujagir, 1993), and chickpea genotypes for resistance to *H. armigera* in India (Srivastava *et al.*, 1996; Yelshetty *et al.*, 1996). Very recently in Australia, Berlandier (1996) studied the effect of alkaloid levels in narrow-leafed lupin on the reproductive performance of the green peach aphid, *M. persicae*, and Thackray *et al.* (1997) assessed the susceptibilities of seedlings of six grain legume species to the redlegged earth mite.

Some researchers have sought to facilitate insect resistance breeding by using capillary gas chromatography-mass spectrometry to detect chemical markers unique to insect-susceptible and -resistant genotypes (Rembold *et al.*, 1990a,b; Clement *et al.*, 1991; Rembold and Schroth, 1993). This research opens opportunities for rapid screening of grain legume germplasm for phytochemicals that correlate with insect resistance under field conditions (Rembold *et al.*, 1990a; Clement *et al.*, 1994).

Transgenic Plants

The first success in genetically engineering a cool season food legume for insect resistance occurred after 1992. This event followed the development of a reproducible transformation and regeneration system for the introduction of foreign genes into cultivars of peas (Schroeder *et al.*, 1993) and involved the transfer of a pest-resistant trait in the seed of common bean (*Phaseolus vulgaris* L.) into pea for resistance to the storage pests *Callosobruchus chinensis* (L.) and *Callosobruchus maculatus* (F.) (Coleoptera: Bruchidae) (Shade *et al.*, 1994). This resistance is based on a gene for a seed protein called α-amylase inhibitor in common bean, which blocks the action of the starch-digesting enzyme α-amylase in weevil larvae, thus preventing digestion of starches in the seeds. Additionally, transgenic peas that expressed the bean α-amylase inhibitor gene in their seeds exhibited resistance to the pea weevil (Schroeder *et al.*, 1995). This seminal research by molecular biologists, plant physiologists, and entomologists places pea at the forefront of biotechnological improvement of legumes for insect resistance. It also opens up possibilities for pyramiding resistance genes with different characteristics into cultivars for more stable insect resistance.

Various other classes of resistance genes are being investigated for reducing grain legume losses to insects, namely the *B. t.* ssp. *tenebrionis cry*IIIA gene and a proteinase inhibitor gene from *Nicotiana alata*

Link et Otto. The latter was found to have toxic effects on *H. punctigera* and *H. armigera* (Heath *et al.*, in press), while *cry*IIIA protein production in pea nodules decreased survival and development of *S. lineatus* (Quinn and Bezdicek, 1996). Much more research is required to understand the future of these and other resistance genes in pest management.

Biological Control

Biological control is "the action of parasites, predators, or pathogens that temporarily control or continually regulate a host population at densities below what they would be in the absence of these natural enemies" (Huffaker and Smith, 1980). The potential role of these natural enemies in regulating pest densities in grain legumes has been increasingly appreciated since the 1980s (Table 2). Evidence summarized by Reed *et al.* (1987), Pawar *et al.* (1986), and Weigand *et al* (1994) shows, for example, that parasitism in *H. armigera* larvae feeding on chickpea in India can be substantial. Although Pimbert (1990) felt that vegetation management and the use of plant diversity may enhance populations of hymenopterous parasitoids for biological control of *H. armigera* in chickpea production zones in India, a recent study suggests that diversification of the chickpea cropping system may not enhance rates of *H. armigera* larval parasitism (Cowgill, 1995). In Syria, parasitoid populations on *L. cicerina* can build up in spring chickpeas and reach high densities late in the season, leading Weigand *et al.* (1994) to conclude that some parasitoids hold promise as biocontrol agents if methods could be developed to increase early-season populations. Moreover, Weigand *et al.* (1994) pointed out that without insecticides in production zones where populations of *H. armigera* are low, but high for *L. cicerina*, natural enemies can be expected to maintain leafminers at subeconomic levels because the EIL is very high.

In Australia, because native parasitoids rarely control the spring generation of *Helicoverpa* spp. infesting chickpea and other crops, two hymenopterous parasitoids were imported and released in Queensland and Western Australia (Murray and Rynne, 1992). Unfortunately, this biological control method is unlikely to be successful because of the unsuitability of chickpea for parasitoid activity and survival (Murray *et al.*, 1995). Biological control of *B. pisorum* in Australia has been considered and preliminary surveys for natural enemies in the weevil's native European distribution have identified potential candidate biocontrol agents (Baker, 1990a,b).

In the USA, the influence of pea genotype was studied on the rates of parasitism in *B. pisorum* (Annis and O'Keeffe, 1987) and the effect of plant morphology on the population growth of *A. pisum* in the presence and absence of lady beetle predators (Kareiva and Sahakian, 1990). Others developed guidelines for sampling *A. pisum* and its predominant predators in lentil fields for optimal control actions (Schotzko and O'Keeffe, 1989b). Current guidelines for insecticidal control of aphids in pea and lentil fields in the USA embrace the potential for biological control by considering the presence of natural enemies during sweepnet sampling (Homan *et al.*, 1992).

Entomopathogens such as nuclear polyhedrosis virus (NPV) kill *Helicoverpa* spp. larvae in India (Reed *et al.*, 1987). This fact plus the need for a biological insecticide to replace or supplement conventional insecticides for gram podborer control has stimulated research into the development of NPV sprays (see section on *Biological Insecticides*).

Cultural Control

The traditional practices, of early sowing of chickpea (October to mid-November) and intercropping with non-host plants help protect crops from *H. armigera* attack in some zones in northern India (Pimbert, 1990; ICRISAT, 1994; Weigand *et al.*, 1994). However, early sowing is ineffective in southern India because high winter temperatures promote faster plant and insect growth (Reed *et al.*, 1987; Wightman *et al.*, 1995). Other control practices may aggravate or ameliorate *H. armigera* problems on chickpea. For example, larval populations increase as plant densities increase, but plowing just before seeding may kill pupae (Reed *et al.*, 1987). In Syria, increased *Sitona* infestations on early sown lentil caused yield reductions compared to yields from late sowing (Weigand *et al.*, 1994).

In Australia, early harvesting of pea seed minimises the losses in seed weight from the feeding of *B. pisorum* larvae and reduces the size of the over-summering population of adults (Baker, 1990b; Smith,

1990a). With the redlegged earth mite in Australia, crops sown into new cropland experience little or no problems compared with crops sown in or near pasture paddocks (Michael *et al.*, 1982; Loss, 1996). Moreover, cropping unacceptable hosts (lupin, chickpea, lentil) of redlegged earth mite before more susceptible and preferred legumes (pea, faba bean) should reduce subsequent infestations and damage by the mite in preferred crops (McDonald *et al.*, 1996).

Behavioral Modifying Chemicals

Of the many plant odorants that permeate the environment, the phytophagous insect must select the few critical signals that stimulate host selection behavior to find food, oviposition sites, and places to rendezvous with receptive mates. In recent years, research has shown that volatile plant kairomones (chemicals that benefit the receiver) have practical use as lures to attract insects to traps for monitoring populations or to baits for control (Metcalf and Metcalf, 1992). While the use of plant kairomones to lure adults of *B. pisorum* (Clement *et al.*, 1991) and *H. armigera* (Rembold *et al.*, 1990a) to toxic baits or field traps has been considered, research has not progressed beyond the exploratory stage. Although Blight and Wadhams (1987) mentioned the potential use of male-produced aggregation pheromone to mass trap *S. lineatus*, research has developed the pheromone to monitor dispersal of adult weevils and to time insecticide sprays (Nielsen and Jensen, 1993; Biddle *et al.*, 1996).

TOWARDS IPM

This chapter is testimony to progress in research on integrated control of grain legume pests, especially in the area of IPM supportive activities such as insect sampling/monitoring and the development and use of economic thresholds. By researching methods beyond conventional insecticides (Table 2), entomologists may have laid the foundation for more biologically based IPM programs in grain legumes. At present, the number of established IPM programs based on the strategy's founding principles of biologically-based pest control methods and a combination of control tactics (Stern *et al.*, 1959) is scarce. We mentioned earlier 1) the early sowing of short-duration chickpeas and intercropping with non-host plants to protect the crop from gram podborer attack in parts of northern India, and 2) in Australia the timely chemical control of adult pea weevils in field margins and early harvesting to minimize losses in grain weight from the feeding of larvae and to reduce the number of adults entering aestivation.

The prospects of managing more major pests through IPM are good if entomologists and other scientists can build upon the ideas and scientific foundations summarized herein and in the entomology reviews from the 1986 and 1992 IFLRCs (Reed *et al.*, 1988; Clement *et al.*, 1994; Weigand *et al.*, 1994). These 'ideas and scientific foundations' include combinations of methods for individual pests, namely: entomopathogens (NPV), botanical and conventional insecticides, plant resistance, and cultural methods to enhance natural enemies, all for *Helicoverpa* control on chickpea in southern India (ICRISAT, 1995a; Wightman, unpubl. report); plant resistance, variation in time of planting, and enhancement of natural enemies for *L. cicerina* on chickpea in West Asia; conventional insecticides, biological insecticides (NPV sprays), and transgenic plants for *H. punctigera* on grain legumes in Australia; and conventional insecticides, plant resistance, transgenic plants, biological control, and behavioral modifying chemicals (USA only) for *B. pisorum* on field pea in Australia and the USA.

The pressure for farmers to implement IPM programs in grain legumes increases each year. We can look to India and other countries where *H. armigera* is resistant to insecticides (Armes *et al.*, 1996; ICRISAT, 1995b), forcing farmers to seek alternatives (Wightman, unpubl. report), and to the USA where the Environmental Protection Agency has canceled the use of some insecticides (Anonymous, 1992) and where it is becoming more difficult for pulse farmers to use cheap and effective pesticides under provisions of The Food Quality Protection Act of 1996 (Anonymous, 1996). As well, the U.S. Government currently has a goal of IPM implementation on 75% of crop area by the year 2000 (Glasener and Johnson, 1997). These changes are in response to society's fear of pesticide residues in food, however minute, and environmental concerns about pesticide use. The same concerns will likely influence in the future, pest management strategies in other countries. Even in the face of these 'pressures' we cannot predict significant growth in the implementation of IPM programs in grain legumes for the foreseeable future. For the present, conventional insecticides will play a key role in grain legume production because farmers have few economic alternatives. However, more judicious use of conventional insecticides through insect monitoring and use of ETs is needed to help preserve the use of these materials.

Roberts (1994) addressed the issue of national and international budgets for agricultural research at the 1992 IFLRC and lamented the low priority given to research on cool season food legumes. Today, static and declining budgets for grain legume research add uncertainty to future progress in IPM.

Acknowledgments

General assistance from L. Elberson and W. Kaiser in the preparation of this manuscript is greatly acknowledged. We thank F. A. Berlandier, C. Daniels, D. Murray, G. Piper, J. Ridsdill-Smith, and D. Schotzko for help in securing literature and N. Bosque-Pérez, G. Long, P. Michael, L. O'Keefe, D. Schotzko, and G. Strickland for their comments on the manuscript. Research was supported in part by grants from the USDA-FAS-ICD-RSED (AS37) in the USA and the Grains Research and Development Corporation in Australia. This article reports the results of research only. Mention of proprietary products does not constitute an endorsement or recommendation for its use by USDA.

References

Annis, B. and O'Keeffe, L.E. 1987. *Environmental Entomology* 16: 653-655.

Anonymous, 1992. *Pesticide Report, No. 83*. Pullman, USA: Cooperative Extension, Washington State University.

Anonymous, 1996. *Bulletin of the USA Dry Pea and Lentil Council* 47: 10.

Armes, N.J., Jadhav, D.R. and DeSouza, K.R. 1996. *Bulletin of Entomological Research* 86: 499-514.

Baker, G.H. 1990a. In: *Workshop Proceedings: National Pea Weevil Workshop*, pp. 80-83 (ed. A.M. Smith). Melbourne, Australia: Victoria Department of Agriculture and Rural Affairs.

Baker, G.H. 1990b. In: *Workshop Proceedings: National Pea Weevil Workshop*, pp. 84-86 (ed. A.M. Smith). Melbourne, Australia: Victoria Department of Agriculture and Rural Affairs.

Baker, G.J. 1990a. In: *Workshop Proceedings: National Pea Weevil Workshop*, pp. 1-7 (ed. A.M. Smith). Melbourne, Australia: Victoria Department of Agriculture and Rural Affairs.

Baker, G.J. 1990b. In: *Workshop Proceedings: National Pea Weevil Workshop*, pp. 95-100 (ed. A.M. Smith). Melbourne, Australia: Victoria Department of Agriculture and Rural Affairs.

Baker, G.J. 1993. *Final Report (DAS 22F)*. Canberra, Australia: Grains Research and Development Corporation.

Bailey, P. and Comery, J. 1987. *Australian Journal of Experimental Agriculture* 27: 439-43.

Bailey, P. and Baker, G. 1988. *Fact Sheet 21/80*. Adelaide, Australia: Department of Agriculture South Australia.

Berlandier, F.A. 1996. *Entomologia Experimentalis et Applicata* 79: 19-24.

Berlandier, F. and Dadour, I. 1992. *Western Australia Journal of Agriculture* 33:43-46.

Bhatnagar, A., Sehgel, V.K. and Rao, S.S. 1995. *LENS Newsletter* 22: 37-43.

Biddle, A.J., Smart, L.E., Blight, M.M. and Lane, A. 1996. In: *Brighton Crop Protection Conference, Pests & Diseases, Volume 1: Proceedings of an International Conference*, pp. 173-178. Farnham, UK: British Crop Protection Council.

Blight, M.M. and Wadhams, L.J. 1987. *Journal of Chemical Ecology* 13: 733-739.

Bommarco, R. And Ekbom, B. 1995. *International Journal of Pest Management* 41:109-113.

Bushara, A.G. 1988. In: *World Crops: Cool Season Food Legumes*, pp. 367-378 (ed. R.J. Summerfield). Dordrecht, The Netherlands: Kluwer Academic Publishers.

Bwye, A.M., Proudlove, W., Berlandier, F.A. and Jones, R.A.C. 1997. *Australian Journal of Experimental Agriculture* 37: 93-102.

Clement, S.L. 1992. *Entomologia Experimentalis et Applicata* 63: 115-121.

Clement, S.L., Pike, K.S. and Fellman, J.K. 1991. In: *Program and Abstracts, Biennial Meeting of The National Pea Improvement Association*, p. 21. Lincoln, USA: National Pea Improvement Association.

Clement, S.L., El-Din Sharaf El-Din, N., Weigand, S. and Lateef, S.S. 1994. *Euphytica* 73: 41-50.

Clement, S.L., Hardie, D. and Poitras, L. 1995. In: *Program and Abstracts, Biennial Meeting of the National Pea Improvement Association*, p. 15. East Lansing, USA: National Pea Improvement Association.

Clement, S.L., Cristofaro, M., Cowgill, S.E. and Weigand, S. In press. In: *Global Plant Genetic Resources for Insect Resistant Crops* (eds. S. L. Clement and S. S. Quisenberry). Boca Raton, USA: CRC Press.

Cowgill, S.E. 1995. *Entomophaga* 40: 307-315.

478

Cowgill, S.E. and Bhagwat, V.R. 1996. *Crop Protection* 15: 241-246.

Dale, M., Gregg, P.C. and Drake, V.A. 1992. *Report of a Workshop on Developing a Heliothis Forecasting Service in Australia,* 43 pp. Narrabri Agricultural Research Station NSW, Australia:.

Ester, A. and Jeuring, G. 1992. *FABIS Newsletter* 30: 32-41.

Ferguson, A.W. 1994. *Crop Protection* 13: 201-210.

Gardner, G. 1996. *World Watch* 8: 21-27.

Glasener, K.M. and Johnson, A. 1997. In: *Agronomy News*, p. 4, April 1997. Madison, USA: American Society of Agronomy.

Gomez, K.A. and Gomez, A.A. 1984. *Statistical Procedures for Agricultural Research.* New York, USA: John Wiley &Sons.

Gregg, P.C., Fitt, G.P., Zalucki, M.P., and Murray, D.A.H. 1995. In: *Insect Migration: Tracking Resources Through Space and Time*, pp. 151-171 (eds. A.G. Gatehouse and V.A. Drake). Cambridge, UK: Press Syndicate, University of Cambridge.

Hagedorn, D.J. (ed.). 1984. *Compendium of Pea Diseases*, 57 pp. St. Paul, USA: American Phytopathological Society.

Hardie, D. 1989. *Farmnote, Western Australian Department of Agriculture*, No. 127/89.

Hardie, D., Baker, G.J. and Marshall, D.R. 1995. *Euphytica* 84: 155-161.

Heath, R.L., McDonald, G., Christeller, J.T., Lee, M., Batemaan, K., West, J., van Heeswijck, R. and Anderson, M.A. In press. *J. Insect Physiol.*

Hoffman, A.A., Porter, S. and Kovacs, I. In press. *Experimental and Applied Acarology.*

Homan, H.W., Stoltz, R.L. and Schotzko, D.J. 1992. *Current Information Series No. 748*, Moscow, USA: University of Idaho, College of Agriculture,.

Horne, J. and Bailey, P. 1991. *Crop Protection* 10: 53-56.

Huffaker, C.B. and Smith, R.F. 1980. In: *New Technology of Pest Control*, pp. 1-24 (ed. C.B. Huffaker). New York, USA: John Wiley & Sons.

ICARDA. 1989. *Annual Report.* Aleppo, Syria.

ICARDA. 1990. *Annual Report.* Aleppo, Syria.

ICRISAT. 1994. *ICRISAT Now, Sowing for the Future*. Patancheru, Andhra, Pradesh, India.

ICRISAT. 1995a. *Annual Report, Asia Region.* Patancheru, Andhra Pradesh, India.

ICRISAT. 1995b. *Report 1995*, Patancheru, Andhra Pradesh, India.

Kareiva, P. and Sahakian, R. 1990. *Nature* 345: 433-434.

Kaya, N. and Hincal, P. 1987. In: *Türkiye I. Entolmolji Kongresi*, pp. 259-266. Izmir, Turkey: Ege Universitesi.

Klein, R.E., Larsen, R.C. and Kaiser, W.J. 1991. *Plant Disease* 75:1186.

Lateef, S.S. 1990. In: *First Naional Workship on Heliothis Management: Current Status and Future Strategies*, pp. 129-140 (ed. J.N. Sachan). Kanpur, India: Directorate of Pulses Research.

Loss, S. 1996. *Farmnote, Agriculture Western Australia*, No. 54/96.

Luckmann, W.H. and Metcalf, R.L. 1975. In: *Introduction to Insect Pest Management,* pp. 3-35 (eds. R.L. Metcalf and W.H. Luckmann). New York, USA: John Wiley & Sons.

Maelzer, D., Zalucki, M.P. and Laughlin, R. 1996. *Bulletin of Entomological Research* 86: 547-557.

Maiteki, G.A. and Lamb, R.J. 1985. *Journal of Economic Entomology* 78: 1449-1454.

McDonald, G. 1995. *Agnote No AG0417.* Melbourne, Australia: Agriculture Victoria.

McDonald, G. 1996. In: *Workshop Report: Heliothis Management in Australia,* 100 pp. (eds. G.G. White, D.A. H. Murray and M.P. Walton). Brisbane, Australia: Cooperative Research Centre for Tropical Pest Management.

McDonald, G. and Bryant, D. 1994. *25th Scientific Conference and A.G.M. of the Australian Entomological Society.* Adelaide, Australia: University of Adelaide.

McDonald, G., Ballinger, D. and Hoffman, A.A. 1996. *Australian Grain* 6: 6-8.

McIntyre, G.T. and Titmarsh, T. 1989. *Software for Primary Producers.* v1.0 Agdex 168.614. Queensland Department of Primary Industry.

Metcalf, R.L. 1996. *American Entomologist* 42: 216-227.

Metcalf, R.L. and Metcalf, E.R. 1992. *Plant Kairomones in Insect Ecology and Control.* New York, USA: Chapman and Hall.

Michael, P.J., Woods, W.M., Richards, K.T. and Sandow, J.D. 1982. *Western Australia Journal of Agriculture, Fourth Series* 23: 83-85.

Michael, P.J., Dhalival, S.S. and Mangano, G.P. 1990. In: *Workshop Proceedings: National Pea Weevil Workshop*, pp. 34-40 (ed. A. M. Smith). Melbourne, Australia: Victoria Department of Agriculture and Rural Affairs.

Miles, M. and McDonald G. 1995. *Proc. Victorian Cropping Zone Technical Update Conference,* pp 23-24. Victoria, Australia: Longeronong College.

Muehlbauer, F.J. and Kaiser, W.J. (eds.). 1994. *Expanding the Production and Use of Cool Season Food Legumes.* Dordrecht, The Netherlands: Kluwer Academic Publishers.

Murray, D. and Rynne, K. 1992. *Australian Grain* 2(4): 4-5.

Murray, D., Rynne, K.P., Winterton, S.L., Bean, J.A. and Lloyd, R.J. 1995. *Journal of the Australian Entomological Society* 34: 71-73.

Nielsen, B.S. 1990. *Journal of Applied Entomology* 110: 398-407.

Nielsen, B.S. and Jensen, T.S. 1993. *Entomologia Experimentalis et Applicata* 66: 21-30.

Ogg, A.G., Jr., Smiley, R.W., Pike, K.S., McCaffrey, J.P., Thill, D.C. and Quisenberry, S.S. In press. In: *Advances in Conservation Farming* (eds. A.G. Ogg, Jr., E. Michalsen and R.I. Papendick). Ankeny, USA: Soil and Water Conservation Society.

O'Keeffe, L.E., Homan, H.W. and Schotzko, D.J. 1991a. *Current Information Series No. 883*, Moscow, USA: University of Idaho, College of Agriculture.

O'Keeffe, L.E., Homan, H.W. and Schotzko, D.J. 1991b. *Current Information Series No. 894*, Moscow, USA: University of Idaho, College of Agriculture.

O'Keeffe, L.E., Homan, H.W. and Schotzko, D.J. 1992. *Current Information Series No. 885*, Moscow, USA: University of Idaho, College of Agriculture.

Oram, P.A. and Agcaoili, M. 1994. In: *Expanding the Production and Use of Cool Season Food Legumes,* pp. 3-49 (eds. F.J. Muehlbauer and W.J. Kaiser). Dordrecht, The Netherlands: Kluwer Academic Publishers.

Pal, S.K., Das, V.S.R. and Armes, N.J. 1996. *International Chickpea and Pigeonpea Newsletter* 3: 44-46.

Parigi, P. and McDonald, G. 1995. *Technical Report,* 34 pp. Melbourne, Australia: Agriculture Victoria.

Pawar, C.S., Bhatnagar, V.A. and Judhav, D.R. 1986. *Proceedings of the Indian Academy of Sciences* 95: 695-703.

Pedigo, L.P. 1996. *Entomology and Pest Management.* Upper Saddle River, USA:Prentice Hall.Pimbert, M.P., 1990. In: *Chickpea in the Nineties: Proceedings of the Second International Workshop on Chickpea Improvement,* pp. 151-163 (eds. H.A. van Rheenen, M.C. Saxena, B.J. Walby and S.D. Hall). Patancheru, Andhra Pradesh, India: ICRISAT.

Qin, T.-K. 1997. *Bulletin of Entomological Research* 87: 289-298.

Quinn, M. and Bezdicek, D.G. 1996. *Journal of Economic Entomology* 89: 550-557.

Reed, W., Cardona, C., Sithanantham, S. and Lateef, S.S. 1987. In: *The Chickpea,* pp. 283-318 (eds. M.C. Saxena & K.B. Singh). UK: CAB International.

Reed, W., Cardona, C., Lateef, S.S. and Bishara, S.I. 1988. In: *World Crops: Cool Season Food Legumes,* pp. 107-115 (ed. R.J. Summerfield). Dordrecht, The Netherlands:Kluwer Academic Publishers.

Rembold, H., Schroth, A., Lateef, S.S. and Weigner, Ch. 1990a. In: *Proceedings of the First Consultative Group Meeting on Host Selection Behavior of Helicoverpa armigera,* pp. 23-26. Patancheru, Andhra Pradesh, India: ICRISAT.

Rembold, H., Wallner, P., Kohne, A., Lateef, S.S., Grune, M. and Weigner, Ch. 1990b. In: *Chickpea in the Nineties: Proceedings of the Second International Workshop on Chickpea Improvement,* pp.191-194. (eds. H.A. van Rheenen, M.C. Saxena, B.J. Walby and S.D. Hall). Patancheru, Andhra Pradesh, India: ICRISAT.

Rembold, H. and Schroth, A. 1993. In: *Breeding for Stress Tolerance in Cool-Season Food Legumes,* pp. 211-224 (eds. K.B. Singh and M.C. Saxena). London, UK: John Wiley & Sons and Sayce Publishing.

Ridland, P.M., Smith, A.M. and McDonald, G. 1993a. *Final Report (DAV 17G),* 47 pp. Canberra, Australia: Grains Research and Development Corporation.

Ridland, P.M., Smith, A.M. and McDonald, G. 1993b. In: *Pest Control and Sustainable Agriculture,* pp. 229-232. Canberra, Australia: CSIRO.

Roberts, E.H. 1994. In: *Expanding the Production and Use of Cool Season Food Legumes,* pp. 983-988 (eds. F.J. Muehlbauer and W.J. Kaiser). Dordrecht, The Netherlands: Kluwer Academic Publishers.

Sandstrom, J. 1994. *Entomologia Experimentalis et Applicata* 71: 245-256.

Schotzko, D. J. and O'Keeffe, L. E. 1986a. *Journal of Economic Entomology* 79: 224-228.

Schotzko, D. J. and O'Keeffe, L. E. 1986b. *Journal of Economic Entomology* 79: 447-451.

Schotzko, D. J. and O'Keeffe, L. E. 1989a. *Environmental Entomology* 18: 308-314.

Schotzko, D. J. and O'Keeffe, L. E. 1989b. *Journal of Economic Entomology* 82: 491-506.

Schroeder, H.E., Schotz, A.H., Wardley-Richardson, T., Spencer, D. and Higgins, T.J.V. 1993. *Plant Physiology* 101: 751-757.

Schroeder, H.E., Gollasch, S., Moore, A., Tabe, L. M., Craig, S., Hardie, D.C., Chrispeels, M.J., Spencer, D. and Higgins, T.J.V. 1995. *Plant Physiology* 107: 1233-1239.

Shade, R.E., Schroeder, H.E., Pueyo, J.J., Tabe, L.M., Murdock, L.L., Higgins, T.J.V. and Chrispeels, M.J. 1994. *Bio/Technology* 12: 793-796.

480

Sime, A. 1996. Honours Thesis. Bundoora, Australia: La Trobe University, School of Zoology.

Singh, K.B. and Weigand, S. 1996. *Crop Science* 36: 472.

Singh, O., Sethi, S.C., Lateef, S.S. and Gowda, C.L.L. 1997. *Crop Science* 37: 295.

Singh, Y. 1985. *Bulletin of Entomology* 26: 86-91.

Smith, A.M. 1990a. In: *Bruchids and Legumes: Economics, Ecology and Coevolution*, pp. 105-114 (eds. K. Fujii, A.M.R. Gatehouse, C.D. Johnson, R. Mitchel and T. Yoshida). Dordrecht, The Netherlands: Kluwer Academic Publishers.

Smith, A.M. 1990b. In: *Workshop Proceedings: National Pea Weevil Workshop*, pp. 8-20 (ed. A.M. Smith). Melbourne, Australia: Victorian Department of Agriculture and Rural Affairs.

Smith, A.M. 1990c. In: *Workshop Proceedings: National Pea Weevil Workshop*, pp. 104-112. (ed. A.M. Smith). Melbourne, Australia: Victorian Department of Agriculture and Rural Affairs.

Smith, A.M. 1992. *Environmental Entomology* 21: 314-321.

Smith, M., Comery, J. and Chaffey, B. 1987. *Research Report Series No. 44.* Burnley, Australia: Department of Agriculture and Rural Affairs.

Smith, A.M. and Hepworth, G. 1992. *Journal of Economic Entomology* 85: 1791-1796.

Smith, A.M. and Ward, S.A. 1995. *Environmental Entomology* 24: 623-634.

Soroka, J.J. and Mackay, P.A. 1990a. *Canadian Entomologist* 122: 503-513.

Soroka, J.J. and Mackay, P.A. 1990b. *Canadian Entomologist* 122: 1201-1210.

Southwood, T.R.E. 1978. *Ecological Methods*. London, UK: Chapman and Hall.

Srivastava, C.P., Singh, U.P. and Singh, R.M. 1996. *International Chickpea and Pigeonpea Newsletter* 3: 43-44.

Stern, V.M., Smith, R.F., van den Bosch, R. and Hagen, K.S. 1959. *Hilgardia* 29(2): 81-101.

Summerfield, R.J., Short, R.W. and Muehlbauer, F.J. 1994. In: *Expanding the Production and Use of Cool Season Food Legumes*, pp. 859-876 (eds. F.J. Muehlbauer, F.J. and W.J. Kaiser). Dordrecht, The Netherlands: Kluwer Academic Publishers.

Thackray, D. 1996. *Australian Grain*, August-September, pp. 42-43.

Thackray, D.J., Ridsdill-Smith, T.J. and Gillespie, D.J. 1997. *Plant Protection Quarterly* 12:141-144.

Todorov, N.A., Pavlov, D.C. and Kostov, K.D. 1996. In: *Food and Feed from Legumes and Oilseeds*, pp. 113-123 (eds. E. Nwokolo and J. Smartt). London, UK: Chapman and Hall.

Tummala, R.L., Haynes, D.L. and Croft, B.A. (eds.) 1974. In: *Modeling for Pest Management, Concepts, Techniques, and Applications (U.S.A./USSR)*, East Lansing, USA: Michigan State University.

Ujagir, R. 1993. *LENS Newsletter* 20: 34-35.

van Emden, H.F. 1978. In: *Pests of Grain Legumes: Ecology and Control*, pp. 297-307 (eds. S.R. Singh, H.F. van Emden and T.A. Taylor). London, UK: Academic Press.

van Emden, H. F., Ball, S. L. and Rao, M. R. 1988. In: *World Crops: Cool Season Food Legumes*, pp. 519-534 (ed. R. J. Summerfield). Dordrecht, The Netherlands: Kluwer Academic Publishers.

Ward, A. and Morse, S. 1995. *Annals of Applied Biology* 127: 239-249.

Walden, K.J. 1995. In: *Insect Migration: Tracking Resources Through Space and Time*, pp. 172-192. (eds. A.G. Gatehouse and V.A. Drake). Cambridge, UK: Press Syndicate, University of Cambridge.

Walden, K. J. 1992. *Western Australia Journal of Agriculture* 33:109-113.

Weigand, S. and Pimbert, M.P. 1993. In: *Breeding for Stress Tolerance in Cool-Season Food Legumes*, pp. 145-156 (eds. K. B. Singh and M. C. Saxena). London, UK: John Wiley & Sons and Sayce Publishing.

Weigand, S. and Joubi, A. 1994. In: *Germplasm Program, Legumes, Annual Report, ICARDA*, pp. 107-108. Aleppo, Syria:

Weigand, S., Lateef, S.S., El-Din Sharaf El-Din, N., Mahmoud, S.F., Ahmed, K. and Ali, K. 1994. In: *Expanding the Production and Use of Cool Season Food Legumes*, pp. 679-694. (eds. E.J. Muehlbauer and W.J. Kaiser), Dordrecht, The Netherlands: Kluwer Academic Publishers.

White, G.G., Murray, D.A.H. and Walton, M.P. 1996. *Workshop Report: Heliothis Management in Australia*, 100 pp. Brisbane, Australia: Cooperative Research Centre for Tropical Pest Management.

Wightman, J.A., Anders, M.M., Rameshwar Rao, V. and Mohan Reddy, L. 1995. *Crop Protection* 14: 37-46.

Wilson, A.G.I. and Morton, R. 1989. *Bulletin of Entomological Research* 79: 265-273.

Yelshetty, S., Kotikal, Y.K., Shantappanavar, N.B. and Lingappa, S. 1996. *International Chickpea and Pigeonpea Newsletter* 3: 41-43.

Zalucki, M.P., Daglish, G., Firempong, S. and Twine, P. 1986. *Australian Journal of Zoology* 34: 779-814.

Integrated weed management for food legumes and lupins

F.L. Young[1], J.Matthews [2], A. Al-Menoufi [3], J Sauerborn [4], A.H. Pieterse [5] and M.Kharrat [6]

1 Unites States Department of Agriculture, Agriculture Research Service,Nonirrigated Agriculture Weed Science Research, Rm. 161, Johnson Hall,Washington State University, Pullman, WA 99164-6421, USA; 2 Department of Agronomy and Farming Systems, University of Adelaide, Roseworthy 5371, South Australia; 3 Alexandria University, Plant Pathology Dept., El-Shatby Alexandria, Egypt; 4 Justus-Leibig-Universitat Giessen, Schottstrasse 2, 35390 Giessen, Germany; 5 Department of Agriculture Research, Royal Tropical Institute, Mauritskade 63, 1092 AD, Amsterdam, The Netherlands; 6 Laboratoire des Legumineuses Alimentaires, INRAT, Ariana 2080, Tunis, Tunisia

Abstract

The history of weed science has been dominated by an emphasis on weed control using synthetic herbicides. However, pressures from social, economic, and environmental factions have suggested weed scientists develop a systems approach for managing weeds that integrates numerous control methods into crop production systems. Some of these methods include preventive, biological, cultural, mechanical, and chemical practices. When incorporated into a cropping system these methods must be socially acceptable, economically feasible, and environmentally sound. Integrated weed management in food legume production is a future research area requiring additional components that will include bioeconomic/predictive modeling, herbicide resistant crops, and precision agriculture.

INTRODUCTION

Although weeds are a major factor affecting crop production, scientists and administrators world-wide have failed to recognize that weeds are the pests most often limiting production (Elmore, 1996). Weeds dictate most crop production practices (Wyse, 1994). In the United States (U.S.), it has been estimated that crop losses because of weeds, without including the actual cost of weed control, was $4.1 billion in 1992 (Bridges and Anderson, 1992). In Sub-Saharan Africa, at least 50% of the inputs to crop production, much of which is hand labor, is for weed control (Akobundu, 1991). A 6 year integrated pest management (IPM) study identified for the first time a profitable, diversified crop production system grown under conservation tillage in the U.S. Pacific Northwest (PNW), (Boerboom et al., 1993). This system, which included spring dry pea (*Pisum sativum* L.), was profitable because grass weeds were controlled using chemical and cultural methods (Young, D. L. et al., 1994; Young, F. L. et al., 1994b).

Weeds in lentil (*Lens culinaris* Medikus), faba bean (*Vicia faba* L.), field pea, chickpea (*Cicer arietinum* L.), grass pea (*Lathyrus sativus* L.), and lupin (*Lupinus albus* L.) crops reduce yields and quality, interfere with harvest, and increase price dockages. Knott and Halila (1988) reviewed the effect of weeds on food legumes and stated that, in general, food legumes are poor competitors because of their slow initial growth. In addition, Papendick et al. (1988) stated that effective weed control is critical for achieving the yield potential of grain legumes. Rarely, except for broomrape (*Orobanche* spp.) are weeds specific to food legumes (Knott and Halila, 1988). Generally, weeds are often a greater problem in legumes than in other crops in the rotation because the use of post-emergence grass herbicides is limited by their high cost, the ineffectiveness of available broadleaf herbicides, and the poor competitiveness of grain legumes. The damage inflicted by weeds on legumes depends on the weed species present, environmental conditions, weed/crop densities, time of weed emergence and soil fertility. In addition to reducing crop yields through interference, some weeds affect legumes by decreasing harvest efficiency, increasing harvest costs, and contaminating the grain. In the U.S., studies showed that spring dry pea was 3 to 20 times more aggressive than mayweed chamomile (*Anthemis cotula* L.). However, the weed continued to grow after pea senescence and interfered with crop harvest (Ogg, Jr. et al., 1993). Growers must then either swathe the crop to desiccate the mayweed chamomile or obtain permission to spray with paraquat before harvest to chemically

481

R. Knight (ed.), Linking Research and Marketing Opportunities for Pulses in the 21st Century, 481–490.
© 2000 *Kluwer Academic Publishers. Printed in the Netherlands.*

desiccate the weed. In each case, costs are increased. Redroot pigweed (*Amaranthus retroflexus* L.) and common lambsquarters (*Chenopodium album* L.) can reduce pea yields 30% (McCue and Minotti, 1979), increase harvest costs by requiring mechanical and/or chemical desiccation (Young, F. L. et al., 1994a), and lower grain quality by staining the seed.

Lupin has been included in the Third International Food Legume Conference. Early research on lupins was concerned with herbicide efficacy and tolerance (Wiedenhoeft and Ciha, 1987; Penner et al. 1993; Lemerle and Hinkly, 1991). In some countries the biology and management of the crop, have received little attention until recently (Brebaum and Boland, 1995) but in Australia, where lupins are the major grain legume grown many studies have been undertaken. Aspects of the cultivation of the crop in Australia are dealt with elsewhere in these proceedings and weed control has been reviewed recently by Gilbey (1993) and by Gill (1994).

In this chapter, we focus on integrated weed management (IWM) in food legumes, and examine new concepts, technologies, and strategies. We will not address herbicide-based control strategies or the harmful effects of weeds in grain legumes. We will provide information and ideas that will integrate weed management practices to reduce herbicide use in food legume systems.

WEED CONTROL VERSUS WEED MANAGEMENT

In the future, farming systems probably will be operating with fewer inputs of pesticides. Government mandates and research funding are emphasizing food safety. For example, in the U.S. it has been proposed that by the year 2000, IPM will be mandatory on 75% of all crops. In the proposal, a goal has been set to reduce agricultural chemicals by two-thirds from current levels and chemicals will be used only as a last resort. The proposals have two important implications. First, if chemicals can be used only as a last resort, weeds will have already damaged the crop considerably, as competition is most pronounced early in weed/crop life cycles. Second, research has been very limited on integrated weed management and information is lacking on how to combine alternative approaches for effective weed management. In developed countries, 60 to 70% of the agricultural pesticides applied are herbicides (Duke, 1996). Worldwide, scientists do not have the information and techniques to allow a smooth and rapid transition to IPM.

The history of weed science, is a history of control, not of weed management (Zimdahl, 1995). Herbicides, since their advent in the 1940's have been efficient, cost effective, and capable of controlling weeds almost 100%. Of the manuscripts published in *Weed Science* from 1960 to 1981 only 4% pertained to IWM, while about 80% dealt with herbicide efficacy and physiology (Thill et al., 1991).

The concept of IPM for weeds is newer and different from IPM for insects or pathogens which is often focused on a single insect or pathogen and is based on accurate threshold levels (Elmore, 1996). When non-chemical methods fail, a decision to spray can be based on economic thresholds. In contrast, concepts of economic and damage threshold levels for weeds are not well developed. The validity of information on single weed species may be of little importance because crop yield losses can vary with weed density among years (Young, F.L., 1986) and locations (Lindquist et al., 1996). Also, weeds are not distributed uniformly and rarely grow in a monoculture. In the PNW IPM cropping systems study, referred to earlier, a complex of more than 50 weed species were identified of which 10 to 12 were important economically (Young et al., 1996). Recent research has attempted to develop thresholds (Berti and Zanin, 1994) and bio-economic decision aid models (Kwon, 1993) for mixtures of weed species. However, even if they are developed, weed thresholds have numerous problems associated with their implementation (O'Donovan, 1996; Cousens, 1987).

Since the early 1980's many definitions of weed management have been made, most of which have limitations such as not including economic thresholds and yield losses, or emphasizing weed control instead of management. The foundation for weed control has been the use of herbicides and tillage, both of which should be reduced for economic and environmental reasons (Gallandt, E., personal communication). One definition of IWM, that has gained acceptance is described as a "directed agroecosystem approach for the management and control of weed and other pest populations at threshold levels that prevent economic damage in the current and future years" (Shaw, 1982). IWM's goals were to decrease losses caused by weeds, reduce costs of control and the associated energy and labor requirements and to decrease tillage and subsequent soil erosion (Shaw, 1982). The concept of management, is the knowledge and understanding of weed biology and ecology which will lead to a greater reliance upon aspects of cropping systems to suppress weeds (Gallandt, E., personal communication). The major component of Shaw's (1982) concept was the agroecosystem, which included the production of high quality food, fiber and livestock, a safe environment, economically feasible systems, and future technology. This concept may have been the forerunner of today's approach to Integrated Crop Management (ICM) or Integrated Farming Systems. Components of these systems include crops, fertility, pests, animals, machinery, economics, environment and people.

Multiple components of IWM are discussed because very little cropping systems research has been conducted. Scientists do not always agree on which category weed management components fall under. Therefore, in our discussion, particular components may overlap several categories. For example, crop competition may be included in biological or cultural control methods. In our concept, IWM includes prevention, biological, chemical, mechanical and cultural methods for weed management.

INTEGRATED WEED MANAGEMENT

Prevention

Prevention is the most basic of all management methods and is defined as those measures taken to forestall the introduction and/or spread of weeds (Walker, 1995). Prevention has been overshadowed by effective chemical weed control and de-emphasized by the concept of economic thresholds. The threshold concept directs that a control measure should not be implemented until the financial loss from crop damage equals the control cost (Walker, 1995). However, if the newly introduced weed is noxious or if there is no effective management method, then the threshold may be one plant, regardless of the impact on the present crop. The simplest preventive method is to plant clean seed. The cleaning of weed-contaminated seed by a grower does not guarantee that crop seed is free of weeds. Buying certified, clean seed is one of the cheapest inputs to weed management. It has been estimated that planting certified seed of spring pea, lentil and chickpea would be a moderately effective management strategy against wild oat (*Avenafatua* L.), hairy nightside (*Solanum sarrachoides* Sendt) and Canada thistle (*Cirisum arvense* L.Scop.) (Ogg, Jr. et al., in press). But against weeds such as common lambsquarters and mayweed chamomile, only a slight benefit would be realized with planting certified seed.

Producers must prevent weeds from being disseminated throughout their farms by cleaning the equipment used for tillage, planting, and harvesting before moving to new fields. Machines distribute troublesome perennial weeds such as Canada thistle (*Cirisum arvense* L. Scop.), quackgrass (*Elytrigia repens* L. Nevski), and field bindweed (*Convolvulus arvensis* L.). Weeds can also be transported by livestock, manure, animal feed and as contaminants of irrigation systems (Day, 1972). Weeds growing in fence rows and ditches must be controlled to prevent their entry into fields. This is especially important for weeds with wind-transported propagules such as Canada thistle. Weeds must be controlled along borders to prevent further encroachment into the field. It may be necessary to plant a different crop in the border to allow for more effective weed management.

Biological

A definition of biological control has been given in the chapter on integrated insect management in these proceedings (see Clement et al.). The definition of biological weed control differs little from biological insect control. Classical biological weed control uses organisms other than the associated crop to reduce the growth and competitive effects of weeds (Cardina, 1995). The most common organisms are insects and pathogens and the goal, like that with chemical control is to reduce the target weed population. In contrast biological management of weeds seeks to enhance processes that hinder the growth and development of weeds and is oriented toward suppressing weeds (Cardina, 1995). Agents used for inundative biological management are more numerous and more diverse than agents for classical biological control. In addition to insects and pathogens, agents may include crops and allelopathy (Cardina, 1995).

Normally, classical biological strategies are not successful in agroecosystems characterized by diversified annual crops and soil disturbance by tillage. However, opportunities exist and the potential for biological agents has been evaluated in some of the food legume crops. Including lupins in a crop rotation suppressed common lambsquarters and redroot pigweed (Rice, 1983). In Canada, the fungal pathogen

Colletotrichum gloeosporiodes f. sp. *malvae* is used to control round-leafed mallow (*Mava pusilla* Sm.) in lentil (Cardina, 1995). The fungus causes stem lesions, but must be applied during periods of high relative humidity or with irrigation. The potential for biological management of broomrape has been reviewed previously (Pieterse et al., 1994). A fly (*Phytomyza orobanchia* Kalt) appears to be the most promising agent. The larvae feed in the stem and fruit capsules of the weed, however its population is sensitive to chemicals, tillage, and crop rotation. In Syria, two-thirds of 31 faba bean fields were infested by broomrape species, and 95% of the infested fields contained *Phytomyza*. Damage included parasitized fruit capsules and destroyed seed (Linke et al., 1992). In the PNW, isolates from bacteria were obtained from the rhizoplane of winter wheat (*Triticum aestivum* L.) and downy brome (*Bromus tectorum* L.) and were evaluated for their selective suppression of grass weed species in winter wheat (Kennedy et al., 1991) where chemicals for the selective control of winter annual grass weeds were not effective. A similar situation exists in the low-rainfall regions of the PNW where growers and agronomists are attempting to move cool season food legume crops to the semiarid regions. A major limitation of this strategy, however, is poor broadleaf weed control, especially Russian thistle, (*Salsola iberica*, Sennen and Pau) in these crops. Because of this dilemma, the potential of soil bacteria to suppress Russian thistle selectively in cool season food legume crops is being investigated (Kennedy, A.C., personal communication).

Chemical

The chemical control of weeds in food legumes was reviewed during the first International Food Legume Research Conference (Knott and Halila, 1988). Information included the herbicides available, weeds controlled, residual carry-over, and drift injury from spraying cereal crops. Other than imazethapyr and quizalofop-p, very few new herbicides have become available for use in food legumes. Imazethapyr controls a wide array of troublesome broadleaf weeds, however, carry-over into cereal crops is a problem, especially in semiarid regions. Quizalofop-p controls many grass weed species, including volunteer crops and is more affordable than current post-emergence herbicides for grass control in legumes. Recent studies have shown that either soaking or coating pea seeds with imazethapyr controlled broomrape 50 to 80% without injuring the seeds (Jurado-Exposito et al., 1996). Generally, as tillage is reduced to maintain residues on the soil surface, producers use more herbicides to control weeds (Thill et al., 1986). In the PNW, spring dry pea was produced under conservation tillage practices and pea yields were equal to or greater than yields in traditional, conventional tillage systems (Young et al., 1994a). Also, the conservation pea production system did not substantially increase herbicide use and cost compared to the conventional system and it reduced soil erosion an estimated 90%. Further complicating the fact that few herbicides are available for use in food legumes, herbicide resistance in weeds is becoming more of a problem worldwide. As of 1997 there are more than a dozen weed species reported to be resistant to the herbicides used in pea and lentil in nine countries (Table 1).

Mechanical

Conventional tillage is an effective means of controlling weeds; however it increases the risk of soil erosion by wind and water and reduces soil quality. Nevertheless, conventional tillage is an important strategy particularly in developing countries. In areas where farms are small, inputs and profits are low, and labor plentiful, hand weeding and hoeing is the traditional method of weed management. Food legumes are normally planted in narrow rows and therefore, between-row mechanical cultivation is not feasible. Furthermore, crops may not respond favorably to wider rows, e.g., yields of faba bean (Wilson and Cussans, 1972) and pea (Davies, et al., 1985) decreased when row width was increased.

Interest has been renewed in using mechanical tillage (not between-row cultivation) for the management of weeds in the growing crop. In Denmark, a model is being developed to predict the optimum intensity of harrowing on the yield of pea and weed control (Rasmussen, 1992). Weed control (based on dry matter reduction) ranged from 0 to 70% and the corresponding pea yield increased from 0 to 5%. Yield increase was limited because of significant mortality of pea plants. In the past, pea and lentil production in the PNW has not been profitable because of the high insecticide and herbicide costs (Young, F.L. et al., 1994a). The most profitable weed management system in pea cost $85/ha/yr (Young, D.L. et al., 1994). Mechanical tillage in pea and lentil has been examined as an alternative to costly herbicide programs. Although crop injury was minimal, harrowing or rotary hoeing in spring dry pea or lentil in the PNW did not reduce weed

Table 1. . Herbicide resistant weeds in cool season food legumes.[a]

ethalfluralin	*Sorgum halepense*	Pea	USA
imazethapyr	*Amaranthus hybridus*	Pea	USA
imazethapyr	*Amaranthus palmeri*	Pea	USA
imazethapyr	*Amaranthus rudis*	Pea	USA
imazethapyr	*Galeopsis tetrahit*	Pea	Canada
imazethapyr	*Lolium rigidum*	Pea	Australia
imazethapyr	*Raphanus raphanistrum*	Pea	Australia
imazethapyr	*Sisymbrium orientale*	Pea	Australia
imazethapyr	*Xanthium strumarium*	Pea	USA
MCPA amine	*Cirsium arvense*	Pea	Sweden, Hungary
MCPA amine	*Sinapis arvensis*	Pea	Canada
metolachlor	*Lolium rigidum*	Pea	Australia
metribuzin	*Amaranthus spp.*	Lentil	Bulgaria, France, Israel, Switzerland
metribuzin	*Amaranthus spp.*	Pea	Bulgaria, France, Israel, Switzerland
metribuzin	*Arenaria serpyllofolia*	Lentil	France
metribuzin	*Arenaria serpyllofolia*	Pea	France
metribuzin	*Chenopodium album*	Lentil	Bulgaria
metribuzin	*Chenopodium album*	Pea	Bulgaria
metribuzin	*Sinapis arvensis*	Lentil	Canada
pendimethalin	*Sorgum halepense*	Lentil	USA
pendimethalin	*Sorgum halepense*	Pea	USA
pronamide	*Avena fatua*	Pea	USA
sethoxydim	*Avena fauta*	Lentil	USA, Canada, Australia
sethoxydim	*Avena fauta*	Pea	USA, Canada, Australia
sethoxydim	*Lolium rigidum*	Pea	Australia
triallate	*Avena fatua*	Lentil	Canada
triallate	*Avena fatua*	Pea	Canada
trifluralin	*Lolium rigidum*	Pea	Australia
trifluralin	*Seteria viridis*	Pea	Canada
trifluralin	*Sorgum halepense*	Pea	USA

[a]/Information provided by Dr. Steve Seefeldt, Chairperson, Herbicide Resistance Committee, Western Society Weed Science, USA.

populations sufficiently (Boerboom and Young, 1994). In lentil, tillage did not improve or supplement weed control provided by a single application of a broadleaf herbicide. It was concluded that in-crop tillage in pea and lentil would not be effective or feasible because of frequent spring showers, large areas, and lack of equipment and labor for timely operations.

Cultural

The suppression of weeds by cultural methods may be the most common and economically feasible non-chemical technique available to growers. Cultural methods may include crop rotations, planting competitive crops and/or cultivars, intercropping, using proper planting equipment, adjusting row spaces, and increasing seeding rate. Before the use of synthetic herbicides, weeds were managed effectively by using clean seed, cultivation, and crop rotation (Froud-Williams, 1988). Crop rotation uses crop monocultures in a rotational sequence that utilizes the combined weed-suppressive effects of different crops within a single season (Liebman, 1988) and prevents the population increases of any single weed species (Walker and Buchanan, 1982). Growing the same crop, with the same management practices, year after year favors weed species that mimic the life cycles and growth habits of the crop. In the PNW, a rotation of winter wheat-spring barley (*Hordeum vulgare*) - spring pea reduced dramatically the winter annual grass downy brome compared to a rotation of winter wheat-winter wheat-spring wheat (*T. aestivum*) (Young, et al., 1996).

When winter wheat followed winter wheat, downy brome populations increased to more than 150 plants m^{-2} but, in the 3-yr rotation, downy brome populations never exceeded 20 plants m^{-2} in winter wheat.

The characteristics of competitive crops include early emergence and seedling establishment, plant vigor, disease resistance, canopy architecture, relative growth rate, root growth, use of nutrients, and tolerance of environmental extremes (Cardina, 1995; Callaway, 1992). Food legumes tend to be poor competitors because of their short stature, slow rate of early growth, and lodging at maturity (Knott and Halila, 1988; Papendick et al., 1988). In the past, plant breeders have not bred or developed plants specifically to suppress weeds (Callaway, 1992). The breeders cannot be held completely responsible for this shortcoming because weed scientists have rarely provided information on plant traits that would increase crop competitiveness. Most traits today will be developed and bred into varietal programs that increase yield and end use quality or disease and insect resistance - traits that will indirectly improve crop competitiveness. Rarely do breeders assess their varieties in the presence of weeds, unlike their assessments of disease and insect resistance. More weed science/plant breeding interactions must take place to advance breeding for crop competitiveness in the cool season food legumes.

Another cultural practice that is assessed rarely for weed competition is intercropping. Normally, the goal of intercropping is to improve crop yield of the mixture compared to each crop individually, but growers may also benefit indirectly from the suppression of weeds (Liebman, 1988). Crop yields and weed suppression in intercropping systems depend on many factors including total crop density, species richness, crop and weed species, planting arrangement, soil fertility, environment and location (Liebman, 1988). When adequate moisture was available, the yield of barley-pea intercrops was almost 38% more than the highest yielding sole crop, yet weed suppression, as measured by biomass, in the intercrop was never more than the weed suppression in the most competitive sole crop (Liebman, 1986). An additional study indicated that weed growth was suppressed 52% in a pea crop, 86% in a barley crop, and 78% in the intercrop (Mohler and Liebman, 1987). Thus a weed-suppressive crop (e.g. barley) may significantly benefit weed management efforts in less weed-suppressive crops (e.g. pea). The use of intercropping can be an excellent example of studying complete IPM systems. For example, leafless and semi-leafless peas reduce disease (Wall and Townley-Smith, 1996) and insect incidence (Clement, S. personal communication,) but are not as competitive as traditional-leafed peas (Wall and Townley-Smith, 1996). However, there may be a compromise of traits between crops with intercropping and the system might be economically and biologically feasible. Intercropping leafless and semi-leafless pea with barley may increase weed-suppression compared to the sole legume crop while these pea cultivars would require fewer pesticide inputs, for insect and disease management.

Precision planting will also improve weed competition by the grain legumes. Normally however, growers use grain drills designed for planting cereals even though this equipment is inappropriate (Papendick, et al.,1988). Conventional planting with grain drills often leave 5 to 15-cm spaces between legume plants in a row and seeding depth varies greatly. Weeds grow prolifically in open spaces where planting skips occur. Uniform planting depth is important for early, uniform crop emergence and placement of seed below the soil layer where preplant herbicides have been incorporated. Uniform spacing with precision drills may also increase radiation interception (Heath et al., 1994). Even if precision planting and a more uniform stand does not increase yield of some legume crops or cultivars (Heath et al., 1994) this practice may increase competition between the crop and associated weeds (Liebman, 1988).

Many similar principles pertaining to precision planting apply to row spacing and seeding rate. In general, weed suppression is increased when crops are planted in narrow rows compared to wide rows (Liebman, 1988) although there are exceptions. For example, weed competition for pea and lentil was similar for both 17 and 34 cm - rows in Chile (Diaz and Penaloza, 1995). In lupin, however, both crop yield and weed competition was increased in narrow (15 cm) rows compared to wide (76 cm) rows (Putnam et al., 1992). Cool season food legumes exhibit compensatory growth and yield may not increase with increased seeding rates. However, higher seeding rates have the potential to increase interspecific competition. Increased seeding rates of lentil (McKenzie et al., 1989; Boerboom and Young, 1994), and pea (Wall and Townley-Smith, 1996; Boerboom and Young, 1994; Townley-Smith and Wright, 1994) decreased weed competition. Increasing density of faba bean did not affect dry weight of broomrape (O. crenata) but the number of attachments onto one genotype was positively correlated with crop density (Manschadi et al., 1997). Because of high seed prices, the cost of increasing the seeding rate must be considered in the production system. Even though increased pea and lentil populations reduced weed biomass and subsequent competition, it is suggested that herbicides continue to be used for economical and effective weed management (Boerboom and Young, 1994). This suggestion implied using the increased seeding rate for entire fields would not be economically feasible. However, some growers in the PNW double the seeding rate for weed suppression and yield increases only in low-lying areas where soil is moist and weed competition is severe.

INTEGRATED WEED MANAGEMENT: CASE STUDIES

It is estimated that over 16 million ha in West Asia and the Mediterranean region are infested with broomrape (Sauerborn, 1991). In a recent review, Pieterse et al. (1994), discussed the integrated management of broomrape (*Orobanche* spp.) in food legumes focusing on cultural, physical, and chemical methods for managing this parasitic weed. Effective management strategies were delayed seeding, hand weeding, crop rotations, solarization, increased fertility, decreased herbicide rates and use of tolerant hosts. New single component research on broomrape continues, which can then be incorporated into IWM systems. Trap crops are included as part of a rotation, because they stimulate the germination of broomrape seed, are not infected themselves (Musselman, 1980) and cause a significant decline of the weed seed in the soil (Schnell et al. 1994). Schnell et al. (1994) evaluated 14 different trap crops as well as fallow and found that pea growth was greatest following legume crops and least following fallow. Of the legume crops, faba bean was the best trap crop and induced the highest reduction of broomrape seeds. In addition to individual trap crops within a rotation, another strategy is to plant a mixture of plant species (Bouhatous and Jacquard, 1994). As a result of germination inhibition studies, Al-Menoufi et al., (1996) concluded that fenugreek (*Trigonella foenum graecum*), lupin (*Lupinus termis*), coriander (*Coriandrum satifum)*, and turnip (*Brassica rapa*) could be intercropped with susceptible economic host plants to reduce broomrape infection in fields.

To implement successful management strategies, scientists must conduct biological and ecological studies including weed surveys. In northern Tunisia, the area of infestation of the red-flowered *O. foetida* has been determined (Kharrat et al., 1992). This species is highly aggressive and is resistant to crop lines that tolerate *O. crenata*. This is important information because subsequent control measures must be identified and implemented. In this regard, sources of resistance to *O. foetida* have now been found within material distributed by ICARDA. (Kharrat and Halila, 1994).

In southern Australia, annual ryegrass (*Lolium rigidum*, Gaud.) is a winter annual weed of all crops. This weed has developed multiple resistance to commonly used selective herbicides. Resistant annual ryegrass infests 4,000 to 5,000 farms (Matthews and Powles, 1996) and the initial frequency of resistant plants can be as high as 2% (Matthews and Powles, 1992). Because of this herbicide resistance, researchers have been investigating an IWM system for the successful management of annual ryegrass. Strategies include delayed sowing of the crop, growing competitive crops, and catching weed seed at harvest (Matthews and Powles, 1996). Delayed sowing of peas reduced ryegrass plants and seeds by 55% and 25%, respectively, compared to normal sowing times. Even though pea is not as competitive as other crops, including them in a rotation increases the competitiveness of subsequent cereal crops by decreasing cereal diseases. Southern Australian farmers have also initiated crop topping, the practice of in-crop use of paraquat to reduce the development of viable seeds by weeds in pulse crops and they are including legume green manures in rotations (Matthews and Powles, 1996). The use of IWM for annual ryegrass has been cost-effective, successful, and crop yields have been maintained compared to herbicide-treated weed free controls.

In contrast to the first two case studies that integrated strategies for the management of a single weed, a study was conducted in the U.S., PNW, to investigate weed management, tillage, and crop rotation on weed complexes, soil erosion and profit and risk of 12 farming systems (Boerboom et al., 1993; Young, D.L. et al., 1994; Young, F.L. et al., 1994a,b). The study was a long-term, large-scale project that used farm-size machinery, focused on weeds, and was designed experimentally to allow careful and detailed statistical comparisons. Including spring dry pea in a 3 year conservation tillage crop production system, diseases of wheat and pea were reduced (Young, F.L. et al, 1994b), winter annual grass weed populations were decreased (Young, F.L. et al., 1996), and farm income was stabilized (Young, D.L. et al., 1994). In the conservation pea production system, herbicide use and costs were similar to conventional pea production systems (Young, F.L. et al., 1994a). Of utmost importance was the reduction in erosion potential when spring pea was produced. In the mid-1980's growers commonly tilled their pea ground six to eight times. In this study tillage operations were reduced to three or four, with the complete elimination of moldboard plowing (Young, F.L. et al., 1994a). Because of the results of this study, growers and researchers are examining the potential of no-till legumes in crop production systems. When evaluated as an individual crop, pea production is costly and has low returns. More research needs to be conducted to further reduce production costs and enhance profits.

NEW STRATEGIES

The development of herbicide resistant crops (HRC) is the most intensively exploited area of plant biotechnology (Duke, 1996). Normally, the herbicides used in a crop are ones to which the crop is naturally resistant, however with HRC's, crops are bred to tolerate a specific herbicide (Duke, 1996). Because legumes are considered minor crops and fewer herbicides have been developed for use in legumes, HRC's could reduce the cost of weed management. Pre-emergence herbicide applications in legumes are expensive, applied at high rates, and often require tillage for incorporation. Use of an HRC will probably increase the use of the herbicide to which the crop is tolerant, but hopefully this herbicide is environmentally safer and cheaper than the herbicide(s) it will replace (Duke, 1996). HRC crops may have their uses in the present weed management strategies, however, it must be realized that HRC's are just an additional tool to be integrated into a complete management system. At the present time, research is being conducted on herbicide resistant lupin (Schmidt and Pannell, 1996), lentil and chickpea.

In the PNW, acceptable weed control is difficult to obtain in dry peas because herbicides are often ineffective and their costs are high. There is a need to balance herbicide cost with adequate weed control in this crop (Beus et al., 1990). A bioeconomic decision model based on 6 years of data from the IWM/IPM field experiment conducted in the PNW has been developed to identify optimal herbicide treatments in dry pea production systems (Kwon, 1993). Before this model, there was no economic weed management analysis for dry pea or other food legumes. A decision-aid model that optimizes profits through the use of appropriate herbicides and rates could benefit growers (Young, D.L. personal communication) by reducing herbicide usage and cost.

In the future, the use of precision agriculture may also reduce the use of herbicides. Geographic information systems and global positioning systems will be used to map weed infestations in fields and then precisely apply herbicides (Duke, 1996).

SUMMARY

In general, changing only one management practice has very little effect on weed dynamics. However, when we integrate several management methods, we can utilize techniques that individually are considered ineffective. It is virtually impossible to discuss integrated weed management without considering the whole crop production system.

Sufficient funding is a major constraint to research. In developed countries the majority of funding for grain legume research is for advancements in breeding, genetics, and biotechnology. Weed management receives few funds, while production agriculture examining long-term cropping systems receives almost no funds. It is our opinion that significant advances in weed management strategies for cool season food legumes will require the redesigning of present cropping systems. Investment in large-scale, long-term systems studies is imperative.

Acknowledgments

I would like to thank E. Gallandt and S. Clement for their critical and thorough review of the manuscript. I would like to acknowledge the assistance of S. Driessen and K. Hemmesch in the preparation of the manuscript and securing literature, respectively.

References

Akobundu, I.O. 1991. *Weed Technology*. 5:680-690.

Al-Menoufi, O.A., Adam, M.A., and El-Safwani, N.A. 1996. In: *Proceedings of The 6th International Parasitic Plant Symposium*. 417-423. (eds. M.T. Moreno, J.I. Cubero, D. Berner, D. Joel, L.J. Musselman, and C. Parker). Cordoba, Spain.

Berti, A. and Zanin, G. 1994. *Weed Research*. 34:327-332.

Beus, C.E., Bezdicek, D.F., Carlson, J.E., Dillman, D.A., Granatstein, D., Miller, B.C., Mulla, D., Painter, K., and Young, D.L. 1990. *Prospects for Sustainable Agriculture in the Palouse: Farmer Experiences and Viewpoints*. Washington State University Station Bulletin, XB1016. Washington State University, Pullman, Washington.

489

Boerboom, C.M., Young, F.L., Kwon, T., and Feldick, T. 1993. *IPM research project for inland Pacific Northwest wheat production*. Agriculture Research Center Bulletin X131029, Washington State University, Pullman, Washington. 46 pp.

Boerboom, C.M. and Young, F.L. 1995. *Weed Technology*. 9:99-106.

Bouhatous, B. and Jacquard, P. 1994. In: *Proceedings of the Third International Workshop on Orobanche and Related Striga Research*. 320-333. (eds. A.H. Pieterse, J.A.C. Verkley, and S.J. terBorg) Amsterdam, The Netherlands.

Brebaum, S., and Boland, G.J. 1995. *Canadian Journal of Plant Science*. 75(4):841-849.

Bridges, D.C. and R.L. Anderson. 1992. In: *Crop Losses Due to Weeds in the United States*. Weed Science Society of America, Champaign, Illinois.

Callaway, M.B. 1992. *American Journal of Alternative Agriculture*. 7(4):168-179.

Cardina, J. 1995. In: *Handbook of Weed Management Systems*. 279-341. (ed. A.E. Smith) New York, USA: Marcel Dekker, Inc.

Cousens, R. 1987. *Plant Protection Quarterly*. 2(1):13-20.

Davies, D.R., Berry, G.J., Heath, M.C. and Dawkins, T.C.K. 1985. In: *Grain Legume Crops*;. 147-198. (eds. R.J. Summerfield and E.H. Roberts) London: Collins.

Day, B.E. 1972. *Pest Control: Strategies for the Future*.. 330-338. National Academy of Sciences, Washington D.C..

Diaz, J.S. and Penaloza, H.E. 1995. *Agricultura Tecnica*. 55:176-182.

Duke, S.O. 1996. In: *Herbicide Resistant Crops, Agricultural, Environmental, Economic, Regulatory, and Technical Aspects*. 1-10. (ed. S.O. Duke). CRC Press, New York, Lewis Publishers.

Elmore, D.L. 1996. *Weed Science*. 44:409-412.

Froud-Williams, R.J. 1988. In: *Weed Management in Agroecosystems : Ecological Approaches*. 213-236. (ed. M. Liebman and M. Altieri). Boca Raton, Florida. CRC Press, Inc.

Gilbey D.J. 1993 *Management of agricultural weeds in Western Australia* pp106-110 (Eds J Dodd, RJ Martin and KM Howes) Western Australia Department of Agriculture Bulletin 4243

Gill G 1994 *Proc First Australian Lupin Technical Symposium* pp324-326 Western Australian Department of Agriculture.

Heath, M.C., Pilbeam, C.J., McKenzie, B.A. and Hebblethwaite, P.D. 1994. In: *Expanding the Production and Use of Cool Season Food Legumes*. 771-790. (ed. F.J. Muehlbauer and W.J. Kaiser). Dordrecht, The Netherlands: Kluwer Academic Publishers.

Jurado-Exposito, M., Castejon-Munoz, M., and Garcia-Torres, L. 1996. *Weed Technology*. 10:774-780.

Kennedy, A.C., Elliot, L.F., Young, F.L. and Douglas, C.L. 1991. *Soil Science Society of America Journal*. 55:722-727.

Kharrat, M., Halila, H.M., Linke, K.H., and Haddar, T. 1992. *FABIS* Newsletter. 30:46-47.

Kharrat, M. and Halila, H.M. 1994. In: *Third International workshop on Orobanche and Related Striga Research*. 639-643. (eds. A.H. Pieterse, J.A.C. Verkleij and S.J. terBorg). Amsterdam, The Netherlands.

Knott, C.M. and Halila, H.M. 1988. In: *World Crops: Cool Season Food Legumes*. 535-548. (ed. R.J. Summerfield). Dordrecht, The Netherlands:Kluwer Academic Publishers.

Kwon, T.J. 1993. Ph.D. dissertation, Department of Agricultural Economics, Washington State University, Pullman, Washington. 355 pp.

Lemerle, D. and Hinkley, R.B. 1991. *Australian Journal of Experimental Agriculture*. 31:379-386.

Liebman, M. 1986. Ph.D. dissertation, Department of Botany, University of California, Berkeley.

Liebman, M. 1988. In: *Weed Management in Agroecosystems: Ecological Approaches*. 197-212. (eds. M. Liebman and M. Altieri). Boca Raton, Florida. CRC Press, Inc.

Lindquist, J.L., Mortensen, D.A., Clay, S.A., Schmenk, R., Kells, J.J., Howatt, K. and Westra, P. 1996. *Weed Science*. 44:309-13.

Linke, K.H., Sauerborn, J. and Saxena, M.C. 1992. In: *Proceedings of the Eighth International Symposium on Biological Control of Weeds*. (eds. E.S. Delfosse and R.R. Scott). Melbourne, Australia: DSIR/CSIRO.

Manschadi, A.M., Sauerborn, J., Kroschel, J., and Saxena, M.C. 1997. *Weed Research*. 37:39-49.

Matthews, J.M. and Powles, S.B. 1992. *Pesticide Science*. 34:365-367.

Matthews, J.M., Llewellyn, R., Powles, S.B., Reeves, T., Jaeschke R. 1996. *Eighth Australian Agronomy Conference*. 684-685. (ed. M. Asghar). University of Southern Queensland, Australian Society of Agronomy.

Matthews, J.M. and Powles, S.B. 1996. *Tenth Australian Weed Conference*. Victorian Weed Science Society, Melbourne, Australia.

McCue, A.S. and Minotti, P.L. 1979. *Proceedings Northeastern Weed Science Society*. 33:106.

McKenzie, B.A., Miller, M.E. and Hill, G.D. 1989. *Proceedings Agronomy Society*, N.Z. 19:11-16.

Mohler, C.L. and Liebman, M. 1987 *Journal of Applied Ecology*. 24:685.

Musselman, L.J. 1980. *Annual Review of Phytopathology*. 18:463-489.

O'Donovan, J.T. 1996. *Phytoprotection*. 77(1):13-28.

Ogg, Jr., A.G., Stephens, R.H., and Gealy, D.R. 1993. *Weed Science*. 41:394-402.

Ogg, Jr., A.G., Smiley, R.W., Pike, K.S., McCaffrey, J.P., Thill, D.C. and Quisenberry, S.S. In press. In: *Advances in Conservation Farming* (eds. A.G. Ogg, Jr., E. Michalsen, and R.I. Papendick). Ankeny, USA: Soil and Water Conservation Society.

Papendick, R.I., Chowdhury, S.L., and Johansen, C. 1988. In: *World Crops: Cool Season Food Legumes.* 237-255. (ed. R.J. Summerfield). Dordrecht, The Netherlands: Kluwer Academic Publishers.

Penner, D., Leep, R.H., Roggenbuck, F.C. and Lempke, J. R. 1993. *Weed Technology.* 7:42-46.

Pieterse, A.H., Garcia-Torres, L., Al-Menoufi, O.A., Linke, K.H. and Ter Borg, S.J. 1994. In: *Expanding the Production and Use of Cool Season Food Legumes.* 695-702. (eds. F.J. Muehlbauer and W.J. Kaiser). Dordrecht, The Netherlands: Kluwer Academic Publishers.

Putnam, D.H., Wright, J., Field, L.A. and Ayisi, K.K. 1992. *Agronomy Journal.* 84:557-563.

Rasmussen, J. 1992. *Weed Research.* 33:231-240.

Rice, E.L. 1983. *Allelopathy,* 2nd ed., New York, USA : Academic Press.

Sauerborn, J. 1991. *Fifth International Symposium on Parasitic Weeds.* Nairobia, Kenya. 137-143.

Schmidt, C.P. and Pannel, D.J. 1996. *Crop Protection.* 15:539-548.

Schnell, H., Linke, K.H. and Sauerborn, J. 1994. *Trop. Science.* 34:306-314.

Shaw, W. 1982. *Weed Science.* 30 (Suppl.):(2):2-12.

Thill, D.C., Cochran, V.L., Young, F.L., and Ogg, Jr., A.G. 1986. In: *STEEP Conservation Concepts and Accomplishments.* 275-287. (ed. F.L. Elliot). College of Agriculture, Washington State University, Pullman, Washington, USA.

Thill, D.C., Lish, J.M., Callihan, R.H. and Bechinski, E.J. 1991. *Weed Technology.* 5:648-656.

Townley-Smith, L. and Wright, A.T. 1994. *Canadian Journal of Plant Science.* 74:387-393.

Wall, D.A. and Townley-Smith, L. 1996. *Canadian Journal of Plant Science.* 76:907-914.

Walker, R.H. and Buchanan, G.A. 1982. *Weed Science.* 30 (Suppl.):17.

Walker, R.H. 1995. In: *Handbook of Weed Management Systems.* 35-49 (ed. A.E. Smith) New York, USA: Marcel Dekker, Inc.

Wiedenhoeft, M.H. and Ciha, A.J. 1987. *Agronomy Journal.* 79:999-1002.

Wilson, B.J. and Cussans, G.W. 1972. In:*Proceedings 11th British Weed Control Conference.* 573-577. Croydon:BCPC.

Wyse, D.L. 1994. *Weed Technology.* 8:403-407.

Young, D.L., Kwon, T.J., and Young, F.L. 1994. *Journal Soil and Water Conservation.* 49:601-606.

Young, F.L. 1996. *Weed Science.* 34:901-905

Young, F.L., Ogg, Jr., A.G., Boerboom, C.M., Alldredge, J.R. and Papendick, R.I. 1994a. *Agronomy Journal.* 86:868-874.

Young, F.L., Ogg, Jr., A.G., Papendick, R.I., Thill, D.C. and Alldredge, J.R. 1994b. *Agronomy Journal.* 86:147-154.

Young, F.L., Ogg, Jr., A.G., Thill, D.C., Young, D.L. and Papendick, R.I. 1996. *Weed Science.* 44:429-436.

Zimdahl, R.L. 1995. In: *Handbook of Weed Management Systems.* 1-18. (ed. A.E. Smith) New York, USA: Marcel Dekker, Inc.

INTEGRATED CONTROL OF NEMATODES OF COOL SEASON FOOD LEGUMES

J.P. Thompson[1], N. Greco[2], R. Eastwood[3], S.B. Sharma[4] and M. Scurrah[5]

1 Queensland Wheat Research Institute, Department of Primary Industries, P.O. Box 2282, Toowoomba 4350, Australia; 2 Istituto di Nematologia Agraria,CNR, Via Amendola 165/A, 70126 Bari, Italy; 3 Victorian Institute for Dryland Agriculture, Private Bag 266, Horsham 3401, Australia; 4 Legume Program, ICRISAT, Patancheru P.O., Andhra Pradesh 502 324, India; 5 Plant Research Centre, SARDI, P.O. Box 397, Adelaide 5001, Australia.

Abstract.

Production of cool season food legumes can be severely limited by nematode attack. Symptoms are yellowing, wilting, stunting, decreased biomass and seed yield. The most damaging nematodes are root-knot (*Meloidogyne* spp.), cyst (*Heterodera* spp.), root-lesion (*Pratylenchus* spp.) and stem (*Ditylenchus dipsaci*). Integrated control is required where profit margins and environmental considerations preclude the use of nematicides. The main factors for effective integrated control are: correct diagnosis of the nematode problems, use of tolerant and resistant cultivars of the main crops, rotation with resistant cultivars of other crops, fallowing, control of weed hosts, choice of sowing time, soil amendment, and sanitation. Present knowledge and future requirements for effective integrated control of the main nematode diseases of each of the cool season food legume crops are discussed.

INTRODUCTION

Nematodes are severe constraints of many crops all over the world. Attacked plants have inefficient root systems and are much more sensitive to biotic and abiotic stresses, such as drought and infertile soil. They may show symptoms of yellowing, wilting, stunting, decreased biomass and poor seed yield. Nematodes may also interact with other soil-borne pathogens increasing disease severity. They may inhibit root nodulation of legumes by rhizobia, thereby limiting nitrogen fixation in cropping systems, and they may reduce the nutritional quality of grain.

The cool-season food legumes, chickpea (*Cicer arietinum* L), pea (*Pisum sativum* L.), faba bean (*Vicia faba* L.), and lentil (*Lens culinaris* Medik.) are parasitised by many species of nematodes. The most damaging genera are the root-knot (*Meloidogyne* spp.), cyst (*Heterodera* spp.), root-lesion (*Pratylenchus* spp.) and stem nematodes (*Ditylenchus dipsaci*). On the basis of responses by 371 nematologists to a worldwide survey, Sasser and Freckman (1987) reported an average yield loss from nematodes of 12.3% for all crops and 13.7% for chickpea, valued at US $328 million (Sharma et al., 1992). Because control with nematicides is largely uneconomic for field crops, other methods must be sought and applied. Often each method is only partially effective and a combination of methods is required in various farming systems to achieve integrated control. These methods principally involve crop rotation with resistant (non-host) or partially resistant and tolerant crop species or cultivars of the same species. Strategies are adopted so that an intolerant crop cultivar is not grown more frequently than about once every 3 years. Rotations should involve economically viable or otherwise useful crops and this becomes more difficult with polyphagous nematode species with wide

R. Knight (ed.), Linking Research and Marketing Opportunities for Pulses in the 21st Century, 491–566.
© *2000 Kluwer Academic Publishers. Printed in the Netherlands.*

host ranges. Other strategies may be invoked, such as altering the sowing date to when the soil is at a less favourable temperature for reproduction of the particular species or supplying supplementary fertiliser or irrigation to offset effects of root damage. Sanitation to exclude nematode introduction into clean fields via infested soil, seed or runoff water should be practised.

During the last two decades, awareness of the impact of nematodes on cool season food legumes has increased, and a number of investigations have been undertaken at national and international research centres. This has resulted in more insights on nematode problems of food legumes and their management, which have been partly discussed in previous conferences (Greco and Di Vito, 1993; Sharma et al., 1994), book chapters (Sikora and Greco, 1990; Riggs and Niblack, 1993) and a monograph (Ali, 1995). The present knowledge of nematode problems of cool season food legumes and integrated control strategies will be discussed here.

CHICKPEA

Reports on numbers of pathogens, particularly the plant parasitic nematodes, associated with chickpea have markedly increased over the past 17 years (Nene et al., 1996). Sharma (1985) reported 46 nematode species or genera found associated with chickpea all over the world and Ali (1995) listed 97 species mainly from the Indian sub-continent and Mediterranean basin. However, several of the reported nematode species were found in the chickpea rhizosphere and there is no certainty that they feed on chickpea. Therefore, emphasis will be placed only on species with confirmed pathogenicity to chickpea.

Root-knot nematodes (*Meloidogyne* spp.)

These sedentary endoparasitic nematodes are distributed worldwide and are considered the major nematode group economically, because of the magnitude of damage caused to different crops and their wide host ranges. Roots of severely infested plants show large galls, poor nodulation and rotting. Interaction of these nematodes with soil-borne fungi has been demonstrated. Three species of root-knot nematodes are important pathogens of chickpea, *Meloidogyne artiellia* Franklin, *M. incognita* (Kofoid et White) Chitwood and *M. javanica* (Treub) Chitwood.

Meloidogyne artiellia occurs in the Mediterranean region and causes severe damage to chickpea in Italy, Spain and particularly Syria. The nematode develops well in the temperature range 15-25 °C and only one generation per growing season occurs. The female stage is attained after an accumulation of about 240 degree-days above 10 °C. It survives in soil during hot, dry summer months as eggs or anhydrobiotic second-stage juveniles (Di Vito and Greco, 1988b), with an average population decline of only 13% by the next autumn. Galls caused by this species are rather small or absent but egg masses (about 500 eggs in a large gelatinous matrix) are easily visible on the roots.

M. artiellia reproduces very well on cereals, cruciferous and leguminous plants. However, maize and the legumes, cowpea, lupin and sainfoin are non-hosts, while oats, french bean, lentil and soybean are poor hosts. The typical rotation of the Mediterranean area, which alternates cool season food legumes with winter cereals, is inappropriate in fields infested with this nematode.

Tolerance limits to the nematode of 0.1 and 0.01 eggs/cm^3 soil have been reported for winter and spring chickpea, respectively, with no yield in fields infested with 8 or 1 eggs/cm^3 soil respectively. The reproduction factor (ratio of the final to the initial population of the nematode) after growth of wheat, winter chickpea and spring chickpea, was 189, 55 and 3 respectively (Di Vito and Greco, 1988a).

Control

Although no chemical control has been attempted against this nematode, the use of nematicides should be effective. However, their use is not recommended as chemicals are expensive and are environmental hazards. Soil solarization also would be very effective in most of the Mediterranean countries but also too expensive for chickpea. Therefore, the management of the nematode should be based on proper crop rotations. In warm and irrigated areas, rotation of chickpea with summer crops is a useful option. Winter chickpea should be preferred over spring chickpea, which suffers more damage. Fallow can be effective but not economic in many situations.

Cultivars of chickpea, resistant to *M. artiellia,* are not available. However, resistance to the nematode has recently been found in the accessions ILWC 64 of *Cicer bijugum* Rech. and ILWC 92 of *C. pinnatifidum* Jaub et Sp., in ICARDA's collection (Di Vito et al., 1996b), but unfortunately these *Cicer* species cannot be crossed with *C. arietinum.* Therefore, it is necessary to search further for resistance in wild germplasm, compatible with *C. arietinum,* or find ways to transfer resistance from incompatible species.

Meloidogyne javanica and *M. incognita*

These two species are serious pests of chickpea (Sharma and McDonald, 1990). They are widespread in the chickpea growing regions of Asia, Africa and South America (Nene et al., 1996). However, most farmers in these countries are unaware of these insidious pests in their soils and damage to chickpea crops is often confused with declining soil fertility or micronutrient deficiencies. Stunting of plants, uneven crop growth, yellowing and bronzing of leaves, delayed flowering and podding, and reduction in number of pods per plant are the above-ground symptoms associated with nematode infection. Symptom expression varies with nematode population densities and chickpea genotypes. Patches of stunted plants in a nematode-infested soil appear earlier in infertile, moisture deficient sandy soils with low pH (Sharma et al, 1992). The characteristic root-swellings (galls) produced from attack by root-knot nematodes are often mistaken for rhizobia nodules. Galls produced at 25-30°C are 30-35% larger than those produced at 15-20°C. Many developing countries have inadequate human resources, trained in nematology, to help farmers identify and manage the damage caused by these nematodes.

The damage threshold levels of *M. incognita* and *M. javanica* range between 0.2 and 2.0 juveniles per cm^3 soil at the time of sowing. Trials conducted in India between 1989 and 1994 revealed that *Meloidogyne* spp may cause loss in yield from 17% to 60% depending on the nematode population levels at planting time (Ali, 1995).

These species complete their life cycle in about 4 weeks at 25-30°C producing several generations in a crop season. Association of these nematodes with *Fusarium oxysporum f. sp. ciceri* advances the onset of wilt and increases wilt incidence. Co-infection of *F. oxysporum* with *M. javanica* moderates the wilt resistance in chickpea cultivars (Maheswari et al, 1995). These nematode species interact with other species of *Fusarium, Rhizoctonia and Sclerotium.* Formation of rhizobia nodules on chickpea roots is also suppressed by these root-knot nematode species.

Control

Summer fallow, solarisation with transparent polyethylene sheets during hot summer months, and seed treatment with biocides such as aldicarb, carbofuran and phorate reduces the population densities of these nematode species (Sharma et al., 1992). The traditional agronomic practices in some parts of India of keeping the land fallow during summer months, deep ploughing, and addition of organic manure to the soil have suppressive effects on the nematodes.

At the International Crops Research Institute for the Semi-Arid Tropics (ICRISAT) Asia Center, all of the 5,000 accessions of chickpea and 40 accessions of wild species were susceptible to *M. javanica* (Sharma et al, 1993). However, some chickpea accessions (N 31, N 59, ICCC 42 and ICCV 90043) have been identified with tolerance to the nematode (Sharma et al, 1995). Sesame, mustard and winter cereals are poor hosts and a 2-3 year rotation of chickpea with these crops may be useful management of these root-knot nematodes for chickpea production.

Cyst forming nematodes.

The most important cyst nematode attacking chickpea is *Heterodera ciceri* Vovlas, Greco et Di Vito. This species is present in Syria, Turkey and recently has been found in Jordan and Lebanon. *H. ciceri* develops better when soil temperature is in the range 15-25°C and usually completes only one generation per growing season of chickpea after an accumulation of 370 degree days (Kaloshian et al., 1986). A reproduction factor of about 250 on chickpea and a population decline of 35% in the absence of a host for seven months were determined in microplots (Greco et al., 1988). Other good hosts of this nematode besides chickpea, are lentil, pea and grass pea (*Lathyrus sativus*). Reproduction on other leguminous species is poor or nil. The tolerance limit of chickpea to the nematode is about 1 egg/cm^3 soil and complete crop failure must be expected in soil infested with 32 or more eggs/cm^3. Moreover, a 20% reduction of seed protein content occurs at high population densities (Greco et al., 1988).

Control

Split application of 10 kg aldicarb/ha and 6-8 weeks of soil solarization significantly reduced root infestation by *H. ciceri* and increased yield of spring chickpea (Di Vito et al., 1991). Other nematicides are also expected to be effective against *H. ciceri*, but they too are expensive and are pollutants. Because of its narrow host range, the control of this cyst nematode should be based primarily on crop rotations. In Syria, chickpea yields increased by 2.1, 4.5 and 7.9 times in fields where cultivation of host crops was suspended for one, two or three years, respectively, in comparison with continuous chickpea cropping (Saxena et al., 1992). Fallowing is rather common in Turkey and Syria and would reduce the soil population density of the nematode by 35-50% per year.

The use of resistant cultivars would be the easiest, simplest and most cost effective way to control *H. ciceri*. Unfortunately, no resistant chickpea cultivar is available. Screening of about 10,000 lines of *C. arietinum* of the ICARDA's germplasm showed no resistance, but among the annual wild chickpeas most lines of *C. bijugum*, six of *C. pinnatifidum* and one of *C. reticulatum* were highly resistant to the nematode (Singh et al., 1989; Di Vito et al., 1996a). Although lines of the first two *Cicer* spp are incompatible with *C. arietinum*, *C. reticulatum* can be crossed and a resistant line of this species has been registered as ILWC 292 in the ICARDA germplasm collection (Singh et al., 1996). It is being used in a breeding programme at ICARDA to transfer resistance to kabuli cultivars with promising results.

Root lesion nematodes

The most important root-lesion nematodes that attack chickpea are species of *Pratylenchus*. All are migratory endoparasites of the root cortical parenchyma in which they cause large necrotic areas that may coalesce and affect most of the root. These symptoms can be confused with blackening caused by other stresses and, therefore, extraction of the nematodes from the roots is necessary for proper diagnosis.

Above-ground symptoms are similar to those described above for attack by other nematodes. Usually, these symptoms are not as severe as those caused by root-knot or cyst nematodes, but they were so in North Africa, Lebanon and Turkey (N Greco unpublished data). However, *Pratylenchus* spp. are more widespread than the previous nematodes. In many Mediterranean countries all chickpea crops are infested with one or more species of *Pratylenchus*. This indicates that the economic damage to chickpea caused by root-lesion nematode, on a country or regional basis, would be larger than that caused by the other nematodes. Among root-lesion nematodes, *P. thornei* Sher et Allen appears to be cosmopolitan, while *P. mediterraneus* Corbett, *P. penetrans* (Cobb) Filipjev et Schuurmans Stekhoven and *P. neglectus* (Rensch) Filipjev et Schuurmans Stekhoven are common on chickpea crops in the Mediterranean basin (Greco et al., 1992; Di Vito et al., 1994a; Di Vito et al. 1994b). A few other species of *Pratylenchus* and *Pratylenchoides*, and *Zygotylenchus guevarai* (Tobar Jiménez) Braun et Loof have also been reported from the Mediterranean basin, but their impact on chickpea has not been investigated.

Root-lesion nematodes pass through several generations per growing season and large numbers can be extracted from infested roots at the early flowering stage. In the absence of a host crop, soil populations can be low especially making prediction of yield loss difficult from pre-sowing populations. However, when rains occur early in autumn the nematode may complete one or more generations on volunteer plants and the following winter crop can suffer severe damage. These nematodes have rather wide host ranges, which include winter cereals. However, Mediterranean populations appear to be much more adapted to cool season than to summer crops. Field observations in Syria revealed that the tolerance limit of chickpea to *P. thornei* is 0.03 nematodes/cm^3 soil and that yield can drop to about 40% when the nematode population at planting is 2 nematodes /cm^3 soil (Di Vito et al., 1992). Unfortunately, similar information for the other root-lesion nematodes is lacking.

In Australia, chickpea is grown as an alternative crop to wheat and many older wheat fields contain the root-lesion nematodes *P. thornei* and *P. neglectus* which both attack chickpea. *P. thornei* prefers clay soils and is predominant in the northern grain region whereas *P. neglectus* prefers lighter-textured soils and is predominant in the southern grain region. Both species have a similar optimum temperature for reproduction of 22 to 25 C (O'Reilly and Thompson, 1993; Thompson, unpublished data; Vanstone and Nicol, 1993). *P. thornei* is the more damaging to wheat causing losses in grain yield of up to 75% in the most intolerant wheat cultivars. Field experiments with nematicides on chickpea have sometimes shown substantial increases (25-60%) in plant biomass or grain yield on sites heavily infested with *P. thornei* in northern New South Wales (T.R. Klein, pers. comm., 1988; M. Schwinghammer, pers comm., 1989), Queensland (Thompson, 1989) and Victoria (Eastwood and Smith, 1995). Similar responses to nematicide have been obtained with chickpea on land infested with *P. neglectus* in South Australia (Taylor, et al., 1997). Many current cultivars of both wheat and chickpea are susceptible to both species of root-lesion nematodes (Table 1) and wheat-chickpea rotations maintain high nematode populations (20 nematodes/g soil) which can cause loss in both crops. This is a potential problem as chickpea is advocated as a break crop for wheat to improve available soil nitrogen and to reduce cereal cyst nematode (*Heterodera avenae* Woll) and fungal diseases caused by *Fusarium graminearum* Schwabe, take-all caused by *Gaeumannomyces graminis* Sacc., and rhizoctonia root rot caused by *Rhizoctonia solani* Kuhn. Root-lesion nematodes are widespread being present in virtually all grain fields in South Australia (V.A. Vanstone and S.P. Taylor, pers comm., 1995). In Victoria almost all grain fields are infested, of which 20% have populations above the damage threshold level of 2 nematodes/ g soil(Eastwood and Smith, 1995). In northern NSW and Southern Queensland, 50% of fields with putative in-crop symptoms of nematode damage are infested with root-lesion nematodes (JP Thompson unpublished data). Annual loss to grain crops from root-lesion nematodes in the eastern Australian grain belt is estimated to be AUD.$ 50-100 million

Table 1. Resistance and tolerance of crop and pasture species to *Pratylenchus thornei* and *P. neglectus.* (Combined information of R. Eastwood, V.A. Vanstone, S.P. Taylor and J.P. Thompson).

Crop	P. THORNEI		P. NEGLECTUS	
	Resistance[1]	Tolerance[2]	Resistance[1]	Tolerance[2]
Winter				
Wheat	VS	I-VT	VS	I-T
Durum	MR-S	MT-T	MR-S	MI
Barley	MR-S	T-VT	MR	MT
Oat	S	-	S	I-T
Rye	R	-	R	-
Triticale	MR-S	MT	R	I-T
Canaryseed	R	T	-	-
Chickpea	VS	MI	VS	-
Pea	R	T	R	-
Faba bean	MR-S	T	R	-
Lentil	R	T	S	-
Lupin	R	-	MR	-
Vetch	S	I	VS	-
Medic	R	-	VS	I
Sub-clover	VS	-	R	-
Canola	MR-S	-	S	MI
Mustard	-	-	S	-
Safflower	R	-	R	-
Linseed	R	T	-	-
Summer				
Sorghum	R	VT	-	-
Maize	S	-	-	-
Millet	R	-	-	-
Mungbean	VS	-	-	-
Pigeonpea	R	-	-	-
Cowpea	VS	-	-	-
Sunflower	R	-	-	-

1 Resistance: R = resistant, MR = moderately resistant, S = susceptible, VS = very
susceptible, MR-S = cultivars ranging from moderately resistant to susceptible, - = no information
2 Tolerance: T = tolerant, MT = moderately tolerant, MI = moderately intolerant, I = intolerant, I-T = cultivars ranging from intolerant to tolerant, - = no information.

Control

Although nematicides are not used to control these nematodes on chickpea, several trials in the Mediterranean region have demonstrated that soil treatments with different chemicals can increase yield. With non-fumigant nematicides, better control can be achieved with split applications of 10 kg a.i./ha. Seed coating with paste containing nematicides has resulted in contradictory results. Soil solarization would also be effective (Di Vito et al., 1991).

Crop rotations and fallow are the only measures adopted for the management of these nematodes. However, alternation of winter with summer crops, fallow and summer ploughing would result in satisfactory control. Moreover, populations of the same nematode species from different geographical areas may have different host ranges that should be ascertained.

No chickpea cultivar resistant to root lesion nematodes is available. Recently, Di Vito et al. (1996b) found resistance to *P. thornei* in accessions of *C. bijugum*, *C. cuneatum* Hochst., *C. judaicum* Bois and *C. yamashitae* Kitamura, but none of these species can be crossed with *C. arietinum*.

In the eastern Australian grain belt, there are diagnostic services available to farmers specifically for root-lesion nematodes. Fields can be diagnosed before sowing to ascertain the nematode threat by extracting and enumerating nematodes in soil samples. In the northern region and the Victorian Wimmera, *P. thornei* occurs as deep as 90 cm in the subsoil of Vertisols and samples for diagnosis are taken to 30 cm depth. Pre-sowing diagnosis is a very useful management tool in areas where the nematodes are well established. *P. thornei* appears to be spreading into newer cropping lands of the northern region in many cases in runoff and flood water. Diagnosis of new infestations is best based on extraction of in-crop soil and root samples using above-ground symptoms as a guide on where to sample a patchy distribution. Where nematodes are well established in a field, rotation to other crops with partial resistance or tolerance (Table 1) is necessary to avoid losses. The aim is not to grow a susceptible, intolerant crop more often than once every 3-4 years. This is achieved in the northern region by rotations of chickpea or wheat with sorghum and barley. Switching between summer and winter crops is often accomplished via a weed-free fallow period of 11-14 months during which nematode numbers decline somewhat. When rainfall permits, chickpea (winter crop) can also be grown directly after sorghum (summer crop) which is resistant to *P. thornei*. Cropping chickpea after wheat should be avoided in nematode-infested fields. In the Victorian Wimmera region, root-lesion nematodes can be effectively controlled with rotations of winter crops. Research (Eastwood and Smith, 1995) showed that a single cycle of good hosts like chickpea and wheat and of poor hosts like pea or barley resulted in a doubling or a halving respectively of the initial soil populations of *P. thornei*. Targeted breeding in the northern region has produced wheat cultivars (Pelsart and Sunvale) with superior tolerance to *P. thornei*. Similar work is underway to ascertain the current levels of tolerance in Australian chickpea cultivars and to improve them through breeding. Tolerant cultivars yield well when attacked by nematodes but still allow nematodes to multiply in their roots leaving a burden in a field to attack subsequent crops or to disseminate the problem. Programs are underway to breed true resistance into wheat cultivars and to search for sources of resistance in chickpea germplasm and related species.

Reniform nematode

Although *Rotylenchulus reniformis* Linford et Oliveira occurs in the Mediterranean area, it has not been reported to damage chickpea. *R. macrosomus* Dasgupta, Raski et Sher also has often been found in the chickpea rhizosphere but never in chickpea roots. *Rotylenchulus reniformis* has been associated with chickpea decline mainly in irrigated areas in parts of northern India (Ali, 1995). However, pathogenic effects of *R. reniformis* populations on chickpea have not been observed in southern and western India. The nematode damage to chickpea is greatly influenced by the preceding crop; sorghum, pigeonpea, mungbean, and urdbean enhance population build up of the nematode. Options to manage the damage caused by *R. reniformis* on chickpea are similar to those for the tropical root-knot nematodes *M incognita* and *M javanica*.

Other nematodes

Other nematodes have been reported from the rhizosphere of chickpea (Sharma, 1985; Ali, 1995; Castillo et al. 1996) but pathogenicity to this crop has only been demonstrated for *Tylenchorhynchus vulgaris* Upadhyay, Swarup et Sethi. Stem nematode (*Ditylenchus dipsaci*) affects chickpea in South Australia in a similar manner to pea (see fuller coverage later). Chickpea seedlings are susceptible and intolerant to stem nematode but resistant as adult plants.

PEA

Several nematode species have been found in association with pea. The most important are cyst-forming, root-knot, root-lesion and stem nematode *Ditylenchus dipsaci* (Kuhn) Filipjev (Sikora and Greco, 1990; Riggs and Niblack, 1993).

Cyst nematodes

The most important and studied nematode of pea is *Heterodera goettingiana* Liebscher. This cyst nematode is distributed in Europe and in the Mediterranean basin (Di Vito and Greco, 1986). In 1992 *H. goettingiana* was also found in several fields in western Washington state (USA) where it was probably causing damage for several years (Handoo et al., 1994). The nematode may interact with soil borne fungi and greatly suppresses rhizobia nodulation. *H. goettingiana* develops during spring and summer in temperate climates and from autumn to mid spring in the coastal area of the Mediterranean basin. Only one generation per growing season of pea is completed if the nematode does not form egg masses and two to three generations if egg masses are produced, which occurs when rhizosphere temperature remains below 15 °C and soil moisture is optimal (Greco et al., 1986). Large numbers of nematode females can be observed on the roots at flowering, when plants show yellowing, patchy growth and few flowers. *H. goettingiana* also damages faba bean, grass pea and vetch.

In microplots the tolerance limit of pea to *H. goettingiana* was 0.5 egg/g soil. Yield losses of 20 and 50% would occur at 3 and 8 eggs/g soil and complete crop failure at 32 eggs/g soil (Greco et al., 1991). Maximum reproduction factors of the nematode have been recorded as high as 90 in microplots of pea, and an average annual decline of 50% in fallow fields (which may increase in warm and arid areas) was determined.

Pea is also a good host for *H. ciceri*, but no damage by this nematode has been reported in farmers' fields. *H. trifolii* Goffart also reproduces on pea but the nematode does not seem to be a problem for pea crops.

Control

The economics of pea crops vary according to whether they are cultivated for fresh or frozen green grains or for dried grains (pulse). In the first case the use of nematicides and other costly means of control could be economical. In Italy (Di Vito and Lamberti, 1976) and England (Whitehead et al., 1979) satisfactory control of *H. goettingiana* was achieved by incorporating fenamiphos (10 kg/ha) or aldicarb and oxamyl (2.8-11.2 kg/ha) in the top 15 cm of soil before sowing. However, oxamyl would rapidly degrade in soils with pH higher than 6. As observed with other cyst nematodes, fumigant nematicides, such as 1,3D and methyl-isothiocyanate, would also be effective. Soil solarization has not yet been assessed against *H. goettingiana*, but in the Mediterranean area it would be as effective as observed for *H. ciceri*.

Like other cyst nematodes, *H. goettingiana* can be controlled by crop rotations. Although proper crop rotation can be designed on the basis of the nematode population density, in general growing a host crop once in three years in heavily infested soil or twice in three years in lightly infested fields would be satisfactory (Ferris and Greco, 1992).

No cultivar of pea resistant to cyst nematodes is available. Moderate resistance, governed by recessive gene(s) (Di Vito and Greco, 1986), was found in accessions of *Pisum abyssinicum* Brown, *P. arvense* L. and *P. elatius* Ster (Di Vito and Perrino, 1978), but no work is in progress to transfer this resistance into cultivars.

Root-knot nematodes

The root-knot nematode species mentioned above may also damage pea. In a pot experiment (Siddiqui et al., 1995) growth reduction of pea was observed at 500 juveniles of *M. incognita*/kg soil. However, in the Mediterranean basin pea is cultivated from mid autumn through spring, when soil temperature is too low for all root-knot nematodes except *M. artiellia* to affect plants. Therefore, only when pea is sown in early autumn would damage ccur. Moreover, damage by *M. artiellia* to pea has never been reported.

Control

Avoidance of early sowing generally is effective for controlling these nematodes. If pea is to be sown earlier in infested areas, then any means of control known to be effective on other crops would also result in satisfactory and economic protection of pea for green pod production.

Stem and bulb Nematode

Stem and bulb nematode *Ditylenchus dipsaci* is among the most destructive nematode species. Its behaviour is that of an endoparasitic, migratory nematode, which mainly attacks above-ground plant parts. It is a complex species comprising 21 recorded host races, each having a different host range. Some races have a rather limited host range while others may reproduce on a large number of plants including weed species. Identification of the nematode race is a difficult task. However, rather than identifying the nematode race it is important to know the reproduction potential of local populations of the nematode on the annual crops cultivated in the same area. In general, depending on the nematode population, bulbous plants, alfalfa, clover, corn, sugarbeet, oats and strawberry are commonly attacked by *D. dipsaci*. Moreover some particular local populations have been found to damage rye, wheat, carrots and Italian ryegrass.

Its epidemiology is greatly influenced by environmental conditions. Air temperature in the range 15-20°C and wet conditions caused by sprinkler irrigations, rain, fog and dew, which usually occur from autumn to spring in the Mediterranean basin, favour nematode infection and reproduction. In central and northern Europe these conditions may also occur in the summer months. Little or no infection occurs during warm and dry periods. *D. dipsaci* attacks all aerial plant parts and is favored by the prostrate habit of pea. Infected stems show large brown to black necrotic areas that may encompass the entire diameter. Because of concomitant infection of other micro-organisms, the stems may rot, become weak and break during windy days. Leaves present black necrotic areas while flowers and pods may show distortion and irregular growth. A few nematodes can be found in the grains and survive as quiescent fourth stage juveniles. The yield can be negligible where nematode infection is heavy.

In South Australia, the oat race of stem nematode is one of the most destructive nematodes causing loss in oat, faba bean, pea and chickpea crops that varies from 10% to crop failure. This race was first recorded in South Australia in 1973 and has since spread to twenty-seven districts covering an area of 150,000 ha. In 1991, the estimated losses to these industries totalled AUD $6m/year. So far this nematode has not been reported in other states. Its weed hosts include wild oats (*Avena fatua* and *Avena sterilis*) and bedstraw (*Galium tricornutum*) which can carry high nematode populations in non-host crop rotation years. It can be spread in hay and in faba been seed and has the potential to spread further in SA. It can be spread within a farm or district, in machinery, water run-off and plant debris. Crop losses can be expected to increase as areas sown to oats, oaten hay and faba beans increase (Scurrah, 1997).

Peas are very intolerant to stem nematode. Crop damage during cold, wet weather is directly related to initial nematode numbers. Stem nematode reduced seedling emergence in trials by up to 30%, and plants that emerged were stunted and deformed. A further 10-30% died during winter but surviving plants produced new tillers and recovered in spring. Yield loss resulted from reduced plant density and delayed maturity. Pea seedlings in South Australia are susceptible and intolerant, but adult plants of all cultivars are resistant (Scurrah, 1997). Pea seed and straw pose a low risk for nematode spread. The current pea cultivars vary significantly in their ability to tolerate stem nematode: Alma and Glenroy are the most tolerant cultivars available; eight weeks after emergence they harbour 50% fewer nematodes than Dinkum, which is very intolerant. With tolerant cultivars, nematodes leave plants earlier and the plants recuperate faster (Scurrah and Szot, 1996).

Control

The European race seems to attack adult pea plants and infest seed while the South Australian race has not been found in pea so far, Therefore the control strategies for stem nematodes differ slightly. Although there is clear evidence that the nematode can survive in seeds and be transported over a long distance, this aspect has been neglected so far. Therefore, control measures must be adopted to produce seed-stock free of the nematode. This goal can be achieved by growing pea in areas where major host plants of the nematode are not cultivated and in fields free of the nematode. The choice of fields having good sunlight and air-flow, the use of pre- and post-planting nematicide treatments and proper weeding are suggested. Examination of the seed-stocks obtained will also be necessary to ensure freedom from nematodes. For this purpose, preplant soil fumigation with 1,3D (100-200 kg/ha) or dazomet (500 kg/ha), or the use of pre-sowing granular non-fumigant nematicides, such as aldicarb, fenamiphos, oxamyl or prophos, all at the rate of 10-12 kg a.i./ha, are suggested. Combination of soil fumigation with post-emergence split application of 5-10 kg/ha of one of the first three granular nematicides will certainly give better nematode control.

Control of stem nematode with nematicides in pea crops for human consumption would be profitable, but post-emergence applications of non-fumigant nematicides is not recommended to avoid residues in the grain. Usually, control of *D. dipsaci* in pea is overlooked and when rotations are adopted they are not designed just to control this nematode. However, proper rotations should consider the geographical origin of the nematode population (race) and the cropping history of the area. Usually, rotating pea with cereal or summer crops is effective in the Mediterranean basin.

No pea cultivar resistant to *D. dipsaci* is available. In South Australia, 800 pea lines have been screened for tolerance by comparing their yield with Alma and by response to nematicide. A number of lines with good tolerance have been selected. These yielded up to 70% more than Alma and gave small responses to nematicide compared with other cultivars (up to 140% increase)(M Scurrah unpublished data). Current advice to farmers in South Australia is to avoid growing susceptible oat cultivars in succession with pea or faba bean and to control the major weed hosts.

Other nematodes

Riggs and Niblak (1993) reported another 30 nematode species hosted by pea. Several of them interact with soil-borne fungi but their impact on pea growth has not been determined. In Brasil, *Helicotylenchus dihystera* (Cobb) Sher is a severe ectoparasite of wheat and pea and 41% yield loss of pea was observed in pots (Sharma et al., 1993).

LENTIL

Few nematode species have been reported in association with lentil and most of them are found in the rhizosphere. Generally, ectoparasitic nematodes found in the rhizosphere of other annual crops in a given area can also be found in lentil crops, but their pathogenicity to lentil is not known.

Heterodera ciceri is the only nematode for which the pathogenicity has been ascertained. This nematode causes yield losses of lentil in Syria and has been reported on the crop in Turkey (Di Vito et al., 1994b). The tolerance limit to the nematode is 2.5 eggs/cm^3 soil and yield losses of 20 and 50% should be expected in fields infested with 20 and 64 eggs/cm^3 soil (Greco et al., 1988). Reduced protein content has been observed in lentil heavily infested by nematode. The pathogenicity of other cyst nematodes reported associated with lentil has not been investigated.

Root-knot nematodes should not constitute a problem for lentil. In the Mediterranean basin lentil is a winter crop while the major root-knot nematodes develop during the warm season and *M. artiellia* reproduces poorly on this food legume.

Ditylenchus dipsaci has been found in stems of lentil in Syria and Turkey in crops with no obvious damage. However, the nematode could affect yield in rainy years.

Pratylenchus spp. have also been extracted from roots of lentil but never in as large numbers as from chickpea roots.

Control

Control measures suggested for the nematodes mentioned would be similar for lentil as for the other crops. In Europe no lentil cultivar resistant to any of the nematodes mentioned is available. In South Australia, the two lentil cultivars tested were resistant to the local race of stem nematode.

FABA BEAN

Faba bean is cultivated either for green pods or for its dried grain. In both cases cyst, root-knot and root-lesion nematodes and *D. dipsaci* are the major nematodes that can damage the crop.

Heterodera goettingiana is the only cyst nematode damaging faba bean. However, the tolerance limit of this food legume to the nematode is about 0.8 egg/g soil, a little less than that of pea, and yield losses of 20 and 50% are expected in soils infested with 5 and 15 eggs/g soil. Complete crop failure would occur when the nematode population at sowing is about 64 eggs/g soil (Greco et al., 1991). The biology and dynamics of the nematode population are similar to those reported for pea.

Control

Control measures are similar to those reported for pea. No source of resistance to the nematode has been found in faba bean.

Root-knot nematodes

Root-knot nematodes, *Meloidogyne* spp., are potentially capable of damaging faba bean. This food legume is cultivated as a winter crop and therefore the soil should be too cool for root-knot nematodes of warm seasons to cause damage. However, damage by these nematodes has been observed in Egypt and in Italy when faba bean is sown early in autumn after a susceptible summer crop. Faba bean is a good host for *M. artiellia*, but no crop damage by the nematode has been observed in the field.

Control

Sowing late in autumn would prevent infection by the warm season root-knot nematodes in infested fields.

Ditylenchus dipsaci

The stem and bulb nematode is considered a serious problem of faba bean because of its survival in seeds and therefore its implication in quarantine regulations. Extensive coverage of the problems caused by this nematode have been published recently (Greco, 1993; Sharma et al., 1994; Caubel and Esquibet, 1995). Most countries will only permit imports of seed stocks free of the nematodes. Two races of stem nematode attack faba bean. The biology of *D. dipsaci* on broad bean is similar to that on other host crops but some differences may occur according to the nematode race. In the Mediterranean basin and Europe the normal race with $2n = 24$ chromosomes is common in broad bean crops. This race mostly infects the basal leaves and stems, while infection of the upper plant parts, including pods and seeds can be negligible. In South Australia heavy seed infestation occurs during wet springs. Symptoms are not specific and diagnosis depends on nematode extraction. In Europe and the Mediterranean basin, symptoms of the nematode attack are browning of stem bases and leaf necrosis. These symptoms can be confused with those of other diseases, such as chocolate spot caused by *Botrytis fabae*. The giant race of the nematode with $2n = 48$ chromosomes is common in North Africa and much more pathogenic on broad bean. This race attacks the upper parts of the plant also, including pods and seeds in which it survives in large numbers as quiescent fourth stage specimens. In addition to the symptoms already described, this race causes distortion and deformation of stems and pods and swelling of the stems, which are characteristic of the nematode infection. Heavily infected grain shows necrotic areas on the cotyledons. Planting infested seeds in healthy soil could result in more damage than planting healthy seeds in infested soil. In the case of soil-borne infections, the crop presents a few areas showing symptoms of the nematode attack, whereas with seed-borne infection the symptoms appear in several small patches depending on per cent of infected seeds.

Control

Because of the seed borne nature of the nematode, strict quarantine regulations and guidelines must be followed to avoid exporting, importing and sowing of infested seed.. Treating infested seed stocks with methyl bromide in a closed container under partial vacuum conditions, at CTP of 1000 mg hr/l (Powel, 1974), 100 g/m^3 x 18h or 80 g/m^3 x 12h (Caubel and Esquibet., 1995) greatly reduced but did not eradicate the nematode without substantially affecting seed germination. Moreover, hot water treatment of the seeds could also be an effective and easy method at farm level. The nematode *Aphelenchoides besseyi* in rice seeds is controlled by hot water treatment of 15 min. at 52-54 °C (Hollis and Keoboonrueng, 1984), but investigations are necessary to ascertain the best combination of temperature and time to kill nematodes without affecting germination of faba bean seeds. Infested seed should be consigned to human or animal consumption, although there is evidence that some nematodes may survive in the animal intestinal tract. Burning of the plant residues is also recommended. High populations of the oat race of stem nematode develop on faba bean in South Australia and the pods and seed pose the danger of long distance dispersal to other areas.

Resistance to *D. dipsaci* has been found in faba bean lines from Morocco, Syria, Tunisia and France (Sharma et al., 1994). Over 100 accessions of faba bean were tested in the UK but no resistance to the oat race of stem nematode was found (Hooer, 1976). Caubel (1989a, b) first reported resistance to the Giant race in INRA 29H. Resistance has also been found in accessions of *Vicia narbonensis* but no information is available on the genetics of these resistances. One hundred accessions tested in South Australia were highly susceptible to the oat race. To date no cultivar with resistance is available. Faba bean cultivars are considered relatively more tolerant to the oat race of stem nematode than pea cultivars in South Australia and do not show obvious symptoms unless heavily infested. Nevertheless, treatment with nematicide increased yield of faba bean cultivars-Fiord by 13%, Icarus by 18%, Ascot by 23% and Aqualdulce by 30%. Breeding for resistance to the oat strain is hampered by the lack of plant symptoms, but counts of nematodes on dried stems indicate that useful variation (800 to 32,000 nematodes/plant) exists in advanced material from the South Australian faba bean breeding program and that higher levels of resistance may be achieved by recurrent selection.

Other nematodes

Specimens of *Pratylenchus* spp. are often extracted from necrotic roots and many ectoparasitic nematode species have been found in the rhizosphere of faba bean, but information on their impact on the crop is lacking. Faba bean appears to be a poorer host than chickpea for *P. thornei* and *P. neglectus* (Table 1) and may prove a better rotational crop with wheat (K. Moore pers. comm., 1997; G. Holloway and R. Eastwood pers. comm., 1997).

GRASS PEA

Investigation on nematodes of grass pea has received little attention probably because of the limited importance of the crop for human consumption. However, grasspea is a good host for the cyst nematodes *Heterodera goettingiana* and *H. ciceri*. In a trial in Syria a cultivar for animal feed was severely damaged by *H. ciceri*. The root-knot nematode *Meloidogyne artiellia* has been shown to reproduce well on grass pea. Pot experiments showed growth reduction by *M. incognita* when soil population densities were at least 0.75 second stage juveniles/cm^3 soil (Thakar et al., 1986). In India severe damage by *Pratylenchus thornei* was observed on *Lathyrus odoratus* (Mishra and Gupta, 1988). This would indicate that this nematode and other species of *Pratylenchus* might also affect grass pea.

LUPINS

Lupins have been shown to have high levels of immunity to most common nematode species of temperate climates and are often recommended as a rotation crop. However information on nematodes affecting lupins is limited. Species of *Pratylenchus* and *M. hapla* have been found in the roots. *Xiphinema lupini*, several species of root-knot, root-lesion, cyst and ectoparasitic nematodes have been reported associated with different lupins (Riggs and Niblack, 1993), but the pathogenicity of most of them is doubtful and the impact of the others on lupins has not been ascertained. However, lupin is a winter crop and, therefore, the warm season root-knot nematodes and *M. artiellia* would not affect it.

GENERAL CONSIDERATIONS ON CONTROL

Control of plant parasitic nematodes must consider the economics of the control measures, which may vary from country to country according to the use of the legume. Usually when a legume is cultivated for the production of green pods or green grain to be consumed fresh or as canned or frozen food, more expensive means of control can be afforded. In this case nematicides or soil solarization would be very useful. This may not be so when cool season legumes are cultivated for dried grain (pulse) production. Moreover, these legumes are very often cultivated on rather marginal land returning net profits that do not allow the use of expensive control measures. In addition, awareness of nematode problems is very poor in many countries and no action is taken to limit yield loss caused by these hidden enemies. In such situations current crop sequences or fallow are not specifically designed to control nematodes or any other parasites, but just to reduce damage by "soil sickness". Instead, a sound approach to minimize yield losses requires precise information on the nematode species, its population density and the pathotype occurring in a given field. Although this may not be possible at farmer field level, surveys and investigations at regional level, would be useful. In general, if crop rotation can be of help in controlling nematodes, rather than suggesting a particular crop sequence, the goal could be the introduction to the area of other crops having different disease problems. This will automatically lead to the adoption of longer term rotations.

The best way to control nematodes of cool season food legumes and by-pass farmer unawareness is to release cultivars resistant to the nematode(s) pests occurring in the area. The second course of action is to educate farmers about nematodes and alternative strategies. Unfortunately, although resistance to nematodes has been reported in cultivated and wild legume species, not enough resources are being assigned to transferring these resistances into cultivars. A breeding program to transfer resistance to *Heterodera ciceri* from *Cicer reticulatum* to kabuli type cultivars of *C. arietinum* is in progress at ICARDA (Aleppo, Syria) (Di Vito et al., 1996). Similar breeding programs should be undertaken for all cool season legumes. However reported resistance should always be confirmed against local populations of a nematode. Moreover, we must be aware that although nematodes are often the major disease problem they usually occur with other pests and diseases. Therefore, breeding programs should aim to produce cultivars with multiple resistances.

In areas infested with nematodes that have wide host ranges, such as root-knot and root-lesion nematodes, the use of resistant cultivars of other crop plants must be considered. For instance, cultivars of tomato, cowpea and haricot bean resistant to root-knot nematodes are available and their inclusion in a rotation can be of help in reducing nematode populations before sowing a susceptible crop.

Until such cultivars are released, the integrated use of nematicides, soil solarization, crop rotation, soil amendments, choice of sowing time, ploughing the soil in summer, weed control, fallow and sowing nematode-free seed should provide satisfactory nematode control and crop benefit in farming systems.

References

Ali, S.S., 1995 *Nematode problems in chickpea*. Indian Institute of Pulses Research, Kanpur, India 184 pp

Castillo, P., Gomez-Barcina, A. and Jimenez-Diaz., R.M., 1996. *Nematologica* 42: 211-219.

Caubel, G. and Esquibet , M., 1995. *La Défense des Végètaux* N° 476: 25-30.

Caubel, G and Leclerq, D., 1989 *Nematologica* 35: 216-224

Di Vito, M. and Lamberti, F., 1976. *Nematologia Mediterranea* 4: 121-131.

Di Vito, M. and Perrino, P., 1978. *Nematologia Mediterranea* 6: 113-118.

Di Vito, M. and Greco, N., 1986. In *Cyst Nematodes*, pp. 321-332 (eds. Lamberti,F. and Taylor, C.E.). New York, USA: Plenum Press.

Di Vito, M. and Greco N., 1988a. *Nematologia Mediterranea* 16:163-166.

Di Vito, M. and Greco N., 1988b. *Revue de Nématologie* 11: 221-225.

Di Vito, M., Greco, N. and Saxena, M.C., 1991. *Nematologia Mediterranea* 19: 109-111.

Di Vito, M., Greco, N. and Saxena, M.C., 1992. *Nematologia Mediterranea* 20: 71-73.

Di Vito, M., Greco, N., Halila, H.M., Mabsoute, L., Labdi, M., Beniwal, S.P.S., Saxena, M.C., Singh, K.B. and Solh, M.B., 1994a. *Nematologia Mediterranea* 22: 3-10.

Di Vito, M., Greco, N., Oreste, G., Saxena, M.C., Singh, K.B. and Kusmenoglu, I., 1994b. *Nematologia Mediterranea* 22: 245-251.

Di Vito, M., Singh, K.B., Greco, N. and Saxena, M.C., 1996. *Genetic Resources and Crop Evolution* 43: 103-107.

Di Vito M., Zaccheo G. and Catalano F., 1996. Proceedings of *"V Congresso della Società Italiana di Nematologia"*, Martina Franca (TA, Italy), 19-21. October 1995. Supplement to *Nematologia Mediterranea* Vol. 23 (1995): 81-83.

Eastwood, R. and Smith, A. (1995), *National Pulses Workshop*, Adelaide, March, 1995. 225, SARDI.

Ferris, H. and Greco, N., 1992. Fundamental and Applied Nematology 15: 25-33.

Greco, N., Di Vito, M. and Saxena, M.C., 1992. *Nematologia Mediterranea* 20: 37-46.

Greco, N., 1993. *Nematropica* 23: 247-251.

Greco, N., Di Vito, M. and Lamberti, F., 1986. *Nematologia Mediterranea* 14: 23-39.

Greco, N., Di Vito, M., Saxena, M.C. and Reddy, M.V., 1988. *Nematologica* 34:98-114.

Greco, N., Ferris, H. and Brandonisio, A., 1991. *Revue de Nématologie* 14: 619-624.

Greco N. and Di Vito M., 1993. In *Breeding for Stres Tolerance in Cool-Season Food Legumes* pp. 157-166 (eds. Singh K.B. and Saxena M.C.). Chichester, UK: John Wiley & Sons.

Handoo, Z.A., Golden, A.M., Chitwood, D.J., Haglund,W.A., Inglis, D.A., Santo, G.S., Baldwin, J.G. and Williams, D.J., 1994. *Plant Disease* 78: 831.

Hollis, J.P. and Keoboonrueng, S., 1984. In *Plant and Insect Nematodes* pp. 95-146.(ed. W.R. Nickle). New York, USA: Marcel Dekker.

Kaloshian, I., Greco, N., Saad, A.T. and Vovlas, N., 1986. *Nematologia Mediterranea* 14: 135-145.

Maheshwari, T.U., Sharma, S.B., Reddy, D.D.R., and Haware, M.P. 1995. *Journal of Nematology* 27:649-653.

Mishra, S.M. and Gupta, P., 1988. *Indian Journal of Nematology*, 18: 357.

Mishra, S.M. and Gupta, P., 1988 *Indian Journal of Nematology*, 16 357

Nene, Y.L., Sheila, V.K. and Sharma, S.B. 1996. *A world list of chickpea and pigeonpea pathogens*. 5 edn. International Crops Research Institute for the Semi-Arid Tropics, Patancheru 502 324, Andhra Pradesh, India; 27p.

O'Reilly, M.M. and Thompson, J.P., 1993. *Proceedings of the Pratylenchus Workshop*, 9th Biennial Australasian Plant Pathology Society Conference, Hobart. pp.5-9.

Powell, D.F., 1974. Plant Pathology 23: 110-113.

Riggs R.D. and Niblack T.L., 1993. In *Plant Parasitic Nematodes in Temperate Agriculture* pp. 209-258 (eds. Evans K., Trudgill D.L. and Webster J.M.). Wallingford, U.K: CAB International.

Sasser J.N. and Freckman D.W., 1987. *InVistas on Nematology: A Commemoration of the Twenty-fifth Anniversary of the Society of Nematologists* pp. 7-14.(eds.Veech J.A. and Dickson D.W.). Hyattsville, Maryland, USA: Society on Nematologists Inc.

Saxena, M.C., Greco, N. and Di Vito, M., 1992. *Nematologia Mediterranea* 20: 75-80.

Scurrah, M., 1995. *National Pulse Workshop*, Adelaide 26.

Scurrah, M., and Szot D., 1996 *Third International Nematology Congress*, Guadeloupe pp 131-132

Sharma, S.B., and McDonald, D. 1990. *Crop Protection* 9:453-458.

Sharma, S.B., Mohiuddin, M., Reddy, M.V., Singh, O., Rego, T.J., and Singh, U. 1995. *Fundamental and Applied Nematology* 18:197-203.

Sharma, S.B., Singh, O., Pundir, R.P.S., and McDonald, D. 1993. *Nematologia Mediterranea* 21:165-167.

Sharma, S.B., Smith, D.H., and McDonald, D. 1992. *Plant Disease* 76: 868-874.

Sharma, R.D., Da Silva, D.B. and Castro, L.H.R., 1993. *Nematologia Brasileira* 17: 85-95.

Sharma, S.B., 1985. *A world list of nematode pathogens associated with chickpea, groundnut, pearll millet, pigeonpea, and sorghum*. Pulse Pathology, Progress Report 42. ICRISAT, Andhra Pradesh, India, pp. 36.

Sharma S.B., Sikora R.A., Greco N., Di Vito M. and Caubel G., 1994. In *Expanding the Production and Use of Cool Season Food Legumes* pp.346-358. (eds. Muehlbauer F.J. and Kaiser W. J.). Dordrecht, The Netherlands: Kluwer Academic Publishers.

Siddiqui, Z.A., Mahmood, I. and Ansari, M.A., 1995. *Nematologia Mediterranea* 23: 249-251.

Sikora R.A. and Greco N., 1990. In *Plant Parasitic Nematodes in Subtropical and Tropical Agriculture* pp. 181-235.(eds. Luc M., Sikora R.A. and Bridge J.). Wallingford, UK: CAB International.

Singh, K.B., Di Vito, M., Greco, N. and Saxena, M.C., 1989. *Nematologia Mediterranea* 17: 113-114.

Singh, K.B., Di Vito, M., Greco, N. and Saxena, M.C, 1996. *Crop Science* 36.

Taylor, S.P., Vanstone, V.A. and Ware, A. (1997) SARDI Report, *Australian Association of Nematologists, Newsletter.*

Thakar, N.A., Patel, H.R. and Patel, C.C., 1986. *Indian Journal of Nematology* 16: 121-122.

Thompson, J.P. 1989. *Australian Chickpea Workshop Proceedings*, Warwick, Oct., 1987, 127-136, (Eds R.B. Brinsmead and E.J. Knights). Australian Institute of Agricultural Science (Queensland Branch), Brisbane.

Vanstone, V.A. and Nicol, J.M., 1993. *Proceedings of the Pratylenchus Workshop*, 9th Biennial Australiasian Plant Pathology Society Conference, Hobart. pp.11-16.

Whitehead, A.G., Bromilow, R.H., Tite, D.J., Finch P.H., Fraser, J.E and French, E.M., 1979. *Annals of Applied Biology* 92: 81-91.

Potential for Pulses in Aquaculture Systems

Geoff L Allan

NSW Fisheries, Port Stephens Research Centre, Taylors Beach Road,,Taylors Beach, NSW 2316 Australia

Abstract

Aquaculture is expanding rapidly as the demand for seafood increases and productivity intensifies using improved, formulated and cost-effective diets. Demand for fishmeal, to be used in these diets is escalating, while supply is static or declining. The shortage has resulted in a global search for alternative ingredients. Pulses have the potential to replace at least some of the fishmeal in aquaculture diets. This potential is influenced by the biochemical composition of pulses; especially their protein content, amino acid profile, carbohydrate content and profile, digestibility and biological availability to the target species, as well as their price and supply. Research indicates that pulses such as lupins, field peas and faba beans are generally well utilised by fish. Their potential could be improved through plant breeding programs which would reduce the content of anti-nutrients and increase the essential amino acids. Processing could be used to reduce the content of poorly digested carbohydrate fractions, increase starch gelatinization and to add exogenous enzymes to assist digestion. Nutrition research within Australia has identified lupins, field peas and faba beans as having the potential to partially replace fishmeal in diets for silver perch, barramundi (sea bass), salmon and shrimp.

1. INTRODUCTION

The demand for seafood is escalating rapidly as both the global population and the preference for seafood increases. As production from capture fisheries is static or declining (Tacon, 1996), the increase in demand for seafood can only be met from aquaculture.

Aquaculture is one of the fastest growing food industries. It has expanded by approximately 9%/year over the past decade (Tacon, 1996) to the stage where in 1993, 22% of total fishery products were from aquaculture (Liao, 1996).

One reason this increase has been possible is the shift to more intensive production made possible by the availability of cost-effective formulated diets. The market for aquaculture diets in Asia increased more than four-fold from 1986 to 1996 (Akiyama, 1991) and New and Csavas (1993) predicted 2.6 million tonnes will be required in Asia by the year 2000.

Marine based ingredients, especially fishmeal and fish oil, are the preferred protein and energy sources as they are rich in essential amino acids and fatty acids and they are well digested and low in anti-nutritional factors. Unfortunately, production of fishmeal already uses 33% of the total global fish catch and within the next decade, Tacon (1996) predicts the proportion of total fishmeal and fish oil used for aquaculture will rise to 25-30% and 30-50% of total production respectively.

Pulses are grain legumes, which are widely used in stock feeding industries, and may have potential to replace at least some of the fishmeal in aquaculture diets.

Availability and composition of pulses

The major pulses produced in Australia, which are used in stock feeds, include lupins (*L. angustifolius* and *L. albus*), field peas (*Pisum sativum*), chick peas (*Cicer arietinum*) and faba beans (*Vicia faba*) while smaller quantities of mung beans (*Vigna radiata*), navy beans (*Phaseolus vulgaris*) and vetch (*Vicia sativa*) are also grown (ABARE, 1996). (For a full list of grain legumes produced in Australia, see Pettersen and MacIntosh, 1994). In total, more than 2.3 million tonnes of legumes are produced, approximately half of which are exported (ABARE, 1996; Table 1).

R. Knight (ed.), Linking Research and Marketing Opportunities for Pulses in the 21st Century, 507–516.
© 2000 *Kluwer Academic Publishers. Printed in the Netherlands.*

Table 1. Production and value of Australian pulses[1]
(Area '000 ha, Production and Exports kt, unit value AUD$/t, Gross value $ million)

	Lupins		Field Peas		Chick Peas		Total[2]	
	94/95	95/96	94/95	95/96	94/95	95/96	94/95	95/96
Area	1258	1308	407	350	169	207	1982	2033
Production	865	1429	214	465	69	258	1231	2335
Exports	291	674	134	217	36	112	547	1133
Ave. unit value	185	199	300	266	559	368	240	239
Gross value	160	284	64	124	39	95	295	558

1 Data from ABARE, 1996

2 Includes lupins, field peas, chick peas, faba beans, mung beans and vetch.

The biochemical composition of pulses (and any other ingredients) will have a major influence on their potential for use in aquaculture diets. Total protein and energy content and the amino acid profile are of primary importance. Fish do not have a requirement for carbohydrates although carbohydrate can be a relatively inexpensive source of energy and spare requirements for protein and lipid (NRC, 1993). Carbohydrate, especially starch, is also important in binding formulated diets.

The digestion and utilization of carbohydrate depends upon its structure. Complex carbohydrates, including plant cell wall material such as cellulose, hemicellulose, lignin and pentosan offer little nutritional benefit, are poorly digested and ultimately pollute the culture environment. In contrast, glucose, maltose and sucrose contributed to the best growth rates when different carbohydrates were fed to chinook salmon at 10% of the diet, followed by dextrin, fructose and galactose (Buhler and Halver, 1961; cited in NRC, 1993). Starch, in particular gelatinised starch, is relatively well digested and digestibility coefficients for dry matter and energy for silver perch were 58 and 66% for raw wheat starch, 60 and 68% for cooked (autoclaved at 121°C for 15 minutes) wheat starch, and 65 and 73% for pregelatinised maize starch respectively (Stone and Allan, in press). The carbohydrates in five pulses are given in Table 2. Anti-nutrients can also restrict the inclusion contents of some pulses (Table 3).

Table 2. Carbohydrates in grain legumes (whole seed %)

Type	Lupins	Field peas	Chick peas	Faba beans	Vetch
Starch	0.30	34.8-54.1	36.0-47.7	31.4-41.8	32.0
Cellulose	11.4-17.7	2.4-7.9	6.0	8.5-10.0	4.7
Sugar	4.8-5.4	1.0-5.7	4.4	1.5-4.1	-
Crude fibre	5.0-12.9	4.9-6.3	5.2-13.0	6.2-11.4	3.2-5.9
ADF	15.4-15.8	6.0-8.7	10.0-16.7	7.2-10.6	6.9-8.0
NDF	18.7-18.8	10.0-12.0	25.4-30.2	9.6-16.8	11.8-21.9

1 Data from Novus, 1992 and Petterson & Macintosh, 1994.

Table 3. Some anitnutrients in selected Australian pulses[1]

	Lupins		Field Peas	Chick Peas	Faba Beans	Vetch
	L.angustofolius	L. albus				
Alkaloids (%)	0.02	<0.01				
Oligosaccharides %)	5.16	6.69	3.69		2.93	
Phytate (%)	0.58	0.79	0.48	0.63		0.66
Tannins (total) (%)	0.32	0.37	0.25	0.49	0.75	0.64
Trypsin inhibitors activity	0.14	0.13	1.01	4.79	0.39	2.40
Chymotrypsin inhibitor activity		0.08	1.60	7.72	0.40	2.25

1 Data from Pettersen and MacIntosh, 1994.

2.   EVALUATING PULSES FOR AQUACULTURE DIETS

Ultimately, the choice of the ingredients to be used in aquaculture rations will depend on availability, price, composition, and the ability of the aquaculture species to digest and utilise the ingredients. In addition, factors such as storage capacity for ingredients at the feed mill and proven performance of fish under commercial conditions when fed diets containing the ingredients of interest, will influence which ingredients are used.

Measurement of digestibility involves subtracting the amount of energy or nutrients, which are voided in the faeces from the amount of energy, or nutrients provided in the diet. Measuring digestibility does not take into account losses which occur in the production of heat or urine but these are much less for fish than terrestrial animals because fish are cold-blooded and expend far less energy in locomotion and excretion (See Cho et al., 1982 and Cho and Kaushik, 1990 for reviews of fish energetics) For fish, digestibility coefficients for different ingredients are additive (Cho et al., 1982; Allan et al., unpublished data) and account for most of the differences between ingredients (Lovell, 1989). Aquaculture species have varying capacities to digest ingredients, especially the considerable amounts of carbohydrates which can be present in pulses. This capacity depends upon the digestive system of the fish species and the presence and activity of various endogenous enzymes (Wee, 1992; Alex Anderson, unpublished data). Apparent digestibility coefficients (ADCs) for a number of pulses fed to silver perch (*Bidyanus bidyanus*), a native Australian freshwater omnivorous fish which is being cultured in NSW, Queensland and Victoria, and rainbow trout, are presented in Table 4. ADCs for juvenile silver perch (<10 g) were determined after faeces were collected by settlement using similar methods to those described by Cho and Kaushik (1990). Although different methods were used for different studies, when compared with trout, silver perch had higher ADCs for protein but lower ADCs for dry matter. The overall high protein digestibility augurs well for use of protein sources from pulses. For both species, ADCs for dry matter and energy for pulses are much lower than for fishmeal, reflecting the poorer digestibility of carbohydrates in pulses.

Table 4 Comparison of digestibility coefficients (%) for dry matter, energy and protein for rainbow trout (*Oncorhynchus mykiss*) and silver perch (*Bidyanus bidyanus*)

	Lupins (*L. angustofolius*)	Field Peas (*Pisum sativum*)	Faba Beans (*Vicia faba*)	Fishmeal
Dry Matter				
silver perch[1]	50.3±3.0	62.0±0.4	56.9±0.3	91.1±1.1
rainbow trout[2]	63.3±0.7	66.1±0.6	66.1±0.9	89.4±0.02
Energy				
silver perch[1]	59.4±1.0	67.0±0.2	62.3±0.4	102.1±0.6
rainbow trout[2]	61.2±0.6	59.2±0.5	60.2±1.0	93.8±0.02
rainbow trout[3]	66.1±2.7			
rainbow trout[4]	64.0±1.9			
Protein				
silver perch[1]	96.6±0.9	84.1±0.3	91.7±1.3	98.6±1.2
rainbow trout[2]	85.5±0.7	80.4±0.5	80.2±0.9	92.3±0.02
rainbow trout[3]	85.2±1.3			
rainbow trout[4]	85.3±1.1			

[1] Data from Allan et al., unpublished data, fishmeal was Danish, 72% protein

[2] Data from Gomes et al., 1995, fishmeal was from Portuguese sardines.

[3] Data from Hughes, 1988.

[4] Data from Smith et al., 1995.

Positive results have been reported where pulses, most often lupins, have been used to replace fishmeal or other protein sources, for rainbow trout, gilthead sea bream and carp

(Viola et al., 1988; Hughes, 1991; Moyano et al., 1992; Morales et al., 1994; Robaina et al., 1995). Hughes (1991) for rainbow trout, and Viola et al. (1988) for carp, reported that lupin was successful as a complete replacement for soybean meal or full fat soy and diets with over 43% lupins performed as well or better than a fishmeal control diet for rainbow trout (Morales et al., 1994) or a fishmeal/soybean meal control diet for carp (Viola et al., 1988).

Morales et al. (1994) reported a protein retention efficiency of 41.2% for rainbow trout fed a diet containing 43% lupins, which was significantly higher than that recorded for a fishmeal-based control diet. Similarly, Gomes and Kaushik (1989) reported protein retention efficiencies of 40-43% for rainbow trout for diets containing 10-30% lupins.

Gouveia et al. (1993) compared the performance of rainbow trout fed diets where lupins (*L. albus*), field peas or faba beans were used to replace 20% of the dietary protein from fishmeal. Best results were obtained with lupins although all the diets yielded a superior performance to the fishmeal control diet (Gouveia et al., 1993).

3. IMPROVING THE VALUE OF PULSES FOR USE IN AQUACULTURE DIETS

Plant breeding programs to increase essential nutrients and reduce anti-nutrients have been successful for a number of grains, including maize, rapeseed, lupins and faba beans (Farrell, 1992). Lupins and faba beans with lower alkaline or tannin contents have an increased value in stock feeds (Farrell, 1992).

Improving the nutritive value of pulses can also be achieved by processing techniques such as grinding, dehulling, classification, cooking and expansion. Grinding to reduce particle size improves digestibility of grains, especially for crustaceans. Protein concentration, through dehulling and removal of starch or non-starch polysaccharides can increase digestibility of pulses for silver perch (Tables 5 and 6). Silver perch have also grown well in laboratory studies on diets with very high contents of dehulled lupins. In one experiment, eight juvenile silver perch (1.8 g/fish) were stocked in 60 litre aquaria (four were used for each diet) and grown for 75 days on experimental diets comprising a reference diet or that diet diluted with different amounts of dehulled lupins or an inert filler (either diatomaceous earth or cellulose). Growth of silver perch was not significantly different for fish fed the reference diet or that diet diluted with up to 60% dehulled lupins (Figure 1). Compared with the reference diet, protein retention efficiency was not reduced until diets contained >60% dehulled lupins (Figure 2).

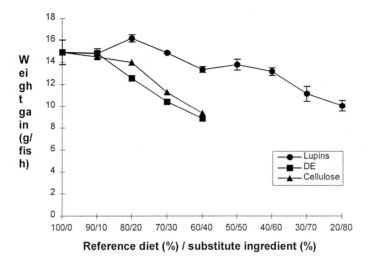

Figure 1. Growth of silver perch (*Bidyanus bidyanus*) fed experimental diets composed of a reference diet and either dehulled lupins, diatomaceous earth (DE) or cellulose.

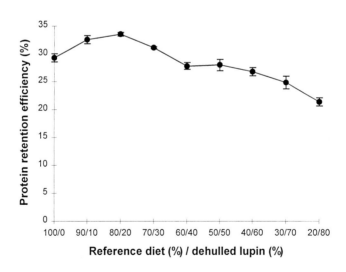

Figure 2. Protein retention efficiency of silver perch (*Bidyanus bidyanus*) fed experimental diets composed of different ratios of a reference diet and dehulled lupins.

Cooking/expansion improved the nutritional value to rainbow trout of field peas and faba beans but not lupins (Gouveia et al., 1991). This is not surprising as lupins contain almost no starch and the improvements, to field peas and faba beans were most likely due to the partial or complete gelatinisation of their starch. A co-extruded product containing rapeseed and field peas, "Colzapro", has been evaluated as a fishmeal replacement in diets for rainbow trout and levels of 45% in the diet did not affect protein efficiency (although lower protein and energy reduced growth) (Gomes and Kaushik, 1989).

There is also a potential to increase the utilisation of pulses in animal feeds through the addition of exogenous enzymes. Such enzymes include proteases, cellulases, pectinases, β-glucanases, lipases and phytases (Batterham, 1992). Riche and Brown (1996) significantly increased phosphorus availability by supplementing plant protein sources with phytase.

4.   FISHMEAL REPLACEMENT RESEARCH IN AUSTRALIA

In Australia, with little fishmeal production but abundant agricultural production, aquaculture will not develop unless aquafeeds based on agricultural proteins are developed. The Australian Fisheries Research and Development Corporation recognised the importance of fishmeal replacement research and aquaculture diet development and created a separate Sub-Program in 1993 to coordinate national research. Additional funding for this research has been provided by the Australian Centre for International Agricultural Research (for collaborative research with the Thailand Department of Fisheries), the Australian Grains Research and Development Corporation, the Australian Meat Research Corporation and the Australian Academy of Grain Technology. The overall aim of the coordinated research is to produce cost-effective aquafeeds, specifically by replacing imported fishmeal with cost-effective, locally-produced, alternative protein sources, defining critical nutritional requirements, evaluating and developing attractants and additives to improve diet palatability, and determining optimum feeding strategies. The research was perceived to be of value to most fish and crustacean species, but for practical purposes it was decided to concentrate efforts on four 'representative' species in Australia; barramundi or sea bass (*Lates calcarifer*), shrimp (*Penaeus monodon*), silver perch (*Bidyanus bidyanus*) and salmon (*Salmo salar*).

A strong focus is on developing highly digestible, "low waste-producing" diets. Results have been very promising: methods for determining digestibility in shrimp (*Penaeus monodon*) and fish have been developed or improved and large data bases of digestibility coefficients for a range of ingredients generated for silver perch (*Bidyanus bidyanus*), barramundi (*Lates calcarifer*) and shrimp (*P. monodon*).

Growth studies have been conducted to measure the contribution of different ingredients, including pulses, oilseeds and animal protein meals. Protein and protein/energy requirements have been investigated. Recently, we formulated, on a least-cost basis, diets for silver perch based on meatmeal, lupins and field peas with only 5 or 10% fishmeal. Fish performance on these diets was similar or superior to that on fishmeal/soybean meal control diets in commercially relevant facilities where fish were grown to market size (Allan, 1996). Composition of the diets did not affect sensory qualities of the fish flesh. Diets for barramundi with no fishmeal have also been developed and for both fish species, results are now being used as the basis of commercial formulations being sold by feed manufacturers.

Table 5 Proximate composition and apparent digestibility coefficients for processed (HO = hulls on; DH = dehulled; PC = protein concentrate) and unprocessed pulses fed to silver perch (Bidyanus bidyanus)

	L. angustifolius[1]			*L. albus*		*Field pea*[2]			*Faba bean*			*Chick pea*		*Vetch*	
Composition[3]	*HO*[4]	*DH*[4]	*PC*[5]	*HO*[4]	*DH*[4]	*HO*[6]	*DH*[6]	*PC*[7]	*HO*[6]	*DH*[6]	*PC*[1]	*HO*[6]	*DH*[6]	*HO*[6]	*DH*[6]
Dry matter %	94.1	95.0	96.3	95.4	95.0	93.5	94.4	90.9	93.9	94.6	91.2	92.5	95.2	94.7	94.9
Energy MJ-kg	17.9	20.7	22.7	20.9	21.4	17.0	17.3	19.8	17.3	17.6	20.6	19.4	19.3	17.9	18.6
Protein %	34.1	43.6	61.4	37.5	42.8	25.5	27.7	42.4	27.7	31.3	48.3	20.8	24.2	30.9	32.3
Apparent Digestibility Coefficients % [8]															
Dry matter %	50.3	67.6	78.4	64.7	77.8	62.0	48.9	85.9	55.9	58.2	66.3	48.7	58.4	41.5	78.3
	3.0	3.2	3.2	0.4	2.0	0.4	2.2	4.0	0.3	1.3	1.8	0.8	0.7	3.2	3.9
Energy %	59.4	74.0	82.0	72.7	85.2	67.0	54.5	91.1	62.2	58.8	73.4	53.6	60.2	55.5	81.8
	1.0	2.3	2.5	1.8	1.5	0.2	2.2	2.8	0.4	0.7	2.0	0.8	0.7	1.0	2.3
Protein %	96.6	100.3	97.4	96.1	101.4	83.3	88.1	98.6	91.6	96.6	95.0	84.8	81.2	74.9	87.7
	0.9	0.4	1.0	0.9	0.3	0.3	1.0	2.0	1.3	0.8	1.4	1.0	3.5	2.6	0.8

1 Gungarru variety; 2 Dunn variety; 3 Analysed composition (see Allan and Frances, 1994 for methods of analysis)
4 Allan et al, unpublished data August 1994; 5 Allan et al, unpublished data October 1994; 6 Allan et al, unpublished data December 1994
7 Allan et al, unpublished data August 1995; 8 Values are means with (± sem) below (n=3 replicates)

Table 6 Amino acid composition and apparent digestibility coefficients for processed (HO=hulls on; DH=dehulled; PC=protein concentrate) and unprocessed pulses fed to silver perch (Bidyanus bidyanus)

	Ingredients														
	L. angustifolius[1]		L. albus			Field pea[2]			Faba bean			Chick pea		Vetch	
% Composition[3]	HO[1]	DH	HO[4]	DH[4]	PC[5]	HO[6]	DH[6]	PC[7]	HO[6]	DH[6]	PC[7]	HO[6]	DH[6]	HO[6]	DH[6]
Arg	4.0	5.2	4.1	4.5		2.5	2.8	5.3	2.8	3.1	5.1	2.0	2.1	2.4	2.7
Hist	0.9	1.2	0.9	1.0		0.6	0.7	1.1	0.6	0.7	1.1	0.5	0.6	0.6	0.8
Iso	1.4	1.8	1.7	1.9		1.1	1.2	2.3	1.1	1.2	2.2	1.0	1.1	1.2	1.4
Leuc	2.4	2.9	2.8	3.2		1.7	1.8	3.5	1.8	2.0	3.5	1.6	1.7	1.9	2.2
Lys	1.4	1.7	1.5	1.7		1.7	1.8	3.4	1.5	1.9	3.0	1.5	1.5	1.7	1.7
Meth	0.2	0.2	0.3	0.3		0.3	0.3	0.4	0.3	0.3	0.4	0.4	0.5	0.3	0.3
Phen	1.3	1.6	1.4	1.6		1.1	1.2	2.3	1.1	1.2	2.0	1.2	1.3	1.1	1.2
Threo	1.3	1.6	1.5	1.7		0.8	0.9	1.7	0.8	1.0	1.6	0.7	0.8	0.8	1.1
Val	1.3	1.6	1.6	1.8		1.2	1.3	2.5	1.2	1.3	2.4	1.0	1.1	1.3	1.5
Apparent Digestibility Coefficients (%)[8]															
Arg	102.9	106.3	101.5	102.6	106.8	88.7	94.0	100.8	94.2	95.4	99.2	81.2	84.3	76.5	87.0
Hist	100.6	101.6	94.2	98.3	101.1	82.4	90.2	98.8	89.3	89.9	96.5	76.8	81.7	72.1	84.5
Iso	95.4	97.5	97.3	91.8	100.8	81.9	82.6	95.3	86.5	87.5	94.0	69.3	71.5	66.3	78.3
Leuc	94.9	96.8	95.8	94.4	99.5	84.5	87.9	96.1	90.7	90.1	95.5	75.8	75.0	71.1	82.2
Lys	98.1	99.5	95.5	96.6	102.5	86.3	88.6	98.2	90.9	94.2	98.2	80.5	83.3	72.7	86.7
Meth	83.9	91.7	91.0	92.2	97.3	87.5	91.2	94.4	93.3	94.2	91.1	85.3	83.2	77.8	88.1
Phen	96.0	98.0	95.6	94.8	100.0	82.9	85.1	96.1	89.2	89.3	94.6	70.8	72.5	62.8	76.7
Threo	95.8	101.3	96.5	97.3	101.8	80.5	83.2	93.7	87.8	86.3	95.2	67.9	75.2	56.6	73.3
Val	94.6	97.2	94.0	91.2	100.4	80.8	82.1	94.6	87.1	87.1	93.5	70.5	71.5	65.8	78.1

1 Gungarru variety; 2 Dunn variety; 3 Analysed composition (see Allan and Frances, 1994 for methods of analysis); 4 Allan et al., unpublished data, August 1994; 5 Allan et al., unpublished data, October 1994; 6 Allan et al., unpublished data, December 1994; 7 Allan et al., unpublished data, August 1995; 8 Values are means (n=3 replicates)

Acknowledgements
Research with silver perch in ponds was conducted under Dr Stuart Rowland at the NSW Fisheries, Grafton Research Centre and assistance from the following colleagues from NSW Fisheries, CSIRO Fisheries, CSIRO Food Science, Queensland Department of Primary Industries and Ridley Agriproducts Pty. Ltd. is gratefully acknowledged: Dave Stone, Jane Frances, Mark Booth, Scott Parkinson, Rebecca Warner-Smith, Helena Heasman, Charlie Mifsud, David Glendenning, Kevin Williams, David Smith, Tony Evans, Vince Gleeson, Zafer Sarac, Chris Barlow and David Overend.

Financial support for Australian aquaculture diet development research has been provided by the Australian Fisheries Research and Development Corporation, Australian Centre for International Aquaculture Research, Australian Grain Research and Development Corporation and the Australian Meat Research Corporation and Australian Native Fish Pty. Ltd.

References
ABARE 1996. Australian Commodities Statistics 1996. *Australian Bureau of Agriculture and Resources Economics.* pp. 364.

Akiyama, D.M. 1991 In: *Proc Aquaculture Feed Processing and Nutrition Workshop, Thailand and Indonesia,* September 19-25 1991. pp 5-9. (Eds E.M. Akiyama and R.K.H. Tan). American Soybean Assoc., Singapore.

Allan, G.L. and Frances, J. 1994 *Proc. Analytical Techniques Workshop* Brisbane 13 April 1993. NSW Fisheries, Port Stephens Research Centre. 102 pp.

Batterham, E.S. 1992 In *Proc. Aquaculture Nutrition Workshop, Salamander Bay,* 15-17 April 1991. pp. 112-117. (Eds G.L. Allan and W. Dall). NSW Fisheries, Brackish Water Fish Culture Research Station, Salamander Bay, Australia.

Buhler, D.R. and Halver, J.E. 1961 *Journal of Nutrition* 74:307-318.

Cho, C.Y. and Kaushik, S.J. 1990. *World Review of Nutrition and Dietetics* 61:132-172.

Cho, C.Y., Slinger, S.J. and Bayley, H.S. 1982 *Comparative Biochemistry and Physiology* 73B(1):25-41.

Farrell, D.J. 1992 In Proc. *Aquaculture Nutrition Workshop, Salamander Bay*, 15-17 April 1991. pp. 105-111. (Eds G.L. Allan and W. Dall). NSW Fisheries, Brackish Water Fish Culture Research Station, Salamander Bay, Australia.

Gomes, E.F. and Kaushik, S.J. 1989 *Proc. Third International Symposium on Feeding and Nutrition in Fish.* Toba. pp. 315-324.

Gouveia, A., Olivia-Teles, A., Gomes, E. and Rema, P. 1993. In *Fish Nutrition in Practice,* Biarritz, (France) 24-27 June 1991. pp. 933-938 (Eds S.J. Kaushik and P. Luquet). INRA, Paris 1993 (Les Colloques, no. 61)/

Hughes, S.G. 1991 *Aquaculture* 93(1):57-62.

Liao, I.C. 1996 In *Development and Sustaining World Fisheries Resources: The State of Science and Management.* Proc. Second World Fisheries Congress, Brisbane, Qld, 1991. pp. 431-436. (Eds D.A. Hancock, D.C. Smith, A. Grant and J.P. Beumer). CSIRO Publishing, Collingwood Vic Australia.

Lovell, T. 1989 *Nutrition and Feeding of Fish.* Van Nostrand Reinhold, New York, 260 pp.

Morales, A.E., Cardenete, G., De la Higuera, M. and Sanz, A. 1994. *Aquaculture* 124:117-126.

Moyano, F-J, Cardenete, G. and De la Higuera, M., 1992 *Aquatic and Living Resources* 5:23-29.

New, M.B. and Csavas, I., 1993.. In *Proc. Regional Expert Consultation on Farm-Made Aquafeeds,* 14-18 December 1992, Bangkok, Thailand. pp. 1-23. (Eds M.B. New, A.G.J. Tacon and I. Csavas). FAO/AADCP.

NRC (National Research Centre), 1993 *Nutrient Requirements of Fish.* National Academy Press. 114 pp.

Petterson, D.S. and Mackintosh, J.B. 1994 *The chemical composition and nutritive value of Australian grain legumes.* Grains Research and Development Corporation. 68 pp.

Riche, M. and Brown, P.B. 1996 *Aquaculture* 142:269-282

Robaina, L., Izquierdo, M.S., Moyano, F.J., Socorro, J., Vergara, J.M., Montero, D. and Fernandez-Palacios, H. 1995 *Aquaculture* 130:219-233.

Smith, R.R., Winfree R.A., Rumsey, G.W., Allred, A. and Peterson, M. 1995 Abstract only. *Journal of the World Aquaculture Society* 26(4):432.

Stone, D.A.J. and Allan, G.L. (in press). *The effects of cooking on the digestibility of a practical diet containing starch products fed to juvenile silver perch (Bidyanus bidyanus).* Abstract only.

Tacon, G.J. 1996 *Global trends in aquaculture and aquafeed production.* FAO/GLOBEFISH Research Programme Report, FAO: Rome (in press).

Viola, S., Arieli, Y. and Zohar, G. 1988 *Israeli Journal of Aquaculture* - Bamidgeh 49(1):29-34.

Wee, K.L. 1992 In *Proc. Aquaculture Nutrition Workshop, Salamander Bay*, 15-17 April 1991. pp. 243-244. (Eds G.L. Allan and W. Dall). NSW Fisheries, Brackish Water Fish Culture Research Station, Salamander Bay, Australia.

Potential and challenges in the marketing of lupins for food and feed

K Swan

The Grain Pool of W.A.,172 St Georges Terrace, Perth, Australia

INTRODUCTION

The Grain Pool of Western Australia (WA) has been marketing lupins internationally for over 25 years, making its first export sale in 1972 with a 1,000 metric tonne (MT) cargo to Europe. This sale followed the establishment of the first voluntary lupin pool in Western Australia that same year, receiving 1,666 MT of lupins. From these humble beginnings, lupin exports by the Grain Pool have grown to an estimated 835,000 MT in 1996-97, generating AUD$190 million in export revenue. It is therefore fitting that the Grain Pool should be asked to address the topic of the potential and challenges in the marketing of lupins for food and feed.

The development of lupin markets since the early 1970's has not been without its challenges. In fact there were times it seemed lupins would not survive as a viable export crop. After the first 1,000 tonne sale, exports from the Grain Pool expanded so rapidly that just four years later in 1976 exports exceeded 50,000 tonnes. However, for the next two years drought devastated the crops and no lupins were exported. In the five years from 1977 to 1981 less than 20,000 tonnes were exported. Despite these set backs the potential for lupins was such that the Grain Pool persevered and was able to rebuild previously established markets in Europe and Taiwan and forge new markets in South East Asia.

INTERNATIONAL MARKETS

Before examining the potential and challenges of marketing lupins, it is appropriate to examine them within the context of international agricultural markets.

According to the Food and Agriculture Organisation of the United Nations (FAO) lupin production world-wide amounted to 1.44 million metric tonnes in 1996. Of these, 1.288 million tonnes or 89% was produced in Australia. By comparison, the second largest producer was Poland with total production of just 45,000 tonnes. The Australian Bureau of Agricultural and Resource Economics (ABARE) estimated Western Australian production for 1996 at 1.1 million metric tonnes, 85% of total Australian production or 76% of total world production.

Unfortunately reliable statistics for world lupin exports are not available, however it is generally accepted that Western Australia alone accounts for around 90% of world exports. Poland, while the world's second largest producer, grows lupins almost exclusively as green manure, exporting only limited volumes of seed into Europe. Eastern Australia and South America are the only other areas known to export lupins, even then only minor volumes.

Lupins continue to be exported almost exclusively as a protein ingredient for stockfeed, with the exception of a small quantity of the Albus lupin for human consumption. Lupins possess a number of attributes that make them an attractive stock feed ingredient. Sweet lupins are low in anti-nutritional factors and thus do not require heat treatment or any other processing, except grinding, to improve palatability and nutritional value to livestock. Consequently the quality is highly consistent and no sophisticated quality control analysis is required to determine the suitability of the product for the ration in which it will be used. Soybean meal by contrast, the main competitor of lupins, must be heat-treated prior to use in order to inactivate a range of nutritionally harmful chemical constituents. Under or over-heating of the soybean meal leads to quality variation and reduces the efficiency with which it is utilised in stockfeed rations. Due to the

R. Knight (ed.), Linking Research and Marketing Opportunities for Pulses in the 21st Century, 517–520.
© *2000 Kluwer Academic Publishers. Printed in the Netherlands.*

nature of their seed coats, lupins are naturally highly resistant to insects, allowing pesticide free delivery. The spherical shape of the seed leads to good flowing attributes and ease of handling.

However, in terms of the international protein market, lupins represent only a small fraction of meals traded. The United States Department of Agriculture (USDA) currently estimates world production of major protein meals at 152 million tonnes, of which around 51 million tonnes is traded in the international market. Lupins therefore represent less than 2% of the major protein meals traded each year.

This brings us to the first and possibly biggest challenge to marketing lupins, which is that they are simply one of a range of stockfeed ingredients that are highly interchangeable. Additionally, most of the alternative protein sources are available in much greater quantities. Soybean meal accounts for over 60% of world's total supply of protein meal, representing some 95 million tonnes per year.

Feed-millers produce livestock feeds with certain specifications, most importantly with respect to energy and protein. Although there are limitations on how specific ingredients can be included in a ration, to a large extent the origin of the energy and protein is not an issue. It can come from a wide range of cereals, vegetable and animal protein sources, by-products and fats. As a mid-level energy and protein source, lupins generally only represent one option in a range of possible ingredients.

While the overall demand for energy and protein in a given market might remain constant the demand for an individual ingredient can vary greatly. The main factor that determines demand for individual ingredients is price.

Feedmillers use a computer-based tool known as "least-cost formulation software" to determine the value of one feed ingredient against another. This takes into account the chemical composition of each ingredient, any limitations to its use, and the specifications set for the ration in arriving at the combination of ingredients which will give the cheapest overall cost for the ration. In this way the maximum price that can be paid for any particular ingredient (relative to the others) is set. Through this method the inclusion rate of ingredients is very sensitive to price.

In the sophisticated compound feed markets of Northern Europe, traditional meal products such as soybeans and lupins have come under increasing competition from cheaper protein sources such as corngluten pellet, palmexpeller, coprapellers, rapemeal and citruspellets. Competition from these lower priced alternatives has been so fierce that consumption of soybean meal this year (1997) is estimated to be down by more than 60 percent. If this trend continues as expected, it is likely lupin market share will also decline. This highlights a constant challenge in marketing lupins of maintaining diversity in marketing options.

NEW MARKETS

The opening of new markets for lupins has always been a high priority in the development of the industry. The rapid growth of Asian markets over the last two decades had made the region an obvious choice for market development. The potential of the South East Asian region was recognised by the Grain Pool as early as the mid 1970's with the first bulk shipment to Taiwan in 1975. Lupin exports to Japan followed in 1985 and to Korea in 1986.

There remains a great potential to develop new markets in Asia, including China, Philippines, Vietnam, Thailand North Korea and Indonesia. Demand for protein sources in Asian continues to grow in response to rapidly rising disposable incomes, associated with strong economic growth and changes in people diets which requires more sophisticated protein sources.

China, with a population of over one billion people with rapidly increasing disposable incomes, is a case in point. The capacity of China to consume large volumes of protein meal was demonstrated during the 1996-97 season when its demand underpinned the international protein market, importing an estimated 3 million tonnes of soymeal. Although China is the fourth largest soymeal producing country in the world, consumption has been surging at annual rate of 30 percent since 1990. China therefore has the potential to consume a large volume of lupins.

A major constraint to establishing new lupin markets in Asia is prohibitively high tariff regimes. In many of these countries an inequitable tariff system discriminates against lupins relative to other competing stockfeed ingredients, particularly soybean meal. A prime example is China where the tariff on lupins is 20 percent but only 5 percent for soybean meal. Lupins simply cannot compete with soymeal with such large tariff disparities. Formal submissions on a government to government level are continuing to push the case for the lowering of restrictive lupin tariffs imposed by China. This comes at an opportune time with China attempting to gain accession to the WTO.

A similar situation exists in Thailand where lupins attract a 30 percent tariff, compared with a zero tariff for soybean meal. This disparity, between closely competing ingredients, is a severe impediment to the development of the lupin trade in Thailand.

The challenge presented by inequitable tariffs is not new in lupin markets. Persistent efforts to reduce tariff levels have seen a number of successes. In May 1985, after concerted efforts by the Grain Pool, the Korean government reduced the import duty on lupins from 20 to 7 percent. This cleared the way for lupins to compete in the Korean market with other protein sources, mainly soymeal. During the 1996-97 season over 200,000 tonnes of lupins were exported to this market.

Similarly, formal approval from the Japanese Ministry of Agriculture and Fisheries for the importation of lupins into Japan in July 1985 came only after 10 years of solid background work. This included intense lobbying and the provision of supporting technical and scientific data. Today the Grain Pool exports around 100,000 tonnes per annum of lupins to Japan.

The past success of the Grain Pool in having prohibitive tariffs reduced gives confidence for further success in new markets. The recent world-wide trend towards trade liberalization and reduction of tariff barriers within the WTO framework is particularly encouraging in this respect.

FURTHER DEVELOPMENTS OF LUPINS FOR FEED

Even in established lupin markets significant opportunities remain, through continuing technical promotion to ensure appropriate and efficient utilisation of lupins, particularly in higher value rations. Technical promotion and reliance on scientific trials to foster market development have always been a mainstay in the marketing process. In some instances the successful use of lupins in one market can be showcased and, together with technical extension, be an effective means of promoting to a new market. In other cases, scientific trials must be part of the process to build customer confidence.

Further opportunities for lupins also exist in the diversification of end uses. The most promising avenue, currently being developed, is the potential for their use in aquaculture. Research conducted by a number of Australian institutions has pointed to the suitability of lupins as a major ingredient in aquafeeds, potentially replacing soybean meal and perhaps fishmeal.

Converting trial results to commercial reality is the next phase of this market development story and is being actively pursued. If successful, potential demand from countries with developed aquaculture industries such as Thailand, Taiwan, Indonesia and the Philippines is enormous. However, it is true that lupin use in aquafeeds is still relatively unknown and successful uptake in international markets will come only after considerable trials and associated market development work.

Increasing the value of lupins to feedmillers, and hence returns to growers, may be achieved by increasing their nutritional value. This can be achieved though plant breeding efforts focused on improving protein content and quality, reduction in seed coat thickness, increasing available energy, and from the use of new processing technologies and feed enzymes.

The discussion of the opportunities and challenges in the marketing of lupins covered so far has been largely in relation to feed applications rather than for human consumption. This is appropriate because the end use of lupins, other than small quantities of Albus lupins, has almost exclusively been as a stockfeed ingredient and probably will continue to be for some time to come.

HUMAN CONSUMPTION OF LUPINS

Australian Sweet Lupins, bred for their low alkaloid content, are a distinct product compared with the bitter varieties traditionally used for food purposes in the Middle East, the Mediterranean region and South America. Approval of sweet lupins for human consumption by the Australian government in 1998 gave encouragement to continue the research on food uses, which began in the early 1980's. A recent breakthrough has been the acceptance of lupins for human consumption by the United Kingdom in 1996 after more than 12 years lobbying by the Grain Pool. This approval came only after many years of scientific studies, including long-term toxicity studies, which proved beyond doubt that lupins were safe for human consumption. The acceptance by the UK is viewed as an important step in obtaining approval in other countries such as Japan.

For at least 15 years lupins have been the subject of laboratory scale investigations which have suggested the potential of the grain and its derivatives to be processed into a range of food products. These include Asian style fermented foods (miso, soy sauce, natto and tofu), flour for pastas and baked goods, a direct substitute for split peas, lentils and other pulses, beverages, sprouts and protein and fibre isolates. Whilst such laboratory and pilot plant scale investigations have been encouraging, it is only through commercial realisation of these products that their market potential will be proved. Unfortunately commercial interest in these products have been very limited and, to date, unsuccessful.

Lupin based food must compete in a market of well established and highly functional products, notably being dominated by foods of soybean origin. Their success will largely be dependent on their functionality relative to other food ingredient choices and relative cost. The competitiveness of lupins in this scenario remains largely untested on a commercial scale.

Given the range of possible food uses suggested by researchers over the many years, the opportunities and challenges of developing food markets for lupin products now lies in the hands of innovated commercial entrepreneurs. Opportunities may exist for exploiting niche markets in the short term, however a conservative view should be adopted in considering the long term potential for widespread adoption of lupin food products.

FUTURE OPPORTUNITIES

In conclusion, despite the many challenges in the international marketing of lupins, the large potential for expanding the utilisation of lupins in established feed markets and developing new markets continues to present exciting and rewarding opportunities. The recent world-wide trend towards trade liberalization, within the GATT framework, gives cause for optimism for the reduction of the prohibitive effects of tariffs, opening the way for new markets. While most of the significant opportunities for lupins are currently limited to feed uses, the longer-term potential for food markets also remain.

The use of pulses in Japan

T. Takabatake
Unicoopjapan Australia Pty Ltd, Suite 603, 25 Bligh St, Sydney, Australia 2000

PULSE FARMERS IN JAPAN

The total population of Japan is about 120 million with a working population of 64 million. The current ratio of farmers to the total population is about 6% and the number of farmers therefore is 4 million. Of these 800,000 grow pulses. The pulses most commonly grown are the Adzuki bean (*Vigna angularis*), peanut (*Arachis hypogea*), field pea (*Pisum sativum*), kidney bean (*Phaseolus vulgaris*), broad bean (*Vicia faba* L), and soybean (*Glycine max*). Other pulses such as chickpea, lentil, small-seeded faba beans (*Vicia faba* L), vetches, lupins (*Lupinus spp*) cowpeas, pigeon peas and mung beans are not produced in Japan.

The domestic demand for the first six pulses is 5.5 million tonnes per year but domestic production is only 200,000 tonnes per year. In other words Japanese self-sufficiency is only 3.6%. About 91% of the domestic demand or 5 million tonnes is for soybeans. Of this quantity 4 million tonnes are used for oil extraction, 900,000 tonnes for food and 100,000 for feed. The domestic production of soybeans is only 100,000 tonnes per year. The rest of the requirement is imported mainly from the USA and Brazil. All the other pulses imported to meet domestic demand come from China, the USA, Myanmar and Thailand

In the last 6 years the main producing areas in Japan and their average production of pulses has been

Adzuki bean –Hokkaido-about 80,000 tonnes.
Peanut- Chiba prefecture-30,000 tonnes.
Field pea- Hokkaido-about 600 tonnes.
Kidney bean- Hokkaido-about 30,000 tonnes.
Soybean-The northern part of Japan including Hokkaido 170,000 tonnes.

Whether produced domestically, or imported, these pulses are used for human consumption. They are either eaten as dry grain, boiled or are boiled and mixed with sugar. Only for soybean do the Japanese have some kind of traditional cooking which will be referred to later.

LUPINS AS FEED IN JAPAN

Currently 90,000 tonnes of lupins are imported each year. All of this quantity is imported by ZENNOH from the Grain Pool of WA (Western Australia). The Grain Pool initiated this import business in 1991. At that time the total demand for compound feeds in Japan was 25 million tonnes per year. Of this, ZENNOH's share was then, and continues to be, over 30%. ZENNOH found that lupins had high contents of protein and calories. As lupins had not been approved at the time a material for compound feeds in Japan, applications were made to the Ministry of Agriculture, Forestry and Fisheries providing them with information on DCP (digestible crude protein) and TDN (total digestible nutrients). At the same time the value of lupins as a feed was tested. They were found to have the same or more value than soybean meal, which is the main plant material used in compound feeds.

Lupins originally contained alkaloids, which meant they were difficult to use as material in compound feeds. But the improved lupins were found to be good material. Their proximate components are moisture 10.8%, protein 29.8%, fat 5.1%, fibre 15.3% and ash 2.7%. Lupins are unusual in having high protein and high fibre. The fat is about 5% less than soybean but more that for

R. Knight (ed.), Linking Research and Marketing Opportunities for Pulses in the 21st Century, 521–523.
© *2000 Kluwer Academic Publishers. Printed in the Netherlands.*

other coarse grains and so we can expect lupins to provide more calories. Due to their high percentage of fibre, lupins have a high value as feed for beef cattle. In addition we found lupins do not contain any urease activity and no trypsin inhibitors and can be used therefore without any treatment.

Digestible component

The digestible component is important in evaluating the value of material for feed. In Japan a digestibility test is required by the Ministry of Agriculture, Forestry and Fisheries before approval is given for material to be used in compound feeds. ZENNOH made, and submitted, the tests in 1991. It was found the digestibility of lupins was high for cattle but low for monogastrics such as pigs and chickens. The high fibre content of lupins was the cause of this effect in pigs and chickens. The values of TDN obtained were 83.2% for cattle, 71.2% for pigs and 44.6% for chickens.

From 1991 Australian lupins have been used as one of the main materials in compound feeds for cattle.

THE WORLD'S PROTEIN CONSUMPTION

About 80% of the world's protein production is plant protein. About 50% is from grains such as rice and wheat, 16% is from soybeans and 7 to 8 % of the rest from other foods such as potatoes, vegetables and fruit. We produce plant protein at four times the rate of animal protein. The present world population is 4.5 billion. It is expected to reach more than six billion by the end of this century. Given this situation we will need to make increasing use of plant protein to maintain our health in the future.

TRADITIONAL USES OF SOYBEANS AS FOOD IN JAPAN

In Japan rice has been the main grain used from early times with the addition of fish and shellfish as animal protein. The old life style was poor but these foods in combination with others gave a suitable protein level. Soybeans were not eaten raw. They were heated to improve their taste and change the chemical components. The trypsin inhibitor, which causes nutritional disorders, was made non-effective by heating. Soybeans also contain essential fatty acids, effective in preventing cholesterol from adhering to the surface of blood vessels.

The Japanese have many kinds of soybean processed foods including tofu, natto, miso and soy sauce. In the author's opinion the reason the Japanese have the highest life expectancy in the world originates from their traditional food culture.

Currently the quantity of soybeans used in Japan for tofu production is 300,000 tonnes per year, 98% of which is imported from USA, China, Brazil and Canada. Tofu is difficult to preserve and transport. There are about 35,000 tofu factories each using on average 40 to 50 kg per day.

Tofu is manufactured by first extracting the soluble components of soybean with boiling water to produce soy-milk. This treatment also denatures anti-nutritional protease inhibitors. The coagulation of the soy-milk is encouraged by the addition of calcium and salt and is promoted by thermal treatment. The curd is then broken, poured into a mould and compressed into a cake mold to set.

Tofu has a high water holding capacity of 90%. It contains 6% protein and 4% fat. Its digestibility is 98% with a good calcium level.

Natto is made by fermentation using Natto bacillus. The particular features of natto are its high viscosity and unique flavour. Natto has been produced in Japan for a very long time using the domestically produced middle to small sized bean. The domestic bean contains sufficient sugar to give it a slightly sweet taste and good flavour as a boiled bean. The bean is suitable for the propagation of Natto bacillus and the sugar is fermentable making the bean good material for the production of natto.

Due to the decrease in production of soybeans in Japan and the increase in their price, imports from China have increased each year. American soybeans are unsuitable and are not used for natto manufacture.

Natto is very popular among the Japanese. It is eaten with rice and soy sauce. It is sometimes used in the same manner as a slice of fish, egg or vegetables used when making sushi. The breakdown of tissue and the changes in the protein that occur during the manufacture of natto make it an improved foodstuff in comparison with normal soybean. Also the fermentation by the Natto bacillus give it a peculiar flavour and a smell attractive to the Japanese.

The current production of Miso in Japan is about 680,000 tonnes per year of which about 580,000 tonnes is industrial production and 100,000 tonnes by the farmer. This converts to 18 g per person per day. The basic materials are soybean, rice, barley and salt. These are mixed with water and then we make malted rice after boiling them. After adding more salt and water the mixture is left to ferment and mature.

The current demand for this brewing is about 1,800 tonnes per year. Of the Japanese market 90% is made in China and the USA, and the remaining 10% in Japan.

Soy sauce

Japan has many kinds of soy sauce but they can be divided roughly into two types. In one wheat is mixed with the soybean. This type shares about 85% of production. The other is that of a little more alcoholic drink made from fermented rice with the above type of soy sauce added. The production of soy sauce is about 1.2 million kilolitres a year in Japan requiring 190,000 tonnes of soybeans a year.

THE JAPANESE POLICY ON IMPORTS

I would like to refer to the Japanese policy regarding imports from foreign countries of pulses such as the kidney bean, broad bean, field peas and the Adzuki bean. Unfortunately there is still a barrier to the importation of these products and sale in Japanese markets. It is the tariff quota system. The Japanese market is liberalised in terms of trade; provided the customs duty is paid anyone can easily import these commodities into Japanese markets. The tariff quota for each year is decided by the Ministry of Agriculture, Forestry and Fisheries. If they judge the level of stock as being too small at the end of the fiscal year they will increase it and if too large decrease it. In other words the quota system is a policy to protect Japanese farmers.

The majority of the total demand for pulses in Japan is dependent on imports from China and other Asian countries. There are about 50 general trading companies, qualified by the Japanese government as importers, who have been operating for a long time. An example of the trade in pulses in Japan is provided by the Chinese Adzuki bean. They are repacked from 50 kg hemp bags to 60 kg hemp bags in order to change the packing to adjust the trade in the futures market. The beans are then delivered to a warehouse for polishing as only polished beans can be traded on the futures market. The cost of polishing is 1,200 to 2,500 yen per tonne and is carried out by the trading companies. The companies profit depends on the situation, sometimes they can make 150,000 yen per tonne, and at other times they will lose money.

Finally some further remarks about ZENNOH which is the National Federation of Agricultural Cooperative Associations in Japan. ZENNOH trades large volumes of grain, including pulses and imports about 1 million tones of US soybean each year and the lupins mentioned above from Australia. ZENNOH is looking for and is pleased to hear about other good materials particularly for use in compound feeds. A current typical target is canola. It handles more than 7 million tonnes of compound feed for Japanese farmers each year out of the total of 25 million tonnes used.

The use of pulses for feed in Australia

A.C. Edwards
ACE Livestock Consulting, P/L. PO Box 108, Cockatoo Valley, SA 5351, Australia

Abstract

Over the last 20 years, pulses have emerged from relative obscurity to become a significant component of the diets of intensively reared and grazing livestock in Australia. The particular pulses employed vary between regions and seasons. The usage is influenced by the geographic separation of supply and demand and competition from alternative markets (human foods and exports). The dominant commodities are lupins and peas but the stockfeed industry employs a wide range of pulses and their milling offals. The suitability of the pulses for each animal species varies with the species/variety of pulses but generally they have wide application and ready acceptance when competitively priced.

INTRODUCTION

The Australian livestock feeding industries are characterised by an ability and a preparedness to utilize a broad range of feed components. Diets for intensively reared livestock are generally based on cereal grains (wheat, barley, sorghum, triticale and oats) and have traditionally been supported by locally produced animal protein-meals (meatmeal, bloodmeal), a range of local vegetable proteinmeals (cottonseed meal, sunflower seed meal, rapeseed/canola meal, soyabean meal) and imported soyabean meal and fishmeal. Table I demonstrates the range of materials used as stockfeed ingredients in Australia.

Table 1. The range of feedstuffs utilized in livestock diets in Australia

Cereal	Milling offals & byproducts	Animal proteins	Vegetable proteins	Legumes	Other
Wheat	Wheat bran	Meatmeal	Soyabean meal	Lupins	Molasses
Barley	Wheat pollard	Bloodmeal	Canola meal	Peas	Salt
Oats	Rice pollard	Fishmeal	Sunflower meal	Faba beans	Limestone
Sorghum	Pea pollard	Poultry meal	Cottonseed meal	Chick peas	Phosphates
Triticale	Oat pollard	Feather meal	Safflower meal	Mung beans	Bentonite
Rye			Linseed meal	Lentils	Bicarb soda
Rice	Whey		Yeast	Vetch	Mg salt
Maize	Milk/Cheese			Cowpeas	Vit/min pax
	Vegetable pomace		F.F. oilseeds	Lablab	Amino acids
	Confectionery		Whole Cottonseeds	Culinary beans	Fats/Oils
	Almond shells			Adzuki beans	Organic acids
	Citrus pulp				
	Brewers grains				
	Potato wastes				
	Malt combings				

Over the last 20 years or so, grain legumes have risen from a position of near obscurity to become significant stockfeed components and export commodities. Table 2 presents the production and disposal statistics for Australian pulses over the last 7 years. This reveals that Australia currently produces of the order of 2.0 million tonnes of pulses annually relative to a total cereal grain production of around 25 million tonnes. Approximately 2/3 of current pulse production is exported and about 700,000 tonnes are utilized as stockfeed (primarily lupins and to a lesser extent peas) which represents about 8% of the total annual stockfeed production.

R. Knight (ed.), Linking Research and Marketing Opportunities for Pulses in the 21st Century, 525–529.
© *2000 Kluwer Academic Publishers. Printed in the Netherlands.*

526

Table 2. . Pulse production and disposal ('000tonnes) in Australia

	90/91	91/92	92/93	Year 93/94	94/95	95/96*	96/97*
Production	1326	1821	1936	2369	1213	2140	2009
Export	781	990	1392	1549	382	1449	1302
Domestic use	414	707	441	741	696	711	744
Food	35	31	32	30	30	18	18
Feed	379	676	409	710	666	693	726
Lupins	267	424	285	503	547	470	493
Peas	87	225	81	165	99	183	191
Faba beans	25	27	43	42	20	40	42

Note: Numbers may not balance due to changes in stocks and the use of grain for seed purposes.
Source: Meyers strategy group - Feed Grains Study (1995).   *Projected.

Grain legumes are generally viewed quite favourably as feedstuffs for livestock (Batterham & Egan, 1986). Their high energy and medium protein content means they can partially replace both the cereal and proteinmeal components of the diet. In the case of peas and lupins the absence of any significant antinutritional factors is also of considerable benefit. Table 3 presents typical values of peas and lupins relative to the major cereal grains and proteinmeals and the nutrient requirements of specific classes of pigs and poultry.

Table 3. Total feed usage by animal species in Australia in 1994/95 with the corresponding grain legume usage (Meyers Strategy Group, 1995)

Animal species	Total Feed* '1000 t	Legumes '1000 t	Proportion of total (%)
Pig	1877	151	8.0
Poultry	2295	156	6.8
Beef cattle	2054	94	4.6
Dairy cattle	1199	176	14.7
Other**	1310	91	6.9
Total	8735	669	7.7

* Not including roughage.
** Includes grazing sheep and cattle, horses, deer, aquaculture, ostrich, emus and petfood.

It is difficult to assign a specific relative value to a commodity as this will vary with changing dietary specifications, the species and age of the animal being fed, and the price and composition of alternative materials. This exercise is best undertaken in the least-cost formulation process via linear programming. Each material does not need to be a near perfect feedstuff in itself. The process of blending allows for the deficiencies and excesses of one to be complemented with reciprocal aspects in other materials to arrive at the optimum nutrient balance in the final diet for a specific feeding purpose. How well various materials complement each other is described as their "nichability", and tends to be high for cereal/grain legume combinations.

As reflected in Table 2 the use of specific pulses for stockfeeds in Australia is quite erratic from year to year and probably more so if analysed on a regional basis. There are numerous factors which contribute to this, some of which are:

Competition from alternative markets

When the export demand for human consumption or stockfeed markets is high, pulses can quickly price themselves out of consideration as domestic stockfeed commodities. In these instances it is usually only the down graded or non-export quality material that is utilized.

Conversely when export markets collapse, the resulting decline in price renders the pulses very competitive as stockfeed ingredients. As there is no absolute requirement for any particular commodity the interchangability of ingredients means that each must compete on its merits in terms of cost per nutrient supplied.

Geographic separation of supply and demand

Livestock production in Australia is spread over a wide area but tends to be concentrated in specific regions. When pulse production and end-users are geographically separated the cost of freight can compromise the competitive value of these commodities and force them into other markets. The classic example is the production of lupins in Western Australian, which far exceeds the local stockfeed demand, but the prohibitive freight costs to markets in the eastern states of Australia means that most of the crop is exported internationally.

Competition from other commodities

When meatmeal is in ready supply due to increased activity in the export beef market or when the international price for oilseed meals is low, pulses will find it harder to compete. Similarly when cereal grain is cheap and in great abundance it is often more economical to use more grain, complemented with proteinmeals and/or synthetic amino acids, than to use pulses.

Proximity of potential pulse end-users to abattoirs or oilseed crushing plants also influences the demand for pulses in specific locations.

Availability, reliability and timing of commitments

The sensitivity of grain production to world market prices and the freedom of growers to sow a range of crops can result in substantial swings in plantings from year to year, in response to anticipated gross margins from each.

Even though there is a general preparedness to use a wide range of alternatives many livestock feeders have their preferred materials and strive to achieve some degree of continuity. Consequently, wild variations in supply of certain commodities between seasons can be considered by some end-users as a negative aspect of using that material.

As livestock production intensifies and becomes less speculative or opportunistic, reliability of feed commodity supply, at fixed or contracted prices, becomes critical for budgeting and profit forecasting. Long term oilseed meal contracts are often written prior to the annual grain harvest and end-users, who perceive the projected pulse market as volatile, may commit to oilseed contracts and largely withdraw from the protein market. As a consequence of these interacting factors it is difficult to predict the volumes of pulses which may be consumed as stockfeed in any one year.

PULSE UTILIZATION BY ANIMAL SPECIES

Peas

Peas are readily accepted and digested by both pigs and poultry and are used freely when cost competitive, up to 25 - 30% inclusion in the diet. They are largely free of antinutritional factors and represent good food value (Davies,1989).

They are not used extensively in ruminant diets since the protein is mostly rumen degradable and hence the high biological value of the protein is not appreciated in ruminants. As a consequence peas tend to be too expensive for ruminant diets in that they can often be more competitively replaced by cereal grain and urea.

Lupins

Most of the lupins available for feed use in Australia are *L. angustifolius* varieties. These are used freely in pig diets but are not well favoured in poultry diets. The storage carbohydrate in lupins is primarily ß-galactan rather than starch. In pigs the digestibility of lupins is high, but about 40% of the digested energy is recovered from the hindgut as volatile fatty acids (Taverner et al.,1983). Since VFA metabolism is energetically less efficient than monosaccharide metabolism this feature reduces the net energy yield from lupins. Despite this, lupins still represent good value for pigs.

Poultry, having no effective hindgut, are not able to recover the nutrients not digested in the small intestine and hence record a relatively low energy value for lupins. As well as this, the passage of undigested non-starch polysaccharides tends to create wet, sticky droppings and for this reason lupins tend to be used sparingly in commercial poultry production.

Protein digestion and amino acid availability have been shown to be quite high for both pigs and poultry (van Barneveld and Hughes, 1994; van Barneveld, pers. comm.) despite some earlier references to the contrary.

The absence of starch and the highly fermentable nature of the carbohydrate fractions in lupins make them particularly attractive for ruminant diets. This allows them to be included at high levels to deliver high metabolisable energy contributions with minimal risk of lactic acidosis. The rumen by-pass characteristics of lupin protein are dependent on particle size and rumen fractional flow rates and, though variable, can be manipulated to make a useful contribution.

Dehulling of lupins can improve their net value by allowing the kernels to be directed to monogastric diets, while the highly fermentable hulls can be employed effectively in ruminant diets.

L. albus varieties have a higher protein and energy value then L. angustifolius varieties and although shown to be quite well utilized by poultry and ruminants they retard the performance of pigs (King, 1981). Factors such as alkaloids, saponins, oligosaccharides, tannins, high manganese levels, low methionine levels, have all been cited as possible causes but are largely disproven. Current theories suggest that delayed transit time through the hindgut may be the prime constraint to intake and subsequent performance.

Faba Beans

Most faba beans are grown primarily to meet an export demand and only become significant as stockfeed components when export demand eases or the product does not meet export standards.

On the basis of their amino acid and energy contributions faba beans are valued by nutritionists below peas and closer to lupins for pigs and poultry. Although they do not demonstrate any overt problems at modest levels of inclusion (up to 15%) faba beans do contain measurable quantities of several antinutritional factors (tannins, trypsin inhibitors, haemagglutinins). These features, plus commercial and scientific evidence of reduced growth and compromised reproductive performance when faba beans are fed at high levels of inclusion, have resulted in most nutritionists adopting a conservative approach to faba bean utilization in pig and poultry diets.

Faba beans are rarely used in ruminant diets unless the price is similar to cereal grains. Although the tannin content has been cited as possibly enhancing the bypass characteristics of the protein, this aspect is not valued to any great extent in commercial formulations.

Faba beans, also known as horse beans or tick beans are more highly valued in horse feeds but the volume employed for this purpose is of minor consequence.

Vetch

Traditionally, native vetches or tares have been viewed as unwelcome contaminants in grain. Varieties such as Popany (*Vicia benghalensis*) and Namoi (*Vicia desycarpa*) which have been used as under-cereal legumes for hay production, have also proven detrimental in that as little as 0.5% seed contamination in cereal grain can result in total feed rejection by pigs.

The introduction of feed varieties of vetch (Blanchfleur and Langedoc) showed considerable promise until concern was raised that they contain potentially neurotoxic compounds (cyanoalanine) if consumed in the raw state. Experimental and commercial evaluation of these vetch varieties revealed quite good performance and no neurotoxic effects but the alarm raised has largely extinguished any interest in growing vetch as a feed legume.

Other pulses

Materials such as chick peas, mung beans, lentils, adzuki beans, Phaseolus spp beans (Navy, Berlotti, White French, Red kidney, etc) cowpeas and *Dolichos lablab* are presented to the stockfeed market when their original target market (human consumption/export) is not reached, either due to depressed demand or inferior quality. As a consequence they are utilized opportunistically at reduced prices. It is generally not economic to grow these pulses for stockfeed purposes at the prices necessary to make them competitive with alternative feedstuffs.

A further complicating factor is that several of these materials contain serious antinutritional factors which require heat treatment to detoxify. This involves additional cost and logistical complications, which further lowers their value. Many of these materials are presented to the stockfeed trade as screenings and as such are not typical of the standard product and contain the added risk of incorporating toxic weed seeds. For these reasons they are often avoided by commercial feedmillers.

CONCLUSIONS

Grain legumes have established themselves as significant stockfeed components for intensive livestock production in Australia. Some are also used extensively as a supplement to grazing livestock.

Feeding value varies with species and variety of pulse and with the species and age of the livestock being fed, but in general pulses are regarded as very useful sources of energy and protein.

Features, desirable in pulses to promote their use, would be a low incidence of antinutritional factors, minimum processing requirements, high digestibility and palatability, and high quality protein (essential amino acid content).

The utilization of pulses as feedstuffs is enhanced where the production sites and end-users are in close proximity and where production is specifically orientated to stockfeed use by contractual commitments.

Future prospects for the use of pulses as stockfeed in Australia are sound with a potential consumption in excess of 1.5 million tonnes per annum where they are price competitive with alternative materials.

References

Batterham, E. S. and Egan A. R. 1986 In *Food Legume Improvement for Asian Fanning Systems*, pp 193-200 (eds E. S. Wallis and D. E. Byth). Canberra:ACIAR.

Davies, R. L. 1989 In *Recent Advances in Animal Nutrition in Australia*, 123-130 (ed D.J. Farrell). University of New England.

King, R. H. 1981 *Animal Feed Science and Technology* 6: 285-296.

Meyers Strategy Group. 1995 *Feed Grains Study.* Commissioned by GRDC, PRDC, DRDC, CMRDC, EIRDC.

Taverner, M. R., Curic, D. M. and Rayner, C. J. 1983 *Journal of the Science of Food and Agriculture* 34: 122-128.

van Barneveld, R. J. and Hughes, R. J. 1994 In *Proceedings of the First Australian Lupin Technical Symposium*, pp 49-57 (eds M. Dracup, and J. Palta). Dept. of Agriculture, Western Australia.

The role of pulses in the Asian livestock industry

Rhonda Treadwell
Agriculture Branch, ABARE GPO Box 1563 Canberra ACT 2601, Australia

Abstract

In this paper, China, South Korea, Indonesia and Thailand are examined to assess the prospects for consumption, production and trade in dairy products, meat and livestock and its effect on feed demand, with particular reference to pulses. Rapid income growth and high population growth in many Asian countries have led to rapid increases in food consumption particularly of livestock products. Some of the increase has been met by expanding domestic production, which has led to significant increases in the demand for feedgrains and protein meals. With an emphasis on production of food crops, their capacity to expand feed production is limited and feed imports are expected to rise.

1. THE ROLE OF PULSES IN THE ASIAN LIVESTOCK INDUSTRY

High income and population growth, in many Asian countries, have led to rapid increases in food consumption in the region. Increased incomes have allowed a change in diets away from cereal based staple foods to dairy and meat products. Despite this, consumption per person of these products is low, with levels varying widely. Consumption of livestock products has generally been highest in countries with higher incomes per person and higher urban populations (table 1).

Table 1. Consumption, incomes, population growth and urbanisation in 1995.

	Consumption/person		GDP/person		Population	Urban
	Dairy products kg	*All meats (a) kg*	*1995 $US*	*Growth rate 1991–95 %*	*Growth rate 1991–95 %*	*Population 1995 %*
China	7.5 b	43.4	560 c	11.0	1.1	30
South Korea	49.2	39.3	8 113 c	6.0	0.9	81
Indonesia	7.3	9.9	772 c	5.2	1.6	35
Thailand	16.0	22.2	2 155 c	6.8	1.2	20
Japan	68.9	43.6	36 174 c	0.4	0.3	78
USA	255.7	117.9	23 061 d	1.6	1.0	76
Australia	263.7	104.4	18 626 e	2.6	1.1	85

a Includes beef, sheep meat, goat, chicken and pigmeat, in dressed carcass weight terms. b China includes Taiwan. c 1990 prices. d 1992 prices. e 1989-90 prices.Sources: UN (1997); FAO (1997); IMF (1997); National Livestock Co-operatives Federation (1997); ASEAN Focus Group Pty Limited (1995).

Some of the increase in consumption has been met by rising domestic production, which has resulted in a significant increase in the demand for feedgrains and protein meals. However, their production capacity is limited and infrastructure is often inadequate in many Asian countries. Consequently, growth in production of dairy products, meats and livestock feeds has not kept pace with rising demand, resulting in rapid increases in imports of these products. In this paper, China, South Korea, Indonesia and Thailand are examined to assess the prospects for consumption, production and trade in dairy products, meat and livestock and the effect of these developments on feed demand, with particular reference to pulses.

R. Knight (ed.), Linking Research and Marketing Opportunities for Pulses in the 21st Century, 531–540.
© 2000 *Kluwer Academic Publishers. Printed in the Netherlands.*

532

1.1 Dairy products

Consumption per person of dairy products has increased in China, South Korea, Indonesia and Thailand, despite the fact these products are not part of the traditional diet. Consumption of dairy products in all four countries has been highest in urban areas because both income and product availability are higher than in rural areas (ADC 1993).

To service the growing consumer market for milk and dairy products in Asia significant dairy industries have developed, in some cases aided by high levels of government support.

China

Although per person consumption of dairy products in China is low, consumption has gradually increased since 1975. China has a small dairy industry relative to its population. However, since 1985 both milk production and cow numbers have more than doubled. Although imports of dairy products have grown rapidly from very low levels in the early 1980s (FAO 1997; ADC 1994) import volumes remain small, with around 80 per cent of the local demand being met from local production (ADC 1994).

South Korea

With rapid growth in consumption in the 1980s, South Korea has become the largest consumer of dairy products among the four countries and has a relatively well-developed dairy industry. Dairy cow numbers and milk production have increased steadily, aided by a favourable climate for dairying and high levels of government support particularly during the 10 year dairy development project 1976–85 (ADC 1993). Average herd size per farm has increased considerably as a result of the trend towards fewer, larger farms (Voboril and Kim 1996). Imports of dairy products are small, with around 93 per cent of local demand being met from domestic production (Voboril and Kim 1996).

Indonesia

Consumption per person of dairy products in Indonesia is low. Indonesia has a relatively underdeveloped dairy industry with most milk being produced on small farm holdings with only three or four cows (ADC 1993). Almost 60 per cent of total domestic demand for dairy products is met by imports (Podbury et.al. 1995).

Thailand

Per person consumption of dairy products in Thailand is high in comparison to Indonesia and China. Prior to 1960, Thailand had no domestic dairy industry and demand was fully supplied by imports. The industry has grown since, aided by restrictions on imports and local content arrangements. Despite this growth, the average herd size on Thailand's 16,000 farms is only around 30 cows. Growth in milk production is likely to be limited by the relatively small size of most farm holdings, inadequate education, a poor feed base and an unfavourable climate. In order to overcome some of these problems the Thai government has announced the Dairy Herd Development Scheme aimed at increasing the national herd by 65,000 cows over five years. Almost 80 per cent of the total domestic demand for dairy products is met from imported inputs (Podbury et.al. 1995).

1.2 Outlook for dairy products

Despite the fact that dairy products are not traditional Asian foods, prospects for increased consumption in these countries are good. Consumption of milk and dairy products in China, South Korea, Indonesia and Thailand is expected to increase at a faster rate than increases in production, and hence a greater volume of imported product is likely to be needed to satisfy domestic demand. Potential for import growth in all countries will be influenced by government policies which act to restrict imports and encourage local production. Such interventions are likely to continue, although the WTO agricultural agreement will constrain support levels.

2.   MEATS

Although the level of meat consumption per person is low in China, South Korea, Indonesia and Thailand it has grown considerably since the mid-1980s (table 2). Consumption preferences however, have remained relatively unchanged, with pork and poultry remaining strongly preferred to beef and sheep meat.

Table 2. Meat consumption ('000 tonnes).

	1986	1987	1988	1989	1990	1991	1992	1993	1994	1995
China										
Pork	17 767	18 149	20 006	21 025	22 573	24 255	26 236	28 394	31 867	36 254
Poultry a	1 851	1 982	2 676	2 795	3 187	3 959	4 556	5 800	7 642	9 582
Beef	563	759	904	1 015	1 101	1 313	1 729	2 184	3 199	4 062
Sheep meat	620	715	799	961	1 065	1 176	1 247	1 363	1 649	2 014
Total	20 801	21 605	24 385	25 796	27 926	30 703	33 768	37 741	44 357	51 912
South Korea										
Pork	401	466	531	665	630	644	731	767	798	838
Poultry a	203	222	235	243	274	292	373	389	398	452
Beef	205	210	196	198	244	308	313	317	372	416
Sheep meat	3	3	1	1	2	2	3	5	0	0
Total	812	901	963	1 107	1 150	1 246	1 420	1 478	1 568	1 706
Indonesia										
Pork	413	418	462	495	545	572	589	622	649	671
Poultry a	341	379	403	437	473	520	529	687	748	748
Beef b	261	240	217	261	260	284	327	442	373	390
Sheep meat c	78	79	77	82	83	84	87	112	114	95
Total	1 092	1 115	1 159	1 274	1 361	1 460	1 531	1 863	1 883	1 903
Thailand										
Pork	276	325	333	334	336	339	342	350	315	301
Poultry a	357	369	402	430	436	466	505	528	627	675
Beef b	226	227	228	234	243	249	252	265	349	353
Sheep meat c	1	1	1	1	1	1	2	1	1	1
Total	860	923	965	1 000	1 017	1 055	1 101	1 144	1 292	1 331

a For China and South Korea poultry includes chicken (or broiler), duck and other poultry meats. For Indonesia and Thailand poultry refers to chicken meat. b Includes buffalo meat for Indonesia and Thailand. c Includes goat meat for Indonesia and Thailand.

Sources: FAO (1997); USDA (1990, 1991, 1993a,b, 1997).

To a large extent, the increase in the consumption of most meats in these countries has been met by rising domestic production, rather than increases in meat imports. While meat production in all four countries has been supported by government policies, the significant growth in some meat sectors, particularly poultry, has been driven largely by the expansion of large-scale commercial farms and intensive livestock operations.

China

Since the mid-1980s China's meat consumption per person has increased by 120 per cent. However, growth in consumption has varied between the different meats and, with the exception of pork, consumption per person of individual meats is relatively low. In addition, growth in consumption is more rapid in some areas such as the coastal provinces which are experiencing greater economic development, urbanisation and income growth than other regions (Heng 1997).

Meat production has been growing by 9 per cent a year on average since 1987. In China there is an emphasis on self-sufficiency in pork. Eighty per cent of China's pig farms are still considered to be small-scale, subsistence-based family farms which use 'low cost technologies' to raise animals. The farmers traditionally use agricultural residues and by-products (such as rice and corn husks and potato vines) as animal feeds, though some use commercial feeds such as maize, barley and soybean meal (Fuell 1997). As the sector is becoming more commercialised, production is becoming more dependent on feedgrains and commercial feeds.

Growth in poultry meat production has averaged 16 per cent a year over the past decade. China currently ranks as the third largest producer in the world, but is projected to become the second largest after the United States (Instate Pty Ltd 1995). Poultry production has grown rapidly under the Chinese government's encouragement of large scale, commercial and intensive poultry farm operations. These tend to be partially owned by foreign companies, are vertically integrated to feed mills and meat processing plants and take advantage of economies of size. Further expansion of the poultry sector will be constrained by the availability and prices of feedgrains, inadequate animal health products and inefficient management practices (Instate Pty Ltd 1995).

In 1996, China became the third largest producer of beef in the world. Although technology is improving, the lack of technical and commercial management skills of cattle operators, poor quality genetics and inadequate feed are constraining the growth of beef production.

South Korea

Seafood (particularly fish) has been the major source of animal protein in South Korea. However, the share of meats (with the exception of sheep and goat meat) in total protein consumption has been increasing (Reynolds *et. al.* 1994). Meat consumption per person increased by 125 per cent over the decade to 1995, mainly driven by rising incomes per person and urbanisation. Pork is the meat mostly consumed, followed equally by beef and poultry.

Pork is the dominant meat produced and is South Korea's most advanced livestock industry. Production is becoming increasingly integrated (Ban 1995) and farm size is increasing. Strong demand for pork has resulted in high prices and record pig numbers, but the industry is still inefficient by world standards.

Beef production in South Korea is characterised by very small-scale enterprises with high costs of production. In 1995, the average herd size was only 4.4 head (Ban 1995). Traditionally, South Korean cattle are intensively fed on forage based rations (rice bran and other vegetable matter). However, due to the limited availability of land suitable for forage production, cattle have been increasingly grainfed in feedlots, largely with imported feedgrains (Reynolds et al. 1994). The rapid growth in demand for beef combined with an inefficient beef industry has led to beef consumption outstripping production in recent years. As a result, beef has accounted for approximately 70 per cent of South Korean meat imports.

Indonesia

Meat makes up a only small proportion of total food consumption in Indonesia. Seafood (particularly fish) is more abundant and relatively cheaper than all types of meat (David McKinna et al. Pty Ltd 1995). In recent years there has been a trend toward higher consumption of meat per person, although the levels are still low. While the rate of consumption growth has been different for each meat, poultry and pork have remained the dominant meats consumed (table 2).

In the past decade meat production has grown by 5 per cent a year on average. Production is dominated by small farmers who own 90 per cent of the animals. However, large scale, commercial, integrated farms are expanding their operations, particularly in the poultry and pork sectors. It is this expansion that has led to the strong growth in poultry and pork production and growing dependence on maize and other feeds such as soybean meal. Further growth will depend on the availability and price of feed.

In contrast to pork and poultry, beef production has not been able to match consumption growth. The availability of land has been a constraint (Dyck 1993). There are a small number of large private feedlots in Indonesia which are increasingly supplied by imported live feeder cattle. Feedlot operations are based on low cost feed in the form of industrial by-products from food processing industries (Rae 1995).

Thailand

Meat consumption patterns in Thailand closely follow those in Indonesia and seafood, which is relatively cheaper, is the dominant source of protein. Nevertheless, meat consumption has been growing rapidly, driven by strong economic growth and urbanisation.

Small scale producers raise 90 per cent of the livestock, although there is an increasing trend towards integrated, large scale operations, particularly in the pig and poultry industries. Total meat production has grown at an average rate of 4.6 per cent a year from 1986 to 1996, with poultry production exhibiting the highest growth. Thailand is largely self sufficient in all meats.

2.1   Outlook for meat

Meat consumption in Asia is expected to increase in the medium term, based on continued income and population growth and increasing urbanisation. The increase in consumption will be met mainly by increased domestic production rather than by significant increases in meat imports. Increased meat production will mean more feed will be imported to meet the demand of the livestock industries. In addition, some countries are expected to increase their imports of live animals to supplement local breeding.

3.   IMPLICATIONS FOR THE FEED SECTOR AND PULSES IN PARTICULAR

The strong growth in commercial production of livestock in China, South Korea, Indonesia and Thailand has increased significantly the demand for feedgrains and protein meals in recent years (table 3). In 1996, 121 million tonnes of grain were used for feed, which is approximately seven times the amount consumed in 1970. The use of protein meals in feed rations has increased by a similar proportion over the same period. Feed use in Asia is dominated by maize as it has higher nutritional value and is cheaper than other feedgrains (Ffoulkes 1997).

Table 3. Feed use ('000 tonnes).

	China		South Korea		Indonesia		Thailand	
	1986	1996	1986	1996	1986	1996	1986	1996
Maize	47575	89250	3540	6650	1500	4500	1600	4100
Wheat	2400	3500	1788	1800	0	140	0	235
Sorghum	3615	3500	45	99	na	na	85	190
Rice	3800	2000a	na	na	795	933a	700	500a
Cassava	1239	2183a	na	na	242	315a	226	212a
Soybean meal	1465	8723	930	1800	313	1025	423	1539
Peanut meal	1004	2187	na	na	35	161	47	126
Fish meal	285	855	172	105	52	100	227	491
Cottonseed meal	85	160	49	297	0	0	5	7
Rapeseed meal	267	354	170	520	50	65	na	na

a 1994 figures. na Not available.   Source: USDA (1997); FAO (1997).

From 1985 to 1995, the use of maize for animal feed in Asia rapidly increased (figure 1). With the high proportion of maize used, the use of high protein soymeal has increased to provide the necessary vegetable protein in feed mixes. Grain by-products, such as wheat pollard, maize bran and rice bran, also constitute important components in feed rations, as they are high protein and energy sources. Although 'by-products' such as rice and maize straw are abundant in the region, they are not widely used for livestock feed, because of their low nutritional value (Ffoulkes 1997). The use of pulses for feed has remained small and relatively constant (figure 1).

536

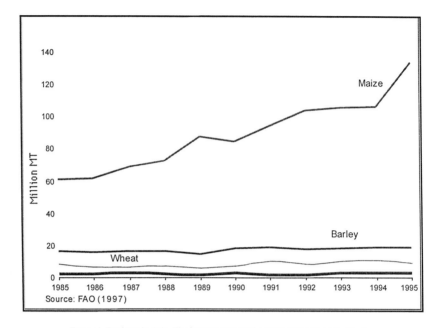

Figure 1. Feed use in Asia. The lowest curve on the graph is the feed use of pulses

As the domestic requirement for feedgrains and protein meals has grown, trade in these commodities has become increasingly important. Maize, wheat and rice dominate the grain trade, although most of the rice traded is used as human food. Maize imports have increased dramatically especially between 1994 and 1995. Soymeal imports also increased in the ten year period. While Asian imports of grains and protein meals over the period 1985 to 1995 have trended upward, imports of pulses have remained constant (figure 2). It must be noted that pulse imports in this figure consist of both food and feed, accordingly imports of feed pulses would account for even a smaller proportion of imports.

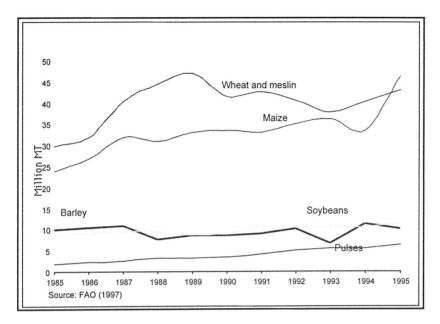

Figure 2. Asian imports

China

Maize is the principal feedgrain in China, accounting for around 70 per cent of total feed use. Accompanying the rise in maize, the use of high protein soybean meal in 1996 was almost twenty times higher than in 1980. In 1996, soybean meal made up almost 70 per cent of China's protein meal use.

In 1996, China produced almost 120 million tonnes of maize, second only to the United States. Driven in part by the country's rapidly expanding livestock sector, maize production has grown at an average annual rate of 5 per cent since 1986. As grain self-sufficiency remains a priority in China, the state trading corporation has kept grain imports to a minimum. In addition, China's grain imports are subject to tariffs and quarantine restrictions.

South Korea

The three main livestock industries in South Korea, poultry, pigs and cattle, are highly dependent on grains, protein meals and industrial by-products due to the limited availability of pasture. Maize is by far the most important feed in South Korea, with over 6 million tonnes used in 1996. However, South Korea's highly price-sensitive feed sector uses feed wheat as a substitute for maize when maize prices are high. Consequently, feed wheat accounted for 47 per cent of total feedgrain consumption in 1993, compared with 4 per cent in 1995, when wheat prices increased markedly. At equal prices, millers prefer feed wheat to maize for cattle and pig mixes because of its higher protein content, while maize is preferred for poultry (Morgan and Raney 1994). Millers also use small quantities of rye in feed mixes. Soybean meal is the most commonly used protein meal in South Korea. However, cottonseed and rapeseed meals make up a large part of feed meal consumption (approximately 30 per cent in 1996.)

Very little maize is grown in South Korea, reflecting the low returns from its production. Soybeans in South Korea are restricted to small plots along the borders and dykes of rice paddies. Virtually all domestic soybean production is used for food. Accordingly, livestock industries are increasingly dependent on importing feed. Maize and wheat dominate grain imports to South Korea, with respectively 63 per cent and 31 per cent of total imports in 1996. In 1995-96, the feed industry used 80 per cent of maize imports, while only 11 per cent of wheat imports were for feed.

Indonesia

The predominant feedgrain in Indonesia is maize, and its use has increased threefold in the last decade to reach 4.5 million tonnes in 1996. Broken rice is also utilised by the feed sector, and its use is increasing, although not as dramatically as maize. Soybean meal dominates protein meal use in Indonesia.

Indonesia is the second largest maize producer of the four countries, producing on average 5 million tonnes a year. However, the rapid increase in demand for maize, relative to production, has increased reliance on imports. To encourage the domestic production of maize, the government has introduced subsidies on farm credits and fertilisers (WTO 1994). In addition, a local content scheme is in place requiring local feed producers to buy a certain amount of their maize from local farmers at a contracted price (AgraFood Asia 1996). Nevertheless, the higher price for rice is limiting the increase in maize production (Alam 1997).

Thailand

The most important feed in Thailand is maize. It accounts for 50–65 per cent of the ration for broilers and 25–30 per cent for pigs (Kurz 1997). Although broken rice can be substituted for maize in pig rations, a practice common in Thailand, the amount of rice used for feed has been decreasing slightly since the early 1990s. This reflects the relatively high prices of rice compared with maize. The Thai livestock sector also uses small but increasing amounts of wheat as feed. Soybean and fishmeal are the most commonly used protein meals in Thailand.

Although maize production in Thailand has fluctuated markedly between years, it has been declining since the mid-1980s. To encourage maize production, the government provides input subsidies for fertiliser and seed. Thailand has largely liberalised its maize and wheat trade as it relies heavily on grain imports. However, the Thai government supports soybean production as part of a quasi-import substitution policy to deter the growth in imports (Giordano and Raney 1993). Imports of lupins, a protein rich livestock feed, are prevented by high import tariffs relative to those on soybeans and soybean meal (40 per cent for whole lupins and 30 per cent for de-hulled lupins).

3.1   Implications for the feed sector

The production of feedgrains is expected to be slowed by the expansion of food grains, reflecting the food self-sufficiency policies of many Asian countries. Furthermore, the availability of suitable land is expected to be a limiting factor. Since 1985, the arable area in these countries has declined, notably in China from 96 million hectares in 1980 to 91 million hectares in 1994. This decline is the result of continued population growth, urbanisation and rising industrial production (Crook and Colby 1996). The fall in South Korea's total arable land has been minimal. However, land is being allocated to other crops, such as vegetables and horticultural crops, resulting in a reduced area for feedgrains. Given these limiting factors, imports are expected to increase in the medium term to meet requirements.

In China, demand for feedgrains and protein meals is expected to increase steadily in response to the rapidly developing livestock industries. China is likely to grow gradually in importance as a grain importer. This is due to the likelihood that demand for feedgrains will outstrip production, and that government policies will accommodate these changes (Crook and Colby 1996).

In South Korea, the removal of import tariff quotas on pig and poultry products in 1997 is expected to result in more imports of these products and a decline in domestic meat production. This is expected to slow the growth in feed imports.

In Indonesia, demand for feed grain (particularly maize) is expected to increase, reflecting the growth in the poultry industry and expansion of feed mills.

Feed demand in Thailand is forecast to increase over the medium term, given the anticipated expansion in livestock production. The Thai Feed Mill Association estimated demand will increase by about 6 per cent a year until 2001 (AgraFood Asia 1997). However, domestic maize production is expected to remain steady in 1997-98, reflecting stable expected prices and a lack of farm labour. Some areas traditionally sown to maize may be planted to crops giving higher returns, such as sugar cane, or to more drought resistant crops, such as sorghum and soybeans. As a result, maize imports are expected to rise.

4. CONCLUSIONS

With continued income growth, Asian consumption of dairy products and meat is expected to increase over the medium term. For dairy products and beef in China, South Korea, Indonesia and Thailand, much of this increased consumption is likely to be met through increased imports. In South Korea, the removal of import tariff quotas on pig and poultry products is expected to result in higher imports and a gradual decline in pork and poultry production, thus reducing the need for feed imports. In contrast, the increased demand for pork and poultry in the other three countries is likely to be met largely through more domestic production. Continued integration and commercialisation of these industries is expected to lead to a significant rise in the use of feedgrains and protein meals, particularly maize and soybean meal. This will require increased imports of these feeds, given the limited arable land and continued population growth, urbanisation and industrialisation of these countries.

References
ADC (Australian Dairy Corporation) 1993, *Dairy Market Briefings: A Report on Selected Countries, Part 1, Asia*, Melbourne, April.
—— 1994, Update to *Dairy Market Briefings: A Report on Selected Countries, Part 1, Asia*, Melbourne.
AgraFood Asia 1996, 'Feedmillers/farmers in maize deal', *AgraFood Asia*, no. 27, Turnbridge Wells, July, p. 13.
—— 1997, 'Feed demand to rise sharply', *AgraFood Asia*, no. 36, Turnbridge Wells, April, p. 18.
Alam, N. 1997, 'Grain and feed', *FAS Online*, Foreign Agricultural Service Attache Report, AGR Number ID7011, US Department of Agriculture, [online], available URL http://www.fas.usda.gov/scriptsw/AttacheRep/attache_frm.idc, 8 April.
ASEAN Focus Group Pty Limited 1995, The Thai dairy industry, Paper prepared for the Australian Dairy Corporation, Sydney, 20 June.
Ban, Y.K. 1995, 'Livestock', *FAS Online*, Foreign Agricultural Service Attache Report, AGR Number KS5041, US Department of Agriculture, [online], available URL http://www.fas.usda.gov/scriptsw/AttacheRep/attache_frm.idc, 17 August.
Crook, F. and Colby, W.H. 1996, *The Future of China's Grain Market*, Economic Research Service, Agricultural Information Bulletin No.730, US Department of Agriculture, Washington DC, October
David McKinna et al. Pty Ltd 1995, South East Asian meat and livestock situation and summary', Paper prepared for Meat Research Corporation, Melbourne, September.
Dyck, J. 1993, 'Asia's diverse food system', *Asia and Pacific Rim*, Situation and Outlook Series, Economic Research Service, RS-93-6, September, US Department of Agriculture, Washington DC, pp. 18–20.
FAO (Food and Agriculture Organisation of the United Nations) 1997, *FAOSTAT Statistics Database*, [online], available URL: http://apps.fao.org/, cited February–July.
Ffoulkes, D. 1997, *Feeding Australian Commercial Cattle in South East Asia*, Northern Territory Department of Primary Industry and Fisheries, Darwin.
Fuell, L.D. 1997, 'Livestock', *FAS Online*, Foreign Agricultural Service Attache Report, AGR Number CH6044, US Department of Agriculture, [online], available URL http://www.fas.usda.gov/scriptsw/AttacheRep/attache_frm.idc, 16 January.
Giordano, M. and Raney, T. 1993, 'Thailand', *Asia*, Situation and Outlook Series, Economic Research Service, RS-92-5, August, US Department of Agriculture, Washington DC, pp. 136–44.
Heng, E. 1997, 'Opportunities in China', *Meat and Livestock Review*, Australian Meat and Live-stock Corporation, Sydney, June, pp.21–6.
IMF (International Monetary Fund) 1997, *International Financial Statistics*, Washington DC, April.
Instate Pty Ltd 1995, Structural change in poultry meat demand and supply in Asia: implications for the international competitiveness of the Australian poultry industry, Report prepared for the Chicken Meat Research and Development Council, June.

540

Kurz, P. 1997, 'Grain and feed', *FAS Online*, Foreign Agricultural Service Attache Report, AGR Number TH7031, US Department of Agriculture, [online], available URL http://www.fas.usda.gov/scriptsw/AttacheRep/attache_frm.idc, 26 March.

Morgan, N. and Raney, T. 1994, *South Korea: Determinants of Corn Import Demand*, ERS Staff Report AGES 9423, Economic Research Service, US Department of Agriculture, Washington DC.

National Livestock Co-operatives Federation 1997, *Materials on Price, Demand and Supply of Livestock Products*, Seoul.

Podbury, T., Ladlow, S., Roberts, I., Felton-Taylor, L. and Chaimun, S. 1995, 'East and South East Asian dairy markets: issues and challenges', *Australian Commodities*, vol. 2, no. 3, pp. 340–57.

Rae, A.N. 1995, 'East Asian food consumption patterns: Projections for animal products', *Agriculture Policy Paper*, no. 18, Massey University, Palmerston North, New Zealand.

Reynolds, R., Shaw, I., Lawson, K., Clark, J., Hamal, K., Bui-Lan, A., and Baskerville, N. 1994, *North Asian Markets for Australian Beef*, ABARE Research Report 94.10, Canberra.

UN (United Nations) 1997, *UN Database*, [online], available URL http://www.un.org/, cited March–July 1997.

USDA (United States Department of Agriculture) 1990, *Dairy, Livestock and Poultry : World Poultry Situation*, Foreign Agricultural Service, FL&P 3–90, Washington DC, August.

—— 1991, *World Livestock Situation*, Foreign Agricultural Service, FL&P 2–91, Washington DC, April.

—— 1993a, *Dairy, Livestock and Poultry: World Livestock Situation*, Foreign Agricultural Service, FL&P 4-93, Washington DC, October.

—— 1993b, *Dairy, Livestock and Poultry: World Poultry Situation*, Foreign Agricultural Service, FL&P 4-93, Washington DC, August.

—— 1997, *Production, Supply and Distribution Database*, Economic Research Services, Washington DC, June.

Voboril, D.B. and Kim, Y.J. 1996 'Dairy', *FAS Online*, Foreign Agricultural Service Attache Report, AGR Number KS6027, US Department of Agriculture, [online], URL http://www.fas.usda.gov/scriptsw/AttacheRep/attache_frm.idc, 28 June.

Warr, P.G. 1996, Poverty and economic growth: the case of Thailand, National Centre for Development Studies Briefing Paper, Australian National University, November 1996, Canberra.

WTO (World Trade Organisation), 1994, *Trade Policy Review Mechansim: Indonesia,* Report by Secretariat, C/RM/S/52, Geneva.

LUPIN BREEDING IN AUSTRALIA

W A Cowling [1] and J S Gladstones [2]
1 Senior Plant Breeder, Agriculture Western Australia, Locked Bag No. 4, Bentley Delivery Centre, WA 6983, Australia, 2 Consultant, 27 Pandora Drive, City Beach, WA 6015, Australia

Abstract

The lupin industry in Australia is based largely on *Lupinus angustifolius*, the narrow-leafed lupin. This species was developed as a new crop in Australia by J S Gladstones in the 1960s from semi-domesticated *L. angustifolius* introduced from Europe. The genes for low alkaloid content (*iuc*) and soft seeds (*moll*) were available in the Swedish forage variety Borre. Spontaneous mutants with white flowers and seeds (*leuc*) and non-shattering pods (*le* and *ta*) were found in the bitter landrace New Zealand Blue. These were recombined with *iuc* and *moll* to release the first commercial sweet varieties Uniwhite (1967) and Uniharvest (1969). However, it was not until early flowering (spontaneous mutant *Ku*) was found in Borre and recombined to produce the variety Unicrop (1973) that the narrow-leafed lupin became a viable crop for the main cropping regions of southern Australia. Breeding since that time has focussed on developing disease resistance, and concurrently yield and quality. Resistance to grey leaf spot (*Stemphylium vesicarium*) in Marri (1976) and Illyarrie (1979) and resistance to Phomopsis stem blight (*Diaporthe toxica*) in Gungurru (1988) and Merrit (1991) were major breakthroughs in the development of the crop. The role of wild lupins in these genetic advances was paramount. Genetic advances in yield of narrow-leafed lupins from Unicrop (1973) to Merrit (1991) occurred at an average rate of approximately 2% per year. Breeding and testing of lupins in Australia is funded partly by industry funds through a nationally co-ordinated project. All new varieties must contain less than 0.02% alkaloids on average. In addition to *L. angustifolius*, *L. albus* cv. Kiev Mutant has been very successful on more fertile loamy soils. A low alkaloid and early flowering variety of *L. luteus*, cv. Wodjil, was released in 1997 and is adapted to very acid sands in low rainfall areas. Breeders are also domesticating new lupin species such as *L. atlanticus* and *L. pilosus* for more alkaline and loamy soils. The greatest challenge facing breeders at present is the urgent need for anthracnose resistance.

INTRODUCTION

Lupins have become an important crop in southern Australia, especially in Western Australia where the the climate and most soils are well suited to the production of the narrow-leafed lupin (*Lupinus angustifolius* L.) (see Nelson and Hawthorne, this volume). However, when lupin breeding began in Australia in the 1950s, the future role of *L. angustifolius* was not obvious and early breeding and testing covered many species including *L. albus* L., *L. luteus* L. and *L. cosentinii* Guss. (Gladstones 1994). The former two were domesticated in Germany before and after World War II (discussed in Cowling *et al.* 1998) and early varieties were introduced to Australia in the 1950s. *L. cosentinii* proliferated as a wild plant after accidental introduction to Western Australia in the mid to late 1800s, where it was later used as a pasture species on poor sandy soils (Gladstones 1994).

Most of the introduced lupin varieties in the 1950s were too late flowering and maturing for successful production in the wheatbelt of Western Australia. Also, *L. albus* was not adapted to the predominant infertile sands of Western Australia and *L. luteus* cv Weiko III suffered from drought stress and severe virus diseases and insect attack. *L. angustifolius* grew well, especially in higher rainfall regions, and when natural mutants were found with non-shattering pods and early flowering, this species became a focus of the breeding effort. The final steps to domestication of *L. angustifolius* and the entire domestication of *L. cosentinii* occurred in the Western Australian breeding program (Gladstones 1994).

This paper describes the breeding of lupins in Australia since the 1950s with the emphasis on the most widely grown species, *L. angustifolius*.

R. Knight (ed.), Linking Research and Marketing Opportunities for Pulses in the 21st Century, 541–547.
© 2000 *Kluwer Academic Publishers. Printed in the Netherlands.*

DOMESTICATION OF *L. ANGUSTIFOLIUS*

The two main parents used in domestication of *L. angustifolius* in Australia were a soft-seeded green-manuring landrace introduced from New Zealand (originally from Europe) named New Zealand Blue, and a soft-seeded sweet forage variety from Sweden cv. Borre (Fig. 1). Borre inherited its low alkaloid gene *iucundis* (*iuc*) from a mutant selected in Germany by von Sengbusch in the late 1920s. The source of the soft-seeded gene *mollis* (*moll*) in both Borre and New Zealand Blue is not known. Von Sengbusch's original sweet variety Müncheberger Blaue Süsslupine II (released in 1944) was hard-seeded (Hackbarth and Troll, 1956). Because of these draw-backs, *L. angustifolius* remained as a minor fodder or forage crop until J S Gladstones in Australia discovered natural mutants in New Zealand Blue with the gene for white flowers and seeds *leucopsermus* (*leuc*) and the genes for reduced pod-shattering *lentus* (*le*) and *tardus* (*ta*) in the late 1950s and early 1960s (Gladstones 1967). The genes *moll, ta* and *leuc* were combined with *iuc* from Borre to develop the first fully domesticated *L. angustifolius* variety Uniwhite in 1967 (Fig. 1).

Pod shattering was reduced by *ta* in Uniwhite, but seed losses before harvest were sometimes high. The second non-shattering gene *le* acted in a complementary fashion to *ta*, and in Uniharvest (1971) shattering was almost eliminated by the presence of both genes (Gladstones 1967).

The lupin industry was still limited by the late flowering and disease susceptibility of Uniwhite and Uniharvest, and would not have expanded into the medium and low rainfall regions but for the discovery of a natural mutant with very early flowering (*Ku*) in Borre (Gladstones and Hill 1969). This mutant was crossed with Uniharvest to produce the early flowering cv. Unicrop which was released in 1973 (Fig. 1). At this stage the lupin industry in Australia seemed set to increase in area dramatically, but a series of setbacks due to drought, disease and poor agronomic practices limited the crop in the latter half of the 1970s. Were it not for the foresight of a few individuals, the breeding program may have been closed at that time (Gladstones 1994).

DISEASE RESISTANCE AND WILD PARENTS

A major disease that concerned lupin growers in Western Australia in the early 1970s was grey leaf spot, which attacked *L. angustifolius* varieties as they approached maturity in high rainfall regions. Following collaboration with a breeding group in the United States Department of Agriculture in Tifton, Georgia, USA, resistance to grey leaf spot and anthracnose was transferred from US breeding lines to Australian *L. angustifolius* varieties (Gladstones 1994). In return, the USDA group was able to make good use of the domestication genes discovered by Gladstones. Resistance to grey leaf spot and moderate resistance to anthracnose in cultivars Marri (1976) and Illyarrie (1979) was derived from the US cultivar Rancher, which in turn derived its resistances from a wild type from Portugal (Forbes *et al.* 1965).

Another hindrance to lupin production was the disease lupinosis which occurred sporadically in animals when they grazed lupin stubbles after harvest. Phomopsis stem blight of lupins was rarely a problem for the plants, but as the plants matured, the causal fungus, now known as *Diaporthe toxica* (Williamson *et al.* 1994), released a mycotoxin which made lupin stubbles toxic to sheep and other farm animals. Resistance to Phomopsis stem blight was found in wild *L. angustifolius* from throughout the natural habitat of the species in Morocco, Portugal, Spain and Italy (Cowling *et al.*, 1987). Some fully domesticated progeny from primary crosses, already made in 1975 between cv. Illyarrie and wild types from Spain and Morocco, proved equally resistant to Phomopsis, leading to the selection and release of cv. Gungurru, Warrah and Yorrel (Fig. 1) (Gladstones 1989). Resistance to Phomopsis stem blight reduced the toxicity to sheep of lupin stubbles (Cowling *et al.* 1988), thereby improving the value of lupins in mixed cropping/grazing farming systems (Warren *et al.* 1989).

The 1975 crosses with wild types were extremely important for the future development of the lupin industry in Australia. Not only did they result in resistance to Phomopsis stem blight, they also improved yields over the domesticated parent. For example, Merrit (1991) yielded about 20% more than its domesticated parent Illyarrie at high plant densities (Cowling and Speijers 1994).

Wild types of *L. angustifolius* from Israel contributed moderate resistance to brown spot (caused by *Pleiochaeta setosa*) in a cv. Myallie (1995) and Tallerack (1997) (Fig. 1). This moderate resistance was found to be highly heritable (Cowling *et al.* 1997) and was rapidly incorporated into new varieties as a result of a recurrent selection program for improved resistance (Cowling 1996). Wild types from Italy were also

Figure 1. Pedigree relationships among narrow-leafed lupin cultivars released in Australia 1967-97

Circles indicate key single crosses; diamonds represent complex crosses. Cultivars in bold were released in Western Australia except for Warrah (South Australia) and Wonga (New South Wales). Year of release follows cultivar name. Bold lines indicate crosses involving Australian cultivars.

Domestication genes: *moll* (soft seeds), *iuc* (sweet), *ta* (non-shattering), *le* (non-shattering), *leuc* (white flowers and seeds), *Ku* (early flowering).

Abbreviations: Gls-R (grey leaf spot resistant), An-R (anthracnose resistant), Ph-R (phomopsis resistant), Bs-MR (brown spot moderately resistant), MRB (moderately restricted branching), Pop. E1.1 (population E1 – first cycle of recurrent selection)

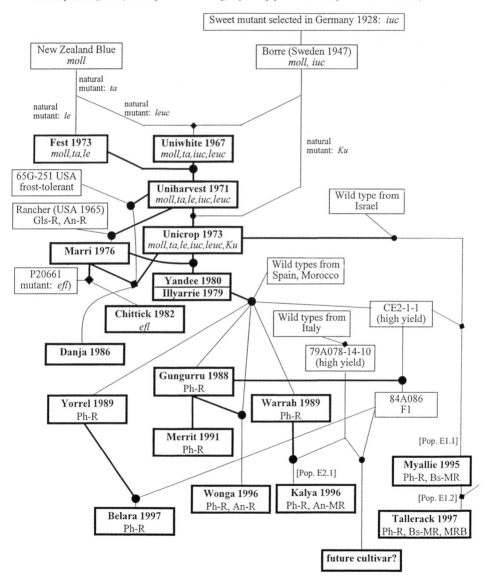

used in the development of breeding line 79A078-14-10, which had a late-season form of resistance to brown spot, and which has become an important parent in the breeding program (Fig. 1). This line is a parent of Kalya (1996), which has moderate resistance to Pleiochaeta root rot, and is a parent of high-yielding varieties for potential release in the near future (Fig. 1). Likewise, breeding line CE2-1-1 (derived from a primary cross between Illyarrie and an unknown wild type, probably from Spain) has been a key parent or grandparent for Myallie (1995) and Belara (1997). Belara is a further advance in both yield and resistance to Phomopsis stem blight. The F_1 and offspring of cross 84A086 (Gungurru/CE2-1-1) have featured in many crosses promising future improved varieties (Fig. 1).

Anthracnose disease became a major threat to the Australian lupin industry following its discovery in 1996. Fortunately, some advanced lines and cultivars such as Wonga and Kalya were found to have greater resistance to anthracnose than any previous cultivars. The genetics of resistance to anthracnose in Wonga is currently being investigated. Resistance most likely originated from the Spanish wild grandparent of Wonga. The other contributing parents are Illyarrie and Gungurru, but neither of these is as resistant to anthracnose as Wonga.

Wild types of *L. angustifolius* will continue to contribute disease resistance and other beneficial traits. Further primary crosses with wild types are made almost every year, although selection of fully domesticated progeny takes several generations (3-4 years from crossing) and the recovery of immediately useful genotypes is infrequent. The choice of wild types may be determined by disease resistance tests or ecogeographical studies of variation in *L. angustifolius* in its natural environment (Clements and Cowling 1994). Best progenies from primary wild by cultivated crosses are ploughed back into the breeding program, even if not suitable for release as varieties.

YIELD AND QUALITY

The success of the lupin industry in Australia is in part due to the quality control imposed by breeders on their material before it is released. Early in the breeding history of *L. angustifolius*, Gladstones made a deliberate decision to associate sweetness (*iuc*) with the genetically recessive gene *leuc* for white flowers and seeds He imposed a zero tolerance for bitter plants in the breeding program, so that released cultivars were all white flowered and pure sweet. Any bitter contaminants or cross-pollinations with bitter wild types were readily identified by their genetically dominant blue flowers and dark seeds (Gladstones 1967). Alkaloids are now controlled by food and feed standards in Australia that specify no more than 0.02% alkaloids in the whole seed. The consistent quality and low alkaloid status of Australian Sweet Lupins (the commercial name for *L. angustifolius* seed exported from Australia) is now recognized around the world.

It is relevant to note that sweet lupins have become a major crop only where such tight controls over alkaloids are possible and have been maintained. In Poland, the yellow lupin industry was kept free of bitter contaminants after World War II and sweet yellow lupins became a major component of cropping systems and animal feeds in that country (Cowling *et al.* 1998). In Mediterranean countries, it is harder (but not impossible) to keep sweet lupins free of bitter contaminants because bitter wild or landrace types are common in or adjacent to sweet crops..

The alkaloid standard of 0.02% imposes some limitations on breeding. The cost of alkaloid testing is substantial – about 10% of the budget for lupin breeding in Western Australia is expended on alkaloid and protein testing in 5000 breeding lines each year. Genetic improvements may be delayed due to unacceptably high seed alkaloids in early selections even though they may be pure for the major sweetness gene *iuc*. For example, moderate resistance to brown spot was first identified in the offspring of a cross between a wild type from Israel and Unicrop (Fig. 1). This breeding line had an alkaloid content in seed more than three times the industry standard and could not be released in its own right. It was therefore crossed with lower-alkaloid lines in an attempt to remove the alkaloid problem.

A recurrent selection program was set up that combined selection for brown spot resistance, improved yield, and lower alkaloids (Cowling 1996). As a result, the variety Myallie was released in 1996 with moderate brown spot resistance and low seed alkaloids (Fig. 1). A second variety Tallerack (1997) from this recurrent selection program had even stronger levels of brown spot resistance and lower seed alkaloids, combined with a mildly restricted branching character. It is therefore possible through recurrent selection to make progress in many economically important characters while meeting the alkaloid standard.

Breeding for lower seed alkaloids may have undesirable side-effects such as increased susceptibility to insects. The low-alkaloid cv. Tallerack is very susceptible to aphid attack. However, the breeding program has always selected for high aphid resistance in the field together with lower seed alkaloids, and in this way is selecting for forms of aphid resistance that are not linked to seed alkaloids. The cultivars Kalya (1996) and Wonga (1996) have high aphid resistance and seed alkaloids less than the industry standard of 0.02%. The mechanism of aphid resistance in these cultivars has yet to be elucidated. The cultivar Belara (1997) and many advanced crossbreds combine higher yields with seed alkaloid contents as low, or even lower, than the sweetest established cultivars.

The yields of lupins in Australia has increased steadily despite the massive increase in area from about 50,000 ha to 1 million ha during the 1980s (see Nelson and Hawthorne, this volume). This is in part due to genetic improvements in yield through breeding. Estimates of genetic improvements in yield were made from historic variety trials in Western Australia. From the release of Unicrop in 1973 to Merrit in 1991, genetic improvements in yield were above 2% per year at high plant density (Cowling and Speijers 1994). This rate of genetic improvement was due to improvements in harvest index from 0.24 in Unicrop to 0.29 in Merrit, and not due to increases in biomass (Tapscott et al. 1994). Compared with other grain legumes, narrow-leafed lupins still have a low harvest index and the potential for improvements through breeding is correspondingly high.

NATIONAL AND INTERDISCIPLINARY COLLABORATION

Lupin breeders in Australia have benefited from close collaboration with other scientific disciplines. A good example is the interdisciplinary work on Phomopsis resistance in lupins. Close working relationships were developed between lupin breeders, plant pathologists, chemists and veterinary pathologists to solve the lupinosis mystery. The link between lupinosis and Phomopsis was elucidated in the early 1970s, and it was shown that the fungus produced a mycotoxin that affected sheep in particular. By the late 1970s it was clear that there were few agronomic options to reduce the risk of lupinosis in lupin stubbles. Wild types were assessed for resistance by plant pathologists in conjunction with lupin breeders in the late 1970s (Cowling et al. 1987) and breeding lines were identified with improved resistance. In joint work with a veterinary toxicologist, these lines were shown to have lower toxicity in stubbles than the older varieties (Cowling et al. 1988). More recent work with chemists has resulted in the detection of very low levels of phomopsin mycotoxin associated with latent infections of D. toxica on green plants. This may be developed as a glasshouse test for resistance at a very early stage of infection (Williamson et al. 1995), as there is a close correlation between the level of stem resistance and the level of phomopsin detected in latently infected stems (Shankar et al. 1998).

Many other research activities have been associated directly or indirectly with lupin breeding in Australia, some of which are summarized by Gladstones (1994) and Cowling (1996).

After 1970, lupin breeding lines were tested by collaborators across southern Australia in an informal arrangement (Gladstones 1994). Late flowering lines did much better in the longer growing seasons of south-eastern Australia than in Western Australia, and the breeding of late flowering L. angustifolius (lacking the Ku gene) was eventually handed over to breeders in Wagga Wagga, New South Wales, in 1989. The program in Western Australia was expanded with the appointment of a second breeder in 1976, focussing first on disease resistance in L. angustifolius with J Hamblin and later combining yield, quality and disease resistance with W A Cowling (Cowling 1996).

The informal arrangements for lupin breeding across southern Australia have been formalized in a national lupin breeding project co-ordinated by W A Cowling and funded by levies on grain production and Government sponsorship through the Grains Research and Development Corporation. This program, known as the Australian Co-ordinated Lupin Improvement Program (ACLIP) co-ordinates the testing and breeding of lupins in Agriculture Western Australia, New South Wales Agriculture, Agriculture Victoria, South Australian Research and Development Institute, the Chemistry Centre of WA, and the Centre for Legumes in Mediterranean Agriculture (CLIMA).

CLIMA has developed technology for producing transgenic lupins, and the world's first transgenic cultivar (a herbicide resistant transgenic plant derived from L. angustifolius cv. Merrit) is currently being field tested in the breeding program and is undergoing seed increase for release by CLIMA. CLIMA supports close collaboration between molecular biologists and lupin breeders, including the development of a molecular map and molecular markers for lupins at Murdoch University in Perth; production of transgenic lupins with high sulphur genes at CSIRO Division of Plant Industry in Canberra; and development of the first transgenic yellow lupin (*L. luteus*) with potential resistance to bean yellow mosaic virus-disease at Murdoch University.

NEW LUPIN SPECIES

In addition to *L. angustifolius*, Gladstones (1994) developed the first domesticated *L. cosentinii* cultivar Erregulla. This cultivar was non-shattering and sweet but had hard seeds. The potential role of this type of plant was as a self-seeding crop that would germinate sufficiently in subsequent years to produce a series of crops with minimal input. A soft-seeded mutant was selected from Erregulla, but this was quite low yielding compared with Erregulla and *L. angustifolius* cultivars. Both Erregulla and its soft-seeded mutant were highly susceptible to aphids and neither was taken up commercially.

An interspecific crossing program was started in the early 1980s to transfer the domestication genes in Erregulla to other rough seeded lupins (Roy and Gladstones 1985, 1988). Gladstones already had selected sweet mutants out of *L. atlanticus*, which were used as a source of low alkaloids. This work was taken up in 1988 (Buirchell and Cowling 1992), and as a result of this long-term effort the first domesticated *L. atlanticus* plant has been developed and is targeted for release in about 2001. Buirchell is now targeting *L. pilosus* which has the highest tolerance of all Mediterranean lupin species to alkaline soils. The objective is to develop a lupin adapted to the large areas of alkaline and heavier soils of southern Australia.

L. albus cv Kiev mutant was grown on 25,000 ha annually in Western Australia until 1996, when the disease anthracnose was found to severely attack this variety. It has not been grown commercially since that time. Breeders in Australia are now collaborating to breed anthracnose resistant albus lupins.

Australia's first *L. luteus* variety Wodjil was released in 1997. It is derived from a single plant selection from the Polish cultivar Teo (released by Poznan Plant Breeders). Wodjil has 0.02% alkaloids whereas Teo has about 0.08% alkaloids in the seed. As a result of close collaboration between the breeders in Poland and Australia, the selection of the low-alkaloid variety Wodjil in Australia has benefited both countries. Wodjil is adapted to the very acidic sands of the eastern wheatbelt of Western Australia and is tolerant of aluminium toxicity on these soils. Like Teo and other *L. luteus* cultivars, it also very resistant to Pleiochaeta root rot and brown spot (Yang *et al.* 1996).

CONCLUSIONS

The development of lupins as a crop plant in Australia demonstrates that successful breeding depends on dedicated and committed individuals as well as strong collaboration (both national and international). Close collaboration among scientists in many disciplines ensured rapid and sound progress in disease resistance breeding while improving yield and quality, and successful crop production was ensured by a strong extension program to farmers who were themselves keen pioneers. Close collaboration with lupin marketers ensured that the high standards and quality of lupins produced in Australia were impressed on potential buyers in international markets.

High quality standards were adopted from the beginning by breeders (for example, adherence to the strict alkaloid standards), and as a result world grain traders have now accepted 'Australian Sweet Lupins' as a valuable source of feed, food and fibre. This represents a major change of view over the past 30 years, from that where all lupins were considered to be bitter and next to useless as a stock feed.

The breeding program has continued to ensure a broad genetic base through crossing with wild lupins, and has continued the development of new species adapted to different soil types despite an economic environment that would prefer not to support such long-term research. The adaptation of lupins to alkaline soils, and soils with a higher proportion of clay, depends on such long-term research.

L. angustifolius has also become a model for recurrent selection for disease resistance and other important traits in self-pollinating crops (Cowling 1996). Recurrent selection in *L. angustifolius* has been successful in the release of new cultivars and in providing improved germplasm for future breeding efforts.

References

Buirchell, B.J. and W.A. Cowling. 1992. *Journal of Agriculture - Western Australia* (4th Series) 33:131-137.

Clements, J.C. and W.A. Cowling. 1994. *Genetic Resources and Crop Evolution* 41:109-122.

Cowling, W.A. 1996. Presidential Address 1994. *Journal of the Royal Society of Western Australia* 79:183-194.

Cowling, W.A. and E.J. Speijers. 1994. In *Proceedings of the First Australian Lupin Technical Symposium.* p 262 (Eds. M. Dracup and J. Palta) Department of Agriculture Western Australia, Perth.

Cowling, W.A., J.G. Allen, and P.McR. Wood. 1988. *Australian Journal of Experimental Agriculture* 28:195-202.

Cowling, W.A., C. Huyghe and W. Swiecicki. 1998. In *Lupins as Crop Plants: Biology, Production and Utilization* (eds J.S. Gladstones, C.A. Atkins and J. Hamblin,.). CAB International, London (in press).

Cowling, W.A., J. Hamblin, P.McR. Wood, and J.S. Gladstones. 1987. *Crop Science* 27:648-652.

Cowling, W.A., M.W. Sweetingham, D. Diepeveen and B.R. Cullis. 1997. *Plant Breeding* 116:341-345.

Forbes, I., H.D. Wells and J.R. Edwardson. 1965. *Phytopathology* 55:627-628.

Gladstones, J.S. 1967. *Australian Journal of Experimental Agriculture and Animal Husbandry* 7:360-366.

Gladstones, J.S. 1989. *Journal of Agriculture - Western Australia* (4th Series) 30:3-7.

Gladstones, J.S. 1994.. In *Proceedings of the First Australian Lupin Technical Symposium* pp 1-38 (eds M. Dracup and J. Palta). Department of Agriculture Western Australia, South Perth.

Gladstones, J.S. and Hill, G.D. 1969. *Australian Journal of Experimental Agriculture and Animal Husbandry* 9:213-220.

Hackbarth, J. and H.-J. Troll. 1956. In *Handbook of Plant Breeding*, Part IV pp. 1-51 [in German] (H. Kappert and W. Rudorf, eds.). 2nd edn. Verlag Paul Parey, Berlin and Hamburg.

Roy, N.N. and J.S. Gladstones. 1985. *Theoretical and Applied Genetics* 71:238-241.

Roy, N.N. and J.S. Gladstones. 1988. *Theoretical and Applied Genetics* 75:606-609.

Shankar, M., W.A. Cowling, M.W. Sweetingham, K.A. Than, J.A. Edgar, and A. Michalewicz. 1998. *Plant Pathology* (in press).

Tapscott, H.L., W.A. Cowling, M. Dracup and E.J. Speijers. 1994. In *Proceedings of the First Australian Lupin Technical Symposium.* p317 (eds. M. Dracup and J. Palta) Department of Agriculture Western Australia, Perth.

Warren, J., J.G. Allen and W.A. Cowling. 1989. *Journal of Agriculture - Western Australia* (4th Series) 30:8-10.

Williamson, P.M., A.S. Highet, W. Gams, K. Sivasithamparam, and W.A. Cowling. 1994. *Mycological Research* 98:1364-1368.

Williamson, P.M., K.A. Than, K. Sivasithamparam, W.A. Cowling, and J.A. Edgar. 1995. *Plant Pathology* 44:95-97.

Yang, H.A., M.W. Sweetingham and W.A. Cowling. 1996. *Australian Journal of Agricultural Research* 47:787-799.

Development of lupins as a crop in Australia

P. Nelson [1] and W.A.Hawthorne [2]

1 The Grain Pool of Western Australia, GPO Box A24, Perth, W Australia; 2 South Australian Research and Development Institut , Naracoorte, S Australia 5271

ABSTRACT

In just twenty years, a new crop plant, *Lupinus angustifolius* has been developed to become a major cropping industry in Australia. This occurred as a result of a successful project of research and development undertaken mainly in Western Australia (WA). It involved plant breeders, agronomists, marketers, farmers and extension personnel. This paper documents the development and highlights the benefits of research and development occurring concurrently.

Before the development of lupins, cropping systems in WA involved either continuous cereal or pasture/cereal rotations. Problems with weeds and diseases were prevalent. Lupins are now the cornerstone of many farming systems in Australia, especially WA where there is a predominance of acidic, sandy soils. Lupins are an economic crop in their own right and provide benefits to other crops in terms of weed management, nutrition and disease breaks which lead to higher yields. Lupins play a vital role in the successful integration of stock and crop enterprises on many farms. Development of varieties, which are resistant to the fungus, *Phomopsis leptostromiformis* has reduced the risk of the disease lupinosis in sheep.

In 1996, 1.5 million hectares of lupins were grown in Australia as a grain crop and earned US$190 million of export income with a further estimated value of US$120 million being generated from sheep feed and increased cereal yields. It is forecast that varieties of *Lupinus angustifolius* will be cultivated on 2 million hectares before the year 2005. A national research and development program continues to develop better varieties of *L.angustifolius*. In addition other lupin species are being developed which are more suited to the heavier and the finer textured soils and also very acid soils.

In 1996 for the first time, the serious disease anthracnose (*Colletotrichum gloeosporoides*) was recorded in commercial lupin crops. Strategies are now in place to eradicate or minimise the effects of this disease through quarantine, plant breeding and agronomic practices. *Fusarium oxysporum* has not been recorded in lupin crops and perhaps the anthracnose experience will strengthen our resolve to prevent this disease entering Australia.

INTRODUCTION

The state of Western Australia (WA) has led the way in the development of the Australian lupin industry, and has consistently sown some 85% of the Australian lupin area. The area lies between latitudes 27°S and 35°S and is subject to a Mediterranean climate with mild wet winters and hot dry summers. Rainfall varies from greater than 750 mm in the south-western corner of the state to less than 300 mm in the eastern and northern agricultural areas. The main cropping areas lie in the 450-mm to 300-mm rainfall zone. In the eastern states (South Australia, Victoria, New South Wales and to a lesser extent Tasmania), the main cropping areas extend to 38°S, and rainfall to 600 mm.

The soil types on which lupins grow well predominate in WA, and are acidic and sandy surfaced. Many are deep sands or sands over gravel or clay at varying depths. Other pulse species grow poorly on these soils, which are infertile in the virgin state and subject to wind erosion when bare of plant cover. Lupins however, have also been found to grow well on heavier textured, acidic soils in eastern Australia.

In Australia, broad scale farming has relied on wheat and wool production for most of its income. The cereal and sheep areas have about 110 million sheep and 19 million hectares of crop. Before 1970, farming systems often involved multiple cropping of wheat on the more loamy or heavier soils. However, on sandy surfaced soils, a ley farming system was practiced, whereby the land was in pasture for three or four years

549

R. Knight (ed.), Linking Research and Marketing Opportunities for Pulses in the 21st Century, 549–559.

and then cropped to cereals. This system integrated the stock and crop enterprises on the farm. During the mid 1970's wheat production became more profitable than sheep and wool and, with the use of nitrogenous fertilisers, the sandy soils were multiple cropped to cereals. After several consecutive cereal crops, grass weed control became difficult and often the land was cultivated 5 or 6 times to control weeds before sowing. Yields gradually became poorer and the financial debt of farmers increased. Continued working of the land led to serious wind erosion and cereal root diseases seriously reduced yields. There was an urgent need to develop new farming systems, preferably legume based, to prevent further deterioration of the sandy soils. Lupin varieties were available, but not widely grown because farmers considered them a high risk crop suited only for areas with annual rainfall greater than 450 mm.

To successfully breed, develop, and then integrate a new crop plant into large scale, established farming systems is rare. This feat was achieved in Australia between 1958 and 1981, led by WA because of its suitable soil types and biological and economic necessity. Uniwhite was the first commercial variety of the sweet narrow-leafed lupin *Lupinus angustifolius* released by John Gladstones to farmers in WA in 1967. Thirty years later, 1.5 million hectares of this species are cultivated by Australian farmers. Farmers of sandy and acidic soils now have a system that is both agronomically and economically diversified and, in economic and ecological terms, promises to remain sustainable.

This paper outlines the contributions made by breeders, plant and animal researchers, marketers, farmers, advisers and funding bodies in accomplishing this task. After the initial release of cv. Uniwhite, breeding, research production and marketing occurred concurrently. The state of WA pioneered the development, but the eastern states of Australia also quickly recognised the role of lupins on specific sandy or acidic, heavy soils, but on a lesser scale than in WA.

LUPIN BREEDING

In the late 1920's the German plant breeder R. von Sengbusch found from field selections, lines of *L. angustifolius* that were low in alkaloid content. Swedish workers using these lines further developed sweet, soft-seeded lines of this species.

In 1954, John Gladstones started lupin evaluation at the University of WA. He established that the narrow-leafed lupin was the species most likely to do well under WA conditions, and in 1958 commenced a breeding program on *L. angustifolius*. Gladstones' task was to develop a crop plant that had sweet seeds, non-shattering pods and early maturity. To readily distinguish it from naturalised types of narrow-leafed lupins, which were blue flowered and bitter, Gladstones selected white flowers and seeds as markers. He selected natural mutants from field populations of the existing cultivars New Zealand Blue and Borre followed by intercrossing to combine all the desirable characteristics into a crop type (Gladstones 1977). Later breeding was based on crosses with wild types selected from native habitats around the Mediterranean. He achieved his first goal with Uniwhite in 1967 and since that time, he and his successors have released 18 varieties with further improvements (Table 1).

A collaborative project with workers from the Coastal Plains Experiment Station at Tifton, Georgia, United States of America - I. Forbes (breeder), H. D. Wells and J. R. Edwardson (plant pathologists), produced lines of narrow-leafed lupins that were resistant to the diseases anthracnose vcg 1 strain (*Colletotrichum gloeosporoides*) and grey leaf spot (*Stemphylium vesicarium*).

Releases since 1988, derived from crosses with wild Mediterranean sources, have included moderate resistance to *Phomopsis leptostromiformis,* the fungus responsible for lupin stubble toxicity and lupinosis disease in sheep (Cowling *et al.* 1986).

Future breeding work is aimed at improving resistance to *Pleiochaeta setosa* root and leaf diseases, yield, insect and virus resistance in *L. angustifolius* (Cowling 1994). *L. atlanticus* and *L. pilosus* (rough seeded lupins) are new species being bred for domestication as crop types for finer textured and alkaline soils (Buirchell and Cowling 1992). The yellow lupin, *L. luteus,* is now being bred and developed for very acid soils containing high aluminium levels. Wodjil was released in 1997 as the first variety, bred for these problem soils, in WA.

National collaboration on lupin variety and breeding line evaluations commenced formally in 1979, but had been occurring since 1972. The recent outbreak of the disease anthracnose has focussed attention on this problem and screening and breeding programs are now under way.

During the late 1970's, Rex Oram, Commonwealth Scientific Research Organisation (CSIRO), Canberra had been breeding late flowering lines for the eastern states. This included *L. angustifolius* and *L. albus* until the early 1980's when funding ceased. Lupin breeding specifically for the eastern states recommenced in 1989 as part of a nationally funded program (Landers 1991). The focus is on heavier soils and the breeding of later and earlier flowering lines. Later flowering lines, which require vernalisation, are needed for areas in the east where early sowing is desirable, but flowering needs to be delayed until the risk of frost has passed. The need for vernalisation is not relevant to the shorter growing seasons of WA. Three narrow leafed lupin varieties have been released specifically for the eastern states (Table 1).

Table 1 Narrow-leafed lupin varieties in Australia (the bold type shows the difference between each variety and its predecessor).

Variety	Year of release	Comments
Uniwhite	1967	Soft-seeded, white-flowered, late-flowering, partially non-shattering and sweet.
Uniharvest	1971	Soft-seeded, white-flowered, late-flowering, **non-shattering**, and sweet.
Unicrop	1973	Soft-seeded, white-flowered, **early-flowering**, non-shattering and sweet.
Marri	1976	Soft-seeded, white-flowered, **late-flowering**, non-shattering, sweet **resistant to grey leaf spot**.
Illyarrie	1979	Soft-seeded, white-flowered, **early-flowering**, non-shattering, sweet and resistant to grey leaf spot. **Adapted to northern farming areas**.
Yandee	1980	Soft-seeded, white-flowered, early-flowering, non-shattering, sweet and resistant to grey leaf spot. **Adapted to southern farming areas**.
Chittick	1982	Soft-seeded, white-flowered, mid-season flowering, non-shattering, sweet and resistant to grey leaf spot. **Adapted to long growing season areas**.
Danja	1986	Soft-seeded, white-flowered, early-flowering, non-shattering, sweet and resistant to grey leaf spot. **Higher yielding replacement for Yandee and Illyarrie**.
Wandoo	1986	Soft-seeded, white-flowered, **mid-season flowering**, non-shattering. Sweet and resistant to grey leaf spot. **Replacement variety for Chittick**.
Geebung	1987	Soft-seeded, white-flowered, **late-flowering**, non-shattering, sweet and resistant to grey leaf spot. **Released for long season areas in eastern Australia**.
Gungurru	1988	Soft-seeded, **white-flowered with purple tinge, seeds have distinctive brown markings, early-flowering**, non-shattering, sweet and resistant to grey leaf spot. Moderate resistance to *Phomopsis leptostromiformis*. **Greater tolerance to the pre-plant herbicide simazine**.
Yorrel	1989	Soft-seeded, white-flowered with purple tinge, **seeds have distinctive brown markings, very early-flowering**, non-shattering, **very sweet**, and resistant to grey leaf spot. Moderate resistance to *Phomopsis leptostromiformis*. **Very susceptible to aphids and brown leaf spot. Released for heavier soil types**.
Warrah	1989	Soft-seeded, white-flowered, white seeded, **early-flowering**, sweet, non-shattering, resistant to grey leaf spot. Moderate resistance to *Phomopsis leptostromiformis*. **Very short therefore needs long seasons and good growth. Released in South Australia**.
Merrit	1992	Soft-seeded, white-flowered with purple tinge, seeds have distinctive brown markings (**2% of seeds lack triangular mark above the hilum**), early-flowering, non-shattering, sweet, and resistant to grey leaf spot. Moderate resistance to *Phomopsis leptostromiformis* **Merrit is a replacement variety for Gungurru in the low, medium and some high rainfall areas**.
Myallie	1995	Soft-seeded, white-flowered with purple tinge, seeds have pale brown markings. Early-flowering, non-shattering, sweet and **resistant to brown leaf spot**. Moderate resistance to *Phomopsis leptostromiformis*. Myallie was released for low and medium rainfall areas where brown leaf spot was a particular problem.
Kalya	1996	Soft-seeded, white-flowered and **white seeded**. Early-flowering, non-shattering, sweet and moderately resistant to *Phomopsis leptostromiformis*. **High yielding** Kalya is a replacement variety for Merrit in the medium to high rainfall areas. **Resistant to anthracnose vcg 2**.
Wonga	1997	Soft-seeded, white-flowered, seeds have brown markings. Early-flowering, non-shattering, sweet and moderately resistant to *Phomopsis leptostromiformis*. Wonga was released for **heavy textured acid soil** in New South Wales . Resistant to anthracnose vcg 2.
Belara	1997	**Has quicker seed filling, and is early maturing for lower rainfall areas**. Early-flowering, non-shattering, sweet and moderately resistant to *Phomopsis leptostromiformis*
Tallerack	1997	**First restricted branching type, short with prolific pod set, very early maturing due to its determinate habit**. Its use in low and very high rainfall areas is being determined. Early-flowering, non-shattering, sweet and moderately resistant to *Phomopsis leptostromiformis*. **Very susceptible to aphids**

L. albus has always been a relatively minor but expanding crop for the heavier, acidic soils. Niche markets have been established. Resistance to cucumber mosaic virus is an added advantage of the species over *L. angustifolius* in north-eastern Australia. The predominant variety has been Kiev Mutant (early-flowering), but Ultra (early-flowering) and Hamburg (late flowering) have also been grown. All have been sweet selections from overseas material. Breeding has progressed to the stage that new releases are likely from 1998 unless anthracnose remains a problem. All *L. albus* lines developed in Australia have the same gene for sweetness to avoid any bitterness developing as a result of outcrossing.

AGRONOMIC CONSIDERATIONS

Many research workers, farmers and advisers were involved in devising the optimum agronomic inputs for lupin cultivation. This work investigated suitable soil types, rotational sequences, seeding rates, times of sowing, seed treatments, fertiliser requirements, soil treatments and insect and disease control measures.

The most important findings were:

the need to plant the crop early (eg Perry 1975; Walton 1976);

some soils were acutely deficient in manganese, which if not corrected resulted in the split seed disorder (eg Hawthorne and Lewis 1982; Gartrell and Walton 1989);

the role of cereal stubble on the soil surface in preventing wind erosion and lessening the effect of the disease brown spot; (eg Sawkins 1981)

the correct use of the pre-sowing herbicide simazine and the benefit to following cereal crops of controlling grass weeds in the lupin crop (eg Gilbey 1982; Nelson 1986);

the development of the lupin/cereal rotation;

the necessity to control insect pests (eg Michael *et al* 1982)

the importance of using high quality seed; and

the use of the fungicides iprodione or procymidone to reduce early infection of brown leaf spot.

This work in WA was summarised for farmers by Nelson and Delane (1990), and in eastern Australia by Lamb and Poddar (1987).

By 1979, the area sown to lupin in WA had fallen to 44% of the national total, having been 83% in 1976 (Figure 1). The technology was available in WA to produce yield of 2 t/ha, and these yields were being achieved on research stations, but farmers' yields were well short of this level (see Table 4).). Graham Walton, who undertook much of the earlier agronomic research in WA, suggested at a seminar involving researchers, farmers and advisers in WA that a production package was available for successful lupin production. His view was that the lack of farmer success was an extension problem, and at that time no further research was needed to make the crop viable (Walton 1979). As well, at that time lupins were being incorporated into farming systems on suitable soil types in eastern Australia, although on a smaller scale than in WA. Eastern state farmers had recognised that lupins were a highly profitable crop. With correct husbandry they could give high yields and have a major impact on the yields of subsequent cereals crops in the rotation (Hawthorne and Lewis 1980, Reeves *et al 1984,* Schultz 1995a, 1995b). There was a need for the known lupin agronomic package to be adopted in WA.

Once this had been achieved, further research became necessary as the industry grew. Research into the 1990's has been focussed on improving the efficiency of phosphate nutrition by deep banding the fertiliser below the seed, examining the effects of wider row spacings so that tines can more easily handle surface trash, examining seed quality (seed size, viable germination percentage, nutritional content), amelioration of non-wetting soils, improving disease and insect control, improving harvest index, and modifying harvest machinery to improve efficiency (eg Blanchard 1992; Jarvis 1992; 1994).

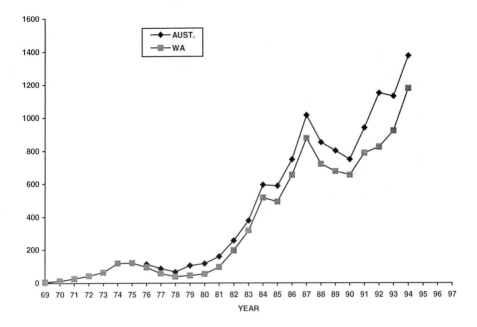

Figure 1. The area ('000 ha) of lupins for grain, by year in the whole of Australia, and in the state of Western Australia.

LUPINS IN ROTATIONAL SYSTEMS

In the Mediterranean type environment of Australia it is feasible to grow only one crop a year as the remainder of the year is too dry for plant growth. The most widely used rotation, involving lupins in Australia, is to grow a cereal or lupin crop in successive years. In some areas where *Pleiochaeta setosa* root rot and brown leaf spot have caused severe yield losses in lupins, the rotation has been extended to include two or more cereal crops (Sweetingham 1986). One of the main benefits of lupins in rotations with cereals is the increased yield of the cereal crop. In general, cereal yields following lupins have been 30% higher than when the cereal followed a cereal crop (Doyle and Herridge 1980, Hawthorne and Lewis 1980, Reeves *et al.* 1984, Rowland *et al.* 1989, Schultz 1995a, 1995b). During the last two years, the reintroduction of canola into farming systems has enabled a lengthening of the rotation to be practiced.

Advantages of the cereal/lupin rotation are:

It is the most profitable and ecological sustainable rotation available to many farmers, especially those with coarse textured sandplain soils.

It is amenable to minimum-tillage and soil-conservation farming practices.

The cereal stubble provides protection for the young lupin plant. This is important on sandy soils where wind erosion may be a problem. The stubble also reduces the rain splash of spores of the brown leaf spot disease.

Root diseases affecting the lupin crop are reduced by the cereal phase of the rotation.

Grassy weeds can be controlled during the lupin phase of the rotation. The absence of grass weeds enables cereal crops to be planted on the first rains and reduces the carry-over of the take-all disease *Gaeumannomyces graminis* which affects cereal crops.

Integrated weed management can be practiced. Broad-leaved weeds can be removed in the cereal phase of the rotation and grassy weeds, such as *Lolium* and *Bromus* species, in the lupin phase.

The cereal benefits from the residual nitrogen left by the lupin crop.

Lupins can extract nutrients, (such as potassium) and water from below the rooting depth of the cereals on deep sandy soils. The greater use of rain, where it falls, reduces the problem of waterlogging and salt accumulation in neighbouring valleys.

Lupins are usually planted and harvested before cereals, which permits an efficient use of farm machinery and the farmer's time.

Lupin stubbles, with spilt seed (often 200 kg/ha), provide excellent nutrition for sheep during the summer when the quality of other feed is low. This enables the integration of crop and livestock enterprises on the farm.

There is no such thing as the perfect system and the lupin/cereal rotation does have some disadvantages:

The interval between lupin crops is only one year and although this rotation has been practised in some areas for 25 years, in other areas, disease has limited yield and forced a longer rotation or a pasture phase.

The increasing use of grass herbicides in both phases has led to herbicide resistance. Again, a pasture phase may be required to assist in solving or avoiding this problem.

There is evidence that lupins acidify the soil at depth. This will require lime treatments to correct.

The rotation increases the water repellence of sandy soils.

Potassium removal under this rotation is high and will need to be replaced on sandy soils.

When cereal cropping has not been as profitable as stock production, some farmers have adopted a lupin/pasture rotation. This rotation requires a greater degree of managerial skill to be successful for the following reasons:

Weed problems in the lupin crop tend to be greater.

There is no stubble to protect the young lupin crop from wind erosion and rain splash of the spores of brown leaf spot.

The soil often requires cultivation to provide a suitable seedbed, increasing the risk of wind erosion.

Diseases may be more prevalent after a clover/grass ley.

Other cropping rotations involving lupins, which are practised to a lesser degree are a lupin/wheat/pasture, lupin/canola/wheat and opportunistic lupin cropping in low rainfall areas when summer rain is received and/or there is a very early start to the season. On some deep, infertile sandy soils in the traditional high rainfall grazing areas of eastern Australia, lupins are sown in clean paddocks and the crop generates some income prior to sowing pastures, especially lucerne. Several lupin crops may be involved, including consecutive crops, and the pasture may be under sown with the last crop of lupins.

The role of lupins in rotations, and having the crop accepted by farmers, is similar to that of other pulse crops, with some unique features associated with the crop's adaptation to specific soil types (eg Hamblin 1987; Delane *et al* 1989).

ANIMAL USE OF LUPINS ON FARM

During the hot dry period of the year, which includes summer, and which may last from October to May in Australia, livestock graze and obtain their nutritional requirements from dry pastures and stubbles. The protein and energy content of this dry feed deteriorates with time and more rapidly if there are any summer thunderstorms, so that grazing stock lose weight and require supplementary feeding. Lupin seed is an important source of feed in Australian farming systems. Sheep, beef cattle, dairy cattle, pigs, poultry and sheep are fed some lupin seed in their rations.

Lupin seed losses during harvesting usually exceed 100 kg/ha, and lupin stubbles are a valuable source of summer feed for stock at a time when other paddock feeds are of low quality. Lupins have enabled farmers to carry more sheep, in better condition, over the summer and early autumn months. Research has shown the successful use of these stubbles requires management that reduces the risks of lupinosis and prevents soil erosion. Lupinosis, caused by the toxins produced by the fungus *Phomopsis leptostromiformis,* which grows on lupin stubble, has caused substantial losses to the sheep industry. These losses have deterred some farmer-graziers from growing lupins if their major income is derived from sheep.

In 1970, the fungus responsible for the disease was isolated (Wood and Brown 1975), which allowed management methods to be developed by advisers and veterinarians to lessen the effects of the disease. These methods involve:

Removing stock from lupin stubbles following summer rain as rain favours production of toxin by the fungus.

Grazing lupin stubbles immediately after harvest. Generally, the risk of toxicity increases as the summer progresses, particularly if moisture is available.

Grazing at conservative stocking rates, certainly not more than 15 sheep per ha.

Providing alternative feed. Stock should never graze lupin stubbles if seed is not present.

Providing more than one, or having a movable watering point for the sheep. This allows a more even grazing and as well as preventing lupinosis, stops wind erosion around the watering points.

Realising that hungry sheep, heavily pregnant ewes and weaner sheep are the most susceptible to lupinosis.

Daily monitoring of the sheep by moving the mob several hundred metres and monitoring the sheep for symptoms of lupinosis.

The development of more resistant varieties such as Gungurru and Merrit, and all the subsequent varieties, has increased the safety margin for grazing (Cowling *et al* 1986; Gladstones 1989; Morecomb and Allen 1990, McDonald and Croker 1991). However, the varieties are not completely free of the fungus and the management practices used to avoid lupinosis need to be continued. Care must be taken to ensure that stubbles, on sandy surfaced soils, are not overgrazed and made susceptible to wind erosion. With the focus on sustainable cropping systems, and a swing to continuous cropping, the lupin stubbles are no longer grazed in some erosion-prone areas.

By 1990 CSIRO scientists had developed a vaccine to prevent lupinosis (Ralph 1990). This vaccine, when it becomes commercially available will further assist the sheep industry and the adoption of lupins as a farm crop in grazing systems.

In addition to the residual seed in lupin stubbles, Australian farmers also use the seed as a supplementary feed (eg Rowe and Ferguson 1986). Research has shown the benefits of including lupins in rations to maintain or increase body weight, increase ram and ewe fertility and prevent weaner losses over summer. When lupins are used as a supplement, and fed to sheep grazing pastures or cereal stubbles, the seeds are no longer fed in a trail, but are now widely distributed once a week, using a mechanical fertiliser spinner. This reduces the labour costs of hand feeding, and the sheep tend to graze the paddock more uniformly as they search for the lupin seeds, which are large, palatable and easily found.

EXTENSION

With any agricultural development, there is a burst of enthusiasm among the innovative farmers. This occurred with the release of the first narrow-leafed lupin variety in WA in 1967. By 1975, there were 122,000 ha under cultivation in WA. However, by 1978, the area had fallen to 39,000 ha (Figure 1.) Problems in production, relating to poor management, coupled with dry seasons, resulted in low yields and a general disenchantment with the crop. A major extension problem existed in having the technology widely adopted (Walton 1979).

In 1978, lupin breeder John Hamblin, suggested that there was a need to demonstrate to farmers in WA the right and the wrong way to produce a lupin crop. He and Peter Nelson, then an agricultural adviser with the Department of Agriculture, proposed they demonstrate to farmers the factors in a "lupin production package" and the yield penalties likely to be incurred by not adopting components of the package.

Before sowing the demonstrations on farms, the mass media were used to inform farmers of the aims of the project. The demonstrations were a great success and again focussed farmer attention on lupins. The results of the demonstrations (Table 2) formed the basis of an extension campaign in the following years. The effects of rhizobia, fertiliser, depth of planting and insect control were also investigated, but at the sites chosen had little effect on yield.

*Table 2.*The effect of complete, partial and nil adoption of the "package approach" on the yield of lupins 1979.

Package approach	Yield (t/ha)
Complete adoption "Do everything right"	1.61
Wrong variety	0.75
Delayed planting	0.56
Low plant stand	1.35
Poor weed control	1.46
Nil adoption "Do everything wrong"	0.21

In 1981, the WA Department of Agriculture leased 250 ha of sandplain soil in the Geraldton area in a 350-mm rainfall zone. This large area was subdivided into 8.5 ha plots. Each plot represented a farming system and received individual management. There were two replications of each system. The demonstration showed to farmers the advantage of including lupins in the rotation compared with continual wheat cropping (Table 3). These large plots had a number of advantages compared with small plots:

The results were more credible in the eyes of the farmer.

They enabled all the interactions such as wind erosion and weed problems to become evident.

They were large enough to allow conventional farming machinery to be used.

They provided a 'laboratory' for research to investigate disease, nutritional or weed problems under different rotational systems within the same environment.

Table 3. The effect of lupins on the yield of wheat, Geraldton 1982-1991

Cropping system	Wheat and lupin yields t/ha
Continuous wheat	0.63
Wheat following lupins	1.32
Lupins following wheat	0.85

Since 1979, the use of lupins in farming systems in WA has been strongly promoted by advisers, consultants and private enterprise (eg Nelson 1985; 1987). An important extension principal emerged. Although the initial demonstration was successful and reawakened farmer awareness, there was a need to continue with field demonstrations and to expand the extension effort to maintain the interest and assist new farmers adopt the system.

The extension campaign is continuing and lupin yields have improved markedly as a result of the campaign. Recent increases in wheat yield can be partially attributed to lupins in the rotation in terms of better weed control. This has enabled one million hectares of wheat to be sown on the first rains of the growing season (Table 4.)

Table 4. Average five-year area (ha) and yields (t/ha) of lupins and wheat in Western Australia 1971-95

Year	Average lupin area	Average lupin yield	Average wheat yield
1971 to 1975	75	0.52	1.15
1976 to 1980	59	0.54	0.93
1981 to 1985	323	0.91	1.11
1986 to 1990	718	0.97	1.38
1991 to 1995	956	1.09	1.61

In 1990, the Grain Pool of WA, the sole marketers of lupins in WA, having successfully established markets for the crop, were concerned that the industry was wavering, and the area sown to lupins was falling (Figure 1). A survey of growers revealed that high wheat prices were causing wheat to be sown at the expense of lupins. Other factors were the inconsistent lupin yields and unsatisfactory returns, diseases in lupin crops and problems with lupinosis in sheep. The Grain Pool also recognised there was a potential to expand the area grown to lupins in the southern parts of WA, and for many new growers to enter the industry. To cater for these developments, they funded a lupin extension specialist to communicate research findings to advisers and growers and provide feedback from the growers as to research and marketing needs. The flagship of this extension effort was a monthly newsletter on lupins called "Lupin Logic" which was sent free to every lupin grower, extension worker and researcher in WA. The service continues with a nominal charge, with leading eastern state farmers and advisers also subscribing. This initiative was a recognition of the need to maintain extension campaigns, especially where technology is changing or when complex issues such as changing farming systems are involved.

In the eastern states of Australia there was no similar campaign, but innovative farmers were stimulated and assisted by local extension programs. They implemented the results of rotation and agronomic trials on lupins conducted in the-mid to late 1970's. Enthusiastic extension workers and farmers worked closely to promote lupins and their benefits as seen in research trials and in farmer's crops. Seeing farmers successfully adopt lupins encouraged other farmers to grow them. The one clear message as was found in WA, was to grow the crop properly or not at all.

FARMER COOPERATION

Farmers have contributed in many ways to the development of the lupin industry in Australia. They have assisted breeders and agronomists by providing land for experimental purposes, conducting large-scale demonstrations, giving feedback to researchers on cultural operations and developing innovations for crop production. In addition, they have contributed levies on harvested grain to fund further research and development.

Farmers fostered the lupin/cereal rotation and some having practised this rotation for 30 years. Framers devised the practices of the pre-sowing application of the herbicide simazine, and its incorporation with the seeding operation, and developed the large-scale seed inoculation of rhizobia. They modified harvesting machinery to improve efficiency.

Farmers have tested research results and modified them to their conditions. Examples are the deep placement of fertilisers and stubble management and retention systems. It is the farmer who integrates all the facets of research and makes the system work so that a profit is made.

MARKETING

The Grain Pool of WA played a significant role in finding and developing markets for the new crop. The Grain Pool is the sole marketeer of WA lupins in bulk, and the world's largest lupin selling organisation. It is grower controlled and funded, and is an important part of WA's rural economy. The Grain Pool played a major role in gaining acceptance for lupins in international markets. This has been a benefit to all of Australia and to potential lupin growing countries (eg Coffey 1988). The majority of the lupins are exported to Europe, Asia and the middle-East. To protect the investment the Grain Pool was making in lupins and to ensure orderly marketing, the WA government, in the 1975/76 season, categorised lupins as a "prescribed grain" which meant that all lupins delivered by growers in WA have to be exported exclusively by the Grain Pool.

The eastern states of Australia do not have a comparable single marketing organisation and marketing was initially a barrier to the growing of lupins in the late 1970's. Growers as groups, tried withholding produce from sale until satisfactory prices were obtained. Cooperatives formed to involve more growers and to control larger parcels of lupins than the individual growers could put together. The participants receiving a pooled priced. The Lupin Growers Association in Victoria, and the SA Pea Growers Cooperative in South Australia established some stability in the relatively small lupin market of eastern Australia. Most lupins were used for domestic stock feed, but eventually some were exported. Both cooperatives were superseded when the Australian Wheat Board, followed by the Australian Barley Board began handling lupins in a 'pool system'. There is now also a predominance of large, often international, trading companies which purchase and market lupins from the eastern Australian states.

Internationally there is still a reluctance by some potential buyers to recognise the qualities of lupins. Unlike wheat, barley, oats and peas where premium prices are paid when they are used for food, lupins are, with few exceptions, yet to be embraced by food manufacturers as an alternative to the soybean. The Grain Pool of WA, and others, are trying to rectify this situation. The Grain Pool has funded research programs following close liaison with growers, agronomists and plant breeders. This includes funding research, in overseas countries, to demonstrate the uses of lupins and to develop new markets.

RECENT THREATS

In 1996, the disease anthracnose vcg strain 2 (*Colletotrichum gloeosporoides*) was detected in commercial crops of *L. albus* cv. Kiev Mutant in northern WA, followed by an unrelated, confined outbreak in *L. angustifolius* on Lower Eyre Peninsula in SA. A massive eradication procedure was implemented in both states, but because of the magnitude of the outbreak in WA, control is now the strategy. Infected ornamental lupins were probably the source of infection in both cases. This devastating disease poses the greatest challenge yet to the Australian lupin industry, and requires the resources of all sectors of the industry to control and eradicate it. Two of the most recently released narrow-leafed lupin varieties are fortuitously resistant. Seed treatment, crop hygiene and management practices are being implemented to manage the disease. The growing of *L. albus* in WA is banned because of its susceptibility to anthracnose.

The development of herbicide resistance in grass weeds, in continuous cropping rotations, is a new challenge. Revised machinery and crop management practices are required. As well, genetically engineered lupins, with resistance to specific knockdown herbicide, are being developed as part of an integrated management program.

FUTURE DEVELOPMENTS

With improved management and new lupin varieties, we expect average yields to increase from the current 1 t/ha to 1.5 t/ha within the next decade. The area sown will increase rapidly in the higher rainfall regions where yield potential is high. This will give a more geographically balanced production base. Lupins will also be grown in drier areas as varieties improve. Australia could have a 1.5m tonnes for export annually by the year 2005.

Lupins have been under selection for less than 40 years and breeders have the potential to produce higher yielding and better varieties for specific environments. Increased resistance to the major diseases will be developed. Finding and incorporating resistance to anthracnose vcg strain 2 has become a priority since 1996. A vaccine to prevent the disease lupinosis in sheep will be used in high risk areas.

The white lupin, *L. albus*, also has a greater role to play on the finer textured soils, especially in the heavier soils of eastern Australia. Screening existing cultivars and an increased breeding effort are in progress. The absence of resistance to anthracnose is however a potential limitation to *L. albus*, especially in WA.

The yellow lupin *L. luteus*, is showing promise as a suitable species for the very acidic soils containing high levels of aluminium.

A breeding project will provide lupins for the finer textured soils. Progress is being made on a rough seeded lupin *L. atlanticus* (Buirchell and Cowling 1992), and this species has the potential to to be grown on the expansive calcareous soils of eastern Australia where the other species will not grow (Egan and Hawthorne 1994).

Agronomists have the opportunity to improve yields through more efficient fertiliser usage, more reliable crop establishment systems, especially on non-wetting sands, and possibly improving the harvest index using chemicals and/or harvesting efficiency. The reintroduction of canola as a crop will enable rotations to be lengthened, thus lessening the disease threat to all crops, including lupins.

The biggest threat to the success of the cereal/lupin rotation is the occurrence of herbicide resistance. The problem of rye-grass resistance is wide spread where herbicides have been the major control agent. Practical solutions are being found, including weed seed collection. Genetically engineered lupins with resistance to specific knock down herbicides are also being developed, and will assist.

A continued effort will be necessary to ensure lupins are accepted worldwide for human consumption and further market opportunities in aquaculture will be developed.

CONCLUSIONS

The narrow-leafed lupin *L. angustifolius* has been bred to become a new crop plant.

Integration of this crop into farming systems was a collaborative effort on the part of breeders, agronomists, farmers and advisers.

International markets for lupins have been established. These exports currently earn about US$190 million per annum. US$120 million is also generated in on-farm use of the lupin grain and stubble and benefits to cereal crops.

Without lupins in farming systems, many of the coarse textured sandy soils would be non-viable as cropping areas.

There is potential for further development of the crop both agronomically and for use in human consumption and aquaculture industries.

Acknowledgments

In addition to the persons named in this paper who have been involved mainly in Western Australia the crop has benefited tremendously in the eastern states of Australia from the enthusiastic research and promotion by Tim Reeves, Wayne Hawthorne, Eric Corbin and Ivan Mock

References

Blanchard, E 1992. *Journal of Agriculture, Western Australia,* 33: 22-25

Buirchell, B. and Cowling, W.A. 1992. *Journal of Agriculture, Western Australia*, 33: 131-137

Coffey, R. 1988. Proc. *World Congress on vegetable protein utilisation* Singapore.

Cowling, W.A. 1994. *Proc. of the First Australian Lupin technical Symposium, Perth, WA.* 129-139.

Cowling, W.A., Allen, J.G., Wood, P.McR. and Hamblin, J. 1986. *Journal of Agriculture, Western Australia,* 27: 43-46

Delane, R.J., Nelson, P. And French, R.J. 1989. *Proc. 5th Australian Society of Agronomy.* 181-191.

Doyle, A.D. and Herridge, D.F. 1980. *Proc. of the Australian agronomy conference.*1:185

Egan, J.P. and Hawthorne, W.A. 1994. *Proc. of the First Australian Lupin technical Symposium, Perth, WA.* 178-181.

Gartrell, J.W. and Walton, G. 1989. *Western Australia Department of Agriculture Farmnote* 89/84.

Gilbey, D.J. 1982. *Journal of Agriculture, Western Australian, 3:* 81-82

Gladstones, J.S. 1977. *Western Australia Department of Agriculture Bulletin* 3990.

Gladstones, J.S. 1989. *Journal of Agriculture, Western Australia, 30:* 3-7.

Hamblin, J. 1987. *Proc. 4th Australian Agronomy Conference.* 67-77.

Hawthorne, W.A. and Lewis, D.C. 1980. *Proc. of the Australian Agronomy Conference* 1 186

Hawthorne, W.A. and Lewis, D.C. 1982. *Proc. of the Australian Agronomy Conference* 2 277.

Jarvis, R. 1992. *Journal of Agriculture, Western Australia,* 33: 8-9

Jarvis, R. 1994. *Proc. of the First Australian Lupin technical Symposium, Perth, WA.*213-220.

Lamb, J. and Poddar, A. 1987. Eds *"The Grain Legume Handbook"*. South Australian South Pea Growers Cooperative, Riverton, South Australia.

Landers, K.L. 1991. *NSW Agriculture technical Bulletin* 44.17

McDonald, C. And Croker, K. 1991. *Journal of Agriculture, Western Australia, 32:* 100-104

Morecombe, P and Allen, J. 1990. *Journal of Agriculture, Western Australia,* 31: 3-4.

Michael, P.J., Woods, W.M., Richards, K.T. and Sandow, J.D. 1982. *Journal of Agriculture, Western Australia, 3:* 83-85.

Nelson, P. 1985 Kondinin and Districts Farm Improvement Group Proceedings "Rethinking Rotations".

Nelson, P 1986. *Western Australia Department of Agriculture Bulletin* 4122.

Nelson, P 1987 *Proc. Australian Agricultural Extension Conference.* Brisbane. 466-481.

Nelson, P. and Delane R.J., 1990. *Western Australia Department of Agriculture Bulletin* 4179 (Revised August 1991).

Perry, M.W. 1975. *Australian Journal of Agricultural Research,* 26: 809-818

Ralph, W. 1990. *Rural Research,* 146: 2-5.

Reeves, T.G., Ellington, A. and Brooke,H.D. 1984. *Australian Journal of Experimental Agriculture and Animal Husbandry,* 24: 595-600

Rowe, J.B. and Ferguson, J. 1986. *Proc. Australian Society of Animal Production,* 16: 343-346.

Rowland, I.C., Mason, M. and Hamblin, J. 1989. *Journal of Agriculture, Western Australia,* 30: 22-25.

Sawkins, D.N., 1981. Unpublished data. Agriculture Western Australia.

Schultz, J.E. 1995a. *South Australian Research and Development Institute Research Report* Series No 3, 52 pp

Schultz, J.E. 1995b. *Australian Journal of Experimental Agriculture,* 35: 865-876

Sweetingham, M.W. 1986. *Journal of Agriculture, Western Australia,* 27: 49-52.

Walton G.H. 1976. *Australian Journal of Experimental Agriculture and Animal Husbandry,* 16: 893-904.

Walton, G.H. 1979. *Proc. Lupin Research Review Workshop* Geraldton.

Wood, P.McR. and Brown, A.G.P. 1975. *Journal of Agriculture, Western Australia,* 16: 31-32.

Lupinus albus as a European crop

Nathalie Harzic1[1], Ian Shield[2], Christian Huyghe[1], George Milford[2].
1 INRA Station d'Amélioration des Plantes Fourragères. 86600 Lusignan. France. ; 2 IACR Rothamsted. Harpenden Herts. AL5 2JQ. UK.

Abstract.

White lupin (*Lupinus albus* L.) is still a minor crop in Europe. The breeding of this species in Western Europe is focused on the autumn-sown type. Detailed studies of the physiology of the plant and the effects of plant architecture on seed yield have shown that a combination of dwarfism and determinate growth is likely to be the most productive type. The development of dwarf determinate genotypes, as the European ideotype, now requires that varieties with high seed yield potential are produced, that they can be grown efficiently and that the seed has a high utilisation value as feed. It has been shown that with this architectural type, biomass accumulation is a major factor limiting yield. Agronomic practices and selection criteria can be defined to increase biomass production. An E.U.-funded study, which started in 1997, of dwarf-determinate genotypes is presented in this paper.

INTRODUCTION.

White lupin (*Lupinus albus* L.) is one of the main grain legumes cultivated in Europe. Others are soybean (classed as an oilseed), the dry food legumes (chickpea, lentil, vetches and beans), and the protein crops (pea, faba bean and lupin). The recent development of protein crops in Europe has been stimulated by political initiatives to reduce the deficit in protein. In 1993, the European Union (E.U.) produced only 32% of the protein it required for animal feed (Carrouée 1995). In 1995, it produced 4,329,103 t of grain legumes and imported 1,777,103 t including 142,103 t of lupin from Australia (UNIP 1997). It has been estimated that the potential use of protein rich material in EU was at least three times greater than the present resources (Carrouée *et al* 1993). The economic context (the high price of soybean cake) also favours the development of white lupin as a protein crop. The white lupin has been preferred to the other lupin species in Western Europe because of its higher protein content (40%). The present paper focuses on this species and especially on its autumn-sown forms. It presents data obtained from collaboration between INRA Lusignan (France) and IACR Rothamsted (UK).

PRODUCTION OF WHITE LUPIN AND CLIMATIC CONSTRAINTS.

The white lupin is largely restricted to the southern and western parts of Europe and is still a minor crop. In 1995 59,103 ha of lupins (white, blue and yellow) were grown in E.U. (UNIP 1997), of which it has been estimated, only one third was the white lupin (Table 1). The average seed yield varies greatly and is higher in Western Europe (3 t/ha in France over the 7 last years) than in Southern Europe (0.8 t/ha in Portugal).

The main climatic constraints for growing the white lupin in these areas are the low winter temperatures, the summer drought and the difficulties of maturation in cool areas. Both spring-sown and autumn-sown types exist. Autumn-sown types are characterised by a high vernalisation requirement, which induces a long over wintering rosette stage. Autumn-sown types have several theoretical advantages over spring-sown types when grown under favourable conditions. The longer growing season, when both winter and summer conditions are compatible for plant growth, induces a greater biomass than in spring-sown types. Therefore they have a higher seed yield potential, assuming a similar harvest index. Autumn-sown types also flower earlier than spring-sown types. This allows them to escape summer drought and high temperature stress during seed formation. For these reasons, lupin breeding in western Europe has focused on the autumn-sown white lupin. Nevertheless, spring-sown white lupins are well adapted to areas with low winter temperatures.

R. Knight (ed.), Linking Research and Marketing Opportunities for Pulses in the 21st Century, 561–567.
© *2000 Kluwer Academic Publishers. Printed in the Netherlands.*

Table 1. Lupins in the E.U (UNIP 1997)

	1993	1994	1995	1996*
France (*L. albus*)				
Area (ha)	3000	4100	3000	2300
Yield (t/ha)	2.00	3.01	3.14	2.60
Italy (*L. albus*)				
Area (ha)	2800	3000	1600	3100
Yield (t/ha)	1.44	1.70	1.70	1.70
Spain (*L. albus, L. angustifolius*)				
Area (ha)	3800	16,100	19,000	17,000
Yield (t/ha)	0.37	0.65	0.53	0.80
Portugal (*L. albus*)				
Area (ha)	2000	1800	3600	3100
Yield (t/ha)	0.80	0.80	0.80	0.80
Germany (*L. albus, L. luteus*)				
Area	0	0	31,800	43,000
Yield (t/ha)			3.00	3.00

* estimated data.

Two breeding strategies existed to adapt the white lupin to European conditions, either to breed early flowering and maturing spring-sown types or taking a longer term view and breed autumn-sown types.

BREEDING OF SPRING SOWN TYPES.

Spring-sown white lupin genotypes have been bred in Europe since the 1930's (Hondelman 1984). Attempts to develop spring-sown cultivars in Germany in the 19th century and in the UK in the late 1970s failed, largely because although the cultivars grew well, they failed to ripen sufficiently early (Gross 1986, Milford *et al* 1991). Breeding program started in the 1970's in France, Germany and Poland. Besides seed yield, the main selection criteria were earliness of flowering and maturity to escape summer drought and to allow early ripening. As a consequence, the current cultivars show little genetic variation for plant structure, i.e. in the number of branches and leaves. That is why, despite being grown at high plant densities, the light interception at flowering is low (Duthion *et al* 1994). The restricted structure during the major part of the vegetative growing cycle prevents high biomass production and reduces seed yield potential. In theory, this could be overcome by shortening the phyllochron to develop plants with a similar flowering date but more leaves, a higher leaf area and an earlier canopy closure. At the present, such genotypes have not been identified.

BREEDING OF AUTUMN-SOWN TYPES.

The first autumn-sown cultivar with an acceptable frost resistance, was released in 1989. The excessive vegetative development of these genotypes was a major limitation to achieving a high seed yield potential. Their vegetative growth is vigorous under cool conditions which provides strong competition for the developing pods, to the detriment of seed yield and ripening. In most cases, these genotypes do not yield well and do not ripen early in the cooler and wetter maritime regions of Northern Europe such as the UK (Julier *et al* 1993a). The high biomass production does not result in high yields because of the low and unstable harvest index (Table 2).

Table 2. Biomass, harvest index and seed yield of cv Lunoble, autumn-sown at Lusignan, France.

Year	Density (plt/m^2)	Biomass (t/ha)	Harvest Index	Seed Yield (t/ha)
1990	13	10.9	0.29	3.16
1990	20	13.4	0.17	2.28
1991	13	10.7	0.38	4.06
1991	20	10.1	0.40	4.04
1992	10	12.0	0.25	3.00
1992	20	11.3	0.31	3.50

A degree of control of vegetative development was obtained through the introduction of a determinate vegetative growth habit (Julier and Huyghe 1993) to produce plants in which all of the vegetative buds became floral at a given time and further vegetative development stopped. The vegetative growth of these genotypes is usually restricted to a mainstem and one order of branches. Moreover, determinacy modifies the profile of the number of leaves on the first-order branches (Figure 1).

Branches of the first determinate genotypes have fewer leaves than those of indeterminate genotypes and the lower first-order branches are longer than the upper ones, so that all the inflorescences are produced at the same height in the canopy at the level of the mainstem inflorescence. The reduction in leaf number on the mainstem and branches also induces near-synchronous flowering on them so that pod growth starts when vegetative growth (stems and leaves) finishes (Julier *et al* 1993b). In contrast, pod growth in indeterminate genotypes is concomitant with the vegetative growth (Figure 2).

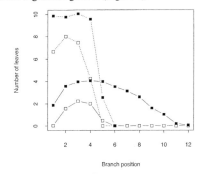

Figure 1. Profile of the number of leaves per first order branch for the indeterminate cultivar Lunoble (·······) and the determinate line CH304-73 (———) for two sowing dates (■ 18/9/90, □ 5/11/90).

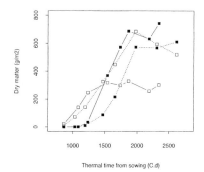

Figure 2. Dry matter partitionning to stems (□) and to pods (■) in the determinate line CH304-70 (———) and in the indeterminate cultivar Lunoble (·······).

564

Because vegetative growth stops early in determinate genotypes, there is less competition between vegetative and reproductive growth during pod development. This partly explains their higher yield stability under a range of seasonal weather and husbandry conditions (Julier *et al* 1993a). In addition, shortening the vegetative growth phase shortens the overall growth cycle. Julier *et al* (1993a) showed that determinate genotypes matured earlier than the indeterminate cultivar Lunoble, ensuring an early harvest in cool areas (Table 3). If sown sufficiently early to produce numerous mainstem leaves and first-order branches, determinate genotypes have a high potential for biomass and seed production, due to a high leaf area index and numerous pod-bearing axes. Unfortunately, they also grow tall and show a high susceptibility to lodging because of the overloading of the mainstem by the large number of pods borne at the top of the canopy.

Table 3. Seed yield, phenology and architectural characters of the dwarf, determinate and dwarf determinate white lupin populations and of the determinate line CH304-73.

| | 1992/93 | | | 1995/96 | |
	Dwarf	Determinate	CH304-73	CH304-73	Dwarf-determinate
Flowering date (°Cd)	994	1081	1053	1139	1316
Mainstem height (cm)	30.0	-	-	56.7	49.7
N° 1st order branches	3.0	4.4	4.4	6.7	7.7
Seed yield (t/ha)	4.96	4.10	5.33	3.61	3.83
Mean seed weight (g)	0.382	0.262	0.228	0.272	0.323

Plant height has been controlled through dwarfism. The first experiments on dwarf lupins were conducted with indeterminate genotypes. It was shown (Harzic and Huyghe 1996) that dwarfism reduces the internode length but not the number of leaves (i.e. internodes), the number of branches, or the leaf area. Dwarf indeterminate genotypes are identical to tall genotypes in all respects except for height. Their seed yield potential is high (Harzic *et al* 1996), and their harvest index higher than that of tall indeterminate lupins (Harzic *et al* 1995).

The combination of dwarfism and determinacy appears to be the most advantageous plant architecture for Western European conditions. Dwarf determinate genotypes are likely to be as stable as the non-dwarf determinate in seed yield, and should mature early and be less susceptible to lodging. Table 3 also shows the average seed yield and architectural characters of dwarf, determinate and dwarf-determinate crops.

The three populations were not grown in the same year or the same location, but comparisons of the dwarf population with the determinate population in Lusignan in 1992/93 and the dwarf determinate population with a determinate standard CH304-73 (cv. Ludet) indicate that the seed potential of the dwarf determinate genotypes was high. The average seed yield of the dwarf determinates was higher than that of CH304-73, one of the best determinate genotypes. The dwarf determinate population flowered late and was thus exposed to summer drought. Breeding for an earlier flowering date is now a major criterion in this material. Lodging susceptibility of the dwarf determinate lupins is linked to the mainstem height (r =0.60 ***). The lodging score of CH304-73 was greater than the average of the dwarf determinate population. This evaluation of the first generation of crosses within the dwarf determinate background showed that the architecture of these genotypes increases lodging resistance. Dwarf determinate genotypes also have the advantage of a greater tolerance, or better avoidance of the deleterious effect of winter frosts, because less stem tissue is exposed to winter frosts (Milford and Huyghe 1996). Dwarf determinate genotypes are likely to have all the advantages of the determinate architecture shown by Julier (1994), in particular for smaller level of competition between vegetative and reproductive growth during pod development (Figure 3).

AUTUMN-SOWN DWARF DETERMINATE WHITE LUPINS AS A EUROPEAN CROP

The development of the dwarf determinate lupin as a European crop requires (1) that genotypes with a high yield potential are available, (2) that they can be grown efficiently and (3) that their seed had a high utilisation value for feed.

The first dwarf and determinate genotypes have been selected in progenies of crosses between a restricted number of determinate progenitors and the dwarf indeterminate population (Figure 4).

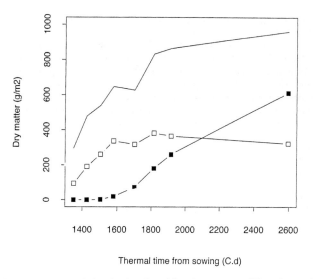

Figure 3. Total dry matter accumulation (——) and partitionning to stems (□) and to pods (■) in the dwarf determinate line dtn12.

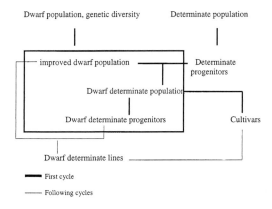

Figure 4. Strategy for breeding dwarf determinate lupin cultivars.

Indeed, it has been found that genetic characters from other genetic sources can easily be introduced in the dwarf indeterminate background in which the architectural characters (number of branches, number of leaves per branch) and phenology (flowering date, maturity) do not present such a bottleneck for manipulation as in the determinate population. Thus, no correlation has been found between seed yield and flowering date in the dwarf population (r=-0.07) (Harzic et al 1996) whereas a positive correlation exists in the determinate population (r=0.61) (Julier et al 1995). These correlations were obtained in two separate multilocational trials, the correlation coefficients were equal to -0.09 (Harzic *et al* 1994) and 0.5-, at Lusignan in 1992/93 (Julier 1994), for the dwarf and the determinate populations, respectively.

Selection is being made within the dwarf determinate background for characters other than seed yield and flowering date, for instance for a high number of leaves per branch to improve light interception during the early stages of growth to improve dry matter production. The positive effect of dry matter production on seed yield is shown on Figure 5 for two dwarf determinate lines (dtn12 and dtn20). The result has been corroborated with 108 genotypes in 1994/95 and 105 in 1995/96 (data not shown). Dry matter production and seed yield can also be improved by changes in crop management. Lower row spacings (0.35 *vs.* 0.50 m) increase light interception early in growth cycle and hence dry matter production (10.0 *vs.* 8.7 t/ha) and seed yield (3.64 *vs.* 3.16 t/ha).

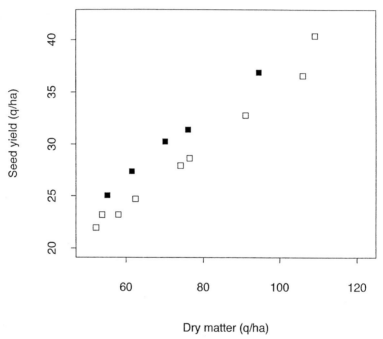

Figure 5. Biomass at maturity and seed yield for two dwarf determinate lines (□dtn12 and ■dtn20) grown under various agronomic practices (density and raw spacing) in 1994/95 and 1995/96 at Lusignan.

A further program of research on selection, agronomy and grain utilisation supported by the European Community, has just started in collaboration with several European groups. The aim of this program is to create varieties and technologies for increasing production and utilisation of high quality protein from the white lupin in Europe. The three main tasks of the project are :

(1) to develop new dwarf-determinate and alkaline-tolerant genotypes capable of sustaining profitable production of protein in the different regions of Europe ;

(2) to evaluate and define the geographic range of these new genotypes and to optimise the agronomic practices needed to grow them profitably in the different European regions, with special attention to anthracnose ;

(3) to develop industrial technologies to improve the utilisation of lupin grain by animal feed industries and for human consumption.

The laboratory or firms involved in this program are: INRA Lusignan-France, IACR Rothamsted-UK, La Cana-France, KVL-Denmark, SWS-Germany, the University of Azores-Portugal and DIAS-Denmark.

The project will produce new and improved dwarf-determinate genotypes of white lupin well adapted for wider cultivation in Europe. Lines with sufficient tolerance to lime to be grown on alkaline soils will also be bred. Simple physiological simulation models will be available as crop management and decision support system for lupin. These will help to determine the suitability of local soils and climates (Siddons *et al* 1994, Milford *et al* 1996), and to optimise husbandry practices for the crop. Technological processes will be developed for improving the utilisation of lupin grain as components of monogastric feeds and as basic ingredients for the human food industry.

CONCLUDING REMARKS

The white lupin is still a minor crop in Europe but new genetic material with higher potential is now available. The utilisation value of the produce will also be increased. By itself, the availability of new dwarf-determinate genotypes does not guarantee the wider commercial uptake. Suitable agronomic practices must be defined and it must be demonstrated that the lupin can be grown widely with little risk of crop failure. The integration of simple physiological models of crop development with soil and climatic databases will help to define where the lupin can be grown and the optimal sowing window. A further step in the white lupin breeding is the modification of the proportion of the pod walls. The modification of the architecture through dwarfism, determinacy and their combination has modified the ratio between reproductive and vegetative biomass with no severe reduction of the total biomass. It thus induced an increased seed yield. The modification of the structure of the reproductive compartment by decreasing the proportion of dry matter in pod walls (currently 35-40% of the pod biomass) would theoretical increase seed yield; i.e. for a given investment in the reproductive compartment, the harvest index of the reproductive organs themselves will be higher. A larger number of pods per unit area should also be produced. A research program on this subject is currently developed at Lusignan.

Acknowledgements : Nathalie HARZIC is grateful to the European Community for financial support (contract FAIR3 CT96-1965 'Creation of varieties and technologies for increasing production and utilisation of high quality protein from the white lupin in Europe).

References.
Carrouée B. 1995. In : *Improving production and utilisation of grain legumes*. 2nd European Conference on Grain Legumes, 2-3 (Ed. AEP)
Carrouée B., Delplancke D., Lapierre O. 1993. *Perspectives Agricoles* 184:10-19
Duthion C., Ney B., Munier-Jolain N..M. 1994. *Agronomy Journal* 86:1039-1045
Gross R. 1986. In : *Proceedings of the 4th International Lupin Conference*, pp 244-77
Hondelman W. 1984. *Theoretical and Applied Genetics* 68:1-9
Harzic N., Huyghe C., Papineau J. 1995. *Canadian Journal of Plant Science* 75:549-555
Harzic N., Huyghe C. 1996. *Journal of Agricultural Science* 127:337-345
Harzic N., Huyghe C., Papineau J., Billot C., Esnault R., Deroo C. 1996. *Agronomie* 16:421-432
Harzic N., Huyghe C., Papineau J. 1994. In : *Proceedings of the 1st Australian Technical Symposium*. 268 (Eds. M. Dracup and J. Palta)
Julier B 1994. Thèse ENSA Rennes. 133p
Julier B., Huyghe C. 1993. *Annals of Botany* 72:493-501
Julier B., Huyghe C., Papineau J., Milford G.F.J., Day J.M., Billot C., Mangin P. 1993a. *Journal of Agricultural Science* 121:177-186
Julier B., Huyghe C., Papineau J. 1993b. *agronomie* 13:877-888
Julier B., Huyghe C., Papineau J., Billot C., Deroo C. 1995. *Euphytica* 81:171-179
Milford G.F.J., Huyghe C. 1996. In : *Problems and Prospects for Winter Sowing of Grain Legumes in Europe*. Pp 59-64 (Ed. AEP)
Milford G.F.J., Day J.M., Leach J.M. Scott T. 1991. *Aspects of Applied Biology* 27:183-188
Milford G.F.J, Shield I.F., Siddons P.A., Jones R.J.A., Huyghe C. 1996. *Aspects of Applied Biology* 46:119-124
Siddons P.A., Jones R.J.A., Hollis J.M., Hallet S.H., Huyghe C., Day J.M., Scott T., Milford G.F.J. 1994. *Journal of Agricultural Science* 121:177-186
UNIP 1997. *Statistiques*.

The Potential of *Lupinus mutabilis* as a crop

P D.S. Caligari[1], P Römer[2], M A Rahim[1], C Huyghe[3], J Neves-Martins[4] and E J. Sawicka-Sienkiewicz[5]

1 Dept. of Agricultural Botany, The University of Reading, Reading RG6 6AS, UK ; 2 Südwestdeutsche Saatzucht, D-76437 Rastatt, Germany ; 3 SAPF-INRA, 86600 Lusignan, France :4 DBEB-ISA, P-1399 Lisboa Codex, Portugal ;5 The Botanical Garden of University of Wroclaw, 50-335 Wroclaw, Poland.

Abstract

Lupinus mutabilis, in its Andean highland centre of origin, has been an important source of protein for human nutrition for more than 2000 years. Today its cultivation in that region is restricted to small, farmers' fields, mostly for their own needs. The valuable seed composition, neutral photoperiodicity, non-shattering pod, large white seeds and its apparent adaptation to moderate climatic conditions made it a potential introduction to Europe. The results of a recently completed research project "Adaptation of *Lupinus mutabilis* to European soil and climate conditions" form the basis of this paper. The results show that, given the present plant architecture, *L mutabilis* will not yield sufficiently highly to be economically feasible as a crop. However, the discovery of a "determinant" mutant opens the potential, *via* further breeding, to produce a crop with a new architecture which might be established as a crop in Europe.

1. THE HISTORY OF LUPINUS MUTABILIS

The idea of using *L. mutabilis* as a crop is not new. In its centre of origin, the Andean highlands (including the present states of Peru, Bolivia and Ecuador), *L. mutabilis* has been known as an important source of protein for human nutrition for more than 2000 years (Gross an Baer, 1975). During the time of the Spanish conquest the percentage of *L. mutabilis* in human diets was estimated to be 5% (Antunez de Mayolo, 1982) which corresponds to an area of 100,000 ha of lupins in Peru (Gross 1983) at that time (16th century). The invasion of the Spanish conquerors caused a change in dietary habits such that up to the present the production of the Andean Lupin ("Tarwi" or "Chocho") in South America is restricted to small, farmers' fields, mostly for their own needs.

The history of *L. mutabilis* in Europe started in the mid 1970's. The valuable composition of the seeds, photoperiodic indifferent behaviour, non-shattering pods, large white seeds and adaptation to moderate climate conditions made it promising as a potential introduction to Europe. The first trials were carried out at the INRA in Lusignan (France) and at The University of Reading (UK). Starting in 1983 the University of Gießen set-up a programme to introduce the Andean Lupin to Germany.

As a result of these activities a research project was funded by the EU (AIR3-CT93-0865): "Adaptation of *L. mutabilis* to European soil and climate conditions". The results from this three-year programme provided the basis for this paper.

2. SEED QUALITY

Of all the lupin species, *L. mutabilis* is the one with the highest percentage of oil and protein in its seeds. In material of direct Peruvian origin, the percentage of protein was as high as 50%, and the oil content reached a maximum value of 24% (Table 1).

R. Knight (ed.), Linking Research and Marketing Opportunities for Pulses in the 21st Century, 569–574.

Table 1. Protein, oil and alkaloid content of 246 ecotypes of *L. mutabilis* of Peruvian origin (% of seed dry matter) (from Römer, 1990).

	mean	minimum	maximum
Alkaloids	2.83	1.66	4.17
Protein	41.92	34.60	50.20
Oil	19.85	14.30	23.60
Protein + Oil	61.77	55.20	71.80

The high percentage of alkaloids has been considered a major disadvantage in terms of seed quality. Due to the work of Erik von Baer in Chile, a very low alkaloid line was selected (Baer and Ibanez, 1986) and named Inti. The alkaloid level of this genotype was reduced to 0.0075% with no reported detrimental effect on the protein (51%) or oil (16%) content (Gross *et al.*, 1988).

Results of seed quality from the EU project are given in Table 2. The mean values for 16 genotypes grown at 5 sites in Europe over 2 years, show a slightly higher percentage of protein but a significantly lower oil content of the grain compared to the Peruvian ecotypes. Our current results reinforce those of earlier workers (Gross *et al.*, 1983) in showing that *L. mutabilis* produces more protein but less oil in its seeds under European conditions than in its centre of origin. The particularly low oil content in our results probably is a reflection of very dry weather conditions in both years.

The low alkaloid line Inti had the highest protein content, thus giving the highest total of protein + oil of all genotypes.

Table 2. Mean values of protein and oil in the seeds of 16 lines of *L. mutabilis* grown at 5 European sites and 2 years (1994, 1995)

Line	Protein	Oil	Protein + Oil
LM 13	44.3	9.2	53.5
LM 18	45.2	8.7	53.9
LM 27	44.8	9.1	53.9
LM 32	45.0	9.0	54.0
LM 34	45.7	8.0	53.7
LM 81	45.2	8.7	53.9
LM 231	45.9	8.4	54.3
CM 157	45.2	8.4	53.6
XM 1-39	45.3	8.4	53.7
XM 5	45.0	8.4	53.4
Potosi	44.3	8.8	53.1
I 82	43.7	9.5	53.2
Mutal	44.7	8.7	53.4
PL 20993	45.2	8.9	54.1
Inti	46.1	8.5	54.6
PT 079	43.6	9.1	52.7
Mean	**44.95**	**8.7**	**53.65**

3.   YIELD

It is therefore clear that seeds of *L. mutabilis* have a valuable composition under European conditions, but the question of the yield potential of the species has yet to be answered.

The yields achieved in the EU project (Table 3) are very poor compared to yields we would expect from other grain legumes in Europe. However, it should be borne in mind that these were two very dry years; it is unlikely that the genotypes chosen for this trial were the optimal ones; and the suitability all the sites for *L. mutabilis* was unknown. Other trials have given higher yields, for example, in England, Masefield (1976) found seed yields between 1.8 and 6.0 t/ha if the crop was sown in March, while the same genotypes produced between 0.5 and 3.5 t/ha when sown in April. Recent experiments in Germany resulted in yields between 1.8 and 6.5 t/ha in 1991 and 0.2 and 2.4 t/ha in 1992 (Rubenschuh, 1997). These results illustrate that yields of *L. mutabilis* are very unstable and strongly influenced by environmental conditions.

Table 3. Seed yield (t/ha) of 16 *L. mutabilis* genotypes at 5 European sites (F, UK, D, P, PL) and in 2 years (1994, 1995)

Mean of 16 Genotypes									
Year	1994					1995			
Site	F	UK	D	P	PL	F	UK	D	P	PL
Yield	1.8	0.9	0.8	2.1	0.4	1.2	1.4	0.5	0.9	0.7

Mean of 5 sites and 2 years									
Mean of 16 Genotypes				Best Genotype			Worst Genotype		
1.1				1.2			0.7		

The following specific reasons for the low yield of *L. mutabilis* were apparent from trials in the EU project:
- response to vernalization
- low potential of dry matter accumulation due to a low leaf area index
- preferential assimilate partitioning to vegetative growth
- high proportion of pod walls.

4. OUTCROSSING RATE

The rate of outcrossing in *L. mutabilis* was investigated in top-cross experiments over 2 years, with low and high anthocyan containing plants. The top-crosses were planted according to the scheme given in Table 4.

Table 4. Scheme of planting low and high anthocyan plants to estimate the outcrossing rate of *L. mutabilis* planted at Rastattin 1994 and 1995

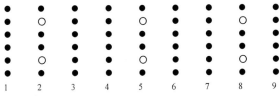

row number

● = Male Line--high in anthocyans
○ = Female Line--low in anthocyans

Table 5 shows the results of these experiments. Outcrossing rates between 16.6 and 58.8% were obtained which indicate that *L. mutabilis* needs to be treated in breeding programmes as a cross-pollinated crop. They also suggest the species is likely to exhibit inbreeding depression on selfing.

Table 5. Percentages of plants with high anthocyan content (A+) in progenies of low anthocyan plants grown with high anthocyan pollinators in the previous year. The results are indicative of the percentage outcrossing in *L. mutabilis* (results from 1994/95 and 1995/96, Rastatt)

1994/95			1995/96		
Genotype	Progeny	%Plants A+	Genotype	Progeny	%Plants A+
1	1	39.4	3	1	38.4
	2	32.7			
Mean		36.05			
2	1	16.6	4	1	25.8
	2	20.5		2	19.2
Mean		18.55		3	25.4
				4	58.8
				5	51.2
				6	32.0
				7	31.0
				8	22.4
				Mean	38.8
Mean	(N = 4)	23.7	Mean	(N = 9)	38.7

5.   EARLY MATURITY

Soon after the first attempts had been initiated to adapt *L. mutabilis* to European climatic conditions, it became obvious that a major problem existed in relation to late maturity (Römer andJahn Deesbach, 1992). The sympodial growth habit results in the continuous formation of lateral, higher order branches.

Selection within the existing variation was not successful because the phenotypes exhibited were poor indicators of the underlying genotypes, the response to selection for early maturity was thus more a reflection of the weather conditions in late summer of the respective year (Table 6) (Römer, 1995a).

Table 6 The relation in *L. mutabilis* between maturity and weather conditions in late summer at Gießen, Germany.

Year	Number of selected "mature" plants	Weather conditions in late summer
1985	45	dry
1986	206	dry
1987	0	very humid

In 1991, within the progenies from a mutation experiment, one completely determinate plant was found in the plant breeding nursery of Südwestdeutsche Saatzucht in Rastatt (Germany). It did not produce lateral branches but instead formed single flowers and pods at the base of the main stem. This mutant was called "KW 1" (Römer, 1995b). The mutant was not considered to be of direct agronomic value because of several disadvantages (tall growth, liability to lodge, low seed production). Therefore a cross-breeding programme was started by crossing the mutant to normal branched plants. The idea was to obtain plants with the determinate character expressed with first order lateral branches. Thus the combination of a good reproductive rate with early maturity might be possible.

Studies of the inheritance of the determinate character by scoring the F_2 progenies of crosses between "normal branched x KW 1" in 1994 and 1995 led to the results given in Table 7.

Table 7 Segregation of normal branched and determinate plants in *L. mutabilis* F_2 progenies of crosses with line "KW 1" at Rastatt in 1994 and 1995

Generation	$F_2'94$	$F_2'95$	Σ
Number of progenies	2	131	133
Total number of plants	244	973	1217
of these: normal branched	200	697	897
determinate	44	276	320
Ratio det.:normal = 1:x	4.5	2.5	2.8

The statistical analysis gave no indication of departure from a 3:1 segregation ratio, thus indicating that the inheritance of the determinate character was monogenic recessive.

In 1995, progenies of crosses "normal branched x KW 1" gave many plants that produced primary branches at the base of the plants. Some plants had a phenotype more or less similar to that which was considered suitable but with the exception that they were not branched completely to the top of the plant. The best-branched plant had 8 primary branches. It is suggested the type be called "semi-determinate".

6. CONCLUSIONS

The seed quality of *L. mutabilis* is not in dispute. The potentially high alkaloid levels do not present an insurmountable barrier since very low alkaloid genotypes already exist. The main hindrance, to the establishment of L. *mutabilis* as a crop is the lack of high yielding, early maturing genotypes.

However, the discovery of the mutant determinate line (KW 1) offers the potential to produce, through breeding, a more suitable ideotype. A semi-determinate plant, i.e. one with the determinate habit as well as possessing additional primary branches is predicted to be the most suitable form. This prediction is based on detailed studies and modelling that have been part of this project. All that is required is further breeding to establish *L. mutabilis* as a crop in Europe.

Acknowledgements
We are grateful to European Union, Directorate General VI (AIR3-CT93-0865) for financial support in carrying out much of the work reported here.

References
Antunez de Mayolo, S. 1982 In *Agricultural and Nutritional Aspects of Lupins* (Eds R. Gross and E.S. Bunting). GTZ Schriftenreihe Nr. 125: 1-11

Baer, E. von and Ibanez, R. 1986 *Proceedings of 4th International Lupin Conference*, Geraldton, Western Australia, 283.

Gross, R. 1983. PhD Thesis, University of Gießen, Germany.

Gross, R. and Baer, E. von 1975. *Zeitschrift fur. Ernährungswissenschaft*. 14: 224-228

Gross, R., Baer, E. von and Rohrmoser, K. 1983. *Zeitschrift fur Acker- und Pflanzenbau* 152:19-31.

Gross, R., Baer, E. von, Koch, F., Marquard, R., Trugo, L. and Wink, M. 1988. *Journal of Food Composition and Analysis* 1: 353-61.

Masefield, G. B. 1976. *Experimental Agriculture* 12: 97-102.

Römer, P. 1990. PhD Thesis, University of Gießen, Germany.

Römer, P. 1995a. In *Fortschritte in der Lupinenforschung und im Lupinen anbau*, pp 195-203 (ed Wink, M) University of Heidelberg, Germany,

Römer, P. 1995b. *Proceedings of the 2nd European Conference on Grain Legumes*, Copenhagen. 254

Römer, P. and Jahn-Deesbach, W. 1992. In *Agrimed research programme Lupinus mutabilis its adaptation and production under European pedoclimatic conditions*, pp 79-85 EUR 14102, Luxembourg,

Rubenschuh, U. 1997. PhD Thesis, University of Gießen, Germany

Research that Overcame the Impediments to Production and Marketing of Lupins in Western Australia

DS Petterson[1], RS Coffey[2], MA Sweetingham[1] and JG Allen[1]

1 Agriculture Western Australia, Locked Bag No 4, Bentley Delivery Centre, Bentley, WA 6983, Australia; 2 Milne Feeds, 103-105 Weshpool Rd, Kewdale, WA 6105, Australia

Abstract

The lupin (*L. cosentinii*) was introduced into the State of Western Australia during the 19th century. By the 1950s it had spread through most of the sandplain country of what is known as the Agricultural Area of the State. Many farmers regarded it as a nuisance because of its self-shattering pods and hard seeds, making it a potential contaminant and reducer of yield in the predominant cereal industries. Others saw it as having potential because of its nitrogen fixation and fodder value for grazing sheep. Some attempts were made to domesticate the species but it was realised that another lupin species was more promising.

Dr John Gladstones of the University of Western Australia, recognised the potential of *L. angustifolius* (narrow-leafed lupin), and released the world's first true lupin crop variety, Uniwhite, in 1967. Since then, others varieties have been released in Western Australia and in other parts of the country. From a base of a few thousand tonnes in the early 1970s the industry has grown to one producing over a million tonnes a year.

This development came about through the dedication of many groups in addition to the plant breeders. Perhaps the most important early contribution came from the Grain Pool of Western Australia, whose marketing staff sought and supported markets world-wide. Their efforts underpinned the work of research and extension groups in giving farmers the confidence to grow a new crop. Plant breeders, agronomists, entomologists and pathologists contributed to increasing yields, overcoming pests and diseases. Animal nutritionists, biochemists, chemists, food scientists and technologists, pharmacologists and toxicologists contributed in developing feed and food uses for lupins, and in demonstrating to industry and legislative bodies they could be used safely.

1. INTRODUCTION

For a new commodity to be accepted by the food or feed industries it must have some intrinsic value.

In the case of the food industries, they need to be convinced a new alternative commodity will result in a lower unit cost of production or a better end-product. Cost advantages could arise from a lower price of raw materials, higher yields or lower processing costs. Quality advantages could include a higher sensory appeal, from appearance, taste, texture or aroma, or a longer shelf-life. The feed industry would need to know potential advantages of either the protein or energy content of the material in terms of relative value per unit cost for a specific formulated diet. In other words, there would have to be a cost advantage when using linear programmes to prepare the feed formulations used in intensive animal industries.

For a crop industry to develop three major groups need to be convinced of its potential. Firstly, the grower will need to know:

- how to grow and harvest the crop;
- how to deal with pests and diseases;
- the benefits of growing the crop;
- that there will be a market for the commodity;
- there will be ongoing research and extension into improving yield, management, value and end-uses for the commodity;
- there will be a financial return from growing the crop.

R. Knight (ed.), Linking Research and Marketing Opportunities for Pulses in the 21st Century, 575–587.

Until these issues are addressed - although not necessarily fully implemented - there can be little expectation that growers will shift from traditional crops to a new one.

The potential cosumers and end-users will want to have information about the crop and its products:

- its chemical and physical characteristics.
- its nutrient content.
- its handling and processing properties (eg splitting, milling and storage).
- any possible limitations to its use.
- the content of antinutritional factors (eg lectins), endogenous toxins (eg alkaloids, neurotoxins) or exogenous toxins (eg mycotoxins).
- the shelf-life of products (eg breakdown, oxidation, predation).
- how safe it is to use (eg whether the content of untoward compounds outweighs the nutrient content).

In between these two groups there are the commodity traders and handlers. In the case of lupins in Western Australia it was the Grain Pool of Western Australia, which had the single-desk selling rights at the time the industry was developing.

The traders' needs are no less complex than the grower's or end-user's. A trader is more likely to invest time and money into developing markets if:

- supply is reasonable;
- handling problems are minimal, or understood and likely to be overcome in the future;
- there are no legal problems with trading the commodity;
- potential buyers have adequate information on which to base their purchases;
- there are no perceived problems with growing, handling or using the crop.

Many of the needs for information are common to all parties, and it is clear that continuing programs of research, development and extension are required for long-term benefits to be sustained.

In this paper we review the chronological development of the industry principally in Western Australia where the major developments took place, and relate how they fitted the requirments of these three groups of interested parties.

2. THE NEED FOR A CROP LEGUME INDUSTRY IN WESTERN AUSTRALIA

Throughout the 1960s and 1970s, government policy in Australia supported efforts to increase the area of land under cultivation. Much of it was on marginal land, where the soils had a low content of nitrogenous matter. Additional burdens were put on the soil when farmers began to grow crops, instead of pasture, on more of their land each year and to introduce rotational cropping. These practices led to an increase in weeds with brome grasses becoming dominant in many areas. To counter the weeds, farmers used more frequent tilling of the soil, leading to a greater risk of soil erosion. During the 1960s, nitrogenous fertilisers became available at economic prices, however their sustained use, combined with these other factors, frequently resulted in a drop in wheat yields. A more sustainable alternative was needed.

2.1 Why grow lupins?

At the time lupins seemed a logical alternative for various reasons.

One was that the Mediterranean climate and poor quality of the soil meant the entire Agricultural Area of Western Australia was best suited to a crop that performed relatively better under adverse conditions than one capable of high yields under the most favourable conditions.

A second reason for considering lupins was that the sandplain lupin (*Lupinus cosentinii*), originally known as the Western Australian blue lupin, had colonised much of the northern part of the Agricultural Area, illustrating lupins could grow well in the area. However, the plants had a high alkaloid content and were not favoured by grazing livestock, although sheep consumed the plant when growing or senescent, if no alternatives were available. The bitter seeds were less acceptable. A breeding program to reduce the alkaloid content, particularly of the seed, and to overcome pod shattering was initiated. It was quickly recognised that the hard seed coat and long trem viability of the seed would make the task of selection and multiplication difficult. Instead, the lupin breeder, Dr John Gladstones of the University of Western Australia, chose to work with the narrow-leafed lupin *L. angustifolius*.

A third reason for considering lupins was that, at the time, there were no other locally adapted varietites of field peas, chickpeas or faba beans suitable for widespread cultivation.

2.2   Benefits of growing lupins

In the development of the industry the main benefits of growing lupins were perceived as:
- providing nitrogen for succeeding cereal crops;
- providing a 'disease break' for cereals in the rotation, particularly when grass weeds were controlled;
- the deeply penetrating tap roots of lupins would recycle leached nutrients back to the soil surface, and provide channels for the roots of succeeding cereals to follow in the search for moisture and nutrients;
- residual stubbles and fallen grain would provide valuable feed for livestock on-farm;
- retained stubble would help minimise wind erosion;
- including lupins would lead to a diversification of farming enterprises.

3.   IMPEDIMENTS OVERCOME IN THE DEVELOPMENT OF THE LUPIN INDUSTRY

3.1   Reducing the alkaloid content of the seed

Historically, lupins have been bitter seeded and toxic. The traditional ways of rendering the seed palatable was to pulverise it into a powder which was washed with copious quantities of water (Gladstones, 1970; Aguilera and Trier, 1978; Uauy et al., 1995). An unfortunate side-effect of these actions was the removal of soluble proteins, free amino acids, carbohydrates and minerals. Nevertheless, the method had been effective for subsistence societies in Mediterranean Europe and the Andean regions of Chile and Peru. In these societies there seems to have been little selection of lupins other than for large seed.

Before lupins could become a major crop it was necessary to breed varieties with low levels of toxic alkaloids. In 1929 Dr Rainer von Sengbusch in Germany announced that after testing seed from 1.5 million plants he had found 'sweet' mutants of *L. albus*, *L. angustifolius* and *L. luteus* (von Sengbusch, 1942). During the next two decades other mutants were identified and these became the starting material for the sweet lupin industry (Gladstones, 1970).

In adapting *L. angustifolius* to the Mediterranean climate of Western Australia, Dr Gladstones selected for reduced alkaloid content in the seed, while at the same time increasing yield and resistance to major diseases. He released the first crop variety of lupin, *L. angustifolius* cv. Uniwhite, in 1967. Since then there has been a steady reduction in the alkaloid content of commercially harvested seed. In a survey of crops in the 1982 growing year, Harris and Jago (1984) reported a mean value of 180 ppm (range 20 to 580 ppm). For 1988 the mean value was 110 ppm (range 20 to 260 ppm) and for 1996 the mean was 100 ppm (range 80 to 120 ppm) (DJ Harris, 1997, personal communication).

The alkaloid screening program started by Dr Gladstones, and now led by Drs Wallace Cowling (*L. angustifolius*) and Bevan Buirchell (*L. albus* and *L. luteus*), required 3-4,000 tests per annum (Cowling et al.,. 1997). Research led by Dr David Harris at the Chemistry Centre (WA) is aimed at increasing the sample throughput, increasing the sensitivity and reducing the cost.

3.2   Improving crop yields

Once low-alkaloid lupins were a reality, the problems of pod-shattering, pests and diseases, establishing seeding and fertiliser protocols, weed control and harvesting practices had to be overcome to make lupins a successful crop (Nelson and Hawthorne 1998). During the early years, the Western Australian Department of Agriculture achieved much higher yields on research stations than did farmers in general. An extension program was needed. Features of this program included on-farm demonstration, where best practices and the consequences of not adopting these practices, were shown. At field days, farmers and agricultural advisers would inspect these demonstrations and discuss the impact of the practices. In many instances farmers were able to tell advisers about techniques that were effective in their circumstances. This information was disseminated. Frequent radio interviews and newspaper articles were features of the program.

Further industry assistance came from modifying conventional cereal harvesters to harvest lupins and to reduce the amount of broken and fallen grain.

Giving farmers the confidence to grow the crop were major factors in exploiting the improvements achieved by the plant breeders. From yields of around 700 kg/ha over only a few thousand hectares in the early 1970s there has been a steady increase in yield across environments and soil types, to about 1100 kg/ha over nearly 1 million hectares in the mid 1990s. Not all environments proved suitable for *L. angustifolius* and breeding programs were initaited on *L. albus, L. atlanticus, L. luteus* and *L. pilosus.*

As the lupin crop went through its exponential expansion phase in the 1980s, problems of disease and weed control emerged. The most profitable system became a continuous lupin-wheat rotation. Without a pasture phase, weeds such as ryegrass and wild radish began to build up and needed to be controlled with knockdown and pre-emergent herbicides. At the time, various grass selective herbicides and diflufenican post-emergent were capable of maintaining lupin crops weed free.

As the rotation bedded in, it became clear that a one year break between lupin crops was insufficient to control brown spot and Pleiochaeta root rot, particularly in the cooler regions. Pathologists and agronomists modified establishment techniques and promoted fungicide seed treatments to minimise disease even in tight rotations (Sweetingham and Jarvis, 1993). In parallel, the breeders continued to breed for improved resistance. Myallie was the first narrow-leafed lupin with significant improvements in brown spot resistance. Kalya has significantly improved resistance to Pleiochaeta root rot.

During the late 1980s, bad aphid years resulted in serious losses from cucumber mosaic virus in the higher rainfall environments. Research showed that if seed stocks could be kept free of the virus the disease could be readily controlled (Jones 1988).

Although improvements in agronomy maintained the rotation from a disease point of view, in the early 1990s the reliance on chemical weed control led to the emergence in ryegrass of resistance to a wide range of herbicides in different chemical groups. Weed scientists, who had predicted this scenario, had to develop an integrated weed management program based on alternative strategies. The capture of weed seeds at harvest and crop-topping (the use of non-selective herbicides late in the season to reduce weed seed set) were developed (Gill, 1994).

3.3   Minimising the risk of lupinosis

One impediment to the development of the lupin industry was the real and apparent threat of lupinosis. This disease is a mycotoxicosis caused by the ingestion of a group of toxins known as phomopsins. They are produced by the fungus *Diaporthe toxica,* previously known as *Phomopsis leptostromiformis* and *P. rossiana,* which colonises lupin plants (van Warmelo et al., 1970; Culvenor et al., 1977; Williamson et al., 1994). Lupinosis is a disease of sheep, but outbreaks have been reported in cattle and other farmed species, and the disease has been produced experimentally in a range of laboratory animal and bird species (Allen, 1986).

Lupinosis is characterised by liver damage, which leads to inappetence, loss of condition, jaundice, lethargy, and in severe cases death. Its most distinctive microscopic feature is an arrested and abnormal mitosis of hepatocytes (Gardiner, 1967; Peterson, 1978). Kidneys, adrenals, pancreas, rumen and muscle tissue may also be damaged in severe cases, and there are reports of abortion, embryotoxicity and reduced lambing rates (Allen, 1986).

The condition was first recognised in Germany, where it virtually destroyed the lupin industry in the 1870s (Zurn, 1879), and has been reported in North America, Poland, New Zealand, South Africa and Spain (Allen, 1986; Rodriguez et al., 1991). In recent times it has only been in Western Australia, where the grazing of lupin stubble after harvest has been extensively practiced, that lupinosis has been a major concern. In this State serious outbreaks had occurred over several decades and memories of these episodes and their impact on farming communities made it difficult for a new lupin industry to be established.

Proof that the disease was a mycotoxicosis (van Warmelo et al., 1970; Gardiner and Petterson, 1972) gave researchers and industry opportunities to overcome the problem:
- Mycologists were able to study how the fungus colonised the plant and the conditions that favoured toxin production.
- Chemists isolated the toxins, and developed assays to detect them.
- Plant breeders selected for resistance to the fungus.

- Scientists developed grazing programs to minimise the risk of the disease.
- Immunologists were able to raise antibodies to the toxins thus assisting chemists with assay development and more importantly to produce a vaccine for livestock.

All of the above gave farmers the confidence to grow lupins. Management procedures to minimise the risk of lupinosis in livestock grazing stubbles were well established by the early 1980s (Allen, 1987; Allen and Chapman, 1988), and extension programs showed farmers how to manage their situations. Surveys conducted by the Grain Pool and the Department of Agriculture found consistently that the risk and problems associated with lupinosis were major concerns to farmers.

A breeding program which started in 1974 culminated in the registration of three *Phomopsis*-resistant (the fungus was still known as *P. leptostromiformis* at this time) cultivars, Gungurru, Yorrel and Warrah, in 1989 (Hawthorne and Gladstones, 1989; Gladstones, 1989a). The program which transferred resistance from wild to domesticated *L. angustifolius* was described by Gladstones (1989b).

The main toxin, phomopsin A, was isolated by Culvenor et al., (1977) and identified as a linear hexapeptide modified by an ether bridge to form a 13-membered cyclic ring incorporating four of the amino acids (Edgar, 1991). Phomopsin B is a chlorinated phomopsin A (Edgar, 1991) and phomopsins C, D and E appear to differ from phomopsin A only in the nature of the terminal amino acid residue (Allen and Hancock, 1989; Cockrum et al., 1994). All of these phomopsins induce the characteristic mitotic abnormalities in hepatocytes (J.G. Allen, 1997, personal communication). An HPLC assay (Hancock etal., 1987) and a sensitive and inexpensive ELISA (Than et al., 1992, 1994a) have been developed for the detection of the phomopsins in a range of substrates.

The availability of the pure toxin enabled the development of an anti-lupinosis vaccine, which has undergone field evaluation (Allen et al., 1994; Than et al., 1994b). Whilst not fully effective to date, expectations are high for a successful commercial vaccine.

A further possible impediment to the industry was the risk of the phomopsins, or any other mycotoxins, being present in the seed of lupin species. There have been only three reports of lupinosis associated with the ingestion of lupin seed by sheep. All three were considered atypical. The first was in a standing crop in which the pods and seeds were heavily discoloured (van Warmelo et al., 1970). The second involved two feeding trials where sheep were fed for 34-45 weeks with lupin seed, of which 7% and 21% were infected with *D. toxica* (Allen et al., 1976, 1983), and the third involved feeding seed, of which 32-46% were discoloured to sheep with a concurrent poisoning from a toxic plant (Links and Walker, 1989). After the first alert by Allen et al., (1976) a series of studies were initiated and it was quickly learned that: infected seed may be either discoloured or asymptomatic (Wood, 1986); virtually all of the toxins produced are in characteristically discoloured seed (Petterson and Hancock, 1986; Wood and Petterson, 1986; Wood et al., 1987; Than et al., 1994a); and the discoloured seeds, being lighter in weight and less dense, can be removed by commercial seed cleaning processes (Wood and Petterson, 1986). Colour-grading using commercial equipment (Sortex) is also effective in removing discoloured seeds (KH Than, PJ Moore and DS Petterson, 1994, unpublished results). Than et al., (1994a) found the concentration of phomopsins in commercial seed lots to be lower in the modern *Phomopsis*-resistant cultivars than in susceptible cultivars.

The combination of breeding for fungal resistance and physical intervention should always ensure that seed destined for human consumption will contain less than 5 µg phomopsins per kg seed (the maximum permitted for human consumption by the National Food Authority in Australia and the Department of Health in the UK).

4.   PROVING THE SAFETY OF LUPINS FOR HUMAN CONSUMPTION

A food might be considered to be 'generally recognised as safe' if there is a long history of use in more than one community and it is approved for human consumption on that basis. If scientific evidence to the contrary were found subsequently, this approval might be withdrawn. Although lupins, *L. albus* in Europe and *L. mutabilis* in South America, have been used as a food for millennia (Gladstones, 1970; Uauy et al., 1995) they could not be regarded as safe because of their bitter alkaloid content. Whilst the communities familiar with these seeds were able to use them safely it could not be assumed that the general community would act in the same way. The case of alkaloid poisoning of a woman in Australia who ate bitter lupini beans imported from Chile illustrates the problem (Lowen et al., 1995). Her symptoms were not life-threatening, but did require intensive care over-night before she fully recovered.

The alternative approach, taken for *L. angustifolius* in Australia and subsequently in the UK and other countries, is to go through a rigorous scientific evaluation that would include a literature search, further toxicological studies, pharmacokinetic studies, and allergy tests. Information on the effects of processing would also help determine possible risks.

4.1 Reported instances of adverse effects

Reported cases of lupin alkaloid poisonings in humans number three deaths in children (Schmidlin-Meszaros, 1973), five acute anticholinergic reactions in adults (Schmidlin-Meszaros, 1973; Smith, 1987; Marquez et al., 1991; Lowen et al., 1995) and a presumed induction of a motor neurone disorder in a single patient (Agid et al., 1988).

The lupin alkaloids showed a low acute oral toxicity for rats and mice (Mazur et al., 1966; Yovo et al., 1984; Petterson et al., 1987) of about 2,000 mg/kg, and no adverse effects were noted on growth, mortality, histology or biochemistry of rats fed lupin-based diets with up to 0.09% alkaloid for 270 days Ballester et al., (1980, 1982). Ballester et al.,. (1984) found fertility in the F1 and F2 generations from a previous study (Ballester et al., 1982) when fed a diet based on *L. albus* seed meal containing 0.013% alkaloids did not differ from controls and all tissues examined were histologically normal.

Production records of the Department of Agriculture, Medina Research Station covering more than eight generations of pigs and a total of over 1100 litters were similar to those of commercial operations that did not use lupins. There were no indications of any teratological effects, and no reports of significant lesions in the sows when slaughtered. During this time, starter, grower and finisher diets contained from 10-40% *L. angustifolius* seed and the sow diets, 10-30%. Alkaloid levels in the diets were estimated to range from 10-80 mg/kg (Petterson, 1985).

Gross (1990) reported a skin prick test on over 200 Chilean children showing the sensitivity to extracts from lupins (*L. albus*) was similar to egg and wheat, and much less than to cow's milk, peanuts and soybeans. Less than 1% of the >3,000 people who had eaten lupin (*L. angustifolius*) - based, foods in Australia showed an immediate allergic reaction (Petterson and Crosbie, 1990), and all of those who reacted had a history of reactions to peanuts, soybeans and others foods. Hefle et al., (1994) reported a typical histamine reaction by a young girl who consumed a lupin (*L. albus*) fortified pasta.

4.2 Additional research conducted

An alkaloid extract from *L. angustifolius* cv. Fest, which has a high alkaloid content ($\approx 2\%$) and a similar profile to modern cultivars, was shown to be devoid of any mutagenic potential in a Salmonella mutation test, a cultured mammalian cells (HGPRT) test and a chromosome aberration test in Chinese hamster ovary cells. Further, the alkaloids were not activated to mutagenic chemicals by a metabolising system derived from chemically activated rat liver (BIBRA, 1986). This cleared the alkaloids of any presumed risk of being (pro-) carcinogenic.

Robbins et al., (1996) conducted a 90-day study in rats using the alkaloids derived from *L. angustifolius* cv. Fest at dose rates of 100, 330, 1,000 and 5,000 mg alkaloid/kg feed. There were no untoward histological or biochemical changes in any of the rats that could have been attributed to the alkaloids in the diet. Grant et al., (1993, 1995) found no adverse effects on appetite or growth of rats when fed raw seed of *L. angustifolius* for up to 800 days.

One of the concerns raised by the Advisory Council on Novel Foods and Processes in the U.K. was whether any groups within the population might have a reduced capacity to metabolise lupin alkaloids in foodstuffs, thus leading to accumulation and the possibility of untoward side-effects. For example, about 5-10% of Caucasians are poor metabolisers of sparteine, which has a similar pharmacological activity profile to lupanine and 13-hydroxylupanine (Mazur et al., 1966; Yovo et al., 1984), although it is considerably more toxic (Yovo et al., 1984; Petterson et al., 1987), and show increased therapeutic response and side effects (reviewed by Cholerton et al., 1992). In a study with 4 poor, and 7 extensive, metabolisers, Petterson et al., (1994) reported no significant difference in the half-life of either lupanine or 13-hydroxylupanine after oral administration (6-7 hr) and the bulk of the two alkaloids studied were excreted unchanged in the urine. They concluded there was little likelihood of any accumulation of the parent molecules and any associated systemic toxicity.

Consumption of any product derived from modern cultivars of 'sweet' lupins is therefore unlikely to have any adverse effects. For example; a standard 100 g serve of pasta fortified with 20% lupin flour would only contain about 4 mg alkaloids. Tempe production has been found to remove about 85% of the alkaloids (Fudiyansyah et al., 1995), and germinating (sprouting) the seeds about 80% (Dagnia et al., 1993). Lupin cell wall (fibre) preparations contain only traces of alkaloids (DJ Harris, 1992, personal communication).

5.   PROCESSING AND FOOD USES

Kernels of *L. albus* and *L. angustifolius* can be fractionated after wet milling, into protein-rich and fibre-rich fractions, each of which can be further purified. Neither protein concentrates nor isolates have any particularly useful functional properties, although a small component of the protein has excellent whipping properties (S.C. Johnson, 1994, personal communication). The purified fibre fraction is colourless, odourless, tasteless and can have a 7 to 8-fold hydration capacity. It has been shown to lower blood cholesterol levels in rats (Evans 1994). Studies in humans have shown the fibre to have good faecal bulking activity and volatile fatty acid production, with an indication of potential for cholesterol-lowering (P Clifton, D Topping and M Noakes 1995, unpublished results). Lupin fibre has been successfully included in breads (Evans 1994) and other foods (Petterson 1997). Lupins have been shown to be an excellent substrate for the production of tempe (Fudijansyah et al., 1995), miso (Coffey 1989) and Oriental sauces (Petterson 1997). They have also been used to supplement breads (Lucisano and Pompeii 1981; Wittig de Penna et al., 1989, Petterson and Crosbie 1990), humitas (Camacho et al., 1989) and pastas (Pompeii et al., 1985, Rayas-Duartes 1993). The use of lupins in food have been reviewed by Birk et al., and Petterson (1987).

6.   ANIMAL NUTRITION STUDIES

6.1   Ruminants

Prior to the development of the lupin industry in Australia there were few references to the use of lupin species (Hill, 1977) but there have been many since on their use as a feed for sheep (Murray, 1994; Godfrey et al., 1993; Kenny and Smith, 1985), and to a lesser extent cattle (Hough and Jacobs, 1994).

Lupins are a good source of metabolisable energy, with Petterson and Mackintosh (1994) quoting 12.5 and 12.0 MJ/kg (as fed) for *L. albus* and 12.2 and 11.7 MJ/kg (as fed) for *L. angustifolius* for sheep and cattle respectively, equivalent to 12.7 and 13.6 MJ/kg DM. Margan (1994) reported ME values between 15 and 15.5 MJ/kg DM for *L. angustifolius*, when fed as whole of diet, well above published values based on calculations or likely from natural variation. Sheep need only a minimal period of introduction to the grain because there is only a low content of quickly fermenting carbohydrates and there is little risk of acidosis unless the animals are poorly nourished (Allen et al., 1997). Cattle also readily consume the grain which are a valuable feed supplement for grazing dairy cows.

The protein is readily degraded in the rumen (Dove, 1984) and is source of nitrogen for microbial digestion. Only 15% to 27% of the protein escapes rumen degradation to become available for absorption in the intestines (Hume, 1974) and attempts to increase the level of bypass protein have met with mixed success Improvements in the reproductive performance of ewes have been variable when their diets have been supplemented with lupins but there are no obvious reasons for the variation (Leury et al., 1990). There are many reports of an improved serving performance of rams fed lupins through increased testis size (see Murray, 1994). The cause of this effect remains unknown, but factors such as energy and protein have been precluded (GB Martin, 1997, personal communication).

6.2   Non-ruminants

6.2.1   Pigs

Grain of *L. angustifolius* is a useful source of protein and energy for pigs (Godfrey and Payne, 1987; Batterham, 1979; Edwards and van Barneveld, 1997). Limitations to its use were found to be alkaloids above 200 mg/kg in the diet and occasional gut distension due to the presence of oligosaccharides. Growth performances vary, depending on the accompanying energy substrate (Edwards and van Barneveld, 1997). In the commercial world, formulations generally contain from 10-30% lupin grain depending on the price and availability of other vegetable protein sources. The grain is widely used in Australia, south-east Asia and parts of Europe.

Grain of *L. albus* varieties grown in Australia contain some feed deterrent- not alkaloids or the common antinutritional factor-that restricts their use to about 10% of the diet (King, 1990; Murphy and Castell, 1995). It is not recommended for use in commercial diets.

Grain of *L. luteus* is presently used in the pig industry in Poland (W Swiecicki, 1997, personal communication), and trial results in Australia suggest in would be a useful (BR Mullan, 1997, unpublished results).

6.2.2   Poultry

Broiler chicken diets can include up to 25% lupin seed meal, when supplemented with lysine and methionine (Brenes et al., 1993). However in commercial practice inclusion rarely exceeds 10% because of the risk of wet-sticky droppings (van Barneveld and Hughes 1994), which pose a health risk due to fly infestations, increased prevalence of coccidiosis and respiratory stress from high ammonia.

The housing and management practices for laying birds differ from those of broilers and this permits a maximum inclusion of 25-35% of *L. albus* or *L. angustifolius* in the meal.

Adding commercial enzymes to poultry diets increases by about 15%, the apparent metabolisable energy of lupins (Annison et al., 1996). Dehulling and autoclaving respectively improve the value of lupins for poultry by deleting indigestible carbohydrates and by a presumed destruction of antinutritional factors (Brenes et al., 1993, Marquardt 1993).

6.2.3   Uses in aquaculture

Lupins are a useful ingredient in diets for marine finfish, estuarine and freshwater species, and for fresh and saltwater crustaceans.

For rainbow trout, Hughes 1988 used 40% *L.albus* seed in their diets and reported that fish grew at a comparable rate to others fed a 40% full fat soybean meal (FFSM) diet or a commercial diet. The calculated value for metabolisable energy of the lupins was 11.7 MJ/kg, with protein digestibility of 85.2% and energy digestibility of 66.1%. Higuera et al., (1988), Gomes and Kaushik (1989) and Moyana et al., (1992) studied other aspects of inclusion and had good results from using lupins. Jenkins et al. (1994) showed that 28% *L. angustifolius* meal inclusion could be successfully substituted for 20% soybean meal in diets for juvenile snapper (red sea bream, *Pagras auratus*), and Petterson et al. (1996) found little difference between the growth rates and FCR, 1.2 to 1.37, for juvenile snapper fed a 44% protein reference diet those fed the same diets containing up to 40% ASL meal (*L. angustifolius*, cv Gungurru). Others have found lupins useful ingredients for gilthead sea bream (*Sparas aurata*) (Robaina et al., 1995), milkfish and tilapia (J Hutabarat, 1997, unpublished results), and carp (Viola et al., 1988).

7.   DEVELOPMENT OF THE LUPIN INDUSTRY

In the development of the industry it was necessary to overcome ignorance of the crop and the commodity before an export market could be established.

Lupins were the first of the then 'unknown' crops to be developed in Australia. Before the 1970's all other crops sold within and out of Australia were familiar to buyer and seller alike. Neither the farming community, the agencies involved with the development of the crop (Department of Agriculture, University of Western Australia, CSIRO), nor the single-desk seller had any concept of the potential for the lupin industry in the 1970s and early 1980s.

Initially a minimum of effort was directed towards customer development. This effectively left the seller in the position of being a price-taker; which is still very much the position today. With world trade in lupins at around 1% of the trade in soybeans, it is difficult to predict any change unless a special application for lupins is discovered.

The export side of the industry followed a tortuous path. Production until the mid 1970s was low, and mostly taken up by the emerging pig and poultry industries or used on-farm for sheep and cattle. A trading house in the Netherlands had an exclusive arrangement for sale into Europe but the principals were not active in market development. In the early 1980s the seller found itself with a surplus to domestic requirments for the first time, and an urgent need to seek an international market. An approach was made to a Japanese trading group, which was a buyer of other grains.

Both of these cases brought home the first lessons to be learned! Trading houses are primarily interested in margins on sales. They are not interested in developing a market for a little-known commodity of doubtful supply, and certainly not in investing resources into the development. There was little point in talking to the company's chief executive officer or purchasing manager when introducing a new commodity. The first hurdle was always the company's nutritionist. Whilst good data were often available from research work under Australian conditions, most company nutritionists were loath to recommend a commodity that was unproven in their own circumstances. Nutritionists and buyers wanted results from work undertaken in their own country. One country would not even allow imports of lupins until samples had been extensively tested by their own agencies.

Another problem was that with limited funding in Australia, market development usually received the smallest allocation. It became incumbent on the Grain Pool and interested researchers to fund the necessary work from their own limited resources. Too often the research was piecemeal and frequently was in response to specific requests from potential customers.

While developing overseas markets during the 1980s, other problems were encountered. One was the Customs classification of lupins in potential markets where they were often listed unfavourably in relation to competing commodities. For example, it took the WA Grain Pool, with the lobbying assistance of the Korean Feed Association and the Australian Government, three years before the Korean Government altered the original classification of lupins, which was one of seed for sowing.

A similar problem occurred with import quotas in Taiwan. These could have been imposed to provide a non-tariff protection of their local oilseed crushing industry. It was not feasible to develop a worthwhile lupin trade. There are neighbouring countries today, which require the tariff on lupin seed or meal to be three to four times higher than for soybeans or soybean meal. No amount of research or rhetoric can overcome these impediments to the growth of an industry.

Another problem arose with nutritionists and others, including health authorities, who were aware of bitter lupins from Europe, North Africa and the Americas and their associated alkaloid problems, particularly the crooked calf disorder associated with rangeland perennial lupins in the US. These people had difficulty accepting low-alkaloid varieties of lupins existed.

Following the development that occurred during the 1980's, interest focused on the use of lupins for food. We were aware of the traditional uses of the lupini bean, large seeded *L. albus*, in Mediterranean countries and of choco or tarwi, *L. mutabilis*, in Andean countries and of the wide range of traditional foods and products derived from soybeans. Information about new food uses for lupins, mostly *L. albus*, was being reported principally from Italy, Germany, Chile and Peru. Work from Australia focussed mainly on *L. angustifolius* (Kidby et al., 1977, Yu 1988). There seemed to be potential for lupins as a fermented food, a protein supplement to breads, pastas, biscuits and similar products and as a source of protein and fibre concentrates.

The first task was to seek approval from the National Health and Medical Research Council (NHMRC) in Australia. The first application made by the Grain Pool was for a maximum of 10% inclusion in any food, on the basis of expected maximum inclusions in breads, sausages etc. This approval was granted in 1978. However, it was not until 1988, after an extensive literature review (Petterson, 1985), a negative result from mutagenicity screening (BIBRA, 1986) and a 90-day feeding study with rats at BIBRA Toxicology International (BIBRA, 1988: Butler et al., 1996), that the NHMRC approved the unrestricted use of lupins provided the alkaloid content did not exceed 200 mg/kg and the phomopsins content did not exceed 5 μg/kg.

After research at the Shinshu Research Institute in Japan found that lupins were an excellent base for miso production (Coffey, 1989), an approach was made to the Japanese Department of Health and Welfare (JDHW) to allow the use of *L. angustifolius* seed for human consumption. This organisation expressed the view that, because of the negligible use of lupins as a food in Australia, acceptance by a third party would be a prerequisite for formal approval by Japan. The JDHW suggested that approval under the guidelines of the Advisory Council on Novel Foods and Processes (ACNFP) in the UK would be appropriate. A formal application was submitted in September 1990. The ACNFP required a further 90-day feeding study with rats to clear an apparent anomaly (Robbins et al., 1996), some evidence of the fate of these alkaloids in humans (Petterson et al., 1994), and other assurances on issues, most of which seemed to arisen from confusion about lupinosis and lupin alkaloid toxicity. In 1996 approval for human consumption was formally granted by the Minister for Health in the UK, with the same two provisos as in Australia. Hopefully this will lead to approval being granted in the rest of Europe, Japan, Canada and the USA.

8.   KEY LESSONS LEARNED

Many lessons have been learnt as a result of the lupin experience. One was that it costs a lot more money, and it takes a lot more time to develop and promote a new commodity than any of the persons involved could have envisaged. For example, it took three years in concert with a very good consumer ally, the Zen-Noh Corporation, to get lupins approved as a feed ingredient in Japan. After 13 years of effort not all the barriers to the use of lupins in food have been lifted, even though there seems to be great potential for their use in miso manufacture (Coffey, 1989).

Another lesson was that is is essential to have a coordinated approach by all the interested parties to generate and disseminate information and to engage in active dialogue with the relevant trade and health groups. For this we would argue for a joint effort by traders, growers, Government agencies and research organisations.

9.   FUTURE RESEARCH AND PROMOTION FOR THE LUPIN INDUSTRY

The bulk of research conducted in Australia to date has been researcher-driven rather than industry-driven which has meant some areas have been well researched and others have not.

The pig industry is one example where an industry-coordinated program would have helped in assessing the value of lupins. Batterham et al., (1984) suggested the availabliity of lysine in lupins was low and variable (0.37 to 0.54), and the current recommended value for lysine availability from lupins in Australia is 0.55 (SCA, 1987), yet from feeding studies in Western Australia, Godfrey and Payne (1987) suggested a value in excess of 0.7 was reasonable and modern commercial practice indicates a range of 0.7 to 0.8. Recent research (Edwards and van Barneveld, 1997), using more precise techniques, indicates a ture availability of between 0.7 and 0.8 for seed and kernel of *L. albus* and *L. angustifolius*.

10.   RECOMMENDATIONS TO ANY GROUP DEVELOPING NEW CROPS

To persons trying to develop a new crop we would suggest from our experience with lupins.

1 Once the main difficulties with growing, harvesting and marketing have been resolved try to estimate the potential size of your crop in the future.

2 Once this forecast has been made consider whether you are likely to be

price takers - accepting the lowest relative value on world markets based on a protein and energy value or

price makers - establishing a relatively higher value for your commodity because of some extra nutritional, functional or hedonic value.

3 Undertake a SWOT (Strengths, Weaknesses, Opportunities and Threats) analysis.

4 Develop a 'business plan' for the industry based on the outcome of this analysis. Input from industry and economists is essential at this stage.

5 Establish the necessary research and development team. Ensure the team has adequate funding. Make sure that no one in the team loses sight of the big picture.

6 Review progress on a regular basis.

7 Be prepared to spend a lot of money on promotion.

8 Listen to your potential customers and work with them as a partners.

References

Agid, Y., Pertuiset, B. and Dubois, M. 1988. *Lancet* 1: 1347.

Aguilera, J.M. and Trier, A. 1978. *Food Technology* 32: 70-76.

Allen, J.G. 1986. In: *Proceedings of the Fourth International Lupin Conference*. International Lupin Association and Western Australian Department of Agriculture, South Perth, pp. 173-187.

Allen, J.G. 1987. In: *Proceedings No. 103, Veterinary Clinical Toxicology*. The Post-Graduate Committee in Veterinary Science, Sydney, pp.113-131.

Allen, J.G. and Chapman, H.M. 1988. In:*Proceedings No. 110, Sheep Health and Production*. The Post-Graduate Committee in Veterinary Science, Sydney, pp. 431-456.

Allen, J.G. and Hancock, G.R. 1989. *Journal of Applied Toxicology* 9: 83-89.

Allen, J.G., Tudor, G.D. and Petterson D.S. 1997. *Proceedings of the 5th International Symposium on Poisonous Plants*. Texas. In press.

Allen, J.G., Moir, R.J. and Mackintosh, J.B. 1983 *Australian Veterinary Journal* 60: 206-208.

Allen, J.G., Than, K.A., Edgar, J.A., Doncon, G.H., Dragicevic, G. and Kosmac, V.H. 1994. In: *Plant-associated Toxins - Agricultural, Phytochemical and Ecological Aspects*. pp. 427-432. (eds Colegate, S.M. and Dorling, P.R.) CAB International, Oxon,

Annison, G., Choct, M. and Hughes, R.J. 1994. In Australian Poultry Science Symposium Editorial Committee (eds.). *Proceedings of the Australian Poultry Science Symposium*. University of Sydney, pp. 92-96.

Annison, G., Hughes, R.J and Choct, M. 1996. *British Poultry Science* 37: 157-172.

Ballester, D., Yanez, E., Garcia, R., Erazo, S., Lopez, F., Haardt, E., Cornijo, S., Lopez, A., Pokniak, J. and Chichester, C.O. 1980. *Journal of Agriculture and Food Chemistry* 28: 402-405.

Ballester, D.R., Brunser, O., Saitau, M.T., Egana, J.I., Yanez, E.O. and Owen, D.F. 1982. *Revista Chilean Nutricion* 10: 177-191.

Ballester, D.R., Brunser, O., Saitau, M.T., Egana, J.I., Yanez, E.D. and Owen, D.F. 1984. *Food Chemistry and Toxicology* 22: 45-48.

Batterham, E.S. 1979. *Australian Journal of Agricultural Research* 30: 369-375.

Batterham, E.S., Murison, R.D. and Andersen, L.M. 1984. *British Journal of Nutrition* 51: 85-99.

BIBRA 1986. *Report on BIBRA Project No. 3.0585*. The British Industrial Biological Research Association, Carshalton, UK. 50 pp.

BIBRA 1988. *Report on BIBRA Project 3.0603*. The British Industrial Biological Research Association, Carshalton, UK. 26 pp.

Brenes, A., Marquardt, R.R., Guenter, W. and Rotter, B.A. 1993. *Poultry Science* 72: 2281-2293.

Brenes, A., Marquardt, R.R., Guenter, W. and Rotter, B.A. 1993. *Poultry Science* 72: 2281-2293.

Butler, W.H., Ford, G.P. and Creasy, D.M. 1996. *Food and Cosmetics Toxicology* 34: 531-536.

Camacho, L., Banados, E., Fernandez, E. 1989. *Archivos Lantinoamericanos de Nutricion* 39: 185-199.

Cholerton, S., Daly A.K. and Idle, J.R. 1992. *Trends in Pharmacological Sciences* 13: 434-439.

Cockrum, P.A., Petterson, D.S. and Edgar, J.A. 1994. In: *Plant-associated Toxins Agricultural, Phytòchemical and Ecological Aspects*. pp. 232-237 (eds Colegate, S.M. and Dorling, P.R.). CAB International, Oxon,.

Coffey, R.S. 1989. In: *Proceedings of the World Congress on Vegetable Protein Utilisation in Human Foods and Animal Feedstuffs*. pp. 410-414. (Ed T.H.Applewhite.). American Oil Chemists' Society, Champaign, Illinois.

Cowling, W.A., Huyghe, C. and Swiecicki, W. 1997. In: *Lupins*. Ch. 4. (eds Gladstones, J.S., Hamblin, J. and Atkins, C.) CAB International, Oxford.

Culvenor, C.C.J., Beck, A.B., Clarke, M., Cockrum, P.A., Edgar, J.A., Frahn, J.L., Jago, M.V., Lanigan, G.W., Payne, A.L., Peterson, J.E., Petterson, D.S., Smith, L.W. and White R.R. 1977. *Australian Journal of Bioogical Sciences* 30: 269-277.

Dixon, R.M. and Hosking, B.J. 1992. *Nutrition Research Reviews* **5**: 19-43.

Edgar, J.A. 1991. In: *Handbook of Natural Toxins*. pp. 371-395. (eds Keeler, R.F. and Tu, A.T.). Marcel Dekker, New York,

Edwards, A.C. and van Barneveld, R.J. 1997. In: *Lupins* Ch. 11. (eds Gladstones, J.S.,. Hamblin, J. and Atkins, C.). CAB International, Oxford.

Evans, A.J. 1994. In: *Proceedings of the First Australian Lupin Technical Symposium*. pp. 110-114. (Eds M.Dracup and J.Palta). Western Australian Department of Agriculture, South Perth.

Fudiyansyah, N., Petterson, D.S., Bell, R.R. and Fairbrother, A.H. 1995. *International Journal of Food Science and Technology* 30: 297-306.

586

Gardiner, M.R. 1967. *Journal of Pathology* 94: 452-455.

Gardiner, M.R. and Petterson, D.S. 1972. *Journal of Comparative Pathology* 82: 5-13.

Gill, G.S. 1994. *Proceedings of the First Lupin Technical Symposium* pp. 324-326.

Gladstones, J.S. 1970. *Field Crop Abstracts* 23: 123-148.

Gladstones, J.S. 1989a. *Australian Journal of Experimental Agriculture* 29: 913-914.

Gladstones, J.S. 1989b. *Journal of Agriculture, Western Australia* 30: 3-7.

Godfrey, N.W. and Payne, H.G. 1987. In *Manipulating Pig Production.* p. 153 APSA Committee (eds.). APSA: Werribee.

Gomes, F.E. and Kaushik, S. 1990. In *The Current Status of Fish Nutrition in Aquaculture.* pp. 315-324. (eds Takeda, M. and Watanabe, T..). *The Proceedings of the Third Internationl Symposium on Feeding and Nutrition in Fish.* Toba, Japan.

Grant, G., Dorward, P.M. and Pustzai, A. 1993. *Journal of Nutrition* 123: 2207-2213.

Grant, G., Dorward, P.M., Buchan, W.C., Armour, J.C. and Pustzai, A. 1995. *British Journal of Nutrition* 73: 17-29.

Gross, R. 1990. In: *Proceedings of the Joint CEC-NCRD Workshop* pp. 164-176. (eds Birk, Y., Dourat, A., Waldman, M. and Uzureau, C.). Israel, 1989,

Hancock, G.R., Vogel, P. and Petterson, D.S. 1987. *Australian Journal of Experimental Agriculture* 27: 73-76.

Harris, D.J. and Jago, J. 1984. Chemical composition of sweet lupinseed. Report of the Government Chemical Laboratories, Perth.

Hawthorne, W.A. and Gladstones, J.S. 1989 *Australian Journal of Experimental Agriculture* 29: 911-12.

Hefle, S.L., Lemanske, R.F. and Bush, R.K. 1994. *Journal of Allergy and Clinical Immunology* 94: 161-172.

Higuera, M. de la, Garcia-Gallego, M., Sanz, A., Cardenete, G., Suarez, M.D. and Moyano, F.J. 1988. *Aquaculture* 71(1-2): 37-50.

Hough, G.M. and Jacobs, J.L. 1994. In *Proceedings of the First Australian Lupin Technical Symposium.* pp. 58-66 (eds Dracup, M. and Palta, J..) Department of Agriculture, West Australia..

Hughes, S.G. 1988. *Aquaculture* 71(4): 379-385.

Jenkins, G.I., Waters, S.P., Hoxey, M.J. and Petterson, D.S. 1994. In *Proceedings of the First Australian Lupin Technical Symposium.* pp. 74-78 (eds Dracup, M. and Palta, J..). Department of Agriculture, West Australia

Jones, R.A.C. 1988. *Annals of Applied Biology* 113: 507-518.

King, R.H. 1990. In *Nontraditional Feed Sources for Use in Swine Production* pp. 237-246. (eds Thacker, P.A. and Kirkwood, R.N) Butterworth Publishers: Stoneham,

Links, I.J. and Walker, R.J. 1989. In: *Proceedings of the Annual Conference of the Australian Society for Veterinary Pathology.* pp. 37-38. Australian Society for Veterinary Pathology, Walkerville,

Lowen, R.J., Alam, F.K.A. and Edgar, J.A. 1995. *The Medical Journal of Australia* 162: 256-257.

Lucisano, M. and Pompei, C. 1981. *Food Science* 14: 323-6.

Margan, D.E. 1994. *Australian Journal of Experimental Agriculture* 34: 331-337.

Marquardt, R.R. 1993. In *Ninth Australian Poultry and Feed Convention: Gold Coast, 1993* p. 47-51.

Marquez, L.R., Guitierrez-Rave, M. and Miranda, F.I. 1991. *Veterinary and Human Toxicology* 33: 265-267.

Mazur, M., Polakowski, P. and Szadowski 1966. *Acta Physiologica Polonica* 17: 299-309.

Moyano, F.J., Cardenete, G. and de-la-Higuera, M. 1992. *Aquaculture and Living Resources* 5(1): 23-29.

Murray, P.J. 1994. In *Proceedings of the First Australian Lupin Technical Symposium.* pp. 67-73 (eds Dracup, M. and Palta, J). Department of Agriculture, West Australia.

Nelson, P. and Hawthorne W.A. 1997. *These proceedings.*

Peterson, J.E. 1978. *Journal of Comparative Pathology* 88: 191-203.

Petterson, D.S. 1985. WA Department of Agriculture, Perth. ISBN 0157-6259.

Petterson, D.S. 1997. In: *Lupins.* Ch. 11 (eds Gladstones, J.S., Hamblin, J. and Atkins, C.) CAB International, Oxford.

Petterson, D.S. and Crosbie, G.B. 1990. *Food Australia* 42: 266-8.

Petterson, D.S. and Mackintosh, J.B. 1994. *The Chemical Composition and Nutritive Value of Australian Grain Legumes.* GRDC: Canberra.

Petterson, D.S., Ellis, Z.L., Harris, D.J. and Spadek, Z.E. 1987. *Journal of Applied Toxicology* 7: 51-53.

Petterson, D.S., Greirson, B.N., Allen, D.G., Harris, D.J., Power, B.M., Dusci, L.J. and Ilett, K.F. 1994. *Xenobiotica* 24: 933-941.

Pompei, C., Lucisano, M. and Ballini, N. 1985. *Sciences des Aliments* 5: 665-668.

Rayas-Duarte, P. 1993. In: *Advances in lupin research. Proceedings of the 7th International Lupin Conference.* pp 537-541 (Eds J.M. Neves-Martins and M.L. Beirao da Costa) Instituto Superior de Agronomia and International Lupin Association, Lisbon.

Robbins, M.C., Brantom, P.G. and Petterson, D.S. 1996. *Food and Chemicals Toxicology* 34: 679-686.

Robiana, L., Izquierdo, M.S., Moyano, F.J., Socorro, J., Vergara, J.M., Montero, D. and Fernandez-Palacios, H. 1995. *Aquaculture* 130(2,3). 219-233.

Rodriguez, F.S., Santiyan, M.P.M., Zamorano, J.D.P. and Cordero, V.R. 1991. *Veterinary and Human Toxicology* 33: 492-94.

SCA 1987. *Feeding Standards for Australian Livestock. Pigs.* Standing Committe on Agriculture. CSIRO. East Melbourne.

Schmidlin-Meszaros, J. 1973. *Mitteilungen aus dem Gebeite der Lebensmitteluntersuchung und Hygiene* 64: 194-205.

Smith, R.A. 1987. *Veterinary and Human Toxicology* 29: 444-445.

Standing Committe on Agriculture 1987. *Feeding Standards for Australian Livestock. Pigs.* East Melbourne: CSIRO.

Sweetingham, M.W. and Jarvis, R.J. 1993. *Australian Grain* 3(1): 48-49.

Than, K.A., Anderton, N., Cockrum, P.A., Payne, A.L., Stewart, P.L. and Edgar, J.A. 1994b In: *Plant-associated Toxins-Agricultural, Phytochemical and Ecological* pp.433-438 (eds Colegate, S.M. and Dorling, P.R.) CAB International, Oxon,

Than, K.A., Payne, A.L. and Edgar, J.A. 1992 In: *Poisonous Plants, Proceedings of the Third International Symposium* pp. 259-263 (eds James, L.F., Keeler, R.F., Bailey, E.M. Jr, Cheeke, P.R. and Hegarty, M.P.). Iowa State University Press, Ames.

Than, K.A., Tan, R.A., Petterson, D.S. and Edgar, J.A. 1994a. In: *Plant-associated Toxins Agricultural, Phytochemical and Ecological Aspects* pp. 62-65 (eds Colegate, S.M. and Dorling, P.R.) CAB International, Oxon

Uauy, R., Gattas, V. and Yáñez, E. 1995. In: *World Review of Nutrition and Dietetics* 77 pp 75-88. (ed. Simopoulos, A.M) Basel, Karger.

van Warmelo, K.T., Marasas, W.F.O., Adelaar, T.F., Kellerman, T.S., van Rensburg, I.B.J. and Minne, J.A. 1970. *Journal of the South African Veterinary Medical Association* 41: 235-247.

Viola, S., Arieli, Y. and Zohar, G. 1988. *Israel Journal of Aquaculture and Bamidgeh* 40: 29-34.

von Sengbusch, R. 1942. *Landwirtsch Jahrb.* 91: 719-880.

Wigan, G.C., Batterham, E.S. and Farrell, D.J. 1994 In *Proceedings of the Fifth Biennial Pig Industry Seminar, WAI.* p. 38-46.

Williamson, P.M., Highet, A.S., Gams, W., Sivasithamparam, K. and Cowling, W.A. 1994. *Mycological Research* 98: 1364-68.

Wood, P. McR, Petterson, D.S., Hancock, G.R. and Brown, G.A. 1987. *Australian Journal Experimental Agriculture.* 27: 77-9.

Wood, P. McR. and Petterson, D.S. 1986. *Australian Journal of Experimental Agriculture* 26: 583-86.

Yovo, K., Huguet, F., Pothier, J., Durand, M., Breteau, M. and Narcisse, G. 1984. *Planta medica* 50: 420-4.

Zurn, F.A. 1879. *Vortrage fur Thierarzte* 7: 251-277.

Are Our Germplasm Collections Museum Items?

N. Maxted [1], W. Erskine [2], L.D. Robertson [2] and A.N. Asthana [3]

1 School of Biological Sciences, University of Birmingham, Edgbaston, Birmingham B15 2TT.; 2International Centre for Agricultural Research in Dry Areas, PO Box 5466, Aleppo, Syria.; 3Institute of Pulse Research, ICAR, Kalyanpur, Kanpur, 208024, Uttar Pradesh, India.

Abstract

Programmes of plant conservation often have a substantial cost, which is met by international or national agencies and indirectly by the general public. To justify this expenditure the products of conservation must have some use to the public. Therefore, the answer to the question in the paper's title is, Yes, unless a clear link is made between conservation and utilisation for the benefit of mankind. The need for this link is established in this paper using models of conservation. The conserved temperate food legume species and potential targets are reviewed, as are the methods of utilisation and the constraints to further exploitation. Conserved germplasm of faba bean, chickpea and lentil has formed the basis of improvement programmes in the Mediterranean region. They provide case studies illustrating the linkage between legume conservation and utilisation.

INTRODUCTION

In the conservation of plant genetic resources the link between conservation and utilisation is fundamental (Maxted et al., 1997a). This point was emphasised in a recently proposed methodology for conservation (Figure 1). The raw materials of conservation are genes within gene pools, covering the total diversity of genetic material of the taxon being conserved. Conservation, is the process that retains the diversity of the gene pool; utilisation is the exploitation of that diversity.

Figure 1

Humankind conserves because it wishes to utilise but conservation does have an economic cost. It would be difficult to persuade society to meet this cost unless the conserved material could be shown to be of 'value'. There is at present no widely agreed method of estimating the value of biodiversity (Flint, 1991; Shands, 1994). Vane-Wright (1996) suggested the value of a species in terms of conservation and utilisation was relative not absolute. It is easier, given a choice of two target taxa to decide which has the higher priority than to provide a monetary value for them separately. For example, major crops, such as the faba bean are likely to be given a higher priority than minor crops and wild relatives such as the narbon bean

It is relatively easy to argue the economic benefit (value) that might accrue from conservation, utilisation and exploitation of land races, or wild relatives of crops in breeding programmes, but it is more difficult to ascribe economic value to truly 'wild' species of limited immediate utilisation. Maxted et al. (1997b) argued that all plants are likely to be of some value, whether in terms of crop potential or for pharmaceutical, recreation, eco-tourism, and educational use, or for less overt utilisation, such as making people feel 'good' or to think that nature was 'safe'. Like all biodiversity, genetic diversity is part of a nation's heritage, alongside its art and culture. Therefore, it is important when considering conservation to make the explicit link to actual or potential utilisation.

THE COST OF CONSERVATION

.Germplasm conservation, where the material is collected and the seed conserved in a gene bank is an expensive activity. Precise estimates of cost are rare and variable. Smith and Linington (1997) estimated that the average cost of collecting an accession from outside one's own country was US $ 597 and incorporating it into a gene bank (including initial germination testing, cleaning, packaging and verification) was an extra US$ 273. Once there, the annual curation costs was US$ 5 and distribution costs US$ 15. The Keystone Center (1991) estimated that the maintenance of an accession was US$50 per year and that this was 43% of

R. Knight (ed.), Linking Research and Marketing Opportunities for Pulses in the 21st Century, 589–602.

Figure 1. Model of Plant Genetic Conservation, taken from Maxted *et al.,* (1997a).

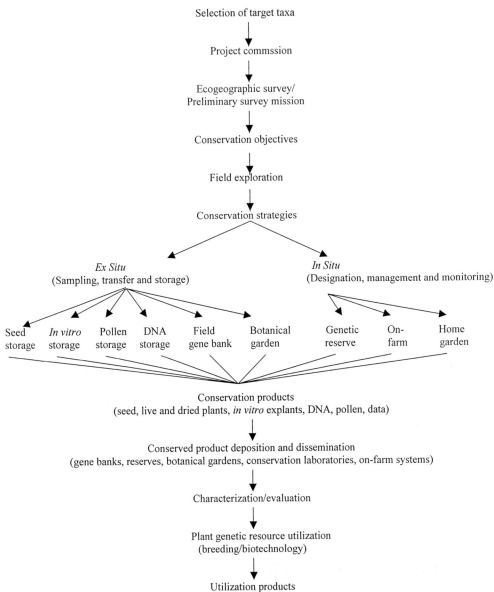

Selection of target taxa

Project commssion

Ecogeographic survey/
Preliminary survey mission

Conservation objectives

Field exploration

Conservation strategies

Ex Situ
(Sampling, transfer and storage)

In Situ
(Designation, management and monitoring)

Seed storage | *In vitro* storage | Pollen storage | DNA storage | Field gene bank | Botanical garden | Genetic reserve | On-farm | Home garden

Conservation products
(seed, live and dried plants, *in vitro* explants, DNA, pollen, data)

Conserved product deposition and dissemination
(gene banks, reserves, botanical gardens, conservation laboratories, on-farm systems)

Characterization/evaluation

Plant genetic resource utilization
(breeding/biotechnology)

Utilization products
(breeding new varieties and crops, pharmaceuticals, pure and applied research, on-farm diversity, recreation, etc)

the total conservation costs. Whichever of these estimates is closer to the truth, we should not lose sight of the economic balance between conservation expenditure and utilisation income. The world's rice collection consists of 215,000 accessions, of which 90,000 distinct genotypes are conserved in 29 collections (Plucknett *et al.*, 1987). The annual cost, to conserve this gene pool, is US$ 20.3 million, but this figure is only 5.1% of the estimated annual global value added to rice yields by the use of rice genetic resources (Brush, 1996). It is important to note that the vast increase in germplasm collections in recent years has not been matched by a corresponding increase in *ex situ* conservation budgets. Clark *et al.* (1997) note that within the National Plant Germplasm System of the USDA, budgets have been stagnant since 1988, but due to inflation and increasing demands, this has lead to a 25% decrease in the *ex situ* funding.

PRIORITIES FOR CONSERVATION OF LEGUME DIVERSITY

Conservation activities are always limited by the financial and technical resources available (Abramovitz, 1994). Because not all taxa have the same utilisation potential, conservationists must prioritise those which are to be the focus of their programmes. The first step in conservation therefore (Maxted *et al.*, 1997a) is the choice of the target taxa. Other authors have assessed priorities for specific legumes (see chapters in this volume), however, we will comment on how priorities should be established. Crucial is the notion of the absolute or comparative "value" of different taxa, an evaluation which has been attempted by several authors (McNeely, 1988; Pearce and Turner, 1990; Groombridge, 1992; Hargrove, 1992; Given, 1994; Pearce and Morgan, 1994; Turner and Postle, 1994; Johnson, 1995; Swanson, 1995; Brush, 1996; Department of the Environment, 1996). Each has stressed the need to allocate value to competing targets, but has failed to outline the factors that might assist in ascribing value. Maxted *et al.* (1997b) suggested these factors should include:

current conservation status,
socio-economic use,
threat of genetic erosion,
genetic distinctiveness,
ecogeographic distribution,
biological importance,
cultural importance,
cost, feasibility and sustainability,
legislation,
ethical and aesthetic considerations,
priorities of the conservation agency.

These factors influencing priorities are summarised with legume examples in Table 1.

Each factor should be weighted according to the mandate of the commissioning agency. A priority list of taxa for conservation should be formulated. The procedure makes the setting of conservation priorities less subjective. Different conservation groups, establishing priorities, are likely to produce different lists depending on the mandate of the commissioning agency and their personal experience. For example breeders of food legumes or of forage legumes will establish different priorities.

ACCESSING CONSERVED PLANT DIVERSITY

Once the plant diversity has been conserved, the first step in utilisation is to find what material is available and what accessions are currently conserved *ex situ* or *in situ*? The material will be in gene banks, botanic gardens, conservation laboratories, genetic reserves and on-farms. The material can be determined by consulting individual gene banks, botanic gardens (or possibly *in situ* genetic reserve) catalogues, directories and databases (Perry and Bettencourt 1995

Searching can be time consuming and to help alleviate the problem national and international directories and databases have been established by various agencies. The International Board for Plant

Table 1. The factors that should be considered when establishing conservation priorities, with examples from temperate legumes.

Contributing Factor	Legume Example
current conservation status	Insufficient material of the genepool is conserved, e.g. *Trifolium* species
socio-economic use	Landraces of faba bean, chickpea, lentil and peas need priority over their wild relatives
threat of genetic erosion	Rare and endemic forage species of Southern Turkey are threatened by tourist development
genetic distinctiveness	Genetically isolated species, like *Vaviloviaformosa* the perennial relative of *Pisum sativum*, need to be conserved to broaden the genetic base of conserved material
ecogeographic distribution	Food and forage species growing at high altitude in Armenia and Georgia because they are likely to be adapted to cold
biological importance	*Medicago littoralis* needs a higher priority because it can be used in sand dune stabilisation
cultural importance	*Sophora toromiro* was used to make wooden scultures on the Easter Islands
cost, feasibility and sustainability	Species that are relatively easy to collect because their collection is cost effective, e.g. weeds of cultivation
legislation	*Percopsis elata* - African teak is listed on CITES App. II
ethical and aesthetic considerations	*Lathyrus mulkak* is an endemic species of Central Asia that has obvious aesthetic appeal (and economic potential) with large flowers. As yet no material is held *ex situ*
priorities of the conservation agency	Food legume breeders, foresters and environmentalists are likely to target different taxa.

Genetic Resources (IBPGR - now the International Plant Genetic Resources Institute, IPGRI) has published catalogues of *ex situ* holdings. The other CG Centres have published breeder's catalogues, which contain passport and characterisation data, but the catalogues usually focus on a single crop. Unfortunately catalogues or even databases rapidly become out of date. It is difficult therefore to compile an up-to-date list of the world's holdings. The collections are housed in national, regional and international gene banks throughout the world and if they are operated efficiently the collections are dynamic, in that new accessions are being added and other accessions sent to users.

Published catalogues and databases are pointers to the significant legume collections. Querying regularly up-dated international databases is a more efficient means of locating current holdings. Example are: the Germplasm Holdings Database of IPGRI, the State of the World database of the Commission on Plant Genetic Resources of FAO, the crop group databases of the European Cooperative Programme on Genetic Resources (ECP/GR), the Genetic Resources Information Network (GRIN) of the USDA and databases associated with national institutes, such as, the Centre for Legumes in Mediterranean Agriculture (CLIMA). The holdings of the CGIAR centres can be queried through the SINGER database, available on CD-Rom or via the Internet (HTTP://www.cgiar.org/singer/index.htm).

As databases become more widely available, so the user will be able to assess the amount of genetic variation and where it is conserved, for any taxon. The user should also check the Plant Genetic Resources literature (e.g. Genetic Resources and Crop Evolution, Plant Genetic Resources Newsletter, Beanbag) to ascertain the results of recent collecting missions.

Care must be taken when interpreting information on gene bank holdings or conserved diversity as the material may have been misidentified. It should be possible to check identifications from voucher material or the material can be grown-out and identified. The size and number of collections may be misleading as gene banks and botanical gardens duplicate their collections in other banks and gardens.. Questions arise, such as are there adequate passport data available? Older collections usually have poor data and the accessions are less likely to be utilised. Are the collections 'good' genetic samples? Were they sampled effectively to maximise genetic variation and has variation been lost since sampling through a) genetic drift when sampling small samples, b) unsuitable regeneration conditions (selection in regeneration plots / out pollination), c) poor gene bank storage resulting in differential erosion, d) human errors, such as mislabelling? The user should consider these factors when choosing the accessions to use.

WHAT LEGUME DIVERSITY IS CONSERVED AND WHERE?

The taxonomic focus of our attention will be the conserved accessions of faba bean, chickpea, lentil, garden pea, their wild relatives (Table 2).

As mentioned above, catalogues may be the first source of information. IBPGR / IPGRI have published catalogues of *ex situ* holdings including volumes on Food Legumes (Bettencourt, Konopka & Damania, 1989). There is information on *Arachis, Cajanus, Cicer, Lens, Lupinus, Phaseolus, Pisum, Psophocarpus, Vicia* and *Vigna* collections. These directories were compiled from data provided by curators of national and international collections and from data held in the international *ex situ* database administrated by IPGRI. IPGRI and ICARDA also maintain databases of germplasm collected during their missions. The data are summarised for temperate food collections (Table 3). The two largest, temperate legume collections are those held by ICARDA and USDA, summarised in Tables 4 and 5 respectively. The ICARDA database is accessible via the SINGER system or the Internet (address given above). The USDA holdings may also be queried via the Internet at (HTTP://www.ars-grin.gov).

UTILISATION OF CONSERVED FOOD LEGUME DIVERSITY

Examples of utilisation are provided by the collections of faba bean, chickpea, lentil held at ICARDA and exploited by breeding programmes in the WANA region and elsewhere for the production of improved cultivars. The first lines distributed were often selected directly from the germplasm collections after initial evaluation. This formed the first flush of releases by national programmes; 56 cultivars were developed from landraces or ecotypes (see Table 6). Many of these were selections stabilised for traits such as seed size, disease resistance, frost tolerance, non-shattering, etc. Today, most lines distributed through the ICARDA International Testing Network are the products of hybridization programmes involving germplasm found resistant or tolerant to one or more biotic or abiotic stresses.

Table 2. Cultivated Food Legume Species with their Wild Relatives.

Crop Species	Crop Latin Name	Crop Relatives	Recent Literature
Faba bean	*Vicia faba* subsp. *faba*	*Vicia* sect. *Narbonensis*	Maxted (1993)
Chickpea	*Cicer arietinum*	*Cicer reticulatum* and *C. echinospermum*	Van der Maesen (1987)
Lentil	*L. culinaris* subsp *culinaris*	*L. culinaris* subsp *orientalis* and other *Lens* sp.	Ferguson *et al.,* (1998)
Pea	*Pisum sativum*	*Pisum fulvum*	Polhill and van der Maesen (1985)
Grass or Indian Pea	*Lathyrus sativus*	*Lathyrus* sect. *Lathyrus*	Kupicha (1983)
	Lathyrus latifolius	*Lathyrus* sect. *Lathyrus*	Kupicha (1983)
	L. sylvestris	*Lathyrus* sect. *Lathyrus*	
	L. annuus	*Lathyrus* sect. *Lathyrus*	
	L. cicera	*Lathyrus* sect. *Lathyrus*	
	L. ochrus	*Lathyrus* sect. *Clymenum*	
White lupin	*Lupinus albus*	Annual *Lupinus* sp.	Plitmann and Heyn (1984)
Blue lupin	*L. angustifolius*		
Yellow lupin	*L. luteus*		
Narbon vetch	*V. narbonensis*	*Vicia* sect. *Narbonensis*	Kupicha (1976)
Common vetch	*V. sativa* subsp. *sativa*	*Vicia* sect. *Vicia*	Maxted (1993)
-	*V. sativa* ssp. *amphicarpa*	*Vicia* sect. *Vicia*	
-	*V. sativa* ssp. *macrocarpa,*	*Vicia* sect. *Vicia*	
Hungarian vetch	*V.pannonica*	*Vicia* sect. *Hypechusa*	
Winter vetch	*V. villosa*	*Vicia* subgen. *Vicilla*	
Bitter vetch	*V. ervilia*	*Vicia* subgen. *Vicilla*	

Table 3. Temperate food germplasm collected during IPGRI and ICARDA missions (the number of accessions, species and countries visited).

Crop Species	Latin Name	ICARDA Acc.	ICARDA Spp	Countries Visited by ICARDA	IPGRI Acc	IPGRI Spp	Countries Visited by IPGRI
Lentil	*L. culinaris* subsp *culinaris*	1131	1	19	297	1	12
Chickpea	*Cicer arietinum*	927	1	16	432	1	11
Faba bean	*Vicia faba*	947	1	18	767	1	11
Pea	*Pisum sativum*	472	1	15	290	1	11
Grass pea	*Lathyrus sativus*	251	1	10	61	1	6
Total		3728	5	-	1847	5	-
Related spp							
Chickpeas	*Cicer* sp.	51	3	5	16	4	1
Vetchlings	*Lathyrus* sp.	847	33	15	282	27	15
Lentils	*Lens* sp.	272	4	8	38	3	6
Lupins	*Lupinus* sp.	159	7	7	1365	12	8
Vetches	*Vicia* sp.	2249	41	21	680	35	11
Total		3578	88	-	2381	81	-

Table 4. The number of temperate food and forage legume accessions held by ICARDA, and those originating from West Asia and North Africa.

Collection	WANA	Total
Chickpea	6753	9586
Faba Bean (ILB)a	1849	4434
Faba Bean (BPL)b	2226	5255
Lentil	4111	7407
Wild *Cicer*	283	291
Wild *Lens*	363	433
Vicia (all species)	2870	5337
sativa ssp. *Sativa*	597	1695
sativa ssp. *amphicarpa*	90	95
villosa ssp. *Dasycarpa*	47	67
ervilia	216	306
narbonensis	205	251
hybrida	138	159
palaestina	97	100
pannonica	15	124
Lathyrus (all species)	1294	1590
Sativus	217	301
Cicera	103	162
Ochrus	71	107
Cilolatus	3	3
Total	19,749	34,333

a:ILB=International Legume Bean accession.

b:BPL=Faba Bean Pure Line.

Table 5 Temperate food legume and related *spp. ex situ* collections held by the USDA. Constructed from the Genetic Resources Information Network (number of accessions, species and countries).

Crop Species	Latin Name	Accessions	Species	Countries
Lentil	*L. culinaris* subsp *culinaris*	2725	1	51
Chickpea	*Cicer arietinum*	4444	1	49
Faba bean	*Vicia faba*	548	1	50
Pea	*Pisum sativum*	4486	1	70
Grass pea	*Lathyrus sativus*	243	1	30
Total		12446	5	-
Related spp				
Chickpeas	*Cicer* sp.	178	21	-
Vetchlings	*Lathyrus* sp.	674	48	54
Lentils	*Lens* sp.	158	5	-
Lupins	*Lupinus* sp.	1273	81	51
Vetches	*Vicia* sp.	1414	54	-
Total		3697	209	-

Table 6 Distribution of legume germplasm from ICARDA collections during 1990-1996. (excluding safety duplications and the GRU's own work).

Crop	1990	1991	1992	1993	1994	1995	1996	Total
Food Legumes	2870	2807	5880	6251	7099	7464	7430	39801
Faba bean	293	60	1278	2705	413	814	221	5784
Chickpea	1103	220	1578	1239	3019	4047	4392	15598
Wild *Cicer* sp.	273	181	613	385	748	568	514	3282
Lentil	1098	2006	1941	1889	2869	1807	2137	13747
Wild *Lens* sp.	103	340	464	33	50	228	166	1384
Lathyrus sp.	228	89	769	854	877	265	569	3651
Pisum sp.	21	56	74	15	93	107	143	509
Vicia sp.	398	778	701	3292	1761	3310	483	10723
Total	3517	3730	7418	10412	9830	11146	8625	54684

As demonstrated in Figure 1, before germplasm can be utilised it must be characterised and evaluated. A lack of such data can seriously limit utilisation. Marshall and Brown (1975) point out that there are two opposing forces in utilisation, the breeder requires as broad a range of genetic variability as possible, but he invariably has limited resources and can only utilise a limited number of samples. The requirement is to provide the breeder with as much characterisation and evaluation data as possible, to enable him to make the best choice of material to use in his crosses. A large number of accessions may be available, 30,000 samples of pulses at the Indian Institute of Pulse Research, but only 2-3% are used in crosses because of the lack of characterisation and evaluation data.

The systematic evaluation of the food legumes for morpho-agronomic characters, based on the IBPGR / ICARDA descriptors, has led to the publication of catalogues and the use of the germplasm in national breeding programmes. Most of the faba bean, kabuli chickpea and lentil material in the collections have been evaluated and the results summarized (Erskine & Witcombe, 1984; Robertson & El-Sherbeeny, 1988; Robertson *et al.*, 1995; Singh *et al.,* 1983, 1991). The descriptors include characters such as seed size, pods per plant, seeds per pod, seed shape, seed colour, testa colour, testa pattern, cotyledon colour, growth habit, and protein quantity. The germplasm has also been scored for biotic stress resistances, including reaction to aphids, Ascochyta blight, and *Orobanche crenata* and for seed and straw yield. These studies have enhanced the value of the conserved germplasm and increased its use in breeding. In the period 1990-1996, an average of 6600 accessions of food legumes was distributed each year (Table 7).

The most important use of the germplasm has been as a source of resistance to biotic stresses. Selections for disease resistance in faba bean for chocolate spot (induced by *Botrytis fabae* Sard.), Ascochyta blight (induced by *Ascochyta fabae* Speg.), rust (induced by *Uromyces fabae* (Pers.) de Bary) and stem nematodes (*Ditylenchus dipsaci* (Khhn) Filipjev) have led to the development of homogeneous resistance

Table 7 Food legume cultivars selected from ICARDA's germplasm collections.

Crop/ Country	Cultivar released	Year of release	Crop/ Country	Cultivar released	Year of release
Kabuli chickpea			Lentil		
Algeria	ILC 482	1988	Algeria	ILL 4400	1988
	ILC 3279	1988	Canada	ILL 481 (Indian head)	1989
China	ILC 202	1988	Chile	ILL 5523 (Centinela)	1989
	ILC 411	1988	Egypt	ILL 4605 (Precoz)	1990
Cyprus	ILC 3279 (Yialousa)	1984	Ethiopia	ILL 358	1984
	ILC 464 (Kyrenia)	1987		NEL 2704	1993
France	ILC 482 (TS1009)	1988	Iraq	ILL 5582	1992
Iraq	ILC 482	1991	Jordan	ILL 5582	1990
	ILC 3279	1991	Libya	ILL 5582	1993
Italy	ILC 72 (Califfo)	1987	Morocco	ILL 4605 (Precoz)	1990
	ILC 3279 (Sultano)	1987		ILL 4606 (Nefza	1986
Jordan	ILC 482 (Jubeiha 1)	1990	Nepal	ILL 4402	1989
	ILC 3279 (Jubeiha 2)	1990	Pakistan	ILL 4605 (Manserha 89)	1990
Lebanon	ILC 482 (Janata 2)	1989	Sudan	ILL 813 (Rubatab1)	1993
Libya	ILC 484	1993	Tunisia	ILL 4400 (Neir)	1986
Morocco	ILC 195	1987	Turkey	ILL 942 (Erzurum '89)	1990
	ILC 482	1987		ILL 1384 (Malazgirt '89)	1990
Oman	ILC 237	1988		ILL 854 (Sazak'91)	1991
Portugal	ILC 5566 (Elmo)	1989	U.S.A.	ILL 784 (Crimson)	1991
Spain	ILC 72 (Fardan)	1985	Faba bean		
	ILC 200 (Zegri)	1985	Australia	ICARUS (BPL 710)	1992
	ILC 2548 (Almena)	1985	Egypt	Giza Blanka (ILB 1270)	1991
	ILC 2555 (Alcaaba)	1985		Giza 461 (ILB 938)	1993
	ILC 200 (Atalaya)	1985	Iran	Barkat (ILB 1269)	1986
Sudan	ILC 1335 (Shendi)	1987			
	ILC 915 (Jeb el Mara 1)	1993	Vetch		
Syria	ILC 482 (Ghab 1)	1986	Jordan	IFVID 683	1993
	ILC 3279 (Ghab 2)	1986		IFVIS 715	1993
Tunisia	ILC 3279 (Chetoui)	1986	Morocco	IFVIS 1812	1993
Turkey	ILC 195	1986			
	ILC 482 (Guney Sarisi 482)	1986	Chickling		
			Jordan	IFLAO 101/185	1993

sources (Table 8). This material was tested internationally and several sources were found to have durable resistances for chocolate spot, rust and Ascochyta blight (Hanounik & Robertson 1987, 1988). Subsequently the resistant sources to chocolate spot and rust have been used by the Egyptian national programme to develop locally adapted lines with resistance to chocolate spot and rust (Nassib*et al.*, 1988), one of which is the recently released cultivar 'Giza 461'.

Intensive efforts have been made to identify Ascochyta blight resistant lines of chickpea at ICARDA since 1978. Screening has included kabuli and desi accessions and over 19,000 have been tested (Singh & Reddy, 1993a). Useful sources have been found (Table 9). The lines released as cultivars were primarily selections for Ascochyta blight resistance (Table 6). Fusarium wilt (induced by *Fusarium oxysporum* Schlecht. emend. Synd. and Hans. f. sp. *ciceri* (Padwck) Synd. and Hans.) causes severe yield loss in chickpea in some countries of WANA. Sources of resistance were identified (Jiménez-Díaz *et al.*, 1991, 1993) (Table 9), and the Tunisian national programme has released a resistant cultivar 'Amdoun 1' (Halila & Harrabi 1990; Halila *et al.*, 1984).

Rust (induced by *Uromyces fabae* (Pers.) de Bary) is the most important foliar disease of lentil. Epiphytotics are common in Chile, Ecuador, Ethiopia, India, Morocco, and Pakistan (Erskine & Saxena, 1993). ICARDA screens collaboratively with the national programmes of Ethiopia, Morocco and Pakistan (Table 10). As a result, rust resistant cultivars have been released in Chile and Ecuador, as well as in Ethiopia, Morocco and Pakistan.

Table. 8 Faba bean inbred resistant sources to various diseases and pests.

Disease	Sources of resistance
Chocolate spot	BPL 110, 112, 261, 266, 710, 1179, 1196, 1278, 1821; ILB 3025, 3026, 2282, 3033, 3034, 3036, 3106, 3107, 2302, 2320,; L82003, L82009
Ascochyta blight	BPL 74, 230, 365, 460, 465, 471, 472, 646, 818, 2485; ILB 752; L83118, L83120, L83124, L183125, L83127, L83129, L183136, L83142, L83149, L83151, L83155, L83156, L82001
Rust	BPL 7, 8, 260, 261, 263, 309, 409, 417, 427, 484, 490, 524, 533, 539; Sel.82 Lat. 15563-1, -2, -3, -4
Stem nematode	BPL 1, 10, 11, 12, 21, 23, 26, 27, 40, 63, 88, 183
Orobanche crenata	18009, 18025, 18035, 18054, 18105, LS 222, LS 225, 8/9-72, -85, -86, -128, -136, -137, -138, -139, -143, -152, -153
BLRV	BPL 756, 757, 758, 769
BYMV	BPL 1351, 1363, 1366, 1371

Table. 9 Sources of resistance in kabuli chickpea germplasm to biotic and abiotic stresses.

Stress	Sources of resistance
Ascochyta blight	ILC 72, 182, 187, 200, 2380, 2506, 2956, 3279, 3856, 4421, 5586, 5902, 5921, 6043, 6090, 6188.
Fusarium wilt*	ILC 54, 240, 256, 336, 487, FLIP 85-29C, FLIP 85-30C, UC 15.
Leaf miner	ILC 316, 992, 1003, 1009, 1216, 2622, 5594, 5901
Cold	ILC 1464, 3287, 3465, 3470, 5638, 5663, 5667, 5947, 5951, 5953, 8262, 8617, 482 (Mut) (M 17033).
Drought	ILC 7191, 7192

*Results from screening by Spanish national program.

Table 10 Sources of resistance to biotic and abiotic stresses in lentil.

Stress	Sources of resistance
Fusarium wilt	ILL 241, 632, 813, 1712, 4403, 5714, 5871, 5883, 6024, 6025, 6410, 6427, 6458, 6461, 6797, 6976, 6991, 7005, 7012, 7180, 7192, 7193, 7199, 7204
Rust	ILL 358, 4605, 5604, 6002, 6209
Ascochyta blight	ILL 358, 2439, 5244, 5480, 5588, 5597, 5684, 5714, 5725, 5755, 6258
Winter hardiness	ILL 52, 465, 468, 590, 662, 780, 857, 975, 1878, 1918

Vascular wilt (induced by *Fusarium oxysporum* f.sp. *lentis* Vasd. and Srin.) is the most important soil-borne disease of lentil in the Mediterranean region. It also causes major yield losses in the Indian sub-continent. Bayaa & Erskine (1990) have developed an efficient screening method and found sources of resistance (Table 10). Ascochyta blight (induced by *Ascochyta lentis* Bond and Vassil.) on lentil can cause considerable losses in WANA, Ethiopia, parts of the Indian sub-continent, and Canada. Losses result from damage to the standing crop, and through reduced seed quality from infection in the swathe. Good sources of resistance have been identified in co-operation with the National Agricultural Research Centre, Islamabad, Pakistan, and are being used in the breeding program (Iqbal *et al.*, 1990) (Table 10).

The WANA region experiences extremes of temperature and moisture supply and there are deficiencies or toxicities of mineral nutrients in the soil. These conditions can limit production. First with chickpea at low elevations (Saxena & Singh, 1984) and then with lentil at high elevations (Sakar *et al.*, 1988), it was recognized yield could be increased with winter planting. For this to be effective with chickpea, Ascochyta blight resistance and cold tolerance were necessary; and with lentil and *L. ochrus,* cold tolerance was needed. Screening for resistance to abiotic stresses started with low temperature. This has lead to the identification of 13 accessions of chickpea, highly tolerant to cold (Table 9), out of nearly 10,000 tested. Recently, emphasis has been placed on screening chickpea and lentil for drought tolerance. Two sources of drought tolerance in chickpea (Table 9) and nine sources of cold tolerance in lentil (Table 10) have been identified. Lack of cold tolerance has hindered the introduction of *L. ochrus* (which has a high degree of *O. crenata* resistance and has high biomass and seed production) to many WANA countries. One accession however, IFLA 109, from Portugal, does have a high level of cold tolerance.

Germplasm collections have also been used to improve the nutritional quality of these crops for human and animal consumption. The faba bean, lentil and kabuli chickpea collections have been evaluated for seed protein and the lentil collection for straw. Seed protein content varied between 18.6% to 30.2% in lentil (Erskine & Witcombe 1984), 18.0% to 31.0% in faba bean (Robertson & El-Sherbeeny, 1992) and 16.0% to 24.8% in chickpea (Singh *et al.*, 1983). Protein content has not generally been a breeding objective for food legumes; the major activity has been a monitoring of new lines to maintain the protein level of existing cultivars.

Special attention has been given to *Lathyrus* spp. and evaluation for low neurotoxin contents (Aletor *et al.*, 1994). *Lathyrus sativus* is a component of human diets and animal feed in times of drought induced famine in Asia and Africa, however excessive consumption can cause "Lathyrism", a nervous disorder. It results in an incurable paralysis of the lower limbs of human and domestic animals, caused by the presence of the free amino acid 3-*N*-Oxalyl-L-2, 3-Diaminopropionic Acid (ODAP) in seed. Screening indicated that none of the *Lathyrus* spp. lines was ODAP-free, although in several the content was low. This also seems to be a problem in related species, since samples of *L. cicera* ranged from 0.01% to 0.22% with a mean of 0.16% (Aletor *et al.*, 1994). *L. sativus* showed the biggest range from 0.16% to 0.74% with a mean of 0.48%, while *L. ochrus* lines were highest in ODAP, ranging from 0.46% to 0.67% with a mean of 0.57%. There appears however to be good potential for breeding *L. sativus* and *L. cicera* lines with low ODAP content.

In addition to cultigens, ICARDA has assembled collections of wild relatives and progenitors of lentil and chickpea. and wild and weedy forms of the forage legume species. It is likely the work with wild species will assume greater importance in coming years, as gene transfer between the wild species and cultigens is improved by new biotechnological techniques. This will drive the need for further collection of these species. The temperate legume crops originated in the 'Fertile Crescent' and is the centre of diversity of their wild relatives (Cubero, 1981; van der Maesen, 1987; Maxted *et al.*, 1990; Maxted, 1995). The progenitor species of chickpea, lentil and pea, *Cicer reticulatum, Lens culinaris* subsp. *orientalis* and *Pisum fulvum* respectively, are endemic to the region. The introgression of wild species of *Vicia,* with *Vicia faba,* has been hampered by the taxonomic distance between the faba bean and its closest relatives and the associated barriers to interspecific hybridization. In the past five years the wild *Lens* and *Cicer* collections at ICARDA have been evaluated for resistances to biotic and abiotic stresses. Sources of resistance have been found in wild *Cicer* species for Ascochyta blight, Fusarium wilt, leaf miner, seed beetle, cyst nematode, cold, and drought (Table 9) (Di Vito *et al.*, 1988; Kaiser *et al.*, 1994; Singh and Reddy, 1993b; Singh and Weigand, 1994; Singh *et al.*, 1989, 1990). Wild species have been the only source of resistance so far found for seed beetle and cyst nematode; and they have a higher level of resistance than the cultivated species for Fusarium wilt, leaf miner, and cold. Accessions of the wild progenitor of lentil, *Lens culinaris* subsp. *orientalis* were found to have higher levels of winter-hardiness than lentil itself (Hamdi *et al.*, 1996). In addition many were more drought resistant and had a lower relative reduction in yield under drought stress (Hamdi and Erskine, 1997). There are also sources of resistance to vascular wilt and Ascochyta blight in wild *Lens* species (Bayaa *et al.*, 1994; 1995).

'Insurance' or 'option' value is also a form of utilisation of conserved germplasm (McNeely, 1988). Germplasm may also be requested for reintroduction of traditional landraces following a disaster situation, either natural (e.g. floods in Bangladesh) or anthropomorphic (e.g. civil disturbance in Rwanda, Afghanistan or Cambodia). Consequently 'insurance' value is also a use of conserved germplasm.

TOWARDS IMPROVING THE CONSERVATION / UTILISATION LINK

The provocative title of this paper has behind it the implication that as conservationists and users of legume germplasm we must ensure we make increased use of collections. Simmonds (1961) was the first to point out that mismanaged collections could be regarded as museum exhibits. If this happens how can we expect the public to continue their financial support? Given (1994) has estimated that 65% of conserved germplasm lacks basic passport data, 80% lacks characterisation and 95% lack evaluation data - which leaves approximately 1% which are correctly catalogued and ready for use. In these circumstances the answer to the question - are our germplasm collections museum items? - must be YES. How can we, as the professionals involved, ensure that conserved germplasm is held in a form suited for utilisation and there is a seamless gradation from conservation to utilisation. The following may assist:

Conservation Planning

Choice of Target Taxa and Area - There are factors, discussed above, to be considered when selecting which taxa to collect. The choice of which areas to target is equally important and will involve an eco-geographic analysis of competing areas. It is often advisable to include potential users in the decision to be made on target taxa and areas.

Field Conservation

Efficient Population Sampling - The better the quality of the sample transferred *ex situ* the more likely it is to be utilised. The collector must ensure that the sample is of sufficient size to avoid the need for immediate regeneration and that it represents the genetic variation in the population sampled.

Collect Associated Materials - Collecting should not be restricted to seed but should include passport data, herbarium voucher specimens and possibly vegetative plants. For legume species, rhizobia cultures also should be obtained.

Passport Data - The more complete the passport data the more useful the accession will be to users. Although the data will vary depending on the taxon and area, a 'good collector' will record: and label a sample (expedition identifier, collector name and number, date, type of material), sample identification (scientific name, vernacular name), sampling information (population estimate - number of plants in population is___, covering ___ m², sampling method), collecting site location (country, province, precise location, latitude, longitude, altitude, farmer's name), site description and context (site disturbance, physiography, soil, biotic factors) and population information (phenology, pests & diseases, uses).

Gene Bank Conservation

Collected Material Processing - The materials require appropriate storage. The passport data require checking and entry into the database. The dried voucher specimen require re-identification, and appropriate treatment before being deposited in the herbarium. Vegetative samples need planting out and rhizobium samples incorporation into collections. The seed may require fumigation, threshing, cleaning and drying to avoid any deterioration in quality.

Collecting Reports and Other Publications - Following the expedition the commissioning agency will require a report, but the existence of novel diversity can be signalled to potential users by publishing collecting reports. The published reports will vary, but may include details of species ecogeography, collection timing, mission personnel, a chronological itinerary, site selection and sampling strategy, material collected, collection processing, and discussion of genetic erosion of the target taxon in the target area and future priorities.

Efficient Storage - Samples from mismanaged collections are less likely to be used. Samples require optimum storage conditions to avoid differential genetic erosion. Human errors through mislabelling or storing the same accession under different accession numbers in the same or different institutes must be avoided. If the sample has to be regenerated it must be done under suitable conditions (e.g. an out-breeding species needs to be in cages to avoid out-crossing).

Duplication - Once collected use of the material can be increased by duplication in national, regional and international collections. This will ensure its longer term safety and make it more widely available to users.

Collection Advertising – Once stored, the gene bank manager must be pro-active and advertise the new accessions in the collection. He or she can draw the attention of potential users by various means, including catalogues, databases (by making the database available on the Internet or by a link via a meta-database, such as IPGRI's holdings database). The SINGER database of CG institute germplasm can be queried at HTTP://www.cgiar.org/singer/index.htm.

Characterisation and Evaluation - Any pre-breeding evaluation will enhance the utilisation of the accessions. This may involve selection based on passport data (e.g. an accession with a high drought or salt tolerance is likely to be useful in breeding tolerant varieties) or characterisation and then selection based on desirable features. More detailed evaluation for drought or salt tolerance, or screening for particular biotic resistance will obviously enhance utilisation. Increasingly there will be evaluation of germplasm for biochemical and molecular markers to assess genetic diversity; and to plan future research in areas such as further *ex situ* collections, *in situ* conservation, core collections, etc. It is often advisable to involve users in the characterisation and evaluation process.

Developing Core Collections - With the increasing size of many collections and limited funds for characterisation and evaluation, it may be necessary to develop a core collection of representative, well characterised and evaluated accessions to assist user select desired traits. As pointed out by Hodgkin *et al.*, 1995) this core should be a point of entry to the entire collection and not a separate entity in itself.

Germplasm Holdings Reports and Other Publication - All results should be published to ensure the potential users are aware of the material available and their attention should be drawn to extreme forms.

Service to Users - The quality of the service provided to users by curators will affect utilisation. If emails remain unanswered or if the user is supplied with dead seed or inadequate passport data he is unlikely to repeat his request and the material will remain under-utilised.

Linking Conservation and Utilisation Sites - Unfortunately the people who conserve and use germplasm are seen as belonging to two distinct professions, often located in separate locations. Utilisation can be improved by bringing the two together, physically and professionally.

Staff training - All of the above actions are only feasible if the gene bank is staffed by appropriately trained personnel. There is a continuing requirement for vocational and specialist PGR courses for custodians of the world's conserved botanical diversity.

The points discussed above are extensive, and it is to be hoped that they are already being applied by the majority of plant collectors, gene bank managers and germplasm users. There are other issues for instance, we have not discussed the somewhat nebulous effects of national and international politics, particularly the Convention on Biological Diversity and Farmers Rights, on the relationship between conservation and utilisation, which we considered beyond the scope of this paper.

CONCLUSION

If we are to have efficient conservation and utilisation, we must follow the example of 'good' germplasm collections and ensure that the material housed has been collected effectively, stored efficiently, advertised thoroughly and disseminated widely. No matter how large the collections, unless they are adequately characterised and evaluated to enable the breeder identify desirable genotypes, they will remain unused. Conservationists have a product available, legume genetic diversity, and like 'good' museums must present their product in an appropriate manner, advertise their 'wares' to ensure support from the general public. No one wants to visit a 'poor' museum where the exhibits are covered in dust and the displays are broken. Therefore, the conservationist, like the 'good' museum director, must make their collections available to meet the varied needs of potential users. This will ensure the general public will continue their long term support for field conservation and germplasm collections, as well as, underwriting conservation and utilisation sustainability for the long term benefit of humankind.

Acknowledgements

The authors would like to thanks ICARDA, IPGRI, and USDA for details of their legume collection holdings and J. Valkoun, M.C. Sawkins and A. Eastwood for their comments on an early draft of this paper.

References

Abramovitz, J.N., 1994. *Trends in biodiversity investments.* World Resources Institute, Washington, D.C.

Aletor, V.A., A. M. Abd El Moneim & A. V. Goodchild, 1994. Journal Sci. and Food Agric. 65: 143-151.

Alinoglu, N. & N. Durlu, 1970. *Journal of Range Management* 23: 61-63.

Bayaa, B. & W. Erskine, 1990. *Arab Journal of Plant Pathology* 8: 30-33.

Bayaa, B., W. Erskine & A. Hamdi, 1994. *Genetic Resources and Crop Evolution,* 41: 61-65.

Bayaa, B., W. Erskine & A. Hamdi, 1995. *Genetic Resources and Crop Evolution,* 42: 231-235.

Bettencourt, E., J. Konopka, & A.B. Damania, 1989. Directory of Crop Gerrmplasm Collections. 1. I. Food Legumes. *Arachis, Cajanus, Cicer, Lens, Lupinus, Phaseolus, Pisum, Psophocarpus, Vicia* and *Vigna.* International Board for Plant Genetic Resources, Rome.

Bettencourt, E., Th. Hazekamp and M.C. Perry, 1992. *Directory of Germplasm Collections. 7. Forages* (Legumes, grasses, browse plants and others). International Board for Plant Genetic Resources, Rome.

Brush, S.B. 1996. *Journal of Environment & Development* 5: 416-433.

Chamberlain, D.F., 1970. *Melilotus* L. *In Flora of Turkey and East Aegean Islands* pp. 3: 448-452. (ed. P.H. Davis). Edinburgh University Press, Edinburgh.

Christiansen, S., A.M. Abd El Moneim, P.S. Cocks & M. Singh, 1998. *Journal of Agricultural Science Cambridge* (In Press).

Clark, R.L., H.L, Shands, P.K. Bretting& S.A. Eberhart, 1997. *Crop Science* 37(1): 1-6.

Cubero, J.I., 1981. In: *Lentils,* pp. 15-38 (Eds C. Webb and G. Hawtin.) CAB/ICARDA, Slough, England.

Cullen, J. 1970. *Anthyllis* L. In *Flora of Turkey and East Aegean Islands* pp. 3: 534 (ed. P.H. Davis) Edinburgh University Press, Edinburgh..

Department of the Environment, 1996. *Towards a methodology for costing biodiversity targets in the UK.* Department of the Environment, London.

Di Vito, M., N. Greco, K.B. Singh & M.C. Saxena, 1988. *Nematologia Mediterranea* 16: 17-18.

Durlu, N. & D.R. Cornelius, 1970. *Agronomy Journal* 62: 55-56.

Erskine, W. & M.C. Saxena, 1993. In: *Breeding for Stress Tolerance in Cool-Season Food Legumes,* pp. 51-62. (Eds K.B. Singh and M.C. Saxena.) John Wiley & Sons, Chichester, U.K.

Erskine, W. & J.R. Witcombe, 1984. *Lentil Germplasm Catalogue.* ICARDA, Aleppo, Syria. 284 pp.

Ferguson, M.E., N. Maxted, L.D., Robertson, H.J., Newbury & B.V. Ford-Lloyd, 1998. *Botanical Journal of The Linnean Society.*

Flint, M., 1991. *Biological Diversity and Developing Countries. Issues and Options.* Overseas Development Administration, London.

Given, D.R., 1994. *Principles and practice of plant conservation.* Chapman & Hall, London.

Groombridge, B., 1992. *Global biodiversity: status of the Earth's living resources.* Chapman & Hall, London.

Halila, M.H. & M.M. Harrabi, 1990. Breeding for dual resistance to Ascochyta and wilt diseases in chickpea. Options Mediterran Jennes Serie Seminaires 9:163-166.

Halila, M.H., H.E. Gridley & P. Houdiard, 1984. *International Chickpea Newsletter* 10: 13-14.

Hamdi, A., I. Kusmeasglu & W. Erskine, 1996. *Genetic Resources and Crop Evolution,* 43: 63-67.

Hamdi, A. & W. Erskine, 1997. *Euphytica* 91: 173-179.

Hanounik, S.B. & L.D. Robertson, 1987. *Plant Disease* 72: 696-698.

Hanounik, S.B. & L.D. Robertson, 1988. *Plant Disease* 73: 202-205.

Hargrove, C. 1992. *The Monist* 75: 183-207.

Hedge, I.C., 1970. *Onobrychis Adans.*In *Flora of Turkey and East Aegean Islands* pp. 3: 560-589. (ed. P.H. Davis). Edinburgh University Press, Edinburgh.

Hodgkin, T., A.H.D., Brown, Th.J.L. van Hintum & E.A.V., Morales, 1995. *Core collections of plant genetic resources.* John Wiley & Sons, Chichester. pp. 1-269.

ICARDA, 1992. *Legume Program Annual Report for 1991.* ICARDA, Aleppo, Syria. pp. 105-7.

ICARDA, 1994. *Legume Program Annual Report for 1993.* ICARDA, Aleppo, Syria. pp. 105-8.

Iqbal, S.M., A. Baksh & B.A. Malik, 1990. *LENS Newsletter* 17:26-27.

Jiménez-Díaz, R.M., K.B. Singh, A. Trapero-Caspas, A. & J.L. Trapero-Casas, 1991. *Plant Disease* 75: 914-918.

Jiménez-Díaz, R.M., P. Crino, M.H. Halila, C. Mosconi & A.T. Trapero-Casas, 1993. In: *Breeding for Stress Tolerance in Cool Season Food Legumes,* pp. 97-106. (Eds K.B. Singh & M.C. Saxena.), John Wiley & Sons, Chichester, UK.

Johnson, N., 1995. *Biodiversity in the balance: approaches to setting geographic conservation priorities.* Biodiversity Support Program, Washington, D.C.

Kaiser, W.J., A.R. Alcal-JimJnez, A. Hervs-Vargas, J.L. Trapero-Casas & R.M. JimJnez-DRaz, 1994. *Plant Disease* 78:962-967.

Keystone Center, 1991. *Final consensus report of the Keystone International Dialogue Series on Plant Genetic Resources* (3rd Plenary Session, Oslo, Norway). Keystone Center, Boulder, CO.

Kupicha, F.K., 1976. *Notes from the Royal Botanic Garden Edinburgh*, 34: 287-326.

Kupicha, F.K., 1983. *Notes From The Royal Botanic Garden Edinburgh* 41: 209-244.

Lassen, P., 1989. *Willdenowia*, 19:49-62.

Lesins, K. A. & I., Lesins, 1979. *Genus Medicago (Leguminosae). A Taxogenetic Study.* Dr. W. Junk., The Hague.

Marshall, D.R. & A.H.D. Brown, 1975. In *Crop genetic resources for today and tomorrow.* pp. 53-80. (Eds. O.H. Frankel & J.G. Hawkes), Cambridge University Press, Cambridge.

Maxted, N., 1993. *Botanical Journal Of The Linnean Society,* 111: 155-182.

Maxted, N., 1995. *Systematic and Ecogeographic Studies in Crop Genepools* 8. IBPGR, Rome. Pp. 184.

Maxted, N., B.V., Ford-Lloyd, & J.G. Hawkes, 1997a. In: *Plant genetic conservation: the in situ approach* pp. 20-55 (eds. N. Maxted, B.V. Ford-Lloyd & J.G. Hawkes) Chapman & Hall, London.

Maxted, N., J.G., Hawkes, L. Guarino & M.C. Sawkins, 1997b. *Genetic Resources and Crop Evolution,* 44: 337-348.

Maxted, N., H. Obari & A. Tan, 1990. *Plant Genetic Resources Newsletter,* 78: 21-26.

McNeely, J.A., 1988. *International Union for Conservation of Nature and Natural Resources,* Gland, Switzerland.

Nassib, A.M., D. Sakar, M. Solh & F.A. Salih, 1988. In: *World Crops: Cool Season Food Legumes,* pp. 1081-94 (Ed R.J. Summerfield.) Kluwer Academic Publishers, The Netherlands.

Pearce, D.W. & R.K. Turner, 1990. *Economics of natural resources and the environment.* Harvester Wheatsheaf, New York, U.S.A.

Pearce, D.W. & D. Morgan, 1994. *The economic value of biodiversity.* Earthscan, London, UK.

Perry, M.C. & E. Bettencourt, 1995. In: *Collecting plant genetic diversity: technical guidelines.* pp 121-129 (eds Guarino, L., V. Ramanatha Rao & R. Reid.) CAB International, Wallingford.

Plitmann, U. and C.C. Heyn, 1984. *Proc. 3rd Int Lupin Conference* La Rochelle, 40-54

Plucknett, D.L., N.H.J., Smith, J.T., Williams and N.M. Anishetty, 1987. *Gene banks and the world's food.* Princeton University Press, Princeton. NJ

Polhill, R.M. and L.J.G., van der Maesen, 1985. In *Grain Legume Crops,* pp. 3-36 (eds. R.J. Summerfield and E.H. Roberts). Collins, London.

Rechinger, K. H., 1984. *Flora Iranica. Papilionaceae II.* Akademische Druck-u. Verlagsanstalt, Graz.

Robertson, L.D. & M. El-Sherbeeny, 1988. *Faba Bean Germplasm Catalogue: Pure Line Collection.* ICARDA, Aleppo, Syria. 140 pp.

Robertson, L.D., K.B. Singh & B. Ocampo, 1995. *A Catalogue of Annual Wild Cicer Species.* Aleppo, Syria, ICARDA. Pp 171

Sakar, D., N. Durutan & K. Meyveci, 1988. In: *World Crops: Cool season food legumes,* pp. 137-146. (Ed R.J. Summerfield.) Kluwer Academic Publishers, The Netherlands.

Saxena, M.C. & K.B. Singh (Eds.) 1984. *Proceedings of the Workshop on Ascochyta Blight and Sowing of Chickpeas.* Martinus Nijhoff/Dr. Junk, The Hague, The Netherlands. 288 pp.

Simmonds, N.W., 1962. *Biological Reviews* 37: 422-465.

Singh, K.B. & M.V. Reddy, 1993a. *Crop Science* 33: 186-189.

Singh, K.B. & M.V. Reddy, 1993b. *Netherlands Journal of Plant Pathology* 99: 163-167.

Singh, K.B. & S. Weigand, 1994. *Genetic Resources and Crop Evolution* 41: 75-79.

Singh, K.B., L. Holly & G. Bejiga, 1991. *A Catalogue of Kabuli Chickpea Germplasm.* ICARDA, Aleppo, Syria. pp.398

Singh, K.B., R.S. Malhotra & M.C. Saxena, 1990. *Crop Science* 30: 1136-1138.

Singh, K.B., R.S. Malhotra & J.R. Witcombe, 1983. *Kabuli Chickpea Germplasm Catalogue.* ICARDA, Aleppo, Syria. pp 284.

Singh, K.B., M. Di Vito, N. Greco & M.C. Saxena, 1989. *Nematologia Mediterranea* 17: 113-114.

Shands, H L, 1994. In: *Conservation of Plant Genetic Resources and the UN Convention on Biological Diversity.* pp. 27-38. (eds D Witmeyer & M S Strauss.) AAAS, Washington DC.

Smith, R.D. & S. Linington, 1977. *Bocconea:* 273-280.

Swanson, T. 1995. *Plant Genetic Resources Newsletter* 105: 1-7.

Turner, P.K. & M. Postle, 1994. In: *Water quality: understanding the benefits and meeting the demands.* (eds Ward, N. & G.D. Garrod.), CPE Research Report, Centre for Rural Economy, Dept. of Agric. Economic and Food Marketing, University of Newcastle upon Tyne

van der Maesen, L.J.G., 1987. In: *The Chickpea,* pp. 11-34. (Eds M.C. Saxena and K.B. Singh.), CAB/ICARDA, Oxon, England.

Vane-Wright, R.I., 1996. In: *Biodiversity: a biology of number and difference.* pp 309-344. (ed Gaston, KT), Blackwell Science, Oxford.

Zohary, M. and D. Heller, 1984. *The Genus Trifolium.* The Israel Academy of Sciences and Humanities.

Cicer species - Conserved Resources, Priorities for Collection and Future Prospects

R S Malhotra [1], R P S Pundir [2], and W J Kaiser [3]

1 ICARDA, P.O. Box 5466, Aleppo, Syria; 2 ICRISAT Asia Center, Patancheru, A.P. 502 324, India; 3 USDA-ARS, WSU-Pullman, WA 99164-6402, USA

Abstract

The genus *Cicer* encompasses 34 wild perennial species, 8 annual wild species, and one annual cultivated species. Most of these species are found in the West Asia and North African region covering Turkey in the north to Ethiopia in the south, and Pakistan in the east to Morocco in the west. Chickpea (*Cicer arietinum*) is the only cultivated species, and is the second most important pulse crop in the world. The two most closely related species to the cultigen, *C. reticulatum*, and *C. echinospermum*, are endemic in southeastern Turkey and adjoining areas of northern Iraq. Good collections have been made and categorized using descriptors. As the level of tolerance to some of the biotic and abiotic stresses is not at a satisfactory level in the cultivated species, limited efforts have been made to collect and evaluate the wild *Cicer* species. For some of the wild annual species namely, *C. yamashitae*, *C. cuneatum*, and *C. chorassanicum*, there are only a few accessions in the collection and more need to be collected.
The annual *Cicer* species are not difficult to grow, and can be conserved and rejuvenated without much difficulty. But the perennial *Cicer* species are extremely difficult to grow. They probably need their original habitats and should be conserved *in situ*. The cultivated species has been extensively developed but it still lacks resistance to many biotic and abiotic stresses. The wild annual species have been evaluated for resistance to these stresses. They provide good prospects for the improvement of chickpea. Desirable genes have been introgressed from the wild species, which are crossable, but not all species are crossable with chickpea and further research is needed. It is hoped biotechnology and tissue culture in future will permit the introgression of their genes into chickpea.

INTRODUCTION

The *Cicer* genus belongs to subfamily Papillionoideae, and tribe Viceae Alef. This genus encompasses 43 species, including 34 wild perennial, 8 wild annual and the cultivated annual chickpea, *Cicer arietinum* L. (Malhotra et al., 1987). All species are diploid and self-pollinating. The chromosome number of all known *Cicer* species is 2n=16 (Ladizinsky and Adler, 1976, Singh and Ocampo, 1993, Pundir et al., 1993). The genus probably originated in the area of Southeast Turkey and adjoining regions of Syria where three species (*C. bijugum* K.H.Rech, *C. echinospermum* P.H.Davis and *C. reticulatum* Lad.) occur naturally. Although there are controversies on the progenitor of chickpea, the studies of Ladizinsky and Adler (1976); Singh and Ocampo (1993); Ocampo et al., (1992); Labdi et al., (1996); indicate that *Cicer reticulatum* is probably the wild progenitor. Remnants of *Cicer* seeds from Hacillar near Burdu (Turkey) dated to 5450 BC (Helbaek 1970) support this view. Chickpea is also known as Bengal gram or Chana in India, Chhola in Pakistan, Hommos in Mediterranean countries, Nohut in Turkey, Shimbra in Ethiopia, and Garbanzo bean in Mexico and the USA.
Chickpea ranks second in area and third in production among the pulses in the world. It is cultivated on about 10 million ha and produces about seven million tones with a yield of 700 kg/ha. Two forms are

R. Knight (ed.), Linking Research and Marketing Opportunities for Pulses in the 21st Century, 603–611.

known, i) kabuli with white flowers, white seed coat and ram-head shaped seeds and ii) desi (indigenous) with dark colored flowers and seeds with an angular shape and various color shades. Kabuli types are grown predominantly in Mediterranean countries, and desi types in countries of South Asia, and eastern and southern Africa. Chickpea is cultivated in moderate winter or spring seasons and occupies a considerable area in over 40 countries. The plant is efficient at fixing nitrogen, the seeds are rich in protein, and the crop can be grown with a minimum of farm inputs.

CONSERVED GENETIC RESOURCES

In recent years the chickpea crop has received increased attention from research workers. The International Crops Research Institute for the Semi-Arid Tropics (ICRISAT) at Hyderabad, India, and The International Center for Agricultural Research in the Dry Areas (ICARDA) at Aleppo Syria, established in

Table 1. The number of cultivated chickpea accessions, classified by country of origin, and held at ICARDA and ICRISAT in Dec 1996.

Country	ICARDA	ICRISAT	Country	ICARDA	ICRISAT
Afghanistan	890	686	Mexico	117	396
Algeria	50	16	Moldavia	6	-
Armenia	3	-	Morocco	225	249
Australia	4	3	Myanmar	-	129
Azerbaijan	16	-	Nepal	6	80
Bangladesh	1	170	Nigeria	1	3
Bulgaria	191	9	Pakistan	265	445
Chile	346	139	Palestine	40	48
China	21	24	Peru	4	3
Columbia	1	1	Portugal	121	84
Cyprus	46	44	Romania	2	-
Czechoslovakia	10	8	Russia	22	-
Ecuador	1	-	Spain	284	121
Egypt	57	53	Sri Lanka	-	3
Ethiopia	124	932	Sudan	11	12
France	18	2	Soviet Union	104	133
Georgia	1	-	Syria	732	203
Germany	1	11	Tanzania	-	97
Greece	18	25	Tadzhikistan	8	-
Hungary	2	4	Tunisia	263	33
India	396	7180	Turkey	864	449
Iran	1737	4856	Uganda	-	1
Iraq	32	18	Ukraine	10	-
Italy	68	45	UK	8	-
Jordan	143	25	USA	121	82
Kazakhstan	1	-	Uzbekistan	11	-
Kenya	-	1	Yugoslavia	6	6
Kyrgyzstan	1	-	Unknown	235	179
Lebanon	28	19	Breeding lines	1444	-
Libya	2	-			
Malawi	3	81	Total	9775	17250

1972 and 1977 respectively, have chickpea as one of their mandate crops. ICRISAT works mainly on desi types and ICARDA on kabuli types. At both centers, efforts are being made to assemble germplasm resources, including land-races, cultivars, genetic stocks and closely related wild species. In addition to donated material, both centres have made collections. ICRISAT and ICARDA have been involved in 41 and 32 collecting missions and have secured 3113 and 2067 accessions respectively. The present holdings at ICRISAT and ICARDA are 17,250 and 9,775 respectively (Table 1)

One of ICRISAT's responsibilities is to serve as a world repository for chickpea germplasm. The collection includes desi and kabuli types and wild species. ICARDA has a regional mandate for the areas where kabulis are grown and its germplasm collection includes only kabuli types. Excepting a few occasional accessions and new acquisitions, the entire ICARDA collection is duplicated at ICRISAT.

The United States Department of Agriculture maintains a collection of 4,237 accessions of cultivated chickpea at their Regional Plant Introduction Station (RPIS) in Pullman, Washington.

CHARACTERIZATION AND EVALUATION

Spring sown and winter sown material from the ICARDA collection has been evaluated for the chickpea descriptors (IBPGR, ICARDA and ICRISAT 1985 and revised 1993) and the results published as catalogues (Singh *et al.*, 1983 and 1991). An indication of the variation observed during spring 1980, and the winter of 1987/88 is given in Table 2. In general, the mean for different traits was much higher for the evaluation of the winter-sown crop than for the spring-sown crop. At ICRISAT, chickpea evaluation work is carried out in the post-rainy season during October to February, which runs through the moderate winter, and spring seasons. Data for 25 morpho-agronomic characters were summarized and published by ICRISAT as a catalogue (Pundir *et al.*, 1988). Parts of the results are given in Table 3. Some of the important traits in chickpea can be related to their country of origin (Table 4). Accessions from Bangladesh were short in height, produced more pods and had a higher resistance to Fusarium wilt. The accessions from Sudan were of relatively short duration, had higher numbers of apical secondary branches and a higher protein content in the seed. Accessions of Indian origin produced the highest average seed yield. The accessions from Jordan were conspicuous by their spreading growth habit, whereas an erect growth habit was common in accessions from Greece and Russia.

Table 2. Variation in chickpea germplasm at ICRISAT.

Trait	Minimum	Maximum	Mean
Time to 50% flowering (day)	28	96	64
Time to maturity (day)	84	169	117
Plant canopy height (cm)	14	96	38
Canopy width (cm)	13	124	40
Pods per plant (no.)	3	238	39
Seeds per pod (no.)	1.0	3.2	1.2
Grain yield (g)	·70	5130	1286
100-seed weight (g)	3.8	59.1	16.1
Seed protein content (%)	12.1	29.6	19.8

Table 3. Variation in kabuli germplasm at ICARDA grown at Tel Hadya, Syria during spring 1980 and winter 1987/88.

Trait	Spring			Winter		
	Min.	Max.	Mean	Min.	Max.	Mean
Time to flower (days)	58	94	81	115	156	137
Flower duration (days)	11	36	23	12	83	29
Time to maturity (days)	114	124	118	174	206	182
Plant height (cm)	15	50	30	25	85	54
Canopy width (cm)	15	60	40	20	96	57
Pods per plant (number)	5	100	25	-	-	-
Seeds per pod (number)	0.1	3.1	1.1	-	-	-
Biological yield (g/m^2)	35	533	204	28	1200	574
Grain yield (g/m^2)	7	292	99	1	567	272
Harvest index (%)	7	84	49	1	78	48
100-seed weight (g)	8.7	59.1	25.1	8.4	70.1	30.0
Protein content (%)	16.0	24.8	20.1	13.5	28.2	23.0

Table 4. Geographical distribution of major chickpea traits

Trait	Country/region
Medium duration	Bangladesh, India, Mexico, Myanmar
Long duration	India, Iran, Morocco, Nepal, Pakistan, Russia, Spain, Syria, Turkey
High branch number	Afghanistan, India, Iraq, Italy
Low branch numbers	Chile, India, Tunisia, Turkey
Erect growth habit	Greece, India, Italy, Russia
Erect (+tall) growth habit	Greece, Italy, Russia
High seed number	Afghanistan, Egypt, India, Mexico, Nepal, Pakistan
Low seed number	Mediterranean countries
High seed mass	Mediterranean countries, India, Mexico
Low seed mass	Indian sub-continent, Ethiopia, Myanmar, Tanzania
Desi (typical) seed	Indian sub-continent, Eastern Africa, Myanmar
Kabuli (typical) seed	Mediterranean countries, West Asia, Chile
Intermediate seed	Ethiopia, Iran
Fusarium wilt resistance	Bangladesh, Ethiopia, India, Iran, Pakistan
Dry root rot resistance	India, Iran
Ascochyta blight resistance	India, Iran, Mexico, Turkey
Gray mold resistance	Iran
Helicoverpa (pod borer) tolerance	India
High seed protein content	Pakistan, Sudan

An evaluation of the ICRISAT chickpea collection, conducted jointly with national programs, was initiated 15 years ago. Accessions were evaluated in India, Nepal and Ethiopia and lines with better regional adaptation were identified. The evaluation of 21,110 lines between 1986 to 1995, undertaken collaboratively with the National Bureau of plant Genetic Resources (NBPGR), India at 5 Indian locations has been very successful (Mathur *et al.*, 1993).

During 1979-80 ICARDA, together with Turkish scientists at Hymana, Turkey (Singh *et al.*, 1981) evaluated 3,158 kabuli chickpea accessions for cold tolerance. Six lines, ILC 410, ILC 2479, ILC 2491, ILC 2636, ILC 2529, and ILC 2406, were highly tolerant to cold under Hymana conditions where the crop was covered with snow for about three months after sowing. Similarly ICARDA with scientists at

Cordoba, Spain identified several sources of resistance to Fusarium wilt, from among 1904 improved lines of chickpea. The lines FLIP 84-43C, FLIP 85-20C, FLIP 85-29C, FLIP 85-30C, ILC 127, ILC 219, ILC 237, ILC 267, and ILC 513 were highly resistant to Fusarium wilt (Jimenez-Diaz *et al.*, 1991).

The joint evaluation of improved sources of tolerance to Ascochyta blight undertaken by the Legume International Testing Program at ICARDA has revealed there are genotypic differences in reaction to the ascochyta blight pathogen present in different areas (Reddy *et al.*, 1992). In another joint evaluation by ICARDA and the Italian national program, 102 accessions of six wild annual *Cicer* species were evaluated for Fusarium wilt resistance under greenhouse conditions in Italy (Infantino *et al.*, 1996). All accessions of *C. bijugum* were highly resistant. Only a few accessions of *C. echinospermum, C. Judaicum, C. pinnatifidum* and *C. reticulatum* were also resistant.

Identification of New Traits
New traits, of value to crop improvement, may occur spontaneously in nature and others may be produced through induced mutation. Some traits identified in recent years include a thick stem, an open flower and short bushy mutants (Dahiya *et al.*, 1984); upright pedicel types (Pundir and van der Maesen, 1977); lobed vexillum (Rao and Pundir, 1983); polycarpy and double pods (Pundir *et al.*, 1988); and the combined occurrence of twin pods and wilt resistance (Pundir and Mengesha, 1988); glabrous stem (Pundir and Reddy, 1989); and determinate growth habit (van Rheenen *et al.*, 1994). The twin pod characteristic occurs in more than 100 accessions in the chickpea collection (mostly Indian origin). It was found this trait could lead to an increase in seed yield of about 6-11% (Sheldrake *et al.*, 1978). Pod size and pod filling percentage are economic traits, but difficult to measure and characterize. These traits were estimated by measuring the replacement of an equivalent volume of water. Accessions were identified with high pod filling percentages (Pundir *et al.*, 1992).

GENETIC RESOURCES OF WILD SPECIES

Wild *Cicer* species were scarce in germplasm collections before 1970 but currently a reasonable number are available. The gene banks of ICRISAT, ICARDA, and RPIS in Pullman hold 135, 268 and 95 accessions of wild annual *Cicer* species, respectively, which consists mostly of the eight wild annual species.

Several efforts to grow and increase seeds of perennial *Cicer* species in ambient conditions at ICRISAT- Patancheru, ICARDA-Aleppo and Izmir-Turkey failed, probably because of unsuitable weather conditions. These species need lower temperatures coupled with drier and longer days. The weather conditions in Pullman, Washington, USA meet these requirements and perennial *Cicers* have been grown successfully. Twelve species (*acanthophyllum, anatolicum, canariense, flexuosum, macracanthum, microphyllum, montbretii, multijugum, nuristanicum, oxyodon, pungens,* and *songaricum*) are maintained at Pullman. Some, such as *C. microphyllum, C. anatolicum* and *C. oxyodon*, grew profusely and produced a large numbers of pods, often with the twin pod characteristic.

Annual species can be raised relatively easily. Accessions of eight wild annual *Cicer* species have been evaluated for various morphological traits at ICARDA and a catalog prepared (Robertson *et al.*, 1995). Some of the accessions have also been evaluated for economic traits at a few locations. Some of the accessions of annual wild *Cicer* species exhibited higher level of expression of economic traits and resistance to biotic and abiotic stresses and the data is summarized in Table 5.

Table 5. Useful traits in wild *Cicer* species.

Trait	*Cicer* species	Reference[1]
Resistance to fusarium wilt	*judaicum*	2,4
	bijugum	2,3,4
	echinospermum	3,4
	canariense, chorassanicum, cuneatum, pinnatifidum	2
	reticulatum	4
Resistance to combined soilborne diseases	*bijugum, cuneatum, judaicum, pinnatifidum*	5
Resistance to gray mold	*bijugum*	3
Resistance to ascochyta blight	*bijugum*	3,7
	judaicum, pinnatifidum	1,3,7
	montbretii	1
Resistance to cyst nematode	*bijugum, pinnatifidum, reticulatum*	8
Tolerance to cold	*bijugum, echinospermum, judaicum, pinnatifidum, reticulatum,*	10
Higher seed protein	*bijugum, reticulatum*	6
Higher biomass	*cuneatum,* most perennial species	-
Resistance to leaf miner	All wild annual species	9
Twin pods	*anatolicum, bijugum, chorassanicum cuneatum, microphyllum, pinnatifidum oxyodon, songaricum*	12,11
Multiple seeds	*cuneatum, montbretii*	12,11

1 Singh *et al.*, 1981; 2 Kaiser *et al.*, 1994; 3 Haware *et al.*, 1992; 4 Infantino *et al.*, 1996; 5 Reddy *et al.*, 1991; 6 Singh and Pundir, 1991; 7 Singh and Reddy, 1993; 8 Di Vito *et al.*, 1996; 9 Singh and Weigand, 1994; 10 Singh *et al.*, 1990; 11 Robertson *et al.*, 1995; 12 van der Maesen, 1987.

Ladizinsky and Adler (1976) classified annual *Cicer* species into three groups, based on their crossability. Crosses between members within a group were successful but not between members of different groups. Group I consisted of *C. arietinum, C. reticulatum* and *C. echinospermum* and Group II of *C. judaicum, C. pinnatifidum* and *C. bijugum.* Group III consisted of *C. cuneatum*, the only other species included in their study. The *Cicer* species were assessed for their value in breeding programs and classified into three gene pools following the scheme of Harlan and De Wet (1971). *C. reticulatum* normally crosses with chickpea and therefore, is a member of the primary gene pool. *C. echinospermum* is in the secondary gene pool because the F_1 hybrids produced from crosses with the cultivated chickpea are highly sterile. Other species, where there is no evidence or possibility of gene exchange with chickpea, were placed in the tertiary gene pool. Subsequent results reported by Singh and Ocampo (1993) and Pundir and Mengesha (1995), showed that *C. echinospermum* crosses normally with chickpea and the F_1 hybrids produced 50% of the normal pod number revealing good prospects for gene exchange. At ICARDA we have been successful in making crosses between *C. arietinum* and *C. pinnatifidum* but the F_1 hybrids survived only up to 3-5 leaf stage and then died after becoming chlorotic (Personal communication R.S. Malhotra).

A recent study of phylogenetic relations, based on isozyme polymorphism among eight wild annual *Cicer* species (Labdi *et al.*, 1996), revealed that levels of polymorphism were high and greater than in the cultivated species. The nine annual *Cicer* species formed four phylogenetic groups based on the neighbor-joining method given by Saitou and Nei (1987). The first group consisted of *C. arietinum, C. reticulatum* and *C. echinospermum*; the second group *C. bijugum, C. pinnatifidum* and *C. judaicum*; the third *C. chorassanicum* and *C. yamashitae*, and the fourth only the one species, *C. cuneatum*.

Germplasm Conservation

Chickpea has seeds that can be stored for long periods with a minimum loss of viability. In the gene banks at ICRISAT and ICARDA, seeds have been stored in medium-term storage (4°C, 20% RH) for 15 years with full seed viability. We are continuing, set by set, to transfer the entire germplasm collection to the long-term (-20°C) storage facility to increase security and minimize the possible loss of genetic diversity. Ellis (1988) has given practical advice on seed viability in storage. For example, chickpea seed having a 99% initial viability and 10% moisture content, will have a viability after 20 years of storage at 4°C, of about 80%. Having a duplicate backup collection elsewhere will further ensure security. The entire ICRISAT chickpea collection is being duplicated at ICARDA and vice versa.

In Situ Conservation

The *in situ* conservation of wild Cicer species is very important especially for the species that are poorly represented in the germplasm pool and are difficult to raise at sites other than their natural habitats. To date very little has been done on this aspect of conservation. Concerted efforts should be made to save these species before these are lost.

Germplasm Distribution and Use

Chickpea research gained momentum with the availability of germplasm from the ICRISAT and ICARDA gene banks. The centers have distributed a large number of samples (Table 6). This includes repatriation of germplasm to countries that have lost their own collections. For example, 4800 chickpea accessions of Iranian origin were sent to Iran and 82 accessions to Nepal. Besides their use in research, many accessions (land races) were found promising and worth cultivating in specific areas of adaptation. Ten accessions supplied from the germplasm collections at ICRISAT and ICARDA were found to be superior and released for cultivation in Algeria, China, Cyprus, Egypt, India, Iran, Iraq, Italy, Jordan, Lebanon, Morocco, Myanmar, Nepal, Oman, Sudan, Syria, Turkey, Tunisia and the USA.

Table 6. Distribution of seed samples from chickpea germplasm collections at ICRISAT and ICARDA, from 1974 to 1996.[1]

Year	ICRISAT	ICARDA	Year	ICRISAT	ICARDA
1974[2]	3070	-	1986	3104	-
1975	7020	-	1987	6268	-
1976	2687	-	1988	5095	-
1977	800	-	1989	8825	-
1978	2318	-	1990	2860	462
1979	1454	-	1991	4745	83
1980	8336	-	1992	1945	714
1981	10202	-	1993	2624	165
1982	5861	-	1994	1166	4481
1983	10548	-	1995	1300	9809
1984	6596	-	1996	5879	4899
1985	4808	-	Total	107511	20613

1 Germplasm samples from ICRISAT, were supplied to various institutes in 80 countries and from ICARDA, to 24 countries.

2 Germplasm samples from ICARDA were supplied from 1977 to 1989, but proper records were not maintained.

A large number of lines identified as sources of resistance to biotic (Ascochyta blight, Fusarium wilt, and leaf miner) and abiotic (cold and drought) stresses have been shared with the national programs through the Legume International Testing Program at ICARDA and ICRISAT.

Priorities for Germplasm Collection

The chickpea collections at ICRISAT and ICARDA are fairly well represented from most chickpea growing countries, except northern Ethiopia, Eritrea, Colombia, Peru, Russia and the Central Asian Independent States. Chickpea germplasm needs to be collected from these countries. A large number of germplasm lines we have received as donations are known only by country of origin and precise passport data are not available. In such cases we need to have germplasm representing diverse agroclimatic-regions of those countries. Our wild *Cicer* collections are far from optimum. Of the wild annuals, *C. yamashitae*, and *C. chorassanicum* are represented by only three and 2 accessions respectively. *C. cuneatum* is represented by only one accession. Representation of perennial *Cicer* species is still very poor. Of the 34 perennial species, live seeds of 42 accessions of only 12 *Cicer* species are available in the collection. We need to secure germplasm of these species as early as possible. Additional collections should also be made from south east Turkey and adjacent areas of Iraq, of the useful wild species namely, *C. reticulatum* and *C. echinospermum* that are crossable with the cultigen.

Future Prospects

Future research and conservation activities relating to the genetic resources of the genus *Cicer* should consider the following:

i) The status of the germplasm needs to be reviewed periodically and new material collected from priority areas. This would include the annual wild species, namely *C. cuneatum* from Ethiopia, *C. chorassanicum* and *C. yamashitae* from Afghanistan, and *C. reticulatum* and *C. echinospermum* from Southeast Turkey and adjacent areas of Iraq and the germplasm of all the perennial *Cicer* species.

ii) Efforts should continue to identify new traits and to collect genetic information and evaluate the germplasm for these traits

iii) Where possible, germplasm sets should be jointly evaluated with National Agricultural Research Systems (NARSs).

iv) the long-term storage of the germplasm should be undertaken at more than one place to increase its security and minimize the possible loss of genetic diversity.

v) A core subset (core collection) of the base collection should be developed to enhance work efficiency and economy in operation.

vi) Passport as well as evaluation details need to be made available to users through (SINGER) the System-wide Information Network for Genetic Resources.

vii) Seeds of *Cicer* germplasm should be rejuvenated at frequent intervals and made available to researchers on demand.

viii) Pre-breeding work on interspecific hybridization needs to be strengthened to facilitate introgression of desirable traits from the wild species to the cultigen. Biotechnological techniques should be used to introgress desirable wild genes into the cultigen.

References

Dahiya, B.S., Lather, V.S., Solanki, I.S. and Kumar, R. 1984. *International Chickpea Newsletter* 11: 408.

Di Vito, M., Singh, K.B., Greco, N. and Saxena, M.C. 1996. *Genetic Resources and Crop Evolution* 43(2): 103-107.

Ellis, R.H. 1988. *Seed Science and Technology* 16: 29-50.

Harlan, J.R. and de Wet, J.M.J. 1971.*Taxon* 20(4): 509-517.

Haware, M.P., Narayan Rao, J. and Pundir, R.P.S. 1992. *International Chickpea Newsletter* 27: 16-18.

Helbaek, H. 1970. In: *Excavation at Hacillar*, pp. 189-244 (ed. J. Mellaart). Edinburg University Press, Gerald Duckworth and Co., London.

IBPGR/ICARDA/ICRISAT. 1985. *Descriptors for chickpea (Cicer arietinum L.).* IBPGR, Rome, Italy.

IBPGR/ICRISAT/ICARDA. 1993. *Descriptors for chickpea (Cicer arietinum L.).* ICRISAT, Patancheru, India.

Infantino, A., Porta-Puglia, A. and Singh, K.B. 1996. *Plant Disease* 80: 42-44.

Jimenez-Diaz, R.M., Singh, K.B., Trapero-Casas, A. and Trapero-Casas, J.L. 1991. *Plant Disease* 75: 914-918.

Kaiser, W.J., Alcala - Jiménez, A.R., Hervás-Vargas, A., Trapero-Casas, J.L. and Jiménez-Díaz, R.M. 1994. *Plant Disease* 78: 962-967.

Labdi, M., Robertson, L.D., Singh, K.B. and Charrier, A. 1996. *Euphytica* 88: 181-188.

Ladizinsky, G. and Adler, A. 1976 . *Theoretical and Applied Genetics* 48: 197-203.

Malhotra, R.S., Pundir, R.P.S. and Slinkard, A.E. 1987. In: *The chickpea*, pp. 67-81 (eds. M.C. Saxena and K.B. Singh). CAB International, Wallingford, Oxon, OX10 8DE, UK.

Mathur, P.N., Pundir, R.P.S., Patel, D.P., Rana, R.S., and Mengesha, M.H. 1993.
 Part 1 (NBPGR-ICRISAT Collaborative Programme). New Delhi, India: National Bureau of Plant Genetic Resources. 194 pp.

Ocampo, B., Venora, G., Errico, A., Singh, K.B. and Saccardo, F., 1992. *Journal of Genetics and Breeding* 46: 229-240.

Pundir, R.P.S. and Mengesha, M.H. 1988. *International Chickpea Newsletter* 18: 3-4.

Pundir, R.P.S. and Mengesha, M.H. 1995. *Euphytica* 83: 241-245.

Pundir, R.P.S., Mengesha, M.H. and Reddy, G.V. 1993 *Euphytica* 69: 73-75.

Pundir, R.P.S., Mengesha, M.H. and Reddy, K.N. 1988. *Journal of Heredity* 79: 479-481.

Pundir, R.P.S. and Reddy, K.N. 1989. *Euphytica* 42: 141-144.

Pundir, R.P.S., Reddy, K.N. and Mengesha, M.H. 1988. *ICRISAT Chickpea Germplasm Catalog: evaluation and analysis*: ICRISAT, Patancheru, A.P. 502 324, India. 99 pp.

Pundir, R.P.S., Reddy, K.N. and Mengesha, M.H. 1992. *International Chickpea Newsletter* 17: 18-20.

Pundir, R.P.S. and van der Maesen, L.J.G. 1977. *Tropical Grain Legume Bulletin* No.10: 26.

Rao, N.K. and Pundir, R.P.S. 1983. *Journal of Heredity* 74: 300.

Reddy, M.V., Raju, T.N. and Pundir, R.P.S. 1991. *Indian Phytopathology* 44: 389-391.

Reddy, M.V., Singh, K.B. and Malhotra, R.S. 1992. *Phytopathologia Mediterranea* 31: 59-66.

Robertson, L.D., Singh, K.B. and Ocampo, B. 1995. *A catalog of annual wild Cicer species*, 171 pp. ICARDA, Aleppo, Syria.

Saitou, N. and Nei, M. 1987. *Molecular Biology and Evolution* 406-425

Sheldrake, A.R., Saxena, N.P. and Krishnamurthy L. 1978. *Field Crops Research* 1: 243-253.

Singh, K.B., Hawtin, G.C., Nene, Y.L. and Reddy, M.V. 1981. *Plant Disease* 65: 586-587.

Singh, K.B., Holly, L. and Bejiga, G. 1991. *Catalogue of kabuli chickpea germplasm* 398 pp. ICARDA P.O. Box 5466, Aleppo, Syria.

Singh, K.B., Malhotra, R.S. and Saxena, M.C. 1990. *Crop Science* 30: 1136-1138.

Singh, K.B., Malhotra, R.S. and Witcombe, J.E. 1983. *Kabuli chickpea germplasm catalog*, 284 pp. ICARDA P.O. Box 5466, Aleppo, Syria.

Singh, K.B., Meyoeci, K., Izgin, N. and Tuwafe, S. 1981. *International Chickpea Newsletter* 4: 11-12.

Singh, K.B. and Ocampo, B. 1993. *Journal of Genetics and Breeding* 47: 199-204.

Singh, K.B. and Reddy, M.V. 1993. *Netherland Journal of Plant Pathology* 99: 163-167.

Singh, K.B. and Weigand, S. 1994. *Genetic Resources and Crop Evolution* 41: 75-79.

Singh, U. and Pundir, R.P.S. 1991. *International Chickpea Newsletter* 25: 19-20.

van der Maesen, L.J.G. 1987. In: *The Chickpea* pp. 11-34. (eds. M.C. Saxena and K.B. Singh). CAB International, Wallingford, Oxon, OX10 8DE, U.K.

van Rheenen, H.A., Pundir, R.P.S. and Miranda, J.H. 1994. *Euphytica* 78: 137-141.

Lens spp: Conserved resources, priorities for collection and future prospects

Morag Ferguson

International Centre for Agricultural Research in the Dry areas (ICARDA), P.O. Box 5466, Aleppo, Syria.
Current address: ICARDA (Arabian Peninsula Regional Program), P.O. Box 13979, Dubai, United Arab Emirates.

Abstract

Information regarding the genetic variation within *Lens* germplasm collections is now emerging. It is hoped that a review of the literature will increase its use for the effective collection, conservation and utilization of lentil germplasm. Two additional species, *L. tomentosus* and *L. lamottei*, have recently been recognized in the genus *Lens* and a new key to the genus has been presented. Studies of agro-morphological and phenological characters as well as stress tolerance responses have revealed potentially useful germplasm. The geographical distribution of genetic variation in both cultivated lentil and its wild relatives has been mapped and areas of high diversity and low diversity identified. This information has been used to prioritize areas for future collection and conservation.

INTRODUCTION

Substantial *ex-situ* collections of *Lens* germplasm already exist. The collection at the International Center for Agricultural Research in the Dry Areas (ICARDA) alone comprises 7758 lentil landrace accessions from 64 countries and 475 wild *Lens* accessions from 16 countries. Partly as a result of recent advances in molecular techniques, information regarding the genetic variation held in these collections is emerging. This information can aid in the identification of potentially useful germplasm as well as form the basis for effective collection and conservation strategies. A review of the literature on genetic variation within *Lens* should aid the exploitation of *Lens* germplasm for these purposes. Here, taxonomic hierarchy is used as a structure to review the information, starting with genetic variation among species, followed by variation within species and finally variation within populations. The implications of this information for future collection, conservation and utilization are discussed.

VARIATION AMONG SPECIES

Recent taxonomic developments
In 1997, two new species were recognized in *Lens*. *Lens tomentosus* was separated from *L. culinaris* subsp. *orientalis* on the basis of its tomentose, as opposed to puberulent, pods and a relatively small, asymmetrical chromosome which bears a minute satellite (Ladizinsky, 1997). It is currently represented by three populations from the Mardin area of south-east Turkey. *Lens lamottei*, originally described by Czefranova (1971) was found to be the same as a differentiated cytotype identified within *L. nigricans* by Ladizinsky *et al.* (1983, 1984) and is now recognized as a separate taxon (Van Oss *et al.*, 1997). The full geographical range of these species still needs to be established. The genus thus consists of the following six species, with the cultivated lentil containing the wild, putative ancestor as a subspecies (Ferguson, 1998):

1.   *L. culinaris* Medikus
a.   subsp. *culinaris*
b.   subsp. *orientalis* (Boiss.) Ponert

613

R. Knight (ed.), Linking Research and Marketing Opportunities for Pulses in the 21st Century, 613–620.
© *2000 Kluwer Academic Publishers. Printed in the Netherlands.*

2.   *L. tomentosus* Ladizinsky
3.   *L. odemensis* Ladizinsky
4.   *L. ervoides* (Brign.) Grande
5.   *L. nigricans* (M.Bieb.) Godron
6.   *L. lamottei* Czefranova

Morphologically *Lens* taxa are closely related and few characters exist to discriminate between them. Recently several additional characters have been reported to aid in the identification of taxa and a new key to the genus has been presented (Ferguson, 1998). An obovate first bifoliate leaflet is indicative of *L. nigricans* and a distinct W-pattern on the seed testa of *L. odemensis* can be used to identify this species.

Crossability relations

Lens culinaris subsp. *orientalis* is readily crossable with the domesticated lentil, although the fertility of the hybrids depends on the chromosome arrangement of the wild parent (Ladizinsky, 1979; Ladizinsky *et al.*, 1984). Ladizinsky and Abbo (1993) identified three distinct karyotypes within *L. culinaris* subsp. *orientalis* with different crossability relations. *Lens odemensis* (referred to as *L. nigricans*, but defined by Ladizinsky *et al.* (1984) as *L. odemensis*) produces partially fertile hybrids with both subspecies of *L. culinaris* (Goshen *et al.*, 1982; Ladizinsky *et al.*, 1984). Pod abortion takes place when the cultivated lentil is crossed with either *L. ervoides* or *L. nigricans*. In addition, no viable hybrids are obtained from crosses between these taxa and *L. culinaris* subsp. *orientalis* or *L. odemensis* (Ladizinsky *et al.*, 1984). Hybrids between *L. ervoides* and *L. nigricans* produce seed-bearing pods at low frequency. Ahmad *et al.* (1995) reported viable hybrids between the cultivated lentil and *L. culinaris* subsp. *orientalis*, *L. odemensis*, *L. ervoides* and *L. nigricans* with the use of GA$_3$ hormone. *Lens tomentosus* was found to be cross-compatible with *L. culinaris* subsp. *orientalis* of a particular karyotype, called the unique crossability group, but not with members of a slightly different karyotype, known as the standard or common crossability group (Ladizinsky and Abbo, 1993). In addition, the two existing *L. tomentosus* accessions were not crossable with each other (Ladizinsky, 1997). Two accessions of *L. lamottei* were inter-fertile with each other, produced partially fertile hybrids with *L. ervoides,* but were reproductively isolated from *L. nigricans* (Ladizinsky *et al.*, 1984). Thus three crossability groups exist in *Lens*: (1) *L. culinaris* and *L. odemensis,* (2) *L. ervoides, L. nigricans* and *L. lamottei,* and (3) *L. tomentosus,* although the extent of crossability varies according to the accessions used. Many studies concerning crossability relations within the genus are based on a small number of accessions. In order to confirm the above crossability groups, further studies, which take into consideration within taxon variation, should be undertaken, particularly involving *L. odemensis, L. lamottei* and *L. tomentosus*.

Phenetic relations

Several studies have been carried out to evaluate phenetic relations within the genus. Conflicting results have emerged which appear to depend on the germplasm and technique used to measure genetic variation (Ahmad and Mc Neil, 1996; Ferguson, 1998). Morphologically *L. lamottei* is closely related to *L. odemensis* (Fig.1) (Ferguson, 1998) and is practically equally associated with *L. odemensis* and *L. culinaris* on the basis of isozyme evidence (Fig. 2) (Hoffman *et al.*, 1986; Ferguson and Robertson, 1996); it does, however, appear to be the taxon most distantly related to all other taxa according to RAPD (Fig. 3) (Ferguson, 1998). These RAPD results are supported by several other studies in which *L. lamottei* was inadvertently included: AFLP (Sharma *et al.*,1996, genotype No. 41); RAPD (Sharma *et al.*, 1995, genotype No. 41; Abo-elwafa *et al.*, 1995, genotype No. 29). Species relations of *L. tomentosus* have not knowingly been evaluated by biochemical or molecular techniques.

Genetic diversity

High genetic diversity has been reported within *L. nigricans*, *L. odemensis* and *L. culinaris* subsp. *orientalis* relative to the cultivated lentil. *Lens lamottei* and *L. ervoides* are the only species reported as having a similar or more restricted genetic base than the cultivated lentil (isozyme: Ferguson and Robertson, 1996; RAPD: Abo-elwafa *et al.* 1995; Sharma *et al.* 1995; Ferguson 1998). Other studies, based on a few accessions, have produced conflicting results (isozyme: Pinkas *et al.* 1985; Hoffman *et al.* 1986; de la Rosa and Jouve, 1992; RFLP: Havey and Muehlbauer, 1989; cpDNA: Mayer and Soltis, 1994).

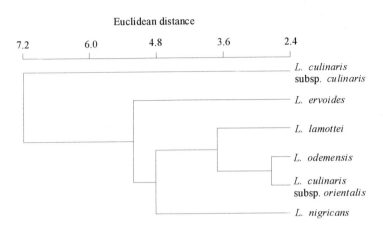

Figure 1. Relationships between taxa according to Euclidean distance based on quantitative morphological characters.

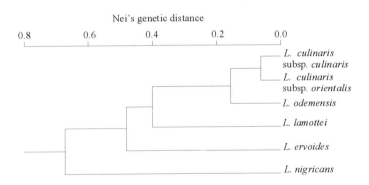

Figure 2. Genetic relationship between taxa according to Nei's genetic distance based on isozyme polymorphisms

616

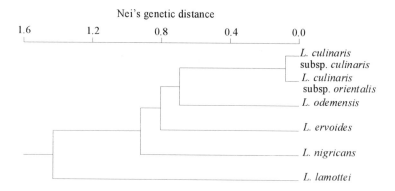

Nei's genetic distance

| 1.6 | 1.2 | 0.8 | 0.4 | 0.0 |

    L. culinaris
    subsp. *culinaris*
    L. culinaris
    subsp. *orientalis*
    L. odemensis
    L. ervoides
    L. nigricans
    L. lamottei

Figure 3. Genetic relationships between *Lens* taxa according to Nei's genetic distance based on RAPD data

Variation within species
Morphological and phenological variation
Agro-morphological and phenological characters have been evaluated in the ICARDA wild and cultivated collections (Erskine and Witcombe, 1984; ICARDA, 1993; Ferguson and Robertson, 1998). In all taxa, adaptation to an appropriate phenology appears to be a key factor in their evolution and clear differences in response to temperature and photoperiod have been observed in cultivated lentils from different geographical regions (Erskine *et al.* 1989; Erskine *et al.* 1990; Ferguson and Robertson 1998). In addition, adaptation to climatic factors such as cold (Erskine *et al.* 1981) and edaphic factors such as iron deficiency (Erskine *et al.* 1993) appear to be important.

The greatest diversity in quantitative descriptors in lentil is for seed yield, biomass, straw yield and seed size, although the greatest gap in the collection is germplasm with a large biomass. Certain *L. culinaris* subsp. *orientalis* accessions have substantially more leaves per plant, peduncles per plant, pods per plant and seeds per plant, and greater leaf area than two cultivated lentil checks (the Argentinian cultivar, Precoz, and a Syrian small-seeded landrace). The total biomass obtained from the best *L. culinaris* subsp. *orientalis* accessions (ILWL 4, 230 and 231) was comparable with the checks. The harvest index of one check was comparable with that of the two best *L. culinaris* subsp. *orientalis* accessions (ILWL 137 and 171). Of the wild taxa, *L. lamottei* had the highest average 100-seed weight (1.12 g). Accessions of *L. culinaris* subsp. *orientalis* from Uzbekistan, Turkmenistan and Tajikistan performed well despite being grown in an environment different from that in which they originated. They had among the largest biomass, the most peduncles per plant, and many pods and seeds per plant, yet seed size did not decline as expected with increased number of seeds. Accessions from this region have a restricted genetic base (Ferguson, 1998) and evidence suggests that they were either carried as a weed to central Asia with the cultivated lentil or that they are in fact relics of previous domestication.
Stress tolerance
Sources of resistance to the major foliar disease of lentil, rust (induced by *Uromyces fabae* (Pers.) de Barry), the most important soil-borne disease of lentil, vascular wilt (induced by *Fusarium oxysporum* Schlecht. em. Snyd. & Hans. f.sp. *lentis* (Vasudeva & Srinivasan) Gordon) and ascochyta blight (induced by *Ascochyta fabae* Speg. f.sp. *lentis* (Bond & Vassil.) Gossen et al.) have been identified.

Resistance to vascular wilt and Ascochyta blight have also been found in *L. culinaris* subsp. *orientalis* (Bayaa *et al.*1994, 1995). No useful source of resistance to the parasitic weed *Orobanche* spp. has yet been found in either the wild or cultivated lentil (Erskine and Saxena, 1993). There is also a lack of resistance to pea leaf weevil (*Sitona* sp.).

Greater resistance to cold tolerance has been found in L. *culinaris* subsp. *orientalis* than in the cultivated lentil (Hamdi *et al.* 1996). The wild genepool, particularly *L. odemensis* and *L. ervoides*, also shows greater resistance to drought in terms of low relative reduction in yield with drought stress (Hamdi and Erskine, 1996).

Geographical distribution of genetic variation

A strong selection pressure in cultivated lentils for adaptation to an appropriate phenology has led to the evolution of distinct ecotypes in different geographical regions, each characterized by a suite of morphological features (Barulina, 1930; Erskine *et al.* 1989). Lentils from South Asia, which are exclusively of the *pilosae* ecotype, exhibit a low diversity and discordance with landraces from other regions according to a combination of qualitative morphological characters (short or rudimentary tendrils and marked pubescence), and are similar to Ethiopian germplasm in terms of quantitative agro-morphological characters, such as early maturity and low biological yield (Erskine *et al.* 1989, 1990, 1998). The restricted diversity in South Asian germplasm is evident from isozyme and RAPD studies (Ferguson *et al.* 1998). These studies, however, also reveal the similarity of South Asian germplasm to that from Afghanistan, and the striking divergence of this germplasm from germplasm from 12 other countries (Figs. 4 and 5). Low diversity and discordance with germplasm from other countries indicates a possible bottleneck when lentils were first introduced into South Asia around 2000 BC (Erskine *et al.* 1998). The divergence of this material is particularly evident in phenological responses. West Asian germplasm, grown in South Asia, flowers as indigenous material matures. Until recently the asynchrony in flowering has reproductively isolated *pilosae* lentils; however, ILL4605, an early large-seeded line, released in Pakistan as 'Manserha 89', has been able to bridge the gap and is now used widely as a parent in breeding programmes in the region (Erskine *et al.* 1998).

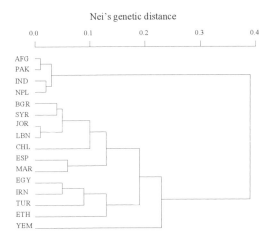

Figure 4. Relationships of lentil landraces from 16 countries based on isozyme variation

The geographical distribution of genetic variation, as revealed by RAPDs, has been mapped in four wild *Lens* taxa. Centres of diversity as well as areas of low diversity have been identified (Ferguson, 1998). For *L. culinaris* subsp. *orientalis,* two centres of diversity exist, one in south-eastern Turkey and north-western Syria, the other in southern Syria and northern Jordan. *Lens culinaris* subsp. *orientalis* accessions from Iran, central Asia and northern Turkey are genetically all very similar and

618

correspond to the common cytotype identified by Ladizinsky *et al.* (1984). The centre of diversity of *L. odemensis* overlaps with the southern centre of diversity of *L. culinaris* subsp. *orientalis* in southern Syria and northern Jordan. A region of high diversity exists for *L. ervoides* along the eastern Mediterranean coast, but the populations from the coastal region of the former Yugoslavia have a particularly narrow genetic base. A clear centre of diversity exists for *L. nigricans* in south-west Turkey with areas of low diversity along the coast of former Yugoslavia, and France and Spain. Centres of diversity for *Lens* are also characterized by a high population density. Maps illustrating the distribution of variation are given in Ferguson (1998).

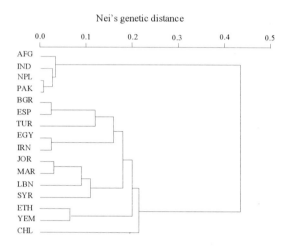

Figure 5. Relationship of lentil landraces from 16 countries based on RAPD variation

Within population variation
Isozyme and RAPD studies have revealed that the genetic structure of wild populations is typical of highly inbreeding species with low levels of heterozygosity (Hoffman *et al.*, 1986; Ferguson and Robertson, 1996) and one or a few discrete genotypes which tend to retain their integrity. Outcrossing is usually less than 1% in cultivated lentil (Wilson and Law, 1972), but has been reported to reach 6% (Erskine and Muehlbauer, 1991). A study on the genetic variation within populations of the wild species using isozymes and RAPDs has in general revealed limited variation within population, with between 78% and 99% of the variation attributable to between-population differences (except from isozyme evidence of *L. lamottei*). Some populations have revealed surprisingly high levels of variation (Ferguson, 1998).

Implications for future collection, conservation and utilization
Studies of phenetic relations as well as genetic diversity and crossability studies indicate that substantial genetic variation exists within the wild species of *Lens* to broaden the genetic base of lentil. Information regarding both species relations and genetic diversity should be used when planning collection missions aimed at increasing genetic variation in genebanks. Agro-morphological, phenological and stress-tolerance data suggest that the wild species may possess variation which could be useful in the improvement of cultivated lentil. *Lens culinaris* subsp. *orientalis* accessions from central Asia appear particularly promising.

Information regarding the geographical distribution of genetic variation, measured using RAPDs, is being used at ICARDA to target areas for future collection missions and for 'core collection' formation. Areas for future collection are targeted by plotting Nei's mean genetic diversity in each of a series of equally sized geographical regions against the number of accessions per region, and calculating the regression line and 95% confidence limits. Those regions, which fall above the 95% confidence limits, are the regions where further collection or sampling is likely to yield novel variation. Priority collection for all species was in south-west Turkey. A collection mission in that region in 1995 added 37 accessions (Ferguson *et al.*, 1996). However, due to the importance of this region for wild lentil, further missions should be conducted, particularly in and around the provinces of Burdur, Isparta and Afyon. In addition, for *L. culinaris* subsp. *orientalis*, collection priorities lie in two broad regions, first southern Syria, north and west Jordan and the West Bank, and second south Turkey across to south-east Turkey. For *L. odemensis* additional priorities lie in the eastern Mediterranean region and parts of North Africa. The region surrounding the coastal junction of Turkey and Syria is a priority for *L. ervoides*, and, although of a lower priority, the Ukraine may yield valuable *L. nigricans* germplasm. Purely on the number of wild lentil accessions available, germplasm from North African countries such as Algeria, Libya, Sudan and Tunisia appear to be under-represented in the collection, as does germplasm from the central and west Asian republics of the former USSR.

Studies of variation within populations show, that, if the conservation objective is to conserve genetic variation, priority should be given to *ex-situ* as opposed to *in-situ* methods. Studies have, however, also revealed that some populations harbour far more variation than others, which implies that levels of diversity within populations must be measured prior to targeting specific populations for *in-situ* conservation.

References

Abo-elwafa, A., K. Murai and T. Shimada. 1995. *Theoretical and Applied Genetics* 90: 335-340.

Ahmad, M., A. G. Fautrier, D. L. McNeil, D. J. Burritt and G. D. Hill. 1995. *Plant Breeding* 114: 558-560.

Ahmad, M. and D. L. McNeil. 1996. *Theoretical and Applied Genetics* 93: 788-793.

Barulina, H. 1930. *Bulletin of Applied Botany, Genetics and Plant Breeding,* Leningrad supplement 40: 265-304.(English summary).

Bayaa, B., W. Erskine and A. Hamdi. 1994. *Genetic Resources and Crop Evolu*tion 41: 61-65.

Bayaa, B., W. Erskine and A. Hamdi. 1995. *Genetic Resources and Crop Evolution* 42: 231-235.

Czefranova, Z. 1971. *Novosti Systematischeski Vyssich Rastenii* 8: 184-191.

de la Rosa, L. and N. Jouve. 1992. *Euphytica* 59: 181-189.

Erskine, W., Y. Adham and L. Holly. 1989. *Euphytica* 43: 97-103.

Erskine, W., S. Chandra, M. Chaudhry, I. A. Malik, A. Sarker, B. Sharma, M. Tufail and M. C. Tyagi. 1998. *Euphytica* (in press).

Erskine, W., R. H. Ellis, R. J. Summerfield, E. H. Roberts and A. Hussain. 1990. *Theoretical and Applied Gen*etics 80: 193-199.

Erskine, W. and F. J. Muehlbauer. 1991. *Theoretical and Applied Genetics* 83: 119-125.

Erskine, W., K. Myveci and N. Izgin. 1981. *LENS Newsletter* 8: 5-8.

Erskine, W. and M. C. Saxena. 1993.. *In: Breeding for Stress Tolerance in Cool-Season Food Legumes.* (eds K. B. Singh and M. C. Saxena). John Wiley *and* Sons. Chichester, U.K.

Erskine, W., N. Saxena and M. Saxena. 1993. *Plant and Soil* 151: 249-254.

Erskine, W. and J. R. Witcombe. 1984. *Lentil Germplasm Catalog.* ICARDA, Aleppo, Syria.

Ferguson, M. 1998. PhD Thesis. School of Biological Sciences, University of Birmingham, Birmingham, UK.

Ferguson, M., N. Acikgoz, A. Ismail and A. Cinsoy. 1996. *Anadolu* 6: 159-166.

Ferguson, M., L. Robertson, B. Ford-Lloyd, H. J. Newbury and N. Maxted. 1998. *Euphytica* (in press).

Ferguson, M. E. and L. D. Robertson. 1996. *Euphytica* 91: 163-172.

Ferguson, M. and L. D. Robertson. 1998. *Genetic Resources and Crop Evolution.* (in press)

Goshen, D., G. Ladizinsky and F. J. Muehlbauer. 1982. *Euphytica* 31: 795-799.

Hamdi, A. and W. Erskine. 1996. *Euphytica* 91: 173-179.

Hamdi, A., I. Küsmenoglu and W. Erskine. 1996. *Genetic Resources and Crop Evolution* 43: 63-67.

Havey, M. and F. Muehlbauer. 1989. *Theoretical and Applied Genetics* 77: 839-843.

Hoffman, D., D. Soltis, F. Muehlbauer and G. Ladizinsky. 1986. *Systematic Botany* 11: 392-402.

ICARDA. 1993. Legume Program; Annual Report for 1992. ICARDA, Aleppo, Syria.

Ladizinsky, G. 1979. *Euphytica* 28: 179-187.

Ladizinsky, G. 1997. *Botanical Journal of the Linnean Society* 123: 257-260.

Ladizinsky, G. and S. Abbo. 1993. *Genetic Resources and Crop Evolution* 40: 1-5.

Ladizinsky, G., D. Braun, D. Goshen and F. Muehlbauer. 1984. *Botanical Gazette* 145: 253-261.

Ladizinsky, G., D. Braun and F. J. Muehlbauer. 1983. *Botanical Journal of the Linnean Society* 87: 169-176.

Mayer, M. S. and P. S. Soltis. 1994. *Theoretical and Applied Genetics* 87: 773-81.

Pinkas, R., D. Zamir and G. Ladizinsky. 1985. *Plant Systematics and Evolution* 151: 131-50.

Sharma, S. K., I. K. Dawson and R. Waugh. 1995. *Theoretical and Applied Genetics* 91: 647-4.

Sharma, S. K., M. X. Knox and T. H. N. Ellis. 1996. *Theoretical and Applied Genetics* 93: 751-8.

Van Oss, H., Y. Aron and G. Ladizinsky. 1997. *Theoretical and Applied Genetics* 94: 452-7.

Wilson, V. E. and A. G. Law. 1972. *Journal of the American Society of Horticultural Science* 97: 142-3.

Lens species, conserved resources in India

B Sharma
Division of Genetics, Indian Agricultural Research Institute, New Delhi, 110012, India

INTRODUCTION

Lentil is among the cool season food legumes cultivated in many countries. Several attempts have been made in the past to collect and evaluate lentil germplasm. In this respect the contribution of ICARDA has been of great significance. Representative samples of germplasm have been collected from all the major lentil growing areas of the world. Although the search for new genes will continue it seems the majority of gene combinations, which survived in nature, are already represented in the germplasm collections.

Our knowledge of single genes governing precise traits is limited. Little effort has been made to induce mutations and discover genes not detected in the natural germplasm. Thus there is scope to collect more germplasm and enhance genetic variability through induced mutation.

Taxonomically the cultivated lentil species *Lens culinaris* has been divide into two subspecies: *microsperma* and *macrosperma*. Besides seed size, the two groups differ in many other properties including intensity of green colour, leaf and leaflet size, presence of pubescence, testa colouration, duration of flowering/maturity, tolerance to stress of various kinds, susceptibility to diseases and cotyledoncolour. The *macrosperma* lentils evolved in the cooler regions of subtropical Europe and are called Mediterranean lentils. Genotypes belonging to this group are usually late in maturity and require better crop management for survival. The *microsperma* lentils are called Indian lentils as they evolved in the harsh conditions of short-day winters in the Indian subcontinent. They have orange (red) cotyledons, small seeds, black spotted testas, deep green foliage, pubescent plants and they flower and mature much earlier than *macrosperma* lentils. They are also tolerant to drought and disease.

As expected, there is a range of genotypes with different character combinations in each of the two groups of lentils. Their subspecies status is also questionable as they are freely crossable and character combinations from both groups are easily achieved.

Germplasm evaluation

A research project was initiated at the Indian Agricultural Research Institute, New Delhi to evaluate and classify the germplasm collection being maintained. Each germplasm line has bee observed in detail for the major characteristics, which can be distinguished on the basis of a property or a plant part. The characters assessed were plant growth, leaf morphology, plant height, green pigmentation, anthocyanin pigmentation, days to flowering and maturity, flower colour, seeds per pod, seed size, testa colour, testa pattern and cotyledon colour.

The entire collection comprising 907 accessions was sown in an augmented design with three commercial varieties as checks, which were repeated every 20 accessions. The mean values of the distinguishing characters of the check varieties are as follows

Character	Lens 4076	Lens 4147	PKVL 1
Plant height (cm)	36	37	35
Days to flowering	81	81	71
Days to maturity	132	132	122
Seeds per pod	1.4	1.7	1.7
1000 grain wt (g)	31.4	23.1	18.7

The entire collection can be evaluated against these values for the check varieties. The results already available reveal a fairly large variability in the germplasm, which can be conveniently maintained under Indian conditions. A summary of the range of characters is presented below with the number of accessions in parentheses.

R. Knight (ed.), Linking Research and Marketing Opportunities for Pulses in the 21st Century, 621–622.
© 2000 *Kluwer Academic Publishers. Printed in the Netherlands.*

1 Growth vigour; Poor (2), medium (1), good (42), very good (526).

2 Growth habit: Erect (208), semi-erect (607), prostrate (126).

3 Plant height: Dwarf <20 cm (6), medium 21-30 cm (207), tall 31-40cm (482), very tall 41-50 cm (108).

4 Green pigmentation: Light green (129), normal green (576), dark green (213).

5 Anthocyanin pigmentation: Stem green (194), light red (674), and deep red (42); leaf green (134), light red (756) and deep red (14); immature pods green (892)) and red (15).

6 Leaflet: Small (210), medium (467), large (220), extra large (21), medium and narrow (4), and long and narrow (2).

7 Leaf pubescence: Sparse (58), medium (540), dense (355).

8 Tendril: Absent (49), rudimentary (256), medium long (495), very long (113).

9 Days to flowering: Extra early< 50 days (22), early 51-75 days (284), medium 76-85 days (532), late 86-110 days (109), very late (3).

10 Flower colour: White (94), white with blue streaks (185), blue (105), light violet (62), medium violet (453), dark violet (5).

11 Seeds per pod: < 1 (53), 1.1-1.5 (285), 1.6-2.0 (390), 2.1-2.5 (4).

12 Seed size (1000 grain wt): 10-20g (337), 21-30g (253), 31-40g (74), 41-50g (10).

13 Testa colour (background): Colourless (73), green (25), gray (3), brown (65), black (5).

14 Testa pattern: Mottled (324), speckled (96). marbled-mottled + speckled (214).

15 Cotyledon colour: Orange (704), yellow (135), green (7).

16 Disease occurrence: Yellow virus (9).

17 Insect damage: < 10% (704), 11-20% (136), 21-30% (41), 31-40% (10), 41-50% (1).

As can be seen from the above groupings of accessions, variability exists for all the traits. This germplasm collection includes mainly *microsperma* genotypes and derivatives from *microsperma* x *macrosperma* crosses. Earliness, seed size, cotyledon colour and disease resistance are the most important economic traits for which new genetic variability has been created from such crosses. For example the earliest lentil genotypes in both *microsperma* and *macrosperma* groups flower (under Delhi conditions) in 72-75 days. Transgressive segregation from crosses between the earliest *macrosperma* (Lens 830, PKVL) and *macrosperma* (Precoz) lentils has produced strains flowering as early as 45-48 days. Such genotypes did not exist in nature.

Similarly, 1000-grain weight in *microsperma* lentils is always below 35g. Their hybridisation with *macrosperma* lentils has generated strains with typical *microsperma* plant morphology but 1000 seed weights reaching 50g. It is expected that Indian lentils with the boldest seed possible (60 g /1000seeds) will be craeted in the near future.

The *microsperma* lentils are generally less affected by disease. Earliness is frequently associated with rust susceptibility. However the Precoz variety of *macrosperma* lentil was among the earliest genotypes in the germplasm and also resistant to rust. It has served as a remarkable donor of three important economic traits: earliness, rust resistance and large seed.

Cotyledon colour of lentil needs special mention. The lentils consumed in the Indian subcontinent have orange (red) cotyledons. Varieties with yellow cotyledon are not cultivated and it may take some time before the consumer in this region becomes accustomed to yellow lentils. The cotyledon colour has been found to be controlled by a system of three genes in which the genes Y and B produce two different kinds of yellow cotyledon, the double dominant YB state results in orange and the double recessive yb light green cotyledons. Another gene, Dg in homozygous recessive condition (dg dg) produces dark green cotyledons. Apparently the Dg,Y and B genes form links in a common pathway of pigment synthesis in the developing seed.

It is evident from the above presentation that a great majority of *macrosperma* lentil genotypes cannot be maintained under field conditions in the countries of South and South-East Asia. But they could be a source of many genes and gene systems of high economic value. It is therefore necessary to accumulate as much germplasm possible of early and medium *macrosperma* lentils and evaluate them under diverse conditions in the tropics and subtropics. Hybridisation of the *macrosperma* strains with unique characters with the promising local genotypes will generate genetic variability never witnessed in nature.

Exhaustive collections have been made for the *microsperma* lentils in the Indian subcontinent. *Macrosperma* lentils are spread over a much larger area in Europe, North and South America. Genetic variability is expected to be greater among *macrosperma* lentils. Introgression between genotypes from the two groups is expected to create entirely new gene combinations and will open possibilities for the release of compartmentalised genetic variability.

Vicia spp: Conserved resources, priorities for collection and future prospects

L. D. Robertson[1], M. Sadiki[2], R. Matic[3] and Lang Li-juan[4]

1 International Center for Agricultural Research in the Dry Areas (ICARDA), Aleppo, Syria; 2 South Australian Research and Development Institute (SARDI), Adelaide, South Australia, Australia; 3 Hassan II Institute of Agronomy and Veterinary Medicine (IAV Hassan II), Rabat, Morocco; 4 Zhejiang Academy of Agricultural Sciences, Hangzhou, China

Abstract

The genus *Vicia* comprises approximately 160 annual and perennial species. The most economically important is *Vicia faba*, which is the only species used currently for human food. Several species have been used as food in the past to a limited extent, usually in times of famine. Other species are important as forage and grain crops. Chief among these are common vetch (*V. sativa* subsp. *sativa*), narbon bean (*V. narbonensis*), woolly pod vetch (*V. villosa*) and bitter vetch (*V. ervilia*). Faba bean grain is also used for animal feed, especially in North Africa and Europe.

ICARDA has the largest collection of this species as well as a large regional (West Asia and North Africa) collection of other *Vicia* species. Other significant collections are maintained in Australia, China, Italy, Spain and Russia. *Vicia faba* and the forage *Vicia* species have orthodox seeds and large collections are available. Some of the other species of *Vicia* are being investigated for their potential for grain production. The partial outcrossing of faba bean causes problems in maintaining the identity of accessions in germplasm collections and methods have been proposed to overcome the difficulty, including pure-line collections and self-pollinated bulks.

Germplasm collections of *Vicia faba* and forage vetch have yielded good sources of resistance to biotic stresses such as chocolate spot, Ascochyta blight, stem nematodes, root-knot nematodes and rust. These sources have been used to develop improved cultivars of faba bean in the UK, France, Egypt, Iran and Australia. Much effort is still needed to overcome other biotic stresses, especially *Orobanche crenata*, a major limitation to faba bean production in Mediterranean areas. Good sources of autofertility have been found in the collections of faba bean.

There is still a need to collect landraces of faba bean in China, South America and to some extent in West Asia and North Africa before they are replaced by modern cultivars.

INTRODUCTION

The genus *Vicia* is divided into two subgenera that comprise approximately 160 annual and perennial species (Allkin et al. 1983). The most economically important species is *Vicia faba* L. It is the only species used currently for human food. Other species have been used as food in the past to a limited extent, usually in times of famine (Enneking 1995) including *Vicia ervilia* L. in Spain and Morocco and *Vicia articulata* Hornem. (black lentil) in Spain and Latin America. *Vicia sativa* L. subsp. *sativa* has been used recently as a lentil substitute in South Asia (Tate & Enneking 1992). Several species of *Vicia* are important as forage and grain crops. Chief among these are common vetch (*V. sativa* subsp. *sativa*), narbon bean (*V. narbonensis* L.), woolly pod

R. Knight (ed.), Linking Research and Marketing Opportunities for Pulses in the 21st Century, 623–633.
© 2000 *Kluwer Academic Publishers. Printed in the Netherlands.*

624

vetch (*V. villosa* Roth) and bitter vetch (*V. ervilia*). Faba bean grain is also used for animal feed, especially in North Africa and Europe.

World faba bean production in 1995 was 3.5 million tonnes from an area of 2.85 million hectares (FAO 1996). Faba bean production has declined in the period 1979 to 1993, mainly due to a reduction in area in China, Italy and France. China, is still the largest producer with 58% of the area and 62% of the production in the world. Next is Ethiopia with 8.3% of the area and 7.0% of the production. Other important countries are Morocco, Egypt, Brazil, Italy and Australia. Area and production statistics for the forage vetches are not readily available. A recent review (Rees 1993) showed the largest production to be in Europe, the former Soviet Union and the former Yugoslavia. Other areas with significant production were Morocco, Turkey, Spain and Sweden.

Faba bean is an Old World legume, referred to frequently by ancient writers (Hawtin & Hebblethwaite 1983). It is cultivated in Europe and the Mediterranean region. Throughout North Africa and West Asia (WANA) faba beans are an important component of the diet. It is sometimes referred to as the "poor people's meat." Faba beans were introduced to South America with European colonization and are mostly grown at high elevations where *Phaseolus* beans cannot be grown. Faba bean was more recently introduced to North America where it is not of significance and to Australia, where the area has increased dramatically in the past 10 years.

The breeding programs of faba bean and forage vetches are built upon the foundation of the germplasm collections. The first years of the program at ICARDA were devoted to collecting and assembling a large germplasm base. Selection was first among, and then within, locally adapted landraces. To date, four cultivars of faba bean and six cultivars of forage vetch have been released from landraces or landrace selections from the germplasm collections of ICARDA. Many of these are selections that have been purified for such traits as seed size, disease resistance, frost tolerance, and non-shattering.

DOMESTICATION, CENTERS OF ORIGIN AND DISTRIBUTION

The immediate ancestor of *V. faba* is not known, and Schafer (1973) suggested it originated from an extinct ancestor. *Vicia narbonensis* (2n=14) has received the greatest attention as the putative ancestor. However, with a different karyotype and chromosome number, *V. narbonensis* has never been crossed successfully with *V. faba*, and cannot be regarded as a direct ancestor.

Vicia faba was most likely domesticated in the region between Afghanistan and the eastern Mediterranean during the period 7000-4000 BC (Zohary & Hopf 1988). Cubero (1984) concluded its culture spread in four directions north to central Europe, northwest to Western Europe, west to the Mediterranean, and east to the Far East (India, China and Japan). The *minor* type was introduced to China in 100 BC (Tao 1981) and the *major* type in AD 1200 (Hanelt 1972)

CURRENT STATUS AND ORIGIN OF GERMPLASM COLLECTIONS

The major collections of faba bean in the world are listed in Table 1. ICARDA has the global mandate from the Consultative Group of International Agricultural Research Centers for crop improvement and genetic resources of faba bean. The largest germplasm collection is maintained at ICARDA, with other large collections in Australia, China, Italy and Russia. The country of origin of the germplasm held in ICARDA is presented in Table 2. Major holdings of other *Vicia* species are listed in Table 3. Most collections contain a large number of species, 34-78 (except for Spain, with only 5). The largest collections are at ICARDA and the Vavilov Institute in St.

Petersburg, Russia. The number of accessions per species of the ICARDA collection is given in Table 4. More than half the collection is *Vicia sativa*, with 2833 accessions. The majority are common vetch (*V. sativa* subsp. *sativa*, 2053 accessions). Most of the other species with large numbers of accessions are species for which ICARDA has breeding programs (*V. ervilia*, *V. hybrida*, *V. narbonensis*, *V. palaestina* Boiss., *V. pannonica* Crantz and *V. ervilia*).

Table 1 .Important faba bean collections.

Country	Organization	No. accessions
Syria	ICARDA	10,086
Germany	IPK	1,507
China	CAAS	1,500
Italy	IDG	1,461
Australia	ATFCC	1,393
Russia	VIR	1,270
Spain	CRF-INIA	1,080
USA	USDA	413

Table 2. Country of origin of the ICARDA faba bean ILB (International Legume Faba Bean) and BPL (Bean Pure Line) collections. The definition of , and method of producing BPLs is given in the section on Conservation.

Country of origin	ILB	BPL	Country of origin	ILB	BPL
Afghanistan	96	111	Latvia	1	1
Algeria	41	41	Morocco	741	182
Argentina	1	4	Mexico	7	14
Australia	9	29	The Netherlands	30	10
Austria	1	2	Nepal	98	1
Bangladesh	36	0	Oman	4	0
Belgium	1	0	Pakistan	34	50
Bolivia	0	2	Palestine	5	10
Bulgaria	14	2	Peru	49	47
Canada	294	9	Poland	42	21
Chile	2	6	Portugal	109	15
China	227	342	Romania	33	14
Colombia	43	95	Russia	3	2
Czech Republic	29	0	Spain	339	782
Cyprus	104	251	Sri Lanka	2	4
Germany	83	65	Soviet Union (Former)*	42	27
Ecuador	205	127	Sudan	115	44
Egypt	93	236	Sweden	6	11
Ethiopia	408	683	Syria	515	176
France	15	15	Tunisia	29	65
Greece	29	54	Turkey	209	212
Hungary	15	12	Ukraine	6	10
Indonesia	1	1	Uruguay	1	2
India	10	10	United Kingdom	83	98
Lebanon	36	88	United States	6	11
Libya	11	0	Unknown		
Lithuania	1	2			

*Specific Republic unknown

Fifteen percent of the faba bean accessions held at ICARDA are landraces collected by ICARDA and ALAD (Arid Lands Agricultural Development Program). However, almost the entire *Vicia* forage collection is of naturally occurring (wild) populations. Most accessions of faba bean and forage vetches in the collections are from the West Asia and North Africa (WANA) region (Tables 2 and 5). This is to be expected, as the WANA region is the center of origin and primary diversity for these temperate cool-season legumes (Zohary & Hopf 1988). This has led to the successful use of faba bean landraces from the collections and forage vetch ecotypes for direct release as cultivars for the WANA region.

Table 3. Major holdings of *Vicia* species (not including faba bean)

Country	Institution	No. species	No. accessions
Russia	VIR	56	5770
Syria	ICARDA	63	5526
Italy	IDG	78	2643
Spain	CRF-INIA	5	1390
Australia	ATFCC	70	1373
USA	USDA	34	1147
Germany	IPK	71	665

Table 4. Number of accessions of *Vicia* species in the ICARDA germplasm collection

Species	No. Acc.	Species	No. Acc.
Vicia aintabensis	9	*Vicia latifolia*	2
Vicia alpestris	1	*Vicia laxiflora*	6
Vicia altissima	3	*Vicia lunata*	1
Vicia anatolica	82	*Vicia lutea*	162
Vicia angustifolia	3	*Vicia melanops*	16
Vicia articulata	21	*Vicia michauxii*	34
Vicia assyriaca	1	*Vicia mollis*	22
Vicia balansae	2	*Vicia monantha*	197
Vicia barbazitae	1	*Vicia montbretii*	3
Vicia benghalensis	19	*Vicia montevidensis*	1
Vicia benthamiana	1	*Vicia multijuga*	1
Vicia bithynica	53	*Vicia narbonensis*	259
Vicia canescens	1	*Vicia noeana*	14
Vicia cassia	5	*Vicia onobrychioides*	4
Vicia cracca	20	*Vicia palaestina*	100
Vicia crocea	1	*Vicia pannonica*	136
Vicia cuspidata	54	*Vicia peregrina*	252
Vicia dichroantha	1	*Vicia pisiformis*	1
Vicia dionysiensis	4	*Vicia pulchella*	14
Vicia eristalioides	1	*Vicia qatmensis*	7
Vicia ervilia	324	*Vicia sativa*	2833
Vicia galeata	1	*Vicia sepium*	5
Vicia galilaea	4	*Vicia sericocarpa*	67
Vicia glareosa	6	*Vicia serratifolia*	4
Vicia grandiflora	15	*Vicia tenuissima*	1
Vicia hirsuta	39	*Vicia tetrasperma*	43
Vicia hyaeniscyamus	9	*Vicia tigridis*	1
Vicia hybrida	187	*Vicia truncatula*	1
Vicia hyrcanica	18	*Vicia unijuga*	1
Vicia johannis	58	*Vicia villosa*	342
Vicia kalakhensis	7	*Vicia* spp.	50
Vicia kokanica	1		
Vicia lathyroides	28	Total	5558

Table 5. Countries of origin for *Vicia* accessions held at ICARDA

Country	No. Acc.	Country	No.Acc.	Country	No.Acc.
Afghanistan	25	Georgia	17	Portugal	71
Albania	12	Greece	128	Paraguay	1
Algeria	221	Hungary	121	Romania	3
Armenia	23	India	1	Russia	34
Australia	34	Iran	50	Saudi Arabia	1
Austria	1	Iraq	16	South Africa	1
Azerbaijan	20	Italy	691	Soviet Union (Former)*	27
Belgium	18	Jordan	11	Spain	60
Bangladesh	2	Japan	41	Sweden	13
Bulgaria	56	Lebanon	42	Syria	1013
Canada	5	Libya	1	Tajikistan	36
Czech Republic	22	Lithuania	3	Turkmenistan	21
Cyprus	97	Latvia	2	Tunisia	18
German	91	Morocco	513	Turkey	1486
Denmark	2	Malta	18	Ukraine	11
Ecuador	9	Mongolia	4	United States	9
Egypt	15	The Netherlands	2	Great Britain	2
Ethiopia	1	Nepal	52	Uzbekistan)	38
Finland	2	Pakistan	38	Yugoslavia (Former)	9
France	63	Palestine	9	Unknown	28
Gabon	1	Poland	16		

*Specific Republic unknown

Recent collections of faba bean by ICARDA have been targeted to fill gaps in geographical areas where the crop is important and to search for germplasm with important traits, such as disease resistance. Disease resistance has been found in material from Ecuador and for heat resistance from Bangladesh (Table 6); see the section "Diversity for Stress Resistance." Also, a number of accessions have been added recently from China, which is not well represented in collections outside of China. Since 1991 over 1200 accessions of forage vetches have been collected by ICARDA, mostly in the WANA region.

Table 6. Recent collections of faba bean by ICARDA

Year	Country	No. accessions
1991	Portugal	95
1992	Libya	12
1993	Morocco	2
	Pakistan	11
1994	Morocco	15
1995	Bangladesh	6
	Nepal	93
	Morocco	2
1996	China	68
	Ecuador	108
1997	Ethiopia	23
	Bangladesh	30
	Spain	1
1991-1997	All	466

Conservation

Vicia species have orthodox seeds that are easily maintained in the active collection at 0°C at a relative humidity of 25%. Seeds are dried to a moisture content of about 6% before storage. The relatively large seed size of faba bean and to a lesser extent *Vicia narbonensis*, creates a problem when compared with other crops because of the large space needed for storage. The large size also results in a low multiplication rate that makes the rate of rejuvenation low. This is the reason for the slow progress in developing a base collection for faba beans at ICARDA.

Clean, healthy and viable seeds are needed for storage and distribution. At ICARDA a schedule of seed testing ensures that only pathogen- and virus-free seeds of faba bean are kept in the collection on a longer-term basis. Also, during multiplication of *Vicia* species in the field, there is routine monitoring for viruses and pathogens.

Another problem in the multiplication of the faba bean collection results from the species being partial out-crossed by insects. The cost of multiplication is increased by the measures that need to be taken when using insect-proof screenhouses, such as tripping to ensure seed set. A pure line faba bean collection has been developed at ICARDA (the bean pure lines, abbreviated to BPLs). These were developed through a "pre-breeding" process by taking randomly selected single plants to progeny rows in a cyclic manner, using insect-proof screenhouses to ensure selfing (Robertson, 1985).

ICARDA has signed an agreement with the Federal Institute of Agrobiology (FIA) in Linz, Austria for the duplication of ICARDA's International Legume Bean collection (ILBs), the BPL collection and for the forage *Vicia* species. To date, approximately 1500 accessions of faba bean and 1910 accessions of forage *Vicia* species have been duplicated.

Diversity for Stress Resistance

Selection in faba bean for resistance to chocolate spot (induced by *Botrytis fabae* Sard.), Ascochyta blight (induced by *Ascochyta fabae* Speg.), rust (induced by *Uromyces fabae* (Pers.) de Bary) and stem nematodes (*Ditylenchus dipsaci* (KÝhn) Filipjev) has followed a parallel course to that described for the development of homogeneous BPL accessions. The same cyclic single-plant progeny row system was used, but with selection for resistance after artificial inoculation with the disease, instead of taking random plants, as was done for the normal BPL accessions. Many useful sources of resistance were developed (Table 7). This material was tested internationally and several selections were found to have durable resistances for chocolate spot and Ascochyta blight (Hanounik & Robertson, 1987, 1988). The best sources of resistance to chocolate spot have come from the Colombian-Ecuador Andean Mountain region. Many sources of resistance to chocolate spot are also resistant to rust especially from this region. Recently (1996), a mission to Ecuador was targeted at collecting faba beans resistant to chocolate spot. This area is a "hot spot" with natural selection occurring for chocolate spot resistance. The germplasm collected shows high levels of resistance compared with germplasm that was collected from the provinces of Yunnan and Sichuan China, a collection made to fill a gap in geographical representation (Fig. 1).

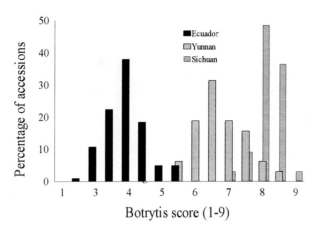

Figure 1. Chocolate spot resistance of recent collections of faba bean from Ecuador and China (1=least susceptible, 9=killed chocolate spot)

Table 7. Biotic stress resistances in faba bean

Stress	Accession
Chocolate Spot	BPL 110, 112, 261, 266, 710, 1179, 1196, 1278, 1821; ILB 2282, 2302, 2320, 3025, 3026, 3033, 3034, 3036, 3106, 3107
Ascochyta Blight	BPL 74, 230, 365, 460, 465, 471, 472, 646, 818, 2485; ILB 752
Rust	BPL 7, 8, 260, 261, 263, 309, 409, 417, 427, 484, 490, 524, 533, 539, 710, 1179
BLRV	BPL 756, 757, 758, 769, 1351, 1363, 1366, 1371
BYMV	BPL 756, 758, 1181, 1185, 1236, 1245

Screening of faba bean for Bean Leaf Roll Virus (BLRV) and Bean Yellow Mosaic Virus (BYMV) was initiated in 1976. Approximately 800 BPLs were screened for BLRV and 450 BPLs for BYMV. Eight accessions were resistant to BYMV, with a long latent period responsible for the resistance (Table 7). Additionally, high levels of resistance to BLRV were present in six BPLs. Two BPLs (BPL 756 and 758) from Afghanistan were resistant to both viruses.

To develop productive forage vetches, with stable performance, resistance to major stem and leaf diseases is necessary. Host-plant screening in *Vicia* species for resistance to powdery mildew (induced by *Erysiphe pisi* Syn. (syn. E. poylgoni D.C.)), Botrytis blight (induced by *Botrytis cinerea* Pers. ex Fr.) and Ascochyta blight (induced by *Ascochyta pisi* Lib.) have started recently. Several sources for resistance to these diseases appear to have been found but the results need to be confirmed. Root-knot nematode (*Meloidogyne artiellia* Frankland) can be a factor limiting the introduction of forage vetches in WANA (ICARDA 1985). At ICARDA, a screening program for this nematode was initiated in 1985. Field and laboratory techniques were developed (Abd El Moneim & Bellar 1993) and accessions with resistance identified (Table 8). Woolly-pod vetch, Selection 683 was highly resistant in the field and laboratory.

Table 8. Abiotic and biotic stress tolerance in forage *Vicia* species

Stress/Species	Accession
Cold	
Vicia sativa subsp. *sativa*	IFVIS 64, 288, 294, 303, 309, 313, 314, 315, 316, 317, 325, 326, 327, 328, 333, 334, 335, 336,339, 372, 377, 385, 391, 399, 403, 406, 407, 408, 416, 447, 3815, 3841
V. ervilia	IFVIE 225, 228, 240, 263, 541, 654
V. narbonensis	IFVIN 67, 749, 1142, 1143, 1147, 1148, 1150, 1152, 2623, 2644, 2663, 2696
Root knot nematode	
V. dasycarpa	IFVID 683, 2095, 1934
V. narbonensis	IFVIN 71
Orobanche crenata	
V. narbonensis	IFVIN 578, 577, 574, 573
V. sativa subsp. *sativa*	IFVIS 708, 715, 1361, 2541

Orobanche crenata Forsk. (an obligate parasitic weed) is the most limiting factor to faba bean production in WANA. Its seriousness is often hidden, because many times faba bean is not grown, since complete crop failure can result in areas of severe infestation. The breeding program at ICARDA, in collaboration with the Egyptian and Spanish national programs, has developed lines with resistance by selecting within progeny of crosses of Family 402 (from Egypt, later used to develop Giza 402) to a Spanish landrace (INIA 06), back-crossed to the Spanish landrace. The selected lines are adapted to the Mediterranean environment (Table 7). Within *V. narbonensis* and *V. sativa* there is considerable variation in the reaction to *O. crenata* with several lines being free of the parasite and others highly susceptible. Lines of *V. villosa* subsp. *dasycarpa* Tan with high levels of resistance were also found. Forage vetch lines resistant to *O. crenata* are listed in Table 8.

Drought can be a constraint to faba bean production in the Mediterranean region. Research at ICARDA has identified lines ILB 1814, 80S43856, and 80L90121 as more efficient in water use (Robertson & Saxena, 1993). Autumn cultivars of faba bean, grown in the Mediterranean region, are less tolerant of frost than the most resistant European genotype Cote d'Or (Lawes et al., 1983), which tolerates temperatures below -15° C. Forage vetch species are mostly resistant to cold and high levels of tolerance have been found. A breeding program with *V. sativa* subsp. *amphicarpa* has been started recently at ICARDA because of its persistence under severe climatic and grazing conditions (Christiansen et al. 1996).

Diversity for agronomic and quality descriptors

The BPL germplasm collection of faba bean was partially evaluated for the IBPGR/ICARDA descriptors (Anonymous 1985) and the results for 840 BPLs summarized in a catalogue (Robertson & El-Sherbeeny 1988). The descriptors evaluated included seed size, pods per plant, seeds per pod, seed shape, seed color, testa color, test pattern, cotyledon color, growth habit, seed yield, straw yield and protein quantity. The catalog gives lists of useful accessions for various descriptors (Table 9) and a statistical summary.

Table 9. Accessions with desirable values for various descriptors

Descriptor/desired state   Accessions (BPL)

Flowering earliness 180, 237, 247, 271, 280, 281, 286, 298, 548, 558, 561, 570, 589, 600, 609, 719, 726, 729, 1230

Lodging/resistant   185, 218, 376, 385, 545, 643

Height of pod/ tall   39, 40, 42, 48, 50, 53, 58, 60, 65, 68, 69, 70, 71, 72, 76, 85, 828, 836, 844, 894, 969, 972, 1262, 1351, 2443

Pods per plant/high 704, 787, 805, 978, 1703, 1732, 1739, 1758, 1766, 1827

Pod length/long   250, 251, 255, 318, 339, 415, 416, 455, 632, 890, 1089, 1102, 1106, 1129, 1157. 1159, 1868, 1872, 1876

Seed yield/high   22, 24, 46, 47, 52, 53, 65, 77, 133, 348, 379, 393, 442, 444, 451, 596, 725, 787, 936, 999, 1015, 1043, 1046, 1120, 1160

Autofertility/high   17, 42, 43, 44, 50, 69, 70, 101, 296, 543, 555, 568, 569, 579, 580, 591, 614, 698, 703, 722, 1029, 1145, 1151, 1342, 1815

Protein content/high 61, 171, 491, 495, 526, 527, 528, 603, 737, 976, 1014, 1386

Pod-shattering is common in *Vicia sativa* subsp. *sativa* and restricts the use of this legume for producing feed. The germplasm collection has been screened to provide non-shattering genotypes (Abd El Moneim 1993). Three accessions were found and selections were made from them to be used as genetic stocks in breeding programs (IFVI 1361, 1416 and 2014). The accessions had undesirable traits such as late flowering, late maturity, and low herbage yield. Crosses to desirable genotypes, followed by selection, has resulted in five non-shattering genotypes (Sel NS 2565, 2558, 2557, 2914, and 1448) with earliness and an erect plant habit (Abd El Moneim 1993).

The faba bean pure-line collection was screened for autofertility (Robertson & El-Sherbeeny, 1995) defined as the ability to self-pollinate in the absence of pollinating insects (Bond & Pope 1987). The mean seed index (SI, measurement for autofertility) among the 840 BPL accessions was 0.54 with a standard error of 0.01. A large number of BPL accessions failed to set seeds without tripping and are described as autosterile. There were many accessions with high autofertility (Table 9). There were marked differences among countries of origin for autofertility, with Egypt having the highest and the former USSR the lowest autofertility. However, most countries expressed a large range, suggesting that breeders should first look in their own material for this trait before searching the exotic germplasm. In general, there was a higher level of autofertility in Middle Eastern countries that may be due to natural selection for non-dependency on pollinators for seed set.

The germplasm collection of faba bean was also evaluated for seed protein. Seed protein content varied from 18.0 to 31.0% with a mean of 24.0% (Robertson & El-Sherbeeny, 1992). Accessions with a high protein content are listed in Table 9. There were no correlations between protein content and seed yield or its components. There was moderate variation among countries for mean protein content: the former Yugoslavia had the lowest content with 22.5% and Japan the highest with 26.6%. However, both had comparable ranges. Regional groupings of countries also had similar means and ranges for protein content.

CURRENT GAPS AND PRIORITIES FOR COLLECTING IN FUTURE

One of the major deficiencies in the faba bean and forage vetch collections is that for many of the accessions there are no passport data. Even the country of origin is not known for more than 2000 ILB accessions, approximately 21% of the collection. For many other accessions there is little information available. Efforts are being made to rectify as much as possible the lack of passport data in order to make the use of the germplasm more efficient.

A major gap that needs to be addressed in the ICARDA faba bean collections is the relatively small number of accessions from China, especially from the more remote areas. China, is a major producer. Other areas that are under-represented are in India, Pakistan and Nepal, although collections were made in Nepal in 1995. In the WANA region there is a need for more collecting in Algeria, Egypt and Yemen.

In regard to desirable traits there is a major gap in the faba bean and forage vetch germplasm collections in terms of sources of resistance to *Orobanche crenata*. Another objective is to broaden the genetic base of the sources of resistance to chocolate spot. Because of the importance of this disease and of rust in faba bean, there is a need for further collecting in the Andean region. There is also a need to find faba bean germplasm with virus resistance and heat tolerance.

FUTURE PROSPECTS

There is considerable genetic variation in the *Vicia faba* germplasm and breeding programs in WANA are beginning to make use of this variation as they become established. In China there is renewed interest in the crop and in restoring to previous levels the area of the crop. In Australia there has been a dramatic increase in faba bean production with the full potential not yet reached.

There are new plant types, which may alleviate the problem of excessive vegetative growth- with the accompaning drop of flowers and young pods- and the high lodging rate. The determinate growth genes and independent vascular supply types (non-branched vascular traces in the raceme) are now available in Mediterranean-type germplasm following introgression with landraces from the region

Disease-resistant cultivars are available from breeding programs that have made use of the resistant sources in germplasm collections. Sources of resistance for two or more diseases are now becoming available. Recent results in breeding for resistance to *Orobanche crenata* offer the hope of controlling this noxious weed. Biotechnological techniques offer the possibility of inserting genes for herbicide resistance into faba bean with the possibility of controlling *Orobanche crenata* with herbicides.

The genetic potential of faba bean is high and yields of 5 to 6 t ha^{-1} are reported. By using the variation present in the collected germplasm, solutions can be found to many of the problems associated with growing faba beans. They offer the promise of increasing yields and overcoming the instability of yield caused by biotic and abiotic stresses.

In the past five years there has been evaluations of the forage vetch germplasm and catalogues will be produced in the next few years. In future biochemical and molecular markers will be used to assess genetic diversity in germplasm collections, to optimize the collection and conservation strategy, and to plan future research in areas such as *ex situ* collections, *in situ* conservation and core collections.

References

Abd El Moneim, A.M. 1993. *Plant Breeding* 110: 168-171.

Abd El Moneim, A.M. & M. Bellar. 1993. *Nematologia Mediterranea* 21: 67-70.

Allkin, R., T.D. McFarlane, R.J. White, F.A. Bisby & M.E. Adey. 1983. *Database Project. Publication* No. 6.

Anonymous. 1985. *Faba Bean Descriptors.* AGPG: IBPGR/85/116. IPGRI, Rome, Italy.

Bond, D.A. and M. Pope. 1987. *Journal of Agricultural Science, Cambridge* 108: 103-108.

Christiansen, S.A., Abd El Moneim, P.S. Cocks & M. Singh. 1996. *Journal of Agricultural Science, Cambridge* 126: 421-427.

Cubero, J. I. 1984. In: *Genetic Resources and Their Exploitation-Chickpeas, Faba Beans and Lentils* pp. 131-144 (Witcombe, J.R. & W. Erskine, eds.). Martinus Nijhoff, The Hague, The Netherlands.

Enneking, D. 1995. *Centre for Legumes in Mediterranean Agriculture* (CLIMA) Occasional Publication No.6. University of Western Australia, Nedlands, WA.

FAO. 1996. FAO *Agricultural Production Yearbook 1995*: Vol. 49. FAO, Rome, Italy.

Hanelt, P. 1972. *Kulturpflanze* 20:209-223.

Hanounik, S.B. & L.D. Robertson. 1987. *Plant Disease* 72: 696-698.

Hanounik, S.B. & L.D. Robertson. 1988. *Plant Disease* 73: 202-205.

Hawtin, G.C. & P.D. Hebblethwaite. 1983. In: *The Faba Bean (Vicia faba L.): A Basis for Improvement,* pp 3-22 (P.D. Hebblethwaite, ed.). Butterworths, London, UK.

ICARDA, 1985. *Annual Report 1985.* pp. 293-295 ICARDA, Aleppo, Syria.

Lawes, D.A., D.A. Bond, & M.H. Poulsen. 1983. In: *The Faba Bean (Vicia faba L.): A Basis for Improvement* pp. 23-76 (P.D. Hebblethwaite, Ed.). Butterworths, London, UK.

Rees, R. 1993. *Proceedings of the Vicia/Lathyrus Workshop*, 22-23 September 1992, Perth, Western Australia. Cooperative Centre for Legumes in Mediterranean Agriculture, Occasional Publication No. 1, Perth, WA.

Robertson, L.D. 1985. In: *Faba Beans, Kabuli Chickpeas, and Lentils in the 1980s, Proceedings of an International Workshop* pp. 15-21 (M.C. Saxena and S. Verma, eds.), 16-21 May 1983, Aleppo, Syria. ICARDA, Aleppo, Syria.

Robertson, L.D. & M.C. Saxena. 1993. In *Breeding for Stress Tolerance in Cool-Season Food Legumes* pp. 37-50 (eds.K.B. Singh and M.C. Saxena,). John Wiley & Sons, Chichester, UK.

Robertson, L.D. & M. El-Sherbeeny, 1988. *Faba Bean Germplasm Catalog: Pure Line Collection.* ICARDA, Aleppo, Syria.

Robertson, L.D. and M. El-Sherbeeny. 1991. *Euphytica* 57: 83-92.

Robertson, L.D. & M. El-Sherbeeny. 1992. *Journal of the Science of Food and Agriculture* 58: 193-196.

Robertson, L.D. and M.H. El-Sherbeeny. 1995. *Genetic Resources and Crop Evolution* 42: 157-163.

Schafer, H.I. 1973. *Kulturpflanze* 21: 211-273.

Singh, K.B., W. Erskine, L.D. Robertson, H. Nakkoul & P.C. Williams. 1988. *Journal of the Science of Food and Agriculture* 44: 135-142.

Tao, Z.H. 1981. *FABIS Newsletter* 3: 24.

Tate, M.E. & D. Enneking. 1992. *Nature* 359: 357-358.

Zohary, D. & M. Hopf. 1988. *Domestication of plants in the Old World.* Clarendon Press, Oxford.

Lupinus ssp: conserved resources, priorities for collection and future prospects

W.K.Swiecicki[1], B.J. Buirchell[2] and W.A. Cowling[3]

1 Institute of Plant Genetics, Polish Academy of Sciences, 60-469 Pozna , Poland; 2 Agriculture Western Australia, 3 Baron-Hay Court, South Perth, Western Australia, Australia 6151; 3 Centre for Legumes in Mediterranean Agriculture, University of Western Australia, Nedlands, Western Australia, Australia 6907

Abstract

Genetic resources of lupins are discussed on a basis of information obtained from institutions with active collections and the taxonomy of the genus *Lupinus* is considered (including two centers of origin, the New and Old World). In drawing any conclusions on resources we have taken account of the agricultural importance of the four lupin crops.

The paper deals with lupin distribution and environmental conditions, collections/gene banks and databases, collected genetic resources, genetic diversity in lupins and future collecting missions.

The four most important issues were listed as:
- the identification and analysis of duplicates in collections,
- free access to South American lupin germplasm,
- mapping of collected genetic resources (when collection site information is available) on a map of a species' area of distribution,
- supplemental collecting missions

INTRODUCTION

The species of the genus *Lupinus* are distributed in two centers of origin, widely separated geographically and in terms of their evolution and taxonomy. One centre is in the Mediterranean basin (the Old World lupins) the other extends through north and south America (the New World lupins).

Over 200 wild species of the New World group - annual and perennial - have small seeds and 2n=48 as the most common chromosome number (exceptionally 2n=96 or 36). A big-seeded exception is *L. mutabilis* (2n=48), a crop in some South American countries. It has a long history of cultivation which parallels the domestication of *L. albus* by early farmers in the Mediterranean region (Hondelmann 1984). The high oil (18-20%) and protein content (40%) of the seeds have created an interest in adapting the species to European conditions (Römer 1984, Sawicka 1994). The ornamental Russel lupin is derived from the interspecific crosses between North American species and is of international importance in horticulture. Several perennial species such as *L. arboreus* and Russel lupins have been introduced to other countries for erosion control and soil improvement.

There are two sub-regions in South America: the Andean and Atlantic sub-regions. There are reports on the similarity of the lupins from the Atlantic sub-region to Old World lupins (eg *L.gibertianus* complex to *L. angustifolius* or *L. bandelierae* and *L. paranensis* to *L. albus* and *L. luteus*, respectively) (Dunn 1984). Unfortunately, these suggestions are not well documented and the limited availability of germplasm has restricted studies. Basic problems are the difficulty of access to South American resources and the need to update the taxonomy of these lupins. For example, for about 500 taxa (species and subspecies) over 1700 different names have been used in the literature (Dunn 1984, Sholars - unpublished information in the Lupinet). The taxa have been poorly studied and collected for conservation in gene banks. This does not imply they have no potential value.

New World lupins have a wide range of distribution (from the Andes to the Arctic) and different plant habits (from small plants - e.g. *L. bracteolaris*, to trees 4.5 m high). Some are adapted to extreme

635

R. Knight (ed.), Linking Research and Marketing Opportunities for Pulses in the 21st Century, 635–644.

environmental conditions. For example, two weeks after the explosion of the Mount St. Helens volcano in 1981, among three plants germinating from the ash layer was a lupin (Findley 1981). In permanently frozen areas of Alaska seeds of *L. arcticus* 10,000 years old were found, which were capable of germination - (Porsild et al. 1967).

In the Mediterranean basin 12 lupin species with different distributions and numbers of chromosomes have been described (Gladstones 1974) (table 1). They were divided in two groups - rough-seeded (7 species) and smooth-seeded (5 species). Two new, smooth-seeded species have been included: an artificial hybrid between *L. hispanicus* subsp. *hispanicus* and *L. luteus*, i.e. *L. (x) eurohybridus* (Swiecicki 1985) and *L. anatolicus* - the result of a collecting mission (Swiecicki et al. 1996). Some doubts exist about *L. anatolicus*. A number of distinguishing characters were defined but some characters fit within the range of variation present in the rough-seeded *L. pilosus* (Clements et al 1996). Differences in characters, crossing abilities and cyto-systematic relationships have been described among the rough-seeded species (Carstairs et al. 1992).

Table 1. The Old World centres of Lupin distribution

SMOOTH-SEEDED:	DISTRIBUTION AREAS	FUTURE EXPEDITIONS
L. hispanicus (2n=52)	Inland Spain and Portugal	
L. micranthus (2n=52)	Mediterranean basin	Any area
L. anatolicus	Western Turkey	
L. (x) eurohybridus	artificial hybrid	
L. albus (2n=50)	Mediterranean basin, northern and eastern African countries, Azores, Madiera, Canary Islands, Greece	Albania, Turkey, Georgia, Greece, Mediterranean islands with alkaline soils eg Corsica, Malta, Sicily (Palermo), the Atlantic islands eg Verde Islands
L. angustifolius (2n=40)	Spain, Portugal, Morocco, Tunisia, Italy and Sicily	Some more of the Greek and Turkish Islands; Morocco around Demnate, Amizmiz, the Anti-Atlas Mountains and north-west of the Atlas Mountains; Algeria, Libya, Tunisia, Israel and Lebanon *L.linifolius* types from Sicily
L. luteus (2n=52)	Iberian Peninsula, coastal Israel and Lebanon, Madiera, Azores Islands	Sardinia, Sicily, south Italy, Rossano Cosenzo- Regio di Calabria (improved yield). Also Israel, Lebanon, Syria (short vegetation) - so called subsp. *orientalis*
Rough-seeded		
L. atlanticus (2n=38)	Morocco	The Anti-Atlas Mountains north west of Tafraoute, south from the Kerdous Plateau towards Ifrane; around Amizmiz in the Atlas Mountains, south west of Amizmiz, area around Demnate, into the mountains between Demnate and Beni Mellal
L. cosentinii (2n=32)	coastal lowlands of Western Morocco, south-west of Spain and Portugal, eastern Tunisia, Sicily, Corsica, Sardinia	Any area
L. digitatus (2n=36)	mountainous of southern Sahara	The Senegal River Valley. Also southern Algeria
L. palaestinus (2n=42)	central and southern plains of Sharon and Philistia, Coast of Israel, Sinai Peninsula	Israel and the Sinai Peninsula mountains
L. pilosus (2n=42)	mountainous regions of Crete Turkey, Greek Isles, Israel, Syria	Lebanon, Israel and Jordan. The Mediterranean coastal areas of Turkey and the nearby islands
L. princei (2n=38)	highlands of Kenya, southern Ethiopia, northern Tanganyika	Highlands of Kenya, Ethiopia Tanzania
L. somaliensis	herbarium specimen	The mountain range in northern Somalia

Three lupin crops: *L. albus, L. angustifolius* and *L. luteus* have smooth seed coats. The white lupin was used by farmers in ancient Greece and became well established in the Roman Empire (Gladstones 1976). At present, there are breeding programs and large areas of these crops in some countries. They are valuable in crop rotations and the grain plays a role in human and animal nutrition (Gladstones 1986, Swiecicki 1986). Cultivars of these species are well adapted to modern agriculture, but further improvements are necessary if they are to compete economically with other crops. There are many examples of germplasm collections and their use in lupin breeding. The importance of collection, documentation, evaluation, conservation and utilization of wild and landrace lupin species was recognized by early breeders in Germany, Holland, Poland and Russia. Early maturity in *L. albus* and *L. luteus* was found in landraces from Palestine and Egypt and resistance to powdery mildew and Fusarium wilt in *L. luteus* from Portugal (Hackbarth and Troll 1957). Very important for lupin domestication and use were genes controlling low alkaloid content and thermo-neutrality (Swiecicki and Swiecicki 1995). Lupin production in the 1990s in Australia, Belarus, Chile, Poland, Russia, South Africa, Ukraine and the USA is based on improved varieties arising from crosses and shared genetic material. For further lupin improvement to occur, for example to overcome the anthracnose problem, well analysed and described collections will be needed,. The research and breeding interest in the four lupin crops is the reason these species dominate in collections.

RANGE OF DISTRIBUTION AND ENVIRONMENTAL CONDITIONS

The distribution of the Old World lupins is well documented - table 1 (Gladstones 1974; Buirchell and Cowling in press). Material from the Mediterranean countries was collected in the 1930s and 1940s (Hackbarth and Troll 1957) 1960s and 1970s (Gladstones1973, 1976; Lehman and Hammer 1978) and 1980s (Mota et al. 1982; Simpson and Mc Gibbon 1982 a,b; Cowling 1986; Papineau and Huyghe 1989; Buirchell 1992; Hammer and Perrino 1995). Collected resources have been exchanged freely among researchers and breeders internationally and duplicates could exist in different institutions.

The term "Mediterranean lupins" covers a broad diversity considering the range of distribution of the different species. *L. albus*, *L. angustifolius* and *L. micranthus* occur almost continuously around the Mediterranean Sea. *L. luteus*, *L. hispanicus*, *L. digitatus* and *L. pilosus* are less widely distributed while *L. cosentinii*, *L. princei*, *L. palestinus* and *L. atlanticus* occur on restricted areas, sometimes widely separated (see *L. cosentinii*, Gladstones 1974). Considering these distributions and cyto-systematic differences, the possibility of natural crossing among the species is interesting. Artificial crossing of the crop species has been attempted. Some have not been successful, for example those made to transfer earliness from *L. angustifolius* to *L. albus*. But others between *L. hispanicus* and *L. luteus* have produced fertile hybrids depending on the direction of the cross and the particular parents used (Swiecicki 1985). Interesting work on this subject was carried out on rough-seeded lupins (Carstairs et al. 1992). The rough seeded lupins were divided into three groups based on their chromosome number and size and inter-specific crossing ability. The groups were Princei - in equatorial Africa (*L. princei*), Atlanticus - in north-western Africa (*L. atlanticus*, *L. digitatus* and *L. cosentinii*) and Pilosus - in the eastern Mediterranean basin (*L. pilosus* and *L. palestinus*).

The lupin species grow in clearly differentiated environmental conditions and the environmental preferences of the material in the Australian Lupin collection has been documented for parameters such as altitude, pH and rainfall:

Collection site altitude

L. angustifolius and *L. albus* grow from sea level near the Mediterranean coast to inland mountain ranges (Table 2). Landraces of *L. albus* were selected by early farmers to grow in diverse environments, including the Canary and Azores islands in the Atlantic ocean and in north and east Africa. The highest elevation recorded for landraces of *L. albus* was 2460 m in the highlands of Kenya. *L. luteus* is restricted to altitudes below 1000 m, while *L. hispanicus* and *L. micranthus* extend into the mountains to 1350 m.

Table 2. Altitude of collection sites for wild and landrace *Lupinus spp.* in the Australian Lupin Collection

Species	Min	Mean	Max	SD	n
L. albus	1	480	2460	518	94
L.angustifolius	1	427	1800	323	610
L. atlanticus	460	985	1630	248	37
L. cosentinii	10	138	1015	206	45
L. hispanicus	149	745	1250	313	47
L. luteus	10	166	500	163	25
L. micranthus	10	499	1350	474	16
L. mutabilis	2200	2807	3500	399	7
L. pilosus	3	231	1100	258	44
L. princei	2100	2280	2460	255	2

L. cosentinii is predominantly found at lower altitudes and on coastal plains. There have been a few collections from higher altitudes in Morocco, and these may be valuable as sources of frost tolerance. *L. pilosus* is found from the low coastal regions to high altitudes (1100m) in the mountains, especially in Crete. *L. atlanticus* only grows in the mountainous areas of Morocco between 450 and 1630 m. *L. princei* is restricted to the eastern highlands of Africa at higher altitudes, on average, than for any other wild lupin. It is

638

found on Mt Kenya at 2200 to 2600 m. *L. digitatus* occurs in the mountainous regions of the southern Sahara and the valleys of the Nile and Senegal Rivers over a range of altitudes (Gladstones 1974). *L. palaestinus* grows in coastal plains and into the mountains of Israel at altitudes mostly less than 1000m.

Collection site pH

L. angustifolius, *L. luteus* and *L.hispanicus* are generally found on acidic soils (Table 3) although some accessions of *L. angustifolius* and *L. luteus* have been found on alkaline soils that do not contain free lime. *L. albus* has been found on a range of soils from pH 5.0 to 9.5. The *L. albus* collection is biased by the large number of Egyptian accessions from high pH soils. The soils in the Nile Valley are very alkaline and yet *L. albus* grows well in this region. It is also found on limestone-derived soils in Greece (Simpson and McGibbon 1982). *L. micranthus* is found across a range of pH from 5.5 to 7.5.

Table 3. Soil pH of collection sites for wild and landrace *Lupinus spp.* in the Australian Lupin Collection

Species	Min	Mean	Max	SD	n
L. albus	5.0	8.25	9.50	1.24	172
L. angustifolius	4.20	6.73	9.00	0.72	445
L. atlanticus	6.50	7.82	9.50	0.78	35
L. cosentinii	6.20	7.64	9.00	0.70	37
L. hispanicus	4.50	5.90	7.00	0.49	26
L. luteus	5.50	6.46	8.5	0.85	22
L. micranthus	5.50	6.31	7.50	0.55	27
L. mutabilis	6.50	6.75	7.00	0.35	2
L. pilosus	5.50	7.60	9.50	1.01	46
L. princei	-	5.7	-	-	1

L. pilosus, *L. atlanticus* and *L. cosentinii* are found on soils ranging from slightly acid (pH 5.7) to highly alkaline (pH 9.5). The only accession of *L. princei*, for which a soil pH was recorded, came from soils of pH 5.7 in the highlands of Kenya. No soil pH data are available for *L. digitatus* although it has been reported to come from calcareous fluvisol type soils. The soils of the Nile Valley, where *L. digitatus* has been reported to grow in the past (Gladstones 1974), are highly alkaline (pH 9.0). *L. palestinus* comes from the alkaline soils of coastal Israel. It is clear that rough-seeded lupins are generally more adapted to alkaline soils than the smooth-seeded lupins.

Collection site rainfall

Most lupin species grow in environments with at least 300 mm average annual rainfall, with the exception of *L. angustifolius*, *L. cosentinii* and *L. atlanticus* which extend to lower rainfall sites (Table 4). The latter two species, along with *L. palestinus*, and presumably *L. digitatus*, occur mainly where rainfall is less than 600 mm. The remaining species are normally found in rainfall regions above 600 mm with some collections of *L. pilosus* at sites with up to 2,000 mm annual rainfall.

Table 4. Rainfall of collection sites for wild and landrace *Lupinus spp.* in the Australian Lupin Collection (from Cowling 1994)

Species	Min	Mean	Max	SD	n
L. albus	350	562	1200	189	36
L. angustifolius	200	643	1500	222	351
L. atlanticus	200	421	650	116	35
L. cosentinii	100	412	700	135	36
L. hispanicus	450	766	1500	215	44
L. luteus	300	645	1100	229	18
L. micranthus	525	701	1000	181	11
L. mutabilis	500	1150	1700	493	4
L. palaestinus	500	566	600	58	3
L. pilosus	350	675	2000	331	33
L. princei	584	792	1000	294	2

LUPIN COLLECTIONS/GENE BANKS AND DATA BASES

Most collecting of crop genetic resources is initiated by breeding institutions interested in using the plant material. Lupin resources are not as large as those for some cereals, or even peas, and there is a need for improvement in the documentation, evaluation and conservation procedures to increase the value of the collections to users. The International Plant Genetic Resources Institute and the International Lupin Association have played an important role in the coordination of lupin collections. At the Vth International Lupin Conference in Pozna, Poland collection data bases were initiated and managers for 4 lupin crops were selected (Swiecicki 1988; Swiecicki and Leraczyk 1994). Work on the European Legume Collection Databases (including lupins) was started at the First Meeting of the Legume Working Group, organized by IPGRI - ECP/GR (1995, Copenhagen/Denmark). Workshops were arranged with "Collection Databases on the Internet" as the main subject (Lipman et al. 1996).

Information on the European Lupin Collection Databases was presented at the VIIIth International Lupin Conference in 1996 and updated version on EUCARPIA Gene Resources Section Symposium. 1996) (Swiecicki and Leraczyk, in press). To represent the World Lupin Gene Resources in working collections it is sufficient to add to the European Database just two centers: from Australia and the USA. In table 5 a list of 16 collections in 14 countries is given together with the collection acronyms (Serwinski 1987) Some of these collections are in government institutions, and could be considered national collections.

Table 5. Institutions and curators of *Lupinus* collections & passport data. Listed are the Country and Acronym, the Institution (Curator) & the number of accessions.

AUS ARGIWA - Agriculture Western Australia, 3 Baron - Hay Court, South Perth, Australia 6151. E-mail: wcowling@infotech.agric.wa.gov.au (Dr. Wallace Cowling) - **3516 accessions**

BGR IPGRSAD - Institute of Plant Genetic Resources, BG-4122 Sadovo, Bulgaria.
Fax: +359 32 270270 (Siyka Angelova) - **221 accessions**

CZE SUMPERK - Research, Breeding and Services (AGRITEC, Ltd.), Zemedelska 16, 787 01, Sumperk-Temenice, The Czech Republic. Fax: +42-649-5975 (Miroslav Hybl) - **60 accessions**

DEU BGRC - Institut f r Pflanzenbau (FAL), Bundesallee 50, D-38116 Braunschweig, Germany.
Fax: (0531) 596 365 (Dr. Lothar Seidewitz) - **1983 accessions**

DEU GAT - Institut f r Pflanzengenetik und Kulturpflanzenforschung, Corrensstrasse 3, D-06466 Gatersleben, Germany.
E-mail: knupffer@ipk-gatersleben.de (Dr. Helmut Knupffer) - **829 accessions**

ESP INIACRF - Instituto National de Investigacion y Tecnologia Agraria y Alimentaria (INIA), Centro de Recursos Fitogeneticos (CRF), Autovia de Arag n, Km. 36 - Finca La Canaleja 28800 Alcal de Henares, Madrit, Spain (Federico Valera Nieto) - **1621 accessions**

GBR RNG - University of Reading, Dept. of Agri. Botany, PO Box 221, Reading RG6 6AS, UK.
E-mail: P.A.Mushtaq@reading.ac.uk (Dr. A.R. Mushtaq) - **1162 accessions**

HUN RCA - Institute of Agrobotany, H-2766 Tapioszele, Hungary.
Fax: (36)-53-380-072, E-mail: holly@mars.iif.hu (Dr. L szl Holly) - **255 accessions**

ISR IGB - Israeli Gene Bank, The Volcani Center, P.O.Box 6, Bet Dagan, Israel.
Fax: 972-3-9669642 (Myra Manoah) - **186 accessions**

ITA PLAPORT - Universita' Degli Studi Di Napoli Federico II, Departimento di Scienze Agronomiche e Genetica Vegetale, Via Universit , 100-800055, Portici, Italia.
Fax: (081) 27 6254 (Prof. Luigi Postiglione) - **39 accessions**

NLD CGN - Centre for Plant Breeding and Reproduction Research (CPRO-DLO), Centre for Genetic Resources, Droevendaalsesteeg 1, P.O.Box 16, NL-6700 AA Wageningen, The Netherlands.
Fax: 31. 83 70. 18094 (Loek J.M. van Soest) - **69 accessions**

POL WTD - Plant Breeding Station, Wiatrowo, 62-100 W growiec, Poland. Fax: 48/67/621 875
E-mail: wswi@igr.poznan.pl (Dr. Wiktor Swiecicki) - **1024 accessions**

PRT BPGV - BPGV/DRAEDM, Quinta dos Peoes, Gualtar 4710 Braga, Portugal (Rena Martins Farias) - **71 accessions**

PRT OEIRAS - Estacao Agronomica National, Quinta do Marques, 2780 Oeiras, Portugal.
Fax: 01/442 0867 (E. Bettencourt) - **951 accessions**

SLO PIEST - Research Institute of Plant Production, Bratislavska 122, 921 68 Pie tany, Slovakia.
Fax: 0042/838/263 06, 237 69 (Dr. Lubor Labaj) - **19 accessions**

USA DAPUL - USDA-ARS, Plant Introduction, RL 194, 59 Johuson Hall, WSU Pullman, WA 99164-6402, USA E-mail:w6cs@ars-grin.gov (Dr Chuck Simon)

The most common descriptors used in collection data bases are: accession number, name, origin country and botanical classification. Less common is information on origin, donor, breeder and pedigree. Not all collections describe a sample's status (wild line, land race, cultivar, breeder's material, mutants, etc.) although this would be useful information

LUPIN GENETIC RESOURCES

Table 6 lists the lupin collections and gene banks. More than 90% of all accessions are protected in 6 institutions. About 84% of resources belong to 4 lupin crops: *L. albus, L. angustifolius, L. luteus* and *L. mutabilis*; the remainder are wild taxa. This indicates that most institutions are interested in lupins as crops. The dominant species in an institution usually reflects the major breeding objectives of that country. Many accessions of the crop species are wild or landrace types (above 80%) but some European institutions have collected cultivars and breeding material almost exclusively.

Table 6 Lupin collections

Lupinus	AUS AGRI WA	BUL IPGR SAD	CZE SUMPE RK	DEU BGRC	DEU GAT	ESP INIAC RF	GBR RNG	HUN RCA	ISR IGB	ITA PLA PORT	NLD CGN	POL WTD	PRT BPGV	PRT OEI RAS	SLO PIEST DA	USA PUL	Total
albus	775	102	28	137	207	571	736	52	45	27	13	302	62	295	10	317	3679
angustifolius	1756	17	9	400	275	553	76	44	42	11		299		246	4	182	3914
luteus	207	18	22	252	125	255	94	68	59	1	56	345	9	262	5	68	1846
anatolicus												1					1
atlanticus	71			18								8				4	101
cosentinii	239			32	6	14	2		1			7		23		5	329
digitatus	3											2			3		8
(x)eurohybridus												3					3
hispanicus	74			90	63	196	5					14		112		50	604
micranthus	43			3	19	12	6	7				2		10			102
palaestinus	10			3					1			4					18
pilosus	146			6	11		14		6			7				11	201
princei	6																6
mutabilis	123	57		998	26	20	194	7	4			14				79	1522
New World (other)		27		37	68			76	26			16		2		285	537
Lupinus sp. (unident.)	63		1	7	29		35	1	2					1		46	185
Total	3516	221	60	1983	829	1621	1162	255	186	39	69	1024	71	951	19	1050	13056

Just 4% of accessions belong to the New World wild species. First of all they were collected in DAPUL (285 accessions for 51 different species; the information is available from http://www.ars.grin.gov/npgs/) but also in RCA, GAT, BGRC, IPGRSAD and WTD. The worst represented in collections is material of the atlantic subregion, e.g. single entries of *L. bracteolaris*. Lupins with simple leaves as well as from *L. gibertianus* - *L. bandeliere* - *L. paranensis* group are not protected. It should be taken into account that many of the New World species are perennial and regeneration of their genetic resources is difficult in countries with Mediterranean climates.

Among wild taxa of the Old World the worst represented is *L.digitatus* (8 accessions in AGRIWA, WTD and DAPUL), *L. princei* (6 accessions in AGRIWA) and *L. palaestinus* (18 accessions in AGRIWA, BGRC, IGB and WTD). There are no entries of *L. somaliensis* in the World Lupin Resources

Some taxa, particularly of the crops appear well represented but passport data suggest a high level of duplication among centers. An analysis of these duplicates could be the next step in the development of a common data base of collections. Some centers have passport information on the collection sites. This information could be used to relate the genetic resources to maps of the species' distribution (Gladstones 1974)

GENETIC DIVERSITY IN LUPINS AND FUTURE COLLECTING MISSIONS

L. albus

White lupin is divided in two taxa: var. *albus* (cultivated forms) and var. *graecus* (the wild ancestor) (Gladstones 1974). *L. albus* var. *graecus* occurs on mainland Greece and the Greek islands. But Perrino et al. (1984) also described this taxon in Calabria, southern Italy. It is a late flowering annual, with small, non-permeable seeds, coloured red or mottled red. Pods are freely shattering with bitter seeds. Its capacity to set pods on the mainstem, could be used in breeding programs to increase this character in cultivars.

For wild and cultivated collections of *L. albus* (from the Iberian Peninsula, the Balkans, the Nile Valley and Turkey) four geographical races were distinguished (Buirchell and Cowling, in press).

The Iberian race has a wide variation of morphological features. Two types of local population could be distinguished. The first, from the Azores, Catalonia, Cadiz and Central Portugal, is characterised by large leaves, pods and seeds. The second, cultivated in the northern provinces of Palencia and Leon, is distinguished by small leaves, pods, seeds and a short time to flowering,

The Nile Valley race has smaller and fewer main-stem leaves, as well as a shorter rosette stage. The synonymous name *L. termis* Forskal was used in the past,

The Turkish race seems to form a homogenous group with small pods, seeds and leaves, few main-stem leaves, few primary pods, more secondary and numerous tertiary and higher order level pods. The earliest flowering types are from this region and they lack a vernalization response. Populations with pink flowers are also characteristic of this group,

The Balkan race includes both wild and cultivated populations with short stature and abundant primary and secondary pods. Wild populations of var. *graecus* are characterized by shorter stems, deep blue flowers, shattering pods and coloured seeds.

Recent collections (Huyghe et al. 1990) have also shown wide genetic variation in characters of possible value for cultivar improvement, eg in flowering time, yield components of pod setting and seed size as well as disease resistance (rust and anthracnose). Additionally there could be variation among mutants for alkaloid content, earliness etc. In Table 1 suggestions have been made for further collecting missions including areas with alkaline soils. Accessions from these areas could be used in breeding programs to produce cultivars adapted to soils with high pH.

L. angustifolius

The largest and most diverse collection of this species is held in Australia (Buirchell and Cowling, in press). The material has been evaluated for agronomic traits, disease susceptibility and chemical composition. Wide variation was found in characters such as time-to-flower (from 90 to 120 days), tallness and branching. Early flowering (90 - 95 days) was observed in accessions from low altitudes of coastal Israel and Morocco, and from the Anti Atlas mountains of southern Morocco. Seed number per pod varied from 4 to 6-7, with the larger numbers being associated with smaller seeds. Large-seeded types, found in Iberia and Morocco, may have been used in cultivation. Large-seeded Iberian types were found below 900 m and the Moroccan exclusively in northern Morocco between 300 and 1000 m. The smallest seeded types were from the central Spanish highlands, the sandy coastal lowlands of Morocco, the Anti Atlas mountains and southern Spain. *L. angustifolius* is the first lupin species to have its polymorphism analysed using molecular techniques. 17 linkage groups with 60 RAPD and 4 isozyme markers were present (Wolko and Weeden 1994). Further work on this subject should result in an estimation of DNA variation in the genus *Lupinus*.

L. angustifolius collections from the Aegean region were analysed recently by Clements and Cowling (1994). Some accessions from the Dodecanese Islands appeared to have desirable agronomic characteristics such as large seeds and many pods on the main stem. The earliest flowering (73 days) wild *L. angustifolius* line was found on the Aegina Island, south of Athens. Geographical areas, important for further collecting are mentioned in table 1.

L. luteus

Regions, with the natural distribution of the species have been well investigated. For future collecting, areas with semi-naturalised populations and semi-domesticated varieties are suggested (table1). Genetic variation seems to be rather broad for many agronomic characters like earliness, seed size and plant architecture. But in breeding some artificially induced characters (not found in natural variation) were used, e.g. unbranched types (Swiecicki 1986, 1988). Also important from a distinctness, homogeneity and stability point of view appeared to be a variation in seed coat color (used as cultivars markers).

L. atlanticus

Fifty-four accessions collected in Morocco were grown in Perth (Western Australia). Four groups of lines were distinguished according to their distribution (Buirchell and Cowling, in press). The most distinct group came from the north of the distribution. They had rapid growth, were tall, late flowering and had large seeds. Total seed alkaloid content varied from 0.29 to 0.50% of dry weight. Suggestions for further collecting missions were made by local farmers (table 1) as previous expeditions had been made during the dry season.

L. cosentinii

The species has not been extensively collected or documented and further collections are needed from all areas. In addition to its regions of natural distribution, this species was introduced to Western Australia over 100 years ago (Gladstones 1974) and became naturalized on the coastal plains.

L. digitatus

Eight accessions are listed in world collections but they are probably "multiplications" of a single collection from the Wadi Karit in south east Egypt. This situation underlines the urgent need for the collection and protection of this species.

L. hispanicus

Two subspecies are described for *L. hispanicus*, distinguished mostly by flower colour and distribution: *subsp. hispanicus* has violet flowers and grows in south and central Spain. Flowers of *subsp.bicolor* are cream, with the corolla becoming lilac when older. It is a little later in flowering and extends into the high altitudes of central and northwestern Spain and central and northern Portugal. The barriers to crossing the species to *L. luteus* are not very strong, (Swiecicki 1985).

L. micranthus

The species is wild, weedy and late flowering and has hairy leaves and pods. The plants are small and prostrate with no potential for domestication. But if it had a lower level of alkaloids in its seeds it could be useful as a natural regenerating pasture on alkaline fine-textured soils.

L. palaestinus

Plants of this species have long inflorescences, many branches and are probably cross-pollinated with insects as the vector. It makes an excellent garden plant.

L. pilosus

Considerable variation has been found within this species for flowering time, seed and leaf size, pod-setting and vernalization requirement. The alkaloid content ranges from 0.13% to 0.85% with multiflorine as the major alkaloid. Rough seeds are a characteristic feature, but an accession with smooth seeds was found in southern Syria. This indicates that the rough/smooth seed character (determined by a single gene and with possible mutations) is not useful for taxonomic purposes. Some regions of the distribution of *L. pilosus* have not been inspected thoroughly (table 1).

L. princei

Only 6 accessions are present in world collections (all in AGRIWA) indicating the need for further collecting. Some areas of its natural distribution have not been thoroughly collected. For example, the extent of its natural habitat in Ethiopia is not known.. Large seeds are characteristic of *L. princei* and distinguish it from other species. Flowers are blue but one accession has white flowers. The species is probably the most late-flowering among lupins.

L. somaliensis

This species is the least well known as the description is based on a single herbarium specimen. Its area of origin, suggested in table 1, should be revisited if possible.

L. mutabilis

Hondelmann (1984) suggested an ancient history of domestication of the Andean lupin in South America, dating back as far as 600-700 BC. It was drawn on ceramic vessels of the Pre-Inca era (Tapia et al. 1988). Unfortunately, there are no known wild relatives of *L. mutabilis* but according to Tapia and Vargas (1982) it is closely related to *L. praestabilis*. Also relevant is that fertile hybrids can be produced from the cross *L. mutabilis* x *L. hartwegi* (Kazimierski, unpublished). *L. mutabilis* occurs in different forms: 1 Annual or biannual plants adapted to long growing seasons of the Andes (240-300 days) with long branches, or 2 Annual types from high altitudes in the central Andes, with a shorter form at 3,800m around Lake Titicaca which has a dominant main stem, few branches and a short growing season (about 155 days).

L. mutabilis flowers in both short tropical days and long summer days in the Temperate Zone. The slow and non-uniform ripening is the main disadvantage from the point of view of a mechanical harvest. The only "sweet" line, cv. Inti still has about 0.06% alkaloids with flower abortion (at least under European conditions) as an additional disadvantage. The species is cross-pollinated and as a result studies on a genetic variation have not been conducted using pure lines. Variation has been observed for characters such as earliness, flower and seed colour, seed size and anthocyanin synthesis. An artificially induced, unbranched mutant is also known (Römer 1994).

As for other South American lupins the main problem for collecting is to have free access to the gathered resources.

References

B.J. 1992. FAO/IBPGR *Plant Genetic Resources Newsletter*, 90: 36-38.

Buirchell, B.J. and Cowling, W.A. *Genetic Resources in Lupins*. (in press),

Carstairs, S.A., Buirchell B.J. and Cowling ,W.A. 1992. *J. Royal. Society. Western Australia*, 75: 78-88.

Clements, J.C., Buirchell, B.J. and Cowling, W.A. (1996) *Plant Breeding*, 115: 16-22.

Cowling, W.A. 1986. FAO/IBPGR *Plant Genetic Resources Newsletter*, 65: 20-22.

Dunn, D.B. 1984 Proc. 3rd *International. Lupin Conference*, La Rochelle, France: 67-85.

Findley, R. 1981 *National Geographic*, 159: 50-65.

Gladstones, J.S. 1973. *Australian Plant Introduction Review*, 9: 11-29.

Gladstones, J.S. 1974. *Technical Bulletin No. 26*, Department. of Agriculture, Western Australia: 1-48.

Gladstones, J.S. 1976 *Australian Plant Introduction Review*, 11: 9-23.

Hacbarth, J. and Troll, H.J. 1956 In:Kappert, H. and Rudorf, W. (eds.). *Handbook of Plant Breeding*, Part IV, 2nd end., P. Parey Verlag, Berlin and Hamburg [in German].

Hammer, K. and Perrino, P. 1995. FAO/IBPGR *Plant Genetic Resources Newsletter*, 103: 19-23.

Hondelmann, W.1984. *Theoretical and Applied Genetics*, 68: 1-9.

Lehman C.O., and Hammer, K. 1983 *Kulturpflanze*, 31: 185-206

Lipman, E., Jongen, M.W.M., Hintum van, Th.J.L., Gass, T. and Maggioni, L. 1996 *Tools for Plant Genetic Resources Management*. IPGRI, CGN and ECP/GR, Rome.

Mota, M., Gusmao, L. and Bettencourt, E. 1982. FAO/IBPGR *Plant Genetic Resources Newsletter*, 50: 22-23.

Papineau, J. and Huyghe, C. 1989 FAO/IBPGR *Plant Genetic Resources Newsletter*, 88/89: 77-78.

Porsild, A.F., Harrington C.R. and Mulligan, G.A. 1967. *Science*, 158: 113-114.

Römer, P. 1994 Advances in Lupin Research. Proc. 7th *International. Lupin Conference*, Evora, Portugal: 90-91.

Sawicka, E.J. 1994. Advances in Lupin Research. Proc. 7th *International. Lupin Conference* Evora, Portugal: 92-94.

Sholars, T.A. (published in Lupinet: lupin - mg@ucdavis.edu).

Simpson, M.J.A. and McGibbon, R.1982a. FAO/IBPGR *Plant Genetic Resources Newsletter*, 50: 14-19.

Simpson, M.J.A. and McGibbon, R.1982b. FAO/IBPGR *Plant Genetic Resources Newsletter*, 52: 28-30.

Swiecicki, W. 1985. *Lupin Newsletter*, 8: 24-25.

Swiecicki, W. 1986. Proc. 4th *International. Lupin Conference* Geraldton, W.A.: 20-24.

Swiecicki, W. 1988. Proc.Vth *International. Lupin Conference* Pozna , Poland: 2-14.

Swiecicki, W.K. and Leraczyk, K. 1994. Proc.7th *International. Lupin Conference* Evora, Portugal: 70-72.

Swiecicki, W.K. and Leraczyk, K. *Proc. Genetic. Resources. Section Meeting*, EUCARPIA, Budapest, Hungary (in press).

Swiecicki, W. and Swiecicki W.K. 1995 *Journal Applied Genetics*. 36 (2): 155-167.

Swiecicki, W., Swiecicki, W.K. and Wolko, B. 1996 *Genetic Resources and Crop Evolution,* 43: 109-117.

Tapia, M.E., Riva, E.A. and Hernandez-Bravo, G. 1988. In: *World Crops: Cool Season Food Legumes* pp 1051-57 (ed. R.J. Summerfield) Kluwer Academic Publishers:.

Tapia, M.E. and Vargas, C. 1982. Proc. *Ist Int. Lupine Workshop.* Lima, Cuzco, Peru.

Wolko, B. and Weeden, N. F. 1995. In: Proc. 7th *International. Lupin Conference.* Evora, Portugal

Lathyrus spp: Conserved Resources, Priorities for Collection and Future Prospects.

A. Sarker', L. D. Robertson' and C. G. Campbell[2]
1 International Center for Agricultural Research in the Dry Areas (ICARDA),P. O. Box 5466, Aleppo, Syria; .2 Agriculture Canada Research Station, Morden, Manitoba, ROG 1JO Canada

Abstract

The center of origin and distribution of the *Lathyrus* gene pool is in the Mediterranean region. The genus contains approximately 150 species, of which *L. sativus* is important for food, feed and fodder while *L. ochrus* and *L. cicera* are important for feed and fodder. *L. cicera* is used for food to a small extent in South America. Some other species have an ornamental value such as *L. odoratus* and *L. latifolius*. Efforts to collect and conserve wild relatives, land races and cultivars of this under-utilized genus have only recently begun in some countries. Collections and conservation of *Lathyrus* germplasm has been made in Bangladesh, Canada, Chile, China, Ecuador, Ethiopia, France, Germany, India, Jordan, Nepal, Morocco, Pakistan, Russia, Syria, Turkey and USA. ICARDA, under the auspices of the FAO, holds "in-trust" germplasm of about 43 species from more than 45 countries

INTRODUCTION

The genetic diversity of *Lathyrus* is of immense significance for rainfed cropping systems in many countries. The importance of the genus is due to its resistance to several biotic and abiotic stresses. Several species are cultivated for food, feed, and fodder, and for ornamental purposes. Therefore, collection, conservation and characterization of the important species of this genus are of significance. The urgency of conserving the resources requires both *ex-situ* (gene banks) and *in-situ* (natural habitats) methods, which offer a holistic approach. There is need to critically assess genetic diversity, evolution, decline of genetic variability at the farm level, and dangers of genetic erosion.

The genus *Lathyrus* contains about 150 species (Kupicha, 1983), of which *L. sativus* L., *L. cicera* L. *and L. ochrus* (L.) DC are important for food, feed and fodder. *L. sativus* is widely cultivated for human consumption, particularly in Bangladesh, China, Ethiopia, India, Nepal and Pakistan. At present, national programs are involved in the development of low or toxin-free *L. sativus*. Additionally, the primary, secondary and tertiary gene pools may play a role in improvement. For example, a toxin-free gene has been identified in *L. tingitanus* L. that is being used to develop toxin-free grass pea varieties in China (Zhou and Arora, 1996).

To prevent genetic erosion and the total extinction of some species of *Lathyrus*, many national and international bodies have launched germplasm collection and conservation activities. However, an extensive and systematic approach has not so far been adopted and requires further attention.

DOMESTICATION AND DISTRIBUTION

The earliest archaeological remains of *Lathyrus* appear in the Neolithic in the Balkans and Near East of Bulgaria, Cyprus, Iraq, Iran, and Turkey (Erskine et al., 1994). A single *Lathyrus* seed, presumed to be a field weed was found in Cayono in Turkey dated around 7200 BC, where bitter vetch was the prevalent pulse (van Zeist, 1972). Compared to the early domesticates of lentil, pea and bitter vetch; *Lathyrus* is only found in small quantities in Turkey, Cyprus, Iraq, Iran and Bulgaria for dates between 6750 to 4770 BC. However a different picture appears from late Neolithic finds at Dimini in Greece (c. 4000-3500 BC), where grass pea is as frequent as pea and lentil (Kroll, 1979). This increased frequency is suggestive of domestication. Grass pea was the chief crop component mixed with lentil (dated 2100-1800), providing stronger evidence of domestication by the Middle Bronze Age (Helback, 1965). It was also found mixed in substantial quantities

645

R. Knight (ed.), Linking Research and Marketing Opportunities for Pulses in the 21st Century, 645–654.

with other leguminous crops in later finds. *L. cicera* is believed to have been domesticated in southwestern Europe by 4000- 3000 BC (Kislev, 1989). Written records provide little knowledge about the origin of grass pea. *Lathyrus* is an ancient Greek plant name probably used for a pulse and possibly for *L. sativus,* but it is not identifiable with certainty and it gives no additional evidence of domestication (Westphal, 1974). The Romans also do not mention *Lathyrus,* which reflects little importance or a lack of knowledge of the crop. Thus, the archaeological evidence suggests that domestication of *Lathyrus* occurred possibly during the late Neolithic and surely by the Bronze Age. Prior to that time it was probably a tolerated weed of other pulses (Erskine et al., 1964). The natural distribution of the genus *Lathyrus* has been completely obscured by cultivation, even in southwest and central Asia (Townsend and Guest, 1974). The progenitor of widely cultivated *L. sativus* is unknown but several Mediterranean species resemble the cultigen morphologically, namely *L. cicera, L. marmoratus* Boiss, *L. blephrocarpus* Boiss and *L. pseudocicera* Pamp The current distribution of the genus covers the Mediterranean region, Indian subcontinent and Ethiopia. It has been observed that *L. ochrus* and *L. sativus* are mostly distributed in coastal, lowland sites, while *L. cicera* is the most common species in highland and cold temperate sites.

MAJOR EFFORTS IN COLLECTION

Expeditions to collect genetic resources of *Lathyrus* have been made in the past by national programs individually and with international organizations but most of the old collections have been lost or destroyed. Unlike other legumes, systematic and intensive collection of the genus started only recently, and is restricted to a few countries. The Bangladesh Agricultural Research Institute (BARI) collected more than 300 landraces of *L. sativus* in the early 1980s. However, most have been destroyed or lost due to lack of proper storage. Later, BARI made collections (1992-1994) in collaboration with the Department of Agricultural Extension and 2078 landraces of *L. sativus* were collected from all grass pea growing areas of Bangladesh (Sarwar et al.,1996). In 1995, in collaboration with ICARDA and the Co-operative Research Centre for Legumes in Mediterranean Agriculture (CLIMA), 62 accessions were collected as part of the project 'Development and Conservation of Plant Genetic Resources in the Mediterranean Region' Thus, BARI now holds 2,130 landraces (Table 1) and maintains passport information in d-Base format under the name 'Grass-BDB'. BARI has provided over 1,153 accessions to ICARDA for back up storage and 200 to CLIMA for evaluation in Western Australia.

In Nepal, a collection of *L. sativus* germplasm was made in 1987 in collaboration with the International Development Research Centre (IDRC) in the *terai* and inner *terai* regions and some parts of sub-tropical basins. A total of 76 samples were collected and preserved at Rampur, Nepal. A second mission in 1995 involved the Nepal Agricultural Research Centre, ICARDA and CLIMA, and 90 accessions were obtained (Neupane, 1996).

Table 1. Major collections and conservation of *Lathyrus* species.

Country/Org.	No. of accessions	No. of species
Bangladesh	2130	1
Canada	953	2
Ethiopia	368	1
ICARDA	3001	45
India	3754	34
Jordan	36	8
Pakistan	138	9
Turkey	600	31
USDA	330	13

The first collection in India of about 600 accessions was made in 1967 (Mehra et al., 1996). They are maintained at the Indian Agricultural Research Institute (IARI). Up to 1977, IARI scientists have collected about 1500 accessions from different grass pea growing areas of India. The *Lathyrus* improvement program in India, now operating from Raipur, Madhya Pradesh and all available collections have been centralized at the National Bureau of Plant Genetic Resources (NBPGR) Regional Station, Raipur. Nearly 2600 accessions are being maintained. Additionally, 1154 exotic accessions of several species have been introduced by NBPGR from Bangladesh, Canada, France, Germany, ICARDA and USA. Eight species, *L. sativus, L.*

aphaca L., *L. pratensis* L., *L. sphaericus* Retz, *L. inconspicuus* L., *L. odoratus* L., *L. altiacus* and *L. luteus* Baker occur in India (Tiwari, 1979). Collections of these species have been made, the largest number being *L. sativus*. Some of the species were collected at altitudes up to 3000 m in the western Himalayas.

Collecting in Pakistan, organized by the Plant Genetic Resources Institute (PGRI), Islamabad has been mostly restricted to Sindh Province. Haqqani and Arshad (1996) reported that 97 landraces of *L. sativus* were collected from different grass pea growing areas. Additionally, 18 accessions of *L. odoratus*, 3 of *L. doratuso* and 1 of *L aphaca* have been collected from Sindh and Punjab Provinces. The PGRI has also collected and conserved 41 accessions of different species from exotic sources, mostly from Canada.

Jordan is a small country but is a rich source of *Lathyrus* diversity. Fifteen species were reported from Palestine of which 10 *(L. digitatus* (M.B) Fiori, *L. hierosolymitanus* Bioss, *L. aphaca, L. gorgonei* Parl., *L. cicera, L. inconspicuus* L. *blepharicarpus, L. ochrus, L. gloeospermus* Warb ct Eig., and *L. pseudocicera)* were reported from Jordan (Syouf, 1996). The national collection program and the Genetic Resource Unit (GRU) of ICARDA made collections in 1981, 1989, 1990 and 1993. A total of 36 accessions of 8 species of *Lathyrus* have been collected from the country. The most abundant species was *L. aphaca*, collected from 15 locations. The entire distribution of the genus ranged from 50 m below sea level to 1125 m above sea level with annual rainfall between 50 to 650 mm with soil pH of 4.0 to 7.9. It was also reported that serious genetic erosion has occurred in the genus, especially in species such as *L. ochrus, L. gorgonei, L. cicera* and *L. digitatus*.

Turkey is the richest source of *Lathyrus* diversity. Davis (1970) reported the existence of about 61 species, some of them endemic at a local or regional level. Over 500 accessions comprising 32 species have been collected from different regions and altitudes ranging from sea level to 2000 m (Table 2). Missions before 1987 collected forage genetic resources in general and *Lathyrus* species were gathered when encountered. In 1987, 1988, and 1995, expeditions focused on specific legumes from 9 agricultural regions (Sabanci, 1996). The number of species collected is nearly half of that found by Davis and the absence of some endemic species indicates enormous genetic erosion has occurred in Turkey.

Table 2. *Lathyrus* species collected in Turkey. The number of accessions and the range in altitude of the accessions.

Species	Number	Altitude (m)	Species	Number	Altitude (m)
L. amoenus	3	165 –1600	*L. laxiflorus*	17	50 – 910
L. annuus	44	s.l.-2 000	*L. marmoratus*	4	s.l.- 100
L. aphaca	119	s.l.-2 000	*L. nissolia*	9	100 – 910
L. blepharicarpus	1	100	*L. ochrus*	1	70
L. chloranthus	4	1500-1800	*L. odoratus*	2	1-500
L. chrysanthus	1	600	*L. pseudoaphaca*	12	100-1 650
L. cicera	90	s.l. 1-600	*L. pseudocicera*	8	100 – 500
L. cilicicus	2	600	*L. rotundifolius*	1	500
L. clymenum	1	s.l	*L. sativus*	17	20 -1 100
L. digitatus	9	150 –1060	*L. saxatilis*	2	1-650
L. gorgonii	27	s.l.- 1100	*L setifolius*	5	60 –100
L. graminoides	4	100-300	*L. sphaericus*	19	125-350
L. hierosolyminatus	22	300-800	*L. spharucatus*	5	350
L. hirsutus	2	200-350	*L. stenophyllus*	2	1-960
L. inconspicus	79	60-1 600	*L. tuberosus*	2	1-960
L. lathyroides	1	100	*L. undulatus*	4	200-1 060

(Sabanchi, 1996)

About 300 accessions of *L. sativus* were collected by breeders at the Adet Research Station in the dense growing areas of Gozam and Gender in Ethiopia in 1987 and 1994. These are not included in collections registered at the Biodiversity Institute in Addis Ababa. In 1997, scientists from the Biodiversity Institute, CLIMA and ICARDA made an expedition in several regions in the country and 68 accessions were collected.

The USDA-ARS *Lathyrus* collection maintained at Pullman, WA, USA contains 330 accessions of 13 species (Muehlbauer and Kaiser, 1994).

ICARDA has the most extensive collection of germplasm for the Mediterranean region as well as for North Africa. ICARDA is concerned with *Lathyrus* species form the West Asia and North Africa (WANA) region, Ethiopia and South Asia. In WANA, *L. sativus, L. cicera* and *L. ochrus* are mainly used as feed legumes. *L. cicera* is common in Greece, Cyprus, Iran, Iraq, Jordan, Spain and Syria; *L. sativus* is used in

Jordan and *L. ochrus* in Cyprus and Greece (Saxena et al., 1993). ICARDA holds 'in-trust' *Lathyrus* germplasm from more than 45 countries under the auspices of the FAO. While the emphasis at ICARDA is for the improvement of *L. sativus, L. cicera* and *L. ochrus*, a collection of 42 other species is maintained (Table 3).

Table 3. Lathyrus species conserved at ICARDA and the number of accessions.

Species	Number	Species	Number	Species	Number
L. amphicarpos	2	*L. clymenum*	2	*L. odoratus*	3
L. angulatus	1	*L. digitatus*	2	*L. pallescens*	1
L. annuus	68	*L. gloeospermus*	2	*L. pratensis*	2
L. aphaca	253	*L. gorgonii*	60	*L. pseudocicera*	65
L. articulatus	104	*L. herticarpus*	2	*L. rotundifolius*	2
L. aureus	1	*L. hierosolymitanus*	104	*L. sativus*	1627
L. basalaticus	4	*L. hirsutus*	17	*L. saxatilis*	2
L. belinensis	1	*L. inconspicuus*	149	*L. setifolius*	7
L. blepharicarpus	33	*L.incurvus*	2	*L. sphaericus*	21
L. cassius	8	*L. latifolius*	1	*L. stenophyllus*	2
L. chloranthus	1	*L.laxiflorus*	12	*L. Sylvestris*	1
L. chrysanthus	3	*L.marmoratus*	33	*L. tingitanus*	18
L. cicera	182	*L. nissolia*	9	*L. tuberosus*	4
L. cilicicus	10	*L. occidentalis*	1	*L. vineali*	4
L. ciliolatus	3	*L. ochrus*	136	*L. sp.*	36

The majority of accessions held in ICARDA are from the WANA region except *L. sativus*. The WANA material includes cultivated and wild forms of naturally occurring populations found mostly in crop fields and orchards (Fig. 1- 4). The *L. sativus* accessions from Ethiopia and the Indian subcontinent are local landraces. A standard collecting form for passport data developed by GRU, ICARDA is based on IPGRT standards.

CHARACTERIZATION AND DOCUMENTATION

Proper evaluation, characterization and documentation are important to utilizing *Lathyrus* genetic resources. However, in-depth evaluation for phenological, morphological, agronomical and quality characters has yet to be made at the national and global level. The Germplasm Resources, Crop Improvement and Agronomy Committee of the International Network for the Improvement of *Lathyrus sativus* and the Eradication of Lathyrism (INILSEL) proposed a list of 16 descriptors to characterize the *Lathyrus* genetic resources (Campbell, 1994). The GRU, ICARDA uses 21 descriptors of which many are common to that of INILSEL's descriptors. However, a detailed list of descriptors, which we propose, is as follows:

1) Growth habit(GRH)	15) Internode length in cm (LINT)
2) Flower colour (FLCO)	16) Seeds per pod(SPD)
3) Anthocyanin present (ANTH)	17) Plant type (indeterminate or determinate) (PLT)
4) Leaf shape(LFSH)	18) Seed coat colour (SCOL)
5) Leaf width (narrow<0.5 , medium<1.0, wide>1.0 cm) (LFWD)	19) Pod shattering at maturity (PSH)
6) Days to 50% flowering (DFLR)	20) 1000 seed weight in g (W1000)
7) Days to 90% podding (DPOD)	21) Downy mildew resistance (DMR)
8) Days to 90% maturity (DMAT)	22) Insect resistance (INR)
9) Plant height in cm (PTHT)	23) Seed yield in kg/ha (SYLD)
10) Height to the first flower in cm (HTFF)	24) Biomass in kg/ha (BYLD)
11) Leaf length in cm (LLF)	25) Straw yield in kg/ha (STYLD)
12) Pod length in cm (LPD)	26) Harvest index in % (HI)
13) Pod width in cm (WPD)	27) B-ODAP concentration of the seeds in % (SOD)
14) Peduncle length in cm (PEDL)	

Evaluation in most countries has been for agronomic characters to meet breeder's immediate needs without a detailed characterization. In Bangladesh 788 landraces were evaluated for 5 characters (Table 4). Characterization of 76 accessions for 13 agro- morphological traits was done in Nepal (Furman and Bharati, 1989). Exotic germplasm and 84 landraces of *L. sativus* were assessed mainly for B-ODAP content. In India,

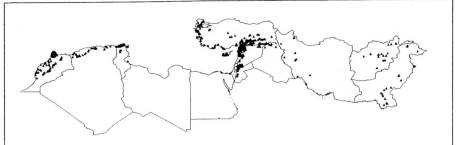

Figure 1. Distribution of all *Lathyrus* species in the ICARDA germplasm collection in WANA

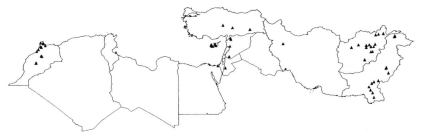

Figure 2. Distribution of *Lathyrus sativus* in the ICARDA germplasm collection in WANA

Figure 3. Distribution of *Lathyrus cicera* in the ICARDA germplasm collection in WANA

Figure 4. Distribution of *Lathyrus ochrus* in the ICARDA germplasm collection in WANA

1,187 accessions have been characterized for phenological, agro-morphological and quality characters. Marked variability was noticed for plant type, leaf size, anthocyanin pigmentation, flower colour etc. (Pandey et al., 1996). About 300 Ethiopian landraces of *L. sativus* have been evaluated for agronomic characters (Campbell, 1994). Canadian evaluation of *Lathyrus* germplasm began in 1967 at the Agriculture Canada Research Station, Morden, Manitoba and 700 accessions were classified for B-ODAP content and other traits. Over 600 accessions of *L. sativus, L. cicera* and *L. ochrus* introduced from the Mediterranean region, Pakistan, India and Bangladesh were evaluated in Western Australia for 9 characters including B-ODAP content (Siddique et al., 1996).

Table 4. Attributes of 788 *Lathyrus sativus* accessions evaluated in Bangladesh during 1994.

Characters	Range	Mean	CV(%)
Days to 50% flowering	57-91	79	6.5
Days to maturity	117-128	125	2.2
Pod length (cm)	2.7-3.5	2.91	4.2
Seeds/pod	3.0-5.3	3.9	10.8
Yield (kg/ha)	960-2502	1577	21.4

Evaluation at ICARDA
Agro-morphological

An evaluation of 1082 accessions of 30 species was performed in 1992 at ICARDA in a series of augmented nurseries using one systematic check (IFLA 347, *L. sativus*) and two random checks (IFLA 101, *L. ochrus* and IFLA 536, *L. cicera).* Results are presented only for the three economically important species, *L. sativus, L. cicera* and *L. ochrus.* Most of the accessions of *L. sativus* and *L. ochrus* had semi-erect GRH, while *L. cicera* also had prostrate and erect plant types (Table 5). The FLCO of almost all accessions of *L sativus* were violet, most accessions of *L cicera* were red and *L. ochrus* were white. Most accessions of all three species had weak anthocyanin pigmentation. *L. sativus* and *L. cicera* had mostly narrow LFSH and *L. ochrus* mostly medium LFSH. These descriptors are used for taxonomic identification in *Lathyrus.*

The accessions of *L cicera* were earlier than the *L. cicera* check, IFLA536, by up to 11 days (Table 6). The SPD was higher for the tested accessions, but the W1000 was smaller and the HI was similar to the check. The BYLD was reduced, which resulted in a lower SYLD and STYLD for the accessions tested. The *L. ochrus* accessions were also earlier than the *L. ochrus* check (Table 7). The vegetative descriptors were similar to the check for *L. ochrus.* The W1000 was smaller for the tested entries as was the SPD. However, for *L. ochrus,* the mean of the tested accessions for BYLD was the same as the check, which resulted in similar values for SnD and STYLD. The HI, BYLD, SYLD and STYLD were much lower for the tested *L. sativus* germplasm accessions than the *L. sativus* check (Table 8). Unlike the other two species, the tested accessions of *L. sativus* were later than the *L. sativus* check.

Table 5. Frequencies (%) for Growth habit GRH, Flower colour FLCO, Anthocyanin pigmentation ANTH, and Leaf shape LFSH for accessions of Lathyrus spp. evaluated at Tel Hadya, Syria, 1992-93

Descriptors/score	L.cicera	L. ochrus	L.sativus	Descriptors/score	L.cicera	L. ochrus	L.sativus
GRH				ANTH			
Prostrate	24.0	3.4	0.7	Weak	83.3	100.0	83.8
Semi-erect	51.0	96.6	96.0	Fair	8.3	0.0	9.9
Erect	25.0	0.0	3.3	Strong	7.3	0.0	6.2
FLCO				Very Strong	1.0	0.0	0.0
White	0.0	53.4	1.5	LESH			
Cream	1.0	12.1	2.2	Narrow	99.0	8.6	99.6
Brick	11.5	0.0	0.4	Medium	1.0	91.4	0.0
Pink	5.2	0.0	0.0	Oval	0.0	0.0	0.4
Violet	4.2	13.8	95.6				
Yellow	2.1	0.0	0.0				
Blue	0.0	0.0	0.0				
Red	76.0	20.7	0.4				

Table 6 Values obtained for a check (IFLA 536) and 96 *Lathyrus cicera* accessions at ICARDA, Syria, during 1992-93.

Descriptors	Check mean	Accessions			
		Mean	Min	Max	CV (%)
DFLR (days)	126.2	123.9	115.0	136.0	3.3
DMAT (days)	161.7	163.9	156.0	181.0	2.8
DPOD (days)	133.7	128.3	122.0	148.0	4.5
PTHT (cm)	36.7	35.4	24.1	49.8	12.5
HTFF (cm)	10.7	8.1	2.4	13.2	18.5
SPD	3.61	3.8	2.3	9.6	24.2
HI (%)	36.6	33.8	12.7	52.0	26.1
W1000 (g)	91.8	83.1	13.9	116.7	24.2
SYLD (kg/ha)	1,237	1,120	117	2,030	52.5
BYLD (kg/ha)	3,350	3,101	635	4,972	37.7
STYLD (kg/ha)	2,113	2,578	488	3,067	26.1

Table 7 Values obtained for a check (IFLA 101) and 58 *Lathyrus ochrus* accessions at ICARDA, during 1992-93.

Descriptors	Check mean	Accessions			
		Mean	Min	Max	CV (%)
DFLR (days)	124.3	120.4	115.0	145.0	4.1
DMAT (days)	160.0	157.0	149.0	184.0	3.2
DPOD (days)	128.0	124.0	118.0	154.0	4.6
PTHT (cm)	33.1	34.7	23	48	15.4
HTFF (cm)	15.8	13.0	7	19	15.5
SPD	4.8	4.6	3.32	5.7	11.9
HI (%)	38.5	36.2	12.7	48.6	20.1
W1000 (g)	130.2	121.3	57.2	156.3	17.7
SYLD (kg/ha)	853	815	105	1,454	38.0
BYLD (kg/ha)	2,214	2,221	726	3,741	32.0
STYLD (kg/ha)	1,362	1,406	564	2,499	32.5

Stress Resistance

Resistance to major biotic and abiotic stresses have been noticed during evaluation. *L. sativus* accessions resistant to powdery mildew caused by *Erysiphe polygoni* DC. and Botrytis blight caused by *Botrytis cinerea* Pers. Ex Fr, have been found in ICARDA collections. Resistant sources for other diseases in other species have also been found but not confirmed. The broomrape, *Orobanche crenata* Forsk. causes major yield losses in *Lathyrus* species particularly in the WANA region. *L. ochrus* accessions remain free of emerged *O. crenata* shoots (Linke et al.,1993) but accessions of *L. sativus* and *L. cicera* were highly susceptible. Lines of *L. ochrus* resistant to *O. crenata* are IFLAO 84, 94, 95 and 101. Among the species evaluated, *L. cicera* is resistant to cold, whereas *L. sativus* and *L. ochrus* accessions are generally susceptible. However, one accession of *L. ochrus*, IFLAO 109, from Portugal has a high level of cold tolerance (Robertson et al., 1996). This can be explained from the eco-distribution of *Lathyrus* species, where most *L. sativus* and *L. ochrus* accessions are from low altitude, mild winter environments while *L. cicera* is adapted to high altitude, continental environments with severe winters.

Quality Factors

Assessment for quality has mostly focused on ß-N-oxalyl-L-a,ß- diaminopropionic acid (B-ODAP), the free amino acid which causes Lathyrism. Preliminary results indicate that none of the *Lathyrus* species lines was free, although in several lines the B- ODAP content was very low. *L. cicera* had a low mean B-ODAP content (0.16%) followed by *L. sativus* (0.48%) and *L. ochrus* (0.57%). Similar observations were made at CLIMA, Australia, where *L. sativus* lines gave an intermediate score.

Conservation of Genetic Resources

Conservation of genes for future use is an important aspect of genetic resources. At present all national programs and ICARDA are following *ex-situ* conservation. Seeds are stored for short, medium and long-term periods. The working collections (breeding lines) are kept in short-term (15°C) storage and the active collections (frequently used lines) in medium- term storage (0° to -2°C) ready for immediate supply. The base collection is kept in long- term storage at -20°C. There is *ex-situ* conservation of *Lathyrus* germplasm in Bangladesh, Chile, China, Canada, Ethiopia, Ecuador, France, Germany, India, Nepal, Pakistan, Russia,

Syria (ICARDA), USA and UK. Many of these are small collections and do not represent a true sample of the genetic variability present in the local germplasm. Co-ordination of *Lathyrus* genetic resources in INILSEL had been attempted by Dr D. Combes, Pau, France in a more organized storage system. Storage at ICARDA has been for long and medium-term. However, regeneration of germplasm is an important part of a conservation program.

In-situ conservation, which is considered as complementary to *ex-situ* initiatives has not been adopted for *Lathyrus*, except for an initial attempt in Turkey. *In-situ* conservation allows the maintenance of a plant population within the ecosystem and environment to which it is adjusted. Arguments in support of *in-situ* conservation of land races and wild relatives focus on allowing evolutionary processes to continue. The *in-situ* Project in Turkey is an important part of the National Plant Genetic Resources Program. Initial surveys have started in Anatolia Diagonal, where *Lathyrus, Vicia* and some other forage legumes will probably be the target species.

GAPS IN COLLECTION OF THE SPECIES

Comprehensive collections of *L. sativus* have been made in Bangladesh and in Madhya Pradesh in India with conservation of the collected germplasm. However, many collections may not be comprehensive, they may not have included related species or in some cases may have been lost after collection. There is a need to determine the status of present collections and determine which areas have not been sampled. This need was expressed in a workshop held at Raipur, India in 1995.

Vast areas in the eastern and northern plains and the Himalayan region in India have not been explored intensively. The *Lathyrus* species *aphaca* and *ochrus* are found as weeds in those areas where *L. sativus* has been collected, however, they were not included in the collections. In addition to the 8 species collected so far, there is the possibility that other species of *Lathyrus* may exist in India, especially in the mountainous regions of the country. Those unexplored areas should be surveyed for the presence of wild relatives as well as for the major species.

Apart from Sindh Province, other areas have not been sampled in Pakistan. There has been no emphasis on collecting related species. Although two missions were organized in Nepal, they were restricted to the plain areas. The north and northwestern hilly areas have not been explored, where it is possible wild relatives occur. Production of *Lathyrus* is declining in Nepal, leading to genetic erosion. In Ethiopia, future collection should cover unexplored areas. Systematic collection of landraces in Gansu and Shaanxi provinces of China has not been undertaken. The representation of these areas in the collection of the Institute of Crop Germplasm Resources, China, needs to be determined.

There are reports of decreased production of *L. sativus* and *L. cicera* in Spain. These species are grown in the more mountainous regions. At present they are often found only at the edges of fields or in gardens for personal use. It appears that there is a threat of this germplasm being lost if a comprehensive collection is not made soon. The status of landraces and collections needs also to be established for southern Europe, including southern France, Greece, Italy and Turkey as the same threat may be present for their germplasm. The need for collection and conservation of *Lathyrus* genetic resources in the central Asian countries and Russia should also be determined.

Although ICARDA has an extensive collection of germplasm from the WANA region, there are still many unexplored areas. The extinction of rare species has been reported from Jordan. The national programs of this region and ICARDA should pay further attention to sampling this large area. Intensive collection needs to be carried out in Chile and Ecuador, where the landraces are also subject to genetic erosion.

INTERNATIONAL EFFORTS IN COLLECTION AND CONSERVATION

The International Plant Genetic Resources Institute (IPGRI) established jointly with ICARDA and FAO a West Asia and North Africa Plant Genetic Resources Network (WANANET), operating from ICARDA headquarters and representing 13 countries. The GRU, ICARDA and the WANANET work in collaboration with national programs to collect and conserve *Lathyrus* germplasm. In the region of Asia, Pacific and Oceania (APO), the IPGRI-APO project is directed to work on the conservation of under-utilized crops and *Lathyrus sativus* is one of the target crops. IPGRI-APO is collaborating with national programs in south Asia and China, on *Lathyrus* genetic resources; linking utilisation with conservation and improving technologies for conservation. INILSEL is mainly involved in conservation. This network operates from the University of Pau, France, and coordinates list of accessions, passport data and descriptors for members of the Network. Coordination of germplasm storage had also been undertaken. In 1995, a regional workshop was organized on *Lathyrus* Genetic Resources in Asia at Raipur, India by IPGRI. The workshop recommended to establish a network on *Lathyrus* Genetic Resources. The proposed network includes countries from South Asia and the WANA region including Ethiopia. CLIMA and other research organizations interested in *Lathyrus* may also be associated as interested research groups. It was also agreed that activities will be coordinated by the IPGRI coordinator for south Asia, New Delhi, India. There will be a representative from each country to act as a country coordinator in the network. ICARDA will take up creation of a database for germplasm from the WANA region and Ethiopia while IPGRI-APO will prepare a database for south Asian countries.

FUTURE THRUSTS

Many of the needs for further exploration, collection, and conservation and for documentation and evaluation have been described above. A global data base network needs to be established under the control of IPGRI for easy access of all the information. Once the databases are prepared, they should be supplied to all persons engaged in *Lathyrus* improvement.

The search for genes for quantitatively inherited traits, such as yield in landraces and weedy and wild relatives should be emphasized. The rigorous assessment of all available germplasm should be intensified to identify sources of resistance to diseases (powdery mildew, Botrytis blight), insects (thrips, aphids), *Orobanche* and cold.

Genetic studies need to be undertaken on traits such as flower colour, B- ODAP content leading to linkage maps and on the reproductive biology of *Lathyrus* species, out crossing mechanisms and interspecific hybridization. Genetical and cytogenetical techniques, molecular markers and other biotechnological approaches should be used to resolve the inter-relations between different *Lathyrus* species;

In any future development of *Lathyrus* for human consumption, zero or very low levels of (B-ODAP) content will be essential and all the germplasm needs to be assessed for toxin- free gene(s) Methods have already been developed for the production of transgenic *Lathyrus* plants. Transfer of this technology and the sharing of toxin-free transgenic plants among countries needs to be carried out. This development needs to be coupled with strategies for the maintenance of genetic purity and proper isolation.

Another important aspect will be to evaluate the adaptability of different species to agro-geographical areas for their dry matter/ biological yield/ grain yield in comparison with other legume crops.

CONCLUSION

Cultivated *Lathyrus* species and their wild relatives are either threatened now, or will be in the near future by severe genetic erosion. Safeguarding their diversity is possible only through planned conservation efforts. A high priority needs to be placed on establishing core collections, conserving germplasm as duplicate sets in different gene banks and the establishment of a global passport and descriptor database.

References

Campbell, C. G, Mehra R. B., Agarwal, S. K., Chen, Y. Z., Abd El Moneim, A. M., Khawaja,Yadov, C. R., Tay, J. W, Araya, W. A. 1994. In *Expanding the Production and Use of Cool Season Food Legumes*, pp 617-630 (Eds F. J. Muehlbauer and W. J. Kaiser). Dordrecht: Kluwer Academic Publishers.

Davis, P. H. 1970. *Flora of Turkey and East Aegean Islands, 3: Edinburgh.*

Erskine, W., Smartt, J. Muehlbauer, F. J. 1994. *Economic Botany* 48: 326-332.

Furman, B. J.and Bharati, M. P. 1989. *National Grain Legumes Imp. Frog.* Rampur, Nepal.

Haqqani, A. M. and Arshad, M. 1996. In *Lathyrus Genetic Resources in Asia*, pp 59-68 (Eds R. K. Arora P. N. Mathur, K. W. Riley and Y. Adham). New Delhi: IPGRJ.

Helback, H. 1965. *Sumer* 19: 27-35.

Kislev, M. E. 1989. *Economic Botany* 43: 262-270.

654

Kroll, H. 1979. *Archaeo-Physika* 8: 173-189.

Kupicha, F. K. 1983. *Notes from the Royal Botanic Garden Edinburgh* 41: 287-326.

Linke, K. H. Moneim A. M. A., and Saxena, M. C. 1993. *Field Crops Research* 32: 277-285.

Mehra, R. B., Raju, D. B. And Himabindu, K. 1996 In *Lathyrus Genetic Resources in Asia*,pp 37-44 (Eds R. K. Arora P. N. Mathur, K. W. Riley and Y. Adham). New Delhi: IPGRI.

Muehlbauer, F. J. and Kaiser, W. J. 1994. In *Expanding the Production and Use of Cool Season Food Legumes*, pp 617-630 (Eds F. J. Muehlbauer and W. J. Kaiser). Dordrecht: Kluwer Academic Publishers.

Neupane R. K. 1996. In *Lathyrus Genetic Resources in Asia*, pp 29-35 (Eds R. K. Arora P. N. Mathur, K. W. Riley and Y. Adham). New Delhi: IPGRI.

Pandey, R. L., Chitale, M. W., Sharma, R. N. And Rastogi, N. 1996. In *Lathyrus* Genetic Resources in Asia, 45-52 (Eds R. K. Arora P. N. Mathur, K. W. Riley and Y.Adham) New Delhi: IFGRI.

Robertson, L D., Singh, K. B., Erskine, W. And Abd El Moneim, A. M. 1996. *Genetic Resources and crop Evolution* 43: 447-60.

Sabanci, C. O. 1996. In *Lathyrus Genetic Resources in Asia*, pp 77-86 (Eds R. K. Arora P. N. Mathur) K. W. Riley and Y.Adham). New Delhi: IPGRI.

Sarwar, C, D. M., Malek, M. A., Sarker, A. and Hassan M. S. 1996. In *Lathyrus Genetic Resources in Asia*, pp 13-20 (Eds R. K. Arora P. N. Mathur, K. W. Riley and Y.Adham). New Delhi: IPGRI.

Saxena M. C., Abd El Moneim, A. M. and Raninam, M. 1993. In *Potential for Vicia and Lathyrus Species as New Grain and Fodder Legumes for Southern Australia*, 2-9 (Eds J. R. Garlinge and M. W. Perry). CLIMA.

Siddique,K. H. M., Loss, S. P., Hervig, S. P. and Wilson, J. M. 1996. *Australian Journal of Experimental Agriculture* 36: 209-18.

Syouf, M. Q. 1996. In *Lathyrus Genetic Resources in Asia*, pp 67-76 (Eds R. K. Arora P. N. Mathur, K. W. Riley and Y.Adham). New Delhi: IPGRI.

Tiwari, S. D. B. 1979. *The Phytogeography of Legumes of Madhya Pradesh*, pp 45-78 (Eds B. Singh and M. P. Singh): Dehradun, India.

Townsend, C.C. and Guest, E. 1974. *Flora of Iraq*, Volume 3 Leguminales. Min. of Agriculture and Agrarian Reform, Bahgdad, Iraq.

van Zeist, W. 1972. *Helinium* 12: 3-19.

Westphal, W. 1974. *Pulses in Ethiopia, their taxonomy and agricultural significance*. Haile Sellassie University, Ethiopia, Agricultural University of Wageningen, the Netherlands, Center for Agricultural publishing and documentation, the Netherlands.

Zhou, M. and Arora, R. K. 1996. In *Lathyrus Genetic Resources in Asia*, pp 91-96 (Eds R. K. Arora P. N. Mathur, K. W. Riley and Y.Adham). New Delhi: IPGRI.

Food legumes in Human Nutrition

G.H. McIntosh and D. L. Topping
CSIRO Division of Human Nutrition, Adelaide, South Australia 5000

Abstract

Grain legumes (pulses) are valuable nutritional resources, providing proteins, complex carbohydrates, unsaturated fats, minerals and vitamins for the human diet. They also contain varying concentrations of non nutrient phytochemicals with functions which are gaining increasing attention given recent claims with regard to heart disease and cancer prevention. Some of these factors are also described as antinutrient factors, being associated with significant growth inhibition in animal production. Processing is aimed at diminishing or eliminating such factors before they are ingested. However some of these factors are considered to be anticancer agents and may be better preserved. This is being examined in current studies. Pulses are an important option for humans seeking a healthy disease-free lifestyle. They can be presented in ways which will see increasing acceptance of their place in a balanced nutritious diet.

INTRODUCTION

Grain legumes or pulses are important traditional foodstuffs in many cultures and are valued as relatively inexpensive sources of protein and energy. Energy is mostly stored as complex carbohydrates (starch/ non starch polysaccharides (NSP)) and/or as lipid. Most grain legumes are low in fat eg faba beans and lentils where the lipid content is 2% or less. A few grain legumes are high in oils e.g. groundnuts and soybeans. Soybeans have lipids and not starch as their main energy store and contain in excess of 19% lipid by weight.

Food systems in the Indian, African and South American subcontinents use legumes successfully either to augment their protein supply or as substitutes for the more expensive animal products. Sgarbieri (1989) reported legume intakes of 57-70g per day (providing 17-18g of protein) in South America. FAO figures (1984) showed that pre-civil war Rwanda had one of the highest recorded intakes (116g per day) providing 47% daily protein needs. By comparison, data for Australia showed intakes of 8-9 grams per person daily (Australian Food Survey, 1993). Nevertheless, large quantities and many species of legumes are grown here: (Table 1). At this stage most is used in animal production or exported overseas.

Table 1. Production in tonnes ('000) of Grain Legumes in Australia (ABARE 1996)

Lupins	1235
Field peas	386
Chickpeas	266
Faba beans	115
Soybeans	73
Navy beans	7.6

NUTRITIONAL CONSIDERATIONS

Significant reviews in this area have been provided by Salunkhe and Kadam (1989), Sgarbieri (1989) and Bressani (1993) in the past, and are comprehensive in their treatment. Composition of some of the main nutrients for significant legumes are shown in table 2.

R. Knight (ed.), Linking Research and Marketing Opportunities for Pulses in the 21st Century, 655–660.
© 2000 *Kluwer Academic Publishers. Printed in the Netherlands.*

Table 2. Composition of some common grain legumes (% Dry Weight)*

	Moisture	Carbohydrate	Protein	Dietary Fibre	Fat
Lupins	8.4	39	32.2	8.1	7
Peas	9.2	62	26.3	6.3	2
Chickpeas	8.7	62	21.7	10.6	4.7
Faba beans	10.0	60	24.3	9.7	1.6
Soybean	8.5	36	36.5	12.5	19.9
Navy beans	10.2	64	23.9	9.7	1.6

* "The chemical composition and nutritive value of Australian Grain Legumes", GRDC, Canberra 1994.

Protein

Legumes vary greatly in protein quality and quantity. Soybeans and lupins have some of the highest concentrations (30% or higher) while common beans (*Phaseolus vulgaris* L.) have 23% or less. Nevertheless, these values are high relative to other plant foods such as cereals (1.6 to 2 times). Protein quality varies considerably both in terms of albumins/globulins present and their amino acid content. Generally, legume proteins are relatively poor in the essential sulphur amino acids (SAA) but are good sources of lysine. However cereals can provide an important complementary source of protein, generally being rich in SAA. Cereals have suboptimal levels of lysine - a deficit which can be remedied conveniently by combining them with legumes. This is a traditional (and highly successful) practice in Indian, African and South American cultures. Research in Australia and elsewhere has shown that foods of high protein quality can be generated by combining cereals and legumes in particular ratios based on knowledge of the individual amino acid profiles. These combinations have protein quality equivalent to dairy proteins. Quality of protein is influenced also by growing, handling and storing conditions. Poor availability of amino acids and/or difficulties in cooking can result from inadequate or prolonged storage of grain legumes.

A significant aspect of legume digestibility is the high level of faecal endogenous nitrogen excretion associated with legume consumption. This can create an impression of poor digestibility (50-70%). Careful balance studies in animals (Sgarbieri 1989) and humans using legumes in which protein amino acids were labeled with [15]N (Gausseres et al 1996) have shown digestibility values of ~90%, i.e. equivalent to animal proteins. The discrepancy between protein digestibility and faecal nitrogen seems to arise from the fact that legume consumption is associated with greater large bowel bacterial fermentation. This leads to a larger faecal biomass and excretion of bacterial protein.

Carbohydrates

Most grain legumes store energy as starch/NSP, oligosaccharides, and/or simple sugars (sucrose). Soybeans are an important exception, where energy is stored as lipid. Generally, legume starches are high in amylose, which has implications for digestion (Annison & Topping, 1994). High amylose starches resist small intestinal amylolysis to a greater degree than amylopectins and escape into the large bowel. This resistant starch (RS) offers some bowel health benefits, as it is fermented by the colonic microflorae to some potentially protective metabolites e.g. butyrate. Numerous animal studies have shown that the presence of RS can add considerably to the apparent fibre content of legumes (Topping et al 1993).

The oligosaccharides (raffinose, verbascose and stachyose) are present in reasonable concentrations. They are viewed as problematic, because of their resistance to small intestinal digestion coupled with rapid colonic fermentation and gas production. They are held responsible for some of the symptoms of discomfort associated with legume consumption. Techniques aimed at removing them involve processing by fermentation and/or germination, and are effective, but time consuming and expensive. Soaking, with or without bicarbonate, is easier and achieves some reduction. However, removal of oligosaccharides may not be advantageous. These compounds function as prebiotics and may stimulate proliferation of bacteria which promote health and provide protection against colonisation by potential pathogens (Gibson & Roberfroid, 1995). Non starch polysaccharides (NSP, major components of dietary fibre) are well represented in grain legumes both as soluble (SNSP) and insoluble (INSP) forms. Cotyledons are higher in SNSP while the hull is largely INSP and some lignin. Because of the associated high levels of tannins in the outer coat of some grain legumes, the hull is removed before they are used for human food.

Minerals Trace Elements And Vitamins

Generally, legumes are good sources of minerals, trace elements and vitamins, particularly those of the B group (e.g. thiamin, riboflavin, niacin, and folate). Mineral availability can be influenced by the presence of phytate, which binds cations such as calcium, iron and zinc and render them less available. However, processing techniques such as cooking (soaking/boiling) and germinating reduce phytate concentrations and increase mineral bioavailability.

Some common legumes (e.g. beans) contain antivitamin factors, such as an antivitamin E compound (Sgarbieri 1989). Their presence has been shown by their tendency to induce signs of E deficiency in farm animals when fed under controlled diet regimes. Their relevance to human nutrition is not known.

Antinutritive Factors

Maximal nutrient availability necessitates the removal (or at least reduction to a safe level) of the antinutrient factors (ANF) present in raw legumes. ANF include phytate, saponins, tannins and lignans, pectins, oligosaccharides, NSP, protease and amylase inhibitors and non-nutritive amino acids. Processing techniques include dehulling and milling, soaking and boiling, roasting, extruding and autoclaving, germinating (and malting) and fermenting. The latter two techniques are particularly useful in lowering oligosaccharides, if the processing is allowed sufficient time. Significant modification of phytate, protease inhibitors and non starch polysaccharides also occurs under these circumstances. Also heating (especially with water) increases starch digestibility.

Germination improves palatability and digestibility of legume/cereal mixes for weaning foods (Marero et al., 1988). We have examined malting and extrusion for the preparation of chickpeas and compared them with the (more traditional) boiling (Wang and McIntosh, 1996; McIntosh et al., 1996). All techniques tested were effective at unmasking cholesterol lowering ability, the boiling approach being best (See Figure 1).

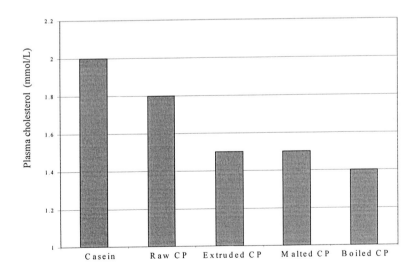

Figure 1 Effect of feeding processed chickpeas on rat plasma cholesterol concentration.

Soaking and boiling were most effective in reducing phytate, oligosaccharides and trypsin inhibitors. Extrusion offers great potential, but needs to be controlled carefully to avoid lowering of methionine availability (Bressani 1993). In instances where toxic agents are present (e.g. vicine/convicine in faba beans), it becomes essential that processing techniques offer maximum protection, e.g. discarding the water after soaking. Other ANF may present less risk and can, in fact, offer some health benefits. An example is phytochemicals with special properties, e.g. phytate, phyto-oestrogens, saponins and lignans which can function as anticancer or cholesterol-lowering agents.

GRAIN LEGUMES FOR HEALTH

Clearly, legumes offer numerous health advantages, not least of which is the low fat /high complex carbohydrate content of nearly all of those used for human foods. This is an important attribute for dieting/weight control. Other important properties include reduction of plasma cholesterol (cardiovascular disease), low glycaemic index (diabetes), colonic bacterial fermentation (bowel health) and phytochemical content (anticancer agents). Of these areas the last is least researched, and in need of more investigation. Significant progress is being made with soybeans in these areas of human health benefits (Messina and Erdman, 1995). For example, interest in phyto-oestrogens and their potentially protective role in hormone sensitive cancers such as breast cancer has raised international interest.

Plasma Cholesterol

Lowering of plasma low-density lipoprotein (LDL) cholesterol has been shown with a number of legumes and legume products (Anderson and Gustafson, 1988; Jenkins et al., 1993) (see Figure 2). Generally the lowering has been shown in animals and humans with preparations high in SNSP, although saponins may contribute substantially in some legumes. It is believed that SNSP (and saponins, if present) increase faecal bile acid and neutral sterol excretion leading to reduction in LDL cholesterol. However, caution needs to be exercised, as some processing (e.g. canning of navy beans) leads to loss of effectiveness (Cobiac et al., 1990). In addition to NSP, there is good evidence that legume proteins contribute to reduction of plasma cholesterol. Significant influences of legume foods on circulating hormone levels (e.g. insulin) have been shown recently and may contribute to some of the observed changes in plasma lipids. Certainly the muted insulin/glucagon response to legume ingestion is associated with minimal hyperglycaemia following a meal, and this feature is seen as beneficial to health, particularly for maturity onset diabetics, for whom glucose control is mandatory.

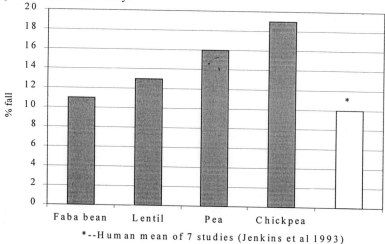

Figure 2 Influence of grain legumes on percent fall of plasma cholesterol concentration in rats and humans.

Large Bowel Health

Both the NSP and RS content of legumes contribute to faecal bulk, and also to colonic bacterial fermentation. This activity leads to production of short chain fatty acids (SCFA) which are important metabolic fuels for the large bowel and liver. In the colon SCFA have a number of important health-related actions including stimulation of fluid and electrolyte transport (prevention of diarrhoea) and promotion of muscular activity and colonic blood flow. In addition, one of the major acids (butyrate) is believed to promote a normal cell phenotype and prevent conditions such as colitis and cancer. Although legumes increase colonic fermentation, a potentially protective role in bowel cancer is disputed. The available epidemiological data are equivocal. Potter (1996), in reviewing the evidence, claimed that legumes (and potatoes) "show no direct benefit in colon cancer prevention". Animal experimentation has provided mixed evidence. There has been one study (Barnes, 1995) reporting fewer aberrant crypts (AC) in rats with genistein (a soy phyto- oestrogen) administration, but this evidence is not supported by other studies (Rao et al, 1997). Secondly, AC numbers do not predict tumor outcomes well and must be regarded as markers only of early mutagenic events. Most animal studies show that soybeans and/or processed byproducts are not protective against colon cancer, but lead to increased fermentation, generation of high concentrations of secondary bile acids and free fatty acids and colonocyte proliferation, probably resulting from high concentrations of soluble dietary fibre, fat, bile acids and cholesterol being delivered to the large bowel (Govers et al 1994, McIntosh et al 1995). Wasan and Goodlad (1996) have pointed out some of the apparent inconsistencies in fibre supplementation in this respect. One aspect not previously adequately appreciated has been the role of starch in the diet. Epidemiological evidence supports a protective role of starch in colonic cancer (Cassidy et al., 1994). In this respect soy must be regarded as an anomaly compared with other starch rich legumes and cereal foods; Japanese eat soy products as part of a high starch diet. Anticancer effects of minor legume components have also been reported (Messina and Erdman 1995). The substantial content of phytoestrogens (eg. genistein, diadzein) in soybeans are considered to contribute to inhibition of development of hormonally sensitive tumours such as prostate and breast cancers. Experimental animal studies are supportive of soybean and soy derived foods having an inhibitory influence on chemically induced breast cancer in rodents (Hawrylewicz et al., 1995). Other grain legumes are probably very variable products in this respect, and need further detailed examination as to how they compare compositionally, with respect to phyto-oestrogens and/or other protective factors. Clearly, more work is needed to help develop our use and understanding of these important food crops.

CONCLUSIONS

Grain legumes provide an important economical protein source, particularly in combination with cereals, which raise protein values to those of high quality sources. The content of NSP and (with the exception of soy) starch and RS has substantial implications for lowering of disease risk. Non-nutritive phytochemicals (saponins, oligosaccharides, protease inhibitors, phytate, phytoestrogens, lignans, flavonoids, etc.) with functional significance in the area of human disease prevention are a new dimension to the health potential of these traditional foods. New processing techniques are available to enhance their palatability and remove some of the factors inhibiting their wider acceptance. Coupled with their agronomic advantages, grain legumes are a valuable dietary option for a health conscious population.

References

Anderson JW and Gustafson NJ 1988. *American Journal of Clinical Nutrition* 48: 749- 53.

Annison G and Topping DL 1994. *Annual Review of Nutrition* 14: 297-320

Australian Food Survey 1993. *Edgell Birdseye* -CSIRO Publication.

Barnes S 1995. *Journal of Nutrition.* 125:777S- 783S

Bressani R 1993. *Food Review International* 9: 237-97

Cassidy A Bingham S and Cummings JH 1994. *British Journal of Cancer* 69: 937-942

Cobiac L, McArthur R and Nestel PJ 1990. *European Journal of Clinical Nutrition* 44: 819-22

FAO Statistics., Rome 1984.

Gausseres N, Mahe S, Benamouzig R and Tome D 1996. *Conference on Plant Proteins from European Crops*. INRA, France.

Gibson GR and Roberfroid MB 1995. *Journal of Nutrition*. 125: 1401-12.

Govers MJAP, Lapre JA, De Vries HT and Van Der Meer R 1993. *Journal of Nutrition*. 123: 1709-13

Hawrlewicz EJ, Zapata JJ and Blair WH 1995. *Journal of Nutrition*. 125: 698S-708S

Jenkins DJA, Spadafora PJ, Jenkins AL and Rainey-Macdonald CG 1993. In *CRC Handbook of Dietary Fiber in Human Nutrition* pp 419-38 2nd Edn. Ed. Gene A Spiller. CRC Press, Boca Raton.

Marero LM, Payumo EM, Librando EC et al 1988. *Journal of Food Science* 53: 1391-95

McIntosh GH, Regester GO , Le Leu RK, Royle PJ and Smithers GW 1995. *Journal of Nutrition*. 125: 809-16.

McIntosh GH, Wang YHA, Hughes G and Le Leu RK 1996. *Conference on Plant Proteins from European Crops*. INRA Nantes, France.

Messina M and Erdman JW 1995. *Journal of Nutrition*. 125:3S

Potter JD 1996. In *Second International Symposium on the Role of Soy in Preventing and Treating Chronic Disease*. pp14 Brussels Belgium Sept 15-18, 1996.

Rao CV, Wang CX, Simi B, Lubet R, Kelloff G , SteeleV and Reddy BS 1997. *Cancer Research* 57: 3717-22

Sgarbieri VC 1989. *World Review of Nutrition and Dietetics* 60:132-198

Salunkhe DK and Kadam SS 1989. *CRC Handbook of World Food Legumes: Nutritional Chemistry, Processing Technology, and Utilisation*. Vol. 1 CRC Press Inc. Boca Raton, Florida.

Topping DL, Illman RJ ,Clarke JM et al 1993. *Journal of Nutrition*. 123; 133-43

Wang YHA and McIntosh GH 1996. *Journal of Nutrition*. 126: 3054-62

Wasan HS and Goodlad RA 1996. *Lancet* 348 (Aug 3) 319-20

Chemical and physical factors influencing the nutritional value and subsequent utilisation of food legumes by livestock

R. J. van Barneveld[1], A. C. Edwards[2] and J Huisman[3]

1 SARDI Pig and Poultry Production Institute, Nutrition Research Laboratory, University of Adelaide, Roseworthy Campus, Roseworthy, SA 5371 Australia.; 2 ACE Livestock Consulting, PO Box 108, Cockatoo Valley, South Australia, 5351 Australia; 3 TNO Nutrition and Food Research, Animal Nutrition and Meat technology (ILOB), PO Box 15 6700 AA Wageningen, The Netherlands.

Abstract

There is significant potential for an increased use of food legumes in livestock feeds, especially considering the recent exclusion of animal proteins from ruminant diets in many countries. Improved nutritional definition of peas, lupins and beans, which are already widely used in livestock feeds, and other legumes such as vetch and *Lathyrus* spp has increased their value and utilisation. Research into the nutritional role of oligosaccharides, soluble and insoluble non-starch polysaccharides, starches, the mode of action of trypsin inhibitors, lectins and tannins and factors that influence amino acid and energy availability is increasing the efficiency of use of food legumes by livestock. Research is also identifying cost-effective mechanisms for improving their nutritional value. In addition, a better understanding of the interactions between food legumes and other feed ingredient is enhancing animal production.

INTRODUCTION

An understanding of the nutritional value of feed ingredients is essential if livestock production is to be optimised. Food legumes such as peas, lupins and faba beans represent valuable protein and energy sources for livestock. Their comparatively low starch and high non-starch polysaccharide content make them ideal fermentation substrates and energy sources for ruminants without the risk of lactic acidosis associated with feeding cereals. High levels of protein and amino acids also make them valuable additions to monogastric diets. Despite this, it is important to define animal responses to legume components including anti-nutritional factors, non-starch polysaccharides and starch types, and to establish the available nutrient content of these legumes. It is also necessary to establish the interactions between food legumes and other feed ingredients when they are combined in manufactured diets. The aim of this paper is to discuss 1) the current an potential consumption of food legumes by livestock, 2) the nutritional value of food legumes used in animal production systems, 3) chemical and physical factors influencing the nutritional value of legumes for livestock and 4) improving the utilisation and nutritional value of food legumes for livestock.

CURRENT AND POTENTIAL CONSUMPTION OF FOOD LEGUMES BY LIVESTOCK

The consumption of grain legumes by livestock varies widely between years and regions depending on prevailing seasonal and market conditions. As a result, it is very difficult to predict legume usage in stockfeeds with any degree of precision. It is also difficult to categorise legume usage by animal species because of the diverse range of raw materials used in livestock diets. The current consumption of food legumes by livestock in Australia will be used to demonstrate the potential of legumes for use in livestock feeds in comparison with other ingredients.

In the five years from 1991-95 the average total consumption of food legumes (lupins, field peas, faba beans) for stockfeed purposes in Australia was 568,000 tonnes/annum being comprised of 405,000 tonnes of lupins, 131,000 tonnes of peas and 32,000 tonnes faba beans (Meyers Strategy Group, 1995). Other food legumes such as chick peas, mung beans, cow peas, lentils, dried culinary beans and adzuki beans have been used occasionally but the volumes involved are minimal. Most of the material in this latter category would

R. Knight (ed.), Linking Research and Marketing Opportunities for Pulses in the 21st Century, 661–670.
© *2000 Kluwer Academic Publishers. Printed in the Netherlands.*

have been offered to stockfeed manufacturers as screenings or material failing to meet export standards. There are also minor volumes of legume milling offals obtained from splitting or seed cleaning activities used in stockfeeds. An estimate of grain legume use, relative to total feed usage for the major livestock species, is presented in Table 1. When grain legumes are readily available and competitively priced they are used in stockfeeds at considerably higher levels than indicated in Table 1.

Table 1. Total feed usage by animal species in Australia 1994/95 with the corresponding grain legume usage (Meyers Strategy Group, 1995)

Animal species	Total feed (not including roughage; '000 tonnes)	Legumes ('000 tonnes)	Legume use as a percentage of the total
Pigs	1,877	151	8.0
Poultry	2,295	156	6.8
Beef cattle	2,054	94	4.6
Dairy cattle	1,199	176	14.7
Other*	1,310	91	6.9
Total	8,735	669	7.7

*Includes grazing sheep and cattle, horses, deer, aquaculture, ostriches, emus and petfood

When estimating the potential for grain legume use in Australian stockfeeds, based on a possible average inclusion for pigs, broiler chickens and dairy cattle of 20 %, laying hens 15 % and feedlot beef 10%, Edwards (1994) suggested a figure of 1.36 million tonnes of legumes in a total feed usage of 7.85 million tonnes (or 17.3 %). Based on this estimate, there appears to be a potential for up to twice the current use in stockfeeds in Australia. Accounting for the omission of animal protein meals from ruminant diets in Australia and many European countries, the demand for protein rich crops, including grain legumes will increase.

NUTRITIONAL VALUE OF LEGUMES USED IN ANIMAL PRODUCTION SYSTEMS

The nutritional value of many food legumes has been widely reported for both ruminants and monogastrics (Gatel and Grosjean, 1990; Dixon and Hosking, 1992; Gdala et al., 1992; Gatel, 1994; Petterson and Mackintosh, 1994; Castell et al., 1996). A summary of this data is presented in Table 2. Note that the chemical composition of a stockfeed ingredient such as a legume is only the starting point when defining the nutritional value for livestock. Where possible, it is important to define the proportion of nutrients in a feed ingredient that are "available" for use by the animal. Accounting for the inevitable losses that occur during digestion and absorption of nutrients is a common method used to define the available nutrient content of a feed ingredient. It is important to note, however, that digestibility can frequently overestimate the nutritive value of feed ingredients depending on the measurement technique and the sample being analysed. For example, the apparent ileal digestibility of amino acids in heat-processed field peas fed to pigs significantly overestimates amino acid availability (van Barneveld et al. 1994).

Table 2. Nutritional value of major Australian grain legumes used in livestock feeds (as received from Petterson and Mackintosh 1994)

Legume	Pigs		Poultry		Cattle	Sheep
	Available lysine(g/Kg)	DE (MJ/kg)	Available lysine(g/Kg)	ME (MJ/kg)	ME (MJ/kg)	ME (MJ/kg)
Pisum sativum	14.3	14.4	14.2	11.5	11.7	12.0
Lupinus angustifolius	11.4	14.2	13.5	10.0	12.0	12.2
Lupinus albus	10.9	15.6	13.7	9.7	11.9	12.5
Vicia faba	12.9	14.5	12.9	11.2	13.4	11.5

DE, digestible energy;   ME, metabolisable energy

* van Barneveld et al (1997); ** J Noblet and R. J. van Barneveld unpublished data

The nutritional value of a number of food legumes for livestock has only recently been evaluated. The vetch *Vicia sativa* cv Blanchefleur contains the neurotoxins L-ß-cyanoalanine (0.1%) and g-L-glutamyl-L- ß -cyanoalanine (0.6%). Van Barneveld et al. (1997) studied the growth response of pigs fed graded levels of this legume. The vetch was included in grower diets (25-55 kg body-weight) at levels of 100 (diet 2), 200 (diet 3) and 300 (diet 4) g/kg, respectively, and in finisher diets (55-100 kg body-weight) at levels of 70 (diet 6), 140 (diet 7) and 210 (diet 8) g/kg, respectively. Diets were formulated using a wheat and barley base to contain equal levels of apparent ileal digestible (ID) lysine/MJ digestible energy (DE) with 0.70 and 0.55 g digestible lysine/MJ DE in the grower and finisher diets, respectively. DE content was also equalised in the grower (14.02 MJ/kg) and finisher (13.6 MJ/kg) diets. Graded dietary inclusion of vetch resulted in a linear decrease (P<0.001) in average daily gain (ADG) over the 25-55 kg growth phase but had no effect over the 55-100 kg growth phase. There was a significant linear decrease (P<0.01) in feed intake over both growth phases. No neurotoxic effects were evident and no L- ß cyanoalanine, or its derivatives, were detected in muscle tissue, suggesting this compound has little impact on pig growth and health. Van Barneveld et al (1997) concluded that *Vicia sativa* cv Blanchefleur is unsuitable for use with pigs during the 25-55 kg growth phase, however, some potential exists for inclusion in pig diets up to levels of 140 g/kg when fed from 55-100 kg. In addition, Szarvas et al (1997) reported that the apparent ileal digestibility of amino acids in vetch fed to growing pigs was equivalent to soybean meal. In poultry, inclusion of *Vicia sativa* cv. Blanchefleur at 100 g/kg resulted in an immediate drop in feed intake and rate of lay of hens, despite selective avoidance during feeding (Glatz et al., 1992).

Lathyrus cicera and *Lathyrus sativus* are two food legumes that have received little attention as a livestock feed ingredient. *Lathyrus* spp. have many agronomic advantages including resistance to drought, high forage and grain yield, and good adaptation to alkaline soils (Farhangi, 1996). For this reason, they are cultivated in many areas of the world including India, Bangladesh and Ethiopia (Chowdhury, 1988). Rotter et al. (1991) reported that *Lathyrus sativus* could be included in the diets of growing chicks at rates of 400 g/kg with no negative effect on weight gain or fat or protein digestibility. In this instance, the *Lathyrus sativus* was contributing no more than 2.2 grams of ß-N-oxalyl-amino-L-alanine per kg of *Lathyrus* seed. In contrast, Castell et al. (1994) suggested that trypsin inhibitors and chymotrypsin inhibitors limit the potential of *Lathyrus* seed as a feedstuff for young pigs with dietary inclusions of *Lathyrus sativus* above 10% reducing average daily gain and feed intake. Farhangi (1996) reported that *Lathyrus* has significant potential as a ruminant feed with *Lathyrus cicera* having a significantly higher dry matter degradability than *Lathyrus sativus*.

CHEMICAL AND PHYSICAL FACTORS INFLUENCING THE NUTRITIONAL VALUE OF LEGUMES FOR LIVESTOCK

All chemical and physical characteristics of a legume will influence their nutritional value for livestock to some extent, and when mixed with other feed ingredients some of these influences will be amplified. The following discusses those characteristics that have the greatest influence on nutritional value for either ruminants or monogastrics.

Protein

The comparatively high crude protein and available amino acid content of grain legumes makes them a highly valuable addition to monogastric diets. The low sulphur amino acid content of legumes such as lupins must be noted, but it is of little consequence when formulating monogastric diets as this deficiency can be made up from other feed ingredients or the addition of synthetic methionine to the diet at minimal cost.

The high biological value of grain legume proteins is of limited relevance in ruminants as most of the protein is degraded in the rumen and reconstituted into microbial protein or lost as ammonia. Despite this, there is some evidence that legume supplements are more effective than cereal grains supported by urea, particularly when used to supplement low to medium quality pastures or conserved forage (Valentine and Bartsch, 1990). Conversely, their role as a supplement to high quality pasture or silage may be limited due to the redundancy of their protein contribution.

The rumen degradability of legume protein varies between the different legume species and with factors such as grain processing (particle size/heat treatment), feeding level and the overall rate of rumen turnover (dwell time). When fed in a finely ground form, especially at low levels of feeding, the rumen by-pass characteristics of legume protein are minimal. Higgins et al (1985) studied the ruminal degradation of the individual proteins of legumes and found that they differed widely in their susceptibility to rumen breakdown. In peas, the high molecular weight polypeptides such as convicilin, vicilin and legumin were degraded rapidly, whereas ovalbumin and albumins were fairly stable in the rumen. Higgins et al. (1985) concluded that since pea albumin contains high levels of cysteine and methionine, essential for wool growth, the relative resistance of this protein to degradation in the rumen could be useful. Clearly, the implications are that despite having low undegradable protein, different legumes differ in the types of undegradable protein, still making them useful protein supplements.

Starch

One advantage the legumes confer on ruminant diets is a lowered risk of lactic acidosis due to a starch content less than cereal grains and the orderly fermentation rate of this and their non-starch carbohydrate components. Most prominent in this regard are lupins which are practically devoid of starch and instead have ß-(1,4)-galactan as their primary storage carbohydrate (Hill, 1977). Valentine and Bartsch, (1987) measured the rumen fermentation parameters for dairy cows fed lupins, peas, faba beans or barley as supplements to oaten hay. They found that the legume grains did not lower rumen pH as markedly as barley grain, despite no significant differences in volatile fatty acid concentration and proportions between cereal grains and legumes. Furthermore, rumen ammonia nitrogen was higher for legume grains, despite the concentrates being isonitrogenous.

In contrast, the lower starch content of legume grains reduces their net energy contributions to monogastric diets (ie a high proportion of the energy derived from legumes fed to monogastrics is derived from fermentation compared to cereals). In addition, the complex non-starch polysaccharide structure of legumes can induce anti-nutritional effects when fed to monogastrics.

Oligosaccharides

In Australia, *L. angustifolius* and *L. albus* are the predominant lupin species available for use in commercial pig diets. Inclusion of *L. albus* in pig diets at levels above 15 % results in a significant reduction in feed intake and subsequent growth rates, a phenomenon not observed with *L. angustifolius*. Oligosaccharides have been implicated in the poor performance of pigs fed diets containing *L.albus* due to their comparatively high levels and their indigestible nature in the small intestine. Van Barneveld et al (1996, 1997) used ethanol extraction to reduce the levels of oligosaccharides in *L. angustifolius* and *L. albus* by approximately 70% and examined the subsequent digestion of energy and amino acids in the small intestine and whole digestive tract of growing pigs fed diets containing these lupins. Ethanol extraction did not influence the ileal energy digestibility of diets containing *L. angustifolius*, but improved the ileal (P<0.05) and faecal (P<0.01) energy digestibility of diets containing *L. albus* and the faecal energy digestibility (P<0.01) of those containing *L. angustifolius* (Table 3). Ethanol extraction improved the DE of diets containing *L. angustifolius* and *L. albus* by 0.5 and 0.7 MJ/kg, respectively. Ethanol extraction significantly improved (P<0.05) the digestion of all reported amino acids in both *L. angustifolius* and *L. albus* (Table 3). The average amino acid digestibility was improved by 9.6% for *L. angustifolius* , by 7.6% for *L. albus*.

Lupin oligosaccharides appear to hinder the digestion of amino acids in the small intestine of pigs. This is in contrast to the findings of Gabert et al. (1995) and Zuo et al. (1996) and suggests that the osmotic effects of these oligosaccharides on the gut dynamics in pigs may change nutrient digestion in the small intestine. On this basis, these results also suggest that the increase in lupin DE consistent with oligosaccharide extraction observed by van Barneveld et al. (1996), was due to more than a digestible energy dilution when oligosaccharides were present.

Non-starch polysaccharides

As discussed previously, the high non-starch polysaccharide (NSP) content of legumes makes them a highly desirable ruminant feed ingredient, yet these non-starch polysaccharides can induce anti-nutritive effects in monogastrics.

Table 3 Energy digestibility coefficients at the end of the small intestine and whole digestive tract and apparent ileal amino acid digestibility coefficients of pigs fed whole dehulled or ethanol extracted *L .angustifolius* or *L. albus*

Treatment.	Dehulled *L. angustifolius*		Dehulled *L. albus*		Statistics	
	Nil	Extracted[1]	Nil	Extracted[1]	Diet	SEM
Energy digestibility (ileum)	0.63a	0.67a	0.68b	0.76c	*	0.017
Energy digestibility (faeces)	0.85a	0.87b	0.87b	0.90c	**	0.005
Diet DE (MJ/kg DM)	13 la	13.6b	14.1c	14.8 d	***	0.077
Amino acids					**	
Threonine	0.7a	0.81bc	0.78b	0.86c	*	0.018
Valine	0.77a	0.84bc	0.81ab	0.88c	*	0.017
Isoleucine	0.8la	0.88bc	0.85ab	0.91c	*	0.014
Leucine	0.78a	0.88b	0.84b	0.90b	*	0.016
Phenylalanine	0.80a	0.88bc	0.84ab	0.90c	*	0.014
Lysine	0.80a	0.86b	0.84ab	0.89 b	*	0.015
Histidine	0.80a	0.85bc	0.8lab	0.87c	*	0.012

Values within a row with different letters differ; *, P<0.05; **, P< 0.01; ***, P < 0.001

DM, dry matter; DE, digestible energy; SEM, standard error of the mean

[1] modified ethanol extraction (Coon et al., 1994) [2] Sum of raffinose and verbascose

Van Barneveld et al (1995) reported that growing pigs fitted with ileal cannulas and fed diets containing graded levels of a lupin NSP isolate experienced a significant linear decrease in lysine and energy digestibility (Table 4). This coincided with a significant linear increase in digesta viscosity, but there was no significant effect on the microbial activity in the small intestine, the large intestine or the caecum. The observed response is likely to be due to a combination of soluble and insoluble NSP components from the lupins. Hansen et al (1991) and van Barneveld (1997) suggested that combinations of ingredients with high levels of soluble NSP, such as legumes and cereals, may result in a further reduction in the digestibility of specific nutrients, including protein and energy.

Table 4. The digesta viscosity, lysine and energy digestibility and ATP activity of digesta from the small intestine, caecum and large intestine of growing pigs fed diets containing graded levels of isolated lupin non-starch polysaccharides

Diet	Viscosity (mPa.s)	Digestibility			ATP (mg/g digesta)		
		Energy	Lysine	S.I.	Caecum	L.I.	
1	1.43	0.85	0.91	0.72	4.78	1.97	
2	2.28	0.71	0.90	0.61	5.73	1.66	
3	2.33	0.50	0.82	0.88	7.14	1.98	
4	4.89	0.58	0.85	0.47	5.45	3.38	
Diet	***	***	**	NS	NS	NS	
SEM	0.304	0.031	0.014	1.185	1.185	1.185	

SEM,mean standard error; NS,not significant; **,P<0.01; ***,P<0.001; Diets 1, 2, 3 and 4 were 0%, 5%, 10% and 15% lupin NSP respectively. (After van Barneveld et al, 1995)

The source of NSP can also influence the response of monogastrics to legumes in diets. Van Barneveld (1996 and unpublished data) determined the apparent ileal amino acid digestibility and digestible energy of wheat, barley, triticale, *Lupinus angustifolius* (cv Gungurru) and *Cicer arietinum* (Desi; chick peas) and then formulated diets to contain 50% of each cereal, respectively, and 35% of each legume, respectively. Diets were equalised for ileal digestible amino acids with lysine limiting at 0.40 g/MJ DE and the growth rates of pigs fed these diets determined (Table 5). Pigs fed combinations of chickpeas and cereals exhibited no significant differences in their empty-body weight gains. In contrast, a highly significant difference was observed in the empty-body-weight gain of pigs fed the diet containing lupins and barley, compared to lupins and wheat and lupins and triticale, respectively. Based on the original diet formulations, all pigs should have grown at the same rate if the apparent lysine digestibility and digestible energy values were additive, when the lupins and cereals were combined in a mixed diet. It appears that the anti-nutritive effects of soluble and insoluble NSP from lupins and barley are amplified when these feed ingredients are combined.

Table 5. Daily liveweight gain, daily empty-body-weight-gain and empty-body-weight feed conversion ratio of growing pigs (25-55 kg) fed diets formulated to equal levels of ileal digestible lysine and containing specific combinations of lupins (*L. angustifolius* cv Gungurru) and either wheat, barley or triticale or chickpeas (*C. arietinum* desi) and either wheat, barley or triticale.

	Average gain (g/d)	EBW gain (g/d)	EBW FCR
Lupinus angustifolius			
+ wheat	677	620a	2.71a
+ barley	662	590b	2.99b
+ triticale	681	630a	2.60a
Cicer arietinum			
+ wheat	692	645	2.45
+ barley	676	624	2.55
+ triticale	671	629	2.58

Values in a column with different letters differ significantly (P < 0. 05)
EBW, empty-body-weight; FCR, feed conversion ratio

Anti-nutritional factors.

Anti-nutritional factors (ANF's) are described as non-fibrous naturally occurring substances exerting negative effects on performance or health of animals. Most of these substances provide the plant a natural protection against attacks by moulds, bacteria, insects and birds (Birk, 1987, 1989; Ryan 1983; Liener & Kakade, 1980). Contaminants, such as mycotoxins and factors attributable to processing, which can also have anti-nutritional effects, are excluded from the definition of ANFs (Yannai, 1980). Grain legumes contain various ANF's (Liener, 1980; Savage 1988; Savage and Deo, 1989; Huisman, 1990, 1991; Huisman and Jansman, 1991, Huisman and Tolman, 1992; Huisman and van der Poel, 1992; Jansman, 1993; Sissons and Tolman, 1991; Gatel 1994).

Huisman and Tolman (1992) clearly outlined that protease inhibitors, trypsin inhibitors, lectins and tannins are the most important ANF's in animal nutrition. For this reason, the following discussion will focus on the latest research associated with these ANF's only. Trypsin inhibitors and lectins are mainly present in peas, lentils and beans. Tannins can be relevant in faba beans and some pea varieties. Other ANF's which can be relevant are vicine and convicine in faba beans and alkaloids in lupins. Less important ANF's in grain legumes are a-amylase inhibitors, saponins and phytates. The negative influences of these ANF's on livestock production are far less pronounced in ruminants due to rumen microorganisms breaking down many of these factors before they affect digestion or are absorbed by the animal. In fact, factors such as tannins, which exert antinutritional effects in monogastrics, may be potentially beneficial in ruminants by inhibiting the digestion of protein and carbohydrate fractions in the rumen (Dixon and Hosking, 1992). As a result, more protein and carbohydrate is likely to reach the small intestine where digestion is more efficient. For this reason, the following discussion will be restricted to monogastrics.

Trypsin inhibitors.

The primary effects of trypsin inhibitors in the animal are the inactivation of trypsin and chymotrypsin produced by the pancreas. Due to this inactivation, the activity of the chymotrypsin may be reduced and, via a negative feedback mechanism, the production of chymotrypsin by the pancreas may be stimulated. Higher secretion of pancreatic enzymes will lead to an increased loss of endogenous proteins.

A study with piglets was carried out to determine the effect of different levels of trypsin inhibitors on the excretion of endogenous and exogenous (feed) protein at the. terminal ileum (Schulze, 1994). Twelve piglets with an average body weight of 13 kg were fitted with a post-valvular T-caecum (PVTC) cannula. The piglets were fed a basal diet containing 160g protein/kg from a soya protein concentrate low in ANFs supplemented with 0, 1.9 and 5.7 mg of soya trypsin inhibitors per kg diet.[15] N-isotope dilution was used to establish endogenous N flow. The results from this experiment clearly showed that trypsin inhibitors decrease the apparent ileal protein digestibility (Table 6). At a trypsin inhibitor activity (TIA) of 2.5, the apparent ileal protein digestibility was 13 units lower than the control. This decrease could be attributed to increased excretion of endogenous protein (9 units) and to increased excretion of undigested feed protein (4 units). The effects were even greater at a TIA of 5.8, when the apparent ileal protein digestibility was decreased by 31 units. This difference could be attributed to 15 units of increased excretion of endogenous protein and 16 units of increased excretion of exogenous protein. The increased excretion of endogenous protein is due to the stimulating effect of trypsin inhibitors on the secretion of the pancreatic enzymes trypsin and chymotrypsin which pass partly undigested to the terminal ileum. At lower trypsin inhibitor levels the decrease in apparent ileal protein digestibility is mainly due to endogenous protein, while at higher trypsin inhibitor levels decreases in apparent ileal protein digestibility is due to both endogenous and exogenous protein.

Table 6. Effect of soya trypsin inhibitors on the excretion of endogenous and exogenous protein in the ileal digesta of piglets with an average body weight of 13 kg (adapted from Schulze, 1994) and the effect of condensed tannins from faba beans on the excretion of endogenous and exogenous protein in ileal digesta of piglets with an average body weight of 14 kg (adapted from Jansman, 1993)

	Trypsin inhibitor activity (TIA) expressed as mg inhibited trypsin per g feed			Condensed tannins		
	0.2	2.5	5.8	< 0.1	<0.15	0.69
Flow of endogenous protein (g/l00g protein intake)	13.2a	22.6b	27.5b	9.0a	11.7 b	16.4c
Flow of exogenous protein (g/l00g protein intake)	0.1	0.4	17.0	2.8a	5.6b	9.5c
Apparent ileal protein digestibility	86a	73b	55c	88.2a	82.7b	74.1c
True ileal protein digestibility	99a	96a	83b	97.2a	94.4b	90.5c

Values with different letters in the same row differ significantly ($P < 0.05$)

The above results demonstrate that trypsin inhibitor activity is an important indicator of the nutritional quality of food legumes for livestock. Modern varieties of high quality peas contain a TIA level of between 1 and 2, yet it is not uncommon for some widely used varieties to have TIA levels of 3 to 4 with some reaching 8 to 11 (Gatel, 1994). The results of Schulze (1994) show that even in the modem pea varieties, TIA can have an impact on livestock production.

Lectins.

Lectins are proteins which bind to the carbohydrate moiety of the glycoproteins in the small intestinal brush border membrane (Kik, 1991). If they bind, the intestinal epithelial cells may be damaged which results in a disturbed digestion and absorption process.

Lectins are predominantly important in *Phaseolus* beans, because they are exceptionally toxic (Pusztai, 1989; Kik et al, 1990). Huisman et al. (1993) showed that lectins from *Phaseolus* beans decreased apparent protein digestibility considerably. This effect was due to lectins increasing the excretion of endogenous protein and reducing true protein digestibility. In contrast, pea lectins have been reported to be non-toxic in piglets and rats (Bertrand et al 1988; Grant et al., 1983; Aubry and Boucrot, 1986; Kik, 1991) although Jindal et al. (1982) reported that pea lectins had a damaging effect on the gut wall. Information on the toxicity of faba bean lectins for livestock is limited. Kik (1991) reported that lectins in faba beans may be toxic, however the levels in faba beans are low (Gatel, 1994).

Tannins.

Tannins are present in the coloured seeds of faba beans and other pulses. Tannins complex with proteins easily, but binding with other nutrients (carbohydrates, minerals) can also take place (Jansman, 1993). As a result, tannins interfere with different aspects of digestion and absorption, however, the mode of action is not entirely clear. Jansman (1993) conducted an experiment with two groups of 4 piglets of 10 kg liveweight each fitted with PVTC cannula. The flow of the endogenous and exogenous protein was measured using the 15 N-dilution method as described by De Lange et al.(1990) and Huisman et al. (1992). A basal diet was formulated to contain 160g crude protein/kg from 250 g dehulled faba beans/kg diet and 90g casein/kg diet. Two test diets were formulated to contain 200 g/kg of low tannin faba bean hulls or high tannin faba bean hulls, respectively, at the expense of maize starch. The content of condensed tannins in the two diets was < 0. 1 % and 0. 69 %, respectively (Table 6). The response observed with the addition of low tannin hulls indicates that other dietary components, such as non-starch polysaccharides. may be influencing endogenous ileal flow and exogenous protein digestion. Condensed tannins increased the excretion of endogenous and exogenous protein. The apparent ileal protein digestibility was decreased by 8 units, four units of which could be attributed to the increased excretion of endogenous protein and 4 units to the increased flow of exogenous feed protein. It was concluded that condensed tannins in faba beans interact with both dietary and endogenous proteins in the digestive tract of pigs reducing the true ileal protein digestibility.

The above discussion demonstrates that the major ANF's in legumes have a significant influence on the flow of endogenous protein, and hence, true ileal amino acid or protein digestibility measurements should be used when assessing the nutritional value of food legumes for livestock.

IMPROVING THE UTILISATION AND NUTRITIONAL VALUE OF FOOD LEGUMES FOR LIVESTOCK

The main legumes (lupins, peas, faba beans) used in livestock production are quite useful commodities in their current state. Although they all contain factors that can limit animal production, the principal constraint determining their level of use in livestock feeds is "cost relative to achieving the same nutrient contribution from other sources". Consequently, when setting breeding objectives for grain legumes, yield as the principal means to reduce cost should probably remain the primary breeding objective, with the manipulation of specific nutrient components being given secondary priority.

Several physical processes can be employed to improve the utilisation of food legumes in the diets of livestock including grinding, dehulling, air classification and soaking/leaching. Before using any of these methods to improve the nutritional value of legumes, the cost-benefit should be carefully assessed.

Grinding

In monogastrics the digestibility of grain legumes is directly related to the fineness of the grind. Smaller particle sizes expose a greater surface area for acid and enzymic hydrolysis. In ruminants particle size is less critical. As long as the seed testa is broken, microbial fermentation will utilise the seed nutrients quite effectively. In fact, no processing is required for sheep as they thoroughly masticate the seeds prior to swallowing. In cattle, however, it is generally recommended that grain legumes be at least coarsely crushed.

Dehulling

The utilisation of some grain legumes can be improved by segregating the seeds into hulls and kernels. In the case of lupins, the largely cellulosic hull is of limited value to monogastrics, with the exception of sows, which can extract more than 14 MJ of digestible energy/kg dry matter from lupin hulls compared to growing pigs which can extract only 7 MJ of digestible energy/kg dry matter (J.Noblet and R.J. van Barneveld, unpublished data). The recovery of some energy by pigs from the hull fraction is largely via hindgut fermentation. In poultry, however, legume hulls act largely as an inert diluent, due to the absence of any effective fermentation function in the avian gut.

Having removed the hull, the residual kernel has a higher energy and protein content than the whole seed and is of increased commercial value to monogastrics. The hull fraction being highly fermentable is of good value to ruminants, so in situations where lupin kernels can be directed to monogastric diets and the hull fraction to ruminant diets there is considerable value in dehulling. In the case of faba beans, most of the tannins are found in the hull and hence dehulling can be an effective means of reducing the negative effects of tannins in monogastric diets (Fowler, 1976). This is practised commercially in Europe but not to any significant extent in Australia.

Air classification

High protein concentrates can be derived from grain legumes by air classification. This allows the high quality protein fractions to be specifically employed in isolation to the usually attendant non-starch polysaccharide fractions and other anti-nutritional components. The cost of this process, however, largely precludes its use in commercial livestock feeding. It is more orientated to human and petfood applications.

Soaking/leaching

The water soluble anti nutritional factors in food legumes can be reduced by traditional soaking or leaching procedures but these are not logistically feasible in commercial feed milling.

Heating

Phaseolus beans can only be used after an intensive toasting procedure (Van der Poel, 1990). Among the toasting procedures, pressure toasting seems to be the most adequate method, because other harmful components in the storage proteins are also inactivated (Van der Poel, 1990). It must be remembered, however, that the application of heat to any protein source can have a negative effect on the availability of the remaining protein (van Barneveld et al., 1994).

CONCLUSIONS

Food legume use in livestock diets will increase and the efficiency of livestock production will improve with an increased understanding of the nutritional role of these legumes. To date, the nutritional value of legumes such as field peas, lupins and faba beans has been well defined, and through this process our understanding of the nutritional role of non-starch polysaccharides, oligosaccharides and anti-nutritional factors such as trypsin inhibitors, tannins and lectins has improved. Further work is required to improve our knowledge of other legumes such as vetch and *Lathyrus* which have the potential to be valuable ingredients for use in livestock diets.

References

Aubry, M.and Boucrot, P. 1986. *Annales de Nutrition et Metabolism*, 30: 175-182.

Bertrand, G., Seve, B., Gallant, B. and Tome, R. 1988. *Sciences des Aliments* 8: 187-212.

Birk, Y. 1987. *Hydrolytic Enzymes*, pp. 257-305 (ed. A. Neuburger and K. Brocklehurst) Elsevier, Amsterdam, Netherlands.

Birk, Y.1989. In: *Recent Advances of Research in Anti-nutritional Factors in Legume seeds* pp 239-250. (ed: J. Huisman, A.F.B. Van der Poel and I.E. Liener) Pudoc, Wageningen, Netherlands

Castell, A.G., Cliplef, R.L., Briggs, C.J., Campbell, C.G. and Bruni, J.E. 1994. *Canadian Journal of Animal Science.* 74:529-39.

Castell, A.G., Guenter, W. and Igbasan, F.A. 1996. *Animal Feed Science and Technology.* 60:209-227.

Chowdhury, S.D. 1988. *World's Poultry Science.* 44: 7-16.

Dixon, R.M. and Hosking, B.J. 1992. *Nutrition Research Reviews.* 5:19-43.

Edwards, A.C. 1994. In *Proceedings of the First Australian Lupin Technical Symposium*, 238-243 (eds M. Dracup and J. Palta) Department of Agriculture: Western Australia.

Farhangi, M. 1996. PhD Thesis. The University of Adelaide.

Fowler, V.R. 1979. In *Vicia faba: Feeding value, processing and viruses.* 31-43 (ed D.A. Bond)

Gabert, V.M., Sauer, W.C., Mosenthin, R., Schmitz, M. and Ahrens, F 1995. *Canadian Journal of Animal Science* 75: 99-107.

Gatel, F. 1994. *Animal Feed Science and Technology.* 45: 317-348.

Gatel, F. and Grosjean, F. 1990. *Livestock Production Science.* 26 (3):155-175.

Gdala, J., Buraczewska, L., and Grala, W. 1 992. *Journal of Animal and Feed Sciences* l (l): 71-79.

Glatz, P.C., Hughes, R.J. and Woodford, R.C. 1992. In *Proceedings of the Australian Poultry Science Symposium*, p. 142, Sydney:Australia.

Grant, G., More, L.J., McKenzie, J.C., Stewart, A. and Pusztai, A.1983. *British Journal of Nutrition,* 50: 207-214.

Hansen, I., Larsen, T., Bach Knudsen, K.E., and Eggum, B.O. 1991. *British Journal of Nutrition.* 66: 27-35

Hill, G.D. 1977. *Nutrition Abstracts and Reviews B.* 47:511-529.

Huisman, J. 1990. Ph D Thesis, Agricultural University Wageningen, Netherlands.

Huisman, J 1991. *Proceedings 8th European Symposium on Poultry Nutrition*, 42-59.

Huisman, A.J.M. 1991 *Nutrition Abstracts and Reviews. Series B:Livestock Feeds and Feeding* 61: 901-921.

Huisman, J. and Tolman, G.H. 1992. In:*Recent Advances in Animal Nutrition*, p. 3-31 (ed. P.C.Garnsworthy, W.Haresign,W. and D.J.A. Cole) Butterworths Heineman, Oxford, UK

Huisman J. and van der Poel, A.F.B. 1994. In: *Expanding the Production and Use of Cool Season Food Legumes.* pp 53-77 (eds F.J. Muehlbauer and W.J. Kaiser) Kluwer Academic Publishers.

Huisman, J., Verstegen, M.W.A., van Leeuwen, P. and Tamminga, S. 1993. *In Proceedings of the First International Symposium on Nitrogen Flow in Pig Production and Environmental Consequenses.* pp 55-61 Pudoc, Wageningen, Netherlands.

Jansman, A.J.M. 1993. *Nutrition Research Reviews.* 6: 209-236.

Jindal, S., Soni, G.L. and Singh, R. 1982. *Journal of Plant Foods.* 4: 95-103.

Kik, M.J.L., Huisman, J.,van der Poel, A.F.B. and Mouwen, J.M.V.M. 1990. *Veterinarian Pathology,* 27: 329-334.

Kik, M.J.L. 1991. Ph D Thesis, State University, Utrecht, Netherlands.

Liener, I.E. 1980. *Toxic constituents of plant foodstuffs*, Academic Press, New York, United States. 502 pp.

Liener I.E. and Kakade M.L. 1980. In: *Toxic constituents of plant foods*, pp.7-71. Academic Press, New York, United States.

Meyers Strategy Group 1995. *"Feed Grains Study Report",* Commissioned by the Grains Research and Development Corporation, Dairy Research and Development Corporation, Pig Research and development Corporation, Chicken Meat Research and Development Corporation and Egg Industry research and Development Corporation.

Petterson D.S. and Mackintosh, J.B. 1994. *The Chemical Composition and Nutritive Value of Australian Grain Legumes.* Grains Research and Development Corporation:Canberra.

Pusztai, A. 1989 In: *Recent Advances of Research in Anti-nutritional Factors in Legume seeds* pp 17-29 (ed: J. Huisman,A.F.B. Van der Poel and I.E. Liener) Pudoc, Wageningen, Netherlands

Rotter, R.G., Marquardt, R.R. and Campbell, C.G. 1991. *British Poultry Science* 32: 1055-67.

Ryan, C.A. 1983. In: *Variable Plant and Herbivores in Natural and Managed Systems.*p 43-60. (ed. R.F. Denno and M.S. McClure) Academic Press, New York, United States

Savage, G.P. 1988. *Nutrition Abstracts and Reviews, Series A,* 58:320-344

Savage, G.P. and Deo, S. 1989. *Nutrition Abstracts and Reviews, Series A,* 59 (2) 66-68.

Schulze,H. 1994. Ph D Thesis Agricultural University Wageningen, Netherlands.

Sissons, J.W. and Tolman, G.H. 1991. In: *Toxic factors in Crop plants*, pp 62-85 (ed. J.P.F.D. Mello and C.M. Duffus) Scottish Agricultural College, Edinburgh, UK.

Szarvas, S.R., Rathjen, J.M. and van Barneveld, R.J. 1997. In *Manipulating Pig Production VI.* (eds. D.P. Hennessy and P. Cranwell) Australasian Pig Science Association: Canberra. (in press)

Valentine, S.C., and Bartsch, B.D. 1987. *Animal Feed Science and Technology.* 16: 261-71.

Valentine S.C. and Bartsch, B.D. 1990. *Australian Journal of Experimental Agriculture.* 30: 7-10.

van Barneveld, R.J. 1996. Rhone-Poulenc Animal Nutrition Seminar *"New Technologies for Livestock Industries",* Sydney Opera House, 1996.

van Barneveld, R.J., Batterham, E.S., and Norton, B.W. 1994. *British Journal of Nutrition.* 72: 257-75.

van Barneveld, R.J., Baker, J., Szarvas, S.R. and Choct, M. 1995. *Proceedings of the Nutrition Society of Australia.* 19: 43.

van Barneveld, R.J., Campbell, R.G., King, R.H., Dunshea, F.R. and Mullan, B.P. 1997. *Proceedings of the Nutrition Society of Australia* (in press)

van Barneveld, R.J., Olsen, L.E. and Choct, M. 1996. *Proceedings of the Nutrition Society of Australia* 20: 114

van Barneveld, R.J., Olsen, L.E. and Choct, M. 1997. In *Manipulating Pig Production VI.* (eds. D.P. Hennessy and P. Cranwell) Australasian Pig Science Association: Canberra. (in press)

van Barneveld, R.J., Davis, B.J., Szarvas, S.R. and Wyatt, G.F. 1997. *In Manipulating Pig Production VI.* (eds. D.P. Hennessy and P. Cranwell) Australasian Pig Science Association:Canberra. (in press)

van Barneveld, R.J., Szarvas, S.R., Wyatt, G.F. and Rathjen, J.R. 1997. In *Manipulating Pig Production VI.* (eds. D.P. Hennessy and P. Cranwell) Australasian Pig Science Association: Canberra. (in press)

Van der Poel, A.F.B. 1990. *Advances in Feed Technology.* Verlag Moritz, Schafer, Detmold p 22-34.

Yannai, S, 1980. In: *Toxic Constituents of Plant Foodstuffs*, pp 371-418 (ed. I.E. Liener) Academic Press, New York, United States

Zuo, Y., Fahey, G.C., Merchen, N.R. and Bajjalieh, N.L. 1996. *Journal of Animal Science* 74: 2441-49.

Towards the elimination of anti-nutritional factors in grain legumes

Enneking, D.[1] and Wink, M.[2]

1. Centre for Legumes in Mediterranean Agriculture, University of Western Australia, Nedlands WA 6907, email: enneking@cyllene.uwa.edu.au ; 2. University Heidelberg, Institute for Pharmaceutical Biology, Im Neuenheimer Feld 364, D-69120 Heidelberg, email: Michael.Wink@urz.uni-heidelberg.de

Abstract

Anti-nutritional factors (ANFs) in grain legumes can be divided into several groups based on their chemical and physical properties such as non protein amino acids, quinolizidine alkaloids, cyanogenic glycosides, pyrimidine glycosides, isoflavones, tannins, oligosaccharides, saponins, phytates, lectins or protease inhibitors. Their elimination can be achieved either by selection of plant genotypes with low levels of such factors, or through post-harvest processing (germination, boiling, leaching, fermentation, extraction etc.). The development of new food crops from *Lupinus*, *Vicia* and *Lathyrus* species is used to illustrate the problems associated with heat-stable low molecular weight ANFs. The relative merits of the various strategies for the elimination of ANFs are discussed in relation to the feasibility of making the elimination and to the ecological function of ANFs. ANFs can be important to the plants producing them, as they can function as defence compounds against herbivores and microorganisms.

1. INTRODUCTION

Plants which produce seeds, rich in energy supplies (carbohydrates, lipids, proteins), usually accumulate potent chemical defence compounds. This occurs in many grain legumes with comparably large and protein-rich seeds. They often contain substantial amounts of "anti-nutritive" factors (ANF), such as lectins, protease inhibitors, non-protein amino acids (NPAAs), alkaloids, cyanogenic glycosides, pyrimidine glycosides, saponins, tannins, isoflavones, oligosaccharides, erucic acid, or phytates (Bell and Charlwood, 1980; Conn, 1981; Rosenthal and Berenbaum, 1991; Rosenthal, 1982; Bardocz et al., 1996; Bardocs and Pusztai, 1996).

Since many of the ANFs are toxic, unpalatable or indigestible, elimination strategies are probably as old as mankind. Grain legumes have been cultivated for many centuries and were a part of Neolithic farming systems (Zohary & Hopf, 1988). The early means of elimination must have consisted mainly of the thermal destruction of heat labile ANFs and leaching out of others by cooking. Today more strategies are available to minimize the impact of ANFs including:

Reduction of ANFs through processing (germination, boiling, leaching, fermentation, extraction, etc)

Reduction of ANFs through genetic manipulation (selection from natural or artificial diversity, genetic engineering of biosynthetic pathways or of the toxic protein itself)

Improvement of tolerance to ANFs through supplementation of diets with protective factors eg, methionine, threonine, feed enzymes.

But before making a decision, whether to breed for ANF free crops or to opt for processing, it is important to discuss the biological consequences and economic constraints of changing ANFs. We will not give a comprehensive review covering all aspects or all relevant publications. Our paper is focussed on two groups of ANFs (lupin alkaloids and NPAAs) with which we have personal experience and references are often restricted to reviews instead of the original papers.

2. ANF BIOLOGY

In order to survive during evolution, plants have developed defences against herbivorous animals, microorganisms and viruses. Furthermore, plants compete with other plants for light, water, and nutrients.

671

R. Knight (ed.), Linking Research and Marketing Opportunities for Pulses in the 21st Century, 671–683.

The production of secondary metabolites (including lectins and toxic peptides) is of ultimate importance as a defence strategy in this context (Rosenthal, 1982; Harborne, 1993; Wink, 1988, 1992, 1993a). The seedling is the most vulnerable stage in the plants' life cycle and it is not suprising that many species are equipped with anti-nutritional factors and other substances in their seeds. Plants are often economic in that they are able to utilize defence compounds (especially the N containing ones) as a nitrogen source during germination. These compounds can be either acutely toxic (such as some lectins, cyanogenic glycosides, NPAAs or alkaloids), unpalatable (such as saponins, tannins, NPAAs, or bitter alkaloids) or "anti-nutritive" reducing growth and fitness of the consumer by nutrient complexation (eg, by phytates), metabolic inhibition (eg, NPAAs, cyanogenic glycosides, isoflavones, alkaloids) or reduction of digestion (eg, through protease inhibitors, lectins, or oligosaccharides).

Considering the evolutionary background of ANF as either antiherbivore or antimicrobial, ANFs have been selected during evolution as biologically active compounds. If isolated and processed they even could be useful in medicine or in agriculture (increasing resistance to pests and pathogens; as natural plant protectants "biorational pesticide" or phytomedicines). The biological background and properties of legume ANFs is briefly summarized in the following. Our results on quinolizidine alkaloids, which constitute the main anti-nutrients of lupins (Wink, 1985, 1988, 1992, 1993a, 1993b), and recent work with non-protein amino acids in Vicia and Lathyrus (Lambein et al., 1990, 1992; Enneking et al., 1993, 1997; Enneking, 1994, 1995) exemplify this view and are covered in more detail.

2.1   . Lectins and protease inhibitors

Those legumes which do not accumulate low molecular weight toxins in their seeds, often store toxic peptides (eg protease inhibitors), or lectins (Bisby et al., 1994). Whereas protease inhibitors block the function of digestive enzymes (proteases) in animals (leading to malnutrition and other disturbances), lectins either bind to receptors in the intestinal tract and related organs (mimicking the activity of other signal compounds) or are taken up by cells and inhibit protein biosynthesis (eg, the lectin abrine from the legume Abrus precatorius is one of the most potent toxins known). These compounds also serve a double purpose for the plants producing them: they are both N storage compounds (which are broken down during germination and seedling development) and defence chemicals against animals. Since lectins are gene encoded proteins, they are direct targets for genetic engineering. For example, the genetic approach will allow the downregulation of lectin synthesis (by "antisense strategies"), eventually leading to lectin-free crops. On the other hand, since lectins mediate insect toxicity, a transfer and expression of lectin genes in other crop plants would be a means of enhancing their resistance to insects.

2.2   Saponins

Triterpene saponins have been detected in soybean, lupins and several other legumes (Price et al., 1987; Bisby et al., 1994). Saponins are amphiphilous compounds which interact with biomembranes of animals, fungi and even bacteria. The hydrophobic part of the molecule complex cholesterol inside the membrane and their hydrophilic sugar side-chain bind to external membrane proteins. Thus fluidity of biomembranes is disturbed leading to holes and pores. As a consequence, cells become leaky and die. This rather unspecific membrane disturbance is responsible for the wide effects of saponins against animals, fungi, bacteria and competing plants. Thus saponins can be considered as a resistance factor in legumes against microbial infection and herbivory. Legume saponins have only moderate toxicity and present a problem only when present in the diet at high concentrations.

2.3   Phytate

Inositols with 4, 5 or 6 phosphate groups are common in the seeds of many of our grain legumes and can reach concentrations higher than 10% of dry matter (Bisby et al., 1994). They can be regarded as stores for phosphate and mineral nutrients that are important for plant nutrition and especially during germination. Since phytates complex iron, zinc, magnesium and calcium ions in the digestive tract, they can cause mineral ion deficiency in animals and humans. Again, these compounds seem to serve a double purpose, ie a defence and a phosphate & mineral store. Phytate content of food can be lowered by the addition of enzymes which hydrolyse them, ie of phytases (Bardocz et al., 1996; Bardocs and Pusztai 1996)

2.4 Oligosaccharides and isoflavonoids

Legume seeds are generally rich in oligosaccharides (up to 20%), such as stachyose and raffinose (Bisby et al., 1994). In animals, they produce flatulence and other disturbances. Since these symptoms are not pleasant, these compounds will contribute to the general deterrence of legume seeds to herbivores. These compounds serve as carbon sources during germination. Therefore their contents can be reduced in legumes through germination which is a common practice, eg in soybeans. Isoflavonoids have been detected in soybean, lupins and several other legumes (Bisby et al., 1994). They are involved in plant defence against fungi, bacteria, viruses and nematodes (phytoalexins, phytoanticipins), act as signals in legume-Rhizobium interactions and exhibit estrogenic activities (Dakora and Phillips, 1996).

These hormonal effects are not necessarily welcomed in humans or livestock. Recently it was found that the isoflavone genistein inhibited tyrosine kinases. Since these enzymes are often stimulated in cancer cells, the lower incidence of some kinds of cancers in people who consume isoflavone rich food, such as soybean products, has stimulated the hypothesis, that some legumes rich in isoflavones can prevent cancer.

2.5 Cyanogenic glycosides

Cyanogens are glycosides of 2-hydroxynitriles and are widely distributed among plants, eg the *Rosaceae, Leguminosae* (eg, in *Phaseolus lunatus* and *Vicia sativa*), Gramineae, and Araceae (Bisby et al., 1994; Bell and Charlwood, 1980; Conn, 1981; Rosenthal and Berenbaum, 1991). In case of emergency, ie. when plants are wounded by herbivores or other organisms, the cellular compartmentation breaks down and cyanogenic glycosides come into contact with an active ß-glucosidase, which hydrolyses them to yield 2-hydroxynitrile. This is further cleaved into the corresponding aldehyde or ketone and HCN by hydroxynitrile lyase. HCN is highly toxic to animals or microorganisms because it inhibits enzymes of the mitochondrial respiratory chain (ie. cytochrome oxidases). In addition to the toxic effects cyanogens can serve as mobile nitrogen storage compounds in seeds, important during germination.

Cassava is a crop plant rich in cyanogenic glycosides. Since farmers in marginal areas rely on "bitter" varieties for food production it has been suggested that the prevention of cyanate poisonings through improved detoxification procedures may be more effective than the development of "low cyanide" cultivars (Tylleskär et al., 1992).

Hanelt and Tschiersch (1967) found that the majority of *Vicia sativa sensu lato* accessions from the Balkans and Turkey contained the cyanogenic glycoside vicianine, whereas the material from Iran was predominantly HCN-free. In *Trifolium repens* a similar polymorphism had been observed. The distribution of cyanogenic *Trifolium repens* genotypes in Europe correlated with the occurrence of snails which feed on this legume and low winter temperatures which can disrupt cyanogenic plant tissue function (Jones, 1972).

2.6 Pyrimidine glycosides

Vicine and convicine are ß-glycosides of the pyrimidines divicine and isouramil (Fig 1). High levels of vicine are present in the seeds of *V. sativa* and *V. faba* (which also contains significant levels of convicine). The ingestion of meals prepared from the seeds of *V. faba* can trigger the onset of Favism, an acute haemolytic disease which affects individuals lacking sufficient activity of the NADPH producing enzyme glucose-6-P-dehydrogenase (G6PD) in their red blood cells (Mager et al., 1980; Marquardt, 1989). Several genetic variants of this enzyme deficiency, which is thought to confer an adaptive advantage against malaria, are known to occur worldwide, with Favism being an extreme manifestation of this trait (Vulliamy et al., 1992). It is particularly prevalent in some Mediterranean and South-West Asian populations (Belsey, 1973). Vicine and convicine have been implicated in favism because their hydrolysis products are unstable and form radicals which can cause a depletion of reduced glutathione (GSH) in G6PD deficient red blood cells. A lack of sufficient NADPH due to G6PD deficiency impedes GSH replenishment and predisposes the red blood cells to oxidative damage which can ultimately result in a haemolytic crisis. Antioxidants have been successfully employed to reduce the effects of vicine in animal diets. (Mager et al., 1980; Marquardt, 1989; and references therein).

Efforts are under way to develop cultivars of *V. faba* with zero levels of vicine in their seeds and some promising material has already been identified (Griffiths and Ramsay, 1992). The work of Bjerg et al. (1984) suggests that these compounds play a role in the resistance of pods and leaves to fungal pathogens, so deletion of these factors from the seeds only would be desirable. However their physiological and biological role during seedling establishment is still unresolved (Griffiths and Ramsay, 1992) and seed specific deletion may affect useful levels of these compounds in young seedlings.

2.7   Lupin alkaloids

The genus *Lupinus* contains several hundred species, 12 in the Old and the rest in the New World. Their evolution has been described recently by Käss & Wink (1997) using nucleotide sequences of marker genes. Since some species have large protein rich seeds (up to 50% protein), these lupins are of considerable agricultural interest. Cultivated species include *L. albus*, *L. luteus*, *L. angustifolius*, *L. mutabilis* and *L. atlanticus*. Reviews of the biochemistry, utilisation and agronomy of lupins can be found in the proceedings of the International lupin conferences (Gross and Bünting, 1982; ILA 1982, 1984, 1986, 1990; Neves-Martin & Beirao da Costa, 1994).

The main secondary metabolites of lupins are quinolizidine alkaloids which are sometimes accompanied by piperidine alkaloids, such as ammodendrine or indole alkaloids such as gramine (Bisby et al., 1994). After synthesis in leaf chloroplasts, alkaloids are exported via the phloem to other parts of the plant and are accumulated in the epidermis of leaves and stems. Reproductive organs, such as flowers and seeds are especially rich in alkaloids (2-6% dry weight). In germinating lupins, alkaloids serve as a source of N (Wink 1993a, b, c). Lupin alkaloids were shown to be feeding deterrents and lethal for a number of insects, especially aphids (Berlandier, 1996; Wink 1992), but also moth- and butterfly larvae, beetles, grasshoppers, flies, bees and ants, other invertebrates and vertebrates, ie alkaloids are active over a wide range of animal orders (Wink, 1985, 1988, 1992, 1993a, b). Furthermore, these alkaloids are inhibitory for competing plants, viruses, bacteria and fungi (Wink, 1985, 1988, 1992, 1993a, c). Consequently, alkaloids appear to be important for the resistance of lupins to insects and other pests, an assumption which could be tested experimentally. Since alkaloids are toxic to humans and animals, breeders have selected varieties with very low alkaloid contents, the "sweet" lupins. When planted in the field sweet lupins suffer substantially from herbivores, such as rabbits and hares. A similar picture was observed for a number of insects, such as aphids, beetles, thrips and leaf mining flies, ie sweet lupins were attacked, whereas the alkaloid-rich ones were largely protected (Wink, 1985, 1988, 1992, 1993a, b). Plant breeders have also observed that bacterial, fungal and viral diseases are more abundant in the sweet forms.

The toxic effects of quinolizidine alkaloids on insects and vertebrates can be explained through the interaction of these alkaloids with acetylcholine receptors (sparteine activates the nAChR, lupanine the nAChR), with Na+, K+-channels and protein biosynthesis (Wink, 1992, 1993a, b; Schmeller et al., 1994). Since synaptic signal transduction and protein biosynthesis are important and vulnerable processes in most animals, it is not surprising that alkaloids have toxic properties over a wide range of animals. These data clearly support the importance of alkaloids for chemical protection.

With our present knowledge of the ecological importance of àlkaloids for the fitness of lupins, it seems doubtful whether the selection of alkaloid free lupins with a decreased resistance is the only possible solution. Similar strategies, ie to eliminate unwanted chemical traits, appear to have been chosen with other agricultural crops (such as cabbage, turnip, rape seed, tomato, potato, cassava or barley) with the consequence, that fitness was much reduced. In these cases the loss of natural protection had to be substituted by man made synthetic agrochemicals with their known problems (Wink, 1993b). For lupin breeding we have proposed two alternatives (Wink, 1993b):

To select mutants which do not translocate the alkaloids to the seeds, since seeds do not produce alkaloids but store them. The plant would retain its chemical resistance in its vegetative parts but would also provide the valuable alkaloid free seeds.

To grow alkaloid rich plants but to process the seeds and to simultaneously produce pure protein, lipids, amino acids, and dietary fibres from bitter seeds. A spin-off product would be alkaloids, which could be either used in medicine or in agriculture (Wink, 1993b). Currently this strategy is being developed by the Mittex Company with aid of the European Commission.

Considering their insecticidal and fungicidal properties, it seems most promising to exploit the alkaloids as natural plant protectants. Alkaloids do not accumulate in the soil, and plant cell culture experiments indicate that alkaloids can be metabolised rapidly. Since alkaloids will be produced as a by-product of a normal crop they could be a comparatively cheap natural ("biorational") pesticide. There are shortages of proteins, insecticides and fungicides in many developing countries and lupin cultivation and processing could be of value to those countries (Wink, 1993b).

In future, the genetic engineering of the pathways leading to secondary metabolites may provide another option. For example, if we could transfer to other crop plants, such as cotton, the genes which encode the biosynthesis, transport and storage of lupin alkaloids, it is likely that we would thereby transfer a novel resistance to insects. When successful, the genetic approach will be even more environmentally friendly than using "biopesticides".

A successful example for seed specific deletion (as suggested for lupin breeding) is provided by cultivars of oilseed rape. Some cultivars, with low glucosinolate (00) in their seeds, have been shown to possess levels of leaf glucosinolates similar to those of high seed glucosinolate (0) cultivars and were not more susceptible to pests and diseases than the more toxic genotypes (Mithen, 1992). Thus, selection for low levels of seed toxins can be achieved while the beneficial protective effects are maintained in other tissues of the plant. However, it has also been demonstrated that the susceptibility of rape seedlings to slugs is related to the seed glucosinolate content. It has been found, as was predicted, that the more widespread cultivation of 00 cultivars required an increased use of molluscides, and thus increases in costs (Moens et al., 1992).

2.8   Non-protein amino acids

More than 900 non-protein amino acids (NPAAs) which are especially abundant in certain legumes (*Vicieae, Phaseoleae, Mimosoideae,* and *Caesalpinioideae*) have been detected which resemble protein amino acids (structural analogues) (Bell and Charlwood, 1980; Bisby et al., 1994; Conn, 1981; Rosenthal, 1982). Concentrations in seeds can exceed 10% of dry weight and since non-protein amino acids are remobilised during germination they function as N-storage compounds (Rosenthal, 1982).

If non-protein amino acids are taken up by herbivores, microorganisms or other plants, they may interfere with several processes (Bell, 1977; Rosenthal, 1982, 1991). They can be accepted in ribosomal protein biosynthesis in place of the normal amino acid leading to defective proteins (example:canavanine, azetidine-2-carboxylic acid). They can inhibit the charging of aminoacyl-tRNA synthetases or other steps of protein biosynthesis or they may competitively inhibit uptake systems for amino acids. In vertebrates, effects may include foetal malformations, neurotoxic disturbances, hallucinogenic effects, hair loss, diarrhoea, paralysis, liver cirrhosis, hypoglycemia, and arrhythmia. In plants, microorganisms and insects non-protein amino acids cause reduced growth or even death. Since non-protein amino acids affect a basic target present in all organisms, they are important in plant to plant, plant to microbe and plant to herbivore interactions.

Holt and Birch (1984) correlated the presence of NPAAs in *Vicia* to resistance to aphids (*Aphis fabae, Acyrtosiphon pisum* (Harr.), *Megoura viciae* (Buckt.)) and found that the most domesticated species were also the least resistant. ß-Cyanoalanine has been shown to be active against *Locusta migratoria* where it exhibited diuretic effects and led to an inhibition of moulting (Schlesinger et al., 1976). It has also been documented as an effective feeding inhibitor in three species of locust (Navon and Bernays, 1978) and one species of bruchid (Janzen et al., 1977).

The biology of canavanine is probably one of the best documented examples of the protective role of non-protein amino acids in plants, although its benefits have not yet been demonstrated in isogenic lines differing only in the canavanine biosynthesis trait (Rosenthal, 1991). It has been shown that canavanine can have toxic effects on a wide range of organisms (Rosenthal, 1977, 1986, 1991; Miersch et al., 1992; Enneking et al., 1993; Enneking, 1994).

The insecticidal properties of these compounds alone, without considering their edaphic, allelopathic (Wright and Srb, 1950; Weaks and Hunt, 1973; Weaks, 1974; Miersch et al., 1992) and possible drought tolerance functions, suggest that it may be prudent to aim for seed specific deletions of these factors, whilst preserving and enhancing their beneficial role in other plant parts and in the phenological growth stages of the plant.

The work of Lambein et al. (1990, 1992; and references therein) has shown that seed NPAA in *Lathyrus* species act as precursors for compounds synthesised during germination. Depletion of seed NPAA toxins would also deplete the reserves from which root exudates (Kuo et al., 1982) and protective chemicals are formed during early seedling establishment. Increased levels of γ-glutamyl-ß-cyanoalanine in young seedlings of *V. sativa* (Ressler et al., 1969) and for canavanine in seedlings of *Canavalia ensiformis* (L.) (D.C.) (Rosenthal, 1972b), *Medicago sativa* L. (Gorski et al., 1991; Miersch et al., 1992) have been documented, suggesting a similar function for these compounds as for those isolated from *Lathyrus*.

2.8.1   Development of new grain legume crops from the genera *Vicia* and *Lathyrus* in Australia

Grain legume species will only be domesticated when there is some easy means of overcoming their ANFs. Peas, lentils and chickpeas have already been developed as food crops because their mainly heat labile ANFs are easily eliminated through processing. The more difficult crops, with heat-stable low molecular weight ANFs in their seeds, remain. These include *Vicia, Lathyrus* and *Lupinus* spp. Several species from these genera have received attention from agronomists, plant breeders and plant collectors because of their ability to grow under adverse environmental conditions and on marginal soils. In most instances the relevant ANFs still need to be identified before they can be effectively eliminated. In recent studies in Australia, a porcine feed intake bioassay was used to identify the principal in *Vicia* species which leads to their unpalatability (Enneking, 1994).

V. villosa seeds are extremely unpalatable to pigs and were found to be a serious problem when present as contaminants of cereals used in feed formulations. Using a series of fractionations and bioassays, the NPAA canavanine, a structural analogue of arginine was identified as the major unpalatability principal in the seed. The same study also found that under alkaline conditions canavanine degrades to the inactive deamino-canavanine (Enneking et al., 1993). *V. ervilia* and *V. articulata* seeds also contain canavanine but at much lower concentrations (0.05-0.2, 0.1-0.4% DW, respectively) (Enneking, unpublished).

The unpalatability of *V. narbonensis* seed was found to be due to the considerably less inhibitory dipeptide γ-glutamyl-S-ethenyl-cysteine (GEC) (Enneking, 1994; Enneking et al., 1997) which can be inactivated by mild acid hydrolysis. This compound renders the seeds unpalatable to pigs, poultry and humans. The limited information about the toxicity of *V. narbonensis* in pigs suggests that it may cause haemolysis, and kidney damage through the formation of crystalline precipitates (The late R. L.Davies, pers. comm.). Further studies of its chemistry, pharmacology and toxicology are clearly needed.

The safe use of *V. sativa* cv. Blanchefleur for human consumption has been the subject of controversy. It is well known, that *V. sativa* contains the Favism toxin, vicine (Pitz el al., 1980) and the neurotoxic peptide γ-glutamyl-ß-cyanoalanine. This compound inhibits the metabolic conversion of methionine to cysteine, leading to raised urinary levels of cystathionine, thus ultimately reducing the supply of vital glutathione. In addition, γ-glutamyl-ß-cyanoalanine-glycine, a structural analogue of glutathione is formed (for a review see Ressler, 1975). Cysteine malnutrition and oxidative stress are likely to increase sensitivity to this toxin. *V. sativa* cv. Blanchefleur is still being exported for human consumption from Australia to developing countries (Steene, 1997, Tate, 1996) despite prior warnings (Tate and Enneking 1992) about its toxicity. Independent reports (Ressler et al.,1997; Roy et al., 1996) also emphasise the unacceptable risk to the health of consumers by this trade.

In a feeding experiment with laying hens the effects on feed intake, weight gain and laying performance of cooked, hydrolysed (4% acetic acid, 30 mins, 121° C) and raw Blanchefleur vetch were compared with raw or cooked red lentils, with all treatments fed at a 10% inclusion level (Enneking, 1994). Cooking did not lead to a reduction of the anti-nutritional activity while acid hydrolysis was effective. The chemical data were in agreement with feed intake and growth performance while no significant differences on laying performance could be detected between treatments.

This assessment of three *Vicia* species, which are representative of the genus, demonstrated that in addition to the pyrimidine glycosides, the major ANFs in the seeds of *Vicia* spp. are NPAAs (Fig 1) which, in principle, can be inactivated by acidic or alkaline hydrolysis.

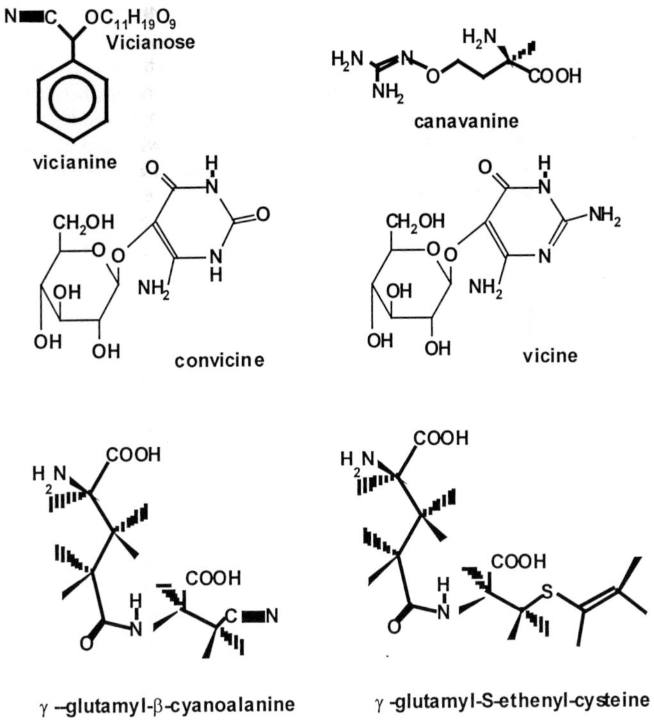

Figure 1. Known *Vicia* toxins

Lathyrus species which are presently of agricultural interest as grain legumes all contain the phagostimulant and neurotoxic glutamate analogue 3-(N-oxalyl)-L-2,3-diaminopropionic acid (beta-ODAP). A major problem associated with the development of *Lathyrus sativus* for human consumption is the risk of lathyrism. There is no acceptable experimental protocol to provoke neurolathyrism in experimental animals and decisions regarding safe intake levels for humans cannot be made. There is no suitable bioassay to test the safety of the newly developed low-toxin cultivars and there is a lack of understanding about the pre-disposing factors eg the malnutrition-poverty complex or genetic susceptibility, which lead to the onset of this nutritional disease. Despite these concerns, millions of people consume *L. sativus* on a regular basis with no apparent ill effect, except when there are famines and consumption is excessive.

Caution is needed with the marketing of the recently released low beta-ODAP *L. cicera* cultivar from South Australia for human consumption in developing countries, since very little information is available about the suitability of this species for monogastric consumption.

The following species are being bred in Australia for reduced grain toxicity with the major institutes responsible for their breeding indicated in parentheses. *Lathyrus sativus* (Ag WA-CLIMA), *L. cicera* (SARDI, Ag WA-CLIMA), *V. sativa* (Waite Institute, Univ. Adelaide-SARDI), *V. ervilia* (CLIMA), *V. narbonensis* (VIDA, CLIMA). Three other species *L. ochrus, L. clymenum,* and *V. articulata*, with a history of human consumption would be of interest but there are no specific programs in place for them. For further details on *Vicia* spp. grain legumes see (Francis et al., this conference).

V. ervilia genotypes with low levels of canavanine (< 0.05% DW) and good yields have been identified recently in Australia. Interestingly, no genotype with canavanine levels approaching those found in *V. villosa* or *V. benghalensis* seeds, has been found after screening 200 accessions of this species, including wild ecotypes from the ICARDA gene bank. This suggests that other unidentified ANFs are present in the seeds of this species. The bitter principle in the seed remains to be isolated and chemically characterised.

Sufficient variation for CN-containing compounds exists in *V. sativa* to make the selection of genotypes with very low levels of these substances a possibility (Delaere, Rathjen and Tate, pers. comm.). Variation in total sulfur levels of the Australian *V. narbonensis* collection suggests that a low GEC genotype can be selected from this material, although screening seems to be confounded by soil sulphur status. Low vicine and convicine genotypes of *V. faba* have already been developed. These genotypes, together with genotypes selected for high toxin levels provide an opportunity for the study of the ecological biochemistry of *Vicia* and *Lathyrus* species. Such genotypes could be evaluated in screening trials for their resistance to diseases, pests and other stress factors to assess the resilience of the low toxin. Conversely, sources of disease, pest and other stress resistances already identified in the two genera could be assessed for their NPAAs and related metabolites eg polyamines. For *Vicia*, information concerning the biosynthesis, transport, subcellular locations, and relative concentrations of the individual NPAAs and other low molecular weight toxins during the life cycle of representative species eg *V. faba* (pyrimidine glycosides and L-Dopa-glycoside), *V. sativa* (pyrimidine glycosides, cyanogenic glycosides, cyano-amino acids), *V. ervilia* (canavanine), *V. villosa* (canavanine, GEC) *V. narbonensis* (pyrimidine glycosides, GEC) would permit predictions and correlations with anti-predator activity and toxicity to grazing animals. The influence of various factors such as grazing, insect or disease attack, cold, heat and water, salt stress or edaphic stresses may also influence the levels of these compounds.

The most attractive feature of *Vicia* and *Lathyrus* spp. as grain and forage legumes for marginal areas is the tolerance of various species or genotypes to stress factors such as drought, cold, temporary waterlogging, pests, diseases and infertile soils. Clearly it would be undesirable to convert such low-input crops into ones with higher management requirements.

Based on the available knowledge of the biology of NPAAs in plants, it is likely that one consequence of their genetic removal would be a reduced fitness as the compounds have a role in the plants' biology, which could be linked to resilient traits enabling them to grow under adverse conditions. It will be feasible to elucidate the nature of this role when genotypes with low levels of seed toxins have been selected and are available for study.

3. ELIMINATION STRATEGIES

3.1 Post harvest processing

Legume seeds are usually cooked when they are to be consumed by humans and cooking inactivates the lectins and protease inhibitors. Low molecular weight compounds are leached out into the cooking water, which is often discarded. These simple techniques were "invented" by man without an understanding of the underlying toxicology, to make legume seeds more palatable and digestible. Today, knowledge of the chemical structure of the anti-nutrients involved can be a help when devising strategies to process legume seeds to obtain toxin free products. Since processing increases the economic value of the grain, food technology and rational processing are an alternative to breeding ANF free plants which can be more susceptible to pests and pathogens.

Using the food technology procedures of separation, filtration etc, pure and nutritionally valuable dietary products, such as protein, dietary fibres, oil and other fine chemicals can be generated. The remaining fractions containing the anti-nutrients need not be discarded: some are useful for the pharmaceutical industry (see below), others in agriculture as biorational pesticides. Because of the costs of these procedures, they are more likely to be used when the grain is destined for human consumption (an example is the processing of soybeans for which a specialized industry has been developed).

The situation in the feed industry is different. This industry is a major user of legume seeds in Europe where pulses are simply ground or pelleted. For this purpose we need varieties which are low in heat stable nutritional factors (eg., alkaloids, saponins, phytates, isoflavones, non-protein amino acids). If heat labile compounds (protease inhibitors, lectins) can be denatured by heat treatment during grinding and pelleting these compounds might be maintained since they confer resistance to the plants.

3.2   Chemical detoxification

Deaminocanavanine is a well known non-toxic deamination product of canavanine (Rosenthal, 1972a; Enneking et al., 1993). The degradation of canavanine to deaminocanavanine under alkaline conditions provides therefore a strategy for detoxification and has been employed for the processing of the canavanine containing seeds of *Canavalia ensiformis* (Obizoba and Obiano, 1988). γ-gutamyl-S-ethenyl-cysteine, γ-gutamyl-β-cyanoalanine the major antinutrient in the seeds of *V. sativa* (besides vicine) can be rendered inactive by mild acid hydrolysis. It is reasonable to propose that in principle post-harvest detoxification procedures can be developed for these anti-nutritional factors.

3.3   Fermentation

Fermentation is widely used in food detoxification processes (Ochse, 1931; Horsfall, 1987; Salih et al., 1991) and a variety of fermented foods are eaten around the world (Yokutsuka, 1991; Campbell-Platt, 1987; Reddy and Salunkhe, 1989). The further development of fermented foods has been advocated by nutritionists because of the nutritional benefits of such products (Hesseltine, 1983). Fermentation is also an effective means for food preservation (Nout and Rombouts, 1992). Fermented foods can be prepared on an industrial or household scale. Indeed, many fermented foods are prepared by very simple techniques a fact which facilitates their adoption in underdeveloped countries for the detoxification of alternative food sources.

Ayyagari et al. (1989) compared Indian households for their effectiveness in detoxifying *L. sativus*. Methods, which included a fermentation step, were the most effective in reducing ODAP levels, eliminating 95% of this toxin. Further improvements are likely to be made with selection for better ODAP degradation. In principle such methods can also be used for the post-harvest detoxification of *Vicia* seeds, thus providing an alternative approach to the wider utilisation of these grains without the need for genetic removal of their low molecular weight antinutritive and unpalatability factors. The incorporation of fermentation processes into other simple food technologies also offers good prospects for a detoxification of food sources while simultaneously giving flexibility in the manipulation of flavour, texture and colour of the raw material.

3.4   Germination

Pea and lentil sprouts (even lupin sprouts in Germany) have gained popularity in recent years. Traditionally, Mediterranean grain legumes have not been used as sprouts. The potential toxicity of beta-isoxazolin-5-one-alanine (BIA), the biosynthetic precursor for the lathyrism toxin beta-ODAP may be a risk factor if consumption of such sprouts is excessive. The concentration of this compound increases during the germination of lentils and peas. This kind of processing, however, which reduces the contents of oligosaccharides and of other N-containing ANFs, has a long history in Asia, where it has served to improve the palatability of soybeans. Pea sprouts are a very recent addition to Chinese cuisine (since ca. 1970).

3.5   Detoxification by ruminants

Vicia and Lathyrus grain legumes, viz. *V faba, V. articulata, V.sativa and V.narbonensis, L. sativus L. cicera, L. ochrus, L. clymenum* can be used as supplemental feeds for ruminant production, which is another form of post-harvest detoxification. These grains have been used for millennia as a ruminant feed in the Mediterranean, Middle-East and West-Asia. Traditional practices are rapidly disappearing with the introduction of more profitable crops, while traditional crops often remain neglected by plant breeders. It is therefore important to preserve this knowledge and transfer it combined with plant introduction to regions where such crops can be grown with profit for ruminant production.

3.6   Can "anti-nutrients" even be useful?

There can be no doubt that the so called "anti-nutrients" of legumes have a biological function. They are certainly important in the physiology of seedlings as N or C storage compounds and to facilitate nutrient uptake and rhizosphere establishment. Since they are also toxic to animals and sometimes even to microorganisms, viruses and other plants, they exhibit defence functions at the same time. As shown for lupin alkaloids, ANFs can be important for the fitness of plants and constitute relevant resistance factors.

An important question arising in the context of selection for low-toxin lines is whether the genetic reduction of anti-feedant and anti-nutritional factors is going to have a negative effect on the ecological fitness of the resulting cultivars? In the words of Bell (1977), who reviewed the ecological function of non-protein amino acids, "The development of a toxin-free crop would be totally impractical if the reduction in toxicity to man or domestic animals was accompanied by an equal or greater reduction in toxicity to predatory insects which might destroy the crop before it could be harvested". Conversely, the breeding of crops for improved resistance and cases where this has led to increased toxicity of the resulting cultivars have been discussed by Breider (1973) and Fenwick et al., (1990).

There are also examples which show that certain levels of ANFs in legumes might be even beneficial on human health. The negative dietary impact of affluence can be held at bay through inclusion of grain legumes (cholesterol lowering properties, dietary fibres, trypsin inhibitors).

The phenylalanine derivative, ß-(3,4-dihydroxyphenyl)-L-alanine (L-Dopa) can be found in free form and bound as a glycoside in some *Vicia* species (incl. *V. faba* and *V. narbon*ensis) and was at one time implicated as a causal agent of Favism, however, it is no longer considered to be important in this disease (Mager et al., 1980). *V. faba* pods might be a cheap dietary source of L-Dopa for the therapy of Parkinson's disease (Kempster et al., 1993).

Canavanine is an inhibitor of soluble nitric oxide synthesis and has recently been found to be useful in treating experimental endotoxaemic shock in rats. Domesticated *Vicia ervilia* and *V. articulata* seeds contain very low to moderately low levels of canavanine (0.05-0.2, 0.1-0.4%), respectively and might be even beneficial.

Lupin alkaloids, such as sparteine, have been used as antiarrhythmic and uterotonic therapeutics. Also other lupin alkaloids exhibit similar activities which could be used either as pure compounds or in mixtures (Wink, 1993b).

Isoflavanoids, which inhibit tyrosine protein kinases, might be interesting as anticancer agents.

3.7   Elimination of ANFs through genetic modification

The identification of alkaloids and NPAAs as the major anti-nutritional principles present in Lupinus and Vicia seeds, respectively, now allows for the selection of genotypes with low levels of these factors, thus enabling the development of more palatable and less toxic cultivars. The selection of genetic material with contrasting levels of anti-nutritional factors is also ideally suited for the elucidation of their biological functions (as shown for the example of bitter versus sweet lupins; see above). Because of space limitations this section will be a brief outline only. The general aim is a selection of non-toxic and palatable genotypes requiring efficient screening techniques to expedite the quantitative detection of individual ANFs for the selection of improved grain legume cultivars. Such techniques should be suitable for testing large numbers of samples to facilitate the screening of the available germplasm and material generated through breeding or artificial mutagenesis. Sometimes simple colour reagents might work for an initial test, such as Reifers reagent for quinolizidine alkaloids in lupins (it produces a brown precipitate with alkaloids; Wink, 1993c). Immunological methods such as ELISA to detect specific proteins can also be established for low molecular weight compounds such as alkaloids (Allen et al., 1990; Wink 1993c). Substantial progress has recently been achieved for *Vicia* and *Lathyrus* employing DRUID-IR for the screening of γ-glutamyl-β-cyanoalanine, and Capillary zone electrophoresis for ODAP, GEC, canavanine and vicine (Eichinger et al., this conference). Furthermore, an enzyme-based method has been shown to be highly effective for ODAP screening. ICARDA routinely screens genotypes by Near-IR; TLC is used to screen *Trifolium subterraneum* for isoflavones. Living systems can also be useful for selection: Whereas alkaloid rich bitter lupins are avoided by rabbits and aphids, sweet alkaloid poor varieties are readily accepted (Wink 1988, 1992). A careful selection of plants attacked by generalist herbivores might provide a clue to plants with lower ANF levels. A snail (*Helix aspersa*) bioassay on 24 well tissue culture plates may prove useful for the detection of phenolic compounds (Enneking, unpublished).

Mass screening is still a labour and capital intensive strategy. With an increased understanding of ANF biology new opportunities, no doubt, will arise for the use of natural selection pressures in screening for ANFs, either to improve crop resistance or to eliminate a particular factor and replace it with ANFs not detrimental to a particular end-use. Tolerance levels for individual ANFs and applications need to be known to enable plant breeders define target levels in their programs.

3.8   Seed specific deletion of ANFs

As discussed already under the section "lupin alkaloids", it might be possible to select strains of legumes which no longer accumulate ANFs in their seed, but still maintain their synthesis in the rest of the plant. In this case only the seeds and seedlings need additional protection, not the whole plant. This selection would work in instances in which ANFs are produced in the leaves but transported to the seeds (such as lupin alkaloids) or in which all parts of a plant produce a certain ANF but not for a seed specific synthesis. In the first instance, selection is directed towards plants in which the translocation via the phloem is blocked and the other instances towards an organ specific inhibition of biosynthesis.

3.9   Genetic engineering options

If the gene is known which encodes a toxic protein (eg, a lectin) or the key enzyme of a biosynthetic pathway leading to alkaloids, saponins, NPAAs etc, genetic engineering offers a set of methodologies at present to downregulate or to knock out the respective activity. Strategies include the expression of antisense mRNA, of gene targetting, and of synthetic oligonucleotides or ribozymes.

Also the introduction of new traits into a crop plant, such as new lectins or proteins rich in methionine/cysteine can be achieved by appropriate molecular techniques. Although these strategies look straightforward in theory, their utilisation and application for a specific problem is often more complicated and tedious. Obstacles are often encountered in that relevant genes have not been detected so far which is usually the situation for biosynthetic enzymes of ANFs. If a time-, developmental and organ specific expression is required then promotor sequences need to be known for a particular plant - and again these data are usually not available for the crop of interest. Transformed plants need to be regenerated which is a severe problem in most legumes.

Several target enzymes could be envisaged for NPAAs or alkaloids, but these genes are still unknown

The economics of producing low toxin *Vicia* or *Lupinus* varieties (with their added-value through improved palatability, reduced toxicity and hence marketability), versus the benefits derived from the protective and adaptive functions of these compounds are going to determine whether genetic deletion of toxins or post-harvest detoxification (itself a costly process) could become viable options for the further development and utilisation of these crops. Their sustainable development, especially for resource-poor farmers in underdeveloped countries may favour post-harvest processing, which has a long tradition in many parts of the world and is of especial importance in traditional cultures that still subsist on a variety of otherwise toxic or unpalatable food stuffs.

Australian industry also has an opportunity to develop post-harvest processes for *Vicia* grain because the inclusion of a simple hydrolysis step into a splitting or compounding operation would allow the destruction of GEC in *V. narbonensis*, while preserving the high levels of sulfur in the grain. The near complete elimination of β-cyanoalanine and γ-glutamyl-β-cyanoalanine from *V. sativa* seed has been achieved by a hydrolytic process incorporated into a commercial splitting operation (Delaere and Tate, pers. comm.). However the lack of cost-effective and pragmatic measures to test the safety of the resultant product and the prohibitive costs incurred by the Australian lupin industry in a similar exercise, have stifled this development.

The development of human food from otherwise toxic or unpalatable grain may justify the added costs for detoxification. This option is not likely to be viable under the current economic conditions, but may become more attractive in the future, especially if the rise in the human population and thus the corresponding rise in demand for food continues.

References

Allen, D.G., Greirson, B.N. and Harris, D.J. 1990. *Abstract 6th intl Lupin Conf. Temuco*, p. 17.

Ayyagari R., Narasinga Rao B. S. and Roy D. N. 1989 *Food Chemistry* 34:229-238.

Bardocz, S., Gelencser, E. and Pusztai, A. 1996 *Effects of antinutrients on the nutritional value of legume diets. Vol. 1*, Brussels:ESSE-EC-EAEC.

Bardocz, S., Pusztai, A. 1996 *Effects of antinutrients on the nutritional value of legume diets. Vol. 2*, Brussels:ESSE-EC-EAEC.

Bell, E. A. 1977 In: *Natural products & the protection of plants*. Proc. of a study week at the Pontificial Academy of Sciences, Oct 18-23, 1976. 571-595 (Ed G. B. Marini-Bettòlo) Amsterdam: Elsevier.

Bell, E.A. and Charlwood, B.V. 1980 *Secondary plant products. Encyclopedia of plant physiology*. Vol. 8 Heidelberg: Springer

Belsey, M. A. 1973. *Bulletin of the World Health Organisation* 48: 1-13

Berlandier, F. A. 1996 *Entomologia Experimentalis et Applicata* 79:19-24

Bisby, F.A., Buckinham, J., Harborne, J.B. (eds) 1994. *Phytochemical dictionary of the Leguminosae*. Vol. 1, Vol. 2; London: Chapman & Hall

Bjerg, B., Heide, M., Norgaad Knudsen, J. C., and Sorensen, H. 1984. *Zeitschrift für Pflanzenkrankheiten und Pflanzenschutz* 91: 483-487

Breider, H. 1973. *Theoretical and Applied Genetics* 43: 66-74.

Campbell-Platt, G. 1987. *Fermented foods of the world. A dictionary and guide*. London: Butterworth

Conn, E.E. 1981 *Secondary plant products. The Biochemistry of plants*, Vol. 7, New York: Academic Press

Dakora, F. D. and Phillips, D. A. 1996. *Physiological and Molecular Plant Pathology* 49:1-20

Enneking. D. 1995. In: *Lathyrus sativus and Human Lathyrism: Progress and Prospects*, 85-92 (Eds H. K. M. Yusuf and F. Lambein) Dhaka: University of Dhaka.

Enneking, D. 1994. Ph. D. thesis, University of Adelaide, South Australia. Republished 1995 as Occasional Publication 6. Nedlands, Western Australia: Centre for Legumes in Mediterranean Agriculture (CLIMA).

Enneking, D., Delaere, I. M.. and Tate, M. E. 1997. *Phytochemistry* (submitted)

Enneking, D., Giles, L. C., Tate, M. E., Davies, R. L. 1993 *Journal of the Science of Food and Agriculture* 61: 315-325

Fenwick, G. R., Johnson, I. T., and Hedley. C. L. 1990 *Trends in Food Science and Technology* July: 23-25.

Gorski, P.M., Miersch, J., and Ploszynski, M. 1991 *Journal of Chemical Ecology* 17: 1135-1144

Griffiths, D.W. and Ramsay, G. 1992 *Journal of the Science of Food and Agriculture* 59: 463-468

Gross, R. and Bunting, E. S. 1982 *Agricultural and nutritional aspects of lupines*. Eschborn: GTZ.

Hanelt, P. and Tschiersch, B. 1967 *Kulturpflanze* 15:85-96

Harborne, J. B. 1993 *Introduction to ecological biochemistry*. 4th ed. New York:Academic Press.

Hesseltine C. W. 1983 *Nutrition Reviews* 41:293-301.

Holt, J. and Birch, N. 1984. *Annals of Applied Biology* 105: 547-556.

Horsfall, N. 1994. In: *Toxic Plants and Animals. A guide for Australia*. (Eds J. Covacevich, P. Davie and J. Pearn) Brisbane: Queensland Museum.

ILA 1982. Proceedings *II International lupin conference*; ILA Madrid, Spain

ILA 1984. Proceedings *III International lupine conference*. ILA, La Rochelle, France

ILA 1986. Proceedings *IV International lupin conference*, ILA Geraldton, Australia

ILA 1990. Proceedings *VI International lupine conference*, ILA Temuco, Chile

Janzen D. H., Juster H. B. and Bell A. E. 1977. *Phytochemistry* 16:223-227.

Jones, D.A. 1972. In *Phytochemical ecology* 103-124 (Ed J.B. Harborne) London: Academic Press.

Käss, E. and Wink, M. 1997. *Plant Systematics and Evolution* (in press)

Kempster, P. A., Bogetic, Z., Secombe, J. W., Martin, H. D., Balazs, N. D. H., Wahlqvist, M. L. 1993. *Asia Pacific Journal Clinical Nutrition* 2, 85-89

Kuo, Y.H., Lambein, F., Ikegami, F., and Parijs, R.V. 1982. *Plant Physiology* 70: 1283-1289

Lambein, F., Kuo, Y.-H., Ikegami, F., and Murakoshi, I. 1990. In *Amino Acids: Chemistry, Biology and Medicine*, 21-28 (Eds G. and Rosenthal G. Lubec: ESCOM Science Publ.B.V.

Lambein, F., Kuo, Y.-H., Ongena, G., Ikegami, F., and Murakoshi, I. 1992. In: *Frontiers and new horizons in amino acid research*, 99-107 (Ed. T. Takai): Elsevier Science Publishers B. V.

Mager, J., Chevion, M. and Glaser, G. 1980. In: *Toxic constituents of plant foodstuffs*. 265-294 (Ed. I. E. Liener) New York: Academic Press.

Marquardt, R. R. 1989. Chapter 6 In: *Toxicants of Plant Origin Vol. II. Glycosides* 161-200 Boca Raton, Florida: CRC Press

Miersch, J., Jühlke, C., Sternkopf, G., and Krauss, G.J. 1992. *Journal of Chemical Ecology* 18: 2117-2129

Mithen, R. 1992. *Euphytica* 63: 71-83.

Moens, R., Couvreur, R., and Cors, F. 1992 *Bulletin des Recherches Agronomiques de Gembloux* 27: 289-307.

Navon A. and Bernays E. A. 1978 *Comparative Biochemistry and Physiology* 59A:161-164.

Neves-Martins, J.M. and Beirao da Costa, M.L. 1994 *Advances in lupin research*. ISA Press, Lisboa

Nout M. J. R. and Rombouts F. M. 1992 *Journal of Applied Bacteriology* Symposium Supplement 73:136S-147S.

Obizoba, I. C. and Obiano, N. 1988 *Ecology of Food and Nutrition* 21: 265-270.

Ochse, J. J. 1931 *Vegetables of the Dutch East Indies* (Eds J. J. Ochse and R. C. Bakhuizen van den Brink) Department of Agriculture, Industry and Commerce of the Netherlands East Indies. Buitenzorg, Java: Archipel Drukkerij

Pitz W. J., Sosulski F. W. and Hogge L. R. 1980 *Journal of the Canandian Institute for Food Science and Technology* 13: 35-39.

Price, K. R., Johnson, I. T. and Fenwick, G. R. 1987 *CRC Critical Reviews in Food Science and Nutrition* 26: 27-135

Reddy N. R., Salunkhe D. K. 1989 In: *CRC handbook of world food legumes: Nutritional chemistry, processing technology, and utilisation* Vol. 3. 177-217 (Eds D. K. Salunkhe and S. S. Kadam) Boca Raton, Florida: CRC Press

Ressler, C. 1975 Chapter 7 In: *Recent Advances in Phytochemistry* 151-166 (Ed. V. C. Runeckles) New York: Plenum Press

Ressler, C., Nigam, S.N., and Giza, Y.-H. 1969 *Journal of the American Chemical Society* 91: 2758-2765.

Ressler, C, Tatake, J.G.,Kaizer, E., and Putman, D.H., 1997 *Journal of Agriculture and Food* 45:189-194

Rosenthal, G. A. 1972a. *Phytochemistry* 11: 2827-2832

Rosenthal, G. A. 1972b. *Plant Physiology* 50: 328-331

Rosenthal, G. A. 1991. In:*Herbivores: Their interaction with secondary plant metabolites Vol I: The chemical participants*. pp. 1-34, Academic Press, San Diego, California.

Rosenthal, G. A. 1977. *The Quarterly Review of Biology* 52: 155-78

Rosenthal, G. A. 1982. *Plant nonprotein amino acids and imino acids*. London & New York: Academic Press.

Rosenthal, G. A. 1986 *Journal of Chemical Ecology* 12: 1145-1156

Rosenthal, G. A. and Berenbaum, M.R. 1991. *Herbivores- Their interactions with secondary plant metabolites*. New York: Academic Press.

Roy, D.N., Sabri, M.I., Kayton, R.J. and Spencer, P.S. 1996. *Natural Toxins* 4: 247-253

Salih, O. M., Nour, A. M. and Harper, D. B. 1991. *Journal of the Science of Food and Agriculture* 57:367-377.

Schlesinger, H. M., Applebaum, S.W., and Birk, Y. 1976. *Journal Insect Physiology* 22: 1421-1425.

Schmeller, T., Sauerwein, M., Sporer, F. and Wink, M. 1994. *Journal of Natural Products* 57: 1316-19.

Steene, M. 1997. *Adelaide Advertiser* 4 April

Tate, M.E. 1996. *Chemistry in Australia* December issue 549-550

Tate, M. E. and Enneking, D. 1992. *Nature* 359: 357-358

Tylleskär, T., Banea, M., Bikangi, N., Cooke, R. D., Poulter, N. H., and Rosling, H. 1992. *The Lancet* 339: 208-211.

Vulliamy, T., Mason, P. and Luzzatto, L. 1992. *Trends in Genetics* 8:138-143.

Weaks, T. E. 1974. *Physiologia Plantarum* 31: 144-148

Weaks, T. E. and Hunt, G. E. 1973. *Physiologia Plantarum* 29: 421-424

Wink, M. 1985. *Plant Systematics and Evolution* 150: 65-81.

Wink, M. 1987. *Planta Medica* 53: 509-514.

Wink, M. 1988. *Theoretical and Applied Genetics* 75: 225-233.

Wink, M. 1992. In: *Insect plant interactions* 4:131-166 (Ed. E. A. Bernays, E.A.) Boca Raton, Florida: CRC-Press

Wink, M. 1993a. In: *The Alkaloids* 43: 1-117 (Ed. G. Cordell), pp. New York: Academic Press.

Wink, M. 1993b. In: *Phytochemistry and agriculture, Proceedings of the Phytochemical Society of Europe* 34: 171-213 (Eds T. A. van Beek and H. Breteler) Oxford: Oxford University Press.

Wink, M. 1993c. In: *Methods in plant biochemistry. Alkaloids and sulphur compounds* 8: 197-239 (Ed P. Waterman).

Wright, J. E. and Srb, A. M. 1950. *Botanical Gazette* 112: 52-57.

Yasui T., Endo Y. and Ohashi H. 1987. *Botanical Magazine Tokyo* 100: 255- 272.

Yokutsuka T. 1991. In: *Handbook of Applied Mycology Vol. 3: Foods and Feeds*. (Eds D. K. Arora, K. G. Mukerji and E. H. Marth) New York: Marcel Dekker.

Zohary D. and Hopf M. 1988. *Domestication of Plants in the Old World*. Oxford: Clarendon.

New Technologies for Toxin Analyses in Food Legumes.

Peter C.H. Eichinger[1] ,Neil E. Rothnie[1], Ian Delaere[2] and Max E. Tate[2]
1. Agricultural Chemistry Laboratory, Chemistry Centre (WA), 125 Hay St., East Perth, WA, 6004; 2. Department of Plant Science, University of Adelaide, Waite Campus, P.O. Box 1,Glen Osmond, South Australia, 5064

Abstract

A key requirement for agricultural research is access to rapid, accurate and inexpensive analyses. This is particularly the case, once desirable crop traits are identified and plant breeders attempt to translate these results into high yielding crops having favourable characteristics. The utility of capillary electrophoresis and diffuse reflectance mid-infrared spectroscopy as tools for rapid analyses of value to plant breeding projects will be described.

INTRODUCTION

In conjunction with the Cooperative Research Centre for Legumes in Mediterranean Agriculture (CLIMA), and the University of Adelaide, we have been involved in screening biologically active components of grain legumes. A range of *Vicia* and *Lathyrus* species has been evaluated for anti-nutritional factors (ANF's), with particular emphasis on those crops likely to have potential in the dry marginal areas of the Australian wheat belt which typically has poor soils. *Vicia* and *Lathyrus* species have evolved a wide range of antipredator ANF's in their seeds. The low Mwt amphiphilic analytes discussed in this paper are:

β-Oxalyl-diamino-propionic acid (β-ODAP), [synonym β-Oxalyl-amino-L-alanine (BOAA)]
L-Homoarginine, a homologue of L-arginine
Lathyrine, an aromatic guanidine, structurally related to L-homoarginine
γ Glutamyl-S-Ethenyl-cysteine (GEC) - a dipeptide
L-Canavanine, an analogue of L-arginine
γ-glutamyl-β-cyano-L-alanine and β-cyano-L-alanine
Vicine and Convicine (pyrimidine glycosides)

These polar analytes, usually analysed by high performance liquid chromatography (HPLC), are also amenable to quantification using a range of capillary electrophoretic techniques. The advantages of capillary electrophoresis for the quantification of these *Lathyrus* and *Vicia* analytes, over HPLC, will be described. The application diffuse reflectance mid-infrared spectroscopy to the rapid quantification of nitrile toxins in *Vicia sativa* will be outlined.

Lathyrus Species

Lathyrus species have been consumed for thousands of years in many areas of the world, being considered in early times as an aphrodisiac - its Latin name being '*a stimulant*'. Unfortunately, many Lathyrus species are by no means harmless and contain the known neurotoxin, β-Oxalyl-diamino-propionic acid (β-ODAP).
We have now used CE analysis for a range of Lathyrus species, most notably *Lathyrus sativus* (grass pea), *Lathyrus ochrus* (cypress vetch), *Lathyrus cicera* (flat pod pea vine) and *Lathyrus clymenum* (Spanish vetch). All these species are able to thrive under conditions of waterlogging

R. Knight (ed.), Linking Research and Marketing Opportunities for Pulses in the 21st Century, 685–692.
© 2000 Kluwer Academic Publishers. Printed in the Netherlands.

such as rice paddies or duplex soils and are also drought tolerant. They are resistant to insects, fungal diseases and in general they are exceedingly hardy plants. However, prolonged ingestion of the seed by humans, especially when it is the sole food source under famine situations, has been associated with the degenerative neurological disease known as Lathyrism. This disease is characterised by a type of paralysis, which leaves the individual (usually male) crippled, with devastating consequences for both the victim and the immediate family. The problem of lathyrism is well documented. The analytical data of bodily fluids, obtained from a group of volunteers consuming cooked *Lathyrus,* was reported by Nunn *et al* (1995). Breeding progams require a simple, accurate and rapid analytical protocol if they are to produce locally adapted cultivars of *Lathyrus sativus* without significant levels of the lathyrogenic agent β-ODAP (**1**) in their seed.

(**1**)

In the past, the most convenient method has been to derivatise a milled seed extract with fluorenylmethoxy chloroformate (FMOC-Cl) and quantify the β-ODAP (**1**) adduct using HPLC. At CLIMA, the method of Geda *et al.*1993, required a total run time of 50 minutes. Furthermore, re-equilibration of the column, required an additional 15-20 minutes. So although the procedure was accurate, such a slow throughput time was unacceptable for a preliminary screening of the large ICARDA *Lathyrus* germplasm collection, as also was the cost of the acetonitrile solvent. Because β-ODAP is a polar non-protein amino acid derivative, whose charge is pH dependent, it was recognised (Arentoft et al, 1995) that it was ideally suited for capillary electrophoresis (Figure 1). Injection to injection time was reduced to 10 minutes and between batch reproducibility was 3.2% (*c.f.* 8.2% reported for HPLC). With CE it was possible to increase the sample throughput six-fold! Most importantly, the labour intensive FMOC derivatisation was no longer required.

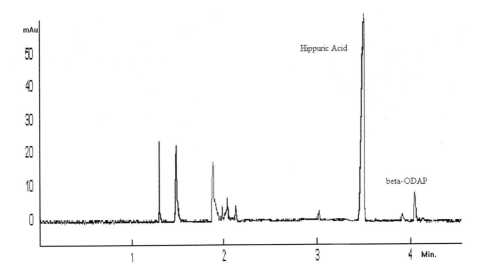

Figure 1. Electropherogram of *Lathyrus sativus.*

Conditions: pH = 7.80 ± 0.05 ; 20mM sodium phosphate, bare fused silica 46mm x 50mm (effective length 40 cm), Voltage = +25kV, Inject 3.0sec (50mBar). The pH conditions used for the capillary electrophoresis are important and poor peak shape is encountered if the pH is not buffered within narrow limits. Furthermore, in this mode, preconditioning of the capillary with 1.0N sodium hydroxide should be avoided, because the peak shape is distorted.

Variation in β-ODAP amongst Lathyrus Accessions (ICARDA)

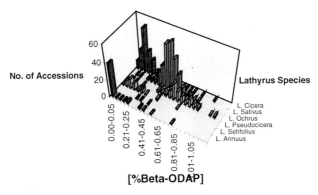

Figure 2. Frequency distribution of β-ODAP (**1**) concentrations amongst *Lathyrus* species ex ICARDA collection (Aleppo, Syria)

L-Homoarginine

Yusuf et al (1995) have reported that L-homoarginine (**2**), a homologue of L-arginine confers some protection against the debilitating effects of Lathyrism in chickens.

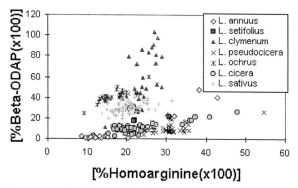

(2)

The amino acids in *Lathyrus* spp. have previously been separated by conventional ion exchange chromatography HPLC and then derivatised on-line using ninhydrin. However, in both speed and resolution, CE in the micellar electrokinetic chromatography (MEKC) mode has distinct advantages over HPLC. In this technique for L-homoarginine, very similar conditions are used to the electrophoretic method described for β-ODAP, but a surfactant, namely sodium cholate is added (data not shown).

The results of the homoarginine analyses are plotted against the results for β-ODAP (Figure 3) for the Lathyrus species from the ICARDA germplasm.

Figure 3. The Variation of % β-ODAP (x100) against % L-Homoarginine (x100) for selected *Lathyrus* species from the ICARDA Germplasm Bank.

While we have yet to compare the data for all ICARDA accessions, it is clear from Fig 3, that in addition to having lower mean levels of β-ODAP than *L. clymenum*, *L. ochrus*,and *L. setifolius*, the mean levels of homoarginine are particularly high for *L. cicera* and higher still for *L. pseudocicera*. In light of the suggestions of Yusuf et al (1995) concerning the protective effect of L- homoarginine against the toxicity of β-ODAP, it would seem that the potential of *L. cicera* and *L. pseudocicera* for animal nutrition are worthy of further investigation.

What is the source of L-homoarginine? In *L. tingitanus*, biochemical pathways have been identified (Brown and Al-Baldawi 1977), which confirm the reversible interconversion of lathyrine (**3**) and homoarginine (**2**) and its 4-hydroxy derivative. However, the evidence provided by these workers suggests that homoarginine arises from the ring opening of lathyrine derived via the orotate pathway from a preformed pyrimidine rather than L-homoarginine acting as a precursor for lathyrine (**3**)

(3)

Vicia Species

Narbon bean (*Vicia narbonensis*) is unique amongst grain legumes in being high in sulfur. Unfortunately, much of this sulfur is bound in an anti-palatibility factor recently identified as γ-glutamyl-S-ethenylcysteine (GEC: **4**) Enneking et al (1998).

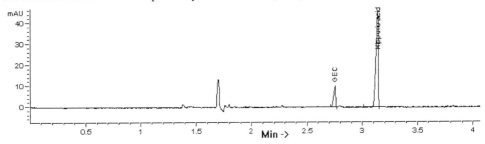

(**4**)

The narbon bean suppresses feed intake of both pigs and poultry , and when it is fed to dairy cows the milk becomes tainted. The purified crystalline γ-glutamyl-S-ethenylcysteine dipeptide (**4**) produces a most disagreeable taste in the mouth. Furthermore, the content of γ-glutamyl-S-ethenylcysteine in the narbon bean has recently been correlated with feed intake inhibition in pigs (Enneking et al 1998, personal communication). Currently, genetic and environmental factors contributing to the partitioning of sulphur into storage proteins and GEC are being examined at CLIMA. Studies of sulfur pathways in transformed *V. narbonensis* have also been reported by Saalbach et al (1995).

Figure 4. CE fractionation of GEC(**4**) and Hippuric acid standards

The availability of crystalline GEC (**4**) has permitted the development of a rapid capillary electrophoresis method (Fig 4). The procedure for extracting the dipeptide, and the electrophoretic conditions are essentially identical to those used for β-ODAP, except that a much wider range of pH can be tolerated during the capillary electrophoresis, without loss of peak shape. Under optimal conditions, the resolution efficiencies for GEC and hippuric acid are 185,000 and 109,000 theoretical plates respectively. The UV spectrum of the S-ethenyl moiety of GEC shows a characteristic shoulder at 220nm (Enneking et al 1998) and with a diode array detector, this feature can be used to confirm its elution time. During a typical batch, the migration time varies < 3%. The hydrolytic stability of pure GEC in aqueous ethanol is quite high, with <2% decomposition occurring over 36 hours, in a carousel at >28°C. [Crude GEC is somewhat less stable]. However, GEC is unstable under acid conditions (0.1N HCl). Observed GEC concentrations in the Narbon bean seed samples have varied from 1.7 - 3.2% DW. For a given narbon bean accession, at a particular site, the amount of GEC produced shows a good correlation with the amount of added total sulfur (as measured by ICP-MS).

L-Canavanine:
Several vetches, including *Vicia ervilia*, *V. benghalensis* and *V. villosa* have been used as either forage or grain for livestock. However under a variety of circumstances, livestock losses through vetch toxicosis or vetch associated diseases have occurred and the presence of L-canavanine in the ingested green matter or grain has been implicated. L-Canavanine (2.2%) in *V. villosa* grain is a potent feed intake inhibitor for pigs (Enneking et al 1993).

L-Canavanine (5) is an analogue of arginine, in which the δ methylene group has been replaced by an oxygen atom. Its close similarity to arginine makes it a suitable substrate for the enzyme arginyl *t*-RNA synthase and consequently numerous proteins have been identified where this residue has been introduced. However, replacing the methylene group with oxygen causes a dramatic change in the pKa of the guanidine moiety, from pKa = 12.48 in arginine to 7.04 in canavanine (Boyar and Marsh 1982) with significant consequences for the shape and properties of proteins incorporating canavanine in place of arginine. Canavanine(5) undergoes a facile base catalysed cyclisation to deaminocanavanine (6).

(5)   (6)

This creates difficulties in derivatisation under basic conditions (*eg* dansylation and FMOC-Cl). In HPLC underivatised L-canavanine cannot be quantified by UV absorption in most commercial instruments, because its UV absorption at or above 205nm is negligible. In contrast, fused silica capillaries used for electrophoresis are transparent to 195nm. Quantification of underivatised L-canavanine has been accomplished down to 0.01% using untreated capillaries, with migration times for canavanine (1.05 min) and benzylamine standard (1.35 min). This method again meets the plant breeder's goals of being rapid, robust and accurate.

Faba Bean (*Vicia Faba*)

Favism is associated with the ingestion of faba beans. It only affects a small proportion of the population, principly amongst peoples of Mediterranean origin who carry a defective gene for glucose-6-phosphate dehydrogenase. Vicine and convicine (pyrimidine glucosides) in the faba have been associated with the development of the haemolytic disease known as Favism and the HPLC analysis has been well documented by Marquadt and Fröhlich (1981). In the current work, ethanol (60%) extracts of faba bean, with added hippuric acid as an internal standard, have proved satisfactory for separating the pyrimidine glucosides by CE. These components can be readily quantified because of their intense and characteristic UV spectra near 270 nm.

Common Vetch (*Vicia sativa*)

Vicia sativa extracts contain the neurotoxins β-cyanoalanine (7) usually <0.1% and its γ-glutamyl derivative (8) near 1.1%. These can readily be identified either by HPLC (Ressler et al 1997) or by capillary electrophoresis in borate buffer as their FMOC derivatives in the MEKC mode. However for screening seeds for toxin content, the most effective quantitative procedure takes advantage of the nitrile (-CN) absorbance band at $2251cm^{-1}$, which can readily be detected using diffuse reflectance mid infrared spectroscopy. As little as 3 mg from the drillings of a single seed is all that is required. Using this technique we have screened over 3000 accessions including the entire *Vicia sativa* collection from ICARDA. It has also been used to demonstrate the unconscionable substitution of *Blanche fleur* vetch for lentils (fig 5) in the markets of Bangladesh. Full details of these techniques for *Vicia sativa*, will be published elsewhere.

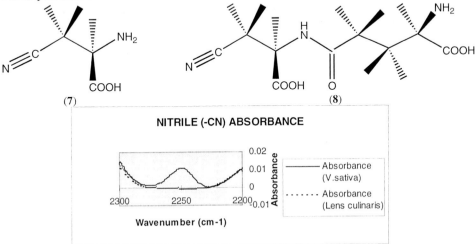

Fig 5 Comparison of the infrared nitrile absorbance at $2251cm^{-1}$ from a sample (morphologically indistinguishable fom Australian Blanche fleur vetch) containing 1.34% γ-glutamyl β-cyanoalanine(**8**) which was sold as "Brazil lentils" in a market (Dec 1996) in Bangladesh and a genuine lentil sample.

CONCLUSIONS

Capillary Electrophoresis and diffuse reflectance mid infrared spectroscopy (not to be confused with NIR), have proven to be powerful techniques which provide rapid analyses for a wide variety of biologically important components (NPAA's, ANF's) in legumes. The analyses are robust, require only limited sample preparation and time consuming derivatisation procedures are usually not required. The cost of consumables is low and the use of hazardous HPLC solvents is avoided, making these "green" technologies. For the plant physiologist, we are able to measure analytes in small volumes (<0.2mL) of phloem (or xylem) sap thereby aiding an understanding of ion transport processes. Potential future, applications include single-seed varietal identification and various food "quality" indicators for correlation of analyte profiles with cooking quality and/or metabolisable energy, etc. Clearly, the power of these techniques remains to be fully exploited by the plant breeding community.

Acknowledgements: We wish to acknowledge financial support (in part) from CLIMA, GRDC and Agriculture (WA). B.N. Grierson, J.M. Dimasi and M.L. Brown have also contributed their technical expertise to this work. MET is indebted to Professor Anisul Haque for the sample of "Brazil lentils"

References:

Arentoft, A. M. and Greirson, B.N. 1995. *Journal of Agriculture and Food Chemistry* 43: 942-945

692

Boyar, A.; and Marsh, R.E. 1982. *Journal American Chemical Society* 104: 1995-8

Enneking, D.; Giles, L.C.; Tate,M.E.; and Davies, R.L 1993. *Journal of the Science of Food and Agriculture* 61: 315

Enneking, D.; Delaere, I.M.; and Tate, M.E. 1998 *Phytochemistry*. In Press

Geda, A.; Briggs C.J. and Venkataram S. 1993. *Journal Chromatography* 635: 338 -341

Marquadt, R.R.; and Fröhlich, A.A. 1981. *Journal Chromatography* 208(2): 373-379

Nunn, P.B.; Perera, K.P.W.C.; Bell, E.A.; Campbell, C.G.; Lambein, F. 1995. *International Conference on Lathyrus and Lathyrism: A Decade of Progress,* Addis Ababa, Ethiopia, Nov 27-29 pp 13 - 17

Ressler, C.; Tatake, J.G.; Kaiser, E.; Putnam, D.H. 1997. *Journal of Agriculture and Food Chemistry* 45: 3311-12

Saalbach, I.; Waddel, D.; Pickard,T.; Schieder,O.; Müntz,K. 1995. *Journal of Plant Physiology* 145: 674-81

Yusuf, H.K.M.; Haque, M.K.; Uddin M.A.; Roy, B.C. and Lambein, F. 1995. *International Conference on Lathyrus and Lathyrism: A Decade of Progress,* Addis Ababa, Ethiopia, Nov 27-29 pp 9-12

Evidence of gains in nutritional quality in *Lathyrus sativus*

S.L. Mehta[1] , I.M. Santha[2] and A.M. Abd El Moneim[3]

1 Indian Council of Agricultural Research, KAB, Pusa, New Delhi 110012.; 2 Division of Biochemistry, Indian Agricultural Research Institute, Pusa, New Delhi 110012.; 3 ICARDA, Aleppo, Syria.

Abstract

By exploiting somaclonal variation a large number of stable low ODAP containing *L. sativus* lines have been developed. Some of the low ODAP strains also have high yields. One such line, Bio L212 has been released. A few of the somaclones have been DNA finger printed. An ODAP degrading gene has been isolated and characterised from a soil microbe and has been cloned into an Agrobacterium strain for transforming *L. sativus*.

INTRODUCTION

Lathyrus sativus, grass pea or Khesari is an important grain legume of India, Bangladesh, Nepal and Ethiopia. It is cultivated over an area of about 1.5 million ha with the annual production of about 0.8 millions tonnes in India. In Bangladesh it is the most important legume crop. It is a hardy crop and is resistant to adverse conditions of drought and water logging and requires minimum inputs for its cultivation. Being a grain legume it also enriches the soil. It has a protein content varying between 28-34%. But the full potential of this crop could not be exploited because of the development of neurolathyrism characterised by lower limb paralysis in people who consume the grain over a prolonged period as a staple of their diet. The causative agent for neurolathyrism has been identified as· (N-oxalyl) amino alanine (BOAA) or N-oxalyl) –2,3-diaminopropionic acid ·- (ODAP) by Sarma and Padmanaban (1969). Malathi et al (1967, 1970) proposed a two step reaction for the synthesis of ODAP   Oxalyl CoA Synthetase 1. Oxalate + CoA+ATP≡Oxalyl-CoA+AMP+Pi. ODAP Synthase 2. Oxalyl CoA+DAP≡ODAP+CoA It has been reported that irrespective of varieties and stages of development, ODAP is detected in all tissues of *Lathyrus* plants but maximum content was observed in leaves during the vegetative stage and in the embryo during the reproductive stage (Prakash et al 1977). Despite a ban on its sale, *Lathyrus* is grown along with other rabi crops to provide an alternative crop during times of adverse conditions. In the eastern part of India it is extensively grown as "Para" or "Utera" with rice, by broadcasting *Lathyrus* seeds in rice fields about 2-4 weeks before the rice harvest. This practice not only minimizes the work but helps to harvest a second crop without any input.

DETOXIFICATION OF *L. SATIVUS* SEEDS.

Physical Methods

Attempts have been made to prevent the onset of the disease neuorlathyrism. Many workers have tried to develop methods which would detoxify the grain and remove the major part of the ODAP. Some of the procedures suggested to remove ODAP by 70-90% were: 1) boiling or steeping the seed splits in hot water followed by draining (Mohan et al 1966) 2) roasting the seeds for about 15-20 minutes at 140 °C (Rao et al 1969) and degerming the cotyledons (Prakash et al 1977).

An attempt was made by Ramchand et al (1981) to use fermentation with *Bacillus* species to detoxify the seeds. They observed that the neurotoxin is broken down by this bacterium without affecting the nutritional quality of the pulse. The utility of such procedures was limited since it could not be adopted on a national scale because of practical difficulties.

R. Knight (ed.), Linking Research and Marketing Opportunities for Pulses in the 21st Century, 693–697.

As an alternative, efforts were made to develop new strains of *L. sativus* using a genetical approach. An analysis of a large collection of *L. sativus* seeds showed wide variation in ODAP levels ranging from 0.1 to 2.5% (Nagarajan and Gopalan 1968). Somayajalu et al (1975) also found variation in ODAP content in a population varying between 0.1 to 1.0%. They reported a strain P-24 having ODAP between 0.19 to 0.32 % and a satisfactory yield.

Mutational breeding has also been tried to develop low ODAP lines. Recurrent irradiation and subsequent selection led to the development of *L. sativus* strains with an ODAP content of 0.375 to 0.117% in the M3 generation (Swaminathan et al 1970). Other methods like spraying with the micro-nutrients cobalt nitrate (0.5 ppm) and ammonium molybdate (20 ppm) at the maximum flowering stage was also found to reduce the ODAP content in seeds by 33% and 11% respectively (Misra and Barat 1981).

The toxin content is influenced by the environment and the toxin content of low ODAP varieties varied between environments (Ramanujam et al 1980). Efforts by several plant breeders to develop low toxin lines were unable to achieve much, because the lines were unstable. A negative correlation has also been found between protein and ODAP content. Four low toxin-containing cultivars of *L. sativus* have been developed in China. They were selected from 73 lines with ODAP concentration ranging from 0.075 to 0.993%. In addition to a low ODAP they have good agronomic characters and the ODAP content has been found to be stable over several years. Toxicological studies of these lines were carried out by feeding the seeds to donkeys, pigs and sheep. These animals did not develop the disease lathyrism (Campbell et al 1994).

Screening of 81 pure lines of *L. sativus* at ICARDA helped in selecting lines varying between 0.02 to 0.74% in ODAP content. Four lines 1FLS521, 943, 519 and 536 originating from Syria, Cyprus and Turkey had an ODAP content between 0.02 to 0.07%. A breeding programme at ICARDA has helped in the development of lines having an ODAP content varying between 0.02 to 0.17%, having white cream coloured seeds and white flowers. It was also found there was no location effect on ODAP content and yield varied between 0.9 and 1.5 t/ha.

In Bangladesh two varieties 8603 (BARI-Khesari-2) and 8612 (BARI-Khesari-1) with ODAP contents of 0.0137 mg/g and 0.006 mg/g respectively have been released. The average yield over a period of 5 years has been 1727 Kg/ha and 1720 Kg/ha respectively against an average yield of 1296 Kg/ha for the check. Crossing programmes are continuing in Bangladesh to develop low ODAP lines (Malek et al 1993)

Biotechnological methods are being used to develop plants with desired characteristics. The methods used in our laboratory include a) Exploitation of somaclonal variation b) Isolation of microbes that can degrade ODAP and to then isolate the gene for ODAP degradation and transfer it to *L. sativus* through Agrobacterium-mediated transformation or by particle gun bombardment. c) Antisense nucleic acid technology to block the synthesis of enzymes involved in ODAP synthesis.

Somaclonal variation:

Somaclonal variation has been used as a novel source of variability in plant improvement (Karp 1991). It has been reported in a wide range of crops including sugarcane, rice, maize, potato, sugarbeet, wheat, brassica, strawberry and carnations. In view of these achievements, somaclonal variation was exploited in an attempt to develop neurotoxin free/low cultivars of *L. sativus* in the Division of Biochemistry, Indian Agricultural Research Institute. We have been successful with *in vitro* regeneration using the leaf, root and internode as explants from the variety P-24 (Roy et al 1991, 1992, 1993). Out of the 300 regenerated plants, grown in the field, about 102 survived and produced viable seeds. The seeds were analysed for ODAP content and were found to range from 0.03 % to 0.89%. An analysis of single seeds in some lines showed a complete absence of ODAP, in some seeds.

In addition to variation in ODAP there was tremendous variability with respect to leaf size and morphology, flower and seed colour, pigmentations of pods, seed yield and seed weight (Mehta et al 1994). Although parent P-24 has blue flowers and grey seeds, somaclones have been obtained with white, pink and red coloured flowers and white, red and varying shades of grey seed coat. White flowered plants also produced seeds with white seed coat. Seeds of low ODAP somaclones were advanced to subsequent generations and individual plants were analysed for ODAP content and selections made. The nomenclature of somaclones has been described earlier (Mehta et al 1994). In the R1 generation the plants having seeds with ODAP content between 0.03 to 0.08%

were selected and advanced further. The low ODAP containing lines when analysed for ODAP in the R2 generation showed segregation. Most of the plants had ODAP content less than 0.10%, but in some had more than 0.1% and others as high as 0.6%. This we observed was due to cross pollination caused by honey bees. This was evident because some of the white flowered and white seeded somaclones, during subsequent generations, had some blue flowered plants despite white colour being recessive. Hence in the subsequent generation the plants were covered with nets to prevent cross pollination and selections were made for low ODAP. Other physical characteristics observed also remained unchanged over generations. The low ODAP somaclones in the R4 generation had contents varying between 0.03 to 0.1% as compared to 0.32% in the parent P-24 (Table 1).

Table 1. ODAP content (%) in the seeds of *Lathyrus sativus* grown in IARI fields during 3 consecutive rabi seasons

SOMACLONES	1994	1995	1996	YIELD
		% ODAP		Kg/10 sq m
BIO 158	0.044	0.055	0.070	2.47
BIO 164	0.034	0.031	0.089	2.45
BIO L203	0.063	0.065	0.081	2.42
BIO L207	0.050	0.044	0.095	4.50
BIO L208	0.028	0.047	0.125	4.50
BIO L212	0.037	0.038	0.087	2.34
BIO L254	0.056	0.062	0.095	2.75
BIO L256	0.065	0.050	0.133	2.25
BIO L257	0.063	0.059	0.105	1.96
BIO R202	0.034	0.056	0.117	2.90
BIO R215	0.041	0.100	0.150	3.12
BIOR224	0.046	0.056	0.097	2.19
BIOR229	0.063	0.068	0.111	2.66
BIOR231	0.051	0.046	0.128	2.91
BIOR233	0.069	0.075	0.132	3.62
BIO1218	0.059	0.056	0.103	2.30
BIO1222	0.031	0.040	0.076	3.90
BIO1230	0.044	0.050	0.130	1.96
P24(ch)	0.321	0.320	0.592	1.70

Some of these somaclones were also analysed at CLIMA, Western Australia by capillary zone electrophoresis and have very low ODAP contents (Table 2). Some also had high yields.

Table 2 . ODAP content of somaclone samples analysed at the Chemistry Centre, University of Western Australia, Perth

Somaclone	ODAP (%)	Somaclone	ODAP (%)
Bio 164	<0.01	Bio I-222	0.06
Bio R-202 (Ratan)	0.06	Bio R-231	<0.01
Bio L-208 (Moti)	0.03	Bio L-254	<0.01
Bio R-209	0.02	P-24	0.34
Bio L-212	0.01		

Under Delhi conditions the yield varied between 19-45 q/ha as compared to 17.0 q/ha by parent P-24 in the 1994-95 rabi season. Five of the somaclones from R2 generation were tested in multilocations by the Indian Institute of Pulse Research, Kanpur under the All India Varietal Trial Programme. These somaclones were Bio L208, Bio R231, Bio L-212, Bio R-202 and Bio L203. The multilocation trials also showed that our somaclones had the lowest ODAP content compared to all the other entries and the check P-24. In the 1994-95 seasons the somaclones had better yields than the others and the first four positions out of the first five were occupied by them. At one location, Bharari, one of the somaclones Bio L 212 yielded 35 q/ha. These trials were repeated in the next generation and based on the observations on ODAP and yield the Central Varietal release committee recommended the release of Bio L 212 for cultivation in the north eastern plain zone and central zone and in March, 1997. Bio L 212 has broad leaves, bold seeds, an ODAP content below 0.05% and an average yield of 25 q/ha under Delhi conditions.

Molecular analysis of somaclones

Molecular analysis of a few of the somaclones varying in different characteristics has been made. The analysis included southern hybridisation pattern with cDN probes and RAPD analysis. The somaclones studied with their characteristics were:

Parent P-24	High ODAP, grey seed, blue flower
Bio I22	Extremely low ODAP, early maturing
Bio 15-8	Extremely low ODAP
Bio R-27	Fairly low ODAP, white flower, white seed coat
Bio-R02	Low ODAP, broad leaf, highest plant height
Bio-L08	Low ODAP, white flower, white seed coat
Bio-16-4	Low ODAP, narrow leaves
Bio-L-19	High ODAP, pink flower

Southern Hybridisation:

Genomic DNA from these somaclones and parent P24 were isolated and restricted with restriction enzymes Eco R1, Pst I, Bam HI, XbaI, Kpn I, HaeIII, MspI, HpaII and fractionated on agarose gel. The fractionated DNA was transferred to nylon membranes and hybridised with γ32P labelled cDNA insert from a randomly chosen cDNA clone (No:29) from cDNA library of *L. sativus*. The hybridisation patterns showed differences (Mandal, 1995). In Eco RI restricted DNA one single band of 4.0 Kb was present in all somaclones and parent P24, but the intensity was very high in Bio L08 and Bio R02. Blots with Bam HI and PstI fragments hybridised with a common 4.3 Kb band in all but with an extra band of 4.5 Kb in BioLo8 and Bio- R02. In XbaI restricts DNA 4.3 Kb band was common in all but BioL08 and Bio R02 had higher intensity and in addition to this Bio R02 had 3 more extra bands of size about 6Kb, 9Kb and 21 Kb of which 6 kb had the highest intensity. MspI and HpaII combination is usually used for detecting methylation in DNA. MspI restricted DNA blot showed hybridastion with a 4.3 Kb with same intensity in all somaclones and parent P-24 and in Bio R02 and Bio L08 there were 3 extra bands of 0.6, 0.8 and 1.2 Kb. HpaI isoschizomer of MspI showed hybridisation with a 4.3 Kb band in all as in the case of MspI but BioL08 and Bio R02 showed a different banding pattern as compared to that of MspI. In BioR02, it hybridised with 3 bands of 5.0 Kb, 10Kb and 15 kb which were absent in all other somaclones and the parent P24.

RAPD analysis

RAPD analyses were conducted to see if there were similarities among the somaclones and the parent as well as to get finger prints of somaclones at DNA level, which could be used to identify somaclones (Mandal et al 1996). Out of 81 oligodeca nucleotide primers used for RAPD analysis only with 24 primers polymorphism could be observed. A similarity matrix, calculated on the basis of the total number of bands and number of showed bands, revealed more than 90% similarity between any two somaclones and between somaclones and the parent. It was not possible to identify a particular somaclone by using any single primer. Two, three or more primers would be required. An RAPD analysis of the remaining low ODAP somaclones is being undertaken.

Isolation of ODAP degrading genes from soil microbes

In the environment, large numbers of micro-organisms exists capable of degrading a wide spectrum of chemical compounds. An attempt was made to isolate microbes from the soil and from a sewage canal near the IARI campus that can degrade ODAP. Serial dilutions and growth in minimal medium containing ODAP as the sole source of carbon and nitrogen was used to isolate organisms which can utilise ODAP for growth. After the 8th serial dilutions they were plated on media containing different antibiotics. Three individual strains BYAI, BYK1 and BYT1 depending on their resistance towards antibiotics ampicillin, kanamycin and tetracycline respectively were isolated. BYAI has been characterized as an *Enterobacter cloacae* (Yadav et al 1992) and other two as *Pseudomonas stutzeri* (Shelly Praveen et al 1994, Sachdev et al 1995). The property for degradation of ODAP was found to be present on respective plasmid present in all these 3 strains. The plasmid from BYAI (pBYAI) has been subjected to detailed studies. It has a size of over 50 Kb and an overlapping circular map with respect to different enzymes namely. Not I, Pst I, KpnI and HindIII have been prepared. The gene responsible for ODAP degradations has been isolated from a library of pBYAI, and sequencing done. Analysis of the sequences showed the largest reading frame of 630 nucleotide in one frame only. The gene coded for a polypeptide of 199 amino acids (Sukanya et al 1993). An homology search did not show any sequence homology with any of the *E coli* decarboxylase or deaminase. This fragment has been cloned to an *E coli* expression vector p Mal c2 and was expressed as a fusion protein with maltose binding protein (mbp). The fusion protein was cleaved with Xa factor, an endoprotease and purified. Molecular weight by SDS-PAGE showed it to be about 20.9 KD which corresponded to that revealed from the sequence analysis and is expressed as a single polypeptide (Nair et al 1994). This 1.8 Kb fragment has also been cloned into *Agrobacterim tumefaciens* for transforming *L. sativus*.

References:

Campbell CG, Mehra RB, Agrawal, SK, Chen YZ, Abd El Moneim AM, Khawaja, HIT, Yadov C R, Tay J U and Arya WA 1994. *Euphytica* 73: 167-175.

Karp A 1991. *Oxford Surveys of Plant Molecular and Cell Biology* 7: 1-58.

Malathi K, Padmanaban G., Rao SLN and Sarma PS 1967. *Biochimica et Biophysica. Acta* 141: 71-78.

Malathi K, Padmanaban G and Sarma PS 1970. *Phytochemistry* 9: 1603-10.

Malek MA, Sarwar CDM, Sarker A and Hassan MS 1995. In *Lathyrus Genetic Resources in Asia*: Proceedings of a Regional Workshop Raipur, India Dec 1995 pp 7-12. (Eds R.K. Arora, PN Mathur, KW Riley and Y. Adhm)

Mandal PK 1995. Ph.D. Thesis submitted to PG school, IARI, New Delhi.

Mandal PK, Santha I.M. and Mehta S.L. 1996. *J Plant Biochem Biotech* 5: 83-86.

Mehta SL, Ali K and Barna KS 1994. *J Plant Biochem Biotech* 3: 73-77.

Misra BK and Barat GK 1981. *Plant Nutrition* 3: 997-1003.

Mohan VS, Nagarajan V and Gopalan C 1966. *Indian Journal of Medical Research* 54: 410-414.

Nagarajan V and Gopalan C 1968. *Indian Journal of Medical Research* 56: 95-97.

Nair AJ, Khatri GS, Santha IM and Mehta SL 1994. *J Plant Biochem Biotech* 3: 103-106.

Prakash S, Misra BK, Adsule RN and Barat GK 1977. *Biochemie und Physiologie Pflanzen* 171: 369-374.

Ranaanujam S, Sethi KL and Rao SLN 1980. *Indian Journal of Genetics* 40: 300-304.

Ramchand, CN, Parekh L J and Ramakrishnan C.V. 1981. *Indian J Biochemistry and.Biophysics* 18 suppl, 140.

Rao SLN, Matalhi K and Sarma PS 1969. Lathyrism *World Review of Nutrition and Dietetics* 10: 214-238.

Roy PK, Singh B, Mehta SL, Barat GK, Gupta N, Kirti PB and Chopra VL 1991 *Indian Journal of Experimental Biology* 29: 327-330.

Roy PK, Barat GK and Mehta SL 1992. *Plant Cell Tissue and Organ Culture*. 29: 135-138.

Roy P.K., Ali K, Gupta A, Barat GK and Mehta SL, 1993. *J Plant Biochem Biotech* 2: 9-13.

Sachdev A, Sharma M, Johari RP and Mehta SL 1995. *J Plant Biochem Biotech* 4: 33-36.

Sarma P.S. and Padmanaban G 1969. In: *Toxic constitutents of Plant Food stuffs*, pp 267-291 (ed IE Liener) Academic Press New York.

Shelly Praveen, Johari RP and Mehta SL 1994. J Plant Biochem Biotech 3: 25-29.

Somayajulu PLN, Barat GK, Shiv Prakash, Misra BK and Srivastava YC 1975. *Proc. Nutrition. Sci India* 19:35-39.

Sukanya R, Santha I M and Mehta SL 1993. *J Plant Biochem Biotech* 2: 77-82.

Swaminathan MS, Naik MS, Kaul AK, and Austin A 1970. *Improving Plant proteins by Nuclear Techniques* IAEA, Vienna pp 165-183.

Yadav V K, Santha I M, Timko MP, Mehta SL 1992. *J Plant Biochem Biotech* 1: 87-92.

The Ever Changing World of Food Legumes

A.E. Slinkard
Crop Development Centre, University of Saskatchewan, 51 Campus Drive, Saskatoon, SK S7N 5A8 Canada

Major changes are occurring in every facet of our lives. These changes have profound effects on the production, utilization and marketing of food legumes. We must keep abreast of these changes and how they affect the food legume industry. In this way we can take an active role in leading research, development and extension efforts in food legumes into the 21st century.

1.1   The Changes

These changes include increased demand for traditional foods, new foods based on various components (protein, oil and starch) and functional foods (soluble fibre) and nutriceuticals for our aging population. Value-added processing for new food or industrial products will become increasingly important. Biotechnology will be used to increase yield, quality and net returns by incorporating selected traits, such as insect resistance, disease resistance or some special nutritive or health-related trait, into the latest cultivars.

A second major change is globalization and all its ramifications. "Bigger appears better" and this trend is permeating society. Farmers are mechanizing and farms are becoming fewer and larger in the developed countries. Urbanization is occurring in both developed and developing countries. Businesses are merging nationally and internationally, including mergers between the crop protection and seed industries. More companies are becoming vertically integrated. Computerization of communications, transportation and financial transactions has helped facilitate the move to "free trade".

A third major change relates to increased emphasis on environmental sustainability and maintenance of biological diversity. Related issues are intellectual property rights, access to germplasm and perceived hazards of transgenic plants.

A fourth major change relates to the sources, adequacy and continuity of funding for food legume research. Traditionally, agricultural research has been funded by the federal government, but most governments have recently reduced their level of investment in agricultural research. This has occurred in spite of numerous studies (e.g., Ruttan, 1982) reporting that agricultural research has provided an internal rate of return to the economy of between 20 and 50% per year (after inflation). The question then becomes, can private industry or the producers achieve this high rate of return on their investment in food legume research? Private companies will support specific high return research projects with the idea that they should be able to make a profit after taxes. Obviously, individual producers cannot expect a return on their investment in food legume research, but groups of producers through a production levy can support an effective research program that will provide a net return to the producers collectively. More and more agricultural research is being funded in this manner, often with matching funds from the government.

1.2   Review of the Previous Two Conferences

The First International Food Legume Research Conference (IFLRC) was held at Spokane, Washington, USA in 1986. The four featured food legumes were pea (*Pisum sativum*), lentil (*Lens culinaris*), faba bean (*Vicia faba*) and chickpea (*Cicer arietinum*). Bunting (1988) reviewed the conference and suggested that other food legumes be included in future conferences. He also emphasized that more research, development and extension efforts are needed on all aspects of food legume production, utilization and marketing. In this manner, net returns from food legume production should become more competitive

R. Knight (ed.), Linking Research and Marketing Opportunities for Pulses in the 21st Century, 699–701.
© 2000 *Kluwer Academic Publishers. Printed in the Netherlands.*

with net returns from cereal production. Then, food legume production and per capita consumption should increase.

As a result of Bunting's (1988) suggestions, the Second IFLRC in Cairo, Egypt (1992) included grasspea (*Lathyrus sativus*) and a separate session on biotechnology and gene mapping. Roberts (1994) reviewed the Second IFLRC and emphasized many of the changes occurring in our lives as noted above. The crisis in research funding has now extended to the International Agricultural Research Centers. Roberts (1994) also emphasized the need for greater stability in yield of food legumes and thus the need for a higher level of resistance to biotic and abiotic stresses including a better understanding of N_2 fixation and nitrogen cycling in drought-prone systems. Consequently, the Third IFLRC emphasized research progress and needs related to increasing yield stability plus several concurrent sessions on marketing.

1.3   The Challenges

In his presentation, Dr. Byerlee (See Chapter 2) concluded that the demand for use of food legumes as feed is increasing, but they are not competitive with soybean meal. Now that is a real challenge. Food legume researchers and extension specialists must increase their efforts and prove that food legume production and utilization in feeds is competitive with soybean meal in many parts of the world, but obviously not in the heart of the soybean producing and processing areas. For example, locally produced pea is competitive with imported soybean meal for use in hog and beef rations in much of western Canada. Pea is a valuable source of both protein and energy in these rations. We need only to reduce the cost of production (increase the yield with no increase in cash inputs) for pea to become competitive with soybean meal in some other areas of the world. Perhaps, this will be better documented by the time of the Fourth IFLRC.

Research on food legume production is designed to reduce risk, increase efficiency and make net returns from food legume production competitive with those from cereal production. Unfortunately, support for food legume research has not kept up with that for cereal research on a per tonne or a per hectare basis. The net result is that food legume production is becoming less competitive with cereal production in many countries. The notable exception is Australia where food legume research has recently been increased as they try to diversify their crop production and crop export mix for a more sustainable crop production system. Thus, a second challenge is to increase the amount and efficiency of food legume research worldwide so that we can achieve a more nearly sustainable crop production system. The benefits and synergy derived from a cereal-legume rotation will help maintain mankind well into the future.

Food legumes fix nitrogen symbiotically in association with *Rhizobium* whereas cereals do not. Nitrogen fixation requires energy (4-10 g carbon is used by nodulated roots per gram of atmospheric nitrogen reduced, Schubert, 1982). Thus, N_2 – fixing plants will have a reduced yield relative to non-N_2-fixing plants, if the content of N in the two plants is the same. Food legumes also have a higher protein concentration than cereals, typically 200-250 g/kg (20-25%) vs. 90-130 g/kg (9-13%) for cereals. Plants require more energy to produce one gram of protein than one gram of carbohydrate. Thus the yield of food legumes will rarely equal that of cereals. Thus, yield breakthroughs, like those in semidwarf wheat, semidwarf rice, hybrid corn and hybrid sorghum (all cereals high in carbohydrates and low in protein and oil) will never occur in food legumes. I challenge you to prove me wrong.

1.4   The Future

My crystal ball indicates that all of the above changes will continue and the big will get bigger, but not necessarily better. Specific changes I see are:
1. Food legume production will continue to increase in areas where specific food legumes have a comparative advantage over other crops – pea in Canada and chickpea and narrow-leaf lupin in Australia at the present time and potentially several food legumes in Argentina, Russia and the Ukraine,
2. Food legume exports will continue to increase,
3. Value-added processing of food legumes will increase, e.g., instantized and/or extruded foods and soluble fibre,

4. The Fourth IFLRC will develop into a larger, more inclusive, international organization with more representation from China, Russia, Ukraine, Argentina and other countries that were poorly represented or absent from the previous conferences,

5. Additional local and regional food legume research coordinating organizations will be organized,

6. Food legume research and production will increasingly be concentrated in those regions that have a comparative advantages for food legume production,

7. More research, development and extension will be done by the private sector as public sector support continues to decline,

8. Producer organizations will become more heavily involved in funding and prioritizing food legume research, and

9. Advances will continually be made in all aspects of food legume research, development, utilization, processing and marketing.

References

Bunting, A.H. 1988. In: *World Crops: Cool Season Food Legumes.* pp. 1155-1167 (ed. R.J. Summerfield). Dordrecht: Kluwer Academic Publishers.

Roberts, E.H. 1994. In: *Expanding the Production and Use of Cool Season Food Legumes.* pp. 983-988 (eds. F.J. Muehlbauer and W.J. Kaiser). Dordrecht: Kluwer Academic Publishers.

Ruttan, V.W. 1982. *Agricultural Research Policy.* Minneapolis: University of Minnesota Press.

Schubert, K.R. 1982. *Workshop Summaries, Vol. 1* Rockville, Maryland: American Society of Plant Physiologists.

CONFERENCE SUMMARY: A REVIEW OF THE MARKETING PAPERS

Robert Rees

Agribusiness and Policy Consultant, Department of Primary Industries and Resources, South Australia.

Three economists set the global picture on Monday of the Conference and the trade scene on Tuesday. As economists, it was difficult for them to reach the same conclusions. It was ever thus.

However, the good news from my point of view was that Indian consumers are behaving rationally and responding strongly to price.

Much has been said by alarmist commentators such as Lester Brown that there is a food security problem. A number of papers went some way to dispelling that myth particularly with pulses.

India has an increasing deficit between annual production and consumption of pulses. It has been a puzzle to me for some time why this is not a major issue for a population which is predominantly vegetarian. I was gratified to learn that instead, consumption of cereals and animal products such as milk and fruit and vegetables have risen strongly.

At the same time, income is rising in many countries, including India, and people are responding by purchasing higher priced goods such as eggs, poultry and milk and shopping at the new supermarkets, which are springing up everywhere. It is clear also that prices of pulses have been rising in real terms. This has led to a great deal of substitution of cheaper pulses such as bulk shipments of dry peas from Australia and Canada, desi chickpeas from Australia and black gram and pigeon pea from Myanmar.

As a consequence of the higher world prices there has been an increase in the area sown to pulses in the last two years as farmers in many countries respond to the higher prices. For supplies to equal demand in the longer term there has to be minimal interference by governments to allow their farmers to respond to higher prices and plant larger areas or substitute less profitable crops for more profitable ones.

Yields of pulses have not increased very much since IFLRC1 which is alarming. As a consequence there have been periodic production shortfalls in a number of countries, highly volatile prices, and increases in imports.

But trade has not increased to anywhere near the projections made by FAO in 1986 with exports rising from 4.6 million tonnes in 1986 to 8 million tonnes in 1996. More importantly it is the relatively small volumes of trade, which the highly aggressive exporting countries are impatient about. The trade is not big enough for all that energy from all their researchers to satisfy their talents.

In 1996 world exports of dry beans totalled 2.5 million tonnes and peas 3.7million tonnes or nearly 80 per cent of the total. Exports of broad beans 0.5 million tonnes, chickpeas 0.3 million tonnes, and lentils 0.7million tonnes are insufficient to warrant large research investments by themselves.

It became clear also that dry beans were an integral part of the product mix for many countries. I was pleased to participate in the debate on the inclusion of dry beans for the next conference. The meeting resolved to recommend to the new IFLRC committee that dry bean, with the exception of soybeans, is included. However, the program needs to be balanced so that it does not lose the focus of this conference.

Our discussion then took us beyond the farm gate and concluded that research into value adding or, as someone said, cost-increasing activities were explored and what an exciting vista they painted. At the present time, the selling of pulses for food is stagnating. Marketing must take a higher profile. By marketing we are talking about branded products, quality products designed for discriminating markets. This is the case on the Indian subcontinent where the revolution in eating habits is just gathering pace particularly as the supermarket trend takes hold.

The developed world needs to concentrate on producing value-added products for itself. There is an enormous need for product development research to take advantage of the obvious health qualities

R. Knight (ed.), Linking Research and Marketing Opportunities for Pulses in the 21st Century, 703–704.
© 2000 *Kluwer Academic Publishers. Printed in the Netherlands.*

704

of pulses. Quality and safety issues are becoming more critical in many countries. Quality needs to take into account the changing consumer buying habits and preferences for size, colour and importantly taste.

Tim Morris outlined the changing shopping habits being adopting worldwide. For example, in Hong Kong 70 per cent of the family income is spent away from home, in New York 80 per cent.

The new vogue is HMR or home meal replacement meals, which are fully prepared away from home. With pulse products there is little excitement in current packaging technology. Who wants to buy pulse seed and soak it for 24 hours anymore?

Tim Morris outlined a number of innovations and opportunities of which researchers need to be acutely. They include:

-the continuous increase in the number of vegetarian and vegan consumers who are highly responsive to health scares in meat products
-fresh prepared salads in supermarkets
-fresh and prepared soups are becoming more popular for lunch in soup tureens
-Japanese style deep fried green peas and single serve soup snacks
-fresh green sprouts for supermarkets and restaurants
-functional foods such as neutraceuticals which have health benefits, fibre, vitamins, nutrients, and health enhancing compounds
-conventional bean based packaged foods
-ethnic specific pulses-Indian, Mexican

Then along came the ball of energy, Susanna Czuchajowska, with her water fractionation technology and the development of gels from pulses for use in the noodle and snack markets of Japan and Korea. South East Asia is not currently a normal consumer of cool season food legumes to any extent.

For the present that leaves us with a large feed overhang in the market. I would prefer to say that in the case of peas and faba bean any quantities of these products going to the feed industry are in fact residual food products not being demanded by food importers. We heard of some very promising developments in the fish industry to use pulses, particularly lupins, faba bean and peas.

For lupins, Kevin Swan of the Grain Pool painted a positive picture for their role in feed use and a list of research areas to advance lupins. They include lupins, with less fibre for the Asian poultry industry and for aquaculture; lupins with 5 −10% oil, with some South American varieties much higher, a hulless lupin, better quality protein, and better cooking properties.

Tony Edwards pointed out the precision required for compounding pulses in animal feeds and the positive role they play in any livestock development. Pulses will also play a part in the enormous growth taking place in the Asian livestock industry but we should be careful of too much expansion, as prices are not expected to rise in real terms because of competition from many substitutes.

A plea for a resolution from this conference. If we want to continue to know what the trends are around the world we need access to information. The decision by FAO to not have anyone working on pulses is dismaying to say the least and I believe the conference must urge FAO to reconsider. There is no other organisation that attempts to pull all the data together on the world's pulse industry.

CONCLUSION

Removal of the supply-side blockages is a key issue for many pulse consuming countries. There is strong demand potential as incomes rise and investment in pulse production leads to lower prices. For the big five exporters the major challenge is to excite investment in value-added products suitable for the supermarket shelves which are in tune with today's cultural imperatives of being easily prepared tasty, timely and interesting foods.

Current Plant Science and Biotechnology in Agriculture

Current Plant Science and Biotechnology in Agriculture

Current Plant Science and Biotechnology in Agriculture

KLUWER ACADEMIC PUBLISHERS – DORDRECHT / BOSTON / LONDON